Lecture Notes of the Institute for Computer Sciences, Social and Telecommunications E

T0092317

Jie Zhou (Ed.)

Complex Sciences

First International Conference, Complex 2009
Shanghai, China, February 23-25, 2009
Revised Papers, Part 1

 Springer

Volume Editor

Jie Zhou
Nanyang Technology University
Network Technology Research Centre
Research Techno Plaza, Block 50 Nanyang Drive
Singapore 637553

E-mail: zjie@ntu.edu.sg

Library of Congress Control Number: Applied for

CR Subject Classification (1998): J.3, J.2, J.6, K.4, J.5

ISSN   1867-8211
ISBN-10   3-642-02465-3 Springer Berlin Heidelberg New York
ISBN-13   978-3-642-02465-8 Springer Berlin Heidelberg New York

springer.com

© ICST Institute for Computer Science, Social Informatics and Telecommunications Engineering 2009
Printed in Germany

Typesetting: Camera-ready by author, data conversion by Scientific Publishing Services, Chennai, India
Printed on acid-free paper   SPIN: 12649990   06/3180   5 4 3 2 1 0

Preface

I was invited to join the Organizing Committee of the First International Conference on Complex Sciences: Theory and Applications (Complex 2009) as its ninth member. At that moment, eight distinguished colleagues, General Co-chairs Eugene Stanley and Gaoxi Xiao, Technical Co-chairs János Kertész and Bing-Hong Wang, Local Co-chairs Hengshan Wang and Hong-An Che, Publicity Team Shi Xiao and Yubo Wang, had spent hundreds of hours pushing the conference half way to its birth. Ever since then, I have been amazed to see hundreds of papers flooding in, reviewed and commented on by the TPC members. Finally, more than 200 contributions were selected for the proceedings currently in your hands. They include about 200 papers from the main conference (selected from more than 320 submissions) and about 33 papers from the five collated workshops:

Complexity Theory of Art and Music (COART)
Causality in Complex Systems (ComplexCCS)
Complex Engineering Networks (ComplexEN)
Modeling and Analysis of Human Dynamics (MANDYN)
Social Physics and its Applications (SPA)

Complex sciences are expanding their colonies at such a dazzling speed that it becomes literally impossible for any conference to cover all the frontiers. We decided to mainly cover the following seven topics, which is already a major challenge for a conference:

Structure and Dynamics of Complex Networks
Complex Biological Systems
Complex Economic Systems
Complex Social Systems
Complex Engineering Systems
Complex Systems Methods
Other Complex Systems

It is our hope that the conference can serve as a bridge for accelerating communication and cooperation between the participants. It is certainly also our hope that more researchers will respond to our invitation in future.

On behalf of all the Organizing Committee members, we thank all the TPC members and reviewers who have carefully helped review and select the contributions. We thank all the local helpers for their endless patience and priceless help. The efforts of the ICST and Springer staff are also gratefully acknowledged. Above all, we thank all the authors for submitting their research results to us. Without their support, there would be no conference.

Last but not least, I would like to take this opportunity to express my personal thanks to all the other Organizing Committee members. Team, it has been amazing and totally enjoyable to work with you.

February 2009   Jie Zhou

Organization

Steering Committee Chair

Imrich Chlamtac   Create-Net, Italy

General Co-chairs

Eugene Stanley   Boston University, USA
Gaoxi Xiao   NTU, Singapore

Technical Co-chairs

Bing-Hong Wang   USTC, China
János Kertész   BUTE, Hungary

Local Co-chairs

Heng-Shan Wang   USST, China
Hong-An Che   SASS, China

Sponsorship Co-chairs

Zhicheng Li   BUPT, China
Sheng Liang   Beihang University, China

Publication Chair

Jie Zhou   NTU, Singapore
email: zjie@ntu.edu.sg

Publicity Chair

Shi Xiao   NTU, Singapore

Web Chair

Yubo Wang   NTU, Singapore

Conference Coordinator

Karen Decker   ICST

Technical Program Committee

Mikko Alava	Helsinki University of Technology, Finland
Azucena Alvarez	Universidad de Sevilla, Spain
Tomasz Arodz	AGH University of Science and Technology, Poland
David K. Arrowsmith	University of London, UK
Alan Baker	Swarthmore College, USA
Juan Gonzalo Barajas	Ramirez IPICYT, Mexico
Patrick Beautement	The abaci Partnership LLP, UK
Mark A. Bedau	Reed College, Portland
Mirza Beg	University of Waterloo, Canada
Jean Botev	University of Trier, Germany
Markus Brede	CSIRO Marine and Atmospheric Research, Australia
Fabricio A Breve	Universidade de São Paulo, Brazil
Christine Broenner	The abaci Partnership LLP, UK
Zdzislaw Burda	Jagellonian University, Poland
John Burns	IT Tallgaht
Yiwei Cao	RWTH Aachen University, Germany
Jean Cavailhes	INRA-CESAER, France
Peilong Chen	National Central University, Taiwan
Yin Jie Chen	University College Cork, Ireland
Yi-Jen Chiu	Diwan College, Taiwan
Wai-Ki Ching	The University of Hong Kong, Hong Kong
Yang Cong	The University of Hong Kong, Hong Kong
Vittoria Colizza	ISI Foundation, Italy
Philip Cordes	University of Bremen, Germany
Michel l Cotsaftis	ECE, France
Jon Crowcroft	University of Cambridge, UK
Matthias Dehmer	University of Coimbra, Portugal
Sahin Delipinar	Bogazici University, Turkey
Zhenyu Dong	Dalian University of Technology, China
Stanislaw Drozdz	Polish Academy of Science, Poland
Chongwei Du	Shanghai Jiao Tong University, China
Zhisheng Duan	Peking University, China
Bohdan Durnota	DeciSci Co. Ltd., China
Schahram Dustdar	Vienna University of Technology
Frank Emmert-Streib	Queen's University Belfast, UK
rasul Enayatifar	Azad University, Iran
Markus Esch	University of Luxemburg, Germany
Maryam Esmaeili	Informatics Faculty of University of Lugano, Switzerland
Jin Fan	The Australian National University, Australia
Mohammad Fassihi	Amir-Kabir, Iran
Philip Vos Fellman	Southern New Hampshire University, USA
Xiang Feng	East China University of Science and Technology, China
John Frazer	Queensland University of Technology, Australia

Rosane Riera Freire   Pontificia Universidade Cató lica do Rio de Janeiro,
                             Brazil
Mauro Gallegati   Polytechnic University of Marche, Italy
Caixia Gao   Inner Mongolia University, China
Jianbo Gao   University of Florida, USA
Lei Gao   DeciSci Co. Ltd., China
Diego Garlaschelli   University of Siena, Italy
Domenico Delli Gatti   UniversitMa Cattolica, Italy
Virendra Gomase   Padmashree Dr. D.Y. Patil University (T.K.I.E.T.,
                             Warananagar), India
Antonio Gómez-Iglesias   CIEMAT, Spain
Yan Gu   University of Melbourne, Autralia
Zhi-Hong Guan   Huazhong University of Science and Technology, China
David Hales   Technical University of Delft, Netherlands
Zhangang Han   Beijing Normal University, China
Da-Ren He   Yangzhou University, China
Jari Saramaki Helsinki   University of Technology, Finland
Jean-Claude Heudin   Pole Universitaire Leonard de Vinci, France
Janusz Holyst   Warsaw University of Technology, Poland
Ping-Nan Hsiao   RCHSS, Academia Sinica, Taiwan
Arthur Huang   University of Minnesota, USA
Chung-Yuan Huang   Chang Gung University, Taiwan
Pan Hui   University of Cambridge, UK
Bin Jiang   University of Gävle, Sweden
LuoLuo Jiang   University of Science and Technology of China, China
Rui Jiang   University of Science and Technology of China, China
Jeff Johnson   The Open University, UK
Shahab Kamali   University of Waterloo, Canada
Beom Jun Kim   Sungkyunkwan University, Korea
Ki-Il Kim   Gyeongsang National University, Korea
Ralf Klamma   RWTH Aachen University, Germany
Ljupco Kocarev   University of California, San Diego, USA
Xiangxing Kong   Central South University, China
Victor Korotkikh   Central Queensland University, Australia
Ondrej Krejcar   VSB Technical University of Ostrava, Czech Republic
Francis Lau   Hong Kong Polytechnic University, Hong Kong
Anna Lawniczak   University of Guelph, Canada
Jae Woo Lee   Inha University, Korea
Wei-Po Lee   National Sun Yat-sen University, Taiwan
Ho-fung Leung   The Chinese University of Hong Kong, Hong Kong
Yu-jian LI   University of Science and Technology of China, China
Sy-Sang Liaw   National Chung-Hsing University, Taiwan
Jijun Lin   Massachusetts Institute of Technology, USA
Nelly Litvak   University of Twente, Netherlands
Jiming Liu   Hong Kong Baptist University, Hong Kong
Ruey-Tarng Liu   National Chung-Hsing University, Taiwan

Wolfgang Loehr	Max Planck Institute for Mathematics in the Sciences, Germany
Eduardo Lopez	University of Oxford, UK
Jianquan Lu	City University of Hong Kong, Hong Kong
Jinhu Lu	Chinese Academy of Sciences, China
Heinz Luediger	IMST GmbH, Germany
Qiang Luo	National University of Defense Technology, China
Amin R Mazloom	Mt. Sinai School of Medicine, USA
Gianluca Mazzini	University of Ferrara, Italy
Bernhard K Meister	Renmin University, China
Telmo Menezes	University of Coimbra, Portugal
Yu Song Meng	Nanyang Technological University, Singapore
Czeslaw Mesjasz	Cracow University of Economics, Poland
Panayotis Michaelides	National Technical University of Athens, Greece
Kevin Mills	NIST, USA
Juergen E Mimkes	Physics Department, Paderborn University, Germany
Jose Nacher	Future University-Hakodate, Japan
Ingve Simonsen Norwegian	University of Science and Technology, Norway
Juan G. Diaz Ochoa	Max-Planck-Institut Dynamik komplexer Technischer Systeme, Germany
Jukka-Pekka Onnela	Harvard University, USA
Spirakis Pavlos	University of Patras, Greece
Matti Peltomaki	Helsinki University of Technology, Finland
Danilo Pescia	ETH Zurich, Switzerland
Manh Cuong Pham	RWTH Aachen University, Germany
Gregory Provan	University College Cork, Ireland
Marcos Quiles	University of São Paulo, Brazil
Jose J Ramasco	ISI Foundation, Italy
Chuanjun Ren	National University of Defense Technology, China
Jie Ren	National University of Singapore, Singapore
Karim Mohammed Rezaul	University of Wales, UK
Colin L Richardson	Imperial College London, UK
Manuel Beltran del Rio	Instituto de Fisica, UNAM, Mexico
Suzanne Sadedin	Monash University, Australia
Sattar B. Sadkhan	University of Babylon, Iraq
Ingo Scholtes	University of Trier, Germany
Caterina Maria Scoglio	Kansas State University, USA
Parongama Sen	University of Calcutta, India
Yingni She	The Chinese University of Hong Kong, Hong Kong
Jingbo Shen	University of Science and Technology of China, China
Paul Sheridan	Tokyo Institute of Technology, Japan
Chuan Shi	Beijing University of Posts and Telecommunications, China
Theodore Simos	University of Peloponnese, Greece

Changsong Zhou   Hong Kong Baptist University, Hong Kong
Jie Zhou   Nanyang Technological University, Singapore
Jin Zhou   Shanghai University, China
Tao Zhou   University of Science and Technology of China, China
Zicong Zhou   Tamkang University, Taiwan

Table of Contents – Part I

Table of Contents – Part II

Part I

Return Intervals Approach to Financial Fluctuations

Fengzhong Wang[1], Kazuko Yamasaki[1,2], Shlomo Havlin[1,3],
and H. Eugene Stanley[1]

[1] Center for Polymer Studies and Department of Physics,
Boston University, Boston, MA 02215, USA
[2] Department of Environmental Sciences,
Tokyo University of Information Sciences, Chiba 265-8501, Japan
[3] Minerva Center and Department of Physics,
Bar-Ilan University, Ramat-Gan 52900, Israel

Abstract. Financial fluctuations play a key role for financial markets studies. A new approach focusing on properties of return intervals can help to get better understanding of the fluctuations. A return interval is defined as the time between two successive volatilities above a given threshold. We review recent studies and analyze the 1000 most traded stocks in the US stock markets. We find that the distribution of the return intervals has a well approximated scaling over a wide range of thresholds. The scaling is also valid for various time windows from one minute up to one trading day. Moreover, these results are universal for stocks of different countries, commodities, interest rates as well as currencies. Further analysis shows some systematic deviations from a scaling law, which are due to the nonlinear correlations in the volatility sequence. We also examine the memory in return intervals for different time scales, which are related to the long-term correlations in the volatility. Furthermore, we test two popular models, FIGARCH and fractional Brownian motion (fBm). Both models can catch the memory effect but only fBm shows a good scaling in the return interval distribution.

Keywords: Financial marekts, Econophysics, Volatility, Return interval, Scaling, Long-term correlation.

1 Introduction

Large and unpredictable fluctuations constitute risk for investments as well as the whole economy. For instance, the credit crisis nowadays is along with turmoil in financial markets, which causes huge losses for many investors and likely initiates a recession worldwide. Moreover, significant risk could be inherent not only in market crashes, but also in less hazardous fluctuations if they are unexpected and investments are not well protected against them. Banks have to properly estimate the risk of their investments and make provisions in order to be able to withstand large fluctuations without going bankrupt. The importance of financial markets attract many researchers and in particular, collaborative work

J. Zhou (Ed.): Complex 2009, Part I, LNICST 4, pp. 3–27, 2009.

joining economists and physicists (which created a new interdisciplinary field of econophysics [1,2,3,4,5,6,7,8,9,10,11,12,13,14,15,16,17,18,20,19,21,22,23,24,25] [26,27,28,29,30,31,32,33,34,35]) has resulted in a better understanding of economic fluctuations. Until relatively recently, theories of economic fluctuations invoked the label of "outliers" (bubbles and crashes) to describe fluctuations that do not agree with the existing theory. However, econophysics research found evidence that the probability distribution of price fluctuations can be described by a power law [27,28,29,30,31,32,33,34,35]. There are no "outliers" since this law also holds for extremely large and unpredictable changes of magnitude sufficient to wreak havoc.

Statistical physics deals with systems comprising a very large number of interacting subunits, for which predicting the exact behavior of the individual subunit would be impossible. Hence, one is limited to making statistical predictions regarding the collective behavior of the subunits. Recently, it has come to be appreciated that many such systems consisting of a large number of interacting subunits obey universal laws, therefore they are independent of the microscopic details. The finding, in physical systems, of universal properties that do not depend on the specific form of the interactions gives rise to the intriguing hypothesis that universal laws or behavior may also be present in economic and social systems [34,35]. An often-expressed concern regarding the application of physics methods to the social sciences is that physical laws are applied to systems with a very large number of subunits (at the order of Avogadro's number, 10^{23}), while social systems comprise a much smaller number of elements. Fortunately, due to the rapid development of electronic trading and data storing in the last few decades, financial data bases have become available with a huge amount of data points (say 10^8), enabling physicists to analyze them as dynamic systems. The data size becomes comparable to nano systems and the "thermodynamic limit" is reached so that methods from statistical physics can be applied. It is worth to note that there is only a small amount of extremely large events even in very huge data bases. To understand these devastating events, it is of great importance to find laws describing the entire data set in order to approach extreme events by extensive analysis on small fluctuations.

Two important conceptual advances on universal laws are *scaling* and *universality*. A system obeys a scaling law if its relation is characterized by the same functional form and exponent over a certain range of scales ("scale invariance"). The typical behavior for scaling is *data collapse*, all curves can be "collapsed" onto a single curve, after a certain scale transformation on the measure. The general principles of scale invariance used here have proved useful in interpreting a number of other phenomena, ranging from elementary particle physics and galaxy structure to finance [35,36,37]. At one time, many imagined that the "scale-free" phenomena are relevant to only a fairly narrow slice of physical phenomena [38,39]. However, the range of systems that apparently display power law and scale-invariant correlations has increased dramatically in recent years, ranging from base pair correlations in noncoding DNA [40], lung inflation [41] and interbeat intervals of the human heart [42] to complex systems involving large numbers of interacting subunits that display "free will," such as city growth [43], university research budgets [44], and even bird populations [45]. In many of these diverse systems, the *same* scaling function exists for a significant range which is remarkable, apparently suggesting the universality of laws. Moreover, many systems share the same scaling functions and characteristic exponents and therefore belong to one universality class. This connection provides the people a comprehensive view over these diverse systems.

Scaling and universality are important properties of a data set describing the global behavior of the probability distribution. This usually does not fully characterize a sequence of data points which also depends on the time organization of the sequence. Only if it is *uncorrelated*, the data points are independent of each other and the sequence is totally determined by the distribution. In most cases, the records is *correlated*, it will affect the order in the data set. This behavior is called "memory", as the data points "remember" previous values. Trivially the memory decays with the time lag. The decay of memory, which could be characterized by the autocorrelation function, may follows different types of function. One typical function is exponential, and the existing of memory is described by a characteristic time scale. The memory almost disappears at the scales above the characteristic time and thus it only exists for a short-term. Such kind of time series is called *short-term correlated*. Another typical function for the autocorrelation is a power law. In this case there is no finite characteristic scale and the correlation exists for a much longer time, therefore it is called *long-term correlated*. Note that short-term memory always exists in a long-term correlated time series. As for the study of financial markets, the temporal structure in a time series is of great importance since it influences the performance of any movement. Many studies show that price change ("return") does not exhibit any linear correlations extending over more than a couple of minutes, but their absolute value, which is a measure of volatility, exhibits long-term correlations (see Ref [34] and references therein). This leads to long periods of high volatility as well as other periods where the volatility is low ("volatility clustering").

Extreme events do not only occur in economics, but also appear in very different fields like climate or earthquakes. For instance, Gutenberg and Richter related huge earthquakes to everyday tremors in one single power law curve [46,47]. If one wants to prepare for a dangerous earthquake, it might be less important to exactly know how strong the next shock will be, but rather to know when a large shock will occur. A good approach is to study the time ("return interval") between two successive shocks larger than a threshold above which a shock would damage a building. This way one can gather information on the temporal structure of the fluctuations. Recently Bunde et al. [48,49,50,51] studied the return intervals for climate records and found that the long-term memory leads to a stretched exponential distribution and clustering of extreme events. They also suggested that these phenomena should therefore also occur in heartbeat records, internet traffic and stock volatility where long-term correlations occur. For financial data, a first effort was conducted by Yamasaki et al. who studied the daily data of currencies and US stocks and showed the scaling in the distribution and long-term memory in the sequence [52]. Following this, Wang et al. studied the intraday data of 30 stocks which constitute of Dow Jones Industrial Average (DJIA) index, Standard and Poor's 500 (S&P 500) index, currencies, interest rates as well as oil and gold commodities and found similar behaviors [53,54]. Similar analysis have been done for the Japanese [55] and Chinese [56,57] stock markets. To compare with the empirical data, Vodenska-Chitkushev et al. examined return intervals from two known models, FIGARCH and fractional Brownian motion (fBm) and showed that both models simulate the memory effects but only fBm yield the scaling feature [58]. Bogachev et al. related the nonlinear correlations to the multiscaling behavior in return intervals [59], they also showed that the return interval distribution follows a power law function for multifractal data sets [63]. Recently, Wang et al. studied systematically 500 components of S&P 500 index and demonstrated a systematic deviation from the scaling. They showed that this multiscaling behavior is related to the nonlinear correlations in volatility sequence [60]. Further, Wang et al. analyzed the relation between multiscaling and several essential factors, such as capitalization and number of trades, and found certain systematic dependence [61]. The multiscaling behavior is also found in the Chinese stock market [62]. These studies help us to better understand the volatility and therefore may lead to better risk estimation and portfolio management [64,65,66,67]. Return

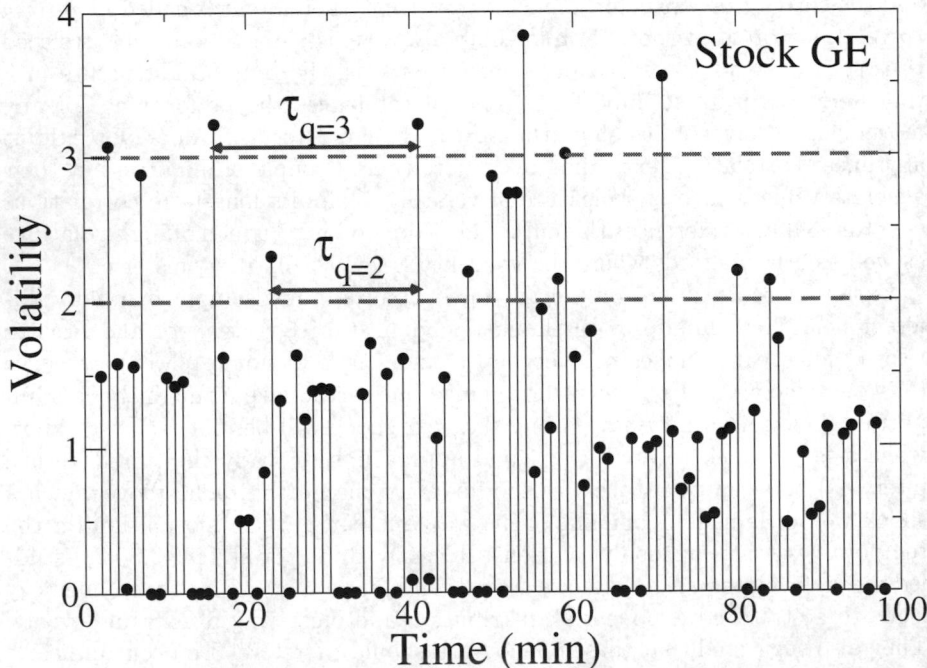

Fig. 1. (Color online) Illustration of volatility return intervals. The volatility is in units of its standard deviation. The solid circles are volatility values of the GE stock on Jan 8, 2001. Return intervals $\tau_{q=2}$ and $\tau_{q=3}$ for two typical thresholds q are displayed.

intervals have also been studied in many other fields (see Ref [68] and references therein). It is calculated in similar ways but with different names, like waiting time, interocurrence time or interspike interval.

In this paper we analyze the volatility return intervals of the entire US stock markets. The database analyzed is the Trades And Quotes (TAQ) from New York Stock Exchange (NYSE). The period studied is from Jan 2, 2001 to Dec 31, 2002, totally 500 trading days. TAQ records every trade for all securities in the US markets. To avoid many missing points in 1 min resolution, we choose to analyze only the 1000 most traded stocks. Their numbers of trades range from 600 to 60,000 times per day. The volatility is defined the same as in Ref [53]. First, we compute the absolute value of the logarithmic change of the minute price, then remove the intraday U-shape pattern, and finally normalize the series with its standard deviation. Therefore the volatility is in units of standard deviations. With 1-min sampling interval, a trading day has 390 points (after removing the market closing hours), and each stock has about 195,000 records. We also examine the S&P 500 index, a benchmark of US stock markets. The data is from Jan 2, 1984 to Dec 31, 1996, totally 130,000 points with 10-min sampling interval. For a typical stock, General Electric (GE), we find volatilities above a certain threshold q and calculate time intervals between them, as illustrated in Fig. 1. These time intervals consist the return interval series and the only free parameter is the threshold q.

2 Distribution of Return Intervals

We begin by analyzing the distribution, one of most important statistical properties for a time series. The distribution can be characterized by probability density function (PDF) or cumulative distribution function (CDF). Previous studies [52,53,54,55,56,57,58,59,60,61,62] showed that PDF for the return interval τ, $P(\tau)$, can be well approximated by a scaling law if τ is scaled by its average $\langle \tau \rangle$ ($\langle ... \rangle$ stands for the average over a data set), i.e.,

$$P(\tau) = 1/\langle \tau \rangle \cdot f(\tau/\langle \tau \rangle). \tag{1}$$

The scaling function f does not depend explicitly on q, but only through the mean interval $\langle \tau \rangle$. If $P(\tau)$ is known for one value of q, Eq. (1) can make predictions for other values of q—in particular for very large q (extreme events), which are difficult to study due to the lack of statistics.

2.1 Stretched Exponential Distribution

An important question is, what is the form of scaling function f? For many markets, the function was suggested to be in a good approximation to a stretched exponential (SE) [52,53,54,55,56,57,58,59,60,61,62],

$$f(x) \sim e^{-(x/x^*)^\gamma}. \tag{2}$$

Here x^* is the characteristic scale and γ is the shape parameter, which is related to the correlations in the volatility sequence and thus called "correlation exponent" [49]. For an uncorrelated series, f reduces to the regular exponential function and $\gamma = 1$. From Eq. (2), the PDF function can be rewritten as

$$P(\tau) \sim e^{-(\tau/a)^\gamma}. \tag{3}$$

Then a is the characteristic scale. From the definition of PDF and $\langle \tau \rangle$, one may find that the parameter a depends exclusively on γ [68,60],

$$a = \langle \tau \rangle \cdot \Gamma(1/\gamma)/\Gamma(2/\gamma). \tag{4}$$

Here $\Gamma(a) \equiv \int_0^\infty t^{a-1} e^{-t} dt$ is the Gamma function. However, due to the discreteness and finite size effects, there are some systematic deviations from the scaling law [51,60]. To avoid them, we will also use a as a free parameter in the SE fit. To simplify the calculation and without loss of generality, we assume τ/a is continuous, then the corresponding CDF, $C(\tau)$, is the integral of the PDF,

$$C(\tau) \equiv \int_x^\infty P(\tau)d\tau \sim \Gamma(1/\gamma, (\tau/a)^\gamma). \tag{5}$$

where $\Gamma(a,x) \equiv \int_x^\infty t^{a-1} e^{-t} dt$ is the incomplete Gamma function. Since CDF accumulates the information of the series and has a better statistics than PDF, in the following we obtain the correlation exponent γ by fitting the CDF with Eq. (5).

As an example, we plot three CDFs (for $q = 2$, 4 and 6 respectively) of the GE stock in Fig. 2. The three curves are distant from the other, due to the difference in $\langle \tau \rangle$. The least-square fits with Eq. (5) are illustrated by the solid lines. We use the classical method, Kolmogorov-Smirnov (KS) Statistic D, to test the goodness-of-fit [69,70]. D is defined as the maximum absolute difference between the cumulative distribution of the original data $C(\tau)$ and that of the fit $F(\tau)$,.

$$D \equiv max(|C(\tau) - F(\tau)|). \tag{6}$$

When D is larger than a certain value, which is called critical value (CV), the SE distribution is rejected. CV is decided by the significance level and data size. In this paper we choose 1% significance level and

$$CV = 1.63/\sqrt{N}, \tag{7}$$

where N is the number of data points.

We fit CDF with SE function for the 1000 most traded stocks [71]. The range of threshold is from $q = 1$ to 6, and the number of fit that is not rejected ("good fit") is listed in Table 1. We can see that most of the cases have a good fit by a SE function. A question naturally arises, for different thresholds, how similar are these correlation exponents? Previous research show that the scaling in distribution is well approximated [52,53,54,55,56,57,58,59,60,61,62]. Trivially, γ for different thresholds are strongly related, and their discrepancy should be

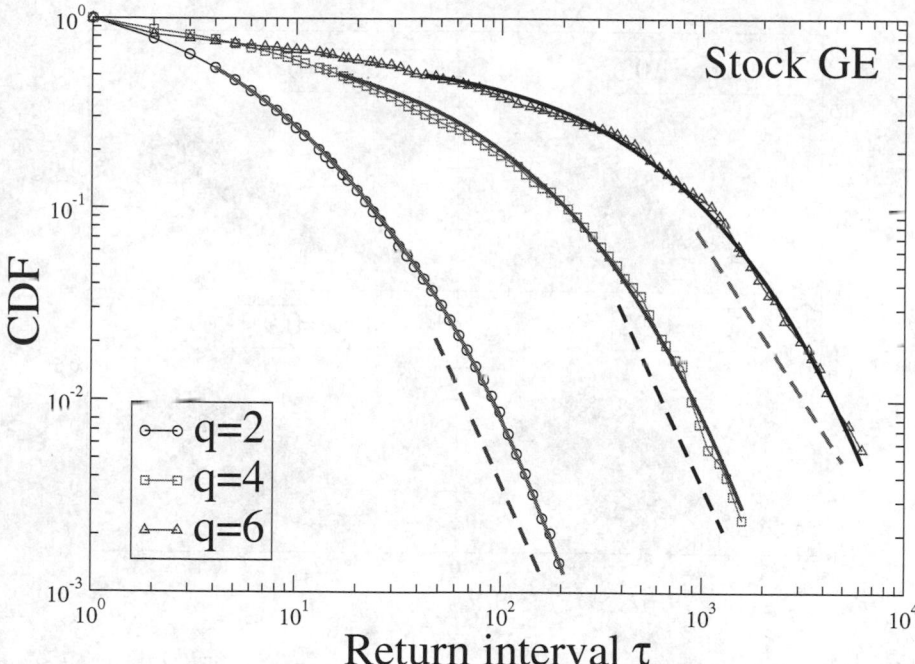

Fig. 2. (Color online) Cumulative distribution function (CDF) of return interval τ. CDF of three typical thresholds $q = 2$, 4 and 6 for the GE stock are plotted. Examples of two types of fit, the dashed lines (left shifted for better visibility) are the power law fit for the distribution tails and the solid lines on symbols are the stretched exponential fit for the whole distributions.

Table 1. Number of good fit on return interval CDF of the 1000 stocks. If KS statistics D (Eq. (6)) is smaller than the critical value CV (Eq. (7)), the corresponding distribution is not rejected. Two types of distribution, stretched exponential (for the whole range) and power law (for the tail), are tested.

Threshold q	1	2	3	4	5	6
Stretched exponential fit	791	795	815	933	977	986
Power law fit	31	349	626	826	839	710

small. To test this assumption we plot in Fig. 3 the dependence of the γ for other thresholds on the γ obtained for $q = 2$. Remarkably, all four cases show significant tendency and the slopes of linear fit are very close to 1. This result supports the well-approximated scaling in the distribution of return intervals. Note that the fluctuation is larger for a higher q, and the slope slightly decreases, which may be due to the limited data size of return intervals for large thresholds. We also test the dependence of other pairs of thresholds and observe similar behaviors. All these behaviors are consistent with Ref [61]. Moreover, we compare the value of the parameter a with Eq. (4) and find that a from the fit is in the same order

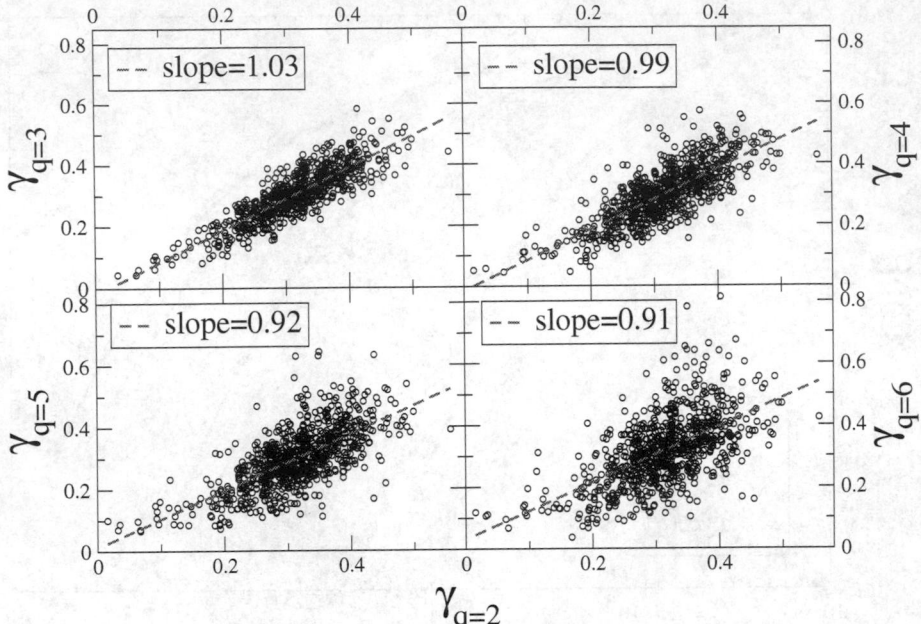

Fig. 3. (Color online) Relation between correlation exponent γ (Eq. (3)) of different thresholds. γ for four thresholds, $q = 3$ to 6 strongly depend on γ for $q = 2$, as indicated by dashed lines from the linear fit. All slopes of fit are quite close to 1, which suggests a good scaling in the distribution of return interval. Note that the fluctuation becomes stronger for a larger q, which relates to the smaller data size for the return interval with a larger q.

as that from Eq. (4), and usually the former is smaller. The ratio between two a is centered from 0.4 (for $q = 1$) to 0.8 (for $q = 6$) for the 1000 stocks [72].

2.2 Power Law Tail

For financial time series, the distribution tail usually is characterized by a power law function [27,28,29,30,31,32,33,34,35]. As for the return interval, Yamasaki et al. suggested that the scaling function is also consistent with a power law tail for large intervals, where the tail exponent is around 1 for both stock and currency data [52]. Moreover, Bogachev and Bunde have shown that the distributions of return intervals are governed by power laws [63]. Then CDF of return intervals would follow

$$C(\tau) \sim \tau^{-\zeta}, \tag{8}$$

where ζ is the tail exponent. To test this hypothesis we examine the distribution tail for the 1000 stocks. A popular way to fit the tail is using the Maximum Likelihood Estimator, specifically, it also called Hill estimator for a power law

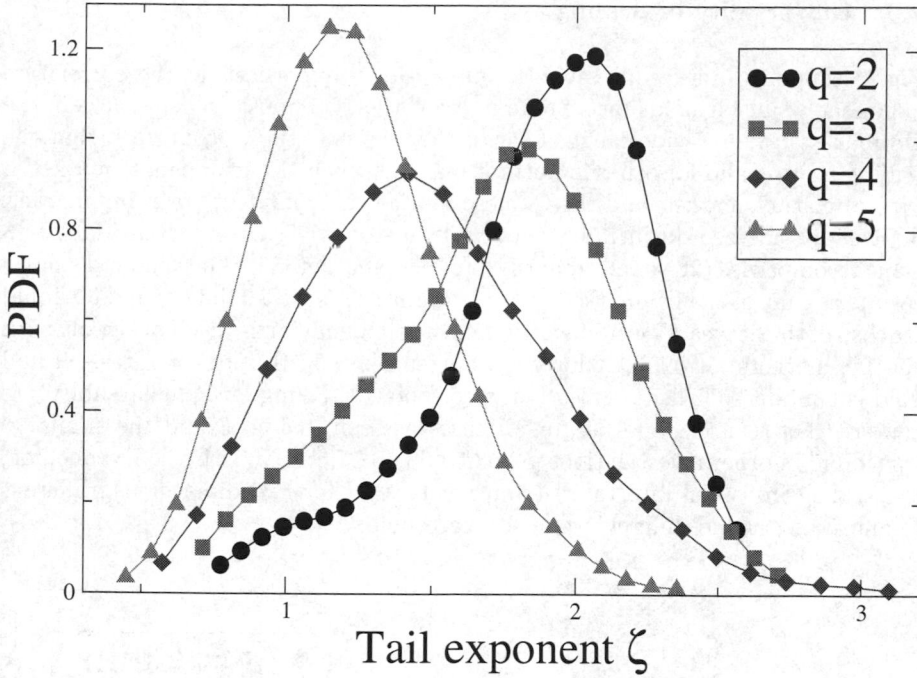

Fig. 4. (Color online) Probability density function (PDF) of tail exponent ζ from power law fit on the cumulative distribution of return intervals. The distribution systematically shifts from right to left, with increasing of the threshold.

tail [32,33,73]. The range of fit is not fixed by the Hill estimator [32,33], thus we examine the entire tail and choose the range that has the minimum KS statistics [33]. Examples of power law fits are demonstrated by the dashed lines in Fig. 2. We still use KS statistics to test the goodness-of-fit. For threshold $q = 1$ to 6, the numbers of good fit are listed in Table 1. For return intervals of $q = 1$ and 2, only for a small portion of the 1000 stocks, the power law distribution is not ruled out. However, for other cases, the power law distribution is not ruled out for a significant portion of stocks. In Fig. 4 we plot the PDF for tail exponent ζ. Interestingly, all PDFs are centered around a certain value which systematically shift from large value to small, with increasing the threshold. For $q = 2$, ζ is centered around 2, and for $q = 5$, ζ is centered around 1. The latter is consistent with Ref [52], which suggests that the difference may due to the limited size of data points. Ref [52] was using daily data, which is about $1/20$ of the intraday data in the current paper ($\sim 10,000$ points for the daily data vs. $\sim 195,000$ points for the intraday data). Similarly, the number of return intervals for $q = 5$ is only about $1/14$ of that for $q = 2$ (average over the 1000 stocks, ~ 850 points for $q = 5$ vs. $\sim 11,800$ points for $q = 2$). We also must note that, for $q = 2$, only about $1/3$ of the 1000 stocks have a good power law fit (Table 1).

2.3   Universality of Scaling

Fig. 3 supports quite impressive the universality hypothesis of the correlation exponent γ since it holds for a broad market, the 1000 most traded stocks in the US markets, with a wide range of thresholds. Recent studies confirmed that the scaling is also valid for other important markets, such as the Japanese market, a typical mature market, and the Chinese market, a prominent emerging market. Jung et al. analyzed the intraday data for 1817 stocks (1 year) and daily data for 3 typical companies (28 years) from the Japanese market [55]. They showed similar results as that of the US markets. For the Chinese market, 2 indices and 30 liquid stocks (both 2.5 years) were investigated, their behavior is also consistent with the US markets [56,57,62]. Moreover, currencies [52,54], interest rates, oil and gold commodities [54] were also found to follow a scaling law. Remarkably, γ is centered between 0.3 and 0.4 for all cases as seen in Fig. 3 and the similar γ was found in other investigations [52,53,54,55,56,57,58,59,60,61,62]. To conclude, the scaling in return interval distribution is valid for two dimensions, different financial assets and different volatility thresholds.

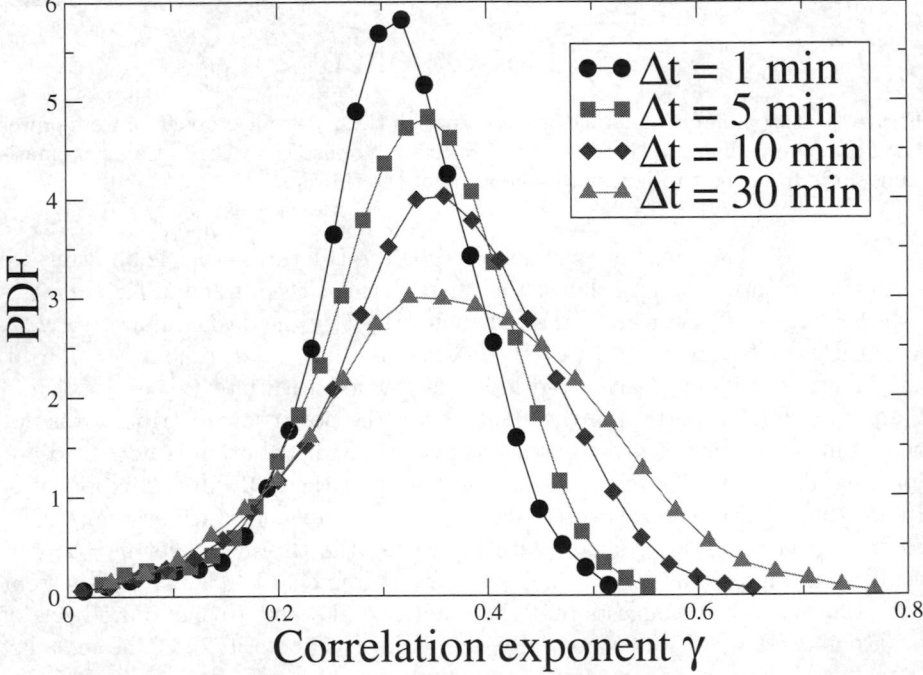

Fig. 5. (Color online) Distribution of correlation exponent γ for four sampling intervals, $\Delta t = 1$, 5, 10 and 30 minutes. With increasing of the sampling interval, the distribution tends to be wider. However, their centers are still close, changing from 0.31 for $\Delta t = 1$ minute to 0.37 for $\Delta t = 30$ minutes, which suggests that scaling is a good approximation for this range of sampling intervals. The broader distribution for lower resolution may be related to its smaller data size.

For statistical analysis, the time resolution of the records is an important aspect since the system may exhibit diverse behaviors in different time windows Δt. This is the third dimension for testing the universality of scaling. In Ref [54], Wang et al. have shown that the scaling is valid even to a sampling interval of 1 trading day. Here we change the volatility sampling interval from 1 minute to 5, 10 and 30 minutes, and then examine its return interval CDF. For 1-day resolution, there is only 500 points for a stock and the statistics is poor, we do not test it here. Also for a good statistics we focus on return intervals of a typical threshold, $q = 2$. Similar to the 1-min resolution, most of cases can be well fit by Eq. (5). For instance, with 5-min resolution, the SE hypothesis for 812 of 1000 stocks are not rejected under 1% significance level. In Fig. 5 we show the PDF of γ for $\Delta t = 1$, 5, 10 and 30 minutes. The shape of PDF systematically changes with increasing the sampling interval, the center shifts to right slightly and the width increases, which is consistent with the change of data size. For a lower resolution, we have fewer data points and consequently stronger fluctuations for γ values. Therefore, these curves show the persistence of the scaling for a broad range of sampling intervals.

2.4 Multiscaling

Financial time series are known to show complex behavior and are not of uniscaling nature [74]. The distribution of activity measure such as the intertrade time has multiscaling behavior [75,76]. From the previous sections we also see some weak but systematic tendencies, which indicate possible multiscaling (Fig. 3). Thus, a detailed analysis of the scaling properties of the volatility return intervals is of interest. Moment μ_m, which is defined as

$$\mu_m \equiv \langle (\tau/\langle \tau \rangle)^m \rangle^{1/m}, \tag{9}$$

accumulates the information over the entire data set and therefore provides a good way for testing the deviations from a scaling law. Pure scaling yields that μ_m should be independent on $\langle \tau \rangle$. Here m is the order of moment. Wang et al. studied the moments for 500 component stocks of S&P 500 index and found that μ_m has a certain tendency with $\langle \tau \rangle$, indicating multiscaling in the distribution of τ [60]. As shown in Fig. 6, the four moments of GE have similar tendencies, they increase to a certain value in the small $\langle \tau \rangle$ regime and then start to decrease. To quantify the tendency, Wang et al. suggested to fit the moments with a power-law [60],

$$\mu_m \sim \langle \tau \rangle^\delta. \tag{10}$$

If the distribution of return intervals follows a scaling law, the exponent δ should be close to 0. In other words, a significant non-zero δ suggests multiscaling. Here we call δ multiscaling exponent since it characterizes the multiscaling behavior [60]. The power law fit is demonstrated by dashed lines in Fig. 6. For very small and very large values of $\langle \tau \rangle$, Wang et al. identified the discreteness and finite size effects respectively [60], which was also recognized for the general case by Eichner et al. [51]. To avoid these effects, we fit Eq. (10) only in the medium

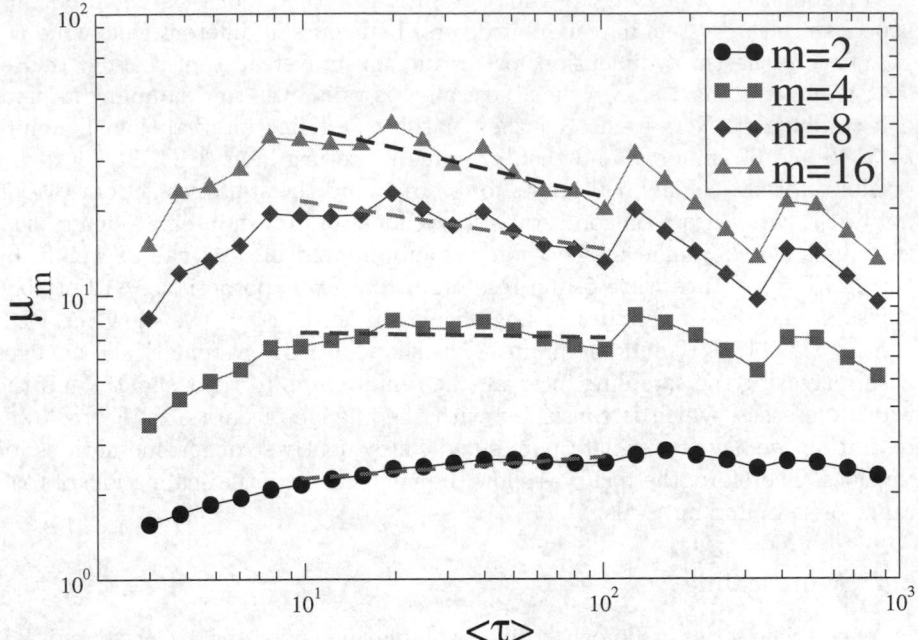

Fig. 6. (Color online) Dependence of moments μ_m on mean interval $\langle\tau\rangle$ for the GE stock. Four orders, $m = 2$, 4, 8 and 16 are showed. Dashed lines are power law fits in the range of $10 < \langle\tau\rangle \leq 100$. Adapted from [61].

range, $10 < \langle\tau\rangle \leq 100$. As shown in Fig. 7, over the 500 component stocks of S&P 500, the average δ systematically changes with m, which supports the existing of multiscaling features in the return interval distribution. Furthermore, we can see that δ are centered around 0 for the surrogate data, suggesting that the multiscaling behavior in the original records is related to the nonlinear correlations in volatility sequence. The surrogate records are generated by the Schreiber method [77,78] where nonlinearities are removed, and the corresponding μ_m is independent with $\langle\tau\rangle$. Ren and Zhou also employed moment analysis on two Chinese indices and confirmed the multiscaling behavior in the return interval distribution [62].

A second way to test the multiscaling is by examining the relation between the correlation exponent γ and threshold q. Wang et al. have shown that γ has a certain dependence on the threshold q for the broad market, especially for small thresholds [61], which is consistent with Fig. 3. A third method for testing is using KS statistics to test compare return interval distributions of two thresholds. If $D > CV$, the null hypothesis that two distributions are same is rejected. For the Japanese market, Jung et al. have shown a good scaling by the KS test [55]. However, for the Chinese market, Ren and Zhou found that the null hypothesis is not rejected only for 12 of 30 liquid stocks. For other 18 stocks,

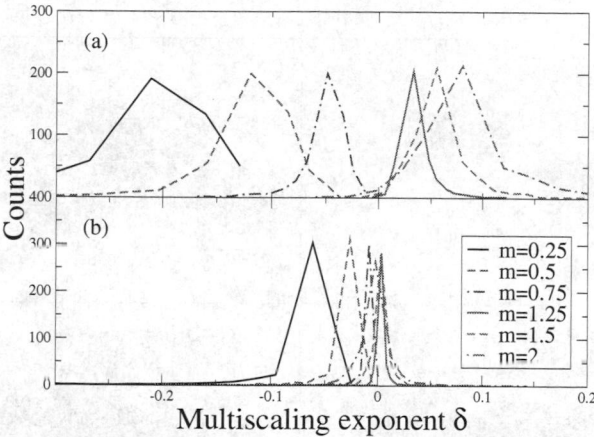

Fig. 7. (Color online) Distribution of multiscaling exponent α for S&P 500 constituents. The exponent α is obtained from the power-law fit for moments in the medium range $10 < \langle \tau \rangle \leq 100$. (a) Histogram of α for the original volatility and (b) for surrogate. The distributions have a systematic shift with m in (a) while all of them almost collapse in (b). This suggests that the multiscaling behavior in the original records dues to the nonlinear correlations in the volatility sequence. Adapted from [60].

the distributions are significantly different for different thresholds therefore they don't obey a single scaling law [62].

2.5   Size Effect

The following question arises, what is the origin for the multiscaling behavior in the return interval distribution? Recently Wang et al. carried out a multi-factor analysis and found similar relations over the factors [61]. Here we focus on the most popular measure, the market capitalization or the size of a stock, which is clearly related to the market activity [76]. Fig. 8 is the scatter plot of the relationship between γ and capitalization for the 1000 stocks. For all the four thresholds, the points are distributed in a wide area, which indicates an insignificant dependence. To better view a possible tendency, we group points according to their logarithmic value of capitalization and plot the average and standard deviation (as the error bar) of γ in each bin, as shown by the triangles in Fig. 8. An increasing trend for most of the range and a drop for very large capitalization is noticed. Interestingly, this behavior is consistent for all four thresholds. Note that the change of average γ is almost in the range of the error bar. Thus, γ systematically depends on the capitalization but the dependence is not strong, which suggests that there is a certain underline nonlinear mechanism and some sort of filtering maybe needed to identify it. Wang et al. also analyzed the dependence on the number of trades, risk and return. They found consistent relation for the risk and return. They also showed that γ is independent on the

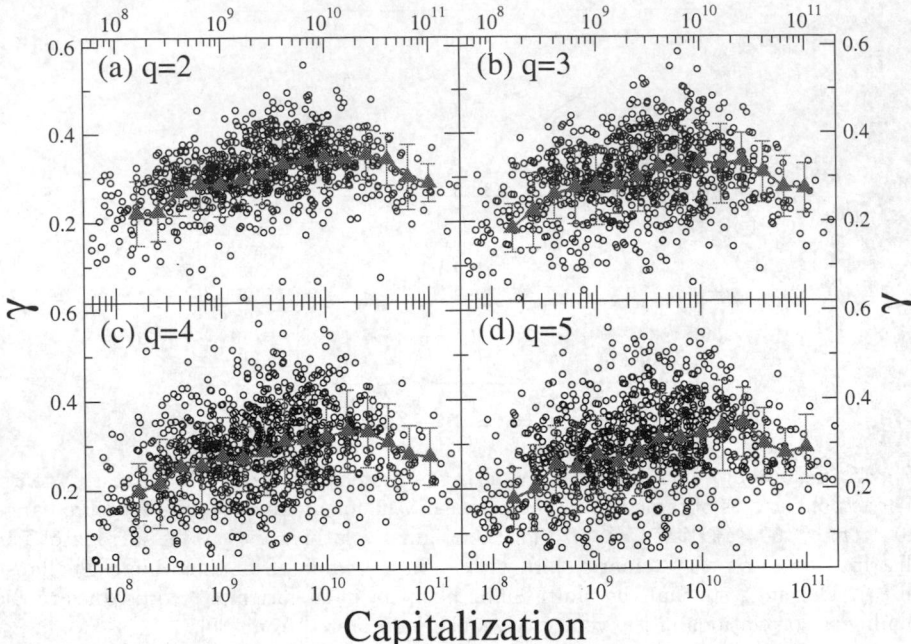

Fig. 8. (Color online) Size effect of correlation exponent γ. Scatter plot of four thresholds, $q = 2$ to 5 are displayed. To better view the tendency, we calculate the average and standard deviation in logarithmic bins of capitalization, as shown by the triangle (average) and error bar (standard deviation). For the four thresholds, average γ increases with the capitalization for most of the range.

number of trades. Similarly, they found a certain dependence on these factors for the multiscaling exponent δ [61].

3   Memory Effects in the Return Interval Sequence

The temporal structure is an essential feature to characterize a time series. It can be examined in different time scales. Here we analyze it in three scales, short, medium and long term.

3.1   Short-Term Memory

The short-term memory can be measured by the conditional PDF, $P(\tau|\tau_0)$, which is the probability of finding a return interval τ immediately after a return interval of size τ_0 [49,50,51,52,53,54]. In records without memory, $P(\tau|\tau_0)$ should be identical to $P(\tau)$ and independent of τ_0. When memory exists, it should depends on the choice of τ_0. Due to the poor statistics for a single value of return interval, a binning of τ_0 is needed. Yamasaki et al. split the entire database into 8 equal-size subsets, Q_1, Q_2, ..., Q_8, with intervals in increasing length [52,53,54].

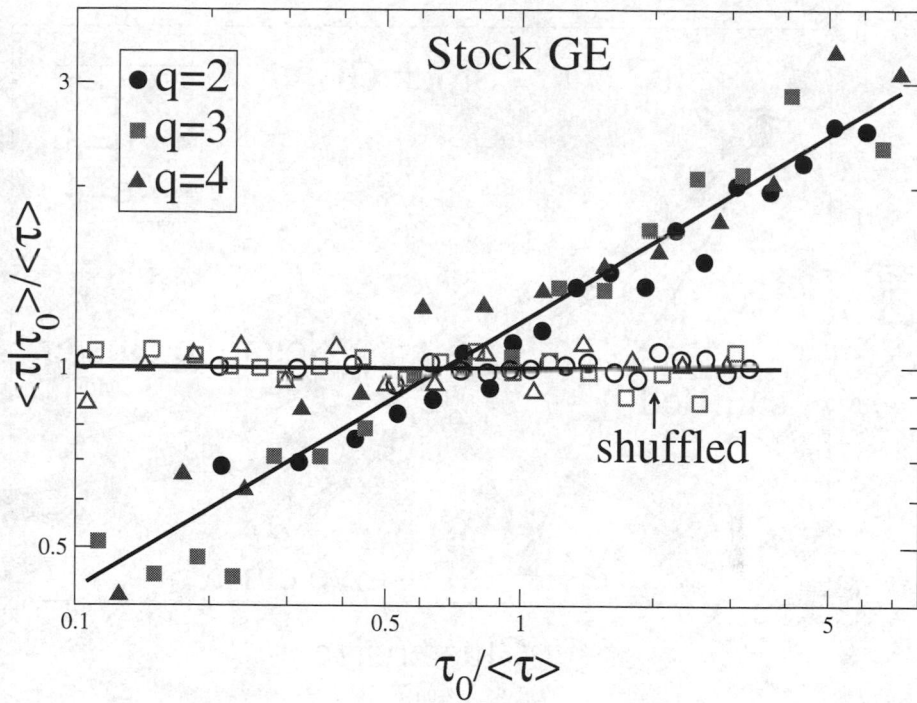

Fig. 9. (Color online) Mean conditional return interval $\langle\tau|\tau_0\rangle/\langle\tau\rangle$ vs $\tau_0/\langle\tau\rangle$ for the GE stock. Symbols are for three different thresholds $q = 2$, 3 and 4. To compare with the real data results (filled symbols), we also plot the corresponding results for shuffled records (open symbols). The distinct difference between the two records implies the memory effect in the original interval sequence. Adapted from [54].

It is found that for τ_0 in Q_1, the probability is higher for small τ, while for τ_0 in Q_8, the probability is higher for large τ. Thus, large (small) τ_0 tends to be followed by large (small) τ ("clustering"), which indicates memory in the sequence. Note that for all thresholds $P(\tau|\tau_0)$ seems to collapse onto a single scaling function for each of the τ_0 subsets, and they can be well fit by a SE function according to Eq. (3). These results are consistent for the US markets, currencies, interest rates and commodities [52,53,54]. Similar results have been found for the Japanese market [55] and Chinese market [56,57].

Further, the short-term memory is also seen clearly in the mean conditional return interval immediately after a given τ_0 subset, $\langle\tau|\tau_0\rangle$, which is the first moment of $P(\tau|\tau_0)$. A power law dependence of $\langle\tau|\tau_0\rangle$ on τ_0 for the GE stock is showed in Fig. 9, as an example. We can see that large (small) τ tend to follow large (small) τ_0, similar to the clustering in $P(\tau|\tau_0)$. Correspondingly, shuffled data (open symbols in Fig. 9) are almost constant as expected, demonstrating that the value of τ is independent of the previous interval τ_0.

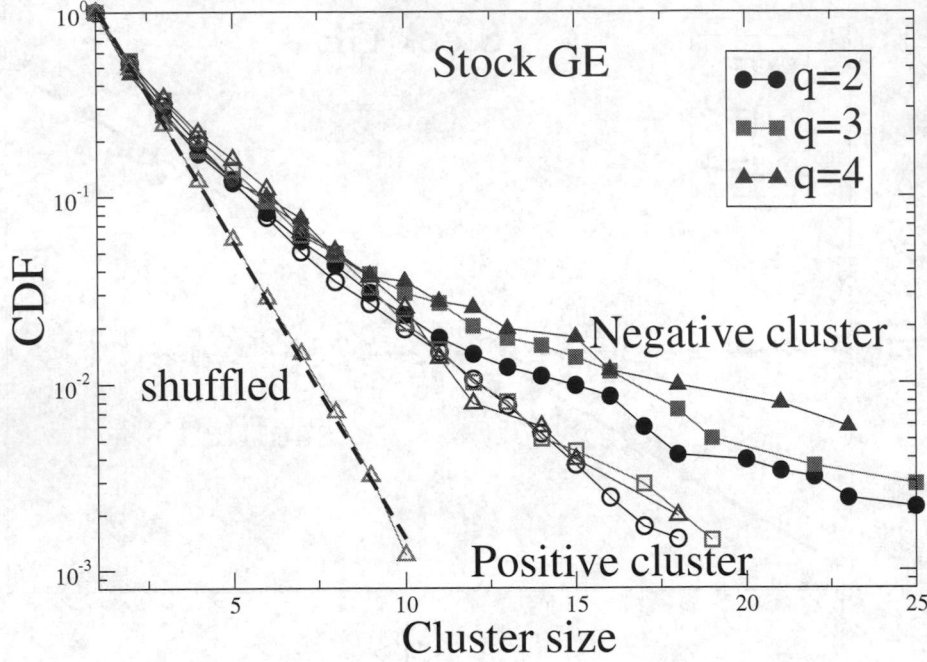

Fig. 10. (Color online) Cumulative distribution of size for return interval clusters. The cluster consists of consecutive return intervals that are all above ("positive cluster", open symbols) or below ("negative cluster", filled symbols) the median of return intervals. For the shuffled records, the distribution follows an exponential function. However, for the original records, their distributions for both positive and negative clusters have much longer tails, suggesting a significant memory in return intervals. Adapted from [54].

3.2   Clustering

Clustering phenomena are displayed by $P(\tau|\tau_0)$ and $\langle\tau|\tau_0\rangle$, indicating the memory in the return intervals. However, both functions measure the intervals that immediately follow an interval τ_0. In order to investigate longer clustering in a straighter way, we analyze "clusters" of return intervals, which are composed by successive intervals with similar size [53,54,55,56,58]. To obtain good statistics we divide the sequence of return intervals into two bins, separated by the median of the entire database. We denote intervals that are above the median by sign "+", and the ones below the median by "–". Accordingly, consecutive "+" or "–" intervals form a positive or negative cluster.

The distribution of cluster sizes n reveal the memory information in the sequence. Fig. 10 shows the cumulative distribution of the cluster size for the GE stock. Both positive and negative clusters have quite long tails, compared to that for the shuffled records which follows an exponential function and shows a much

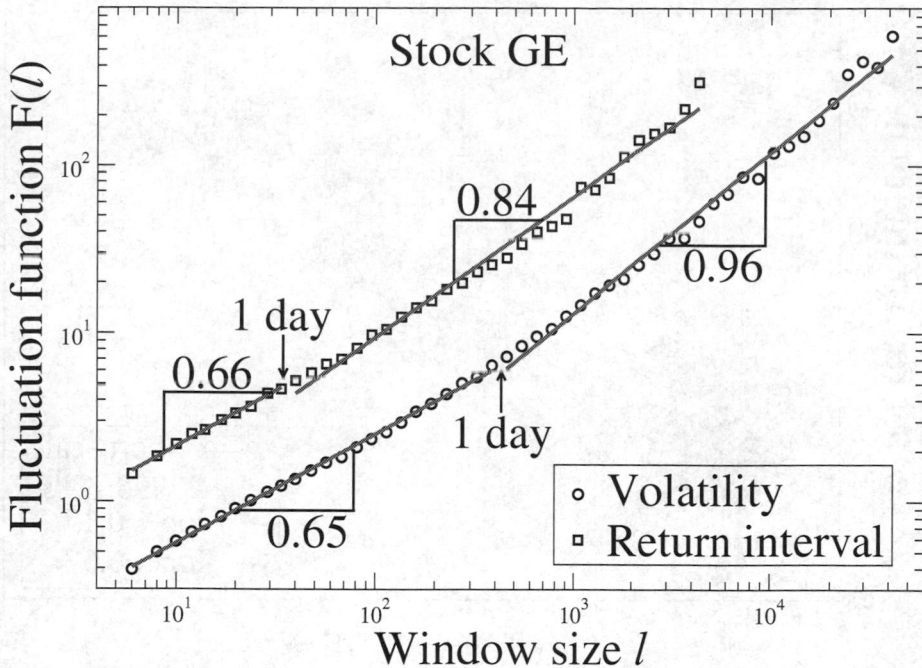

Fig. 11. (Color online) Detrended fluctuation analysis (DFA) on the volatility and return interval ($q = 2$) for the GE stock. Two curves are similar and their crossovers are around 1 trading day. Solid lines are for power law fits on the two regimes. For short scale, two α (slopes in the plot) are almost same. For long scale, two α are different but they are strongly related.

faster decay. For the positive clusters, the distribution still has good statistics even for size $n = 18$, while the negative clusters extend to $n = 25$. Thus, the memory effects persist for quite long times (e.g., the average return interval for GE with threshold $q = 2$ is about 9 minutes, so there are still some clusters corresponding to even 200 minutes in the time scale). Note that the distribution of positive clusters is very similar for different thresholds $q = 2, 3, 4$, while the negative clusters show the same effect only for $n \leq 10$. Similar clustering has been found also in earthquake and climate data [50,79].

3.3 Long-Term Correlations

The volatility is known to have long-term correlations [31], thus an examination of long-term correlations in the return interval is needed. We apply the Detrended Fluctuation Analysis (DFA) method [80,81,82] to the volatility and their return interval sequence. Without loss of generality, we investigate the return interval for a typical threshold $q = 2$. After removing trends, DFA computes the

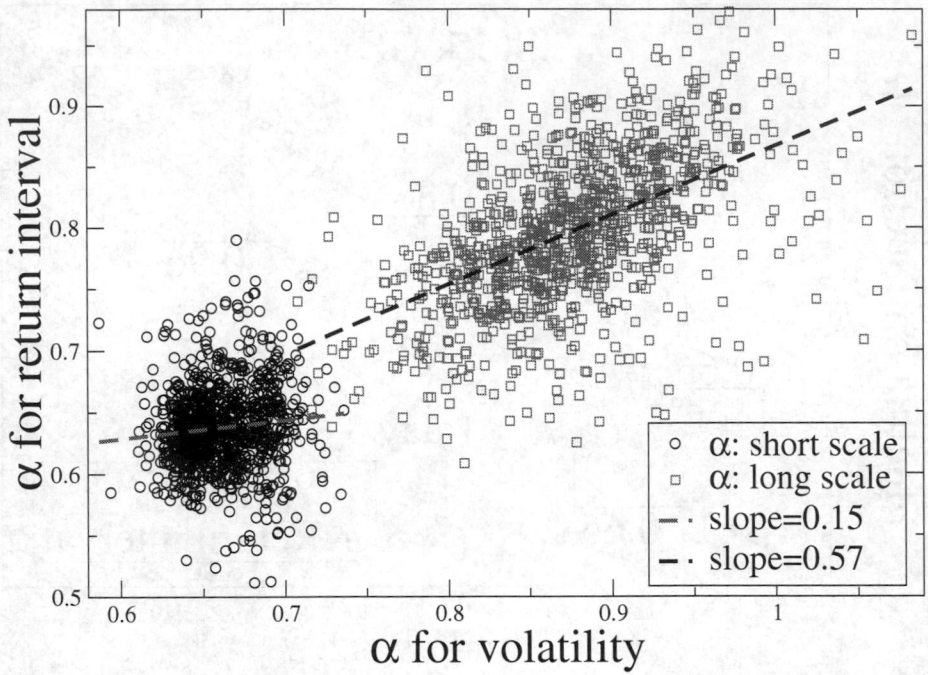

Fig. 12. (Color online) Dependence of long-term correlations in the volatility and the return interval ($q = 2$) sequence. The results for two scale regimes are showed. As indicated by the two linear fits on the symbols (dashed lines), the dependence is not strong for the short scale but it is significant for the long scale. The weak relation for short scale is related to the small range of their α.

root-mean-square fluctuation $F(\ell)$ of a time series within windows of ℓ points, and determines the exponent α from the scaling function,

$$F(\ell) \sim \ell^{\alpha}. \tag{11}$$

The correlation in the time series is characterized by the exponent $\alpha \in (0, 1)$. If $\alpha > 0.5$, the records has positive correlations If $\alpha = 0.5$, it has no correlation (white noise). If $\alpha < 0.5$, it has negative correlations.

Similar to the volatility [31], there is a crossover in the DFA curve for return interval thus the entire regime can be split into two sub-regimes $\ell < \ell^*$ and $\ell > \ell^*$ (ℓ^* is chose for that the corresponding time spanned is 390 minutes or 1 trading day) [31,53]. As an example, we show DFA curves for volatility and return interval ($q = 2$) of the GE stock in Fig. 11. We see that the corresponding values for α are distinctly different in the two regimes. However, both α are significantly larger than 0.5, suggesting long-term correlations in return intervals. In the short scale regime ($\ell < \ell^*$), we find $\alpha = 0.64 \pm 0.04$ for the return interval of the 1000 stocks, while $\alpha = 0.66 \pm 0.02$ for the volatility. The two cases are almost the same. In the long scale regime ($\ell > \ell^*$), we find $\alpha = 0.80 \pm 0.06$ for the return interval and $\alpha = 0.88 \pm 0.06$ for the volatility. and the discrepancy is slightly larger but in the range of the error bars. Here error bar refers to the standard deviation of the 1000 stocks. Such behavior suggests a common origin for the strong persistence of correlations in both volatility and return interval records, and in fact the clustering in return intervals is related to the known effect of volatility clustering [19]. To further examine the relation between two types of α, we draw the scatter plot for the dependence of two α in two regimes respectively, as shown in Fig. 12. We can see a significant dependence for α in the long scale. However, α for the short scale are crowded together so that there is no strong tendency.

4   Models

To further understand the financial fluctuations, Vodenska-Chitkushev et al. simulated models for the volatility series and tested the corresponding return intervals. Two popular long-term memory models, FIGARCH [83] and fractional Brownian motion (fBm) [84] are examined (see Ref [58] and references therein).

4.1   FIGARCH

Fractional integrated generalized autoregressive conditional heteroscedasticity (FIGARCH) [83] is a popular model for the return simulation. In this model the return r_t can be generated by the following process,

$$r_t = \mu + a(L) \cdot \epsilon_t. \tag{12}$$

Here μ is the mean value of return, L is the lag operator, $a(L)$ is the coefficient from the autoregressive moving average (ARMA) procedure, and ϵ_t is the disturbance term,

$$\epsilon_t \equiv z_t \cdot \sigma_t. \tag{13}$$

z_t is an $i.i.d.$ process with zero mean and unit variance, and the conditional variance σ_t^2 is determined by the following process,

$$\sigma_t^2 = \sigma^2 + \lambda(L) \cdot (\epsilon_t^2 - \sigma^2). \tag{14}$$

Here σ^2 is the unconditional variance of ϵ_t, and $\lambda(L)$ is from ARCH and GARCH coefficients which follows $\lambda(L) \sim (1-L)^d$. $d \in (0, 1)$ is the fractional differencing parameter and $\lambda(L)$ can be expanded into an infinite polynomial of L. FIGARCH process can captures the long-term dependence in volatilities, which is connected to the parameter d. When d increases, the long-term memory will gradually vanish.

After extracting parameters from the S&P 500 index data, we simulate returns from which volatilities are derived and analyze their return intervals properties [58]. First, we test the scaling of return intervals distribution, as shown in Fig. 13. There are significant deviations from the scaling for both small and large intervals. This result manifests that FIGARCH does not show good scaling in

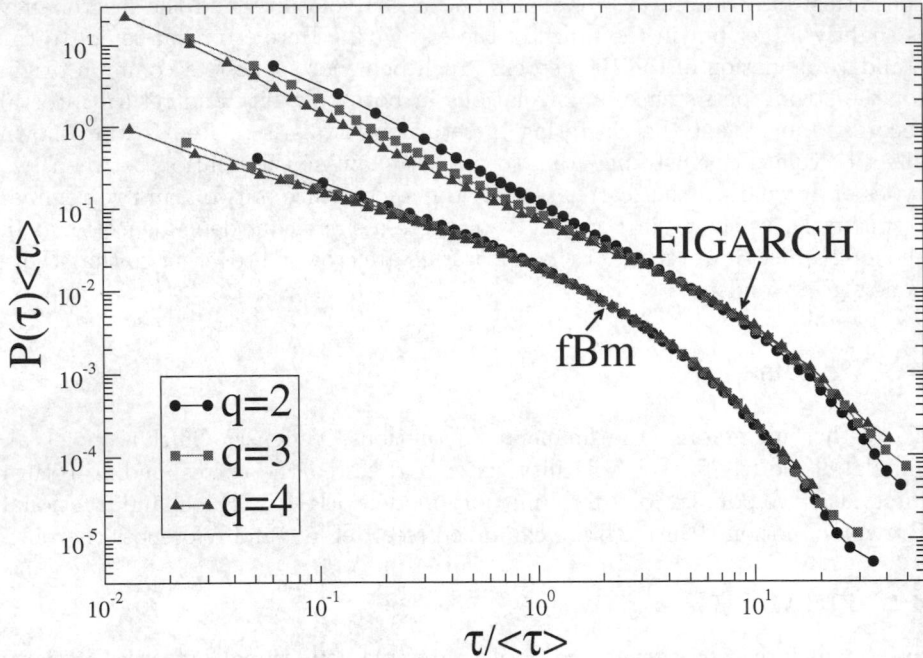

Fig. 13. (Color online) Scaling in the distribution of return intervals for two models, FIGARCH and fBm. Curves for fBm are vertically shifted down for better visibility. For FIGARCH, the scaled PDF, $P(\tau) \cdot \langle\tau\rangle$, does not collapse onto a single curve, especially for small and large scaled interval $\tau/\langle\tau\rangle$, which suggests no good scaling. For fBm, the three scaled PDFs collapse for most of range (the small deviations at very small or very large scaled intervals correspond to discreteness and finite size effects respectively). This indicates a good scaling in the distribution for the fBm model.

the return interval distribution. Further, we examine the cluster size distribution, which is demonstrated in Fig. 14. We can see that FIGARCH captures the memory effects for both positive and negative clusters. Their effects are slightly stronger than the empirical memory.

4.2   Fractional Brownian Motion

Fractional Brownian motion (fBm) [84] is a generalization of Brownian motion. The only difference from a regular Brownian motion is that the increments of fBm are correlated. The long-range dependence of the increments can be characterized by the Hurst parameter $H \in (0, 1)$, which is the only parameter to index a fBm process $B_H(t)$. Note that $B_H(t)$ reduces to a regular Brownian motion when $H = 1/2$, while $H > 1/2$ ($H < 1/2$) corresponds to positive (negative) correlation. An important feature of fBm is the scale invariance,

$$B_H(c \cdot t) = c^H \cdot B_H(t) \tag{15}$$

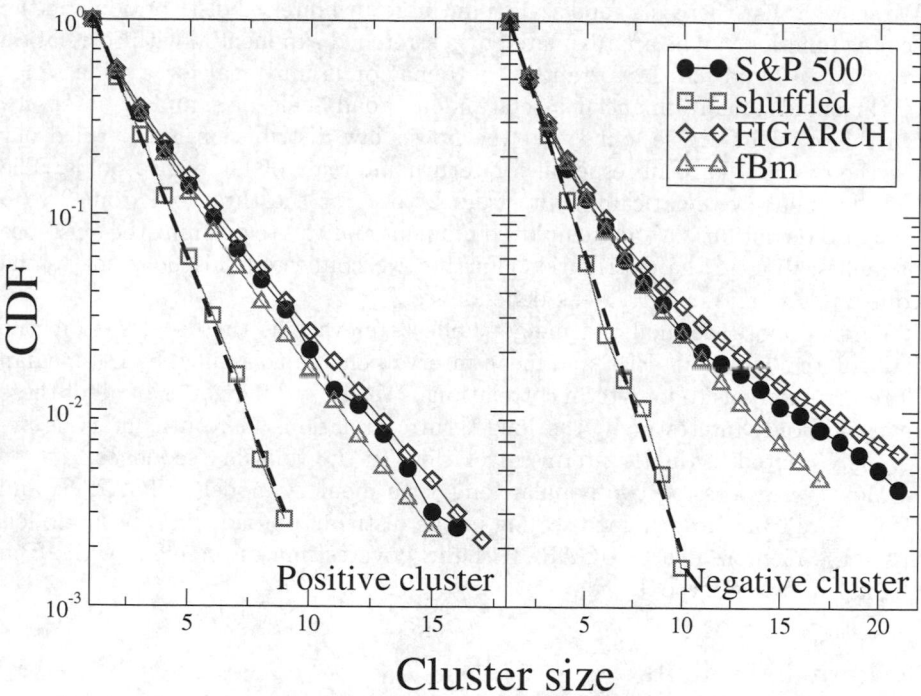

Fig. 14. (Color online) Cluster size distribution for the FIGARCH and fBm models. The output of two models are very close to that of S&P 500 index and significantly away for the shuffled data, which suggests that the memory in the empirical data can be repeated by both FIGARCH and fBm. The figure shows that FIGARCH slightly overestimates the memory while fBm slightly underestimates it.

for all $c > 0$. Since the return only has short-term correlations while the volatility has long-term correlations, we simulate the return by

$$r_t = e^{B_H(t+1) - B_H(t)} \cdot \eta_t \tag{16}$$

where η_t is an i.i.d. process with zero mean and unite variance.

We simulate return intervals with fBm process and calculated their PDF and distributions of cluster size [85]. PDF of return intervals is showed in Fig. 13, which has a well-approximated scaling. In Fig. 14, the cluster size distributions of fBm process is quite close to the empirical data. The two curves are only slightly smaller for both positive and negative cluster.

5   Conclusions

We analyzed the properties of the return intervals for the 1000 most traded stocks in the US markets, as well as reviewed recent studies on return interval analysis.

We showed that there is a good scaling in the return interval distribution and the scaling function can be approximated by a stretched exponential with correlation exponent around 0.4. Importantly, the behavior is universal for a wide range of thresholds, many financial assets and a broad scale of sampling intervals. On the other hand, we found that the power law distribution is not ruled out for the distribution tail, especial for return intervals of large thresholds. The tail exponent systematically shifts from 2 to 1 for the threshold from 2 to 5 standard deviations. We also employed moment analysis to examine the existence of multiscaling in the distribution. Further we connected this behavior to the company size and found a weak dependence.

Further more we analyzed memory effects in various time scales, from the immediate conditional PDF and mean interval, clusters classified by the median of return intervals to long-term correlations. We showed memories in all of these investigations. Interestingly, the long-term correlations in return intervals are strongly related to the long-term correlations in the volatility sequence.

Moreover, we tested two popular long-term memory models, FIGARCH and fBm. Only fBm shows a good scaling in the distribution. However, both models catch the memory effect. FIGARCH slightly overestimates the effect while fBm slightly underestimate it.

Acknowledgments

We thank S.-J. Shieh, X. Gabaix, P. Gopikrishnan, V. Plerou, B. Rosenow, J. Nagler, F. Pammolli and especially A. Bunde, L. Muchnik, P. Weber, W.-S. Jung and I. Vodenska-Chitkushev for collaboration on many aspects of this research, and the NSF and Merck Foundation for financial support.

References

1. Pareto, V.: Cours d'Economie Politique, Lausanne and Paris (1897)
2. Bachelier, L.: Théorie de la spéculation [Ph.D. thesis in mathematics]. Annales Scientifiques de l'Ecole Normale Supérieure III-17, 21 (1900); translated and Reprinted in, Cootner, P. (ed.). The Random Character of Stock Market Prices, p. 17. MIT Press, Cambridge (1967)
3. Lévy, P.: Théorie de l'Addition des Variables Aléatoires, Gauthier-Villars, Paris (1937)
4. Fama, E.F.: J. Business 36, 420 (1963)
5. Officer, R.R.: J. Amer. Statistical Assoc. 67, 807 (1972)
6. Clark, P.K.: Econometrica 41, 135 (1973)
7. Wood, R.A., McInish, T.H., Ord, J.K.: J. Finance 40, 723 (1985)
8. Harris, L.: J. Financ. Econ. 16, 99 (1986)
9. Admati, A., Pfleiderer, P.: Rev. Financ. Stud. 1, 3 (1988)
10. Schwert, G.W.: J. Finance 44, 1115 (1989); Chan, K., Chan, K.C., Karolyi, G.A.: Rev. Financ. Stud. 4, 657 (1991); Bollerslev, T., Chou, R.Y., Kroner, K.F.: J. Econometr. 52, 5 (1992); Gallant, A.R., Rossi, P.E., Tauchen, G.: Rev. Financ. Stud. 5, 199 (1992); Le Baron, B.: J. Business 65, 199 (1992)

11. Ding, Z., Granger, C.W.J., Engle, R.F.: J. Empirical Finance 1, 83 (1993)
12. Dacorogna, M.M., Muller, U.A., Nagler, R.J., Olsen, R.B., Pictet, O.V.: J. Int. Money Finance 12, 413 (1993)
13. Loretan, M., Phillips, P.C.B.: J. Empirical Finance 1, 211 (1994)
14. Pagan, A.: J. Empirical Finance 3, 15 (1996)
15. Mantegna, R.N., Stanley, H.E.: Nature 383, 587 (1996)
16. Cizeau, P., Liu, Y., Meyer, M., Peng, C.-K., Stanley, H.E.: Physica A 245, 441 (1997)
17. Cont, R.: Ph.D. thesis, Universite de Paris XI (1998) (unpublished); see also e-print cond-mat/9705075
18. Pasquini, M., Serva, M.: Econ. Lett. 65, 275 (1999)
19. Lux, T., Marchesi, M.: Int. J. Theor. Appl. Finance 3, 675 (2000); Giardina, I., Bouchaud, J.-P.: Physica A 299, 28 (2001); Lux, T., Ausloos, M · In: Bunde, A., Kropp, J., Schellnhuber, H.J. (eds.) The Science of Disasters: Climate Disruptions, Heart Attacks, and Market Crashes, p. 373. Springer, Berlin (2002)
20. Rosenow, B.: Int. J. Mod. Phys. C 13, 419 (2002)
21. Gabaix, X., Gopikrishnan, P., Plerou, V., Stanley, H.E.: Nature 423, 267 (2003); Gabaix, X., Gopikrishnan, P., Plerou, V., Stanley, H.E.: Quart. J. Econ. 121, 461 (2006)
22. Farmer, J.D., Gillemot, L., Lillo, F., Mike, S., Sen, A.: Quant. Finance 4, 383 (2004); Farmer, J.D., Lillo, F.: ibid 4, C8, (2004); Plerou, V., Gopikrishnan, P., Gabaix, X., Stanley, H.E.: ibid 4, C11 (2004)
23. Lillo, F., Farmer, J.D., Mantegna, R.N.: Nature 421, 129 (2003)
24. Weber, P., Wang, F., Vodenska-Chitkushev, I., Havlin, S., Stanley, H.E.: Phys. Rev. E 76, 016109 (2007)
25. Eisler, Z., Bartos, I., Kertész, J.: Adv. Phys. 57, 89 (2008)
26. Jung, W.-S., Kwon, O., Wang, F., Kaizoji, T., Moon, H.-T., Stanley, H.E.: Physica A 387, 537 (2008)
27. Lux, T.: Appl. Finan. Econ. 6, 463 (1996)
28. Gopikrishnan, P., Meyer, M., Amaral, L.A.N., Stanley, H.E.: Eur. Phys. J. B 3, 139 (1998)
29. Muller, U.A., Dacorogna, M.M., Pictet, O.V.: Heavy Tails in High-Frequency Financial Data. In: Adler, R.J., Feldman, R.E., Taqqu, M.S. (eds.) A Practical Guide to Heavy Tails, p. 83. Birkhäuser Publishers, Basel (1998)
30. Plerou, V., Gopikrishnan, P., Amaral, L.A.N., Meyer, M., Stanley, H.E.: Phys. Rev. E 60, 6519 (1999); Plerou, V., Gopikrishnan, P., Amaral, L.A.N., Gabaix, X., Stanley, H.E.: ibid 62, 3023 (2000); Gopikrishnan, P., Plerou, V., Gabaix, X., Stanley, H.E.: ibid 62, 4493 (2000)
31. Liu, Y., Gopikrishnan, P., Cizeau, P., Meyer, M., Peng, C.-K., Stanley, H.E.: Phys. Rev. E 60, 1390, (1999); Plerou, V., Gopikrishnan, P., Gabaix, X., Amaral, L.A.N., Stanley, H.E.: Quant. Finance 1, 262, (2001); Plerou, V., Gopikrishnan, P., Stanley, H.E.: Phys. Rev. E 71, 046131 (2005); for application to heartbeat intervals see, Ashkenazy, Y., Ivanov, P.C., Havlin, S., Peng, C.-K., Goldberger, A.L., Stanley, H.E.: Phys. Rev. Lett. 86, 1900 (2001)
32. Plerou, V., Stanley, H.E.: Phys. Rev. E 76, 046109 (2007)
33. Clauset, A., Shalizi, C.R., Newman, M.E.J.: http://arxiv.org/abs/0706.1062v1
34. Mantegna, R., Stanley, H.E.: Introduction to Econophysics: Correlations and Complexity in Finance. Cambridge Univ. Press, Cambridge (2000)
35. Mandelbrot, B.B.: J. Business 36, 394 (1963)
36. Bunde, A., Havlin, S. (eds.): Fractals in Science. Springer, Heidelberg (1994)

37. Mantegna, R.N., Stanley, H.E.: Nature 376, 46 (1995)
38. Stanley, H.E.: Introduction to Phase Transitions and Critical Phenomena. Oxford University Press, Oxford (1971)
39. Stanley, H.E.: Rev. Mod. Phys. 71, 358 (1999)
40. Peng, C.K., Buldyrev, S., Goldberger, A., Havlin, S., Sciortino, F., Simons, M., Stanley, H.E.: Nature 356, 168 (1992); Buldyrev, S.V., Goldberger, A.L., Havlin, S., Peng, C.-K., Stanley, H.E., Stanley, M.H.R., Simons, M.: Biophys. J. 65, 2673 (1993); Buldyrev, S.V., Goldberger, A.L., Havlin, S., Peng, C.-K., Simons, M., Stanley, H.E.: Phys. Rev. E 47, 4514 (1993); Mantegna, R.N., Buldyrev, S.V., Goldberger, A.L., Havlin, S., Peng, C.-K., Simons, M., Stanley, H.E.: Phys. Rev. E 52, 2939 (1995)
41. Suki, B., Barabási, A.-L., Hantos, Z., Peták, F., Stanley, H.E.: Nature 368, 615 (1994)
42. Peng, C.K., Mietus, J., Hausdorff, J., Havlin, S., Stanley, H.E., Goldberger, A.L.: Phys. Rev. Lett. 70, 1343 (1993)
43. Makse, H.A., Havlin, S., Stanley, H.E.: Nature 377, 608 (1995); Makse, H.A., Andrade, J.S., Batty, M., Havlin, S., Stanley, H.E.: Phys. Rev. E 58, 7054 (1998)
44. Plerou, V., Amaral, L.A.N., Gopikrishnan, P., Meyer, M., Stanley, H.E.: Nature 400, 433 (1999)
45. Keitt, T.H., Stanley, H.E.: Nature 393, 257 (1998); Keitt, T.H., Amaral, L.A.N., Buldyrev, S.V., Stanley, H.E.: Scaling in the Growth of Geographically Subdivided Populations: Scale-Invariant Patterns from a Continent-Wide Biological Survey. Focus issue: The Biosphere as a Complex Adaptive System, Phil. Trans. Royal Soc. B: Biological Sciences 357, 627 (2002)
46. Gutenberg, B., Richter, C.F.: Seismicity of the Earth and Associated Phenomenon, 2nd edn. Princeton University Press, Princeton (1954)
47. Turcotte, D.L.: Fractals and Chaos in Geology and Geophysics. Cambridge University Press, Cambridge (1992)
48. Bunde, A., Eichner, J.F., Havlin, S., Kantelhardt, J.W.: Physica A, 342, 308 (2004)
49. Bunde, A., Eichner, J.F., Kantelhardt, J.W., Havlin, S.: Phys. Rev. Lett. 94, 048701 (2005)
50. Livina, V.N., Havlin, S., Bunde, A.: Phys. Rev. Lett. 95, 208501 (2005)
51. Eichner, J.F., Kantelhardt, J.W., Bunde, A., Havlin, S.: Phys. Rev. E 75, 011128 (2007)
52. Yamasaki, K., Muchnik, L., Havlin, S., Bunde, A., Stanley, H.E.: Proc. Natl. Acad. Sci. U.S.A. 102, 9424, (2005); Yamasaki, K., Muchnik, L., Havlin, S., Bunde, A., Stanley, H.E.: In: Takayasu, H. (ed.) Proceedings of the Third Nikkei Econophysics Research Workshop and Symposium, The Fruits of Econophysics, Tokyo, p. 43. Springer, Berlin (2005)
53. Wang, F., Yamasaki, K., Havlin, S., Stanley, H.E.: Phys. Rev. E 73, 026117 (2006)
54. Wang, F., Weber, P., Yamasaki, K., Havlin, S., Stanley, H.E.: Eur. Phys. J. B 55, 123 (2007)
55. Jung, W.-S., Wang, F.Z., Havlin, S., Kaizoji, T., Moon, H.-T., Stanley, H.E.: Eur. Phys. J. B 62, 113 (2008)
56. Qiu, T., Guo, L., Chen, G.: Physica A 387, 6812 (2008)
57. Ren, F., Guo, L., Zhou, W.-X.: http://arxiv.org/abs/0807.1818v1
58. Vodenska-Chitkushev, I., Wang, F.Z., Weber, P., Yamasaki, K., Havlin, S., Stanley, H.E.: Eur. Phys. J. B 61, 217 (2008)
59. Bogachev, M.I., Eichner, J.F., Bunde, A.: Phys. Rev. Lett. 99, 240601 (2007)
60. Wang, F., Yamasaki, K., Havlin, S., Stanley, H.E.: Phys. Rev. E 77, 016109 (2008)

61. Wang, F., Yamasaki, K., Havlin, S., Stanley, H.E.:
    http://arxiv.org/abs/0808.3200v1
62. Ren, F., Zhou, W.-X.: http://arxiv.org/abs/0809.0250v1
63. Bogachev, M.I., Bunde, A.: Phys. Rev. E 78, 036114 (2008)
64. Black, F., Scholes, M.: J. Polit. Econ. 81, 637 (1973)
65. Cox, J.C., Ross, S.A.: J. Financ. Econ. 3, 145 (1976); Cox, J.C., Ross, S.A., Ru-
    binstein, M.: J. Financ. Econ. 7, 229 (1979)
66. Bouchaud, J.-P., Potters, M.: Theory of Financial Risk and Derivative Pricing:
    From Statistical Physics to Risk Management. Cambridge Univ. Press, Cambridge
    (2003)
67. Johnson, N.F., Jefferies, P., Hui, P.M.: Financial Market Complexity. Oxford Univ.
    Press, New York (2003)
68. Altmann, E.G., Kantz, H.: Phys. Rev. E 71, 056106 (2005)
69. Stephens, M.A.: J. Am. Stat. Assoc. 69, 730 (1974)
70. Engle, R., Russel, J.: Econometrica 66, 1127 (1998)
71. To avoid the discreteness for small τ (Ref [51] suggested a power law function
    for this range) and large fluctuations for very large τ, we choose the range of
    $0.01 \leq CDF \leq 0.50$ to perform the stretched exponential fit
72. It sounds that the ratio between two a significantly deviates from 1. However, it
    is not that huge if we test the sensitivity on γ for Eq. (4). For instance, when γ
    changes from 0.30 to 0.31, the value a from Eq. (4) increases more than 30% with
    a constant $\langle \tau \rangle$
73. Hill, B.M.: Ann. Stat. 3, 1163 (1975)
74. Di Matteo, T.: Quant. Finan. 7, 21 (2007)
75. Ivanov, P.C., Yuen, A., Podobnik, B., Lee, Y.: Phys. Rev. E 69, 056107 (2004)
76. Eisler, Z., Kertész, J.: Phys. Rev. E 73, 046109 (2006); Eisler, Z., Kertész, J.: Eur.
    Phys. J. B 51, 145 (2006)
77. Schreiber, T., Schmitz, A.: Phys. Rev. Lett. 77, 635 (1996); Schreiber, T., Schmitz,
    A.: Physica D 142, 346 (2000)
78. Makse, H.A., Havlin, S., Schwartz, M., Stanley, H.E.: Phys. Rev. E 53, 5445 (1996)
79. Eichner, J.F., Kantelhardt, J.W., Bunde, A., Havlin, S.: Phys. Rev. E 73, 016130
    (2006)
80. Peng, C.-K., Buldyrev, S.V., Havlin, S., Simons, M., Stanley, H.E., Goldberger,
    A.L.: Phys. Rev. E 49, 1685 (1994); Peng, C.-K., Havlin, S., Stanley, H.E., Gold-
    berger, A.L.: Chaos 5, 82 (1995)
81. Hu, K., Ivanov, P.C., Chen, Z., Carpena, P., Stanley, H.E.: Phys. Rev. E 64, 011114
    (2001); Chen, Z., Ivanov, P.C., Hu, K., Stanley, H.E.: ibid 65, 041107 (2002); Xu,
    L., Ivanov, P.C., Hu, K., Chen, Z., Carbone, A., Stanley, H.E.: ibid 71, 051101
    (2005); Chen, Z., Hu, K., Carpena, P., Bernaola-Galvan, P., Stanley, H.E., Ivanov,
    P.C.: ibid 71, 011104 (2005); Kantelhardt, J.W., Zschiegner, S., Koscielny-Bunde,
    E., Havlin, S., Bunde, A., Stanley, H.E.: Physica A 316, 87 (2002)
82. Bunde, A., Havlin, S., Kantelhardt, J.W., Penzel, T., Peter, J.-H., Voigt, K.: Phys.
    Rev. Lett. 85, 3736 (2000)
83. Baillie, R.T., Bollerslev, T., Mikkelsen, H.O.: J. Econometrics 74, 3 (1996)
84. Mandelbrot, B.B., Van Ness, J.W.: SIAM rev. 10, 442 (1968)
85. To simplify the simulation and without loss of generality, we neglect the crossover
    in the DFA curve and obtain H = 0.86 for the S&P 500 index

Organizational Adaptative Behavior: The Complex Perspective of Individuals-Tasks Interaction

Jiang Wu[1], Duoyong Sun[2], Bin Hu[1], and Yu Zhang[3]

[1] School of Management, Huazhong University of Science and Technology
1037, Luoyu Road, Wuhan, China, 430074
jiangwu.john@gmail.com, bin_hu@mail.hust.edu.cn
[2] College of Information System and Management, National University of Defense
Technology, Changsha, Hunan, China, 410073
duoyongsun@gmail.com
[3] Computer Science Department, Trinity University
One Trinity Place San Antonio, TX, USA, 78212-7200
yzhang@cs.trinity.edu

Abstract. Organizations with different organizational structures have different organizational behaviors when responding environmental changes. In this paper, we use a computational model to examine organizational adaptation on four dimensions: Agility, Robustness, Resilience, and Survivability. We analyze the dynamics of organizational adaptation by a simulation study from a complex perspective of the interaction between tasks and individuals in a sales enterprise. The simulation studies in different scenarios show that more flexible communication between employees and less hierarchy level with the suitable centralization can improve organizational adaptation.

Keywords: Organizational Adaptation, Simulation.

1 Introduction

Nowadays, rapid technological changes, extensive globalization, and intense competition have created significant pressures on organizations. To build an agile organization, one of the most essential issues is how to build an organization that can respond rapidly to the changing business environment. Organizational researchers have long recognized the value of formal models—mathematical, logical, computational—for examining organizational behavior in general and organizational adaptation in particular [2-7]. However, it is still challenging to identify the organizational adaptation and examine the suitable organizational structure that adapts to the environment change.

Research into organizational adaptation is a complex issue because the relationships are intricate, multiple participants are involved during interacting over time and interactions are nonlinear. To study the nonlinear phenomena, there are difficulties to uncover by the classical inductive case methods and to examine them by standard statistical techniques. Computer-based simulation can be used for theory development and hypothesis generation [10]. Simple, but non-linear processes often underlie the team and group behavior. Computational analysis enables the theorist to think through

J. Zhou (Ed.): Complex 2009, Part I, LNICST 4, pp. 28–38, 2009.

the possible ramifications of such non-linear processes and to develop a series of consistent predictions. Simulations are also particularly valuable when we seek to explain longitudinal phenomena that are challenging to study using empirical methods because of their time and data demands [4].

Particularly, agent-based simulations can give insights into the "emergence" of macro level phenomena from micro level actions and enable examination of how sensitively the simple rules affect the final results using sensitivity analysis[11].OrgAhead [12] is one of the first computational models used to study organization adaptation, and it focuses on organizational learning designed to test different forms of organizations under a common task representation. Subsequently, Epstein built an organizational adaptation model in which individual agents endogenously generate internal organizational structure to adapt optimally to dynamic environment [13].

In this paper, we propose a new computational model of organizational adaptation that expands Epstein's model so as to be suitable to explore the different dimensions of adaptation—Agility, Robustness, Resilience and Survivability. In this paper, an organization, termed adaptive, can generate new strategies and/or reconfigure its structure to potentially achieve even higher performance [3]. However, a robust organization is able to sustain high levels of performance in dynamic environments without having to change its structures. Robustness involves the ability of the system to survive variations in structural/internal parameters without disrupting its behavior. This would relate to loss of nodes in the company, cost of loss and cost/delay in replacing the nodes. This does have an effect on the agility as a secondary consideration. In this paper, we use an agent-based computational model to address the issue of organizational adaptation. We consider the hierarchy, span of control and communication culture among employees as the representation of organizational structure. The internal environment changes with the hiring or firing employees, and the external environment changes with increasing or decreasing sales opportunities. We assume the enterprise in this paper is a sales company whose major tasks are to recognize and grasp sales opportunities. Thus, the central issue is to build a simulation model on the interaction between individuals and tasks so as to regard it as a test-bed to examine the "fast", "stable", "large" and "long" dimensions of organizational adaptation. Also, we use cumulative bank balance related to revenue, cost and profit as the main measurement to observe the dynamics of organizational adaptation. Our objective is to find insights about organizational adaptation from the simulation study on interaction between organization and task.

The paper is structured as follows. In Section 2, we propose a new computational model that is used to examine the different dimensions of organizational adaptation. Section 3 specifies the design of our virtual experiments. In Section 4, through a series of virtual experiments (simulations), we observe the organizational agility, robustness, resilience and survivability. Section 5 concludes this paper.

2   The Computational Model

Modeling Agility, Robustness, Resilience and Survivability of Organizational Adaptation System (MARRS) is fundamentally expanded from Epstein's organizational adaptation model[13]; But our computational model (MARRS) is designed especially for examining four dimensions of organizational adaptation. MARRS is an agent-based simulation model and developed using Netlogo[1]. MARRS provides a design/analysis toolkit in which the policies (or traditions) governing the generation of structure and allocation of resources within a self adaptive self organizing multi-agent system of systems can be explored, modeled and/or designed. The tool provides to the designer/analyst the ability to both models how well a particular organization's management policy will fare in one or more alternative environmental dynamics contexts.

As shown in Fig. 1, in the output panel, the red dots represent sales opportunities that move from left to right and can be controlled by the "Opportunity" parameter set. When the organization faces sales opportunities, the managers in organization assign workers to recognize and grasp sales opportunities that are represented by white dots. The opportunities that cannot be intercepted are represented by yellow dots. Enterprise organization is represented by squares in Fig. 1, and it is a structure of hierarchy. The workers are located in the zero level in the hierarchy and in charge of recognizing sales opportunities; the managers are located in the 1~5 levels, and assign tasks to workers and manage task schedules; the managers on the higher level will lead the managers on the lower level. The blue links between employees in Fig. 1 represent the span of control for each manager, and the managers on the higher level have larger span of control. The managers on the first level have span of control $2^1=2$, the managers on the second level have span of control $2^2=4$ and so on. When the organization responds to the change of environment, managers can communicate with the managers on the same level, and exchange workers to schedule tasks; or turn to the managers on the higher level who will be in charge of the tasks assignment. Like other agile enterprises, this organization's structure will adapt based on hierarchy level, span of control and communication mechanisms.

To respond to sales opportunities, workers cannot move to other locations by themselves and must be led by the direct managers on the higher level. Managers can take measures according to the management thresholds from T_{min} to T_{max}. The following four mechanisms are employed to adapt to the changes of the business environment:

(1) **Task Scheduling Mechanism:** managers can manage the workers in his own span of control. They can place the workers who are free at time t into "Free List" and the workers who will meet sales opportunities at time $t+1$ into "Anticipation List". Managers will randomly assign the workers in "Free List" into "Anticipation List" to adapt to the change of environment at time $t+1$.

(2) **One-level Communication Mechanism:** in the range of their memory length, managers can calculate the lost sales opportunities percentage P_i in their span of control: $P_i=lost\_opportunities / memory\_length$. If $P_i>T_{max}$, the manager i demands workers to recognize sales opportunities: $D_i=(P_i-T_{max})\times(1-demand\_inertia$); if $P_j<T_{min}$, the manager j supplies workers to managers who demand workers: $S_j=(T_{min}-P_j)\times(1-supply\_inertia)$. The managers who have demands communicate with other managers on the same level to obtain the supplied workers.

(3) **Upward Communication Mechanism:** if manager i has exhausted all the supplies from the managers on the same level, and the manager i still holds the status of

---

[1] Epstein's model is based on Ascape(http://ascape.sourceforge.net/), but our new model for different dimensions of organizational adaptation is based on Netlogo(http://ccl.northwestern.edu/netlogo/) . They are different agent-based platforms.

$P_i > T_{max}$ after "*upward_inertia*" time period, the manager i will turn to the direct manager on the higher level. The maximum upward level is controlled by "Max Level" parameter setting.

(4) **Downward Communication Mechanism:** if manager j still holds the status of $P_j < T_{min}$ after "*downward_inertia*" time period, the manager j will authorize the managers on the lower level to take in charge of task scheduling between workers in the range of his span of control.

The bank accounting balance is used to generate the measurements that examine the four dimensions of organizational adaptation. As shown in the output graph of Fig. 1, "Salary" and "Accounting" parameter settings can control budget and cost. The cost includes salary and transaction cost. The salary of busy employees and salary of free employees are different. Managers who actively manage workers at time t obtain more salary than managers who are free at time t. The busy managers are paid a bonus in addition to the basic salary.

The parameters of this model include four categories: opportunity generation parameters, worker and manager parameters, coordination parameters and accounting parameters. The category of coordination parameters is very important because these define the organizational structure. Demand Inertia is the willingness to ask for help from your trading partner workers and ranges from 0 (never willing) to 1 (always willing). Supply Inertia is the willingness to provide help to your trading partner workers and ranges from 0 (never willing) to 1 (always willing). Max-Hierarchy determines how many levels of the hierarchy on maximum and also includes the different "span of control" on the different level.

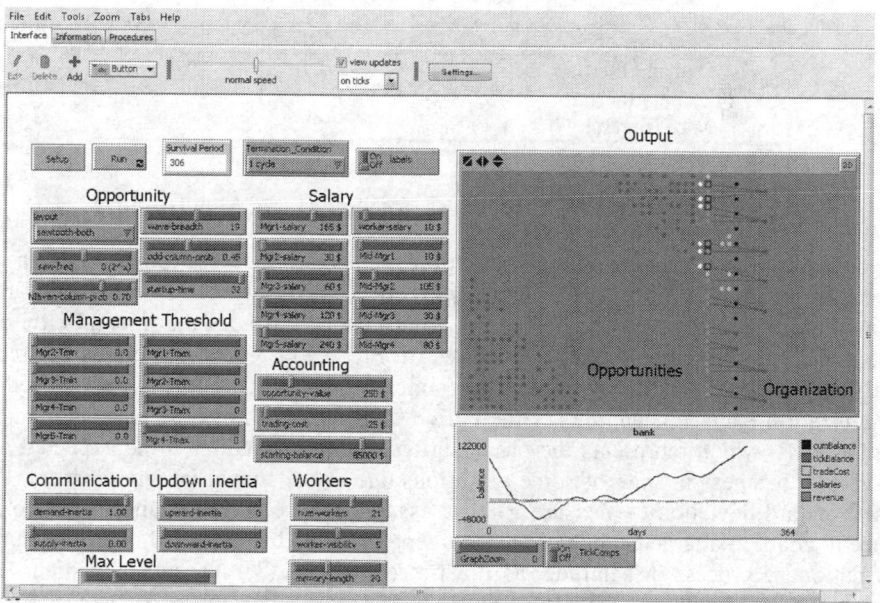

Fig. 1. MARRS—Organizational Adaptation Simulation System

3 Simulations Design

The simulation model can be regarded as the test-bed to alter parameters according to our requirements and to observe the resulting behaviors. From the findings of virtual experiments, potentially useful implications for practitioners can be revealed. As shown in Table 1, we design a series of virtual experiments in order to examine the Agility, Robustness, Resilience and Survivability of organizational adaptation. The different combinations of Demand Inertia and Supply Inertia represent different communication cultures among workers. Specially, we focus on the three different conditions of communication: free trade, medium trade and no trade. In the condition of free trade (Demand Inertia=1, Supply Inertia=1), workers can communicate with each other very freely and do not have inertia between their interactions. In the condition of medium trade (Demand Inertia=0.5, Supply Inertia=0.5), workers faces some pressures to supply or demand help from other workers, and they should keep balance in the communication. In the condition of no trade (Demand Inertia=0, Supply Inertia=0), the culture is very dull, people don't communicate with each other, there is no chance to exchange the labors, and workers have the highest priority of finishing their own tasks (standing always in the same department and finishing the assignments from his/her manager). There are 3×3×5 = 45 (see Table 1) conditions for the virtual experiments; we ran every virtual experiment independently for 100 runs by Monte-Carlo simulation procedure (Carley 1995). During simulations, we keep other parameters at their default values, as described in.

Table 1. Simulations design of organizational structure change for adaptation

Parameters	Range
Demand Inertia	0, 0.5, 1
Supply Inertia	0, 0.5, 1
Max-Hierarchy	1, 2, 3, 4, 5

3.1 Measurements of Agility, Robustness, Resilience and Survivability

In the evolutionary process during simulation, we need to measure four dimensions of organizational adaptation. Agility is measured by velocity, which represents how fast the organization adapts to changes in the environment (includes the internal environment and external environment). As shown in Fig. 2 (1), y axis represents the bank balance, and x axis is the tick time during simulation. Velocity is the average speed of organization balance changing for the period T. Fig. 2(2) shows the measurement of Robustness, which represents the organization's ability to maintain the expected objective in business in spite of some vibrations due to the change of environment. We use standard deviation to measure Robustness. Resilience is the organization's tolerance to adapt to a large change of environment and keep the expected objective in the evolutionary process. Magnitude, as illustrated in Fig. 2(3), measures Resilience of organizational adaptation. This magnitude is the maximum number of workers who were fared when organization can still obtain positive profit, or the maximum opportunities were reduced when organization can still accumulate positive profit. We have

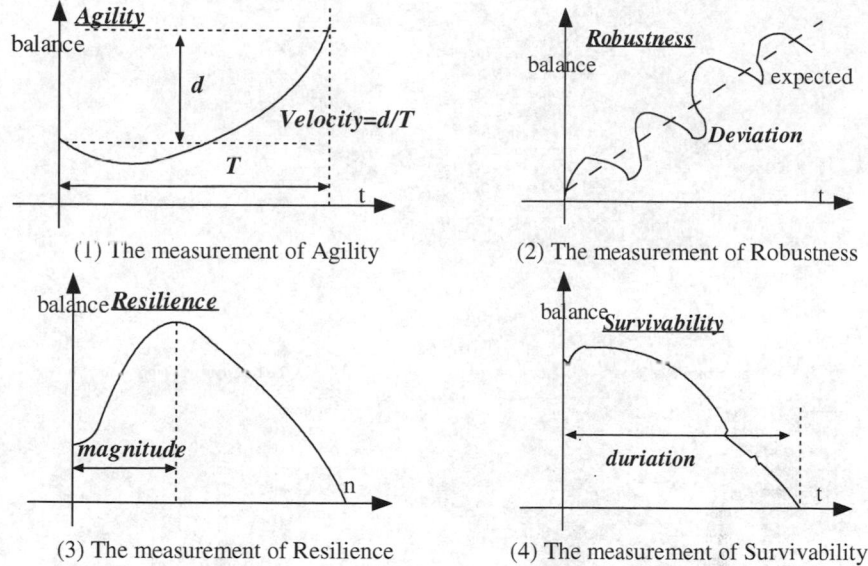

(1) The measurement of Agility

(2) The measurement of Robustness

(3) The measurement of Resilience

(4) The measurement of Survivability

Fig. 2. The measurements of different dimensions of organizational adaptation

assumed that organization will bankrupt if its bank balance is negative. Thus, as shown in Fig. 2(4), we use duration to measure the survivability. In practice, some companies have a wonderful visionary approach which enables them to adapt remarkably well, continue to survive and possibly never die [14].

4 Simulation Studies

We first run the simulations according to Table 1 under the different dimensions of organizational adaptation. Through simulations, we examine the genome of organizational structure (the value combination of max-hierarchy, demand-inertia and supply-inertia) of Robustness, Agility, Survivability and Resilience of the organization. Because every condition is needed to run independently for 100 repetitions, the results presented in the following are the average cumulative balance of all the repetitions.

4.1 The Study of Robustness

To examine the Robustness of organizational adaptation, we design a scenario that the wave-breadth of opportunities can change from 16 to 8 for one time cycle of 100 ticks, which represent the external environment change periodically. This scenario describes situations in which the sales environment changes according to the needs of the market, which usually changes periodically due to the ebb and flow of supply-and-demand. We keep the number of workers a constant value 16 and examine the robustness under different organizational structures (the combination of hierarchy,

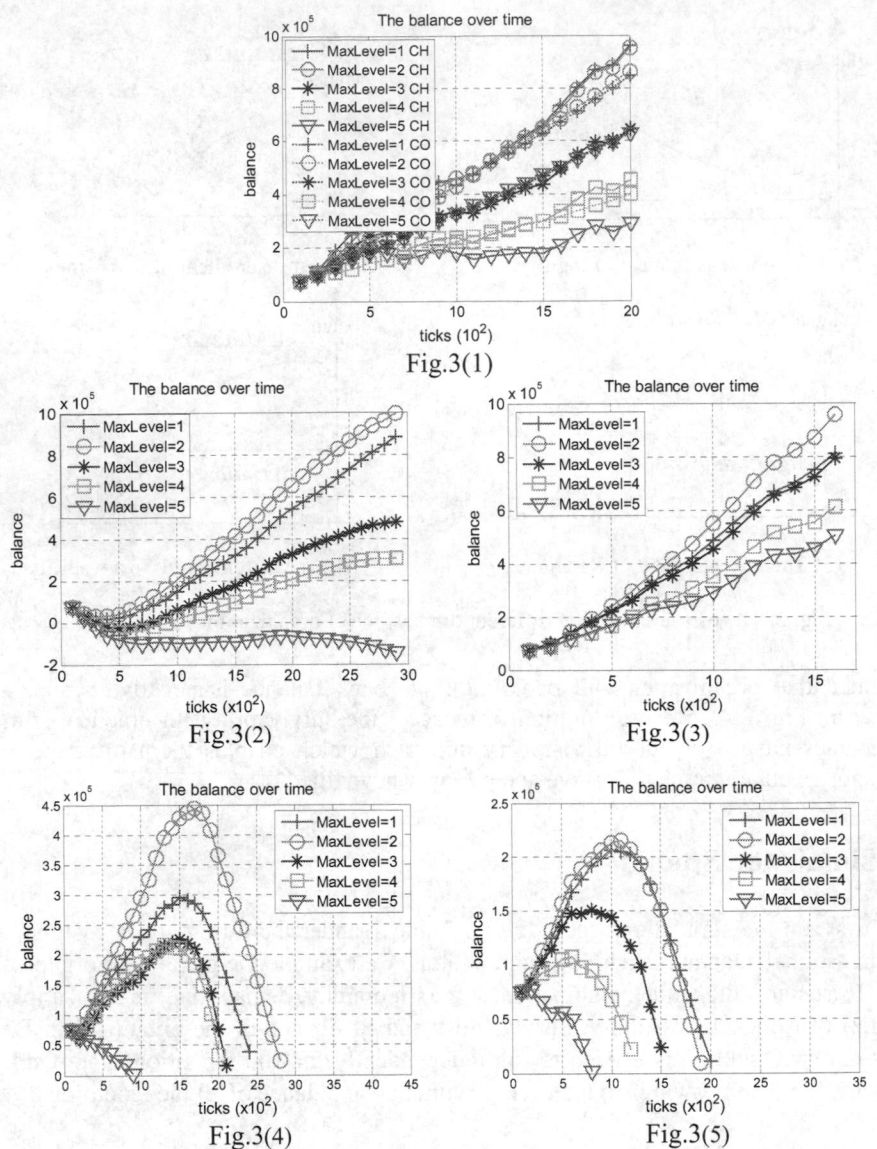

Fig. 3. Results of different dimensions of organizational adaptation in medium trade

demand inertia and supply inertia). As shown in Fig. 3(1), the curves with the legend "CH" represent the scenarios of opportunities change, and these scenarios compare to the other category of scenarios with the legend "CO" (the wave-breadth is a constant 16 for all the time). The Monte Carlo simulation results of the "CO" scenario are expected output and its deviations between output results of "CH" scenarios are organizational Robustness. The smaller deviation represents the higher Robustness.

In practice, more communication and more hierarchy means the organization is more robust. This conclusion has real world implications: to appropriately respond to the change of environment, timely communication can improve the organizational ability of the risk-averse. The communication among workers needs to be spontaneous, but meanwhile this communication must be limited to achieving the unitive purpose generated by the manager in the hierarchy, and if not, this communication is may be counterproductive. Also, greater hierarchy centralization produces a larger span of control and the command transferred to workers is more authoritative and direct. In this way, the organization can avoid the information distortion that occurs during the process of transferring command across a broader, more decentralized hierarchy. In conclusion, the centralization (high hierarchy and large span of control) and flexible communication mechanism (less demand-supply inertia) are significant to improve the organizational Robustness.

4.2 The Study of Agility

To examine the Agility dimension of Organizational Adaptation, we design two types of scenarios to represent the change of the internal environment and the external environment respectively. *(1) Number of Workers* increase from 3 to 32 every 100 ticks (hire new employee gradually) and set the wave-breath=16 as a constant value when running the business. Other parameters are set according to default values. *(2) Wave-Breadth* increase from 16 to 32 every 100 ticks (the opportunities increase gradually) and set number of workers = 12 as a constant value when running the business. Other parameters are set according to default values. The Agility is the average velocity that is equal to the final cumulativeBalance divided by the ticks period, we examine the impact of hierarchy level and communication (trade) on the Agility. The results are the average values of numerous independent Monte Carlo simulations. The dynamics of the output result are shown in Fig. 3(2) and Fig. 3(3).

As shown in Table 2, by analysis of SPSS, we obtain results: under the change of the internal environment, demand-inertia, supply-inertia and the level of hierarchy affect the Agility negatively; under the change of the external environment, demand-inertia and supply-inertia affect the Agility negatively, but the maximum level of hierarchy affects the Agility positively.

Table 2. The correlation analysis result by SPSS

Measurements	demand-inertia	supply-inertia	max-level
Agility(In)	-.478**	-.458**	-.298*
Agility(Out)	.074	-.465**	-.499**
Robustness	.065	-.347*	.354*
Survivability(In)	-.503**	492**	-.134**
Survivability(Out)	-.484**	-.434**	-.254**
Resilience(In)	-.507**	-.336*	-.086
Resilience(Out)	-.476**	-.458**	-.318*

* Correlation is significant at the 0.05 level.
** Correlation is significant at the 0.01 level.

In practice, in order to accommodate the quick change of internal and external environment, the enterprise needs to maintain a kind of flat organizational structure. More flexibility is needed by the managers on the low level, because these managers don't have a large span of control and they can guide the workers more carefully, and meanwhile the flat organizational structure can avoid the unnecessary transaction cost when communicating between the members in the hierarchy. However, we must notice another point that more hierarchy can keep a more stable business running which prevents bankruptcy in turbulent business environments. When the external environment is rapidly changing, enterprises need additional hierarchy to address attrition and more emphasis needs to be placed on communication. Demand-supply inertia should be minimized, and with the free trade atmosphere information will flow smoothly to improve the quality of responding to environmental demands. In conclusion, flat organizational structure (lower hierarchy and smaller span of control) and flexible communication mechanisms (less demand-supply inertia) are significant to improve the organizational Agility.

4.3 The Study of Survivability

Assessment of survivability of the organization was based on financial viability which is measured by the bank balance over a time period of ticks. If the bank balance fell at or below zero then the organization did not survive, as it was out of business. If the organization maintained a positive bank balance throughout the timeframe then it was designated robust for this study. The period duration is used to measure survivability. To examine, we designed two scenarios: (1) *Number of Workers* decreasing from 20 to 2 every 100 ticks (fire the employees gradually), which represents the change of internal environment. We keep external opportunities environment unchanged (wave-breadth = 12); (2) *Wave-Breadth* decreasing from 16 to 2 every 100 ticks (the opportunities decrease gradually), which represents the change of the external environment. We keep the internal human resource environment unchanged (Number of Workers=8). The dynamics of cumulative bank balance for the above two scenarios are shown in Fig. 3(4) and Fig. 3(5) respectively.

The analysis result by SPSS in Table 2 shows that maximum level of hierarchy and supply, demand inertia affect survivability positively no matter what kind of the change of environment. Thus, the organization with more communication among workers and less hierarchy (smaller span of control) has stronger survivability. It is obvious that flexible communication and flat organizational structure allows organizations to survive longer.

4.4 The Study of Resilience

To examine the Resilience of organizational adaptation, we also design two scenarios that are the same as previous section. The first scenario (see Fig. 3(4)) is used to examine the Workers Resilience to internal environmental change: How many persons can leave and the organization still be profitable and survive? The second scenario (see Fig. 3(5)) is used to examine Opportunities Resilience to external environmental change: How much can the opportunities decrease and the organization still be profitable and survive? To calculate the resilience, we first smooth the curves firstly and then find the inflection of the curve by calculating the gradients along the curve. The correlation analysis result by SPSS for worker resilience and opportunity resilience are shown in and respectively. The analysis results show that demand, supply inertia and hierarchy all affect organizational resilience negatively, which means that more flexible communication and less hierarchy can improve the resilience of organizational adaptation.

In practice, the flat organizational structure and flexible communication among workers allows the organization to withstand larger changes in the external and inter nal environments.

5 Conclusions

In this paper, we have expanded Epstein's organizational adaptation computational model [13] to propose a new computation model to examine four dimensions of organizational adaptation: Robustness, Agility, Resilience and Survivability. Using a Netlogo agent-based simulation platform, we ran Monte Carlo simulations on this computational model and designed five types of scenarios to run virtual experiments of examining the four dimensions of organizational adaptation.

Through simulation study, we concluded that besides improving Robustness, more flexible communication and lower hierarchy (smaller span of control) can promote the Agility, Resilience and Survivability of organizational adaptation. That is, flat organizational structure and flexible communication mechanism (less demand-supply inertia) are significant to improve the organizational Agility, Resilience and Survivability. However, for improving Robustness, the centralization (high hierarchy and large span of control) and flexible communication mechanism (less demand-supply inertia) are significant. Thus, in agile enterprise, the flat organizational structure and flexible communication is necessary, but meanwhile the organization need to keep a suitable centralized command transferring structure to run business under the right vision, strategic direction of organization designed to maintain ongoing competitiveness [14].

Acknowledgments

We specially thank Joshua Epstein for his lecture on Organizational Adaptability in the Santa Fe Institute's 2008 Complex Systems Summer School. We will thank Kathleen M. Carley (CASOS, CMU) for her support to our research. For valuable comments and discussions, we thank Catherine Spence, Steve Hall, Brad Jones and Brian Hirshman. The authors gratefully acknowledge China Scholarship Council to provide scholarship for my visiting Carnegie Mellon University. This work was funded by the China National Natural Science Fund (No. 70671048). This work was also partially supported by the Santa Fe Institute whose research and education programs are supported by core funding from the USA National Science Foundation and by gifts and grants from individuals, corporations, other foundations, and members of the Institute's Business Network for Complex Systems Research.

References

[1] Anderson, P.: Complexity Theory and Organization Science. Organization Science 10(3), 216–232 (1999)

[2] Foisel, R., Chevrier, V., Haton, J.P.: Modeling adaptive organizations. In: Proceedings of International Conference on Multi Agent Systems, 1998 (1998)

[3] Levchuk, G.M., Meirina, C., Pattipati, K.R., Kleinman, D.L.: Design and analysis of robust and adaptive organizations. In: 2001 IEEE International Conference on Systems, Man, and Cybernetics (2001)

[4] Carley, K.M.: Computational organizational science and organizational engineering. Simulation Modelling Practice and Theory 10(5-7), 253–269 (2002)

[5] Carley, K.M., Svoboda, D.M.: Modeling Organizational Adaptation as a Simulated Annealing Process. Sociological Methods Research Sociological Methods & Research 25(1), 138–168 (1996)

[6] Dessein, W., Santos, T.: Adaptive Organizations. Journal of Political Economy 114(5), 956–995 (2006)

[7] Handley, H.A.H., Levis, A.H.: A Model to Evaluate the Effect of Organizational Adaptation. Computational & Mathematical Organization Theory 7(1), 5–44 (2001)

[8] Davis, J.P., Eisenhardt, K.M., Bingham, C.B.: Developing theory through simulation methods. Academy of Management Review 32(2), 480–499 (2007)

[9] Baligh, H.H., Burton, R.M., Obel, B. (eds.): Devising Expert Systems in Organization Theory: The Organizational Consultant (1990)

[10] Carley, K.M.: On generating hypotheses using computer simulations. Systems Engineering 2(2), 69–77 (1999)

[11] Nigel, G.: How to build and use agent-based models in social science. Mind and Society 1(1), 57–72 (2000)

[12] Lee, J.-S., Carley, K.M.: OrgAhead: A Computational Model of Organizational Learning and Decision Making. Carnegie Mellon University, School of Computer Science, Institute for Software Research International, Technical Report CMU-ISRI-04-117 (2004)

[13] Epstein, J.M.: Generative social science: studies in agent-based computational modeling. Princeton University Press, Princeton (2006)

[14] Collins, J.C., Porras, J.I.: Built to Last: Successful Habits of Visionary Companies. HarperCollins Publishers (1997)

Optimization Using a New Bio-inspired Approach

Xiang Feng[1], Francis C.M. Lau[2], and Daqi Gao[1]

[1] East China University of Science and Technology, Shanghai 200237, China
{xfeng,gaodaqi}@ecust.edu.cn
[2] The University of Hong Kong, Hong Kong
fcmlau@cs.hku.hk

Abstract. There is growing interest in bio(logy)-inspired approaches that are inspired by the principles of biology and that can solve difficult problems. In this paper, we propose a new computational algorithm that is inspired by molecular mechanics for the solution of complex problems. There is a deep and useful connection between mechanics mechanics and combinatorial optimization. This connection exposes new information and allows an unfamiliar perspective on traditional optimization problems and approaches. The alternative of *molecular mechanics algorithm* (MMA) to traditional approaches has the advantages of inherent parallelism and the ability to deal with a variety of complicated social interactions, autonomous behaviors and multiple objectives.

Keywords: Bio-inspired algorithm, multi-objective optimization, molecular mechanics algorithm (MMA), molecular dynamics.

1 Introduction

Many artifacts have been built throughout history, and many of which obtained their inspiration from phenomena in the natural world. It is noted that "progress often occurs at the boundaries between disciplines." [1] In the field of computer science, especially in artificial intelligence, there is growing interest in parallel-distributed intelligent theories and approaches that are inspired by the principles of nature and that can solve difficult problems. Bio(logy)-inspired approaches are probably the best known example of such nature-inspired approaches. Successful bio-inspired approaches include Genetic Algorithm (1975) [2], Ant Colony Optimization (1991) [3,4], and Particle Swarm Optimization (1995) [5]. There were also physics-inspired approaches such as Simulated Annealing Algorithm (1983) [6]. The fields of biology and physics have flourished in a rich soil for many years. We believe there exist many opportunities for the application of the principles in biology and physics to computing. This paper proposes a brand new bio-inspired approach for solving difficult computational problems.

J. Zhou (Ed.): Complex 2009, Part I, LNICST 4, pp. 39–51, 2009.

2 Molecular Mechanics Algorithm (MMA)

The distribution problem is one of those difficult computational problems. The mathematical structure of the distribution problem is simple. The distribution problem defined below is a typical NP-hard combinatorial optimization problem.

Definition 1. *In a multi-objective framework, the distribution problem can be formulated as*

$$
\begin{cases}
\max: \\
\quad z^q(R) = (C^q)^T X(R)^T = \displaystyle\sum_{i=1}^{I}\sum_{j=1}^{J} c_{ij}^q x_{ij} r_{ij} \quad q = 1, \cdots, Q \\
\quad\quad\quad s.t. \quad \displaystyle\sum_{j=1}^{J} r_{ij} = 1 \quad i = 1, \cdots, I \\
\quad\quad\quad\quad x_{ij} = 0, 1
\end{cases}
\tag{1}
$$

where q represents the objective, R is a two-dimensional distribution vector, C^q is a two-dimensional weight vector $(q = 1, 2, \cdots, Q)$, and X is a two-dimensional Boolean vector.

With this problem model, we can now examine the bio-inspired multi-objective model which can mathematically describe an MMA. The theory of evolution behind the model is a dynamical theory.

The bio-inspired dynamics will drive the MMA to its equilibrium state.

Definition 2. *The distribution and weight dynamic equations of the MMA are defined, respectively, by*

$$
r_{ij}(t+1) = r_{ij}(t) + \Delta r_{ij}(t)
\tag{2}
$$

$$
c_{ij}^q(t+1) = c_{ij}^q(t) + \Delta c_{ij}^q(t)
\tag{3}
$$

The two dynamic equations are seen as the "MMA evolution" by fictitious agents (molecules) which manipulate the distribution and weight vectors until an equilibrium is reached. In the MMA, every entry of distribution vector R is treated as a fictitious agent (molecule). In fact, the weight vector is invariable, and the evolution of the weight vector only occurs in the computing process in order to obtain efficient solutions for the distribution vector.

For the fictitious agents—the molecules—there are four factors related to the evolutionary distribution vector (R) and the weight vector (C):

– personal utility (u) (to realize the multiple objectives);
– whole utility (J) (to increase the overall utility);
– minimal personal utility (P) (to realize max-min fair distribution);
– interaction among the molecules (Q) (to satisfy the restrictions and to describe high-dimensional, highly nonlinear, random behaviors and dynamics).

We try to solve the distribution problem defined in Definition 1 by subdividing this hard problem with respect to the four factors. In molecular mechanics, "energy" makes molecules move. When "energy" is equal to zero, the molecules will stop moving, being in an equilibrium state. For the MMA, we will define the "energy" function, which makes fictitious agents (molecules) evolve towards the optimum or until an equilibrium is reached. The "energy" function is based on the four factors each of which corresponds to a component of the "energy" function.

Definition 3. *The "energy" function is defined by*

$$E_{ij}^q(t) = \lambda_1 u_{ij}^q(t) + \lambda_2 J^q(t) - \lambda_3 P^q(t) - \lambda_4 Q^q(t) \tag{4}$$

where $0 < \lambda_1, \lambda_2, \lambda_3, \lambda_4 < 1$. The larger the "energy" function, the faster the fictitious agent (molecule) would evolve towards the optimum.

According to "differential equation theory", a variable's increment to make it minimum is equal to the sum of negative items from related factors differentiating the variable. Because our defined problem is a "maximum" problem, a variable's increment to make it maximum is equal to the sum of the items from related factors differentiating the variable. Thus we have the following definitions.

Definition 4. *The increments of distribution and weight are defined, respectively, by*

$$\Delta r_{ij}(t+1) \approx \frac{dr_{ij}(t)}{dt} = \frac{\partial E_{ij}^q(t)}{\partial r_{ij}(t)} = \sum_{q=1}^{Q} (\lambda_1 \frac{\partial u_{ij}^q(t)}{\partial r_{ij}(t)} + \lambda_2 \frac{\partial J^q(t)}{\partial r_{ij}(t)} - \lambda_3 \frac{\partial P^q(t)}{\partial r_{ij}(t)} - \lambda_4 \frac{\partial Q^q(t)}{\partial r_{ij}(t)}) \tag{5}$$

$$\Delta c_{ij}^q(t+1) \approx \frac{dc_{ij}^q(t)}{dt} = \frac{\partial E_{ij}^q(t)}{\partial c_{ij}^q(t)} = \lambda_1 \frac{\partial u_{ij}^q(t)}{\partial c_{ij}^q(t)} + \lambda_2 \frac{\partial J^q(t)}{\partial c_{ij}^q(t)} - \lambda_3 \frac{\partial P^q(t)}{\partial c_{ij}^q(t)} - \lambda_4 \frac{\partial Q^q(t)}{\partial c_{ij}^q(t)} \tag{6}$$

$$q = 1, 2, \cdots, Q$$

Four kinds of factor functions in the "energy" function will be defined here, respectively.

Definition 5. *The individual personal utility function for every agent (molecule) is defined by*

$$u_{ij}^q(t) = 1 - \exp(-c_{ij}^q(t) r_{ij}(t) x_{ij}(t)) \quad q = 1, 2, \cdots, Q \tag{7}$$

Definition 6. *The whole utility function for every agent (molecule) is defined by*

$$J^q(t) = \sum_{i=1}^{I} \sum_{j=1}^{J} u_{ij}^q(t) \quad q = 1, 2, \cdots, Q \tag{8}$$

Definition 7. *The gravitational potential energy function that makes minimal personal utility increase for every agent (molecule) is defined by*

$$P^q(t) = k^2 \ln \sum_{i=1}^{I} \sum_{j=1}^{J} \exp[-(u_{ij}^q)^2(t)/2k^2] - k^2 \ln IJ \qquad (9)$$

$$q = 1, 2, \cdots, Q$$

Definition 8. *The interaction energy function is defined by*

$$Q^q(t) = \sum_{i=1}^{I} | \sum_{j=1}^{J} r_{ij}(t)x_{ij}(t) - 1 |^2 - \sum_{i,j} \int_0^{u_{ij}^{q}} \{[1 + \exp(-\zeta_{ij}x)]^{-1} - 0.5\}dx \qquad (10)$$

where ζ_{ij} represents the intention strength of social coordination.

Now, we explain why the four kinds of functions are chosen.

1. Personal utility function. For the q-th objective, the larger the value of $c_{ij}^q(t)r_{ij}(t)x_{ij}(t)$ in Eq. (7), the more profit the (i,j)-th molecule gets. The optimization problem here is posed as a maximization problem. And we use the exponential function in order that $u_{ij}^q(t)$ would be between 0 and 1. $u_{ij}^q(t)$ can be regarded as the q-th dimensional utility of molecule. The larger $u_{ij}^q(t)$ is, the more profit the (i,j)-th molecule gets. Schematically, the q-th dimensional utility function u_{ij}^q of a molecule corresponds to the q-th dimensional coordinate of the (i,j)-th molecule's q-th dimensional force field. We define the distance from the bottom boundary to the upper boundary of all of the molecule's q dimensional force fields to be 1. The biological meaning of the MMA will be discussed in Section IV. $1 - e^{-x}$ is chosen as the definition of u_{ij}^q because $1 - e^{-x}$ is a monotonically increasing function and is between 0 and 1. Obviously, the larger $u_{ij}^q(t)$ the better.
2. Whole utility function. For this definition, we assume that the individual personal utilities are additive. Obviously, the larger $J^q(t)$ the better.
3. The potential energy function. For Eq. (9), $0 < k < 1$ is a parameter to be tuned in the implementation. The smaller P^q is, the better. With Eq. (9), we attempt to construct a potential energy function, P^q, such that the decrease of its value would imply the increase of the minimal utility of all the molecules. We prove that in Theorem 1. This way we can optimize the distribution problem in the sense that we consider not only the individual personal utility, but also the aggregate utilities, by increasing the minimum utility of all the molecules again and again. In fact, k represents the strength of the upward gravitational force in the q-th dimensional force field. The bigger k is, the faster the molecules would move up; hence, k influences the convergence speed of the distribution problem. k needs to be carefully adjusted in order to maximize the q objectives.
4. The interaction energy function. For Eq. (10), the first term of $Q^q(t)$ is related to the constraints on capability; the second term involves social coordinations, with ζ_{ij} coming from Eqs. (11) − (13). If social coordinations are not involved in the system, ζ_{ij} will be a constant (e.g., 6 or 8). The first

term of $Q^q(t)$ corresponds to a penalty function with respect to the constraint on the utilization of resources. The second term of $Q^q(t)$ is chosen as shown because we want $\frac{\partial Q^q}{\partial u_{ij}^q}$ to be a monotonically decreasing sigmoid function. $-\{[1 + \exp(-\zeta_{ij} u_{ij}^q)]^{-1} - 0.5\}$ is such a function. Therefore we let $\frac{\partial Q^q}{\partial u_{ij}^q}$ equal to $-\{[1 + \exp(-\zeta_{ij} u_{ij}^q)]^{-1} - 0.5\}$. Then $\frac{\partial Q^q}{\partial u_{ij}^q}$ is integrated to be Q^q.

$$\zeta_{ij}(t) \uparrow \Rightarrow \frac{\partial Q^q}{\partial u_{ij}^q} \downarrow \Rightarrow -\frac{\partial Q^q}{\partial u_{ij}^q} \uparrow \overset{(5)}{\Rightarrow} \Delta r_{ij}(t+1) \uparrow \overset{a}{\Rightarrow} r_{ij}(t+1) \uparrow \overset{(7)}{\underset{b}{\Rightarrow}} u_{ij}^q \uparrow$$

a: $r_{ij}(t+1) = r_{ij}(t) + \Delta r_{ij}(t+1)$;
b: u_{ij}^q is a monotonically increasing function.

$$\zeta_{ij}(t) = \sum_{l=1}^{I} \zeta_{ilj}(t) + \sum_{l=1}^{I} \zeta_{lij}(t) \tag{11}$$

$$\zeta_{ilj}(t) = \begin{cases} 1 & if \ \beta_{ilj} \in (II) \cup (IV) \\ -1 & if \ \beta_{ilj} \in (I) \cup (III) \end{cases} \tag{12}$$

$$\zeta_{lij}(t) = \begin{cases} 1 & if \ \beta_{lij} \in (I) \cup (IV) \\ -1 & if \ \beta_{lij} \in (II) \cup (III) \end{cases} \tag{13}$$

β_{ilj} is the social coordination of agent i with respect to agent l for object j, which gives rise to the change $\zeta_{ilj}(t)$ of intention strength $\zeta_{ij}(t)$. The social coordination (β_{ilj}) can be divided into four main categories as follows.
 - I: Unilateral adaptive coordination
 - II: Unilateral inducing coordination
 - III:Bilateral adaptive coordination
 - IV: Bilateral inducing coordination

We can therefore obtain the iteration speed by the following equation.

$$v_{ij}^q = du_{ij}^q/dt = \frac{\partial u_{ij}^q}{\partial r_{ij}} \frac{dr_{ij}}{dt} + \frac{\partial u_{ij}^q}{\partial c_{ij}^q} \frac{dc_{ij}^q}{dt}$$

$$= [\lambda_1 + \lambda_2 \frac{\partial J^q(t)}{\partial u_{ij}^q(t)} - \lambda_3 \frac{\partial P^q(t)}{\partial u_{ij}^q(t)} - \lambda_4 \frac{\partial Q^q(t)}{\partial u_{ij}^q(t)}]\{[\frac{\partial u_{ij}^q(t)}{\partial r_{ij}(t)}]^2 + [\frac{\partial u_{ij}^q(t)}{\partial c_{ij}^q(t)}]^2\}$$

v_{ij}^q represents the iteration speed of molecule (i,j) (the (i,j)-th entry of distribution vector) with respect to the q-th objective. Meanwhile, v_{ij}^q represents the speed of the upward movement of molecule (i,j) in the q-th dimensional force field.

After having proposed the mathematical model of MMA, we give the parallel MMA in Table 1.

Proving the mathematical model of MMA

We now discuss the properties of the mathematical model of MMA as proposed above.

Theorem 1. *If k is very small, decreasing the potential energy function $P^q(t)$ of Eq. (9) amounts to increasing the minimal utility of molecules (entries in distribution vector R), minimized over R.*

Table 1. Algorithm MMA

Input: c_{ij}^q, x_{ij}, ζ_{ij}

Output:

1. Initialization:

   $t \leftarrow 0$

   $r_{ij}(t)$ —Initialize in parallel

   2. **While** $(du_{ij}^q/dt \neq 0)$ **do**

   $t \leftarrow t+1$

   $u_{ij}^q(t)$ —Compute in parallel according to Eq.(7)

   du_{ij}^q/dt —Compute in parallel according to Eq.(14)

   $dr_{ij}(t)/dt$ —Compute in parallel according to Eq.(5)

   $r_{ij}(t) \leftarrow r_{ij}(t-1) + dr_{ij}(t)/dt$

   $dc_{ij}^q(t)/dt$ —Compute in parallel according to Eq.(6)

   $c_{ij}^q(t) \leftarrow c_{ij}(t-1) + dc_{ij}(t)/dt$

Proof. Supposing that $H(t) = \max\limits_{i,j}\{-(u_{ij}^q(t))^2\}$, we have

$$[\exp(H(t)/2k^2)]^{2k^2} \leq \{\sum_{i=1}^{I}\sum_{j=1}^{J}\exp[-(u_{ij}^q(t))^2/2k^2]\}^{2k^2} \leq [IJ\exp(H(t)/2k^2)]^{2k^2}.$$

Taking the logarithm of both sides of the above inequalities gives

$$H(t) \leq 2k^2\ln\sum_{i=1}^{I}\sum_{j=1}^{J}\exp[-(u_{ij}^q(t))^2/2k^2] \leq H(t) + 2k^2\ln IJ.$$

Since IJ is constant and k is very small, we have

$$H(t) \approx 2k^2\ln\sum_{i=1}^{I}\sum_{j=1}^{J}\exp[-(u_{ij}^q(t))^2/2k^2] - 2k^2\ln IJ = 2P^q(t).$$

It turns out that the potential energy $P^q(t)$ at the time t represents the maximum of $-(u_{ij}^q(t))^2$ among all the molecules, which is the minimal personal utility of entries with respect to an objective q at time t. Hence the decrease of potential energy $P^q(t)$ will result in the increase of the minimum of $u_{ij}^q(t)$.   □

Theorem 2. *Updating the allotted entries r_{ij} and weights c_{ij}^q by Eq. (5) and Eq. (6) respectively amounts to changing the speed of molecule by $v_{ij}^q(t)$ of Eq. (14).*

Proof. Denote the k-th terms of Eq. (5) and Eq. (6) by $\langle \frac{dr_{ij}(t)}{dt} \rangle_k$ and $\langle \frac{dc_{ij}^q(t)}{dt} \rangle_k$, respectively. When allotted entry r_{ij} is updated according to (5), the first and second terms of (5) will cause the following speed increments of the iteration, respectively:

$$\langle du_{ij}^q(t)/dt \rangle_1^r = \frac{\partial u_{ij}^q(t)}{\partial r_{ij}(t)} \langle \frac{dr_{ij}(t)}{dt} \rangle_1 = \lambda_1 [\frac{\partial u_{ij}^q(t)}{\partial r_{ij}(t)}]^2 \tag{14}$$

$$\langle du_{ij}^q(t)/dt \rangle_2^r = \frac{\partial u_{ij}^q(t)}{\partial r_{ij}(t)} \langle \frac{dr_{ij}(t)}{dt} \rangle_2 = \lambda_2 \frac{\partial u_{ij}^q(t)}{\partial r_{ij}(t)} \frac{\partial J^q(t)}{\partial r_{ij}(t)}$$

$$= \lambda_2 \frac{\partial u_{ij}^q(t)}{\partial r_{ij}(t)} \frac{\partial J^q(t)}{\partial u_{ij}^q(t)} \frac{\partial u_{ij}^q(t)}{\partial r_{ij}(t)} = \lambda_2 \frac{\partial J^q(t)}{\partial u_{ij}^q(t)} [\frac{\partial u_{ij}^q(t)}{\partial r_{ij}(t)}]^2 \tag{15}$$

Similarly, the third and the fourth term of Eq. (5) will cause the following speed increments of the iteration:

$$\langle du_{ij}^q(t)/dt \rangle_3^r = -\lambda_3 \frac{\partial P^q(t)}{\partial u_{ij}^q(t)} [\frac{\partial u_{ij}^q(t)}{\partial r_{ij}(t)}]^2$$

$$\langle du_{ij}^q(t)/dt \rangle_4^r = -\lambda_4 \frac{\partial Q^q(t)}{\partial u_{ij}^q(t)} [\frac{\partial u_{ij}^q(t)}{\partial r_{ij}(t)}]^2$$

Similarly, for Eq. (6), we have $\langle du_{ij}^q(t)/dt \rangle_k^{c^q}$, $k = 1, 2, 3, 4$. We thus obtain

$$\sum_{k=1}^{4} [\langle du_{ij}^q(t)/dt \rangle_j^{c^q} + \langle du_{ij}^q(t)/dt \rangle_j^r]$$

$$= [\lambda_1 + \lambda_2 \frac{\partial J^q(t)}{\partial u_{ij}^q(t)} - \lambda_3 \frac{\partial P^q(t)}{\partial u_{ij}^q(t)} - \lambda_4 \frac{\partial Q^q(t)}{\partial u_{ij}^q(t)}] \{ [\frac{\partial u_{ij}^q(t)}{\partial r_{ij}(t)}]^2 + [\frac{\partial u_{ij}^q(t)}{\partial c_{ij}^q(t)}]^2 \} = v_{ij}^q(t)$$

Therefore, updating $r_{ij}^{(k)}$ and $(c_{ij}^q)^{(k)}$ by (5) and (6), respectively, gives rise to the speed increment of the iteration that is exactly equal to $v_{ij}^q(t)$ of Eq. (14). $\square$

Theorem 3. *Updating the allotted entries r_{ij} and weights c_{ij}^q by Eq. (5) and Eq. (6) respectively amounts to increasing the minimal utility of an entry with respect to an objective q in direct proportion to the value of λ_3.*

Proof. The speed increment of the iteration, which is related to potential energy $P^q(t)$, is given by

$$\langle \frac{du_{ij}^q(t)}{dt} \rangle_3 = \langle du_{ij}^q(t)/dt \rangle_3^r + \langle du_{ij}^q(t)/dt \rangle_3^{c^q} = -\lambda_3 \frac{\partial P^q(t)}{\partial u_{ij}^q(t)} \{ [\frac{\partial u_{ij}^q(t)}{\partial r_{ij}(t)}]^2 + [\frac{\partial u_{ij}^q(t)}{\partial c_{ij}^q(t)}]^2 \}.$$

Denote by $\langle \frac{dP^q(t)}{dt} \rangle$ the differentiation of the potential energy function $P^q(t)$ with respect to time t arising from using Eqs. (5), (6). We have

$$\langle \frac{dP^q(t)}{dt} \rangle = \frac{\partial P^q(t)}{\partial u_{ij}^q(t)} \langle \frac{du_{ij}^q(t)}{dt} \rangle_3 = -\lambda_3 [\frac{\partial P^q(t)}{\partial u_{ij}^q(t)}]^2 \{ [\frac{\partial u_{ij}^q(t)}{\partial r_{ij}(t)}]^2 + [\frac{\partial u_{ij}^q(t)}{\partial c_{ij}^q(t)}]^2 \}$$

$$= -\lambda_3 \omega_{ij}^2(t) (u_{ij}^q(t))^2 x_{ij}^2(t) [r_{ij}^2(t) + (c_{ij}^q(t))^2] [u_{ij}^q(t)]^2 \leq 0.$$

where, $\omega_{ij}(t) = \exp[-(u_{ij}^q(t))^2/2k^2] / \sum_{i=1}^{I} \sum_{j=1}^{J} \exp[-(u_{ij}^q(t))^2/2k^2]$.

It can be seen that using Eqs. (5) and (6) gives rise to monotonic decrease of $P^q(t)$. Then by Theorem 1, the decrease of $P^q(t)$ will result in the increase of the minimal utility of entries in distribution vector R, in direct proportion to the value of λ_3. $\square$

Theorem 4. *The first and second terms of Eqs.* (5) *and* (6) *will enable the personal utility of every entry of distribution R to increase, in direct proportion to the value of* $(\lambda_1 + \lambda_2)$.

Proof. According to Eqs. (15) and (16), the sum of the first and second terms of Eq. (5) and (6) will be

$$\langle du^q_{ij}(t)/dt\rangle^r_1 + \langle du^q_{ij}(t)/dt\rangle^r_2 + \langle du^q_{ij}(t)/dt\rangle^{c^q}_1 + \langle du^q_{ij}(t)/dt\rangle^{c^q}_2$$
$$= [\lambda_1 + \lambda_2 \frac{\partial J^q(t)}{\partial u^q_{ij}(t)}]\{[\frac{\partial u^q_{ij}(t)}{\partial r_{ij}(t)}]^2 + [\frac{\partial u^q_{ij}(t)}{\partial c^q_{ij}(t)}]^2\} = (\lambda_1 + \lambda_2)x^2_{ij}(t)[r^2_{ij}(t) + (c^q_{ij}(t))^2][-u^q_{ij}(t)]^2$$
$$\geq 0.$$

Therefore, the first and second terms of (5) and (6) will cause $u^q_{ij}(t)$ to monotonically increase. □

Theorem 5. *Updating r_{ij} and c^q_{ij} by Eqs.* (5) *and* (6) *gives rise to monotonic increase of the whole utility of all the entries of distribution R, in direct proportion to the value of λ_2.*

Proof. Similar to Theorem 4, it follows that when an entry (i,j) modifies its r_{ij} and c^q_{ij} by Eqs. (5) and (6), differentiation of $J^q(t)$ with respect to time t will not be negative—i.e., $\langle \frac{dJ^q(t)}{dt}\rangle \geq 0$, and it is directly proportional to the value of λ_2. □

Definition 9. *(Max-min Fairness) A feasible distribution R is max-min fair if and only if an increase of any entries of distribution vector r within the domain of feasible distributions must be at the cost of an decrease of some already smaller entries r. Formally, for any other feasible distribution Y, if $y_{ij} > r_{ij}$ then there must exist some (i',j') such that $r_{i'j'} \leq r_{ij}$ and $y_{i'j'} < r_{i'j'}$.*

Theorem 6. *(Max-min fair allocation) Max-min fair solution with multi- objective can be obtained by updating the allotted entries r_{ij} and weights c^q_{ij} by Eq.* (5) *and Eq.* (6) *respectively.*

Proof. It is straightforward from Theorems 1–5 and Definition 9. □

Theorem 7. *Updating r_{ij} and c^q_{ij} by Eqs.* (5) *and* (6) *gives rise to monotonic decrease of the interaction energy $Q^q(t)$, in direct proportion to the value of λ_4.*

Proof. As in the above, we have

$$\langle \frac{du^q_{ij}(t)}{dt}\rangle_4 = -\lambda_4 \frac{\partial Q^q(t)}{\partial u^q_{ij}(t)}\{[\frac{\partial u^q_{ij}(t)}{\partial r_{ij}(t)}]^2 + [\frac{\partial u^q_{ij}(t)}{\partial c^q_{ij}(t)}]^2\}; \text{ and}$$

$$\langle \frac{dQ^q(t)}{dt}\rangle = \frac{\partial Q^q(t)}{\partial u^q_{ij}(t)}\langle \frac{du^q_{ij}(t)}{dt}\rangle_4 = -\lambda_4 [\frac{\partial Q^q(t)}{\partial u^q_{ij}(t)}]^2\{[\frac{\partial u^q_{ij}(t)}{\partial r_{ij}(t)}]^2 + [\frac{\partial u^q_{ij}(t)}{\partial c^q_{ij}(t)}]^2\} \leq 0.$$

Updating r_{ij} and c^q_{ij} by Eqs. (5) and (6) makes the interaction energy $Q^q(t)$ smaller and smaller. Thus it is possible to satisfy the restrictions. □

Theorem 8. *The MMA can solve the distribution problem defined in Definition 1.*

Proof. It is straightforward from Theorems 6–7. □

3   Simulation

Because the problem-related matrices are too large to be listed in this paper, we go directly to the results of these two problems. The experimental evolutionary results for z from $t = 1$ to $t = 1000$ are depicted in Fig. 1.

As shown in Fig. 1, Curve A, at $t = 983$, z reaches its maximum, and stays unchanged in the remainder of the iterations; for Curve B, at $t = 968$, z reaches its maximum, and stays unchanged in the remainder of the iterations. At the two points, the MMA converges to a stable equilibrium state and produces the optimum solutions. The results confirm the usefulness of MMA for large-scale NP-hard combinatorial optimization problems.

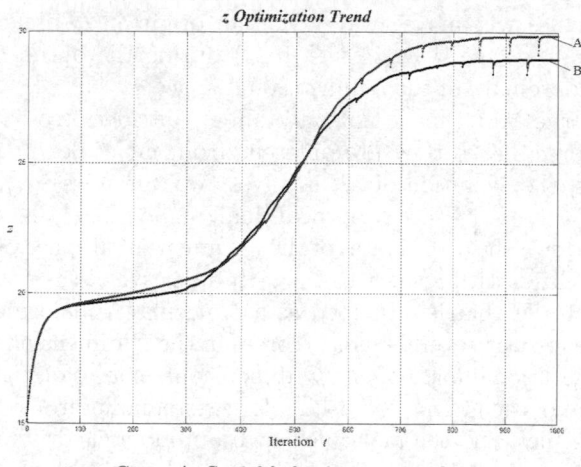

Curve A: Social behaviors not involved

Curve B: Social behaviors involved

Fig. 1. Comparison of z optimization in large-scale problems with social behaviors involved and not involved

4   Motivation and Biological Meaning of MMA

In the MMA, all the entries in distribution vector R are treated as molecules which are located in their own force fields. This transforms the distribution problem into the kinematics and dynamics of the molecules in a set of force fields. In molecular mechanics, Newtonian mechanics is used to model molecular systems and force fields are used to calculate the potential energy.

MMA puts emphasis on

- providing a view of individual and whole optimization (with one to two objectives);
- parallelization with reasonably low time complexity;
- all multiple objectives being optimized individually as well as collectively;
- the ability to deal with social interactions;
- the biological meaning of the model.

In the biological model of MMA, there are $I \times J$ molecules (every entry in distribution vector R is treated as a molecule), and the same number $(I \times J)$ of force fields. These force fields are evenly distributed in the horizontal plane. Every molecule moves in its own force field. When the number of optimization objectives is Q, the force fields of all molecules form Q dimensions where the coordinates in the space are in $[0, 1]$. In the biological model of MMA, s_{ij} represents molecule (i, j).

If the number of maximum objectives is 1, the molecules will move upwards on one-dimensional spaces (lines) $(u_{ij}^1 \in [0, 1])$ during the optimization process. If the number of maximum objectives is 2, the molecules will move away from the origin in two-dimensional spaces (planes) $(u_{ij}^1 \in [0, 1], u_{ij}^2 \in [0, 1])$ during the optimization process. Analogously, if the number of maximum objectives is Q, the molecules will move away from the origin in Q-dimensional spaces $(u_{ij}^1 \in [0, 1], \cdots, u_{ij}^q \in [0, 1], \cdots, u_{ij}^Q \in [0, 1])$ during the optimization process, where u_{ij}^q is a coordinate of the q-dimensional space.

Molecules in the MMA move not only under outside forces, but also under their internal force; hence they are different from molecules in molecular mechanics. In fact, the evolution of s_{ij} involves two variables—r_{ij}, c_{ij}^q. r_{ij} and anyone of $c_{ij}^q (q = 1, \cdots, Q)$ are reciprocal dual.

In a force field F_{ij}, the coordinates of the Q-dimensional space of the molecules represent the utilities with respect to the q-th objective of the entry (i, j) of the distribution vector R that is described as a molecule. A molecule will be influenced simultaneously by several kinds of forces in the Q-dimensional space, which include the gravitational force of the Q-dimensional space force field where the molecule is located, the pulling or pushing forces stemming from the interactions with other molecules, and the molecule's own autonomous driving force.

When the number of maximum objectives is 1, that is, a single objective optimization problem, all the above-mentioned forces that are exerted on a molecule are dealt with as forces along a vertical direction (along a line). Thus a molecule will be driven by the resultant force of all the forces that act on it upwardly or downwardly, and move along a vertical direction. The larger the upward resultant force on a molecule, the faster the upward motion of the molecule. When the upward resultant force on a molecule is equal to zero, the molecule will stop moving, being at an equilibrium status. As shown in Fig. 2, the molecules move in their own one-dimensional force fields F_{ij} (lines).

The upward gravitational force of a force field on a molecule causes an upward component of the motion of the molecule, which represents the tendency that the molecule pursues the common benefit of the whole. The upward or downward component of the motion of a molecule, which is related to the interactions with other molecules, depends upon the strengths and categories of the interactions. The molecule's own autonomous driving force is proportional to the degree the molecule tries to move upwards in its own force field where it is located, i.e., the molecule tries to acquire its own maximum utility.

When the number of maximum objectives is 2, each molecule moves away from the origin in its own force field (a unit plane), as shown in Fig. 2.

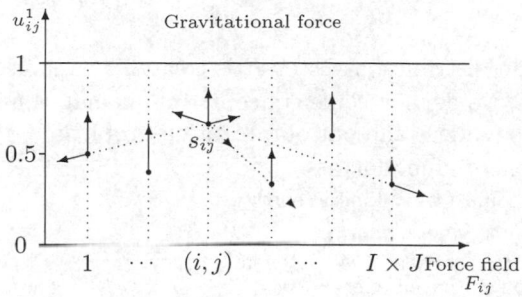

Fig. 2. The biological model of MMA for single objective optimization

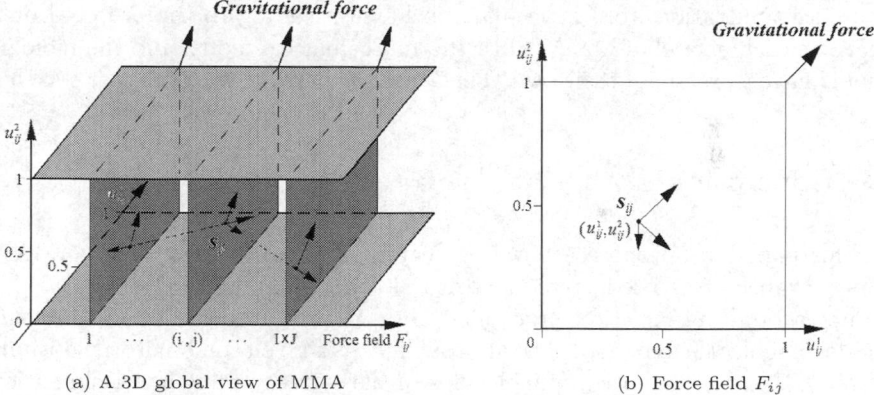

(a) A 3D global view of MMA   (b) Force field F_{ij}

Fig. 3. The biological model of MMA for two objectives optimization

When the number of maximum objectives is $Q > 2$, each molecule moves away from the origin in its own Q-dimensional force field. Q-dimensional force field and $Q+1$-dimensional biological model of MMA are abstract mathematical spaces.

One major difference between the molecule of the proposed MMA and the molecule of a classical biological model is that the molecule in MMA has its own driving force which depends upon the autonomy of the molecule. All the molecules, in their own Q-dimensional force fields, simultaneously, evolve under their exerted forces; as long as they gradually reach their equilibrium positions from their initial positions which are set at random, we can obtain a feasible solution to the multiple objectives distribution problem.

There are three main components of the MMA biological model. We call them as "skeleton", "muscle" and "blood".

Molecules motion————————————————skeleton
How to move (mathematical model of MMA)————muscle
Energy (energy function) ————————————-blood

In summary

- Every entry in distribution vector R is treated as a molecule.
- In Section II, we defined some concepts in kinematics and dynamics based on both biology ideas and our optimization problem.
- We defined energy function as
  Energy = personal utility kinetic energy
  + whole utility kinetic energy
  + max-min fair gravitational potential energy
  + interaction potential energy

It is difficult to say which one occurred to us earlier: the idea of the mathematical model of MMA or the idea of the biological model. We found inspiration in and a foundation from molecular mechanics, leading to the proposal of the mathematical model of MMA. Both the mathematical model and the biological model have a system of their own. There are the correspondences between them.

5   Conclusion

In this paper, we propose a novel Molecular Mechanics Algorithm (MMA) as a new branch of nature-inspired algorithms for solving multi-objective NP-hard combinatorial optimization problems. The MMA is inspired by the biological model of molecule dynamics. The approach maps a given combinatorial optimization problem to the motion of molecules in the corresponding multi-dimensional force fields. The molecules move according to certain rules defined by a mathematical model until arriving at a stable state; subsequently, the solution of the multi-objective combinatorial optimization problem is obtained by anti-mapping the stable state. We have discussed the mathematical model, algorithm, motivation, biological model, and the experiments of the MMA in detail.

The MMA can work out the theoretical optimum solution, which is important and exciting. We have given the theoretical proofs and experiments to verify this key point.

Although there are many differences between molecules in molecular mechanics and those in the MMA, we have shown that being inspired by molecular mechanics, the MMA enables feasible multi-objective optimization in very large scales. The MM approach can work out the theoretical optimum solution and has a low computational complexity, which is crucial for the functioning of large-scale NP-hard combinatorial optimization problems.

Acknowledgements. This project is supported by the Hong Kong General Research Fund under Grant No. HKU 713708E, the National Science Foundation of China (NSFC) under Grant No. 60575027, the High-Tech Development Program of China (863) under Grant No. 2006AA10Z315, and the Specialized Research Fund for the Doctoral Program of Higher Education under Grant No. 20060251013.

References

1. Shadbolt, N.: Nature-Inspired Computing. IEEE Intelligent Systems 19, 1–2 (2004)
2. Holland, J.: Adaptation in Natural and Artificial Systems. MIT Press, Cambridge (1976)
3. Dorigo, M.: Optimization, Learning and Natural Algorithms. Ph.D. Thesis, Dipartimento di Elettronica, Politecnico di Milano, Italy (1992) (in Italian)
4. Dorigo, M., Maniezzo, V., Colorni, A.: Ant System: Optimization by a Colony of Cooperating Agents. IEEE Trans. Systems, Man, Cybernet., Part B 26(1), 29–41 (1996)
5. Kennedy, J., Eberhart, R.: Particle Swarm Optimization. In: IEEE Conf. Neural Networks, pp. 1942–1948. IEEE Press, Los Alamitos (1995)
6. Kirkpatrick, S., Gelatt, C., Vecchi, M.: Optimization by Simulated Annealing. Science 220(4598), 671–680 (1983)

Optimality Conditions of a Three-Dimension Non-smooth Thermodynamic System of Sea Ice

Wei Lv[1], Hong Bao[2], and Enmin Feng[3]

[1] Shanghai University, Department of Mathematics, Shanghai 200444, China
lvwei7809@yahoo.com.cn
[2] Liaoning University of Traditional Chinese Medicine,
Information and Engineering College, Shenyang 110032, China
[3] Dalian University of Technology, Department of Applied Mathematics,
Dalian 116024, China

Abstract. This study is intended to provide the mathematical foundation for the numerical computation of the parameter identification problems of the three-dimensional two-layer thermodynamic system of sea ice. The non-smooth thermodynamic system with mixed boundary conditions is established, its properties are obtained and the first-order necessary conditions of the parameter identification problem of the non-smooth system are derived.

Keywords: coupled 3D non-smooth thermodynamic system, parameter identification, necessary condition for optimality.

1 Introduction

Sea ice plays an important role in the global climate system [1,2,3]. The thin sea ice, sometimes having a snow cover, forms a new interface between the lower ocean and the upper atmosphere, reducing the transfer of moisture, heat and momentum between the atmosphere and the ocean. The freezing and melting processes of sea ice are influenced by the temperature distribution, thus the numerical simulation for ice temperature distribution have attracted great attention [4-8]. However, the physical parameters such as the density, the specific heat, the thermal conductivity and the heat exchange coefficient, and so on, are crucial for exactly describing the sea ice temperature profile. Therefore, accurately estimating these physical parameters can improve the sea ice thermodynamic modelling.

Until now, these physical parameters in the sea ice thermodynamic system are mainly estimated by field data [4-8]. However, the field data are spare and unsatisfactory due to the difficulties associated with fieldwork, especially during the polar winter. Some parameters could not be detected continuously and automatically up to now, such as the ice salinity; some could not be detected directly, such as the ice thermal conductivity. Thus it could not help us to thoroughly understand the physical evolution of sea ice just by field data. The parameter identification method is effective to solve this problem. Parameter identification

J. Zhou (Ed.): Complex 2009, Part I, LNICST 4, pp. 52–65, 2009.

refers to the determination of the physical parameters that could be detected discontinuously or difficultly from the physical parameters which can be detected continuously in the system model such that the predicted response of the model is close, in some well-defined sense, to the process observations. In recent years, there are many researchers devoting to the parameter identification problems of thermodynamic systems. Some researchers [9-12] considered the determination of source terms, some researchers [13-15] considered the determination of thermal conductivities. In [16], the properties and the optimality conditions for a one-dimension non-smooth thermodynamic system of sea ice were provided. The coefficients describing the sea ice salinity and other two parameters in a one-dimension nonlinear and non-smooth thermodynamic system of sea ice were identified using the parameter identification method in [17]. In this paper, we will deal with a three-dimension two-layer thermodynamic system of sea ice, and identify the densities, the specific heats, the thermal conductivities and the heat exchange coefficients of the snow and the sea ice using the parameter identification method. The properties of the non-smooth thermodynamic system are discussed, and the first-order necessary conditions of the parameter identification problem of the non-smooth system are derived. Therefore, the parameter identification theories of the three-dimension non-smooth distributed parameter system are applied to the actual sea ice problems, and the mathematical foundation for the numerical computation of the parameter identification problems of the sea ice thermodynamic system is provided.

The rest of this paper is organized as follows. In Section 2, we describe the three-dimension two-layer thermodynamic system. Section 3 derives the properties of the system. In Section 4, we establish an identification model, prove the existence of optimal control, and derive the necessary conditions for optimality. Section 5 concludes this research.

2   The Coupled 3D Thermodynamic System of Sea Ice

In this section, according to the distribution characteristics of the sea ice temperature field, we will describe a three-dimension two-layer thermodynamic system coupled by the snow and the sea ice, (as in Fig.1 and Fig.2), which is denoted by SIS.

SIS is from longitude W_1 degrees west to longitude W_2 degrees west, latitude N_1 degrees north to latitude N_2 degrees north, and L meters depths, where W_1, W_2, N_1, N_2 and L are positive constants, and $W_1 < W_2$, $N_1 < N_2$. Take the position at longitude W_1 degrees west and latitude N_1 degrees north on the surface of SIS as the origin denoted by A_1. Let x_1 represent the distance coordinate in the latitude direction taken as positive north, x_2 the distance coordinate in the longitude direction taken as positive east, x_3 the depth coordinate of SIS taken as positive downward, and their units are meters; set $x = (x_1, x_2, x_3)$ be the spatial coordinate. For convenience of our analysis, set $I_n = \{1, 2, \cdots, n\}$ and $\overline{S}$ express the closure of the set S, where n is any positive integer. Let A_i, A_i'' $(i \in I_4)$ denote the eight vertexes of SIS. The plane $A_1' A_2' A_3' A_4'$ denotes the

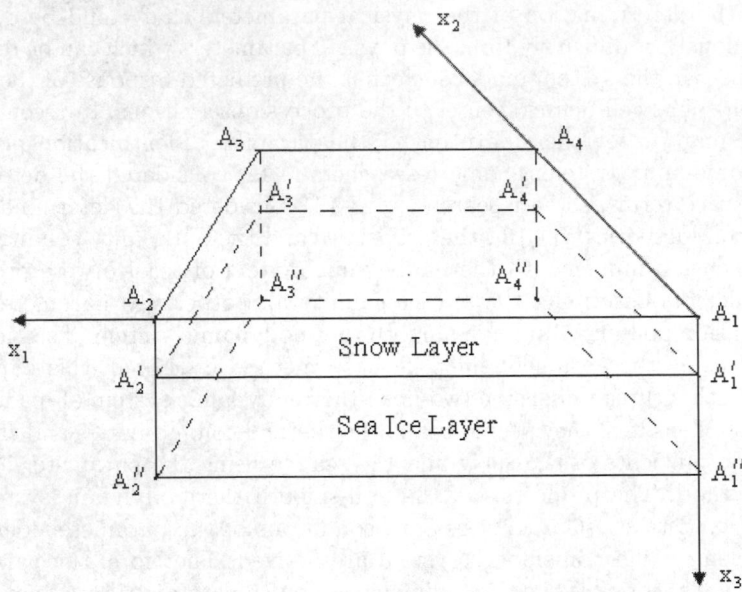

Fig. 1. The configuration of the 3D two-layer model

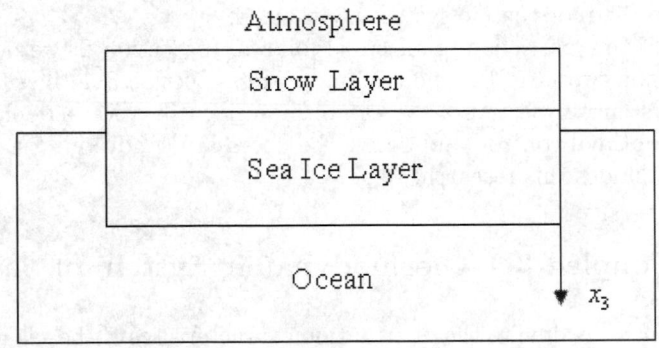

Fig. 2. The schematic diagram of the 3D two-layer model in profile

interface between the snow layer and the sea ice layer, where A_i' ($i \in I_4$) denote the four vertexes. Let L_{sn}, L_{si} and L_{ia} be the depths of the snow layer, the sea ice layer and the sea ice layer in the atmosphere. Let t denote time, t_f the final time, $I = (0, t_f)$, and their units are seconds. Let $T(x,t)$ be the temperature of SIS at position x and time t, $T_0(x)$ the initial temperature, and their units are Kelvins. Let Γ_1 denote the plane $A_1 A_2 A_2'' A_1''$, Γ_2 the plane $A_2 A_3 A_3'' A_2''$, Γ_3 the plane $A_3 A_4 A_4'' A_3''$, Γ_4 the plane $A_4 A_1 A_1'' A_4''$, Γ_5 the plane $A_1'' A_2'' A_3'' A_4''$, Γ_6 the plane $A_1 A_2 A_3 A_4$, Γ_7 the plane $A_1' A_2' A_3' A_4'$, $\Gamma = \bigcup_{j=1}^{6} \Gamma_j$ the boundary of SIS. Let Ω_{sn} be the interior area of the snow layer, Ω_{si} the interior area of the sea ice layer, $\Gamma_{sn} = \partial \overline{\Omega}_{sn} \backslash ri\Gamma_7$ the boundary of the snow layer, $\Gamma_{si} = \partial \overline{\Omega}_{si} \backslash (\Gamma_7 \cup ri\Gamma_5)$

the boundary of the sea ice layer, Γ_{ia} the boundary of the sea ice layer in the atmosphere. $\Omega = \Omega_{sn} \cup \Omega_{si} \cup ri\Gamma_7$ is the interior area of SIS, $Q_{sn} = (\Omega_{sn} \cup ri\Gamma_7) \times I$, $Q_{si} = (\Omega_{si} \cup ri\Gamma_5) \times I$, $Q = \Omega \times I$. Take note that for any $(x,t) \in Q$, $T = T(x,t)$, $T_0 = T_0(x)$. Then according to the energy conservation law and the Fourier's law of heat conduction, the unsteady coupled three-dimensional two-layer thermodynamic system is described by the following heat equations

$$(\rho c)_{sn} \frac{\partial T}{\partial t} = \nabla \cdot (k_{sn} \nabla T) + g_{sn}(x,t), \qquad (x,t) \in Q_{sn}, \qquad (1)$$

$$k_{sn} \nabla T = k_{si} \nabla T, \qquad (x,t) \in Q_{ni}, \qquad (2)$$

$$(\rho c)_{si} \frac{\partial T}{\partial t} = \nabla \cdot (k_{si} \nabla T) + g_{si}(x,t), \qquad (x,t) \in Q_{si}. \qquad (3)$$

The boundary conditions are

$$-k_{sn} \frac{\partial T}{\partial_{su}} = h_{sn}(T - T_{at}(x,t)), \qquad (x,t) \in \Gamma_6 \times I, \qquad (4)$$

$$-k_{sn} \frac{\partial T}{\partial \boldsymbol{n}_{sn}} = h_{sn}(T - T_{at}(x,t)), \qquad (x,t) \in (\Gamma_{sn} \backslash \Gamma_6) \times I, \qquad (5)$$

$$-k_{si} \frac{\partial T}{\partial \boldsymbol{n}_{si}} = h_{si}(T - T_{at}(x,t)), \qquad (x,t) \in \Gamma_{ia} \times I, \qquad (6)$$

$$-k_{si} \frac{\partial T}{\partial \boldsymbol{n}_{si}} = h_{si}(T - T_{oc}(x,t)), \qquad (x,t) \in (\Gamma_{si} \backslash \Gamma_{ia}) \times I, \qquad (7)$$

$$-k_{si} \frac{\partial T}{\partial \boldsymbol{n}_{io}} = h_{si}(T - T_{oc}(x,t)), \qquad (x,t) \in \Gamma_5 \times I. \qquad (8)$$

The initial condition is

$$T(x,0) = T_0(x), \qquad x \in \Omega. \qquad (9)$$

Where ∇ is gradient operator; ρ, c, k, h and g are the density, the specific heat, the thermal conductivity, the heat exchange coefficient on the boundary and the source term, respectively; the subscripts sn, si and oc denote snow, sea ice and ocean, respectively; $\boldsymbol{n}_{su}$, $\boldsymbol{n}_{sn}$, $\boldsymbol{n}_{si}$ and $\boldsymbol{n}_{io}$ are the unit exterior normal vectors of the boundaries Γ_6, $\Gamma_{sn} \backslash \Gamma_6$, Γ_{si} and Γ_5, respectively; $T_{at}(x,t)$ and $T_{oc}(x,t)$ are the temperature functions of the atmosphere and the ocean which are adjacent to SIS.

Since the densities, the specific heats, the thermal conductivities, the heat exchange coefficients and the source terms are different in the three layers of SIS, the thermodynamic system is non-smooth.

According to the physical properties of SIS, we give the following assumptions.

(A1) The temperature $T(x,t)$ of SIS is continuous on $\overline{Q}$.
(A2) The densities, the specific heats, the thermal conductivities and the heat exchange coefficients on the boundaries of snow, sea ice and ocean are bounded, so there exist $q_{Lj} > 0$, $q_{Uj} > 0$, and $q_{Lj} \leq q_{Uj} (j \in I_6)$, such that $q_{L1} \leq (\rho c)_{sn} \leq q_{U1}$,

$q_{L2} \leq (\rho c)_{si} \leq q_{U2}$, $q_{L3} \leq k_{sn} \leq q_{U3}$, $q_{L4} \leq k_{si} \leq q_{U4}$, $q_{L5} \leq h_{sn} \leq q_{U5}$, $q_{L6} \leq h_{si} \leq q_{U6}$.

(A3) $T_0(x) \in C^2(\Omega)$; $T_{at}(x,t) \in C^0((\Gamma_{sn} \cup \Gamma_{ia}) \times \bar{I})$, $T_{oc}(x,t) \in C^0((\Gamma \backslash \Gamma_{sn} \backslash \Gamma_{ia}) \times \bar{I})$; $g_{sn}(x,t) \in C^0(\bar{Q}_{sn})$, $g_{si}(x,t) \in C^0(\bar{Q}_{si})$; and for any $t \in \bar{I}$, $T_{at}(L_{sn} + L_{ia}, t) = T_{oc}(L_{sn} + L_{ia}, t)$, $g_{sn}(L_{sn}, t) = g_{si}(L_{sn}, t)$.

For convenience of our analysis, next we will simplify the thermodynamic system. Let

$$\alpha(x,t) = \begin{cases} \alpha_1 = (\rho c)_{sn}^{-1}, & (x,t) \in \overline{\Omega}_{sn} \times I, \\ \alpha_2 = (\rho c)_{si}^{-1}, & (x,t) \in (\overline{\Omega}_{si} \backslash \Gamma_7) \times I. \end{cases}$$

$$\beta(x,t) = \begin{cases} \beta_1 = k_{sn}, & (x,t) \in \overline{\Omega}_{sn} \times I, \\ \beta_2 = k_{si}, & (x,t) \in (\overline{\Omega}_{si} \backslash \Gamma_7) \times I. \end{cases}$$

$$g(x,t) = \begin{cases} g_{sn}(x,t), & (x,t) \in \overline{\Omega}_{sn} \times I, \\ g_{si}(x,t), & (x,t) \in (\overline{\Omega}_{si} \backslash \Gamma_7) \times I. \end{cases}$$

$$\boldsymbol{n} = \begin{cases} \boldsymbol{n}_{su}, & (x,t) \in \Gamma_6 \times I, \\ \boldsymbol{n}_{sn}, & (x,t) \in (\Gamma_{sn} \backslash \Gamma_6) \times I, \\ \boldsymbol{n}_{si}, & (x,t) \in \Gamma_{si} \times I, \\ \boldsymbol{n}_{io}, & (x,t) \in \Gamma_5 \times I. \end{cases}$$

$$\gamma(x,t) = \begin{cases} \gamma_1 = h_{sn}, & (x,t) \in \Gamma_{sn} \times I, \\ \gamma_2 = h_{si}, & (x,t) \in \Gamma_{si} \times I. \end{cases}$$

$$T_c(x,t) = \begin{cases} T_{at}(x,t), & (x,t) \in (\Gamma_{sn} \cup \Gamma_{ia}) \times \bar{I}, \\ T_{oc}(x,t), & (x,t) \in ((\Gamma_{si} \backslash \Gamma_{ia}) \cup (\Gamma_{oc} \backslash \Gamma_5)) \times \bar{I}. \end{cases}$$

Then (1)-(9) can be simplified to the following form denoted by U3D2LTS.

$$\frac{\partial T}{\partial t} = \alpha(x,t) \nabla \cdot (\beta(x,t) \nabla T) + \alpha(x,t) g(x,t), \quad (x,t) \in Q_{sn} \cup Q_{si}, \quad (10)$$

$$k_{sn} \nabla T = k_{si} \nabla T, \quad (x,t) \in Q_{ni}, \quad (11)$$

$$-\beta(x,t) \frac{\partial T}{\partial \boldsymbol{n}} = \gamma(x)(T - T_c(x,t)), \quad (x,t) \in \Gamma \times \bar{I}, \quad (12)$$

$$T(x,0) = T_0(x), \quad x \in \overline{\Omega}. \quad (13)$$

Since $g_{sn}(x,t)$, $g_{si}(x,t)$, $T_{at}(x,t)$ and $T_{oc}(x,t)$ are given, the temperature T of U3D2LTS is dependent on the parameters ρ_{sn}, ρ_{si}, c_{sn}, c_{si}, k_{sn}, k_{si}, h_{sn} and h_{si}, that is $T = T(x,t; \rho_{sn}, \rho_{si}, c_{sn}, c_{si}, k_{sn}, k_{si}, h_{sn}, h_{si})$. For convenience, let $q = (\alpha_1, \alpha_2, \beta_1, \beta_2, \gamma_1, \gamma_2) = (q_1, q_2, q_3, q_4, q_5, q_6)$. We often use the notation X_1 to denote $X_1(x)$ or X_2 to denote $X_2(x,t)$, where $x \in \overline{\Omega}$ or $(x,t) \in \overline{Q}$.

From the assumption (A2), set $q_L = \min\{q_{U1}^{-1}, q_{U2}^{-1}, q_{L3}, q_{L4}, q_{L5}, q_{L6}\}$, $q_U = \max\{q_{L1}^{-1}, q_{L2}^{-1}, q_{U3}, q_{U4}, q_{U5}, q_{U6}\}$, then $q_L \leq \alpha_j, \beta_j, \gamma_j \leq q_U$, $j \in I_2$. Let $Q_{ad} = \{q = (\alpha_1, \alpha_2, \beta_1, \beta_2, \gamma_1, \gamma_2) | 0 < q_L \leq \alpha_j, \beta_j, \gamma_j \leq q_U, j \in I_2\}$ be the parameter set, then Q_{ad} is a bounded closed convex subset of $\mathbb{R}^6$.

3   Properties of U3D2LTS

In this section, we will derive the properties of U3D2LTS.

Let $V = H^1(\Omega)$ be a Sobolev space, for any $\varphi \in V$, multiply the two sides of (10) by φ and integrate them over Ω, we have that

$$\int_\Omega \varphi \frac{\partial T}{\partial t} dx = \int_\Omega \alpha\varphi\nabla \cdot (\beta\nabla T) dx + \int_\Omega \alpha\varphi g dx, \tag{14}$$

for the first term on the right of (14),

$$\int_\Omega \alpha\varphi\nabla \cdot (\beta\nabla T) dx = \int_\Omega \nabla \cdot (\alpha\varphi\beta\nabla T) dx - \int_\Omega \beta\nabla T \cdot \nabla(\alpha\varphi) dx$$

$$= \int_\Gamma \alpha\gamma\varphi(T_c - T) d\Gamma - \int_\Omega \beta\nabla T \cdot \nabla(\alpha\varphi) dx. \tag{15}$$

Substitute (15) into (14), and we get that

$$\int_\Omega \varphi \frac{\partial T}{\partial t} dx + \int_\Omega \beta\nabla T \cdot \nabla(\alpha\varphi) dx + \int_\Gamma \alpha\gamma\varphi T d\Gamma$$

$$= \int_\Omega \alpha\varphi g dx + \int_\Gamma \alpha\varphi\gamma T_c d\Gamma. \tag{16}$$

For any $(x, t) \in Q$, let $T(x, t) = e^{rt} P(x, t)$, where r is a nonnegative constant, then (16) is transformed into

$$\int_\Omega \varphi \frac{\partial P}{\partial t} dx + \int_\Omega \beta\nabla P \cdot \nabla(\alpha\varphi) dx + r \int_\Omega P\varphi dx + \int_\Gamma \alpha\gamma\varphi P d\Gamma$$

$$= \int_\Omega \alpha\varphi g e^{-rt} dx + \int_\Gamma \alpha\gamma\varphi T_c e^{-rt} d\Gamma.$$

Let $U_r(\alpha, \beta; u, v) = \int_\Omega \beta\nabla u \cdot \nabla(\alpha v) dx + r \int_\Omega uv dx, \forall\ u, v \in V$, then for any $\alpha, \beta \in Q_{ad}$, $U_r(\alpha, \beta; u, v)$ is a continuous linear functional on $V \times V$, so there exists $A_r(\alpha, \beta) \in \mathcal{L}(V, V')$, such that

$$U_r(\alpha, \beta; u, v) = \langle A_r(\alpha, \beta)u, v\rangle, \quad \forall\ u, v \in V.$$

Let $U_\Gamma(\alpha, \gamma; u, v) = \int_\Gamma \alpha\gamma uv d\Gamma, \forall\ u, v \in V$, then $U_\Gamma(\alpha, \gamma; u, v)$ is a bilinear functional on $V \times V$, the embedding operator $H^1(\Omega) \hookrightarrow L^2(\Gamma)$ is linear continuous, and there exists $A_\Gamma(\alpha, \gamma) \in \mathcal{L}(V, V')$, such that

$$U_\Gamma(\alpha, \gamma; u, v) = \langle A_\Gamma(\alpha, \gamma)u, v\rangle, \quad \forall\ u, v \in V.$$

Let $\mathcal{U}_r(q; u, v) = U_r(\alpha, \beta; u, v) + U_\Gamma(\alpha, \gamma; u, v), \forall\ u, v \in V, \forall\ q \in Q_{ad}$, then

$$\mathcal{U}_r(q; u, v) = \int_\Omega \beta\nabla u \cdot \nabla(\alpha v) dx + r \int_\Omega uv dx + \int_\Gamma \alpha\gamma uv d\Gamma. \tag{17}$$

Property 1. *Suppose that assumptions (A1)-(A3) are valid, then for any $q \in Q_{ad}$, $u, v \in V$, $\mathcal{U}_r(q; u, v)$ defined by (17) is a bilinear functional on $V \times V$.*

Proof. For any $q \in Q_{ad}$, any $\lambda_1, \lambda_2 \in \mathbb{R}^1$, any $u, u_1, u_2, v, v_1, v_2 \in V$, we have that

$$\mathcal{U}_r(q; \lambda_1 u_1 + \lambda_2 u_2, v)$$
$$= U_r(\alpha, \beta; \lambda_1 u_1 + \lambda_2 u_2, v) + U_\Gamma(\alpha, \gamma; \lambda_1 u_1 + \lambda_2 u_2, v)$$
$$= \lambda_1(U_r(\alpha, \beta; u_1, v) + U_\Gamma(\alpha, \gamma; u_1, v)) + \lambda_2(U_r(\alpha, \beta; u_2, v) + U_\Gamma(\alpha, \gamma; u_2, v))$$
$$= \lambda_1 \mathcal{U}_r(q; u_1, v) + \lambda_2 \mathcal{U}_r(q; u_2, v),$$

and $\mathcal{U}_r(q; u, \lambda_1 v_1 + \lambda_2 v_2)$
$$= U_r(\alpha, \beta; u, \lambda_1 v_1 + \lambda_2 v_2) + U_\Gamma(\alpha, \gamma; u, \lambda_1 v_1 + \lambda_2 v_2)$$
$$= \lambda_1(U_r(\alpha, \beta; u, v_1) + U_\Gamma(\alpha, \gamma; u, v_1)) + \lambda_2(U_r(\alpha, \beta; u, v_2) + U_\Gamma(\alpha, \gamma; u, v_2))$$
$$= \lambda_1 \mathcal{U}_r(q; u, v_1) + \lambda_2 \mathcal{U}_r(q; u, v_2),$$

which completes our proof. □

Property 2. *Suppose that assumptions (A1)-(A3) are valid, then for any $q \in Q_{ad}$, there exist $M_1 > 0$, $M_2 > 0$, such that*

$$\mathcal{U}_r(q; u, u) \geq M_1 \|u\|_V^2, \quad \forall\, u \in V, \tag{18}$$
$$|\mathcal{U}_r(q; u, v)| \leq M_2 \|u\|_V \|v\|_V, \quad \forall\, u, v \in V. \tag{19}$$

Proof. For any $u \in V$, we obtain that

$$U_r(q; u, u) = \int_\Omega \beta \nabla u \cdot \nabla(\alpha u) dx + r \int_\Omega u^2 dx$$
$$= \int_{\Omega_{sn}} \alpha_1 \beta_1 (\nabla u)^2 dx + \int_{\Omega_{si}} \alpha_2 \beta_2 (\nabla u)^2 dx + r \int_\Omega u^2 dx$$
$$\geq q_L^2 \|\nabla u\|_{L^2(\Omega)}^2 + r \|u\|_{L^2(\Omega)}^2, \tag{20}$$

and

$$U_\Gamma(q; u, u) = \int_\Gamma \alpha \gamma u^2 d\Gamma$$
$$\geq q_L^2 \|u\|_{L^2(\Gamma)}^2$$
$$\geq q_L^2 \left(\frac{1}{m} \|\nabla u\|_{L^2(\Omega)}^2 + 3(1+m) \|u\|_{L^2(\Omega)}^2 \right). \tag{21}$$

From the inequalities (20) and (21), we have that

$$\mathcal{U}_r(q; u, u) = U_r(\alpha, \beta; u, u) + U_\Gamma(\alpha, \gamma; u, u)$$
$$\geq (r + 3(1+m)q_L^2) \|u\|_{L^2(\Omega)}^2 + (q_L^2 + \frac{q_L^2}{m}) \|\nabla u\|_{L^2(\Omega)}^2, \tag{22}$$

take $m = 1$, then for any $r \geq 0$, $r + 3(1+m)q_L^2 > 0$, $q_L^2 + \frac{q_L^2}{m} > 0$, $M_1 = \min\{r + 3(1+m)q_L^2, q_L^2 + \frac{q_L^2}{m}\} > 0$, then $\mathcal{U}_r(q; u, u) \geq M_1 \|u\|_V^2$, $\forall\, u \in V$.

Next we prove the second inequality. According to Poincaré inequality, we can obtain that $\|u\|_{L^2(\Omega)}^2 \leq C \|\nabla u\|_{L^2(\Omega)}^2$, where C is a positive constant, then

$$|\mathcal{U}_r(q; u, v)|$$
$$= |U_r(\alpha, \beta; u, v) + U_\Gamma(\alpha, \gamma; u, v)|$$

$$= |\int_{\Omega} \beta \nabla u \cdot \nabla(\alpha v) dx + r \int_{\Omega} uv dx + \int_{\Gamma} \alpha \gamma uv d\Gamma|$$

$$\leq (r + q_U^2) \|u\|_{L^2(\Omega)} \|v\|_{L^2(\Omega)} + q_U^2 \|\nabla u\|_{L^2(\Omega)} \|\nabla v\|_{L^2(\Omega)}$$

$$\leq B(\|u\|_{L^2(\Omega)} \|v\|_{L^2(\Omega)} + \|\nabla u\|_{L^2(\Omega)} \|\nabla v\|_{L^2(\Omega)})$$

$$\leq B(C + 1) \|\nabla u\|_{L^2(\Omega)} \|\nabla v\|_{L^2(\Omega)}$$

$$\leq M_2 \|u\|_V \|v\|_V,$$

where $B = r + q_U^2$, $M_2 = 2B^2(C + 1)^2$, which completes the proof. $\qquad\square$

From (17), Property 1 and Property 2, we can get that $\mathcal{U}_r(q; u, v)$ is a bilinear continuous functional on $V \times V$, so there exists an $\mathcal{A}_r(q) \in \mathcal{L}(V, V')$, such that

$$\mathcal{U}_r(q; u, v) = \langle \mathcal{A}_r(q)u, v \rangle_{V', V}, \quad \forall \ u, v \in V, \tag{23}$$

that is $\mathcal{A}_r(q) = A_r(\alpha, \beta) + A_\Gamma(\alpha, \gamma)$.

Property 3. *Suppose that assumptions (A1)-(A3) hold, then for any $q \in Q_{ad}$, the following inequalities*

$$\langle \mathcal{A}_r(q)u, u \rangle_{V', V} \geq M_1 \|u\|_V^2, \quad \forall \ u \in V, \tag{24}$$

$$|\langle \mathcal{A}_r(q)u, v \rangle_{V', V}| \leq M_2 \|u\|_V \|v\|_V, \quad \forall \ u, v \in V, \tag{25}$$

are true, where M_1 and M_2 are defined in Property 2.

Let $\mathcal{F}_r(t, \alpha, \gamma; v) = \int_{\Omega} \alpha v g e^{-rt} dx + \int_{\Gamma} \alpha \gamma v T_c e^{-rt} d\Gamma$, $\forall \ v \in V$, then for any $t \in (0, t_f)$, any $\alpha, \gamma \in Q_{ad}$, $\mathcal{F}_r(t, \alpha, \gamma; v)$ is a linear continuous functional on V, so there exists $f_r(t, \alpha, \gamma) \in V'$, that is $f_r(t, \alpha, \gamma) \in \mathcal{L}(V, V')$, such that

$$\mathcal{F}_r(t, \alpha, \gamma; v) = \langle f_r(t, \alpha, \gamma), v \rangle, \quad \forall \ v \in V. \tag{26}$$

From (23) and (26), then U3D2LTS can be written as an evolution equation on V' denoted by EESS.

$$\frac{dP}{dt} + \mathcal{A}_r(q)P = f_r(t, \alpha, \gamma), \tag{27}$$

$$P(0) = P_0. \tag{28}$$

Where $P_0 = T_0$. Next we will present the existence and uniqueness of solution of EESS.

Property 4. *Suppose that assumptions (A1)-(A3) hold, then for any $q \in Q_{ad}$, EESS has a unique weak solution $P(t; q) \in L^2(0, t_f; V) \cap C(0, t_f; V)$.*

Proof. The proof is as Theorem 3.16 in Ref.[18]. $\qquad\square$

Property 5. *Under the assumptions of Property 4, then U3D2LTS also has a unique weak solution $T(x, t; q) \in L^2(Q; V) \cap C(Q; V)$, for any $q \in Q_{ad}$.*

Define $S(Q_{ad}) = \{T(x,t;q) \in L^2(Q;V) \cap C(Q;V) | T(x,t;q)$ is the weak solution of U3D2LTS corresponding to $q \in Q_{ad}\}$ be the solution set of U3D2LTS.

4   Optimality Conditions of U3D2LTS

In this section, we will consider the parameter identification problem of U3D2LTS. Let $\tilde{T}_{x,t}$ be the observed temperature data in $\overline{Q}$, $g_{sn}(x,t)$, $g_{si}(x,t)$, $T_0(x)$, $T_{at}(x,t)$ and $T_{io}(x,t)$ are all given. Define the performance criterion as follows

$$J(q) = \|T(x,t;q) - T_{mea}(x,t)\|^2_{L^2(Q)}, \qquad (29)$$

where $T(x,t;q)$ is the temperature obtained from U3D2LTS, $T_{mea}(x,t)$ is a fitted continuous temperature function of $\tilde{T}_{x,t}$ in the selected area $\overline{\Omega}$ and the time interval $[0, t_f]$, where $q \in Q_{ad}$. Our goal is to obtain the value of the parameter $q \in Q_{ad}$ to make $T(x,t;q)$ approach $T_{mea}(x,t)$, that is

$$PIP: \quad min \quad J(q)$$
$$s.t. \quad T(x,t;q) \in S(Q_{ad}), q \in Q_{ad}. \qquad (30)$$

Where $J(q)$ is defined by (29), $S(Q_{ad})$ is the solution set of U3D2LTS.

4.1   Existence of Optimal Parameter of PIP

To get the existence of optimal solution of PIP, first we will consider the strong continuity of the mapping $\prod : q \to T(x,t;q)$.

Theorem 1. *Suppose that assumptions (A1)-(A3) hold, then the mapping $\prod :$ $q \to T(x,t;q)$ is strongly continuous, where $q \in Q_{ad}$, $T(x,t;q) \in S(Q_{ad})$.*

Proof. For any fixed parameter $q_0 = (\alpha_1^0, \alpha_2^0, \beta_1^0, \beta_2^0, \gamma_1^0, \gamma_2^0) \in Q_{ad}$, let $\{q_n\} \subset Q_{ad}$ be a sequence, such that $\|q_n - q_0\|_{Q_{ad}} \to 0$, as $n \to \infty$, where $q_n = (\alpha_1^n, \alpha_2^n, \beta_1^n, \beta_2^n, \gamma_1^n, \gamma_2^n)$; let $T_n = T_n(x,t;q_n)$ and $T^0 = T^0(x,t;q_0)$ be the weak solutions of U3D2LTS corresponding to q_n and q_0 respectively; set $\omega_n = T_n - T^0$, $\alpha_0 = \{\alpha_1^0, \alpha_2^0\}$, $\alpha_n = \{\alpha_1^n, \alpha_2^n\}$, $\beta_0 = \{\beta_1^0, \beta_2^0\}$, $\beta_n = \{\beta_1^n, \beta_2^n\}$, $\gamma_0 = \{\gamma_1^0, \gamma_2^0,$ $\gamma_n = \{\gamma_1^n, \gamma_2^n\}$, then we have

$$\int_\Omega \omega_n \frac{\partial \omega_n}{\partial t} dx - \int_\Omega \omega_n \alpha_n \nabla \cdot (\beta_n \nabla \omega_n) dx$$
$$= \int_\Omega \omega_n \alpha_n \nabla \cdot (\beta_n \nabla T^0) dx - \int_\Omega \omega_n \alpha_0 \nabla \cdot (\beta_0 \nabla T^0) dx + \int_\Omega \omega_n (\alpha_n - \alpha_0) g dx,$$
$$\qquad (31)$$

$$-\beta_n \frac{\partial \omega_n}{\partial \boldsymbol{n}} - \gamma_n \omega_n = (\beta_n - \beta_0) \frac{\partial T^0}{\partial \boldsymbol{n}} + (\gamma_n - \gamma_0)(T^0 - T_c), \qquad (32)$$
$$\omega_n(x,0) = 0. \qquad (33)$$

For the equation (31), integrate over $[0, t]$, we have

$$\int_0^t \int_\Omega \omega_n \frac{\partial \omega_n}{\partial s} dx ds - \int_0^t \int_\Omega \alpha_n \omega_n \nabla \cdot (\beta_n \nabla \omega_n) dx ds$$
$$= \int_0^t \int_\Omega \alpha_n \omega_n \nabla \cdot (\beta_n \nabla T^0) dx ds - \int_0^t \int_\Omega \alpha_0 \omega_n \nabla \cdot (\beta_0 \nabla T^0) dx ds$$
$$+ \int_0^t \int_\Omega (\alpha_n - \alpha_0) \omega_n g dx ds, \tag{34}$$

then (34) can be simplified as

$$\int_0^t \int_\Omega \omega_n \frac{\partial \omega_n}{\partial s} dx ds + \int_0^t \int_\Omega \beta_n \nabla \omega_n \cdot \nabla(\alpha_n \omega_n) dx ds + \int_0^t \int_\Gamma \alpha_n \gamma_n \omega_n^2 d\Gamma ds$$
$$= \int_0^t \int_\Omega \omega_n ((\alpha_n - \alpha_0) \nabla (\beta_n \nabla T^0) + \alpha_0 \nabla ((\beta_n - \beta_0) \nabla T^0)) dx ds$$
$$+ \int_0^t \int_\Gamma \omega_n (\alpha_n (\beta_0 - \beta_n) \frac{\partial T^0}{\partial n} + \alpha_n (\gamma_0 - \gamma_n)(T^0 - T_c)) dx ds$$
$$+ \int_0^t \int_\Omega \omega_n (\alpha_n - \alpha_0) g dx ds. \tag{35}$$

For the left term of (35), from (24), we get that

$$LEFT = \frac{1}{2} \|\omega_n(t)\|_{L^2(\Omega)}^2 + \int_0^t \langle A_0(\alpha_n, \beta_n) \omega_n, \omega_n \rangle_{V', V} ds$$
$$+ \int_0^t \langle A_\Gamma(\alpha_n, \gamma_n) \omega_n, \omega_n \rangle_{V', V} ds$$
$$= \frac{1}{2} \|\omega_n(t)\|_{L^2(\Omega)}^2 + \int_0^t \langle A_0(q_n) \omega_n, \omega_n \rangle_{V', V} ds$$
$$\geq \frac{1}{2} \|\omega_n(t)\|_{L^2(\Omega)}^2 + M_1 \int_0^t \|\omega_n(s)\|_V^2 ds, \tag{36}$$

where $A_0(q_n) = A_0(\alpha_n, \beta_n) + A_\Gamma(\alpha_n, \gamma_n)$.

For the right term of (35),

$$RIGHT \leq \frac{M_1}{4} \int_0^t \|\omega_n(s)\|_{L^2(\Omega)}^2 ds$$
$$+ \frac{3}{M_1} (\int_0^t \|(\alpha_n - \alpha_0) \nabla \cdot (\beta_n \nabla T^0) + \alpha_0 \nabla ((\beta_n - \beta_0) \nabla T^0)\|_{L^2(\Omega)}^2 ds$$
$$+ \int_0^t \|\alpha_n (\beta_0 - \beta_n) \frac{\partial T^0}{\partial n} + \alpha_n (\gamma_0 - \gamma_n)(T^0 - T_c)\|_{L^2(\Gamma)}^2 ds$$
$$+ \int_0^t \|(\alpha_n - \alpha_0) g\|_{L^2(\Omega)}^2 ds). \tag{37}$$

Substitute (36) and (37) into (35), we can obtain that

$$\|\omega_n(t)\|_{L^2(\Omega)}^2 + \frac{3M_1}{2}\int_0^t \|\omega_n(s)\|_V^2 ds$$

$$\leq \frac{6}{M_1}\Big(\int_0^t \|(\alpha_n - \alpha_0)\nabla\cdot(\beta_n\nabla T^0) + \alpha_0\nabla((\beta_n - \beta_0)\nabla T^0)\|_{L^2(\Omega)}^2 ds$$

$$+ \int_0^t \|\alpha_n(\beta_0 - \beta_n)\frac{\partial T^0}{\partial \boldsymbol{n}} + \alpha_n(\gamma_0 - \gamma_n)(T^0 - T_c)\|_{L^2(\Gamma)}^2 ds$$

$$+ \int_0^t \|(\alpha_n - \alpha_0)g\|_{L^2(\Omega)}^2 ds\Big)$$

$$+ \int_0^t (\|\omega_n(s)\|_{L^2(\Omega)}^2 + \frac{3M_1}{2}\int_0^s \|\omega_n(\theta)\|_V^2 d\theta)ds.$$

Set $Y_n(t) = \|\omega_n(t)\|_{L^2(\Omega)}^2 + \frac{3M_1}{2}\int_0^t \|\omega_n(s)\|_V^2 ds$, using Gronwall's lemma, we have that

$$Y_n(t) \leq e^{t_f}\frac{6}{M_1}\Big(\int_0^t \|(\alpha_n - \alpha_0)\nabla\cdot(\beta_n\nabla T^0) + \alpha_0\nabla((\beta_n - \beta_0)\nabla T^0)\|_{L^2(\Omega)}^2 ds$$

$$+ \int_0^t \|\alpha_n(\beta_0 - \beta_n)\frac{\partial T^0}{\partial \boldsymbol{n}} + \alpha_n(\gamma_0 - \gamma_n)(T^0 - T_c)\|_{L^2(\Gamma)}^2 ds$$

$$+ \int_0^t \|(\alpha_n - \alpha_0)g\|_{L^2(\Omega)}^2 ds\Big).$$

Since $\|q_n - q_0\|_{Q_{ad}} \to 0$, as $n \to \infty$, so we can get that $\Pi : q \to T(x,t;q)$ is strongly continuous, and obtain the desired result. □

Theorem 2. *Suppose the assumptions (A1)-(A3) hold, then there exists at least one optimal parameter $q^* = (\alpha_1^*, \alpha_2^*, \beta_1^*, \beta_2^*, \gamma_1^*, \gamma_2^*) \in Q_{ad}$ satisfying (31).*

Proof. For any $q \in Q_{ad}$, let $X(T;q) = T(x,t;q) - T_{mea}(x,t)$, then $J(q) = \|X(T;q)\|_{L^2(Q)}^2 \geq 0$. Obviously, $X(T;q)$ is continuous on V. Hence, by Theorem 1, the mapping $\Pi : q \to T(x,t;q)$ is continuous for all $q \in Q_{ad}$, so $q \to J(q)$ is continuous on Q_{ad}. Since Q_{ad} is a nonempty bounded closed set, there exists $q^* \in Q_{ad}$, such that for all $q \in Q_{ad}$, $J(q^*) \leq J(q)$, i.e. $q^* \in Q_{ad}$ is an optimal parameter, thus we obtain the desired result. □

4.2   Necessary Conditions for Optimality

Here we will derive the necessary conditions for optimality. For this purpose, we will give a theorem as follows.

Theorem 3. *Assume assumption (A1)-(A3) hold, let $P(q)$ be the solution of EESS corresponding to $q \in Q_{ad}$, then at each point $q^0 \in Q_{ad}$ the function $q \to P(q)$ has a weak Gâteaux differential in the direction $q - q^0$, denoted by $P'(q^0; q - q^0)$, and it is the solution of the Cauchy problem*

$$\varphi_t + \mathcal{A}_r(t, q^0)\varphi = -\mathcal{A}_r'(t, q^0; q - q^0)P(q_0) + f_r'(t, q^0; q - q^0), \quad (38)$$

$$\varphi(0) = 0, t \in I, \quad (39)$$

where $P(q^0)$ is the solution of EESS corresponding to $q = q^0$, $\mathcal{A}'_r(q^0; q - q^0)$ and $f'_r(t, q^0; q - q^0)$ denote the weak Gâteaux differential of $\mathcal{A}_r(q)$ and $f_r(t, q)$ in the direction $q - q^0$.

Proof. Since Q_{ad} is a closed convex set, set $q_\delta = q^0 + \delta(q - q^0) \in Q_{ad}$ for $0 \leq \delta \leq 1$, $\varphi^\delta = \frac{P(q_\delta) - P(q^0)}{\delta}$. Then from (23), we obtain

$$\varphi^\delta_t + \mathcal{A}_r(q_\delta)\varphi^\delta = \frac{f_r(t, q_\delta) - f_r(t, q^0)}{\delta} + \frac{\mathcal{A}_r(q^0) - \mathcal{A}_r(q^\delta)}{\delta} P(q^0).$$

Let $\delta \to 0$, then we obtain the desired result.   □

With the help of the above Theorem 3, we get the following necessary conditions for optimality.

Theorem 4. *Suppose that assumptions (A1)-(A3) are valid, let $q^* \in Q_{ad}$ is an optimal solution of PIP, and $T(q^*) \in S(Q_{ad})$ satisfies the following equations,*

$$\frac{dT(q^*)}{dt} + \mathcal{A}_0(q^*)T(q^*) = f_0(t, q^*), \tag{40}$$

$$T(0, q^*) = T_0, \tag{41}$$

then the adjoint state $\Psi(q^*)$ of $T(q^*)$ is determined by the adjoint equations

$$-\frac{d\Psi(q^*)}{dt} + \mathcal{A}^*_0(q^*)\Psi(q^*) = 2(T(q^*) - T_{mea}), \tag{42}$$

$$\Psi(t_f, q^*) = 0, \tag{43}$$

and the inequality

$$\int_0^{t_f} \langle -\mathcal{A}'_0(q^*; q - q^*)T(q^*), \Psi(q^*) \rangle_{V', V} dt + \int_0^{t_f} \langle f'_0(t, q^*; q - q^*), \Psi(q^*) \rangle_{V', V} dt \geq 0. \tag{44}$$

is true, where $\mathcal{A}^*_0(q^*)$ is the conjugate operator of $\mathcal{A}_0(q^*)$, $\mathcal{A}'_0(q^*; q - q^*)$ and $f'_0(t, q^*; q - q^*)$ are the weak Gâteaux differentials of $\mathcal{A}_0(q^*)$ and $f_0(t, q^*)$ in the direction $q - q^*$, $T_{mea} = T_{mea}(x, t)$ is the fitted temperature function of the observed data, $\mathcal{A}_0(q^*) = A_0(\alpha^*, \beta^*) + A_\Gamma(\alpha^*, \gamma^*)$.

Proof. Since $\Pi : q \to T(q)$ has a weak Gâteaux differential, it follows that $J(q)$ as defined by (25) also has a Gâteaux differential. Then in order that $J(q)$ attains its minimum at $q^* \in Q_{ad}$, it is necessary that

$$J'(q^*)(q - q^*) = \lim_{\sigma \to 0} \frac{J(q^* + \sigma(q - q^*)) - J(q^*)}{\sigma} \geq 0, \tag{45}$$

for all $q \in Q_{ad}$. Using the result of Theorem 3, it follows from the above that

$$J'(q^*)(q - q^*) = 2 \int_0^{t_f} \langle T'(q^*; q - q^*), T(q^*) - T_{mea} \rangle_{L^2(\Omega)} dt \geq 0. \tag{46}$$

Let $\Psi(q^*)$ is the adjoint function of $T(q^*)$, it is the solution of the following equation

$$-\frac{d\Psi(q^*)}{dt} + \mathcal{A}_0^*(q^*)\Psi(q^*) = 2(T(q^*) - T_{mea}), \tag{47}$$

$$\Psi(t_f, q^*) = 0. \tag{48}$$

Then, since $T(q^*) - T_{mea} \in L^2(\Omega)$, reversing the flow of time $t \to t_f - t$, it follows from Property 5 that the adjoint system also has a unique solution $\Psi(q^*)$.

Next we will prove the inequality (44).

Multiplying both sides of (47) by $T(q) - T(q^*)$ and integrating over $[0, t_f]$, we have

$$\int_0^{t_f} \langle \Psi(q^*), \frac{d(T(q) - T(q^*))}{dt} + \mathcal{A}_0(q^*)(T(q) - T(q^*)) \rangle_{V,V'} dt$$

$$= 2 \int_0^{t_f} \langle T(q^*) - T_{mea}, T(q) - T(q^*) \rangle_{L^2(\Omega)} dt. \tag{49}$$

Let $q_\theta = q^* + \theta(q - q^*)$, $0 \le \theta \le 1$, and multiply (49) by $\frac{1}{\theta}$,

$$\int_0^{t_f} \langle \Psi(q^*), \frac{d}{dt}(\frac{T(q) - T(q^*)}{\theta}) + \mathcal{A}_0(q^*)\frac{T(q) - T(q^*)}{\theta} \rangle_{V,V'} dt$$

$$= 2 \int_0^{t_f} \langle T(q^*) - T_{mea}, \frac{T(q) - T(q^*)}{\theta} \rangle_{L^2(\Omega)} dt.$$

Let $\theta \to 0$, we obtain

$$\int_0^{t_f} \langle \Psi(q^*), \frac{dT'(q^*; q - q^*)}{dt} + \mathcal{A}_0(q^*)T'(q^*; q - q^*) \rangle_{V,V'} dt$$

$$= 2 \int_0^{t_f} \langle T(q^*) - T_{mea}, T'(q^*; q - q^*) \rangle_{L^2(\Omega)} dt.$$

Since $T'(q^*; q - q^*)$ is the solution of (38) by Theorem 3,

$$\int_0^{t_f} \langle -\mathcal{A}_0'(q^*; q - q^*)T(q^*) + f_0'(q^*; q - q^*), \Psi(q^*) \rangle_{V',V} dt$$

$$= 2 \int_0^{t_f} \langle T'(q^*; q - q^*), T(q^*) - T_{mea} \rangle_{L^2(\Omega)} dt.$$

Then from (46), we obtain the desired result.   □

5   Conclusions

In this paper, we have considered an unsteady three-dimension two-layer thermodynamic system coupled by the snow, the sea ice, presented the properties and derived the optimality conditions of the system. Thus we provide the mathematical foundation for the numerical computation of the parameter identification problems of the three-dimensional two-layer thermodynamic system of sea ice. The optimization algorithm and numerical results will be presented in a forthcoming paper.

Acknowledgments

This work was supported by Innovation Fund Project of Shanghai University under Grant No. A.10-0101-08-413, Shanghai Outstanding Young Teachers special Research Project under Grant No. B.37- 0101-08-004, and Shanghai Leading Academic Discipline Project under Grant No. S30104.

References

1. Haskell, T.: What's so important about sea ice? Water & Atmosphere 11(3), 28–29 (2003)
2. Wolff, E.: Whithor Antarctic sea ice? Science 302, 1164 (2003)
3. Laxon, S., Peacock, N., Smith, D.: High interannual variability of sea ice thickness in the Arctic region. Nature 425, 947–950 (2003)
4. Maykut, G.A., Untersteiner, N.: Some results from a time-dependent thermodynamic model of sea ice. J. Geophys. Res. 76, 1550–1575 (1971)
5. Gabison, R.: A thermodynamic model of the formation growth and decay of first-year sea ice. J. Glaciol. 33, 105–109 (1987)
6. Ebert, E.E., Curry, J.A.: An intermediate one-dimensional thermodynamic sea-ice model for investigating ice-atmosphere interaction. J. Geophys. Res. 98(C6), 10085–10109 (1993)
7. Cheng, B.: On the numerical resolution in a thermodynamic sea-ice model. J. Glaciol. 48(161), 301–311 (2002)
8. Reid, T., Crout, N.: A thermodynamic model of freshwater Antarctic lake ice. Ecol. Model. 210, 231–241 (2008)
9. Shidfar, A., Karamali, G.R.: Numerical solution of inverse heat conduction problem with nonstationary measurements. Appl. Math. Comput. 168(1), 540–548 (2005)
10. Li, G.S.: Data compatibility and conditional stability for an inverse source problem in the heat equation. Appl. Math. Comput. 173, 566–581 (2006)
11. Shidfar, A., Karamali, G.R., Damirchi, J.: An inverse heat conduction problem with a nonliear source term. Nonlinear Anal. 65, 615–621 (2006)
12. Ikehata, M.: An inverse source problem for the heat equation and the enclosure method. Inverse Probl. 23, 183–202 (2007)
13. Christov, C.I., Marinov, T.: Identification of heat-conduction coefficient via method of variational imbedding. Math. Comput. Modell. 27(3), 109–116 (1998)
14. Engl, H.W., Zou, J.: A new approach to convergence rate analysis of Tikhonov regularization for parameter identification in heat conduction. Inverse Probl. 16, 1907–1923 (2000)
15. Telejko, T., Malinowski, A.: Application of an inverse solution to the thermal conductivity identification using the finite element method. J. Mater. Process. Technol. 146, 145–155 (2004)
16. Lv, W., Feng, E., Li, Z.: A coupled thermodynamic system of sea ice and its parameter identification. Appl. Math. Modell. 32, 1198–1207 (2008)
17. Lv, W., Feng, E., Lei, R.: Parameter Identification for a Nonlinear Thermodynamic System of Sea Ice. International Journal of Thermal Sciences 48, 195–203 (2009)
18. Wang, Y.: L^2 Theories on Patial Differential Equation. Beijing University Press, Beijing (1989)

Optimal Service Capacities in a Competitive Multiple-Server Queueing Environment

Wai-Ki Ching[1], Sin-Man Choi[1], and Min Huang[2]

[1] Advanced Modeling and Applied Computing Laboratory,
Department of Mathematics,
The University of Hong Kong
{wching,kellyci}@hkusua.hku.hk
[2] College of Information Science and Engineering,
Northeastern University, Shenyang, 110004, China
Key Laboratory of Integrated Automation of Process Industry,
Ministry of Education, Shenyang, 110004, China
mhuang@mail.neu.edu.cn

Abstract. The study of economic behavior of service providers in a competition environment is an important and interesting research issue. A two-server queueing model has been proposed in Kalai et al. [11] for this purpose. Their model aims at studying the role and impact of service capacity in capturing larger market share so as to maximize the long-run expected profit. They formulate the problem as a two-person strategic game and analyze the equilibrium solutions. The main aim of this paper is to extend the results of the two-server queueing model in [11] to the case of multiple servers. We will only focus on the case when the queueing system is stable.

Keywords: Markovian Queueing Systems, n-server Queue, Nash Equilibrium, Competition.

1 Introduction

The problem of finding the optimal strategy and control policy of a queueing system is a traditional mathematical problem and has been well studied in the literature, see for instance [2,9,10,11,12,17]. In an optimal control problem, it usually involves making decisions on system parameters such as the system service capacity and number of servers in the system under a specified cost structure (convex or concave). Here service capacity is an important competitive factor in the design of a service system, for example, in the areas of telecommunication networks [6] data transmission systems [11] and Vendor-Managed Inventory (VMI) system [3,16]. In particular, the current development in supply chain management emphasizes the coordination and integration of inventory and transportation logistics [4,18]. VMI is a supply chain initiative where the distributor is responsible for all decisions regarding the selection of retailers or agents. This creates a competitive environment for the agents and retailers to compete in the market [14].

J. Zhou (Ed.): Complex 2009, Part I, LNICST 4, pp. 66–77, 2009.

Kalai et al. [11] studied a strategic game of two servers competing for their market shares through determining their service capacities. A Markovian queueing system of two servers is used in their model and analysis. Markovian queueing systems are popular tools for modeling servicing systems as they are mathematically tractable [6,7] when compared to the non-Markovian queueing systems. The problem is then analyzed using game theory [15]. Game theory is a popular and promising approach [1,5] for the captured problem. They classified the Nash equilibria into three different cases concerning the cost function and the revenue per customer. The waiting time is finite in one of these cases and there is a unique symmetric equilibrium. Although their model is simple, it brings in two important concepts. The first one is the "competitive game of servers" and the second one is "the market share of a server in a multi-server facility". Furthermore, they also report that when the marginal cost of providing service is "high", there is a unique symmetric equilibrium and the total service capacity is less than the mean demand rate. In such a case, each server actually behaves as if it were a monopolist. Competition therefore has no effect and this leads to an undesirable situation. On the other hand, when the marginal cost of providing service is "low", a unique symmetric equilibrium exists and the total service capacity is greater than the mean demand rate. In this paper, we will extend the model in [11] by allowing the number of servers to be more than two. In particular, we are interested in the case when the total service capacity is greater than the mean demand rate.

The remainder of the paper is structured as follows. In Section 2, we will give a brief review on the two-server queueing system discussed in [11] and the analytic results therein. We then present our multiple-server queueing system and also our analysis on the system performance in Section 3. A numerical demonstration is given in Section 4 for the case of a 3-server queueing systems. Finally concluding remarks are given to address further research issues in Section 5.

2 A Review on the Two-Server Queueing System

The service system studied in Kalai et al [11] consists of two independently operated servers. Customers arrive according to a Poisson process of rate λ and the service times are assumed to follow the exponential distribution. Each of the server i operates independently and determines its own service capacity μ_i so as to maximize its own profits. The cost to operate at service capacity μ is $c(\mu)$. Here the operating cost function $c(.)$ is assumed to an increasing and strictly convex function, i.e., both $c'(\mu)$ and $c''(\mu)$ are both positive and an example of such a function is $c(\mu) = \mu^2$.

The servers earn a fixed amount R for each unit of service rendered. The queueing system consists of a single First-In-First-Out queue. If a customer arrives when both servers are idle, the customer will be assigned to either server with equal likelihood. No server is allowed to be idle when at least one customer in the system. If a customer arrives when one server is idle and the other is busy, he/she will be assigned to the idle server. In the following subsections, we will present briefly the main results obtained in [11] concerning the two-server queueing system.

2.1   The System Steady-State Probability Distribution

If Server i $(i = 1, 2)$ chooses service capacity μ_i and such that

$$\mu_1 + \mu_2 > \lambda \tag{1}$$

the system has a steady-state probability distribution. We remark that condition (1) is a necessary and sufficient condition for the Markovian queueing system to be stable or to have steady-state probability distribution. Let P_n be the probability that there are n customers in the system; P_{10} be the probability that server 1 is busy and server 2 is idle; P_{01} be the probability that server 2 is busy and server 1 is idle. By studying the balanced equations of the queueing system, we have the following results:

$$P_0 = \frac{1 - \rho}{1 - \rho + \frac{\lambda(\mu_1 + \mu_2)}{2\mu_1\mu_2}} \quad \text{and} \quad P_{10} = \frac{\lambda P_0}{2\mu_1} \quad \text{and} \quad P_{01} = \frac{\lambda P_0}{2\mu_2} \tag{2}$$

where

$$\rho = \frac{\lambda}{(\mu_1 + \mu_2)} \tag{3}$$

is the system load. Moreover, we also have

$$P_1 = P_{10} + P_{01} \quad \text{and} \quad P_n = \rho^{n-1} P_1 \quad n = 2, 3, \ldots \tag{4}$$

2.2   The Market Share

Computing the market share of Server i is equivalent to computing the mean number of customers per time unit that enter service with Server i. Using the results in Section 2.1, if $\mu_1 + \mu_2 > \lambda$, the mean number of customers per time unit that enter service with Server 1 is

$$P_0 \frac{\lambda}{2} + P_{01}\lambda + P_3\mu_1 + P_4\mu_1 + \ldots \tag{5}$$

and that with Server 2 is

$$P_0 \frac{\lambda}{2} + P_{10}\lambda + P_3\mu_2 + P_4\mu_2 + \ldots \tag{6}$$

We then divide by the mean number of customers per time unit that enter service, i.e., λ, to obtain the *market share* of Server i. Thus the fraction of all customers served by Server $i(i = 1, 2)$, is given by

$$\alpha_i(\mu_1, \mu_2) = \frac{\lambda \mu_i^2 + \mu_1 \mu_2 (\mu_1 + \mu_2)}{\lambda(\mu_1 + \mu_2)^2 + 2\mu_1 \mu_2 (\mu_1 + \mu_2 - \lambda)}. \tag{7}$$

2.3 The Profit Function

Given the market shares of the servers in Section 2.2, the profit function $\pi_i(\mu_1, \mu_2)$ for Server $i \in \{1, 2\}$, the expected profit per time unit earned by Server i, is then given by

$$\pi_i(\mu_1, \mu_2) = \begin{cases} R\lambda \alpha_i(\mu_1, \mu_2) - c(\mu_i) & \text{if } \mu_1 + \mu_2 > \lambda \\ R\mu_i - c(\mu_i) & \text{if } \mu_1 + \mu_2 \leq \lambda. \end{cases} \tag{8}$$

Here $c(\mu)$ is the cost of providing service at capacity μ and R is the revenue per customer served.

2.4 The Nash Equilibrium of the Queueing System

Kalai et al. [11] considered the situation as a two-person strategic game and they found that finite waiting times exist at equilibrium if and only if

$$c'(\frac{\lambda}{2}) < \frac{R}{2}. \tag{9}$$

Moreover, if this condition is satisfied, then a unique equilibrium exists in which both servers select the same service capacity $\mu_c = \mu_1 = \mu_2$ such that

$$c'(\mu_c) = \frac{R\lambda^2}{2\mu_c(2\mu_c + \lambda)}. \tag{10}$$

3   The General Multiple-Server Queueing System

In this section, we extend the two-server queueing system studied in [11] to a general n-server queueing system. The arrival process of customers is assumed to be a Poisson process. In this queueing system, arriving customers wait in a single First-In-First-Out (FIFO) queue if all servers are busy. No server is allowed to be idle when there is at least one customer in the queueing system. If a customer arrives when more than one server is idle, the customer is assigned to any of the idle servers with equal likelihood. Once a server completes the service of a customer, the first customer in the queue, if any, is assigned to the server. Each server i may choose its own service capacity μ_i, and its service time follows the exponential distribution with mean $1/\mu_i$. The servers earn a revenue of R per customer served, and each of them incurs a cost of $c(\mu)$ to operate at service capacity μ, where $c(.)$ is an increasing and strictly convex function, i.e., both $c'(.)$ and $c''(.)$ are both positive.

In the following subsections, we present some important properties of the multiple-server queueing system through the propositions. The proofs of the propositions are omitted but can be found in [8].

3.1   The Steady-State Distribution of the Queueing System

Given the service capacities $\mu_1, \ldots, \mu_n$ and the mean demand rate λ, suppose $\sum_{i=1}^{n} \mu_i > \lambda$. This condition is to guarantee that the queueing system is stable and the system steady-state probability distribution exists. We would like to obtain the steady-state probability distribution of the number of customers in the system. Let us give the following definitions. Let P_i be the steady-state probability of having i customers in the system, where $i = 0, 1, 2, \ldots$. Also let $P_{\mathbf{s}}$, where $\mathbf{s} = (s_1, s_2, \ldots, s_n)$ and $s_i = 0$ or 1, be the steady-state probability of having s_i customers at Server i. We note that by definition

$$P_k = \sum_{\{\mathbf{s}|s_1+\ldots+s_n=k\}} P_{\mathbf{s}} \quad \text{for} \quad k = 0, 1, \ldots, n. \tag{11}$$

We establish the equations governing the steady-state probabilities. The equations can be obtained by equating the incoming rate and outgoing rate at each of the state. For $s_i = 0, 1$ and $\sum_{i=1}^{n} s_i \neq n$, we have

$$(\sum_{\{i|s_i=1\}} \mu_i + \lambda) P_{(s_1,s_2,\ldots,s_n)} = \sum_{\{i|s_i=0\}} \mu_i P_{(s_{-i},s_i=1)} + \sum_{\{i|s_i=1\}} \frac{\lambda P_{(s_{-i},s_i=0)}}{|\{j|s_j=0\}|+1}. \tag{12}$$

where (s_{-i}, s_i') denotes $(s_1, \ldots, s_{i-1}, s_i', s_{i+1}, \ldots, s_n)$. When $s_i = 0$ for all i this gives

$$\lambda P_{(0,0,\ldots,0)} = \mu_1 P_{(1,0,\ldots,0)} + \mu_2 P_{(0,1,0,\ldots,0)} + \cdots + \mu_n P_{(0,\ldots,0,1)}. \tag{13}$$

For the states with at least n customers we have

$$(\sum_{i=1}^{n} \mu_i + \lambda) P_{(1,1,\ldots,1)} = \sum_{i=1}^{n} \mu_i P_{n+1} + \sum_{i=1}^{n} \lambda P_{(s_{-i}=\mathbf{1},s_i=0)} \tag{14}$$

and

$$(\sum_{i=1}^{n} \mu_i + \lambda) P_k = (\sum_{i=1}^{n} \mu_i) P_{k+1} + \lambda P_{k-1} \text{ for } k = n+1, n+2, \ldots. \tag{15}$$

We note that these two equations together are equivalent to

$$(\sum_{i=1}^{n} \mu_i + \lambda) P_k = (\sum_{i=1}^{n} \mu_i) P_{k+1} + \lambda P_{k-1} \text{ for } k = n, n+1, \ldots. \tag{16}$$

We also have the normalization equation

$$\sum_{i=0}^{\infty} P_i = 1. \tag{17}$$

It can be shown by direct verification that the solution is given by the following proposition.

Proposition 1. *We have*

$$P_{(s_1,s_2,\dots,s_n)} = \frac{(n-k)!\lambda^k P_0}{n! \prod_{\{i|s_i=1\}} \mu_i} \qquad where \quad k = s_1 + s_2 + \dots + s_n > 0 \qquad (18)$$

and

$$P_k = \rho_{k-n} P_n \qquad for \quad k > n \qquad (19)$$

and

$$P_0 = \left(1 + \sum_{k=1}^{n-1} \frac{(n-k)!\lambda^k (\sum_{i_1 < i_2 < \dots < i_{n-k}} \mu_{i_1}\mu_{i_2}\cdots\mu_{i_{n-k}})}{n!\mu_1\mu_2\cdots\mu_n} + \left(\frac{1}{1-\rho}\right)\frac{\lambda^n}{n!\mu_1\mu_2\cdots\mu_n}\right)^{-1}.$$
$$(20)$$

The steady-state probability distribution describes the long run behavior of the system. Each of these probabilities P_k represents the long-run proportion of time that there are k customers in the system. They are essential in studying how each server determines its strategy to maximize its long-run profit. In the next subsection, we will write the market share of each server in terms of these probabilities and obtain an expression for the market share.

3.2 The Market Share of Each Server

We derive the market share of each server from the steady-state distribution. We note that when $\sum_{j=1}^{n} \mu_j \leq \lambda$, i.e., customers arrive at least as fast as the servers can serve them, the steady-state probability distribution does not exist and the queue is infinite. In this case, each server receives customers at its service capacity in the long run. Otherwise, $\sum_{j=1}^{n} \mu_j > \lambda$ and all customers will be served. Each server only receives a fraction of the arriving customers, at a rate lower than its service capacity. The server's profit thus depends on the fraction of all customers it serves, i.e. its market share.

When $k(1 \leq k \leq n)$ servers are idle, customers arrive at a rate of λ and an arriving customer is served by any one of the k idle servers with equal likelihood. Each of these idle servers therefore receives customers at a rate of λ/k. On the other hand, when all servers are busy with at least one customer waiting in the system, each of the busy servers i receives a new customer when it completes the service for a customer, i.e. at a rate of its service capacity μ_i.

To obtain the market share, we find the expected value of the server's rate of receiving customers in different states of the systems, taking expectation over the steady-state probabilities. In the following, we give the formula for the market share for an individual server.

Proposition 2. *If $\sum_{j=1}^{n} \mu_j > \lambda$, the market share of Server i, $\alpha_i(\mu_1,\mu_2,\dots,\mu_n)$ is given by*

$$\frac{\mu_i \left[\sum_{k=0}^{n-1} k!\lambda^{n-k-1} \left(\sum_{j_1<j_2<\dots<j_k, j_p \neq i \,\forall p} \mu_{j_1}\mu_{j_2}\cdots\mu_{j_k} \right) + \lambda^{n-1}\left(\frac{\rho}{1-\rho}\right) \right]}{\sum_{k=1}^{n} k!\lambda^{n-k}\left(\sum_{j_1<j_2<\dots<j_k} \mu_{j_1}\mu_{j_2}\cdots\mu_{j_k} \right) + \frac{\lambda^n}{1-\rho}}. \qquad (21)$$

As we focus on the case when the mean demand rate is less than the total service rate, the market share is directly tied to the profit of a server. Before formulating the profit function of a server, we state the following two propositions related to the partial derivatives of the market share α_i with respect to μ_i. These will be useful in determining the Nash equilibrium of the system when we considered the system as a n-player strategic game.

Proposition 3. *Suppose that* $\sum_{j=1}^{n} \mu_j > \lambda$ *then* $\frac{\partial \alpha_i(\mu_1,\mu_2,...,\mu_n)}{\partial \mu_i} > 0$. *Furthermore, when* $\mu_i \to \infty$, *we have* $\frac{\partial \alpha_i(\mu_1,...,\mu_n)}{\partial \mu_i} \to 0$.

Proposition 4. *Suppose that* $\sum_{j=1}^{n} \mu_j > \lambda$, *then* $\frac{\partial^2 \alpha_i(\mu_1,\mu_2,...,\mu_n)}{\partial \mu_i^2} < 0$.

Propositions 3 and 4 together mean that the market share α_i is increasing and concave with respect to μ_i $(i = 1, 2, \ldots, n)$.

3.3   The Profit Function

Here we proceed to find out the profit function of an individual server, which represents the server's profit per time unit in the long run. There are two cases to be considered. Suppose that $\sum_{j=1}^{n} \mu_j > \lambda$, Server i receives customers at a rate of $\lambda\alpha_i(\mu_1, \mu_2, \ldots, \mu_n)$. When $\sum_{j=1}^{n} \mu_j \leq \lambda$, Server i receives customer at a rate of μ_i. In both cases, Server i incurs a cost of $c(\mu_i)$. Therefore similar to [11], the profit function of Server i is given by

$$\pi_i(\mu_1,\mu_2,\ldots,\mu_n) = \begin{cases} R\lambda\alpha_i(\mu_1,\mu_2,\ldots,\mu_n) - c(\mu_i) & \text{if } \sum_{j=1}^{n} \mu_j > \lambda \\ R\mu_i - c(\mu_i) & \text{if } \sum_{j=1}^{n} \mu_j \leq \lambda \end{cases} \quad (22)$$

Each of the servers aims to maximize its long-run profit when determining its service capacity. Therefore, how a server's profit changes with its service capacity (when other servers' capacities remain unchanged) is important in characterizing the server's decision. By proposition 3 and 4, we readily obtain the following proposition describing the properties of the profit function π_i with respect to μ_i.

Proposition 5. *For* $i = 1, 2, \ldots, n$, *for each fixed* $\lambda > 0$ *and* $\mu_j > 0$ *where* $j \neq i$, *the function* $\pi_i(\mu_1, \mu_2, \ldots, \mu_n)$ *is continuous and strictly concave in* μ_i.

The continuity and concavity of the profit function ensure that the first-order condition is a sufficient condition for a value of μ_i to maximize the profit function.

3.4   The Nash Equilibrium of the Queueing System

Since servers' decisions of their service capacities would affect the profit of each other, we model the situation as an n-player strategic game, in which each server i chooses its service capacity μ_i to maximize its profit π_i. Here we discuss the Nash equilibrium of the system. In the two-server model in [11], a unique symmetric equilibrium is found in the case when the total demand rate is less than the total service rate. In our analysis, we will show that, similar to the two-server case, when the marginal cost is low enough, there is a unique equilibrium, in which all servers choose the same service capacities. In the following, we will first look at how the profit of Server i changes with its service capacity when all other servers choose the same service capacities.

Proposition 6. *For $\mu_c > \lambda/n$,*

$$\frac{\partial}{\partial \mu_i} \alpha_i(\mu_1, \mu_2, \ldots, \mu_n)\bigg|_{\mu_1=\mu_2=\ldots=\mu_n=\mu_c} = \frac{\lambda}{n^2 \mu_c^2} \left[1 - \frac{\lambda^{n-1}}{\displaystyle\sum_{k=0}^{n-1}(k+1)! \binom{n-1}{k} \lambda^{n-k-1} \mu_c^k} \right]$$

which is decreasing in μ_c. Also, we have

$$\lim_{\mu_c \to (\lambda/n)^+} \frac{\partial}{\partial \mu_i} \alpha_i(\mu_1, \mu_2, \ldots, \mu_n)\bigg|_{\mu_1=\mu_2=\ldots=\mu_n=\mu_c} = \frac{n-1}{n\lambda}$$

and

$$\lim_{\mu_c \to \infty} \frac{\partial}{\partial \mu_i} \alpha_i(\mu_1, \mu_2, \ldots, \mu_n)\bigg|_{\mu_1=\mu_2=\ldots=\mu_n=\mu_c} = 0.$$

It should be noted that proposition 6 implies that for $\mu_c > \lambda/n$, we have

$$\frac{\partial}{\partial \mu_i} \alpha_i(\mu_1, \mu_2, \ldots, \mu_n)\bigg|_{\mu_1=\mu_2=\ldots=\mu_n=\mu_c} < \frac{n-1}{n\lambda}.$$

We also note that the partial derivative in proposition 6 gives the marginal benefit Server i gets by unilaterally deviating from a service capacity μ_c commonly chosen by all servers.

The following proposition gives the Nash equilibrium of the game, which represents the decision of the servers on their service capacities in the long run.

Proposition 7. *If $(n-1)R/n > c'(\lambda/n)$ then there is a unique equilibrium where $\mu_1 = \mu_2 = \ldots = \mu_n = \mu_c$ and μ_c is the unique solution that satisfies $\mu_c > \lambda/n$ and*

$$R\lambda \frac{\partial}{\partial \mu_i} \alpha_i(\mu_1, \mu_2, \ldots, \mu_n)\bigg|_{\mu_1=\mu_2=\ldots=\mu_n=\mu_c} = c'(\mu_c). \tag{23}$$

i.e.,

$$R\left(\frac{\lambda}{n\mu_c}\right)^2 \left[1 - \frac{\lambda^{n-1}}{\displaystyle\sum_{k=0}^{n-1}(k+1)! \binom{n-1}{k} \lambda^{n-k-1} \mu_c^k} \right] = c'(\mu_c). \tag{24}$$

If $(n-1)R/n \le c'(\lambda/n)$ then the system has no equilibrium in which the expected waiting time is finite.

We note that from the proposition, we have $\mu_c > \lambda/n$ and so the expected waiting times are finite. This means that we know that if the marginal cost of serving $1/n$ of all customers is less than $(n-1)/n$ of the revenue received per customer, there is a unique symmetric equilibrium with finite waiting times.

For equation (23) to hold, it means that the marginal benefit Server i gets by unilaterally deviating from a service capacity μ_c commonly chosen by all servers must be equal to the marginal cost to do so. In this case, Server i does not benefit from changing its service capacity. Mathematically, the first-order condition for π_i holds. From the concavity of π_i obtained in proposition 5, we know that choosing μ_c as the service capacity maximizes the profit for Server i.

Since the servers share the same cost function and the same profit function with respect to their own service capacities, the condition for which the marginal benefit equals the marginal cost is identical for all servers when they choose the same service capacities. The proposition asserts that there is only one value of μ_c which satisfies the condition, and that this symmetric equilibrium is the unique equilibrium of the system.

This proposition shows that, given the arrival rate of customer λ, the number of servers n and the revenue per customer R, all servers will choose the same service capacity given by equation (24) in the long run if the condition

$$\frac{(n-1)R}{n} > c'(\frac{\lambda}{n}) \tag{25}$$

is satisfied. The proposition is useful for determining the minimum value of revenue per customer R for which the system will have a finite-waiting time equilibrium.

When $n = 2$, Propositions 6 and 7 reduce to the results in [11]. It is worth noting that as n increases, $(n-1)R/n$ increases and $c'(\lambda/n)$ decreases. Therefore, the minimum value of R required for the existence of a finite waiting-time equilibrium decreases as n increases. An increase in the number of servers causes competition to become more intense. Thus the minimum revenue per customer needed to achieve an equilibrium with finite waiting times becomes lower.

4   A Numerical Example on Three-Server Queueing System

In this section, we present a numerical example for the case of a three-server queueing system, i.e., $n = 3$. Here we assume the cost function takes the following form:

$$c(\mu) = \mu^2 \tag{26}$$

and the condition for the queueing system to be stable

$$\mu_1 + \mu_2 + \mu_3 > \lambda. \tag{27}$$

We note that $c'(\mu) > 0$ and $c''(\mu) > 0$ for $\mu > 0$. Thus $c(\mu)$ is strictly increasing and strictly convex.

We first give the steady-state probability distribution of the system. The following result comes from Proposition 1 in Section 3.1. We have

$$P_0 = \frac{1-\rho}{(1-\rho)\left(1 + \frac{\lambda(\mu_1\mu_2 + \mu_1\mu_3 + \mu_2\mu_3)}{2\mu_1\mu_2\mu_3}\right) + \frac{\lambda^2(\mu_1 + \mu_2 + \mu_3)}{6\mu_1\mu_2\mu_3}},$$

$$P_{(0,0,1)} = \frac{\lambda P_0}{3\mu_3}, \quad P_{(0,1,0)} = \frac{\lambda P_0}{3\mu_2}, \quad P_{(1,0,0)} = \frac{\lambda P_0}{3\mu_1},$$

$$P_{(0,1,1)} = \frac{\lambda^2 P_0}{6\mu_2\mu_3}, \quad P_{(1,0,1)} = \frac{\lambda^2 P_0}{6\mu_1\mu_3}, \quad P_{(1,1,0)} = \frac{\lambda^2 P_0}{6\mu_1\mu_2},$$

and

$$P_k = \rho^{k-2} P_2 \quad \text{for} \quad k > 2$$

where

$$P_2 = P_{(0,1,1)} + P_{(1,0,1)} + P_{(1,1,0)}.$$

Moreover, we have

$$\alpha_i(\mu_1, \mu_2, \mu_3) = \frac{\mu_i\left[\lambda^2 + \lambda(\mu_j + \mu_l) + 2\mu_j\mu_l + \frac{\lambda^3}{\mu_i + \mu_j + \mu_l - \lambda}\right]}{\lambda^2(\mu_i + \mu_j + \mu_l) + 2\lambda(\mu_i\mu_j + \mu_i\mu_l + \mu_j\mu_l) + 6\mu_i\mu_j\mu_l + \frac{\lambda^3(\mu_i + \mu_j + \mu_l)}{\mu_i + \mu_j + \mu_l - \lambda}}.$$

where $j, l \in \{1, 2, 3\}$ and i, j, l are distinct. Now we have

$$\left.\frac{\partial}{\partial \mu_i}\alpha_i(\mu_1, \mu_2, \mu_3)\right|_{\mu_1 = \mu_2 = \mu_3 = \mu_c} = \frac{2\lambda(2\lambda + 3\mu_c)}{9\mu_c(\lambda^2 + 4\mu_c\lambda + 6\mu_c^2)}.$$

If $2R/3 > c'(\lambda/n) = 2\lambda/3$, i.e., $R > \lambda$ then there is a unique symmetric equilibrium where $\mu_1 = \mu_2 = \mu_3 = \mu_c$ and μ_c is the unique solution that satisfies $\mu_c > \lambda/3$ and

$$\left[\frac{2\lambda^2(2\lambda + 3\mu_c)}{9\mu_c(\lambda^2 + 4\mu_c\lambda + 6\mu_c^2)}\right] R = c'(\mu_c) = 2\mu_c$$

i.e.,

$$54\mu_c^4 + 36\lambda\mu_c^3 + 9\lambda^2\mu_c^2 - 3R\lambda^2\mu_c - 2R\lambda^3 = 0.$$

5   Concluding Remarks

In this paper, we extend the analytic results of the two-server queueing system discussed in [11] to an n-server queueing system. To extend our study to the incentive aspect of the queueing system is our future work.

In fact, a service system of two servers coordinated by one central agency was studied by Gilbert and Weng [12]. The principal-agent relationship [13] between the central agency and the servers was studied, from the principal's perspective. It is of interest whether the allocation policy with a *separate queue* or that with a *common queue* would allow the coordinator to control waiting times at a lower cost. The service system studied in [12] consists of two independently operated servers coordinated by one central agency. Again customers arrive according to a Poisson process and the service times are assumed to follow an exponential distribution. Each of the server operates independently and determines its own service capacity so as to maximize its individual profits. The coordinating agency determines a fixed amount R, the compensation to the servers for each unit of service rendered, to induce a desirable service capacity. The coordinating agency's goal is to minimize its cost to maintain expected sojourn time below a given level. It was found that the servers have a weaker incentives to increase their service capacities in common queue systems than in separate queue systems. In many cases, the competition incentive effects can more than offset the risk-pooling benefits of a common queue. In particular, cases with small permissible waiting times or not severe diseconomies on increasing capacity favor the separate queue system.

The queueing system discussed in this paper corresponds to the common queue with n servers. Therefore the results obtained here are ready to apply to generalize the models and conclusions addressed in [12].

Acknowledgment. Ching is supported in part by Hong Kong RGC Grant No. 7017/07P and the HKU Strategic Research Funding on Computational Sciences and CRCG Grants of the University of Hong Kong. Huang is supported in part by the National Natural Science Foundation of China (Project no. 70671020, 70721001, 70431003, 60673159), the Program for New Century Excellent Talents in University (Project no.NCET-05-0295,NCET-05-0289), Specialized Research Fund for the Doctoral Program of Higher Education (20070145017, ,20060145012). The authors would like to thank the anonymous referees for their helpful comments and corrections.

References

1. Altman, E.: Non-zero-sum Stochastic Games in Admission, Service and Routing Control in Queueing Systems. Queueing Systems Theory Appl. 23, 259–279 (1996)
2. Andradotir, S., Ayhan, H., Down, D.: Server Assignment Policies for Maximizing the Steady-State Throughput of Finite Queueing Systems. Manag. Sci. 47, 1421–1439 (2001)
3. Ben-Daya, M., Hariga, M.: Integrated Single Vendor Single Buyer Model with Stochastic Demand and Variable Lead Time. International Journal of Production Economics 92, 75–80 (2004)
4. Bernstein, F., Chen, F., Federgruen, A.: Coordinating Supply Chains with Simple Pricing Schemes: The Role of Vendor-Managed Inventories. Manag. Sci. 52, 1483–1492 (2006)
5. Ching, W.: On Convergence of Asynchronous Greedy Algorithm with Relaxation in Multiclass Queueing Environment. IEEE Communication Letters 3, 34–36 (1999)
6. Ching, W.: Iterative Methods for Queuing and Manufacturing Systems. Springer Monographs in Mathematics. Springer, London (2001)
7. Ching, W., Ng, M.: Markov Chains: Models, Algorithms and Applications. International Series on Operations Research and Management Science. Springer, New York (2006)

8. Ching, W., Choi, S., Huang, M.: Optimal Service Capacity in a Multiple-server Queueing System: A Game Theory Approach (preprint) (2008), http://hkumath.hku.hk/papers/~wkc/cchpaper1.pdf

9. Crabill, C., Gross, D., Magazine, M.: A Classified Bibliography of Research on Optimal Control of Queues. Oper. Res. 25, 219–232 (1977)

10. El-Taha, M., Maddah, B.: Allocation of Service Time in a Multiserver System. Manag. Sci. 52, 623–637 (2006)

11. Kalai, E., Kamien, M., Rubinovitch, M.: Optimal Service Speeds in a Competitive Environment. Manag. Sci. 38(8), 1154–1163 (1992)

12. Gilbert, S., Weng, Z.: Incentive Effects Favor Nonconsolidating Queues in a Service System: The Principal-Agent Perspective. Manag. Sci. 44(12), 1662–1669 (1998)

13. Laffont, J., Martimort, D.: The Theory of Incentives: the Principal-agent Model. Princeton University Press, Princeton (2002)

14. Mishra, B., Raghunathan, S.: Retailer vs. Vendor-Managed Inventory and Brand Competition. Manag. Sci. 50, 445–457 (2004)

15. Morries, P.: Introduction to Game Theory. Springer, New York (1994)

16. Tai, A., Ching, W.: A Quantity-time-based Dispatching Policy for a VMI System. In: Gervasi, O., Gavrilova, M.L., Kumar, V., Laganá, A., Lee, H.P., Mun, Y., Taniar, D., Tan, C.J.K. (eds.) ICCSA 2005. LNCS, vol. 3483, pp. 342–349. Springer, Heidelberg (2005)

17. Teghem, J.: Control of the Service Process in a Queueing System. Euro. J. of Oper. Res. 23, 141–158 (1986)

18. Thomas, D.: Coordinated Supply Chain Management. European Journal of Operational Research 94, 1–15 (1996)

One Kind of Network Complexity Pyramid with Universality and Diversity

Jin-Qing Fang and Yong Li

China Institute of Atomic Energy, Beijing 102413
fjq96@126.com

Abstract. It is based on well-known network models Euler graph, Erdös and Renyi random graph, Watts-Strogatz small-world model and Barabási-Albert scale-free networks, and combined the unified hybrid network theoretical frame. One kind of network complexity pyramid with universality and diversity is constructed, described and reviewed. It is found that most unweighted and weighted models of network science can be investigated in a unification form using four hybrid ratios (dr, fd, gr, vg). As a number of hybrid ratios increase, from the top level to the bottom level complexity and diversity of the pyramid is increasing but universality and simplicity is decreasing. The network complexity pyramid may have preferable understanding in complicated transition relationship between complexity-diversity and simplicity-universality.

1 Introduction

Pyramid architecture can be widely found in nature and most social fields. For example, Zoltvai and Barabási firstly proposed the life's complexity pyramid in biology science [1], and it was found that "the topologic properties of cellular networks share surprising similarities with those of natural and social networks. This suggests that universal organizing principles apply to all networks, from the cell to the World Wide Web." Based on profound analysis for network science development history [2-15] and the unified hybrid network theory frame proposed by Fang's group [15-26], we suggest and investigate one of kind network complexity pyramid with seven levels, as shown in Fig.1, so-called the network model's complexity pyramid(NMCP). The top three levels of the NMCP are Euler graph(EG, level-7)[1], Erdös and Rényi random graph (ERRG, level-6)[2], Watts-Strogatz (WS) small-world model[3] and Barabási-Albert (BA) scale-free networks (level-5)[3], respectively. These network models mark the three milestones in network science development history. The level-4 of the NMCP is the weighted evolution networks(WENM) [27,28]. The top four levels have grabbed main intrinsic quality of complex network respectively. As in depth study of network science, however, how exactly depict and fully mirror all characteristics of most real-world networks is still challenging subject because the real-world is one harmonious and unification world with both determinacy and randomness. Therefore, we have put forwarded the unified hybrid network theoretical frame with three unified hybrid network models [15-26] , which can be constructed as following three levels of the NMCP. The level-3 is the harmonious unification hybrid preferential network model (HUHPM,), the level-2 is the large unified hybrid network model(LUHNM)

J. Zhou (Ed.): Complex 2009, Part I, LNICST 4, pp. 78–89, 2009.

and the level-1 is the unified hybrid network model with variable speed growth (UHNM-VSG). From the top level-7 to the bottom level-1 complexity and diversity is increased but universality and simplicity is decreasing. The NMCP may have preferable understanding in complicated transformation relationship between complexity-diversity and simplicity-universality.

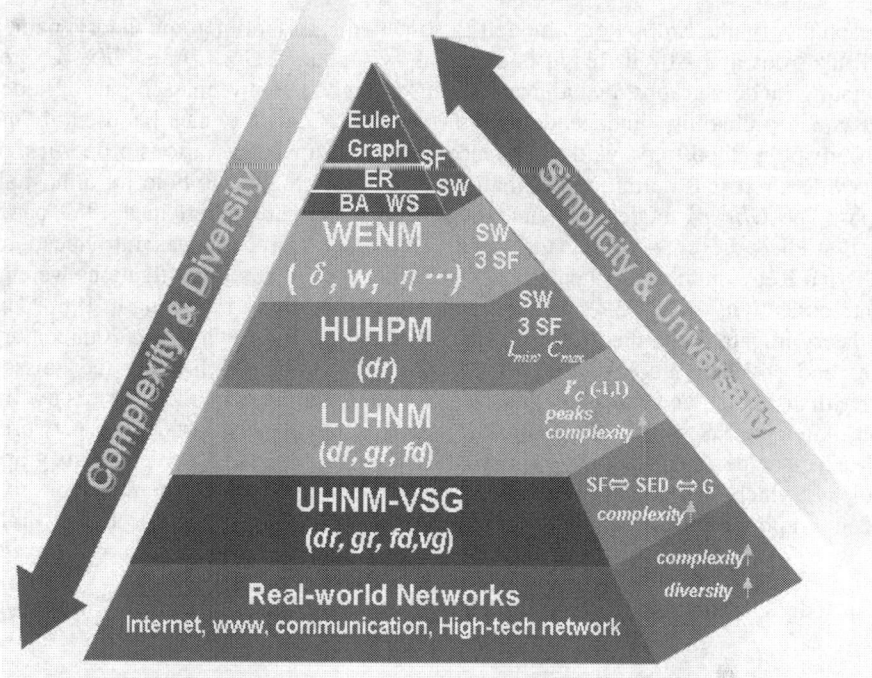

Fig. 1. Complex network model pyramid diagrams

2   The Top Three Levels of the NMCP

Retrospective network in the footsteps of scientific development, network theoretical model research has been one of the most significant issues in the network sciences. So far the history of this area has gone through three milestones which all breakthroughs from the theoretical model.

2.1   The Level-7: Euler (Regular) Graphs

A graph that has an Euler circuit is called an Eulerian graph. The first milestone was Euler graphs born in the 1736 [1], which attributed to the graph father. Euler has done many pioneering work, such as he first solved the famous Konigsberg Seven Bridge problem and the many facets of the Euler theorem [1]. The Euler's theorem is that (a)If a graph has more than two vertices of odd degree then it cannot have an Euler path. (b)If a graph is connected and has just two vertices of odd degree, then it at least has one Euler path. Any such path must start at one of the odd-vertices and end at the other odd vertex.The Euler graphs have been studied for longest period since then. The regular EG theory has laid the foundation of the graph theory development and should be at the top level-7 of the pyramid.

2.2 The Level-6: ER Random Graph

In graph theory, the Erdös-Rényi model[2], so-called ER random graph theory, named for Paul Erdős and Alfréd Rényi, is either of two models, $G(n, p)$ and $G(n, M)$, for generating random graphs, including one that sets an edge between each pair of nodes with equal probability, independently of the other edges. It can be used in the probabilistic method to prove the existence of graphs satisfying various properties, or to provide a rigorous definition of what it means for a property to hold for almost all graphs. The $G(n, p)$ model was first introduced by Edgar Gilbert in a 1959 paper which studied the connectivity threshold. The $G(n, M)$ model was introduced by Erdös and Rényi in their 1959 paper. As with Gilbert, their first investigations were as to the connectivity of $G(n,M)$, with the more detailed analysis following in 1960. The ER theory impact graph theory for 40 years long. Erdös is known as the 20th century Euler, and obtained Wolf Award in 1984. The ER random graph obeys the Poisson degree distribution, and has a smaller average path length and smaller clusters coefficient. After the ER model, from the late 1950s to late 1990s, large-scale networks with no clear design principles primarily uses this simple and easy random graph topology, which is accepted by the majority of people. Many mathematicians give random graph theory strict mathematical proof, and obtain many similar and accurate results. Properties of $G(n, p)$ are as follows. A graph from $G(n, p)$ has on average $\binom{n}{2} p$ edges. The distribution of the degree of any particular vertex is binomial: $P(\deg(v) = k) = \binom{n-1}{k} p^k (1-p)^{n-1-k}$, where n is the total number of vertices in the graph. In a 1960 paper, Erdös and Rényi described the behavior of $G(n,p)$ very precisely for various values of p. Their results included that:

(a) If $np < 1$, then a graph in $G(n,p)$ will almost surely have no connected components of size larger than O(logn).

(b) If $np = 1$, then a graph in $G(n,p)$ will almost surely have largest component whose size is of order $n^{2/3}$.

(c) If np tends to a constant $c > 1$, then a graph in $G(n, p)$ will almost surely have a unique "giant" component containing a positive fraction of the vertices. No other component will contain more than O(logn) vertices.

(d) If $p < \dfrac{(1-\varepsilon)\ln n}{n}$, then a graph in $G(n, p)$ will almost surely not be connected.

(e) If $p > \dfrac{(1-\varepsilon)\ln n}{n}$, then a graph in $G(n, p)$ will almost surely be connected.

Thus $\dfrac{\ln n}{n}$ is a sharp threshold for the connectivity of $G(n, p)$.

So far the ER random graph has succeeded in revealing the emergence of certain structural properties and multi threshold function and so on. Thus it should be at the level-6 of the pyramid.

2.3 The Level-5: Small-World Network and Scale-Free Models

In the level-6, both of the two major assumptions of the $G(n, p)$ model (that edges are independent and that each edge is equally likely) may be unrealistic in modeling real situations. In particular, an Erdös-Rényi graph will likely not be scale-free like many real networks. Therefore the Watts and Strogatz model attempts to correct this limitation.

In 1998, Watts and Strogatz proposed small world (SW) network mode[3].They revealed that the SW effect of the complex network is a kind of hybrid results of determinacy and randomness. Soon Newman and Watts and others made some improvements for the SW models [6-12]. The degree distribution of ER random model and the WS model are not completely in line with many networks in reality and have certain limitations. Many empirical graphs are well modeled by small-world networks. Social networks, the connectivity of the Internet, and gene networks all exhibit small-world network characteristics. A certain category of small-world networks were identified as a class of random graphs by Watts and Strogatz. They noted that graphs could be classified according to two independent structural features, namely the clustering coefficient and average node-to-node distance, the latter also known as average shortest path length. Purely random graphs, built according to the ER model, exhibit a small average shortest path length (varying typically as the logarithm of the number of nodes) along with a small clustering coefficient. Watts and Strogatz measured that in fact many real-world networks have a small average shortest path length, but also a clustering coefficient significantly higher than expected by random chance. Watts and Strogatz then proposed a novel graph model, now currently named the WS model, with (i) a small average shortest path length, and (ii) a large clustering coefficient. The first description of the crossover in the WS model between a "large world" (such as a lattice) and a small-world was described by Barthelemy and Amaral in 1999. This work was followed by a large number of studies including exact results.

In 1999, Barabási and Albert (BA) proposed a scale-free (SF) network model[4] and found the power-law nature of the complex networks, i.e. degree distribution follows $p(k) \sim k^{-\gamma}$. Two discoveries of the SW and the SF networks mark the third milestone of network development[5] and network sciences was born [5-12]. The formation mechanism of the SF network is based on two rules: growth and preferential attachment in accordance with the degree of nodes. The BA model is the first model of a random network with the SF property. Further, network with complex topology describe as diverse as the cell, the WWW or society. One of the most surprising finding is that despite their apparent differences and sharing the same large-scale topology, each having a SF structure. Subsequently, it was found that the formation mechanisms of the SF are also as diverse as replication, nearest neighbour connections, hybrid preferential linking and local connective information. In summary, main feature is the evolution of complex network is driven by self-organizing processes that are governed by simple but generic scaling laws. Many subsequent empirical research of real-world networks(RWN) have demonstrated that the RWNs is neither regular nor random, but they belong to a large class of hybrid network both determinacy and randomness, and commonly possess both the SW and the SF properties, as well as the statistical property which is completely different from the level-7 regular graph and the level-6 random graph.

3   The Level-4: Weighted Evolving Network Models

Up to now three milestones above from the level-7 to the level-5 are all un-weighted networks. They reflect most of topological properties and dynamical behavior between network nodes and connectivity but they could not describe different role of nodes and all characteristics of the RWNs completely since almost RWNs belong to weighted networks. Only weighted evolving networks can carefully portray the nodes connection and mutual interaction. Thus it is natural boost that from the un-weighted network models above toward weighted evolving network models (WENM) , which has became the level-4 of the NMCP. Along with more and more empirical studies on weighted networks, fresh properties related link weight are obtained by some typical WENMs [27,28]. In the level-4 there are a lot of preferential driving mechanisms: (1) Node strength; (2)Edged weight; (3)Both strength and edged weight; (4)Both weight and fitness; (5) Both topological growth and strength driving; (6) Geographical link of position neighborhood; (7) Local information or both local world and weight driving; (8) Topological growth with strengths' driving, and so on. In the level-4 the WENMs have revealed some common characteristics: the SW as well as the three SF (node degree, strength and weight distributions), i.e., all obey the power-law property with different exponents.

4   The Level-3: HUHPNM

It is noted that all WENMs in the level-4 belong to generalized random networks, which always ignored deterministic linking. They are useful for theoretical analysis easily and reproduce main topological properties for the RWNs. But based on the foundational observation fact for a unifying world in natural and social networks, one cannot ignore anyone of order and random since their interactions in real world are neither completely regular nor completely random and lying between the extremes of order and randomness.

To overcome weak point of the level-4, the unified hybrid network model frame [15-26, 31-32] with trilogy was proposed and can be constructed as following three levels of the NMCP. The level-3 is the harmonious unifying hybrid preferential network model (HUHPNM), in which one total hybrid ratio is introduced by

$$dr = \frac{d}{r} = \frac{DPA}{RPA} \qquad (1)$$

where d is a number of time intervals (step) for deterministic preferential attachment (DPA), and r is a number of random preferential attachment (RPA). It was found in the level-3 that some universal topological properties, including the exponents γ of the three power-laws (degree, node strength, and edged weight) are highly sensitive to total hybrid ratio the dr. A threshold of the exponent is at $d/r = 1/1$.

Through theoretical analysis for the HUHPNM, we obtain the complicated function relationship of power exponent γ with d/r for some weighted HUHPNM which are quite coincide with the numerical curves. Moreover, for all three BA, BBV and TDE models their γ has quite complicated relation with the weighted parameters (δ, w) and the total hybrid ratio d/r. Their complicated function relationship of power exponent γ with d/r for the BA and BBV are [16]:

$$\gamma_{BA} = \frac{1}{\beta} + 1 = A_1 e^{-\left(\frac{d/r}{A_2}\right)^{A_3}} + A_4 .$$ (2)

where $\gamma_0 = 3$.

$$\gamma_{BBV} = \frac{4\delta + A_1 e^{-\left(\frac{d/r}{A_2}\right)^{A_3}} + A_4}{2\delta + 1} .$$ (3)

where A_i is parameter, i=1, 2, 3, 4.

This reflects both mutual competition and harmonious unification. The level-3 has both the SF and the SW properties. It was found that the HUHPNM-BA is of the shortest average path length(APL) and largest average clustering coefficient (ACC).

5   The Level-2 : LUHNM

In fact, the level-3 HUHPNM does not completely reflect the actual network links of the diversity and complexity. To describe diverse complex networks and improve the HUHPNM, we have extended the HUHPNM toward a large unifying hybrid network model (LUHNM)[19-26], which become the level-2 of the NMCP. Two new hybrid ratios: determinist hybrid ratio fd and random hybrid ratio gr are introduced respectively by

$$fd = \frac{f}{d} = \frac{HPA}{DA}$$ (4)

and

$$gr = \frac{g}{r} = \frac{GRA}{RA}$$ (5)

where HPA is helping poverty attachment, and GRA is general random attachment, thus we have DA = HPA + DPA; RA = GRA + RPA. In the level-2, it is found that much more complex relation of topological properties depending on three hybrid ratios (dr, fd, gr). The degree-degree correlation r_c (the assortative coefficient) is one of interesting quantity. The level-2 exhibits two fresh transition features of the r_c from negative 1 to positive 1 in the both un-weighted and weighted LUHNM. Firstly, only if the $fd \geq 0.9/1$, whatever the gr value is, the r_c curves appear multiple peaks phenomena as (*dr, fd, gr*) change. As *dr* increases, the r_c increases and can reach largest positive 1. The *fd* =0.9/1 plays a key role for the transition features of the r_c depending on the matched sense of 3 different hybrid ratios (*dr, fd, gr*). The LUHNM can have a better understanding the r_c change in different hybrid rations. Obviously, the results in the level-2 are more closer to the RWNs and can give a reasonable answer concerned question: why social networks are mostly positive degree-degree correlation but biological and technological networks tend to be negative degree-degree correlation. The LUHNM can further increase additional hybrid ratio according to actual need, and makes it more flexible and potential application.

6   The Level-1: UHNM-VSG

Further comparison to the RWNs and in-depth analysis, it is obvious that, even so, in the level-2 is still not fully reflect the actual network growth situation, because actual networks usually display variable speed growing process, such as high-tech network, the Internet, the WWW, human social networks, communication networks and so on. Therefore it is necessary to introduce a variable growth hybrid ratio, *vg,* which is defined by

$$vg = \frac{DVG}{RVG} \tag{6}$$

where *DVG* is time intervals of deterministic variable speed growth, and *RVG* is time intervals of random variable speed growth. Thus we propose and construct the unified hybrid network model with various speed growing (UHNM-VSG)[31-32] as the level-1 of the NMCP[31-33]. The level-1 has two variable growth pictures: deterministic and random growth, for example, one may take a growing format as follows [13-14]:

$$m(t) = p(N(t))^\alpha \tag{7}$$

where *m(t)* is *t* time to increase the number of nodes connected edge, *N(t)* is the number of nodes at the network at *t* time, α is growth index, *p* is a constant for deterministic growth; but for random growth the linking probability is $0 < p(t) < 1$. According to the value of the variable speed index α we have normal(α=0), deceleration(α<0), acceleration($0 < \alpha < 1$) and super-accelerated situation(α>1). Therefore the UHNM-VSG is of flexible and includes most current important kinds of network models. The level-1 has rich fresh features as follows.

6.1   Transition of *P(k)* from Single Scale to Broad Scale as α Changes

It is found in the level-1 that transition of the cumulative degree distribution *P(k)* can be changed from single SF to double stretched exponential (SED) as growth index α increases. Fig.2(a)-(c) shows comparison of the *P(k)* under different α for fixed *fd*=0/1and *gr*=0/1 case. In Fig.2(a) α =0, no matter *dr* how change, *P(k)* follows the power-law distribution:

$$p(k) \approx k^{-\gamma} \tag{8}$$

The power exponent γ increases in nonlinear way as α increases, the change of γ is sensitive to dr increase, this is consistency with the level-3. However, if α is not equal to zero, for instance, when $\alpha =0.3$ and 0.6 under different dr, the $P(k)$ not only emerges the SF but can change to double SED, which is expressed by [33] :

$$P(k) = e^{-(\frac{k}{k_0})^c} \tag{9}$$

where k_0 is a parameters, c is stretched exponent. It is found that $P(k)$ obeys the SED and there exits a transition point near at $dr=1/1$, at two sides of the transition point $P(k)$ has two different SED, named as first SED and second SED. The topological properties can be changed from the SF to double SED as $\alpha >0.3$, as shown in Fig.2, and depend on the hybrid ratios. It implies the relation of stretched exponent c with dr and α is much more complex.

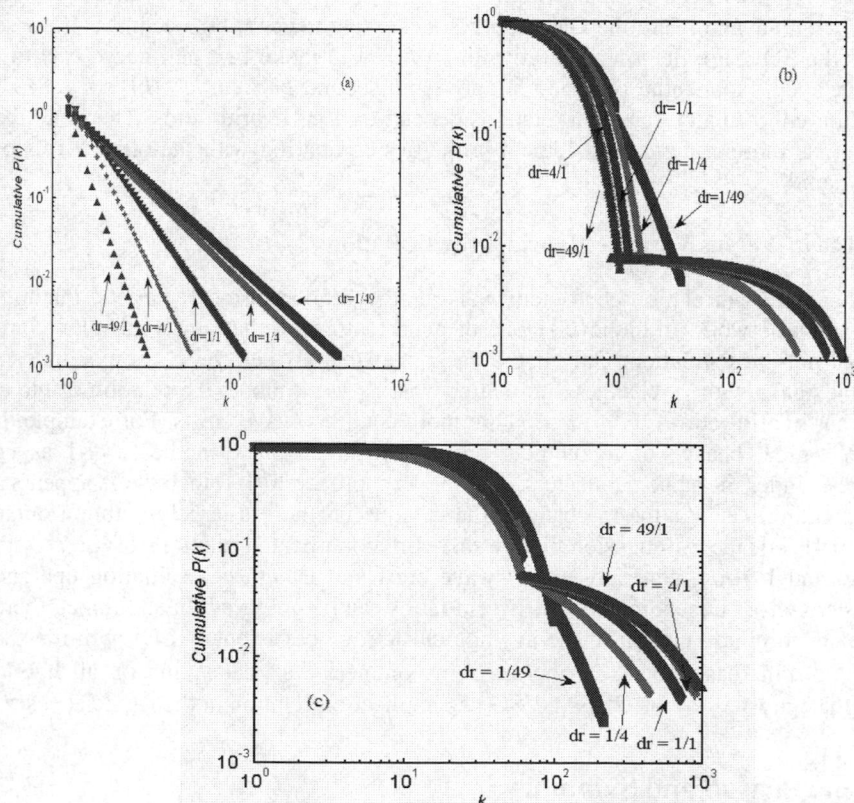

Fig. 2. For $fd=0/1$ and $gr=0/1$ cumulative degree distribution $P(k)$VS k. (a) $\alpha =0$; (b) $\alpha =0.3$, (c) $\alpha =0.6$.

6.2 Transition of *P(k)* from Single Scale to Broad Scale as *vg* Changes

The hybrid growth ratio *vg* is another key control parameter in the level-1 and effects on topological properties largely. It is found that all cumulative degree distribution *p(k)* display two kinds of distribution. For random prevailing (*dr*=1/49) case first half curve *p(k)* follows delayed exponential distribution:

$$p(k) = A_1 e^{-\frac{k}{t_1}} + y_0 \qquad (10)$$

where A_1, y_0 and t_1 are three parameters but second half curved *p(k)* obeyed the power-law $p(k) \approx k^{-\gamma}$.

However, for *dr*=1/1 case first half curve obeys the Gaussian distribution defined by

$$P(x) = y_0 + \frac{A}{w\sqrt{\frac{\pi}{2}}} e^{-2(\frac{x-x_c}{w})^2} \qquad (11)$$

where y_0, x_c, w and A are the Gaussian associated parameters, but second half curve *p(k)* is the SDE. For determinacy prevailing (*dr*=49/1) mode first half curve *p(k)* follows delayed exponential distribution above but second half curve *p(k)* is the SED. The different *p(k)* and transition relations depend on four hybrid ratios. Therefore the level-1 has more complex topological properties comparing with the other levels of the pyramid.

6.3 The r_c Versus Variable Hybrid Growth Ratio *vg*

Degree-degree correlation coefficient r_c is another important characteristic quantity of complex network. In the level-1 one of main features is that always display complicated nonlinear relation. Only if the *vg* greater than 1 then the r_c always appears multiple peaks with *vg* changes,, no matter what *dr* work mode, the relation of the r_c with four hybrid ratios (*vg, dr, fd, gr*) are more complex wave crests. For example, if *dr*=49/1, great change of topological property is taken place in the level-1 as *vg* changes. Under the same parameters, if the *vg* is approach to 1/1, then the r_c appears a r_c maximum value. As the *vg* changes, the r_c value is changed largely within a range of [-1,0.4]. For the 3-dimension picture there are complicated relations of the r_c with log(*vg*) and log(*dr*), generally, many wave crest and trough of oscillation ups and downs crisscross depending on four hybrid ratios can be observed. In additional, there also exits complex relation of the average clustering coefficient (ACC) with (*fd, gr, dr, α*), as *α* increases the ACC increases in nonlinear way depending on all hybrid ratios. The mystery of the UHNM-VSG has implication relation in their special cases.

7   Comparison and Summary

It is seen from the level-1 the UHNM-VSG model can provide much more information about topological properties and their transition relations of the cumulative degree distribution $P(k)$ between the SF and the SED by controlling four hybrid ratios (vg, dr, fd, gr) for two kinds of growth pictures in Eq.(7). Three-dimensional relationships for the ACC and the r_c with the hybrid ratios (dr, fd, gr, vg, α) are much more complicated. The ACC can be changed from 0 to 1 in nonlinear fashion and appear a variety of wave crests and troughs. The r_c is changed between +1 and -1 only depending various matching of the 4 hybrid ratios. Compared with the level-3 the HUHPM and the level-2 the LUHNM, the level-1 the UHNM-VSG can include most current network models and approaches to real-world networks in a close range as demon strated in Ref. [14]. The SED can provide a better description for economical networks [14] and high technology networks [19-22].

In short, we suggest, describe and review one kind of network model complexity pyramid with seven levels. Table-1 gives a comparison and summary for various levels of the pyramid under the different hybrid ratios.

It is seen from Fig.1 and Table-1 that all models of the pyramid levels can be very well studied in unification form by the 4 hybrid ratios (dr, fd, gr, vg). It is found that from the bottom level-1 to the top level-7 of the pyramid universality-simplicity is increasing but complexity-diversity is decreasing. On the other hand, from the top level-7 to the bottom level-1 of the pyramid universality-simplicity is reducing but complexity-diversity is strengthening. All properties and changes between seven levels of the pyramid depend on matching of four hybrid ratios (dr, fd, gr, vg)

Table 1. Comparison of the pyramid levels under the different hybrid ratios

Model \ Hybrid ratios	dr	gr	fd	Vg	Properties	Pyramid Level
EG	1/0	0/0	0/0	0/0	Simple	7
ER	0/1	1/0	0/0	0/0	Emergence	6
WS	1/0.1 (a few)	1/0	0/1	0/0	Small World Simplicity↑	5
BA	0/1	0/1	0/0	0/0	SF, universality↑	
	0/1	0/1	0/0	0/0	Scale-free(SF)	
BB,BBP,BBV, , TDE etc	The same as BA model above				3-power-law SF,SW	4
HUHPM	tunable	0/1	0/1	0/0	Complexity Diversity SF,SW	3
LUHNM	tunable	tunable	tunable	0/0	Complexity↑ Diversity↑ simplicity↓	2
LUHNM-VSG	tunable	tunable	tunable	tunable	Complexity↑↑ simplicity↓↓ SF↔SED	1

strongly. The network model complexity pyramid has preferable understanding in complicated transition relationship among various characteristics as well as the change features. The NMCP is of universal, self-adapting and flexibility and can be extended to study many real-world networks. However, more exact theoretical work of complex network pyramid is still open and a very challenging for researchers.

Acknowledgment. This work is supported by Nature Science Foundation of China (Grand Nos. 60874087, 70431002, 60773120 and 10647001). Nature Science Foundation of Beijing (Grand No. 4092040).

References

1. Leonhard Euler(1707-1783), http://www2.zzu.edu.cn/math/
2. Erdös, P., Rényi, A.: On the evolution of random graphs. Publ. Math. Inst. Hung. Acad. Aci 5, 17–61 (1960)
3. Watts, D.J., Strogatz, S.H.: Collective dynamics of "small-world" networks. Nature 393, 440–442 (1998)
4. Barabási, A.L., Albert, R.: Emergence of scaling in random networks. Science 286, 509–512 (1999)
5. Watts, D.J.: The "New" Science of Networks. Annu. Rev. Sociol. 30, 243–270 (2004)
6. Newman, M.E.J., Watts, D.J.: Renormalization group analysis of the small-world network model. Phys. Lett. A. 263, 341–346 (1999)
7. Dorogovtsev, S.N., Mendes, J.F.F.: Evolution of Networks: From Biological Mets to the Internet and WWW. Oxford University Press, Oxford (2003)
8. Kleinberg, J.: Navigation in a small world. Nature 406, 845 (2000)
9. Albert, R., Barabási, A.L.: Statistical mechanics of complex networks. Rev. Mod. Phys. 74, 47–97 (2002)
10. Strogatz, S.H.: Exploring complex networks. Nature (London) 410, 266–275 (2001)
11. Mattick, J.S., Gagen, M.J.: Accelerating networks. Science 307, 856–858 (2005)
12. Gagen, G.M., Mattick, J.S.: Accelerating, hyperaccelerating and decelerating probabilistic networks. Phys. Rev. E. 72, 016123 (2005)
13. Sen, P.: Accelerated growth in outgoing links in evolving networks: Deterministic versus stochastic picture. Phys. Rev. E. 69, 046107 (2004)
14. Laherrère, J., Sornette, D.: Stretched exponential distributions in nature and economy: "fat tails" with characteristic scales. Eur. Phys. J. B. 2, 525–539 (1998)
15. Fang, J.Q., Liang, Y.: Topological properties and transition features generated by a new hybrid preferential model. Chin. Phys. Lett. 22, 2719–2722 (2005)
16. Fang, J.Q., Bi, Q., Li, Y., et al.: A harmonious unifying preferential network model and its universal properties for complex dynamical network. Science in China Series G 50(3), 379–396 (2007)
17. Fang, J.Q., Bi, Q., Li, Y., et al.: Sensitivity of exponents of three-power-laws to hybrid ratio in weighted HUHPM. Chi. Phys. Lett. 24(1), 279–282 (2007)
18. Lu, X.B., Wang, X.F., Li, X., et al.: Topological transition features and synchronizability of a weighted hybrid preferential network. Physica A 370, 381–389 (2006)
19. Fang, J.Q.: Exploring and advances in theoretical model of network science. Review of Science and Technology 24(12), 67–72 (2006)
20. Fang, J.Q., Wang, X.F., Zheng, Z.G., et al.: A New interdisciplinary Science— Network Science(I). Progress in Physics 27(3), 239–343 (2007)

21. Fang, J.Q., Wang, X.F., Zheng, Z.G., et al.: A New interdisciplinary Science— Network Science(II). Progress in Physics 27(4), 361–448 (2007)
22. Fang, J.Q., Bi, Q., Li, Y.: Advances in Theoretical models of network science. Front. Phys. China. 1, 109–124 (2007)
23. Fang, J.Q., Li, Y., Bi, Q., et.al.: From the harmonious unifying hybrid preferential attachment model toward a large unifying hybrid network model. J. Modern Phys., 21(30), 5121–5142 (2007)
24. Li, Y., Fang, J.Q., Liu, Q.: New transition features of associatativity in large unified hybrid network. Review of Science and Technology 25(11), 23–29 (2007)
25. Fang, J.Q.: Evolution Features of Large Unifying Hybrid Network Model with a Variable Growing Speeds. In: The 4th International Workshop Hangzhou 2007 on Simulational Physics, invited talk, November 9-12, 2007, Hangzhou, Zhejiang (2007)
26. Fang, J.Q.: Some Progresses in Theoretical Model of Nonlinear Dynamical Complex Networks. In: Invited talk, 2008 National Physics Conference, September 18, Nanjing (2008)
27. Barrat, A., Barthélemy, M., Vespignani, V.: Weighted evolving networks: Coupling topology and weight dynamics. Phys. Rev. Lett. 92, 228701 (2004)
28. Wang, W.X., Wang, B.H., Hu, B., et al.: General Dynamics of Topology and Traffic on Weighted Technological Networks. Phys. Rev. Lett. 94, 188702 (2005)
29. Fang, J.Q.: Investigating High-Tech Networks with Four Levels From Developing Viewpoint of Network Science. World SCI – Tech R&D 30(5), 667–674 (2008) (in Chinese)
30. Wang, W.X., Wang, B.H., Hu, B., et al.: Mutual attraction model for both assortative and disassortative weighted networks. Phys. Rev. E. 73, 016133 (2006)
31. Fang, J.Q., Li, Y.: Advances in unified hybrid theoretical model of network science. Advances in Mechanics 6, 663–678 (2008)
32. Fang, J.Q., Li, Y., BI, Q.: Unified Hybrid Variable Speed Growth Model and Transition of Topology Property. Complex Systems and Complexity Scinece 5(4), 56–65 (2008) (in Chinese)
33. Laherrère, J., Sornette, D.: Stretched exponential distributions in nature and economy: "fat tails" with characteristic scales [J]. Eur. Phys. J.B. 2, 525–539 (1998)

On Traveling Diameter of an Instance of Complex Networks – Internet

Ye Xu[*], Zhuo Wang, and Wen-bo Zhang

College of Information Science and Engineering, Shenyang Ligong University,
Shenyang 110168, China
{xuy.mail,zhuow,wenbozhang}@gmail.com

Abstract. As an instance of complex networks, Internet has been a hot topic for both complex networks and traditional networks research fields. Internet Traveling Diameter (ITD) is an important property defined in this paper representing the dynamic flow of Internet performance, and was mainly discussed. Short-term forecast model of ITD was firstly studied, then after it was proved that the short-term one was not good enough for long term forecast due to the complexity of Internet, the long-term model was studied. Both short-term and long-term model were given their mathematical descriptions at last.

Keywords: Internet traveling diameter; Logistic Model; GA; chaos; correlation dimension.

1 Introduction

With great improvements in the Internet research fields recently, more and more emerging research approaches, either from new research viewpoints or by a crossover study with other subjects, are applied to Internet related studies.

Studies in reference [1-4] made use of many new research approaches on Internet from complex networks point of view. In these approaches, Internet was regarded as an example of complex network due to its large scale and complicated variations, and definitions such as power law distribution [1,3,4], spectrum density [5], scale-free [2] and so on were utilized to depict qualitatively or compute quantitatively properties of Internet.

Referring to the idea of these research approaches, together with considering the fact that the hops of Internet datagram traveling from one node (router) to another are closely related to Internet performance, a definition of Internet traveling diameter, short for Internet diameter, would be discussed. Besides, the computing approaches of Internet diameter in both short-term type and long-term type would be discussed detailedly in this paper.

1.1 Definition of Internet Diameter

Assume that one datagram's transferring from one node (router) to a direct link node counts for 1 hop in Internet, and the datagram transferred from one source node to a destination

---

[*] Corresponding author. Ye XU, ph.D in computer application technology, associate professor, his current research interests include complex networks modeling, adaptive signal processing and pattern recognition.

J. Zhou (Ed.): Complex 2009, Part I, LNICST 4, pp. 90–99, 2009.

counts for J hops, where J is a number greater than or equal to 1. A definition of Internet traveling diameter is brought forward to represent the average hops, or statistical hops, of millions of datagram transferred from any source to any end at any time in Internet.

Definition 1. Assume J_i represents the hops of No. i datagram in Internet, the size of total datagram sample is N, and the frequency of No. i datagram in the sample is F_i, its probability is p_i, then Internet Diameter is

$$D = \frac{1}{N} \sum_{i=1}^{N} J_i F_i = \sum_{i=1}^{N} J_i p_i \tag{1}$$

1.2 Datagram Samples

As is well known, the validity of statistical results are entirely dependent on the size of statistical samples, the larger the sample is, the more accurate results would be.

Sample in this paper comes from three CAIDA[1] monitor nodes -- riseling node in SanDiego, CA, US, k-peer node in Amsterdam, NorthHolland, NL and apan-jp node in Tokyo, Kanto, JP[2]. Sampling time is from July 1999 to Jun. 2004[3]. The sample comprises more than 75,000,000 items in all and is of large scale. What's more, several other CAIDA monitors are also referred to sometime for a better accuracy.

There are datagram that could not reach the destination and are discarded by routers in Internet, and these datagram were depicted as "incomplete" or "I" in our sample, and those that could reach the target were depicted as "complete" or "C". Table 1 gives the detail.

Table 1. Sample details

Monitor node	Size of C	Size of I	$C\%$	Total
riseling	16448329	13669728	**54.6%**	30118057
k-peer	9479028	9555955	**49.8%**	19034983
apan-jp	15172408	10423192	**59.3%**	25595600
Total				74748640

Since the unreachable datagram is simply discarded by routers in Internet, we ignore them and focus on the reachable ones in the sample. We could see from table 1 that the reachable ones account for larger percentage of the whole sample, and could still be regarded as a sample with great size.

1.3 A Quick Look at Hops in the Sample

Basically there are two techniques for us to have a quick look at properties of Internet Diameter. The first method was called "space-dimension analysis", by which differences

---

[1] CAIDA, the Cooperative Association for Internet Data Analysis, is a worldwide research center on Internet-related research fields. CAIDA has more than thirty monitor nodes distributed throughout the whole world, measuring and monitoring the variations of Internet.

[2] The reason selecting these monitors is that the nodes are separately located in three different continents on earth. This way of selection might provide a more general view of Internet throughout the whole world.

[3] Data items of one day out of one month are drawn out to build up the sample in this paper.

between different monitor nodes from different continents was illustrated in figure 1(a). And the other method is "time-dimension analysis", by which features of data items from only one node but at different time were illustrated in Fig. 1(b).

Another three monitors, a-root node in Herndon, VA, US, cdg-rssac node in Paris, France and nrt node in Tokyo, Kanto, JP were added to current sample for space-dimension analysis. And for "time-dimension analysis", a twelve months dataset (from 2003 to 2004) from riseling was selected, this selection does not lose generality because samples from all six nodes represent great consistency in figure 1(a).

We can see From fig. 1 that summits of six curves lie between 10 and 18 hops, which indicates that, although the data items in the sample were drawn out from different monitors at different time, they presents highly similar characters.

From figure 1(a) and (b), we see that all hops lie in an interval of [2, 32]. Taking errors into account, we enlarge the interval from [2, 32] to [1, 36], by which we could ensure that the entire sample was included.

Then for Traveling Diameter, according to equation (1), D is an average hops, so

$$D \in [1,36] \tag{2}$$

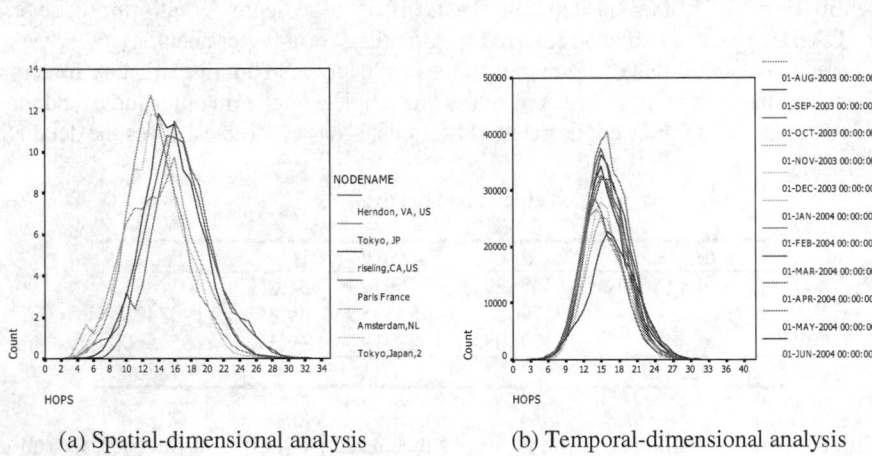

(a) Spatial-dimensional analysis   (b) Temporal-dimensional analysis

Fig. 1. A quick look at Internet Diameter

1.4 A Quick Look at Internet Diameter

We then begin to look at properties of Internet Diameter (D) by constructing a Cartesian coordinate system with D as Y axis and time as X axis, which is illustrated in Fig. 2.

It's obvious that all three curves have consistent variations in Fig. 2. With t growing longer, D is decreasing gradually. Though there is much difference among $D_{riseling}$, D_{kpeer} and D_{apanjp} at the beginning, the difference gets narrowed rapidly. Three Ds reach at a very small difference interval when t is around 60.

From Fig. 2, we can conclude that $D_{riseling}$, D_{kpeer} and D_{apanjp} represent great consistency when t is getting longer. Then, for simplicity of calculation, we average the three Ds and illustrate it in Fig. 3.

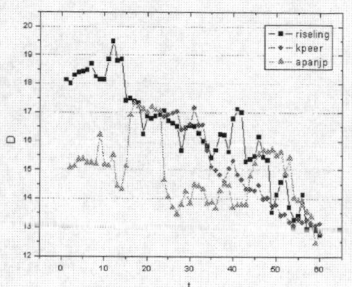 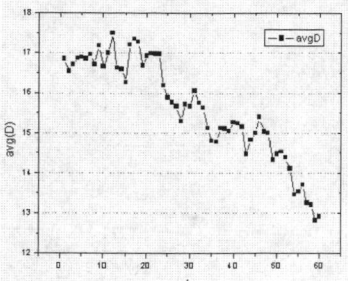

Fig. 2. A quick look at Internet Diameter (*D*). There are three monitor nodes selected in the graph, riseling, k-peer, apan-jp node, and the time is from 1999.07 to 2004.06. Y axis is value of *D* calculated by equation (1). X axis is time with a resolution of month, there are totally sixty months from 1999.07 to 2004.06.

Fig. 3. Average *D* from 1999.07 to 2004.06.

AvgD in Fig. 3 is the basic element for short-term and long-term computing approaches.

In short-term part, variations of avgD in Fig. 3 is much similar to a Logistic Model, so a Logistic Model is selected as a fundamental framework model, but additional adjustment models are necessary [6].

2 Short-Term Model of Traveling Diameter

2.1 Improving Logistic Model

After applying avgD sample to Logistic Model [7,8,11], we get a nonlinear differential equation:

$$\frac{dD}{dt} = rD(1 - \frac{D}{k}) \tag{3}$$

where $D(\geq 0)$ means avgD at t, $t(>0)$ is time(month), $k(>0)$ means upper bound of avgD and $r(>0)$ means growth rate. Solving equation (3), we get an equation

$$D = \frac{k}{1 + \dfrac{k}{D_0 - 1}e^{-rt}} = \frac{k}{1 + me^{-rt}} \tag{4}$$

where $D_0 = D(t = 0), m = \dfrac{k}{D_0 - 1}$.

(1) Transform one: from increasing function to decreasing one
Standard Logistic Function is an increasing function, but avgD is decreasing. Transform is needed.

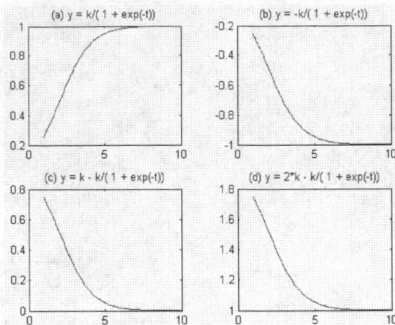

Fig. 4. Transform standard Logistic function into decreasing one through four steps. Step 1, standard Logistic function, $y=k/(1+e^{\wedge}-t)$, k is 1. Step 2, add a minus sign to the function, then $y=-k/(1+e^{\wedge}-t)$. Step 3, $y=y+k$, then $y=k-k/(1+e^{\wedge}-t)$. And step 4, add another k to function because avgD is greater than 1, then $y=2k-k/(1+e^{\wedge}-t)$.

Take a standard Logistic Function, as in equation (5), for an example,

$$y = \frac{k}{1+e^{-t}}. \quad \text{where } k = 1 \tag{5}$$

It is transformed into a decreasing one through four steps illustrated in Fig. 4(a)~(d).

In Fig. 4(d), the transition amount along Y axis is $2*k$, but this is not necessary. It could be any real number greater than k since avgD belong to an interval of [1, 36] and its lower limit is 1 (according to Fig. 4(c)). Then the improved Logistic Function is

$$D = d - \frac{k}{1+me^{-rt}} \tag{6}$$

(2) Transform two: additional adjustment models

There is a kind of quasi-periodic vibrations in the variations of avgD in figure 3, and according to reference [6], this kind of vibration could be simulated by adjustment models composed of sine and cosine functions. In reference [6], only sine function is referred to. However, we import both sine and cosine functions for a better efficiency in this paper.

After two transforms, we get a composite Logistic Function as:

$$D=d-\frac{k}{1+me^{-(v+p_1e^{-g_1t}\sin[\pi(\frac{t}{h1}+u1)]+p_2e^{-g_2t}\cos[\pi(\frac{t}{h2}+u2)])t}} \tag{7}$$

where $D(\geq0)$ is avgD at t, d is adjustment parameter, p_1,p_2 is amplitude of vibration of avgD, h_1,h_2 is half period of vibration in month, u_1,u_2 is the initial phase and g_1,g_2 are parameters of modulus decay.

2.2 Curve Fitting

Float-point Genetic Algorithm [6][9][10] listed in table 2 is used for curve fitting of equation (7).

Table 2. An implementation of GA

procedures	information	equations and algorithms
(1) Definition of genes in GA.	Randomly initializing a gene group comprising 100 genes.	$x = (d, k, m, v, p_1, p_2,$ $g_1, g_2, h_1, h_2, u_1, u_2)$
(2) Definition of evaluation function	Assume $D(t)$ is the value of avgD out of the composite model at t, and $D^*(t)$ means real value at t from sample. Then evaluation function $f(x)$ and its score function in GA are set up to be:	$f(x) = \sum_{i=1}^{60} \lvert D(t_i) - D^*(t_i) \rvert$ $score(x) = 1/f(x)$
(3) Selection	Genes were sorted by scores from high to low in the gene group, and the first $m*N$ genes, m is a random number ($0<m<1$), were selected for the next round calculation by GA. Then we duplicate the selected genes, and deleted the last m genes to keep the group size remaining the same.	
(4) Crossover	Randomly select two genes, $xi(vi...)$ 、 $xj(vj...)$ out of the group to perform the crossover	$v_i' = v_i(1-\alpha) + \beta v_j$ $v_j' = v_j(1-\alpha) + \beta v_i$
(5) Mutation	Randomly select two genes, $xi(vi...)$ 、 $xj(vj...)$ out of the group to perform the crossover.	$v_i = v_i(1+\alpha)$ if $\gamma \geq 0.5$ $v_i = v_i(1-\alpha)$ if $\gamma < 0.5$
(6) Termination conditions	Basically there are two termination conditions in GA in this paper. The first condition is when score of the best gene in the group is greater than a threshold s, s is set to be 0.1 in the algorithm. The other condition is when the number of calculation rounds in GA gets greater than a threshold n, and n is set to be 50000.	

After the curve fitting of GA, the result, i.e., the short-term model of equation (7) is:

$$D = 16.9495 - \frac{6.8262}{1 + 58.2418 \times e^{-(A+B)t}} \qquad (8)$$

where $t>0$, $t \in N$, A and B are:

$$A = 0.007844 + 0.001218\, e^{-0.001191\, t} \times \sin[\,\pi(\frac{t}{0.000866} + 0.014911\,)]$$

$$B = 0.082215\, e^{-0.004712\, t} \times \cos[\,\pi(\frac{t}{0.001} + 0.003242\,)]$$

2.3 Problem of Long-Term Computing with the Short-Term Model

Experiments indicate that the accuracy degree of the short-term model is getting worse with time getting longer, i.e., the short-term model is not suitable for long-term computation. So study on long-term computing approaches is to be introduced.

3 Long-Term Model of Traveling Diameter

3.1 Long-Term Forecast Principle in Chaos System

According to reference [12], it's very difficult to perform long-term forecast or computing a Chaos system due to the initial sensitivity. Mr. Lorenz E. N. had proved during his study on weather forecast that it is impossible to perform a long-term forecast in a Chaos system [17].

However, if "odd attractor, OA" exists in the Chaos system, long-term forecast of the system during a limited space and a limited period of time, is computable [12,13]. Though the forecast time is limited into a period of time, it's still much longer than that in short-term part, so we call the model under such conditions as long-term forecast model.

Since OA only exists in Chaos system, what we ought to do now is to prove whether variations system of Internet Diameter is a Chaos or not.

3.2 Chaos Proof

To prove whether a system is a Chaos, a new notion "correlation dimension, D_2" has to be introduced. If D_2 could be obtained out of the target system, the system is confirmed to be a Chaos, and OA is proved to exist [12].

Table 3. Algorithm for $D2$

Algorithm: Correlation dimension solving algorithm
Input: list /*Time sequence sample of Internet Diameter (1999.07~2004.06)*/

/* Initialization */
tao = 3; length = list.length;
/* calculate D_2 with an increaing m */
loop when m=4,8,10,12 … until D_2 is convergent
/*Constructing a coordinate system of length-(m-1)*tao dimensions */
vecgroup = zeros(m, length-(m-1)*tao);
vecgroup = getValue(list);
/* Calculate distances between vector i and vector j in the coordinate system */
rij = calcRIJ(vecgroup(:,i), vecgroup (:,j));
/* Get a Matrix r*/
r = [maxRij:(maxRij-minRij)/15:minRij]
/* Calculating correlation integral, cr */
cr = calcCR(r, rij);
/* plot */
plot(log(r), log(cr));
/* get $D2$, D2 is the slope of curve in plot. If $D2$ exists, the curve in the plot would be similar to a straight line and the slope of curves would increase till a limited bound with the increasing m. */
calcD2();
/* end of loop while m=4,8,10,12,14,16,18 */
end loop

For calculating D_2, a special sample called "time sequence" would have to be drawn out first from the sample ranging from 1999.07 to 2004.06.

With the time sequence sample, we then construct an m-dimensional coordinate system, where m is an integer and is usually less than 100. After this, different m in an increasing order would be input into the algorithm [12,14~16] to construct different m-dimensional system. The output of the algorithm D_2 is the slope of curve in the output plot, and it would increase very fast when m increases. If D_2 increases to an infinite number such as $\tan(pi/2)$ finally, the system is proved to not be a Chaos. On the contrary, if D_2 increases till a limited upper bound, the system is confirmed to be a Chaos with OA. The algorithm is listed in table 3 [12][14][15][16].

Reference [12] suggests that tao should lie between (4, 8), but since the size of the time sequence sample of Internet Diameter is rather small (only sixty months), so we set tao to be 3 in the algorithm. Result of D_2 solution is illustrated in Fig. 5.

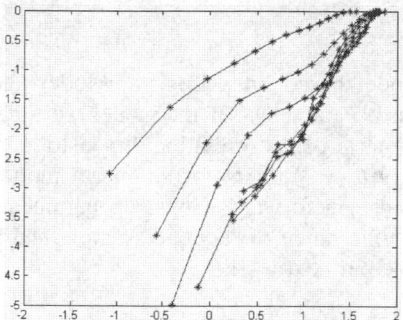

Fig. 5. D2 solutions diagram with $\ln C2(r,m)$ as Y axis and $\ln(r)$ as X axis. Algorithm parameters are: $\tau=3$, $m=2,5,8,11,13,15,16$. Seven curves represent seven m from left to right, when $m=2,5,8, 11,13,15,16$.

From Fig. 5, there is good convergence of curves when $m=11,13,15,16$ -- the last four curves from left to right. The reason is that, firstly, the four curves are much similar to straight lines. Secondly, the slopes of the lines remain stable with m getting larger.

So D_2 in the variations system of Internet Diameter exists and is:

$$D_2(m_c = 16, \tau = 3) = 2.2444 \qquad (9)$$

3.3 Chaos Forecast Model for Long Time System

The existence of D_2 proves that the system is a Chaos with OA, and could provide further useful information for setting up long-term models.

Study in reference [12] indicates that only a model with at least $\lceil D_2 \rceil$ dimensions, i.e., a function set with $\lceil D_2 \rceil$ functions, could perform a long-term forecast with relatively acceptable accuracies. According to equation (9), $\lceil D_2 \rceil$ is three, then the model should be:

$$\begin{cases} x = f_1(x, y, z, x^{(n)}, y^{(n)}, z^{(n)}) \\ y = f_2(x, y, z, x^{(n)}, y^{(n)}, z^{(n)}) \\ z = f_3(x, y, z, x^{(n)}, y^{(n)}, z^{(n)}) \end{cases} \tag{10}$$

where $x \sim z$ represents important physical variables in target system, and $x^{(n)} \sim z^{(n)}$ means n orders of differential coefficient of the corresponding variable.

Equation (10) is only a framework of a Chaos model. It's still very difficult to confirm the parameters of the model because Chaos is a system with great complexity and the researches on it are still not very clear nowadays, as well as the time range of the time sequence sample is rather small (only from 1999 to 2004). And this would be our future work.

4 Conclusions

In short-term part, a model based on a Logistic Model and additional adjustment models is brought forward. Parameters of the model were finally confirmed through GA experiments. In long-term part, correlation dimension D_2 of Internet Diameter is calculated out of an algorithm. With D_2, a long-term model with three differential coefficient functions is brought forward. However, parameters of the model are still uncertain due to a short time range of sequence sample and complexity of the target system. And this would be our next work.

References

1. Floyd, S., Paxson, V.: Difficulties in simulating the Internet. IEEE/ACM Trans. on Networking 9(4), 392–403 (2001)
2. Watts, D., Strogatz, S.: Collective dynamics of 'small-world' networks. Nature 393(6684), 440–442 (1998)
3. Aiello, W., Chung, F., Lu, L.Y.: A random graph model for massive graphs. In: Proc. of the ACM STOC 2000, pp. 171–180. ACM Press, Portland (2000)
4. Zhang, Y., Zhang, H.-L., Fang, B.-X.: A Survey on Internet Topology Modeling. Journal of Software 15(8), 1221–1226 (2004)
5. Farkas, I.J., Derényi, I., Barabási, A., Vicsek, T.: Spectra of 'real-world' graphs: Beyond the semicircle law. Physical Review E 64(2), 1–12 (2001)
6. Yin, C.Q., Yin, H.: Artificial Intelligence and Expert System, pp. 291–295. China Waterpower Publishing House (2002)
7. Yang, Z.J., Xu, Z.R.: Forecast of the Population Growth in the Country of HEILONGJIANG by the Forecast Method of Dynamic Logistic. Journal of Agriculture University of HEILONGJIANG 9(2), 23–28 (1997)
8. Wu, S.L.: Forecast of development of China Numerical Library by Logistic Model. Journal of Information 4 (2004)
9. Wang, J.M., Xu, Z.L.: New crossover operator in float-point genetic algorithms. Control Theory And Applications 19(6) (December 2002)
10. Rudolph, G.: Covergence properties of canonical genetic algorithms. IEEE Trans. on Neural Networks 5(1), 96–101 (1994)

11. Zhang, H.: Two New Population Growth Equation. Journal of Bimathematics 10(4), 78–82 (1995)
12. Huang, R.S.: Chaos and its application, vol. 1.1, pp. 191–239. WU HAN University Press (2000)
13. Zhang, Q.C., Wang, H.L., Zhu, Z.W.: Split and Chaos, vol. 1.1, pp. 256–281. TIAN JIN University Press (2005)
14. Kenneth, J.: Falconer. Fractal Geometry – Mathematical Foundations and Applications, vol. 8, pp. 41–75. Northeastern University Press (1991)
15. Xu, P.: Analysis of attractor of dam observations. Journal of Dam Observations 10 (1992)
16. Kang, Z., Huang, J.-W., Li, Y., Kang, L.-S.: A Multi-Scale Mixed Algorithm for Data Mining of Complex System. Journal of Software 14(7), 1229–1237 (2003)
17. Lorenz, E.N.: Deterministic nonperiodic flow. J. Atoms, Sci. 20 (1963)

On the Approximation Solution of a Cellular Automaton Traffic Flow Model and Its Relationship with Synchronized Flow

R. Jiang[1,2], Y.M. Yuan[1], and K. Nishinari[2]

[1] School of Engineering Science, University of Science and Technology of China,
Hefei 230026, China
[2] Department of Aeronautics and Astronautics, School of Engineering,
The University of Tokyo, Hongo, Bunkyo-ku, Tokyo 113-8656, Japan

Abstract. This paper studies approximation solution of a cellular automaton model. In the model, the finite size effect is trivial because the congested flow is quite homogeneous. Thus, the approximation solution of a small sized system can be regarded as solution of large system. We have investigated the approximation solution of a small traffic system with two vehicles. The analytical result is in good agreement with simulation. Finally, it is demonstrated that the homogeneous congested flow is closely related to synchronized flow.

Keywords: traffic flow, cellular automaton, synchronized flow.

1   Introduction

Recently, the vehicular traffic problem has attracted the interests of physicists [1-6]. On the one hand, the researches contribute to traffic planning, design, control and implementation of transportation network by revealing the basic principles governing traffic congestion. On the other hand, vehicular traffic, as a system of interacting particles driven far from equilibrium, offers the possibility to study various fundamental aspects of nonequilibrium systems.

To simulate the various traffic phenomena, many traffic flow models are proposed. These models could be generally classified into macroscopic continuum models, mesoscopic kinetic models, microscopic car-following models and cellular automaton (CA) models. We focus on CA models in this paper.

In 1992, Nagel and Schreckenberg proposed the well known NaSch model [7]. They extended the CA-184 rules, in which the maximum velocity $v_{max} = 1$, to case $v_{max} > 1$. Together with the randomization effect, the NaSch model could reproduce spontaneous formation of jams in congested traffic. Since then, a large number of papers that improve NaSch models have been published. Analytical theories of the NaSch model are also developed (see Ref.[2] and references therein).

In recent years, Kerner and his colleagues found that congested flow can be further classified into synchronized flow and wide moving jams [1,8-12]. Different

J. Zhou (Ed.): Complex 2009, Part I, LNICST 4, pp. 100–109, 2009.

from wide moving jams, which exhibit the characteristic to maintain a constant mean velocity of the downstream jam front, synchronized flow is usually fixed at bottlenecks. Empirical observations show that a two-dimensional scattering of empirical data on flow-density plane is identified in synchronized flow.

On an open road with an isolated on-ramp, it is found that when the on-ramp flow rate is high, the onset of congestion is usually associated with the transition from free flow to synchronized flow, and then moving jams emerge spontaneously in synchronized flow, usually at different location. However, widening synchronized flow pattern (WSP) could be observed when the on-ramp flow rate is low. No spontaneous jam occurs in WSP [1].

The NaSch model and the existing NaSch-based models fail to reproduce the synchronized flow, because randomization effect always triggers avalanche-like braking of vehicles and finally leads to jams in congested flow.

In this paper, we present a simple CA model, in which homogeneous congested flow is reproduced. It is demonstrated that the homogeneous congested flow is closely related to synchronized flow. Furthermore, we present an approximation solution of the model. This is achieved because the finite size effect is trivial in the model.

The paper is organized as follows. In section II, the model is introduced. Section III presents the approximation solution of the model. Section IV discusses the relationship between homogeneous congested flow and synchronized flow. The conclusion is given in Section V.

2   Model

In this paper, we present a simple CA model, the approximation solution of which could be obtained. The parallel update rules of our model are as follows.

- Velocity adjustment:
$$v_n(t+1) = \min(v_n(t) + 1, v_{\max}, d_n(t), v'),$$
where $v' = \left\lfloor \dfrac{-T + \sqrt{T^2 + \frac{2}{D}\left(d_n(t) + \frac{v_{n-1}^2(t)}{2D}\right)}}{1/D} \right\rfloor$
- Randomization:
$$v_n(t+1) = \max(v_n(t+1) - 1, 0) \text{ with probability } p_d$$
- Movement:
$$x_n(t+1) = x_n(t) + v_n(t+1)$$

Here $v_{\max}$ is maximum velocity, D is a comfortable deceleration, T is reaction time, $\lfloor x \rfloor$ is the maximum integer that is not larger than x, $d_n = x_{n-1} - x_n - l$ is spatial gap in front of vehicle n, l is vehicle length.

The only difference between our model and NaSch model is introduction of v', which is obtained from

$$\frac{v'^2}{2D} + v'T = \frac{v_{n-1}^2(t)}{2D} + d_n(t). \tag{1}$$

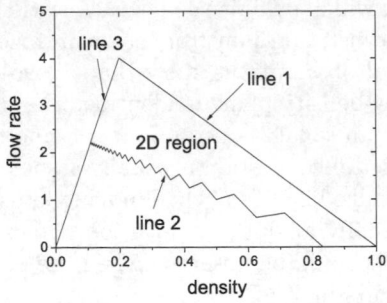

Fig. 1. The steady states of the model, which cover a two-dimensional region in the flow-density plane in the noiseless limit (i.e., $p_d = 0$). Line 1 is determined by $v = d$, line 2 is determined by $v = \min(v')$ and v' is determined by $v' = \lfloor \frac{-T + \sqrt{T^2 + \frac{2}{D}\left(d + \frac{v'^2}{2D}\right)}}{1/D} \rfloor$, line 3 is determined by $v = v_{\max}$. Here v is velocity and d is spatial gap of the steady state. The parameters $T = 1, D = 1, v_{\max} = 20$. Each cell is 1.5 m and one vehicle occupies five cells.

Here $\frac{v'^2}{2D} + v'T$ is braking distance of a car travelling with velocity v' and $\frac{v_{n-1}^2(t)}{2D}$ is braking distance of the car $n - 1$. Note that T is not necessarily to be one and D is not necessarily to be an integer in the model. It is obvious that when D is extremely large and $T \leq 1$, the model reduces to NaSch model.

In our model, the vehicles brake with comfortable deceleration if possible. However, if the comfortable deceleration could not guarantee safety (i.e., Eq. (1) is not met), the vehicles will brake with much larger deceleration.

In the noiseless limit, the steady states of the model cover a two-dimensional region as shown in Fig. 1. Therefore, the model is within the framework of Kerner's three-phase traffic theory [1].

3   Approximation Solution

Fig. 2 shows fundamental diagram of the model with parameters $D = 1, T = 1$, $p_d = 0.1$. A critical density ρ_c is identified. When $\rho < \rho_c$, free flow exists. When $\rho > \rho_c$, homogeneous congested flow is identified (see Fig. 3). It can be seen that the velocities of the vehicles fluctuate around 12 and 13. Velocities smaller than 10 or larger than 14 are seldom observed. Due to homogeneity of the congested flow, the finite size effect is trivial (Fig. 2).

Based on this fact, if we could obtain approximation solution of small sized traffic system with two vehicles, then the solution can also be regarded as approximation solution of large system.

Now we investigate such a small system with two vehicles by using car-oriented mean-field theory [13]. Let us consider a specific density $\rho = 1/6$, which corresponds to system size $L = 60$. Since the traffic flow is quite homogeneous, we

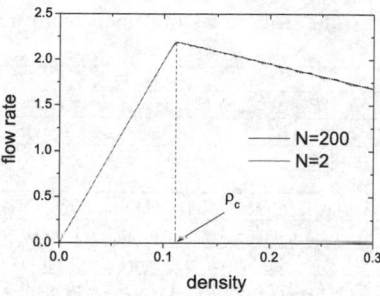

Fig. 2. Fundamental diagram of the model with parameters $D = 1$, $T = 1$, $p_d = 0.1$. The system size $L = N(d + l)$, N is the number of vehicles, d is average spatial gap of vehicles. The flow rate of system consisting of two vehicles ($N = 2$) is almost identical to that of large system.

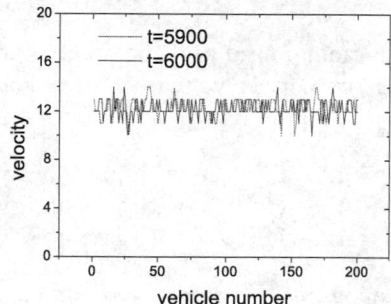

Fig. 3. Snapshots of velocities of the vehicles in congested flow. The density $\rho = 1/6$, i.e., $d = 25$.

only consider the following configurations of the two vehicles and assume the appearance probabilities of other configurations could be neglected [1]:

$$12{-}11{-}26, 13{-}12{-}25, 12{-}12{-}25, 12{-}11{-}25, 12{-}10{-}25, 11{-}11{-}25, 13{-}12{-}24,$$

$$13{-}11{-}24, 12{-}12{-}24, 12{-}11{-}24, 11{-}11{-}24, 13{-}11{-}23, 12{-}11{-}23, 11{-}11{-}23.$$

Here the first number denotes the velocity of one vehicle, the third number denotes the spatial gap in front of the vehicle, and the second number denotes the velocity of the other vehicles. Note due to the symmetry, $x - y - z$ and $y - x - (L - z)$ denote the same configuration.

We study the evolution of the configurations. For example, consider configuration 11-11-25. Without considering randomization, this configuration will evolve into 12-12-25. When considering randomization, 12-12-25 will turn into 11-11-25 with probability p_d^2 (both vehicles are randomized), 12-11-24 with probability

---

[1] Generally speaking, considering more configurations will lead to more accurate approximation solution.

Table 1. The table shows other equations obtained from mean field approximation

From	we have
$\frac{dP_{13-12-25}}{dt} = 0$	$(P_{12-12-24} + P_{13-12-24})(1 - p_d)^2 \approx P_{13-12-25}$
$\frac{dP_{12-12-25}}{dt} = 0$	$(P_{11-11-25} + P_{12-11-25})(1 - p_d)^2 + P_{13-12-25}(1 - p_d)p_d \approx P_{12-12-25}(1 - (1 - p_d)^2)$
$\frac{dP_{12-11-25}}{dt} = 0$	$(P_{12-12-24} + P_{13-12-24})p_d^2 + (P_{12-11-26} + P_{11-11-24} + P_{12-11-24} + P_{13-11-24})(1 - p_d)p_d$ $\approx P_{12-11-25}$
$\frac{dP_{12-10-25}}{dt} = 0$	$(P_{11-11-23} + P_{12-11-23} + P_{13-11-23})(1 - p_d)p_d \approx P_{12-10-25}$
$\frac{dP_{11-11-25}}{dt} = 0$	$(P_{12-12-25} + P_{12-11-25})p_d^2 + P_{12-10-25}(1 - p_d)^2 \approx P_{11-11-25}(1 - p_d^2)$
$\frac{dP_{13-12-24}}{dt} = 0$	$P_{13-12-25}(1 - p_d)^2 \approx P_{13-12-24}$
$\frac{dP_{13-11-24}}{dt} = 0$	$(P_{12-12-24} + P_{13-12-24})(1 - p_d)p_d \approx P_{13-11-24}$
$\frac{dP_{12-12-24}}{dt} = 0$	$P_{13-12-24}(1 - p_d)p_d + (P_{12-11-26} + P_{11-11-24} + P_{12-11-24} + P_{13-11-24})(1 - p_d)^2$ $\approx P_{12-12-24}(1 - (1 - p_d)p_d)$
$\frac{dP_{12-11-24}}{dt} = 0$	$(P_{11-11-25} + P_{12-12-25} + P_{12-11-25})2(1 - p_d)p_d + P_{13-12-25}p_d^2 \approx P_{12-11-24}$
$\frac{dP_{11-11-24}}{dt} = 0$	$(P_{12-11-26} + P_{12-11-24} + P_{13-11-24})p_d^2 \approx P_{11-11-24}(1 - p_d^2)$
$\frac{dP_{13-11-23}}{dt} = 0$	$P_{13-12-25}(1 - p_d)p_d \approx P_{13-11-23}$
$\frac{dP_{12-11-23}}{dt} = 0$	$(P_{12-11-26} + P_{11-11-24} + P_{12-11-24} + P_{13-11-24})(1 - p_d)p_d \approx P_{12-11-23}$

$2(1 - p_d)p_d$ (one vehicle is randomized and the other is not) and remains 12-12-25 with probability $(1 - p_d)^2$ (neither vehicles are randomized). This means the configuration

$$11 - 11 - 25 \rightarrow \begin{cases} 12 - 12 - 25 & \text{with probability } (1 - p_d)^2 \\ 11 - 11 - 25 & \text{with probability } p_d^2 \\ 12 - 11 - 24 & \text{with probability } 2(1 - p_d)p_d \end{cases}$$

The evolution of other configurations could be obtained similarly.

Based on these, the master equation of the evolution of the configurations could be written out. The master equation of configuration 12-11-26 is

$$\frac{dP_{12-11-26}}{dt} \approx (P_{11-11-23} + P_{12-11-23} + P_{13-11-23})(1 - p_d)^2 - P_{12-11-26}. \quad (2)$$

Here " $\approx$ " is used because we have neglected many possible configurations, whose appearance probabilities are very small. In the steady state, $\frac{dP_{12-11-26}}{dt} = 0$. Thus, we have

$$(P_{11-11-23} + P_{12-11-23} + P_{13-11-23})(1 - p_d)^2 \approx P_{12-11-26}. \quad (3)$$

Similarly, other equations could be obtained as shown in Table 1.

Finally, our assumption gives

$$P_{12-11-26} + P_{13-12-25} + P_{12-12-25} + P_{12-11-25} + P_{12-10-25} + P_{11-11-25} + P_{13-12-24}$$
$$P_{13-11-24} + P_{12-12-24} + P_{12-11-24} + P_{11-11-24} + P_{13-11-23} + P_{12-11-23} + P_{11-11-23} \approx 1$$
$$(4)$$

Now we have 14 variables and 14 linear equations [2]. The equations could be solved. The average velocity of the vehicles is then calculated by

---

[2] Note from $\frac{dP_{11-11-23}}{dt} = 0$, we can obtain another equation. However, this is not an independent equation.

$$\begin{aligned}
\bar{v} = &\; 11.5P_{12-11-26} + 12.5P_{13-12-25} + 12P_{12-12-25} + 11.5P_{12-11-25} \\
&+ 11P_{12-10-25} + 11P_{11-11-25} + 12.5P_{13-12-24} + 12P_{13-11-24} \\
&+ 12P_{12-12-24} + 11.5P_{12-11-24} + 11P_{11-11-24} + 12P_{13-11-23} \\
&+ 11.5P_{12-11-23} + 11P_{11-11-23}
\end{aligned} \tag{5}$$

and the result is $\bar{v} = 12.188$, which is in very good agreement with simulation result $\bar{v} = 12.19$.

If the density is changed, then one needs to study the master equations of different configurations. It can be verified that the approximation solution is always in good agreement with simulation result at any density.

4   Discussion

In this section, we discuss the possible relationship between homogeneous congested flow and the synchronized flow. To this end, we firstly study the synchronized flow in three CA models, i.e., the model proposed by Lee et al. (model A) [14], the Kerner-Klenov-Wolf model (model B) [15], and the model proposed by Jiang and Wu (Model C) [16].

Fig. 4(a) shows the fundamental diagram of model A. Two critical densities ρ_{c1} and ρ_{c2} are identified. When $\rho < \rho_{c1}$, the traffic is in free flow. When $\rho_{c1} < \rho < \rho_{c2}$, synchronized flow appears in free flow. Fig. 5(a) shows the typical snapshot of velocities corresponding to the coexistence phenomenon. With the increase of density, more and more vehicles are involved in the synchronized flow state (Fig. 5(b)). When $\rho > \rho_{c2}$, the free flow disappears and all vehicles are in synchronized flow state (Fig. 5(c)). With the further increase of density, the average velocity of the synchronized flow decreases (Fig. 5(d)).

Fig. 4(b) shows the fundamental diagram of model B. When ρ is smaller than ρ_c, which corresponds to maximum flow rate, light synchronized flow has already appeared (Fig. 6(a)). With the increase of density, the coexistence of light synchronized flow and heavy synchronized flow is identified (Fig. 6(b)). With the further increase of density, the average velocity of synchronized flow

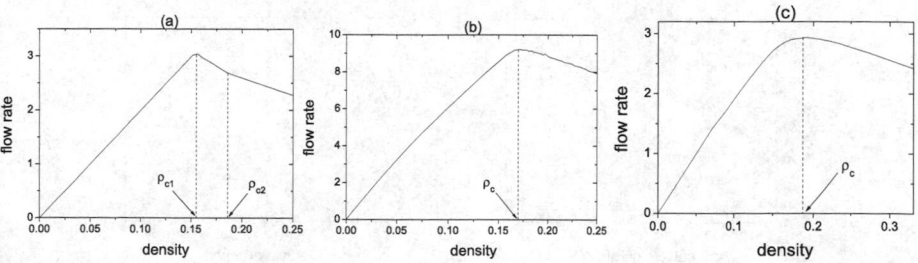

Fig. 4. Fundamental diagram of (a) model A, (b) model B, (c) model C. In (a), the parameters are the same as in Ref.[13]. In (b), $p_{a1} = p_{a2} = 0.052$, other parameters are the same as in parameter set I in Ref.[14]. In (c), $h = 3$, other parameters are the same as in Ref.[15]. Note that in model B, each cell is 0.5 m and one vehicle occupied 15 cells.

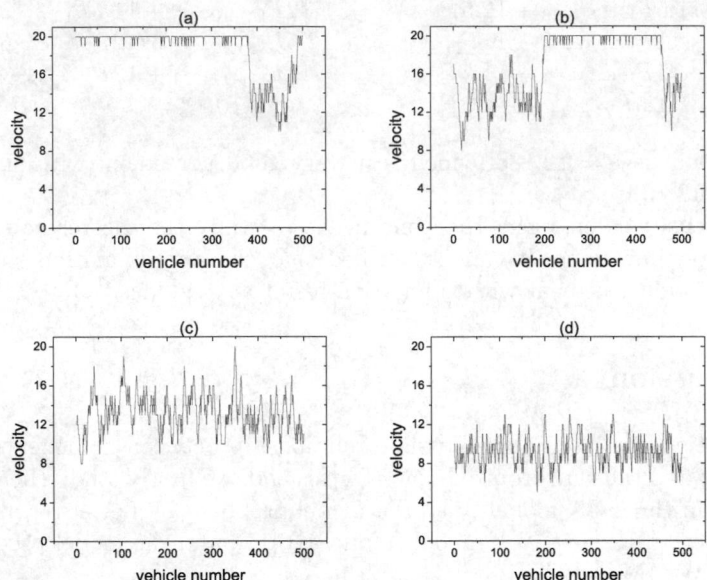

Fig. 5. Snapshots of velocities in model A. (a) $\rho = 0.1613$, (b) $\rho = 0.1724$, (c) $\rho = 0.2$, (d) $\rho = 0.25$.

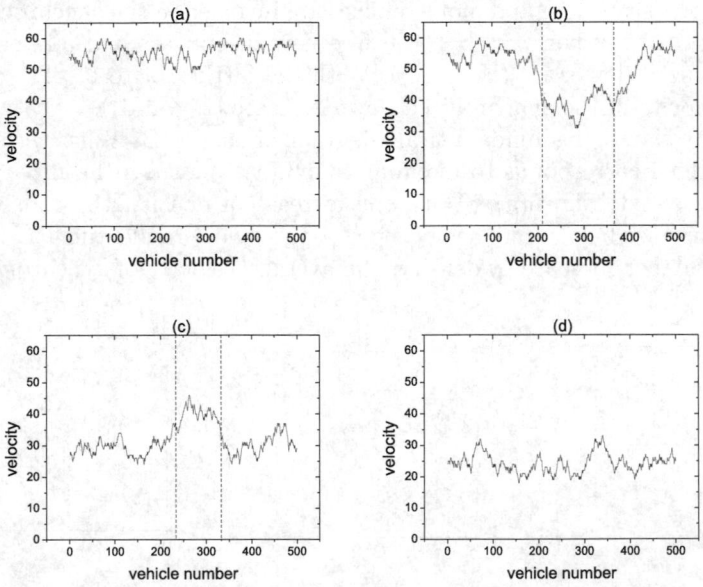

Fig. 6. Snapshots of velocities in model B. (a) $\rho = 0.1667$, (b) $\rho = 0.1875$, (c) $\rho = 0.25$, (d) $\rho = 0.3$. In (b) and (c), the dashed lines show the boundaries of the coexisting states.

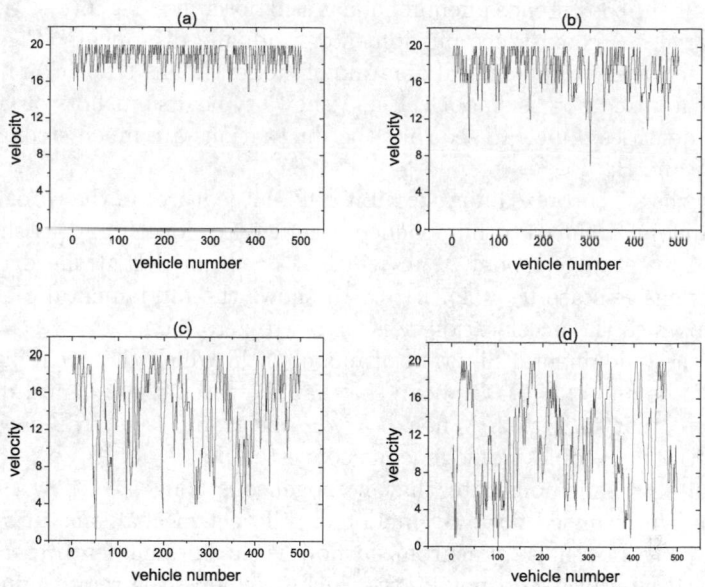

Fig. 7. Snapshots of velocities in model C. (a) $\rho = 0.125$, (b) $\rho = 0.1667$, (c) $\rho = 0.2$, (d) $\rho = 0.25$.

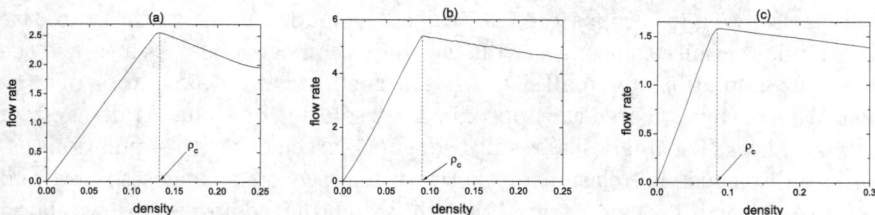

Fig. 8. Fundamental diagram of slightly modified (a) model A, (b) model B, (c) model C

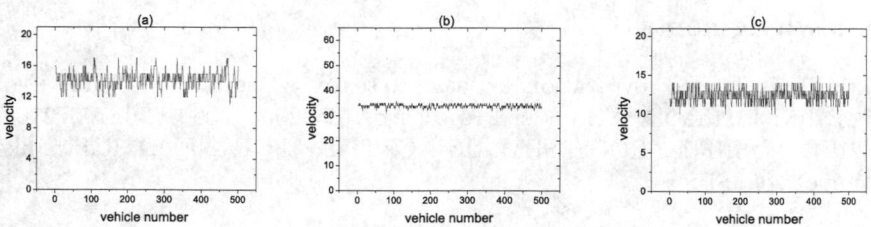

Fig. 9. Snapshots of velocities in (a) modified model A, $\rho = 0.1667$, (b) modified model B, $\rho = 0.15$, (c) modified model C, $\rho = 0.125$

decreases, but the coexistence phenomenon is still observed (Fig. 6(c)). When the density is large, the coexistence phenomenon gradually disappears (Fig. 6(d)).

Fig. 4(c) shows the fundamental diagram of model C. The transition from free flow to synchronized flow is smooth. Fig. 7 shows typical snapshots of velocities at different densities. One can see that the fluctuations are much stronger than in models A and B.

Although the synchronized flows exhibit different features in the three models, they are all closely related to homogeneous congested flow. Fig. 8(a) shows the fundamental diagram of model A, in which it is assumed that the drivers are always in defensive state ($\gamma = 1$). Fig. 8(b) shows the fundamental diagram of model B, in which the acceleration noise is set to zero, i.e., $p_{a1} = p_{a2} = 0$. Fig. 8(c) shows the fundamental diagram of model C, in which the brake lights are always set to be on ($b_n = 1$). One can see that the three fundamental diagrams are similar to that in Fig. 2. When $\rho < \rho_c$, the traffic is in free flow. When $\rho > \rho_c$, the traffic becomes homogeneous congested flow (Fig. 9).

Our results thus demonstrate that homogeneous congested flow might be backbone of synchronized flow. By including different mechanisms into the homogeneous congested flow, synchronized flow with different features could be reproduced. Presently we do not know which mechanism is real origin of synchronized flow, which needs to be further investigated.

5   Conclusion

To summarize, we have proposed a CA traffic flow model, in which the finite size effect is trivial because the congested flow is quite homogeneous. As a result, the approximation solution of a small sized system can be regarded as solution of large system. We have investigated the approximation solution of a small traffic system with two vehicles. The analytical result is in good agreement with simulation.

We also discussed the relationship between homogeneous congested flow and synchronized flow. It is demonstrated that the synchronized flow is closely related with homogeneous congested flow. By introducing different mechanisms into the homogeneous congested flow, synchronized flow with different features could be reproduced.

Acknowledgments

This work is funded by National Basic Research Program of China (No. 2006CB705500), the NNSFC under Project Nos.10532060, 70601026, 10672160, 10872194, the SRF for ROCS, SEM, the NCET and the FANEDD. R.J.thanks the support of JSPS.

References

1. Kerner, B.S.: The Physics of Traffic. Springer, New York (2004)
2. Chowdhury, D., Santen, L., Schadschneider, A.: Statistical physics of vehicular traffic and some related systems. Phys. Rep. 329, 199–329 (2000)

3. Helbing, D.: Traffic and related self-driven many-particle systems. Rev. Mod. Phys. 73, 1067–1141 (2001)
4. Nagatani, T.: The physics of traffic jams. Rep. Prog. Phys. 65, 1331–1386 (2002)
5. Nagel, K., Wagner, P., Woesler, R.: Still flowing: Approaches to traffic flow and traffic jam modeling. Oper. Res. 51, 681–710 (2003)
6. Maerivoet, S., De Moor, B.: Cellular automata models of road traffic. Phys. Rep. 419, 1–64 (2005)
7. Nagel, K., Schreckenberg, M.: A cellular automaton model for freeway traffic. J. Physique I 2, 2221–2229 (1992)
8. Kerner, B.S., Rehborn, H.: Experimental properties of phase transitions in traffic flow. Phys. Rev. Lett. 79, 4030–4033 (1997)
9. Kerner, B.S., Rehborn, H.: Experimental features and characteristics of traffic jams. Phys. Rev. E 53, R1297–R1300 (1996)
10. Kerner, B.S., Rehborn, H.: Experimental properties of complexity in traffic flow. Phys. Rev. E 53, R4275–R4278 (1996)
11. Kerner, B.S.: Experimental features of self-organization in traffic flow. Phys. Rev. Lett. 81, 3797–3800 (1998)
12. Kerner, B.S.: Empirical macroscopic features of spatial-temporal traffic patterns at highway bottlenecks. Phys. Rev. E 65, 046138 (2002)
13. Schadschneider, A., Schreckenberg, M.: Car-oriented mean-field theory for traffic flow models. J. Phys. A 30, L69–L75 (1997)
14. Lee, H.K., Barlovic, R., Schreckenberg, M., Kim, D.: Mechanical restriction versus human overreaction triggering congested traffic states. Phys. Rev. Lett. 92, 238702 (2004)
15. Kerner, B.S., Klenov, S.L., Wolf, D.E.: Cellular automata approach to three-phase traffic theory. J. Phys. A 35, 9971–10013 (2002)
16. Jiang, R., Wu, Q.S.: Cellular automata models for synchronized traffic flow. J. Phys. A 36, 381–390 (2003)

On Scale-Free Prior Distributions and Their Applicability in Large-Scale Network Inference with Gaussian Graphical Models

Paul Sheridan, Takeshi Kamimura, and Hidetoshi Shimodaira

Tokyo Institute of Technology, Meguro-ku, Tokyo 152-8552, Japan
sherida6@is.titech.ac.jp

Abstract. This paper concerns the specification, and performance, of scale-free prior distributions with a view toward large-scale network inference from small-sample data sets. We devise three scale-free priors and implement them in the framework of Gaussian graphical models. Gaussian graphical models are used in gene network inference where high-throughput data describing a large number of variables with comparatively few samples are frequently analyzed by practitioners. And, although there is a consensus that many such networks are scale-free, the *modus operandi* is to assign a random network prior. Simulations demonstrate that the scale-free priors outperform the random network prior at recovering scale-free trees with degree exponents near 2, such as are characteristic of many real-world systems. On the other hand, the random network prior compares favorably at recovering scale-free trees characterized by larger degree exponents.

Keywords: Bayesian inference, complex networks, Gaussian graphical model, Markov chain Monte Carlo, prior distribution, scale-free, "small n, large p" problem, small-sample inference.

1 Motivation

Gaussian graphical models (GGMs) are commonly used to estimate a gene network from microarray data [17]. In this framework, p genes are represented by an undirected network $G = (V, E)$ where the node set $V = \{1, \ldots, p\}$ indexes the Gaussian random vector $X = (X_1, \ldots, X_p)$. A data matrix D of n microarray experiments is taken as a random sample from the multivariate Gaussian $X \sim N(\mu, \Sigma)$. The edge set E is defined by the conditional independence structure of X so that the edge $\{i, j\}$ is in E if, and only if, X_i and X_j are conditionally dependent given the remaining variables in X.

When Σ is nonsingular, conditional independence (a missing edge) between two variables X_i and X_j is equivalent to $\omega_{ij} = 0$ in the precision matrix $\Omega = \Sigma^{-1}$. Thus, model fitting over GGMs, known as *covariance selection*, amounts to identifying zero entries in Ω.

In classical GGM theory $n > p$ is necessary [3]; however, with genomic data it is frequently the case that $n \ll p$: the so called "small n, large p" problem.

J. Zhou (Ed.): Complex 2009, Part I, LNICST 4, pp. 110–117, 2009.

Many authors have taken to finding estimates for Ω when $n < p$ using either a full Bayasian approach [4] [7] [18], or via an empirical Bayes manner [10] [11].

Covariance selection, then, is accomplished either by heuristic searches, or by sampling the posterior distribution

$$\pi(G, \theta | D) \propto P(D|G)\pi(G|\theta)\pi(\theta) \tag{1}$$

over the space of undirected networks G on p nodes where $\pi(G|\theta)$ is a prior over networks that may depend on a set of parameters θ; $P(D|G)$ is the likelihood. To our knowledge, current methodologies assume G is sparse, and make inference on that basis. In particular, the approach to prior specification adopted in [7] assigns an inclusion probability $\beta = 2/(p-1)$ to each edge in G. This choice of β encourages sparse networks as the expected number of edges is p. In effect, $\pi(G|\theta)$ is the formula for the probability of a network under the random network model of [5], and we will refer to this as the *random prior*.

However, over the past decade, numerous examples of large-scale biological, technological, and sociological networks have been reported to be *scale-free*: that is, the *degree distribution* $p(k)$—the fraction of nodes in the network with degree k—closely follows a *power-law* $p(k) \propto k^{-\gamma}$ with exponent γ typically between 2 and 3 [9]. In particular, this property is thought to be a feature of gene networks [16] and $\gamma \approx 2.2$ has been verified for the known interactions in *S. cerevisiae* [6]. A further example comes from finance where it has been shown that cross-correlation between stock prices for companies on certain stock exchanges follows a power-law [15].

In this paper, we provide specifications for the prior $\pi(G|\theta)$ based on the formula for the probability of a network under three different scale-free models (Section 2). Our approach to employing scale-free network models in statistical inference to estimate a network G from data D is original insomuch as previous research has focused on estimating θ for a particular network, G. In Section 3, we give the results of a simulation study comparing these priors to the random prior at estimating scale-free trees from synthetic data. Finally, in Section 4 we muse on the practicality and possible future applications of our methodology.

2   Scale-Free Priors over Network Structures

In this section, we propose three scale-free assignments for $\pi(G|\theta)$. Each prior is defined by the formula for the probability of a network under a simple scale-free model. We selected the static model [8], the Poisson-growth (PG) model [13], and the proteome growth model [14].

Random network model: The random network, or Erdös-Rényi, model gives rise to a network G by connecting each pair of nodes in G with specified probability, β. Consequently, the probability of a network with $|E|$ edges is

$$\pi(G|\theta) = \beta^{|E|}(1-\beta)^{T-|E|}$$

where $\theta = (\beta)$ and $T = p(p-1)/2$ is the number of possible edges.

Static model: This model relies on *node fitness* as a generating mechanism, meaning that nodes in a network are assigned weights; edges are added to the network such that nodes with higher weight get more edges. Specifically, the static model is defined by $\theta = (\gamma, K)$ and generates a network as follows: Each node $1, \ldots, p$ is assigned a weight $P_i \propto i^{-1/(1-\gamma)}$ where $2 < \gamma < 3$ is the degree exponent. For $p \times K$ steps:

1. Select nodes i and j with probabilities P_i and P_j, respectively.
2. Connect i and j with an edge, unless they are already connected.

Unlike for the random network model, the probability of a network under any of the scale-free models depends on the order of the nodes in G. Therefore it is necessary to include the extra parameter $\sigma = (\sigma_1, \ldots, \sigma_p)$, a permutation of the nodes in V. Thus the posterior in Equation (1) becomes $\pi(G, \theta, \sigma | D)$ with accompanying prior $\pi(G|\theta, \sigma)$. The static model prior, as described in [8], is given by

$$\pi(G|\theta, \sigma) = e^{pK(1-M)} \prod_{e_{\sigma_i \sigma_j} \in G} \left(e^{2pK P_{\sigma_i} P_{\sigma_j}} - 1 \right)$$

where M is the sum of squares $\sum_{i=1}^{p} P_{\sigma_i}^2$.

PG model: The PG model is an offshoot of the Barabási-Albert (BA) model, which is based on two simple mechanisms: *growth*, where a network is built iteratively, over a series of steps $t = 1, \ldots, p$, by introducing a new node with m (fixed) edges at each step, and *preferential attachment* where the m edges are connected to exactly m nodes already in the system such that the probability a node of degree k gets an edge is proportional to $r(k) = k$, the *attachment function*. When $m = 1$, the BA model is known to have degree exponent $\gamma = 3$ [1].

In the PG model, m is assigned according to a Poisson random variable with parameter λ so that number of edges added at each step can vary. In addition, the attachment function is defined by $r(k) = k + a, k \geq 1$, and $r(0) = b$ where $a \geq -1$ is a small offset and $b = \geq 0$ is a threshold parameter. The PG model, then, is defined by $\theta(\lambda, a, b)$, and has been shown in [13] to follow a power-law with degree exponent ranging $\gamma > 2$. The formula for the associated prior is

$$P(G|\theta, \sigma) = \prod_{t=1}^{p-1} \left(\prod_{i=1}^{t} e^{-\lambda q_t(k_{\sigma_i, t})} \frac{(\lambda q_t(k_{\sigma_i, t}))^{s_{\sigma_i, t}}}{s_{\sigma_i, t}!} \right).$$

where $k_{\sigma_i, t}$ is the degree of node σ_i at step t, q_t is the normalized attachment function at step t, and $s_{\sigma_i, t}$ is the number of edges connecting nodes σ_i and t.

Proteome growth: This model is based on growth and *node duplication*. In particular, At each step $t = 1, \ldots, p$, a node is selected from the network at random and duplicated so that the new node inherits its edge structure. Edges emanating from the new node are deleted with probability q; new edges are added between the new node and all other nodes in the system with a small probability β/t. Thus the model is specified by the parameter $\theta = (\beta, q)$, and have been show to exhibit scale-free-like properties [14].

The probability of a network under this model depends on the duplication history $\psi = \{\psi_2, \ldots, \psi_p\}$ in addition to σ, where ψ_t is the node in σ from which σ_t was duplicated. It follows that $\pi(G|\theta, \sigma, \phi)$ is simply the product of the edge inclusion probabilities, which are defined by

$$P(e_{\sigma_i, \sigma_t}) = \begin{cases} 1 - q(1 - \beta/t) & \text{if } \sigma_i \text{ is a neighbor of } \psi_t, \\ \beta/t & \text{otherwise.} \end{cases}$$

Unlike for the random network model, the probability of a network under any of the scale-free models depends on the order of the nodes in G. Therefore it is necessary to include the extra parameter $\sigma = (\sigma_1, \ldots, \sigma_p)$, a permutation of the nodes in V. Thus the posterior in Equation (1) becomes $\pi(G, \theta, \sigma|D)$ with accompanying prior $\pi(G|\theta, \sigma)$.

3 Simulation

We conduct a simulation study to compare the scale-free priors against the random prior at recovering a range of tree topologies from small-sample synthetic data sets.

Tree generation: We generated $p = 100$ node trees from given degree distributions using the stochastic algorithm described in [2], and Table 2 summarizes the trees generated in each case.

Data generation: Trees are used because it is simple to generate multivariate normal data satisfying their conditional independence structures. For each tree in Table 2 we generated ten small-sample data sets, each with $n = 10$ observations.

MCMC implementation: MCMC algorithms are commonly used for sampling from high-dimensional probability distributions such as those encountered in GGMs. We ran the ready-to-use MCMC software from [7] to explore the space of decomposable GGMs for each tree, under each simulated data set. To transition from a decomposable model G to another G' they add or delete an edge from G at random to obtain G'.

In order to accommodate our scale-free priors we modified their sampler to include both the node permutation σ and the duplication history ψ for the proteome growth model. We update the node permutation by choosing σ' in a "neighborhood" of σ. Specifically, we select a node $i \in \{1, \ldots, p\}$ at random, and

Table 1. A summary of scale-free network models used as the basis for prior distributions

Model	Mechanism	Parameters	γ	Ref.
Static	Node fitness	$\theta = (\gamma, K)$	$2 - 3$	[8]
PG	Pref. attach	$\theta = (\lambda, a, b)$	> 2.0	[13]
Prot. gr.	Node dup.	$\theta = (\beta, q)$	$2 - 3$	[14]

Table 2. A wide range of tree topologies were generated for used in the simulation: Generic trees having a binomial degree distribution, scale-free trees, and a star tree in which all nodes are connected to a central hub. The *two parameter model* of citeburda (with parameters α and β) was used as the model for generating most of the scale-free trees. γ is the value of the degree exponent as predicted by the generating model. APL stands for average path length, which, is smaller for more highly centralized trees.

Topology	Model Name	Parameters	γ	APL
Generic	Erdös-Rényi	—	—	10.46
Generic	Two param.	$\alpha = 7.0, \beta = 1.5$	—	8.54
Scale-free	Two param.	$\alpha = 0.24, \beta = 3.0$	3.0	5.86
Scale-free	Two param.	$\alpha = 0.93, \beta = 2.5$	2.5	4.87
Scale-free	Barabási-Albert	$m = 1$	3.0	4.37
Scale-free	Two param.	$\alpha = 8.0, \beta = 2.0$	≈ 2.0	3.34
Scale-free	Two param.	$\alpha = 2.5, \beta = 2.5$	≈ 2.5	3.18
Scale-free	Two param.	$\alpha = 3.6, \beta = 2.2$	2.2	2.96
Scale-free	Two param.	$\alpha = 8.5, \beta = 2.1$	2.1	2.32
Star	—	—	—	1.98

proceed to transpose σ_i and σ_{i+1} to obtain σ'. In the case of the proteome growth model, the duplication history ψ is updated conditionally on σ' be rewiring the node represented by σ'_i to $\sigma'_i - 1$, if $\psi_{\sigma'_i} = \sigma'_i - 1$; otherwise σ'_i is rewired randomly to another node $\sigma'_j < \sigma'_i$. Additionally, we assigned the uniform distribution over each model parameter to compute $\pi(\theta)$, the prior distribution over the network model parameters.

We ran each MCMC chain for 5×10^6 steps after a burn-in of 10^5 steps. Each chain was started from the empty network, and for the scale-free priors with node σ and ψ were taken at random. We obtained similar results when starting from a variety of different initial conditions.

Results: Simulation results are summarized in Table 3. To estimate a network from a chain we took the $p - 1$ (i.e. the number of edges in a tree) edges of highest frequency over all networks in a chain. The model parameter estimate $\hat{\theta}$ was obtained by taking mean of the parameter values from a chain.

- The random network prior did quite well at recovering the generic trees as well as the scale-free trees with (large) $\gamma = 2.5$ to 3.0.
- The proteome growth prior exhibited poor performance overall. This can likely be attributed to model misspecification insomuch as this model may not be able to capture the tree structures that it was employed to estimate.
- The static and PG model priors outperformed the random prior at recovering the scale-free trees with underlying γ values near 2. This behavior likely reflects that as γ in a network tends to smaller values, the connectivity to the hub, the node with highest degree, tends to be higher, resulting in a more highly centralized network. In the pathological case of a star topology

Table 3. Results of the MCMC simulation ($p = 100$) with the numerical values in the table averaged over ten runs. PPV is the positive predictive value defined by $TP/(TP + FP)$, where TP stands for true positive, and FP for false negative.

Network Prior	Generated Tree Topology	γ	Estimated Parameters θ	$\hat{\gamma}$	PPV
Random	Generic	—	$\hat{\beta} = 0.015$	—	0.36
	Generic	—	$\hat{\beta} = 0.016$	—	0.35
	Scale-free	3.0	$\hat{\beta} = 0.016$	—	0.27
	Scale-free	2.5	$\hat{\beta} = 0.017$	—	0.28
	Scale-free	3.0	$\hat{\beta} = 0.016$	—	0.26
	Scale-free	≈ 2.0	$\hat{\beta} = 0.019$	—	0.19
	Scale-free	≈ 2.5	$\hat{\beta} = 0.016$	—	0.20
	Scale-free	2.2	$\hat{\beta} = 0.017$	—	0.12
	Scale-free	2.1	$\hat{\beta} = 0.017$	—	0.10
	Star	—	$\hat{\beta} = 0.018$	—	0.06
Static	Generic	—	$\hat{\gamma} = 2.82, \hat{K} = 0.63$	2.82	0.33
	Generic	—	$\hat{\gamma} = 2.80, \hat{K} = 0.70$	2.80	0.32
	Scale-free	3.0	$\hat{\gamma} = 2.70, \hat{K} = 0.76$	2.70	0.28
	Scale-free	2.5	$\hat{\gamma} = 2.58, \hat{K} = 0.90$	2.58	0.30
	Scale-free	3.0	$\hat{\gamma} = 2.76, \hat{K} = 0.70$	2.76	0.28
	Scale-free	≈ 2.0	$\hat{\gamma} = 2.39, \hat{K} = 1.18$	2.39	0.37
	Scale-free	≈ 2.5	$\hat{\gamma} = 2.54, \hat{K} = 0.88$	2.54	0.40
	Scale-free	2.2	$\hat{\gamma} = 2.25, \hat{K} = 1.27$	2.25	0.51
	Scale-free	2.1	$\hat{\gamma} = 2.30, \hat{K} = 1.16$	2.30	0.57
	Star	—	$\hat{\gamma} = 2.25, \hat{K} = 1.28$	2.25	0.78
PG	Generic	—	$\hat{\lambda} = 0.66, \hat{a} = -0.57, \hat{b} = 0.76$	2.88	0.27
	Generic	—	$\hat{\lambda} = 0.83, \hat{a} = -0.83, \hat{b} = 0.72$	2.56	0.17
	Scale-free	3.0	$\hat{\lambda} = 0.92, \hat{a} = -0.88, \hat{b} = 0.70$	2.51	0.23
	Scale-free	2.5	$\hat{\lambda} = 1.02, \hat{a} = -0.89, \hat{b} = 0.70$	2.51	0.28
	Scale-free	3.0	$\hat{\lambda} = 0.82, \hat{a} = -0.80, \hat{b} = 0.71$	2.58	0.22
	Scale-free	≈ 2.0	$\hat{\lambda} = 1.19, \hat{a} = -0.94, \hat{b} = 0.69$	2.50	0.41
	Scale-free	≈ 2.5	$\hat{\lambda} = 0.96, \hat{a} = -0.91, \hat{b} = 0.69$	2.49	0.45
	Scale-free	2.2	$\hat{\lambda} = 1.21, \hat{a} = -0.95, \hat{b} = 0.68$	2.49	0.62
	Scale-free	2.1	$\hat{\lambda} = 1.13, \hat{a} = -0.95, \hat{b} = 0.68$	2.48	0.62
	Star	—	$\hat{\lambda} = 1.20, \hat{a} = -0.95, \hat{b} = 0.70$	2.49	0.94
Prot. gr.	Generic	—	$\hat{\beta} = 0.64, \hat{q} = 0.91$	—	0.24
	Generic	—	$\hat{\beta} = 0.71, \hat{q} = 0.92$	—	0.22
	Scale-free	3.0	$\hat{\beta} = 0.72, \hat{q} = 0.92$	—	0.19
	Scale-free	2.5	$\hat{\beta} = 0.77, \hat{q} = 0.95$	—	0.23
	Scale-free	3.0	$\hat{\beta} = 0.68, \hat{q} = 0.92$	—	0.17
	Scale-free	≈ 2.0	$\hat{\beta} = 0.81, \hat{q} = 0.94$	—	0.02
	Scale-free	≈ 2.5	$\hat{\beta} = 0.72, \hat{q} = 0.92$	—	0.02
	Scale-free	2.2	$\hat{\beta} = 0.80, \hat{q} = 0.92$	—	0.10
	Scale-free	2.1	$\hat{\beta} = 0.75, \hat{q} = 0.94$	—	0.10
	Star	—	$\hat{\beta} = 0.83, \hat{q} = 0.94$	—	0.02

Note: For the proteome growth model, $\hat{\gamma}$ cannot be obtained analytically.

the scale-free priors perform exceptionally well in comparison to the random prior.

- The static model prior was able to produce was able to estimate the γ values with some accuracy, while the PG model prior did not discriminate between different degree exponents.

Out of the three scale-free prior distributions, the static model was the most accurate at estimating a network. In addition, it produced reasonable estimates for the degree exponent, γ.

4   Discussion

In summary, GGMs provide a framework for making inference about the conditional independence structure of a set a Gaussian variables when the number of observations are small compared to the number of variables. And this methodology, in various forms, is commonly applied in gene network inference. Our contribution in this paper has been to study the *a priori* inclusion of scale-free network topology into GGM inference via the specification of scale-free prior distributions. Our simulation study suggests that scale-free priors outperform the random prior at recovering scale-free trees, from small-sample data sets, with degree exponent γ near 2—a property thought to be typical of a wide variety of real-world networks.

We used trees, not networks, in our simulation in order to facilitate data generation. It would be beneficial to further this work by conducting a simulation over general network structures.

The implementation of [7], on which our results are based, can comfortably work for networks of a few hundred nodes; however, gene network inference can often involve networks with nodes numbering in the thousands. Therefore, scalability is a major issue, and integrating scale-free topological features using a different approach to GGM inference is of interest [10] [11].

References

1. Barabási, A.-L., Albert, R.: Emergence of scaling in random networks. Science 286, 509–512 (1999)
2. Burda, Z., Correia, J.D., Krzywicki, A.: Statistical Ensemble of Scale-Free Random Graphs. Phys. Rev. E 64, 046118+ (2001)
3. Dempster, A.P.: Covariance selection. Biometrics 28, 95–108 (1972)
4. Dorba, A., Hans, C., Jones, B., Nevins, J.R., West, M.: Sparse graphical models for exploring gene expression data. J. Multiv. Analysis 90, 196–212 (2004)
5. Erdös, P., Rényi, A.: On Random Graphs I. Pub. Math. Debrecen 6, 290–297 (1959)
6. Jeong, H., Mason, S.P., Barabasi, A.L., Oltvai, Z.N.: Lethality and Centrality in Protein Networks. Nature 411, 41–42 (2001)
7. Jones, B., Dobra, C., Carvalho, C., Hans, C., Carter, C., West, M.: Experiments in Stochastic Computation for High-Dimensional Graphical Models. Statistical Science 20(4), 388–400 (2005)

8. Lee, D.-S., Goh, H.-I., Kahng, B., Kim, D.: Scale-Free Random Graphs and Potts Model. Pramana J. Phys. 64, 1149–1159 (2005)
9. Newman, M., Barabási, A.-L., Watts, D.J.: The Structure and Dynamics of Networks. Princeton University Press, Princeton (2006)
10. Schäfer, J., Strimmer, K.: An Empirical Bayes Approach to Inferring Large-Scale Gene Association Networks. Bioinformatics 21, 754–764 (2005a)
11. Schäfer, J., Strimmer, K.: A Shrinkage Approach to Large-Scale Covariance Matrix Estimation and Implications for Functional Genomics. Stat. Appl. Genet. Mol. Biol. 4, article 32 (2005b)
12. Sheridan, P., Kamimura, T., Shimodaira, H.: Scale-Free Networks in Bayesian Inference with Applications to Bioinformatics. In: Proceedings of The International Workshop on Data-Mining and Statistical Science (DMSS 2007), Tokyo, pp. 1–16 (2007)
13. Sheridan, P., Yagahara, Y., Shimodaira, H.: A Preferential Attachment Model with Poisson Growth for Scale-Free Networks. Ann. Inst. Stat. Math. 60 (2008)
14. Solé, R.V., Pastor-Satorras, R., Smith, E., Kepler, T.B.: A Model of Large-Scale Proteome Evolution. Adv. Complex Systems 5, 43–54 (2002)
15. Vandewalle, N., Brisbois, F., Tordoir, X.: Non-Random Topology of Stock Markets. Quant. Finance 1, 372–374 (2001)
16. Wagner, A.: How the Global Structure of Protein Interaction Networks Evolve. Proc. R. Soc. B 270, 457–466 (2003)
17. Werhli, A.V.V., Grzegorczyk, M., Husmeier, D.: Comparative Evaluation of Reverse Engineering Gene Regulatory Networks with Relevance Networks, Graphical Gaussian Models and Bayesian Networks. Bioinformatics 22, 2523–2531 (2006)
18. Wong, K., Carter, C., Kohn, R.: Efficient Estimation of Covariance Selection Models. Biometrika 90, 809–830 (2004)

On General Laws of Complex Networks

Wenjun Xiao[1], Limin Peng[2], and Behrooz Parhami[3]

[1] School of Software Engineering, South China University of Technology,
Guangzhou 510641, P.R. China
wjxiao@scut.edu.cn
[2] Department of Computer Science, South China University of Technology,
Guangzhou 510641, P.R. China
penglm86@126.com
[3] Dept. Electrical & Computer Eng., University of California,
Santa Barbara, CA 93106-9560, USA
parhami@ece.ucsb.edu

Abstract. By introducing and analyzing a renormalization procedure, Song et al. [1] draw the conclusion that many complex networks exhibit self-repeating patterns on all length scales. First, we aim to demonstrate that the aforementioned conclusion is inadequately justified, mainly because their equation (7) on the invariance of degree distribution under renormalization does not hold in general. Secondly, Barabási and Albert [2] find that many large networks exhibit a scale-free power-law distribution of vertex degrees. They show this common feature to be a consequence of two generic mechanisms: (i) networks expand continuously by the addition of new vertices, and (ii) new vertices attach preferentially to those that are already well connected. We show that when vertex degrees of large networks follow a scale-free power-law distribution with the exponent $\gamma \geq 2$, the number of degree-1 vertices, when nonzero, is of the same order as the network size N and that the average degree is of order less than log N. Given that many real networks satisfy these two conditions, our results add another necessary characteristic of the scale-free power-law distribution of vertex degrees in such networks. Our method has the benefit of relying on conditions that are static and easily verified. They are verified by many experimental results of diverse real networks.

Keywords: complex network, self-similarity, scale-free, computer network, computer communication.

1 Introduction

Complex systems with many components and associated interactions arise in nature, society, and many human artifacts. Interactions in such systems can be modeled by networks composed of vertices and links, which are in turn abstracted as undirected or directed graphs. A graph G, denoted as $G = (V, E)$, has a set V of vertices or nodes and a set of edges or links, where each edge is defined by a pair of vertices (ordered pair, for directed graphs) [7]. Complex systems in the three categories of natural, societal, and synthetic include:

J. Zhou (Ed.): Complex 2009, Part I, LNICST 4, pp. 118–124, 2009.

1. Protein interactions, metabolic systems, contagious diseases
2. Acquaintances, movie-actor peer group, research collaborators
3. Power grid, Internet connectivity, Worldwide Web linkages.

Two models of complex networks have been studied extensively [1-10]: the small-world model and the scale-free one. The small-world model features localized clusters that are connected by occasional long-range links, leading to an average distance between vertices that grows logarithmically with the network size N. Watts and Strogatz [3] investigated mechanisms via which a regular network can be transformed into a small-world network, without significantly modifying the vertex-degree distribution, and quantified the parameters that characterize the resulting structures.

Scale-free networks, on the other hand, tend to have uneven vertex connectivities, so that a certain fraction of vertices, independent of network size, are highly connected (the hubs). Barabási and Albert [2] demonstrated that the scale-free power-law distribution of vertex degrees in many large networks is a direct consequence of two generic mechanisms that govern network formation: (i) Networks expand over time through the addition of new vertices, and (ii) New vertices attach preferentially to those that are already well connected. It is well-known that scale-freedom of a network has significant implications for its diffusion properties and its robustness.

In this paper, we focus on scale-free networks. After reviewing the parameters and key attributes of such networks in Section 2, we provide necessary characteristic conditions for scale-free complex networks and show that these conditions are both easy to verify and satisfied by many natural and man-made scale-free networks (Section 3). Section 4 contains our conclusions and some directions for further research.

2 On Self-similarity of Complex Networks

We begin by reviewing some relevant properties of complex networks [1-6]. Two models of real complex networks have been studied extensively: the small-world model and the scale-free one. We shall focus first on scale-free networks. To avoid confusion with equations in this paper, equation numbers in our references will be enclosed in square brackets.

For many complex networks, the probability distribution $P(k)$ of the number of degree-k vertices, also known as the degree distribution, can be represented (independent of scale) by a power law with characteristic exponent γ ([2] means that this is equation [2] in ref. 1):

$$P(k) \approx k^{-\gamma}. \tag{[2] (1)}$$

The renormalization method, introduced in ref. 1, is as follows. Let G be a network of degree distribution $P(k)$ satisfying equation (1). Vertices of G are covered by N_B boxes of linear size l_B. Boxes are then viewed as vertices of a renormalized network G', with two such vertices (boxes) connected in G' if and only if there exists at least one link between their constituent vertices in G. Song et al. [1] guided only by experimental results for WWW, then proceed with the assumption that the degree distribution is invariant under renormalization:

$$P(k) \to P(k') \approx (k')^{-\gamma}. \tag{[7] (2)}$$

And here lies the problem: equation (2) is neither proved mathematically nor verified adequately by experimentation with different complex networks. Hence, the validity of equation (2) is suspect. We shall endeavor to show equation (2) invalid by examining both the theoretical and experimental analyses of Song et al [1].

In the theoretical analysis that follows, we refer to, and use, four equations from ref. 1, reproduced below for completeness. Here the network G is called self-similarity if the equation (3) holds.

$$\langle M_B(l_B) \rangle \equiv N/N_B(l_B) \approx l_B^{d_B}. \tag{[5] (3)}$$

$$k \to k' = s(l_B)k. \tag{[6] (4)}$$

$$s(l_B) \approx l_B^{-d_k}. \tag{[8] (5)}$$

$$\gamma = 1 + d_B/d_k. \tag{[9] (6)}$$

In the equations above, $\langle M_B(l_B) \rangle$ is the average mass of (or the number of vertices in) a box, d_B is the fractal dimension (or box dimension) derived from $N_B \approx l_B^{-d_B}$, $s(l_B) < 1$ is the scaling of vertex degrees owing to renormalization (i.e., the ratio k'/k), and d_k is a new exponent characterizing the variation of s with l_B.

We prove that equations (4)-(6) can be derived directly from equation (3), without using the suspect equation (2). For this, we set $s = (N/N')^{1/(1-\gamma)}$, where N' is the size of G'. By equation (3), we have $s \approx l_B^{d_B/(1-\gamma)}$. Let $d_k = d_B/(\gamma-1)$, we obtain equation (6), which leads to $s \approx l_B^{-d_k}$; viz., equation (5). Finally, we obtain equation (4) from setting $k' = s\,k$, complete our proof. However, we cannot obtain equation (2) from equation (3). On the other hand, it is easily verified that equation (2) is equivalent to $n'(k') \approx n(k)$, where $n(k)$ and $n'(k')$ represent the number of vertices of degrees k and k' in networks G and G', respectively.

It is easily shown that multiple renormalizations at fixed l_B can be achieved by an equivalent renormalization. In fact, defining

$$N^{(i)}/N^{(i+1)} \approx l_B^{d_B^{(i)}}. \tag{7}$$

for $i = 0, 1, \ldots, t$ and $N = N^{(0)}$, we have

$$N/N^{(t)} \approx l_B^{d_B^{(0)}+\ldots+d_B^{(t-1)}}. \tag{8}$$

Next, setting $s^{(i)} = (N^{(i)}/N^{(i+1)})^{1/(1-\gamma)}$, we obtain $s^{(i)} \approx l_B^{d_B^{(i)}/(1-\gamma)}$. Finally, from $d_k^{(i)} = d_B^{(i)}/(\gamma-1)$, we get $s^{(i)} \approx l_B^{-d_k^{(i)}}$. Taking $k^{(0)} = k$ and $k^{(i+1)} = s^{(i)}k^{(i)}$ leads to the following two equations:

$$k^{(t)} = s^{(0)} \cdots s^{(t-1)} k .$$ (9)

$$s^{(0)} \cdots s^{(t-1)} \approx l_B^{-d_k^{(0)} - \cdots - d_k^{(t-1)}} .$$ (10)

We have thus demonstrated that the hypothesis which Song et al. [1] derive experimentally (namely, that based on Fig. 2d of ref. 1, the degree distribution of the WWW of special sizes is invariant under renormalization for different box sizes) is not supported by theoretical analysis, leading to serious doubts regarding its correctness in general. As a consequence, it is not surprising that many complex networks (Internet, protein interactions, and some random networks) lack self-similarity, as indicated in the supplementary materials of ref. 1. This can be explained theoretically as follows. By equation (7), we may set

$$N^{(i)} / N^{(i+1)} \approx C_i l_B^{d_B^{(i)}} .$$ (11)

thus, leading to

$$N / N^{(t)} \approx C_0 \cdots C_{t-1} l_B^{d_B^{(0)} + \ldots + d_B^{(t-1)}} .$$ (12)

In equation (12), the product $C_0 \cdots C_{t-1}$ may be an exponential function of the average distance $\bar{l}$ when $C_i > 1$ and $t \approx \bar{l}$. This would imply that the network size is an exponential function of the average distance $\bar{l}$, thus, the network lacks self-similarity in this case.

3 Necessary Conditions for Scale-Free Networks

In the following, we study the conditions for vertex degrees of complex networks having scale-free power-law distribution. We assume that the network is connected; similar arguments apply to disconnected networks. Let $P(k)$ be the probability distribution of the number of vertices of degree k, as previously defined. Let A denote the average vertex degree and n_k the number of degree-k vertices. We have $M = \frac{1}{2}NA$ and $n_k = NP(k)$, where N and M are numbers of vertices and edges, respectively. The preceding definitions imply

$$\sum_{k=1}^{N-1} n_k = N .$$ (13a)

$$\sum_{k=1}^{N-1} kn_k = 2M .$$ (13b)

Supposing that $n_1 \neq 0$, we have $n_k = n_1 P(k) / P(1)$ and

$$\sum_{k=1}^{N-1} [P(k) / P(1)] = N / n_1 .$$ (14a)

$$\sum_{k=1}^{N-1} [k P(k) / P(1)] = 2M / n_1 .$$ (14b)

For scale-free networks, we have $P(k) = P(1) k^{-\gamma}$, which leads to $\sum_{k=1}^{N-1} k^{-\gamma} = N/n_1$. Therefore, assuming $\gamma \geq 2$, which is known to hold for many scale-free networks [4,5], we have

$$N/n_1 \leq \sum_{k=1}^{\infty} k^{-\gamma} \leq \sum_{k=1}^{\infty} k^{-2} = \pi^2/6 . \tag{15}$$

This leads to the conclusion $N \approx n_1$. Let $f(k) = P(k)/P(1)$. When $N \approx n_1$, equations (14) yield

$$\sum_{k=1}^{N-1} f(k) \approx 1 . \tag{16a}$$

$$\sum_{k=1}^{N-1} k f(k) \approx A . \tag{16b}$$

Assuming that all logarithms are in base 2, equations (16) yield $1 < A \leq \log N$, given that $\sum_{k=1}^{N} k^{-1} \approx \log N$. Thus, we have proved that for many real complex networks of scale-free power-law distribution with $\gamma \geq 2$, the number of degree-1 vertices, when nonzero, is of the same order as the network size N and that the average degree is of order less than $\log N$.

On the other hand, if $\log \log N \leq A \leq \log N$, then equations (16) imply

$$f(k) \approx k^{-\gamma}. \tag{17}$$

We now elaborate on equation (17). There are three canonical cases for the function $f(k)$ when $\log \log N \leq A \leq \log N$: (a) $f(1) = 1$ and $f(k)$ a constant for $k \neq 1$; (b) $f(k) = e^{1-k}$; and (c) $f(k) \approx k^{-\gamma}$. For case (c), equations (16) hold when $2 \leq \gamma < 3$. However, for cases (a) and (b), equation (16b) is not satisfied if equation (16a) holds. Note that more complex functions satisfying equations (16) exist. For example, one can define the function

$$f(k) \approx e^{1-k}, \text{ if } k \leq \log N; \ f(k) \approx k^{-\gamma}, \text{ otherwise} . \tag{18}$$

which satisfies equations (16). As a consequence of our results above, the scale-free property of complex networks must be viewed as an approximate or fuzzy property.

The preceding leads to a model for scale-free networks, known to satisfy $k^{\gamma} = n_1/n_k$. Taking this equality to be exact, and noting that the right-hand side is a rational number, we can readily prove that k^{γ} must be an integer that divides n_1. Then, n_1 must be divisible by the least common multiple of $k_1^{\gamma}, k_2^{\gamma}, \ldots, k_l^{\gamma}$, where $1 = k_1 < k_2 < \cdots < k_l$ is the degree sequence of the network. This is a general model for scale-free networks that we now aim to study further.

Let $n_1 = c[k_1^{\gamma}, k_2^{\gamma}, \ldots, k_l^{\gamma}] \neq 0$, where c is constant. If $\gamma \geq 2$ then

$$N = n_1 \sum_{i=1}^{l} k_i^{-\gamma} \leq \tfrac{\pi^2}{6} n_1 < \tfrac{5}{3} n_1 \text{ and } M = \tfrac{1}{2} n_1 \sum_{i=1}^{l} k_i^{1-\gamma} \leq \tfrac{1}{2} n_1 \leq N \log \log N.$$

Now we give a sufficient and necessary condition that a connected graph is a tree.

Proposition. Assume that G is connected and scale-free, then G is a tree

$$\Leftrightarrow \sum_k k^{1-\gamma} = \frac{2N-2}{n_1}.$$

Proof. G is a tree $\Leftrightarrow 2N - 2 = N\sum_k P(1)k^{1-\gamma} = N\sum_k \frac{n_1}{N}k^{1-\gamma} = n_1\sum_k k^{1-\gamma}$.

Table 1. Number of vertices (N), number of edges (M), average degree (A), and characteristic exponent (γ) in some complex networks

Network	N	M	A	γ
Internet	10,687	31,992	5.98	2.5
Film actors	449,913	25,516,482	113.43	2.3
Metabolic network	765	3686	9.64	2.2
Protein interactions	2115	2240	2.12	2.4

Table 1 lists the parameters N, M, A, and γ for several real scale-free networks [4,5]. We note that $A \approx \frac{1}{2} \log N$ (respectively, $5 \log N$, $\log N$, and $\log \log N$) for the Internet (film actors, metabolic, and protein interaction) network.

4 Conclusion

We have shown that for many real networks of scale-free power-law degree distribution with the exponent $\gamma \geq 2$, the number of degree-1 vertices, when nonzero, is of the same order as the network size N and that the average degree is of order less than $\log N$. Our method has the benefit of relying on conditions that are static and easily verified for any network. Such distributions are known to be applicable to diverse fields of study, including computer communication and software architecture. However it is worth to further research networks of scale-free power-law degree distribution with the exponent $\gamma < 2$.

Such extensions and variations will further broaden the applications of our results in diverse subfields within computing, communication, biology, and the social sciences.

Acknowledgements. Research of W. Xiao is supported by Guangdong Key Laboratory of Computer Network (CCNL200705) and Laboratory of Basic Software and Applied Construction Technology of Guangdong(2006B80407001).

References

1. Song, C., Havlin, S., Makse, H.A.: Self-similarity of complex networks. Nature 433, 392–395 (2005)
2. Barabási, A.-L., Albert, R.: Emergence of scaling in random networks. Science 286, 509–512 (1999)

3. Watts, D.J., Strogatz, S.H.: Collective dynamics of 'small-world' networks. Nature 393, 440–442 (1998)
4. Albert, R., Barabási, A.-L.: Statistical mechanics of complex networks. Rev. Mod. Phys. 74, 47–97 (2002)
5. Newman, M.E.J.: The structure and function of complex networks. SIAM Rev. 45, 167–256 (2003)
6. Xiao, W.J., Parhami, B.: Cayley Graphs as Models of Deterministic Small-World Networks. Information Processing Letters 97, 115–117 (2006)
7. Biggs, N.: Algebraic Graph Theory. Cambridge University Press, Cambridge (1993)
8. Myers, C.R.: Software systems as complex networks: Structure, function, and evolvability of software collaboration graphs. Physical Review E 68, 046116 (2003)
9. Montoya, J.M., Sole, R.V.: Small World Patterns in Food Webs. Journal of Theoretical Biology 214, 405–412 (2002)

On Distributed Multi-Point Concurrent Test System and Its Implementation

Hao Luo and Huaxin Zeng

School of Information Science and Technology,
Southwest Jiaotong University, Chengdu 610031, P.R. China
jackfrued@gmail.com, huaxinzeng1@yahoo.com.cn

Abstract. As the rapid expansion of the Internet, novel network applications are constantly emerging accompanied by the increasingly growing of the complexity for network protocols. As a result, IP routers are becoming more and more important in today's Internet. However, test methodology and test systems for IP routers often fall behind the state of the art. Based on the deficiency of current stand-alone test systems, this paper analyzes the necessity for distributed test architecture and introduces a new test system called Distributed Multi-point Concurrent Test System (DMC-TS) which can mirror real-world networks in the test experiments and conduct conformance testing, performance testing and interoperability testing for the routers under test. Key issues for the implementation of DMC-TS are discussed, especially on test synchronization problem. Consequently, two types of test synchronization problem are pointed out with the corresponding solutions. Some experiments are carried out to illustrate the feasibility and practicability of DMC-TS.

Keywords: distributed, concurrent test manager, test agent, synchronization.

1 Introduction

With world-wide applications of the Internet, IP routers are playing more and more important roles in computer networks. Whether these routers can reliably work or not has direct impact on today's Internet; also, the Quality of Service (QoS) as well as customer satisfaction with service provider networks are directly influenced by the performance of these routers. Consequently, product testing for IP routers becomes the prerequisite to ensure their ability to process routing information and to forward data and thus must be carried out before deployment in mission critical networks.

Since routing protocol is one of the most important factors for routers, testing of routing protocols has drawn great interest of academic and industrial communities worldwide since late 1970's. Pioneer work on conformance testing resulted in a framework on conformance testing as specified in ISO/IEC standard 9646 [1]. The document provided a framework on methodology for conformance testing of OSI (Open System Interconnection) protocols. As a result, four abstract test methods for end systems and two test methods for relay systems were proposed. In addition, a semi-formal test definition language called TTCN (Tree and Tabular Combined Notation) was also defined which

J. Zhou (Ed.): Complex 2009, Part I, LNICST 4, pp. 125–139, 2009.

put the history of conformance testing using non-formal methods such as natural languages to an end.

Two test methods for relay systems defined in ISO/IEC standard 9646 are Loopback Test Method (LTM) and Transverse Test Method (TTM). LTM executes test procedures by sending test data to the router under test and then receiving the loop back messages from the same port. It had been the most popular test method for relay systems until the emergence of TTM in early 1980's. TTM enables the testing of routers through a pair of ports and thus more applicable in practice and superior over LTM. At present, TTM is still the most popular test method for routers as it is simple to apply and capable to test the behavior of a router between a pair of ports with predefined test data. Generally, both of the two test methods can be implemented by stand-alone systems, however only one port or a pair of ports of the routers can be tested simultaneously.

The rapid development of network technology makes the port number and the forwarding rate of a single port increase constantly. Nowadays the core routers in the backbone of the Internet generally have dozens or even hundreds of ports and the throughput of each port have reached the magnitude of Gbps or more. Meanwhile the demands on security, robustness as well as QoS capability for IP routers are being put forward. By and large, traditional test systems based on LTM and TTM are no longer feasible and realistic as the routers under test have become increasingly complex and the related protocols that need to be tested are widely distributed. It's necessary to update the traditional test methodology and to develop new test system that can apply test control and observation to all ports of the routers under test so as to mirror real-world networks in the test experiments.

The deficiency of current stand-alone test systems as stated above stimulates the activity of research work on Distributed Multi-point Concurrent Test System (DMC-TS) and the relevant issues presented in this paper and the paper is organized in the following manner. The necessity for distributed test architecture and corresponding system is discussed in section 2. In section 3, the prototype of DMC-TS is presented with the detailed discussion on its implementation. In section 4, we deal with test synchronization problem which is very common in distributed test architecture. Some experiments are carried out to illustrate our current work as well as the feasibility and practicability of DMC-TS in section 5. Finally, conclusions are drawn in section 6, summarizing our contributions and discussing some future research issues.

2   The Necessity for Distributed Test System

IP routers have been traditionally perceived as a network device with a 3-layer architecture adopting the in-band signaling point of view. However, this point of view cannot explain why the routing information protocols such as RIP and BGP and the network management protocol such as SNMP in a router are actually implemented on the top of a transport protocol (TCP or UDP). This contradiction can be avoided by adopting out-band signaling concept [2]. At this point of view, a router is composed of two sets of protocol stacks, one for user data transfer (up to network layer, usually being referred to as User Plane, or U-Plane for short), and the other for constructing a network capable of transporting data traffic (up to application layer, usually being referred to as Control Plane, or C-Plane for short) [3]. However, current performance testing for IP routers often takes only U-Plane into account but ignores the influence of C-Plane. Indeed, the activities of C-Plane have crucial impact on the performance of U-Plane. As a result, performance testing that considers the interaction between U-Plane and C-Plane usually yields more accurate and reliable results than the ones that only take the performance of U-Plane into account, since the former provides the routers under test with the circumstances that is more like the real-world networks when putting these routers into practice.

As stated previously, the rapid development of network technology resulted in the growing complexity of the protocols that run inside IP routers. The running of some protocols often involves a number of communication ports simultaneously. For example, if test operator wants to duplicate real-world conditions for the router under test, routing simulation must be done beforehand. Therefore a large routing table must be created automatically by the test system through multiple communication ports in limited time. Also, normal routing information including routing requests and updates must be transmitted from the test system to the router under test periodically or randomly during the test process as well as flapping routes for simulation of abnormal conditions. In this case, traditional test system based on LTM and TTM that involves only one port or a pair of ports becomes impotent. With distributed test system and concurrent testing approach, all possible conditions can be simulated by means of injecting all kinds of test data including routing information, network management queries and so on into the router under test from a number of test components in a controllable manner. This is also the main benefit for developing a distributed test system.

Another important reason for developing a distributed test system is modeling and generation of reasonable workload. The performance of an IP router depends on the characteristics of the workload it must server. As a result, what kind of test traffic should be used becomes an important issue that must be considered, especially when carrying out performance testing or QoS testing. In order to acquire the reliable test results, it is ideal to replicate Internet traffic or to use a workload model that can reveal various characteristics of Internet's traffic. However, it is very difficult to replicate Internet traffic or to put forward a feasible workload model due to the complexity and uncertainty of the network system. Although some research work has pointed out the "self-similar" nature of network traffic [4], it is hard to make the best of this model in test practice.

However, by making full use of distributed test architecture, the problem of workload modeling and generation can be expediently resolved in a simple manner by generating packet streams with different source/destination addresses, different length, different types of service and different priorities from a number of test components. In addition, by superposition of the packet streams from different test components, workload that fits to a specific arriving process and a special traffic pattern can be generated and in consequence workload that a router would serve when being deployed in real-world networks can be simulated.

Last but not least, the scalability and extensibility of the test system are considered. Although some commercial test systems for IP routers based on stand-alone test systems can be configured with a number of test ports so that they can also carry out extensive testing for routers under test, the scalability and extensibility is inevitably limited. Employing a distributed test system is the key factor to surmount the problem of scalability and extensibility in testing. As new test components can be added up into existing test system dynamically as needed, the capability of the test system can therefore be promoted without making great changes to existing system. In addition, a distributed test system can make the best of all potential resources and thus provides the test system with high performance/price ratio.

At present, researchers have reached an agreement of developing test systems based on distributed architecture [5, 6], but the prerequisite is to design a reasonable architecture for distributed testing and actualize an efficient approach for concurrent test control. Also some other issues on the implementation must be solved beforehand. These issues are the major motivation of the research work presented in this paper.

3 The Prototype and Implementation of DMC-TS

In order to duplicate real-world conditions for routers under test and satisfy the requirements for scalability and extensibility of the test system itself, a new test system called Distributed Multi-point Concurrent Test System (DMC-TS) is introduced in this paper, which is composed of one Concurrent Test Manager (CTM) and several Test Agents (TAs). The CTM itself does not have the test ability but fulfills the specified tests by sending test requests and control requests to one or more TA(s). Each TA listens to the test and control requests and then executes the test procedures according the requests from CTM by injecting test data into the Router Under Test (RUT) as well as observing the external behavior of the RUT. A test verdict will be generated when the assigned test is accomplished and it will be sent back to CTM as a response to the former test request.

The test and control requests sending from CTM to TAs during the test process are non-blocking ones, which means that CTM is able to send requests successively to several TAs without blocking for the responses of the former requests. Also, there are some requests that need to be sent to several TAs at the same time and this can be done by connecting CTM and TAs with a fast switch and communicating through multicast or broadcast. As several TAs are involved, a number of points of control and observation (PCOs) can be established and the RUT can be controlled and observed exhaustively from various PCOs.

Both of the functional testing (such as conformance testing) and the non-functional testing (such as performance testing) can be carried out with DMC-TS using the test configuration as illustrated in Fig. 1. Under this configuration, each TA can represent an adjacent router or a subnet connected to the RUT and thus background traffic, routing information update messages and network management queries can be generated by these TAs through the test ports connected to the RUT during the test process. In this way, real-world conditions can be simulated for the RUT so that the test results of conformance testing and performance testing can be more accurate and reliable. Of course, more than one RUT can be involved in the test. These routers can form a system that can be treated as a whole and usually be referred to as the System Under Test (SUT).

Experience has also shown that two implementations of the same protocol are not necessarily interoperable even both of the implementations have previously passed the conformance testing. Therefore, sometimes there is a need for arbitration. A complete test system for IP routers should encompass interoperability testing as well as conformance testing and performance testing and this can be done with DMC-TS by employing a special TA (as the shadowed TA in Fig. 1) with the ability to fulfill the passive monitoring or active intervention of the communication process between two RUTs as well as tracing the potential faults.

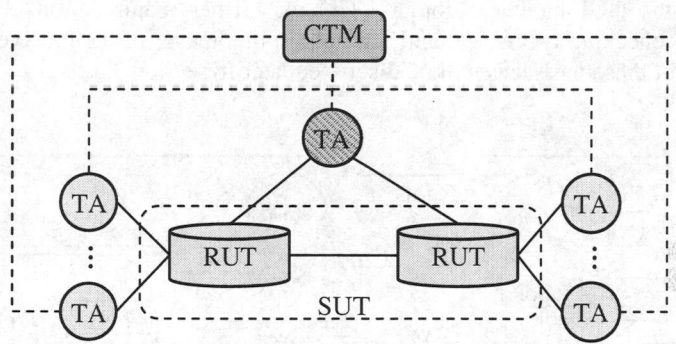

Fig. 1. Conformance and performance testing configuration of the DMC-TS

In the architecture of DMC-TS, TA is loosely coupled with CTM (using dotted lines in Fig. 1), i.e. each TA can perform some simple test without the participation of CTM. In this way, we can make TA into a stand-alone portable tester in the future. However, the most important feature of TA is to play the role of a test component in DMC-TS. Therefore, on one hand the number of TAs can be increased in order to promote the test ability of the system; on the other hand the connection between the RUTs and TAs can be modified for the purpose of establishing some possible network topologies for testing.

3.1 Design of Concurrent Test Manager

As show in Fig. 2, CTM is composed of the Graphical User Interface (GUI) and the Execution Engine (EE). The GUI module is mainly used to set test configuration through dialog boxes or forms, implement user control by buttons or menus and display the test results with line, pie, or bar graphs. Among these functionalities, test configuration setting is usually done by selecting and editing the Concurrent Test Control Script (CTCS). The CTCS is a simple script language dedicated to implement concurrent test control by providing test operator with the core language features including basic operators, loops and branch statements as well as some macros that are used to accomplish a series of complicated operations by a single line command. CTM keeps a mapping table to maintain the relationship between macros and their related implementation functions. As a matter of course, the table can be modified by test operators in order to make some extension to existing macros. Apparently, CTCS is very convenient to test operators as they want to add new functionalities to the script language according to the test requirements.

The EE module is the most important module to achieve multi-point concurrent test control. It interprets the CTCS by converting the code into test request or control request messages and then sending them to one or more TA(s). The EE module can also communicate with the GUI module through Data and Control Channel (DCC) and it converts the control operations initiated by test operator into control request messages and sends them to one or more TA(s). On receiving a response message from TA, the EE module will decide whether to handle the message itself or to forward it to the GUI module through DCC for further manipulation. This can be achieved by checking type and identifier field in the header of the received message. The format of these messages will be discussed later in section 3.3.

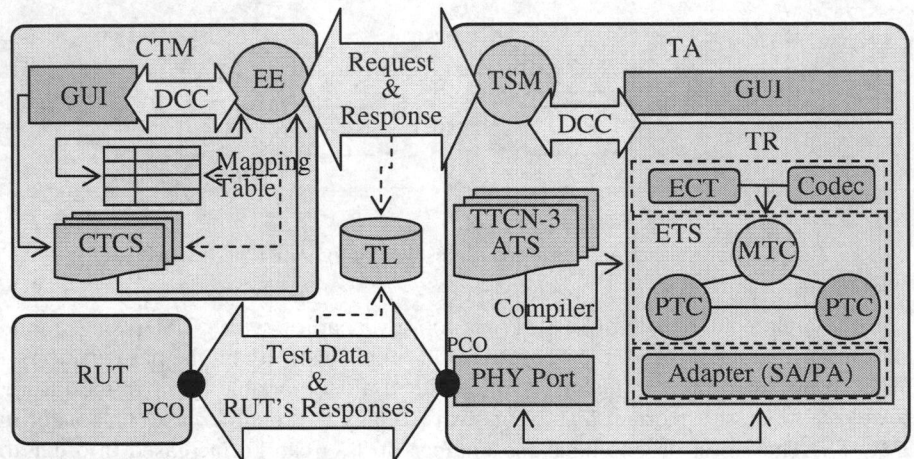

Fig. 2. Details on implementation of concurrent test manager and test agent

3.2 Design of Test Agent

As show in Fig. 2, TA is mainly composed of three parts, including the Test Service Module (TSM), the Test Runner (TR), and the GUI module. The TSM is responsible for listening to the test requests and control requests from CTM, scheduling these messages and then deciding what should be done next. The TR is the most important part in the implementation of TA. It is responsible for executing the actual test jobs by integrating the Execution Control Thread (ECT), the Codec (Encoder/Decoder), the Executable Test Suites (ETS), and the Adapter. When TA receives a request from CTM, TSM queues the message into its buffer and then do some scheduling for the message according to its priority. ECT is initiated by TSM on receiving the first test request message and it is responsible for controlling the execution of ETS. ECT will not stop its execution until the test is accomplished by normal termination, abnormal termination or user compulsory termination. For normal termination and abnormal termination, the ECT needs to send the test verdict or the exception status to TSM through DCC. TSM will encapsulate the information reported by ECT into a specific response message using pre-defined message format and after that send it back to CTM as the corresponding response to the former request. CTM will make further manipulation of the response message.

The TTCN-3 [7] (Testing and Test Control Notation version 3) is selected as the test description language for DMC-TS. TTCN-3 has been put forward and maintained by ETSI as a general purpose test specification and implementation language that is dedicated to black-box testing of a wide range of computer and telecommunication systems. Typical areas of application are protocol testing, service testing, module testing, testing of CORBA based platform, APIs, etc. TTCN-3 introduces lots of advanced features such as template matching mechanism and dynamic test control which make the description of test data and test control very flexible and convenient. One of the most important reasons to select TTCN-3 is that it defines test cases on abstract level so that test developers can concentrate on the development of test logic instead of worrying about how to implement them on a given platform or operating system. A collection of the test cases on the abstract level is called Abstract Test Suites (ATS) which can be converted into ETS by specific TTCN-3 compiler or interpreter. Together with the Codec, System Adapter (SA) and Platform Adapter (PA), ETS can run on different test platform and Implementation Under Test (IUT) and thus the implementation independent test can be achieved.

The GUI module for TA has the similar functionality as it is in CTM. As stated above, TA can conduct some simple test that involves only one test port or a pair of test ports without the participation of CTM. In this case, the GUI module can provide a man-machine interface for the test operators. PHY Port in Fig. 2 is the physical port of TA for sending test data and receiving responses from the RUT. PCO stands for the point of control and observation during the test process.

3.3 Some Relevant Issues

Other issues relevant to the implementation of DMC-TS that haven't been discussed above are list below:

- **Test Log.** In order to carry out some off-line analysis and debugging for the test system itself, Test Log (TL) module is needed for DMC-TS. As illustrated in Fig. 2, TL provides the storage function for test data and the responses of RUT as well as request and response messages between CTM and TA.
- **Message Format.** In order to ensure unambiguous communication between CTM and TA, the format of the request and response message must be clearly defined. The message is composed of 1-byte header, 1-byte reserved field and message data as shown in Fig. 3. The detail of fields in message header is outlined in Table 1.

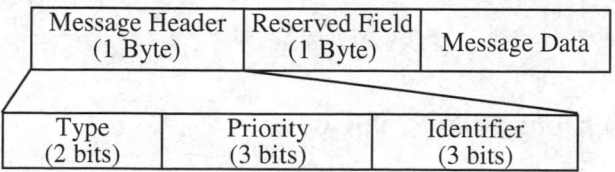

Fig. 3. Message Format

Table 1. Detail of message header fields

Message Type	Priority		Identifier
00	000	High Priority (for control message)	000 – STATE CHECK
Test Request	001		001 – INITIALIZE
01	010		010 – START
Control Request	011	Normal Priority	011 – STOP
10	100	(for test message)	100 – PAUSE
Test Response	101	Low Priority (for date transmission)	101 – RESUME
11	110		110 – DATA
Control Response	111		111 – UPDATE

- **State Diagram of CTM and TA.** The working state of CTM and TA can be depicted by state diagram as illustrated in Fig. 4 and Fig. 5 respectively.

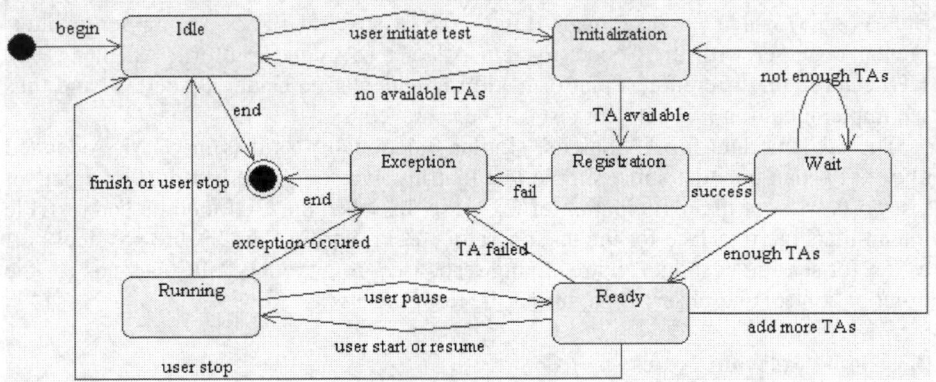

Fig. 4. State diagram of CTM

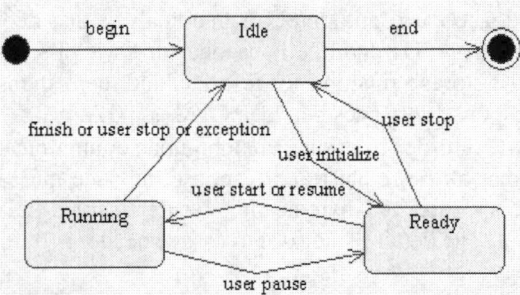

Fig. 5. State diagram of TA

4   Test Synchronization Problem

The significant difference between the DMC-TS and current test system is that the DMC-TS is constructed using a distributed architecture and thus may encounter the synchronization problem due to the lack of global state knowledge and global timer. The synchronization problem is referred to as the test sequences executed on multiple test components are not full ordered and thus has adverse impact on the test results, especially to conformance testing. So this problem must be solved before the DMC-TS can be used to conduct some test experiments. Typically, multi-port finite state machine is adopted to describe the synchronization problem [8].

A multi-port finite state machine with n ports (np-FSM) is a 6-tuple $M=(S, \Sigma, \Gamma, \delta, \lambda, s_0)$. S is a finite set of states and $s_0 \in S$ is the initial state. $\Sigma=(\Sigma_1, \Sigma_2, \dots, \Sigma_n)$, where Σ_k is the input alphabet of port k, and $\Sigma_i \cap \Sigma_j=\Phi$, for $i \neq j$. Let $I=\Sigma_1 \cup \Sigma_2 \cup \dots \cup \Sigma_n$. $\Gamma=(\Gamma_1, \Gamma_2, \dots, \Gamma_n)$, where Γ_k is the output alphabet of port k, and $\Gamma_i \cap \Gamma_j=\Phi$, for $i \neq j$. Let $O=(\Gamma_1 \cup \{\varepsilon\}) \times (\Gamma_2 \cup \{\varepsilon\}) \times \dots \times (\Gamma_n \cup \{\varepsilon\})$, where ε stands for the null output. δ is the transition function $S \times I \to S$, and λ is the output function $S \times I \to O$.

A transition of an np-FSM M is a 4-tuple (s, σ, γ, s') where $s, s' \in S, \sigma \in I, \gamma \in O$, such that $\delta(s, \sigma)=s'$ and $\lambda(s, \sigma)=\gamma$. An np-FSM M can be represented by a directed graph $G=(V, E)$ where V represents the set S of states of M and E presents all specified transitions of M. An example of 3p-FSM is given in Fig. 6, where $S=\{S_0, S_1, S_2, S_3\}$, $\Sigma_1=\{X\}, \Sigma_2=\{Y\}, \Sigma_3=\{Z\}, \Gamma_1=\{a, b\}, \Gamma_2=\{c\}, \Gamma_3=\{d, e\}$. In this figure, the transition t_1 denotes that if S_0 is the current state and the input X is received, the state changes to S_1 and the output d is sent at ports 3. ε means no output occurs at port 1 and port 2.

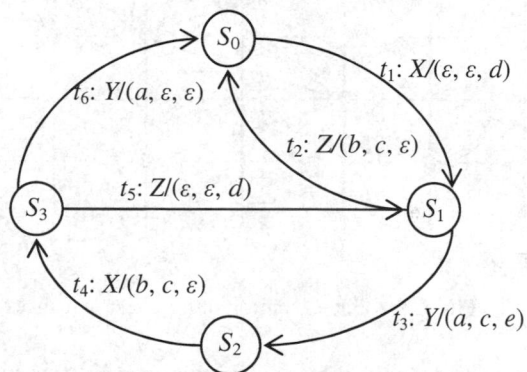

Fig. 6. An example of 3p-FSM

A global test sequence of np-FSM is a sequence in the form: $!x_1?y_1!x_2?y_2 \dots !x_t?y_t$ where, for $i=1,2,\dots,t$, $x_i \in I$ and $y_i \in O$, for each port k, $|y_i \cap \Gamma_k| \leq 1$, i.e. y_i contains at most one symbol from the output alphabet of each port. '$!x_i$' means sending message x_i to the RUT and '$?y_i$' means receiving the messages belonging to y_i from the RUT.

Generally, the global test sequence sare generated from the specification of the IUT and characterized by their fault coverage (output faults and transfer faults) [9, 10]. A possible global test sequence for 3p-FSM of the example illustrated in Fig. 6 is:

$$!X?\{d\}!Y?\{a, c, e\}!X?\{b, c\}!Z?\{d\}!Z?\{b, c\} \qquad (1)$$

In the DMC-TS, each TA executes a local test sequence constructed from the complete test sequence of the IUT. We can get the local test sequence for each TA from the global test sequence (1), and recorded as Seq_1, Seq_2 and Seq_3 with the corresponding TA recorded as TA_1, TA_2 and TA_3.

$$Seq_1 = !X?a!X?b?b, Seq_2 = !Y?c?c?c, Seq_3 = ?d?e!Z?d!Z \qquad (2)$$

Apparently, the conform execution of local test sequences Seq_1, Seq_2 and Seq_3 must give the result shown in Fig. 7(a), but if TA_1, TA_2 and TA_3 execute their local test sequences separately, the execution gives the failed result shown in Fig. 7(b). This kind of failure is called Synchronization Problem of Type-I (short for SPT-I) which is formally defined as follows.

Definition 1. For any two consecutive transitions $t_i=(s_i, \sigma_i, \gamma_i, s_i')$ and $t_j=(s_j, \sigma_j, \gamma_j, s_j')$, a SPT-I occurs if $port(\sigma_j) \notin ports(\gamma_i) \cup \{port(\sigma_i)\}$, where $port(\sigma_j)$ and $port(\sigma_i)$ denotes the port associated with input σ_j and input σ_i respectively; $ports(\gamma_i)$ denotes the set of ports associated with values from γ_i that are not null.

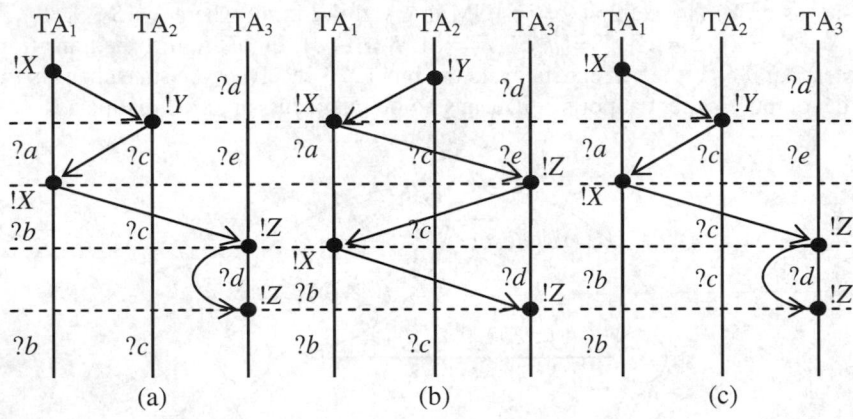

Fig. 7. Possible execution of local test sequences

The SPT-I can be easily solved by exchanging coordination messages among TAs. In [11], an algorithm is introduced to generate send and receive operations for coordination messages and insert them into the proper position of the local test sequences. Applying the proposed algorithm to the global test sequence (1), we get the following local test sequences, where '$!C_i$' means sending a coordination message to port i and '$?C_j$' means receiving a coordination message from port j.

$$Seq_1 = !X?a!C_2!X?b!C_3?b, Seq_2 = ?C_1!Y?c?c?c, Seq_3 = ?d?e?C_1!Z?d!Z \qquad (3)$$

However, the proposed algorithm still cannot overcome the errors as illustrated in Fig. 7(c). Although the input operations on each TA conform to (1) and the output operations on each TA conform to (2), the position of some output operations are not the same as show in Fig. 7(a). This kind of failure is called the Synchronization Problem of Type-II (short for SPT-II) which is formally defined as follows.

Definition 2. Given two consecutive transitions $t_i=(s_i, \sigma_i, \gamma_i, s_i')$ and $t_j=(s_j, \sigma_j, \gamma_j, s_j')$, a SPT-II occurs at port k if $(k \neq port(\sigma_j) \wedge (\gamma_i^k \neq \varepsilon$ XOR $\gamma_j^k \neq \varepsilon))$, where $port(\sigma_j)$ denotes the port associated with input σ_j; γ_i^k and γ_j^k denotes the output at port k on t_i and t_j respectively.

The existing of the SPT-II will probably cause the test system fails to detect the potential errors of the IUT. In [12], some algorithms for overcoming the SPT-II have been proposed but made a simplifying assumption that the time required for a coordination message to travel from a tester to another is greater than the reaction time of the IUT. Apparently, this assumption cannot be satisfied at all times. For DMC-TS, the SPT-II can be thoroughly overcome if and only if the position of output operations of the IUT can be checked without the consideration of time constraints and we follow the steps outlined below:

1. Record the position value of all output operations for each TA according to the global test sequence and input/output alphabets, where the position value means at which transition the output operation has occurred.
2. Add '!T' operation before all input operations in local test sequences, where '!T' stands for sending a special message to CTM. When CTM receives T message from TA, it broadcasts the message to all TAs including the one that initiates the message.
3. Define two variables ot and nt with the initial value of 0 for each TA respectively. Block all input operations in the local test sequence until the condition $nt > ot$ can be satisfied and after that, let $ot := nt$. These operations can be easily actualized in the test cases written in TTCN-3.
4. When TA receives the T message from CTM, let $nt := nt + 1$. For all output operations, record the current value of nt as its position value on each TA respectively.

We assume that the sending and receiving of T message is based on the reliable transmission and this assumption can be easily satisfied if taking TCP as transport layer protocol. For conformance testing, if the output of the IUT is not the expected one, the corresponding TA will yield a test verdict of "fail". If all TAs finished the execution of their local test sequences, the position value of output operations will be checked by comparing the values recorded by step 1 and step 4 to verify whether the SPT-II has occurred. If the SPT-II has indeed taken place, the fault of the IUT will be revealed by the unmatched position value.

5 Test Experiments

Conformance testing, performance testing and interoperability testing can be carried out for IP routers with DMC-TS. The following is an experiment of route flap testing. Route flap testing is a procedure that repeatedly changes routing tables with route withdrawals and updates. Core Internet routers must be able to handle many gigabits or terabits of traffic flawlessly during constant changes to route tables of 100,000 entries or more. Obviously, route flap testing involves both the U-Plane testing and the C-Plane testing of the RUT simultaneously. Following is the definition of U-Plane testing and C-Plane testing:

– **U-Plane Testing.** U-Plane testing helps determine how rapidly and accurately routers can manipulate very large volumes of data traffic. This type of testing includes generating realistic Internet traffic, including multiple types and classes of service, and stressing a router's forwarding ability and QoS performance.

– **C-Plane Testing.** As C-Plane uses various routing and signaling protocols to establish routes or path for U-Plane traffic, C-Plane testing usually includes route and link flapping, path establishment and destruction, failover and policy testing.

5.1 Testing of BGP Route Flapping

The objective of BGP route flapping test is to record the data performance through the RUT in the presence of an unstable BGP peer. This test is important to characterize the stability of the RUT and verify its ability to continue to provide data transport while being subjected to a chronic network failure resulting in an unstable BGP peer.

The test will establish two BGP peers on two different interfaces. As illustrated in Fig. 8. TA_1 plays the role of a BGP peer that advertises routes and forwards the UDP traffic from the RUT and TA_2 plays the role of another BGP peer that advertises and withdraws its routes (flapped routes) at a set rate for a set time. The rest of the TAs plays the role of ordinary routers that send UDP traffic to the RUT at specific rate during the test process.

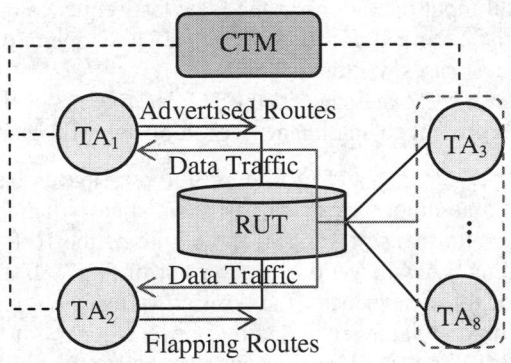

Fig. 8. Test BGP route flapping using the DMC-TS

The CTM will execute a pre-written CTCS to control the execution of the test and we follow the steps outlined below:

1. Configure BGP on the RUT.
2. Advertise a set number of routes to the RUT from TA_1.
3. Send UDP traffic from TA_3 – TA_8 to the RUT and record traffic statistics.
4. Perform route flapping on a set number of flapped routes from TA_2 for a set amount of time and record traffic statistics at a set period.
5. To characterize router behavior, vary one of the following and continue at step 3.
   – Number of traffic routes.
   – Number of flapped routes.
   – Data packet length.
   – Data packet distribution.

5.2 Test Results

In our experiment, two routers from different manufacturers marked as RUT-1 and RUT-2 are selected to make some comparison. 50,000 of BGP routes have been established and 10,000 of them are unstable ones. We use UDP packets with six different lengths from 64 bytes to 1518 bytes as the forwarding traffic and record the throughput and the latency statistics of the RUT-1 and the RUT-2 as shown in Fig. 9 and Fig. 10 respectively.

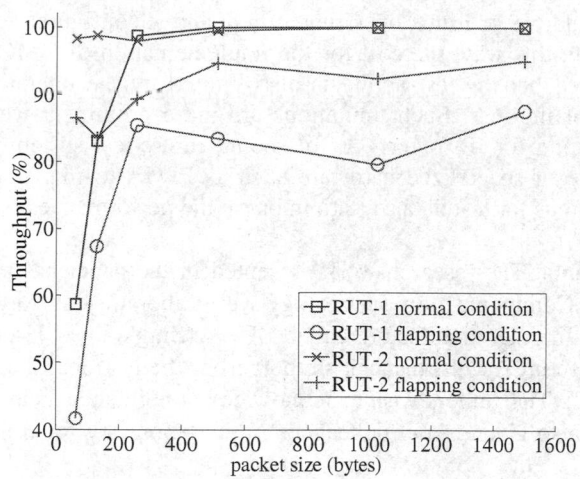

Fig. 9. Throughput under normal and flapping condition

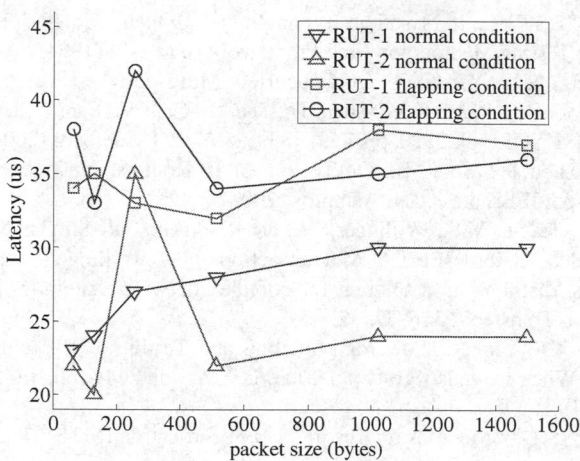

Fig. 10. Latency under normal and flapping condition

6   Conclusion and Future Works

The work presented in this paper is dedicated to construct a reasonable distributed architecture for concurrent testing and resolve the difficulty of test synchronization problem. Preliminary experiments have confirmed the feasibility and practicality of DMC-TS. However, using T message for checking the position of output operations will bring about additional overhead, and therefore how to reduce the overhead for sending and receiving of T message as well as coordination message is the next step of our research work. In addition, the current implementation of TA is based on the PC that runs Windows or Linux operating system and is configured with two or more network cards. In this way, the cost for the implementation of DMC-TS can be significantly reduced but the test ability is also limited by the operating systems and network cards of the PCs. Such limitations are more prominent when carrying out performance testing for IP routers. At this point, future test system should consider taking advantage of specialized hardware such as FPGA to implement sending and receiving operations for test data so as to improve the performance of the test system.

Acknowledgments. The research work presented in the paper has sponsored by Sichuan Network Communication Technology Key Laboratory as part of the research work on a new Internet architecture called SUPA (Single-layer User-data switching Platform Architecture), with financial support from the National Science Foundation of China (NSFC). The authors wish to acknowledge the financial support from NSFC. Our thanks are also given to our colleagues in the laboratory for fruitful cooperation in the past.

References

1. ISO/IEC JTC1/SC21.: Information Technology – Open Systems Interconnection – Conformance Testing Methodology and Framework (part 1-7) (1994)
2. Song, B., Zeng, H.X., Yue, L.Q.: On Concurrent Multi-port Test System for Routers and Its Support Tools. In: Liew, K.M., Shen, H., See, S., Cai, W., Fan, P., Horiguchi, S. (eds.) PDCAT 2004. LNCS, vol. 3320, pp. 469–483. Springer, Heidelberg (2004)
3. Zeng, H.X., Zhou, X., Song, B.: On Testing of IP Routers. In: Proceedings of PDCAT 2003, pp. 61–65. IEEE press, Los Alamitos (2003)
4. Leland, W.E., Taqqu, M.S., Willinger, W., et al.: On the Self-Similar Nature of Ethernet traffic (extended version). IEEE/ACM Transaction on Networking 2(1), 1–5 (1994)
5. Viho, C.: Test distribution: a solution for complex network system testing. Int. J. Softw. Tools Technol. Transfer 7, 316–325 (2005)
6. Wu, J.P., Li, Z.J., Yin, X.: Towards Modeling and Testing of IP Routing Protocols. In: Hogrefe, D., Wiles, A. (eds.) TestCom 2003. LNCS, vol. 2644, pp. 49–62. Springer, Heidelberg (2003)
7. ETSI ES201873 1-7.: Methods for Testing and Specification (MTS); The Testing and Test Control Notation version 3 (2008)
8. Khoumsi, A.: A Temporal Approach for Testing Distributed Systems. IEEE Transactions on Software Engineering 28(11), 1085–1103 (2002)

9. Petrenko, A., Bochmann, G., Yayo, M.: On Fault Coverage of Tests for Finite State Speci-
    fication. Computer Network and ISDN Systems 29, 81–106 (1996)
10. Pap, Z., Subramaniam, M., Kovacs, G., Nemeth, G.A.: A Bounded Incremental Test Gen-
    eration Algorithm for Finite State Machines. In: Petrenko, A., Veanes, M., Tretmans, J.,
    Grieskamp, W. (eds.) TestCom/FATES 2007. LNCS, vol. 4581, pp. 244–259. Springer,
    Heidelberg (2007)
11. Rafiq, O., Cacciari, L., Benattou, M.: Coordination Issues in Distributed Testing. In: Pro-
    ceeding of the International Conference on Parallel and Distributed Processing Techniques
    and Applications (PDPTA 1999), Las Vegas (Nevada), USA. CSREA Press (1999)
12. Ural, H., Whittier, D.: Distributed testing without encountering controllability and ob-
    servability Problems. Inf. Processing Lett. 88(3), 133–141 (2003)

Organizational Structure of the Transcriptional Regulatory Network of Yeast: Periodic Genes

Frank Emmert-Streib[1] and Matthias Dehmer[2]

[1] Queen's University Belfast, Computational Biology and Machine Learning,
Center for Cancer Research and Cell Biology, School of Medicine,
Dentistry and Biomedical Sciences, 97 Lisburn Road, Belfast BT9 7BL, UK
v@bio-complexity.com
[2] Institute for Bioinformatics and Translational Research, UMIT,
Eduard Wallnoefer Zentrum 1, 6060, Hall in Tyrol, Austria
Matthias.Dehmer@umit.at

Abstract. In this paper we investigate the organizational structure of the transcriptional regulatory network of *S. cerevisiae* with respect to the connectivity structure of periodic genes. We demonstrate that the giant strongly connected component plays a prominent role serving as central connector for genes experimentally found to be periodically expressed during the cell cycle of yeast. Numerically, we find by randomization of the gene labels that this organizational structure is unlikely to be formed by chance.

Keywords: graph theory, transcriptional regulatory network, causality, randomization, periodic genes.

1 Introduction

The analysis of complex network has gain much attention during the last decade [2,5,14,20,19]. This interest comes in part from the fact that many natural phenomena can be cast into a network framework that enables an analysis of the problem. Especially, in molecular biology such approaches have been used frequently [8,15,18]. In contrast to the theoretical analysis of general complex networks and their properties in biology that major interest consists in understanding the functional organization of gene networks [4]. So far, however, it is largely unknown how to connect, e.g., graph theoretical network properties meaningfully to the biological function of a molecular biological system.

In this paper we use the transcriptional regulatory network of yeast to investigate the organizational structure of periodic genes. Genes are called periodic if they are expressed periodically during the cell cycle [1,12,21,23] that means if they are not just switched on or off but alternate periodically between activation states. Traditional approaches studying periodic genes use, e.g., time series data from DNA microarray experiments trying to identify periodic pattern. Here we do not aim to identify periodic genes but are interested instead in their structural organization in the transcriptional regulatory network. That

J. Zhou (Ed.): Complex 2009, Part I, LNICST 4, pp. 140–148, 2009.

means, we use the transcriptional regulatory network [13,22], which is a directed, unweighted network, and a list of genes known to be periodic [23] to perform a structural analysis of this network. Our analysis is based on the observation that a (general) network may contain one or more strongly connected components. A strongly connected component is a subnetwork connecting each pair of nodes in this subnetwork bidirectionally. With other words, the strongly connected component has a cyclic structure allowing to connect nodes on closed paths (cycles). The transcriptional regulatory network of yeast contains such strongly connected components [9]. Due to the fact, that only nodes in the strongly connected component can occur on cycles we hypothesis that only these genes can be directly activated periodically [9]. All other gene that are periodic need to be triggered by these genes. For this reason we hypothesis that the strongly connected component plays a prominent role in the organization of the cell cycle and the activation of periodic genes. We calculate the shortest paths [6] from the strongly connected component to all periodic genes in the network (if possible) and investigate the observed structure. More strictly, due to the fact that the transcriptional regulatory network is a curated network its structure can be considered as *causal* representing molecular interactions instead of just some form of association between the genes. This implies that our graph theoretical approach is causality based because a path connects only genes that (potentially) influence each other causally.

This paper is organized as follows. In the next two sections we present the method we apply to the transcriptional regulatory network and the data we use for our analysis. In section 4 we present numerical results and this paper finishes in section 5 with conclusions.

2 Methods

We use a graph theoretical approach to study the structural organization of periodic genes in the transcriptional regulatory network G of yeast. In addition to the transcriptional regulatory network we use a list of genes known to be periodically expressed during the cell cycle.

The transcriptional regulatory network can be partitioned by the presence or absence of cycles connecting genes. In mathematical terms a part of the network that is cyclic is also called a strongly connected component (SCC) [7]. For example, for a SCC containing at least three genes, A_i, A_j, A_k there exists a cycle $A_i \rightarrow \ldots \rightarrow A_j \rightarrow \ldots \rightarrow A_k \rightarrow \cdots \rightarrow A_i$. The dots indicate that there are possibly other genes involved. However, the important point is that there exists a cycle on which all three genes appear. This observation is important because the presence of a cycle in a network is a necessary condition that truly periodic behavior can be observed because these genes have the ability to interact (activate/inhibit) each other consecutively and, hence, can form a limit cycle [17]. This leads us to the separation of the genes in two classes. The first class consists of genes that belong to the SCC. The genes in the second class do not belong to the SCC. Further the two classes are not equal but the information should flow in one direction namely from $SCC \rightarrow G/SCC$. The reason is that only genes in the SCC can establish a periodic behavior, as explained above, while genes in G/SCC can not. Based on this classification and hierarchy we raise the following hypothesis [9].

Hypothesis 1. *Given a causal path from a gene in the SCC to a gene in G/SCC, obtained from the transcriptional regulatory network, connecting two genes known to be periodic than all genes on this path are periodic if: First, the connecting path is a shortest path. Second, there is just one shortest path connecting the periodic genes.*

In this paper we will not analyze the predictions of our hypothesis but instead we focus on the structural organization of the obtained subnetwork. More precisely, we will analyze if properties of the observed subnetwork are formed by chance or significant with respect to gene label randomization.

3 Data

For our analysis we use the transcriptional regulatory network (TRN) of yeast [13,22] which is a directed, unweighted network. From this network we extract the weakly connected component (WCC) consisting of 3357 genes and 7230 interactions. The weakly connected component of a network is defined as the subnetwork that connects every pair of nodes by at least one directed path [7]. In contrast, the strongly connected component (SCC) is defined as subnetwork that connects each pair of genes in both directions. That means there exists a path connecting, e.g., gene A with gene B but there exists also a path connecting gene B with gene A. The TRN consists of two strongly connected components. One consists of 36 and the other of just 2 genes. When we speak in the following of the SCC of the TRN we speak always about the larger subnetwork also called the giant strongly connected component [24]. The strongly connected component is part of the weakly connected component, $SCC \subseteq WCC$. We use a list of ZHAO et al. as reference for periodic genes [23]. They categorized 260 genes as periodic from which 179 are in the subnetwork (WCC) considered in our analysis.

4 Results

In Fig. 1 we show a subnetwork of the transcriptional regulatory network. This subnetwork contains the strongly connected component and all periodic genes that can be reached from there. The SCC is shown as one node only (red) because we are here not interested in the connectivity of the SCC but the connectivity from the SCC to periodic genes. The periodic genes are shown in orange and genes that are non periodic are shown in blue. We want to emphasize that we included only edges that occur on shortest paths from the SCC to periodic genes. This does not only simplify the situation but corresponds also to an assumption frequently employed in the context of gene networks in general [3,11,16] assuming that interactions follow shortest paths. The resulting network looks

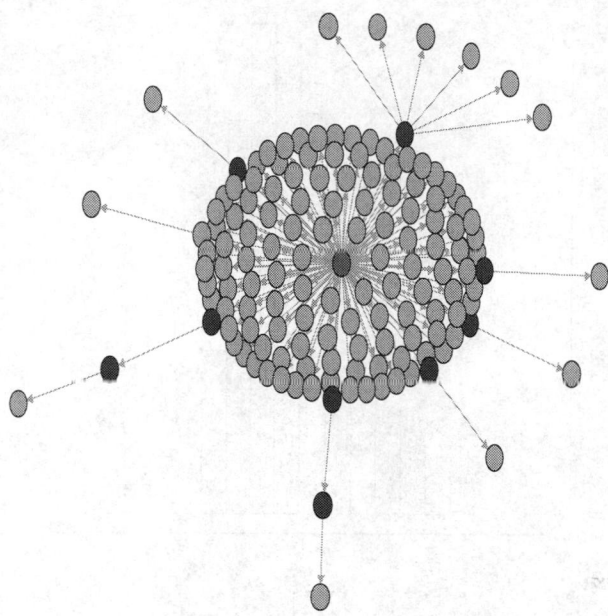

Fig. 1. Subnetwork of the TRN of yeast. Shown are 141 genes and the strongly connected component represented as one red node. Nodes in orange correspond to periodic genes [23], blue nodes are genes not categorized as periodic. The connections shown are shortest paths connecting the periodic genes to the strongly connected component. All other connections are omitted.

remarkably simple containing many periodic genes and only very few non periodic genes. More precisely, we find among all 179 periodic genes in the WCC 132 are connected to the SCC. This corresponds to 73% of all periodic genes. Considering the fact that the SCC contains nine more periodic genes our model view covers 78% of all periodic genes. Another interesting result from Fig. 1 is that only 9 non periodic genes (blue nodes) are necessary to accomplish the shown connected subnetwork.

The crucial question arsing from these observations is if these results are an effect caused by evolution or if these results are merely random structures. To investigate this we randomize the transcriptional regulatory network in the following way. We keep all genes from the SCC fixed. All other node labels, which correspond to gene names, are randomized by permuting these node labels. According to this randomization we generate an ensemble of $N_E = 1000$ networks and repeat our analysis for Fig. 1. The results of these randomizations are shown in Figs. 2–4.

In Fig. 2 we show histograms of the number of periodic genes N_{sc}^p that are directly connected to the SCC (top) and of the number of periodic genes N_{tot}^p that are reachable via a shortest path from the SCC (bottom). It is interesting to see that in these randomized networks the number of periodic genes is much smaller compared to the results for the (normal) TRN shown in Fig. 1.

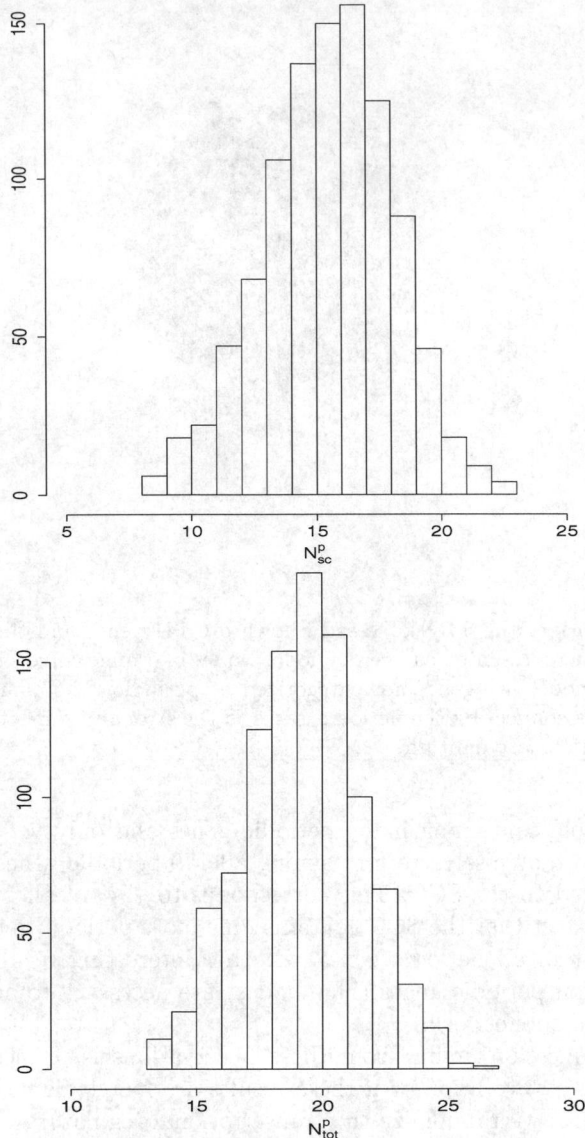

Fig. 2. Top: Histogram of the number of periodic genes N_{sc}^p that are directly connected to the SCC as a result of gene label randomization. The mean value of N_{sc}^p is 16.02. Bottom: Histogram of the number of periodic genes N_{tot}^p that are reachable from the SCC as a result of gene label randomization. The mean value of N_{tot}^p is 19.75.

In Fig. 3 we show the percentage of non-periodic genes involved to connect the SCC to N_{tot}^p periodic genes,

$$p_{np}^p = \frac{N_{np}}{N_{tot}^p}. \tag{1}$$

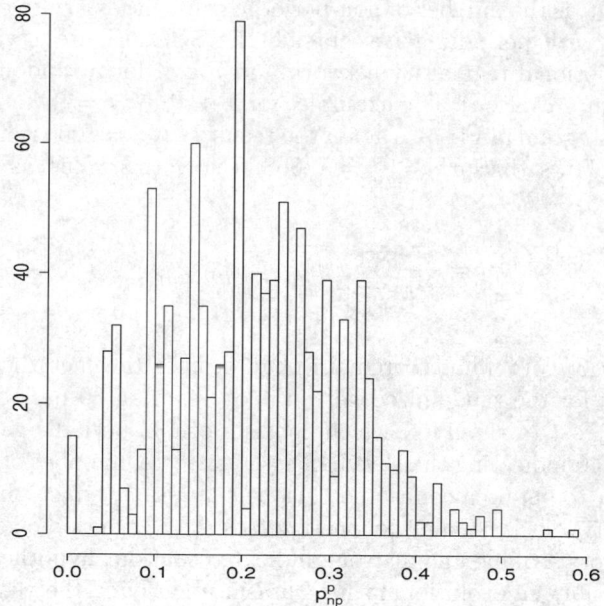

Fig. 3. Histogram of p_{np}^p, the percentage of non-periodic genes involved to connect the SCC to N_{tot}^p periodic genes

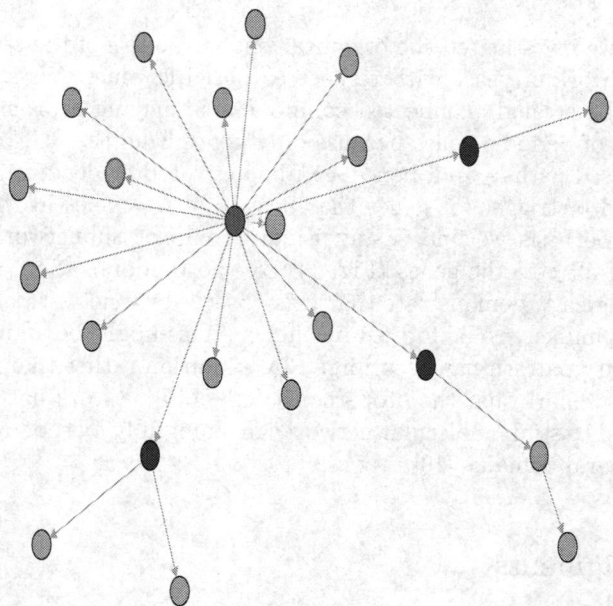

Fig. 4. Same as Fig. 1, however, for a gene label randomized network

That means, N_{np} is the number of non-periodic genes necessary to connect genes from the SCC with periodic genes outside the SCC. Figure 1 visualizes this for the transcriptional regulatory network and Fig. 4 for a randomized version thereof (see figure caption). For example, for Fig. 1 $N_{np} = 9$ (number of blue nodes). The histogram in Fig. 3 is again the result of $N_E = 1000$ randomizations. From Fig. 1 follows that $p_{TRN,np}^p = 0.068$. Using this value as threshold to calculate

$$p_b = \frac{1}{N_E} \sum_{p_{np}^p} I(p_{np}^p \leq p_{TRN,np}^p) \qquad (2)$$

gives $p_b = 0.083^3$. We want to remark that despite the fact that the largest cluster consists for the randomized networks of less than 30 periodic genes (see Fig. 3), in contrast to the TRN which consists of 132 periodic genes, this is a quite low number indicating that by chance the expected number of non-periodic genes necessary to obtain a connected cluster is larger. The fact, that non of the randomized networks is capable connecting close to 132 periodic genes from the SCC is even more striking and a strong indicator that our hypothesis is sensible unraveling possibly an evolutionary mechanism underlying the yeast cell cycle.

5   Conclusions

In this paper we investigated the organizational structure of the transcriptional regulatory network of yeast with respect to periodic genes. We started by assuming that the strongly connected component is playing a prominent role in the regulation of periodic genes because only genes from the SCC can be found on cycles - closed paths - and, hence, can be activated or deactivated cyclically forming a kind of trigger for genes that occur not on cycles in the TRN. Applying our hypothesis we find a surprisingly compact subnetwork that spans almost 80% of all periodic genes (Fig. 1). By the randomization of gene labels we could numerically demonstrate that this observed subnetwork and its constituting parts is unlikely to be formed by chance. This might be an indicator that this connectivity pattern has been forged by evolution rather than accidentally.

We want to remark that the property *cyclicity* of a network has been already previously used to study molecular networks meaningfully by separating proteins in their structural domains [10].

Acknowledgments

We would like to thank Galina Glazko for fruitful discussions and two anonymous reviewers for their comments that helped to improve this paper.

---

[3] In Eqn. 2 $I()$ is the indicator function which is 1 if the argument is true and 0 else.

References

1. Ahnert, S., Willbrand, K., Brown, F., Fink, T.: Unbiased pattern detection in microarray data series. Bioinformatics 22(12), 1471–1476 (2006)
2. Albert, R., Barabasi, A.: Statistical mechanics of complex networks. Rev. of Modern Physics 74, 47 (2002)
3. Arita, M.: The metabolic world of escherichia coli is not small. Proc. Natl. Acad. Sci. USA 101(6), 1543–1547 (2004)
4. Barabasi, A.L., Oltvai, Z.N.: Network biology: Understanding the cell's functional organization. Nature Reviews 5, 101–113 (2004)
5. Bornholdt, S., Schuster, H. (eds.): Handbook of Graphs and Networks: From the Genome to the Internet. Wiley-VCH, Chichester (2003)
6. Dijkstra, E.: A note on two problems in connection with graphs. Numerische Math. 1, 269–271 (1959)
7. Dorogovtesev, S., Mendes, J.: Evolution of Networks: From Biological Nets to the Internet and WWW. Oxford University Press, Oxford (2003)
8. Emmert-Streib, F., Dehmer, M. (eds.): Analysis of Microarray Data: A Network Based Approach. Wiley-VCH, Chichester (2008)
9. Emmert-Streib, F., Dehmer, M.: Analyzing and predicting periodic genes of the transcriptional regulatory network of yeast: A causality based approach (submitted) (2008)
10. Emmert-Streib, F., Mushegian, A.: A topological algorithm for identification of structural domains of proteins. BMC Bioinformatics 8, 237 (2007)
11. Jeong, H., Tombor, B., Albert, Z.N., Olivai, R., Barabasi, A.L.: The large-scale organization of metabolic networks. Nature 407, 651–654 (2000)
12. Luan, Y., Li, H.: Model-based methods for identifying periodically expressed genes based on time course microarray gene expression data. Bioinformatics 20(3), 332–339 (2004)
13. Luscombe, N., Badu, M., Snyder, Y.H.M., Teichmann, S., Gerstein, M.: Genomic analysis of regulatory network dynamics reveals large topological changes. Nature 431, 308–312 (2004)
14. Newman, M.E.J.: The structure and function of complex networks. SIAM Review 45, 167–256 (2003)
15. Palsson, B.: Systems Biology. Cambridge University Press, Cambridge (2006)
16. Rahman, S., Schomburg, D.: Observing local and global properties of metabolic pathways: load points and choke points in the metabolic networks. Bioinformatics 22(14), 1767–1774 (2006)
17. Schuster, H.G.: Deterministic Chaos. Wiley VCH Publisher, Chichester (1988)
18. Shmulevich, I., Dougherty, E.: Genomic Signal Processing. Princeton University Press, Princeton (2007)
19. Watts, D.: Small Worlds: The Dynamics of Networks between Order and Randomness. Princeton University Press, Princeton (1999)
20. Watts, D., Strogatz, S.: Collective dynamics of 'small-world' networks. Nature 393, 440–442 (1998)
21. Wichert, S., Fokianos, K., Strimmer, K.: Identifying periodically expressed transcripts in microarray time series data. Bioinformatics 20(1), 5–20 (2004)

22. Yu, H., Kim, P., Sprecher, E., V., T., Gerstein, M.: The importance of bottlenecks in protein networks: Correlation with gene essentiality and expression dynamics. PLoS Computational Biology 3(4), e59 (2007)
23. Zhao, L., Prentice, R., Breeden, L.: Statistical modeling of large microarray data sets to identify stimulus-response profiles. Proc. Natl. Acad. Sci. USA 98(10), 5631–5636 (2001)
24. Zhu, D., Qin, Z.: Structural comparison of metabolic networks in selected single cell organisms. BMC Bioinformatics 6, 8 (2005)

Packet-Level Traffic Allocation for Real-Time Streaming over Multipath Networks

Yanfeng Zhang, Cuirong Wang, and Yuan Gao

Northeastern University, Shenyang LN 110004, P.R. China
threewells14@gmail.com

Abstract. We address packet-level traffic allocation problem for real-time media streaming under multipath network environment. Based on an in-depth analysis of multipath real-time streaming model, also considering fluctuation of multipath network status as well as burst of media sending rate, we suggest that traffic load should be allocated to paths in proportion to the paths' available bandwidths, which minimizes the overall bandwidth overload probability. Moreover, due to the smallest transmission unit is packet, in order to execute the traffic allocation policy exactly, weighted size-aware packet distribution algorithm is proposed to avoid the actual traffic deviation due to variance of packet sizes. Simulation results show that the proposed algorithm outperforms other traditional algorithms, especially for reducing packet late arrivals, which has negative impaction in real-time transmission.

Keywords: traffic allocation, multipath, real-time streaming, available bandwidth, path redundance.

1 Introduction

In despite of the development of novel network infrastructures and constantly increasing bandwidth, Internet media streaming applications still suffer from limited and fluctuated bandwidth. Multipath streaming transmission has recently been proposed as a solution to overcome packet networks limitations [1], [2], [3]. It allows to increase the streaming bandwidth by balancing the load over multiple disjoint network paths between media sender and receiver. It also improves the error resilience of the media streaming system by means of redundant paths. Essential to such a multipath streaming system, at sender, is the packet distributor that dispatches media packets to the paths. It is necessary for the sender to distribute workload in a reasonable manner so that the multipath system can achieve its full potential.

How to distribute packets to achieve maximum benefit? Numerous studies [3], [4], [5], have made contributions on this research field. The fundamental concept is to allocate traffic in terms of available bandwidth. While all these works do not consider the fluctuation of network status enough. Unlike these approaches, which rely on UDP for streaming, some researchers focus on exploiting TCP for multipath real-time streaming, imposing TCP's state-awareness ability [6],

J. Zhou (Ed.): Complex 2009, Part I, LNICST 4, pp. 149–162, 2009.
© ICST Institute for Computer Sciences, Social Informatics and Telecommunications Engineering 2009

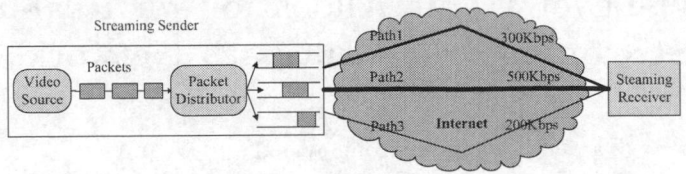

Fig. 1. Multipath streaming framework

[7]. For real-time specific, based on UDP, we try to implement a dynamic traffic allocation mechanism to "sense" the transmission characteristics of each path, and distribute packets fairly over the paths to achieve the designed goal.

In the framework of multipath network as shown in Fig. 1, our work addressed to the problem of streaming packet distribution, which takes into account real-time streaming characteristics. We are aiming at distributing packets fairly in order to achieve efficient utilization of bandwidth resources. Two key challenges are what is the distribution policy and how to execute this policy exactly. By means of analyzing media specific scenario, this paper gives corresponding solutions of these challenges.

In this paper, we make the following three contributions. (i) We analyze end-to-end multipath real-time streaming system in depth, and provide a model of bandwidth overload probability. (ii) Based on the model, we prove that allocating traffic in proportion to paths' available bandwidths respectively helps to reduce the overall overload probability. (iii) Following the traffic allocation policy, a weighted size-aware packet distribution algorithm for multipath real-time streaming is proposed, which is fine grained for its perceiving the smallest data unit (i.e. packet) over packet switching networks.

The rest of the paper is organized as follows. The multipath real-time streaming model is analyzed in Section II. Section III provides our optimal media-driven traffic allocation scheme and proves it. In section IV, we propose weighted size-aware packet distribution algorithm imposing upon traffic allocation policy. Simulation results are presented in Section V. Section VI concludes the paper.

2   Multipath Real-Time Streaming Analysis

2.1   Multipath Real-Time Streaming

We consider an end-to-end transmission framework where the media streaming application uses $M(M \geq 2)$ disjoint paths. Paths are considered to be disjoint if they do not share performance bottlenecks. The set of available loop-free paths between a media sender and a receiver is defined as $P = \{P_1, P_2, \ldots, P_M\}$.

For end-to-end perspective, we do look into the network status from an end-to-end point of view, rather than focus the hop-by-hop process during transmission. The network available bandwidth $b_i(t)$ (i.e., spare bandwidth), that is the bandwidth left unused by idle and non-greedy connections, is hence given by the following expression:

$$b_i(t) = C_i - \sum_{k \in K} \eta_i^k, \ \forall P_i \in P \tag{1}$$

where the first summation represents the total bandwidth of path P_i, while the latter summation of η_i^k represents the bandwidth allocated to other applications K, known as background traffic. Background traffic is always unsteady, and this instability lead to $b_i(t)$'s up and down.

Real-time video streaming is usually captured frame by frame by a video capture device every other fixed time, and the raw video frames are instantly encoded into compressed frames using some video encoder (e.g. H.264/AVC or MPEG-4). These compressed frames are commonly of different sizes in terms of video sequence characteristics and video encoder's configuration. Every encoded frame is then fragmented into network packets under the general rule stating that 1) each network packet contains data relative to at most one video frame, 2) several packets may contains data belong to the same frame. Let $\Pi = \{p_1, p_2, \ldots, p_N\}$ be the chronologically ordered sequence of N network packets, after fragmentation of the encoded frames. Any network packet p_n is characterized by its size s_n in bytes, frame number f_n, and its timestamps t_n. Timestamp is important for video player to play video packet at the right time. For the packets derived from the same frame, their frame numbers f_n, and timestamps t_n are uniform, which can be written as

$$f_n = f_{n+1} = \ldots = f_{n+k} \Longleftrightarrow t_n = t_{n+1} = \ldots = t_{n+k} \tag{2}$$

A packet distributor is set to permit data packets to be dispersed on multiple outgoing paths under a distribution scheme. Steaming application sends data at instantaneous rate of $R(t)$, which is split into many "fractional" rate $r_i(t)$, i.e. $R(t) = \sum_{i=1}^{M} r_i(t)$. $r_i(t)$ is the sending rate allocated to P_i at time instant t.

We denote by $\Phi = (\phi_1, \phi_2, \ldots, \phi_N)$ the distribution policy adopted by the streaming sender, and the ϕ_n represents the path chosen for packet p_n. In the multipath network scenario presented above, the sender can decide to send packet p_n through any path. Therefore, if p_n is distributed to path P_m, the packet p_n's imposed action $\phi_n = m$.

2.2 Packet Loss and Packet Late Arrival

In our streaming model, in order to decrease the video quality distortion, the streaming strategy aims at avoiding allocated bandwidth overload that results in packet losses and late arrivals. Firstly, we consider that the transmission links are lossless, and that packet loss only happens when sending a packet with the sending rate higher than the available bandwidth. Assuming a packet p_n allocated on P_i, i.e. $\phi_n = i$, we have

$$p_n \text{ is lost, if } r_i(t_n^s) > b_i(t_n^s) \tag{3}$$

where t_n^s is packet p_n's send time.

At the same time, even packet p_n is not lost, i.e. $r_i(t_n^s) < b_i(t_n^s)$, it still suffers from the danger of late arrival, which will be dropped too. Note that, time related metric such as packet late arrival and transmission delay is highly important for real-time real-time streaming, which distinguishes real-time streaming traffic from other traffic such as large file transmission.

Based on the previous work [8], we model the bottleneck link of each path as a work conserving queuing system with a service rate b_i, $i = 1, 2, \ldots$. We assume that the source flow is regulated by a σ, ρ leaky bucket (or a token bucket, which is implemented in most commercial routers). Let the real-time traffic's sending rate at t be $r(t)$, which is regulated by a σ, ρ leaky bucket, i.e., $r(t)$ conforms to a deterministic envelope process [9]. Due to this traffic shaping function, the source instantaneous rate on every path is shaped as:

$$r(t) = \rho + \sigma(t) \tag{4}$$

where ρ is the long-term average rate of the process (the rate factor), and $\sigma(t)$ is the burst during a small period of time, which is related to video sequence's characteristics.

Consider a work conserving queue with capacity $b(t)$, i.e. the available bandwidth. If the queue is stable, the queuing delay is upper bounded by the maximum busy period of the system [10]

$$d = \frac{\sigma}{\int_0^t b(u)du - \rho}$$

by means of (4), the the instantaneous fractional delay at t can be computed

$$d(t) = \frac{\sigma(t)}{b(t) - r(t) - \sigma(t)}$$

Given a decoding deadline's upper bound, and a packet p_n allocated on P_i, i.e. $\phi_n = i$, for $\sigma(t)$ is fixed in terms of video sequence, we have

$$p_n \text{ is late, if } b_i(t_n^s) - r_i(t_n^s) < \varepsilon \tag{5}$$

where ε is a positive bound to indicate late packets and t_n^s is p_n's send time.

2.3   Bandwidth Overload Probability

Packet loss or late arrival (i.e. unsuccessfully decoded packet) happens in terms of (3) and (5), which is due to traffic allocated overload. It is clear that lost packets is a subset of late packets, that is $b(t) - r(t) < \varepsilon$ limitation tighter than $r(t) > b(t)$ limitation. So we define the overload situation if $b(t) - r(t) < \varepsilon$ occurs.

Assuming at time η, the network available bandwidth is measured as $b(\eta)$, possibly with feedback of the receiver or other bandwidth detection approaches [11], [12]. However, network available bandwidth usually experiences change abruptly, given instantaneous detected bandwidth $b(\eta)$, during the period between two consecutive bandwidth detections, the actual available bandwidth is

$$b(t) = b(\eta) - X, t \in [\eta, \eta + \tau),$$

where X is the available bandwidth variance (i.e., traffic load variance) from $b(\eta)$, also known as background traffic burst, and τ is the bandwidth detection interval. Therefore, the probability of overload can be written as

$$Pr\left\{[b(\eta) - X] - r(t) < \varepsilon\right\} = Pr\left\{X > b(\eta) - r(t) - \varepsilon\right\}.$$

The burst length X (negative when light load) is commonly considered according to Pareto distribution [13]. Hence, according to Pareto property, we can carry on this consequence

$$Pr\left\{X > b(\eta) - r(t) - \varepsilon\right\} = \left[\frac{b(\eta) - r(t) - \varepsilon}{X_m}\right]^{(-\alpha)},$$

where the burst X converges to X_m in the limit of a large value of the exponent α, and α is a positive parameter (note that, the smaller α is, the greater probability overload occurs). In other words, X_m is the expected value of $b(\eta) - r(t) - \varepsilon$, i.e. $E[b(\eta) - r(t) - \varepsilon]$. For $b(\eta)$, its expected value keeps the same until the next available bandwidth detection, and ε is determined by the streaming application, while for $r(t)$, its expected rate can be computed as the mean rate during time scale $t \in [\eta, \eta + \tau)$, which is

$$E[r(t)] = \frac{\int_\eta^{\eta+\tau} r(t)dt}{\tau}$$

To sum up, given the instantaneous detected available bandwidth $b(\eta)$ at time η and packet late bound ε, during the period of $t \in [\eta, \eta + \tau)$, the overload probability is

$$Pr\left\{b(t) - r(t) > \varepsilon\right\} = \left\{\frac{b(\eta) - r(t) - \varepsilon}{b(\eta) - E[r(t)] - \varepsilon}\right\}^{(-\alpha)} \tag{6}$$

The analysis of multipath streaming as well as bandwidth overload probability, provides an in-depth study of multipath network behavior's character, and help us propose the optimal traffic allocation in the next section.

3 Traffic Allocation: Path Weight Determination

We generalize the previous observations, and derive theorems that guide the design of an optimal traffic allocation strategy. Since sending rate of every path decides the traffic load on that path, traffic allocation problem can be transformed to the problem of allocating rate among multiple paths. This section shows that, in the optimal traffic allocation, sending rate of every path is assigned in proportion to the path's available bandwidth, which minimize the overall bandwidth overload probability. We start from a multipath streaming scenario assuming available bandwidth of paths can be precisely detected periodically.

Theorem 1 (Rate allocation). *Given media application's instantaneous sending rate* $R(t) = \sum_{i=1}^{M} r_i(t)$, *and the detected available bandwidth* $b_i(\eta)$ *over* P_i *at time* η, *the optimal rate allocation* $\boldsymbol{R(t)}^* = [r_1(t), \ldots, r_M(t)]^*$ *during time interval* $t \in [\eta, \eta + \tau)$, *that minimizes the overall bandwidth overload probability based on (6):*

$$\boldsymbol{R(t)}^* = \left[r_1(t), \ldots, r_M(t)\right]^* = \arg\min_{R(t)} \sum_{i=1}^{M} Pr\left\{b_i(t) - r_i(t) > \varepsilon\right\} \qquad (7)$$

is set in proportion to paths' available bandwidths

$$\boldsymbol{R(t)}^* = \left[R(t) \cdot \frac{b_1(t)}{\sum_{i=1}^{M} b_i(t)}, \ldots, R(t) \cdot \frac{b_M(t)}{\sum_{i=1}^{M} b_i(t)}\right] \qquad (8)$$

Proof. Deriving the minimum function given in (7),

$$\sum_{i=1}^{M} \left\{\frac{b_i(\eta) - r_i(t) - \varepsilon}{b_i(\eta) - E\left[r_i(t)\right] - \varepsilon}\right\}^{(-\alpha)}$$

its minimum value is obtained when all the items are equal

$$\frac{b_1(\eta) - r_1(t) - \varepsilon}{b_1(\eta) - E\left[r_1(t)\right] - \varepsilon} = \ldots = \frac{b_M(\eta) - r_M(t) - \varepsilon}{b_M(\eta) - E\left[r_M(t)\right] - \varepsilon}$$

Only focusing on the first two paths, we have the cumulative equation during time period of $(\eta, \eta + \tau]$,

$$\int_{\eta}^{\eta+\tau} \left\{b_1(\eta) \cdot E\left[r_2(t)\right] - b_2(\eta) \cdot r_1(t) + r_1(t) \cdot E\left[r_2(t)\right]\right\} dt$$

$$= \int_{\eta}^{\eta+\tau} \left\{b_2(\eta) \cdot E\left[r_1(t)\right] - b_1(\eta) \cdot r_2(t) + r_2(t) \cdot E\left[r_1(t)\right]\right\} dt$$

Since

$$\int_{\eta}^{\eta+\tau} r_i(t) dt = \int_{\eta}^{\eta+\tau} E\left[r_i(t)\right] dt,$$

we finally obtain

$$\frac{\int_{\eta}^{\eta+\tau} r_1(t) dt}{\int_{\eta}^{\eta+\tau} r_2(t) dt} = \frac{b_1(\eta)}{b_2(\eta)}.$$

Considering instantaneous rate allocation, we let

$$\frac{r_1(t)}{b_1(\eta)} = \frac{r_2(t)}{b_2(\eta)}$$

to get the minimum value. In a similar way, we have

$$\frac{r_1(t)}{b_1(\eta)} = \frac{r_2(t)}{b_2(\eta)} = \ldots = \frac{r_M(t)}{b_M(\eta)},$$

where the path's rate is set in proportion to the available bandwidth.

Considering the constraint $R(t) = \sum_{i=1}^{M} r_i(t)$, to get the optimal rate allocation $\boldsymbol{R(t)^*} = [r_1(t), \ldots, r_M(t)]^*$, we should set path P_j's rate according to P_j's fraction of total available bandwidth, which minimizes the overall overload probability.

$$r_j(t)^* = R(t) \cdot \frac{b_j(t)}{\sum_{i=1}^{M} b_i(t)} \tag{9}$$

The traffic allocation method provides a reasonable way of distributing packets in order to reduce packet loss and late arrival probability. In our packet distribution scheme, path weight vector $(\omega_1, \omega_2, \ldots, \omega_M)$ is introduced, which indicates respective distribution capabilities of paths. By means of all paths' instant available bandwidth acquired by periodic detection, path P_m's weight can be determined

$$\omega_m = \frac{b_m(t)}{\sum_{i=1}^{M} b_i(t)}, \text{ and } \sum_{i=1}^{M} \omega_i = 1, \tag{10}$$

by which we execute packet distribution. A path with larger weight, is more likely to attract media traffic. Actually, $b_m(t) = 0$ is possible, which means no available resource can we consume on path P_m, then the path's weight $\omega_m = 0$ allows us to transmit no packet through path P_m, i.e. path P_m is abandoned. Extremely when only one path have available bandwidth, multipath transmission transforms to unipath transmission, which is reasonable in practical environment [17]. In the next section, we describe our complete packet distribution algorithm applying the path weight in detail.

4   Weighted Size-Aware Packet Distribution Algorithm

Suppose real-time streaming application generates a sequence of frames every other capture time interval, and they are encoded by some encoder (e.g. MPEG-4 or H.264/AVC). In practical, if an encoded frame's size is larger than network MTU, it is fragmented into several smaller network packets, each with size s_n in bytes. Then, in multipath streaming, these network packets $\Pi = p_1, p_2, \ldots, p_N$ are distributed to a set of M paths $P = P_1, P_2, \ldots, P_M$. Except this packet distributing thread, another work for available bandwidth detection thread is running. This detection and path weight computation are carried out every other interval τ. The path weight vector $(\omega_1, \omega_2, \ldots, \omega_M)$ is acquired in terms of (10). Actually, path weight indicates that path's expected traffic load proportion and it is updated periodically.

Focus back to the main sending thread, given periodic renewed path weight vector, packets should be exactly distributed according to path weight (i.e., the expected traffic load). Despite the rate allocation approach is an idealized scheme, but the smallest possible data unit in streaming is a packet, differentiated by size. Thus, a more explicit packet distribution scheme aware of packet size is proposed, whose philosophy is to minimize the deviation of actual traffic distribution from the given path weight vector, i.e., from the expected distribution.

Let $T_m(n)$ and $T'_m(n)$, respectively, be the expected traffic load in bytes (determined by ω_m), and the actual traffic load in bytes to be sent on path P_m, just after the packet p_n's distributing decision has been made. For an idealized packet distributor, we have

$$T_m(n) = \omega_m \cdot \sum_{j=1}^{n} s_j$$

where s_j is the size of packet p_j and $j = 1, 2, \ldots, N$.

The main idea of packet distribution is to simulate optimal rate allocation as closely as possible. However, the assignment of a complete packet to a path may cause a transient load imbalance with respect to the targeted traffic allocation, that is some paths may be fed more traffic than expected temporarily while other paths may have less, after the distribution for a certain packet. Those paths fed with more traffic than expected have the tendency of not having the next packet assigned to them. Therefore, the current level of load imbalance as well as the size of the next successive packet is required for the traffic distributor to make the next distribution decision.

To quantify the above selection criterion, a metric is introduced to measure the traffic underload on a path. The residual traffic load of every path, just before distributing the packet p_n, $R_m(n)$, is defined as the amount of traffic load in bytes that should be fed on path P_m in order to achieve the expected traffic load. In other words,

$$R_m(n) = T_m(n) - T'_m(n-1), \quad \sum_{i=1}^{M} R_i(n) = s_n.$$

We use $R_m(n)$ to measure the streaming traffic underload on P_m, just before distributing p_n. If $R_m(n) > 0$, path P_m has been injected with less traffic than expected and, hence, p_n can be sent on this path. On the other hand, if $R_m(n) < 0$, there is too much streaming traffic being assigned on it and, hence, packet p_n should not be transmitted on this path. Briefly, $R_m(n)$ provides an indicator to the packet distributor for deciding which path p_n should be transmitted on.

Algorithm 1 presents the sketch of the main distributting process, where, for clarity, we bring up again $\phi_n = m$, if packet p_n is sent on path P_m. After running this algorithm, we can determine the optimal distribution policy Φ^*.

Concerning the distributing packet procedure's time and space complexities, it takes $O(N)$ time for processing each packet as it searches for a path P_m such that $R_m(n)$ is maximized. Also, it needs $O(N)$ counters to store its working variables. As the number of paths is generally small and fixed, we consider that the computational and storage costs are minimal. For the path weight computation, due to its simplicity and executed not very soon, its complexities are neglectable. At the same time, we argue that the packet distribution is fair and explicit. For any sequences of packets to be dispersed, the variance between the actual traffic load and the expected traffic load allocated to each path is always bounded by a finite constant.

Algorithm 1. Weighted Size-Aware Packet distribution

Require: $p_n, s_n, M, P_m, 1 \le m \le M$
Ensure: Optimal packet distribution $\Phi^* = [\phi_1, \phi_2, \ldots, \phi_n]^*$

1: Initialize the variables
2: **while** frame capture time comes **do**
3:     capture frame
4:     **while** bandwidth detection time comes **do**
5:         invoke Update_Pathweight()
6:     **end while**
7:     encode frame
8:     split frame into a packet sequence A_p with n' packets
9:     **for all** packets p_n in A_p **do**
10:         invoke Distribute_Packet(p_n)
11:     **end for**
12: **end while**

13: **procedure** DISTRIBUTE_PACKET(p_n)
14:     $S \leftarrow s_n$
15:     **for all** each $m, m \in 1, 2, \ldots, M$ **do**
16:         $R_m(n) \leftarrow R_m(n-1) + \omega_m \cdot S$
17:     **end for**
18:     choose a path $P_{m'}$ such that $R_{m'}(n)$ is maximized
19:     $\phi_n \leftarrow m'$
20:     $R_{m'}(n) \leftarrow R_{m'}(n-1) - S$
21: **end procedure**

22: **procedure** UPDATE_PATHWEIGHT
23:     **for all** each $m, m \in 1, 2, \ldots, M$ **do**
24:         detect P_m's available bandwidth b_m
25:         update $\omega_m = b_m(t) / \sum_{i=1}^{M} b_i(t)$
26:     **end for**
27: **end procedure**

In summary, our packet distribution algorithm guarantees the variance between the actual traffic and the expected traffic under a limit bound. It is deployed at the media sender side, usually working for just one media flow, thus its complexity is acceptable for practical streaming applications.

5   Simulation Results

5.1   Simulation Setup and Relate Algorithms

We use ns-2 [16] to simulate multipath network scenarios. Two disjoint paths are selected between video sender (source) and video receiver (sink), with bandwidths of 1Mbps and 500Kbps respectively, and with the same end-to-end transmission delay of 100ms. A background traffic flow is generated according to the

On/Off Pareto distribution on the first path (namely path1) and on the second path (namely path2). The available bandwidth for our streaming application is considered to be the background traffic's rate subtracts from the total link bandwidth, which is detected every other 1 second.

Four packet distribution algorithms are studied, namely, weighted size-aware (WSA), weighted round robin (WRR), additive increase and multiplicative decrease (AIMD), and greedy (Greedy) [4], while the WSA approach is described in Section IV. WRR distributes packets to each path in a weighted cyclical fashion, where the weight is determined in terms of the total bandwidth of each path. AIMD focuses on a particular path, and utilizes this path in a probe manner. A initial threshold working as traffic load indicator is set at first, and media applications allocate traffic load lighter than the threshold. When the allocated traffic load does not exceed the available bandwidth, this threshold increases additively, otherwise, it decreased multiplicatively. The following packets after threshold hitting are distributed to the next path, where another instance of AIMD is running. Greedy method is based on [4], it will not chose another path for transmission unless all other available paths with higher available bandwidth have been chosen. Moreover, the chosen paths should be used at their maximum available bandwidth. Certainly, this available bandwidth is detected periodically by video streaming applications. Excluding WRR, all the other three algorithms are working by means of detected available bandwidth. Except the difference between traffic allocation schemes, an extraordinary of WSA from other schemes is its fine-grained property resulted from packet size awareness.

5.2   Comparison of Performance

We evaluate these algorithms introducing standard CIF sequences *foremancif* under different background traffic load levels, which are set as presented in Table 1. Fig. 2a compares the number of lost packets achieved by the four packet distribution schemes. Greedy as well as WSA performs better even under high background traffic load level. On the other hand, in order to test the late arrivals under different background traffic load levels, packet's maximum endurable transmission delay is set to 500ms, all the packets arrive later than this deadline are late arrivals. Fig. 2b gives the comparison of late arrivals over four algorithms. As expected,

Table 1. Background traffic load setup

Path	Param	L1	L2	L3	L4	L5
	burst time (ms)	200	200	250	250	250
Path1	idle time (ms)	50	50	30	30	30
	mean rate (Kbps)	750	800	850	900	950
	burst time (ms)	100	100	200	200	200
Path2	idle time (ms)	50	50	30	30	30
	mean rate (Kbps)	200	250	350	400	420

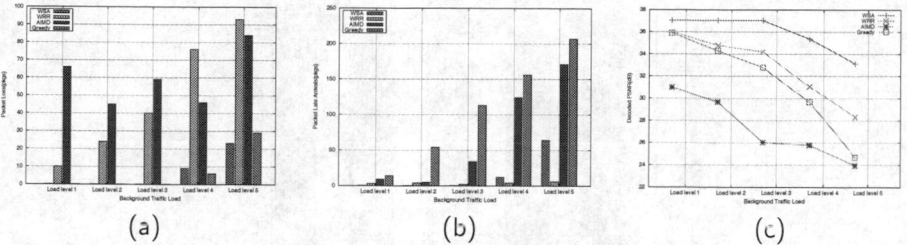

Fig. 2. Comparison of performance from four packet distribution algorithms under different load levels. (a) Lost packets. (b) Late packets. (c) PSNR.

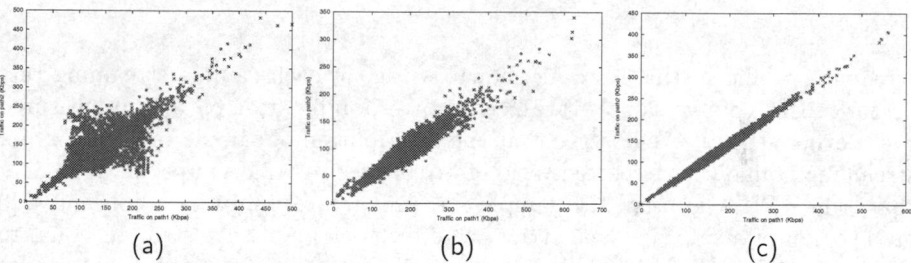

Fig. 3. Sample streaming rate on two paths with weigh ratio 3 to 2. (a) Round robin distributing packets. (b) Weighted round robin distributing packets. (c) Weighted size-aware (WSA) distributing packets.

Greedy generates a much larger number of late arrivals than other schemes, and AIMD also produces amount of late arrivals. Interestingly, WRR seems to have late arrivals avoidance, but we take notice that it has lost numerous packets, that have already deteriorated video quality.

As an approach of comprehensively considering packet loss and late arrival, we evaluate received video's quality measured by PSNR metric, as depicted in Fig. 2c. It demonstrates that, WSA always has the highest PSNR in all background traffic load levels. Another observation is that, Greedy's performance degrades faster than other schemes with the increasing of background traffic load.

5.3 Packet Size Aware

We now elucidate that WSA distributes streaming traffic load fairly by means of packet size awareness, which is subsequent upon path weight determination (i.e., rate allocation). Since encoded packets are of different sizes, even though every path's rate has been determined, distributing packets without considering packet size may lead to actual traffic load deviation from expected.

Fig. 3 plots a set of sample streaming rate vectors to demonstrate this by contrasting simple round robin and weighted round robin distribution with our weighted size aware distribution (i.e., WSA). *StarWarsIV* is used to generate

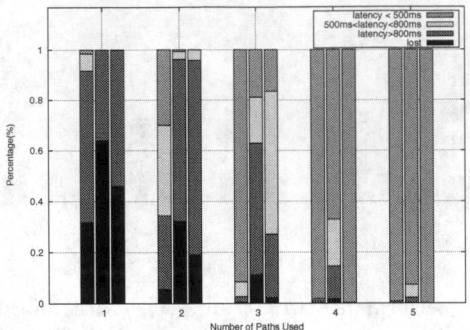

Fig. 4. Effect of path number

streaming traffic in this test. As shown is Fig. 3, each sample streaming rate vector consists of two sample rates, each corresponding to a path, and measured every other second. We observe that the distribution of sample rate vectors for streaming traffic with expected ratio of 3:2 between path1 and path2. It is clear that, when WSA is used, the sample streaming rate vectors are concentrated on a region of a shin diagonal stripe, where the slope of that stripe is equal to the expected ratio between path1 and path2. There is a much thicker stripe for using weighted round robin packet distribution, and even a worse sample rate vectors when simple round robin is employed. Anyway, WSA distributes packets to paths in a fine-grained manner according to expected traffic allocation.

5.4   Effect of Path Number

In the above simulation, we have focused on two paths transmission scenario to test weighted size-aware packet distribution algorithm. The effect of path number used for real-time streaming based on our WSA scheme is also evaluated. To survey this effect accurately, three long-term video trace files (i.e., *StarWarsIV*, *SouthPark*, and *OfficeCam*) is used as video source. Each path's bandwidth is set to 1Mbps, and background traffic of mean bitrate 800Kbps with On/Off exponential distribution is running on every path. All the three sequences are streamed using 1 to 5 paths.

The results are presented in Fig. 4, three columns of each path number represent the situations when introducing different source files. With the increase of path number, multipath's benefit is gained significantly. Interestingly, by increasing only one path improves the performance more than double times, and the effect of multipath streaming is quite tremendous, which implies our excellent packet distribution scheme. Additionally, it shows that, transmission latency is minimized greatly, which contributes to late arrivals avoidance. The simulation results prove that, by using weighted size-aware packet distribution for multipath real-time streaming, we make efficient utilization of network resources taking into account real-time streaming characteristics.

In summary, WSA packet distribution scheme performs better for multipath real-time streaming, it distributes packets through multiple paths to avoid bandwidth overload of a single path. The similar method aiming at balancing traffic load between different paths is WRR, which generates no late arrived packet as well as WSA. On the other hand, path with higher available bandwidth is preferred to other paths with lower available bandwidth in Greedy, and this strategy bears less packet losses than other strategies. However, packets distributor with Greedy algorithm brings a great number of late packets, which will be dropped by real-time streaming applications.

6 Conclusions

In this paper, we provide an in-depth analysis of multipath real-time streaming system considering media characteristics. These analyses point that by splitting traffic in proportion to the path's available bandwidth, streaming applications experience minimal bandwidth overload probability, which results in packet losses and packet late arrivals. And based on the distribution policy, a novel weighted size-aware packet distribution algorithm (i.e., WSA) for multipath real-time streaming is described, which ensures actual load distribution with a small deviation from expected. Our simulation results demonstrate the effectiveness of WSA in reducing overall packet loss rate and packet late arrivals as well as in improving video quality. Due to its satisfied effect and low complexity, the weighted size-aware packet distribution algorithm provides a very practical solution to efficient real-time streaming over multipath networks.

References

1. Golubchik, L., et al.: Multi-path continuous media streaming: What are the benefits? ACM J. Perform. Eval. 49, 429–449 (2002)
2. Apostolopoulos, J.G., Trott, M.D.: Path diversity for enhanced media streaming. IEEE Communication Magazine 42, 80–87 (2004); Nagel, W.E., Walter, W.V., Lehner, W. (eds.) Euro-Par 2006. LNCS, vol. 4128, pp. 1148–1158. Springer, Heidelberg (2006)
3. Frossard, P., de Martin, J.C., Civanlar, M.R.: Media Streaming With Network Diversity. Proc. IEEE. 96, 39–53 (2008)
4. Jurca, D., Frossard, P.: Media Flow Rate Allocation in Multipath Networks. IEEE Trans. on Multimedia 9, 1227–1240 (2007)
5. Bui, V., Zhu, W.: Improving multipath live streaming performance with Markov Decision Processes. In: Proc. ISCIT 2007, pp. 580–585. IEEE Press, New York (2007)
6. Wang, B., Wei, W., Guo, Z., Towsley, D.: Multipath Live Streaming via TCP: Scheme, Performance and Benefits. In: Proc. ACM SIGCOMM CoNEXT 2007. ACM, New York (2007)
7. Fang, C., Fu, X.: Probe-Aided MulTCP: an Aggregate congestion control mechanism. ACM SIGCOMM Computer Communication Review 38, 17–28 (2008)
8. Shiwen, M., Panwar, S.S., Hou, Y.T.: On minimizing end-to-end delay with optimal traffic partitioning. IEEE trans. on Vehicular Technology 55, 681–690 (2006)

9. Cruz, R.L.: A calculus for network delay. I. Network elements in isolation. IEEE Trans. on Information Theory 37, 114–131 (1991)
10. Chang, C.S.: Stability, queue length, and delay of deterministic and stochastic queueing networks. IEEE Trans. on Automatic Control 39, 913–931 (1994)
11. Qi, F., Wei, J., Wu, J.: Available bandwidth detection with improved transport control algorithm for heterogeneous networks. In: Proc. IEEE Workshops on Distributed Computing Systems, pp. 656–659. IEEE Press, New York (2005)
12. Li, Y., Munro, A., Kaleshi, D.: Multi-rate congestion control using packet-pair bandwidth detection with session and layer changing manager. In: Proc. IEEE IPCCC, pp. 485–490. IEEE Press, New York (2005)
13. Zukerman, M., Neame, T.D., Addie, R.G.: Internet traffic modeling and future technology implications. INFOCOM 2003 Review 1, 587–596 (2003)
14. Bauke, H.: Parameter estimation for power-law distributions by maximum likelihood methods. The European Physical Journal B 58, 167–173 (2007)
15. MPEG-4 Video Traces for Network Performance Evaluation,
    http://www.tkn.tu-berlin.de/research/trace/trace.html
16. Network Simulator, http://www-mash.cs.berkeley.edu/ns/
17. Andersen, D., Snoeren, A., Balakrishnan, H.: Best-path vs.multi-path overlay routing. In: Proc. ACM SIGCOMM Internet Measurement Conference, pp. 91–100. ACM Press, New York (2003)

Particle Competition in Complex Networks for Semi-supervised Classification

Fabricio Breve, Liang Zhao, and Marcos Quiles

Institute of Mathematics and Computer Science, University of São Paulo,
São Carlos SP 13560-970, Brazil
{fabricio,zhao,quiles}@icmc.usp.br

Abstract. Semi-supervised learning is an important topic in machine learning. In this paper, a network-based semi-supervised classification method is proposed. Class labels are propagated by combined random-deterministic walking of particles and competition among them. Different from other graph-based methods, our model does not rely on loss function or regularizer. Computer simulations were performed with synthetic and real data, which show that the proposed method can classify arbitrarily distributed data, including linear non-separable data. Moreover, it is much faster due to lower order of complexity and it can achieve better results with few pre-labeled data than other graph based methods.

Keywords: semi-supervised learning, particle competition, complex networks, community detection.

1 Introduction

Complex networks is a recent and active area of scientific research, which studies large scale networks with non-trivial topological structures, such as computer networks, telecommunication networks, transportation networks, social networks and biological networks [1,2,3]. Many of these networks are found to be divided naturally into communities or modules, thus discovering of these communities structure became one of the main issues in complex network study [4,5,6,7,8]. Recently, a particle competition approach was successfully applied to detect communities modeled in non-weighted networks [9].

The problem of community detection is also related to the machine learning field, which is concerned with the design and development of algorithms and techniques that allow computers to "learn", or improve their performance through experience [10]. Machine learning algorithms usually falls in one of these two categories: *supervised learning* and *unsupervised learning*. In supervised learning, the algorithm learns a function from the training data, which consists of pairs of samples and their respective labels, so after having seen a number of training examples the algorithm can predict the labels of unseen data. On the other hand, in unsupervised learning the samples are unlabeled and the objective is to determine how the samples are organized. One form of unsupervised learning is clustering, which is the partitioning of a data set into subsets (clusters), so

J. Zhou (Ed.): Complex 2009, Part I, LNICST 4, pp. 163–174, 2009.

that the data in each cluster share some characteristics. The algorithm proposed in [9] to detect communities belongs to the unsupervised learning category.

With the emergence of the complex networks field and the study of larger networks, it is common to have large data sets in which only a small subset of samples are labeled. That happens because unlabeled data is relatively easy to collect, but labeling samples is often an expensive, difficult or time consuming task, since it often requires the work of humans specialists. Supervised learning techniques cannot handle this kind of problem because they require all samples labeled before the training process. Unsupervised learning techniques cannot be applied to solve this kind of problem either because they ignore label information of samples. In order to solve these problems a new class of machine learning algorithms arose, the semi-supervised class. *Semi-supervised* learning is halfway between supervised and unsupervised learning, it address these problems by combining a few labeled samples with a lot of unlabeled samples to produce better classifiers while requiring less human effort [11,12]. For example, consider Fig. 2a, it is a toy data set with 2000 samples, but only 20 of them are labeled (red circles and blue squares), a supervised algorithm would learn from only these 20 samples and it would probably misclassify a lot of unseen samples, while an unsupervised algorithm would consider all the 2000 samples without any distinction, thus not taking advantage of the labeled ones. On the other hand, a semi-supervised algorithm can learn from both labeled and unlabeled samples, probably producing a better classifier. Semi-supervised methods include generative models [13,14], cluster-and-label techniques [15,16], co-training techniques [17,18], low-density separation models, like Transductive Support Vector Machines (TSVM) [19], and graph-based methods, like Mincut [20] and Local and Global Consistency [21].

Traditional semi-supervised techniques, such as Transductive Support Vector Machine (TSVM) [19], can identify data classes of well defined form, but usually fail to identify classes of irregular form. Thus, assumptions on class distribution have to be made and unfortunately it is usually unknown a priori. On the other hand, since most graph based methods have high order of computational complexity $(O(n^3))$, it makes their use limited to small data sets [11]. This is considered as a serious shortage because semi-supervised learning techniques are usually applied to data sets with large amount of unlabeled data. Also, many graph based methods can be viewed as regularization frameworks, they are similar to each other, basically differing only in the particular choice of the loss function and the regularizer [22,20,21,23,24,25].

In this paper we present a new kind of network-based semi-supervised classification technique, by using particle walking and competition. We extend the model proposed in [9] to handle weighted networks and to take advantage of pre-labeled data. The main contributions of this new method are:

- unlike most other graph-based models, it does not rely on loss functions or regularizers;
- it can classify arbitrarily distributed data, including linear non-separable data;

- it is much faster than other graph-based methods due to its low order of complexity and thus it can be used to classify large data sets;
- it can achieve better results than other graph-based methods when a small number of data samples is labeled.

This paper is organized as follows: Section 2 describes the model in details. Section 3 shows some experimental results from computer simulations, and in Section 4 we draw some conclusions.

2 Model Description

Our model is an extension of the particle competition approach proposed by [9]. Their model is used to detect communities in networks, represented by non-weighted networks. There are several particles walking in a network, competing with each other for the possession of network nodes, and rejecting intruder particles. In this way, after a number of iterations, each particle will be confined within a community of the network, so the communities can be divided by examining the nodes ownership.

In this paper, we have changed the nodes and particles dynamics, and some other details that will follow, so the new model is not only suitable to conduct semi-supervised learning, but also can represent weighted networks (weights represent pair-wise similarity between data samples). The model is described as follows:

Given a data set $X = \{x_1, x_2, \ldots, x_n\} \subset \mathbb{R}^m$ and a label set $L = \{1, 2, \ldots, c\}$, some samples x_i are labeled as $y_i \in L$ and some are unlabeled as $y_i = \emptyset$. The goal is to provide a label to these unlabeled samples.

First, we define a graph $G = (V, E)$, with $V = \{v_1, v_2, \ldots, v_n\}$, and each node v_i corresponds to a sample x_i. An affinity matrix W [21,26] defines the weight between the edges in E as follows:

$$W_{ij} = \exp{-\|x_i - x_j\|^2 / 2\sigma^2} \quad \text{if} \quad i \neq j, \tag{1}$$
$$W_{ii} = 0, \tag{2}$$

where W_{ij} defines the pair-wise relationship between x_i and x_j, with the diagonal being zero, and σ is a scaling parameter which controls how quickly the affinity W_{ij} falls off with the distance between x_i and x_j.

Then, we create a set of particles $P = (\rho_1, \rho_2, \ldots, \rho_c)$, in which each particle corresponds to a label in L. Each particle ρ_j has three variables $\rho_j^v(t)$, $\rho_j^\omega(t)$ and $\rho_j^\tau(t)$. The first variable, $\rho_j^v(t) \in V$, is used to represent the node v_i being visited by particle ρ_j at time t. The second variable, $\rho_j^\omega \in [\omega_{\min} \quad \omega_{\max}]$ is the particle potential characterizing how much the particle can affect a node at time t, in this paper we set the constants $\omega_{\min} = 0$ and $\omega_{\max} = 1$. The third variable, $\rho_j^\tau(t) \in V$, represents the target node by particle ρ_j at time t, sometimes the particle will be accepted by the node, so $\rho_j^\tau(t) = \rho_j^v(t)$, and some times it will be reject, thus $\rho_j^\tau(t) \neq \rho_j^v(t)$, as we will explain later.

Each node v_i have two variables: $v_i^\rho(t)$ and $v_i^\omega(t)$. The first, $v_i^\rho(t) \in P$ register the particle that owns node v_i at time t. The second variable is a vector $v_i^\omega(t) = \{v_i^{\omega_1}(t), v_i^{\omega_2}(t), \ldots, v_i^{\omega_c}(t)\}$ of the same size of L, where each element $v_i^{\omega_j}(t) \in [\omega_{\min} \quad \omega_{\max}]$ corresponds to the level of ownership by particle ρ_j over node v_i. So, at any given time t the particle ρ_j that owns v_i is defined as:

$$v_i^\rho(t) = \arg\max_j v_i^{\omega_j}(t). \tag{3}$$

Also, the following equations always holds:

$$\sum_{j=1}^c v_i^{\omega_j} = \omega_{max} + \omega_{\min}(c-1). \tag{4}$$

We begin the algorithm by setting the initial level of ownership vector v_i^ω by each particle ρ_j as follows:

$$v_i^{\omega_j}(0) = \begin{cases} \omega_{\max} & \text{if } y_i = j \\ \omega_{\min} & \text{if } y_i \neq j \text{ and } y_i \neq \emptyset \\ \omega_{\min} + \left(\frac{\omega_{\max} - \omega_{\min}}{c}\right) & \text{if } y_i = \emptyset \end{cases}, \tag{5}$$

which means the nodes corresponding to labeled samples already starts with their ownership set to the corresponding particle with maximum strength, while the other nodes starts with all particles ownership levels equally set.

The initial position of each particle $\rho_j^v(0)$ is set to one of the nodes that they already owns (corresponding to pre-labeled samples) as set in Eq. 5, so the following holds:

$$\rho_j^v(0) = \{v_i | y_i = j\}. \tag{6}$$

The initial potential of each particle is set as:

$$\rho_j^\omega(0) = \omega_{\max}. \tag{7}$$

We kept the concept of *random moving* and *deterministic moving* from the original model, where *random moving* means the particle will try to move to any neighbor randomly chosen, and *deterministic moving* means the particle will visit a node that it already owns. Here we extended these rules to handle weighted networks as follows: in *random moving* the particle ρ_j will try to move to any neighbor v_i randomly chosen with probability defined by:

$$p(v_i | \rho_j^v) = \frac{W_{ki}}{\sum_{q=1}^n W_{qi}}, \tag{8}$$

where k is the index of the node stored in ρ_j^v, so W_{ki} represents the weight of the edge connecting nodes ρ_j^v and v_i. In *deterministic moving* the particle ρ_j will try to move to any neighbor v_i randomly chosen with probability defined by:

$$p(v_i | \rho_j^v) = \frac{W_{ki} v_i^{\omega_j}}{\sum_{q=1}^n W_{qi} v_i^{\omega_j}}, \tag{9}$$

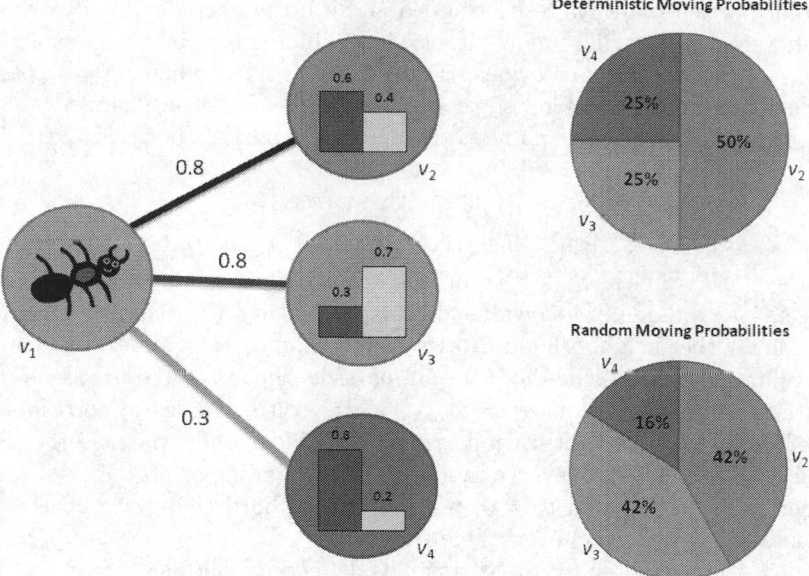

Fig. 1. Deterministic Moving and Random Moving Example Illustration. On the left, the particle ρ_1 is going to choose its target ρ_1^τ among v_2, v_3 and v_4. The graphics inside each node denotes their respective ownership levels $v_i^{\omega_1}$ (red) and $v_i^{\omega_2}$ (yellow), which corresponds to ρ_1 and ρ_2 respectively. On the right, the graphs show the probabilities of choosing each node for either deterministic or random moving.

and again, k is the index of the node stored in ρ_j^v. At each iteration, each particle has probability $p_{\det}$ of taking deterministic moving and probability $1 - p_{\det}$ of taking random moving, and $0 \le p_{\det} \le 1$. Once the deterministic moving or random moving is chosen, the target neighbor $\rho_j^\tau(t)$ will be randomly chosen with probabilities defined by Eq. 8 or Eq. 9 respectively. Figure 1 shows an example situation where a particle is going to choose among three neighbors with different edges weights and ownership levels, the graphics show the probabilities of each node being chosen when using either deterministic or random moving.

Regarding the node dynamics, at time t, each ownership level $v_i^{\omega_k}(t)$ of each node v_i, which was chosen by a particle ρ_j as its target $\rho_j^\tau(t)$, is defined as follows:

$$
v_i^{\omega_k}(t+1) = \begin{cases} v_i^{\omega_k}(t) & \text{if } y_i \neq \emptyset \\ \max\{\omega_{\min}, v_i^{\omega_k}(t) - \frac{\Delta_v \rho_j^\omega(t)}{c-1}\} & \text{if } y_i = \emptyset \text{ and } k \neq j \\ v_i^{\omega_k}(t) + \sum_{q \neq k} v_i^{\omega_q}(t) - v_i^{\omega_q}(t+1) & \text{if } y_i = \emptyset \text{ and } k = j \end{cases}, \quad (10)
$$

where $0 < \Delta_v \le 1$ is a parameter to control the ownership levels changing speed. If Δ_v takes a low value, the node ownership levels change slowly, while if it takes a high value, the node ownership levels change quickly. Each particle ρ_j will increase their corresponding ownership level $v_i^{\omega_j}$ of the node v_i they are targeting while decreasing the ownership levels (of this same node) that corresponds to the other particles, always respecting Eq. 4. So if the particle already owns the node its targeting, it will reinforce it, else it will increase its own ownership level, possibly becoming the new owner. However, v_i^ω is fixed when $y_i \neq \emptyset$ (particle visiting a pre-labeled sample), so their ownership levels never change.

Regarding the particle dynamics, at time t, each particle potential $\rho_j^\omega(t)$ is set as:

$$\rho_j^\omega(t+1) = v_i^{\omega_j}(t+1) \quad \text{with} \quad v_i(t+1) = \rho_j^\tau(t+1), \qquad (11)$$

which means every particle ρ_j have their potential ρ_j^ω set to the value of its ownership level $v_i^{\omega_j}$ from the node it is currently targeting. This way, a particle will be strong as it is walking in its own neighborhood, but it will become weak if it try to invade another neighborhood. Notice that to handle semi-supervised learning we have made v_i^ω fixed when $y_i \neq \emptyset$, so a particle ρ_j can always "recharge" their potential to the maximum ($\rho_j^\omega = \omega_{\max}$) when visiting nodes v_i corresponding to pre-labeled samples of its own class ($y_i = j$). Meanwhile, particles ρ_j cannot visit or change ownership levels of nodes v_i corresponding to pre-labeled samples of other class ($y_i \neq j$ and $y_i \neq \emptyset$) no matter how hard they try, and they will become weak ($\rho_j^\omega = \omega_{\min}$) every time they try.

Finally, the particle position at time t is defined as follows:

$$\rho_j^v(t+1) = \begin{cases} \rho_j^\tau(t+1) & \text{if } v_i^\rho(t+1) = \rho_j \\ \rho_j^v(t) & \text{if } v_i^\rho(t+1) \neq \rho_j \end{cases}, \qquad (12)$$

with $v_i = \rho_j^\tau(t+1)$, which means that after raising its own ownership level on the target node ρ_j^τ, the particle ρ_j will move to it if it already owned it or if it became the new owner after that ownership level increase, else, it will stay where it was. Notice that the node owner v_i^ρ at any time is defined by Eq. 3.

We have also introduced a "reset" mechanism to take the particles back to one of the nodes corresponding to pre-labeled samples after a pre-defined amount of steps, so the particles could walk around all the pre-labeled nodes. This way, after each r steps, the particles are reset using Eq. 6, and their respective potentials are set to the maximum by Eq. 7.

So, in summary, our algorithm works as follows:

1. Build the affinity matrix W by using Eq. 1,
2. Set nodes ownership levels by using Eq. 5,
3. Set particles initial positions and potentials by using Eq. 6 and Eq. 7 respectively,
4. Repeat steps 5 to 11 until convergence or for a pre-defined number of steps,
5. Select between deterministic moving or random moving,
6. Select the target node for each particle by using Eq. 8 or Eq. 9 for deterministic moving or random moving respectively,
7. Update nodes ownership levels by using Eq. 10,
8. Update nodes ownership flags by using Eq. 3,
9. Update particles potentials by using Eq. 11,
10. Update the particles positions by using Eq. 12,
11. If r steps are reached, reset particles positions and potentials using Eq. 6 and Eq. 7 respectively.

3   Computer Simulations

In this section, we present the simulation results of some semi-supervised classification tasks by using the proposed model with synthetic and real data sets. We also compare our method with the Global and Local Consistency Method [21] for both results accuracy and execution time. For our method, the following parameters were held constant: $p_{det} = 0.6$ and $\Delta_v = 0.1$. The other parameters (σ and r) were set to their optimal values for each experiment. For the Consistency Method, the parameter $\alpha = 0.99$ was held constant, as the authors did in their original article [21], and σ was set to its optimal value for each experiment.

The first experiment was carried by using the artificial image shown in Fig. 2a, a toy data set with 2000 samples divided in two linearly non-separable classes (1000 samples per class), 20 of these samples (1%) are pre-labeled (10 from each class). Our algorithm was able to accurately classify all the unlabeled data as shown in Fig. 2b.

In the second experiment, we have used the Iris data set from the UCI Machine Learning Repository [27], which contains 4 attributes and 3 classes of 50 instances each, where each class refers to a type of iris plant. Both our Particle Method and the Consistency Method were used to perform semi-supervised classification in this data set. At each set of experiments, some samples (10% to 2%) were randomly chosen as the pre-labeled samples, and the remaining ones were presented unlabeled to both algorithms for classification. The classification results from the algorithms are shown in Figure 3. As our algorithm is non-deterministic there is small differences in the results obtained from different runs, so all the results presented here are the average of 100 runs with the same parameters and the same pre-labeled samples.

Our third experiment was performed using another real database, the Wine data set, also from the UCI Machine Learning Repository [27]. This data set results from a chemical analysis of wines grown in the same region in Italy, but derived from three different cultivars. The analysis determined the quantities

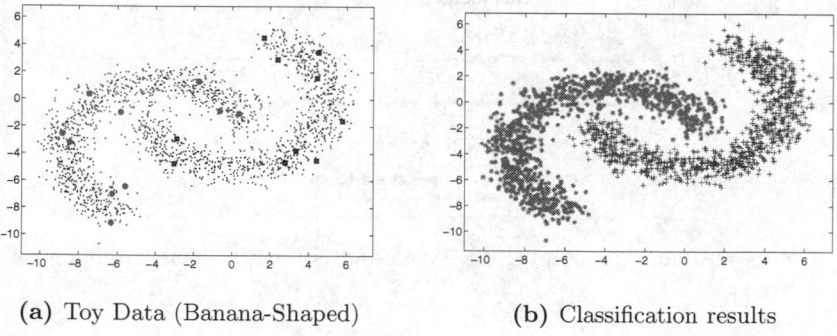

(a) Toy Data (Banana-Shaped)   (b) Classification results

Fig. 2. Classification of the banana-shaped patterns. (a) toy data set with 2000 samples divided in two classes, 20 samples are pre-labeled (red circles and blue squares). (b) classification achieved by the proposed method.

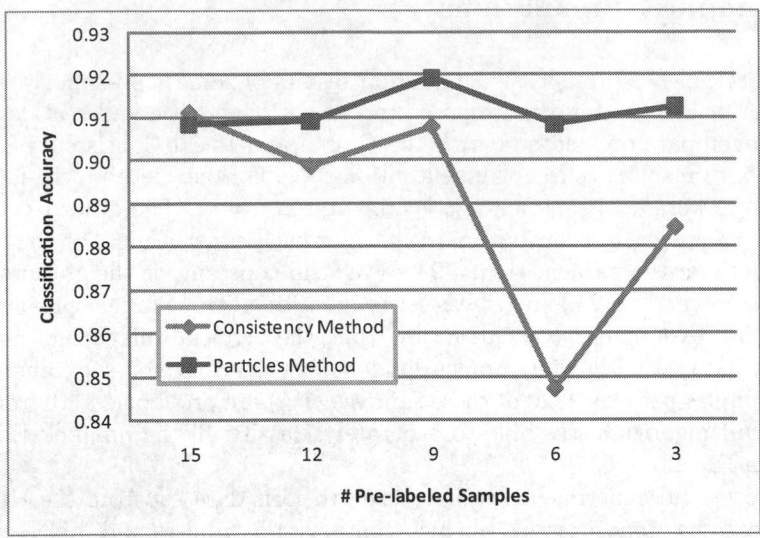

Fig. 3. Classification Accuracy in the Iris data set with different number of pre-labeled samples

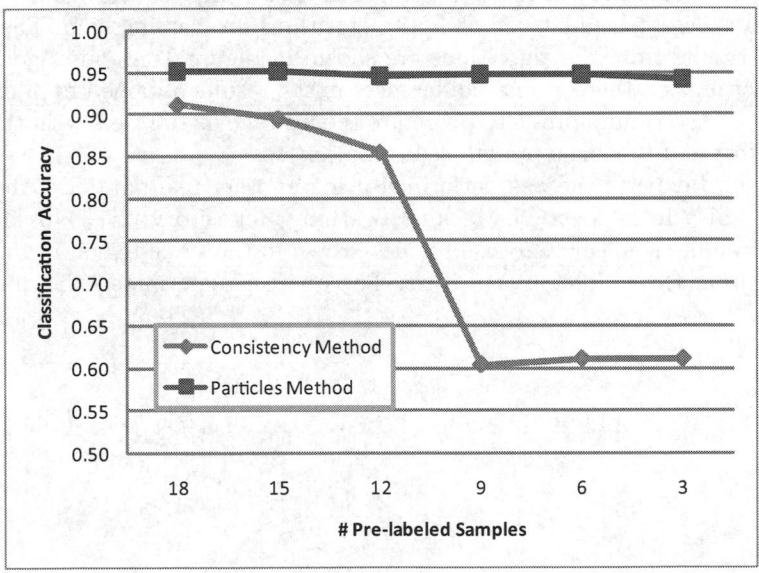

Fig. 4. Classification Accuracy in the Wine data set with different number of pre-labeled samples

of 13 constituents (attributes) found in these three types of wines, there are 178 samples in total. Again, for each set of experiments, some samples were randomly chosen as the pre-labeled samples while the others were presented to

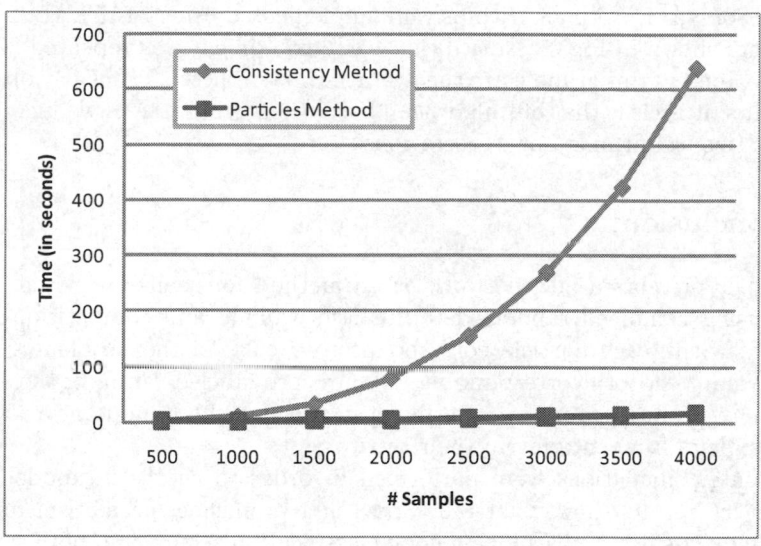

Fig. 5. Time elapsed to reach 95% correct classification with data sets of different sizes

the algorithms unlabeled. The classification results from both algorithms are shown in Figure 4. The presented results, once more, are the average of 100 runs with the same parameters and pre-labeled samples.

By observing Figures 3 and 4 we can notice that the Particle Method outperformed the Consistency Method in most cases, specially with the Wine data set. The major gains are observed when there are fewer pre-labeled nodes, which is another advantage of our method.

We also expect our method to be faster than other graph-based methods, because most of them have order of complexity $O(n^3)$ [11]. For instance, consider the Consistency Method [21], although it usually requires few iterations to converge, each iteration is $O(n^3)$ as they include $n \times n$ matrix multiplications. Also, there is a single step before the iterations, the laplacian normalization, which has high computational cost, for small data sets the cost of this step is even higher than all the iterations together. On the other hand, although our method usually requires thousands of iterations even for small data sets, each iteration is only $O(n \times c)$, so we expect our method to escalate well and to be quite fast for larger data sets.

In order to verify our expectations we have generated some banana-shaped data sets with increasing number of samples, using PRTools [28] function `gendatb` with the variance parameter fixed to 0.8. Then, we let both our method and the Consistency Method classify each of these data sets, running with their optimal parameters. For each data set we randomly chosen 10% of the samples (half from each class) to be the pre-labeled samples input for both algorithms, and finally we have measured the time each algorithm takes to achieve at least 95% correct classification of the remaining samples. All these tests were ran in a regular desktop computer with an Intel Core 2 Quad Processor model Q9450 and 4GB of RAM. Both algorithms were implemented using MATLAB [29]. The results are shown in Fig. 5. Notice that each experiment was repeated 20 times and the values in this graphic are the average time in these 20 runs. By observing the results, it is clear that our algorithm becomes much faster as we increase the data set size, confirming our expectations.

4   Conclusions

This paper presents a new network-based method for semi-supervised classification using combined random-deterministic walking and competition among particles, where each particle corresponds to a class of the problem. Starting from a small territory corresponding to a few pre-labeled samples, these particles can expand their domain by walking in their neighborhood and preventing other particles from entering in their territory.

Computer simulations were performed in order to check the model viability, and the results shows that our model is a promising mechanism for semi-supervised classification, achieving good classification accuracy in both synthetic and real data. It is important to notice that as the data set size grows our technique becomes much faster than other graph-based methods. This is a desirable feature since semi-supervised classification techniques are usually applied to data sets with large number of unlabeled samples. Also, our method seems to be less affected by the size of pre-labeled set, the ability to learn from less pre-labeled samples is another desirable feature in semi-supervised classification, as pre-labeling samples is the expensive or time consuming task to be avoided.

Another advantage of our method is that it can incorporate unseen data without any modifications in the algorithm, just insert the new nodes in the graph, calculate their connection weights and after some iterations each of these new node will belong to a particle, probably the particle that owns its neighbors. The new nodes can even affect the classification of older nodes, as they can become a new strong link between different neighborhoods.

Acknowledgements

This work is supported by the State of São Paulo Research Foundation (FAPESP) and the Brazilian National Council of Technological and Scientific Development (CNPq).

References

1. Newman, M.: The structure and function of complex networks. SIAM Review 45, 167–256 (2003)
2. Dorogovtsev, S., Mendes, F.: Evolution of Networks: From Biological Nets to the Internet and WWW. Oxford University Press, Oxford (2003)
3. Bornholdt, S., Schuster, H.: Handbook of Graphs and Networks: From the Genome to the Internet. Wiley-VCH, Chichester (2006)

4. Newman, M., Girvan, M.: Finding and evaluating community structure in networks. Physical Review E 69, 026113 (2004)
5. Newman, M.: Modularity and community structure in networks. Proceedings of the National Academy of Science of the United States of America 103, 8577–8582 (2006)
6. Duch, J., Arenas, A.: Community detection in complex networks using extremal optimization. Physical Review E 72, 027104 (2005)
7. Reichardt, J., Bornholdt, S.: Detecting Fuzzy Community Structures in Complex Networks with a Potts Model. Physical Review Letters 93(21), 218701 (2004)
8. Danon, L., Díaz-Guilera, A., Duch, J., Arenas, A.: Comparing community structure identification. Journal of Statistical Mechanics: Theory and Experiment 9, P09008 (2005)
9. Quiles, M., Zhao, L., Alonso, R., Romero, R.: Particle competition for complex network community detection. Chaos 18, 033107 (2008)
10. Mitchell, T.: Machine Learning. McGraw-Hill, New York (1997)
11. Zhu, X.: Semi-Supervised Learning Literature Survey. Technical Report, Computer Sciences, University of Wisconsin-Madison (2005)
12. Chapelle, O., Schölkopf, B., Zien, A.: Semi-Supervised Learning. MIT Press, Cambridge (2006)
13. Nigam, K., McCallum, A., Thrun, S., Mitchell, T.: Text classification from labeled and unlabeled documents using EM. Machine Learning 39, 103–134 (2000)
14. Fujino, A., Ueda, N., Saito, K.: A hybrid generative/discriminative approach to semi-supervised classifier design. In: AAAI 2005, Proceedings of the Twentieth National Conference on Artificial Intelligence, pp. 764–769 (2005)
15. Demiriz, A., Bennet, K., Embrechts, M.: Semi-supervised clustering using genetic algorithms. In: Proceedings of Artificial Neural Networks in Engineering, pp. 809–814. ASME Press (1999)
16. Dara, R., Kremer, S., Stacey, D.: Clustering unlabeled data with SOMs improves classification of labeled real-world data. In: Proceedings of the World Congress on Computational Intelligence (WCCI), pp. 2237–2242 (2002)
17. Blum, A., Mitchell, T.: Combining labeled and unlabeled data with co-training. In: COLT: Proceedings of the Workshop on Computational Learning Theory, pp. 92–100 (1998)
18. Mitchell, T.: The role of unlabeled data in supervised learning. In: Proceedings of the Sixth International Colloquium on Cognitive Science (1999)
19. Vapnik, V.: Statistical Learning Theory. Wiley, New York (2008)
20. Zhu, X., Ghahramani, Z., Lafferty, J.: Semi-supervised learning using Gaussian fields and harmonic functions. In: Proceedings of the Twentieth International Conference on Machine Learning, pp. 912–919. Morgan Kaufmann, San Francisco (2003)
21. Zhou, D., Bousquet, O., Lal, T., Weston, J., Schölkopf, B.: Learning with local and global consistency. In: Advances in Neural Information Processing Systems, vol. 16, pp. 321–328. MIT Press, Cambdridge (2004)
22. Blum, A., Chawla, S.: Learning from labeled and unlabeled data using graph mincuts. In: Proceedings of the Eighteenth International Conference on Machine Learning, pp. 19–26. Morgan Kaufmann, San Francisco (2001)
23. Belkin, M., Matveeva, I., Niyogi, P.: Regularization and semisupervised learning on large graphs. In: Conference on Learning Theory, pp. 624–638. Springer, Heidelberg (2004)

24. Belkin, M., Niyogi, P., Sindhwani, V.: On manifold regularization. In: Proceedings of the Tenth International Workshop on Artificial Intelligence and Statistics (AISTAT 2005), pp. 17–24. Society for Artificial Intelligence and Statistics, New Jersey (2005)
25. Joachims, T.: Transductive learning via spectral graph partitioning. In: Proceedings of International Conference on Machine Learning 2003, pp. 290–297. AAAI Press, Menlo Park (2003)
26. Ng, A., Jordan, M., Weiss, Y.: On spectral clustering: analysis and an algorithm. In: Advances in Neural Information Processing Systems, vol. 14, pp. 849–856. MIT Press, Cambridge (2001)
27. Asuncion, A., Newman, D.: UCI Machine Learning Repository. University of California, Irvine, School of Information and Computer Sciences (2007), http://www.ics.uci.edu/~mlearn/MLRepository.html
28. PRTools: The Matlab Toolbox for Pattern Recognition, http://prtools.org
29. MATLAB, http://www.mathworks.com/products/matlab/

Retail Location Choice with Complementary Goods: An Agent-Based Model

Arthur Huang and David Levinson

Department of Civil Engineering, University of Minnesota
500 Pillsbury Drive SE., Minneapolis, MN 55455
{huang284,dlevinson}@umn.edu

Abstract. This paper models the emergence of retail clusters on a supply chain network comprised of suppliers, retailers, and consumers. Firstly, an agent-based model is proposed to investigate retail location distribution in a market of two complementary goods. The methodology controls for supplier locales and unit sales prices of retailers and suppliers, and a consumer's willingness to patronize a retailer depends on the total travel distance of buying both goods. On a circle comprised of discrete locations, retailers play a non-cooperative game of location choice to maximize individual profits. Our findings suggest that the probability distribution of the number of clusters in equilibrium follows power law and that hierarchical distribution patterns are much more likely to occur than the spread-out ones. In addition, retailers of complementary goods tend to co-locate at supplier locales. Sensitivity tests on the number of retailers are also performed. Secondly, based on the County Business Patterns (CBP) data of Minneapolis-St. Paul from US Census 2000 database, we find that the number of clothing stores and the distribution of food stores at the zip code level follows power-law distribution.

Keywords: clustering, agent-based model, location choice, distribution pattern.

1 Introduction

In economic geography, clusters are geographical agglomerations of firms with similar or complementary capabilities [1]. Geographical clusters of business locations have been prominent phenomena in almost all countries and regions. Global integration increasingly contributes to regional specialization, with decreasing transportation costs and trade barriers enabling firms to closely interact with other firms to benefit from local economies of scale [2,3].

An early investigation of clustering was performed by Marshall [4], who argued that while firms are directly connected through business exchange, they are also indirectly linked through competition for labor and production factors, and that clustering of locations represented the distribution of economic activities. Weber [5] proposed a theory of industrial location where industrial organizations locate to minimize transportation costs of raw materials and final product. Christaller,

J. Zhou (Ed.): Complex 2009, Part I, LNICST 4, pp. 175–187, 2009.

in his central place theory [6], indicated that the effects of market threshold and consumers' preferences in terms of range of patronizing lead to a system of central places, wherein each center supplies certain types of products forming levels of hierarchy. Krugman [7] asserted that the geographical distribution of firms is balanced by centripetal and centrifugal forces. The centripetal forces entice firms to cluster, and the centrifugal forces cause firms to scatter.

The mechanism of business clustering has gained increasing attention. Porter [8] formulated a diamond model to identify the mechanism of fostering industrial dynamism and long-term development. Levinson and Krizek [9] proposed four factors impacting a firm's decision of where to locate on a spatially-structured supply chain networks: complementors, competitors, connectors, and customers, which comprised the diamond of exchange. Huang and Levinson [10] studied retail location choice on a supply chain network of one product and found that different numbers of competitors and transportation can lead to different retail distribution patterns.

Hierarchical distributions of resources and economic activities have been widespread in almost every city, region, and nation. Power-law distributions have been found to well fit many natural and social phenomena, such as the population of cities [12], distribution of land uses [13], and the number of citations received by published academic papers [14]. Zipf [15] proposed that city sizes follow a special form of the hierarchical distribution which is latter named as the Zipf's law. Gabaix [16] and Ioannide et al. [17] indicated that the rank-distribution of US cities follows the Zipf's law. Zipf's law was also found to fit the distribution of US firm sizes and the sales of US manufacturing firms [18,19]. Similar results are also found for European and Japanese firms [20,21,22].

Yet human beings are still lacking understanding about the micro-foundations of the agglomeration of human activities and resources; quantitative theoretical models that can properly answer the question of how and why clusters emerge and prosper need to be formulated. Microscopically, clusters form by experiencing a complex and self-organized process in developmental stages, where business agents are constantly learning and adapting [23]. It is of interest to examine how individual agents' seemingly random and chaotic decisions and interactions as a whole lead to clusters of firms.

As an extension of Huang and Levinson [10], this research builds an agent-based model to examine retail location choice on a simplified supply chain network of suppliers, retailers, and consumers. Retailers maximize profits by locating, which is modeled as a repetitive game. We are interested in the retail distribution pattern in equilibrium. It should be noted that this study analyzes a pure model of agglomeration of firms without forms of co-operation and other inter-organizational linkages. The basic assumptions of this model are: (1) Two categories of products exist in the market; one retailer only sells one category of products. (2) Each consumer needs both products. (3) Each consumer buys all needed products of one category from one retail=cr in one trip. (4) Consumers share the same utility function, suggesting that they have the same taste when patronizing retailers. (5) Suppliers of the same product offer the same unit sales price and keep their price and locations fixed at all times. (6) Retailers of the same product have the same fixed unit sales prices. (7) Retailers' moving is costless.

The rest of the paper is organized as follows. Section 2 describes autonomous players and defines the concepts of cluster and average cluster density to measure retail spatial patterns. Section 3 depicts and analyzes the simulation results. Section 4 performs sensitivity tests on the number of retailers. Section 5 discusses the principles of retail location choice. Section 6 analyzes the distribution patterns of clothing stores and food stores in Minneapolis–St. Paul in the state of Minnesota, United States. Finally, Section 7 concludes the paper.

2   The Model

In this section, a multi-agent paradigm is adopted to model a repetitive non-cooperative game of retail location choice. All players sit on the circle of a finite number of uniform locations. This section introduces agents in this model: suppliers, retailers, and consumers. Variables and constants used in this research are listed in Table 1.

2.1   Consumers

The market has two categories of products x and y, which are sold by two kinds of retailers. Let R_{xi} indicate retailer i of product x, and R_{yj} indicate retailer j of product y. Consumers hope to buy both products with minimum cost, which, in this research, implies minimum total travel distance. A trip is defined as a round-trip from home to visit R_{xi} and R_{yj}. Trips are assumed not to have fixed costs; only total distance matters. In the scenario of W_x number of R_{xi} and W_y number of R_{yj}, there are in total $W_x \cdot W_y$ trip candidates, from which a consumer chooses the shortest trip.

Let d_{pi}, d_{pj}, and d_{ij} respectively denote the shortest distance between consumer p and retailer R_{xi}, between consumer p and R_{yj}, and between R_{xi} and R_{yj}. Since players locate on a circle, it can be easily deduced that d_{pi}, d_{pj}, and d_{ij} are larger than or equal to the half of the circle perimeter. Given retailer R_{xi} and retailer R_{yj}, the shortest trip distance d_t for consumer p can be calculated by the following method: if the summation of d_{pi}, d_{pj}, and d_{ij} is larger than or equal to the perimeter of the circle where retailers are located, the shortest trip is the perimeter of the circle. Otherwise, the shortest trip distance equals twice of the largest value of d_{pi}, d_{pj}, and d_{ij}.

Consumers are assumed to be homogeneous, sharing the same utility function. The utility for consumer p to patronize retailer R_{xi} equals:

$$U_{pi} = \sum_{t=1}^{W_x \cdot W_y} k_1 \cdot d_t^{\beta} \cdot \pi_{ti} \tag{1}$$

Where β is expected to be negative because longer travel distance generally diminishes consumers' willingness of patronizing. To account for preferences

Table 1. List of Variables

Variables	Description
d_{pi}	shortest distance between consumer p and retailer R_{xi}
d_{pj}	shortest distance between consumer p and retailer R_{yj}
U_{pi}	consumer p's utility of partronizing retailer R_{xi}
π_{ti}	dummy variable, equaling 1 if Retailer R_{xi} is included in trip t
d_t	total travel distance of trip t
b_{pi}	binary variable, which equals 1 if consumer p patronizes retailer R_{xi}
ρ_{pi}	probability for consumer p to patronize retailer R_{xi}
ρ_{pm}	probability for consumer p to patronize a retailer sitting at locale m
m_{ik}	shortest distance between retailer R_{xi} and supplier k of product x
Ω_{im}	expected profit for retailer R_{xi} when locating at m
Π_i	actual profit of retailer R_{xi}
σ_{mk}	shortest distance between supplier k and locale m on the circle
l_{mk}	binary variable, which equals 1 if a retailer in location m patronizes supplier k
l_{ik}	binary variable, which equals 1 if retailer R_{xi} patronizes supplier k
ϵ_i	number of retailers in cluster i
τ_i	number of locations in cluster i
φ_n	mean cluster density of the distribution pattern of n retailers

Constants	Description
k_1	a constant in consumers' uitlity function
θ	unit retail sales price of product x
χ	unit retail y sales price of product y
λ_x	individual customer's demand on product x
λ_y	individual customer's demand on product y
u	retailers' unit shipping cost per product
δ	unit sales price of suppliers of x
υ	unit sales price of suppliers of y
N	number of consumers
K	number of suppliers of product x
L	number of suppliers of product y
W_x	number of retailers of product x
W_y	number of retailers of product y
C	total number of locales on the circle

associated with factors other than traffic cost (and thus avoid a deterministic model, which is a special case where travel cost dominates), we use a logit model in which the probability for a consumer to patronize a retailer depends on travel cost but has a random component. The probability for consumer p to patronize retailer R_{xi} is formulated as:

$$\rho_{pi} = \frac{e^{U_{pi}}}{\sum_{j \in W_x} e_j^{U_{pi}}} \tag{2}$$

Similar formulas can be established for retailer R_{yj}. It should be noted that when a consumer's locale contains both retailers of x and retailers of y, the total travel distance becomes zero. In reality, there is always some distance between a consumer's home and a retail store. Thereby in this case, we set this intra-zonal distance to be 0.25 of the distance between two adjacent locations on the circle, which is a typical empirical value used in regional transportation planning models.

The *Roulette Wheel Selection* method is adopted for a consumer to select a retailer. This approach suggests that a retailer with a larger ρ_{pi} for consumer p has a greater chance to be selected by this consumer. A consumer's probabilities of patronizing all retailers comprise his *wheel of selection*, which is updated in every round; and a spin of the wheel selects a retailer. The sequence for consumers to patronize retailers is randomly decided for each round.

2.2 Suppliers

There are two kinds of suppliers which sell product x and product y, who are evenly distributed on the circle and are co-located. The model assumes that all suppliers offer the same unit sales price and can always produce enough goods to meet market demand. Suppliers locations are fixed in all rounds.

2.3 Retailers

Retailers connect suppliers and consumers on supply chains. Retailers' initial locations are randomly assigned. In the beginning of each round, retailers evaluate expected profits of all locations. For example, retailer R_{xi}'s expected profit in locale m, Ω_m, is calculated as:

$$\Omega_m = (\sum_{p=1}^{N} \lambda \cdot \rho_{pm}) \cdot [\theta - \sum_{k=1}^{K} (\delta + u \cdot \sigma_{mk}) l_{mk}] \tag{3}$$

Where $\sum_{p=1}^{N} \lambda \cdot \rho_{im}$ represents total expected sales of products in locale m. The following part in brackets refers to expected profit per product, equaling sales price minus cost. A retailer's cost includes the purchasing cost of products from a supplier and the shipping cost which is proportional to shipping distance and quantity of products. Here we assume a retailer patronizes the closest supplier. After evaluating profits of all localities, a retailer moves to the locale that has the highest expected profit (revenue - cost), given others are geographically fixed at that time. Each retailer can only move once per round; the sequence for retailers to move is randomly decided, no matter what products they sell.

After all retailers choose locations, consumers begin to patronize retailers; the method is introduced in Section 2.1. Retailers' *actual* profits are calculated at the end of one round. A typical formula of retailer R_{xi}'s actual profit, Π_{xi}, is as follows:

$$\Pi_i = (\sum_{p=1}^{N} \lambda \cdot b_{pi}) \cdot [\theta - \sum_{k=1}^{K} (\delta + u \cdot m_{ik}) l_{ik}] \tag{4}$$

Compared with function (3), the main difference is the way total sales amount is calculated, which in this function equals R_{xi}'s sales price times actual sales amount at the end of a round. The actual profit and expected profit are different for two reasons: First, actual profit is calculated when all retailers have had an opportunity to relocate, that is at the end of a round, whereas expected profit may be estimated before other retailers find new locations. Second, consumers patronize retailers stochastically.

2.4 Measuring Spatial Distribution

We measure the number of discrete retail clusters and average cluster density when a game reaches equilibrium. A cluster is defined as an agglomeration of retailers that are geographically adjacent or in the same locale of the circle. Cluster density is calculated as the number of retailers in a cluster divided by the number of locations in the cluster. The mean average cluster density of n retailers, φ_n, is formulated as:

$$\varphi_n = \frac{1}{M} \sum_{i=1}^{M} \frac{\epsilon_i}{\tau_i} \tag{5}$$

where τ_i is the number of locations in cluster i; ϵ_i is the number of retailers in cluster i; M is the total number of clusters.

3   Experiments and Results

In our first basic experiment, all agents sit on a circle of 100 uniform discrete locations, where 5000 consumers are evenly distributed on these locations. A consumer's demand on product x is 20 and on product y is 10. First, we examine the scenario of 5 retailers of x, 5 retailers of product y, 5 suppliers of x, and 5 suppliers of y. Table 2 shows the values of the parameters used in this experiment. A game is believed to have achieved an equilibrium when all retailers stay at their current locales for three consecutive rounds. Since multiple equilibria may exist in this game, we test 400 different retail initial location patterns.

Typically a stable pattern emerges after the first round. Our results find five retail location distribution patterns, which can be grouped into four categories by the number of clusters (it should be noted that here we consider the geographical patterns, although each individual retailers' final location may vary in each game depending on their initial patterns and the sequence of moving). The

Table 2. Value of parameters in the scenario of 10 retailers and 10 suppliers

Parameters	value	Parameters	value
N	5000	k_1	1
C	100	χ	2.5
u	0.08 (\$)	δ	1.5 (\$)
υ	1.0 (\$)	K	5
θ	3.5	L	5
λ_x	20	W_x	5
λ_y	10	W_y	5

Table 3. Probability distribution of the number of clusters and cluster density for the case of 10 retailers and 10 suppliers

Number of clusters	Probability	Cluster density	Probability
1	0.861	10	0.741
2	0.128	6	0.120
3	0.011	5	0.117
		3.3	0.011
		3	0.011

probabilities for different numbers of clusters and cluster densities are shown in Table 3. The most common pattern is only one cluster, where all retailers accumulate at a supplier locale. All the retail distribution patterns share two features: (1) retailers only stay at supplier locales; (2) the same number of retailers of x and retailers of y co-locate, indicating that they constitute pairs. It is interesting to notice that the evenly distributed pattern of retailers—every one retailer of x and every one retailer of y double at a supplier locale—does not appear in our experiments. To further explore its possibility, we purposely set the initial distribution pattern to be very similar to the evenly distributed one, the result of which is that the evenly distributed pattern emerges.

It can be summarized that the hierarchical pattern of one cluster is most common, while the spread-out patterns with a larger number of clusters are very rare, which is the feature of power-law distribution. According to the definition of power-law distribution, a quantity x follows a power law if its distribution obeys

$$p(x) \propto x^{-\alpha}, \tag{6}$$

where α is the scaling parameter. The logarithmic scale of $p(x)$ and x have the form of a linear relationship with slope $-\alpha$. Figure 1 shows log-log plots of

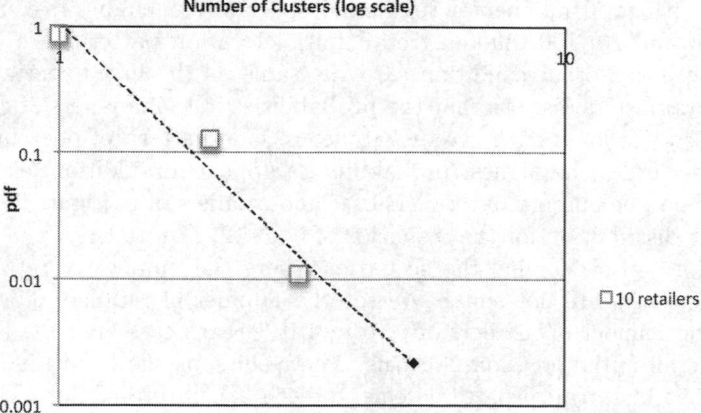

Fig. 1. Log-log plot showing power-law distributions in the number of clusters for 10 retailers (5 retailers of x and 5 retailers of y), where $\alpha = 3.83$

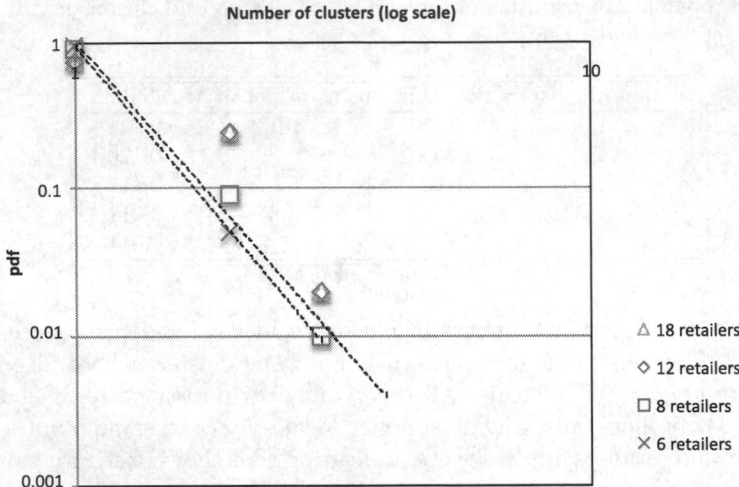

Fig. 2. Log-log plot showing power-law distributions in the number of clusters for the scenarios of 6, 8, 12, and 18 retailers in total, where the number of retailers of x equals the number of retailers of y

probability distribution of the number of clusters in our result. The model fit test confirms our hypothesis of power law; α is estimated to be 3.83 through maximum likelihood estimation (MLE).

4   Sensitivity Tests on the Number of Retailers

Our sensitivity tests investigate different numbers of retailers. First, presumably having the same number of retailers of x and retailers of y, we examine the scenarios with total number of retailers respectively equaling 4, 6, 8, 10, 12, 14, 16, 18, and 20. 100 different retail initial location patterns are examined for each scenario; other conditions are the same as the first experiment. The experimental results disclose that the probabilities of the numbers of clusters in all these cases follow power-law distributions. The pattern of one cluster at a supplier locale has the highest probability to appear; in addition, each cluster contains the same number of retailers of x and retailers of y. Figure 2 shows the probability distribution for the scenarios of 6, 8, 12, 18 retailers.

Second, we also examine the scenarios where the number of retailers of x and retailers of y are not equal. We set the number of retailers of x to be 5, and run the number of retailers of y from 4 to 9; each case are tested with 100 different retail initial location patterns. The results, as shown in Figure 3, also indicate that hierarchical patterns emerge with a high probability. The probability distributions of the numbers of resultant clusters are all found to well fit the power-law distribution. Table 4 displays the estimated exponent α for the scenarios of various numbers of retailers. It is disclosed that controlling for the

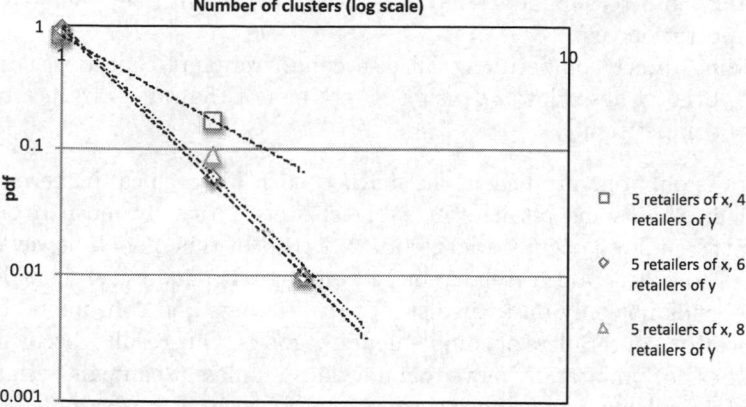

Fig. 3. Log-log plot showing power-law distributions in the number of clusters for the scenarios where the number of retailers of x equals 5 and the number of retailers of y respectively equals 4, 6, and 8. When the number of retailers of y equals 6, $\alpha = 2.27$; when the number of retailers of y equals 6, $\alpha = 4.11$; when the number of retailers of y equals 6, $\alpha = 4.03$.

Table 4. Estimated exponent α for different numbers of retailers

# of retailers of x	# of retailers of y	α
5	4	2.27
5	5	3.81
5	6	4.11
5	7	4.58
5	8	4.01
5	9	4.01
6	6	3.59
7	7	3.77
8	8	2.00
9	9	3.10
10	10	3.49

number of retailers of x, as the number of retailers of y increases, the scaling parameter becomes larger, meaning that the slope of the power-law trend line gets steeper. Additionally, the values of the scaling parameter in the cases of different numbers of retailers of x and y tend to be larger than those of the scenarios with the same number of retailers of x and y.

5   Discussion

Our experiments illustrate some principles for retail location choice in our model:

1. Pairing of retailers of complementary goods. *Ceteris paribus*, a retailer is more likely to move to a locale with more retailers of complementary goods.

2. Staying close to suppliers. Retailers all co-locate with suppliers to reduce transportation cost.
3. Avoiding direct competition. All else equal, retailers choose places whose neighborhood has a lower density of retailers of the same product to avoid direct competition.

If there is only one product in the market, such hierarchical patterns cannot be stable in that some retailers in a big cluster can easily move to an open space on the circle to occupy a larger market [10]. In this model, however, since consumers consider total travel distance of buying both goods, retailers' location choice depends not only on their distance to suppliers and consumers, but also on the locations of retailers of complementary goods. Our results discover that a retailer does not unilaterally move to a new locale unless it can pair with another retailer of complementary goods there.

In the central place theory, Christaller [6] claimed that in the areas with evenly-distributed population and resources, settlements have equidistant spacing between centers of the same order; high-order services are farther away from low-order services. Yet this research reveals that even in a market of two equally important products, hierarchical distribution patterns can autonomously emerge. This comports with the notion of retail districts found in many cities [9], such as the Kappabashi district of Tokyo specializing in kitchen equipment (and plastic sushi) along with similar examples of clustered competitors. In this model, although the even distribution pattern of retailers can occur under certain circumstances, to acheive this each cluster needs almost the same timing to emerge, which has a high requirement for retail initial distribution conditions and the sequence of location choice. Thereby it is much more difficult to emerge than the hierarchical ones.

6   Retail Geographical Distribution in the Twin Cities

Based on the County Business Patterns (CBP) data from the US Census 2000 database, we further examine the retail geographical distribution patterns in the Twin Cities. Each zip code area is considered as a cluster. Two categories of retailers are selected according to the 6-digit NAICS code. One category is food and beverage stores, including supermarkets and other grocery (except convenience) stores (445110) and convenience stores (445120); we believe products in this category are complementary to each other. The other category is clothing and clothing accessories stores, which include: mens clothing stores (448110), womens clothing stores (448120), childrens and infants' clothing stores (448130), family clothing stores (448140), other clothing stores (448190), and shoe stores (448210); the commodities from these stores are also complementary.

Fig.4 presents the cumulative distribution for the number of food and beverage store establishments. The distribution fits the power law with an R-square of 0.94, and the scaling parameter is estimated to be 1.10. Fig. 5 shows the

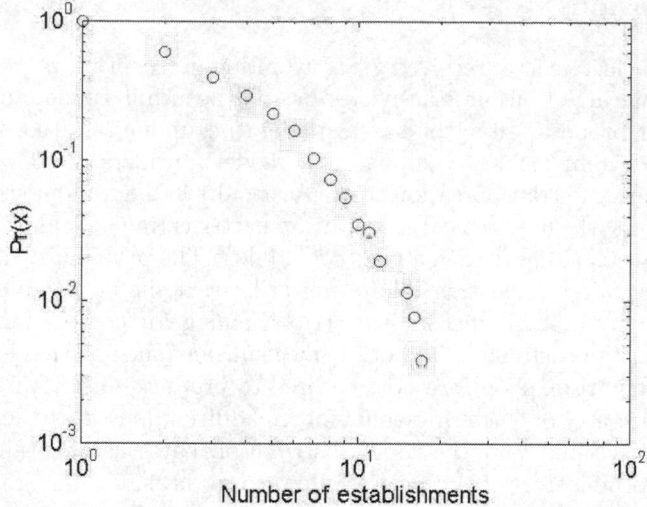

Fig. 4. The cumulative distribution for the number of food and beverage stores by zip code in the Twin Cities, MN in 2000. The scaling parameter is estimated to be 1.10.

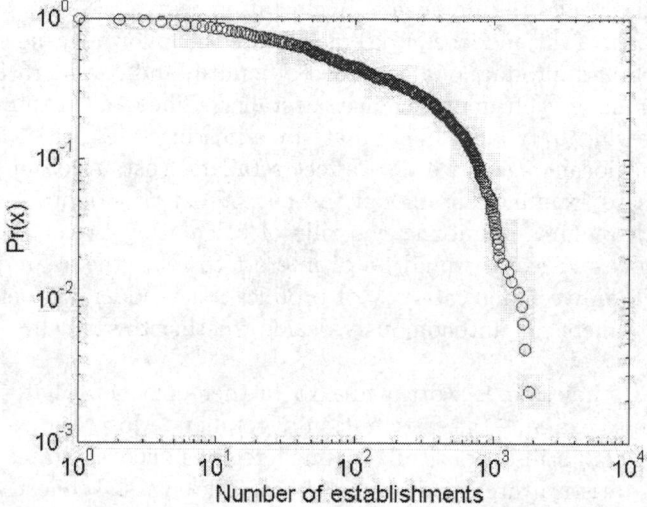

Fig. 5. The cumulative distribution for the number of clothing and accessories stores by zip code in the Twin Cities, MN in 2000. The scaling parameter is estimated to be 0.45.

cumulative distribution for the number of clothing and clothing accessories stores. Its distribution is power law with a kink in the curve; the MLE method estimates the scaling parameter to be 0.45. Such findings can be seen as empirical evidence of hierarchical distribution patterns of retailers of complementary goods.

7   Conclusions

Geographical clusters have received great attention in recent years, yet the mechanism of clustering of business activities has not been fully understood.

This paper proposes an agent-based model to examine retail location choice in a market of complementary goods. This model considers the impact of both market demand and transportation cost. Our results find autonomous emergence of retail clusters; the hierarchical distribution patterns (in particular, the pattern of only one cluster) appear with a high probability. The probability distributions of the number of retail clusters follows power laws. It should be noted that this occurs with only a single mechanism (trip chaining for complementary goods on the part of the customer), and other mechanisms (such as the desire of customers to comparison shop) are not required to produce clusters, but may also be additional source of clustering behavior. In addition, based on the US Census data in 2000, we find that the retail distribution patterns of retailers of food stores and clothing stores by zip code follow power laws.

Retail location choice is a process involving balancing many different factors, such as distance to suppliers, distance to its complementary goods, and distance to direct competition. This research discloses that co-locating of retailers of complementary goods is a striking phenomenon. In addition, retailers settle down at their supplier locales to minimize transportation cost.

The framework of this model can be extended in the following aspects. Factors such as price, brand, product quality, word of mouth, and considerations about scheduling can impact consumers' choice of retailers. The model can incorporate mechanism by which retailers can compete by adjusting sales price and quality. Land price and local wage price also affect retailers' cost. Another intriguing future work is to examine this model in the context of a more general grid network, which requires modifying the rule of calculating travel distance and transportation cost. Also, it would be of interest to relax the assumption that one retailer can only sell one category of products; the model will better reflect the reality if retailers can autonomously decide whether to specialize or expand their scope.

In empirical studies, it is worthwhile to further consider what the proper level should be to examine the distribution of retailers. Most studies categorize industries and plants into politically defined regions such as states, counties, Metropolitan statistical areas (MSA), and zip code areas. While they can be meaningful indicators of clusters, clusters that are on the edges of several adjacent counties or zip code areas are not properly measured by looking from the county or zip code level.

References

1. Richardson, G.B.: The Organization of Industry. Economic Journal 82, 883–896 (1972)
2. Brakmann, S., Garretsen, H., van Marrewijk, C.: An Introduction to Geographical Economics. Cambridge University Press, Cambridge (2001)

3. Baldwin, R., Forslid, R., Martin, P., Ottaviano, G., Robert-Nicoud, F.: Economic Geography and Public Policy. Princeton University Press, Princeton (2003)
4. Marshall, A.: Principles of Economics. Macmillan, London (1890)
5. Weber, A.: Theory of the Location of Industries. University of Chicago Press (1957)
6. Christaller, W.: Central Places in Southern Germany (Translated by C.W. Baskin). Prentice-Hall, NY (1966)
7. Krugman, P.: Urban concentration: The Role of Increasing Returns and Transport Costs. International Regional Science Review 19, 5–30 (1996)
8. Porter, M.: The Role of Location in Competition. International Journal of the Economics of Business 1, 35–40 (1994)
9. Levinson, D., Krizek, K.: Planning for Place and Plexus: Metropolitan Land Use and Transport. Routledge, New York (2008)
10. Huang, A., Levinson, D.: An Agent-based Retail Location Model on a Supply Chain Network. University of Minnesota. Working paper (2008)
11. Simon, H.A., Bonini, C.P.: The Size distribution of business firms. American Economic Review 48(4), 607–617 (1958)
12. Clauset, A., Shalizi, C., Newman, M.E.J.: Power-law Distributions in Empirical Data. SFI Working Paper (2007)
13. Andersson, C., Hellervik, A., Lindgren, K.: A Spatial Network Explanation for a Hierarchy of Urban Power Laws. Physica A 345, 227–244 (2005)
14. Redner, S.: How Popular is Your Paper? An Empirical Study of the Citation Distribution. European Physical Journal B 4, 131–134 (1998)
15. Zipf, G.: Human Behavior and the Principle of Least Effort: An Introduction to Human Ecology. Addison-Wesley, Reading (1949)
16. Gabaix, X.: Zipf's Law For Cities: An Explanation. Quarterly Journal of Economics 114, 739–767 (1999)
17. Ioannides, Y., Overman, H.: Zipf's Law for Cities: An Empirical Examination. Regional Science and Urban Economics 33, 127–137 (2003)
18. Axtell, R.: Zipf Distribution of U.S. Firm Sizes. Science 293, 1818–1820 (2001)
19. Stanley, M.: Zipf Plots and the Size Distribution of Firms. Economics Letters 49, 453–457 (1995)
20. Fujiwara, Y., Di Guilmi, C., Aoyama, H., Gallegati, M., Souma, W.: Do Pareto-Zipf and Gibrat laws Hold True? An analysis with European rms. Physica A: Statistical Mechanics and its Applications 335, 197–216 (2004)
21. Okuyama, K., Takayasu, M., Takayasu, H.: Zipf's Law in Income Distribution of Companies. Physica A 269, 125–131 (1999)
22. Mori, T., Nishikimi, K., Smith, T.: The Number-Average Size Rule: A New Empirical Relationship Between Industrial Location and City Size. Journal of Regional Science 48, 165–211 (2008)
23. Maskell, P., Malmberg, A.: Localised Learning and Industrial Competitiveness. Cambridge Journal of Economics 23, 167–185 (1999)

Research on Web2.0 System Design Based on CAS Theory

Kai Chen and Hengshan Wang

University of Shanghai for Science and Technology,
Shanghai 200093, P.R.China
kookia.chen@gmail.com, wanghs@usst.edu.cn

Abstract. According to complexity theory, this paper analyses several characteristics of some present Web2.0 systems, such as Blog, Wiki, SNS and social tags. It also summarizes the disadvantages of current information system design methods and finally re-designs it based on CAS and DSDM.

Keywords: Complex adaptive systems (CAS) Web2.0.

1 Introduction

The concept of "Web 2.0" began with a conference brainstorming session between O'Reilly and MediaLive International. Dale Dougherty, web pioneer and O'Reilly VP, noted that far from having "crashed", the web was more important than ever, with exciting new applications and sites popping up with surprising regularity [1]. Though there is no common answer to what is Web 2.0, in the next few years the core ideas unknowingly and continued to infiltrate into the Internet. Users gradually changed from the recipient of information into the producers and disseminators, from the audience to their own social group who have the right to speak, and gradually shift from an individual to a social group with a common concern, the Internet services model also gradually change.

With the idea of Web2.0 widely understanding, and information system evolves from low-level to advanced, simple to complex, close to open, and isolated to cooperative, The components of information system become more dependent, the coupling degree among components gets lower, while the components interact and collaborate more flexibly. In a word, information system shows more characters of complex system than before. The most prominent ones are web2.0 systems. Therefore, it becomes a growing real need that using some theories and methods of related with system complexity research to study and design the information system.

2 CAS Theory

2.1 The Core of CAS

The core of CAS theory is that adaptation makes complexity [2]. Complex adaptive systems are special cases of complex systems. They are complex in that they are

J. Zhou (Ed.): Complex 2009, Part I, LNICST 4, pp. 188–195, 2009.
© ICST Institute for Computer Sciences, Social Informatics and Telecommunications Engineering 2009

diverse and made up of multiple interconnected elements and adaptive in that they have the capacity to change and learn from experience [3]. A Complex Adaptive System (CAS) is a dynamic network of many agents (which may represent cells, species, individuals, firms, nations) acting in parallel, constantly acting and reacting to what the other agents are doing. The control of a CAS tends to be highly dispersed and decentralized. If there is to be any coherent behavior in the system, it has to arise from competition and cooperation among the agents themselves. The overall behavior of the system is the result of a huge number of decisions made every moment by many individual agents. The term complex adaptive system (CAS) was coined at the interdisciplinary Santa Fe Institute (SFI), by John H. Holland, Murray Gell-Mann and others.

2.2 General Properties

Despite substantial different systems center on them in detail, CAS share four major features [4]:

1. Parallelism. CAS consists of large numbers of agents that interact by sending and receiving signals. Moreover, the agents interact simultaneously, producing large numbers of simultaneous signals.

2. Conditional action. The actions of agents in a CAS usually depend on the signals they receive. That is, the agents have an IF/THEN structure: IF [signal vector x is present] THEN [execute act y]. The act may itself be a signal, allowing quite complicated feedbacks, or the act may be an overt action in the agent's environment.

Interlocking sequences of signal-processing rules become programs that are executed in parallel, with all that implies for flexibility and breadth of repertoire.

3. Modularity. In an agent, groups of rules often combine to act as "subroutines". For example, the agent can react to the current situation by executing a sequence of rules. These "subroutines" act as building blocks that can be combined to handle novel situations, rather than trying to anticipate each possible situation with a distinct rule. Because potentially useful building blocks are tested frequently, in a wide range of situations, their usefulness is rapidly confirmed or disconfirmed.

4. Adaptation and evolution. The agents in a CAS change over time. These changes are usually adaptations that improve performance, rather than random variations.

3 Research on Present Characteristic Web2.0 Systems

In recent years, thousands of new web information systems appeared on the internet which based on the conception of Web2.0. Blog, Wiki, Social Network Service, and Social Tags are the four main kinds of Social Software [5] which are recognized as applications of Web 2.0. From the view of CAS (Complex Adaptive System) theory, Web 2.0 system has characteristics of aggregation, nonlinearity, flows and diversity, mechanisms of tagging, internal model and building blocks [6]. These characteristics and mechanisms are recognized by Holland [2] as the seven basic points of CAS.

Different from traditional software and information system, web2.0 systems have some typical characters as follows:

1. They use more participative-architecture and open-architecture; the most important is that people become a part of design of system;

2. There are plenty of nonlinear and self-organized mechanisms in system, these mechanisms can make the system adjust its function and structure continually to adapt to the changing environment;

3. The interactions and inter-operations among these systems become very frequent and complex than ordinary ones, the complex relation of the interactions and inter-operations facilitate a dynamic complex network coming into being, The complex network, which is quite similar to ecology network;

4. Social network analysis has been applied to information system design as embedded algorithms. In these systems, spontaneous cooperation among users allows many kinds of social networks to grow up from the bottom.

According to observation, user' enthusiasm of creativity started to descend, and the interaction among participators is increasing. How can we find the change when we design an information system? This problem has been raised in the paper [4]. So in the new method of information system design will be concerned, one is that keep the enthusiasm of participators, the other is that how to find the rule of system's adaptation and evolution. And we never deem these types of web2.0 systems as isolated ones, they are a whole.

4   The New Method of Web2.0 Information System Design

4.1   The Trend of Information System Design and Development

With the rapid development of information technology, the design and development of information systems also took place change time and again. From the evolution and development patterns of information systems, they were the earliest single-user, stand-alone system (or a host with multiple terminals), then developed the client/server model (C/S) systems which were used in the local area network or a closed network environment, after that, browser/server (B/S) model systems were used in the open environment of the Internet, as well as ,the expansion of the three-tier and multi-layer structure based on B/S model. Recently, developed to the so-called fat client model and decentralized peer-to-peer (P2P) model, information systems were toward more personalized, intelligent, distributed and decentralized. We can see from the Figure 1.

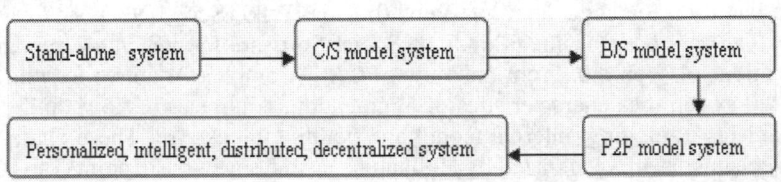

Fig. 1. The evolution and development of information system

4.2 Present Method of Development and Design

In the software engineering domain, software or information system development process is a structure imposed on the development of a software product. Synonyms include software life cycle and software process. There are several models for such processes, each describing approaches to a variety of tasks or activities that take place during the process.

The best-known and oldest process is the waterfall model, where developers are to follow these steps in order: Requirements specification, Design, Construction (implementation or coding), Integration, Testing and debugging (validation), Installation (deployment), Maintenance. After each step is finished, the process proceeds to the next step, just as builders don't revise the foundation of a house after the framing has been erected.

Iterative development is a cyclic software development process developed in response to the weaknesses of the waterfall model. The basic idea behind iterative enhancement is to develop a software system incrementally, allowing the developer to take advantage of what was being learned during the development of earlier, incremental, deliverable versions of the system. Learning comes from both the development and use of the system, where possible. Key steps in the process were to start with a simple implementation of a subset of the software requirements and iteratively enhance the evolving sequence of versions until the full system is implemented. At each iteration, design modifications are made and new functional capabilities are added.

The traditional approaches of system design have a common feature, it's top-down, according to needs design and analysis system. All of the design, structural order and classification are pre-determined by the designer, once the demand changing, it is difficult to make rapid and adaptive responding. Designer doesn't take into account that users' involvement is a key part of information system design. To methodology, they are a product of dualism, the relationship of user and system is use and used, with the change of requirements, traditional approaches are difficult to adapt themselves.

Agile development and dynamic systems development method (DSDM) bring us a new hope. DSDM is one of a number of agile methods for developing software. It is an iterative and incremental approach that emphasizes continuous user involvement. Its goal is to deliver software systems on time and on budget while adjusting for changing requirements along the development process. There are 9 underlying principles consisting of four foundations and five starting-points [7].

4.3 The Method of Web2.0 Information System Design

In a paper called "The Dynamic Business Applications Imperative," John R. Rymer, a senior analyst with Forrester, points to a fundamental shortcoming of today's applications: Today's applications force people to figure out how to map isolated pools of information and functions to their tasks and processes, and they force IT pros to spend too much budget to keep up with evolving markets, policies, regulations, and business models. IT's primary goal during the next five years should be to invent a new generation of enterprise software that adapts to the business and its work and evolves with it. Forrester calls this new generation Dynamic Business Applications, emphasizing close alignment with business processes and work (design for people) and adaptability to business change (build for change).

In the book of "The Fifth Discipline", Peter Senge said: From a very early age, we are taught to break apart problems, to fragment the world. This apparently makes complex tasks and subjects more manageable, but we pay a hidden, enormous price. We can no longer see the consequences of our actions; we lose our intrinsic sense of connection to a larger whole. When we then try to "see the big picture," we try to reassemble the fragments in our minds, to list and organize all the pieces. But, as physicist David Bohm says, the task is futile—similar to trying to reassemble the fragments of a broken mirror to see a true reflection. Thus, after a while we give up trying to see the whole altogether [8].

After analyzing the features of present Web2.0 system, the trend of information system and the characters of present approaches of development, now we will put these fragments together, design a method of Web2.0 information system development combined with CAS theory and DSDM. Figure 2.

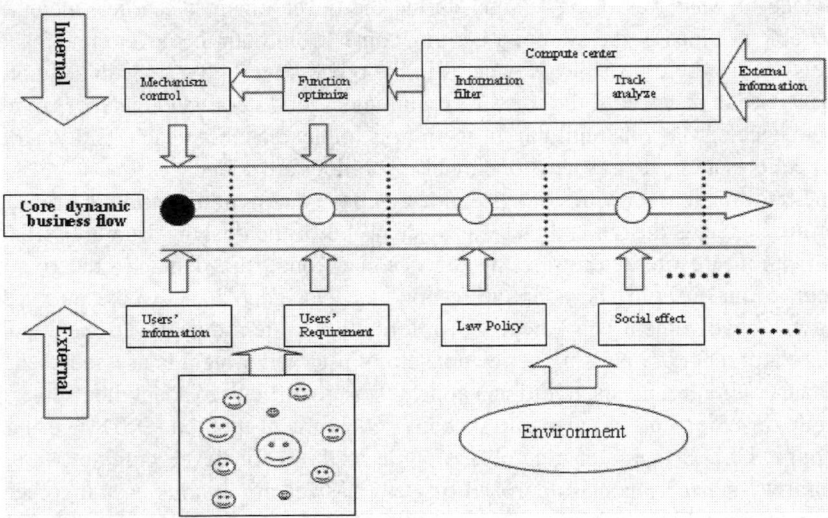

Fig. 2. The method of Web2.0 information system design

1. Core dynamic business flow

According to different circumstances, core business, as an engine of system, must be designed the fundamental business and service. Allow users to participate in the entire process of dynamic business and applications, in the process of participation exert their creativity, encourage them to cooperate with their friends or partners. With incentives make their participation more meaningful, and share the joy of success. In this way user will maintain long-term enthusiasm for participation. And unlike most of the SNS is made only for entertainment with your partners. To reach adaptability of system's business flow, every state is open, and can accept information from internal control and external feedback. Own the adaptability, and system can supply better services and applications.

2. External Feedback

Based on the theory of CAS, the constitution of a complex adaptive system has agent and environment. In the approach of web2.0 information system development and design, people serve as an important part; they are not only the creators of system's information, but also the foundation of the self-organization and creators of emergence. So user involvement is a prominent character. So-called environment is laws, policies and social effects and others. The information will be collected and transmitted to compute center of internal control for system's improvement. At the same time, the maintainer will realize dynamic reflection of user's requirements and environment's change, in order to optimize the program in compute center.

3. Internal control

Just agents and environment, we can not get our goal. We need another one, it is mechanism. What is mechanism? It is the rule of user's behavior in the system, there is no system can run without regulation. Agent, environment and regulation can create complexity, and then the system can evolve. We don't know where the system will go, but we can control the regulation of the system, via these mechanism lead and help the evolution of system.

Compute center. Its main work is to find the rule of system evolution. The rule is not permanent, and can change with the external feedback information. Filtering information, analyze and track record of users' behavior is the specific ways. Creators' information and behavior record are collected by compute center, and then through the calculation of algorithm and statistic analysis get the related data. At this time, maintainer of the system will give a report about suggestion of optimizing function and the situation of users' behavior and interests based on external environment and related data. Compute center will be optimized by programmer accordance with the report. In order to realize better individual service, we record users' behavior. With constantly improved, users also feel more in line with individual needs and preferences.

Functions optimize. A regulation is a function component, it is needed to realize by programmer. The main work of this part is to manage, realize and optimize these components based on report of compute center. In order to realize the adaptability of the information system, it supplies the interface for the application components which used in other systems. Of course, if you are professional user, the system would accept your own application through API. Anther important work is to optimize the core dynamic business flow.

Mechanism control. According to real requirement, design, add or delete the regulation of information system. If the regulation is good for the evolution or system, we will add it, or not.

Through users' participation, external feedback, dynamic control and dynamic management, new generation information system has its life. The more run, the better will be. With the user involvement, information system has the complexity. As designer, only can provide a platform for services and applications, as to how the system develop, we can not predict. If only an information system has the characters of complex adaptive system, it has its own evolution; only can we guide its development through the mechanism. So we must do a rapid response when requirement changed from users' involvement information, and gradually improve the mechanism to guide the development of the system's evolution. As a single user in the system, even if you asked him what good or not, what does not need or not, it is very difficult to get an unequivocal answer, because they are not sure. The solution of this problem depends on the analysis of the external feedback information. To realize this process, on the one hand, directly listen to the suggestions of users, on the other hand, complete through statistical analysis and psychological analysis of users' behavior, it has a big different with traditional design of information system which only focuses on technology. So at the beginning of design, system will be designed the tables to record users' behavior data. After analyzing these data, one is that designer can understand the needs of individual users, provide better personalized service in the future; the other is that can know group behaviors, improve the quality of service in system.

5 Conclusions

In this paper, analyzing the characters of present Web2.0 system, the trend of information system development and the situation of traditional design approaches, then build a new method of Web2.0 information system development and design based on CAS theory and DSDM. In order to realize the convenient management, the whole design is divided into three parts, including core dynamic business flow, external feedback and internal control, which are dynamically controlled and managed by users' participation and information feedback to make the system work effectively. The system may like getting a real life and having an evolution process and finally realize the adaptation for users' personality. That's the Web2.0 adaptive information system we want to get.

Acknowledgement

This paper is supported by Shanghai Key project Management Science and Engineering under Grant Number S30504.

References

1. O'Reilly, T.: What Is Web 2.0: Design Patterns and Business Models for the Next Generation of Software. O'Reilly Media, Inc., Sebastopol (2007),
   http://www.oreillynet.com/pub/a/oreilly/tim/news/2005/09/30/what-is-web-20.html
2. Holland, J.H.: Hidden Order: How Adaptation Builds Complexity. Addison-Wesley, Massachusetts (1995)
3. Complex adaptive system,
   http://en.wikipedia.org/wiki/Complex_adaptive_system
4. Holland, J.H.: Studying complex adaptive systems. Journal of Systems of Science and complexity 19(1) (2006)

5. Zhang, S.: Study from Social Software, Web 2.0 to Complex Adaptive Information System (in Chinese). Ph.D Thesis, Renmin University of China (2006)
6. Wang, R., Fang, M., Chen, Y.: A Model of Collaborative Knowledge-Building Based on Web2.0. In: IFIP International Federation for Information Processing, vol. 255., Xu, L., Tjoa, A., Chaudhry, S.: Research and Practical Issues of Enterprise Information Systems II, vol. 2, pp. 1347–1351. Springer, Boston (2007)
7. Dynamic Systems Development Method, http://en.wikipedia.org/wiki/DSDM
8. Peter, M.S.: The Fifth Discipline: The art and practice of the learning organization. Doubleday, New York (1994)

Reconstructing Gene Networks from Microarray Time-Series Data via Granger Causality*

Qiang Luo, Xu Liu, and Dongyun Yi

Department of Mathematics and Systems Science
National University of Defense Technology
Changsha, Hunan 410073, China
dongyun.yi@gmail.com

Abstract. Reconstructing gene network structure from Microarray time-series data is a basic problem in Systems Biology. In gene regulation networks, the time delays and the combination effects which are not considered by most existent models are key factors to understand the genetic regulatory networks. To address these problems, this paper proposed a fast algorithm to learn initial network structures for gene networks from time-series data by employing the Granger causality model to analyze the time delays and the combination effects for gene regulation. The simulation results on a synthetic network and the ethylene pathway in *Arabidopsis* show that the proposed algorithm is a promise tool for learning network structures from time-series data.

Keywords: partial Granger causality, gene regulatory networks, time series data, projection pursuit.

1 Introduction

Gene networks controlling how genes are up and down regulated in response to signals play a key role in the life phenomena [1], such as development, metabolizability, adaptability, immunity, etc. Earlier, the genetic networks are investigated by constructing the mathematical model of few genes, and the characteristics of the biology networks are analyzed through simulation [2]. These approaches work well in the small scale networks. Nowadays, the invention and application of high-throughput technologies make it possible to study the genes in the genome scale enabling the quantitative understanding of large gene networks [3]. For analysis in the genome scale, it is more suitable to reconstruct the network models of the genetic regulatory networks from data [4]. Recently, as more and more biology databases available, reverse engineering cellular networks has become a hot issue in biology, computer science, as well as mathematics, and the results of these researches are fruitful [5, 6, 7]. Methods for gene network reconstruction have

---

* This work is partially supported by National Basic Research Program of China (No. 2005CB321800) , and the Graduate Innovation Foundation of National University of Defense Technology (No. B060203).

J. Zhou (Ed.): Complex 2009, Part I, LNICST 4, pp. 196–209, 2009.

been proposed on the basis of statistical analysis such as Boolean models [8], differential equation models [9], Beyesian networks [10], and so on.

In recent years, learning the structures of Bayesian networks from massive data to reconstruct the gene networks from Microarray data attracted many scholars' attention [11, 12, 13, 14, 15]. Actually, the data used by Bayesian networks is produced by the perturbation experiments which usually knock out a gene and study the downstream effects, i.e., these data reflect the stationary status of the gene networks response to a stimulus. However, the use of perturbation experiments is limited due to technical and biological reasons [16]. Time-series expression data which imply a number of regulatory interactions have been widely used to study biological systems in many different species [17]. To model the time-series Microarray data, the dynamic Bayesian network (DBN) which has been shown to be appropriate for representing complex stochastic non-linear relationships among multiple random variables has been employed [18].

The initial structure of gene networks plays a key role in the present structure learning algorithms [19, 20, 21, 22] for DBN, since the learning algorithms usually start from prior network structures which are constructed by expert according to the background knowledge, and perform heuristic searches in the space of directed acyclic graphs to improve the network structure in the light of the information contained in data. Unfortunately, the complexity of the prior network structure avoids structure missing, but exponentially increases the computation complexity of the learning algorithm. Considering the combination effects in the regulation, one has to do the permutation test $m2^{m-1}$ times for the gene network with m genes. Besides, the published learning algorithms of DBN are almost based on the first order Markov assumption of the variables, but many researches on the practical data sets challenge this assumption [23, 24].

To address these problems, we proposed a learning algorithm for DBN from Microarray time-series data without the first order Markov assumption. This algorithm consists of three steps: first, the Granger causality model are employed for pairs of genes with their time sequence expression data to build an initial network; second, the false regulations in the initial network are deleted by partial-Granger causality; third, if necessary, it depends on the practical data, we could apply the partial-Granger causality to further filter out the false regulation between genes by computing the combination effects of the conditional candidate genes. The application results of the proposed algorithm on both the simulation data and the practical data show that it is a promising method of dynamical network structure learning.

2   Method

Consider a gene network with n genes, denoted by $\mathbf{G} = (G_1, G_2, \cdots, G_n)$ and the expression of the genes are jointly stationary. The expression time series with length T of the genes in this network are available, $g_i(t) \in R^+$ ($i = 1, 2, \cdots, n$, $t = 1, 2, \cdots, T$) is the stochastic time process for each gene. The main aim of this paper is to reconstruct the structures of the gene networks from the Microarray time-series data. Since the most concern of this study is the relationship between changes of genes, we can assume that $EG_i(t) = 0$ for all i.

2.1 Granger Causality

Assume that each process $G_i(t)$ $(i = 1, 2, \cdots, n)$ admits an autoregressive representation

$$G_i(t) = \sum_{p=1}^{\infty} a_p^{(i)} G_i(t-p) + \varepsilon_i(t), \text{var}(\varepsilon_i(t)) = \sigma_{\varepsilon_i}^2. \tag{1}$$

Jointly, they are represented as

$$G_i(t) = \sum_{p=1}^{\infty} a_p^{(i|j)} G_i(t-p) + \sum_{q=1}^{\infty} b_q^{(i|j)} G_j(t-q) + \varepsilon_{i|j}(t), \text{var}(\varepsilon_{i|j}(t)) = \sigma_{\varepsilon_{i|j}}^2. \tag{2}$$

The intensity of causal influence of G_j on G_i can be measured by

$$F_{G_j \to G_i} = \frac{\sigma_{\varepsilon_i}^2}{\sigma_{\varepsilon_{i|j}}^2} - 1. \tag{3}$$

If G_i and G_j are independent, then b_q are uniformly zero and $\sigma_{\varepsilon_i} = \sigma_{\varepsilon_{i|j}}$; otherwise, the expression of G_j will be helpful for the prediction of G_i, i.e., $\sigma_{\varepsilon_i} > \sigma_{\varepsilon_{i|j}}$. Hence, it is clear that $F_{G_j \to G_i} = 0$ when there is no causal influence from G_j to G_i, and the greater the value of $F_{G_j \to G_i}$ is the more strong the causal influence will be. With the expression data, if the orders in this model (2) are determined by some criterion (e.g., Akaike Information Criterion, AIC) as P_i and Q_j for G_i and G_j, respectively, the variances can be estimated as follows:

$$\hat{\sigma}_{\varepsilon_i} = \frac{1}{T - 2P_i} \sum_{i=P_i+1}^{T} \hat{\varepsilon}_i^2, \tag{4}$$

$$\hat{\sigma}_{\varepsilon_{i|j}} = \frac{1}{T - 2P_i - 2Q_j} \sum_{i=P_i+Q_j+1}^{T} \hat{\varepsilon}_{i|j}^2, \tag{5}$$

where $\hat{\varepsilon}_i$ and $\hat{\varepsilon}_{i|j}$ are the residuals of models (1) and (2), respectively. Then, we have

$$\hat{F}_{G_j \to G_i} = \frac{\hat{\sigma}_{\varepsilon_i}}{\hat{\sigma}_{\varepsilon_{i|j}}} - 1 \sim F(2Q_j, T - 2P_i - 2Q_j). \tag{6}$$

Given a significance level F_1, an F test of the null hypothesis that G_j does not have causality influence on G_i. Now, the dynamic network structure can be given below:

$$M^{(1)} = \left((m_{ij}^{(1)}) \right), \tag{7}$$

where

$$m_{ij}^{(1)} = \begin{cases} 1, & \hat{F}_{G_j \to G_i} > F_1; \\ 0, & \text{otherwise.} \end{cases} \tag{8}$$

2.2   Partial-Granger Causality

The conditional independence confuses pairwise algorithms, for instance, gene Y is correlated with gene Z when $X \to Y$ and $X \to Z$, i.e., given X, Y and Z are conditional independence. To filter out the fake edges from the learning structure, the partial-Granger causality (P-G-C) is induced.

When $m_{ij}^{(1)} = 1$ $(i \neq j, i, j = 1, 2, \cdots, n)$, the conditional independence between G_i and G_j given G_k will be examined for each $k = 1, 2, \cdots, n$ and $k \neq i, k \neq j$. Consider the following models

$$G_i(t) = \sum_{p=1}^{\infty} a_p^{(i|k)} G_i(t-p) + \sum_{q=1}^{\infty} c_q^{(i|k)} G_k(t-q) + \varepsilon_{i|k}(t), \ \mathrm{var}(\varepsilon_{i|k}(t)) = \sigma^2_{\varepsilon_{i|k}}, \quad (9)$$

$$G_i(t) = \sum_{p=1}^{\infty} a_p^{(i|j,k)} G_i(t-p) + \sum_{r=1}^{\infty} b_r^{(i|j,k)} G_j(t-r) + \sum_{q=1}^{\infty} c_q^{(i|j,k)} G_k(t-q) + \varepsilon_{i|j,k}(t),$$

$$\tag{10}$$

where

$$\mathrm{var}(\varepsilon_{i|j,k}(t)) = \sigma^2_{\varepsilon_{i|j,k}}.$$

Similarly, we assume that the noise terms in these models are white noise, and the null hypothesis that G_j is conditional independence with G_i given G_k, i.e., $b_r^{(i|j,k)}$ in (10) are uniformly zero, could be tested with the following statistic:

$$F_{G_j \to G_i | G_k} = \frac{\sigma^2_{\varepsilon_{i|k}}}{\sigma^2_{\varepsilon_{i|k,j}}} - 1, \tag{11}$$

and its estimation with the observation data is similar to (6):

$$\hat{F}_{G_j \to G_i | G_k} \sim F(2R_{j|k}, T - 2P_{i|k} - 2Q_k - 2R_{j|k}), \tag{12}$$

where $P_{i|k}$, $R_{j|k}$ and Q_k are orders given by AIC for G_i, G_j and G_k, respectively. Given the significant level F_2, the F test can be performed to filter out the fake correlations in $M^{(1)}$ to get the initial dynamical network structure $M^{(2)}$ as follows:

$$m_{ij}^{(2)} = \begin{cases} 0, & \text{for } m_{ij}^{(1)} = 1 \text{ and } \hat{F}_{G_j \to G_i | G_k} \leq F_2; \\ m_{ij}^{(1)}, & \text{otherwise.} \end{cases} \tag{13}$$

In gene networks, the number of conditional genes is $n-2$ for each regulation pair indicated by $M^{(1)}$, i.e., we need to compute the P-G-C $n^2(n-2)$ times in the worst case. Besides, the combination effect of the conditional genes can not be exclude in this one by one version of P-G-C. Let the set of the conditional genes denoted by $\mathbf{G}(i,j) = \mathbf{G} \backslash \{G_i, G_j\}$, and the causality model can be given below:

$$G_i(t) = \sum_{p=1}^{\infty} a_p^{i|\mathbf{G}(i,j)} G_i(t-p) + \sum_{k \neq i,j} \sum_{q=1}^{\infty} c_q^{(k)} G_k(t-q) + \varepsilon_{i|\mathbf{G}(i,j)}(t), \quad (14)$$

$$G_i(t) = \sum_{p=1}^{\infty} a_p^{(i|\mathbf{X}(i,j),j)} G_i(t-p) + \sum_{r=1}^{\infty} b_r G_j(t-r)$$

$$+ \sum_{k \neq i,j} \sum_{q=1}^{\infty} c_q^{(k,j)} G_k(t-q) + \varepsilon_{i|\mathbf{G}(i,j),j}(t). \tag{15}$$

Generally, it is not practical to perform F test for this model, since the number of parameters in this model is most likely to be much bigger than the sample size of the data, i.e., the parameters can not be well estimated with limited sample size. Next, let's give a fast algorithm to compute the multivariate partial Granger causality in a projection pursuit manner.

For $G_j \rightarrow G_i$, let $G_{k_s} \in \mathbf{G}(i,j)(s = 1, \cdots, n-2)$, then define

$$H(t) = I - B^T(t)(B^T(t)B(t))^{(-1)}B(t), \tag{16}$$

where

$$B(t) = (G_{k_1}(t), G_{k_2}(t), \cdots, G_{k_{n-2}}(t)), t = 1, 2, \cdots, T.$$

Then, the effects of the conditional variables could be excluded from variable pairs of (G_i, G_j) by

$$G_i'(t) = H(t)G_i(t), \tag{17}$$
$$G_j'(t) = H(t)G_j(t), \tag{18}$$

and the partial-Granger causality model can be defined as follows:

$$G_i'(t) = \sum_{p=1}^{\infty} a_p^{(i)} G_i'(t-p) + \varepsilon_i(t), \tag{19}$$

$$G_i'(t) = \sum_{p=1}^{\infty} a_p^{(i|j)} G_i'(t-p) + \sum_{q=1}^{\infty} b_q G_j'(t-q) + \varepsilon_{i|j}(t). \tag{20}$$

Performing the F test with the statistic

$$\hat{F}_{G_j \rightarrow G_i|\mathbf{G}(i,j)} \sim F(2Q_j, T - 2P_i - 2Q_j), \tag{21}$$

the network structure $M^{(2')} = (m_{ij}^{2'})$ is established by

$$m_{ij}^{(2')} = \begin{cases} 0, & \text{for } m_{ij}^{(1)} = 1 \text{ and } \hat{F}_{G_j \rightarrow G_i|\mathbf{G}(i,j)} < F_{2'}; \\ m_{ij}^{(1)}, & \text{otherwise.} \end{cases} \tag{22}$$

Now the multivariate Granger causality needs to be computed for a pair of regulation genes only once. The steps of the main algorithm of this paper are described as follows:

MAIN ALGORITHM

Step 1. Data Centralizing: G_i is replaced by $G_i - \frac{1}{T}\sum_{t=1}^{T} G_i(t)$ for $i = 1, 2, \cdots, n$;

Step 2. G-C analysis: The network $M^{(1)}$ is given by calculating the statistics $\hat{F}_{G_j \to G_i}$ as (6);

Step 3. P-G-C analysis: Univariate P-G-C analysis is performed to build network structure $M^{(2)}$ according to (13);

Step 3'. P-G-C analysis: Multivariate P-G-C analysis is carried out to exclude false relationships in the network structure (22).

In this algorithm, you can choose to use Step 3 or Step 3' in your application: when the prior knowledge about the conditional gene of a pair of regulation genes is available, Step 3 needs to be run only for the candidate conditional gene; If no prior information is available or the combination effects is noticeable, Step 3' is preferred.

3   Experimental Results

3.1   Synthetic Dynamical Networks

We first test our algorithm of structure learning for the synthetic network of 5 genes consisting 5 regulations [25]. As presented on Fig. 1(a), the directed edges represent the regulations between genes which can also be formulated by a dynamical system:

$$x_1(t) = 0.95\sqrt{2}x_1(t-1) - 0.9025x_1(t-2) + \varepsilon_1(t),$$
$$x_2(t) = 0.5x_1(t-2) + \varepsilon_2(t),$$
$$x_3(t) = -0.4x_1(t-3) + \varepsilon_3(t),$$
$$x_4(t) = -0.5x_1(t-1) + 0.25\sqrt{2}x_4(t-1) + 0.25\sqrt{2}x_5(t-1) + \varepsilon_4(t),$$
$$x_5(t) = -0.25\sqrt{2}x_4(t-1) + 0.25\sqrt{2}x_5(t-1) + \varepsilon_5(t).$$

Without loss of generality, we may set $\varepsilon_1 \sim N(0, 0.6), \varepsilon_2 \sim N(0, 0.5), \varepsilon_3 \sim N(0, 0.3), \varepsilon_4 \sim N(0, 0.3), \varepsilon_5 \sim N(0, 0.6)$ and, for simplicity, we assume that $\forall i \neq j, Cov(\varepsilon_i, \varepsilon_j) = 0$.

As described in last section, our algorithm learns the network structure from data by many steps, and thereby the learning results given by Step 2 and Step 3' are portrayed in Fig. 1(b) and Fig. 1(c), respectively. Fig. 1(b) shows many fake relationships, such as $gene1 \to gene5$, an indirect casual interaction generated by $gene1 \to gene4$ together with $gene4 \to gene5$. From Fig. 1(c), we can see that except for the edge $gene4 \to gene5$ the edges of the original synthetic network have all be reconstructed from the simulation data, meanwhile no false edge have been learned from the data. The result presented in Fig. 1 is one of the results given by the proposed algorithm with threshold values 0.9 and 0.9 with no repeat, and the length of the time series, i.e., the sample size, generated by the synthetic network is 500. Clearly, the sample size and the threshold values

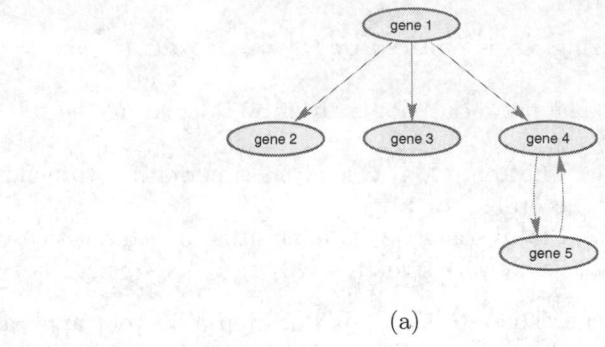

(a)

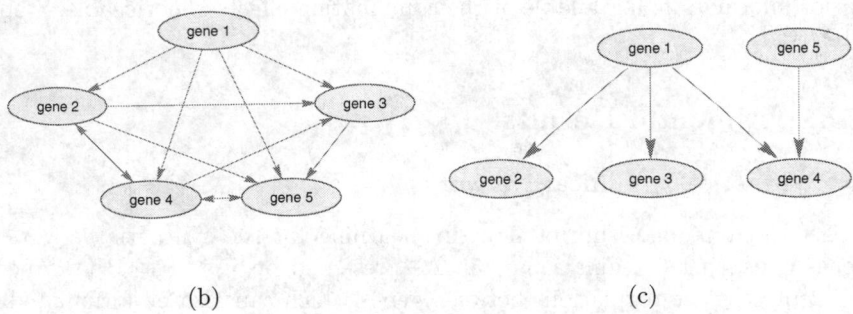

(b)   (c)

Fig. 1. The numerical experimental results of our algorithm on the synthetic network. (a) the original synthetic network; (b) the learned structure by G-C analysis; (c) the learned structure by Step 3′

of the F-tests, which are key factors for the performance of our algorithm, need further investigation.

Practically, the sample size are often limited, so the performance of our algorithm, as a function of sample size is also instructive. The sensitivity (SE) and the specificity (SP) have been employed to quantitatively compare the performances of our algorithm over different sample sizes. The SE of an approach to learn the gene regulations from expression data is a measure of the probability to detect the regulation by this approach but does not say whether any candidate regulation is truly a transcriptional regulation, and it can be computed by the number of true positives (TP) and the number of true regulations in gene network (NG) as follows

$$SE = \frac{TP}{NG} \times 100\%.$$ (23)

The SP measures the accuracy of a given approach, and can be estimated from the percentage of the predicted regulation of this approach that are present in the reference gene network by the following equation:

$$SP = \frac{TP}{NL} \times 100\%,$$ (24)

where NL stands for the number of edges in the learned network structure. For a given sample size, the data are iteratively generated from the synthetic network, and the network structure is learned from these data by our algorithm with some threshold values. A total of 50 repeated for each experimental setting are conducted and the averages of the sensitivity and the specificity are computed. The numerical experimental results are shown in Fig. 2 by our algorithm with different threshold values vary from 70 to 700. Clearly, no matter which threshold values are chosen, the SE and the SP totally increase with the sample size.

Note that the results in Fig. 2 also show that the proposed algorithms with different threshold values exhibit different performances. Therefore, the two threshold values, in which one for the F-test in G-C analysis and the other for the F-test in P-G-C analysis, deserve to be discussed in details. In this paper, we set the two threshold to be the same value. As the threshold values specify the confidence level of the F-test in the algorithm, the higher value of the threshold will get higher specificity and lower sensitivity. For a given sample size 90, Fig. 3(a) shows the box plots for the performance comparison between the proposed methods with different threshold values ranging from 0.5 to 0.98. The p-values of the one-way analysis of variance are both 0 for the recall percentage and the precision percentage given by the proposed algorithms with different threshold values, which means the effect of threshold values on the performance of our algorithm is statistically significant. Fig. 3(b) illustrates the average performance level variation against the threshold value and the threshold value somewhere between 0.7 and 0.8 reaches the balance between sensitivity and specificity.

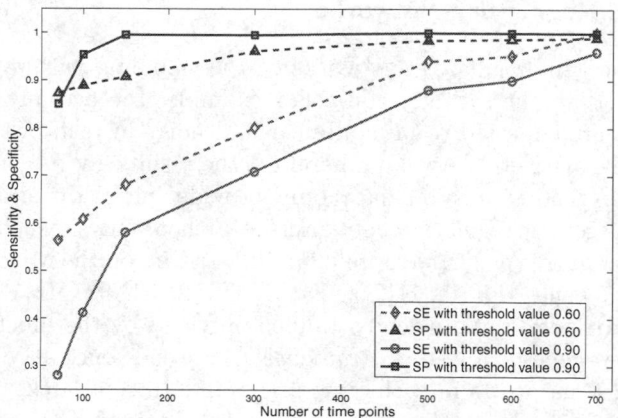

Fig. 2. Performance comparison of our methods on the data sets with different sample sizes

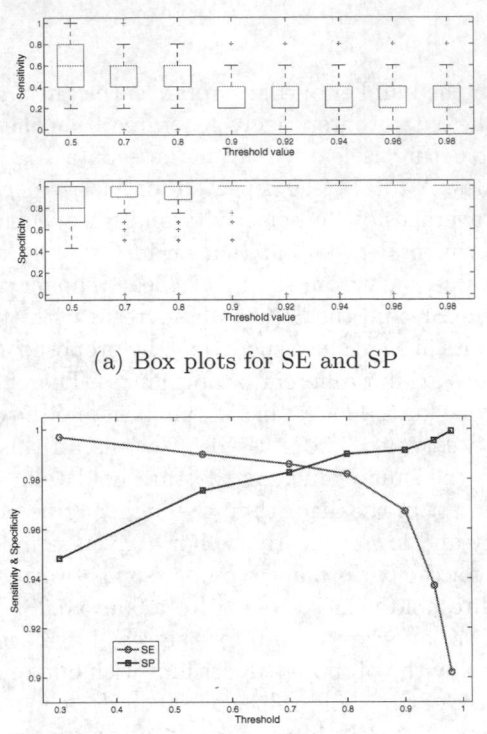

(a) Box plots for SE and SP

(b) Average performances against threshold values

Fig. 3. Performance comparison of our methods with different threshold values

3.2   Genetic Regulatory Networks

In practice, the gene regulatory networks or the signaling pathways are much more complex than the synthetic one. For example, the actually interactions between genes are expected to be nonlinear and noisy instead of the linear interactions in the synthetic network. Therefore, the results given by the proposed algorithm with gene expression data only provide initial dynamical network structure of the genetic regulatory networks for further analysis, such as Bayesian networks. Here, in this paper, we applied our algorithm on the Microarray time-series data of 7 genes (ETR1, ETR2, ERS1, ERS2, EIN4, CTR1 and MPK6) related to the detection of ethylene stimulus provided by the functional analysis of regulatory genes involved in *Arabidopsis* leaf senescence at Warwick HRI, which have 22 time points in each time series and 16 replicates. Table. 1 lists the gene catma_id [1]. Fig. 4 illustrates the results obtained by the proposed algorithm. Since the space is limited, the description details of these genes can be

---

[1] The aim of the Complete *Arabidopsis* Transcriptome MicroArray (CATMA) project (http://www.catma.org/) was the design and production of high quality Gene-specific Sequence Tags (GSTs) covering most *Arabidopsis* genes.

Table 1. Gene name and catma_id discussed in this paper

Gene	Name	Catma_id
ETR1	At1g66340	CATMA1a55610
ETR2	At3g23150	CATMA3a23140
ERS1	At2g40940	CATMA2a39280
ERS2	At1g04310	CATMA1a03150
EIN4	At3g04580	CATMA3a03560
CTR1	At5g03730	CATMA5a02913
MPK6	At2g43790	CATMA2a42185

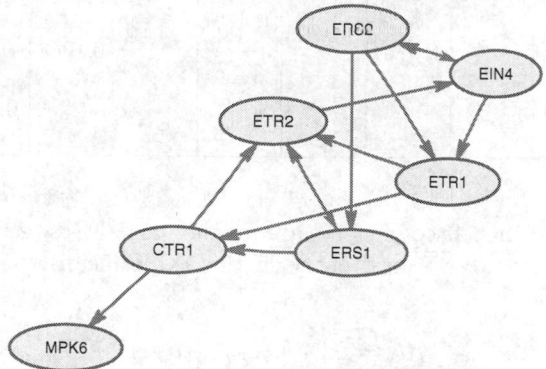

Fig. 4. Learned network structure by the proposed algorithm

found on the web site of the Complete *Arabidopsis* Transcriptome MicroArray (CATMA) project (http://www.catma.org/).

During the past decade, the reference plant *Arabidopsis* have been well studied and the ethylene is a gaseous plant hormone involved in many life process [26]. We've already known that the ethylene is perceived by a family of five membrane-associated receptor (ETR1, ETR2, ERS1, ERS2 and EIN4) in *Arabidopsis*, and the ethylene binding leads to functional inactivation of the receptors which are negative regulators of ethylene responses. In the presence of ethylene, CTR1 which is another negative regulator of the pathway loses its ability to repress the downstream genes. That's the early events of ethylene perception and signaling. However, the detailed network structure of these genes needs to be further investigated, and a MAPK pathway involving MPK6 in *Arabidopsis* has recently been proposed in operating downstream of CTR1 as a positive regulator in the pathway. From the result given by the proposed algorithm, we can see that except for EIN4 and ERS2 the other ethylene receptors are all have directed relationship with CTR1. The existence of the directed edge from CTR1 to MPK6 supports that the MAPK cascade is involved in this pathway.

The coefficients in the partial Granger casuality model have been listed in Table 2. We set the maximum order for delay to be 5 and the AIC is used to select the optimal order. On the diagonal, the coefficients are for the autoregulatory

Table 2. Coefficients given by partial Granger Casuality model

	ETR1	ETR2	ERS1	ERS2	EIN4	CTR1	MPK6
ETR1	0.2479	-0.0638				0.4304	
		-1.0446				-1.1091	
ETR2		0.15	0.7018		-0.0798		
			-0.4192		-0.1611		
ERS1		-0.7355	0.1045			-0.0224	
		1.1327				0.6591	
ERS2	-0.1398		0.1535	0.0239	0.1793		
	0.2747		0.2808				
EIN4	0.1822			-0.087	-0.0056		
	-0.2256			-0.4767			
CTR1		-0.4751				0.1063	-0.251
		0.4915					0.2538
MPK6							0.243
							-0.477

equation (19); and the left coefficients are for the partial Granger casuality equation (20). For example, the first entry in this table means the autoregulatory equation for ETR1 is

$$G'_{ETR1}(t) = 0.2479 G'_{ETR1}(t-1) + \varepsilon. \tag{25}$$

We may say it is the positive autoregulation that works behind the expression behavior of ETR1. However, for MPK6, there is a second order delay, which makes some confusion:

$$G'_{MPK6}(t) = 0.243 G'_{MPK6}(t-1) - 0.477 G'_{MPK6}(t-2) + \varepsilon. \tag{26}$$

This can be explained biologically that MPK6 not only has a directly positive autoregulation but also has some negative regulation which might be works in some negative loop instead of the directly regulation. Therefore, there will be a delay in this negative effect, which means that some genes other than the 7 genes we studied here also have great effects on the regulation of MPK6, i.e., more genes need to be included in this network for further understanding.

In Table 2, the second entry is the coefficients of the partial Granger model which describes how ETR1 regulates ETR2 excluding the influence of the other 5 genes. And it can be written as follows:

$$G'_{ETR2}(t) = -0.0638 G'_{ETR2}(t-1) - 1.0446 G'_{ETR1}(t-1) + \varepsilon. \tag{27}$$

Similarly, the above equation may help us to understand the negative regulation from ETR1 to ETR2, which indicates that the receptors are not function independently. As listed in Table 2, the autoregulation for ETR2 is a positive one, but in the partial Granger model the first coefficient in (27) is -0.0638, which represents a negative autoregulation. It can be explained as follows: the autoregulation coefficients on the diagonal of the result table describe regulations that

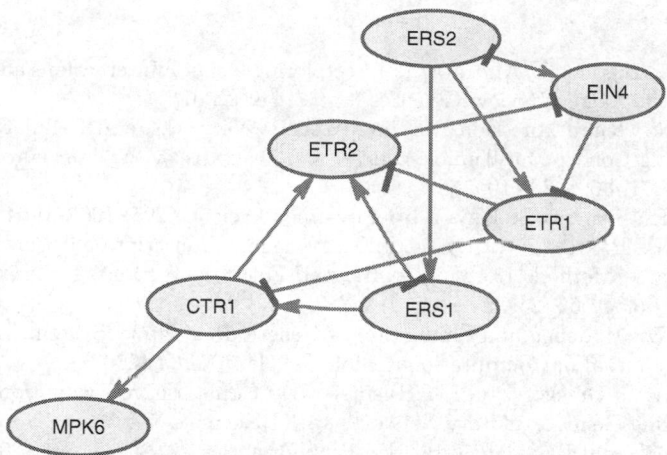

Fig. 5. Reconstructed gene network for the ethylene response pathway in *Arabidopsis*

include all genes influences. Instead of the effects contributed to by the whole gene set, the partial Granger causality model excludes the regulation effect on ETR2 given by the genes other than ETR1. That's to say, the autoregulation can not be determined at the current stage. Therefore, the combination effects must be considered during the reconstruction of genetic regulatory networks, and partial Granger causality model provides us a promise tool to address this problem. At last, the result shown on Fig. 4 can be redrawn with more details about the negative or positive regulations between the genes, which is shown in Fig. 5.

4 Conclusion

Reconstructing gene regulatory networks from Microarray time-series data has been a big challenge in systems biology due to the complex regulatory relationships among genes and the limit of available data. Many algorithms have been developed to deal with this reverse engineering problem, and most of them need to start from the initial network structures which affect the sensitivity and specificity of these algorithms. In this paper, an algorithm for learning structure from expression time-series data has been proposed by applying the partial Granger causality model. Instead of the first order Markov assumption, the proposed algorithm considers the multiple delays of the regulation effects by optimally determining the orders in the model with AIC. The present learning algorithms for Bayesian networks are embarrassed by the exponential computation complexity. However, in our algorithm, the combination effects of the candidate conditional genes need to be computed only once for each pair of regulation genes in the projection pursuit manner. Despite the linear assumptions inherent in the Granger causality models, the results reported above indicate that it is a promise tool for reconstructing gene networks. Therefore, one of the further investigation direction of this algorithm is its nonlinear extension.

References

1. Boone, C., Bussey, H., Andrews, B.J.: Exploring Genetic Interactions and Networks with Yeast. Nature Review Genetics 8, 437–449 (2007)
2. Huang, S.: Gene Expression Profiling, Genetic Networks, and Cellular States: An Integrating Concept for Tumori-genesis and Drug discovery. Journal of Molecular Medicine 77, 469–480 (1999)
3. Kitano, H.: Systems Biology: a brief overview. Science 295, 1662–1664 (2002)
4. Smolen, P., Baxter, D.A., Byrne, J.H.: Modeling Transcriptional Control in Gene Networks — Methods Recent Results, and Future Directions. Bulletin of Mathematical Biology 62, 247–292 (2000)
5. Jong, H.D.: Modeling and Simulation of Genetic Regulatory System: A Literature Review. Journal of Computational Biology 9(1), 67–163 (2002)
6. Werhli, A.V., Grzegorczyk, M., Husmeier, D.: Comparative Evaluation of Reverse Engineering Gene Regulatory Networks with Relevance Networks, Graphical Gaussian Models and Bayesian Networks. Bioinformatics 22, 2523–2531 (2006)
7. Schlitt, T., Brazma, A.: Current Approaches to Gene Regulatory Network Modelling. BMC Bioinformatics, 8(suppl. 6), S9 (2007)
8. Kauffman, S., Peterson, C., Samuelsson, B., Troein, C.: Random Boolean Network Models and the Yeast Transcriptional Network. Proc. Natl. Acad. Sci. USA 100, 14796–14799 (2003)
9. Chen, K.C., Wang, T.Y., Tseng, H.H., Huang, C.Y.F., Kao, C.Y.: A Stochastic Differential Equation Model for Quantifying Transcriptional Regulatory Network In Saccharomyces Cerevisiae. Bioinformatics 21(12), 2883–2890 (2005)
10. Friedman, N., Nachman, I., Pe'er, D.: Learning Bayesian Network Structure from Massive Datasets: the Sparse Candidate Algorithm. In: Laskey, K.B., Prade, H. (eds.) UAI 1999. Proceedings of the Fifteenth Conference on Uncertainty in Artificial Intelligence, pp. 206–215. Morgan Kaufmann, Stockholm (1999)
11. Friedman, N., Linial, M., Nachman, I., Pe'er, D.: Using Bayesian Networks to Analyze Expression Data. In: Proceeding of the Fourth Annual International Conference on Computational Molecular Biology (RECOMB), Tokyo, Japan, pp. 127–135 (2000)
12. Pe'er, D., Regev, A., Elidan, G., Friedman, N.: Inferring Subnetworks from Perturbed Expression Profiles. Bioinformatics 17 (suppl. 1), S215–S224 (2001)
13. Friedman, N.: Inferring Cellular Networks Using Probabilistic Graphical Models. Science 303(5659), 799–807 (2004)
14. Sachs, K., Perez, O., Pe'er, D.: Casual Protein-Signaling Networks Derived from Multi-Parameter Single-Cell Data. Science 308(5721), 523–529 (2005)
15. Gevaert, O., Smet, F.D., Timmerman, D., Moreau, Y., Moor, B.D.: Predicting the Prognosis of Breast Cancer by Integrating Clinical and Microarray Data with Bayesian Networks. Bioinformatics 22, e184–e190 (2006)
16. Tong, A., Evangelista, M., Parsons, A.B., et al.: Systematic Genetic Analysis with Ordered Arrays of Yeast Deletion Mutants. Science 294, 2364–2368 (2001)
17. Ernst, J., Nau, G.J., Bar-Joseph, Z.: Clustering Short Time Series Gene Expression Data. Bioinformatics 21(suppl. 1), I159–I168 (2005)
18. Murphy, K., Mian, S.: Modelling Gene Expression Data Using Dynamic Bayesian Networks. Technical report, Computer Science Division, University of California, Berkeley, CA (1999)
19. Perrin, B.E., Ralaivola, L., Mazurie, A., et al.: Gene Networks Inference Using Dynamic Baysian Networks. Bioinformatics 19(suppl. 2), ii138–ii148 (2003)

20. Kim, S., Imoto, S., Miyano, S.: Dynamic Bayesian Network and Nonparametric Regression for Nonlinear Modeling of Gene Networks from Time Series Gene Expression Data. Biosystems 75, 57–65 (2004)
21. Zou, M., Conzen, S.D.: A New Dynamic Bayesian Network Approach for Identifying Gene Regulatory Networks from Time Course Microarray Data. Bioinformatics 21(1), 71–79 (2005)
22. Dojer, N., Gambin, A., Mizera, A., Wilczyski, B., Tiuryn, J.: Applying Dynamic Bayesian Networks to Perturbed Gene Expression Data. BMC Bioinformatics 7, 249 (2006)
23. Li, X., Rao, S., Jiang, W., Li, C., Yun, X.: Discovery of Time-Delayed Gene Regulatory Networks Based on Temporal Gene Expression Profiling. BMC Bioinformatics 7, 26 (2006)
24. Shi, Y., Mitchell, T., Bar-Joseph, Z.: Inferring Pairwise Regulatory Relationships from Multiple Time Series Datasets. Bioinformatics 23(6), 755–763 (2007)
25. Schelter, B., Winterhalder, M., Timmer, J.: Handbook of Time Series Analysis: Recent Theoretical Developments. Wiley-VCH, Weinheim (2006)
26. Li, H., Guo, H.: Molecular basis of the ethylene signaling and response pathway in Arabidopsis. Journal of Plant Growth Regulation 26(2), 106–117 (2007)

Recognition of Important Subgraphs in Collaboration Networks

Chun-Hua Fu, Yue-Ping Zhou, Xiu-Lian Xu, Hui Chang, Ai-Xia Feng,
Jian-Jun Shi, and Da-Ren He*

College of Physics Science and Technology, Yangzhou University,
Yangzhou, 225002, China
darendo10@yahoo.com.cn

Abstract. We propose a method for recognition of most important subgraphs in collaboration networks. The networks can be described by bipartite graphs, where basic elements, named actors, are taking part in events, organizations or activities, named acts. It is suggested that the subgraphs can be described by so-called k-cliques, which are defined as complete subgraphs of two or more vertices. The k-clique act degree is defined as the number of acts, in which a k-clique takes part. The k-clique act degree distribution in collaboration networks is investigated via a simplified model. The analytic treatment on the model leads to a conclusion that the distribution obeys a so-called shifted power law $P(q) \propto (q + \alpha)^{-\gamma}$ where α and γ are constants. This is a very uneven distribution. Numerical simulations have been performed, which show that the model analytic conclusion remains qualitatively correct when the model is revised to approach the real world evolution situation. Some empirical investigation results are presented, which support the model conclusion. We consider the cliques, which take part in the largest number of acts, as the most important ones. With this understanding we are able to distinguish some most important cliques in the real world networks.

Keywords: subgraph, collaboration network, bipartite graph, clique, shifted power law.

1 Introduction

Complex networks have attracted attentions in recent years [1,2]. Among the studies, the investigations on social networks become increasingly more attractive. An obvious feature of social networks is the community structure of the basic elements ("vertices" or "actors"). Inside communities the connections ("edges" or "ties" or "arcs") between actors are much denser than the connections between the communities [3]. This indicates that the active or self-determining actors get together and form groups. In addition to social networks, Zhang, Chang, Su, Fu and cooperators suggested considering the networks, in

---

* Corresponding author.

J. Zhou (Ed.): Complex 2009, Part I, LNICST 4, pp. 210–219, 2009.
© ICST Institute for Computer Sciences, Social Informatics and Telecommunications Engineering 2009

which the basic elements were not active, however, they might be influenced by some "manipulators" and showed indirect activity [4-7]. We may call such networks "quasi-social". Examples of quasi-social networks may be mentioned as transportation networks (with stations as the quasi-actors), language networks (with languages as the quasi-actors), and other man-made or man-collected systems.

One of the most interesting questions in social network studies is whether communities can be divided into some basic subgraphs. The basic subgraphs are composed of several (2 to k) actors, which are always together in performing a network function. For example, if a movie actor and an actress always perform sweet heart couple in many famous movies, the audience thinks them as a fixed unit. They will strongly unsatisfy if one of them suddenly performs couple with a different partner. In social networks, the pair of actors can be defined as a "dyad", which is "an unordered pair of actors and the arcs that exist between the two actors in the pair" (Ref. [3], page 510). In this paper we shall only discuss mutual dyads, in which both of the actors has a directed tie to the other (Ref. [3], page 511), and use an ordinary un-directed tie to express the two directed ties. A "triad" is similarly defined as a triple of actors with the ties between them (Ref. [3], page 559). We also only consider mutual triads where all the three pairs of actors are connected by un-directed ties. In social networks "a clique is defined as a maximal complete subgraph of three or more vertices" (Ref. [3], page 254). If we revise the definition to "a complete subgraph of two or more vertices", we can address the mutual dyads as "2-cliques" and mutual triads as "3-cliques". We can similarly define "k-cliques".

Interesting research achievements have been published about social collaboration networks, including Hollywood actor collaboration network and scientist collaboration network [8-11]. The collaboration networks can be described by bipartite graphs. In these graphs the vertices can be divided into two sets [3]. One type of the vertices is "actors" taking part in some activities, organizations or events. The other type of vertices is the activity, organization or event named "acts". When we note only the collaboration relationship between actors, we may project the bi-graph onto the actor-vertices, and obtain a unipartite graph. In the projected graph each act is represented by a complete subgraph where edge is connected between every pair of vertices. Each act complete subgraph can be divided into some smaller complete subgraphs, or cliques. Different act complete subgraphs may share some cliques. A clique is more important if it takes part in more acts. The number of vertices in an act complete subgraph is addressed as "act size" and denoted by T. The number of acts, in which an actor takes part, is addressed as "act degree" of the actor vertex and denoted by h [4-7]. Zhang, Chang, Su and Fu and cooperators investigated some quasi-social collaboration networks [4-7].

In this article we shall concentrate on the recognition of most important cliques in social and quasi-social collaboration networks. The most important cliques must take part in the largest number of acts, thus they must show the largest k-clique act degree (defined as the number of acts, in which the k-clique takes part). The k-clique act degree distribution then becomes the most important property in the current study. The article will be organized as follows. In section 2 we shall develop a model describing the evolution of social and quasi-social collaboration networks in a very ideal and simplified situation. By analysis of the model we can show the general function form of the k-clique act degree distribution in the simplified case. For the situations nearer to practical, we shall show, by some numerical investigation, that the general function form probably is qualitatively correct. In section 3 we shall present empirical investigation results in three real world quasi-social collaboration networks as proofs, which show good agreement with the model conclusion. We also shall recognize most important 2-cliques and 3-cliques in these real world networks. The empirical investigation results on eight different real world collaboration networks shall also be mentioned. In the last section the text will be summarized and some discussions will be presented.

2   The Model

2.1   The Definition of k-Clique Act Degree

Now we introduce accurate definition of k-clique act degree. In a bipartite graph the act degree of a 2-clique (a mutual dyad) is defined as $D_{i,j} = \sum_m a_{i,m} a_{j,m}$, where i, j denotes two different actors, and m denotes an act. $a_{i,m}$ is the element of the bipartite graph adjacency matrix. It is defined as $a_{i,m} = 1$ if actor i takes part in act m (they are connected by a bipartite graph edge); and $a_{i,m} = 0$ otherwise. Similarly, the act degree of a 3-clique (a mutual triad) is defined as $Tr_{i,j,k} = \sum_m a_{im} a_{jm} a_{km}$, where i, j, k denotes three different actors, m denotes an act. One can then write the general definition of k-clique act degree as $q_{i_1,i_2,\cdots,i_k} = \sum_m a_{i_1,m} a_{i_2,m} \cdots a_{i_k,m}$. A k-clique act degree distribution $P(q)$ is defined as the probability of a k-clique with act degree q, stands for the number of k-cliques with act degree q in the network [12].

2.2   The Simplified Model

The ideal and simplified situation we consider firstly is that the act size (the number of vertices in an act), T, is a constant, also, it takes a value, $k \times n$ where n is an integer, so that each act can just include n "legal k-cliques" where k is a constant. All the vertex actors unite together and form legal k-cliques. There are just $k \times N$ vertex actors in the network therefore every actor is in a legal k-clique. There is no single vertex remaining in the network. During the evolution of the network, the legal k-cliques will not disband, and each vertex actor can only join one of them. The legal k-cliques perform as a fixed unit in the network evolution. At each time step, a new legal k-clique joins network and select $n - 1$ old legal k-cliques by certain rule to form a new act. Of course, when a new act is formed, because all the edges between every pair of vertex actors must be connected, certainly some "illegal k-cliques", which share actor vertices with legal k-cliques, should appear. However, for the simplified model considered in this subsection we shall only count the act degree of legal k-cliques.

Firstly we consider the rule of selecting $n-1$ old legal k-cliques with a probability proportional to the k-clique act degree q of each old legal k-clique. This can be addressed as a "k-clique act-degree linear preference rule". We can, similar to what BA did [1,9], get a conclusion that the legal k-clique act degree distribution $P(q)$ takes exact power functions.

Secondly, we consider the rule of selecting $n-1$ old legal k-cliques randomly. We can write down, following the references [1,4,5,9], the evolution equation of the legal k-clique act degree and analytically obtain the conclusion that the legal k-clique act degree distribution $P(q)$ takes exact exponential functions.

To interpolate between the above two extreme cases [4,5,13,14], we consider the rule of selecting the $n-1$ old legal k-clique randomly with a probability p, and using legal k-clique act degree linear preference rule with probability $1-p$. Similarly, we have (when t is large)

$$\frac{\partial q_i}{\partial t} = p\frac{n-1}{t} + (1-p)\frac{(n-1)q_i}{nt}. \tag{1}$$

This equation can be written as

$$\frac{\partial q_i}{\partial \ln(t)} = p\frac{T-k}{k} + (1-p)\frac{T-k}{T}q_i. \tag{2}$$

This can be solved to give

$$q_i = C_i t^{(T-k)(1-p)/T} - \frac{Tp}{k(1-p)}, \tag{3}$$

where C_i is the integration constant, which can be determined using the condition $q_i(t=t_i)=1$. Let $\alpha = Tp/[k(1-p)]$ and $\eta = T/[(T-k)(1-p)]$. Now we have

$$P(q_i < q) = P(t_i > t(\frac{q+\alpha}{1+\alpha}))^{-\eta}. \tag{4}$$

The legal k-clique act degree distribution is then given by

$$P(q) = \frac{dP(q_i < q)}{dq} = \frac{\eta}{1+\alpha}(\frac{q+\alpha}{1+\alpha})^{-\eta-1}. \tag{5}$$

The legal k-clique act degree distribution function is called "Shifted Power Law" (SPL) [5]. Now let's check the limiting case. For $p=0$, $\alpha=0$ and $\eta = T/(T-k)$; it is easy to see that

$$P(q) \propto q^{-\frac{kT-k}{T-k}}. \tag{6}$$

For $p \to 1$, $\alpha \to \infty$, and $\eta \to k\alpha/(T-k)$,

$$P(q) \propto e^{k(1-q)/(T-k)}. \tag{7}$$

So the distribution we obtained for $0 < p < 1$ interpolates between the power-law distribution and the exponential distribution. When the parameter p continuously changes from 0 to 1, $P(q)$ continuously varies from a power-law distribution to an exponential distribution.

2.3   The Models Approaching the Real World Network Evolution

In the real world network evolution one usually cannot distinguish the legal k-cliques and illegal k-cliques, therefore it is unreasonable to ignore the illegal k-clique act degree. Also, the act size (the number of vertices in an act), T, in general cannot just take the value $k \times n$ therefore the acts may include some "isolated vertices". In this subsection we shall consider both the situations. However, the act size, T, will be considered still as a constant since our previous investigation results showed that this simplification could be accepted for many real world network studies [4,5].

It is difficult to solve the more complex model analytically, we have to discuss it numerically. The results show that the conclusions are qualitatively the same. Several figures showing the numerical investigation results for $k = 2$ have been already published in Ref. [15]. Some numerical investigations have been done for $k = 3$, which show similar results. We can therefore expect a model conclusion that in collaboration networks 2-clique and 3-clique act degree distribution obeys SPL functions. Considering the simplified model prediction we believe that at least in a large potion of real world collaboration networks all the k-clique act degree distributions obey SPL functions. SPL means a very uneven k-clique act degree distribution. Only a few k-cliques take part in a lot of acts, they can be viewed as fixed units and the most important subgraphs in the network. Most of the k-cliques take part only in a few acts. The vertices in the cliques join different k-cliques with different partners when perform cooperation function in different acts. They cannot be thought of as fixed units, neither as important subgraphs.

3   Empirical Investigations on Some Real World Collaboration Networks

The above mentioned model analysis and simulations only discuss one fixed k value although the discussion is rather general for all k. In the real world collaboration networks, however, many k-cliques (with different k) coexist. If we could get empirical investigation results, which show that the k-clique act degree distributions for several values of k obey SPL functions, we can believe that the model conclusion is correct at least for a large portion of real collaboration networks. In this section we present such empirical investigations.

3.1   Empirical Investigation on World Language Distribution Network

World language distribution has been of research interests [16]. We propose a network description on world language distribution. The actor vertices of the network are defined as languages. The acts are defined as countries or regions where the languages are spoken. Two vertices are connected by an edge if they are coexisting in a common region. The data were downloaded from the fifteenth

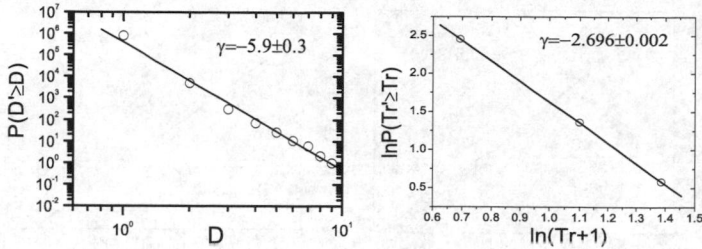

Fig. 1. Left: The cumulative dyad (2-clique) act degree distribution of the language network; Right: The cumulative triad (3-clique) act degree distribution of the language network.

edition of Ethnologue [17], published in 2005, which lists 6142 languages and 228 countries plus the 8 regions.

Figure 1 shows that the cumulative dyad (2-clique) act degree distribution of the language network, which can be described with a power law, $P(D' \geq D) \propto D^{-5.9}$. With an approximation that the data number is large and quasi-continuous, one can easily prove that the original dyad act degree distribution obeys $P(D) \propto D^{-4.9}$. This is an extreme case of a SPL distribution. Figure 1 also shows the cumulative triad (3-clique) act degree distribution of the language network, which can be described with a SPL function, $P(Tr' \geq Tr) \propto (Tr + 1)^{-2.696}$.

The most important 2-cliques can be listed as: 1) Izora and Mangas (appear together in 9 regions); 2) Mbulungish and Pular (appear together in 8 regions); 3) Mann and Mbulungish (appear together in 7 regions); 4) Mbulungish and Bainouk-Gunyu Adobe (appear together in 7 regions); 5) Mann and Pular (appear together in 7 regions).

The most important 3-cliques can be listed as: Biali, Gude and Majera; Biali, Gyele and Wawa; Giziga(South), Majera and Mofu(North); Giziga(South), Majera and Wawa; and Giziga(South), Mofu(North) and Wawa. All these 3-cliques take part in 3 acts.

3.2 Empirical Investigation on Mixed Drink Network

The acts of mixed drink network are defined as mixed drinks. The actor vertices are defined as drink ingredients. Two vertices are connected by an edge if they are coexisting in a common act. The data were downloaded from the website http://www.drinknation.com, which lists 1501 drink ingredients and 7804 mixed drinks.

Figure 2 shows the cumulative dyad (2-clique) act degree distribution of the mixed drink network, which can be described with a SPL, $P(D' \geq D) \propto (D + 20)^{-2.81}$. Figure 2 also shows the cumulative triad (3-clique) act degree distribution of the mixed drink network, which can be described with a SPL, $P(Tr' \geq Tr) \propto (Tr + 5)^{-8.4}$.

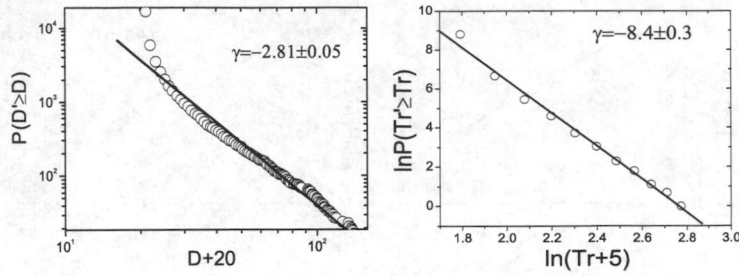

Fig. 2. Left: The cumulative dyad (2-clique) act degree distribution of the mixed drink network; Right: The cumulative triad (3-clique) act degree distribution of the mixed drink network

The most important 2-cliques can be listed as: 1) Vodka and OrangeJuice (appear together in 273 drinks); 2) OrangeJuice and PineappleJuice (appear together in 201 drinks); 3) OrangeJuice and Grenadine (appear together in 195 drinks); 4) Vodka and TripleSec (appear together in 191 drinks); 5) Vodka and GranberryJuice (appear together in 177 drinks).

The most important 3-cliques can be listed as: 1) PineappleJuice, Rumdark and Rum; 2) OrangeJuice, PineappleJuice and Grenadine; 3) OrangeJuice, PineappleJuice and Grenadine; 4) OrangeJuice, Amaretto and SouthernComfort; 5) IrishCreamBaileys, Kahlua and Vodka; 6) Amaretto, SouthernComfort and SloeGin.

3.3   Empirical Investigation on Information Technology Product Network in China

The data are downloaded from the websites: http://www.pcpop.com/ and http://www.it168.com/. In the IT product network the companies are defined as the actor vertices, and the IT products are defined as the acts. An edge between two actor vertices represents that the two companies produce at least one common IT product and thus compete in the market. This network includes 265 IT products and 2121 IT companies.

Figure 3 shows the cumulative dyad (2-clique) act degree distribution of the IT product network, which can be described with a power law, $P(D' \geq D) \propto D^{-3.7}$. Figure 3 also shows the cumulative triad (3-clique) act degree distribution of the mixed drink network, which can be described with a SPL, $P(Tr' \geq Tr) \propto (Tr + 3)^{-10.2}$.

The most important 2-cliques can be listed as: 1) BIGBUFFALO (HEDY company) and MASTER (produce 16 common products); 2) Kangguan ViewTech Center and Unika (produce 15 common products); 3) SAMSUNG and NEWGRAND (produce 14 common products); 4) SAMSUNG and Adobe (produce 13 common products); 5) MASTER and SAMSUNG (produce 13 common products).

The most important 3-cliques can be listed as: 1) BIGBUFFALO, MASTER and TESSM (produce 10 common products); 2) BIGBUFFALO, MASTER and

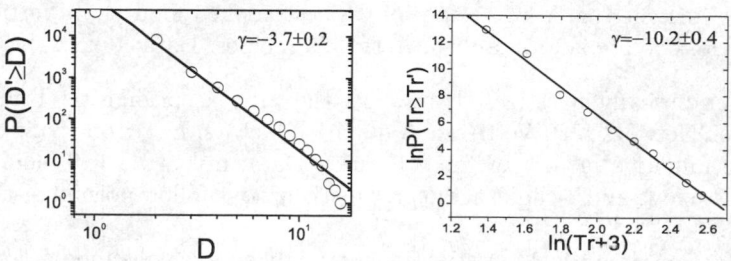

Fig. 3. Left: The cumulative dyad (2-clique) act degree distribution of the IT product network; Right: The cumulative triad (3-clique) act degree distribution of the IT product network

SAMSUNG (produce 10 common products); 3) BIGBUFFALO, MASTER and NEWGRAND (produce 9 common products); 4) Kangguan ViewTech Center, Unika and Oneforall (produce 9 common products); 5) MASTER, SAMSUNG and NEWGRAND (produce 9 common products).

3.4   Some Other Empirical Investigated Collaboration Networks

We have investigated another 9 collaboration networks. 8 of them also show SPL 2-clique and 3-clique act degree distributions:

(1) The undergraduate elective network of Yangzhou University (YZU) [7]: Vertices: 121 general support courses, Edge: two vertices are belong to a common scientific subject, Acts: 78 scientific subjects, Data: provided by University Academic Affairs office.

(2) The Chinese professional training organization network: Vertices: 2674 training courses, Edge: two vertices are provided by a common training organization, Acts: 398 training organizations, Data: http://www.ot51.com/, http://www. 00100.cc/, http://philosophy.cass.cn/, and http://www.people.com.cn/ et al..

(3) China mainland movie network [18]: Vertices: 3085 movies, Edge: a common movie actor performs in these two movies, Acts: 920 movie actors, Data: http://soft6.com/, http://www.mtime.com/movie/.

(4) 2004 Olympic game network: Vertices: 4496 athletes, Edge: two athletes take part in a common sports item, Acts: 229 sports items,
Data: http://2004.sina.com.cn/results/summary/.

(5) Traditional Chinese herb prescription formulation network [4,5]: Vertices: 681 herbs, Edge: two herbs included in a prescription, Acts: 1536 prescriptions, Data: Refs. [19,20].

(6) Beijing bus route network [4,5]: Vertices: 4199 bus stations, Edge: two stations in a common route, Acts: 1572 bus routes, Data: http://www.bjbus.com/.

(7) The Travel Route Network of China [4,5]: Vertices: 171 scenic spots, Edge: two scenic spots in a travel route, Acts: 240 routes, Data: http://www.cnta.com/ 8-ssls/lyqd.asp.

(8) Network of Huai-Yang recipes of Chinese cooked food [4,5]: Vertices: 242 foods, Edge: two foods form a dish, Acts: 329 recipes, Data: Ref. [21].

In our empirically investigated networks, the only exception is the Fruit Nutritive Factor Network [22]: Vertices: 45 nutritive factors, Edge: one fruit contains these two nutritive factors, Acts: 151 fruits, Data: http://www.fumuqin.com/. Both the 2-clique and 3-clique act degree distributions follow normal distribution functions.

Therefore our empirical investigations strongly support the model prediction.

4 Conclusion and Discussion

We show, with a very simplified network evolution model, that k-clique act degree distribution probably always obey SPL functions in collaboration networks. Some empirical proofs are presented, which have been obtained in some real world systems and show SPL 2-clique and 3-clique act degree distributions. This indicates that small complete subgraphs do widely exist in collaboration networks, which may include some non-social networks. However, only a few of them take part in many collaboration acts so that they can be considered as important basic units of the networks. It is worth noting that some other real world collaboration networks show k-clique act degree distributions, which do not obey SPL distribution. The evolution mechanism in these collaboration networks must be basically different and thus deserves a further investigation.

Acknowledgement

The research is supported by the Chinese National Natural Science Foundation under the grant numbers 10635040 and 70671089.

References

1. Albert, R., Barabasi, A.-L.: Statistical mechanics of complex network. Rev. Mod. Phys. 74, 47–97 (2002)
2. Newman, M.E.J.: The structure and function of complex networks. SIAM Review 45, 167–225 (2003)
3. Wasserman, S., Faust, K.: Social Network Analysis: Methods and Applications. Cambridge Univ. Press, Cambridge (1994)
4. Zhang, P.P., Chen, K., He, Y., et al.: Model and empirical study on some collaboration networks. Physica A 360, 599–616 (2006)
5. Chang, H., Su, B.-B., Zhou, Y.-P., He, D.-R.: Assortativity and act degree distribution of some collaboration networks. Physica A 383, 687–702 (2007)
6. Su, B.-B., Chang, H., Chen, Y.-Z., He, D.-R.: A game theory model of urban public traffic networks. Physica A 379, 291–297 (2007)
7. Fu, C.-H., Zhang, Z.-P., Chang, H., et al.: A kind of collaboration-competition networks. Physica A 387, 1411–1420 (2008)

8. Watts, D.J., Strogatz, S.H.: Collective dynamics of 'small-world' networks. Nature 393, 440–442 (1998)
9. Barabasi, A.-L., Albert, R.: Emergence of scaling in random networks. Science 286, 509–512 (1999)
10. Newman, M.E.J.: Scientific collaboration networks. I. Network construction and fundamental results. Phys. Rev. E 64, 016131 (2001); Newman, M.E.J.: Scientific collaboration networks. II. Shortest paths. weighted networks and Centrality. Phys. Rev. E 64, 016132 (2001)
11. Ramasco, J.J., Dorogovtsev, S.N., Pastot-Satorras, R.: Self organization of collaboration networks. Phys. Rev. E 70, 036106 (2004)
12. Krapivsky, P.L., Redner, S.: Rate equation approach for growing networks. In: P-Satorras, R., Rubi, M., D-Guilera, A. (eds.) Statistical Mechanics of Complex networks, p. 4. Springer, Heidelberg (2003)
13. Liu, Z., Lai, Y.-C., et al.: Connectivity distribution and attack tolerance of general networks with both preferential and random attachments. Phys. Lett. A 303, 337–344 (2002)
14. Li, X., Chen, G.: A local world evolving network model. Physica A 328, 274–286 (2003)
15. Chang, H., He, D.-R.: General collaboration networks. In: Guo, L., Xu, X.-M. (eds.) Complex Networks, pp. 166–186. Shanghai Scientific and Educational press, Shanghai (2006) (in Chinese)
16. Gomes, M.A.F., Vasconcelos, G.L., Tsang, I.J., Tsang, I.R.: Scaling relations for diversity of languages. Physica A 271, 489–495 (1999)
17. http://www.ethnologue.com/
18. Liu, A.-F., Fu, C.-H., Chang, H., He, D.-R.: An empirical statistical investigation on Chinese mainland movie network. Complex Sys. and Complexity Sci. 4, 10–16 (2007) (in Chinese)
19. Zhu, Y.-X.: Chinese Herb Prescription Manual, 2nd edn. Jindun Publishing House, Beijing (1996) (in Chinese)
20. Liu, D.-L., et al.: Most Frequently Used Chinese Medication Handbook. People's Military Medication Publisher, Beijing (1996) (in Chinese)
21. Compiling group of Beijing Nationality Restaurant.: Huai-Yang Bill of Fare. Chinese Travel Publisher, Beijing (1993) (in Chinese)
22. Qu, Y.Q., Jiang, Y.M., He, D.-R.: Fruit Nutritive Factor Network. Jrl. Syst. Sci. and Complexity 22, 150–158 (2009)

Queueing Transition of Directed Polymer in Random Media with a Defect

Jae Hwan Lee and Jin Min Kim

Department of Physics, Soongsil University, Seoul 156-743, Korea
jmkim@ssu.ac.kr

Abstract. We study a queueing transition of directed polymer in random media with an attractive defect at the center of the one dimensional substrate. The end to end distance Δx of the polymer follows $\Delta x \sim t^{1/z}$ with $z = 3/2$, for weak defect strength ϵ where t is the polymer length. If $\epsilon \geq \epsilon_c$ then the polymer is localized with finite Δx in long t limit. The transition is related to the queueing phenomena of the asymmetric simple exclusion process.

Keywords: directed polymer in random media, queueing transition.

1   Introduction

A traffic jam occuring in a bottleneck, or near a road under construction is a typical example of queueing phenomena. The asymmetric simple exclusion process (ASEP) [1,2,3,4] have been introduced to explain the relation between the queueing phenomena and nonequilibrium driven dynamic process. In the ASEP, most particles jump into a vacant neighboring site in one direction with hopping probability one, but a selected site of hopping rate r with $r < 1$ plays a role of the bottleneck. Many studies have suggested that the critical hopping rate is one, $r_c = 1$, i.e., the nonzero hopping rate always gives rise to queueing phenomena [5,6,7,8,9]. However, some recent studies on the ASEP with a slow bond insist that a queueing transition of a jamming state at $r_c < 1$. The overall flux passing through the defect site is rarely influenced if the defect strength is not strong enough [10,11,12,13,14].

The ASEP can be interpreted as surface roughening problem of crystal growth in the body-centered solid-on-solid interface model [15,16,17] where the height difference between nearest neighbors is restricted by ±1. An increase (decrease) of surface height is equivalent to the presence (absence) of a particle in the ASEP. Both models belongs to the Kardar-Parisi-Zhang (KPZ) [18] universality class of two dimensional problems.

The directed polymer problem in random media (DPRM) [19,20,21,22] is well described by the KPZ equation and can be mapped into the ASEP. An attractive line defect in two-dimensional DPRM is related to the slow bond in the ASEP. Some arguments on the DPRM have proposed $r_c = 1$ [5,6,7,8,9], for the queueing transition. If a queueing transition exists at $r_c < 1$ in the ASEP problem, it would

J. Zhou (Ed.): Complex 2009, Part I, LNICST 4, pp. 220–224, 2009.

be interesting to find the critical point and phase transtion in the DPRM. Here, we study the DPRM with an attractive defect and measure various physical quantities as a function of in order to observe whether there exist the transition or not.

The Hamiltonian of DPRM with a defect at $\mathbf{x} = 0$ is

$$\mathcal{H} = \int dt \left[\gamma \left(\frac{d\mathbf{x}}{dt} \right)^2 + \mu(\mathbf{x}, t) - \epsilon\delta(\mathbf{x}) \right], \tag{1}$$

where $\mathbf{x}$ is the $d - 1$ dimensional transverse vector, t is the polymer length perpendicular to the substrate, and $-\epsilon\delta(\mathbf{x})$ means a time-independent defect at $\mathbf{x} = 0$. There are three competing terms: a bending energy γ forcing the polymer straight against a transverse bending, the impurity $\mu(\mathbf{x}, t)$ assigned to each point $(\mathbf{x}, t)$ preferring the polymer to be deformed through the minimum impurities, and the attractive defect potential at $\mathbf{x} = 0$ making the polymer return to the origin. The random potential $\mu(\mathbf{x}, t)$ is a white noise satisfying

$$\langle \mu(\mathbf{x}, t)\mu(\mathbf{x}', t') \rangle = 2D\delta(t - t')\delta^{d-1}(\mathbf{x} - \mathbf{x}'). \tag{2}$$

The partition function $Z(\mathbf{x}, t)$ for the polymer, starting from $(0, 0)$, and ending at $(\mathbf{x}, t)$, can be written as the path integral

$$Z(\mathbf{x}, t) = \int_{(0,0)}^{(\mathbf{x},t)} \mathcal{D}\mathbf{x}'(t')$$

$$\times \exp\left\{ -\frac{1}{T} \int_0^t dt' \left[\gamma \left(\frac{d\mathbf{x}'}{dt'} \right)^2 + \mu(\mathbf{x}', t') - \epsilon\delta(\mathbf{x}') \right] \right\}, \tag{3}$$

where T is temperature.

2   Discrete Model and Numerical Results

At zero temperature, entropy is ignored and the problem in Eq. (1) becomes much simplified by finding the optimal path and its energy $E(x, t)$ among all the pathes arriving at (x, t). The initial energy $E(x, 0) = 0$ is given at $t = 0$. A continuous random number between 0 and 1 with uniform distribution is assigned for the randomness $\mu(x, t)$ on a discrete structure. In addition, the attractive defect potential $-\epsilon$ is given at the center site $x = 0$.

First we consider a triangular structure. The polymer starts from $x = 0$ and its path is restricted by $|x(t) - x(t+1)| = 0$ or 1. There is a bending energy γ against a transverse jump $|x(t) - x(t+1)| = 1$. The minimum energy $E(x, t)$ for the polymer ending at (x, t) can be obtained recursively [20,21,22]: in $d = 1+1$,

$$E(x, t + 1) = \min\{ E(x, t) + \mu(x, t) - \epsilon\delta_{x,0},$$
$$E(x - 1, t) + \mu(x - 1, t) + \gamma - \epsilon\delta_{x-1,0}, \tag{4}$$
$$E(x + 1, t) + \mu(x + 1, t) + \gamma - \epsilon\delta_{x+1,0}\},$$

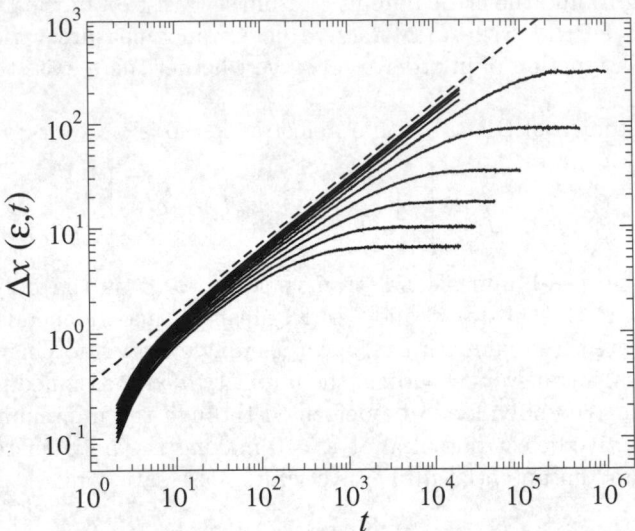

Fig. 1. $\Delta x(\epsilon, t)$ as a function of t on a discrete triangular structure with $\epsilon = 0.00, 0.01,$ 0.02, 0.03, 0.04, 0.05, $\cdots$, 0.09 from top to bottom. The guideline has the slope of 2/3.

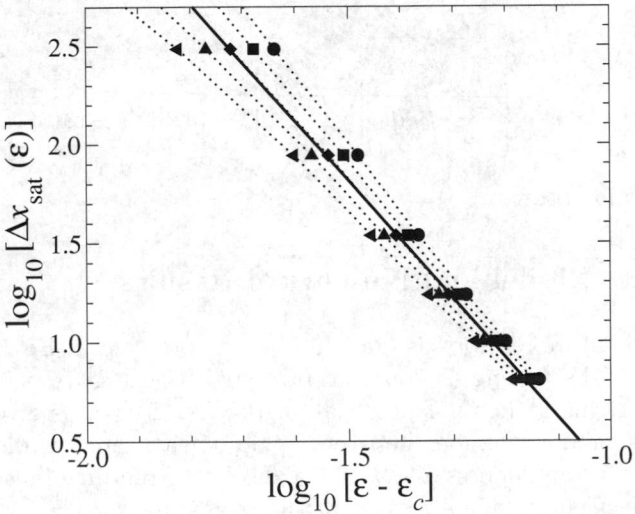

Fig. 2. $\Delta x_{sat}(\epsilon)$ as a function of $\epsilon - \epsilon_c$ for arbitrary critical values $\epsilon_c = 0.011, 0.016,$ 0.021, 0.026, and 0.031 from right to left. The most straight line is obtained at $\epsilon_c = 0.021$, where $\Delta x_{sat} \sim (\epsilon - \epsilon_c)^{-\delta}$ with $\delta = 3.00$.

where $\min\{A, B, C\}$ takes the minimum value among A, B, and C. We shall write $d = 1 + 1$ to indicate that there is one transverse and one longitudinal direction. Following Eq. (4) the polymer energy at each site is updated in parallel.

We monitor the end to end distance Δx of the polymer as a function of the polymer length t. In general it increases with t. Without defect, Δx follows

$$\Delta x \sim t^{1/z}, \tag{5}$$

with $z = 3/2$ [23]. The optimal path is affected by the strength of the defect ϵ. Its contribution to Δx is negligible as long as ϵ is smaller than ϵ_c as shown in Fig. 1 where Δx still shows the power law behavior with $z = 3/2$. For $\epsilon > \epsilon_c$, Δx increases with t and then approaches a finite value Δx_{sat} in long time limit. Actually Δx_{sat} depends on ϵ. We assume that Δx_{sat} diverges as ϵ approaches ϵ_c following a scaling law

$$\Delta x_{sat} \sim (\epsilon - \epsilon_c)^{-\delta}. \tag{6}$$

The log-log plot of Δx_{sat} against $(\epsilon - \epsilon_c)$ for various values of ϵ_c is given in Fig. 2, where we estimate $\epsilon_c \approx 0.021$ with $\delta \approx 3.0$.

3 Summary

A directed polymer in random media with an attractive defect in the middle of one dimensional substrate is studied. The end to end distance Δx of the polymer is monitored as a function of the polymer length t and ϵ. We find that there is a phase transition at $\epsilon_c \approx 0.021$. We measure Δx in a triangular structure to avoid the finite system size effect. Due to the triangular structure, the simulation is limited up to $t_{max} \leq 10^6$ depending on ϵ. For $\epsilon < \epsilon_c$, the contribution of the defect to Δx is negligible. For $\epsilon > \epsilon_c$, Δx becomes finite. This behavior supports the queueing transition at finite ϵ_c [13]. Analytic works and larger simulations are required to get more accurate values of the critical exponents.

Acknowledgments

This work was supported by the Korea Research Foundation Grant funded by the Korean Government(MOEHRD) (KRF-2008-314-C00143) and also partially by the Soongsil University Research Fund (JMK).

References

1. Gwa, L.-H., Spohn, H.: Six-vertex model, roughened surfaces, and an asymmetric spin Hamiltonian. Phys. Rev. Lett. 68, 725–728 (1992)
2. Derrida, B., Evans, M.R., Hakim, V., Pasquier, V.: Exact solution of a 1D asymmetric exclusion model using a matrix formulation. J. Phys. A 26, 1493–1517 (1993)
3. Blythe, R.A., Janke, W., Johnston, D.A., Kenna, R.: The grand-canonical asymmetric exclusion process and the one-transit walk. J. Stat. Mech., P06001 (2004)
4. Rosini, M., Reggiani, L.: A Monte Carlo investigation of noise and diffusion of particles exhibiting asymmetric exclusion processes. J. Phys., Condens. Matter 19, 036226 (2007)

5. Tang, L.-H., Lyuksyutov, I.F.: Directed polymer localization in a disordered medium. Phys. Rev. Lett. 71, 2745–2748 (1993)
6. Balents, M., Kardar, M.: Disorder-induced unbinding of a flux line from an extended defect. Phys. Rev. B 49, 13030–13048 (1994)
7. Kinzelbach, H., Lässig, M.: Depinning in a random medium. J. Phys. A 28, 6535–6541 (1995)
8. Hwa, T., Nattermann, T.: Disorder-induced depinning transition. Phys. Rev. B 51, 455–469 (1995)
9. Lässig, M.: On growth, disorder, and field theory. J. Phys. Condens. Matter 10, 9905–9950 (1998)
10. Kandel, D., Mukamel, D.: Defects, Interface Profile and Phase Transitions in Growth Models. Europhys. Lett. 20, 325–329 (1992)
11. Slanina, F., Kotrla, M.: Weak pinning: surface growth in the presence of a defect. Physica A 256, 1–17 (1998)
12. Myllys, M., Maunuksela, J., Merikoski, J., Timonen, J., Horváth, V.K., Ha, M., den Nijs, M.: Effect of a columnar defect on the shape of slow-combustion fronts. Phys. Rev. E 68, 051103 (2003)
13. Ha, M., Timonen, J., den Nijs, M.: Queuing transitions in the asymmetric simple exclusion process. Phys. Rev. E 68, 056122 (2003)
14. Song, H.S., Kim, J.M.: Faceting Transition of a Restricted Solid-on-Solid Growth Model with a Defect Site. J. Korean Phys. Soc. 48, S245–S248 (2006)
15. van Beijeren, H.: Exactly Solvable Model for the Roughening Transition of a Crystal Surface. Phys. Rev. Lett. 38, 993–996 (1977)
16. Kotrla, M., Levi, A.C.: Kinetic six-vertex model as model of bcc crystal growth. J. Stat. Phys. 64, 579–604 (1991)
17. Kotrla, M., Levi, A.C.: Kinetic roughness in the BCSOS model. J. Phys. A 25, 3121–3132 (1992)
18. Kardar, M., Parisi, G., Zhang, Y.-C.: Dynamic Scaling of Growing Interfaces. Phys. Rev. Lett. 56, 889–892 (1986)
19. Kardar, M., Zhang, Y.-C.: Scaling of Directed Polymers in Random Media. Phys. Rev. Lett. 58, 2087–2090 (1987)
20. Kim, J.M., Moore, M.A., Bray, A.J.: Zero-temperature directed polymers in a random potential. Phys. Rev. A 44, 2345–2351 (1991)
21. Kim, J.M.: Phase transition of directed polymer in random potentials on 4+1 dimensions. Physica A 270, 335–341 (1999)
22. Kim, J.M.: Restricted Solid-on-solid Model and a Directed Polymer in Random Potentials. J. Korean Phys. Soc. 45, 1413–1419 (2004)
23. Gelfand, M.P.: Random walks in random media with random signs. Physica A 177, 67–72 (1991)

Pollution Modeling and Simulation with Multi-Agent and Pretopology

Murat Ahat[1], Sofiane Ben Amor[1], Marc Bui[1],
Michel Lamure[2], and Marie-Françoise Courel[1]

[1] Ecole Pratique Des Hates Etudes, 46 rue de Lille, 75007 Paris, France
Murat.Ahat@etu.ephe.sorbonne.fr, {s.benamor,m.bui}@ephe.sorbonne.fr,
marie-francoise.courel@cnrs-dir.fr
[2] Université Lyon 1, 43 bd du 11 novembre 1918, 69622 Villeurbanne cedex, France
lamure@mac.com

Abstract. Pollution in metropolitan cities has become a serious problem, resulting in poor living conditions and serious health problems. Pollution being qualified as a complex system, we propose a multi-agent approach to model and simulate it, so that we could study, analyze and predict it better. As in the early stage of the project, we have some successful experiments and attempts to integrate the mathematical theory of Pretopology in the modeling and simulation levels. In addition, these interesting results shades some light on our future direction.

Keywords: Complex System, Pollution, Pretopology, Multi-Agent Simulation, Repast Simphony.

1 Introduction

Pollution, as a by product of development and industrialization, is posing threat to the health of millions of people, environment and ecology. The situation is even worse in the developing metropolitan cities. The case of Ouagadougou [8], a young African city, is the topic of Project Mousson [7]. In this project, scientists and experts from different disciplines (geographers, meteorologists, sociologists, anthropologists, mathematicians and computer scientists) gather to study and analyze the air pollution in Ouagadougou, which is a complex system, in order to build a pollution alarm and alert system.

One of the problem in complex system modeling and simulation is that it is difficult to model and express the qualitative variables and factors. The advantage of multi-agent modeling and simulation is being able to integrate these qualitative information much more easily [9]. Our collaboration with the experts from different domains may yield a better understanding and modeling of the system, in turn results a system that can express the pollution in different levels, and from different angles.

Being in the early stage of modeling, we have conducted some interesting experiments. We have successfully integrated the pretopology method into a classical heat bugs model to simulate the basic pollution diffusion process. Though

J. Zhou (Ed.): Complex 2009, Part I, LNICST 4, pp. 225–231, 2009.

the model is simple, it has shown some interesting results. Thus, in the following sections we will discuss the pollution diffusion model based on the classical heat bugs model, pretopology and pretopology integrated diffusion model.

2   Pollution Diffusion Model and Pretopology

The diffusion process of this simple model is realized according to the heat diffusion in heat bugs model. However, we integrate the pretopological method into this diffusion process to have more insights into the diffusion process. The agents in the model are very primitive, as they just produce pollution into its environment with its pollution strength. Gradual discussion in the following subsections shows how pretopology is used in the model and why.

2.1   Diffusion Process in Heat Bugs Model

Heat bugs model [4] is a famous classical multi-agent model that exists as a example model in various multi-agent simulation toolkits: repast [3], swarm [5], netlogo [6], to name a few. We take its diffusion process and combine it with polluter agents instead of heat bugs. And below is a brief description of how the combination works.

We discretize the space into a two dimensional grid (torus) $G(X, Y)$, as X stands for width, and Y for height. Also, we discretize the time into small time steps t_i, while i is from 0 to n. A location in the grid is denoted by its coordinate $c(x, y)$, at the same time it represents the pollution level at this location. A polluter agent is denoted by pa, whilst its pollution strength is denoted by s. Agent location in the grid is determined from the beginning of the simulation and does not change afterwards. The simulation takes place in the following steps:

1. Simulation preparation: let all $c(i, j)$ in the $G(X, Y)$ be 0; Populate the grid with n polluter agents with pollution strengths s, and randomly scatter them over it.
2. Let every polluter agent pollutes its location.
3. Diffuse the pollution over the grid.
4. Go to step 2.

Now we will introduce the corresponding pseudo code to the step 3.

```
Procedure Diffuse ()
   Constant
      ERate = 0.99; DRate = 1.0;
   Var
      G2(X,Y) as temporary grid;
      c2(x,y) as cells of G2(X,Y);
   Begin
      For (i:=1 to X)
```

```
    For (j:=1 to Y)
      avg := mooreNeighbourAverage();
      c2(i,j) := ERate*(c(i,j)+DRate*(avg-c(i,j)));
    End For; //j
  End For; //i
  copy(G2(X,Y) to G(X,Y));
End Procedure //Diffuse()
```

In the above code, there are two important constants: evaporation and diffusion rate. Evaporation rate defines how much of the pollution evaporates, while diffusion rate determines the pollution spread speed. They both are in [0..1].

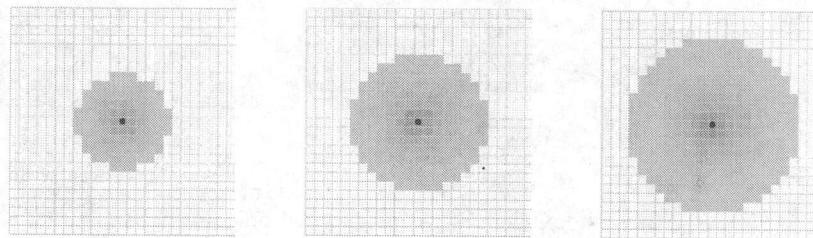

Fig. 1. Pollution diffusion process at different time steps: from left to right at 40, 100 and 250. G(25,25), EvaporationRate = 0.99, Diffusion Rate = 1.0. A polluter agent (red dot) is placed at the center of the grid.

Figure 1 shows diffusion results at different time steps in the simulation. Red color is used to indicate the pollution in the grid. The deeper the color is, the stronger the pollution is. This means around the polluter agent the pollution is stronger. Pollution spreads as time passes.

Besides this fact, experiments with different evaporation and diffusion rates show that by this means different diffusion processes can be achieved.

2.2   The Basic Pretopological Concepts

Developments in hardware technologies facilitate the application of pretopology theory in real-world scenarios. Recently a pretopology library based on java is developed to provide ready to use pretopological data types and algorithms [10]. And there are also attempts and suggestions for using pretopology in the complex system modeling [11] [12] [13]. Our motivation is using pretopological concept of *pseudoclosure* to simulate the pollution diffusion, and meanwhile to observer the evolution of this diffusion process. Here is a brief mathematical introduction to pretopology.

Let E be a non-empty set and let $\mathcal{P}(E)$ designate all of the subsets of E.

Definition 1. *A pseudoclosure $a(.)$ is a mapping from $\mathcal{P}(E)$ to $\mathcal{P}(E)$, which satisfies following two conditions:*

$$- \ a(\emptyset) = \emptyset \qquad\qquad (P1)$$
$$- \ A \subseteq a(A) \qquad\qquad (P2)$$

A pretopological space is a pair (E, a), where E is endowed with a pseudoclosure $a(.)$.

Subset $a(A)$ is called the *pseudoclosure* of A. As $a(a(A))$ is not necessarily equal to $a(A)$, a sequential appliance of pseudoclosure on A can be used to model expansions: $A \subseteq a(A) \subseteq a(a(A))...$, for example,as in pollution diffusion. Figure 2 illustrates the pseudoclosure and its expansion process.

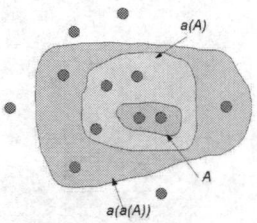

Fig. 2. A and sequential pseudoclosure appliance on A

In our model we will use a type V_D pretopology space, as defined below.

Definition 2. *A V_D pretopology space is a pretopology space that satisfies: $\forall A$, B, $A \subseteq E, B \subseteq E$ $a(A \cup B) = a(A) \cup a(B)$* $\qquad (P3)$

The notion of closure is also important in the pretopological space.

Definition 3. *$A \in P(E)$ is a closure only and only if : $A = a(A)$* $\qquad (P4)$

2.3   Integration of Pretopology into the Model

From the above discussions, one can easily relate the *pseudoclosure* with expansion of the pollution. Before discussing the *pseudoclosure* integration , we need to introduce the definition of pollution cluster. A pollution cluster is a connected subset of the grid $G(X, Y)$, whose pollution level is above a *threshold*, in our case threshold is 0. This means in the beginning of the simulation, there will be more pollution clusters. With the diffusion process, pollution clusters merge and result bigger ones. Given enough polluters, after certain steps in the simulation, only one pollution cluster will be left in the grid. Now we discuss how we use *pseudoclosure* to simulate the pollution diffusion. Here are the steps:

1. Simulation preparation, let every $c(i, j)$ in the $G(X, Y)$ be 0; Populate the grid with n polluter agents with pollution strength s, and randomly scatter them over it. Create one pollution cluster for every location that has a polluter agent.
2. Let every polluter agent pollutes its location.

3. Calculate the pseudoclosure of every pollution cluster.
4. Merge clusters if they are mergable.
5. Go to step 2.

The pseudo code of step 3 and 4 are as shown below:

```
Procedure PseudoClosure ()   //Step 3
  Var
    PollutionClusterSet: Set;
    G2(X,Y) as temporary grid;
    c2(x,y) as cells of G2(X,Y);
  Begin
    For (every pollutionCluster in PollutionClusterSet)
        For (every c(i,j) in pollutionCluster) //diffuse
            average:=mooreNeighbourAverage();
            c2(i,j):=ERate*(c(i,j)+DRate*(average-c(i,j)));
        End For;
        For (every c(i,j) neighbour of pollutionCluster)
            avg:=mooreNeighbourAverage();
            c2(i,j):=ERate*(c(i,j)+DRate*(avg-c(i,j)));
            If (c2(i,j)>0) Then
                Add c2(i,j) to pollutionCluster;
        End For;
    End For;
    copy(G2(X,Y) to G(X,Y));
  End Procedure //PseudoClosure()

Procedure MergePollutionClusters ()   //Step 4
  Var
    PollutionClusterSet: Set;
  Begin
    LABEL: DEBUT;
    If (sizeof(PollutionClusterSet) > 1)
      For (every pc1 in PollutionClusterSet)
        For (every pc2 in PollutionClusterSet)
            If ((pc1<>pc2)and(pc1,pc2 has common area)) Then
                pc1 = merge(pc1,pc2);
                remove(pc2, PollutionClusterSet);
                goto DEBUT;
            End If;
        End For;
  End Procedure //MergePollutionClusters()
```

This pretopology integration gives us the advantage of tracking the evolution of pollution clusters.

3   Simulation Results

In the simulation, evaporation rate is fixed to 0.99, and diffusion rate is to 1.0. Our pollution environment is a 200x200 torus grid. We have populated the grid with 40 to 220 polluter agents in ten runs, adding 20 agents in each subsequent run. The pollution strength of the agent is fixed to 10.

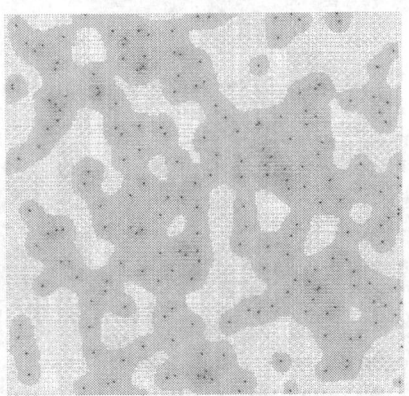 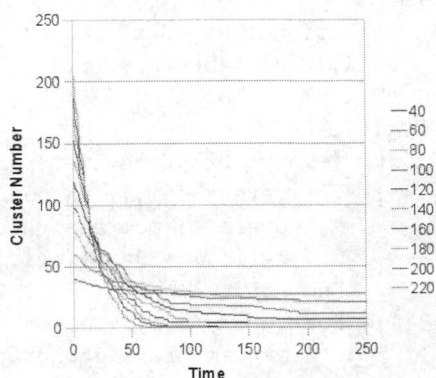

Fig. 3. Left: visualization of pollution diffusion, showing merging of pollution clusters. Right: Pollution cluster number plot in ten run.

Figure 3, illustrates our simulation results. On the left, the pollution clusters are merging to form a single big cluster. So, our question is what kind of conditions trigger the big cluster emergence? The chart on the right answers the question: big cluster emerges when there are more than 180 polluter agents with pollution strength 10, or when the polluter agent density is higher than 0.45%. When pollution strength is set to 5 and 20, the emergence has occurred with 0.9% and 0.25% of polluter agent density. This strongly indicates the emergence occurs when the overall pollution amount given off at every time step reaches a point, in the case this example, roughly around 1800 per time step.

We are now able to observer the diffusion process in the terms of pollution clusters, thanks to pretopology integration. That is why we obtained results that show there is a link between emergence of a single cluster and overall pollution amount. Though at this level, we are not able to validate simulation results, we hope that in the future to validate the results using the data available from the Project Mousson.

The model is developed with Repast Simphony [3], which considerably decreased the development time, eased the visualization [1] and analysis steps [2].

4   Conclusion

Though a simple one, the model paves the road for the more complex and complete pollution simulation system. This also shows a successful pretopology integration into simulation, and its usefulness in the complex system modeling and analysis.

Our work does not stop here. Our goal in the near future is to keep collaborates with other experts from different domains, and complete the model in a gradual manner: integration of GIS, real world agents, complex environment, etc. In the long run, through collaboration and gradual development, we aim a simulation system that is usable in the real world scenarios.

References

1. North, M.J., Tatara, E., Collier, N.T., Ozik, J.: Visual Agent-based Model Development with Repast Simphony. In: Proceedings of the Agent 2007 Conference on Complex Interaction and Social Emergence, Argonne National Laboratory, Argonne, USA (November 2007)
2. Tatara, E., North, M.J., Howe, T.R., Collier, N.T., Vos, J.R.: An Introduction to Repast Modeling by Using a Simple Predator-Prey Example. In: Proceedings of the Agent 2006 Conference on Social Agents: Results and Prospects, Argonne National Laboratory, Argonne, IL USA (September 2006)
3. Repast Simphony (visited October 2008), http://repast.sourceforge.net/
4. Swarm Heat Bugs example (visited October 2008),
   http://www.swarm.org/examples-heatbugs.html
5. Swarm MAS toolkit (visited October 2008), http://www.swarm.org
6. Netlogo MAS toolkit (visited, October 2008),
   http://ccl.northwestern.edu/netlogo/
7. Mousson Project (french) (visited October 2008),
   http://mousson.csregistry.org
8. Cachier, H., Sciare, J., Favez, O., Guinot, B.: Factors Influencing Urban Carbonaceous Aerosols: Comparison of Beijing (RP China), Paris (France), Cairo (Egypt) and Ouagadougou (Burkina Fasso). In: 9th International Conference on Carbonaceous Particles in the Atmosphere, Berkeley, US (August 2009)
9. Crooks, A., Castle, C., Batty, M.: Key Challenges in Agent-Based Modelling for Geo-Spatial Simulation. Geocomputation 2007, NUI Maynooth, Ireland (September 2007)
10. Levorato, V., Bui, M.: Data Structures and Algorithms for Pretopology: The JAVA based software library PretopoLib. In: 8th International Conference on Innovative Internet Community Systems, Schoelcher, Martinique (June 2008)
11. Ben Amor, S., Lavallee, I., Bui, M.: Percolation, Pretopology and Complex Systems Modeling. Models and Simulations, Paris, France (June 2006)
12. Ben Amor, S., Levorato, V., Lavallee, I.: Generalized Percolation Processes Using Pretopology Theory. In: 2007 IEEE International Conference on Research, Innovation and Vision for the Future, March 5-6, 2007, pp. 130–134 (2007)
13. Ben Amor, S., Ahat, M., Bui, M., Lavallee, I.: Modlisation et simulation des phnomnes complexes dans le cadre des SIG. In: WaterID Conference. Urumqi, China (October 2006)
14. Levorato, V., Ahat, M.: Modlisation de la Dynamique des Rseaux Complexes associe la Protopologie. In: Conference ROADEF 2008. Clermont-Ferrand, France (February 2008)

Policy, Design and Management: The *in-vivo* Laboratory for the Science of Complex System

Jeffrey Johnson

Faculty of Mathematics, Computing and Technology, The Open University
Walton Hall, Milton Keynes, MK7 6AA, United Kingdom
j.h.johnson@open.ac.uk

Abstract. Complex systems scientists cannot by themselves perform experiments on complex socio-technical systems. The best they can do is to perform experiments alongside policy makers who are constantly engaged in experiments as they design and manage the systems the systems for which they are responsible. In this context the nature of *prediction* in the implementation of real systems is much more complicated than it is in traditional science. The *goals* identified by policymakers change through time, and this is usually managed through the design and management processes. The combination of policy and design is the opportunity – the only opportunity – for complex systems scientists to engage and to be allowed to be involved in *in-vivo* experiments in large socio-technical systems. In turn this opens up new methodological approaches and questions for the science of complex systems.

Keywords: Complex Systems, Policy, Design, Management, Prediction.

1 Introduction

The science of complex systems differs from traditional science in many respects. One of the most important is that complex systems scientists cannot perform active experiments on many of the systems they study. For example, a scientist cannot decide to build a bridge to test predictions of traffic flows, a scientist cannot implement a stay-at-home policy to investigate the consequent impact on an epidemic, a scientist cannot implement a radical energy policy to investigate its impact on climate change, and a scientist cannot declare war or take measures to keep the peace. Initiating change assumes purpose and a view on how future systems ought to be (Simon, 1969). In democracies deciding what ought to be is the prerogative of policymakers, not scientists.

Large complex socio-technical systems have strong political dimensions, and changing them often involves huge resources and complicated planning procedures with delivery through many agents. Most policies are experiments with unpredictable outcomes, but they are generally not treated as scientific experiments and not instrumented. Put simply, only policy makers have the mandate and the money to conduct experiments on large complex socio-technical systems.

J. Zhou (Ed.): Complex 2009, Part I, LNICST 4, pp. 232–241, 2009.

The best that experimental scientists can do is to align themselves with policy makers, ensuring that the scientific basis of policy is sound and persuading the policy makers to tolerate the collection of data for scientific purposes. This is necessary but not sufficient for scientific experiments, since the whole nature of scientific prediction has to be rethought for complex systems.

2 Performing Experiments on Complex Socio-technical Systems

Complex systems scientists are seeking new methods of understanding how systems might evolve from current states to future states. *Experiment* generally involves putting a system in a particular state, making an intervention and observing the consequences. These may or may not support hypotheses on the consequences of the intervention. Since only policy makers can legitimately and practically make interventions, the best that complex systems scientists can do is make precise the desired future state(s) of the system (what policy says it *ought* to be), make precise which system states might evolve, and suggest interventions that may best achieve the desired state(s): "tell us what the target is and we'll tell you how to hit it".

current state now   target state at t

Intervention kick

Fig. 1. Experiments: predicting that given interventions will result in future system states

This is illustrated in Figure 1. Somehow the system must be *represented*. For complex systems such as a cancer or a city this representation cannot just be a few words or symbols. The representation will be complicated and generally have many levels with many relationships and numbers characterizing particular states. Characterizing the system as it is now and how it ought to be in the future involves complicated data structures holding a lot of data and computing state transitions and trajectories from these data at many scales.

The idea of a policy-oriented *prediction* is that if one makes an intervention one 'kicks' the system off its current trajectory onto another trajectory that will hit the target. Let s_0 be the state of system at now, time zero, and s_t be the state of the system at time t. Suppose the system is governed by *transition dynamics* with $f: s_0 \rightarrow s_t$. Let k_{NULL} be the *null kick* meaning that one does nothing to the system. Then let us write $f: (s_0, k_{NULL}) \rightarrow s_{t,NULL}$ to mean the transition dynamics of the system when no intervention is made. Let us write $f: (s_0, k_i) \rightarrow s_{0,i,t}$ to mean the state of the system at time t when intervention k_i is applied to the system in state s_0 at time t_0.

The cooperation between scientists and policy makers assumes that the former know something about the dynamics of the system that the latter find useful. It is assumed that scientists can bring to bear better ways of representing that states of systems, s_t, their transition dynamics, f, and that this enables predictions to be made on the consequences of any particular policy kick, k_i. It also assumes that scientists are *practical*, proposing data structures that enable data collection or, more often, use of data that already exists. Also the transition dynamics must be *computable*. Policy makers are not interested in abstract predictions from scientists that do not give precise statements of what might be.

Paraphrasing the argument so far, scientists cannot do *in-vivo* experiments on most complex socio-technical systems. Generally they have neither the mandate nor the money to make interventions in large complex systems. Therefore the best they can do is sit alongside policy makers who do have the mandate and the money to make interventions. The best that a scientist can do is persuade policy makers that they know how to *predict* the outcome of any particular policy intervention, and possibly suggest particular interventions that the policy makers might find helpful.

3 Prediction and Control in Complex Systems

As an example of prediction, consider an ancient warship. The captain's policy is that an enemy ship *ought* to be hit by a cannon ball. Assuming the scientist will take the commission, their job is to suggest a way of setting up the cannon in a way that can be predicted to hit the target ship when fired.

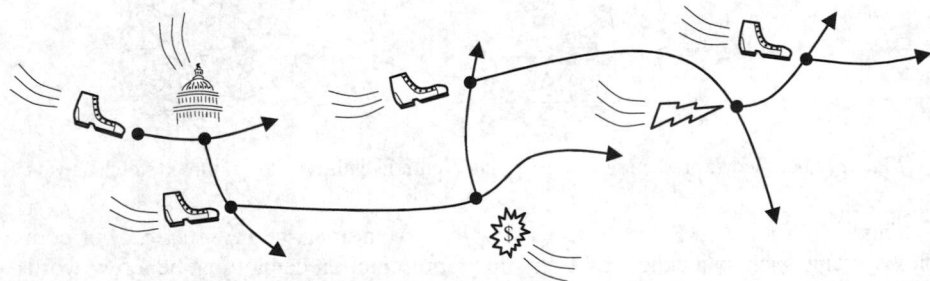

Fig. 2. Policy is subject to many forces from many external sources

For complex socio-technical systems the reality is much more complicated that this since no-one can control all their aspects. In Figure 2 it is supposed that the original policy kick sends the system on a trajectory towards the goal, but legislation knocks it off course. For example, the policy of one Government department can be knocked off course by policy changes made by another department, *e.g.* planning regulations may become more severe, or fiscal rules may change. Figure 2 shows intervention kicks to put the system back on course towards the goal, but being knocked off trajectory by an external financial force, such as the severe global financial crisis emanating from the USA in 2007 and 2008. Again interventions attempt to put the systems back on track, and but other extreme events knock it off trajectory from the goal.

Traditional science assumes that the laboratory can be separated from its environment. Let S be the system under investigation and U the *universal system*. Then let the *environment*, E, be defined to be $E = U - S$. Thus anything not in the system is by definition in the environment. Traditional science progressed by being able to isolate

S in the laboratory with *E* having negligible effect. For example, Galileo did not have to worry about bank rate or the weather when studying motion down an inclined plane. Often *E* does not have much effect on complex systems when they are behaving normally. The problem with complex systems is that they are subject to *extreme event* when they do not behave normally, and *the occurrence of extreme events is itself normal.* Understanding the nature of extreme events and the interaction and coevolution between systems and environment is part of the challenge in complex systems science.

To continue our earlier example, the cannon may be on a ship in the middle of a storm, with huge waves tossing it about. Or the regulations may change so that cannon must be fired in a different way. Or the cost of powder may change meaning that less can be used. Or a wheel may have fallen off the gun carriage. Predicting how the target might be hit becomes much more difficult as the system and its environment change. Figure 2 is much closer to feedback control than open-ended prediction. The former is essential in real systems while the latter is almost unattainable in complex social technical systems. *Thus whole concept of prediction has to change relative to traditional science.*

4 Prediction Fans and Prediction Horizons

Figure 2 is a simplification, since the outcome of any particular intervention kick may not be unique. In other words, an intervention may have a number of outcomes as illustrated in Figure 3. Thus the possible future states of systems fan out towards a *prediction horizon* beyond which predictions are meaningless. For example, what will the bank rate be ten years from today? Or how many cars will there be in a hundred years time? As the clock ticks the horizon moves forward, allowing some predictions to become meaningful and even useful.

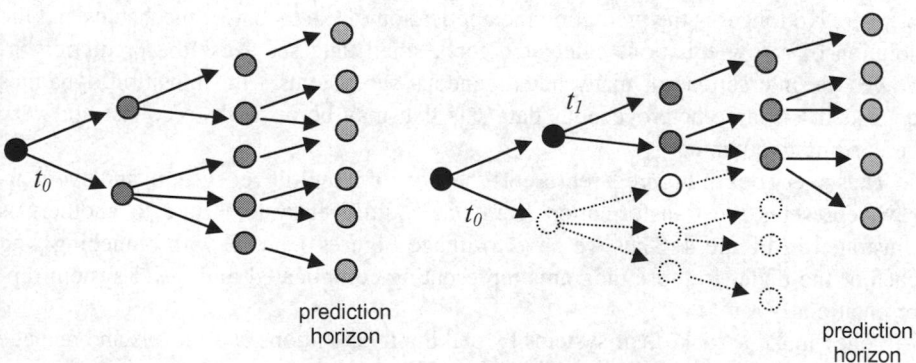

(a) Future trajectories fans out from the present   (b) The horizon moves forward with time

Fig. 3. Predictions in complex socio-technical systems fan out and have horizons

To continue the example of the cannon and target, the analogy now is that the target ship is over the horizon and part of the prediction becomes knowing when, if ever, it will become visible and what to do until it comes in range.

5 Prediction and Policy in Multilevel Systems

The story gets even more complicated. Most complex systems have many levels, from micro-level to macro-level. When one speaks of an intervention, it may matter at what level the intervention occurs. For example, national novernments can make interventions by changing laws or making funds available for their objectives. They cannot intervene at the level of the individual person, as some agencies can at a much more disaggregate level. Generally national governments do not build particular schools, this is done by intermediate level administrations. Whatever the level of an intervention, it is possible that its consequences will be felt at higher and lowest levels of aggregation.

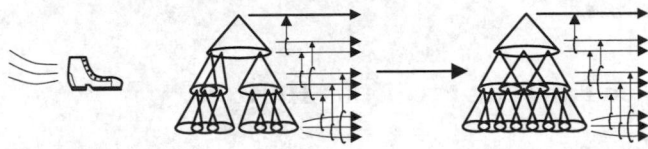

Fig. 4. Predicting the consequences of a kick to a multilevel system, $k:S_t \rightarrow S_{t+\Delta t}\, k$.

To continue the analogy of the cannon, it might be that to fire the cannon requires the permission of the admiral and the captains of all the other ships, and that this permission depends on the detailed states of all the other ships and the configuration of the fleet, etc. Alongside all this are considerations of the cost of gunpowder and the lurching of the ship in the gale, and the possible position of the target beyond the horizon. No longer is the prediction the application of the theory of mechanics and the solution of a few equations calculated for a small data set. Now the prediction involves the interactions of many heterogeneous subsystems with transition dynamics dependent on large heterogeneous data sets that may be difficult to collect and very demanding to compute.

Thus every dot in Figure 3 represents the state of a multilevel system, and every arrow represents the transformation from one multilevel system state to another, as illustrated in Figure 4. Thus we have to image Figures 1 and 2 with branching, and each of the nodes in those diagrams representing complicated multilevel system representations.

Policy makers 'kick' their systems by making interventions at all levels and manage the consequences as best they can (Figure 4). What can complex systems scientists say about this? What does it mean to make a scientific prediction in these circumstances? Does it include the consequences at all levels?

The idea of the scientist using predictive methods to advise the policy maker how to achieve their goals is getting very complicated. But there is one final twist:

6 The Goalposts Move!

In the case of the cannon, it is as if the target ship changes, or the target becomes assisting rather than destroying the ship, because a more beneficial target emerges from the process of trying to hit the original target.

Consider policy makers identifying and trying to solve problems. There is the normative aspect that the system *ought* to be better, and there is the practical aspect of what could make it better. Such situations invariably have many competing dimensions. Ideally they would all be optimized, but usually the constraints compete with each other for any given solution. They have to be satisficed, meaning an acceptable compromise must be found. Sometimes the problem of creating the system that *ought* to be is over-constrained and there is no solution. For progress to be made, one or more constraint must be relaxed. Changing the constraints leads to a different problem. To be useful, any prediction made by a complex systems scientist must address the problem as it exists now, not as it was previously.

7 Design and Management

By now we have entered the world of *design* which is well known to policy makers. It will be argued that the design and management of real systems is the *in-vivo* laboratory of complex systems science.

Figure 5 shows a simple diagram of the way people solve practical problems. On the left is the process of establishing what is required, which comes from those who make policy. On the right is the *generate-evaluate cycle* that characterises the design process. In this cycle designers generate new systems and evaluate them against the requirements. If the new system satisfies all the requirements it is a 'solution' to the design problem, and the process passes to implementation. If the proposed system does not satisfy the requirements then alternative systems are generated themselves to be tested against requirements.

The *design cycle* is a spiral process in time, since the design rarely returns to a previous state. The reason is that the designer *learns* about the problem during each iteration around the cycle. Each time a potential solution is generated the designer is making

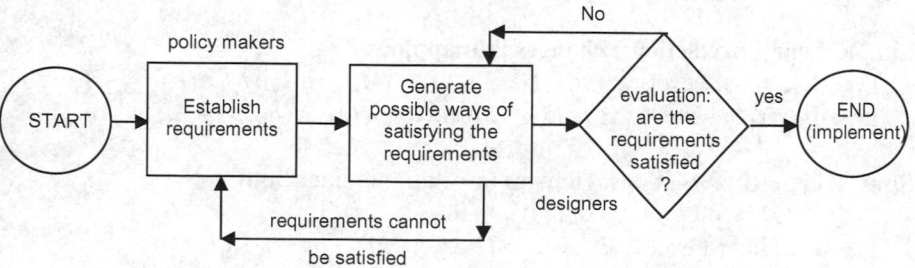

Fig. 5. The simplified requirements-generate-evaluate model of the design process

hypotheses as to what the parts of the system might and how they might fit together. Each time the potential solution is evaluated these hypotheses are reviewed and possibly tested. Each time a potential solution is rejected the designer learns something new about the system.

Anyone who has been involved in a large complicated project will recognise the process of 'working up' the definition of the project, and will be familiar with the interplay between constraints and possible ways forward. Policy makers know this from experience. Many scientists experience a similar process in their own projects. In this paper we are not just saying that the messy business of managing projects is necessary to administer science, we are saying that engaging in messy design and management projects *is* science, as far as complex socio-technical systems are concerned. Design and management are intimately tied up with the emerging new scientific method, with data collection, with prediction, and any possibility of testing predictions using new statistical methods.

The argument is that complex systems scientists cannot do experiments on large socio-technical systems but must align themselves with policy makers (who do experiments all the time). Policy makers execute their experiments through the design process which is a systematic way of allowing a coevolution between what *ought* to be and what *can* be.

This process involves many predictions at many levels, and the process of making predictions evolves with the requirements and proposed solutions.

8   Prediction and Testing Predictions *in-vivo*

Arising from the previous discussion, many kinds of predictions that can be made about complex systems. Let the experimental multilevel system be written as (B, M, H), where B is the relational *backcloth* of the system and M is the class of mappings representing the *traffic* of system activity over the backcloth. We write $B = B_N \oplus B_{N+1} \oplus B_{N+2} \ldots \oplus B_{N+L}$ where B_{N+j} is the relational backcloth of the system at level $N+j$, which can be represented by networks and hypernetworks. The mappings on the backcloth can be written $M = M_N \oplus M_{N+1} \ldots \oplus M_{N+L}$ where $M_{N+i} = \{m_{N+i,j} \mid m_{N+i,j} : B_{N+i} \rightarrow Z\}$ where Z is a number system such as the real or rational numbers[1]. H is a class of mappings that aggregate the mappings M_{N+i} over the backcloth, $h_{ij} : M_{N+i} \rightarrow M_{N+j}$.

Simple Type-I predictions: changes in mappings
    Type-I-1, Fixed Level. $k : (M_{N+i}(B_{N+i}), t) \rightarrow (M_{N+i}(B_{N+i}), t + \Delta t)$
    Type-I-2, Inter-Level. $k : (h_{ij} M_{N+i}(B_{N+i}), t) \rightarrow (M_{N+j}(B_{N+j}), t + \Delta t)$

Simple Type-II predictions: changes in relational backcloth
    Type II-1, Fixed Level. $k : (B_{N+i}, t) \rightarrow (B_{N+i}, t + \Delta t)$
    Type II-2, Inter-Level $k : (B_{N+i}, t) \rightarrow (B_{N+j}, t + \Delta t)$

---

[1] It can be shown that any system can be written this way, where the levels are relationally defined by parts and wholes (Type-α aggregation) and classification (Type-β aggregation) (Johnson, 2008).

These predictions correspond to a single arrow in Figure 1. They include *time series* predictions as the some of the simplest examples of Type I-1 predictions, where the future value of a mapping on a fixed backcloth is determined by its previous values. Type I-2 mappings include the aggregation of numbers over the levels of a fixed backcloth, such as the cost of building a system from components bottom-up, or the top-down distribution of resource. The employment of a new person is an example of a Type II-1 change. The impact that new person might have on the team is an example of a Type II-2 change.

Simple Stochastic Predictions
Both Type-I and Type-II predictions can be stochastic. This corresponds to putting transition probabilities on the arrows in Figure 3. For complex systems the fans may spread out very rapidly, meaning that the transition probabilities rapidly become small so that they may carry no useful information. But they may do.

Compound Predictions: multi-kick control
As Figure 2 suggests, the reality of keeping a system on a trajectory that will hit a given goal requires constant monitoring of the systems state, and constantly making new predictions on which to base new kicks to keep the system on trajectory to the target. In engineering systems feedback control is achieved by a few sensors and, usually, a few fixed equations that are used to compute changes to the control action (*e.g.* applying more or less power to the actuators). In such systems one does not predict that any particular control action will ensure that the system hits the target, but one can predict that the multi-kick control regime will ensure that the system will hit the target.

For complex socio-technical systems this means the empirical scientist must propose a control regime which combines Type-I and Type-II predictions across multilevel systems. To my knowledge no complex systems experiment has ever been conducted in these terms.

The reason for this is probably because complex systems are not designed and implemented in this way. The evolution of the control system occurs during the design process, which is not widely seen as part of the scientific method, even though this view is well known in design theory (e.g. Herbert Simon, Ross Ashby).

9   Design Predictions

As we have seen, design is the ultimate test of prediction since it requires an understanding of the system dynamics of the specification-design process. On the UK Embracing Complexity in Design project (Johnson, *et al*, 2007) it emerged that

- designing complex systems requires a scientific understanding of their dynamics
- design processes can be complex, e.g. manufacturing processes, supply chains
- the environment of design can be complex , e.g. regulation, fashion, economy
- design is a complex collaborative cognitive process

Complex systems scientists can contribute to the design and implementation of complex systems by providing system models allowing simple predictions to be made. To implement systems requires an understanding of the processes involved, since these play an important role in the selection of the target(s) at any time.

There are no quantitative models of complex systems prediction that take into account the environment within which systems are design, but many descriptive models of the design process.

Design as a complex collaborative process involves iterating around the coevolutionary cycles of Figure 5. Usually the 'client' occupies the left part of the diagram, deciding what they want and don't want, and what they like or don't like. Here the client is shown as policy makers – the people that have the mandate to decide what *ought* to be and the money to commission its design and implementation.

Design is characterised by juggling constraints and making compromises, from the sketch stage all the way through to the blueprint and implementation. Even when new systems are being constructed constraints may change as new problems are discovered. The design process implicitly involves designing the *management* of the systems when it has been built.

From conception to delivery and day-to-day performance, the design process involves predictions of many kinds at many levels. This is the opportunity – the only opportunity – for complex systems scientists to engage and to be allowed to be involved in *in-vivo* experiments.

10   The New Statistics

As presented here, prediction in policy and design is much more complicated than in conventional experiments, which are generally contrived to be as simple as possible. What can it mean to *test* a prediction in this context? Certainly conventional statistical techniques can be applied to local predictions, but how can the 'correctness' of the design process be tested in a rigorous statistical way? Of the many complications, what does it mean to make a prediction of a multilevel systems? In physics we are content that the gas laws give highly reliable predictions at the macrolevel, while the states of individuals are unknowable at the microlevel. In complex socio-technical systems the behaviour of individuals at the microlevel can have massive effects at meso and macro levels. This suggests that isolated single-level predictions will not do for complex systems, and that statistical tests will themselves have to be multilevel.

Conventional statistical methods were not developed for the kind of large complex multilevel systems discussed in this paper. For example, in the UK the major North-South M1 Motorway has been redeveloped considerably over the last few years, increasing the number of lanes in each direction to four, redesigning many intersections, and replacing a number of bridges. Such a project involves hundreds if not thousands of interacting predictions. What methods could be developed for testing any or all of those predictions? The discussion in this paper suggests that there is a completely new approach to statistical analysis waiting to be discovered and developed.

11   Conclusion

It has been argued that complex systems scientists cannot by themselves perform experiments on complex socio-technical systems, and that the best they can do is to perform experiments alongside policy makers who are constantly undertaking large and small experiments. In this context it has been shown that the nature of *prediction* in the implementation of real systems is much more complicated than it is in traditional science. In particular, the *goals* identified by policymakers change through time, and this is usually managed through the design and management processes. It is suggested that the combination of policy and design is the opportunity – the only opportunity – for complex systems scientists to be involved in *in-vivo* experiments. In turn this opens up the need for new methodological and statistical approaches for the science of complex systems.

References

1. Johnson, J.H., Alexiou, K., Creigh-Tyte, A., Chase, S., Duffy, A., Eckert, C., Gasgoine, D., Kumar, B., Mitleton-Kelly, E., Petry, M., Qin, S.-F., Robertson, A., Rzevski, G., Teymur, N., Thompson, A., Young, R., Willis, M., Zamenopoulos, T.: Embracing Complexity in Design. In: Inns, T. (ed.) Designing for the 21st Century: Interdisciplinary Questions and Insights, pp. 129–149. Design Council, London (2007)
2. Ashby, W.R.: Cybernetics. Chapman & Hall Ltd., London (1956)
3. Johnson, J.H.: Science and policy in designing complex futures. Futures 40, 520–536 (2008)
4. Johnson, J.H.: Hypernetworks in the science of complex systems. Imperial College Press, London (2009)
5. Simon, H.: The sciences of the artificial. MIT Press, Boston (1969)

Phase Transition of Active Rotators in Complex Networks

Seung-Woo Son[1], Hawoong Jeong[1], and Hyunsuk Hong[2]

[1] Department of Physics, Institute for the BioCentury, KAIST,
373-1 Guseong-dong, Yuseong-gu, Daejeon 305-701, Korea
[2] Department of Physics, Research Institute of Physics and Chemistry,
Chonbuk National University, Jeonju 561-756, Korea

Abstract. We study the nonequilibrium phenomena of a coupled active rotator model in complex networks. From a numerical Langevin simulation, we find the peculiar phase transition not only on globally connected network but also on other complex networks and reveal the corresponding phase diagram. In this model, two phases — stationary and quasi-periodic moving phases — are observed, in which microscopic dynamics are thoroughly investigated. We extend our study to the non-identical oscillators and the more heterogeneous degree distribution of complex networks.

Keywords: active rotator model, phase transitions, complex networks.

1 Introduction

Various coupled oscillatory systems in nature have been known to exhibit many interesting behaviors including synchronization. Collective synchronization has attracted much interest due to the beauty of simultaneousness and the spontaneous emergence in such phenomena as the synchronous flashing of fireflies, the chorusing of crickets, and the clapping of hands after an astonishing orchestral performance [1]. In order to understand such synchronized behaviors, nonlinear coupled oscillators have been studied extensively with various models. Among them, the Kuramoto model is one of the most studied models due to its simplicity and analytical tractability [2,3]. The Kuramoto model has been extended with many variations for applications in diverse systems [3]. One natural extension is to add external fields, which implies the external current applied to a neuron to describe an excitable systems. This is also known to be an *active rotator* model when each oscillator has the constant natural frequency [3,4].

Most studies of the active rotator model have assumed that all oscillators are connected to each other, i.e., globally connected network, or sometimes 2 and 3-dimensional regular lattice is used [4,5,6]. However, such a type of interaction has a limitation when applied to most real systems. Therefore we need to consider such nontrivial connectivity and extend the study of synchronization to complex networks. Thus, in the present paper, we report our study of active rotator model in complex networks.

J. Zhou (Ed.): Complex 2009, Part I, LNICST 4, pp. 242–246, 2009.

2   Model System

The dynamics of N coupled limit-cycle oscillators having the phase $\{\phi_i(t)|i = 1, 2, \ldots, N\}$ is described by the set of equations

$$\frac{d\phi_i}{dt} = \omega_i - b\sin\phi_i - \frac{K}{\langle k \rangle}\sum_{j=1}^{N}a_{ij}\sin(\phi_i - \phi_j) + \eta_i(t). \tag{1}$$

The first term ω_i represents the natural frequency of the ith oscillator, which is assumed the random normal distribution having the correlation $\langle \omega_i\omega_j \rangle = \sigma^2\delta_{ij}$ with the variance σ^2 and the mean $\langle \omega_i \rangle = \omega_0$. The second and third terms indicate the pinning force and the coupling between the oscillators respectively; the coupling strength K is set to be a positive one ($K > 0$), so the interacting oscillators favor their phase difference minimized. The adjacency matrix element $a_{ij} = 1(0)$ if oscillators i and j are connected (disconnected), and $\langle k \rangle$ denotes the mean degree given by $\sum_i k_i/N$, where the degree $k_i = \sum_j a_{ij}$. In the last term of Eq. (1), $\eta_i(t)$ is the Gaussian white noise with properties $\langle \eta_i(t) \rangle = 0, \langle \eta_i(t)\eta_j(t') \rangle = 2D\delta(t - t')\delta_{ij}$.

When all oscillators are connected to each other, i.e., $a_{ij} = 1$ for all $i \neq j$, and $b = 0$, $D = 0$, the model corresponds to the original Kuramoto model [2]. If all oscillators are identical and $b = 0$, it describes the thermodynamic system of classical XY spins, where D plays role of the temperature of the spin systems [6]. When all oscillators have the same frequency, we call the system as active rotators.

Collective phase synchronization is conveniently described by the order parameter defined by

$$r(t)e^{i\theta(t)} \equiv \frac{1}{N}\sum_{j=1}^{N}e^{i\phi_j(t)}, \tag{2}$$

where $r > 0$ implies emergence of the phase synchronization. Then we take the time average of $r(t)$ such as $r \equiv \overline{r(t)} = (2/T)\sum_{t=T/2+1}^{T}r(t)$, where the over line represents the time averaging and we set T to enough large number after confirming the state passes over the transient period. In the case of the original Kuramoto model, the time averaged r delivers most information since $r(t)$ saturates to a value r. However, active rotators do not always go to the stationary phase but show periodic behavior. Therefore, Shinomoto et $al.$ [4] introduced another order parameter σ and a kind of fluctuation measure $\tilde{\chi}$ defined by

$$\sigma e^{i\varphi} \equiv \overline{r(t)e^{i\theta(t)}} = \frac{2}{T}\sum_{t=T/2+1}^{T}r(t)e^{i\theta(t)}, \tag{3}$$

$$\tilde{\chi} \equiv N \cdot \overline{|r(t)e^{i\theta(t)} - \sigma e^{i\varphi}|^2}. \tag{4}$$

One can easily show that $\tilde{\chi}$ is equivalent to $N \cdot [\overline{r^2(t)} - \sigma^2]$, which measures the difference between r and σ.

In order to investigate phase transition, we have performed a numerical simulation of Eq. (1). We use the second-order Runge-Kutta method [7] with discrete time step $\Delta t = 0.01$. For given b and D, we get the total simulation time $T = 2 \times 10^4$ steps so the first 10^4 steps are discarded as a transient period to achieve steady state and 10^4 steps are used to compute the order parameters.

3   Phase Transition in Active Rotator Model

First of all, we fix the natural frequency $\omega_i = 1$ for all i in order to study the active rotator model on Erdös-Rényi random networks [8]. To generate the random network, we visit each node and connect to other nodes with the probability $p = \langle k \rangle / (N - 1)$, where we fix $\langle k \rangle = 5$ for convenience. Then we perform a numerical simulation on the Eq. (1), and investigate phase transition.

Figure 1 shows the phase diagram with $K = 5$. When $D = 0$, the active rotators show a transition at $b = 1$, which corresponds to the natural frequency $\omega = 1$. For $b > 1$, the system becomes a steady state and rotators are fixed to specific angle, otherwise rotators are synchronized and move periodically. And when $b = 0$, Eq. (1) becomes simply coupled identical oscillators without any external fields. As D increases, the order parameter r becomes smaller since the noise disturbs the oscillators to be synchronized. Finally, the system becomes a desynchronized state at $D = 2.5$, which corresponds to the half of coupling strength $K/2$. This transition point $D = K/2$ well agrees with the result of globally coupling case and overall features of phase diagram are not much different from the mean-field expectation.

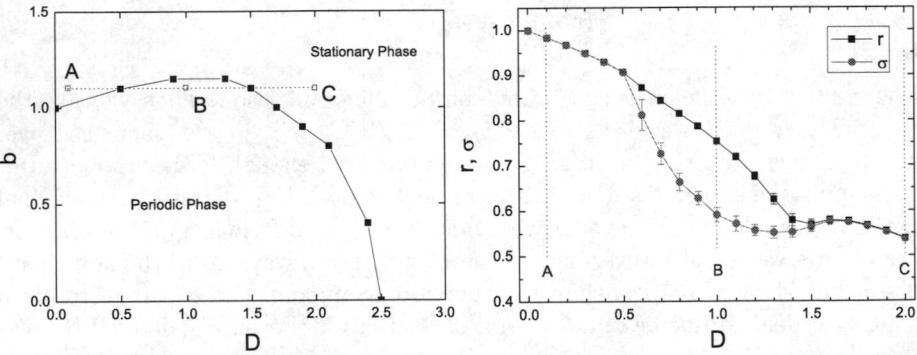

Fig. 1. Phase diagram for the random network with fixed $\omega_i = 1$ for every node i and $K = 5$ (left). When b and D is small enough, the active rotator behaves periodic motion. However, the active rotator goes to the stationary phase if the external field strength b or noise strength D gets strong. On the right: Order parameter behavior for $b = 1.1$. When the noise amplitude is small, oscillators fixed to the external field potential. However, if noise becomes a proper level, oscillators show the periodic motion. For the strong noise, each oscillator scatters and overall behavior shows stationary state.

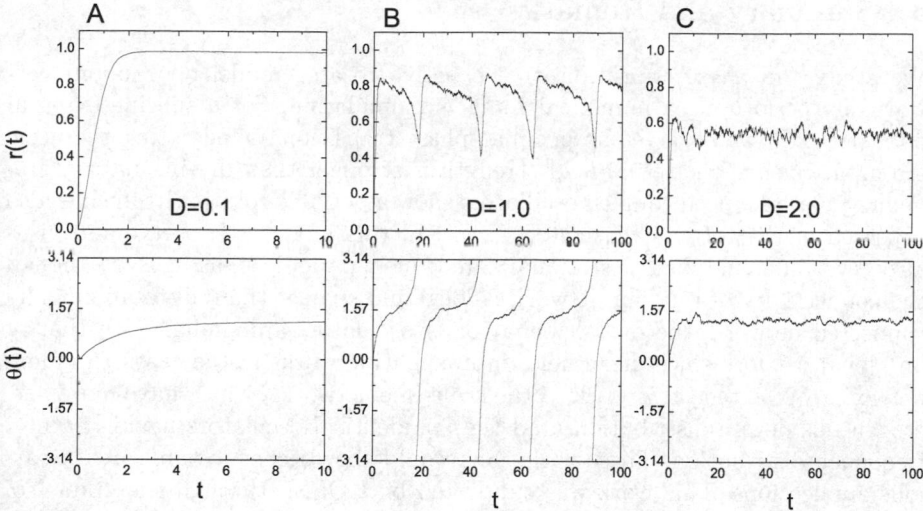

Fig. 2. Three phase states for $b = 1.1$. From the phase diagram, we observe the behaviors of order parameter at the three different points. The label A, B, and C correspond with the three points in Fig. 1.

Noise doesn't always make synchronization bad. For instance, when $b = 1.1$, since the rotator natural frequency ω is smaller than the external fields strength, rotators cannot overcome the external field and stay at a fixed point if the coupling and noise are very weak. However, if the coupling is strong enough and moderate strength of noises are added, the rotators shows a periodic motion. First several oscillators excited by noises try to overcome the external potential and these oscillators pull other oscillators to upward. Average phase gradually moves up to the peak of the potential and then slides down fast. This transition is shown in the left panel of Fig. 1. After $D = 0.5$ the system shows a periodic phase, then it becomes a stationary state again above $D = 1.5$. These three different phases are shown in Fig. 2. The label A, B, and C corresponds to that of Fig. 1. As shown in the middle panels in Fig. 2, the average angle θ rotates quasi-periodically, which means oscillators are rotating together. Therefore, the order parameter σ becomes small since σ is calculated from averaging over complex order parameters including phase information also. Each different phases cancel each other. In this reason, we can find the transition points by observing the difference between order parameters r and σ or the divergent behaviors of the susceptibility $\widetilde{\chi}$.

We extend this study to the non-identical oscillators, i.e. oscillators having the natural frequency distribution [5], and the complex networks having more heterogeneous degree distribution such as *scale-free* networks [9]. As a primary result, we observe that the area of phase diagram is enlarged as networks' degree distribution becomes more heterogeneous. And if oscillators have the natural frequency distribution, the dynamics become more complicated since oscillators make several clusters which have a similar effective frequency.

4   Summary and Remarks

We study the phase transition of the active rotator model on random networks by performing a numerical Langevin simulation. For a specific external field strength, we observe the peculiar phase transition as increasing the noise strength. Even the external field strength is stronger than driving natural frequency, noise-induced coupled oscillators show a periodic rotations, which is also observed in that of globally coupled case. In this model we observe and visualize two different phases, stationary and quasi-periodic phases. Even though the connectivity of random network is local and sparser than all-to-all globally connected network, the overall behavior of oscillators are similar with that of on global network since mean-field approximation works in the case of random networks. We extend this work to the scale-free networks, which have more heterogeneous degree distribution, and the nonidentical oscillators having natural frequency distribution. These variations would show richer dynamics with various applications. This work was supported by KOSEF through the grant No. R17-2007-073-01001-0 and R01-2007-000-20084-0 (H.H).

References

1. Winfree, A.T.: The Geometry of Biological Time. Springer, New York (1980); Pikovsky, A.S., Rosenblum, M.G., Kurths, J.: Synchronization: A Universal Concept in Nonlinear Sciences. Cambridge University Press, Cambridge (2001)
2. Kuramoto, Y.: Chemical Oscillations, Waves, and Turbulence. Springer, Berlin (1984); Kuramoto, Y., Nishikawa, I.: Statistical Macrodynamics of Large Dynamical Systems. Case of a Phase Transition in Oscillator Communities. J. Stat. Phys. 49, 569–605 (1987)
3. Acebrón, J.A., Bonilla, L.L., Pérez Vicente, C.J., Ritort, F., Spigler, R.: The Kuramoto Model: A Simple Paradigm for Synchronization Phenomena. Rev. Mod. Phys. 77, 137–185 (2005)
4. Shinomoto, S., Kuramoto, Y.: Phase Transition in Active Rotator Systems. Prog. Theor. Phys. 75, 1105–1110 (1986); Cooperative Phenomena in Two-Dimensional Active Rotator Systems. ibid. 75, 1319–1327 (1986)
5. Sakaguchi, H.: Cooperative Phenomena in Coupled Oscillator Systems under External Fields. Prog. Theor. Phys. 79, 39–46 (1988)
6. Arenas, A., Pérez Vicente, C.J.: Phase Diagram of a Planar XY Model with Random Field. Physica A 201, 614–625 (1993); Exact Long-Time Behavior of a Network of Phase Oscillators under Random Fields. Phys. Rev. E 50, 949–956 (1994)
7. Kim, S., Park, S.H., Ryu, C.S.: Nonequilibrium Phenomena in Globally Coupled Active Rotators with Multiplicative and Additive Noises. ETRI Journal 18, 147–160 (1996); Newton, N.J.: Asymptotically Efficient Runge-Kutta Methods for a Class of Itô and Stratonovich equations. SIAM J. Appl. Math. 51, 542–567 (1991)
8. Erdös, P., Rényi, A.: On Random Graphs. Publ. Math. (Debrecen) 6, 290 (1959); Bollobás, B.: Random Graphs. Academic Press, London (1985)
9. Goh, K.-I., Kahng, B., Kim, D.: Universal Behavior of Load Distribution in Scale Free Networks. Phys. Rev. Lett. 87, 278701 (2001)

Personal Recommendation in User-Object Networks

Tao Zhou

Department of Physics, University of Fribourg, Chemin du Muse 3, CH-1700
Fribourg, Switzerland
Department of Modern Physics and Nonlinear Science Center,
University of Science and Technology of China, Hefei Anhui, 230026, P.R. China
zhutou@ustc.edu

Abstract. Thanks to the Internet and the World Wide Web, we live in
a world of many possibilities we can choose from thousands of movies,
millions of books, and billions of web pages. Far exceeding our personal
processing capacity, this excessive freedom of choice calls for automated
ways to find the relevant information. As a result, the field of information
filtering is very active and rich with unanswered challenges. In this short
paper, I will give a brief introduction on the design of recommender systems, which recommend objects to users based on the historical records of
users' activities. A diffusion-based recommendation algorithm, as well as
two improved algorithms are investigated. Numerical results on a benchmark data set have demonstrated the advantages in algorithmic accuracy.

Keywords: Infophysics, Personal Recommendation, Bipartite Networks,
User-Object Networks, Diffusion.

1 Introduction

The last few years have witnessed an explosion of information that the exponential growth of the Internet and World Wide Web confronts us with an information overload: We face too much data and sources to be able to find out those
most relevant for us. Indeed, we have to make choices from thousands of movies,
millions of books, billions of web pages, and so on. Evaluating all these alternatives by ourselves is not feasible at all. As a consequence, an urgent problem is
how to automatically find out the relevant objects for us, namely information
filtering. A landmark for information filtering is the use of search engine, by
which users could find the relevant web pages with the help of properly chosen
keywords. However, the search engine has three essential disadvantages. First,
it does not take into account personalization and returns the same results for
people with far different habits. Therefore, if a user's habits are different from
the mainstream, even with some *right keywords*, it is hard for him to find out
what he likes from the countless searching results. Secondly, the search engine is
a tool helping users to find out the web pages at least containing some content
known to them. Many web pages, having potentialities to match a user's tastes,

J. Zhou (Ed.): Complex 2009, Part I, LNICST 4, pp. 247–253, 2009.

are, however, completely out of his horizon. In a word, the search engine is only helpful to find *what you know* instead of *what you like*, since you may have no idea of the latter. Thirdly, some tastes, such as the feelings of music and poem, can not be expressed by keywords, even language. The search engine, based on keyword matching, will lose its effectiveness in those cases.

To our knowledge, the most promising way to efficiently filter the overload information is to automatically provide personal recommendations based on the historical record of a user's activities [1,2]. For example, *Amazon.com* uses one's purchase record to recommend books, *AdaptiveInfo.com* uses one's reading history to recommend news, and *Recipefinder.com* uses one's stated interests to recommend restaurants. In a web-based serving system, a recommendation engine could improve loyalty by creating a value-added relationship between the site and the user. Actually, the more a user uses the recommendation engine – teaching it what he wants – the more loyal he is to the site. Recommendation engines also improve cross-sell for E-commerce systems by suggesting additional products for the customer to purchase. For example, the statistical investigation by *VentureBeat.com* shows that the recommendation engine in *Amazon.com* contributes about 35% of sales. Motivated by its significance in economy and society, the design of an efficient recommendation algorithm becomes a joint focus from engineering science to marketing practice, from mathematical analysis to physics community. Various kinds of recommendation algorithms have been proposed, including collaborative filtering [3], content-based analysis [4], spectral analysis [5], iteratively self-consistent refinement [6], principle component analysis [7], and so on.

In this short paper, I will introduce a diffusion-based algorithm for personal recommendation in bipartite user-object networks. Numerical results on a benchmark data set have demonstrated its advantage in algorithmic accuracy. Two improved algorithms are also introduced, which perform even better.

2   Diffusion-Based Algorithm

A recommendation system consists of users and objects, and each user has collected some objects. Denoting the object set as $O = \{o_1, o_2, \cdots, o_n\}$ and the user set as $U = \{u_1, u_2, \cdots, u_m\}$, the recommendation system can be fully described by a bipartite user-object network with $n + m$ nodes, where an object is connected with a user if and only if this object has been collected by this user. Connection between two users or two objects is not allowed. A Reasonable assumption is that the objects a user has collected are what he likes, and a recommendation algorithm aims at predicting his personal opinions (to what extent he likes or hate them) on those objects he has not yet collected. That is to say, given a target user, a recommendation algorithm should provide an ordered list of all the objects having not been collected by this user. Those objects in the top of this list are recommended to this user.

Based on the bipartite user-object network, an object-object network can be constructed, where each node represents an object, and two objects are connected if and only if they have been collected simultaneously by at least one user. We assume a certain amount of resource (i.e., recommendation power) is associated with each object, and the weight w_{ij} represents the proportion of the resource o_j would like to distribute to o_i. For example, in the book-selling system, the weight w_{ij} contributes to the strength of recommending the book o_i to a customer provided he has already bought the book o_j. The weight w_{ij} can be determined following a network-based diffusion process [8,9] where each object distributes its initial resource equally to all the users who have collected it, and then each user sends back what he has received equally to all the objects he has collected. For a general user-object network, the weighted projection onto the object-object network reads [9]:

$$w_{ij} = \frac{1}{k(o_j)} \sum_{l=1}^{m} \frac{a_{il}a_{jl}}{k(u_l)}, \qquad (1)$$

where $k(o_j) = \sum_{i=1}^{n} a_{ji}$ and $k(u_l) = \sum_{i=1}^{m} a_{il}$ denote the degrees of object o_j and user u_l, and $\{a_{il}\}$ is the $n \times m$ adjacent matrix of the bipartite user-object network.

For a given user u_i, we assign some resource (i.e., recommendation power) on those objects already been collected by u_i. In the simplest case, the initial resource vector $\mathbf{f}$ can be set as

$$f_j = a_{ji}. \qquad (2)$$

That is to say, if the object o_j has been collected by u_i, then its initial resource is unit, otherwise it is zero. After the resource-allocation process, the final resource vector is

$$\mathbf{f'} = W\mathbf{f}. \qquad (3)$$

Accordingly, all u_i's uncollected objects o_j $(1 \le j \le n, a_{ji} = 0)$ are sorted in the descending order of f_j', and those objects with highest values of final resource are recommended.

To test the algorithmic accuracy, we use a benchmark data-set, namely *Movie-Lens*. The data consists of 1682 movies (objects) and 943 users, and users vote movies using discrete ratings 1-5. We therefore applied a coarse-graining method similar to that used in Ref. [10]: a movie has been collected by a user if and only if the giving rating is at least 3 (i.e. the user at least likes this movie). The original data contains 10^5 ratings, 85.25% of which are ≥ 3, thus after coarse gaining the data contains 85250 user-object pairs. To test the recommendation algorithms, the data set is randomly divided into two parts: The training set contains 90% of the data, and the remaining 10% of data constitutes the probe. The training set is treated as known information, while no information in the probe set is allowed to be used for prediction.

A recommendation algorithm should provide each user with an ordered queue of all its uncollected objects. For an arbitrary user u_i, if the relation $u_i - o_j$ is in the probe set (according to the training set, o_j is an uncollected object for u_i), we measure the position of o_j in the ordered queue. For example, if there are 1000 uncollected movies for u_i, and o_j is the 10th from the top, we say the position of o_j is 10/1000, denoted by $r_{ij} = 0.01$. Since the probe entries are actually collected by users, a good algorithm is expected to give high recommendations to them, thus leading to small r. Therefore, the mean value of the position value $\langle r \rangle$ (called *ranking score*, which approximately equals one minus the area under the receiver operating characteristic (ROC) curve [11]), averaged over all the entries in the probe, can be used to evaluate the algorithmic accuracy: the smaller the ranking score, the higher the algorithmic accuracy, and vice verse. The average values of ranking scores over 10 independent runs (one run here means an independently random division of data set) are 0.106, 0.122, and 0.140 for the present algorithm, the collaborative filtering[1], and the global ranking method[2], respectively. Clearly, the present diffusion-based algorithm performs the best.

3   Two Improved Algorithms

3.1   Diffusion-Based Algorithm with Tunable Initial Recommendation Power

Consider the initial resource located on object o_i as its assigned recommendation power. In the whole recommendation process, the total power given to o_i is $p_i = \sum_j f_i^j$, where the superscript j runs over all the users u_j. In the above mentioned algorithm, the total power of o_i is $p_i = \sum_j f_i^j = \sum_j a_{ij} = k(o_i)$. That is to say, the total recommendation power assigned to an object is proportional to its degree, thus the impact of high-degree objects (e.g., popular movies) is enhanced. Although it already has a good algorithmic accuracy, this uniform configuration may be oversimplified, and depressing the impact of high-degree objects in an appropriate way could, perhaps, further improve the accuracy. Motivated by this, we propose a more complicated distribution of initial resource to replace Eq. (2):

$$f_j^i = a_{ji} k^\beta (o_j), \tag{4}$$

where β is a tunable parameter. Compared with the original case, $\beta = 0$, a positive β strengthens the influence of large-degree objects, while a negative β weakens the influence of large-degree objects. In particular, the case $\beta = -1$ corresponds to an identical allocation of recommendation power ($p_i = 1$) for each object o_i.

---

[1] The collaborative filtering is based on measuring the similarity between users. For two users u_i and u_j, their similarity can be simply determined by $s_{ij} = \sum_{l=1}^n a_{li} a_{lj} / \min\{k(u_i), k(u_j)\}$. For any user-object pair $u_i - o_j$, if u_i has not yet collected o_j (i.e., $a_{ji} = 0$), the predicted score, v_{ij} (to what extent u_i likes o_j), is given as $v_{ij} = \sum_{l=1, l \neq i}^m s_{li} a_{jl} / \sum_{l=1, l \neq i}^m s_{li}$. For any user u_i, all the nonzero v_{ij} with $a_{ji} = 0$ are sorted in descending order, and those objects in the top are recommended.

[2] The global ranking method sorts all the objects in the descending order of degree and recommends those with highest degrees.

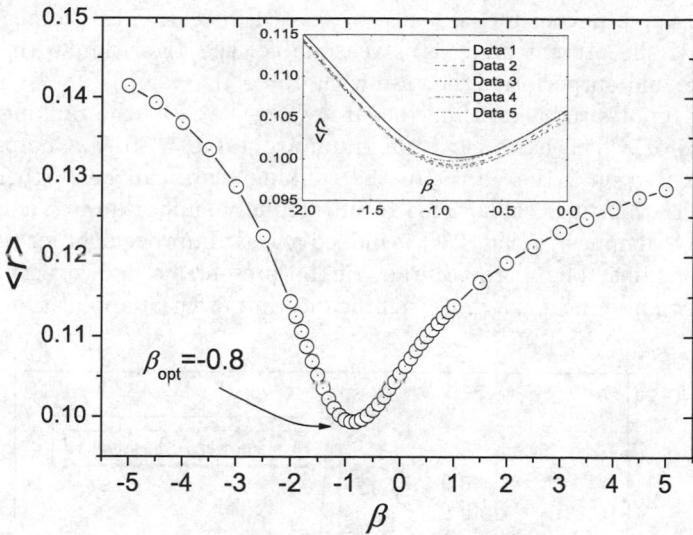

Fig. 1. (Color online) The ranking score $\langle r \rangle$ vs. β. The optimal β, corresponding to the minimal $\langle r \rangle \approx 0.098$, is $\beta_{opt} \approx -0.8$. All the data points shown in the main plot is obtained by averaging over five independent runs with different data-set divisions. The inset shows the numerical results of every separate run, where each curve represents one random division of data-set. After Ref. [12].

Ref. [12] reported the algorithmic accuracy as a function of β. As shown in Fig. 1, the curve has a clear minimum around $\beta = -0.8$. Compared with the uniform case, the ranking score can be further reduced by 9% at the optimal value. It is indeed a nice improvement for recommendation algorithms. Note that β_{opt} is close to -1, which indicates that the more homogeneous distribution of recommendation power among objects may lead to a more accurate prediction.

3.2   Redundant-Eliminated Algorithm

In the diffusion-based algorithm mentioned in Section 2, for any user u_i, the recommendation value of an uncollected object o_j is contributed by all u_i's collected object, as

$$f'_j = \sum_l w_{jl} a_{li}. \tag{5}$$

Those contributions, $w_{jl} a_{li}$, may result from the similarities in same attributes, thus lead to heavy redundance. Generally speaking, if the correlation between o_i and o_k and the correlation between o_j and o_k contain some redundance to each other, then the two-step correlation between o_i and o_k, as well as that between o_j and o_k should be strong. Accordingly, subtracting the higher order correlations in an appropriate way could, perhaps, further improve the algorithmic accuracy. Motivated by this idea, we replace Eq. (5) by

$$\mathbf{f}' = (W + aW^2)\mathbf{f}, \tag{6}$$

where a is a free parameter. When $a = 0$, it degenerates to the algorithm in Section 2. If the present analysis is reasonable, the algorithm with a certain negative a could outperforms the case with $a = 0$.

Figure 2 reports the algorithmic accuracy, measured by the ranking score, as a function of a, which has a clear minimum around $a = -0.75$. Compared with the case in Section 2 (i.e., $a = 0$), the ranking score can be further reduced by 23% at the optimal value. This result strongly supports our analysis. It is worthwhile to emphasize that, 23% is indeed a great improvement for recommendation algorithms. The ultra accuracy of the present method, even far beyond our expectation, indicates a great significance in potential applications.

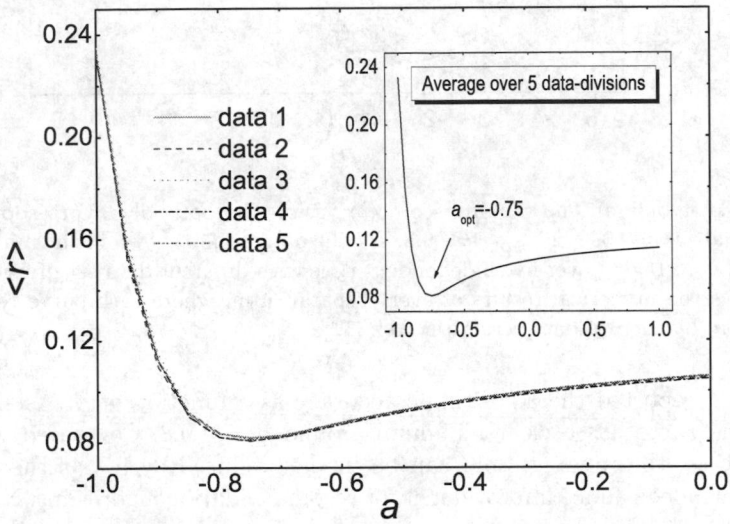

Fig. 2. The ranking score $\langle r \rangle$ *vs.* a. The main plot shows the numerical results of five independent runs, where each run corresponds to a random division of data set. The relation between $\langle r \rangle$ and a is very stable, and the fluctuation induced by the randomness in data division can be neglected. The curve shown in the inset is obtained by averaging over those five independent runs. The optimal a, corresponding to the minimal $\langle r \rangle \approx 0.0822$, is $a_{opt} \approx -0.75$.

4   Conclusion

In this short paper, I introduced a diffusion-based personal recommendation algorithm, which performs obviously better than the commonly used collaborative filtering and global ranking method. In addition, I discussed two improved algorithms with remarkably higher accuracies. The former one has the same computation complexity as the original diffusion-based algorithm, while the latter one is ultra accurate. Those advantages are of significance in potentially real applications.

Acknowledgement

I acknowledge *GroupLens Research Group* for data (http://www.grouplens.org). This work is supported by the National Natural Science Foundation of China under Grant Nos. 60744003 and 10635040.

References

1. Konstan, J.A.: Introduction to recommender systems: Algorithms and Evaluation. ACM Trans. Inf. Syst. 22, 1–4 (2004)
2. Adomavicius, G., Tuzhilin, A.: Toward the next generation of recommender systems: A survey of the state-of-the-art and possible extensions. IEEE Trans. Knowl. Data Eng. 17, 734–749 (2005)
3. Herlocker, J.L., Konstan, J.A., Terveen, K., Riedl, J.T.: Evaluating Collaborative Filtering Recommender Systems. ACM Trans. Inform. Syst. 22, 5–53 (2004)
4. Pazzani, M.J., Billsus, D.: Content-Based Recommendation Systems. Lect. Notes Comput. Sci. 4321, 325–341 (2007)
5. Maslov, S., Zhang, Y.C.: Extracting Hidden Information from Knowledge Networks. Phys. Rev. Lett. 87, 248701 (2001)
6. Ren, J., Zhou, T., Zhang, Y.C.: Information Filtering via Self-Consisent Refinement. Europhys. Lett. 82, 58007 (2008)
7. Goldberg, K., Roeder, T., Gupta, D., Perkins, C.: Eigentaste: A Constant Time Collaborative Filtering Algorithm. Inf. Retr. 4, 133–151 (2001)
8. Ou, Q., Jin, Y.D., Zhou, T., Wang, B.H., Yin, B.Q.: Power-law strength-degree correlation from resource-allocation dynamics on weighted networks. Phys. Rev. E 75, 021102 (2007)
9. Zhou, T., Ren, J., Medo, M., Zhang, Y.C.: Bipartite network projection and personal recommendation. Phys. Rev. E 76, 046115 (2007)
10. Blattner, M., Zhang, Y.C., Maslov, S.: Exploring an opinion network for taste prediction: An empirical study. Physica A 373, 753–758 (2007)
11. Hanely, J.A., McNeil, B.J.: The meaning and use of the area under a receiver operating characteristic (ROC) curve. Radiology 143, 29–36 (1982)
12. Zhou, T., Jiang, L.L., Su, R.Q., Zhang, Y.C.: Effect of initial configuration on network-based recommendation. Europhys. Lett. 81, 58004 (2008)

Performance Analysis of Public Transport Systems in Nanjing Based on Network Topology

Ping Li[1], Zhen-Tao Zhu[2,3], Jing Zhou[2], Jin-Yuan Ding[3], Hong-Wei Wang[3], and Shan-Sen Wei[3]

[1] Department of Basic Sciences, Nanjing Institute of Technology,
Nanjing 211167, China
liping@njit.edu.cn
[2] School of Management and Engineering, Nanjing University,
Nanjing 210093, China
jzhou@nju.edu.cn
[3] School of Economics and Management, Nanjing Institute of Technology,
Nanjing 211167, China
zztnit@yahoo.com.cn

Abstract. The urban public transport network (UPTN) in Nanjing is characterized by a complex network with topological pedestals. The empirical data indicates that it is a small-world network. Under malicious attack to the high connectivity nodes of the network, the average path-length will increase 2.5 times, the reliability and traffic capacity of the UPTN will greatly decline, and the travel expenditure will distinctively increase. The topological significance of stations and routes are redefined to help assess the small-world property of UPTNs, so as to improve city transportation. It is also found that if the urban rail transit, such as metro, is introduced to the UPTN, then the topological diameter of the network is reduced, and its structure is optimized.

Keywords: complex network, network topology, public transport system, urban rail transit.

1 Introduction

One of the common challenges that urban cities are confronted with is traffic congestion. According to the 2005 annual report on the development of urban road traffic of Nanjing, the average waiting time of citizens in Nanjing City at a bus station is 5.80 min, which is the longest time since 1999. Most of roads in Nanjing are almost saturated with traffic flow. On the other hand, the passenger volume of 2005 in the public transportation is less than that of 2003, the year when China suffered from SARS outbreaks. Facing the increasingly serious traffic problem, one needs a new approach to analyze the characteristics of urban public traffic systems (UPTS) from new perspective, so as to deal with the traffic problem in practice; a complex network approach fits to this requirement [1,2,3]. The UPTS is a complex system consisting of thousands of vehicles and passengers, and its performance is directly related to the topological structure of the

J. Zhou (Ed.): Complex 2009, Part I, LNICST 4, pp. 254–264, 2009.

network [4,6,5,7,8,9,10,11]. However, contrasting with the similar findings of the small-world and scale-free properties in other complex networks [1,3], the recent studies of 14 major cities of the world [12] have shown the diversity in statistical and topological properties of UPTNs due to diverging historical evolutions and other external factors such as wartime destruction and political constraints. In this paper, we analyze the characteristics of the UPTN in Nanjing from the novel perspective of topology.

2   Statistic Parameters of Network Topology

In order to quantitatively describe the characteristics of the topological structure of networks, a series of characteristic parameters have been introduced into network analyses [3], such as node degree k, degree distribution $P(k)$, node strength, s, distribution of nodes' strength, $p(s)$, characteristic path length, L, clustering coefficient, C, and betweenness centrality, BC.

2.1   Node Degree and Degree Distribution

The degree of a node, k, means that this node is connected with k edges. In the case of UPTNs, it can be described as "a bus station has k neighbor stations". Here, two stations are defined as neighbor only if one station is the successor of the other in the series serviced by a route. One of the important statistical attributes is the distribution of node degree, for not all nodes have the same number of incident edges. The distribution of nodes' degree can be depicted by density distribution function, $p(k)$, (or cumulative distribution function, $P(k) = \int_k^\infty p(\xi)d\xi$), which represents the probability that a randomly selected node in a network has k neighbor nodes, and is also equivalent to the ratio of the number of the nodes with degree k in the network to the total number of nodes of the network. In a scale-free network, both $p(k)$ and $P(k)$ take power law function forms with respect to node degree. In a UPTN, the distribution of node degree has much impact on the accessibility of the public traffic system. Carefully analyzing the degree distribution in a UPTN and figuring out which function form it satisfies can contribute to understanding the topological characteristics of the UPTN. In the case of weighted networks, another meaningful parameter of topological characteristics is the strength of a node, which is defined as the sum of the weight of all its neighbor nodes, i.e. $s_i = \sum_{j \in V_i} \omega_{ij}$. Here, ω_{ij} denotes the weight of the link from node i to its neighbor node j. In a UPTN, the strength of a node can be described as the total frequency with which different bus routes serve the bus station.

2.2   Characteristic Path Length

Characteristic path length, L, is defined as the average value of all shortest path lengths over all pairs of nodes in a network; it is a characteristic parameter describing the distance between two arbitrary nodes in a whole sense. In a UPTN, it reflects the average number of bus stations one may traverse from one place to another place using the UPTS. It is straight forward that the smaller L the more convenience one travels by bus and the better accessibility the public traffic network has.

2.3 Clustering Coefficient

Assuming that node i has k_i neighbor nodes, there may exist at most $C_{k_i}^2$ edges among k_i nodes; in fact, there are only t edges among them, so the clustering coefficient C_i of node i , $C_i = t/C_{k_i}^2 = 2t/k_i(k_i - 1)$. The clustering coefficient of the whole network, C, is the average of the clustering coefficients over all the nodes in the network, namely, $C = \sum_{i=1}^{n} C_i/n$. The clustering coefficient of a network describes the local clustering characteristics of the network, i.e. measures the tendency that nodes in the network form cliques. In a UPTN, it reflects the local connectivity and intensive degree of public traffic routes. The larger the C, the higher the local connectivity, and the more intensive the urban public traffic routes. It ensures that there will be little impact on the accessibility between any other directly connected stations when a certain bus station is congested. Therefore it contributes to the robustness of public traffic networks.

2.4 Betweenness

Node betweenness reflects the centrality of a node in a network. It can be used to identify the hub nodes in a network. Much information and many other resource flows from a node must traverse these hub nodes to rapidest reach other nodes through shortest paths. The betweenness centrality of node i, denoted by BC_i, can be obtained by counting the sum of the fraction of shortest paths between all pairs of nodes passing through node i in the network. In a UPTN, the betweenness of a node reflects the capability that this bus station acts as transfer station. In this paper, aforementioned network topological statistic parameters will be adopted as major criteria to investigate the topological characteristics and its evolution of the UPTN.

3 Statistic and Topological Properties of the UPTN

3.1 The Construction of UPTNs

The bus station network is defined in a natural way with nodes representing bus stations, such as A and B, and if there is at less one route passing through A and B without any other stations between them, then the two stations are linked. If there are multiple links between A and B, then the number of these routes is assigned to the link (edge) between node A and node B as its weight. Obviously the modeled network is a directed weight network, which depicts the fundamental topological characteristics of UPTNs. Using its basic topological parameters such as average shortest path, shortest path distribution, degree distribution etc, one can study the topological property of the network. For the case study of Nanjing, the UPTN is modeled in this way. This network consists of 224 routes and 1542 stations [13](Fig.1).

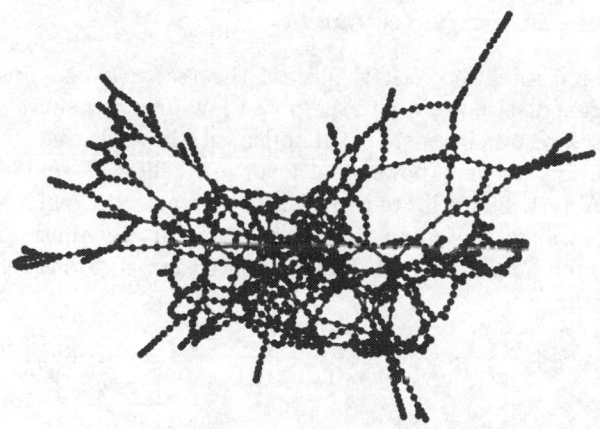

Fig. 1. The topologic structure of the UPTN in Nanjing

3.2 Small World Property Testament

To describe the small world property of the complex networks, Watts et al. [1] introduced two characteristic parameters: characteristic path length (L) and clustering coefficient (C). The small world property is mathematically characterized by the average shortest path length that depends at most logarithmically on the network size n and the clustering coefficient in such a way that

$$C \gg C_{random} \sim <k>/n \tag{1}$$

$$L \gg L_{random} \sim \ln n/\ln <k> \tag{2}$$

where, $<k>$ is the average degree over all nodes in the network, and n is the total number of nodes. The results of the average degree $<k>$, characteristic path length L, and clustering coefficient C of the UPTN in Nanjing are listed in Table 1(L_{random} and C_{random} in Table 1 are the characteristic path length and clustering coefficient of the corresponding random network with same size, respectively).

Table 1. Topological statistical parameters of the UPTN in Nanjing

n	$<k>$	L	L_{random}	C	C_{random}
1542	5.856	17.00	4.15	0.111	0.0038

From Table 1, $C/C_{random} \sim 29.25$ and $L/L_{random} \sim 4.10$ indicate the small-world property of the network. Therefore it can be concluded that the UPTN in Nanjing is a small-world network.

3.3   Scale-Free Property Testament

Many empirical studies show that most of the real networks display a power law shaped degree distribution, and the power law function curve decreases relatively slow, which results in existence of nodes with large degree. Networks whose degree distribution obeys a power-law form are called as scale-free networks. Barabási and Albert [2] attributed the self-organization of real systems into the scale-free structure to two major factors: growth and preferential linking. Scale-free networks often have the small-world property as well. However, according to

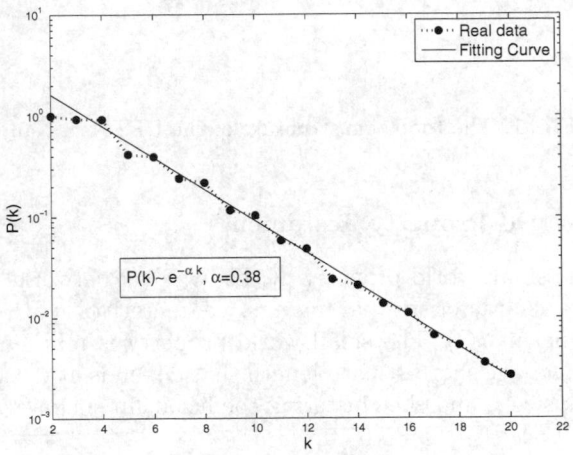

Fig. 2. The degree cumulative distribution of the UPTN in Nanjing

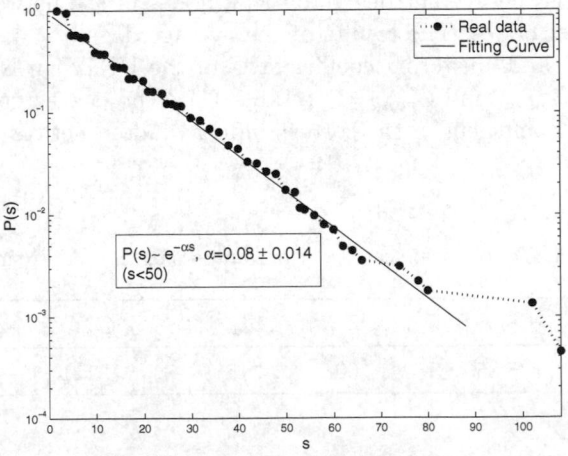

Fig. 3. Semi-logarithmic plot of the strength cumulative distribution of the UPTN in Nanjing

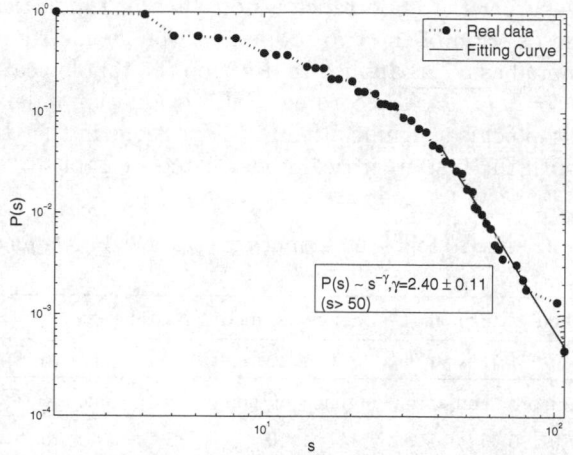

Fig. 4. Log-log scale plot of the strength cumulative distribution of the UPTN in Nanjing

our empirical data of the UPTN in Nanjing, not all the degree distribution and the strength distribution have a typical power law form. The degree distribution (Fig. 2) more likely takes an exponential form, while the strength distribution takes a certain form in between a power law form and an exponential one. The nodes whose strengths are less than 50 seem to take an exponential distribution (Fig. 3), while the nodes with a larger strength appear to take a power law distribution (Fig. 4); which is consistent with some research reports by other scholars [12,14], who have also found that the degree distributions of UPTNs in some cities such as Berlin, Hamburg, Moscow, Hong Kong, Beijing do not take a power law form. Obviously, the finite-size and spatial constraint of the UPTN are two major factors which hinder the formation of the scale-free property. On the other hand, one can see from Fig. 3. and Fig. 4. that the fat-tail distribution results in the coexistence of a few hub nodes and a larger number of poor connected nodes.

4   Effectiveness Analysis of the UPTN Based on Topological Statistics

4.1   Improvement of City Transportation by Enhancing the Small-World Property of UPTN

A few of shortcuts should be built between critical nodes, which may shorten the average path length and improve the reliability of the whole network. To do this, one shall first identify the critical nodes in the network. One possible way is to identify these critical nodes according to the role that nodes play in topological structure. It is known that node degree reflects the total number of a given station's connections, while BC represents the capability that the station acts as a transfer station. The two parameters are both the important characteristic parameters measuring the importance a node has in topological structure. To integrate the description of the two properties, the geometric mean of the two parameters denoted as SI, is applied to describe the topological importance of a node, where, $SI = \sqrt{K \cdot BC}$. According to this evaluation index, the sequencing analysis of the topological significance of 1542 stations in the UPTN in Nanjing is performed, with the first 10 critical nodes listed in Table 2.

Table 2. First 10 topologic significance stations in Nanjing UPTN

Station	Chalukou	Xinzhuang	Zhongyang men	Nanjing railway station	Huamugongsi
SI value	1.26×10^{-2}	1.21×10^{-2}	1.19×10^{-2}	0.91×10^{-2}	0.87×10^{-2}
Station	Hedingqiao	Yuhuatai	Gongjiaozong gongsi	Changlelu	Xinjiekou
SI value	0.79×10^{-2}	0.76×10^{-2}	0.75×10^{-2}	0.74×10^{-2}	0.69×10^{-2}

Aiming to improve the traffic capability between the main city and new towns of Nanjing, the Xincheng bus company offers a No.101 bus service, which starts from Dongshanzongzhan, passes through Jingfashichang, Dajiedongzhan, Xinyilu, Chengzhong, Zhushanlu, Fuqianlu, Gongxiaoshangxia, Wuyihuayuan, Hedingqiao, Shijiali, Chalukou, Yanhuihongcun, and ends at Zhonghuamennei. According to the company's estimation, bus route No.101 will run efficiently. In our analysis (Table 2), Chalukou, Hedingqiao, and Zhonghuamennei rank the first, sixth, fifteenth place in the significance index of 1542 stations, respectively, which provides the theoretic basis for running this route. In the light of station significance index, one can define the significance index of a route, LI, by the average of SI over all stations of the route. Assuming that a route has m stations and the significance index of node i is SI_i, then the significance index of the route is $LI = \sum_{i=1}^{m} SI_i / m$.

4.2   Robustness and Vulnerability Analysis of UPTNs

Small-world networks have a common feature that they are robust against random attacks, yet vulnerable to malicious attacks. The robustness (vulnerability) of a network can be measured to find out whether the network will still connected after some of nodes have been deleted. Owing to the inequality of the role that different nodes in a network play, a few of critical nodes play a key role in the running of the whole network; which makes the network to be highly vulnerable when these nodes are attacked deliberately. That is to say, if only those small amount of nodes (not more than 5%) with largest connectivity are halted or become congested, which may cause the entire network fail to work. So the small-world scale-free property brings about the advantage of effectiveness and fast communication of the network, but might result in the quick spread of congestion as well. For this reason, we should attach much attention to the construction of the key stations from the following aspects: protecting key nodes and attaching much importance to "long-range link". Protecting the transport capability of key stations is able to enhance the transfer capability of the whole network as well as to prevent traffic jam from spreading quickly. As learned from the study on small-world networks research, a few shortcuts will shorten the average distance L significantly, which improves the transit capability of the network. This has been demonstrated by the example of Nanjing Xincheng bus company's bus route No.101 operation. To assess the robustness of Nanjing UPTN, all the stations have been sequenced by their node degree and betweenness (BC), respectively. After the removal of the 1% stations with the largest degree and betweenness, and the 1% randomly selected stations separately, the average degree $< k >$, characteristic path length L, clustering coefficient C, and other topological characteristic parameters are recalculated with the results from both cases listed in Table 3.

Table 3. Changes in characteristic parameters of Nanjing UPTN after the removal of 1% stations

Type of removal	1% largest degree	1% largest BC	1% random selected	No removal
$< k >$	5.452	5.633	5.774	5.856
L	42.8959	40.3965	26.5748	17.0056
L_{random}	4.32	4.24	4.18	4.15
C	0.0973	0.1053	0.1078	0.111
C_{random}	0.00357	0.00278	0.00378	0.00380

It can be learned from the Table 3 that the small-world property of the network do not change in whatever way one delete the 1% stations. However, under the deliberate attacks the average path length increases 2.5 times relative to the original one, which suggests that the reliability and transit capacity of the network are heavily declined and the travel cost of citizens will increase remarkably. In comparison, change in the average path length due to the removal of 1% randomly selected vertexes is much less than that caused by malicious attacks. This is accordant with the robustness and vulnerability of small-world networks.

5   Synergetic Relation between Urban Rail Transit and UPTN

Urban rapid rail transit systems enjoy the advantage of large volume, high speed, less pollution and energy consumption. Vigorously development of urban rail transit has a great significance in mitigating the congestion of urban transportation and in improving urban atmosphere environment. Meanwhile, rail transit promotes the optimization of the spatial structure, and quickens the communication between the city center and sub-centers of a city as well.

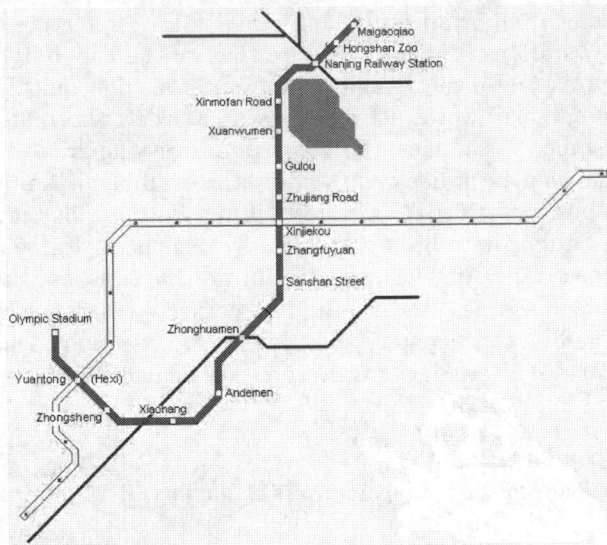

Fig. 5. Map of Nanjing Metro Line 1 and 2 (source:www.urbanrail.net)

Table 4. Changes in characteristic parameters of Nanjing UPTN before and after the joint of Metro Line 1 and 2 (R is the diameter of the network, i.e. the longest one of all shortest paths in the network)

Traffic mode	Normal bus	Bus and Metro Line 1	Bus and Metro Line1,2
$<k>$	5.856	5.869	5.885
L	17.00	16.59	16.59
R^*	56	53	53
L_{random}	4.15	4.14	4.14
C	0.111	0.111	0.111
C_{random}	0.0038	0.00278	0.00378

In the case of Nanjing rail transit system, Nanjing Metro Line 1 runs across main city from north to south, while Line 2 traverses main city from east to west, which effectively shortens the spatiotemporal distance between main city and three new towns (Xianlin, Hexi, Jiangning). This conclusion can be verified by the analysis of the diameter of Nanjing UPTN. Table 4 demonstrates the variation of the topological characteristics of the UPTN before and after the joint of the rail transits.

It can be seen from Table 4 that despite of the long length of the two Metro Lines the small-world property of the network is strengthened after their joint. Not only the average path length L but also the network diameter R reduce as well. The latter is more suggestive of that the spatiotemporal distances between the center and sub-centers of the city have been optimized effectively.

6   Conclusion

Topological structure of the UPTN determines the connectivity and accessibility of public transportation service in such a kind of complex networks. Therefore, it is meaningful to study the topological feature such as the small-world and scale-free properties of the networks. The case study of the UPTN in Nanjing shows that it does have such properties. Meanwhile, it is demonstrated theoretically and practically that the "long-range link" of a network system takes a very important role in enhancing the accessibility of the public transportation. It should be mentioned that the development of urban rail transit shortens the topological diameter of the whole network, which contributes to optimization of city's spatial structure and communication between the center and sub-centers of the city. The complex network model can be used to analyze dynamic behaviors of urban public transport systems, and furthermore to forecast and control the systems, therefore it may possess engineering meaning and potential application value.

Acknowledgments. This work was partly supported by the State Key Development Program for Basic Research of China (Grant No 2006CB705500), and the National Natural Science Foundation of China (Grant No 70571033) and the Student Foundation (S20080803) of Nanjing Institute of Technology.

References

1. Watts, D.J., Strogatz, S.H.: Collective dynamics of "small-world" networks. Nature 393, 440–442 (1998)
2. Barabasi, A.L., Albert, R.: Emergence of Scaling in Random Networks. Science 286, 509–512 (1999)
3. Albert, R., Barabasi, A.L.: Statistical mechanics of complex networks. Reviews of Modern Physics 74, 47–97 (2002)
4. Guimera, R., Mossa, S., Turtschi, A., et al.: The worldwide air transportation network: Anomalous centrality, community structure, and cities' global roles. Proceedings of the National Academy of Sciences 102, 7794–7799 (2005)
5. Kurant, M., Thiran, P.: Extraction and analysis of traffic and topologies of transportation networks. Physical Review E 74, 36114 (2006)
6. Lee, K., Jung, W.S., Park, J.S., et al.: Statistical analysis of the Metropolitan Seoul Subway System: Network structure and passenger flows. Physica A: Statistical Mechanics and its Applications 387, 6231–6234 (2008)
7. Sienkiewicz, J., Holyst, J.A.: Statistical analysis of 22 public transport networks in Poland. Physical Review E 72(4), 46127 (2005)
8. Kalapala, V., Sanwalani, V., Clauset, A., et al.: Scale invariance in road networks. Physical Review E 73, 26130 (2006)
9. Zheng, J.F., Gao, Z.Y.: A weighted network evolution with traffic flow. Physica A: Statistical Mechanics and its Applications 387, 6177–6182 (2008)
10. Wu, J.J., Gao, Z.Y., Sun, H.J., Huang, H.J.: Urban transit as a scale-free network. Mod. Phys. Lett. B 18, 1043 (2004a)

11. Xu, X., Hu, J., Liu, F., et al.: Scaling and correlations in three bus-transport networks of China. Physica A: Statistical Mechanics and its Applications 374, 441–448 (2007)
12. Von, F.C., Holovatch, T., Holovatch, Y., et al.: Network harness: Metropolis public transport. Physica A: Statistical Mechanics and its Applications 380, 585–591 (2007)
13. Nanjing passenger transport management office, http://www.njkgc.cn
14. Chen, Y.Z., Li, N.: The randomly organized structure of urban ground bus-transport networks in China. Physica A: Statistical Mechanics and its Applications 386, 388–396 (2007)

Non-sufficient Memories That Are Sufficient for Prediction

Wolfgang Löhr[1] and Nihat Ay[1,2]

[1] Max Planck Institute for Mathematics in the Sciences,
Inselstraße 22, D-04103 Leipzig, Germany
{Wolfgang.Loehr,Nihat.Ay}@mis.mpg.de
[2] Santa Fe Institute, 1399 Hyde Park Road, Santa Fe, New Mexico 87501, USA

Abstract. The causal states of computational mechanics define the minimal sufficient (prescient) memory for a given stationary stochastic process. They induce the ε-machine which is a hidden Markov model (HMM) generating the process. The ε-machine is, however, not the minimal generative HMM and minimal internal state entropy of a generative HMM is a tighter upper bound for excess entropy than provided by statistical complexity. We propose a notion of prediction that does not require sufficiency. The corresponding models can be substantially smaller than the ε-machine and are closely related to generative HMMs.

Keywords: hidden Markov models, HMM, computational mechanics, causal states, ε-machine, prediction.

1 Introduction

Computational mechanics is a theory developed by Crutchfield, Young, Shalizi and others ([1,2]). It tackles the problem of building predictive models of stationary stochastic processes[1] and finding the minimal such model. This problem is solved by the so-called ε-machine which operates on the causal states. Although the ε-machine is a hidden Markov model (HMM) and minimal under the assumptions of computational mechanics, it is (in general) distinct from and can be much larger than the minimal HMM capable of generating the process. In the literature, this distinction is not always clear. Also, minimal entropy of a generative HMM provides a tighter upper bound for excess entropy than statistical complexity does (see Example 7).

In the present paper, we compare and highlight the difference between the approach of computational mechanics, which is based on the fundamental concept of sufficient statistics, and the construction of the minimal generative HMM. We propose a notion of predictive model that is weaker than sufficiency and thereby allows for smaller models. More specifically, we require our models to be able to generate a prediction of the future that follows the same conditional distribution as the real future (Section 4). It turns out that if a process is generated by

---

[1] Extensions to spatio-temporal systems exist, but we do not consider them here.

J. Zhou (Ed.): Complex 2009, Part I, LNICST 4, pp. 265–276, 2009.

an HMM, the minimal predictive model in our sense cannot be larger than the original HMM. We have already presented main idea and results of the present paper in [3]. Therefore, we omit the proofs of the propositions in this less technical review; they can be found in the appendix of [3]. Complementary to [3], we discuss the relation between excess entropy, statistical complexity and the size of generative HMMs (Corollary 6 and Example 7).

2   Sufficient Statistics and Causal States

Consider a stationary stochastic process $X_{\mathbb{Z}} = (\ldots, X_{-1}, X_0, X_1, \ldots)$ on a discrete alphabet D. We interpret $X_{-\mathbb{N}_0}$ as the observed past and $X_{\mathbb{N}}$ as the future, which we want to predict. Not all information of $X_{-\mathbb{N}_0}$ is necessary for predicting $X_{\mathbb{N}}$. Therefore, one tries to compress the relevant information in a memory variable M, which assumes values in a set M of memory states, via a memory kernel (transition probability) mem. This is illustrated as

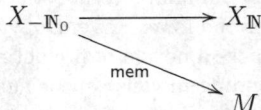

Sometimes, we call both the memory variable M and the memory kernel mem simply *memory*. No confusion arises, as one determines the other. For technical simplicity, we restrict to countable M, although this restriction is not necessary (see the appendix of [3]).

The usual approach in computational mechanics is to consider the special case of deterministic functions instead of memory kernels mem, but recently an extension to stochastic maps has been considered by Still and Crutchfield ([4]). We adopt this extension and do not require mem to be deterministic, allowing for a stochastic assignment. That is

$$\text{mem} \colon D^{-\mathbb{N}_0} \to \mathcal{P}(M) \quad \text{measurable,}$$

where $\mathcal{P}(M)$ denotes the set of probability measures on M. Note that M is embedded in $\mathcal{P}(M)$ via Dirac measures and thus a (measurable) deterministic memory function $f \colon D^{-\mathbb{N}_0} \to M$ induces a memory kernel $\text{mem}_f(x_{-\mathbb{N}_0}) = \delta_{f(x_{-\mathbb{N}_0})}$, where δ_m is the Dirac measure in m. In general, mem reduces the information about the future, which is expressed by the following inequality:

$$I(M : X_{\mathbb{N}}) \leq I(X_{-\mathbb{N}_0} : X_{\mathbb{N}}) =: E(X_{\mathbb{Z}}).$$

where I denotes the mutual information between two random variables[2] and E is the *excess entropy*, an important complexity measure also known as *effective measure complexity* and *predictive information* ([5,6]). In computational

---

[2] $X_{-\mathbb{N}_0}$ and $X_{\mathbb{N}}$ are not discrete-valued. Their mutual information is defined by the limit $I(X_{-\mathbb{N}_0} : X_{\mathbb{N}}) := \sup_{n,m} I(X_{[-n,0]} : X_{[1,m]}) = \lim_{n \to \infty} I(X_{[-n,0]} : X_{[1,n]})$.

mechanics, one requires that the memory preserves all information about the future. This property is called *prescient* ([2]) and formalized by

$$I(M : X_{\mathbb{N}}) = E(X_{\mathbb{Z}}).\tag{1}$$

It is this central requirement that ensures minimality of causal states (Proposition 1) and ε-machine while ruling out smaller hidden Markov models. We will relax it in Section 4 to a different notion of "predictive". Requirement (1) is equivalent to conditional independence of past and future given the memory:

$$X_{-\mathbb{N}_0} \perp\!\!\!\perp X_{\mathbb{N}} \mid M.$$

Using the language of statistics, we say that such a memory is *sufficient* for the future, or simply that M is a **sufficient memory**. Sufficient memories are the candidates for *predictive models* proposed by computational mechanics. It is natural to ask how big a sufficient memory has to be and how to obtain a minimal one. There are mainly two possibilities to measure the size of a memory: cardinality $|M|$ of the set of memory states and Shannon entropy $H(M)$ of the memory variable. Both notions of size, however, yield the same notion of minimality and the unique solution is given by the causal states, which are constructed in the following way: We identify two history trajectories, $x_{-\mathbb{N}_0}, \hat{x}_{-\mathbb{N}_0} \in D^{-\mathbb{N}_0}$, if they induce the same conditional probability on the future, i.e.

$$x_{-\mathbb{N}_0} \sim \hat{x}_{-\mathbb{N}_0} \quad :\Leftrightarrow \quad P(X_{\mathbb{N}} \mid X_{-\mathbb{N}_0} = x_{-\mathbb{N}_0}) = P(X_{\mathbb{N}} \mid X_{-\mathbb{N}_0} = \hat{x}_{-\mathbb{N}_0}).^3$$

The *causal state* $\mathfrak{C}(x_{-\mathbb{N}_0})$ of $x_{-\mathbb{N}_0}$ is its equivalence class,

$$\mathfrak{C}(x_{-\mathbb{N}_0}) := \{\, \hat{x}_{-\mathbb{N}_0} \mid x_{-\mathbb{N}_0} \sim \hat{x}_{-\mathbb{N}_0} \,\},$$

and the function $\mathfrak{C}$ defines a deterministic sufficient memory (see [2]).[4] Its set of memory states is the set of causal states,[5]

$$M_{\mathfrak{C}} := \operatorname{Im}(\mathfrak{C}) = \{\, \mathfrak{C}(x_{-\mathbb{N}_0}) \mid x_{-\mathbb{N}_0} \in D^{-\mathbb{N}_0} \,\},$$

and the memory kernel $\mathrm{mem}_{\mathfrak{C}}$ is defined by $\mathrm{mem}_{\mathfrak{C}}(x_{-\mathbb{N}_0}) = \delta_{\mathfrak{C}(x_{-\mathbb{N}_0})}$, the Dirac measure in the corresponding causal state. It is well-known that the set $M_{\mathfrak{C}}$ of causal states is the minimal prescient partition of $D^{-\mathbb{N}_0}$. Consequently, $\mathrm{mem}_{\mathfrak{C}}$ is the minimal sufficient deterministic memory. This property easily extends to the non-deterministic case:

Proposition 1 (minimality of causal states). *Any sufficient memory with set* M *of memory states and memory variable* M *satisfies*

$$|M| \geq |M_{\mathfrak{C}}| \quad and \quad H(M) \geq H(M_{\mathfrak{C}}).$$

---

[3] $P(X \mid Y = y) = P(X \mid Y = \hat{y})$ means that $P(X \in B \mid Y = y) = P(X \in B \mid Y = \hat{y})$ for every measurable set (event) B.

[4] We fix a regular version of conditional probability $P(X_{\mathbb{N}} \mid X_{-\mathbb{N}_0})$. Therefore, the function $\mathfrak{C}$ is measurable and the causal states are measurable subsets of $D^{-\mathbb{N}_0}$.

[5] In general, $M_{\mathfrak{C}}$ need not be countable. Here, we restrict to processes with a countable number of causal states. For the more general case, see the appendix of [3].

Due to the minimality of the causal states, their entropy

$$C_{\mathfrak{C}}(X_{\mathbb{Z}}) := H(M_{\mathfrak{C}})$$

is an important complexity measure called *statistical complexity*. It is evident from (1) that statistical complexity is lower bounded by excess entropy.

A memory kernel mem does not only induce a (random) memory state $M = M_0$ at time zero, but a whole stationary process $M_{\mathbb{Z}}$ of memory states. The conditional distribution of $M_{\mathbb{Z}}$ is computed as

$$P(M_{[0,T]} = m_{[0,T]} \mid X_{\mathbb{Z}} = x_{\mathbb{Z}}) = \prod_{k=0}^{T} \mathsf{mem}(x_{]-\infty,k]}; m_k), \qquad T \in \mathbb{N}_0,$$

where we use the notation $[0,T]$ for the discrete interval $\{0, \dots, T\}$ and $M_{[0,T]} = m_{[0,T]}$ for $M_0 = m_0, \dots, M_T = m_T$. Note that the process $M_{\mathbb{Z}}$ of a sufficient memory need not be Markovian. However, the memory process of the *minimal sufficient memory*, i.e. the process of causal states, is always Markovian ([2]).

3   Hidden Markov Models (HMMs) and ε-Machine

Sufficient memories, such as given by the causal states, contain all information about the future that is available in the past. How do we actually extract this information and justify the term "model" for sufficient memories? In computational mechanics, the ε-machine describes the mechanism of prediction. It is defined as a *stochastic output automaton*, i.e. a "machine" with the following components: It has a set S of internal states and is initialized by one of these states according to some initial probability distribution $\mu \in \mathcal{P}(\mathsf{S})$. We assume S to be countable for technical simplicity. At each time step t, depending on the current internal state S_t, an output symbol Y_{t+1} from the finite alphabet D and a new internal state S_{t+1} are (stochastically) generated. This is modeled by a joint transition probability gen from the internal states to output symbols and internal states:

$$\mathsf{gen} \colon \mathsf{S} \to \mathcal{P}(\mathsf{D} \times \mathsf{S}).$$

Thus the pair (gen, μ) of generating mechanism and initial distribution induces processes $S_{\mathbb{N}_0}$, $Y_{\mathbb{N}_0}$ of internal states and output symbols. The situation is illustrated as

$$S_0 \longrightarrow S_1 \longrightarrow S_2 \ \cdots \ S_{T-1} \longrightarrow S_T$$

$$Y_1 \qquad Y_2 \ \cdots \ Y_{T-1} \qquad Y_T$$

The joint distribution of internal- and output process is computed according to

$$P(S_{[0,T]} = s_{[0,T]}, \ Y_{[1,T]} = y_{[1,T]}) = \mu(s_0) \prod_{k=1}^{T} \mathsf{gen}(s_{k-1}; y_k, s_k), \qquad T \in \mathbb{N}.^6$$

---

[6] $\mathsf{gen}(s; y, \hat{s})$ denotes the probability of the pair $(y, \hat{s})$ w.r.t. the measure $\mathsf{gen}(s)$.

Stochastic automata are also widely known as *edge-emitting hidden Markov models*.[7] We use this terminology, but for brevity we call the pair (gen, μ) hidden Markov model (*HMM*) and always mean edge-emitting HMM.

If the initial distribution μ is gen-invariant, i.e. if

$$\mu(s) = \sum_{\hat{s} \in \mathsf{S}, d \in \mathsf{D}} \mu(\hat{s}) \, \mathsf{gen}(\hat{s}; d, s) \qquad \forall s \in \mathsf{S},$$

then the processes $S_{\mathbb{N}_0}$ and $Y_{\mathbb{N}_0}$ are (jointly) stationary and uniquely extended to processes $S_{\mathbb{Z}}$ and $Y_{\mathbb{Z}}$ respectively. We then call the HMM *stationary*. Because our aim is to investigate given processes $X_{\mathbb{Z}}$ with time set $\mathbb{Z}$, we assume in this section that μ is gen-invariant. If the law of the output process $Y_{\mathbb{Z}}$ of a stationary HMM (gen, μ) coincides with the law of the given process $X_{\mathbb{Z}}$, the HMM is a *generative model* for the process of interest: we can easily simulate and investigate statistical properties of $X_{\mathbb{Z}}$ by means of the HMM. We call such an HMM **generative**. A generative HMM is a possibility, how the process $X_{\mathbb{Z}}$ might have been produced, although it is of course highly non-unique. The question about a *minimal* generative HMM suggests itself. With "minimal" we either mean minimal cardinality $|\mathsf{S}|$ of the set of internal states or minimal entropy $H(\mu) = H(S_0)$ of the invariant initial distribution. Unlike in the situation of sufficient memories, these two notions do not coincide (see Example 7).

Finding the minimal generative HMM is intrinsically difficult,[8] but every sufficient memory M induces an HMM, thus providing an upper bound. In general, the process of internal states of the associated HMM cannot have the same distribution as the process $M_{\mathbb{Z}}$ of memory states, because the latter process need not be Markovian. The (first order) Markov approximation of the joint process $(M_{\mathbb{Z}}, X_{\mathbb{Z}})$, however, yields the desired HMM:

Proposition 2 (sufficient memories induce generative HMMs). *Let* mem *be a sufficient memory kernel and* $M_{\mathbb{Z}}$ *its process of memory states. Then a generative HMM is given by* $\mathsf{S} := \mathsf{M}$, $\mu(s) := P(M_0 = s)$ *and*

$$\mathsf{gen}(s; d, \hat{s}) := P(X_1 = d, M_1 = \hat{s} \mid M_0 = s).$$

Example 3 (ε-machine). If we take the causal states as sufficient memory, the HMM $(\mathsf{gen}_{\mathfrak{C}}, \mu_{\mathfrak{C}})$ constructed in Proposition 2 is the ε-machine of computational mechanics. As the process of causal states is already Markovian, the ε-machine fully describes the (statistics of the) time evolution of the causal states. $\Diamond$

The causal states provide the minimal sufficient memory and induce the ε-machine. But is the latter also the minimal generative HMM? In general, the answer is "no". The ε-machine may be arbitrarily much bigger than the minimal HMM. It can be infinite or even uncountable, while there is a generative HMM

---

[7] "Edge-emitting" means that in visualizations as transition graphs the output symbols appear as edge labels.

[8] A geometric condition for minimality in terms of cardinality was specified by Heller in [7], but no constructive algorithm is known to us.

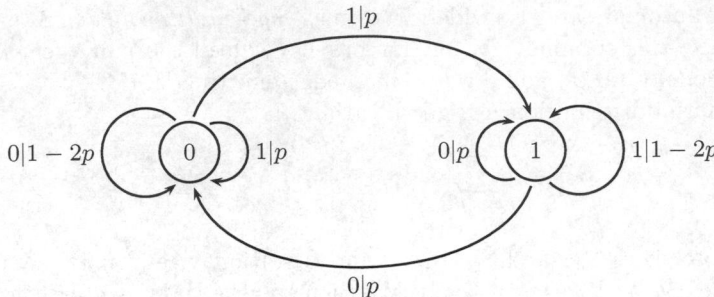

Fig. 1. Transition graph of the generator defined by (2). Circled nodes are internal states and edges are transitions, labeled with output symbol x and transition probability q as "$x|q$".

with only two internal states. This was already mentioned by Crutchfield in [8], but not everyone who applies computational mechanics seems to be aware of the fact. In the following, we give an example of this phenomenon.

Example 4 (uncountable ε-machine). We define the observable process $X_{\mathbb{Z}}$ by a stationary HMM with $\mathsf{D} := \mathsf{S} := \{0, 1\}$. It is clear that there is a generative HMM with two internal states, namely the original one. Nevertheless, the number of causal states (and thus the ε-machine) turns out to be uncountable. The initial distribution μ of the HMM is the uniform distribution. With a parameter $0 < p < \frac{1}{4}$, we define the generator by

$$\mathsf{gen}(s; x, \hat{s}) := \begin{cases} 1 - 2p, & \text{if } \hat{s} = x = s \\ p, & \text{if } x \neq s \\ 0, & \text{otherwise} \end{cases} \qquad (2)$$

See Figure 1 for an illustration of the transition graph. It is easy to check that μ is gen-invariant. One can show (see [3]) that the conditional probability for the internal state given a finite history behaves as follows: There exist intervals $I_n(x_{[-n,0]})$, disjoint for fixed n, such that

$$P(S_0 = 1 \mid X_{[-n,0]} = x_{[-n,0]}) \in I_n(x_{[-n,0]})$$

and the intervals are nested for increasing n, i.e.

$$I_{n+1}(x_{[-n-1,0]}) \subset I_n(x_{[-n,0]}).$$

Note that $P(S_0 \mid X_{-\mathbb{N}_0}) = \lim_{n \to \infty} P(S_0 \mid X_{[-n,0]})$ (a.s.) and that histories inducing different expectations on S_0 also induce different expectations on $X_{\mathbb{N}}$. Consequently, every causal state contains at most two (infinite) histories.[9] $\diamond$

---

[9] This is true for the canonical version of conditional probability $P(X_{\mathbb{N}} \mid X_{-\mathbb{N}_0} = x_{-\mathbb{N}_0}) = \lim_{n \to \infty} \sum_{s=0}^{1} P(S_0 = s \mid X_{[-n,0]} = x_{[-n,0]}) P(X_{\mathbb{N}} \mid S_0 = s)$. Note that this limit always (not only a.s.) exists. Other choices may produce identifications on sets of measure zero, but still lead to uncountably many causal states.

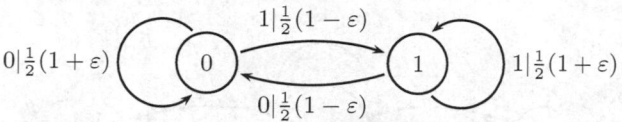

Fig. 2. ε-machine for a "nearly i.i.d." Markov process

Analogously to statistical complexity, one can consider the minimal internal-state entropy of a generative HMM:

Definition 5. Let $X_{\mathbb{Z}}$ be a stationary process. We call the quantity

$$C_{\mathrm{hmm}}(X_{\mathbb{Z}}) := \inf_{(\mathrm{gen},\mu)} H(\mu),$$

where the infimum is taken over all generative HMMs, **generative complexity**.

For any generative HMM, $Y_{-\mathbb{N}_0} \to S_0 \to Y_{\mathbb{N}}$ is a Markov chain and $I(Y_{-\mathbb{N}_0} : Y_{\mathbb{N}}) = I(X_{-\mathbb{N}_0} : X_{\mathbb{N}})$. Therefore, the internal state entropy $H(\mu) = H(S_0)$ is lower-bounded by the excess entropy. Together with Proposition 2 we obtain

Corollary 6. $E(X_{\mathbb{Z}}) \leq C_{\mathrm{hmm}}(X_{\mathbb{Z}}) \leq C_{\mathfrak{C}}(X_{\mathbb{Z}})$

The following example demonstrates that for some processes both inequalities in Corollary 6 are strict. It also illustrates that HMMs with minimal entropy need not have the minimal number of internal states.

Example 7. Let $\mathsf{D} := \{0,1\}$ and consider the stationary Markov process $X_{\mathbb{Z}}^{\varepsilon}$ defined by

$$P(X_0^{\varepsilon} = d) := \tfrac{1}{2} \quad \text{and} \quad P(X_{n+1}^{\varepsilon} = \hat{d} \mid X_n^{\varepsilon} = d) := \begin{cases} \tfrac{1}{2}(1+\varepsilon), & \text{if } d = \hat{d} \\ \tfrac{1}{2}(1-\varepsilon), & \text{if } d \neq \hat{d} \end{cases}.$$

$X_{\mathbb{Z}}^{\varepsilon}$ is a disturbed i.i.d. process with disturbance of magnitude ε towards a constant process: For $\varepsilon = 0$, it is i.i.d. and for $\varepsilon = 1$ it is constantly 0 or 1, each with equal probability. For $0 < \varepsilon \leq 1$, there are two causal states which correspond to the last observed symbol. The ε-machine is visualised in Figure 2 and statistical complexity is given by

$$C_{\mathfrak{C}}(X_{\mathbb{Z}}^{\varepsilon}) = H(X_0^{\varepsilon}) = \ln(2),$$

regardless how small the parameter ε is. At $\varepsilon = 0$, $\varepsilon \mapsto C_{\mathfrak{C}}(X_{\mathbb{Z}}^{\varepsilon})$ has a discontinuity and assumes the value 0. The excess entropy behaves differently: at $\varepsilon = 1$ and $\varepsilon = 0$ it coincides with statistical complexity, but it is continuous in ε and behaves like $\tfrac{1}{2}\varepsilon^2$ for small ε. It can easily be calculated:

$$E(X_{\mathbb{Z}}^{\varepsilon}) = I(X_1^{\varepsilon} : X_0^{\varepsilon}) = \tfrac{1}{2}\big((1+\varepsilon)\ln(1+\varepsilon) + (1-\varepsilon)\ln(1-\varepsilon)\big).$$

Now we show that for sufficiently small $\varepsilon > 0$, the generative complexity is strictly greater than excess entropy and strictly smaller than statistical complexity, i.e. $E(X_{\mathbb{Z}}^{\varepsilon}) < C_{\mathrm{hmm}}(X_{\mathbb{Z}}^{\varepsilon}) < C_{\mathfrak{C}}(X_{\mathbb{Z}}^{\varepsilon})$.

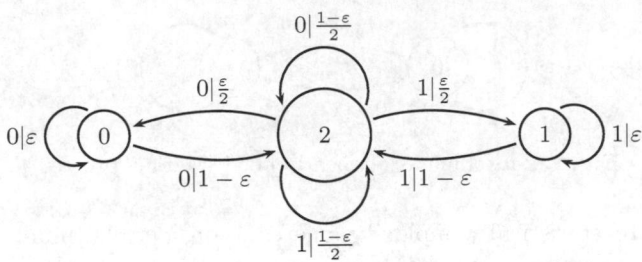

Fig. 3. HMM for the same Markov process as in Figure 2. The internal state entropy is lower, as node 2 carries nearly all weight: $\mu^\varepsilon(2) = 1 - \varepsilon$.

It is clear that no HMM can do with less than two internal states, but we can construct an HMM with lower internal state entropy on three states. The idea is to have one state corresponding to the i.i.d. process and getting most of the invariant measure when ε is small. The other two states correspond to the disturbances towards constantly 0 and 1 respectively. More precisely, let $S := \{0, 1, 2\}$ and consider the stationary HMM given by the generator visualized in Figure 3 together with the invariant initial distribution $\mu^\varepsilon(s) = \begin{cases} 1 - \varepsilon, & \text{if } s = 2 \\ \frac{\varepsilon}{2}, & \text{if } s \in \{0, 1\} \end{cases}$.

It is straightforward to verify that this HMM indeed generates $X_{\mathbb{Z}}^\varepsilon$. The internal state entropy is given by

$$H(\mu^\varepsilon) = -(1-\varepsilon)\ln(1-\varepsilon) - \varepsilon\ln(\tfrac{\varepsilon}{2}) \quad \overset{\varepsilon \to 0}{\longrightarrow} \quad 0.$$

Thus it is smaller than $C_{\mathfrak{C}}(X_{\mathbb{Z}})$ for sufficiently small ε. On the other hand, any generative HMM has to take the "disturbance of magnitude ε" into account: It is easy to see that no single internal state can get greater invariant measure than $1 - \frac{\varepsilon}{2}$. Thus the generative complexity is lower bounded as follows:

$$C_{\mathrm{hmm}}(X_{\mathbb{Z}}^\varepsilon) \geq L := -(1 - \tfrac{\varepsilon}{2})\ln(1 - \tfrac{\varepsilon}{2}) - \tfrac{\varepsilon}{2}\ln(\tfrac{\varepsilon}{2}) \geq -\tfrac{\varepsilon}{2}\ln(\tfrac{\varepsilon}{2}).$$

This bound converges to zero slower than linearly in ε. Consequently, for sufficiently small ε, excess entropy cannot be achieved or approximated by entropies of generative HMMs. The different entropies are plotted in Figure 4.   ◇

4   Predictive Interpretation of HMMs

We have seen that there can be a huge discrepancy between minimal sufficient memory and minimal generative HMM. The requirement of sufficiency is based on a certain understanding of "prediction". Here, we propose an alternative, weaker notion of prediction that allows for a predictive interpretation of all HMMs.

We model prediction by two steps: First the past $X_{-\mathbb{N}_0}$ is processed by a memory kernel mem, like in Section 2 but without the sufficiency assumption.

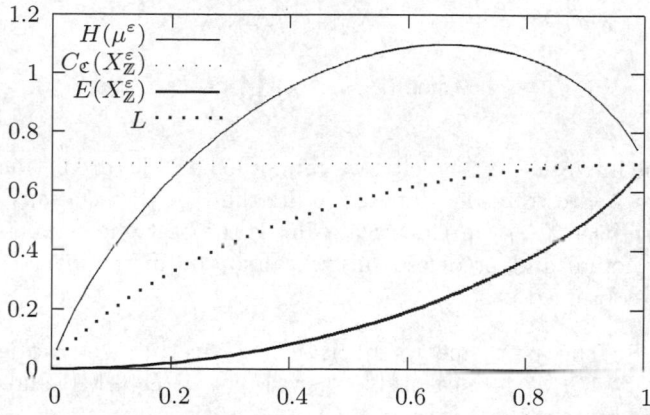

Fig. 4. Internal state entropy of the HMM of Figure 3, statistical complexity, excess entropy and the lower bound L for the generative complexity are plotted against the parameter ε. For $\varepsilon = 0$, all values are 0 and for $\varepsilon = 1$, all values are $\ln(2)$.

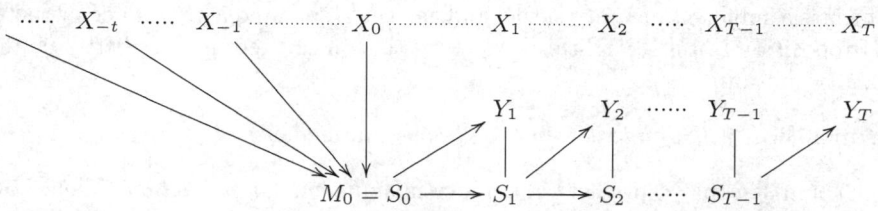

Fig. 5. The process of generating $Y_{\mathbb{N}}$ as prediction of $X_{\mathbb{N}}$. The dotted lines symbolize that $X_{\mathbb{Z}}$ may have arbitrary dependencies and need not be Markovian.

Then the actual prediction is done by generating a predicted future $Y_{\mathbb{N}}$. To this end we assume a generator **gen**, which uses the set of memory states as internal states and is initialized by the random memory state M_0 produced by **mem**. Thus **gen**, or rather the (non-invariant) HMM $\big(\text{gen}, \text{mem}(X_{-\mathbb{N}_0})\big)$, generates the prediction $Y_{\mathbb{N}}$ as in Section 3. The situation is illustrated as

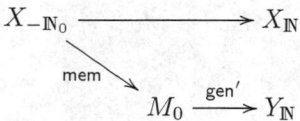

where **gen'** is the kernel from $\mathsf{M} = \mathsf{S}$ to $\mathsf{D}^{\mathbb{N}}$ obtained by iterating **gen** and projecting to the output. Figure 5 shows the situation in more detail. The resulting joint conditional distribution is given by

$$P(S_{[0,T]} = s_{[0,T]},\ Y_{[1,T]} = y_{[1,T]} \mid X_{-\mathbb{N}_0} = x_{-\mathbb{N}_0})$$
$$= \mathsf{mem}(x_{-\mathbb{N}_0};\ s_0) \prod_{k=1}^{T} \mathsf{gen}(s_{k-1};\ y_k, s_k), \qquad T \in \mathbb{N}.$$

Due to the intrinsic stochasticity, we cannot expect the prediction $Y_{\mathbb{N}}$ and the actual future $X_{\mathbb{N}}$ to coincide. But we require that the *distributions*, conditioned on the known past $X_{-\mathbb{N}_0}$, are identical. This is the best one can possibly do and means that actual and predicted future cannot be distinguished statistically, based on the observed past.

Definition 8. The pair $(\mathsf{mem}, \mathsf{gen})$ is called **predictive model** of $X_{\mathbb{Z}}$ if $\mathsf{mem} \colon \mathsf{D}^{-\mathbb{N}_0} \to \mathcal{P}(\mathsf{M})$ is measurable, $\mathsf{gen} \colon \mathsf{M} \to \mathcal{P}(\mathsf{D} \times \mathsf{M})$, and the process $Y_{\mathbb{N}}$ generated by the HMM $(\mathsf{gen}, \mathsf{mem}(X_{-\mathbb{N}_0}))$ satisfies

$$P(Y_{\mathbb{N}} \mid X_{-\mathbb{N}_0}) = P(X_{\mathbb{N}} \mid X_{-\mathbb{N}_0}) \qquad \text{a.s.}$$

A memory kernel mem (resp. generator gen) is called **predictive** if there exists a generator gen (resp. memory mem) such that $(\mathsf{mem}, \mathsf{gen})$ is a predictive model.

If mem is a sufficient memory kernel and gen the associated generator constructed in Proposition 2, it is straightforward to see that $(\mathsf{mem}, \mathsf{gen})$ is a predictive model. Thus we obtain:

Proposition 9. *Sufficient memory kernels are predictive.*

One could say that a predictive memory is *sufficient for prediction*. Then, however, sufficiency for prediction does not imply sufficiency in the sense of statistics. In fact, predictive memories can be much smaller than any sufficient memory: Assume any generative HMM (gen, μ). We know from Example 4 that for certain processes the number of internal states can be substantially smaller than the number of causal states. But now we show that gen is predictive, i.e. the HMM induces a predictive model and thus in particular a predictive memory kernel mem. Of course mem is in general not sufficient but has only as few memory states as the generative HMM.

Proposition 10 (generative HMMs are predictive). *Let* (gen, μ) *be a generative HMM. Then* gen *is predictive, i.e. there is a memory kernel* mem, *such that* $(\mathsf{mem}, \mathsf{gen})$ *is a predictive model. More specifically, we can choose*

$$\mathsf{mem}(x_{-\mathbb{N}_0}) := P(S_0 \mid Y_{-\mathbb{N}_0} = x_{-\mathbb{N}_0}).$$

If $(\mathsf{gen}_{\mathfrak{c}}, \mu_{\mathfrak{c}})$ is the ε-machine, then the memory kernel mem constructed in Proposition 10 recovers the causal state projection, i.e. $\mathsf{mem} = \mathsf{mem}_{\mathfrak{c}}$. In particular, this memory mem is deterministic. Of course, for general generative HMM, the associated memory need not be deterministic. Even more, it cannot be deterministic whenever the HMM is smaller than the corresponding ε-machine: In the following proposition we see that determinism implies sufficiency.

Proposition 11 (determinism implies sufficiency). *If* mem *is a predictive memory kernel and deterministic, i.e.* mem $=$ mem$_f$ *for some measurable* $f \colon \mathsf{D}^{-\mathbb{N}_0} \to \mathsf{M}$, *then* mem *is sufficient. In particular,* $|\mathsf{M}| \geq |\mathsf{M}_{\mathfrak{C}}|$ *and* $H(M) \geq H(M_{\mathfrak{C}})$.

5   Summary and Discussion

There are two aspects of prediction: a memory which compresses the past to a set of memory states (such as the causal states) and an encoding of the mechanism of prediction (such as the ε-machine). Looking at the memory part, sufficiency is a natural requirement, which leads to minimality of causal states and ε-machine. Sufficiency is the central assumption of computational mechanics. It has to be stressed, however, that the ε-machine is not the minimal generative hidden Markov model. Analogously to statistical complexity $C_{\mathfrak{C}}(X_{\mathbb{Z}})$, we defined generative complexity $C_{\mathrm{hmm}}(X_{\mathbb{Z}})$ as size in terms of entropy of the minimal generative HMM and obtained that $E(X_{\mathbb{Z}}) \leq C_{\mathrm{hmm}}(X_{\mathbb{Z}}) \leq C_{\mathfrak{C}}(X_{\mathbb{Z}})$, where $E(X_{\mathbb{Z}})$ is the excess entropy. Furthermore, we gave an example, where both inequalities are strict. We proposed a different notion of "predictive" and compared it to the sufficiency requirement used in computational mechanics. According to our notion, it has to be possible to generate a prediction $Y_{\mathbb{N}}$ for the future with the same statistical properties as the real future $X_{\mathbb{N}}$, conditioned on the observed past, i.e. $P(Y_{\mathbb{N}} \mid X_{-\mathbb{N}_0}) = P(X_{\mathbb{N}} \mid X_{-\mathbb{N}_0})$. It turned out that predictive in this sense is strictly weaker than sufficient and that any generative HMM can be interpreted as predictive in our sense.

Extending the model class from sufficient to predictive includes models that are substantially smaller than the ε-machine. At the same time, it preserves a notion of predictive power which we consider quite natural. Nevertheless, we have to point out two drawbacks of our approach: Firstly, constructing a minimal HMM is intrinsically difficult, whereas efficient algorithms are available for the construction of the ε-machine from data. Secondly, and conceptually more important, the memory state is no longer a complete substitute for the past. Given a sufficient memory, the complete conditional future distribution that corresponds to an observation $x_{-\mathbb{N}_0}$ is encoded in a single memory state $m \in \mathsf{M}$. On the other hand, assume that we have observed a particular past $x_{-\mathbb{N}_0}$ and want to use a predictive model for sampling the conditional future distribution several times. We first choose a memory state m according to mem$(x_{-\mathbb{N}_0})$ and then initialize gen with m for generating a prediction $y_{\mathbb{N}}$. We repeat this sampling procedure and obtain the correct future distribution $P(Y_{\mathbb{N}} \mid X_{-\mathbb{N}_0} = x_{-\mathbb{N}_0}) = P(X_{\mathbb{N}} \mid X_{-\mathbb{N}_0} = x_{-\mathbb{N}_0})$. But if we "forget" the history state $x_{-\mathbb{N}_0}$ and, instead of sampling new m's according to mem$(x_{-\mathbb{N}_0})$, initialize gen always with the same m, the resulting distribution of $Y_{\mathbb{N}}$ can be different from $P(X_{\mathbb{N}} \mid X_{-\mathbb{N}_0} = x_{-\mathbb{N}_0})$. Thus, we have to memorize the *distribution* (the information state) mem$(x_{-\mathbb{N}_0})$ of the initial memory states m. It is easy to show that the number of these information states is lower bounded by the number of causal states, because the map from history to information state defines a predictive deterministic memory, which is sufficient according to Proposition 11.

Currently, we do not know which of the two notions of prediction is more natural in which situations, and further steps towards revealing and comparing operational aspects of prediction are subject of our research.

References

1. Crutchfield, J.P., Young, K.: Inferring statistical complexity. Phys. Rev. Let. 63, 105–108 (1989)
2. Shalizi, C.R., Crutchfield, J.P.: Computational mechanics: Pattern and prediction, structure and simplicity. Journal of Statistical Physics 104, 817–879 (2001)
3. Löhr, W., Ay, N.: On the generative nature of prediction. Accepted for publication in Advances in Complex Systems (preprint, 2008),
   http://www.mis.mpg.de/publications/preprints/2008/prepr2008-8.html
4. Still, S., Crutchfield, J.P.: Optimal causal inference. Informal publication (2007),
   http://arxiv.org/abs/0708.1580
5. Grassberger, P.: Toward a quantitative theory of self-generated complexity. Int. J. Theor. Phys. 25, 907–938 (1986)
6. Bialek, W., Nemenman, I., Tishby, N.: Predictability, complexity, and learning. Neural Computation 13, 2409–2463 (2001)
7. Heller, A.: On stochastic processes derived from Markov chains. Annals of Mathematical Statistics 36, 1286–1291 (1965)
8. Crutchfield, J.P.: The calculi of emergence: Computation, dynamics and induction. Physica D 75, 11–54 (1994)

New Statistics for Testing Differential Expression of Pathways from Microarray Data

Hoicheong Siu[1,2], Hua Dong[1,2,3], Li Jin[1,2], and Momiao Xiong[1,3]

[1] Laboratory of Theoretical Systems Biology, School of Life Science, Fudan University, Shanghai, 200433, China
[2] State Key Laboratory of Genetic Engineering and MOE Key Laboratory of Contemporary Anthropology, School of Life Sciences and Institutes of Biomedical Sciences, Fudan University, Shanghai, 200433, China
[3] Human Genetics Center, University of Texas School of Public Health, Houston, TX 77225
{hoejohn,hdong0425,ljin007,momiao}@gmail.com

Abstract. Exploring biological meaning from microarray data is very important but remains a great challenge. Here, we developed three new statistics: linear combination test, quadratic test and de-correlation test to identify differentially expressed pathways from gene expression profile. We apply our statistics to two rheumatoid arthritis datasets. Notably, our results reveal three significant pathways and 275 genes in common in two datasets. The pathways we found are meaningful to uncover the disease mechanisms of rheumatoid arthritis, which implies that our statistics are a powerful tool in functional analysis of gene expression data.

Keywords: microarray, pathway, linear combination test, quadratic test, de-correlation test, rheumatoid arthritis.

1 Introduction

Understanding biological implication of gene expression profiles is important but challenging. A popular approach to gene expression data analysis is to identify a number of differentially expressed genes. Although such approach is useful for uncovering principles underlying biological processes, it has at least two limitations. First of all, in many cases, after Bonferroni correction, only a few individual genes may meet the threshold for statistical significance, because the relevant biological effects are modest relative to the noise inherent to the microarray technology. Second, one may be left with a long list of statistically significant genes without any unifying biological theme. Genes carry out their functions via intricate pathways of reactions and interactions. Pathways are sets of genes that act together to achieve certain cellular or physiologic functions. Prioritizing pathways relevant to a particular phenotype can help researchers to focus on the subset of most relevant genes, and generate further biological hypotheses. Genes belonging to the same pathway often exhibit subtle, coordinated changes in their expressions. Alternative to using a gene as a unit for analyzing expression profiles is to take a pathway as a unit for gene expression data

J. Zhou (Ed.): Complex 2009, Part I, LNICST 4, pp. 277–285, 2009.

analysis. Gene Set Enrichment Analysis (GSEA) which was developed by examining the overall differences in expression patterns between predefined gene sets and the whole gene list on the array is a useful tool for perform-ing pathway analysis [1]. Most methods for GSEA assume that the expressions of the genes within a pathway are independent [2,3]. However, in reality, the expressions of the genes within a pathway are correlated. Ignoring correlation among genes within a pathway may lead to misleading results.

Purpose of this paper is to develop statistics for GSEA which take correlations among gene expression into account. To accomplish this goal, we first investigated correlations among genes within the pathway and find that correlations among genes cannot be ignored. This motives us to develop three novel statistics which are able to combine dependent P-values of genes within the pathway. Finally, we apply the developed statistics to two rheumatoid arthritis gene expression datasets. Our method revealed pathways involved in rheumatoid arthritis disease, some of which have been independently validated by other microarray studies and by in vivo functional studies.

2 Methods

A pathway-based differential expression analysis is to use a pathway as the basic unit of analysis. Instead of testing differential expression of single gene between normal and abnormal tissues, pathway-based differential expression analysis is to jointly test for differential expressions of all genes within the pathway. Formally, suppose that there are k genes in the pathway. The null hypothesis for testing differential expression of the ith gene in the pathway is represented by:

$$H_{i0} : \theta_i = \theta_{i0} \tag{1}$$

where θ_i denotes the parameter, e. g., the difference of expression value between cases and controls. Then, the null hypothesis for testing differential expression of a pathway between normal and abnormal tissues is defined as testing for the combined null hypothesis:

$$H_{i0} : \theta_i = \theta_{i0}, i = 1,2,...,k. \tag{2}$$

The alternative hypothesis is defined as $H_{ia} : \theta_i \neq \theta_{i0}$, for at least one gene.

In this report, we will focus on combining individual differential expression tests of genes. We developed methods for combining dependent P-values which take correlations among gene expressions into account.

Let n_A be the number of affected tissues and n_G be the number of normal tissues. There are k genes in the pathway, define the mean expressions of i-th gene in cases and controls, respectively, as:

$$\overline{X}_i = \frac{1}{n_A} \sum_{j=1}^{n_A} x_{ij} \text{ and } \overline{Y}_i = \frac{1}{n_G} \sum_{j=1}^{n_G} y_{ij} \tag{3}$$

X_{ij} is the expression level of gene I from the j-th abnormal tissue, and y_{ij} is the expression level of the gene I from the j-th normal tissue. Let

$$\overline{X} = \begin{bmatrix} \overline{X}_1 \\ \vdots \\ \overline{X}_k \end{bmatrix}, \overline{Y} = \begin{bmatrix} \overline{Y}_1 \\ \vdots \\ \overline{Y}_k \end{bmatrix}, X_j = \begin{bmatrix} x_{1j} \\ \vdots \\ x_{kj} \end{bmatrix} \text{ and } Y_j = \begin{bmatrix} y_{1j} \\ \vdots \\ y_{kj} \end{bmatrix} \tag{4}$$

then the sampling covariance matrix of the genes in the pathway defined as:

$$S = \frac{\sum_{j=1}^{n_A}(X_j - \overline{X})(X_j - \overline{X})^T + \sum_{j=1}^{n_G}(Y_j - \overline{Y})(Y_j - \overline{Y})^T}{n_A + n_G - 2} \tag{5}$$

where S has a form like $S = \begin{bmatrix} S_{11} & S_{12} & \cdots & S_{1K} \\ S_{21} & S_{22} & \cdots & S_{2K} \\ \cdots & \cdots & \cdots & \cdots \\ S_{K1} & S_{K2} & \cdots & S_{KK} \end{bmatrix}$. Let $d_{ii} = \dfrac{1}{\sqrt{S_{ii}}}$, then

$$D = \begin{bmatrix} d_{11} & 0 & \cdots & 0 \\ 0 & d_{22} & \cdots & 0 \\ \cdots & \cdots & \cdots & \cdots \\ 0 & 0 & \cdots & d_{KK} \end{bmatrix} \text{ and } R = DSD = \begin{bmatrix} 1 & r_{12} & \cdots & r_{1k} \\ r_{21} & 1 & \cdots & r_{2k} \\ \cdots & \cdots & \cdots & \cdots \\ r_{k1} & r_{k2} & \cdots & 1 \end{bmatrix} \tag{6}$$

R is the correlation matrix of the genes in the pathway.
The statistic for testing differential expression of gene i is defined as:

$$T_i = \frac{\overline{X}_i - \overline{Y}_i}{\sqrt{\left(\dfrac{1}{n_A} + \dfrac{1}{n_G}\right)S_{ii}}} \tag{7}$$

T_i follows a student distribution with degree freedom $n_A + n_G - 2$. Denote its cumulative distribution by $F(T_i)$. Define the transformation as:

$$z_i = \Phi^{-1}(F(T_i)) \tag{8}$$

where Φ is a standard normal cumulative distribution and $z_i \sim N(0,1)$.
We define three statistics for testing differential expression of a pathway. Let

$$Z = \begin{bmatrix} Z_1 \\ \vdots \\ Z_k \end{bmatrix} \text{ and } e = \begin{bmatrix} 1 \\ \vdots \\ 1 \end{bmatrix} \tag{9}$$

(1) Linear Combination Test:
      The first test statistic is linear combination test, which is defined as:

$$T_L = \frac{e^T Z}{\sqrt{e^T Re}} \tag{10}$$

T_L follows a standard normal distribution.

(2) Quadratic Test
      The second statistic is based on the quadratic form of Z and defined as:

$$T_Q = Z^T R^{-1} Z \tag{11}$$

T_Q is asymptotically distributed as a central $\chi^2_{(k)}$ distribution, where k is the number of genes in the pathway.

(3) Decorrelation Test
      We decompose the matrix R as $R = CC^T$ and let $T = [T_1 \ ... \ T_k]^T$, and define the de-correlated statistics $W = C^{-1}T = [W_1, \ \cdots, \ W_k]^T$, which are asymptotically distributed as a vector of independent standard normal variables. For each W_i, we calculate its P-value P_i. Define the statistic:

$$T_D = -2\sum_{i=1}^{k} \log P_i \tag{12}$$

T_D follows a $\chi^2_{(2k)}$ distribution, where k is the number of genes in the pathway.

3   Results

3.1   Data Filtering and Statistical Analysis

We chose two gene expression datasets of rheumatoid arthritis by two standards: 1) case-control study: both rheumatoid arthritis patients and health controls are included; and 2) More than 45 samples are included. Datasets I was downloaded from Gene Expression Omnibus [4]. It contains 46 samples composed of 35 cases and 11 controls, using Affymetrix GeneChip Human Genome U95 Set HG-U95A array on which including 8685 genes [5]. Since it is one color DNA chip, we use the abstract intensity value for calculation. Datasets II is downloaded from Stanford MicroArray Database [6], which is composed of 35 cases and 15 controls, totally 14337 genes. As it is a two-color cDNA array, we extracted the log2 rations of gene expression values for our analysis [7]. 7434 genes shared between the two arrays. Gene represented more than once on the microarrays were averaged. The differential expression of the gene was tested by Mann-Whitney Test as the distribution of gene expression is unknown. Total 2069 genes in dataset I, 2003 genes in dataset II and 275 genes in both datasets showed mild differential expression with P-value < 0.05. However, after Bonferroni correction for multiple tests (P < 5.75705E-06 for dataset I and

Table 1. Significant differentially expressed genes in rheumatoid arthritis studies

	P-value	Number of genes
Dataset I	<0.05	2069
	<5.75705E-6*	7
Dataset II	<0.05	2003
	<3.48748E-6*	0
Combined 2 datasets	<0.05	275
	<6.72585E-6*	0

* indicates Bonferroni correction was applied to adjust for multiple test corrections.

$P < 3.48748E-06$ for dataset II), 7 genes were significant in dataset I but none gene was significant in dataset II.

3.2 Pathway Analysis

Since number of significantly differentially expressed genes identified were not large enough to explore disease mechanisms. Genes belonging to the same pathway often exhibit subtle, coordinated changes in the expression profile. Pathway analysis can detect genes which conferred small disease risk individually, but whose joint actions can be implicated in the development of disease. In this paper, we collected pathways of human and studied with 501 of them included more than 2 genes, 202 from KEGG [8] and 299 from BioCarta [9] (updated until Dec 2008). The three statistics: linear combination test, quadratic test, de-correlation test were used to identify the differentially expressed pathways.

3.2.1 Correlation Structure

To examine whether correlation among genes in a pathway can be ignored or not, for example we calculate the correlation among genes in the Expression Role of PPAR-gamma Coactivators in Obesity and Thermogenesis pathway (biocarta591) which were shown in Table 2. We can see that the correlations between the genes were quite large and cannot be ignored.

Table 2. Correlations among 6 genes within Biocarta591 in dataset I (upper triangular) and dataset II (lower triangular). Absolute value list as below:

	CREBBP	EP300	LPL	RXRA	NCOA1	NCOA2
CREBBP	1	0.541728	0.099420	0. 278448	0.266486	0.225759
EP300	0.751867	1	0.005371	0.349918	0.165928	0.112575
LPL	0.446183	0.362063	1	0.239493	0.063079	0.121320
RXRA	0.552236	0.560826	0.477218	1	0.157935	0.077518
NCOA1	0.576320	0.618891	0.174420	0.486905	1	0.111558
NCOA2	0.493967	0.658855	0.129232	0.236496	0.461819	1

To investigate how the correlations among genes affect the P-values, we present Table 3 to summarize P-values for testing differential expression of the pathway by combining independent P-values and dependent P-values. Independent P-values was got by replacing the correlation coefficient matrix R with identity matrix I. From Table 3, we can see that the P-value for testing pathway by combining independent P-values is much smaller than that by combining dependent P-values. This indicates that the statistic by combining independent P-values of genes may have high false positive rates. Thus we conclude that correlations have big impacts on the P-values of the statistics for testing differential expressions of the pathways and thus cannot be ignored.

Table 3. P-values of the pathways using linear combination of independent and dependent P-values of genes within pathways

Pathway name	independent		dependent	
	$T = \dfrac{e^T Z}{\sqrt{K}}$	P-value	$T_L = \dfrac{e^T Z}{\sqrt{e^T Re}}$	P-value
Five pathways in dataset II				
Glycolysis / Gluconeogenesis	-4.75595	1.98E-06	-1.85403	0.063734
Citrate cycle (TCA cycle)	-1.12204	0.261844	-0.42385	0.671678
Pentose phosphate pathway	-4.56911	4.90E-06	-2.28524	0.022299
Pentose and glucuronate interconversions	-1.41608	0.156753	-1.05021	0.29362
Fructose and mannose metabolism	-4.98735	6.12E-07	-2.00286	0.045192
Five significant pathways in dataset II				
Cell Communication	-11.3023	0	-5.05448	4.32E-07
C21-Steroid hormone metabolism	-4.17168	3.02E-05	-4.57937	4.66E-06
Complement Pathway pathway	-6.78801	1.14E-11	-4.39731	1.10E-05
Complement and coagulation cascades	-10.5662	0	-4.2202	2.44E-05
Lectin Induced Complement Pathway	-4.47688	7.57E-06	-4.05495	5.01E-05

3.2.2 Application to Rheumatoid Arthritis in Two Datasets

The proposed statistics were applied to two RA gene expression datasets. Table 4 showed the P-values of Dentatorubropallidoluysian atrophy (DRPLA) (hsa05050) and genes in the pathway. To evaluate the performance of the proposed statistics for testing differential expressions of the pathway, we also listed the P-value of the TAPPA method for comparison [3]. See supplementary Tables 3 for another pathway with more genes.

Table 5 listed pathways with both P-values < 0.01 which were obtained by novel linear combination test (LCT) in pathway-based gene expression studies of rheumatoid arthritis. We found three pathways showing significance in the two gene expression datasets. Obviously, these pathways are involved in inflammatory process. See supplementary Tables 2 for more pathways with P-values < 0.05.

Table 4. P-values of genes in Dentatorubropallidoluysian atrophy (hsa05050)

Dataset I		Dataset II	
Method	P-value	Method	P-value
LCT	5.76E-04	LCT	0.001985
QT	1.39E-09	QT	1.52E-04
DT	9.27E-10	DT	2.57E-04
TAPPA	0.347244	TAPPA	0.006951
Gene	P-value	Gene	P-value
ATN1	0.273753	ATN1	0.002935
BAIAP2	0.511389	BAIAP2	0.024157
CASP1	0.00022	CASP1	0.147013
CASP3	0.562305	CASP3	0.60399
CASP8	0.652234	CASP7	0.439692
GAPDH	0.334194	INS	0.215546
INSR	0.202407	INSR	0.50486
ITCH	0.231124	ITCH	0.294672
MAGI1	0.708848	MAGI1	0.248591
MAGI2	0.388306	MAGI2	0.223494
RERE	0.000894	RERE	0.799462
WWP1	0.00044	WWP1	0.200264
WWP2	0.000179	WWP2	0.379637

Table 5. Three Pathways with P-values < 0.01 in rheumatoid arthritis studies obtained by novel linear combination test (LCT)

Pathway name	Gene[#1]	Gene[#2]	Dataset I P-value	Gene[#3]	Dataset II P-value
Dentatorubropalliduluysian atrophy (DRPLA)	15	13	0.000576	13	0.001985
Axon guidance	128	95	0.000710	106	0.008328
Tetrachloroethene degradation	3	2	0.001051	3	0.000479

Gene[#1], Gene[#2], Gene[#3] mean in the specific pathway, the total number of genes, the number of genes contained in dataset I and the number of genes contained in dataset II.

4 Discussion

Despite great success in microarray technology, traditional strategies for gene expression analysis have focused on identifying individual genes that exhibit differences in expressions between abnormal and normal samples. Although useful, single gene differential expression analysis will miss many genes with moderate genetic effects and fail to detect biological processes which play an important role in disease development. To overcome these limitations, several pathway-based data analysis methods have been proposed for gene expression data analysis[10-14]. However, most statistical methods for GSEA have ignored correlations among the genes in the pathway.

To investigate whether correlations exist among the genes in the pathway and if there are, whether the correlations have impact on the results of pathway differential expression analysis, we calculated correlations among genes in the pathway and test statistics without consideration of the correlations among genes using real RA gene expression datasets. We found that the correlations among the genes in the pathway were quite large and cannot be ignored when we design test statistics.

In this report, we proposed new statistics for GSEA which take correlations among the genes in the pathway into account. The newly developed statistics were applied to two RA gene expression datasets. We found three common pathways in two datasets which are significantly different between case and control samples. Our results were not very consistent with other published literatures due to the following reasons[5,7,15-17]: 1) Sample sizes in both two datasets are very small. 2) The patients with different rheumatoid arthritis subtypes may fall ill by different disease mechanisms thus we find different related pathways instead of shared common pathways. Dataset I includes 35 patients, 25 of which are polyarticular rheumatoid arthritis and 10 are pauciarticular rheumatoid arthritis. There are gene expression differences in two RA subtypes. Dataset II also contains two rheumatoid arthritis subtypes signature by IFN-induced gene. 3) Different microarray platform definitely would affect the results. One-color affymetrix DNA chip use the abstract intensity value while two-color cDNA chip get the log2 ratio of intensity value of test sample by reference sample. The different design principle in two microarray platform causes noise and variances of results. The merits of our statistic should be further validated in more trusted datasets.

Although in most cases QT and DT are much powerful than the LCT method, the QT and DT methods are not reliable due to singularity of the correlation matrix of the expressions of the genes in the pathway. In the future, we need to design a strategy to ensure that the correlation matrices of the gene expressions are positive definite.

Acknowledgments

Hoicheong Siu, Hua Dong, Li Jin, Momiao Xiong are supported by National Science Foundation of China (30890034), Hi-Tech Research and Development Program of China(863) (2007AA02Z312) and Shanghai Commission of Science and Technology (04dz14003). Li Jin is also supported by Shanghai Leading Academic Discipline Project (B111) and the Center for Evolutionary Biology.

References

1. Subramanian, A., Tamayo, P., Mootha, V.K., Mukherjee, S., Ebert, B.L., Gillette, M.A., Paulovich, A., Pomeroy, S.L., Golub, T.R., Lander, E.S., et al.: Gene set enrichment analysis: a knowledge-based approach for interpreting genome-wide expression profiles. Proc. Natl. Acad. Sci. USA 102, 15545–15550 (2005)
2. Wang, K., Li, M., Bucan, M.: Pathway-Based Approaches for Analysis of Genome-wide Association Studies. Am. J. Hum. Genet. 81, 1278–1283 (2007)
3. Gao, S., Wang, X.: TAPPA: topological analysis of pathway phenotype association. Bioinformatics 23, 3100–3102 (2007)

4. Gene Expression Omnibus, http://www.ncbi.nlm.nih.gov/geo/
5. Barnes, M.G., Aronow, B.J., Luyrink, L.K., Moroldo, M.B., et al.: Gene expression in ju-venile arthritis and spondyloarthropathy: pro-angiogenic ELR+ chemokine genes relate to course of arthritis. Rheumatology (Oxford) 43(8), 973–979 (2004)
6. Stanford MicroArray Database, http://genome-www5.stanford.edu/
7. van der Pouw Kraan, T.C., Wijbrandts, C.A., van Baarsen, L.G., Voskuyl, A.E., Rusten-burg, F., Baggen, J.M., Ibrahim, S.M., Fero, M., Dijkmans, B.A., Tak, P.P., Verweij, C.L.: Rheumatoid arthritis subtypes identified by genomic profiling of peripheral blood cells: as-signment of a type 1 interferon signature in a subpopulation of patients. Ann. Rheum. Dis. 66(8), 1008–1014 (2007)
8. KEGG (Kyoto Encyclopedia of Genes and Genomes), http://www.genome.jp/kegg
9. BioCarta, http://www.biocarta.com
10. Benfey, P.N., Mitchell-Olds, T.: From genotype to phenotype: systems biology meets natural variation. Science 320, 495–497 (2008)
11. Curtis, R.K., Oresic, M., Vidal-Puig, A.: Pathways to the analysis of microarray data. Trends Biotechnol 23, 429–435 (2005)
12. Werner, T.: Bioinformatics applications for pathway analysis of microarray data. Curr. Opin. Biotechnol. 19(1), 50–54 (2008)
13. Curtis, R.K., Oresic, M., Vidal-Puig, A.: Pathways to the analysis of microarray data. Trends Biotechnol. 23(8), 429–435 (2005)
14. Adewale, A.J., Dinu, I., Potter, J.D., Liu, Q., Yasui, Y.: Pathway analysis of microarray data via regression. J. Comput. Biol. 15(3), 269–277 (2008)
15. Olsen, N., Sokka, T., Seehorn, C.L., Kraft, B., Maas, K., Moore, J., Aune, T.M.: A gene expression signature for recent onset rheumatoid arthritis in peripheral blood mononuclear cells. Annals of the Rheumatic Diseases 63, 1387–1392 (2004)
16. Szodoray, P., Alex, P., Frank, M.B., Turner, M., Turner, S., Knowlton, N., Cadwell, C., Dozmorov, I., Tang, Y., Wilson, P.C., Jonsson, R., Centola, M.: A genome-scale assess-ment of peripheral blood B-cell molecular homeostasis in patients with rheumatoid arthri-tis. Rheumatology 45, 1466–1476 (2006)
17. Chang, M., Rowland, C.M., Garcia, V.E., Schrodi, S.J., Catanese, J.J., van der Helm-van Mil, A.H., Ardlie, K.G., Amos, C.I., Criswell, L.A., Kastner, D.L., Gregersen, P.K., Kur-reeman, F.A., Toes, R.E., Huizinga, T.W., Seldin, M.F., Begovich, A.B.: A large-scale rheumatoid arthritis genetic study identifies association at chromosome 9q33.2. PLoS Genet. 27, 4(6), e1000107 (2008)

Multiple Phase Transitions in the Culture Dissemination

Bing Wang[1,2], Yuexing Han[3], Luonan Chen[4], and Kazuyuki Aihara[1,2]

[1] ERATO Aihara Complexity Modelling Project, JST, Institute of Industrial Science,
The University of Tokyo, 4-6-1 Komaba, Meguro-ku, Tokyo, 153-8505, Japan
[2] Institute of Industrial Science, The University of Tokyo, Tokyo, Japan
[3] Graduate School of Information System, The University of Electro-Communications,
1-5-1, Chofugaoka, Chofu-Shi, Tokyo, Japan
[4] Department of Electrical Engineering and Electronics, Osaka Sangyo University,
Daito, Osaka, 574-8530, Japan

Abstract. We study the coevolution process in the Axelrod's model with the consideration of agents' abilities to access to the information. With a parameter to control the ability of communication, we observe two kinds of phase transitions both for cultural domains and network fragments, respectively. With the simulation results, we find the relationship between the critical value and the controlled parameter. The results indicate that the powerful ability to access to the information benefits the dissemination of culture in the system.

Keywords: Self-organized systems, Complex systems, Dynamics of social systems.

1 Introduction

During the last few years, a great deal of efforts have been devoted to the study of social phenomena and social behaviors, such as opinion formation, rumors, disease propagation. In [1], Axelrod proposed one model to describe the dissemination of culture among interacting agents in a society system. With this model it generates a global convergence to a single culture state.

Since the Axelrod model has been proposed, The behavior of this model has been much studied in static networks [3,4], such as the effect of the network structure [5], noise [6], mass media [7,8] and cultural threshold [9,10]. Recently, the study of the relationship between the networks' topologies and the dynamical behavior on them has led to a deep understanding, and showed that the topology of the network or the interaction between agents is not static in time. Therefore, it is emergent to integrate the new framework of the coevolution of network structures and state dynamics [11,12,13,14,15,16].

In this work we investigate the influence of communication ability on the coevolution of agents' interactions and state dynamics in the Axelrod's model. Agents can interact with their neighbors who share common features with them or construct new contacts with others. The candidate agents can be chosen within

J. Zhou (Ed.): Complex 2009, Part I, LNICST 4, pp. 286–290, 2009.

the limited spatial space due to the constraints of transportation condition, technology factor or the limited information sources. With simulations we show that the system can display two kinds of phase transitions, that is, cultural domains and network fragments, respectively. The results indicate that the more powerful the ability to access to the information, the better the ordered phase can be observed.

The rest of the paper is organized as follows. In Sec. 2, we briefly introduce the Axelrod's model and the coevolution process by taking into account the space constraints. In Sec. 3, we show the results on the effects of the spatial constraint on the dissimilation of culture. In Sec. 4, we conclude our results and give a brief discussion.

2   The Model

Assume a system with N agents at the sites of a square lattice $L \times L$ without periodical condition is given, the state S_i of node i is defined as a vector of F components $\sigma_i = (\sigma_{i1}, ... \sigma_{iF})$ representing cultural features such as language, music, and sports. Each component σ_{if} represents the preference for each culture feature. Initially for each node i, σ_{if} is randomly assigned from the set $\{0, 1, ..., q-1\}$ with equal probability $1/q$. The time-discrete dynamics evolves by iterating the following steps.

(1) Select a node i and one of its neighbors j randomly. Let c_{ij} represent the number of common features shared by i and j, denoted by $c_{ij} = \sum_k \delta_{\sigma_{ik}, \sigma_{jk}}$.

(2) Interaction. If $c_{ij} = 0$ or $c_{ij} = F$, nothing happens; otherwise if $0 < c_{ij} < F$, i and j interact with probability c_{ij}/F. The interaction means that one of i's feature σ_{if}, if $\sigma_{if} \neq \sigma_{jf}$, then set $\sigma_{if} = \sigma_{jf}$ [1].

Step (1) and (2) are the basic process of the Axelrod's model and by Step (2) agents become more similar. With the repeat of the above process a *frozen state* can often be reached, where the node's status does not change anymore. That is, neighboring sites have an overlap equal to F or 0. The relative size of the largest component of agents sharing the same culture features S/N is a measure of the cultural diversity.

In the following we consider the coevolution process that agents search new contacts with common features to interact with access to the information.

(3) Rewiring. If $c_{ij} = 0$, agent i will break the link with j and look for new contacts. Agents within the geographical distance $\alpha \times L$ can be the candidate nodes, where $\alpha \in [1/L, \sqrt{2}]$, and no multiple connections are allowed. Candidates sharing common features with i will be given priorities to be chosen. If there are no nodes sharing common features with i, i will connect with a candidate randomly.

The parameter α can be described as agents' abilities of communication constrained by the spatial space and technology factors.

3   Simulation Results

We will measure the ordered parameter, the size of the giant component with the same cultural features S_{max} divided by N for the described model and the Axelrod's model, respectively.

In Fig. 1, we first display S_{max}/N versus q for different values of α. The curve for $\alpha = 0$ corresponds to the results in the Axelrod's model. It is shown that there exists a phase transition for each curve and the critical value q_c increases with α, which means that the powerful ability to obtain the information favors the homocultural state. The critical value q_c can be evaluated by verifying the point at which S_{max}/N achieves the maximal variance. In the inset of Fig. 1, we show the critical value of q_c versus α in a more careful way. It clearly shows that with the increase of α, more agents can be ideal candidates to interact and it becomes possible to find compatible ones to interact, which leads to the homocultural state, and a higher q_c can be achieved.

In the rewiring process, agents will find new contacts and rewire with them, which will induce the broken of the network into several parts and make the size of the giant component for the network structure denoted by S'_{max} also reduce, see Fig. 2. However, if q is very large, since most of the agents do not share any common cultural features and they cannot interact with each other to become more similar. In this case, the rewiring process governs the dynamics and the network reconnects again, which leads to a higher value of S'_{max} again, see Fig. 2.

We can see that for each α, S'_{max}/N first decreases to the lowest value and then sharply increases to approach to the state where almost all the nodes are connected. In order to discover the exact value of q^*, we analyze the number of components for cultural domains and network fragments, denoted by g with black and red curves in Fig. 3. For $q < q^*$, the number of groups for cultural domains and network fragments coincide with each other and agents in the same network component possess the same cultural traits. However, when $q > q^*$, cultural

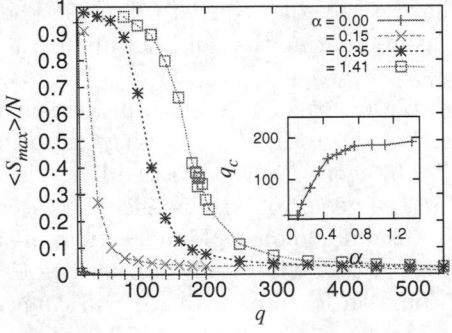

Fig. 1. (Color online) The ordered parameter S_{max}/N for cultural domains versus q for $\alpha = 0.0, 0.15, 0.35, 1.41$ (from left to right). $\alpha - 0$ corresponds to the case of the Axelrod's model. The inset shows the critical value q_c versus α.

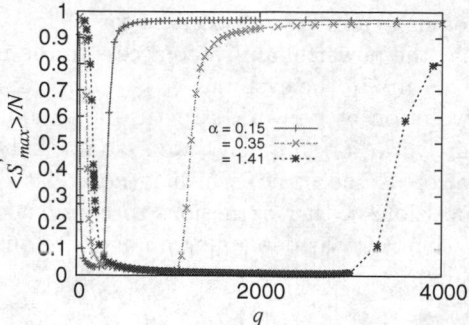

Fig. 2. (Color online) The ordered parameter S'_{max}/N for network structure versus q for $\alpha = 0.15, 0.35$, and 1.41

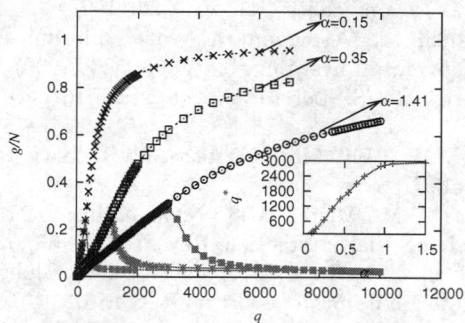

Fig. 3. (Color online) The ordered parameter g/N for network fragments (red) and cultural domains (black) versus q for $\alpha = 0.15, 0.35$ and 1.41. The inset shows the critical value q^*, at which the network fragments and cultural domains do not coincide, versus α.

traits and network components do not coincide anymore and the number of groups for cultural domains increases while the number of groups for network fragments decreases. In other words, each component possesses several cultural states. The inset in Fig. 3 shows the exact value of q^* versus α. We find that q^* grows with α until most of compatible pairs have been found.

4   Conclusion and Discussion

In this work, we studied the coevolution of agents' interactions and the update of status in the Axelrod's model by tuning agents' abilities to access to the information. With the increase of the parameter, the model shows two kinds of phase transitions both for the cultural domains and the network components. The numerical results indicate that the critical value q_c for cultural domains increases with the controlled parameter α soon and then does not change much anymore. This phenomena can also be observed for q^* for the network components. It indicates that the powerful ability to access to the information benefits the dissemination of culture in the system.

The study of coevolution of network structure and nodes' states is still at the beginning. In this work, for simplicity, we assume that all agents' abilities to access to information is the same, which is not related with the network topology or nodes' positions. Other extensions to consider different abilities to access to the information, such as depending on agent's connectivity may bring more interesting results.

References

1. Axelrod, R.: The Dissemination of Culture: A Model with Local Convergence and Global Polarization. J. Conflict Res. 41, 203–226 (1997)
2. Castellano, C., Marsili, M., Vespignani, A.: Nonequilibrium Phase Transition in a Model for Social Influence. Phys. Rev. Lett. 85, 3536 (2000)
3. Castellano, C., Fortunato, S., Loreto, V.: Statistical Physics of Social Dynamics. arXiv:0710.3256
4. Flache, A., Macy, M.W.: Why More Contact May Increase Cultural Polarization. arXiv:physics/0604196
5. Klemm, K., Eguíluz, V.M., Miguel, M.S.: Nonequilibrium Transitions in Complex Networks: A Model of Social Interaction. Phys. Rev. E 67, 026120 (2003)
6. Klemm, K., Eguíluz, V.M., Toral, R., Miguel, M.S.: Global Culture: A Noise-induced Transition in Finite Systems. Phys. Rev. E 67, 045101R (2003)
7. González-Avella, J.C., Eguíluz, V.M., Cosenza, M.G., Klemm, K., Herrera, J.L., Miguel, M.S.: Local Versus Global Interactions in Nonequilibrium Transitions: A Model of Social Dynamics. Phys. Rev. E 73, 046119 (2006)
8. González-Avella, J.C., Cosenza, M.G., Tucci, K.: Nonequilibrium Transition Induced By Mass Media in a Model for Social Influence. Phys. Rev. E 72, 065102R (2005)
9. Flache, A., Macy, M.W.: Local Convergence and Global Diversity: The Robustness of Cultural Homophily. arXiv:physics/0701333
10. De Sanctis, L., Galla, T.: Effects of Noise and Confidence Thresholds in Nominal and Metric Axelrod Dynamics of Social Influence. arXiv:physics/0707.3428
11. Vazquez, F., González-Avella, J.C., Eguíluz, V.M., Miguel, M.S.: Time-scale Competition Leading to Fragmentation and Recombination Transitions in the Coevolution of Network and States. Phys. Rev. E 76, 046120 (2007)
12. Holme, P., Newman, M.E.J.: Nonequilibrium Phase Transition in the Coevolution of Networks and Opinions. Phys. Rev. E 74, 056108 (2006)
13. Vazquez, F., Eguíluz, V.M., Miguel, M.S.: Generic Absorbing Transition in Coevolution Dynamics. Phys. Rev. Lett. 100, 108702 (2008)
14. Grabowski, A., Kosiński, R.A.: Evolution of a Social Network: The Role of Cultural Diversity. Phys. Rev. E 73, 016135 (2006)
15. Gil, S., Zanette, D.H.: Coevolution of Agents and Networks: Opinion Spreading and Community Disconnection. Phys. Lett. A 356, 89–94 (2006)
16. Zanette, D.H., Gil, S.: Opinion Spreading and Agent Segregation on Evolving Networks. Physica D 224, 156–165 (2006)

Joint Channel-Network Coding (JCNC) for Distributed Storage in Wireless Network*

Ning Wang and Jiaru Lin

Key Laboratory of Information Processing and Intelligent Technology,
Beijing University of Posts and Telecommunications
bjxt713@gmail.com, jrlin@bupt.edu.cn

Abstract. We propose to construct a joint channel-network coding (knosswn as Random Linear Coding) scheme based on improved turbo codes for the distributed storage in wireless communication network with k data nodes, s storage nodes (k<s) and a data collector. This framework extends the classical distributed storage with erasure channel to AWGN and fading channel scenario. We investigate the throughput performance of the Joint Channel-Network Coding (JCNC) system benefits from network coding, compared with that of system without network coding based only on store and forward (S-F) approach. Another helpful parameter: node degree (L) indicates how many storage nodes one data packet should fall onto. L characterizes the en/decoding complexity of the system. Moreover, this proposed framework can be extended to ad-hoc and sensor network easily.

Keywords: Joint Channel-Network Coding(JCNC), Distributed Storage, Wireless networks.

1 Introduction

Network coding [1] is a recent field in information theory that breaks with this assumption: instead of simply forwarding data, nodes may recombine several input packets into one output packets. This coding method can cope with the condition that source nodes disappear or inactive, which is the biggest trouble in the distributed communication. In this essay, we bring forward the scheme of distributed storage, where data packets generated by distributed sources are channel-coded first and sent to storage nodes through AWGN or Rayleigh channel, and then combined through network coding and saved at the storage nodes. System maintenance requirements and throughput can be significantly improved while adopting the integrity of stored information.

One of the key processes in distributed network storage is how to reconstruct the k data packets from any k out of s storage nodes, which is essentially the erasure channel coding problem [2]. Many codes used in this scenario have been investigated

---

* This work has been supported by the Doctor Fund of Beijing University of Posts and Telecommunications.

J. Zhou (Ed.): Complex 2009, Part I, LNICST 4, pp. 291–301, 2009.

on erasure channels. Erasure codes have been introduced in distributed storage in the OceanStore project [3]. In [4] the authors provide a random linear coding with a centralized server for distributed storage. Recently, fountain codes, like LT-code [5] and Raptor code [6], have been researched in distributed communication. All the codes introduced above work well for distributed storage in erasure channel. However, the coding structure introduced in this work can be used in wireless broadcast channel, like Gaussian or fading channel. We approximate a digital fountain basing turbo coding [7] [8] to broadcast source data and adopt random network coding to collect the data. Simulation results show that system with joint channel-network coding (JCNC) outperforms the system with only channel or erasure coding. It is the redundancy contained in the transmission of the relay on the storage nodes with JCNC that brings the throughput improvement.

The main innovations and contributions of this paper: we approximate a digital fountain based on turbo code, so the designed distributed storage system can be applied in Gaussian or fading channel instead of only in traditional erasure channel. Digital fountain breaks the tradition that the collector always receives a sequent stream of data packets, instead, the data source can product limitless encoding packets in digital fountain pattern. The simulation results show that the throughput of the system is greatly enhanced. It is network coding that brings the gratifying improvement. In addition, the optimal query intervals for the collector with different simulation conditions are provided in this paper, which reduces the total communication overhead greatly.

This paper is organized as follows: Related work about distributed network storage is discussed in section 2. We present the theory background, system model and assumption in section 3. Section 4 provides the simulation results and performance analysis. Finally, in section V we draw the conclusion of this work and discuss the future work.

2 Background and Related Work

2.1 Distributed Storage System

Distributed storage covers many key technologies about communication, e.g. file distribution, data replication and collection, distributed transaction mechanism, distributed timing and coordination, network security. Nowadays, the research about distributed network storage in the internet-based application and services is relatively interesting and mature. However, distributed media across wide areas and distributed network storage in wireless environment, e.g. sensor and ad-hoc network, pose an interesting challenge for present distributed technologies.

It is may be pretty to view distributed architecture as an idea, not just a technology. The book [9] gives a number of complex tradeoffs to be considered when managing the delivery, storage and collection of distributed data, which can not easily be duplicated at different sites.

2.2 Reed-Solomon (RS) Code

As a typical erasure code, RS code is one of Maximum Distance Separate (MDS) codes. The distinct advantage for RS code over simply replicating codes is that any d out of the s encoded fragments (storage nodes) suffices to recover the original d data fragments. In practice, however, decoding of RS code becomes expensive even infeasible in the network with large-scale nodes, just because the inverse matrix of generation matrix is required for the encoding. Standard algorithms for decoding RS code require exponential time. Maybe RS code is suitable on the condition that small blocks of data need to be encoded. Hard decoding is practically a fatal limitation for RS code applying to distributed network storage where the network scale is usually large or dynamic.

2.3 Low Density Parity Check (LDPC) Code

Low Density Parity Check (LDPC) LDPC (Low Density Parity Check) was proposed as an alternative with random construction [10], a random matrix with large weight. LDPC code relies on a parity check matrix based on bipartite graphs with the data nodes on the left and storage nodes on the right. Linear equations are produced between data blocks and coding blocks. The encoding and decoding is extremely fast [11]. However, since the LDPC codes are not MDS codes, coding blocks are required to reconstruct the original blocks. The overhead of application to distributed storage is too large. So LDPC structure needs to be modified for information distributed applications [10] [12].

2.4 Fountain Code

Fountain code was put forward by M.Luby in 1998. It was not until 2002 that was the feasible coding method raised. LT codes [5] and Raptor codes [6] are classical fountain codes in practice. The encoding of LT-code is XOR operation on d information symbols selected randomly according to the predetermined degree distribution of encoding symbol. For Raptor code, the M original blocks are preencoded into M' symbols firstly, and then only a successive function of M' suffice to recover the original information.

Compared to the limitation of slow encoding and decoding over large block sizes, fountain code breaks the source data into small blocks of packets and encodes over these blocks. The main idea of fountain code is that the encoding symbols can be transmitted limitlessly. An additional encoding block will be retransmitted when the receiver doesn't receive the given symbol correctly, which eliminates the need for retransmission dramatically. The key feature of fountain code is that the receiver can reconstruct the original data from any d out of s encoding blocks (the original data is break into d blocks and encoded into s blocks). From above discussion, fountain code can be well applied in distributed network environment, e.g. internet, satellite networks and wireless sensor or ad-hoc networks.

2.5 Network Coding

The capacity of information stream from end to end in communication networks is determinate by the mini-cut of network digraph. However, traditional Store-Forward

(SF) scheme cannot reach the up-bound of maxflow-mincut theorem raised by Shannon. In 2000, Ahlswede.R and Li.S-Y.R brought forward the idea of network coding: routers can combine different information flows via encoding instead of only storing, thus available network resource can be maximum utilized.

The key feature of network coding is that the receiver encodes what it has received and then forwards the coding packets. There are two main benefits of this approach: potential throughput improvements and a high degree of robustness. For the server, the original data (d blocks) can be decoded as long as the rank of decoding matrix reaches d. The network decoding will be faster when the data nodes get more dispersed, so network coding is being paid much attention and applied to distributed network communication. Fast en/decoding and improvement on throughput promote the application and research of network coding recently, e.g. multicast capacity, cooperation communication, mesh network and sensor network.

3   System Model and Realization

3.1   System Model and Assumption

There are k data nodes, s (s>k) storage nodes with limited memory and one collector in our system shown in Figure.1. We assume one data node generates only one data packet and the data packets are sent to storage nodes independently. One data packet is sent to L storage nodes (selected randomly) uniformly at one time. We emphasize the parameter L and simulate the measurable value with different d (data nodes) and s (storage nodes). L, so-called data node degree, represents the number of edges in Figure.1, which influences the sparsity of the encoding matrix S and decoding complexity directly.

Before sent to storage nodes through AWGN or Rayleigh channel, channel coding is operated on the source data. In this work, we approximate a digital fountain based on turbo code. Thus the encoding message can be sent to the storage nodes limitlessly until the collector retrieves the original data.

The storage nodes play the role of channel decoding, re-encoding (network coding on storage nodes) and saving devices. We assume one storage node has the same limited memory as one data node. The storage nodes decode the message received from data nodes firstly. Only if decoded correctly, the message will be saved. When more than one packet is decoded correctly, they will be compressed into one packet via network coding.

The data collector queries the storage nodes at an interval and reconstructs the original data. The collector is assumed to have enough memory and power for decoding the k original data packets. We define the system throughput: k/n, where n is the number of storage nodes the collector needs to query for reconstructing the original k data blocks. It is assumed that the collector can retrieve the data from any k storage nodes, which is the critical feature and design criteria for distributed storage system, from the moment that k/n reaches 1. So many important parameters in our

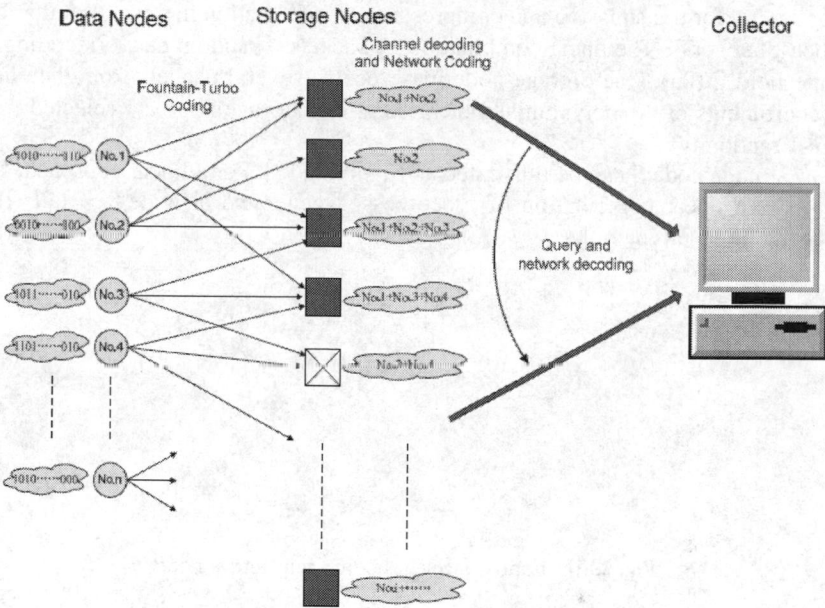

Fig. 1. System model: d data nodes, s storage nodes and 1 collector which can be set anywhere of the network

system are simulated at the condition k/n=1. We calculate the value of k/n whenever the rank of decoding matrix G' gets full. Network decoding at one SNR will be operated until k/n reaches 1. Also, the whole times of decoding matrix G' getting full before k/n reaches 1 is defined as the system communication time or source data reconstruction time (expressed as Time in section 4), representing the communication efficiency of the system.

3.2 System Realization

The main task in distributed storage system is that the k source data packets are saved in a redundant way in the s storage nodes and the collector can retrieve the original data by querying any k storage nodes in its vicinity. The channel encoding data packets fall into storage nodes randomly and random network coding (will be detailed in next section) is operated on the storage nodes.

In this wireless distributed network storage system, turbo coding with good BER performance is operated on the source data through AWGN or fading channel. On the storage nodes, the channel coding packets will be channel-decoded firstly. If more than one packet has been decoded correctly by the storage nodes, the packets will be network encoded, not simple saving and forwarding (SF).

We assume that information is transmitted as vectors of bits which are of length u, represented as elements in the finite field F (2u). In this paper we consider random linear coding for transmission and compression of information in general multi-source multicast networks. The linear combination of packets is random network coding over a finite field F (2u). One Storage node may receive several packets from data nodes. The coefficients of the polynomial generated at one storage node are selected from 0 to 2u -1 randomly.

Any storage node S can be illustrated as Figure.2 [13]. For a linear approach, S j on a link j is a linear combination of processes Xi generated at node v = tail (j) and signals Di on incoming links.

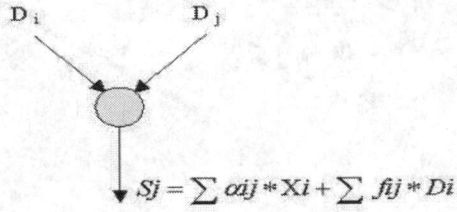

$$Sj = \sum \alpha ij * Xi + \sum fij * Di$$

Fig. 2. Illustration of linear coding at a storage node

The decoding can be presented as:

$$Zi = \sum \eta ij * Si \tag{1}$$

As long as d (the number of source nodes) linear combinations are received, the source signal will be retrieved theoretically. Based on the random linear coding/decoding theory, this paper provides JCNC approach for wireless distributed scenario and analyses its performance.

Correspondingly, the row vectors of generation matrix G_{k*n} are random and independent absolutely and the data vector is D_{1*k}. Nonzero Element in storage encoding vector S_{1*n} represents the edge connecting data and storage nodes of the bipartite graph in Figure.1. This just shows the key property of our design: random and decentralized code structure.

$$S_{1*n} = \begin{bmatrix} d_1 & d_2 & \ldots & d_k \end{bmatrix} * \begin{bmatrix} g_{11} & g_{12} & \cdots & g_{1n} \\ g_{21} & g_{22} & \cdots & g_{2n} \\ \cdot & \cdot & \cdot & \cdot \\ g_{k1} & g_{k2} & \cdots & g_{kn} \end{bmatrix} \tag{2}$$

To retrieve the k data packets, it is clear that successful decoding requires a full rank matrix G'_{k*k} from G_{k*n}. The value of n will plus 1 when the collector queries data from one storage node to another. However, that of k plus 1 only when the rank of G' rises.

The realization process is described in figure 3,

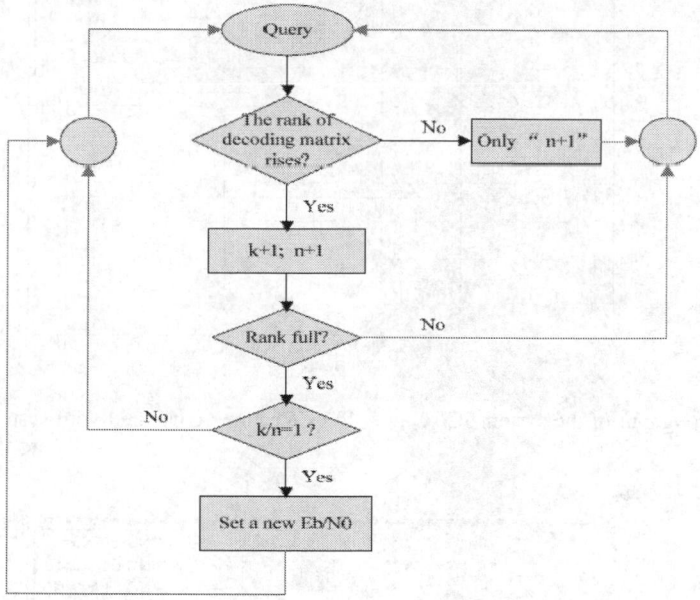

Fig. 3. Simulation flow chart: It shows the method of how to get the throughout of the System

4   Performance Evaluation

In this section, we provide the simulation results and analyze improvement performance of our scheme.

4.1   Throughput (k/n)

In this joint channel and network coding (JCNC) scheme, the performance of channel coding affects the whole capacity of system. So the average collector throughput (k/n) at several Eb/N0 is shown in Figure 4 and Figure 5, for AWGN channel and Rayleigh channel respectively.

Throughput of system with joint channel and network coding (JCNC) is compared with that of system without network coding (only fountain-channel coding). In the system with only channel coding (CC), every storage node saves only one packet. If one packet is decoded correctly and saved, the storage node will not accept new packet until next query. The count of n in only channel coding (CC) system is the same as JCNC system. However, the value of k plus 1 only when the channel decoding correctly (according to BER as shown in section 3.) and the number of packet is not repeated with that of former.

Distinctly from the simulation results, the throughput of JCNC is largely improved at the same SNR, which just benefits from the network coding. The throughput

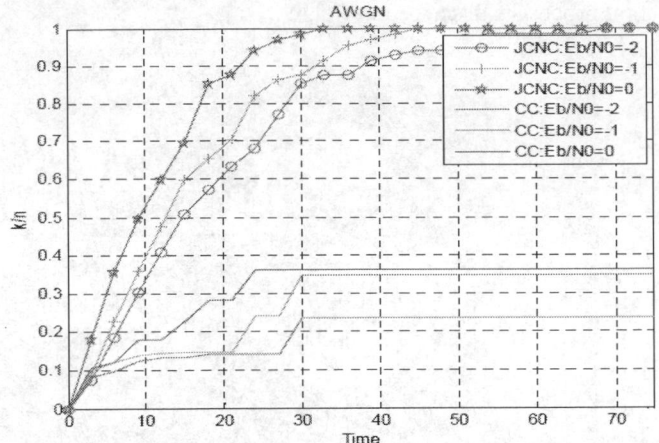

Fig. 4. Throughput of the system JCNC in AWGN Channel, compared with system CC. L=1, k/s=10%.

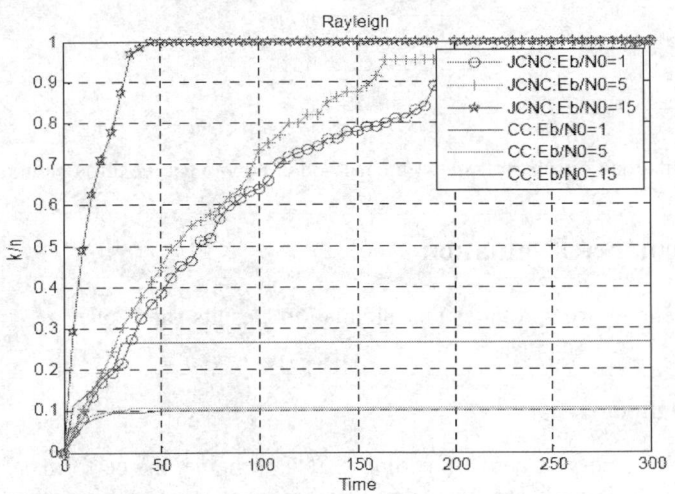

Fig. 5. Throughput of the system JCNC in Rayleigh Channel, compared with system CC. L=1, k/s=10%.

doesn't reach 0.5 even at high SNR (AWGN: Eb/N0= 0; Rayleigh: Eb/N0=15). Moreover, the throughput is related partially with the performance of BER as interpreted in section 3. And the BER is greatly improved as Eb/N0 rises as a result of channel coding. So it is clearly that the throughput gets better with the higher Eb/N0.

4.2 Node Degree (L)

As discussed in section 3, the cost and time delay of data query is minimal when k/n reaches 1. Moreover, some key parameters, the system overhead, query delay and en/decoding complexity, mainly relates with source node degree L. Emphasizing the effects of L in the system, we provide the simulation results at ideal channel condition (AWGN: Eb/N0=1dB, Rayleigh: Eb/N0=12dB).

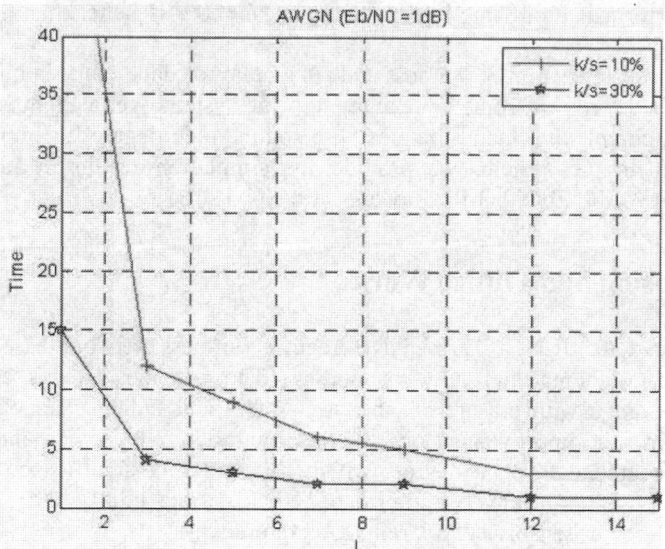

Fig. 6. Time of data reconstruction with different node degree (L) in AWGN Channel

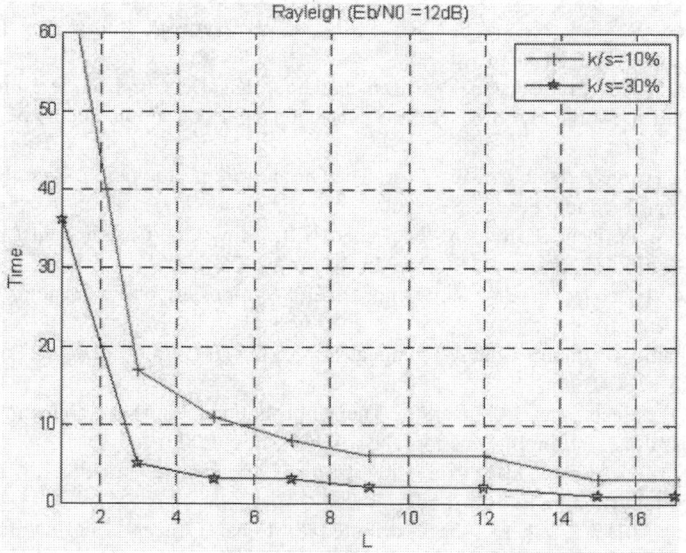

Fig. 7. Time of data reconstruction with different node degree (L) in Rayleigh Channel

Both Figure.6 and Figure.7 present the time of data reconstruction at the collector versus L with different network (k/s=10% and k/s=30%). Communication time decreases sharply with value of L rises from 1 to 4. However, the time of retrieving original data decreases much slowly, even remains constant, with L (L>5) rising. In addition, performance of Time in network with k/s=30% is better than network with k/s=10%.

As discussed in section 3, the cost and time delay of data query is minimal when k/n reaches 1. Moreover, some key parameters, the system overhead, query delay and en/decoding complexity, mainly relates with source node degree L. Emphasizing the effects of L in the system, we provide the simulation results at ideal channel condition. (AWGN: Eb/N0=1dB, Rayleigh: Eb/N0=12dB)

5 Conclusion and Future Work

We introduced the JCNC distributed storage in wireless network based on digital fountain turbo coding and network coding. The scheme extends the classical distributed storage with erasure channel to AWGN and fading channel. Simulation results confirm the improvement on system throughput. We provide another helpful parameter: source data node degree, which have guiding significance in practice. Finding effective algorithm to optimize d, s, L and time jointly and extending the system to ad-hoc or sensor network remain as future work.

References

1. Ahlswede, R., Cai, N., Li, S.-Y.R., Yeung, R.W.: Network Information Flow. IEEE-IT 1(46), 1204–1216 (2000)
2. Dimakis, A.G., Prabhakaran, V., Ramchandran, K.: Decentralized erasure codes for distributed networked storage. IEEE Transaction on Information Theory 52, 2809–2816 (2006)
3. Rhea, S., Eaton, P., Kubiatowicz, P. J.: The OceanStore prototype. In: Proc. USENIX File and Storage Technologies (FAST) (2003)
4. Acedanski, S., Deb, S., Médard, M., Koetter, R.: How Good is Random Linear Coding Based Distributed Networked Storage. In: NetCod 2005 (2005)
5. Luby, M.: LT codes. In: Proc. 43rd Annual IEEE Symposium on Foundations of Computer Science (2002)
6. Shokrollahi, A.: Raptor Codes, Technical Report DR2003-06-001 Digital Fountain (June 2003)
7. Jenkac, H., Hagenauer, J., Mayer, T.: The Turbo-Fountain and its Application to Reliable Wireless Broadcast. In: Information Theory Workshop (2006)
8. http://www.ecse.rpi.edu/Homepages/shivkuma/teaching/sp2001/readings/digital-fountain
9. Coulouris, J.D., Kindberg, T.: Distributed Systems: Concepts and Design, 4th edn. Addison Wesley/Pearson Education (2005)
10. Kameyama, H., Sato, Y.: Erasure Codes with Small Overhead Factor and Their Distributed Storage Applications. In: CISS 2007, 41st Annual Conference on Information Science and System (2007)

11. Plank, J.S., Thomason, M.G.: On the Practical use of LDPC Erasure Codes for Distributed Storage Applications. Technical Report CS-03-510, University of Tennessee (2003)
12. Plank, J.S., Thomason, M.G.: A Practical Analysis of Low-Density Parity-Check Erasure Codes for Wide-Area Storage Applications. The International Conference on Dependable Systems and Networks. IEEE, Los Alamitos (2004)
13. Ho, T., Medard, M., Shi, J., Effros, M., Karger, D.R.: On Randomized Network Coding, http://web.mit.edu

Invariance of the Hybrid System in Microbial Fermentation

Caixia Gao[1] and Enmin Feng[2]

[1] School of Mathematical Sciences
Inner Mongolia University, Hohhot, Inner Mongolia, 010021, China
gaocx0471@163.com
[2] Department of Applied Mathematics
Dalian University of Technology, Dalian, Liaoning, 116024, China

Abstract. In this study, we propose a nonlinear hybrid dynamical system to describe the concentrations of extracellular and intracellular substances in the process of bio-dissimilation of glycerol to 1,3-propanediol. An invariance principle is established for the hybrid dynamical system. At the same time, we state and prove new stability criteria for the nonlinear hybrid system. These results provide less conservative stability conditions for hybrid system as compared to classical results in the literature and allow us to characterize the invariance of a class of nonlinear hybrid dynamical systems.

Keywords: invariance, hybrid system, microbial fermentation.

1 Introduction

1,3-propanediol(1,3-PD) possesses potential applications on a large commercial scale, especially as a monomer of polyesters or polyurethanes, its microbial production is recently paid attention to in the world for its low cost, high production and no pollution, etc. [1]. Among all kinds of microbial production of 1,3-PD, dissimilation of glycerol to 1,3-PD by *Klebsiella pneumoniae* has been widely investigated since 1980s due to its high productivity [2,3]. The experimental investigations showed that the fermentation of glycerol by *K. pneumoniae* is a complex bioprocess, since the microbial growth is subjected to multiple inhibitions of substrate and products. The researches about the fermentation include the quantitative description of the cell growth kinetics of multiple-inhibitions, the metabolic overflow kinetics of substrate consumption and product formation in continuous cultures, feeding strategy of glycerol in fed-batch culture, and so on [4]. In these researches on fed-batch culture, all numerical results are based on the continuous dynamical models and there exist big errors between computational and experimental results. In fact, there exist impulsive phenomena in fed-batch culture, so the process characterized by continuous models is not fit for the actual process any longer. In order to characterize the actual process, the impulsive differential equations are applied to the fed-batch fermentation [5].

J. Zhou (Ed.): Complex 2009, Part I, LNICST 4, pp. 302–309, 2009.

The parameters in continuous system are not fit for the impulsive system, so parameter identification is necessary. Usually, ranges of parameters change in the neighborhood of initial values during the identification process. But we can't ensure the system is stable under the given ranges of parameters. Thus, stability of the system becomes a fundamental issue in system analysis and design, that is necessary for system identification and optimal control.

This paper is organized as follows. In section 2, we formulate the problem of impulsive system. In Section 3, an invariance principle is established for the hybrid dynamical system. At the same time, we state and prove new stability criteria for the nonlinear hybrid system. These results provide less conservative stability conditions for hybrid system as compared to classical results in the literature.

2   Hybrid Nonlinear Dynamical System

In this paper, we consider the effects of some enzymatic catalyses on substrates and products within cells. In this way, the computational load can be reduced greatly.

Based on [6], the hybrid nonlinear dynamical system $S(l, i)$ concerning enzymatic catalyses and transports of glycerol and 1,3-PD can be described as

$$\begin{cases} \dot{x}(t) = f(x(t), l, i, v(l), q(i)), & t \in [t_0, t_f], \\ x(t_0) = x_0, & (l, i) \in L \times G, \end{cases} \tag{1}$$

where $x(t) \in R_+^k$ is the state variable with k components of the ith dynamical system, $f : R_+^k \times L \times G \times Dc \times D_s(i) \to R^k$ is the rate of reactions, $L = \{1, 2, \cdots, l\}$ is the serial number set of experiments. $v(l) = (D(l), c_{s0}(l))^T \in D_c(l) \subset R_+^2$, $D_c(l)$ is the admissible set of dilution rate and initial glycerol concentration in lth experiment. $G = \{1, 2, \cdots, g\}$ is the serial number set of possible metabolic pathways, and g is the total number of possible metabolic pathways. $q(i) = (k_{i,1}, k_{i,2}, \cdots, k_{i,ds(i)})^T \in D_s(i)$ is the kinetic parameter vector in the ith dynamical system, $ds(i)$ is the total number of kinetic parameters. Since each component of the state variable $x(t)$ represents a certain substance concentration, there exists a nonempty bounded closed region $W_a \subset R_+^k$ such that solution $x(t; v(l), q(i))$ of $S(l, i)$ is in W_a. In addition, $D_c(l)$ and $D_s(i), (l, i) \in L \times G$, are nonempty bounded closed sets.

Because the relationships among substrates, intracellular substances and enzymes haven't been fully determined in experiments, the metabolic pathways have 72 possible cases according to mechanism analysis. That is, in the above model the total number (g) of metabolic pathways is 72. For simplicity, we only discuss system $S(1, 1)$ corresponding to the first experiment and the first case. Other models have properties similar to those of $S(1, 1)$. $S(1, 1)$ is expressed as follows.

$$\begin{cases} \dot{x}_1 = (\mu - D)x_1 \\ \dot{x}_2 = D(c_{s0} - x_2) - p_2 x_1 \\ \dot{x}_3 = k_1(x_8 - x_3)x_1 - Dx_3 \\ \dot{x}_4 = p_4 x_1 - Dx_4 \\ \dot{x}_5 = p_5 x_1 - Dx_5 \\ \dot{x}_6 = \frac{1}{k_2}(k_3 \frac{x_2}{x_2 + k_4} + k_5(x_2 - x_6) - p_2) - \mu x_6 \\ \dot{x}_7 = k_6 u_1 \frac{x_6}{k_{m1}^*(1 + \frac{x_7}{k_7}) + x_6} - k_8 u_2 \frac{x_7}{k_{m2}^* + x_7(1 + \frac{x_7}{k_9})} - \mu x_7 \\ \dot{x}_8 = k_8 u_2 \frac{x_7}{k_{m2}^* + x_7(1 + \frac{x_7}{k_9})} - k_{10}(x_8 - x_3) - \mu x_8 \end{cases} \tag{2}$$

In (2), D and c_{s0} are the dilution rate and the initial glycerol concentration of the first experiment. k_{m1}^*, k_{m2}^* are given constants. The specific growth rate of cells μ, specific consumption rate of substrate p_2, specific formation rates of products p_i, $i = 4, 5$, and u_i, $i = 1, 2$, are expressed as follows.

$$\mu = \mu_m \frac{x_2}{x_2 + k_s^*} \prod_{i=2}^{5}(1 - \frac{x_i}{x_i^*}), \tag{3}$$

$$p_2 = m_2 + \frac{\mu}{Y_2} + \Delta_2 \frac{x_2}{x_2 + k_2^*}, \tag{4}$$

$$p_4 = m_4 + \mu Y_4 + \Delta_4 \frac{x_2}{x_2 + k_4^*}, \tag{5}$$

$$p_5 = p_2(\frac{c_1}{b_1 + Dx_2} + \frac{c_2}{b_2 + Dx_2}), \tag{6}$$

$$u_1 = k_{11} + k_{12}\mu + k_{13} \frac{x_2}{x_2 + k_{14}}, \tag{7}$$

$$u_2 = k_{15} + k_{16}\mu + k_{17} \frac{x_2}{x_2 + k_{18}}. \tag{8}$$

In (3)-(8), $\mu_m, k_s^*, k_2^*, k_4^*, \Delta_2, \Delta_4, m_2, m_4, c_i, b_i(i = 1, 2)$ are given constants, respectively. Moreover, $x_i^*(i = 1, 2, \cdots, 5)$ are given critical concentrations. $q = (k_1, k_2, \cdots, k_{18}) \in D_s(1) \subset R^{18}$ is the parameter vector in the first case.

According to the transport mechanisms of glycerol and 1,3-PD across cell membrane, we assume that

$(A1)$ The absolute values of differences between intracellular and extracellular concentrations of glycerol and 1,3-PD have upper bounds M_1 and M_2, respectively.

Under assumption $(A1)$, we can easily obtain the following properties of system (1).

Property 1. For any pair $(l, i) \in L \times G$, $v(l) \in D_c(l)$ and $q(i) \in D_s(i)$, function $f(x(t), l, i, v(l), q(i))$ satisfies that $f \in C([t_0, t_f]; R^k)$ and f is locally Lipschitz continuous in x on R_+^k.

Property 2. For any pair $(l, i) \in L \times G$, $v(l) \in D_c(l)$ and $q(i) \in D_s(i)$, function $f(x(t), l, i, v(l), q(i))$ satisfies linear growth condition, i.e., there exist positive constants $\alpha, \beta > 0$ such that

$$\|f(x(t), l, i, v(l), q(i))\| \leq \alpha + \beta \|x(t)\|, \ \forall \ t \in [t_0, t_f],$$

where $\| \cdot \|$ is Eulcidean norm.

Proof. For $f(x(t), 1, 1, v(1), q(1))$ the linear growth condition is satisfied as the following proof. For any $x(t) \in R_+^k, v(1) \in D_c(1)$ and $q(1) \in D_s(1)$, we know that

$$|f_1(x(t), 1, 1, v(1), q(1))| \leq (|\mu_m| + |D(1)|)|x_1|.$$

Letting $L_1 = |\mu_m| + |D|$, we obtain that $|f_1(x(t), 1, 1, v(1), q(1))| \leq L_1\|x\|$. Furthermore, let $L_2 = \max\{|D|, |m_2| + |\mu_m||Y_2| + |\Delta_2|\}$. Since

$$|f_2(x(t), 1, 1, v(1), q(1))| \leq |D||c_{s0}| + |D||x_2| + (|m_2| + |\mu_m||Y_2| + |\Delta_2|)|x_1|,$$

we must conclude that $|f_2(x(t), 1, 1, v(1), q(1))| \leq |D||C_{s0}| + L_2\|x\|$. Let $L_3 = \max\{|k_1||M_2|, |D|\}$. Then

$$|f_3(x(t), 1, 1, v(1), q)| \leq |k_{1,1}||M_2||x_1| + |D(1)||x_3|,$$

and we have that $|f_3(x(t), 1, 1, v(1), q(1))| \leq L_3\|x\|$. Set $L_4 = \max\{|m_4| + |\mu_m||Y_3| + |\Delta_3|, |D(1)|\}$. Since

$$|f_4(x(t), 1, 1, v(1), q)| \leq (|m_4| + |\mu_m||Y_3| + |\Delta_3|)|x_1| + |D||x_4|,$$

we obtain that $|f_4(x(t), 1, 1, v(1), q(1))| \leq L_4\|x\|$. Let $L_5 = \max\{(|m_2| + |\mu_m||Y_2| + |\Delta_2|)(|\frac{c_1}{b_1}| + |\frac{c_2}{b_2}|), |D(1)|\}$. Since

$$|f_5(x(t), 1, 1, v(1), q(1))| \leq (|q_2|(|\frac{c_1}{b_1}| + |\frac{c_2}{b_2}|))|x_1| + |D(1)||x_5|,$$

we obtain that $|f_5(x(t), 1, 1, v(1), q(1))| \leq L_5\|x\|$. Let $L_6 = |\frac{k_5}{k_2}| + |\mu_m|$. Since

$$|f_6(x(t), 1, 1, v(1), q(1))| \leq |\frac{kk_5}{k_2}|\|x_2\| + (|\frac{k_5}{k_2}| + |\mu_m|)|x_6| + (|\frac{q_2}{k_2}| + |\frac{k_3}{k_2}|),$$

we obtain that $|f_6(x(t), 1, 1, v(1), q(1))| \leq L_6\|x\| + (|\frac{q_2}{k_2}| + |\frac{k_3}{k_2}|)$. Let $L_7 = |\mu_m|$. So

$$|f_7(x(t), 1, 1, v(1), q(1))| \leq |k_6||u_1| + |k_8||u_2| + |\mu_m||x_7|,$$

and we see that $|f_7(x(t), 1, 1, v(1), q(1))| \leq L_7\|x\| + |k_6||u_1| + |k_8||u_2|$. Let $L_8 = |k_9| + |\mu_m|$. Then

$$|f_8(x(t), 1, 1, v(1), q(1))| \leq L_8\|x\| + |k_8||u_2|.$$

Let $\beta = \frac{\sqrt{2}}{2}\max\{L_1, \cdots, L_8\}$ and $\alpha = \frac{\sqrt{2}}{2}\max\{|D(1)||c_{s0}(1)|, |k_8|(|k_{15}| + |k_{16}||\mu_m| + |k_{17}|) + |k_6|(|k_{11}| + |k_{12}||\mu_m| + |k_{13}|), \frac{|m_2| + |\mu_m||Y_2| + |\Delta_2|}{|k_2|} + |\frac{k_3}{k_2}|\}$. In view of the boundedness of $D_c(1)$ and $D_s(1)$, we must conclude that

$$\|f(x(t), 1, 1, v(1), q(1))\| \leq \alpha + \beta \|x(t)\|, \ \forall \ t \in [t_0, t_f].$$

In the same way, it can be proved that for any $(l, i) \in L \times G$ and $q \in D_s(i)$ the function $f(x(t), l, i, v(l), q(i))$ defined in (1) satisfies linear growth condition.

Property 3. For any pair $(l, i) \in L \times G$, $v(l) \in D_c(l)$ and $q(i) \in D_s(i)$, system (1) has a unique solution, denoted by $x(t; v(l), q(i))$. Furthermore, $x(t; v(l), q(i))$ is continuous with respect to $q(i)$.

Proof. Since f is continuous with respect to $q(i) \in D_s(i)$, it follows from Property 1 and Property 2 that system (2) has a unique solution $x(t; v(l), q(i))$. In addition, $x(t; v(l), q(i))$ is continuous in q on $D_s(i)$ in term of the theory of continuous dependence of solution to differential equations on parameters.

3   Stability Criteria

In this section, an invariance principle is established for the hybrid dynamical system. At the same time, we state and prove new stability criteria for the nonlinear hybrid system. These results provide less conservative stability conditions for hybrid system as compared to classical results in the literature.

Definition 1. *A function $V(t, x)$ is said to be*
(i) *decrescent if there exists a function $a : R_+ \to R_+$ such that*

$$V(t, x) \leq a(\|x\|), \quad (t, x) \in R_+ \times s(\rho).$$

(ii) *positive definite if there exists a continuous function $b : R_+ \to R_+$ such that*

$$b(\|x\|) \leq V(t, x), \quad (t, x) \in R_+ \times s(\rho)$$
$$V(t, 0) \equiv 0, \qquad\qquad t \in R_+.$$

Lemma 1. *[7] Assume that*
(i) $V \in \sum$, *there exist $\lambda_k \in R$ and a continuous function $c_k : R_+ \to R_+$ such that*

$$D^+ V(t, x) \leq \frac{\lambda_k}{\Delta t_k} c_k(V(t, x)), \quad (t, x) \in (t_{k-1}, t_k) \times s(\rho);$$

(ii) *there exist $v_k \in R$ and a continuous function $d_k : R_+ \to R_+$ such that*

$$V(t_k^+, x + I_k(x)) \leq V(t_k, x) + v_k d_k(V(t_k, x)), \quad x \in s(\rho);$$

(iii) $\lambda_k + v_k \leq 0$, *for $s \in (0, \rho)$, $c_k(s) \leq d_k(s)$ if $v_k < 0$ and $c_k(s) \geq d_k(s)$ if $u_k < 0$.*
Then system (1) is stable. Suppose further that
(iv) $V(t, x)$ *is decrescent and for any $\eta > 0$, there exists a $\sigma > 0$ such that*

$$s + |v_k| d_k(s) < \eta, \quad \forall s \in (0, \sigma), \quad k = 1, 2, \cdots$$

Then system (1) is uniformly stable.

Theorem 1. *The dynamical system* (2) *is impulsive stable if parameters in system* (2) *satisfy that* $\dfrac{b_1}{c_1} + \dfrac{b_2}{c_2} < 1$ *and* $1 + \dfrac{1}{Y_2}(a+1) + Y_3 + Y_4 > 0$.

Proof.   Define Lyapunov function

$$V(x) = \frac{1}{2}(x_1 - x_2)^2 + x_1 x_3 + x_1 x_4 + x_1 x_5, \tag{9}$$

Then $V(x)$ is positive definite, decrescent. Along solutions of (4), we have

$$D^+V(x) = \frac{\partial}{\partial x}V(x) \cdot f(x) = (x_1 - x_2 + x_3 + x_4 + x_5)\mu x_1$$
$$-q_2 x_1(x_2 - x_1) + q_3 x_1^2 + q_4 x_1^2 + q_5 x_1^2$$

Let $a = \dfrac{b_1}{c_1} + \dfrac{b_2}{c_2}$ in expressing q_5. Since $a < 1, 1 + \dfrac{1}{Y_2}(a+1) + Y_3 + Y_4 > 0$, and

$\dfrac{\Delta_i x_2}{x_2 + k_i}$, $i = 2, 3, 4$, is increasing about x_2, we have

$$D^+V(x) \le [(a+1)m_2 + m_3 + m_4 + (1 + \frac{1}{Y_2}(a+1) + Y_3 + Y_4)\mu_m$$
$$+ (a+1)\frac{\Delta_2 x_2^*}{x_2^* + k_2} + \frac{\Delta_3 x_2^*}{x_2^* + k_3} + \frac{\Delta_4 x_2^*}{x_2^* + k_4}]x_1^2$$
$$+ \mu_m x_1 x_2 + \mu_m x_1 x_3 + \mu_m x_1 x_4 + \mu_m x_1 x_5$$

Let $\lambda_0 = (a+1)m_2 + m_3 + m_4 + (1 + \dfrac{1}{Y_2}(a+1) + Y_3 + Y_4)\mu_m + \dfrac{\Delta_3 x_2^*}{x_2^* + k_3} + \dfrac{\Delta_4 x_2^*}{x_2^* + k_4}$
and $\lambda^* = \max\{2\lambda_0, \mu_m\}$, thus by (6)

$$D^+V(x) \le \lambda^* V(x) \tag{10}$$

This implies, by the definition of $I_i(x(t_i))$, that

$$V(x + I_i(x)) = V(x_1(1 - u_i), cu_i + x_2(1 - u_i), x_3(1 - u_i),$$
$$x_4(1 - u_i), x_5(1 - u_i))$$
$$= (1 - u_i)^2 V(x) - cu_i(1 - u_i)x_1 + \frac{1}{2}c^2 u_i + cu_i x_2(1 - u_i) \tag{11}$$

Hence $V \in \sum$, let $\lambda_i = \Delta t_i \lambda^*$, $c_k(s) = s$, the Condition (i) of Lemma 1 are satisfied from (7).

There are two cases to consider for Condition (ii) of Lemma 1.

Case 1. $x_1 < x_2$. Choose u_i that satisfies $u_i \le \dfrac{2x_1 - 2x_2 - c}{2x_1 - 2x_2}$, which implies, in view of (8), that

$$V(x + I_i(x)) \le V(x) - u_i(2 - u_i)V(x) \tag{12}$$

Case 2. For $x_1 \ge x_2$, we can also choose the controllable variable u_i such that (9) is satisfied.

Let $v_i = -u_i(2 - u_i)$, $d_k(s) = s$, the Condition (ii) of Lemma 1 is satisfied from (9).

We can also choose the controllable variable u_i such that $u_i \geq 1 - \sqrt{1 - \lambda^* \Delta t_i}$, when $\lambda_i + v_i = \Delta t_i \lambda^* - u_i(2 - u_i) \leq 0$, we get $c_i(s) = d_i(s)$ if $v_i < 0$. Thus Condition (iii) of Lemma 1 is satisfied. Hence system (5) is impulsive stable by Lemma 1.

Theorem 2. *Under the conditions of Theorem 1, system* (2) *is uniformly stable.*

Proof.   From Definition 1, the function $V(x)$ of Theorem 1 is decrescent and $s + |v_i| d_i(s) = s + u_i(2 - u_i)s$. Let $\sigma = \eta/2$, we have

$$s + |v_i| d_i(s) \leq 2s < \eta.$$

This implies that the condition (iv) of Lemma 1 is satisfied, and the system (5) is uniformly stable.

4   Conclusion

The conditions of parameters are given by Theorem 1 in the paper. Under these conditions, the parameter identification is realized for the fermentation process. Numerical simulation shows that the impulsive dynamical system presented in this paper can characterize the process of 1,3-propanediol production by fermentation. We conclude that the impulsive system is more fit for formulating fed-batch fermentations.

Acknowledgements

This work was supported by the College Project of Research in Inner Mongolia (grant no.NJZY07012) and 513 Program' Project of Inner Mongolia University.

References

1. Biebl, H., Menzel, K., Zeng, A., Deckwer, W.: Microbial production of 1,3-propanediol. Appl. Microbiol. Biotechnol. 52, 297–298 (1999)
2. Xiu, Z., Song, B., Wang, Z., Sun, L., Feng, E., Zeng, A.: Optimization of biodissimilation of glycerol to 1,3-propanediol by klebsiella pneumoniae in one-stage and two-stage anaerobic cultures. Biochemical Engineering Journal 19, 189–197 (2004)
3. Zeng, A., Rose, A., Biebl, H., Tag, C., Guenzel, B., Deckewer, W.: Multiple product inhibition and growth modeling of Clostridium butyricum and Klebsiella pneumoniae. Biotechnol. Bioeng. 44, 902–911 (1994)
4. Reimann, A., Biebl, H.: Production of 1,3-propanediol by clostridium butyricum DSM5431 and product tolerant mutants in fedbatch culture, feeding strategy for glycerol and ammonium. Biotechol Lett. 18, 827–832 (1996)

5. Caixia, G., Kezan, L., Enmin, F., Zhilong, X.: Nonlinear impulsive system of fed-batch culture in fermentative production and its properties. Chaos Solitons and Fractals 28(1), 271–277 (2006)
6. Enmin, F., Chongyang, L., Zhaohua, G., Yaqin, S.: Identification of Intracellular Kinetic Parameters in Continuous Bioconversion of Glycerol by Klebsiella pneumoniae. In: The Second International Symposium, OSB 2008, Lijiang, China, Proceedings, vol. 37, pp. 93–100 (2008)
7. Xinzhi, L., Yanqun, L., Lay, T.K.: Stability Analysis of Impulsive Control System, Mathematical and Computer Modeling. Mathematical and Computer Modeling 37, 1357–1370 (2003)

Is Self-organization a Rational Expectation?
A Critical Review of Complexity and Emergence

Heinz Luediger

IMST GmbH, Information and Communication Systems
C.F. Gaußstraße 2, Kamp-Lintfort, Germany
luediger@imst.de

Abstract. Over decades and under varying names the study of biology-inspired algorithms applied to non-living systems has been the subject of a small and somewhat exotic research community. Only the recent coincidence of a growing inability to master the design, development and operation of increasingly intertwined systems and processes, and an accelerated trend towards a naïve if not romanticizing view of nature in the sciences, has led to the adoption of biology-inspired algorithmic research by a wider range of sciences. Adaptive systems, as we apparently observe in nature, are meanwhile viewed as a promising way out of the complexity trap and, propelled by a long list of 'self' catchwords, complexity research has become an influential stream in the science community. This paper presents four provocative theses that cast doubt on the strategic potential of complexity research and the viability of large scale deployment of biology-inspired algorithms in an expectation driven world.

Keywords: complexity, emergence, mind, biology-inspired engineering, perception, knowledge, reductionism, holism.

> **"To gaze implies more than to look at – it signifies a psychological relationship of power, in which the gazer is superior to the object of gaze."**
>
> Jonathan E. Schroeder

1 Introduction

The difficulty of the subject starts with the absence of a positive definition of what complexity research is, and more critical, what problem it wants to solve when it talks about self-organisation. Most, if not all involved in its research, would agree that it is anti-reductionist by nature, which implies that it does not conform to the structures and methodologies of 'conventional' science. Likewise its proponents would agree that 'conventional' science cannot possibly explain the processes far off the thermal equilibrium, characterized by abundance of multitude and a tendency to produce some strange kind of order. The reason for failure is assigned to its very methodology of

J. Zhou (Ed.): Complex 2009, Part I, LNICST 4, pp. 310–319, 2009.

breaking structures into well behaved and understood elements before reassembling the parts, which cumulatively equate to the whole. The reductive methodology indeed becomes cumbersome, questionable and eventually useless when the phenomena under study cannot be sufficiently isolated from the environment they are embedded in, e.g. the local weather, traffic jams, the immune system or certain aspects of a next generation ICT system. Not even the solar system, the ultimate instance of reductionist clockworks, can be predicted (in the very long run) by the knowledge conventional science brought about. The bleak prophecy concerning its limits is thus not unjustified.

The methodology of complexity research rests on a holistic view of the world and a qualitative difference between the whole and the sum of its parts is generally accepted by its advocates. What appears to be the perfect balance of natural and in particular biological processes is interpreted as the result of a yet undiscovered distributed control scheme. Since nature is believed to be a system with built-in self-organisation and sustainability, the yet undiscovered mechanisms governing this intricate balance have become a major target of complexity research. Emergence[1], the perplexing effect occurring when cause and effect of a supposedly integral process are observed in different domains of experience[2], has become its earmark and universal solution likewise, while indicating the limits of conventional science and engineering. A novel kind of (soft) knowledge is proclaimed to be its result, but what this realistically means remains entirely obscure. Little more can be added to a generic description of complexity research because, like other holistic movements, it has neither developed rigorous semantics nor a quantitative framework. In this setting, where technological progress is challenged by complexity, and complexity research by the absence of semantics, it may be informative to see what complexity research potentially has in store for us. The following four theses are meant to re-initiate a fundamental debate about complexity research. The often and for good reason voiced reproach of immunization, as e.g. generally and specifically expressed in [2], requires the 'complexity' issue to be discussed on neutral ground; the degree of semantic consistency of complexity research, i.e. the degree to which we understand what we mean when we talk about complexity, is the only practical measure that currently can be applied to judge its goals, methods and results.

2 Self-organization and Expectation Are Incompatible in Principle

The 'self' qualities we believe to find in nature are the causally unintelligible effects of non-conservatory, undirected and unpredictable processes[3]. The occasionally beautiful patterns they create are easily misinterpreted as an engineered or learnt order, whereas they are a suffering and identity-sacrificing response to changing conditions in essence. The absence of stable identity and predictability make those processes incompatible with expectation-based individual and societal pursuit.

---

[1] A broad discussion and extensive bibliography concerning *emergence* can be found in [1].
[2] E.g. in significantly different dimensions of time, space, aggregation or through different modalities.
[3] The assigned non-qualities are only placeholders for the 'nothing' we know about naturalistic processes when viewed through the spectacles of the 'precise' sciences.

The roots of complexity research go back to the fifties of the last century, but only the recent societal unfolding of a romanticizing view of nature enabled the concentration of substantial efforts on the study of the processes 'used' by nature to create, develop and control its forms and processes. The difficulty of the subject has been unequivocally admitted at any time - and may yet have been underestimated in its full extent. Today, the majority of complexity research efforts are clearly directed at its commercial exploitation, i.e. at attempting to solve real world problems, for which next to emergence another monster lies in wait - purpose.

When we desire to make use of the processes that appear to self-arrange the complex interplay of the biosphere, it often goes unnoticed that we insinuate that they have been purposefully created or learnt, namely with the effect of having those 'self' qualities. Purpose and deliberate learning, however, are the domains of humans, possibly with few exceptions. They are the expression of human will to influence the future so as to make expectations more probable. Thus, what we observe in the biosphere is neither an instance of an engineered or controlled order nor a symbiotic equilibrium of 'interests' for the welfare of its 'parts', but rather a non-recurrent status quo that reflects the unique transitional balance of multiple mutual influences which is passé before it can become lawful. This status quo is neither special nor superior to any other as there is no distinguished or expected realization. On this background, biological processes cannot be explained but being predominately reactions or unconditional adaptations to changes in the environment they are inextricably entangled in. They react to changing conditions whereby producing change that causes other processes to react and so forth ad infinitum. The apparently distributed self-control of biological processes actually is a spirit-, direction- and endless chain reaction of volatile identities, rather than a locally orchestrated coexistence of systems[4]. The response to condition on the one hand and expectation on the other is what separates biological processes from machines, reaction from function and complex phenomena from actual systems. In a conscious world, explicit expectations translate into functional requirements that are prior to the system and to be observed over long periods of time and under a wide range of conditions. The ratio of the number of forbidden and granted states of such systems is near infinite - in opposition to the living world, where 'anything goes' seems to be the maxim.

Our systems and infrastructures require continuous maintenance to prevent them from falling victim to entropy, whereas the biosphere appears to be the animated equivalent of entropy and for this reason transformative but hardly destroyable. Life is, by all standards, extremely likely! It survived extreme climates, the impact of asteroids, gigantic volcano explosions and would almost certainly continue to exist after a nuclear disaster. The forms and functions developing in the wake of such a

---

[4] The strength of Darwin's theory of evolution is in its Kantian epistemic conception as it intrinsically prohibits a look behind the mirror where the concepts of the mind are getting embraced by paradox. The 'blind watchmaker' (R. Dawkins) is a suitable metaphor not only at the level of the content of the theory but also at the level of its coming into the world. This higher level 'blind watchmaker' is the mind. Its raw material and touchstone are an independent and incorruptible yet unknowable reality external to our minds, against which it tests 'intelligent designs' like the theory of evolution. This is why the whole is indeed much more than the sum of parts - it contains a big deal of human ingenuity.

disaster might, however, exceed our power of imagination. For biological processes any environment is a sufficient environment and any realization is a 'good' realization since nothing, except some kind of replication, needs to be preserved, and if it isn't - who cares!? The secret of nature's 'sustainability' is the absence of expectation. This very absence, however, is why nature's 'sustainability' is only apparent. When evolution turns a nose into a trunk or larvae convert into butterfly, there is nothing 'sustainability' corresponds to - except nature's metamorphic survivability. Biological processes and purposefully engineered systems represent incompatible extremes, for expectation impairs survivability as 'self-organization' undermines purpose.

3   Complexity and Emergence Are No Objects of Natural Science

A US military observer summarized: '...in our view, complexity is a result and not a cause of confusion...[5]*'. Indeed, complexity and paradox arise when a theory, model or notion is applied beyond its frame of applicability. They can therefore be interpreted as the result of our asking unfit questions, i.e. questions that have no answers under the current scientific paradigm(s). Emergence is the ultimate warning bell indicating that we are transgressing a frame of applicability. Complexity and emergence are neither natural phenomena nor potential causes that can be researched; they are the fabric of a plurality of mutually independent domains of perception shining through the appearances when we confuse our ideas of the world with the world itself.*

In many decades and despite increasing efforts cybernetics, catastrophe, chaos and complexity research have not produced the expected answers, except that complex or chaotic behavior can result from simple interaction of aggregates of simple elements. Rather they have remained at a narrative, gathering and complexity (re)producing level. The chronic lack of a theoretical and quantitative framework is a strong indicator for 'asking wrong questions'. Complexity research is often motivated by the experience of emergence that corresponds to the amazement befalling us when, for instance, we realize that there is nothing of what makes water in hydrogen and oxygen. Water, as we know it, refers to something that is and always has been part of our lives. It is visible, tangible and can be smelled and heard under certain circumstances. In other words, 'water' can be experienced via our natural senses and its basic qualities, e.g. its fluidity at high and solidity at low temperatures, have been derived from the interaction of what we call water with those senses. When, however, we talk about H_2O, we talk about a mental construct that nobody ever has experienced naturally. We obtain knowledge about this artificial domain through the use of a compound of real and virtual prostheses, the latter being logic and mathematics whereas electronic amplifiers, particle accelerators and powerful microscopes are examples of real prostheses. The effects of instrumentation in the sciences are discussed in [3]. What Giere calls 'scientific perspectivism', a selective, partial and not necessary view of the world is exemplary illustrated with regard to color vision as a human investigative activity that cannot produce absolute knowledge, not even as an

---

[5] 'Complexity: a cognitive barrier to defense system acquisition management' Acquisition Review Quarterly, Winter 2001, George H. Perino.

ideal. He consequently takes the step Kant and Popper denied [4] - in holding that human knowledge, despite its speculative nature, asymptotically may converge toward true knowledge - and makes clear why we cannot transcend our role of observers 'however much some may aspire to a God's-eye view of the universe'. While this step does in no way simplify the enterprise of science, it may yet provide novel insight into its origin, use and limitations as well as into the immense responsibility science carries.

While biology is and has developed as a science that roots very much in the world of our natural senses and hence language, physics developed its own prosthetic senses and accordingly emancipated from natural language. On this background, the attempt to superimpose, relate or interchange the knowledge of two (or several) scientific disciplines, as often operated in interdisciplinary complexity research, appears surprising. Biology and physics, for instance, under the current paradigms, exist as independent sciences because their objects of study are complementary, 'orthogonal' or mutually exclusive projections of the same world. As 'red', 'dissonant' and 'soft' uniquely correspond to visual, acoustic and tactile experiences, and have only figurative meaning beyond, the abstract concepts of physics become meaningless when we try to relate them to the naturalistic forms and processes dealt with by biology and vice versa. The ultimate disparateness of apperception through the various natural and artificial senses may well represent the insurmountable cliffs we call emergence, i.e. disconnected clusters of aesthetic or semantic coherence. They can be interpreted as the effect of an evolutionary orthogonalisation of our sensual and semantic theories of the world making experience and talk about the experienced possible in the first place by categorically structuring and ordering our senses and theories such, that, contrary to the very paradigm of complexity research, *not everything is connected to everything*. As there is nothing of smell in vision there is nothing of physics in biology[6]. Complexity research and its networking paradigm agitate these ordering schemes - and harvest complexity.

4 'Complex' Knowledge Is Logically and Economically Unviable

Complexity research cannot escape the paradox of attempting to understand the 'whole' on the basis of its very opposite, i.e. on the basis of notions and concepts that have been stripped off the 'wholeness' in the process of reduction. It therefore has no operational ground to stand on and tries to pull itself out of the water at its own hair. Also its claim to aim at a radically novel kind of knowledge is not convincing, for it would assume revolutionary changes to brain structures genetically bestowed to us.

There are trillions of trees on this planet of which no two are identical. And yet, the notion 'tree' and the associated concept 'tree', that goes with it, allow us to talk about trees without pointing at or making reference to each or any of them. This is the

---

[6] Which does not mean that there are no physical effects in living beings, it means that the world is open to multiple views, each ideally expanding an orthogonal set of categories and therefore adding a 'pure' dimension of experience. Intra-categorical consistency (knowledge), however, is at the expense of the meaninglessness of inter-categorical questions.

process of reduction - or abstraction from the particular. In a creative-deductive[7] feed-back loop it *integrates* a range of phenomena which from a certain point of observation exhibit symmetries and thus invariant properties. Before notion and concept 'tree' were created in the early history of mankind, 'trees' did not exist! The advantage of this reduction is sheer immeasurable as it allows, in conjunction with other abstractions, to manipulate 'trees' in a mental process[8] before actually carrying out or having carried out a manipulation with a high rate of success, because unfit manipulations can be a priori dropped[9]. We can in fact teach others how to successfully solve classes of problems without the involved things being in reach or sight, or even exactly known. What is conveyed in the process is knowledge. Without abstraction from the particular, which is synonymous for abstraction from the whole, knowledge is impossible for economic reasons. In sharp contrast, the know-how we acquire via plain observation and imitation or trial and error cannot be integrated with other knowledge for the reason of absent semantics. Compared to knowledge, know-how is a useful blind alley. The vast number of such blind alleys we are facing today, i.e. the increasing algorithmic determination of the actions we (and machines) take, is a major source of complexity, for they are semantically unrelated, not deducible and thus incomprehensible.

As powerful the process of reduction is, it inevitably comes at the expense of the loss of individuality and hence wholeness. Knowledge is the result of selective and 'theory-laden' observation [5]. Like facts, it appears to be discrete (by notion) and possible only at the necessary renunciation of other potential knowledge, as ultimately suggested by the uncertainty principle of quantum mechanics. That quantum-mechanics' renunciation is not the effect of under-determination is discussed in [6, 7]. Knowledge of the world, involving more than one of a set of 'orthogonal' notions at a time (e.g. particle and wave, matter and life, etc.), appears to become blurred, because it cannot be consistently thought[10]. The particle-wave example shows that asking certain questions is literally unreasonable, but also that the quest for better knowledge is rationally justified as it is the sense-variant semantic illustration of the experience of an underlying stratum the true nature of which is independent of and unobservable by the ideas of the mind. This view naturally explains the anti-absolute argument of *pessimistic induction*, which describes the historical succession of successful but incompatible theories which once were believed true but eventually proved wrong.

---

[7] Kant posited that a 'saltus' of the mind is needed as the first step towards scientific progress. Popper generalized the idea and made the principle, that nothing can be observed without a genetically/logically leading hypothesis the central point of his theory of speculative knowledge (see annex).

[8] In this process imagination deals with semantic systems where the system (the whole) is logically and temporally leading. It is utterly impossible to envision a system bottom-up from (its) parts. The parts of the simplistic system *bicycle,* dropped over virgin tribes land would likely be used for fishing, hunting, child play or be declared cult but not get assembled, leave alone used as a bicycle.

[9] The effects of biological evolution on mankind ceased long ago in response to the much higher efficiency of the predictive qualities of the mind.

[10] A *thing* cannot possibly be e.g. particle and wave at the same time given the classical (archaic) concepts we associate with *particle* and *wave*.

The tree in front of us is a much more complex phenomenon than the notion 'tree' can convey. Actually it is unique. Complexity research deliberately drives at the investigation of unique (holistic) phenomena and thus at the reversal of the process of reduction. The inevitable result is inflation, for the methodological arrow now points from the universal to the particular. Our apparatus of thinking though is based on the finite resource brain and on a categorical and hence reductive scheme of perception for economic reasons. It appears that human knowledge exists in a corridor between holistic inflation and reductive renunciation. We can comprehend only those discrete and restricted aspects which we can talk about in an economic and semantically consistent manner, those which can be explained without the need of being demonstrated, pointed at or made reference to. The pathway of the (yet) un-understood to conscience is not through the logic and semantic centers of reasoning but through the less strictly censored boulevards of emotion and impression. The latter, having served generations of scientists as the motivation to understand, i.e. make semantic, has now been declared the original subject of complexity research. Its dilemma ironically is that any knowledge it would produce would need to be reductive, because it necessarily would need to be semantic. In order to rescue its holistic claim, complexity research evades its 'Copenhagen Convention[11]' at all costs and instead markets the promise of a radically new kind of knowledge while, however, suppressing the unpleasant implication that it would require a novel brain.

If knowledge is out of reach of complexity research, what can realistically be expected its output to be? For weakly complex systems, which are 'systems' with a potentially knowable state space, the answer is anti-know-how. In fact, we, not the 'system', will successively learn how not to do it, since such 'systems' will provide services at the expense of continued ex post elimination of their undesirable states. Once all undesirable 'system' states have been eliminated, or equivalently, once those 'systems' behave according to expectation, they will emulate conventional systems via the inefficient process of state exclusion, and thus become void of any 'self' quality for their then severely restricted state space is no longer containing a sufficient number of 'solutions' to cope with the 'unexpected'. For truly complex 'systems', having (near) unrestricted state space, the answer is simple: Nothing can be expected from 'systems' serving no expectation - by definition. In practice a third world between expectation-less 'self-organization' (nature) and predetermined functionality (man-made systems) may be illusionary. One might argue that humans are an example of something amid the extremes. The argument appears convincing but is nonetheless false; we cannot recycle the concepts of the mind (e.g. molecules, cells, neurons, agents, self-organization etc.) in an explication how these very concepts organize their originator - the mind. What we are ourselves does not figure among the concepts framed by the mind and therefore the argument is wrong.

5   Real Reduction of Complexity Is a Semantic Program

Processes, that are highly individual and do not underlie laws which are significantly less individual than the processes in question, are useless in trying to estimate the

---

[11] The Copenhagen Convention, i.e. Bohr's principle of complementarity, made quantum mechanics empirically relevant because it implies that quantum experiments are consistently describable in classical terms involving either 'particle' **or** 'wave'.

future. Human culture is largely based on the ability to estimate the future by use of a superior, semantics-based process called reasoning, capable of manipulating virtual objects and processes before releasing them to the world. Reasoning, however, needs notions which complexity research denies. 'Experience without notions is blind' (I. Kant).

Despite the high promise of biology-inspired 'self'- research it remains essentially unsatisfactory from a methodological point of view. Operationally it is fundamentalist empiricism for the reason of its self-elected detachment from the wealth of 'conventional' theoretical knowledge and its procedures of acquisition and dissemination. It is industrious with collecting (and increasingly producing) wonderful patterns and strange attractors of which we don't know what they correspond to. A microscopic turn of the wheel of the tap will cause a dramatic change of its dripping pattern - sufficient to catalogue another strange attractor. The activities of complexity research have in common that the encountered effects cannot be communicated but by describing the structure of the 'machine' or its algorithmic elements, i.e. the parts of complexity research relating to conventional science. But the very patterns they produce are not intelligible and render range and qualitative degree of the correlation of objectives and effects undeterminable, thus making complexity research incommunicado.

This muteness breaks the knowledge chain, the way we administer and accumulate knowledge. Conventional knowledge can be made the object of operations; it can be compared, transposed, permuted, integrated, and tested, for it has become knowledge (and a societal good) only after having become semantic. Moreover, the totality of knowledge or parts thereof can be abstracted from in the attempt to achieve a higher economy of reasoning that, if successful, builds theories over theories and represents the ultimate reduction of complexity. Following Kant, the forces behind this process are the regulative ideas of the *unity of mind* and the *unity of reason*. The regulative idea of unity also underlies the reflexive level of perception, provoking the integration of experience to highest degree possible. The foundation of this process are coherent and consistent models and theories. The importance of semantic consistency, enabling the concept of *'information is what creates information'*, is stressed by von Weizsäcker throughout [7]. In the absence of this precondition complexity research is highly divergent and means a different thing to anyone involved in or observing its investigation[12]. What currently remains for its proponents and opponents is a philosophical-methodological debate: The promise of radically new knowledge versus the threat of irrationalism when semantics are sacrificed and the particular is favored over the universal.

6 Conclusions

Emergence has always fascinated scientists but has remained a baffling riddle throughout the centuries. It appears to coincide with structures of apperception and

---

[12] Some branches of complexity research try to escape its muteness by developing hierarchies of 'systems' and rudimentary semantics. The notions they create do, however, not interface with traditional experience in any point, such that they remain without meaning. 'Notions without experience are void' (I. Kant).

reasoning rather than physical phenomena or causes. Complexity arises when we ask questions that have no answers under the current scientific paradigms, i.e. questions exceeding current linguistic potentiality. The history of science has shown that scientific paradigms are not eternal. But it has also shown that their transformation goes with dramatic changes in the meaning of apparently well established notions of apparently well established sciences, which are the shock waves of a major reduction of complexity corresponding to scientific progress. The inflationary methodology of complexity research makes it no candidate to trigger the next scientific transformation. Rather it may further the erosion of coherent and consistent communication in the sciences and the society.

References

[1] Corning, P.A.: The Re-Emergence of 'Emergence',
    http://www.complexsystems.org/publications/pdf/emergence3.pdf
[2] Horgan, J.: The End of Science. Helix Books (1996)
[3] Giere, R.M.: Scientific Perspectivism, p. 15. The University of Chicago Press (2006)
[4] Hahn, R.: Die Theorie der Erfahrung bei Popper und Kant, pp. 46–47. Alber Verlag Freiburg, München (1982)
[5] Popper, K.R.: Objektive Erkenntnis, Ein evolutionärer Entwurf., Hoffmann und Campe, Hamburg, p. 271 (1995)
[6] Heisenberg, W.: Der Teil und das Ganze, Gespräche im Umkreis der Atomphysik, Auflage 2001, pp. 141–149. Piper Verlag, München, (1969)
[7] von Weizsäcker, C.F.: Zeit und Wissen, p. 847. Hanser Verlag (1992)

Appendix: Thoughts about Reductionism

There is evidence from various directions that reductionism is not merely a philosophical preference we can hold at liberty. Similar to the fact that we are not at free will to map our everyday experiences into other than the system of space, time and causality, we may not be at free will to acquire (anticipative) knowledge through other than reductive methods of reasoning. If so, reductionism represents the condition of the possibility of knowledge, i.e. denotes an intrinsic principle of the brain-mind complex. It would then serve a fundamental purpose which in the widest sense could be described as being of economical nature. The physical interpretation of this purpose is associated with the limited resource brain (memory and its organisation), which requires schemes that reduce the infinite number of world states to a set of rules relating objects and subjects in time. The second, psychological, interpretation is that reductionism is the observable effect of the mind's most distinguished enterprise to rescue its unity. Like physical systems involving feed-back loops need to be tightly controlled to avoid chaotic and other undesirable behaviour, the reflexive mind is in constant danger of falling apart. Panic and horror are the transient effects when the unity of the mind is severely challenged, i.e. when our role as rational observers of the world is at stake. Psychological studies speak of the 'Stalinistic Methods' of the brain to prevent this happen and forgery of inter-subjective reality to surprising degrees has been reported. What is 'the unity of the mind'? It is its singular point of view (as a regulative idea rather than an achievable state) from where the world makes sense, from where it *appears* causal, coherent and consistent, or in brief - explainable. Explanations on the one hand need words while on the other hand an increasing number of words imply an exploding quantity of possible relations between these words (and hence meanings), such that a check of the consistency of an explanation quickly becomes difficult or even impossible. Ockham's razor is appealing to us because it is a condition of the possibility of verifying the consistency of a statement about the world. This is when reductionism may get involved as a method of theoretical verification by condensing the astronomical number of states of natural processes to a workable set of hypothetical entities and attributes, i.e. a system, which ideally represents all possible states of selected observable processes under well defined conditions. These conditions are reductions themselves and therefore reductionism presents the world-as-such by way of analogy rather than by way of convergence toward an absolute reality, which explains the paradox of factual scientific progress (in the natural sciences) in a succession of mutually exclusive theories as disconnected domains of stringent semantic coherence.

In section four the procedural elements of the reductive method have been briefly mentioned. According to Kant they consist of two fundamentally different steps that give rise to the world as we know it. First, a 'saltus' of the mind' is needed, corresponding to the fact that the novel is not deducible from current knowledge. It appears to represent its findings in phenomenal space for further elaboration by semantic processes. This first step is tentative and kind of art rather than kind of science. In a second step, the analytic capabilities of the mind come into play by trying to deduce and assess what flows from the 'saltus'. The reductive method thus creates causal *wholes* by a joint effort of the powers of imagination, logical deduction and inference, while the famous 'breaking into parts'[13] is absent in the process. If it were not, the absurdity would occur that the knowledge of the novel would be a necessary condition for its subsequent derivation in the Cartesian process of 'breaking it into parts'.

For the novel cannot be legitimated by experience, it is unlikely that 'truth' in the sense of 'conformance with experience' - as generally accepted in the sciences - is a critical concept in the logical assessment of the 'saltus'. The fact that nevertheless empirically valid theories can be the outcome of a purely mental process suggests that the involved mechanisms and in particular the role of our senses are not well understood[14]. At the same time, if the reductive scientific method is linked to an aesthetic-semantic process instead of a semantic-empirical methodology, it is not unreasonable to assume that its outputs are potentially furnished with validity prior to any empirical test. Otherwise the reductive method would not be of great value in obtaining and securing the unity of the mind. The most delicate process in the world may in fact require more sophisticated means than corporeal trial and error to establish multiple, mutually orthogonal and therefore non-conflicting world perspectives.

---

[13] Descartes mentions in several places [see Bernard Williams, 'Descartes', p.15] that i) his 'method' is not obligatory and ii) the *Discours de la Méthode* is not adequately explaining it.

[14] Einstein's statement that '...*only the theory determines what can be observed*' may set the framework for a better understanding of our world perceptions.

Inter-Profile Similarity (IPS): A Method for Semantic Analysis of Online Social Networks

Matt Spear, Xiaoming Lu, Norman S. Matloff, and S. Felix Wu

University of California, Davis
batman900@gmail.com, lu@ucdavis.edu, matloff@cs.ucdavis.edu,
wu@cs.ucdavis.edu

Abstract. Online Social Networks (OSN) are experiencing an explosive growth rate and are becoming an increasingly important part of people's lives. There is an increasing desire to aid online users in identifying potential friends, interesting groups, and compelling products to users. These networks have offered researchers almost total access to large corpora of data. An interesting goal in utilizing this data is to analyze user profiles and identify how similar subsets of users are. The current techniques for comparing users are limited as they require common terms to be shared by users. We present a simple and novel extension to a word-comparison algorithm [6], entitled Inter-Profile Similarity (IPS), which allows comparison of short text phrases *even if they share no common terms*. The output of IPS is simply a scalar value in $[0, 1]$, with 1 denoting complete similarity and 0 the opposite. Therefore it is easy to understand and can provide a total ordering of users. We, first, evaluated the effectiveness of IPS with a user-study, and then applied it to datasets from Facebook and Orkut verifying and extending earlier results. We show that IPS yields both a larger range for the similarity value and obtains a higher value than intersection-based mechanisms. Both IPS and the output from the analysis of the two OSN should help to predict and classify social links, make recommendations, and annotate friends relations for social network analysis.

Keywords: Online Social Network, Semantic Analysis, Profile Similarity, Natural Language Processing.

1   Introduction

Online Social Networks (OSN) are experiencing an explosive growth rate and are becoming an increasingly important part of people's lives. There is an increasing desire to aid users in identifying potential friends, interesting groups, and compelling products. Furthermore, these networks have offered researchers almost total access to large corpora of data. An interesting goal in utilizing this data is to analyze user profiles and identify how similar subsets of users are. The current techniques for comparing users are limited as they require common terms to be shared.

J. Zhou (Ed.): Complex 2009, Part I, LNICST 4, pp. 320–333, 2009.
© ICST Institute for Computer Sciences, Social Informatics and Telecommunications Engineering 2009

In this paper, we devise a simple, novel method which extends [6] to compare short-text snippets using Natural Language Processing (NLP). Inter-Profile Similarity (IPS) provides an application-independent mechanism to give a total ordering according to the similarity value. This algorithm requires no extra work from the user and provides a comparison of profiles[1], which OSN already have user's input. To our knowledge, this paper is the first to provide an NLP style similarity analysis on social network graph.

The benefits of using NLP to compare the similarity of users are twofold: (1) different words that possess the same meaning will be correctly identified, and (2) the number of terms in common decreases as the size of the vocabulary increases. As an example in [8] the authors show a difference between immediate friends and a random non-friend, but looking at the mean similarity, there is not a large difference between the two groups. Using an algorithm which utilizes NLP, e.g. IPS, should aid researchers in quickly evaluating the similarity of users despite there existing a large vocabulary set. We show that IPS yields both a larger range for the similarity values and obtains higher values than the intersection-based approaches.

IPS lends itself to many practical applications in OSN e.g.: (1) product recommendations to users based on their profiles, (2) personalizing the ordering of search results based on their profiles, and (3) building communities by recommending groups to users who based on their profiles. Sites such as Netflix and Amazon recommend items that they believe the user would enjoy given their prior history, but these algorithms are done in a closed fashion and are not directly compatible. Secondly, when executing searches for potential friends one is generally interested more in people similar to them. Therefore, it is preferable to provide a total ordering utilizing the user's similarity to the people in the result-set. Some current OSN already do this, but in a rather simplistic manner, such as utilizing the group and affiliation information of the users. Finally, being able to quickly find communities with people whom share interests with the user is an impoerant part of OSN.

The core of all of above problems is identifying "similar" users, whether it be due to their history of movie ratings, history of items bought, or activities/interests in common. IPS provides a common front-end to identify similar users utilizing their profiles which already exist in OSN. Once this is accomplished, all of the above problems can be solved.

We evaluated the effectiveness of IPS with a user-study and found that IPS's ordering is closely aligned to what users expect. We also applied IPS to Facebook and Orkut and were able to verify and extend earlier results. We showed, in agreement with [9], that there is a trend of decreasing similarity with increasing distance, and, further, showed that there was no significant between same gender and opposite gender neighbor similarities. Both IPS and the analysis of the two OSN should help to predict and classify social links, make recommendations, annotate friends relations for social network analysis.

---

[1] Where a profile is considered to be a set of short phrases, e.g. "play basketball", "read book".

The remainder of the paper is organized as follows: in Section 2 we describe the related work, in Section 3 we formalize the limitations of using the current approaches and introduce the IPS algorithm, then in Section 4 we provide the details and results of a user-study of IPS, next in Section 5 we apply IPS to analyze datasets from Facebook and Orkut, finally in Section 6 we note future work and present our conclusions.

2   Related Work

OSN have grown very rapidly and have become a popular mechanism for discovering (and rediscovering) friends and relationships [12]. For example, in its January 2004 debut, Orkut had over 50,000 communities, but by May, 2005 had over 1,500,000. Many OSNs now have members numbering in the tens-of-millions, and users are increasingly using them as a recommendation and community building application. Efficient and well-ordered search results are essential to the discovery of new friends in OSN.

There are a number of existing algorithms that provide the similarity of profiles (see [11] for a good survey), but there is no existing approach that utilizes NLP. Current similarity measure used in profile comparisons, documents clustering and search result presentations tend to be based on word intersection [2,3,4]. In [8], the similarity amongst neighboring nodes were compared to values of a node picked at a random in the network. It is interesting to note that the maximum intersection of any two users was below 0.16, which should intuitively imply that users are not related although they might be more related when comparing the semantic meaning of their image tags. User profiles tend to be short snippets and word-intersection algorithms might not achieve satisfactory results due to lack of common words [5]. The approach in [5] treats each snippet as queries, obtains documents related their snippets, and then compare these returned documents using a word/phrase intersection similarity measure such as cosine [3].

IPS extends [6], which provides a word-comparison algorithm using context-vectors in WordNet [10]. WordNet is a lexical dictionary that also has a hierarchy amongst the words. Words which have different meanings (or parts of speech) are designated by *senses*. The IPS similarity measure uses NLP, which allow a more flexible measure of similarity.

3   IPS Algorithm

In this section, first we analytically show why the use of a simpler algorithm which utilizes intersection will not provide good results. Then we present the IPS algorithm and describe its complexity, benefits and limitations.

3.1   Why Not Intersection?

The most obvious problem with intersection is that it requires both users to choose the same keyword. To help illustrate the issue, we will walk through a simple example to show the expected number of items in the intersection of two equally sized sets. We assume the keywords are chosen from a universe U of cardinality N, and are numbered $\{1, \ldots, N\}$ in order of decreasing popularity. We assume that the keyword popularity follows a Zipf distribution[2], i.e. the probability of drawing the kth most popular item is $\Pr(k) \stackrel{\text{def}}{=} Ck^{-1}$, where C is a normalizing constant. We also assume keywords are chosen independently; this is not generally the case, e.g. if "JPOP" is drawn it is more likely "Ai Otsuka" will be drawn even if it is not a popular keyword overall, but this assumption makes the analysis much easier. Furthermore, we assume that the user profiles are all the same size[3].

We now consider the expected number of items in common amongst two sets of items picked from $\mathbf{U}$. Our derivation uses Wallenius' Multivariate Noncentral Hypergeometric Distribution [1], as each keyword has different weight and keywords from $\mathbf{A}$ and $\mathbf{B}$ are picked without replacement from $\mathbf{U}$:

$$
\mathrm{E}[|\mathbf{A} \cap \mathbf{B}|] = \sum_{k=1}^{N} 1 \cdot \Pr(k \in A \wedge k \in B) = \sum_{k=1}^{N} [\Pr(k \in A)]^2
$$

$$
= \sum_{k=1}^{N} \left[\underbrace{\sum_{\mathbf{x} \in \chi} \underbrace{\int_0^1 \prod_{j=1}^{N} \left(1 - t^{w_j/D}\right)^{x_j} dt}_{\Pr(k \in A | \mathbf{x}), \text{ from WMNHD}}}_{\Pr(k \in A)} \right]^2 \tag{1}
$$

where χ is the set of vectors of length N such that $x_k = 1$ and $\sum_j x_j = |A|$, w_j is the weight of the jth item in $\mathbf{U}$ and $D = \mathbf{w} \cdot (1 - \mathbf{x}^i)$. Note that for any fixed $|A|$[4]:

$$
\lim_{N \to \infty} \mathrm{E}[|\mathbf{A} \cap \mathbf{B}|] = 0
$$

In Figure 1, we simulated $\mathrm{E}[|\mathbf{A} \cap \mathbf{B}|]$ where we drew $|A|$ items from the N (number of keys in the universe) possible for various $|A|$ and $|N|$. The simulation shows that the intersection of $\mathrm{E}[|\mathbf{A} \cap \mathbf{B}|]$ is very low with increasing N and decreasing $|A|$.

3.2   IPS

The entirety of the IPS algorithm is shown in Algorithm (1). The main component of IPS is the ProfileSimilarity on line 1. This function takes two profiles A and B (each consisting of a set of phrases) and outputs a similarity value between 0 and 1. The phrases consist of a small set of words and, in general,

---

[2] This assumption is backed up by many studies on Peer-to-Peer (P2P) and our own studies into Facebook and Orkut.

[3] It is trivial to extend the probabilities to cover the case where they are not.

[4] This approaches 0 at a rate of $\log(N)$, so one could *in theory* multiply by $\log(N)$ to achieve a stable value, although calculating N would be extremely cumbersome.

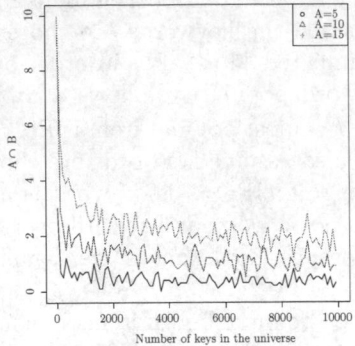

Fig. 1. Simulated $E[|\mathbf{A} \cap \mathbf{B}|]$ for various $|A|$ and varying N

Algorithm 1. IPS

1: **Function** ProfileSimilarity($\mathbf{A}$, $\mathbf{B}$)
2: $\quad(s, c) \leftarrow (0, 0)$
3: $\quad$**ForEach** $a \in \mathbf{A}$ **Do**
4: $\quad\quad(t, b) \leftarrow \max_{b \in \mathbf{B}} PhraseSimilarity(a, b)$; $(s, c) \leftarrow (s + t, c + 1)$; $\mathbf{B} \leftarrow \mathbf{B} - b$
5: $\quad\quad$**Break If** $|\mathbf{B}| = 0$
6: $\quad$**EndFor**
7: $\quad$**Return** $\frac{s}{c}$
8: **End**
9: **Function** PhraseSimilarity($\mathbf{A}, \mathbf{B}$)
10: $\quad$/* a, b from line 4 were split on spaces */
11: $\quad$**If** $|A| = 1$ **AND** $|B| = 1$ **Then**
12: $\quad\quad$**Return** SIMILARITY(A_0, B_0)
13: $\quad$**EndIf**
14: $\quad(\mathbf{adj}_A, \mathbf{noun}_A, \mathbf{verb}_A) \leftarrow WSD(\mathbf{A})$; $(\mathbf{adj}_B, \mathbf{noun}_B, \mathbf{verb}_B) \leftarrow WSD(\mathbf{B})$; $\mathbf{S} \leftarrow \emptyset$
15: $\quad$**ForEach** $T \in \{adj, noun, verb\}$ **Do**
16: $\quad\quad(s, c) \leftarrow (0, 0)$
17: $\quad\quad$**ForEach** $a \in \mathbf{T}_A$ **Do**
18: $\quad\quad\quad(t, b) \leftarrow \max_{b \in \mathbf{T}_B} SIMILARITY(a, b)$; $(s, c) \leftarrow (s + t, c + 1)$; $\mathbf{T}_B \leftarrow \mathbf{T}_B - b$
19: $\quad\quad\quad$**Break If** $|\mathbf{T}_B| = 0$
20: $\quad\quad$**EndFor**
21: $\quad\quad\mathbf{S}_T \leftarrow \frac{s}{c}$
22: $\quad$**EndFor**
23: $\quad(\mathbf{A}, \mathbf{B}) \leftarrow \bigcup_{T \in \{adj, noun, verb\}}(\mathbf{T}_A, \mathbf{T}_B)$; $(s, c) \leftarrow (0, 0)$
24: $\quad$**ForEach** $a \in \mathbf{A}$ **Do**
25: $\quad\quad(t, b) \leftarrow \max_{b \in \mathbf{B}} SIMILARITY(a, b)$; $(s, c) \leftarrow (s + t, c + 1)$; $\mathbf{B} \leftarrow \mathbf{B} - b$
26: $\quad\quad$**Break If** $|\mathbf{B}| = 0$
27: $\quad$**EndFor**
28: $\quad\mathbf{S}_{lo} \leftarrow \frac{s}{c}$
29: $\quad$**Return** weighted_avg($\mathbf{S}$)
30: **End**

---

are not full sentences; instead they are generally short-text snippets describing some activity or interest. In simplest terms, ProfileSimilarity repeatedly finds the maximally similar items from A and B and removes them until either A

or B is empty; then returns the average similarity value. This function utilizes PhraseSimilarity to compare phrases.

In PhraseSimilarity (line 9), the short-text snippets A and B need to be compared. This requires two functions from [6]: (1) WSD, and (2) SIMILARITY. Word Sense Disambiguation (WSD) is a common problem encountered in NLP— it consists of taking a sentence and identifying the sense, i.e. the part of speech (e.g. noun, adjective, verb) and the meaning, of each word. The second operation SIMILARITY provides the similarity of two words given their senses. We do not go into the details of the algorithm and refer the reader to [6] for a detailed explanation.

Now that the two external functions have been described we detail the evaluation of the phrase-similarity. To begin with, each phrase is broken into its sets for each of its parts of speech (line 14). Then, for each part of speech, the maximally similar words from A and B are found and removed until there are no more words in the part of speech. Once all parts of speech have been compared any remaining words are compared without regard to the part of speech; this is done for two reasons: (1) WSD is not a perfect algorithm and will sometimes incorrectly identify a word, and (2) some phrases do not share any parts of speech (e.g. "eat" and "JPOP"), but it is still desirable to know their similarity. Once all the parts of speech and leftover words have been compared a weighted average is returned wherein more weight is placed on nouns than verbs.

3.3   Benefits and Limitation

In this section we discuss some of the benefits and current limitations of the IPS system.

The benefits are:

- **Identify similar concepts despite being expressed with different words.** Previous intersection-based methods do not scale due to the enormity of most languages. Employing IPS overcomes this fundamental limitation.
- **Provides a total ordering over any set of users with regard to a querier.** This is useful in many applications from data mining to recommendation systems, and IPS provides a simple common interface for doing so.
- **Handles phrases of varying length by ignoring words that do not match.** Phrases of differing lengths are fairly common, and IPS handles them by comparing the most similar words first and ignore the rest. For example: let A = "play basketball" and B = "basketball", then IPS will report these users with a high similarity value.

The limitations are:

- **Ignores negation.** Phrases or sentences with negated weight such as "do not like basketball" are ignored to avoid confusion to algorithm computation.

- **The left-over words for phrases of varying length may be important.** To handle sentences of varying lengths IPS simply ignores the leftover words, this may not be the most appropriate course of action.

All of the limitations are currently future work and should be addressed in later iterations of IPS.

4   IPS User Study

To evaluate the effectiveness of IPS, we conducted a user-study[5] While IPS outputs a number between 0 and 1, it is very difficult to evaluate the "correctness" of this number, e.g. it is hard for a user to state that "basketball" and "volleyball" are 0.75 similar. To reduce the user's burden, we asked the users to order a set of phrases, and then compared this ordering with IPS' ordering.

4.1   User-Study Description

The user-study consisted of three sections: (1) word similarity, (2) phrase similarity, and (3) profile similarity. Each section presented the user with a base word, phrase, or profile and then three other words, phrases, or profiles to order with respect to the base word, phrase, or profile. The user-study was implemented as an online interface using JavaScript to allow the users to visually order the words, phrases, or profiles.

The words, phrases, or profiles are in bold and the remaining words, phrases, and profiles are ordered according to IPS' value (shown in parenthesis). In the actual user study, these values were hidden from the user and were presented in a random order.

An example of the word similarity questions asked in the questionnaire is:

basketball: (1) volleyball (0.99), (2) soccer (0.61), (3) California (0.35)
The number in parenthesis is determined by IPS and not presented to the user.
An example of the phrase similarity questions asked is:
play soccer: (1) play football (0.82), (2) watch football (0.52), (3) water garden (0.27).
An example of the profile similarity questions asked is:
educated partner, read book, has patience, stable job, good salary, play sports, enjoy travel, enjoy movie: (1) educated and well read, stable job and finance, nice salary, play sports,watch movies, has patience (0.66) (2) read book, watch movies, go to church, good job, hang out with friends, good dresser (0.58); (3) party and drink, listen to music, dining at fancy restaurant, shopping online, go to concert (0.37);

4.2   Analysis

We had 30 users complete the survey. Overall, the mode of the user responses always matched IPS' ordering. Also, 98% of the users were within one transposition from IPS' ordering, and 79% of the user's orderings agreed with IPS.

---

[5] Available at:
http://wwwcsif.cs.ucdavis.edu/%7Elu/similarity/examples/test.html

To evaluate the each individual questions, we numbered the possible responses according to a gray-code wherein IPS' ordering was in the center. A gray-code has the property that neighboring codes contain only a single transposition, which allows us to consider how close the response was to IPS. For example, if IPS stated the order was *bac* then we would use the ordering {*cab*, *cba*, *bca*, **bac**, *abc*, *acb*} so IPS' output appears at position 4 and each item has only a single transposition between each of its neighbors.

We now present the confidence intervals for each question in each section using a z value chosen for 0.01 confidence, and provide some statistical observations.

For the word similarity, 68% of the user's orderings completely agreed with IPS' ordering, and 97% of the users made at most one transposition from IPS' ordering. The aggregate of word similarity section yielded $\mu_{word} \in [3.98, 4.28]$, where the correcting ordering was numbered 4.

For the phrase similarity 87% of the user's orderings completely agreed with IPS' ordering, and 99% of the users made at most one transposition from IPS' ordering. The aggregate of word similarity section yielded $\mu_{phrase} \in [3.92, 4.06]$, where the correcting ordering was numbered 4.

Finally, for the profile similarity 74% of the user's orderings completely agreed with IPS' ordering, and 96% of the users made at most one transposition from IPS' ordering. The aggregate of word similarity section yielded $\mu_{profile} \in [3.91, 4.27]$, where the correcting ordering was numbered 4.

In all cases the confidence interval is much smaller than 1 (the minimum transposition distance). This in conjunction with the fact that 98% of the responses were within one transposition distance shows that IPS does a good job of providing orderings that would be consistent with what a user would expect.

5   Applying IPS in an OSN Study

We applied IPS to evaluate two popular social networks: Facebook and Orkut. We show how similarity correlated with topological distance with various sub-grouping; part of this is validating the results from [9] using NLP instead of the intersection based approach they utilized and part is extending said work with flow inside affiliations and across genders. Our emphasis is to show that IPS provides more logical results for similarity values than prior work and to aid in developing a deeper insight into OSN growth. As such, we also show how IPS compares with the standard L1 intersection method.

5.1   Data Overview

The overall statistics of the two networks are shown in Table 1. We refer implicitly to this table throughout Section 5.1. Facebook allows users to describe themselves in a number of different categories; however, we concentrate only on the following categories: (1) activities, (2) interests, (3) gender, and (4) networks (affiliations). To help ensure clarity throughout the paper we will refer to the networks category as affiliations. Using the affiliations, users are able to restrict

Table 1. Overall Statistics

Attribute	Facebook	Orkut
Vertices	1265	15329
Edges	7827	61738
Average degree	12.4	8.1
Average number of keywords	5.17	29.5
Average number of words per keyword	5.2	1.99
Fraction of the network within 4 hops	74.4%	32.9%

the set of people that can view their profile, the default policy is that only those in the same affiliation or are immediate friends may view each others profiles. As a result crawling a large sample users in Facebook is considerably harder (hence the smaller graph size).

Similarly, in Orkut we concentrate on the following categories: (1) activities, (2) passions, (3) sex, and (4) communities. We chose to focus on these few categories as they generally use terms that exist in the dictionary (versus, e.g., Movies). Communities and affiliations were not compared using NLP, but instead used L1 intersection as the choices were from a fixed set with no semantic meaning to the keys. Furthermore, these categories allow cross-network comparison to ensure trends are consistent. For the NLP similarity comparison, we created rings for distances between 1 and 4, so we also present the fraction of the graphs that this covers.

5.2   Semantic Analysis

We now turn to investigate the similarity of users using IPS. We are primarily interested in how similarity between pairs of users changes with increasing topological distance. To compute these similarity distributions for every node in the network, we constructed the ring of nodes at a distance 1, then those at a distance 2 but not at a distance of 1 and so on until distance 4. For each set of nodes in the ring we applied the similarity algorithm to get the pairwise similarity with varying distances.

Figures 2(a) and 2(b) shows the average profile similarity versus topological distance and also compares IPS with intersection. Both IPS and intersection show a trend of decreasing similarity with increasing distance, which agrees with [9]. IPS provides much higher similarity values than intersection. Note that the affiliations (communities) was compared using intersection as the values were chosen from a fixed vocabulary set with no semantic meaning in the identifiers.

Figures 2(c) and 2(d) shows the CDF ($\Pr(X \leq x)$) of IPS similarity distribution at different distances, ball of 1, 2, 3 and 4 (where ball of 1 means all the nodes 1-hop away, ball of 2 means all the nodes 2-hop away and not 1-hop away, and so on). Given a similarity value, say 0.4, the CDF of ball 2, 3, 4 are all closer to 1 than the CDF of ball 1 (for Facebook, it is around 0.83, for Orkut, it is around, 0.86. This shows neighboring nodes tend to have a higher similarity value than non-neighboring nodes.

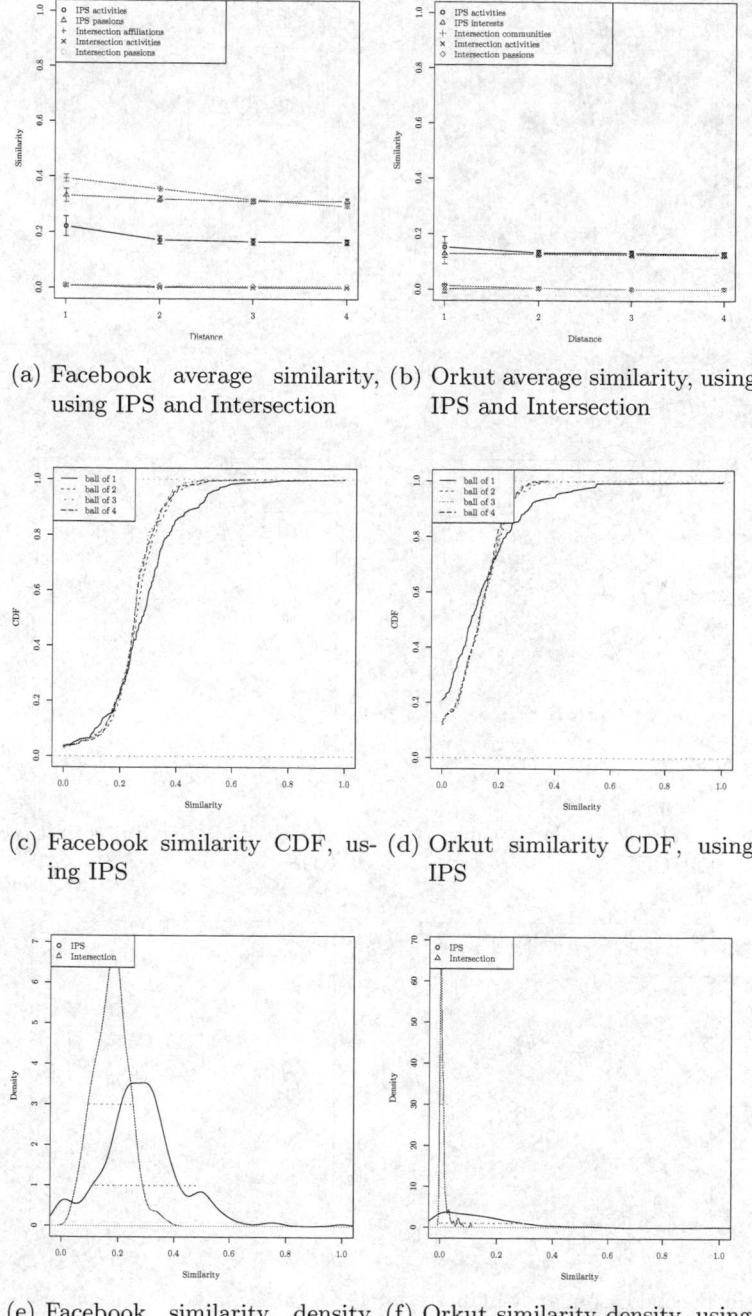

(a) Facebook average similarity, using IPS and Intersection

(b) Orkut average similarity, using IPS and Intersection

(c) Facebook similarity CDF, using IPS

(d) Orkut similarity CDF, using IPS

(e) Facebook similarity density, using IPS and Intersection

(f) Orkut similarity density, using IPS and Intersection

Fig. 2. Similarity comparisons over all profiles using Facebook and Orkut data

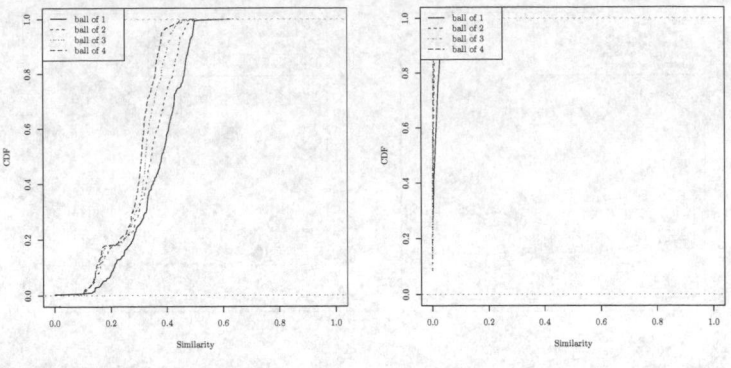

(a) Facebook similarity in affilia- (b) Orkut similarity in communi-
tions CDF   ties CDF

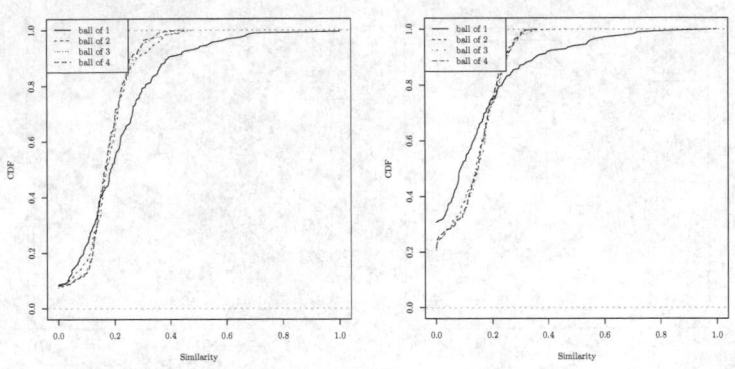

(c) Facebook similarity in activi- (d) Orkut similarity in activities
ties CDF   CDF

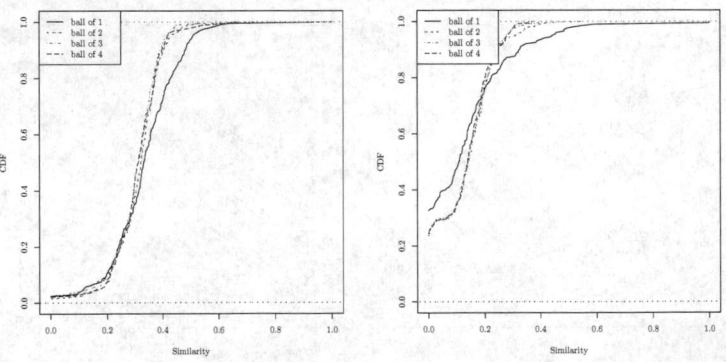

(e) Facebook similarity in interests (f) Orkut similarity in passions
CDF   CDF

Fig. 3. IPS Similarity comparisons in affiliations, activities, and interests/passions using Facebook and Orkut data

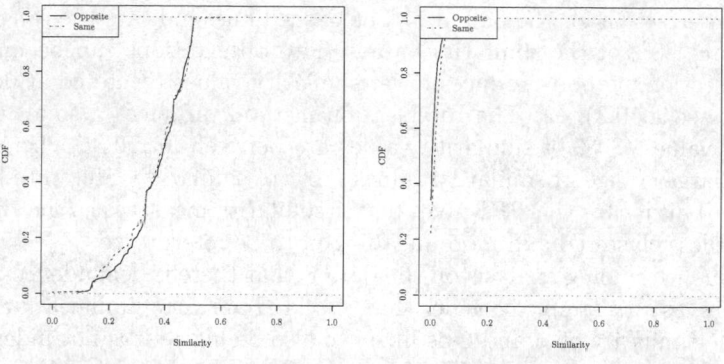

(a) Facebook similarity by gender in affiliations distance of 1

(b) Orkut similarity by gender in communities distance of 1

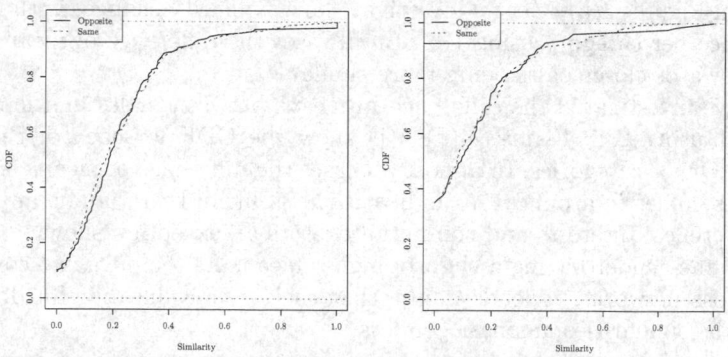

(c) Facebook similarity by gender in activities distance of 1

(d) Orkut similarity by gender in activities distance of 1

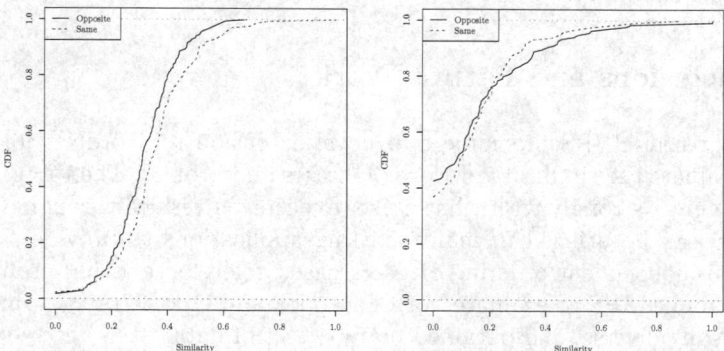

(e) Facebook similarity by gender in interests distance of 1

(f) Orkut similarity by gender in interests distance of 1

Fig. 4. Gender Similarity comparisons

Finally, Figures 2(e) and 2(f) shows the similarity density distribution using IPS and intersection for the ball of 1. The dashed lines indicate the range which encompasses 80% of the similarity values. Two observations can be made: (1) The intersection method produces lower similarity values (centered at 0.2) than IPS (centered at 0.3). (2) The Intersection method produces a small range of similarity values (80% of similarity values are between $[0.1, 0.3]$, where as IPS produces larger range of similarity values (80% of similarity values are between $[0.1, 0.45]$. This means that IPS gives better similarity measures as the similarity values are less clustered and more meaningful to be interpreted.

One issue with doing analysis on the data is that there is dependence between data points; as an example consider that A and B are very similar as are B and C then the similarity of A and C is likely to also be high—it is not independent of the first two measurements. The bootstrap is a useful tool for data analysis arising in nonstandard situations, such as the correlated-data setting we have here [13,14].

Next, in Figures 3(a)–3(f) we investigate how similarity changes with increasing distance per category using IPS. In all cases the difference of means shows a trend towards closer pairs being more similar.

Next, we investigated the difference in means between males and females at distance 1 using IPS. Figures 4(a)–4(f) show the CDF for each of the Facebook and Orkut categories. In almost all cases the difference of means is almost perfectly symmetric around zero indicating it is likely there is not any meaningful difference. Interests and communities were an exception showing a trend towards same-gender having a slightly higher mean. As there was no consistent trend across networks, it lends doubt that there is any meaningful difference between the similarities amongst genders.

Finally, we investigated if there was any correlation between *geographic* distance and similarity. Due to space limitations we do not display the graph nor the means, but we did not see any significant correlation between geographic distance and similarity.

6   Conclusions and Future Work

We have presented IPS, a simple and novel extension to WordNet [6] can be used to evaluate the similarity of words, phrases and profiles. The total ordering produced agrees strongly with what a user expects, as verified through our user-study. IPS can be utilized in many existing applications to provide a simple, application-independent ordering of users based solely on existing profile data. We, also, applied IPS to evaluate both Facebook and Orkut graphs, which were geographically diverse, and obtained many pieces of data.

IPS has a few shortcomings as described in Section 3.3. In the future more study into better mechanisms to address these and optimizing IPS would be important. With the power of IPS, we also want to investigate how network properties change over time and see if similarity helps to explains why OSN growth exceeds the prediction by preferential attachment [7].

References

1. Wallenius' noncentral hypergeometric distribution, http://en.wikipedia.org/wiki/Wallenius_noncentral_hypergeometric_distribution
2. Hammouda, K., Kamel, M.: Phrase-based Document Similarity Based on an Index Graph Model. In: Proceedings of the 2002 IEEE International Conference on Data Mining (ICDM 2002), p. 203. IEEE Computer Society, Washington (2002)
3. Investigating Measures for Pairwise Document Similarity, http://www.ncstrl.org:8900/ncstrl/servlet/search?formname=detail&id=oai
4. Zamir, O., Etzioni, O., Karp, R.: In: Kamel, M.K. (ed.) Knowledge Discovery and Data Mining, pp. 287–290 (1997)
5. Sahami, M., Heilman, T.: A web-based kernel function for measuring the similarity of short text snippets. In: Proceedings of the 15th international conference on World Wide Web, pp. 377–386. ACM, New York (2006)
6. Patwardhan, S., Pedersen, T.: Using WordNet-based context vectors to estimate the semantic relatedness of concepts. In: Proceedings of the EACL 2006 workshop, pp. 1–8 (2006)
7. Mislove, A., Marcon, M., Gummadi, K., Druschel, P., Bhattacharjee, B.: Measurement and analysis of online social networks. In: Proceedings of the 7th ACM SIGCOMM conference on Internet measurement, pp. 29–42. ACM, New York (2007)
8. Marlow, C., Naaman, M., Boyd, D., Davis, M.: HT 2006, tagging paper, taxonomy, Flickr, academic article, to read. In: Proceedings of the 17th conference on Hypertext, pp. 31–40. ACM, New York (2006)
9. Information Flow in Social Groups, http://www.citebase.org/abstract?id=oai:arXiv.org:cond-mat/0305305
10. Fellbaum, C.: WordNet: An Electronic Lexical Database (Language, Speech, and Communication). The MIT Press, Cambridge (1998)
11. Spertus, E., Sahami, M., Buyukkokten, O.: Evaluating similarity measures: a large-scale study in the orkut social network. In: Proceedings of the eleventh ACM SIGKDD international conference on Knowledge discovery in data mining, pp. 678–684. ACM, New York (2005)
12. Wen, Z., Tzerpos, V.: Evaluating Similarity Measures for Software Decompositions. In: Proceedings of the 20th IEEE International Conference on Software Maintenance, pp. 368–377. IEEE Computer Society, Washington (2004)
13. Bradley, E.: Better Bootstrap Confidence Intervals. Journal of the American Statistical Association 82, 171–185 (1987)
14. Tan, P., Steinback, M., Kumar, V.: Introduction to Data Mining. Addison-Wesley, Reading (2005)

Inefficiency in Networks with Multiple Sources and Sinks

Hyejin Youn[1], Michael T. Gastner[2,3,4], and Hawoong Jeong[1]

[1] Department of Physics, Korea Advanced Institute of Science and Technology,
Daejeon 305-701, Korea
[2] Santa Fe Institute, 1399 Hyde Park Road, Santa Fe, NM 87501, USA
[3] Department of Computer Science, University of New Mexico, Albuquerque,
NM 87131, USA
[4] Institute for Chemistry and Biology of the Marine Environment,
Carl von Ossietzky Universität, 26111 Oldenburg, Germany
visang@kaist.ac.kr

Abstract. We study the problem of optimizing traffic in decentralized transportation networks, where the cost of a link depends on its congestion. If users of a transportation network are permitted to choose their own routes, they generally try to minimize their personal travel time. In the absence of centralized coordination, such a behavior can be inefficient for society and even for each individual user. This inefficiency can be quantified by the "price of anarchy", the ratio of the suboptimal total cost to the socially optimal cost. Here we study the price of anarchy in multi-commodity networks, (i.e., networks where traffic simultaneously flows between different origins and destinations).

Keywords: flow optimization, transportation network, Nash equilibrium, multi-commodity flow.

The past few years have witnessed dramatic advances in finding, understanding, and characterizing complex networks [1,2]. One frontier for scientists is now to understand dynamic processes on networks such as the flow of matter or information in technological or social networks, vehicles traveling in transportation networks, electricity exchanged through the power grid, and the spread of diseases in biological networks [3,4,5,6]. If users of a network can decide freely which paths they take, then it is important to understand the users' behaviors and their interactions in order to optimally design the network and control the flows [7,8]. If the users' paths do not interfere with each other, users with even a small amount of local information are able to navigate on the network almost as efficiently as if they possessed global knowledge [9]. However, if the users' decisions are mutually dependent, the flow can in reality still be far from optimal even if all individuals have complete information about the network and other users' behaviors [10].

Consider, for instance, the simple network depicted in Fig. 1 [11]. Suppose that there is a constant flow of travellers F between the nodes s and t which

J. Zhou (Ed.): Complex 2009, Part I, LNICST 4, pp. 334–338, 2009.

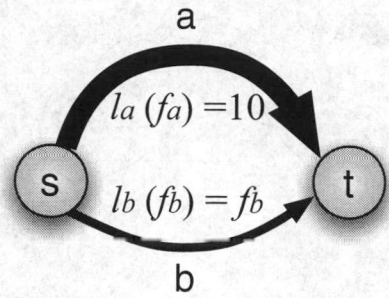

Fig. 1. Schematic depiction of a simple network with an origin s and a destination t. If 10 users travel from s to t, the flows are $f_a^{SO} = f_b^{SO} = 5$ in the social optimum and $f_a^{NE} = 0$ and $f_b^{NE} = 10$ in the Nash equilibrium.

are connected by two different types of links: a short but narrow bridge a where the effective speed becomes slower as more cars travel on it, and a long but broad multi-lane freeway b where delays due to congestion are negligible. Suppose further that the delay on link a is proportional to the flow, $l_a(f_a) = f_a$, while the delay on b is flow-independent, $l_b(f_b) = 10$, where $f_{a(b)}$ is the flow on link $a(b)$. The total time spent by all users is given by the "cost function" $C(f_a) = l_a(f_a) \cdot f_a + l_b(f_b) \cdot f_b$ where the flow on b is equal to $f_b = F - f_a$. The socially optimal flow $f_{a(b)}^{SO}$ is defined as the minimum of C. If, for instance, the total flow $F = 10$, it can be easily calculated that $f_a^{SO} = f_b^{SO} = 5$ and $C = 75$.

On the other hand, every user on link b could reduce his delay from 10 to 6 by switching paths, which poses a social dilemma: as individuals, users would like to reduce their own delays, but this reduction comes at an additional cost to the entire group. In our example, as long as l_a is not equal to l_b, there will be an incentive for the users experiencing longer delays to shift to another link and finally flows reach a Nash equilibrium $f_{a(b)}^{NE}$, where no single user can make any individual gain by changing his own strategy unilaterally. All users take the link a at the total cost of $C = 100$. The *price of anarchy* (POA) is defined as the ratio of the total cost of the Nash equilibrium to the total cost of the social optimum [12],

$$\text{POA} = \frac{\sum_z l_z(f_z^{NE}) \cdot f_z^{NE}}{\sum_z l_z(f_z^{SO}) \cdot f_z^{SO}}, \tag{1}$$

where, for a general network, the sums are over all links. This ratio indicates the relative inefficiency of the decentralized system; in our simple example POA=$100/75 = 1.33$.

There have been a number of theoretical studies about the POA (see [11] and references therein). Recently, we published calculations of the POA in real transportation networks [13]. Like most previous studies, we based our analysis on flows with a single origin and a single destination. This approach was motivated by analogies to electric circuits where even multiple current sources or sinks can be mapped onto a network with a single origin-destination pair by connecting

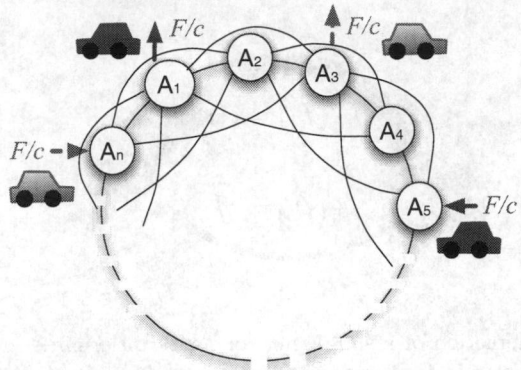

Fig. 2. Schematic depiction of a multi-commodity flow in a one-dimensional lattice with n nodes and a periodic boundary. Each node is connected to its six nearest neighbors and each commodity is assigned a flow F/c where c is the number of commodities. In the figure, red and blue cars symbolize different commodities. Origins and destinations of different commodities are not interchangeable. For example, red cars starting at node A_n are bound for A_3. A_1 is a closer destination, but it is not a destination for red cars.

all sources (sinks) to a fictitious auxiliary source (sink) and solving for the optimal and equilibrium flows. In other words, an electric current can flow to any of the sinks present in the network. However, if the flows are constrained to move from specific sources to specific destinations, the situation becomes more complicated [14]. Unfortunately, flows in human transportation networks or in the Internet fall in this class of so-called "multi-commodity flows". An important question for future research is whether the POA is affected by the number of commodities c.

In order to develop some intuition, we have studied the effect of multiple commodities in regular one-dimensional lattices (Fig. 2). All the networks contain 100 nodes; each node is connected to its nearest, next-nearest, and second-next-nearest neighbors, and periodic boundaries are applied (i.e., the lattice has the topology of a circle). Every link between nodes i and j has a delay of the form $l_{ij} = a_{ij}f_{ij} + b_{ij}$, where $a_{ij} = a_{ji}$ is a random integer equal to 1, 2, or 3, and $b_{ij} = b_{ji}$ between 1 and 100. The coefficient a_{ij} denotes how steeply the delay on the link increases with the flow present on $i \rightarrow j$. The constant b_{ij} implies a given delay regardless of congestion. The delays depend only on the total flows $f_{ij} = \sum_{k=1}^{c} f_{ij,k}$, where $f_{ij,k}$ is the flow of commodity k on the link $i \rightarrow j$ and c is the number of commodities. We then randomly choose c origins s_k and destinations t_k ($k = 1, \ldots, c$), among the nodes. For each commodity, origin and destination must be distinct, i.e., $s_k \neq t_k$, but the origin or destination of one commodity may be the origin or destination of another commodity. To each commodity we assign a traffic flow F/c, i.e., all commodities transport an equal share of the total traffic volume F. Every commodity is permitted to split its

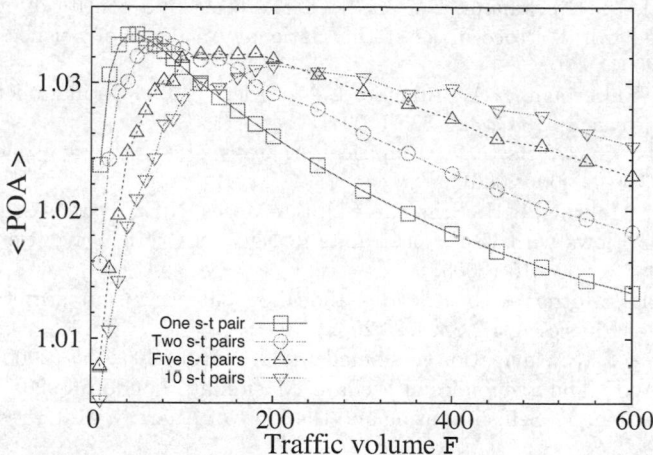

Fig. 3. The price of anarchy in regular lattices with multiple random sources and sinks ("multi-commodity flows") averaged over 100 to 400 networks. Each pair contributes equally to F.

flow on different paths from s_k to t_k. We then calculate the multi-commodity POA

$$\text{POA} = \frac{\sum_{\text{link } i \rightarrow j} \left[l_{ij} \left(\sum_k f_{ij,k}^{NE} \right) \cdot \left(\sum_k f_{ij,k}^{NE} \right) \right]}{\sum_{\text{link } i \rightarrow j} \left[l_{ij} \left(\sum_k f_{ij,k}^{SO} \right) \cdot \left(\sum_k f_{ij,k}^{SO} \right) \right]}, \tag{2}$$

with the straightforward Frank-Wolfe algorithm [15].

The results, averaged over 100 to 400 random instances of such networks, are shown in Fig. 3. Our preliminary results indicate that the POA as a function of F is a unimodal curve for all c, but the values depend on c: the peak becomes broader and shifts to higher values as c increases. We currently hypothesize that it is possible to collapse the POA for all c to a universal scaling function $f(F)$. Unfortunately, the quality of our data is not yet sufficient to test this hypothesis. The convergence of the Frank-Wolfe algorithm is too slow to extend our calculations to $c \gg 1$. We are currently redesigning our code to implement the PARTAN algorithm of Ref. [16] which should significantly reduce the computation time.

References

1. Albert, R., Barabási, A.-L.: Statistical mechanics of complex networks. Reviews of Modern Physics 74, 47–97 (2002)
2. Newman, M.E.J.: The structure and function of complex networks. SIAM Review 45, 167–256 (2003)
3. Argollo de Menezes, M., Barabási, A.-L.: Fluctuations in Network Dynamics. Physical Review Letters 92, 028701 (2004)

4. Hufnagel, L., Brockmann, D., Geisel, T.: Forecast and control of epidemics in a globalized world. Proceedings of the National Academy of Sciences 101, 15124–15129 (2004)
5. Wu, F., Huberman, B.A., Adamic, L.A., Tyler, R.J.: Information flow in social groups. Physica A 337, 327–335 (2004)
6. Kossinets, G., Watts, D.J.: Empirical Analysis of an Evolving Social Network. Science 311, 88–90 (2006)
7. Jahn, O., Möhring, R.H., Schulz, A.S., Stier-Moses, N.E.: System-Optimal Routing of Traffic Flows with User Constraints in Networks with Congestion. Operations Research 53, 600–616 (2005)
8. Kim, D.H., Motter, A.E.: Resource allocation pattern in infrastructure networks. Journal of Physics A 41, 224019 (2008)
9. Kleinberg, J.M.: Navigation in a small world. Nature 406, 845 (2000)
10. Pigou, A.C.: The Economics of Welfare. Macmillan, London (1920)
11. Roughgarden, T.: Selfish Routing and the Price of Anarchy. MIT Press, Cambridge (2005)
12. Papadimitriou, C.: Algorithms, games, and the Internet. In: Proceedings of the 33rd Annual ACM Symposium on Theory of Computing, pp. 749–753 (2001)
13. Youn, H., Gastner, M.T., Jeong, H.: The price of anarchy in transportation networks: Efficiency and optimality control. Physical Review Letters 101, 128701 (2008)
14. Carmi, S., Wu, Z., López, E., Havlin, S., Eugene Stanley, H.: Resource allocation pattern in infrastructure networks. European Physical Journal B 57, 165–174 (2007)
15. Frank, M., Wolfe, P.: An algorithm for quadratic programming. Naval Research Logistics Quarterly 3, 95–110 (1956)
16. Arezki, Y., van Vliet, D.: A Full Analytical Implementation of the PARTAN/Frank Wolfe Algorithm for Equilibrium Assignment. Transportation Science 24, 58–62 (1990)

Impacts of Local Events on Communities and Diseases

Xin-Jian Xu[1], Li-Jie Zhang[1], Guo-Hong Yang[1], and Xun Zhang[2]

[1] College of Science, Shanghai University, 99 Shangda Road, Shanghai 200444, China
xinjxu@shu.edu.cn
[2] Centre for Computational Science and Engineering,
National University of Singapore, Singapore 117542

Abstract. The study of community networks has attracted considerable attention recently. In this paper, we propose an evolving community network model based on local events, the addition of new nodes intra-community and new links intra- or inter-community. Employing growth and preferential attachment mechanisms, we generate the network with a generalized power-law distribution of nodes' degrees. Furthermore, we study epidemic spreading in the resulting network by the simple SIS model to understand the influence of the network structure on the dynamics. We find that the existence of communities in networks causes the critical behavior of the spreading dynamics and keeps epidemics endemic.

Keywords: complex networks, community networks, SIS model.

1 Introduction

Complex networks, evolved from the Erdös-Rényi random graph [1], are powerful models for describing many complex systems in biology, sociology, and technology [2]. In the past decade, the explosion of the general interest in the structure and the evolution of most real-world networks is mainly reflected in two striking characteristics. One is the small-world property [3], which suggests that a network has a highly degree of clustering like regular networks and a small average distance among any two nodes similar to random networks. The small-world phenomenon has been successfully described by network models with some degree of randomness [3,4]. The other is the scale-free behavior [5], which means a power-law distribution of connectivity, $P(k) \sim k^{-\gamma}$, where $P(k)$ is the probability that a node in the network has k connections to other nodes and γ is a positive real number determined by the given network. The origin of the scale-free behavior has been traced back to two mechanisms that are observed in many systems, growing and preferential attachment [5,6].

Recently, with the progress of research in networks, many other statistical characteristics of networks appeared on the stage. Of particular renown is the so-called "community"(or "modularity"). That is to say, a network is composed of many clusters of nodes, where the nodes in the same cluster are highly connected, while there are few links among the nodes belonging to different clusters. For

J. Zhou (Ed.): Complex 2009, Part I, LNICST 4, pp. 339–350, 2009.

instance, groups are formed in scientific collaboration networks [7]. Also, it has been found that dynamical processes on networks are affected by community structures, such as tendencies spread well within communities [8] and diffusion between different communities is slow [9].

In the study of community networks, most research has been directed in two distinct directions. On the one hand, attention has been paid to designing algorithms for detecting community structures in real networks. A pioneering method was made by Girvan and Newman [7], who introduced a quantitative measure for the quality of a partition of a network into communities. Later, a number of algorithms have been proposed in order to find a good optimization with the least computational cost. The fastest available procedures use greedy techniques [10] and extremal optimization [11], which are capable of detecting communities in large networks. On the other hand, research has focused on modeling of networks with community structures. In Ref. [12], a static social network was introduced where individuals belong to groups that in turn belong to groups of groups and so on. In Ref. [13], a networked seceder model was suggested to illustrate group formation in social networks. In Ref. [14], a growing bipartite network for social communities with group structures was proposed. Each of those models is constructed based on one aspect of reality.

In this paper, we introduce a network model with communities that gives a realistic description of local events [15,16,17]. The model incorporates three processes, the addition of new nodes intra-community and new links intra- or inter-community. Using growing and preferential attachment mechanisms, we generate the community network with a good right-skewed distribution of nodes' degrees, which has been observed in many social systems. Then, we investigate the standard SIS model on the generated network. We notice a great influence of the community structure on the epidemic dynamics over such complex networks.

2   Network Model

The Barabási-Albert network [5] only describes a particular type of evolving networks, the addition of new nodes preferential connecting to the nodes already present in the network. Systems in the real world, however, are much richer. For example, in friendship networks, a person usually makes friends with people belonging to different communities besides the community he belongs to. To give a realistic description of the network construction like that, we introduce a growing model of community networks based on local events, the addition of new nodes intra-community and new links intra- or inter-community. The proposed model is defined as follows.

We start with M (≥ 2) isolated communities and each community consists of a small number n of isolated nodes. At each time step, we perform one of the following three operations.

(i) With probability p we add a new node in a randomly chosen community. Here the randomly chosen means that the community is selected according to the uniform distribution. The new node is only connected to one node that already presented in the selected community. We denote it as the uth commnuity. The probability that node i in community u will be selected is proportional to its intra-community degree

$$\prod(k_i^{\text{intra}}) = \frac{k_{u,i}^{\text{intra}} + 1}{\sum_j (k_{u,j}^{\text{intra}} + 1)}, \tag{1}$$

where the sum runs over nodes in community u and $k_{u,i}^{\text{intra}}$ is the intra-community degroe of nodc i in community u.

(ii) With probability q we add a new link in a randomly chosen community. For this we randomly select a node in a randomly chosen community u as the starting point of the new link. The other end of the link is selected in the same community with the probability givcn by Eq. (1).

(iii) With probability r $(= 1 - p - q)$ we add a new link between two communities. For this we randomly select a node in a randomly chosen community u as the starting point of the new link. The other end i of the link selected in the other community v is proportional to its inter-community degree

$$\prod(k_i^{\text{inter}}) = \frac{k_{v,i}^{\text{inter}} + 1}{\sum_{v \neq u;j} (k_{v,j}^{\text{inter}} + 1)}, \tag{2}$$

where the sum runs over nodes in all communities except for community u and $k_{v,i}^{\text{inter}}$ is the inter-community degree of node i in community v.

After t time steps, this scheme generates a network of $Mn + pt$ nodes and t links. The parameters p, q, and r control the network structure. In the case of small r, the generated network will have a strong community structure. Notice that whatever process is chosen at each time step, only one link is added to the system (duplicate and self-connected edges are forbidden), however, this is not essential. We choose link probabilities $\prod(k_i^{\text{intra}})$ and $\prod(k_i^{\text{inter}})$ to be proportional to $k_i^{\text{intra}} + 1$ and $k_i^{\text{inter}} + 1$, respectively, such that there is a nonzero possibility of isolated nodes acquiring new links.

3 Degree Distribution

In our community network, the degree of a node consists of two parts, the intra-community degree and the inter-community degree. Increase in the node's connectivity can be divided into two processes, the increases of the intra-community degree and the inter-community degree. In each process, we assume that k_i^{intra} and k_i^{inter} change continuously, and the probabilities $\prod(k_i^{\text{intra}})$ and $\prod(k_i^{\text{inter}})$ can be interpreted as the rates at which k_i^{intra} and k_i^{inter} change, respectively. Thus, the operations (i)-(iii) all contribute to k_i, each being incorporated in the continuum theory as follows.

(i) Addition of a new node in a randomly chosen community with probability p:

$$\frac{\partial k_{u,i}^{\text{intra}}}{\partial t} = p \frac{1}{M} \frac{k_{u,i}^{\text{intra}} + 1}{\sum_j (k_{u,j}^{\text{intra}} + 1)}. \tag{3}$$

(ii) Addition of a new link in a randomly chosen community with probability q:

$$\frac{\partial k_{u,i}^{\text{intra}}}{\partial t} = q[\frac{1}{N} + \frac{1}{M}\frac{k_{u,i}^{\text{intra}} + 1}{\sum_j (k_{u,j}^{\text{intra}} + 1)}], \tag{4}$$

where N is the number of total nodes. The first term on the right-hand side (rhs) corresponds to the random selection of one end of the new link, while the second term on the rhs reflects the preferential attachment (Eq. (1)) used to select the other end of the link.

(iii) Addition of a new links between two communities with probability r :

$$\frac{\partial k_{v,i}^{\text{inter}}}{\partial t} = r[\frac{1}{N} + (1 - \frac{1}{M})\frac{k_{v,i}^{\text{inter}} + 1}{\sum_{v \neq u;j}(k_{v,j}^{\text{inter}} + 1)}]. \tag{5}$$

The first term on the rhs represents the random selection of one end of the new link, while the second term on the rhs considers the preferential attachment (Eq. (2)) used to select the other end of the link in the other community.

Combing the contribution of above processes, we have

$$\frac{\partial k_{u,i}^{\text{intra}}}{\partial t} = \frac{p+q}{M}\frac{k_{u,i}^{\text{intra}} + 1}{\sum_j (k_{u,j}^{\text{intra}} + 1)} + \frac{q}{N}, \tag{6}$$

$$\frac{\partial k_{v,i}^{\text{inter}}}{\partial t} = \frac{r}{N} + r\frac{M-1}{M}\frac{k_{v,i}^{\text{inter}} + 1}{\sum_{v \neq u;j}(k_{v,j}^{\text{inter}} + 1)}, \tag{7}$$

with

$$\sum_j (k_{u,j}^{\text{intra}} + 1) = \sum_j k_{u,j}^{\text{intra}} + \frac{N}{M}$$

$$= 2t(p\frac{1}{M} + q\frac{1}{M}) + \frac{Mn + pt}{M}$$

$$= \frac{3p + 2q}{M}t + n,$$

$$\sum_{v \neq u;j} (k_{v,j}^{\text{inter}} + 1) = \sum_{v \neq u;j} k_{v,j}^{\text{inter}} + N(1 - \frac{1}{M})$$

$$= 2tr\frac{M-1}{M} + (Mn + pt)\frac{M-1}{M}$$

$$= \frac{(2 - p - 2q)(M-1)}{M}t + (M-1)n.$$

We can simplify Eqs. (6) and (7) for large t, and get

$$\frac{\partial k_{u,i}^{\text{intra}}}{\partial t} \approx \frac{p+q}{3p+2q}\frac{(k_{u,i}^{\text{intra}} + 1)}{t} + \frac{q}{pl}, \tag{8}$$

$$\frac{\partial k_{v,i}^{\text{inter}}}{\partial t} \approx \frac{1 - p - q}{2 - p - 2q}\frac{(k_{v,i}^{\text{inter}} + 1)}{t} + \frac{1 - p - q}{pt}. \tag{9}$$

The boundary conditions of the intra-community degree and the inter-community degree at initial time t_s can be estimated in the sense of mathematical expectations, $k_{u,i}^{intra}(t_s) = p + q$ and $k_{v,i}^{inter}(t_s) = r$, respectively. So we write the solutions of Eqs. (8) and (9)

$$k_{u,i}^{intra}(t) = \frac{p^3 + p^2 + 2p^2q + 4pq + pq^2 + 2q^2}{p(p+q)}(\frac{t}{t_s})^{\frac{p+q}{3p+2q}}$$

$$- \frac{p^2 + 4pq + 2q^2}{p(p+q)}, \tag{10}$$

$$k_{v,i}^{inter}(t) = \frac{2 - 2q - pq + p - p^2}{p}(\frac{t}{t_s})^{\frac{1-p-q}{2-p-2q}}$$

$$- \frac{2 - 2q}{p}. \tag{11}$$

In random networks, the degree distribution can be calculated by

$$P(k) = \frac{1}{t}\sum_{i=1}^{t}\delta(k_i(t) - k), \tag{12}$$

which gives

$$P(k^{intra}) = \frac{3p^2 + 2pq}{p^2 + 2q^2 + 4pq + 2p^2q + pq^2 + p^3}$$

$$\times \left[\frac{p^2 + 4pq + 2q^2 + (p^2 + pq)k^{intra}}{p^2 + 2q^2 + 4pq + 2p^2q + pq^2 + p^3}\right]^{-(3+\frac{p}{p+q})}, \tag{13}$$

$$P(k^{inter}) = \frac{2p - 2pq - p^2}{2 - p - 4q - 2p^2 + 2q^2 + 2p^2q + pq^2 + p^3}$$

$$\times \left[\frac{2 - 2q + pk^{inter}}{2 + p - 2q - pq - p^2}\right]^{-(3+\frac{p}{1-p-q})}. \tag{14}$$

Thus, the degree distribution of our network obeys a generalized power-law form

$$P(k) \sim [A(p,q)k + B(p,q)]^{-\gamma(p,q)}. \tag{15}$$

In Fig. 1 we present numerical results of distributions of the intra-community degree, the inter-community degree, and the total degree of nodes in log-log scale. The experimental network is generated by the proposed scheme with $N = 10^5$, $M = 10$, $n = 5$, $p = 0.4$, and $q = 0.4$, respectively. The distributions of the intra-community degree and the inter-community degree, shown in Figs. 1(a) and 1(b), agree with analytical results of Eqs. (13) and (14), respectively. The small deviations between computer simulations and analytical solutions at both ends of the distributions appears to be the mathematical approximation of the boundary conditions and the finite size effect due to the relatively small network size used in the simulations. According to the evolving rule of our network, nodes with larger intra- (or inter-) degree have higher probabilities to gain new

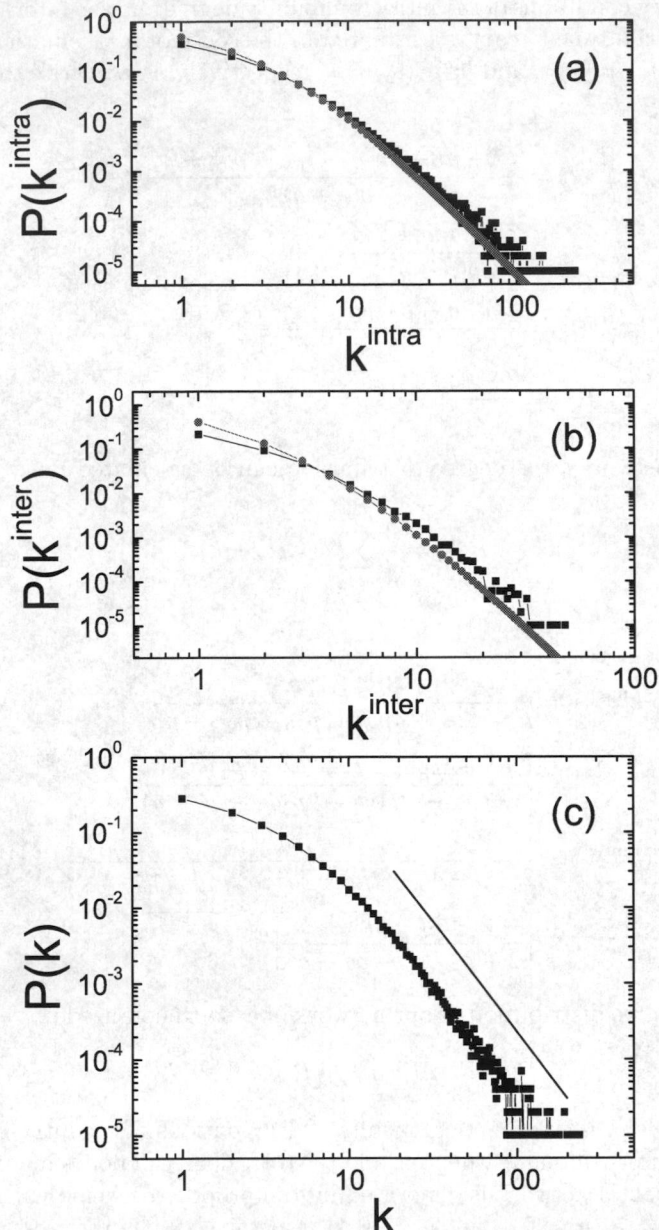

Fig. 1. (Color online) Log-log representation of distributions of intra-community degree (a), inter-community degree (b), and total degree (c) of nodes. All the simulation results (squares) display good right-skewed distributions. The circles in (a) and (b) denote analytical results predicted by Eqs. (13) and (14), respectively. The solid line in (c) is guide to the eye with power-law decay exponent $\gamma = 3$. The experiment network has a total number of nodes $N = 10^5$ with parameters $M = 10$, $n = 5$, $p = 0.4$, and $q = 0.4$, respectively.

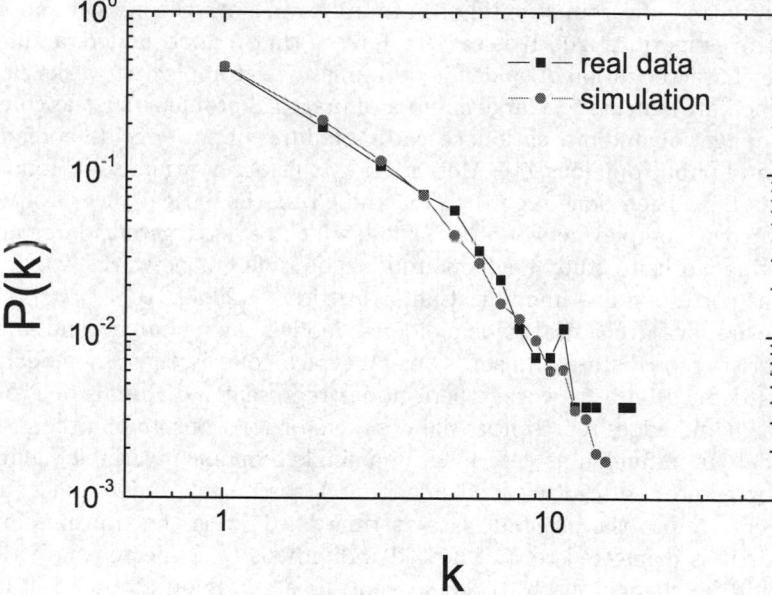

Fig. 2. (Color online) The degree distribution of econophysicists (squares) of an econophysics scientific collaboration network [19]. The circles correspond to computer simulations of our model with parameters $M = 10$, $n = 2$, $p = 0.4$, and $q = 0.4$, respectively.

links, then the usual degree preferential attachment is reasonably kept. This means that the right-skewed character of the network, such as the node's total degree, will retain. As shown in Fig. 1(c), the total degree distribution of nodes is well expected showing a good right-skewed character, which is reasonably in agreement with the condition of many realistic systems [18].

To illustrate the predictive power, we also compare the numerical result of our network with the statistics of an econophysics collaboration network. In the econophysics collaboration network, each node represents one scientist. If two scientists have collaborated one or more papers, they would be connected by an edge. Zhang *et al.* took the largest connected component of this network, which includes 271 nodes and 371 edges, and provided the best division, i.e., $M = 10$ [19]. In Fig. 2 we plot the degree distribution of econophysicists of the econophysics collaboration network which is fitted by computer simulations of our network starting with 10 communities. To gain p and q, we fit the connectivity distribution $P(k)$ obtained from this collaboration network with Eq. (15), obtaining a good overlap for $p = 0.75$ and $q = 0.15$ (Fig. 2).

4 Epidemic Spreading

In the consequent study of complex networks, an important topic is to inspect the effect of their topologies on dynamical behaviors and evolutionary processes. Of particular importance is the spread of infectious diseases, which has attracted increasing attention from scientific communities in both theoretical and experimental investigations [20]. It is easy to foresee that a good understanding and accurate characterization of epidemic dynamics over complex networks can provide immediate benefits to a large number of practical problems such as computer virus propagation and prevention, cascading failures of power grids, spreading of rumors and public opinions, etc. Motivated by this observation, some fundamental works have been done regarding an unified mathematical theory of disease spreading over complex networks [21,22,23], which have triggered a large number of following works to study epidemic models on different networks [24,25,26,27] as well as corresponding immunization strategies [28,29,30,31].

There are several classical epidemiological models taking into account different characteristics of disease transmission. A typical one is the SIS model. This model is defined on a network, where nodes represent individuals or groups of individuals and edges represent social contacts or relations among them. In the SIS model, an individual is described by a single dynamical variable having one of the two states: susceptible and infected. A susceptible individual at time t will be changed to the infected state at time $t + 1$, with the transmission rate $\nu > 0$, if it is connected to an infected individuals. An infected individual at time t will be changed back to the susceptible state again at time $t + 1$, with the transmission rate $\delta > 0$, due to immunity or medical treatment. Here, an effective spreading rate is defined to be $\lambda = \nu/\delta$ and, without loss of generality, one may assume $\delta = 1$ since it only affects the definition of the time scale of epidemic propagation.

In present work, we have performed Monte-Carlo (MC) simulations of the SIS model with synchronously updating after the network has been constructed. The two courses, the evolution of the network and the epidemic dynamics, are independent. Initially, the number of randomly infected nodes is 1 percent of the size of the network. After appropriate relaxation times, the systems stabilize in a steady state. Simulations were averaged over 10 different starting configurations, performed on 10 different realizations of the network. Given a network, an important observable is the prevalence ρ, which is the time average of the fraction of infected individuals in steady states.

Figures 3(a) and 3(b) depict the steady density of infected nodes ρ vs the spreading rate λ and its reciprocal, respectively. Four epidemic plots are presented in Fig. 3(a). Here we choose $p = q$ in simulations, just as an illustration of the influence of the local events on the dynamics. One can easily notice a drop of the epidemic prevalence as p and q become larger. In addition, the critical behavior of the dynamics is implied by Fig. 3(b), namely, there is a transition from the epidemic state to the endemic one as the network gets more modular. This is different from the results obtained in Balabási-Albert scale-free networks, although the connectivity of our community network displays a similar power-law decay. Only in cases that the addition of links among groups are equal to or larger than that inside groups, i.e., the networks are reduced to typical growing scale-free networks, can the disease show the feature of the absence of the threshold. To describe the critical behavior in detail, we plot the epidemic threshold

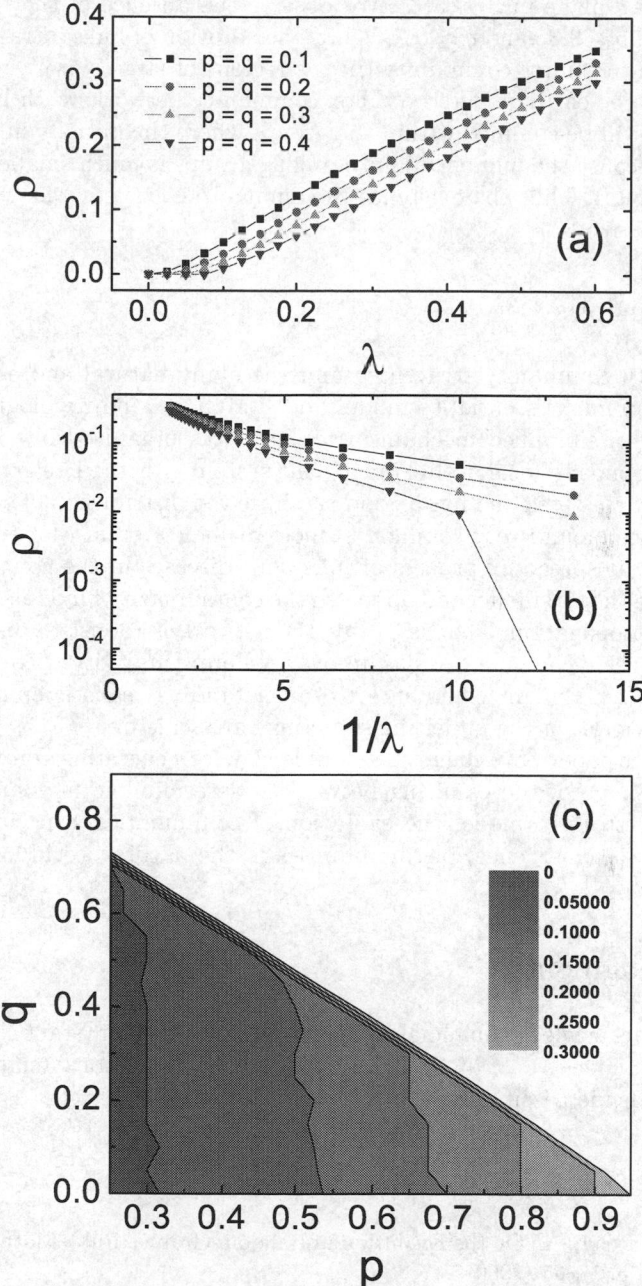

Fig. 3. (Color online) Density of infected nodes ρ vs λ (a) and $1/\lambda$ (b) for different local events. The trend of the network modularity enhances the critical value of the spreading rate. (c) The epidemic threshold λc in the p? q plane. Parameters of the underlying networks are the same as in Fig. 1.

λ_c in the $p - q$ plane in Fig. 3(c). We observe the absence of the threshold in regions about $p \leq 0.35$ and $q \leq 0.7$. While the addition of links intra-community is larger than that inter-community (in the regions of large p or q), we find that $\lambda_c > 0$. It can be easily understood. For community networks with large p or q, the infection will be confined within the groups where the initially infected seeds are chosen because the number of links among groups is much smaller than that inside the groups. Thus the existence of communities in networks will suppress spreading.

5 Conclusions

Networks with community structures underlie many natural and artificial systems. It is becoming essential to model and study this kind topological feature. We presented a simplified mechanism for networks organized in communities, which corresponds to local events during the system growth. The generated network is highly clustered and has a good right-skewed distribution of connectivity, which have been found very common in most realistic systems. Furthermore, we investigated the spreading process of infectious diseases in the community network by using the SIS model and observed the containment of modular structures to epidemic propagation. That is to say, the network is robust to diseases when connections inside communities has an overwhelming majority than that among communities. This might contribute to understanding realistic epidemics with critical behavior even the underline structures are scale free.

The present paper only suggests a simple way for generating community networks. The shape of the resulting network is deterministic in some extent. It is more interesting to model the evolution of communities, especially the self organization (or emergence) of communities in the natural world [32], which is left to future work.

Acknowledgments

The authors acknowledge financial support from NSFC under Grant No. 10805033 and STCSM under Grant No. 08ZR1408000. This work is sponsored by Shanghai Leading Academic Discipline Project.

References

1. Erdös, P., Rényi, A.: On the Evolution of Random Graphs. Publ. Math. Inst. Hung. Acda. Sci. 5, 17–65 (1960)
2. Albert, R., Barabási, A.-L.: Statistical Mechanics of Complex Networks. Rev. Mod. Phys. 74, 47–97 (2002)
3. Watts, D.J., Strogatz, S.H.: Collective Dynamics of "Small-world" Networks. Nature 393, 440–442 (1998)
4. Newman, M.E.J., Watts, D.J.: Scaling and Percolation in the Small-world Network Model. Phys. Rev. E 60, 7332–7342 (1999)

5. Barabási, A.-L., Albert, R.: Emergence of Scaling in Random Networks. Science 286, 509–512 (1999)
6. Krapivsky, P.L., Redner, S., Leyvraz, F.: Connectivity of Growing Random Networks. Phys. Rev. Lett. 85, 4629–4632 (2000)
7. Girvan, M., Newman, M.E.J.: Community Structure in Social and Biological Networks. Proc. Natl. Acad. Sci. USA 99, 7821–7826 (2002)
8. Bettencourt, L.M.A.: Tipping the Balances of a Small World. arXiv:cond-mat/0304321
9. Eriksen, K.A., Simonsen, I., Maslov, S., Sneppen, K.: Modularity and Extreme Edges of the Internet. Phys. Rev. Lett. 90, 148701 (2003)
10. Newman, M.E.J.: Fast Algorithm for Detecting Community Structure in Networks. Phys. Rev. E 69, 066133 (2004)
11. Duch, J., Arenas, A.: Community Detection in Complex Networks Using Extremal Optimization. Phys. Rev. E 72, 027104 (2005)
12. Watts, D.J., Dodds, P.S., Newman, M.E.J.: Identity and Search in Social Networks. Science 296, 1302–1305 (2002)
13. Grönlund, A., Holme, P.: Networking the Seceder Model: Group Formation in Social and Economic Systems. Phys. Rev. E 70, 036108 (2004)
14. Noh, J.D., Jeong, H.-C., Ahn, Y.-Y., Jeong, H.: Growing Network Model for Community with Group Structure. Phys. Rev. E 71, 036131 (2005)
15. Albert, R., Barabási, A.-L.: Topology of Evolving Networks: Local Events and Universality. Phys. Rev. Lett. 85, 5234 (2000)
16. Dorogovtsev, S.N., Mendes, J.F.F.: Scaling Behaviour of Developing and Decaying Networks. Europhys. Lett. 52, 33–39 (2000)
17. Li, X., Chen, G.: A Local-world Evolving Network Model. Physica A 328, 274–286 (2003)
18. Amaral, L.A.N., Scala, A., Barthelemy, M., Stanley, H.E.: Classes of Small-world Networks. Proc. Natl. Acad. Sci. USA 97, 11149–11152 (2000)
19. Zhang, P., Li, M., Wu, J., Di, Z., Fan, Y.: The analysis and Dissimilarity Comparison of Community Structure. Physica A 367, 577–585 (2006)
20. Biley, N.T.J.: The Mathematical Theory of Infectious Diseases and its Applications, 2nd edn. Griffin, London (1975)
21. Pastor-Satorras, R., Vespignani, A.: Epidemic Spreading in Scale-Free Networks. Phys. Rev. Lett. 86, 3200–3203 (2001)
22. Newman, M.E.J.: Spread of Epidemic Disease on Networks. Phys. Rev. E 66, 016128 (2002)
23. Warren, C.P., Sander, L.M., Sokolov, I.M.: Epidemics, Disorder, and Percolation. Physica A 325, 1–8 (2003)
24. Eguíluz, V.M., Klemm, K.: Epidemic Threshold in Structured Scale-free Networks. Phys. Rev. Lett. 89, 108701 (2002)
25. Barthélemy, M., Barrat, A., Pastor-Satorras, R., Vespignani, A.: Dynamical Patterns of Epidemic Outbreaks in Complex Heterogeneous Networks. J. Theor. Biol. 235, 275–288 (2005)
26. Vazquez, A.: Polynomial Growth in Branching Processes with Diverging Reproductive Number. Phys. Rev. Lett. 96, 038702 (2006)
27. Volz, E.: SIR Dynamics in Random Networks with Heterogeneous Connectivity. J. Math. Biol. 56, 293–310 (2008)
28. Pastor-Satorras, R., Vespignani, A.: Immunization of Complex Networks. Phys. Rev. E 65, 036104 (2002)

29. Cohen, R., Havlin, S., ben-Avraham, D.: Efficient Immunization Strategies for Computer Networks and Populations. Phys. Rev. Lett. 91, 247901 (2003)
30. Hufnagel, L., Brockmann, D., Geisel, T.: Forecast and Control of Epidemics in a Globalized World. Proc. Natl. Acad. Sci. USA 101, 15124–15129 (2004)
31. Gross, T., Dommar D'Lima, C.J., Blasius, B.: Epidemic Dynamics on an Adaptive Network. Phys. Rev. Lett. 96, 208701 (2006)
32. Kumpula, J.M., Onnela, J.-P., Saramäki, J., Kaski, K., Kertész, J.: Emergence of Communities in Weighted Networks. Phys. Rev. Lett. 99, 228701 (2007)

Identifying Social Communities in Complex Communications for Network Efficiency

Pan Hui[1,*], Eiko Yoneki[2], Jon Crowcroft[2], and Shu-Yan Chan[2]

[1] Deutsche Telekom Laboratories / TU Berlin,
Ernst-Reuter-Platz 7, 10587 Berlin, Germany
Pan.Hui@telekom.de
[2] Computer Laboratory, University of Cambridge,
15 JJ Thomson Avenue, CB3 0FD Cambridge, UK
lastname.firstname@cl.cam.ac.uk

Abstract. Complex communication networks, more particular Mobile Ad Hoc Networks (MANET) and Pocket Switched Networks (PSN), rely on short range radio and device mobility to transfer data across the network. These kind of mobile networks contain duality in nature: they are radio networks at the same time also human networks, and hence knowledge from social networks can be also applicable here. In this paper, we demonstrate how identifying social communities can significantly improve the forwarding efficiencies in term of delivery ratio and delivery cost. We verify our hypothesis using data from five human mobility experiments and test on two application scenarios, asynchronous messaging and publish/subscribe service.

Keywords: Pocket Switched Networks, Human Mobility, Community, Social Network, Asynchronous Messaging, Publis/Subscribe.

1 Introduction

We envision a future in which a multitude of devices carried by people are dynamically networked, forming Pocket Switched Networks (PSN) [1]: a type of Delay Tolerant Network (DTN) [2] for such environments. A PSN uses contact opportunities to allow humans to communicate without network infrastructure.

An efficient data forwarding mechanism over the temporal graph of the PSN [3] is required that copes with dynamic network topology by human mobility, and repeated disconnection and re-wiring. We believe the traditional approach of building and updating routing tables is not cost effective for a PSN, since mobility patterns are often unpredictable and topology changes can be rapid. Rather than exchanging much control traffic to create unreliable routing structures, we search for characteristics of the network that are less volatile than mobility. A PSN is formed by people, and the social relationships among people may prove to be a more stable network structure. Unicast and multicast

---

* This work was done when Pan Hui was in Cambridge.

"routes" in this system are emergent properties of the community structure and community interests, respectively. Thus, such a social backbone can be used for better forwarding decisions.

Community is an important attribute of PSNs. Cooperation binds, but also divides human society into communities. Human society is structured. For an ecological community, the idea of correlated interaction means that an organism of a given type is more likely to interact with another organism of the same type than with a randomly chosen member of the population [4]. This correlated interaction concept also applies to human, so we can exploit this kind of community information to select forwarding paths. We believe identifying social communities can help to choose next relays for particular destinations, and hence reduce the number of unwanted traffic generated (delivery cost).

In this paper, we use five experimental datasets, which cover a rich diversity of environments from busy metropolitan city to quite university town, with an experimental period from several days to almost one year, to verify our hypothesis that identifying social communities can help to improve forwarding efficiency. We evaluate our results on both single-point communication and multi-point communication to make a more general conclusion. For these two kinds of communication, we use more particularly the *asynchronous messaging* and *publish/subscribe* applications.

2   Experimental Datasets

In this paper, we use four experimental datasets gathered by the Haggle Project [1] over two years, referred to as *Infocom05*, *HongKong*, *Cambridge*, and *Infocom06*; one dataset from the MIT Reality Mining Project [5], referred to as *Reality*. Previously, the characteristics of these datasets such as inter-contact and contact distribution have been explored in several studies [6] [1] [7], to which we refer the reader for further background information. We believe these five datasets cover a rich diversity of environments from busy metropolitan city (*HongKong*) to quite university town (*Cambridge*), with an experimental period from several days (*Infocom06*) to almost one year (*Reality*). Datasets from cellular operators, for example the one used by Gonzalez *et al.* [8], can be much larger in scale but lack of peer-to-peer proximity logging of neighbor devices, hence can not be used for evaluation of PSN applications.

- In *Infocom05*, the devices were distributed to approximately fifty students attending the Infocom student workshop. Participants belong to different social communities (depending on their country of origin, research topic, etc.). However, they all attended the same event for 4 consecutive days and most of them stayed in the same hotel and attended the same sections (note, though, that Infocom is a multi-track conference).
- In *Hong-Kong*, the people carrying the wireless devices were chosen independently in a Hong-Kong bar, to avoid any particular social relationship

---

[1] http://www.haggleproject.org

Table 1. Characteristics of the five experimental data sets

Experimental data set	Infocom05	Hong-Kong	Cambridge	Infocom06	Reality
Device	iMote	iMote	iMote	iMote	Phone
Network type	Bluetooth	Bluetooth	Bluetooth	Bluetooth	Bluetooth
Duration (days)	3	5	11	3	246
Granularity (seconds)	120	120	600	120	300
Number of Experimental Devices	41	37	54	98	97
Number of internal contacts	22,459	560	10,873	191,336	54,667
Average # Contacts/pair/day	4.6	0.084	0.345	6.7	0.024

between them. These people have been invited to come back to the same bar after a week. They are unlikely to see each other during the experiment.

- In *Cambridge*, the iMotes were distributed mainly to two groups of students from University of Cambridge Computer Laboratory, specifically undergraduate year1 and year2 students, and also some PhD and Masters students. This dataset covers 11 days.
- In *Infocom06*, the scenario was very similar to *Infocom05* except that the scale is larger, with 80 participants. Participants were selected so that 34 out of 80 form 4 subgroups by academic affiliations.
- In *Reality*, 100 smart phones were deployed to students and staff at MIT over a period of 9 months. These phones were running software that logged contacts with other Bluetooth enabled devices by doing Bluetooth device discovery every five minutes.

The five experiments are summarised in Table 1.

3   Communities in the Mobility Traces

A social network consists of a set of people forming socially meaningful relationships, where prominent patterns or information flow are observed. In PSN, social networks could map to computer networks since people carry the computer devices. In this section, we implement and apply Newman's weighted network analysis (WNA) for our data analysis [9].

For each community partitioning of a network, one can compute the corresponding modularity value using the following definition of *modularity* (Q):

$$Q = \sum_{vw} \left[\frac{A_{vw}}{2m} - \frac{k_v k_w}{(2m)^2} \right] \delta(c_v, c_w) \tag{1}$$

where A_{vw} is the value of the weight of the edge between vertices v and w, if such an edge exists, and 0 otherwise; the δ-function $\delta(i, j)$ is 1 if $i = j$ and 0 otherwise; $m = \frac{1}{2} \sum_{vw} A_{vw}$; k_v is the degree of vertex v defined as $\sum_w A_{vw}$; and c_i denotes the community of which vertex i belongs to. Therefore the term in the formula $\frac{\sum_{vw} A_{vw}}{2m} \delta(c_v, c_w)$ is equal to $\frac{\sum_{vw} A_{vw} \delta(c_v, c_w)}{\sum_{vw} A_{vw}}$, which is the fraction of the edges that fall within communities. *Modularity* is defined as the difference between this fraction and, the fraction of the edges that would be expected to

Table 2. Communities detected from the four datasets

Dataset	Info06	Camb	Reality	HK
Q_{max}	0.2280	0.4227	0.5682	0.6439
Max. Community Size	13	18	23	139
No. Communities	4	2	8	19
Avg. Community Size	8.000	16.500	9.875	45.684
No. Community Nodes	32	33	73	868
Total No. of Nodes	78	36	97	868

fall within the communities if the edges were assigned randomly but keeping the degrees of the vertices unchanged. The algorithm is essentially a genetic algorithm, using the modularity as the measurement of fitness. Instead of testing on some mutations of the current best solutions, it enumerates all possible merges of any two communities in the current solution, evaluates the relative fitness of the resulting merges, and chooses the best solution as the seed for the next iteration.

Table 2 summarises the communities detected by applying WNA on the four datasets. According to Newman [9], nonzero Q values indicate deviations from randomness; values around 0.3 or more usually indicate good divisions. For the *Infocom06* case, the Q_{max} value is low; this indicates that the community partition is not very good in this case. This also agrees with the fact that in a conference the community boundary becomes blurred. For the *Reality* case, the Q value is high; this reflects the more diverse campus environment. For the *Cambridge* data, the two groups spound by WNA is exactly matched the two groups (1st year and 2nd year) of students selected for the experiment.

4   Single-Point Communication

Here we propose the BUBBLE algorithm, with the intention of bringing in a concise concept of community into PSN forwarding to achieve significant improvement of forwarding efficiency. BUBBLE combines the knowledge of community structure with the knowledge of node centrality to make forwarding decisions. There are two intuitions behind this algorithm. Firstly, people have varying roles and popularities in society, and these should be true also in the network – the first part of the forwarding strategy is to forward messages to nodes which are more popular than the current node. Secondly, people form communities in their social lives, and this should also be observed in the network layer – hence the second part of the forwarding strategy is to identify the members of destination communities, and to use them as relays. Together, we call this BUBBLE forwarding. For this algorithm, we make two assumptions:

- Each node belongs to at least one community. Here we allow single node communities to exist.
- Each node has a global ranking (i.e.global centrality [10]) across the whole system, and also a local ranking within its local community. It may also belong to multiple communities and hence may have multiple local rankings.

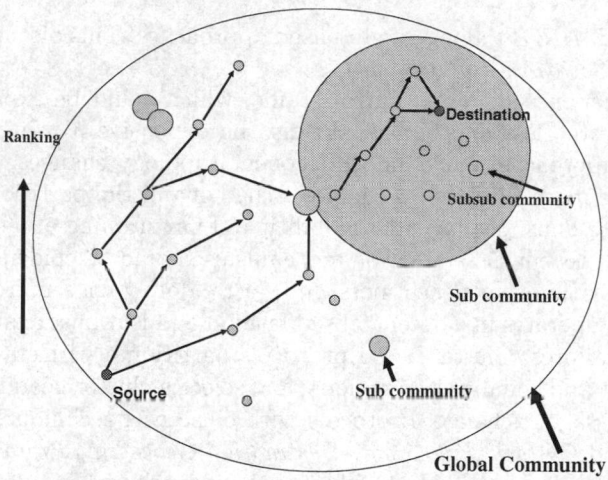

Fig. 1. Illustration of the BUBBLE algorithm

Forwarding is carried out as follows. If a node has a message destined for another node, this node first *bubbles* the message up the hierarchical ranking tree using the global ranking, until it reaches a node which is in the same community as the destination node. Then the local ranking system is used instead of the global ranking, and the message continues to bubble up through the local ranking tree until the destination is reached or the message expires. This method does not require every node to know the ranking of all other nodes in the system, but just to be able to compare ranking with the node encountered, and to push the message using a greedy approach. In order to reduce cost, we also require that whenever a message is delivered to the community, the original carrier can delete this message from its buffer to prevent further dissemination. This assumes that the community member can deliver this message. We call this algorithm BUBBLE, using the metaphor of *bubble* for a community.

The forwarding process fits our intuition and is taken from real life experiences. First you try to forward the message via surrounding people more popular than you, and then you bubble it up to well-known popular people in the wider-community, such as a postman. When the postman meets a member of the destination community, the message will be passed to that community. The first community member who receives the message will try to identify more popular members within the community, and bubble the message up again within the local hierarchy, until the message reaches a very popular member, or the destination itself, or the message expires. Figure 1 illustrates the BUBBLE algorithm.

5   Multi-point Communication

Creating an overlay for message dissemination has been a popular technique for multi-point communication. Below, we describe a brief discussion of existing approaches in MANETs along gossip based approaches. This discussion leads to our proposal: *Socio-Aware Overlay.*

State maintenance requires control traffic, which could be expensive to operate, while a stateless approach could also be expensive if using event flooding. Stateful approaches suffer under frequent topology changes, and stateless approaches are more suitable for topology change and the partitioning and isolation of nodes. Thus, dealing with mobility and partitioning of networks shows that the basis of event dissemination mechanisms should be epidemic. The basic gossip dissemination sends each message to a randomly chosen group of nodes. This approach operates in a decentralised fashion and is robust against node and network link failures. Cluster-based protocols partition a wireless network into several disjoint and equally sized regions, and select a cluster head in each region to operate message exchange. Protocols with clustering techniques include *Ge-oGRID* [11] and *Obstacle-Free Single-Destination Geocasting Protocol* (OFSGP). *Socio-Aware Overlay* takes a clustering-based approach and a membership of the group is dynamically detected through community detection process rather than implicitly defined as the set of nodes within a certain area in geographical or physical casting.

Structured overlays assign identifiers to nodes and control the identifiers of neighbours in overlay networks and the keys of the objects they store. This is effective since lookups can be done with cost $O(logN)$; this is better than a flooding approach. However, the characteristics of MANETs require a significant amount of traffic to maintain the overlay links. Thus, strict layering may not work.

In [12], a structured P2P overlay network is used for a publish-subscribe system. Subscriptions are mapped to keys and sent to a rendezvous node. There is some optimisation such as bundled notification dissemination. The performance of this approach depends on the real mapping between the overlay network and the underlying network topology.

We propose multi-point event dissemination using an overlay constructed by closeness centrality nodes in communities. Detected communities are well connected implying that socially they share the same interests with high chances. Thus, similar subscriptions may coexist within the same community. The fundamental idea of this approach is instead of artificially constructing an overlay based on various contexts (e.g. location, group mobility), the existing structure is detected and mapped to the function. Thus, this approach strongly depends on the dynamic community detection mechanisms. A crucial factor is how good the community detection mechanism is. The current simple detection algorithm detects approximately 60% of communities compared to the centralised approach. We are adding messaging passing at a certain temporal point to improve the community detection.

We currently choose a closeness centrality node for the broker node as the closeness centrality imply the best visibility in the community. Thus, once this node gets the message, delivery to any member of the community has high reliability. Because of the characteristics of human networks (i.e. scale-free networks), many nodes within a community are tightly connected and multiple closeness

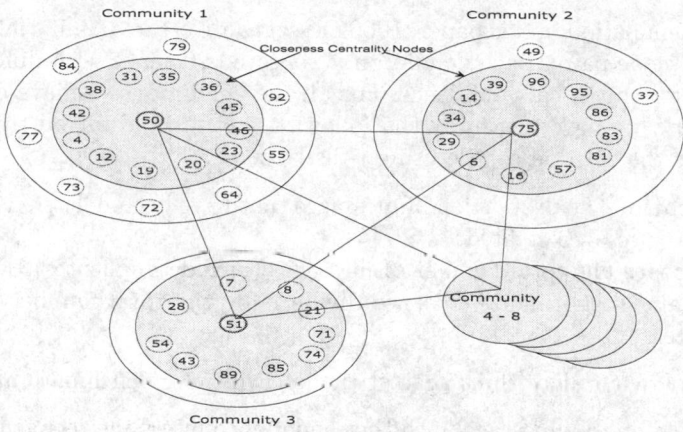

Fig. 2. Community Structure

centrality nodes can coexist. This is an advantage to balance the workload of brokers and will be the subject of future work to add a load balancing mechanism. We are also investigating other criteria to select the broker node, which is work in progress. Subscription propagation occurs as part of gossiping, thus, it does not cause any extra cost. The proposed multi-point communication takes advantage of PSNs, where various communication methods can be used to control delay in DTNs.

Communication between brokers can have two modes: *Unicast* and *Direct*. *Unicast* is based on the underlying unicast algorithms. Thus, it could end up epidemic routing. *Direct* provides more direct communication mechanism such as WiFi access points or GPRS *Direct* approach gives accelerated message delivery with some cost. When *Unicast* is used for the communication between broker nodes, the average hop count follows the distance of the pair nodes (i.e. 1.6 hops for MIT Reality mining trace). Using the betweenness centrality, where a node has dual visibility from and to communities will improve the hop counts, and we are investigating this extension.

6   Results and Evaluations

In this section, we show the results for both single-point and multi-point communication, using asynchronous messaging and publish/subscribe service as the specific applications.

6.1   Single-Point Communication

In order to evaluate different forwarding algorithms, we use a discrete event simulator called *HaggleSim*. The original trace files are divided into discrete sequential contact events, and they are fed into the emulator as inputs. For every discrete encounter event, the emulator makes a forwarding decision based on the forwarding algorithm under study.

For each emulation in this paper, 1000 messages are created, uniformly sourced between all node pairs. Each emulation is repeated 20 times with different random seeds for statistical confidence. For all the emulations we have conducted for this work, we have measured the following metrics and for all the metrics, we compute the 95th percentile using t-distribution.

Delivery ratio: The proportion of messages that have been delivered out of the total unique messages created.

Delivery cost: The total number of messages (include duplicates) transmitted across the air. To normalize this, we divide it by the total number of unique messages created.

We compare our algorithms against the following five benchmark algorithms.

WAIT: Hold on to a message until the sender encounters the recipient directly, which represents the lower bound for delivery and cost.

FLOOD: Messages are flooded throughout the entire system, which represents the upper bound for delivery and cost.

MCP: Multiple-Copy-Multiple-Hop. Multiple Copies are sent subject to a time-to-live hop count limit on the propagation of messages. This is a controlled flooding strategy.

LABEL: A social based forwarding algorithm introduced by Hui *et al.* [13]. Messages are only forwarded to the nodes in the same community (i.e.with the same label) as the destination.

PROPHET: A standard non-oblivious benchmark that has been evaluated against several previous works [14]. It calculates the delivery predictability at each node for each destination by using history of encounters and transitivity. A message is forwarded to a node if it has higher delivery predictability than the current node for that particular destination.

In this paper, we only show the *Reality* dataset as an example due to the limit of space. To evaluate the forwarding algorithm, we extract a 3 week session during term time from the whole 9 month dataset. Emulations were run over this dataset with uniformly generated traffic. There is a total 8 groups within the whole dataset. We observed that within each individual group, the node centralities demonstrate diversity similar to the *Cambridge* case.

From Figure 3(a) and Figure 3(b), we can see that of course flooding achieves the best for delivery ratio, but the cost is 2.5 times that of MCP, and 5 times that of BUBBLE. BUBBLE is very close in performance to MCP in the multiple-group case as well, and even outperforms it when the time TTL of the messages is allowed to be larger than 2 weeks. However, the cost is only 50% that of MCP.

Regarding LABEL forwarding, we can observe from Figure 3 that LABEL only achieves around 55% of the delivery ratio of the MCP strategy and only 45% of the flooding delivery although the cost is also much lower. However it is not an ideal scenario for LABEL. In this environment, people do not mix as well as in a conference [13]. A person in one group may not meet members in another group

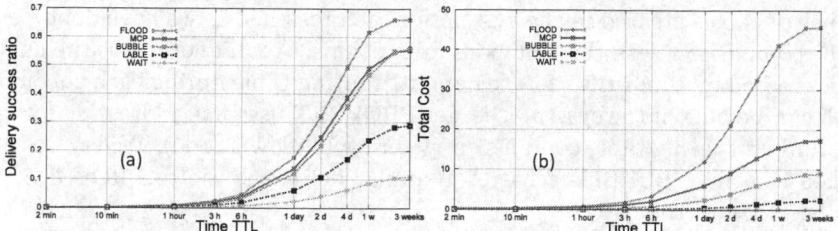

Fig. 3. Comparisons of several algorithms on *Reality* dataset, all groups

so often, waiting to meet a member of the destination group before transmitting is not effective.

In order to further justify the significance of social based forwarding, we also compare BUBBLE with a benchmark 'non-oblivious' forwarding algorithm, PROPHET[14]. PROPHET uses the history of encounters and transitivity to calculate the probability that a node can deliver a message to a particular destination. Since it has been evaluated against other algorithms before and has the same contact-based nature as BUBBLE (i.e. do not need location information), it is a good target to compare with BUBBLE.

PROPHET has four parameters. We use the default PROPHET parameters as recommended in [14]. However, one parameter that should be noted is the time elapsed unit used to age the contact probabilities. The appropriate time unit used differs depending on the application and the expected delays in the network. Here, we age the contact probabilities at every new contact. In a real application, this would be a more practical approach since we do not want to continuously run a thread to monitor each node entry in the table and age them separately at different time.

Figure 4 (a) and (b) shows the comparison of the delivery ratio and delivery cost of BUBBLE and PROPHET. Here, for the delivery cost, we only count the

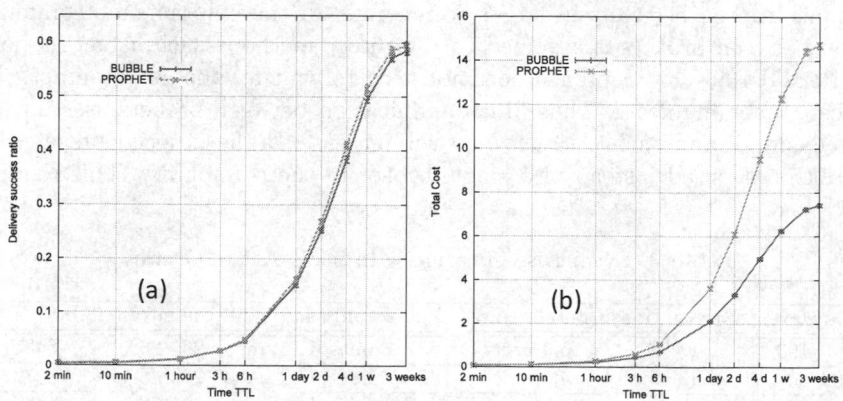

Fig. 4. Comparisons of BUBBLE and PROPHET on *Reality* dataset

number of copies created in the system for each message as we have done before for the comparison with the 'oblivious' algorithms. We did not count the control traffic created by PROPHET for exchanging routing table during each encounter, which can be huge if the system is large (PROPHET uses flat addressing for each node and its routing table contains entry for each known node). We can see that most of the time, BUBBLE achieves a similar delivery ratio to PROPHET, but with only half of the cost.

Considering that BUBBLE does not need to keep and update an routing table for each node pairs, the improvement is significant. Similar significant improvements by using BUBBLE are also observed in other datasets, these demonstrate the generality of the BUBBLE algorithm, but because of page limit, we can not include the results here.

6.2   Multi-point Communication

For validation and evaluation of the proposed approach, we use a discrete event emulator to replay the connectivity traces. The original trace files are divided into discrete sequential contact events and fed into the emulator as inputs. Although the current subscription model is simply topic-based, content-based filtering can be operated in the broker nodes. In the experiments, ten topics are predefined. Randomly selected nodes create 20 to 100 unique subscriptions, and 200 to 1000 publications unless stated otherwise. The message creation times are uniformly distributed throughout the experimental duration. The experiment is performed with MIT (100 devices) traces. Table 3 summarises the results of the *Socio-Aware Overlay* approach. The second column (*Average hops*) is hop counts per publication. The experiment with the MIT trace shows around 1.3 hops regardless of the scale of publication/subscription. The average pair distance of the network is 1.6 hops, which indicates that the *Socio-Aware Overlay* approach performs better than flooding to every subscriber by epidemic approach. The total hop count in the entire operation is shown in the final column (*Total Hops*). A pure epidemic approach results in larger hop counts. In the experiments, communication between brokers is assumed to use direct methods such as access-point WiFi or GPRS. This approach does not need to wait for the next contact with devices to communicate. Thus, if communication between brokers uses unicast or an epidemic approach, *Average hops* will increase. In the experiments, a group of brokers are used instead of a single broker in the community. This requires

Table 3. Event Dissemination with Socio-Aware Overlay

# Pub/Sub	Average Hops	Contact to Sub	Pub to Sub	Latency	Undelivered	Total Hops
1000/100	1.28	5.6 units	631.6 units	5.26 mins	261(26%)	6431
500/50	1.34	4.6 units	828.5 units	6.90 mins	242(48%)	1373
200/20	1.32	4.3 units	831.4 units	6.93 mins	115(58%)	204
1000/100C	1.35	2.7 units	449.4 units	3.75 mins	33(3%)	-

further work for balancing network work load of brokers and increasing reliability by replication of brokers.

Each publication has three stages during the simulation: (i) a publication is created at time unit (A), (ii) a publisher contacts the other devices to inject its publication to the network at time unit (B), and (iii) the publication is delivered to the subscriber at time unit (C). The timeline of a publication's life is depicted below:

A: Publication Created
B: Publisher → First Node Contact
C: Subscriber Received Publication

The third column (*Contact to Sub*) indicates $C - B$ in the number of time units. The fourth column (*Pub to Sub*) shows that total duration of publishing $(C - A)$. A single time unit has a duration of 0.5 seconds, and *Latency* indicates the approximate latency in minutes. Thus, $C - A$ and $C - B$ are indicators of the latency of publications. $C - B$ is much smaller than $C - A$ and $C - A \approx B - A$. On average, it takes over 3 days to get a first contact from when a publication is ready. However, the majority of nodes gets much shorter waiting time until getting a first contact (see Fig. 5). Once the publication is passed to the contacted device, in a few minutes subscribers will receive a publication.

Fig. 5 depicts the distribution of values $C - A$ in three different settings of publication and subscription. The result shows a power law distribution indicating that most event dissemination has short durations. Fig. 6 depicts the distribution of values $C - B$ from publisher's and subscriber's aspects from an experiment with 1000 publications and 100 subscriptions. Certain subscribers (e.g. 70-80) have higher durations, which has various reasons such as that these nodes are away from the centrality nodes in the community (i.e. more than single hop distance), or these nodes may not be part of the community despite them being detected. This will require further investigation.

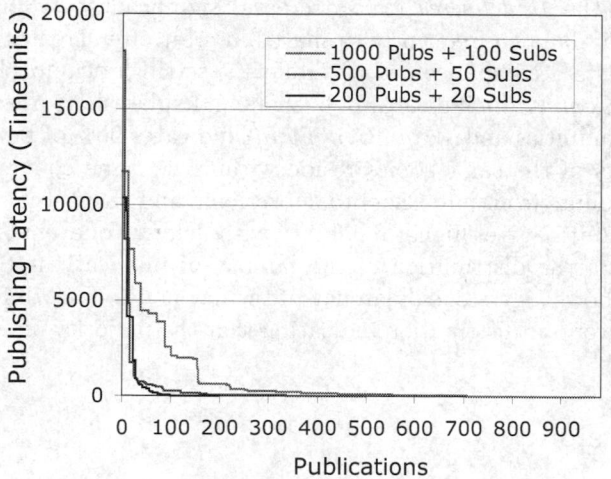

Fig. 5. Latency of Publications I

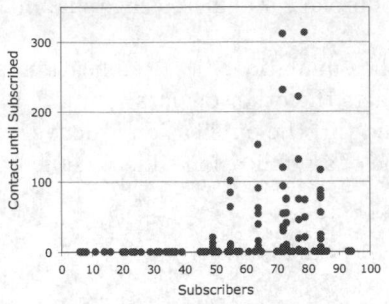 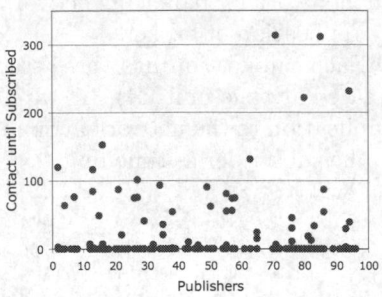

Fig. 6. Latency of Publications II

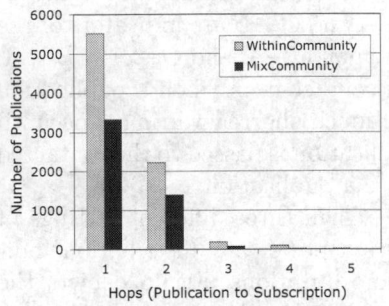 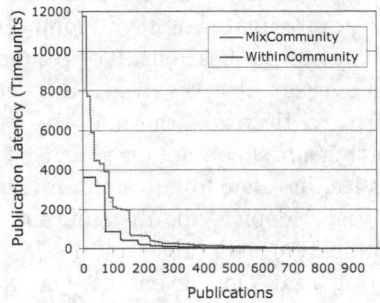

Fig. 7. Latency: Within and Mix Community

The value of *Undelivered* indicates the reliability of delivery. The ratio varies from 26% to 58% in 3 settings. The result shown in the last row of Table 3 has the same setting as the first row except publishers and subscribers are in the same communities. When both publishers and subscribers are in the same communities, the *Undelivered* ratio decreases significantly to 3%. In the real world, this may happen frequently as shared interest often creates communities.

Fig. 7 depicts a comparison of two different settings of publishers and subscribers. *MixCommunity* indicates publishers and subscribers are spread across different communities and *WithinCommunity* indicates 90% of both subscribers and publishers of the same topics reside within the same community. Fig. 7a depicts hop counts from publishers to subscribers and shows that topic sharing within communities gives higher reliability with delivery of events in fewer hops. Fig. 7b depicts the distribution of the latency of publications $(C - A)$. *MixCommunity* shows high value of latency of the few nodes. Fig. 7b fundamentally presents a power law distribution indicating that the majority of nodes have low latency.

7   Conclusion

We have shown empirically that identifying social communities enhances communication efficiency for complex mobile networks for both single-point communication and multi-point communication. Social information can be further explored to provide better applications, for example a city-wide Pocket Switched Network, community-based media sharing application, and social-tagging metadata system for information searching. We are devoting two projects [2] to further characterise and understand social networks, and utilize this knowledge to develop more novel applications.

Acknowledgement. This research is funded in part by the EC IST SOCIAL-NETS - Grant agreement number 217141. We would like also to acknowledge comments from Steven Hand, Brad Karp, Frank Kelly, Richard Mortier, Pietro Lio, Andrew Moore, Nishanth Sastry, Derek Murray, Sid Chau, Andrea Passarella, and Hamed Haddadi.

References

1. Hui, P., Chaintreau, A., et al.: Pocket switched networks and human mobility in conference environments. In: Proc. WDTN (2005)
2. Fall, K.: A delay-tolerant network architecture for challenged internets. In: Proc. SIGCOMM (2003)
3. Kempe, D., et al.: Connectivity and inference problems for temporal networks. J. Comput. Syst. Sci. 64(4), 820–842 (2002)
4. Okasha, S.: Altruism, group selection and correlated interaction. British Journal for the Philosophy of Science 56(4), 703–725 (2005)
5. Eagle, N., Pentland, A.: Reality mining: sensing complex social systems. Personal and Ubiquitous Computing V10(4), 255–268 (2006)
6. Chaintreau, A., Hui, P., et al.: Impact of human mobility on the design of opportunistic forwarding algorithms. In: Proc. INFOCOM (2006)
7. Leguay, J., Lindgren, A., et al.: Opportunistic content distribution in an urban setting. In: ACM CHANTS, pp. 205–212 (2006)
8. Gonzalez, M.C., Hidalgo, C.A., Barabasi, A.L.: Understanding individual human mobility patterns. Nature 453(7196), 779–782 (2008)
9. Newman, M.E.J.: Analysis of weighted networks. Physical Review E 70, 056131 (2004)
10. Freeman, L.C.: A set of measuring centrality based on betweenness. Sociometry 40, 35–41 (1977)
11. Liao, W.H., Tseng, Y.C., Lo, K.L., Sheu, J.P.: GeoGRID: A geocasting protocol for mobile ad hoc networks based on grid. Journal of Internet Technology 1(2) (2000)
12. Baldoni, R., Marchetti, C., Virgillito, A., Vitenberg, R.: Content-based publish/subscribe over structured overlay networks. In: Proc. ICDCS (2005)
13. Hui, P., Crowcroft, J.: How small labels create big improvements. In: Proc. IEEE ICMAN (2007)
14. Lindgren, A., Doria, A., et al.: Probabilistic routing in intermittently connected networks. In: Dini, P., Lorenz, P., de Souza, J.N. (eds.) SAPIR 2004. LNCS, vol. 3126, pp. 239–254. Springer, Heidelberg (2004)

---

[2] http://www.social-nets.eu/, http://www.amillionpeople.net/

Hypernetworks of Complex Systems

Jeffrey Johnson

Faculty of Mathematics, Computing and Technology, The Open University
Walton Hall, Milton Keynes, MK7 6AA, United Kingdom
j.h.johnson@open.ac.uk

Abstract. Hypernetworks generalise the concept of a relation between two things to relations between many things. The notion of *relational simplex* generalises the concept of network edge to relations between many elements. Relational simplices have multi-dimensional connectivity related to hypergraphs and the Galois lattice of maximally connected sets of elements. This structure acts as a kind of *backcloth* for the dynamic system *traffic* represented by numerical mappings, where the topology of the backcloth constrains the dynamics of the traffic. Simplices provide a way of defining multilevel structure. This relates to system time measured by the formation of simplices as system events. Multilevel hypernetworks are classes of sets of relational simplices that represent the system backcloth and the traffic of systems activity it supports. Hypernetworks provide a significant generalisation of network theory, enabling the integration of relational structure, logic, and topological and analytic dynamics. They provide structures that are likely to be necessary if not sufficient for a science of complex multilevel socio-technical systems.

Keywords: Complex Systems, Hypernetworks, Networks, Simplex, Simplicial Complex, Backcloth, Traffic, Multilevel Systems, Dynamics.

1 Introduction

Hypernetwork generalise the concept of a relation between two things to relations between many things. The higher dimensional analogues of the network edge are the triangle, the tetrahedron, the pentahedron, and so on (Figure 1). Thus n-ary relations can be represented by *polyhedra* in multidimensional space. Polyhedra provide a multidimensional generalisation of one-dimensional network edges.

An n-ary relation R between n elements, $x_1, x_2, .., x_n$ is defined by a proposition P_R where it is assumed that $P_R(x_1, x_2, .., x_n)$ is well formed and there is a practical procedure for deciding whether or not $P_R(x_1, x_2, .., x_n)$ is true.

Binary relations yield the usual network edge, written (a, b), where a is related to b under the relation R if and only if there is a proposition P_R with $P_R(a, b)$ = True. The graphical representation of networks uses small solid circles called *vertices* to represent the elements and *lines* between the vertices to represent relationship. These lines are called *edges* or *links*. Generally $P_R(a, b) \neq P_R(b, a)$ and $(a, b) \neq (b, a)$, and the edges of networks are said to be *oriented*. (a, b) is oriented from a to b. Oriented edges are often represented by *arrows*.

J. Zhou (Ed.): Complex 2009, Part I, LNICST 4, pp. 364–375, 2009.

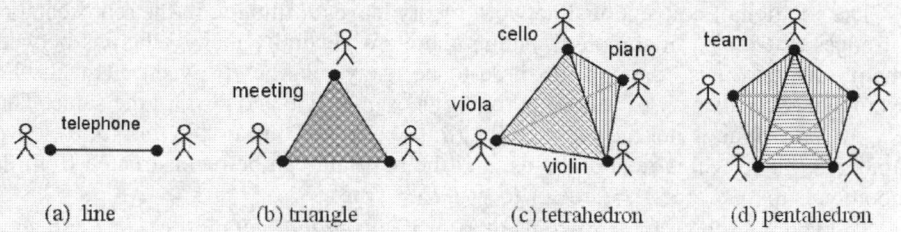

Fig. 1. Polyhedra generalise the concept of binary related to pairs to n-ary relations

Links and arrows are very powerful for representing things in complex systems. Links show that *a* and *b* are related in some way, and arrows can represent ideas such as *flow*, *transformation*, and *entailment* between the vertices *a* and *b*.

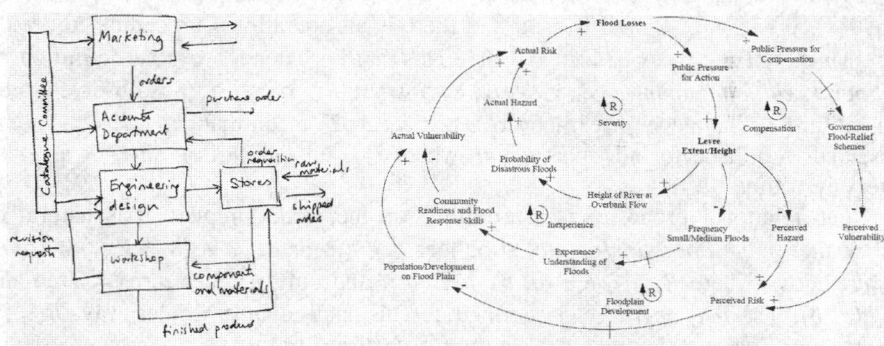

Fig. 2. Network with oriented edges can represent flow and transformations

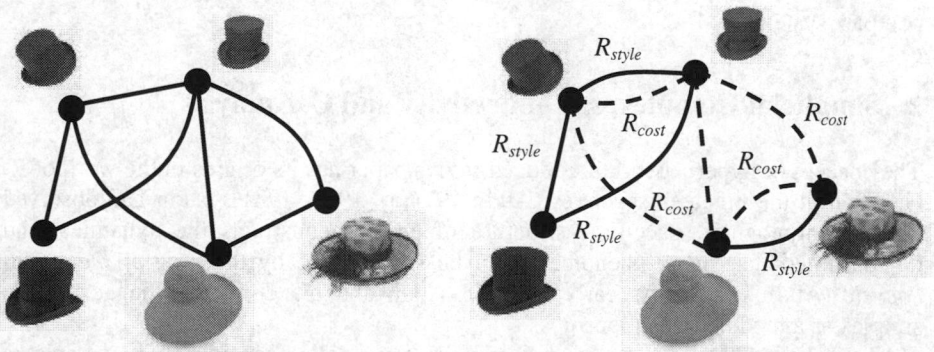

(a) The price-style hat network is ambiguous   (b) real networks have heterogeneous links

Fig. 3. In complex systems the relational structure must be explicit

The notation of conventional network theory leaves R implicit in the representation of edges as (a, b). In complex systems there are generally many relations between many sets, and it is common for there to be many relations between elements. For example, two hats may be related by being the same style and costing the same. Thus $R_{style}(a, b)$ = True is not the same as $R_{cost}(a, b)$ = True. To draw the link (a, b) leaves ambiguity as to which relation holds, and the best that can be reconstructed from the notation is that $R_{style}(a, b)$ = True or $R_{cost}(a, b)$ = True.

To overcome this problem we use the *explicit relation* notation $\langle a, b; R \rangle$ to show that $R(a, b)$ = True. Then $\langle a, b; R_{style} \rangle \neq \langle a, b; R_{cost} \rangle$ is clear from the notation.

Let $X_0, X_1, \ldots, X_n$ be sets. In the usual way let the *Cartesian product* of these sets be defined as $\prod_i X_i = X_0 \times X_1 \times \ldots \times X_n = \{ \langle x_0, x_1, .., x_n \rangle \mid$ for all x_i belongs to X_i for all $i = 0, \ldots, n \}$. The ordered set of elements, $\langle x_0, x_1, .., x_n \rangle$ is called an *abstract n-simplex*. Such a simplex can be represented by a polyhedron on n-dimensional space. Simplices are the natural generalisation of network edges.

A *relational simplex*, $\langle x_0, x_1, .., x_n; R \rangle$ is said to exist if the is proposition P_R such that $P_R \langle x_0, x_1, .., x_n \rangle$ is well formed and there is an operational procedure to decide whether or not $P_R \langle x_0, x_1, .., x_n \rangle$ = True. This can be extended by the definition of *temporal relational simplex* $\langle x_0, x_1, .., x_n; R; t \rangle$ where $P_R \langle x_0, x_1, .., x_n \rangle$ is observed to be true at time t. This generalises: let $\langle x_0, x_1, .., x_n; R; T \rangle$ mean that $P_R \langle x_0, x_1, .., x_n \rangle$ is observed to be true for all times t in T where T is an *interval* of time, or a set of intervals of time.

When relational propositions are defined on the same simplex, it is natural to define the *wedge operation* on two simplices, e.g. $\langle a, b; R_{style} \rangle \wedge \langle a, b; R_{cost} \rangle = \langle a, b; R_{style} \wedge R_{cost} \rangle$, where $R_{style} \wedge R_{cost}(a, b)$ = True if and only if $R_{style} (a, b)$ = True and $R_{cost}(a, b)$. The *vee operation* is defined for disjunction in a similar way, $R_{style} \vee R_{cost}(a, b)$ = True if and only if $R_{style} (a, b)$ = True or $R_{cost}(a, b)$.

In general we use the notation $\underline{X}$ to represent sequences of vertices such as $x_0, x_1, .., x_n$. Then the notation $\langle \underline{X}; R; T \rangle$ provides a way of combining relational structure, logic, and time, all of which are necessary if not sufficient for a science of complex systems.

2 Simplicial Complexes, Connectivity and Q-Analysis

The concept of hypernetwork introduced in this paper has its origins in the work of R. H. Atkin in the nineteen seventies (Atkin 1974(a), 1977, 1981). Atkin had observed that the topological space-time structure of physics constrains the dynamics, and demonstrated that many phenomena can be summarised by the *Law of the trivial cocycle* (Atkin, 1972). As early as 1968 Atkin and his coworkers suggested the simplex as a model for relationships:

"To examine the idea of connectivity in more detail consider, for example, a collection of people and the sociological roles which they are said to be playing. Let the role-set be denoted by Y and let it contain a finite number of roles Y1 , Y2, . . ; similarly let there be a finite number of persons X1, X2, . . . in the collection of people X. An individual person X1 plays, say, roles Y1 , Y2 ,Y3 and a second person X2 plays the roles Y2 Y3,Y4,Y5.

We now define an abstract p-simplex to be a subset of Y containing (p + 1) roles provided that there is at least one individual who plays all these roles. Thus the 2-simplex (Y1 Y2 Y3) exists since X1 plays the three roles represented therein, so also do the 3-simplex (Y2 Y3 Y4 Y5), the 0-simplex (Y5), and many others. The two simplices (Y1 Y2 Y3) and (Y2 Y3 Y4 Y5) are clearly joined by the 1-simplex (Y2 Y3) – which is referred to as a face of both the 2- and 3-simplices. The collection of all such simplices actually forms a complex K(Y) which has the property that a person is represented by one of its simplices together with all the faces of that simplex. We may note that two people who play the same roles are indistinguishable in this model.

If we use the language which has historical connections with geometry we would refer to X1 as an abstract closed triangle, whilst X2 would be an abstract closed tetrahedron: in general, with respect to this particular role-set Y, we would observe a person X as an abstract closed polyhedron. All such polyhedra are connected (if at all) by faces (which are also polyhedra) which are common – the whole structure, or complex, therefore exhibits a connectivity which possesses a natural classification in terms of the dimensionality of the various polyhedra and their common faces. This connectivity seems to be a natural expression of the possibility of communication among the persons in the structure.

Thus our persons X1 and X2 can communicate with each other because they have a connection via their common simplices (Y2 Y3), (Y2) and (Y3). This connection exhibits the fact that X1 "sees" X2 via the common faces of their separate polyhedra. On the other hand X2 has many faces (15 in all, if we include the whole tetrahedron) any one of which might serve as a connecting face between himself and someone else." (Atkin *et al*, 1968).

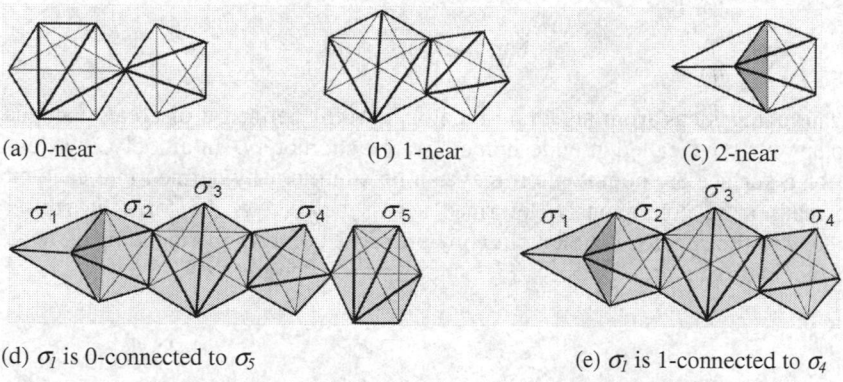

(a) 0-near   (b) 1-near   (c) 2-near

(d) σ_l is 0-connected to σ_5 (e) σ_l is 1-connected to σ_4

Fig. 4. q-nearness and q-connectivity

Let $\sigma_p = \langle x_0, x_1, .., x_p \rangle$ be an abstract p-simplex with vertices $x_0, x_1, .., x_p$. The simplex $\sigma_q = \langle x'_0, x'_1, .., x'_q \rangle$ is a *q-dimensional face*, or q-face, of σ_p if and only if every vertex of σ_q is also a vertex of σ_p, i.e. σ_q is a face of σ_p if and only if $\{x_0, x_1, .., x_q\} \subseteq \{x_0, x_1, .., x_p\}$. Let $\sigma_p = \langle x_0, x_1, .., x_p \rangle$ and $\sigma_{p'} = \langle x'_0, x'_1, .., x'_{p'} \rangle$ be two abstract simplices. Their *shared face* is defined as $\sigma_p \cap \sigma_{p'} = \sigma_{p''} = \langle x''_0, x''_1, .., x''_{p''} \rangle$ where $\{ x''_0, x''_1, .., x''_{p''} \} = \{ x_0, x_1, .., x_p \} \cap \{ x'_0, x'_1, .., x'_{p'} \}$.

In algebraic topology a set of simplices with all its faces is called a *simplicial complex*. Atkin defined two simplices to be *q-near* if they shared a q-dimensional face, and he defined two simplices to be *q-connected* if there was a chain of pairwise q-near simplices between them.

Technically, q-connectivity is the transitive closure of the q-nearness relations, and is an equivalence relation on a set of simplices with dimension q or greater. As such it partitions those simplices into equivalence classes of *q-connected components*. Atkin defined a listing of those components and related statistics to be a *Q-analysis*.

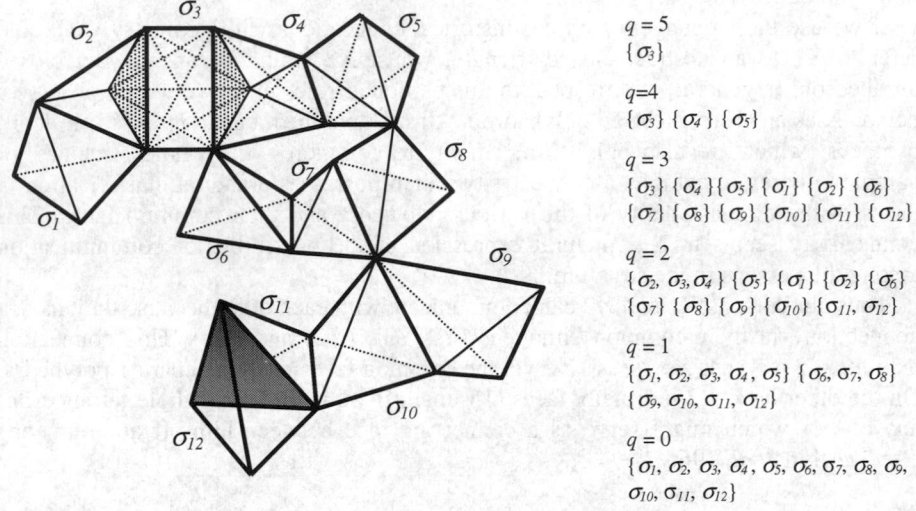

$q = 5$
$\{\sigma_3\}$

$q=4$
$\{\sigma_3\} \{\sigma_4\} \{\sigma_5\}$

$q = 3$
$\{\sigma_3\} \{\sigma_4\} \{\sigma_5\} \{\sigma_1\} \{\sigma_2\} \{\sigma_6\}$
$\{\sigma_7\} \{\sigma_8\} \{\sigma_9\} \{\sigma_{10}\} \{\sigma_{11}\} \{\sigma_{12}\}$

$q = 2$
$\{\sigma_2, \sigma_3, \sigma_4\} \{\sigma_5\} \{\sigma_1\} \{\sigma_2\} \{\sigma_6\}$
$\{\sigma_7\} \{\sigma_8\} \{\sigma_9\} \{\sigma_{10}\} \{\sigma_{11}, \sigma_{12}\}$

$q = 1$
$\{\sigma_1, \sigma_2, \sigma_3, \sigma_4, \sigma_5\} \{\sigma_6, \sigma_7, \sigma_8\}$
$\{\sigma_9, \sigma_{10}, \sigma_{11}, \sigma_{12}\}$

$q = 0$
$\{\sigma_1, \sigma_2, \sigma_3, \sigma_4, \sigma_5, \sigma_6, \sigma_7, \sigma_8, \sigma_9,$
$\sigma_{10}, \sigma_{11}, \sigma_{12}\}$

Fig. 5. A Q-analysis

Again using ideas from algebraic topology, Atkin defined a discrete analogue of homotopy that he called pseudo-homotopy, or shomotpty. Intuitively, two closed loops on a surface are homotopic if they can be continuously defined into each other. Loops cannot be continuously deformed across holes, *e.g.* the torus has different homotopy to the sphere, and the homotopy properties of a simplicial complex relate to its topological structure.

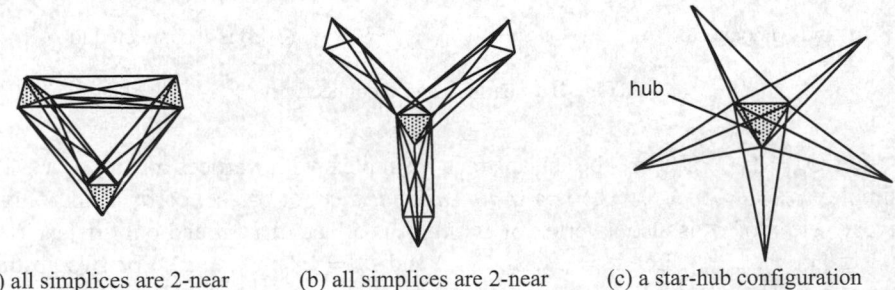

(a) all simplices are 2-near   (b) all simplices are 2-near   (c) a star-hub configuration

Fig. 6. Star-hub configurations

This work highlighted the need to discriminate between the configurations shown in Figures 6 (a) and 6(b), where all the simplices are q-near to each other, but one configuration has a hole while the other does not. This leads to the definition of the star-hub configuration shown in Figure 6(c) (Johnson (1983)).

3 Hypergraphs, Galois Connections, Maximal Rectangles, Star-Hubs

Most binary relations hold between different sets, A and B, giving rise to *bipartite networks* such as that shown in Figure 7(a) with the elements of A arranged in a line, the elements of B arranged in a line, and lines drawn between a and b when a is R-related to B.

A *hypergraph* is a set A with a class of its subsets. Let $\sigma_R(a) = \{ b \mid a$ is R-related to $b\}$. For any $A' \subseteq A$ let $\sigma_R(A') = \{ b \mid a$ is R-related to b for all a belong to $A'\}$. Then let $H_A(B, R) = \{ \sigma_R (A') \mid$ for all $A' \subseteq A\}$ and $H_B(A, R) = \{ \sigma_R (B') \mid$ for all $B' \subseteq B\}$. These are hypergraphs, as illustrated in Figure 7(b).

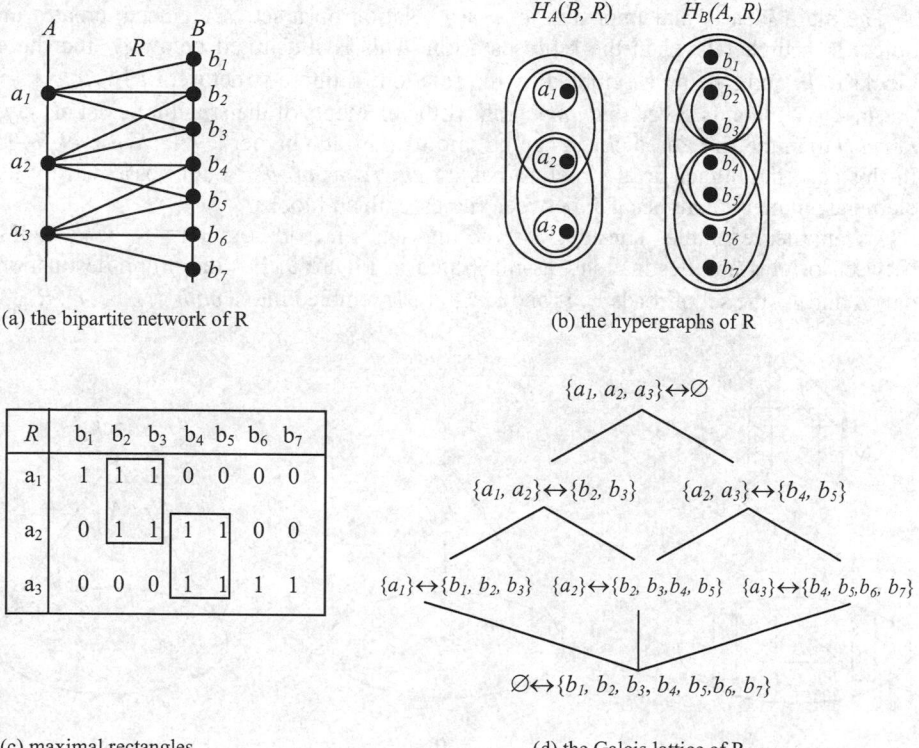

(a) the bipartite network of R

(b) the hypergraphs of R

(c) maximal rectangles

(d) the Galois lattice of R

Fig. 7. Bipartite network, hypergraphs and Galois lattice for a relation R between sets A and B

It can be shown that Then $H_A(B, R)$ and $H_B(A, R)$ are in one-to-one correspondence. Intuitively the subsets of A and B are paired as $A' \leftrightarrow B'$ so that every member of A' is R-related to every member of B'. A' and B' are *maximal* in the sense that no element outside A' is related to all the elements of B', and no element outside B' is related to all the elements of A'. If the relation R is represented by an incidence matrix with entry $m_{ij} = 1$ if $a_i R b_j$ and equals zero otherwise, then the rows and columns can be arranged to show the $A' \leftrightarrow B'$ pairs as blocks of ones in so-called *maximal rectangles*. For example, Figure 7(c) shows the two maximal rectangles corresponding to the pairs $\{a_1, a_2\} \leftrightarrow \{b_2, b_3\}$ and $\{a_2, a_3\} \leftrightarrow \{b_4, b_5\}$. This one-to-one correspondence is called a *Galois connection* and the pairs sets can be arranged as a *Galois lattice* (Barbut and Monjardet, 1970) as shown in Figure 7(d).

4 Relational Simplices and Multilevel Hypernetworks

The ideas sketched in the previous section are essentially set-theoretic. Hypernetworks enrich this set-theoretic approach by making a distinction between sets and structured sets, and this enables a powerful approach to representing the dynamics of complex multilevel systems.

The main idea is that imposing an n-ary relation on a set of elements creates an object at a higher level in the representation. This is illustrated below by the three blocks a, b, and c being assembled by the relation R into a structure, $R: \{a, b, c\} \rightarrow \langle a, b, c; R \rangle$, that is given the *name* arch. If the elements of the structure exist at, say, *Level N* then the structured object can be said to exist at a higher level, say *Level N+1*. In this case the higher level structure has an *emergent property* not possessed by its elements, namely there is a gap between the assembled blocks.

As another example, consider assembling sets of road segments to form paths between origins and destinations, as illustrated in Figure 9. For the origin-destination pair A and A' the set of roads r_1, r_2 and r_3 can be assembled into a *path*, $\langle r_1, r_2, r_3; R_{AA'} \rangle$,

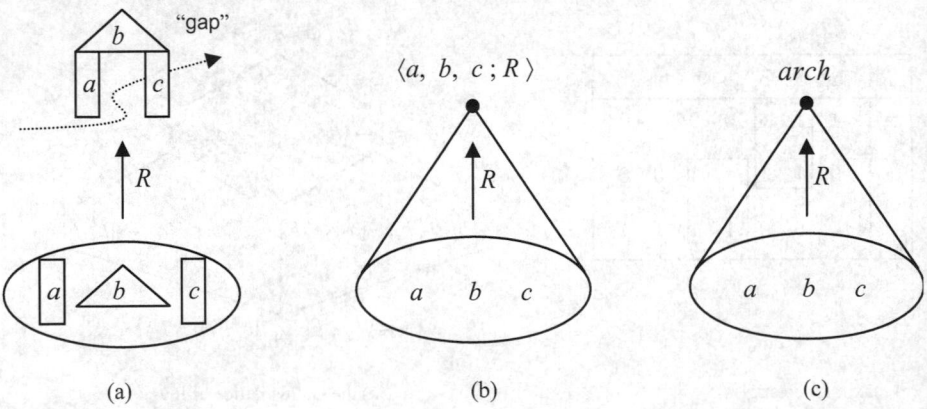

Fig. 8. n-ary relations map objects to higher levels of representation

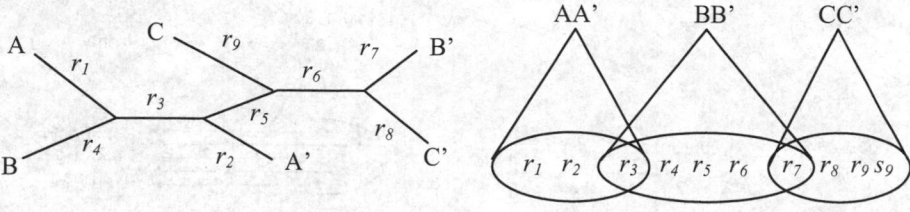

(a) AA', BB' and CC' define paths in the networks   (b) the paths as structured sets of roads

Fig. 9. Links assembled to routes in a road network

enabling vehicles to travel from A to A'. The fact that this path is a structure and not a set can be seen from the necessity for the assembly relation to order the roads as r_1 followed by r_3 (not r_2) followed by r_2. The simplices $\langle r_1, r_2, r_3; R_{AA'} \rangle$ and $\langle r_3, r_4, r_5, r_6, r_7; R_{BB'} \rangle$ are connected through the vertex $\langle r_3 \rangle$ and this is where their traffic interacts, with AA' traffic delaying BB' traffic and *vice-versa*.

5 Backcloth and Traffic

Networks allow a distinction to be made between relatively static infrastructure such as a road or computer network and relatively dynamics flows such as vehicles or information on that infrastructure. Generally the infrastructure is *relational* while the flows are *numerical*. Atkin suggested the metaphor of a structural *backcloth* supporting a traffic of flows measured by numbers. For example, the flow of vehicles through a road network is traffic on the relatively fixed infrastructure of roads. The term can be generalised to mappings related to the flows, such as the travel time on a road network.

Traffic can exist at many levels across a system, and aggregates over the relational structure. For example, the travel time on the path chosen between the origin and destination, $\langle r_1, r_2, r_3; R_{AA'} \rangle$, is the sum of the travel times on the individual roads as vertices, as shown in Figure 10(a). Sometimes the relations themselves carry numerical traffic, *e.g.*.in Figure 10(b) the cost of the arch is the sum of the cost of its components

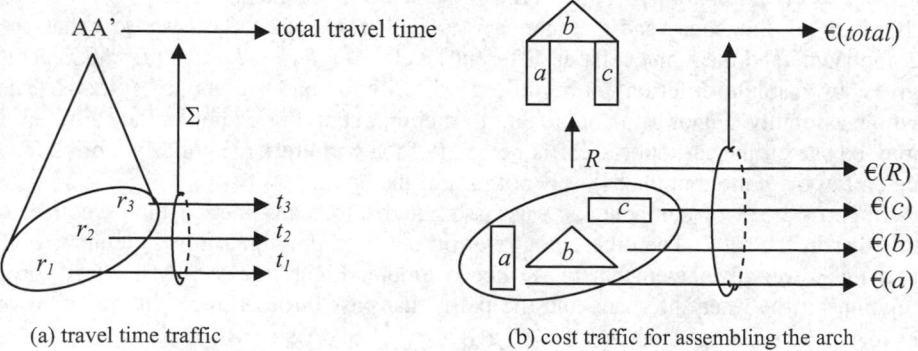

(a) travel time traffic   (b) cost traffic for assembling the arch

Fig. 10. Mappings as traffic on multidimensional structure

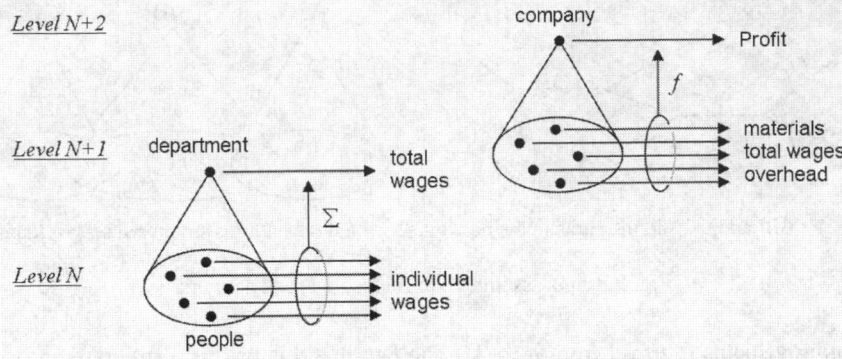

Fig. 11. Aggregating traffic over the backcloth in multilevel systems

plus the assembly cost. It could include other costs such as tax and profit, with a non-linear relationship between lower and higher level traffic.

Figure 11 illustrates the bottom-up aggregation of traffic in a multilevel system, with micro-level traffic aggregating into meso-level traffic, and meso-level traffic aggregating into macro-level traffic. Generally there is both bottom-up and top-down interaction between traffic at all levels. For example, an overall price could be set for a project at a high level of representation, and this price would have a top-down influence on the lower level traffic.

A *multilevel hypernetwork* is any class of sets of relational simplices with sets of mapping on the simplices and their faces. In other words multilevel hypernetworks provide a means for representing the backcloth and the traffic of multilevel systems.

6   Hierarchical Cones in Heterarchical Systems

Given an *n*-ary relation such as $R:\{x_0, x_1, x_2\} \rightarrow \langle x_0, x_1, x_2 ; R \rangle$, the imposition of the relation creates a new object $\langle x_0, x_1, x_2 ; R \rangle$. When modelling complex systems it is common to give this new structure a *name*, say y. For example, the blocks were assembled into an *arch*. We extend the notation for representing simplices to say that $\langle x_0, x_1, x_2 ; y; R \rangle$ is a *hierarchical cone*. The set of components $\{x_0, x_1, x_2\}$ is called the *base* of the cone, and the name y is called its *apex*. The we say that the components and the name exist at different levels, $Level(x_i) < Level(y)$ for all x_i. This gives an absolute criterion for multilevel discrimination in complex systems. The whole assembly cannot be a component of a component. For example, the carburettor may be part of the car, but the car is not part of the carburettor. Similarly a brick may be part of the house but the house is not part of the brick.

Suppose a set of components X is assembled into many named things collected together in the set Y. Then there is a relation between the higher level elements in Y and the lower level elements in X. For example, Figure 12 shows the top down relationship between the roads and the paths that pass through them. In this case we have the 'interesting' cones $\langle r_3 ; AA', BB'; R_{down} \rangle$ and $\langle r_6 ; BB', CC'; R_{down} \rangle$, where the meaning of R_{down} is discussed below.

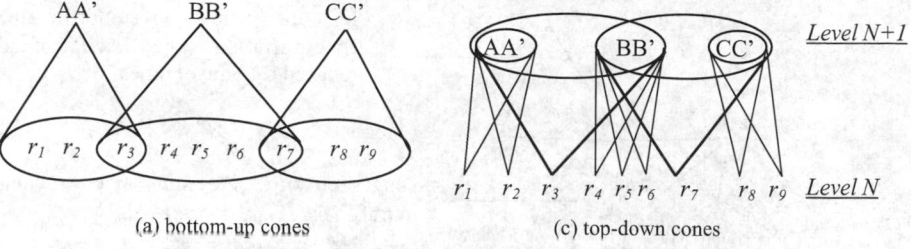

(a) bottom-up cones   (c) top-down cones

Fig. 12. Bottom-up and top-down hierarchical cones

The relation $R_{AA'}$ assembles roads into a path between A and A', and the relation $R_{BB'}$ assembled roads into a path between B and B'. What does it mean to write

$$\langle r_1, r_2, r_3; AA'; R_{AA'}\rangle \cap \langle r_3, r_4, \ r_5, r_6, r_7; BB'; R_{BB'}\rangle \ = \ \langle r_3; AA', BB', R_{down}\rangle$$

or more generally

$$\langle X; y; R\rangle \ \cap \ \langle X'; y'; R'\rangle \ \cap \ \langle X''; y''; R''\rangle \ \overset{def}{=} \ \langle X \cap X' \cap X''; y, y', y''; R \oplus R' \oplus R''\rangle$$

as the combination of cones? The expression $R \oplus R' \oplus R''$ concerns the combination of bottom-up n-ary relations to form new top down relations whose meaning depends on context. It can be noted that $X \cap X' \cap X'' \leftrightarrow \{y, y', y''\}$ is a Galois pair in which all the Xs are related to all the ys. The answer to the questions of what it means to write $\langle r_3; AA', BB', R_{down}\rangle$ is that the paths between AA' and BB' share the vertex write $\langle r_3\rangle$.

7 Q-Transmission

In almost all systems the topological structure of the backcloth constrains the traffic dynamics. This is obviously the case in electrical networks where components are connected together in ways decided by the designer to achieve specified electrical flows. Change the topology, as with a short circuit, and dramatically different behaviour can emerge.

As a simple example consider the road network in Figure 9. This can be represented as three connected simplices as shown in Figure 13. Suppose that there is a large increase in travel demand between A and A' resulting in higher traffic flows along the path $\langle r_1, r_2, r_3; R_{AA'}\rangle$. Then the traffic on $\langle r_3\rangle$ will be heavier than usual, acting a barrier to the traffic on BB'. After $\langle r_3\rangle$ the BB' traffic will be less. This lighter traffic on BB' means there will be lighter traffic than usual on $\langle r_7\rangle$ resulting in freer flow traffic on $\langle r_7, r_8, r_9; R_{CC'}\rangle$ and reduced CC' travel times. In this case the dynamic behaviour on $\langle r_1, r_2, r_3; R_{AA'}\rangle$ is *transmitted* to $\langle r_7, r_8, r_9; R_{CC'}\rangle$, even though the two path simplices share no road links. In other words, for the dynamics of one part of the system to affect the dynamics of another part of the system it is sufficient that they are connected by a chain of connected.

In this case just sharing a single vertex was sufficient for transmission to occur. In general the more highly connected the simplices, the greater is the magnitude of transmission. It can be hypothesised that some processes need two simplices to be

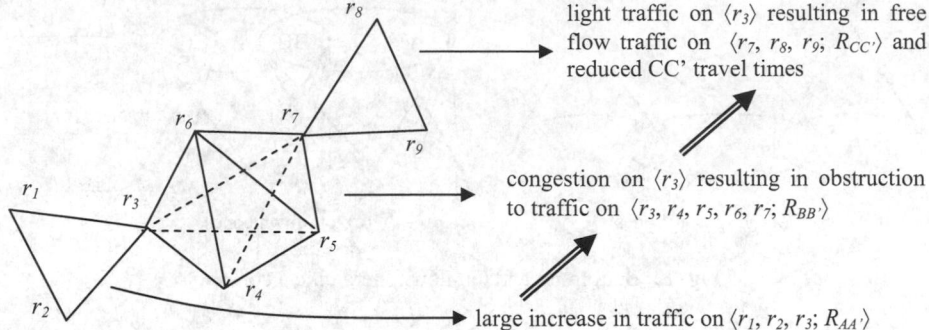

Fig. 13. Transmission of dynamics between connected simplices

q-near for one to affect the other. In this case the q-connectivity of the backcloth constrains what are called the *q-transmission* dynamics of the system. In many cases simply being connected is sufficient for transmission to occur.

8 Time and Structural Events

Apart from giving a way to represent multilevel structure, simplices give a way of measuring time as *system events*. Consider the arch in Figure 8 where there is a transition from a state in which the arch is not built to a state where the arch is built, $R: \{a, b, c\} \rightarrow \langle a, b, c; R \rangle$. The moment that the simplex $\langle a, b, c; R \rangle$ comes into being marks what Atkin (1981) called an *event*. Events in physical space-time are usually measured by physical systems such as pendulums or oscillating crystals. These may or may not be synchronised with events in system time. For example, it may be planned to open the bridge to traffic on 1st June, but if it is not complete it will not be opened. The defining event for opening the bridge is when it is finished, *i.e.* after the many polyhedral events that define the bridge existing in a safe form.

Matching system time to clock time lies at the heart of much design and management. Human beings consume resource in clock time – we eat periodically in clock time, are paid in clock time, pay rent in clock time, and so on. This has to be resolved against system time when human beings are involved creating system events. In large complex systems there are events at every level on different time scales.

Much mathematical modelling has focussed on formulae relating numerical properties of systems. More recently it has become widely acknowledged that the underlying network topology plays a large role in the dynamics of systems. But it is necessary to go beyond this to consider the evolution of the networks and how the relations change. Put simply, when does a link appear and disappear in a network? What causes links to form or to break?

This relates to the possibility of making predictions in complex systems science. Predictions can be classified as

Simple Type-I predictions: changes in mappings
Type-I-1, Single level
Type-I-2, Multiple level

Simple Type-II predictions: changes in relational backcloth
Type II-1, Single Level
Type II-2, Multiple Level

The combination of types I-2 and II-2 is the norm for complex socio-technical systems, and this presents a big challenge in that hypernetworks may help to meet.

9 Conclusions

This paper has briefly introduced the notion of *relational simplex* which generalises the concept of network edge to relations between many elements. Relational simplices have higher dimensional connectivity related to hypergraphs and the Galois lattice of maximally connected sets of elements. This structure acts as a kind of *backcloth* for the dynamic system *traffic* represented by numerical mappings, where the topology of the backcloth constrains the dynamics of the traffic. Simplices provide a way of defining multilevel structure, where this relates to system time measured by the formation of simplices as system events. Multilevel hypernetworks are classes of sets of relational simplices that represent the system backcloth and support the traffic of systems activity. Hypernetworks provide a significant generalisation of network theory and provide structures that are likely to be necessary if not sufficient for a science of complex multilevel socio-technical systems.

References

1. Atkin, R.H., Bray, R.W., Cook, I.T.: A mathematical approach towards a social science, Essex University Review, No. 2, Colchester (Autumn 1968)
2. Atkin, R.H.: Mathematical structure in human affairs. Heinemann Educational Books, London (1974)
3. Atkin, R.H.: From cohomology in physics to q-connectivity in social science. I.J. Man-Machine Studies 4, 139–167 (1972)
4. Atkin, R.H.: Multidimensional Man. Penguin Books, Harmondsworth (1981)
5. Barbut, M., Monjardet, B.: Ordre et Classification. Hachette Université (1970)
6. Johnson, J.H.: Q-analysis: a theory of stars. Environment and Planning B: Planning and Design 10(4), 457–469 (1983)
7. Johnson, J.H.: Hypernetworks in the science of complex systems. Imperial College Press, London (2009)

Less Restrictive Synchronization Criteria in Complex Networks with Coupling Delays

Yun Shang[1] and Maoyin Chen[2]

[1] Institute of Mathematics, AMSS, Academia Sinica, Beijing 100080, P.R. China
shangyun602@163.com
[2] Department of Automation, TNList, Tsinghua University, Beijing 100084, P.R. China
mychen@mail.tsinghua.edu.cn

Abstract. This paper considers the synchronization in complex networks with coupling delays, whose topologies could be be symmetric and asymmetric. Differing from most works on the synchronization in complex networks with coupling delays, this paper only uses a positively-defined function, which is definitely not a Krasovskii-Lyapunov function, to analyze the synchronization criteria. Further, we can derive novel but less restrictive synchronization criteria than those resulting from the Krasovskii-Lyapunov theory. Theoretical analysis and numerical simulations fully verify the main results.

Keywords: complex networks, synchronization, matrix measure.

1 Introduction

Recently, the dynamics of complex networks has been extensively investigated, with special emphasis on the interplay between the overall topology and the local dynamics of coupled nodes. As a typical kind of dynamics, the synchronization in complex networks has been a research topic [1-17]. In 1998, Pecora and Carroll proposed the *master stability function* (MSF) based method to study the synchronization in networks [5]. The Lyapunov's direct method can be also used to study the synchronization in networks by constructing a Lyapunov function, which decreases along trajectories and gives analytical criteria for local and global synchronization [6-9,11,12].

Due to the finite speeds of transmission and spreading as well as traffic congestions, a signal or influence traveling through a complex network is often associated with time delays, and this is very common in biological and physical networks. Complex networks with coupling delays have recently attracted attention in many fields. From works [12,15-17], the Krasovskii-Lyapunov theory [18] is a useful and powerful tool to discuss the synchronization in networks with coupling delays. According to this kind of theory, some sufficient delay-independent and delay-independent conditions are given to ensure synchronization in networks with coupling delays. However, synchronization criteria resulting from the Krasovskii-Lyapunov theory may be too strict since they require that the

J. Zhou (Ed.): Complex 2009, Part I, LNICST 4, pp. 376–388, 2009.

derivative of a positively-defined Krasovskii-Lyapunov function is non-positive for all time [12,15-17].

Without using the Krasovskii-Lyapunov theory, this paper tries to derive less restrictive criteria for the synchronization in complex networks with coupling delays. Differing from most works on the synchronization in complex networks with coupling delays, this paper only uses a positively-defined function, which is definitely not a Krasovskii-Lyapunov function, to analyze the synchronization. We can derive novel but less restrictive synchronization criteria than those resulting from the Krasovskii-Lyapunov theory. The idea in this paper can be applied to complex networks with symmetric and asymmetric topologies.

The rest of this paper is organized as follows. A complex network model with coupling delays is presented, and some preliminaries are introduced in Section 2. In Section 3, we give one novel synchronization criterion for complex networks with symmetric topology. The synchronization criterion in complex networks with asymmetric topology is considered in Section 4. Numerical simulations are illustrated to show the effectiveness of the proposed synchronization criteria in Section 5. The last Section draws our conclusion.

2   A Complex Network Model and Necessary Preliminaries

Consider a complex network consisting of N nodes in the following form

$$\dot{x}_i(t) = f(x_i(t)) + \sum_{j \neq i} g_{ij} \Gamma(x_j(t-\tau) - x_i(t-\tau)) \tag{1}$$

for $1 \leq i \leq N$, where $x_i(t) = (x_{i1}(t), \ldots, x_{in}(t))^T$ is the state vector of node i, $\tau > 0$ is the time delay, the initial states for states $x_i(t)$ are $x_{i0} = \psi_{i0}(t), t \in [-\tau, 0]$, $\Gamma = \text{diag}\{r_1, \cdots, r_n\}$ with r_i being 0 or 1 is the inner coupling matrix, and $f : R^n \to R^n$ is a smooth vector-valued function. Matrix $G = (g_{ij})_{N \times N}$ is the outer coupling matrix representing the topology of networks, and its elements are chosen as follows: if nodes i and j are connected, $g_{ij} = g_{ji} \neq 0$; otherwise $g_{ij} = g_{ji} = 0$, and the diagonal elements are defined by $g_{ii} = - \sum_{j=1, j \neq i}^{N} g_{ij}$.

Network (1) is said to be in a synchronized manifold Ξ: $\{x_1(t) = \cdots = x_N(t) = s(t)\}$ if $\lim_{t \to \infty} (x_i(t) - s(t)) = 0$ for $1 \leq i \leq N$, where $s(t)$ is a solution of an isolated node, denoted by $\dot{s}(t) = f(s(t))$. In this paper, suppose that $s(t)$ is an orbitally stable solution of the isolated node, and the Jacobian matrix $J(t) = Df(s(t))$ satisfies that $\frac{dJ(t)}{dt} = \frac{Df(s(t))}{dt}$ is bounded for all time.

In this paper our main results are based on the concept of matrix measure and one lemma with respect to the stability of time-delayed equations.

Definition 1: The matrix measure of a complex square matrix $C = (c_{ij})$ is defined as follows [20]:

$$\mu_{\cdot}(C) = \lim_{\varepsilon \to 0^+} \frac{||I_n + \varepsilon C|| - 1}{\varepsilon} \tag{2}$$

in which $|| \cdot ||$ is a matrix norm, and I_n is the identity matrix.

When $||C||_1 = \max_j \sum_{i=1}^{n} |c_{ij}|$, $||C||_2 = [\lambda\max(C^T C)]^{1/2}$ and $||C||_\infty = \max_i \sum_{j=1}^{n} |c_{ij}|$, we obtain $\mu_1(C) = \max_j\{\text{Re}(c_{jj}) + \sum_{i=1,i\neq j}^{n} |c_{ij}|\}$, $\mu_2(C) = \frac{1}{2}\lambda_{\max}(C^* + C)$, and $\mu_\infty = \max_i\{\text{Re}(c_{ii}) + \sum_{j=1,j\neq i}^{n} |c_{ij}|\}$ respectively, where C^* is the complex conjugate transpose of a complex matrix.

Lemma 1 [21,22]: Consider the following time-delayed equations

$$\dot{x}(t) = Ax(t) + Bx(t - \tau), x(t) = \phi(t),\ t \in [-h,\ 0] \tag{3}$$

where $A, B \in R^{n \times n}$, $\phi(t)$ is a continuous vector-valued initial function. Eq.(3) is asymptotically stable i.o.d (independent of delay) if and only if, for any given positive definite hermitian matrix $Q(z)$, $\forall |z| = 1$, the solution of $P(z)$ of the complex Lyapunov matrix equation

$$A^*(z)P(z) + P(z)A(z) = -Q(z),\ |z| = 1 \tag{4}$$

is also a positive definite hermitian matrix, where $A(z) = A + zB$, $|z| = 1$, $z = \exp(jw)$, $w \in [0, 2\pi]$, $j = \sqrt{-1}$.

Lemma 1 can be viewed as the asymptotical stability condition of a generalized linear system described by [22]

$$\dot{y}(t) = A(z)y(t),\ |z| = 1 \tag{5}$$

3   Synchronization in Complex Networks with Symmetric Topology

We first give a fundamental lemma for the network (1) with symmetric topology.

Lemma 2: For network (1) with symmetric topology, assume the outer coupling matrix G is a *nonnegative diffusively coupled matrix*, and can be irreducible and diagonalized. The manifold Ξ is asymptotically stable, if the following $N - 1$ systems are asymptotically stable:

$$\dot{w}_i(t) = J(t)w_i(t) + \lambda_i \Gamma w_i(t - \tau),\quad 2 \leq i \leq N \tag{6}$$

where λ_i are nonzero eigenvalues of G.

Proof: Since G is a *nonnegative diffusively coupled matrix*, G has zero-sum rows and nonnegative off-diagonal elements. In addition, 0 is one eigenvalue of multiplicity 1, and there exists a nonsingular matrix $\Phi = (\phi_1, \cdots, \phi_N)$ such that $G^T\phi_k = \lambda_k\phi_k$ for $1 \leq k \leq N$, where $0 = \lambda_1 > \lambda_2 \geq \cdots \geq \lambda_N$ [9]. After a similar procedure [6,7,9,13-15], the manifold Ξ is asymptotically stable if the $N - 1$ linear systems (6) are asymptotically stable.   □

From works [12,15-17], the Krasovskii-Lyapunov theory is very useful and powerful to analyze the stability of Eq. (6). Generally speaking, a Krasovskii-Lyapunov function can be chosen as

$$V^i(t) = w_i(t)^T P w_i(t) + \mu \int_{t-\tau}^{t} w_i(\alpha)^T Q w_i(\alpha)d\alpha \tag{7}$$

where $P = P^T > 0$, $Q = Q^T > 0$, and $\mu > 0$ is an arbitrary positive parameter. The main purpose of the Krasovskii-Lyapunov theory is to find the condition for the negativeness of $\frac{dV^i(t)}{dt}$ when the error $w_i(t)$ is not zero. In this paper, differing from most results with respect to the Krasovskii-Lyapunov theory, we only use a positively-defined function

$$V_i(t) = w_i(t)^T P w_i(t) \tag{8}$$

to analyze the stability of Eq. (6). Our aim in this paper is to make $\lim_{t \to \infty} V_i(t) = 0$, which also leads to $\lim_{t \to \infty} w_i(t) = 0$.

In order to do so, we first introduce a segmentation strategy for Eq. (6). For the time interval $[t_0, \infty)$, we segment it into $[t_0, \infty) = \bigcup_{j \geq 0} [t_j, t_{j+1})$, where $\tau_0 > 0$ is sufficiently small, j is an integer, $t_{j+1} = t_j + \tau_0$, and τ is a multiple of τ_0. If τ_0 is sufficiently small, the isolated dynamics $s(t)$ can be approximated by $s(t_j)$ within the interval $[t_j, t_{j+1})$, which further results in the approximation of $J(t)$ by $J(t_j)$. Therefore, within the interval $[t_j, t_{j+1})$, Eq. (6) can be approximated by

$$\dot{w}_i(t) = J(t_j)w_i + \lambda_i \Gamma w_i(t - \tau), \quad 2 \leq i \leq N \tag{9}$$

For the approximation system (9), we have the following result.

Theorem 1: Eq. (9) is asymptotically stable for the sample time τ_0, if there exists a symmetric positive definite matrix $P = M^T M \in R^{n \times n}$, $||M|| \neq 0$, such that

$$\int_{t_0}^{\infty} [\mu_\theta(MJ(t_j)M^{-1} + M^{-T}J(t_j)^T M^T) + (\frac{k^2}{\varsigma} + \varsigma) + 2|\lambda_i|]dt = -\infty \tag{10}$$

where μ_θ is one of μ_1, μ_2, μ_∞, ς and k are two sufficiently small positive constants.

Proof: Along with the solution of system (9), we get $\dot{V}_i(t) = w_i(t)^T[PJ(t_j) + J(t_j)^T P]w_i(t) + 2\lambda_i w_i^T(t)P\Gamma w_i(t-\tau)$. The second term satisfies $2\lambda_i w_i(t)^T P\Gamma w_i(t-\tau) \leq 2|\lambda_i| \cdot ||Mw_i(t)|| \cdot ||M\Gamma w_i(t-\tau)|| \leq |\lambda_i| \cdot [V_i(t) + V_i(t-\tau)]$. Hence we obtain $\dot{V}_i(t) \leq w_i(t)^T M^T[MJ(t_j)M^{-1} + M^{-T}J(t_j)^T M^T]Mw_i(t) + |\lambda_i| \cdot [V_i(t) + V_i(t-\tau)]$. Inspired by the concept of matrix measure (2), we get $\dot{V}_i(t) \leq (\mu_\theta(MJ(t_j)M^{-1} + M^{-T}J(t_j)^T M^T) + |\lambda_i|)V_i(t) + |\lambda_i|V_i(t-\tau)$. Hence

$$V_i(t) \leq \exp(\int_{t_0}^t (\mu_\theta(MJ(t_j)M^{-1} + M^{-T}J(t_j)^T M^T) + |\lambda_i|)d\alpha)V_i(0)$$
$$+ \int_{t_0}^t \exp(\int_\vartheta^t (\mu_\theta(MJ(t_j)M^{-1} + M^{-T}J(t_j)^T M^T) + |\lambda_i|)d\alpha)|\lambda_i|V_i(\vartheta - \tau)d\vartheta$$

From the comparison theorem [19], the solution $V_i(t)$ satisfies

$$V_i(t) \leq \Gamma_i(t) \tag{11}$$

where $\Gamma_i(t)$ is the maximal solution of $\Gamma_i(t) = \exp(\int_{t_0}^t (\mu_\theta(MJ(t_j)M^{-1} + M^{-T}J(t_j)^T M^T) + |\lambda_i|)d\alpha)V_i(0) + \int_{t_0}^t \exp(\int_\vartheta^t (\mu_\theta(MJ(t_j)M^{-1} + M^{-T}J(t_j)^T M^T) + |\lambda_i|)d\alpha)|\lambda_i|\Gamma_i(\vartheta - \tau)d\vartheta$, or equivalently,

$$\frac{d\Gamma_i(t)}{dt} = (\mu_\theta(MJ(t_j)M^{-1} + M^{-T}J(t_j)^T M^T) + |\lambda_i|)\Gamma_i(t) + |\lambda_i|\Gamma_i(t-\tau) \quad (12)$$

with the same initial condition $V_i(0)$. From Lemma 1, within the interval $[t_j, t_{j+1})$, the stability of Eq. (12) is equivalent to the stability of

$$\frac{d\Gamma_i'(t)}{dt} = [\mu_\theta(MJ(t_j)M^{-1} + M^{-T}J(t_j)^T M^T) + |\lambda_i| + \vartheta|\lambda_i|]\Gamma_i'(t) \quad (13)$$

where $\vartheta = \exp(j\theta)$, $\theta \in [0, 2\pi]$, $j = \sqrt{-1}$. Since the relationship that $\|\exp[(E + \vartheta F)t]\| \leq \exp[\mu(E + \vartheta F)t] \leq \exp[\mu(E) + \|F\|)t]$ for $E, F \in R^{n \times n}$ and $\forall |\vartheta| = 1$ (please refer to Ref. [25] and Lemma 2 in Ref. [26]), we get

$$\begin{aligned}
\|\Gamma_i'(t)\| &= \|\exp(\int_{t_j}^t [\mu_\theta(MJ(t_j)M^{-1} + M^{-T}J(t_j)^T M^T) + |\lambda_i| + \vartheta|\lambda_i|]ds) \cdots \\
&\quad \times \exp(\int_{t_0}^{t_1} [\mu_\theta(MJ(t_0)M^{-1} + M^{-T}J(t_0)^T M^T) + |\lambda_i| + \vartheta|\lambda_i|]ds)V_i(0)\| \\
&\leq \exp(\int_{t_{j-1}}^t [\mu_\theta(MJ(t_j)M^{-1} + M^{-T}J(t_j)^T M^T) + 2|\lambda_i|]ds) \cdots \\
&\quad \times \exp(\int_{t_0}^{t_1} [\mu_\theta(MJ(t_0)M^{-1} + M^{-T}J(t_0)^T M^T) + 2|\lambda_i|]ds)V_i(0) \\
&= \exp(\int_{t_0}^t [\mu_\theta(MJ(t_j)M^{-1} + M^{-T}J(t_j)^T M^T) + 2|\lambda_i|]ds)V_i(0)
\end{aligned}$$

$$(14)$$

when $t \in [t_j, t_{j+1})$. From condition (10), $\int_{t_0}^\infty [\mu_\theta(MJ(t_j)M^{-1} + M^{-T}J(t_j)^T M^T) + 2|\lambda_i|]dt = -\infty$ holds. Hence we get $\lim_{t \to \infty} \Gamma_i'(t) = 0$, which means $\lim_{t \to \infty} \Gamma_i(t) = 0$ and $\lim_{t \to \infty} V_i(t) = 0$. This implies that the approximation system (9) can be asymptotically stable. $\qquad \square$

Note that there exists the term of $\frac{k^2}{\varsigma} + \varsigma$ in condition (10), and this can be approximatively zero if we choose two sufficiently small constants k and ς. In the following we show that this term is very useful for dealing with the approximation error between Eq. (6) and Eq. (9). From Theorem 1, we know that condition (10) only ensures the stability of Eq. (9), but it cannot ensure the stability of Eq. (6). Now we consider the stability condition for Eq. (6).

Theorem 2: Eq. (6) is asymptotically stable, if there exists a symmetric positive definite matrix $P = M^T M \in R^{n \times n}$, $\|M\| \neq 0$, such that

$$\int_{t_0}^\infty [\mu_\theta(MJ(t)M^{-1} + M^{-T}J(t)^T M^T) + (\frac{k^2}{\varsigma} + \varsigma) + 2|\lambda_i|]dt = -\infty \quad (15)$$

Proof: Obviously, the following relationships hold:

$$\begin{aligned}
&\int_{t_0}^t [\mu_\theta(MJ(t)M^{-1} + M^{-T}J(t)^T M^T) + (\frac{k^2}{\varsigma} + \varsigma) + 2|\lambda_i|]dt \\
&= \lim_{\tau_0 \to 0} \{\sum_{j \geq 0}^{n-1} \int_{t_j}^{t_{j+1}} [\mu_\theta(MJ(t_j)M^{-1} + M^{-T}J(t_j)^T M^T) + (\frac{k^2}{\varsigma} + \varsigma) + 2|\lambda_i|]dt \\
&\quad + \int_{t_n}^t [\mu_\theta(MJ(t_n)M^{-1} + M^{-T}J(t_n)^T M^T) + (\frac{k^2}{\varsigma} + \varsigma) + 2|\lambda_i|]dt\} \\
&= \lim_{\tau_0 \to 0} \int_{t_0}^t [\mu_\theta(MJ(t_j)M^{-1} + M^{-T}J(t_j)^T M^T) + (\frac{k^2}{\varsigma} + \varsigma) + 2|\lambda_i|]d\vartheta
\end{aligned}$$

when $t \in [t_n, t_{n+1})$, and

$$\begin{aligned}
&\int_{t_0}^\infty [\mu_\theta(MJ(t)M^{-1} + M^{-T}J(t)^T M^T) + (\frac{k^2}{\varsigma} + \varsigma) + 2|\lambda_i|]dt \\
&= \lim_{\tau_0 \to 0} \int_{t_0}^\infty [\mu_\theta(MJ(t_j)M^{-1} + M^{-T}J(t_j)^T M^T) + (\frac{k^2}{\varsigma} + \varsigma) + 2|\lambda_i|]dt
\end{aligned}$$

This implies that, for arbitrary small positive constant ε, there exists a constant $\delta_1 > 0$ such that

$$| \int_{t_0}^{\infty} [\mu_\theta (MJ(t)M^{-1} + M^{-T}J(t)^T M^T) + (\frac{k^2}{\varsigma} + \varsigma) + 2|\lambda_i|] dt$$
$$- \int_{t_0}^{\infty} [\mu_\theta (MJ(t_j)M^{-1} + M^{-T}J(t_j)^T M^T) + (\frac{k^2}{\varsigma} + \varsigma) + 2|\lambda_i|] dt| < \varepsilon$$

(16)

if $0 < \tau_0 < \delta_1$. From Eq. (14), we have

$$\lim_{t \to \infty} ||\Gamma_i'(t)|| \le \exp(\int_{t_0}^{\infty} [\mu_\theta (MJ(t_j)M^{-1} + M^{-T}J(t_j)^T M^T) + (\frac{k^2}{\varsigma} + \varsigma) + 2|\lambda_i|] dt (V_i(0)$$
$$\le \exp(\int_{t_0}^{\infty} [\mu_\theta (MJ(t)M^{-1} + M^{-T}J(t)^T M^T) + (\frac{k^2}{\varsigma} + \varsigma) + 2|\lambda_i|] dt)(\exp(\varepsilon)V_i(0)$$

(17)

It means that condition (15) can be also one sufficient condition for the asymptotical stability of Eq. (9) if the sample time τ_0 satisfies $0 < \tau_0 < \delta_1$.

Now we prove the stability of Eq. (6) if condition (15) is satisfied. For Eq. (6), if the sample time τ_0 is sufficiently small, we obtain

$$\dot{w}_i(t) = J(t_j)w_i(t) + \lambda_i \Gamma w_i(t - \tau) + O(t, t_j, \tau_0)w_i(t)$$

(18)

where $O(t, t_j, \tau_0) = J(t) - J(t_j)$. Since $O(t, t_j, \tau_0) = \frac{dJ(t_j)}{dt}(t - t_j)$ and the assumption that $\frac{dJ(t_j)}{dt}$ is bounded for all time, we obtain $\lim_{\tau_0 \to 0} ||O(t, t_j, \tau_0)|| = 0$. Therefore, for the constant k, there exists a constant δ_2 satisfying $\delta_1 > \delta_2 > 0$ such that $-kI_n < O(t, t_j, \tau_0) < kI_n$ for $0 < \tau_0 < \delta_2$. From the function $V_i(t)$ given by Eq. (8), we get $\dot{V}_i(t) \le w_i(t)^T M^T [MJ(t_j)M^{-1} + M^{-T}J(t_j)^T M^T + (\frac{k^2}{\varsigma} + \varsigma)]Mw_i(t) + |\lambda_i| \cdot [V_i(t) + V_i(t - \tau)]$ since $2w_i^T(t)O^T(t, t_j, \tau_0)M^T Mw_i(t) \le (\frac{k^2}{\varsigma} + \varsigma)w_i(t)^T M^T Mw_i(t)$. Similar to the proof procedure in Theorem 1, we conclude that $\lim_{t \to \infty} V_i(t) = 0$ if condition (10) is satisfied for $0 < \tau_0 < \delta_2$. This means that Eq. (18), namely Eq. (6), is asymptotically stable for $0 < \tau_0 < \delta_2$ provided that condition (10) holds. Further, from Ineqs. (16,17), we conclude that condition (15) is also a sufficient condition for the stability of Eq. (6).   □

We have several remarks.

Remark 1: From Ref. [15], the stability of Eq. (6) can be analyzed by the Krasovskii-Lyapunov function (7), and a general condition is $PJ(t) + J^T(t)P + Q + \lambda_N^2 c^2 PAQ^{-1}A^T P < 0$ for all time t. Let $P = M^T M$ with $||M|| \ne 0$, and we get $MJ(t)M^{-1} + M^{-T}J^T(t)M^T < -M^{-T}QM^{-1} - \lambda_N^2 c^2 MAQ^{-1}A^T M^T < 0$ for all time. Obviously, this is too strict for all time, and this can not be applied to the case where $MJ(t)M^{-1} + M^{-T}J^T(t)M^T$ is larger than zero during certain time intervals. In this paper condition (15) does not require the condition that $MJ(t)M^{-1} + M^{-T}J^T(t)M^T < 0$ for all time. In this sense condition (15) is less restrictive than Theorem 2 in Ref. [15]. For the case of $s(t) = (1/N) \sum_{k=1}^{N} x_k(t)$, we can also give one less restrictive condition for synchronization in network (1) with coupling delays than Theorem 1 in Ref. [17].

Remark 2: Based on the above idea, we can consider the case where the inner coupling $\Gamma(t) = \text{diag}\{r_1(t), \cdots, r_N(t)\}$ is continuously time-varying. If $\Gamma(t)$ is independent of the node dynamics $x_i(t)$, the stability of the synchronized state is equivalent to the stability of the linear systems $\dot{w}_i(t) = J(t)w_i(t) + \lambda_i \Gamma(t)w_i(t-\tau)$ for $2 \leq i \leq N$. Similar to Theorems 1 and 2, one sufficient stability condition is given as follows

$$\int_{t_0}^{\infty} [\mu_\theta(MJ(t)M^{-1} + M^{-T}J(t)^T M^T) + (\frac{k^2}{\varsigma} + \varsigma) + |\lambda_i|(1 + \frac{||M\Gamma(t)||^2}{\lambda_{\min}(M^T M)})]dt = -\infty \tag{19}$$

where $P = M^T M \in R^{n \times n}$, $||M|| \neq 0$, is a n-dimensional symmetric positive definite matrix, and $\lambda_{\min}(M^T M)$ is the minimum eigenvalue of matrix $M^T M$.

Remark 3: We can also consider the case where a coupling delay occurs when the signals from each of the nodes are transmitted to interconnected nodes. In this case the dynamics of the network is given by $\dot{x}_i(t) = f(x_i(t)) + c\sum_{j \neq i} g_{ij}\Gamma(x_j(t-\tau) - x_i(t))$ where c is the coupling strength. Let $g_i = \sum_{j \neq i} g_{ij}$. Under the condition of $g_1 = g_2 = \cdots = g_N = g$, the synchronized state is given by $\dot{s}(t) = f(s(t)) + cg\Gamma(s(t-\tau) - s(t))$. From Ref. [16], the stability of the synchronized state $x_1(t) = \cdots = x_N(t) = s(t)$ can be transformed into the stability of the following linear systems $\frac{d}{dt}(\varphi(t)) = (J(t) - cg\Gamma)\varphi(t) + c(\lambda_i + g)\Gamma\varphi(t-\tau)$ for $2 \leq i \leq N$. Similar to Theorems 1 and 2, we can obtain the following sufficient synchronization condition

$$\int_{t_0}^{\infty} [\mu_\theta(M(J(t) - cg\Gamma)M^{-1} + M^{-T}(J(t) - cg\Gamma)^T M^T) + (\frac{k^2}{\varsigma} + \varsigma) \\ + 2|c(\lambda_i + g)|]dt = -\infty \tag{20}$$

where $P = M^T M$, $||M|| \neq 0$, is a n-dimensional symmetric positive definite matrix. Compared with Theorem 1 in Ref. [16], condition (20) is also less restrictive.

Remark 4: Now we extend the procedure in Theorems 1 and 2 to the stability of the n-dimensional time-varying linear systems

$$\dot{x}(t) = A(t)x(t) + B(t)x(t-\tau) \tag{21}$$

where $A(t)$ and $B(t)$ are continuously time-varying, and are independent of the dynamics $x(t)$. Similar to Theorems 1 and 2, we can obtain one asymptotical stability condition

$$\int_{t_0}^{\infty} [\mu_\theta(M(A(t)M^{-1} + M^{-T}A(t)M^T) + (\frac{k^2}{\varsigma} + \varsigma) + (1 + \frac{||MB(t)||^2}{\lambda_{\min}(M^T M)})]dt = -\infty \tag{22}$$

where $P = M^T M$, $||M|| \neq 0$, is a n-dimensional symmetric positive definite matrix. Similar to the analysis in Remark 1, condition (22) is less restrictive than conditions from the Krasovskii-Lyapunov theory.

4   Synchronization in Complex Networks with Asymmetric Topology

If the topology in network (1) is symmetric, criteria (10,15) are not be applicable since eigenvalues of G may have the non-zero imaginary part. Hence we further analyze the synchronization criteria for complex networks with asymmetric topology. Note that the procedure developed in this section can be also applied to networks with symmetric topology.

Network (1) can be rewritten in an equivalent form

$$\dot{X}(t) = F(X(t)) + (G \otimes \Gamma)X(t - \tau) \tag{23}$$

where '$\otimes$' is the Kronecker product, $F(X(t)) = (f(x_1(t))^T, \cdots, f(x_N(t))^T)^T$, and $X(t) = (x_1^T(t), \cdots, x_N^T(t))^T$. By choosing a suitable continuously time-varying matrix $K(t) \in R^{n \times n}$, Eq. (23) is equivalent to the following system

$$\dot{X}(t) = F'(X(t)) - (I_N \otimes K(t))X(t) + (G \otimes \Gamma)X(t - \tau) \tag{24}$$

where $F'(X(t)) = ((f(x_1(t)) + K(t)x_1(t))^T, \cdots, (f(x_N(t)) + K(t)x_N(t))^T)^T$.

Let $\eta_j(t) = x_{j+1}(t) - x_1(t)$ for $1 \leq j \leq N-1$, and $\eta(t) = (\eta_1^T(t), \cdots, \eta_{N-1}^T(t))^T$. Then we get

$$\dot{\eta}(t) = \bar{F}(X(t)) - (I_{N-1} \otimes K(t))\eta(t) + (S_G \otimes \Gamma)\eta(t - \tau) \tag{25}$$

where $\bar{F}(X(t)) = ((f(x_2(t)) + K(t)x_2(t) - f(x_1(t)) - K(t)x_1(t))^T, \cdots, (f(x_N(t)) + K(t)x_N(t) - f(x_1(t)) - K(t)x_1(t))^T)^T$, and S_G is described by

$$S_G = \begin{bmatrix} -g_{12} - \sum_{j \neq 2} g_{2j} & g_{23} - g_{13} & \cdots & g_{2N} - g_{1N} \\ g_{32} - g_{12} & -g_{13} - \sum_{j \neq 3} g_{3j} & \cdots & g_{3N} - g_{1N} \\ \vdots & \vdots & \ddots & \vdots \\ g_{N2} - g_{12} & g_{N3} - g_{13} & \cdots & -g_{1N} - \sum_{j \neq N} g_{Nj} \end{bmatrix} \tag{26}$$

The above procedure can be seen in Refs. [10,14]. The procedure given by Eqs. (25,26) is very useful for dealing with the synchronization in networks without coupling delays, and the derived synchronization criteria are less restrictive than many exiting synchronization criteria [14]. In this paper we also utilize the procedure to consider the synchronization in networks with coupling delays.

Suppose that the feedback gain $K(t)$ is not affected by the node dynamics $x_i(t)$. Applying the segmentation strategy developed in the previous section, Eq. (25) can be approximated by the following system

$$\dot{\eta}(t) = \bar{F}(X(t)) - (I_{N-1} \otimes K(t_j))\eta(t) + (S_G \otimes \Gamma)\eta(t - \tau) \tag{27}$$

within the interval $[t_j, t_{j+1})$. Further, stability conditions for Eqs. (25,27) are stated as follows:

Theorem 3: Let $K(t)$ be a suitable feedback gain such that $f(x(t)) + K(t)x(t)$ is V-uniformly decreasing for a symmetric positive definite matrix $V \in R^{n \times n}$. Eq. (27) is asymptotically stable if there exists a positive definite matrix $U = \mathrm{diag}(u_1, \cdots, u_{N-1})$ such that

$$\int_0^{+\infty} [\mu_\theta(-M(I_{N-1} \otimes K(t_j))M^{-1} - M^{-T}(I_{N-1} \otimes K(t_j))^T M^T) \\ + (\frac{k^2}{\varsigma} + \varsigma) + (1 + \frac{\|M(S_G \otimes \Gamma_2)\|}{\lambda_{\min}(M^T M)}) - \frac{c_0 u_0}{\lambda_{\max}(M^T M)}] dt = -\infty \tag{28}$$

where $U \otimes V = M^T M$, $M \in R^{n(N-1) \times n(N-1)}$ is nonsingular, $u_0 = \min\{u_1, \cdots, u_{N-1}\}$, c_0 is a positive constant, and $\lambda_{\max}$ stands for the maximal eigenvalue of $M^T M$.

Proof: We choose a positively-defined function $V_0(t) = \eta^T(t)(U \otimes V)\eta(t)$. Hence its derivative along with the trajectory of Eq. (27) is $\dot{V}_0(t) = 2\eta^T(t)(U \otimes V)\bar{F}(X(t)) + 2\eta^T(t)(U \otimes V)(S_G \otimes \Gamma)\eta(t - \tau) + \eta^T(t)((U \otimes V)(-(I_{N-1} \otimes K(t_j)) - (I_{N-1} \otimes K(t_j))^T(U \otimes V))\eta(t)$. From the V-uniformly decreasing property of $f + K$ [11,12], the first term is of the form $2\eta^T(t)(U \otimes V)\bar{F}(X(t)) \leq -c_0 \sum_{j=2}^{N} u_{j-1} \|x_j - x_1\|^2 \leq -\frac{c_0 u_0}{\lambda_{\max}(M^T M)} V_0(t)$. The second term satisfies $2\eta^T(t)(U \otimes V)(S_G \otimes \Gamma)\eta(t - \tau) \leq V_0(t) + \frac{\|M(S_G \otimes \Gamma)\|}{\lambda_{\min}(M^T M)} V_0(t - \tau)$. The third term satisfies $\eta^T(t)((U \otimes V)(-I_{N-1} \otimes K(t_j)) - (I_{N-1} \otimes K(t_j))^T(U \otimes V))\eta(t) \leq \mu_\theta(-M(I_{N-1} \otimes K(t_j))M^{-1} - M^{-T}(I_{N-1} \otimes K(t_j))^T M^T)V_0(t)$. Similar to the proof of Theorem 1, condition (28) is a sufficient stability condition for the approximation system (27). □

Theorem 4: Assume that $K(t)$ and V satisfy Theorem 3. Eq. (25) is asymptotically stable if there exists a positive definite matrix $U = \text{diag}(u_1, \cdots, u_{N-1})$ such that

$$\int_0^{+\infty} [\mu_\theta(-M(I_{N-1} \otimes K(t))M^{-1} - M^{-T}(I_{N-1} \otimes K(t))^T M^T) + (\frac{k^2}{\varsigma} + \varsigma) + (1 + \frac{\|M(S_G \otimes \Gamma_2)\|}{\lambda_{\min}(M^T M)}) - \frac{c_0 u_0}{\lambda_{\max}(M^T M)}] dt = -\infty \tag{29}$$

where $U \otimes V = M^T M$, $\|M\| \neq 0$, $u_0 = \min\{u_1, \cdots, u_{N-1}\}$, c_0 is a positive constant, and $\lambda_{\max}$ stands for the maximal eigenvalue of $M^T M$.

Proof: This can be easily by the procedure in Theorems 1, 2 and 3. □

Remark 5: Theorems 3 and 4 do not require the linearization strategy (please see Eq.(6)). Moreover, conditions (28,29) can be regarded as global synchronization criteria. If the feedback gain matrix $K(t)$ is chosen as a constant matrix K_0, similar to the proof procedure in Theorem 3, we get $\dot{V}_0(t) \leq (\mu_\theta(-M(I_{N-1} \otimes K_0))M^{-1} - M^{-T}(I_{N-1} \otimes K_0)^T M^T) + 1 - \frac{c_0 u_0}{\lambda_{\max}(M^T M)})V_0(t) + \frac{\|M(S_G \otimes \Gamma)\|}{\lambda_{\min}(M^T M)} V_0(t - \tau)$. From Lemma 1 and Ref. [23], one sufficient stability condition for the case of the time-invariant feedback gain K_0 is

$$(\mu_\theta(-M(I_{N-1} \otimes K_0))M^{-1} - M^{-T}(I_{N-1} \otimes K_0)^T M^T) + 1 - \frac{c_0 u_0}{\lambda_{\max}(M^T M)}) + \frac{\|M(S_G \otimes \Gamma)\|}{\lambda_{\min}(M^T M)} < 0 \tag{30}$$

Remark 6: The idea in Theorems 3 and 4 can be also extended to the synchronization in networks, whose topology $G(t)$ is time-varying. Let $G(t) = (g_{ij}(t))$ have the same definition as matrix G in the network (1) at the t instant. Suppose that the topology $G(t)$ is continuously time-varying, and it is not affected by the node dynamics $x_i(t)$. Inspired by Theorems 3 and 4, we also obtain one sufficient synchronization condition

$$\int_0^{+\infty} [\mu_\theta(-M(I_{N-1} \otimes K(t))M^{-1} - M^{-T}(I_{N-1} \otimes K(t))^T M^T)$$
$$+(\tfrac{k^2}{\varsigma} + \varsigma) + (1 + \tfrac{\|M(S_G(t) \otimes \Gamma_2)\|}{\lambda_{\min}(M^T M)}) - \tfrac{c_0 u_0}{\lambda_{\max}(M^T M)}]dt = -\infty \qquad (31)$$

where $S_G(t)$ has the same structure as Eq. (26) at the t instant.

Remark 7: We further consider the synchronization in the network given by

$$\dot{x}_i(t) = f(x_i(t)) + \sum_{j \neq i} g_{ij}(t)\Gamma_1(x_j(t) - x_i(t)) + \sum_{j \neq i} g_{\tau,ij}(t)\Gamma_2(x_j(t-\tau) - x_i(t-\tau))$$
$$(32)$$

where Γ_i ($i = 1, 2$) are the inner coupling matrices. Let $G(t) = (g_{ij}(t))$ and $G_\tau(t) = (g_{\tau,ij}(t))$ have same definition as matrix G in the network (1) at the t instant. Similar to Theorems 3 and 4, one sufficient synchronization condition for network (32) is

$$\int_0^{+\infty} [\mu_\theta(M(S_G(t) \otimes \Gamma_1 - I_{N-1} \otimes K(t))M^{-1}$$
$$+ M^{-T}(S_G(t) \otimes \Gamma_1 - I_{N-1} \otimes K(t))^T M^T)$$
$$+(\tfrac{k^2}{\varsigma} + \varsigma) + (1 + \tfrac{\|M(S_{G_\tau}(t) \otimes \Gamma_2)\|}{\lambda_{\min}(M^T M)}) - \tfrac{c_0 u_0}{\lambda_{\max}(M^T M)}]dt = -\infty \qquad (33)$$

Further, Theorem 4 can be also generalized to network (32) with continuous time-varying inner coupling matrices $\Gamma_1(t)$ and $\Gamma_2(t)$. From Eq. (2) in the work [12], one necessary condition for the synchronization in network (32) is $(U \otimes V)(G(t) \otimes \Gamma_1(t) - I_n \otimes K) \leq 0$ for all time. Similar to the above discussion, this condition is too strict, and condition (33) for the coupling matrices $\Gamma_1(t)$ and $\Gamma_2(t)$ is less restrictive.

5   Numerical Simulations

In this section we verify the effectiveness of the proposed synchronization criteria by using a three-dimensional system as a node in network (1). Each individual node is described by $\dot{x}_1(t) = (-1 + 1.5\sin(t))x_1(t)$, $\dot{x}_2(t) = -3x_2(t)$, $\dot{x}_3(t) = -3x_3(t)$. Further, its Jacobian is $J(t) = \text{diag}\{-1 + 1.5\sin(t), -3, -3\}$. To begin with, we prove the stability of the above system at its zero solution. From the proofs of Theorems 1 and 2, the stability condition for the isolated node is $\int_{t_0}^{\infty}[\mu_1(J(t) + J(t)^T) + (\tfrac{k^2}{\varsigma} + \varsigma)]dt = -\infty$ for $M = I_3$, and arbitrary small positive constants ς and k. If we choose $\tfrac{k^2}{\varsigma} + \varsigma = 1$, $\int_{t_0}^{\infty}[\mu_1(J(t) + J(t)^T) + 1]dt = \int_{t_0}^{\infty}[-2 + 3\sin(t) + 1]dt = \int_{t_0}^{\infty}[-1]dt + \int_{t_0}^{\infty}[3\sin(t)]dt = -\infty$. Therefore the isolated node can be asymptotically stable at its zero solution.

In this section the star-type coupled network is chosen to be the simulated network. In this network, only one node is a center node with degree $N - 1$, and all the other nodes with degree 1 are connected to this center node. Suppose that all nodes are connected by their first states, namely $\Gamma = \text{diag}\{1, 0, 0\}$. In this case the coupling matrix is

$$G = c \begin{bmatrix} -1 \cdots & 0 & 1 \\ \vdots & \ddots & \vdots & \vdots \\ 0 \cdots & -1 & 1 \\ 1 \cdots & 1 & -(N-1) \end{bmatrix}$$

where c is a positive constant coupling. Let the number of nodes $N = 10$ and $c = 0.08$. Obviously, two different nonzero eigenvalues of G are $\lambda_1 = -0.08$ and $\lambda_2 = -0.8$. From Theorems 1 and 2, the synchronization condition for arbitrary delay time τ is $\int_{t_0}^{\infty} [\mu_1(J(t) + J(t)^T) + (\frac{k^2}{\varsigma} + \varsigma) + |\lambda_i|]dt = -\infty$, which can be easily verified by the the above analysis. Simulations results with respect to three states $x_1(t)$ (the dashed lines), $x_2(t)$ (the solid lines), and $x_3(t)$ (the dashdot lines) are plotted in Figure 1 for the delayed time $\tau = 2$. From this figure, the network can be asymptotically stabilized at the zero solution of the isolated node. Since $\mu_1(J(t) + J(t)^T) = -2 + 3\sin(t)$ are larger than zero during certain time intervals, the Krasovskii-Lyapunov theory can not be successfully used to analyze the stability of the network.

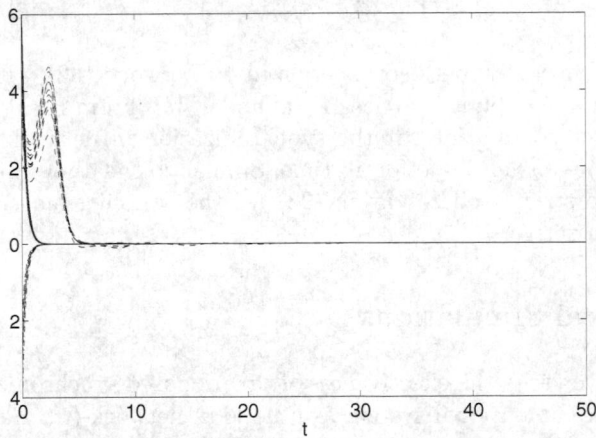

Fig. 1. The history curves of states $x_i(t)$ ($i = 1, 2, 3$)

6   Conclusion

In this paper we propose some novel synchronization criteria in complex networks with coupling delays, in which the topologies in networks can be symmetric and asymmetric. Compared with synchronization criteria resulting from the Krasovskii-Lyapunov theory, the proposed synchronization criteria are less restrictive.

Acknowledgments. Yun Shang thanks the partial support by NSFC projects (No. 60736011 and No. 60603002) and 863 project (No. 2007AA01Z325); Maoyin Chen thanks the partial support by NSFC project (No. 60804046), Special Doctoral Fund in University by Ministry of Education (No. 20070003129) and the Alexander von Humboldt Foundation, Germany.

References

1. Albert, R., Barabasi, A.L.: Statistical mechanics of complex networks. Rev. Mod. Phys. 74, 47–91 (2002)
2. Boccaletti, S., Latora, V., Moreno, Y., et al.: Complex networks: structure and dynamics. Phys. Repor. 424, 175–308 (2006)
3. Watts, D.J., Strogatz, S.H.: Collective dynamics of 'small world' networks. Nature 393, 440–442 (1998)
4. Barabasi, A.L., Albert, R.: Emergence of scaling in random networks. Science 286, 509–512 (1999)
5. Pecora, L.M., Carroll, T.L.: Master stability functions for synchronized coupled systems. Phys. Rev. Lett. 80, 2109 (1998)
6. Wang, X.F., Chen, G.: Synchronization in scale-free dynamical networks: robustness and fragility. IEEE Trans. Circuits Syst. I 49, 54–62 (2002)
7. Lü, J., Yu, X., Chen, G., et al.: Characterizing the synchronizability of small-world dynamical networks. IEEE Trans. Circuits Syst. I 51, 787–796 (2004)
8. Lü, J., Chen, G.: A time-varying complex dynamical network model and its controlled synchronization criteria. IEEE. Trans. Auto. Contr. 50, 841–846 (2005)
9. Lü, J., Yu, X., Chen, G.: Chaos synchronization of genearl complex dynamical networks. Physica A 334, 281–302 (2004)
10. Stefanski, A., Wojewoda, J., Kapitaniak, et al.: Simple estimation of synchronization threshold in ensembles of diffusively coupled chaotic systems. Phys. Rev. E 70, 026217 (2004)
11. Wu, C.W.: Synchronization in coupled arrays of chaotic oscillators with nonreciprocal coupling. IEEE Trans. Circuits Syst. I 50, 294–297 (2003)
12. Wu, C.W.: Synchronization in arrays of coupled nonlinear systems with delay and nonreciprocal time-varying coupling. IEEE Trans. Circuits Syst. I 52, 282–286 (2005)
13. Chen, M.: Some simple synchronization criteria for complex dynamical networks. IEEE Trans. Circuits Syst. II 53, 1185–1189 (2006)
14. Chen, M.: Chaos synchronization in complex networks. IEEE Trans. Circuits Syst. I 55, 1335–1346 (2008)
15. Li, C., Chen, G.: Synchronization in general complex dynamical networks with coupling delays. Physica A 343, 263–278 (2004)
16. Lu, W., Chen, T., Chen, G.: Synchronization analysis of linearly coupled systems described by differential equations with a coupling delay. Physica D 221, 118–134 (2006)
17. Zhou, J., Chen, T.: Synchronization in general complex delayed dynamical networks. IEEE Trans. Circuits Syst. I 53, 733–744 (2006)
18. Kuang, Y.: Delay Differential Equations. Academic Press, London (1993)
19. Lakshmikantham, V., Leela, S.: Differential and Integral Inequalities. Academic Press, New York (1969)
20. Nicykescy, S.I.: Delay effects on stability: a robust control approach. Springer, London (2001)
21. Brierley, S.D., Chiasson, S.D., Lee, J.N., et al.: On stability independent of delay for linear systems. IEEE. Trans. Auto. Contr. 27, 252–254 (1982)
22. Hmamed, A.: Futher results on the robust stability of uncertain time-delay systems. Int. J. Systems Sci. 22, 605–614 (1991)

23. Mori, T.: Criteria for asymptotic stability of linear time-delay dystems. IEEE. Trans. Auto. Contr. 30, 158–161 (1985)
24. Mori, T., Kokame, H.: Stability of $\dot{x}(t)=Ax(t)+Bx(t\text{-}\tau)$. IEEE. Trans. Auto. Contr. 34, 460–463 (1989)
25. Lancaster, P.: Theory of matrices. Academic Press, New York (1969)
26. Chiou, J.S.: Stability analysis for a class of switched large-scale time-delay systems via time-switched method. IEE Pro. Contr. Theor. Appl. 153, 684–688 (2006)

MANIA: A Gene Network Reverse Algorithm for Compounds Mode-of-Action and Genes Interactions Inference

Darong Lai[1,2], Hongtao Lu[2], Mario Lauria[3], Diego di Bernardo[3], and Christine Nardini[1,*]

[1] CAS-MPG PICB, Yue Yuan Road 320, Shanghai, PRC
darong.lai@gmail.com, christine@picb.ac.cn
[2] Dept. of computer science and engineering, Shanghai Jiao Tong University, PRC
lu-ht@co.ojtu.cdu.cn
[3] TIGEM, 111 Via Pietro Castellino, Naples, Italy
{lauria,dibernardo}@tigem.it
www.picb.ac.cn/ClinicalGenomicNTW

Abstract. Understanding the complexity of the cellular machinery represents a grand challenge in molecular biology. To contribute to the deconvolution of this complexity, a novel inference algorithm based on linear ordinary differential equations is proposed, based on high-throughput gene expression data. The algorithm can infer (i) gene-gene interactions from steady state expression profiles *AND* (ii) mode-of-action of the components that can trigger changes in the system. Results demonstrate that the proposed algorithm can identify *both* information with high performances, thus overcoming the limitation of current algorithms that can infer reliably only one.

Keywords: gene network, gene expression, reverse engineering, Ordinary Differential Equations (ODE), compound mode-of-action.

1 Introduction

Thanks to the fast moving and recent advancements in technology, our society is assisting to an unprecedent high-throughput production of information coming from a variety of areas of human activity. This comprises, but is not limited to, economic, social and biological data. In particular, we focus our attention on biomolecular data. To deconvolute the structure underlying such data, cross fertilization from diverse areas of research, and notably the introduction of exact sciences in the realm of biology, has been a fundamental requirement to mine the complex interaction that explains the data we observe. However, the task is far from completed, and although economical, sociological and molecular systems own peculiar characteristics, advances in the deconvolution of the complexity in any of these areas bears the potential to significantly contribute to explain

---

* Corresponding author.

J. Zhou (Ed.): Complex 2009, Part I, LNICST 4, pp. 389–399, 2009.
© ICST Institute for Computer Sciences, Social Informatics and Telecommunications Engineering 2009

the complexity of the global system we live in. In the area of molecular biology several high-throughput platforms are quickly becoming available [1], however, gene expression data represents at the moment the most abundant source of molecular high-throughput information. This work focuses on the identification of networks of interaction among genes. Networks of interactions identify general relationships among the nodes of the network (genes), thus, a link in the network may not represent a physical interaction (carried on by intermediate molecules such as proteins). However, these algorithms can be extremely powerful in the initial characterization of unknown systems, taking advantage of low-cost, high-throughput screens and generating relevant *in silico* hypotheses that can be further and efficiently tested in *wet lab*. Moreover, these algorithms, thanks to their ability to reconstruct networks on genome-wide data, offer a systemic perspective of the interactions. Depending on the model adopted, these methods can infer a causality in the relationship (directed networks) or rather a simple 'connection' among items (undirected networks). Many methods [2] have been proposed to reverse-engineer gene expression data, that can either take advantage of the evolution in time of the state of the system (time series, i.e. [3]), or of different equilibrium states reached by the system (steady state, i.e. [4]). Our approach focuses on the latter, more abundant, steady state data. To achieve different equilibrium states of the system, the system is perturbed in different ways (i.e. knock-out, knock-down, alterations in the growing medium) and the resulting expression data is collected once the system has reached the novel equilibrium. Algorithms that handle these data typically output a representation of the gene network in the form of a graph or an adjacency matrix (here called A, [4,5,6]). These networks represent the relationships occurring among genes, and offer a first impression of the complex pathways that are being activated in the system under study. Alternatively these algorithms offer an estimation, for example in the form of a ranked list, of the genes that were directly affected by the perturbation in the experiments [7] (here called P). When the perturbation is obtained adding a compound in the environment of the cell, the genes identified by P represent the direct targets of the perturbing agents that have been used to alter the equilibrium. This identifies an information extremely valuable in areas such as *chemogenomics*, where the identification of a small molecule's direct target (also called *transcriptional perturbations*) can provide fundamental information on its use as a drug. Because of this, P is also known to represent the *mode-of-action* of the perturbing compound. So far, *a priori* knowledge of the direct targets of perturbation was required for a proper identification of A [4], or alternatively, the identification of an estimate of P was not able to produce a reliable representation of A [7], due to the high sensitivity of the algorithms to errors in P. With our novel approach we aim at the identification of both the gene network (A) and the single direct target matrix (P), overcoming the current limitation, while preserving and improving both performances. Our approach can handle efficiently experiments resulting in single transcriptional perturbations. Single transcriptional perturbations are useful to be quantified when the entity of a single gene knock-down is unknown and when the action of perturbagens is supposed to target predominantly an individual (unknown) transcript or protein, rather than several elements of a pathway. In the following we present related methods (Section 2), details of our algorithm (Section 3), validation results (Section 4) and their interpretation (Section 5).

2 Related Work

Number of approaches are being designed and tested to uncover the complexity of molecular interactions. In the following we briefly describe currently used tools based on Bayesian theory (Banjo, [6]), information science approach (ARACNe, [5]) and ordinary differential equations (ODE, NIR and MNI [4,7]), that have proven to be useful in the identification of gene networks or compounds mode-of-action. All the above methods can handle steady-state data. Banjo [6] generates a network space and screens then the best network structures attributing the most appropriate conditional density function, by optimization of an objective function (Bayesian Dirichlet equivalence, or Bayesian information criterion). Banjo can reconstruct signed directed network indicating regulation among genes, but it cannot infer networks involving cycles (or loops). ARACNe (Algorithm for Reconstruction of Accurate Cellular Networks, [5]) is regarded as an information-theoretic approach to gene network inference. It computes mutual information (MI, [8]) for all pairs of genes profiles to estimate the independence between genes and uses strategies (Data Processing Inequality, DPI) to successfully filter out the number of false-positive interactions. ARACNe cannot reconstruct directed networks. Our approach is strongly rooted in two ODE-based methods previously developed and validated. Namely, we used as a starting point NIR [4] able to infer the network of genes interactions (A), provided the perturbations (P) are known, and MNI [7], able to rank the most likely direct target genes of perturbations (estimate of P). Briefly, these algorithms aim at the identification of the function that describes the variation of gene expression matrix x over time $x' = f(x, p)$, with x representing the steady state expressions of the N genes involved in the network across M experiments, f is a non linear function that models how the expression values x and M experiments provoking external influences p modify the genes' activity. Assuming steady state and small perturbations, these equations can be linearized around the equilibrium state, and become, for a scalar element of the expression matrix $x'_{il} = \sum_j a_{ij} x_{jl} + p_{il} = \underline{a}_i^T \cdot \underline{x}_l + \underline{p}_{il}$ with $i, j = 1..N$ indicating genes and $l = 1..M$ experiments. In matricial form and at equilibrium this becomes $AX = -P$, with A being the N by N network matrix (a_{ij} represents the action of gene j on gene i), X the expression data (x_{il} represents the expression of gene i in experiment l) and P the matrix of transcriptional perturbations (p_{il} represents the transcriptional perturbation of gene i in experiment l), explaining the origin of our notations. These approaches assume that only a limited number of connections among genes are possible, to reflect the structure of the molecular pathways. Based on this sparsity assumption, NIR uses multiple linear regression to infer the connections among genes. Conversely, MNI is trained on the expression data in X to evaluate A and P

through an iterative process based on the minimization of an objective function (Sum of Square Errors, SSE).

3   Method

Our approach aims at identifying A and P based solely from the expression data X, thus overcoming the necessity to have *a priori* information on the direct target of the perturbation (P), which is very often an important unknown of the problem. To achieve this goal, we sought to chain the two algorithms in order to use the prediction of MNI to feed NIR and infer the network. To do so, our algorithm uses iteratively $M-1$ experiments in X to predict the M-th column (experiment) of P, as a ranked list of most likely targets. Due to the intrinsic noise of the data and the limited deterministic predictive power of MNI, the reliable identification of A, P is not trivial, especially when predicting complex data, as it can be shown in Section 4, in the varying performances of MNI+NIR, which represent the trivial chaining of the 2 algorithms (output of MNI used directly as P for NIR, see Figure 1(a)). For this reason, other strategies had to be integrated, schematically shown in Figure 1(b). Based on previous acronyms (and on the obsessive search for the network identification) we call this new approach Mode of Action & Network Identification Approach (MANIA).

In this approach, an estimate of P is produced by MNI, called P_{MNI}, this matrix contains the top ranking perturbations (we tested 1 and 10 top best, parameter $topP$), while all other values of P_{MNI} are set to zero. When choosing the single top perturbation option $(topP = 1)$, the algorithm should perform at its best, provided MNI reliably identifies the correct transcriptional perturbation as the most likely (i.e. the best prediction *is* indeed the gene target). In this case, in fact, no noise is added; however, we also tested the algorithm preserving the top 10 best predictions $(topP = 10)$, to offer backup solutions in case MNI is not able to find the correct perturbation as first choice. The core step of the algorithm consists of the strategy used to *clean* P_{MNI} from the incorrect

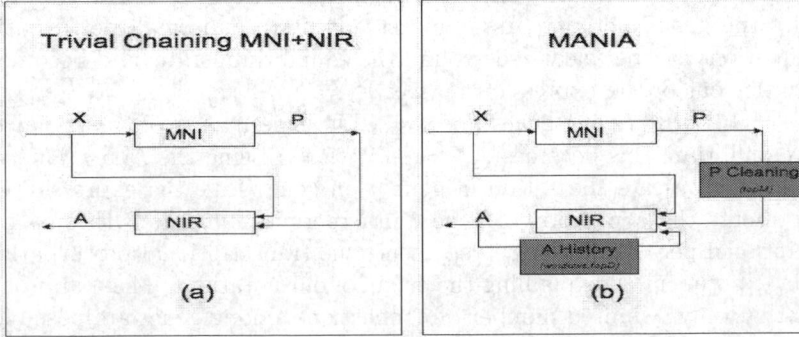

Fig. 1. Schematic view of the trivial chaining of MNI and NIR and of the strategies implemented in MANIA, discussed in Section 3

predictions, so that only the appropriate perturbation is used in NIR to predict A. This strategy consists of two steps. The first is the iterative computation of all the solutions for a given row of A, using all the predictions offered by P_{MNI}. The solutions are then ranked and only the $topM$ best are preserved (along with the corresponding perturbations) while computing the following rows. However, this step alone is not sufficient, since, often, the solution that minimizes the objective function (SSE) produces a local minimum of the objective function. Choosing this solution can result in the identification of a unique minimum for P and thus for all the rows of A. To overcome this issue, information about the previous rows computed in A are used. Thus, another parameter ($windows$) has been introduced to indicate the number of rows used as previous knowledge to calculate and minimize SSE. In particular, SSE is computed on all the $topM$ solutions as $X = -A_{tmp}^{-1}P$, where A_{tmp} is the identity matrix (self-relation is always assumed true) with the corresponding $windows$ rows replaced by the solutions already computed. By construction A is always invertible. For each row of A and P only the best $topM$ solutions are preserved before computing the following rows of A, P. In our simulations, we set $windows = 5$ and $topM = 200$. In our experience, these values represent a good compromise between computation time and accuracy. Pseudocode in Algorithm 1 gives more details about the process.

4   Experimental Results

To validate our approach, we used two known benchmark datasets (here called Dataset 1 [2] and Dataset 2 [9]), and compared our performances to state-of-the-art algorithms briefly summarized in Table 1. These algorithms were used with their default parameters values.

Dataset 1. This dataset consists of 20 instances of expression matrices X with 100 genes and 100 experiments, obtained from 20 instances of network matrices A with sparsity 10 (indicating a maximum of 10 possible interactions for each gene), and single perturbation for P (identity matrix). Gaussian noise (10%) is added to expression data to better mimic real data. Performances are computed using positive predictive value (PPv, also called $accuracy$) defined as $TP/(TP + FP)$ and Sensitivity $TP/(TP + FN)$, where TP, FP and FN stand for True Positive, False Positive and False Negative, respectively. Results were averaged,

Table 1. Network inference algorithms used for performances comparison

Software	Download link	Model
BANJO	www.cs.duke.edu/ amink/ software/banjo	Bayesian method
ARACNe	www.amdec-bioinfo. cu-genome.org/html	Information-theory method
MNI/NIR	http://dibernardo.tigem.it/ wiki/index.php	ODE based method

Input: gene expression profiles matrix X, user defined parameters
Output: Adjacency matrix A of gene network; mode-of-action matrix P
N:number of genes;
M: number of experiments;
$topP$: max number of perturbations proposed by MNI preserved in P_{MNI}
per experiment;
$topM$: max number of solutions preserved per each row of A;
$windows$: max number of previously computed row preserved;
for $i \leftarrow 1$ to M do
   Compute $P_{MNI}(:,i)$ with MNI from $X(:, [1..i-1, i+1..M])$;
end
Sort P_{MNI} columnwise in descending order $\rightarrow pIdx$ sorted index
matrix of P_{MNI};
$arrayA_{1:topM} \leftarrow$ NULL;
$arrayP_{1:topM} \leftarrow$ NULL;
for $i \leftarrow 1$ to N do
   for $j \leftarrow 1$ to $topP$ do
      Create perturbation matrix $Pmat$ in two steps:
      (a). $Pmat \leftarrow O$(zero-matrix);
      (b). $Pmat(pIdx(j,m),m) \leftarrow P_{MNI}(pIdx(j,m),m)$, where
      $1 \leq m \leq M$;
      Get j-th perturbation vector $P_i(j,:) = Pmat(j,:)$;
      for $k \leftarrow 1$ to $Total$ $number$ of all $combinations$ of non-$zero$ $perturbations$
      in $P_i(j,:),$ say P_{ijk} do
         compute i-th row of A, i.e. A_{ijk}, with NIR and perturbation
         P_{ijk};
         compute $A_{tmp,ijk}$ as identity matrix with $windows$ rows of
         $arrayA_h(1 \leq h \leq topM)$ and A_{ijk} as i-th row;
         compute $P_{tmp,ijk}$ as zeros matrix with previous $windows$ rows
         of $arrayP_h(1 \leq h \leq topM)$ and P_{ijk} as i-th row;
         compute $SSE_{ijk} = (X - A_{tmp,ijk}^{-1} \cdot P_{tmp,ijk})^2$;
      end
   end
   Rank SSE and select $topM$ A_{ijk} solutions;
   for $h \leftarrow topM$ do
      $arrayA_h = [arrayA_h(1:i-1,:); A_{tmp,h}(i,:)]$;
      $arrayP_h = [arrayP_h(1:i-1,:); P_{tmp,h}(i,:)]$;
   end
end
$A = arrayA_1$;
$P = arrayP_1$;

Algorithm 1. Pseudocode for MANIA. Matricial notations follow Matlab syntax:
$M(:,i)$ indicates column i in matrix M, $M(i,:)$ indicates row i in matrix M.

and proved to be stable with $st.dev < 0.08$ in all cases (standard deviations of PPv/sensitivity for undirected networks are 0.07/0.07, and 0.06/0.06 for directed networks).

Dataset 2. This dataset comes from the Dream 2 Competition (Heterozygous InSilico 1, Challenge 4) organized by the DREAM (Dialogue for Reverse Engineering Assessments and Methods, [9]) consortium, whose objective is to catalyze the interaction among researchers and improve progresses in the area of cellular network inference. Data was generated using simulations of biological interactions. Namely, the rate of synthesis of the mRNA of each gene is considered to be affected by the level of mRNA of other genes. For these reasons, this represents a valuable and challenging benchmark to test reverse engineering approaches. This dataset contains steady state levels of 50 genes of an hypothetical wild-type organism and 50 heterozygous knock-down strains. All ODE algorithms were tested assuming the number of connections associated with each gene (connectivity of the network, and sparsity of the matrix) is 10, including self-connection. Data were preprocessed with log-transformation of the expression ratio for each gene (knock-down vs wild-type strains). Ratios corresponding to null levels of expression in wild-type were treated as unknown values, and were set to zero as it was done in [4]. Standard deviation of each entry of the data matrix X was computed against the 25-nearest neighbors of the gene of interest, with the approach illustrated in [4]. Finally before computing A, the absolute value of P_{MNI} was normalized column-wise for numeric stability consideration, however this step does not affect the results. Besides evaluating PPv and Sensitivity (ROC curves), the adjacency matrix A was also scored following the procedure adopted in the DREAM 2 Challenge, after scoring the connections of A (normalizing the absolute values). Results were graded using the area under the curve (AUC) for ROC (false positive vs true positive rate) and precision-versus-recall curve (Prec vs Rec)) for the whole set of predictions. For the first

Table 2. Matrix A performance results of Dataset 1. *MNI+NIR* represents the trivial chaining of MNI and NIR, with no strategy to identify the best performances and keeping the single first best and 10 first best predictions of MNI (called respectively MNI+NIR1 and MNI+NIR10). The same values were used for *MANIA*. *Random* refers to the expected performances of an algorithm that selects pairs of genes randomly and then infers an edge between them.

	Directed		Undirected	
Algorithm	PPv	Sensitivity	PPv	Sensitivity
MNI+NIR1	**0.84**	**0.75**	**0.86**	**0.76**
MNI+NIR10	0.18	0.15	0.27	0.23
MANIA1	**0.89**	**0.81**	**0.95**	**0.81**
MANIA10	**0.75**	**0.68**	**0.79**	**0.70**
NIR	**0.96**	**0.86**	**0.97**	**0.87**
ARACNe	-	-	0.56	0.28
BANJO	0.42	-	0.71	0.00
Random	0.10	-	0.19	-

k predictions (ranked by score, and for predictions with the same score, taken in the order they were put in the prediction files), Precision was defined as the fraction of correct predictions to k, and Recall was the proportion of correct predictions out of all the possible true connections.

Figure 2 is the graphical version of the results of Table 3.

Table 4 and Figure 3 give the results produced by the algorithms for the performances on the directed network. Although MANIA always shows good

Table 3. Performance Results on Dataset 2 on A undirected network

| Algorithm | Precision at n^{th} Correct Prediction | | | | AUC | |
	1st	2nd	5th	20th	Prec vs Rec Curve	ROC Curve
MNI+NIR1	1.0000	1.0000	0.8333	0.4651	0.2859	0.6965
MNI+NIR10	0.0909	0.0609	0.1219	0.1639	0.1158	0.6230
MANIA1	**0.5000**	**0.4000**	**0.5556**	**0.5714**	**0.3513**	**0.7957**
MANIA10	**1.0000**	**1.0000**	**0.6250**	**0.5405**	**0.3014**	**0.7191**
NIR	**1.0000**	**1.0000**	**1.0000**	**1.0000**	**0.5968**	**0.8202**
ARACNe	1.0000	1.0000	0.5000	0.3279	0.2143	0.6658
BANJO	1.0000	0.3333	0.3125	0.4167	0.1900	0.5925

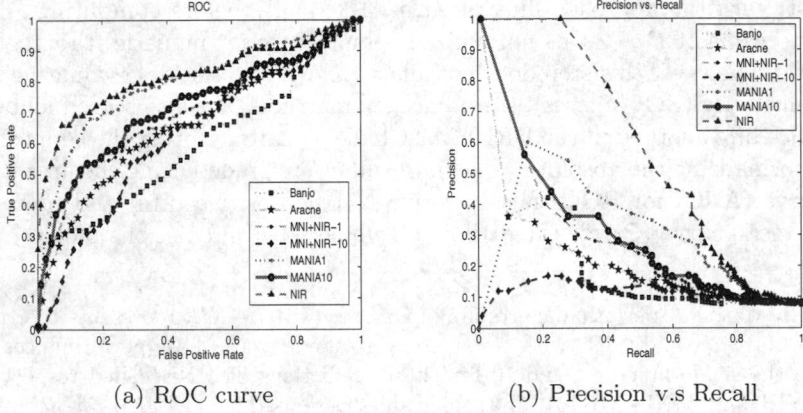

(a) ROC curve   (b) Precision v.s Recall

Fig. 2. Performances for A on Dataset 2. AUC Curves for undirected network.

Table 4. Performance Results on Dataset 2 on A directed network

| Algorithm | Precision at n^{th} Correct Prediction | | | | AUC | |
	1st	2nd	5th	20th	Prec. vs Rec Curve	ROC Curve
MNI+NIR1	0.5000	0.6667	0.5556	0.3279	0.1921	0.6999
MNI+NIR10	0.1429	0.2222	0.1724	0.1835	0.1107	0.6864
MANIA1	**0.5000**	**0.4000**	**0.6250**	**0.3846**	**0.2258**	**0.7877**
MANIA10	**1.0000**	**0.6667**	**0.5556**	**0.3448**	**0.2066**	**0.7232**
NIR	**1.0000**	**1.0000**	**1.0000**	**1.0000**	**0.5781**	**0.8314**
BANJO	0.5000	0.6667	0.2174	0.0559	0.0724	0.5441

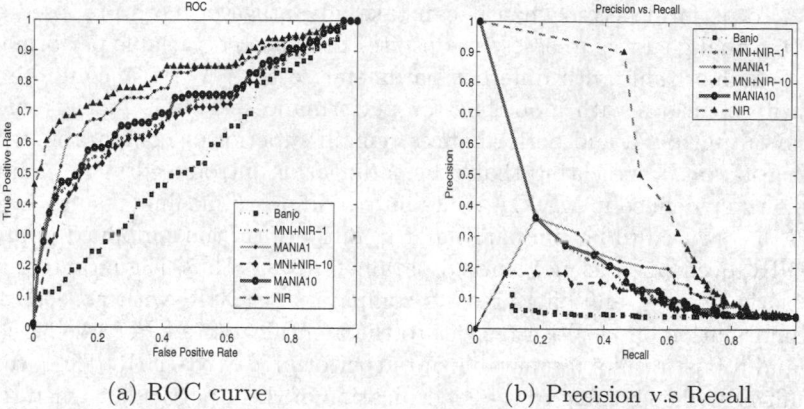

(a) ROC curve   (b) Precision v.s Recall

Fig. 3. Performances for A on Dataset 2. AUC Curves for directed network.

Table 5. Performance Results on Dataset 1 and Dataset 2 for the ODE-based algorithms that can predict P

	Dataset 1		Dataset 2	
	PPv	Sensitivity	PPv	Sensitivity
MNI+NIR1	0.870	0.870	0.540	0.540
MNI+NIR10	0.094	0.920	0.066	0.660
MANIA1	**0.950**	**0.860**	**0.750**	**0.540**
MANIA10	0.798	0.830	0.526	0.600

performances, in this test, interestingly, it also clearly outperforms other ODE-based algorithms. Possible reasons for this are discussed in Section 5.

5   Discussion

We have tested our approach against 4 different algorithms and across 2 datasets interpreted as directed and undirected networks. In general, MANIA can perform better than the state-of-the-art non-ODE approaches listed in Table 1, which were used by setting parameters to their default values, and comparably well or superiorly to ODE approaches as NIR or MNI+NIR. Our objective was to infer A with performances as close as possible to NIR, which we considered as our gold standard [2]. Before discussing further the performances, it is worth noting that comparison between MNI+NIR and MANIA were done with the purpose to assess the validity of the enhancements proposed, compared with our simpler idea of directly chaining the two approaches (MNI and NIR). We figured that, given the reasonably good performances of MNI on the identification of one perturbation, MNI+NIR would be advantaged when used with the parameter $topP = 1$, which offers to NIR the best possible P. In order to perform a fair comparison, MANIA was also tested with this value of the parameter, however, we expected

MANIA to perform better when it can take advantage of more proposed solutions. Our final goal was to assess if, despite the expected variable performances of the two algorithms with different parameter setting, MANIA could be able to identify solutions with global better performances. This is indeed true for the final output of A, and performances remain superior or comparable for the identification of P, indicating that the modification introduced in the multiple regression step defined in MANIA contribute to improve the final results. Overall MANIA has proved to be comparable or to outperform the simplified approach MNI+NIR, even when using identical parameter $topP$, thus, it guarantees more stable performances, and offers results comparable to NIR, with no need for a priori information on P. With respect to the identification of P, MANIA shows stable and robust results comparable or outperforming MNI+NIR. These results are confirmed when coming to the identification of A. In particular, when tested against simple models for simulations (Dataset 1) the performances of MANIA and MNI+NIR are inferior to NIR, however they do not degrade much and both algorithms have the fundamental advantage to infer A from expression data only. In the validation on Dataset 2 a more complex and realistic model, these trends are confirmed, with even less variance in the performances, and highlighting the superiority of MANIA. This supports the introduction of the algorithmic variations peculiar to MANIA. Since NIR was the best algorithm tested on this dataset in the Challenge DREAM 2, the possibility to preserve or degrade little the performances under more difficult conditions, represents an important achievement. In general, compared to the best performing algorithm NIR, MANIA has comparable performances in accuracy and the great advantage of not requiring a priori knowledge on the targets of the perturbations. Compared to NIR+MNI it has higher ability to remove the noise in matrix P, a characteristic that becomes more and more crucial when the network represents more complex interactions. Although at this level of analysis this is only speculation, it is quite reasonable to assume that real network are indeed complex ones. At the other hand, the lack of a priori information about P makes MANIA need to search for an optimal P in a space spanned by P_{MNI}, which increases its computation effort when compared to NIR or MNI+NIR. From the Algorithm in section 3, the main time cost is in the line of the computation of SSE involving inversion which takes $O(N^3)$, thus MANIA scales as $O(N^4)$. Therefore MANIA can infer networks with up to about thousands of vertices in reasonable time. However, MANIA is easily parallelized (row-wise computations of A), and can thus handle larger networks in reasonable time.

6   Conclusion

In this paper, a new reverse-engineering algorithm (MANIA) has been proposed, which effectively couples two assessed approaches MNI and NIR and overcomes their limitations, while preserving their performances. In our simulation experiments, we have shown that MANIA can identify the network of interactions among genes from steady state experiments provided single perturbations are causing the expression variations. This covers several applications, like single gene knock-down or systematic small molecules testing [10], when assuming the perturbation affects a single target. These are two widely used experimental approaches with applications in chemo- and pharmaco-genomics and model organism research. Although MANIA performed encouragingly, it is worth noting that there is only one single gene perturbed in each experiment in the system under study. Our current work consists in the identification of a proper heuristic for extending this application to multiple perturbation targets and apply the validation to a larger variety of cases.

Acknowledgments. Darong Lai and Hongtao Lu are supported by the Specialized Research Fund for the Doctoral Program of Higher Education (SRFDP, No. 20050248048) and the Program for New Century Excellent Talents in University (NCET-05-0397).

References

1. Guiducci, C., Nardini, C.: High parallelism, portability and broad accessibility: Technologies for genomics. ACM J 4(1) Article 3 (2008)
2. Bansal, M., Belcastro, V., Ambesi-Impiombato, A., di Bernardo, D.: How to infer gene networks from expression profiles. Mol. Syst. Biol. 3 (2007)
3. Bansal, M., Gatta, G.D., di Bernardo, D.: Inference of gene regulatory networks and compound mode of action from time course gene expression profiles. Bioinformatics 22(7), 815–822 (2006)
4. Gardner, T.S., di Bernardo, D., Lorenz, D., Collins, J.J.: Inferring genetic networks and identifying compound mode of action via expression profiling. Science 301(5629), 102–105 (2003)
5. Margolin, A.A., Nemenman, I., Basso, K., Klein, U., Wiggins, C., Stolovitzky, G., Favera, R.D., Califano, A.: Aracne: An algorithm for the reconstruction of gene regulatory networks in a mammalian cellular context (2004)
6. Yu, J., Smith, V.A., Wang, P.P., Hartemink, A.J., Jarvis, E.D.: Advances to bayesian network inference for generating causal networks from observational biological data. Bioinformatics 20(18), 3594–3603 (2004)
7. di Bernardo, D., Thompson, M.J., Gardner, T.S., Chobot, S.E., Eastwood, E.L., Wojtovich, A.P., Elliott, S.J., Schaus, S.E., Collins, J.J.: Chemogenomic profiling on a genome-wide scale using reverse-engineered gene networks. Nat. Biotechnol. 23, 377–383 (2005)
8. Cover, T.M., Thomas, J.A.: Elements of Information Theory. John Wiley and Sons, Chichester (2001)
9. http://wiki.c2b2.columbia.edu/dream
10. Lamb, J., Crawford, E.D., Peck, D., Modell, J.W., Wrobel, I.C.M.J., Lerner, J., Brunet, J.-P., Subramanian, A., Ross, K.N., Reich, M., Hieronymus, H., Wei, G., Armstrong, S.A., Haggarty, S.J., Clemons, P.A., Wei, R., Carr, S.A., Lander, E.S., Golub, T.R.: The connectivity map: Using gene-expression signatures to connect small molecules, genes, and disease. Science 313(5795), 1929–1935 (2006)

Measurement and Statistics of Application Business in Complex Internet

Lei Wang[1], Yang Li[2], Yipeng Li[3], Shuhang Wu[3], Shiji Song[1], and Yong Ren[3]

[1] Tsinghua University, Department of Automation,
100084 Beijing, China
leiwang03@mails.tsinghua.edu.cn
[2] Beijing Institute of Petrochemical Technology, School of Information Technology,
102617 Beijing, China
[3] Tsinghua University, Department of Electronic Engineering
100084 Beijing, China

Abstract. Owing to independent topologies and autonomic routing mechanism, the logical networks formed by Internet application business behavior cause the significant influence on the physical networks. In this paper, the backbone traffic of TUNET (Tsinghua University Networks) is measured, further more, the two most important application business: HTTP and P2P are analyzed at IP-packet level. It is shown that uplink HTTP and P2P packets behavior presents spatio-temporal power-law characteristics with exponents 1.25 and 1.53 respectively. Downlink HTTP packets behavior also presents power-law characteristics, but has more little exponents $\gamma = 0.82$ which differs from traditional complex networks research result. Moreover, downlink P2P packets distribution presents an approximate power-law which means that flow equilibrium profits little from distributed peer-to peer mechanism actually.

Keywords: Internet, traffic, application business, peer-to-peer.

1 Introduction

Unlike the early description of the network's topology based on the random graph theory, the rapidly developing theory and methods of complex network have already been widely used in the research of Internet [1-4]. In 1999, the Faloutsos discovered the four kinds of power-law distribution characteristic involved in the topology of Internet AS(autonomous system) [5,6], which led to researches on the large number of topology models and growth mechanisms of Internet that have characteristics of power-law distribution, Inet, BRITE[7,8], etc. represented. The data resource of this kind of researches mainly comes from the measuring results of Internet topology based on Trace-route, provided by research institutions CAIDA and so on, such as Skitter, Oregon, etc[9]. On the other hand, after A.L.Barabási discovering the scale-free characteristic of complex network, a large number of information networks' topological characteristics based on the Internet application have been experimentally researched. A.L.Barabási, as a typical representative, used Robert to follow and catch

J. Zhou (Ed.): Complex 2009, Part I, LNICST 4, pp. 400–410, 2009.

the WWW (World-Wide Web) page links, and proved that the WWW which consists of pages and hypertext links have the characteristic of power-law distribution [10].

Aiming at measurement and improvement of Internet performance, the work above has greatly advanced people's comprehension of the Internet and the topological distribution of business in the Internet application layer. However, the default identity of network nodes in the statistical model leads to the research on application layer separated from that on the network layer. In the early work, we pointed out that the users' behaviors are a kind of logical application behavior which depended on the transmission capacity supplied by the Internet infrastructure, and meanwhile had independent topology and messaging rules [11,12]. The users' behaviors couple the users' nodes with the network nodes, which represent that the users' nodes launch business according to the logic of application layer, and after reaching the network nodes the business drives many changes of the network layer, such as the acceptance, routing, traffic, etc. This network structures which have coupled characteristic are already not able to be exactly described by the simple Internet power-law or WWW power-law.

With the number of internet users growing exponentially, the influence to the distribution of network traffic bring by the distribution of users' behaviors is increasing and can not be neglected any more. The latest statistics from ISC(Internet Society of China) show that the site of number one network traffic in China attracts 31% of the internet users to click it, and have the average click-through rate of 6.8/(day person). The large-scale gathering characteristics of the users' behaviors lead to the serious imbalance in network traffic, so that the load-balancing technology such as CDN (Content Delivery Network) has been proposed. We have already found that the virtual network behaviors had a malignant influence on the whole characteristics of internet, although people have gradually understood the power-law in the internet and the topological distribution of business in the Internet application layer, the distribution characteristics of users' behaviors which joined them have not been paid enough attention to. The scientific measurements and quantitative descriptions of the distribution characteristics of users' behaviors have not been seen in literature.

In order to understand the distribution characteristics of the behaviors of Internet application layer, this paper observes and analyzes the distribution characteristics of behaviors of Tsinghua University campus network, which is a local group of users, raises a topological constituting mechanism which uses bipartite graph model to couple users' nodes with network nodes, and observes the Traces traffic of the total export of Tsinghua University campus network. Through filtering and merging we select certain kinds of typical business which take a large proportion (such as HTTP, P2P), and simplify the users' behaviors to the packet level to form a conclusion of quantitative analysis. The research indicates that for a kind of directed logical network constituted by users' behaviors, The uplink and downlink HTTP packets and uplink P2P packets show a consistent characteristic of power-law distribution in both time and space. The difference from conclusion of the traditional complex network research is that the power index γ of the distribution of business behaviors are both less than 2. Downlink packets of the P2P business satisfy a kind of distribution similar to the power-law. In this paper we calculate and discuss the status when the γ changes, and point out the physical meaning involved.

2 Bipartite Graph Model

In the face of Internet, such a complex network system, we only focus on the whole measure of the users` behavior, so it can be abstracted to a structure with three network layers as shown in Figure 1. The core network takes charge of the main data transmission and exchange and the edge network is adjacent to the core network, which provides business and services. The access network is used to describe the physical topology of users` access nodes. The network business process is abstracted to: users produce business by the connecting network, and visit the edge network; The edge network answer the business request and drive the traffic distribution mainly of the core network, and also of other levels of networks.

In order to measure the distribution of users` behaviors, i.e. the interaction between the edge network and the connecting network, this paper bring the "bipartite graph" in graph theory to further describe the logical connection between the edge network and the connecting network. The logical connection is different from the physical connection in Figure 1, see in [12].

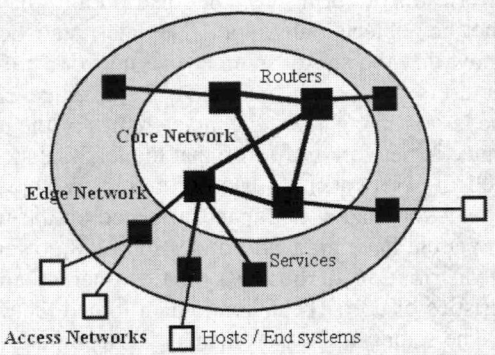

Fig. 1. Network architecture

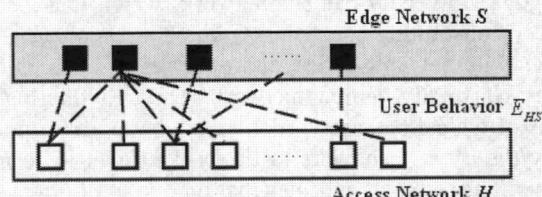

Fig. 2. Bipartite graph model of users' behavior

G is called bipartite graph, if there is a partition of the vertexes V of the undirected graph $G = <V,E>$, $V = V_1 \cup V_2$, $V_1 \cap V_2 = \Phi$, so that any for edge of G, its two ends are in V_1 and V_2 respectively.

Let S (services) to represent the set of nodes of the edge network, H (hosts) to represent the set of nodes of the connecting network, the users` behaviors in the specified time period constitute the one-way set of logical connections E_{HS} between S and H. It is known through the network structure that $S \cap H = \Phi$, it does not matter to let $V = S \cap H$, so the one-way graph $G_{HS} = <V, E_{HS}>$ is a bipartite graph which describes the characteristic of behavior that the connecting network visits the edge network, the topology is shown in Figure 2.

By measuring uplink traffic E_{HS}, we can get the in-degree distribution of S, which can be seen as the distribution of the users behaviors to the network nodes. By the same token, let E_{SH} to represent the behaviors of answers of network to the visit of users, which constitutes the bipartite graph of downlink traffic. The out-degree distribution of S is whole reflected of the feedback behaviors of the application layer.

3 Application Business Packets Processing

As one of the largest campus networks in China, the Tsinghua University campus network owns a ten thousand million bandwidth backbone. To ensure the high enough sampling rate, the original velocity of flow got in the total export is about 250MBps, even if we analyze traffic of 1 minute, the quantity of date is about 15GB, so it will be difficult to measure and process the data in a long period. In fact, each IP datagram header contains a wealth of information, the filed of source address and destination address in the protocol tree can completely define the visiting behaviors of a group. This paper catches all packet headers of group in the experiment, and completes the data statistics of EHS by analyzing the source and destination address, which greatly reduces the data quantity.

For the two typical business which take the largest proportion of network traffic: HTTP and P2P (peer-to-peer), the original packet header files can not used directly for the statistics of users` behaviors, since there are a large number of different protocols, different sorts of business and useless "noise" packet headers, which should be filtered a second time. The experiment uses tools such as tcpdump and Ethereal to do multi-stage filter to the original traffic, and finally obtains the pure set of packet headers of HTTP and P2P.

Take the uplink bipartite graph as example, a visit behavior from connecting network H to the edge network S, is sliced up to several groups in the according IP layer. By classification and integration the set of packet headers obtained above, we can get the set of visit records from H to S, which is shown as E_{HS} in Figure 2. This paper decomposes users` behaviors to the scale of data packet so as to make a more basic analysis.

4 Performance Evaluation

In the experiment we used the uplink and downlink traces traffic in the Tsinghua University campus network. Data are collected in 7 different dates to ensure the statistical stability. The time period is about 1 hour, the number of groups is 10^6 orders, the data quantity of packet headers is more than 25GB, and the magnitude of the connecting network and the edge network are both 104 orders.

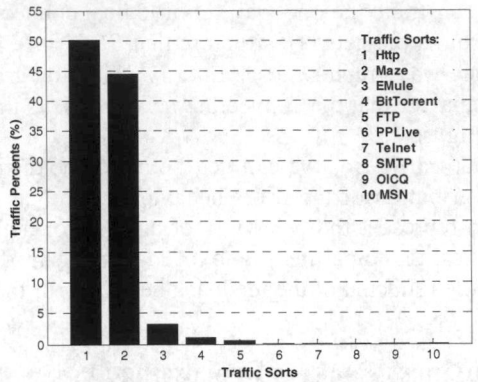

Fig. 3. Proportion of different Internet application business in backbone traffic

Figure 3 counts the uplink proportion distribution of several mainstream business (or application environment). The results indicate that the users' behaviors lead to the big difference between the proportions different business's traffic take. Maze, Emule, BiTTorrant and PPLive are all P2P application, they take the proportion close to HTTP, and the two take the vast majority of network resources. Actually, the experiment simplifies the connecting network H to the local range of Tsinghua University campus network, and the campus network is itself a huge Web Cache, inside which there are widely resource sharing, and the export traffic is reduced. So in actual situation the traffic of P2P business may take a greater proportion. As the mainstream businesses, HTTP and P2P accept the vast majority of users' behaviors, so we will mainly focus on these two businesses to do the statistics and discussions.

4.1 Statistics of HTTP Packets

This section discusses the behaviors of HTTP business between the connecting network and the edge network. This kind of behaviors in the application layer mostly result from the users clicking URL hypertext links to visit the web sites and download information, so it is a kind of two-way interactive behaviors. We first consider the uplink from the connecting network to the edge network. In the experiment, we filter the groups of HTTP businesses from the uplink trace flows and deal with them with relevant methods, and measure the in-degree distribution of the edge network nodes (web site) in different scales of time. It should be noted that the whole characteristic that the edge network accept the users' behaviors directly influence the basic network performance of the physical network such as load balance, quality of service and so on,[12] which is more significant relative to the distribution characteristics of the node degree of connecting network.

Because of the limitation of data quantity and processing ability, it is not able to do the traffic monitoring and analyzing in any large time scale at will. In order to confirm the statistical stability in time, we do the statistics of the grouping behaviors in two time scales. Figure 4 and Figure 5 shows the results of the uplink distribution of the grouping HTTP businesses, with the log-log coordinate system to be easily shown, and in which uplink traffic of 1 minute and 1 hour time intervals are respectively used. In the figure, k represents the degree of the edge network nodes, $p(k)$ is the corresponding distribution function. The results indicate that the visits from HTTP groups to the edge network nods distribute as power law, and have consistent characteristics in different time intervals. Take the results of 1 hour observing time as example (Figure 5), the distribution of k satisfies

$$p(k) = C_l k^{-\gamma} \tag{1}$$

By fitting we can get $C_l = 0.69$, $\gamma_l = 1.25$, and the error is less than 0.1. In the case that the observing time is 1 minute, $\gamma_s = 1.23$, which is a little smaller. The statistics above reveals that the there exists power-law among the behaviors of grouping HTTP business, most of the edge nodes servers accept a relatively small part of business, while the minority of the nodes accept most of the user` behaviors, so the HTTP business have the apparent characteristics of cluster. Not like the scale-free properties pointed out by the complex network research, we confirm this phenomenon from the perspective of application-layer behaviors, and provide a direct explanation about the load imbalance of the actual network. Figure 6 can help us to understand the influence that scale-free HTTP business behaviors bring to the network more profoundly. After descending sorting the edge network nodes by degree, the top 10% node servers accept about 80% of the HTTP business, which leads to a local network peak, but a low efficiency. Therefore, to optimize the popular resource nodes has naturally become the main idea of the technology of Internet load balance at present.

In the statistics of the downlink distribution of HTTP grouping behaviors, the phenomenon of power-law is also discovered (Figure 7), and $\gamma = 0.82$. The resource nodes of the edge network will generally answer the users` uplink visiting behaviors, based on the analysis above, few popular resource nodes will produce the downlink answers which take a relatively large proportion (such as the file download, etc.), which leads to the distribution of scale-free behaviors, and aggravates the entirely load imbalance.

In the bipartite graph shown in Figure 2, N represents the maximal accessing number of the edge network nods in the specified period of time, which is the increasing function of the number of user M. For the sake of a convenient qualitative discussion, we uniformly use N to represent them. It is easy to see that the distribution of degree of the edge network nodes satisfies.

$$\sum_{k=1}^{N} P(k) = 1 \tag{2}$$

From formulas (1)(2) we can get the first-order and the second-order moment of k is

$$D_1 = \sum_{k-1}^{N-1} k P_d(k) = C \sum_{k-1}^{N-1} k^{1-\gamma} \tag{3}$$

$$D_2 = \sum_{k-1}^{N-1} k^2 P_d(k) = C \sum_{k-1}^{N-1} k^{2-\gamma} \tag{4}$$

$$C = \frac{1}{\sum_{k=1}^{N-1} k^{-\gamma}} \tag{5}$$

The mean and variance is $d_\mu = D_1$, $d_\sigma = D_2 - D_1^2$ respectively, and with the increase of γ, the D_1 and D_2 both decrease. When $\gamma \to \infty$, $d_\mu \to 1$, $d_\sigma \to 0$. So the increasing process of γ is a topological changing process that the heterogeneous characteristic of scale-free network gradually fades away, and tends to uniformity.

For the uplink business, $\gamma \in (1,2]$, when N is large enough, we have $D_1 = O(N^{2-\gamma})$, $D_2 = O(N^{3-\gamma})$, $d_\mu \ll N$. The mean value of the degree of the edge network node is significantly lower than the entire network level, while the mean and the variance both diverge, which is the obvious feature of the scale-free property. Compared with the traditional explanation about the flat network from complex network theory ($\gamma \in (2,3]$), the actual measured behaviors of application layer have stronger characteristic of cluster. Since the proportions of the traffic of uplink and downlink businesses are quite asymmetric, large number of downlink groups aggravates the heterogeneous character-istics of network, which results in the smaller power index $\gamma = 0.82$.

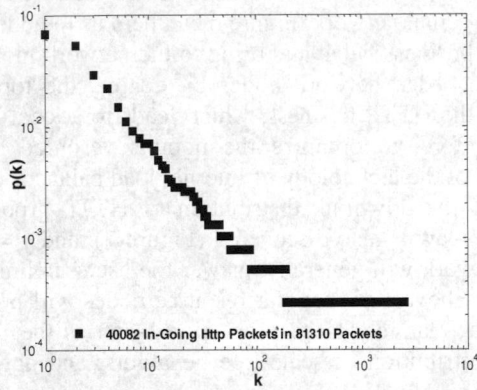

Fig. 4. Statistics of HTTP uplink packets. One minute uplink trace flow is measured.

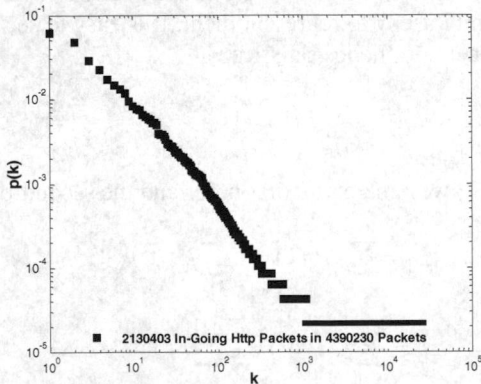

Fig. 5. Statistics of HTTP uplink packets. One hour uplink trace flow is measured.

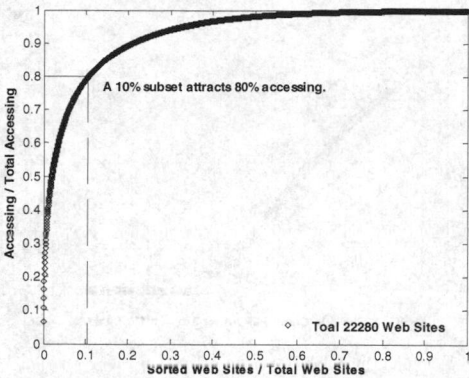

Fig. 6. Uplink HTTP packets collective behavior versus edge network nodes sorted by accessing times

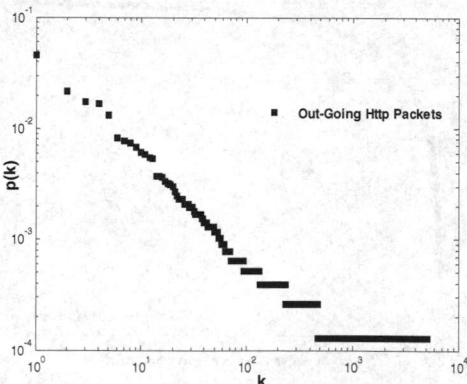

Fig. 7. Statistics of HTTP downlink packets. One hour downlink trace flow is measured.

4.2 Statistics of P2P Packets

With the development of broad band and the increasingly demands for the audio and video files, P2P businesses have gradually become the kind of business which use the most network resources. From the experiments, we can see that the uplink packets of P2P also entirely shows an obvious characteristic of power-law, with the power index $\gamma = 1.53$, a little bigger than that of HTTP grouping status, which indicates that the application layer behaviors diverges more for the assembly of P2P resources. Since the users' uplink behavior is a common reflect of information demands, the same with the characteristics of the HTTP uplink packets, a small number of popular resource nodes attracts the vast majority of grouping visits (Figure 9), represented by the assembly of users to the searching services nodes of P2P.

For the two mainstream businesses, HTTP and P2P, the downlink mechanisms of traffic and the resources using statuses are quite different, which is also reflected in the statistics of the downlink behavior of P2P business. As in Figure 10, it is a similar

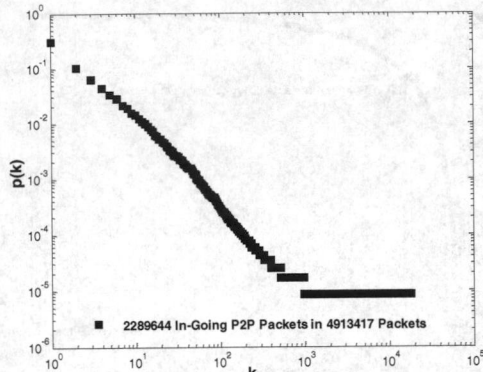

Fig. 8. Statistics of P2P uplink packets. One hour uplink trace flow is measured.

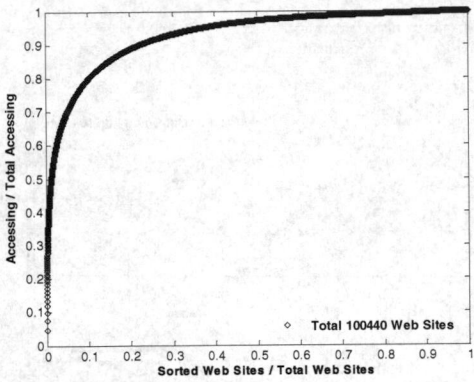

Fig. 9. Uplink P2P packets collective behavior versus edge network nodes sorted by accessing times.

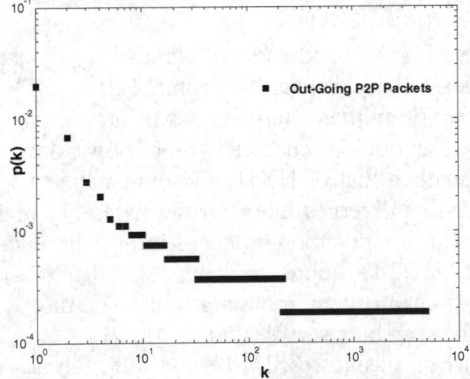

Fig. 10. Statistics of P2P downlink packets. One hour downlink trace flow is measured.

power-law distribution, the difference between nodes decreases, and the number of popular nodes is enlarged. The P2P business itself mainly shares long stream transmission. If we gather the statistics from the point of grouping, each node must have quite big degrees; and the mechanism of distributed peer-to-peer sharing resources, to a certain extent, curbs the flow imbalance and alleviates the power-law characteristic of business. However the grade difference of the network nods also exists obviously, for the edge network S, the node load is quite of imbalance, and the characteristic of assembling (diverging) still exists. The experiment indicates that the P2P mechanism improves the efficiency of resource sharing from the peer-to-peer perspective. However, it is a pity that it contributes very little to the flow balance in the physical network, and itself also occupies large part of network resources and redundantly costs the long stream, which will enlarge the influence of this kind of similar power-law characteristic to the traffic distribution, and this is a major conflict that can not be avoided.

5 Conclusion

In the former work, we use the simulation of abstract model to point out that the behaviors of application behaviors and the coupling of physical networks will influences the overall dynamics of Internet, with the development of network and the increasing of number of users, the influence from the behaviors of application layer to the distribution of the entire network traffic can not be ignored any more. This paper observes and researches the distribution characteristic behaviors of Tsinghua University campus network, such a local users group, based on the topology of bipartite graph which couples users nodes with the network nodes, in the scale of IP grouping we do the statistics and discussion about the two typical businesses, HTTP and P2P businesses, which take the most proportion of network traffic, and obtain the conclusions below: 1) The uplink distributions of HTTP business and the P2P business groups both show a consistent power-law distributed characteristic in time and space, the power index is respectively 1.25 and 1.53. A few popular resource nodes attract the vast majority of users' behaviors, which has a obvious characteristic of cluster. 2) The downlink packets of HTTP business show the power-law characteristic, and $\gamma = 0.82$, which is a little different from the traditional explanation about the flat network from the complex network theory. And the actual measured behaviors of application layer have a stronger characteristic of cluster. 3) Downlink packets of P2P business satisfy a distribution similar to the power-law, and the peer-to-peer mechanism contributes little to the traffic of physical network.

The general power-law characteristic shown in the behaviors of Internet application will heavily leads to the network load imbalance, the declining of the bandwidth utilization rate, and aggravates the rate of deterioration of the local network. The increasing P2P business occupies large quantities of network resources, however it doses not provide the obvious support for the physical network balance at the same time. This kind of actually measured characteristic of behaviors of application layer urges us to not only focus on the single layer of network or technical development in local range, but also entirely observe and research a series of theories and methods of analysis of network traffic, routing balance, topological design and so on.

Acknowledgment

This work is supported in part by the National Nature Science Foundation of China (NSFC) grants 60672142, 60772053, the National Basic Research Program of China (973 Program) grants 2007CB307100 / 2007CB307105, and by the China Postdoctoral Science Foundation grants 20080430400.

References

1. Fuks, H., Lawniczak, A.T.: Mathematics and computers in simulation 51, 101 (1999)
2. Li, Y., Liu, Y., Shan, X., Ren, Y., Jiao, J., Qiu, B.: Chin. Phys. 14, 2153 (2005)
3. Takayasu, M., Fukuda, K., Takayasu, H.: Physica A 274, 140 (1999)
4. Takayasu, M., Takayasu, H., Fukuda, K.: Physica A 277, 248 (2000)
5. Faloutsos, M., Faloutsos, P., Faloutsos, C.: ACM SIGCOMM Computer Communication Review 29, 251 (1999)
6. Siganos, G., Faloutsos, M., Faloutsos, P., Faloutsos, C.: IEEE/ACM Transactions on Networking 11, 514 (2003)
7. Medina, A., Lakhina, A., Matta, I., Byers, J.: Proc. of MASCOTS, Washington, 346 (2001)
8. Winick, J., Jamin, S.: Technical report CSE-TR-456-02, Department of EECS, Universtiy of Michigan (2002)
9. Chen, Q., Chang, H., Govindan, R., Jamin, S.: Proc. of IEEE INFOCOM, New York, 608 (2002)
10. Albert, R., Barabasi, A.-L.: Reviews of Modern Physics 74, 47 (2002)
11. Liu, F., Shan, X., Ren, Y.: Acta Phys. Sin. 53, 273 (2004)
12. Wang, L., Zhou, S., Yuan, J., Ren, Y., Shan, X.: Acta Phys. Sin. 56, 36 (2007)

Moving Breather Collisions in the Peyrard-Bishop DNA Model

A. Alvarez[1], F.R. Romero[1], J. Cuevas[2], and J.F.R. Archilla[2]

[1] Grupo de Física No Lineal. Área de Física Teórica. Facultad de Física. Universidad de Sevilla. Avda. Reina Mercedes, s/n. 41012-Sevilla, Spain
azucena@us.es
http://www.grupo.us.es/gfnl
[2] Grupo de Física No Lineal. Departamento de Fisica Aplicada I. ETSI Informática. Universidad de Sevilla. Avda. Reina Mercedes, s/n. 41012-Sevilla, Spain

Abstract. We consider collisions of moving breathers (MBs) in the Peyrard-Bishop DNA model. Two identical stationary breathers, separated by a fixed number of pair-bases, are perturbed and begin to move approaching to each other with the same module of velocity. The outcome is strongly dependent of both the velocity of the MBs and the number of pair-bases that initially separates the stationary breathers. Some collisions result in the generation of a new stationary trapped breather of larger energy. Other collisions result in the generation of two new MBs. In the DNA molecule, the trapping phenomenon could be part of the complex mechanisms involved in the initiation of the transcription processes.

Keywords: Discrete breathers, intrinsic localized modes, moving breathers, breather collisions, Peyrard-Bishop model.

1 Introduction and Model Set-Up

The DNA molecule is a discrete system consisting of many atoms having a quasi-one-dimensional structure. It can be considered as a complex dynamical system, and, in order to investigate some aspects of the dynamics and the thermodynamics of DNA, several mathematical models have been proposed. Among them, it is worth remarking the Peyrard–Bishop model [1] introduced for the study of DNA thermal denaturation. This model, as well as some variations of it, have also been used extensively for the study of some dynamical properties of DNA.

The study of discrete breathers (DBs) in chains of oscillators is an active research field in nonlinear physics [2,3,4,5]. Under certain conditions, stationary breathers can be put in motion if they experience appropriate perturbations [6], and they are called *moving breathers* (MBs). There are no exact solutions for MBs, but they can be obtained by means of numerical calculations.

In the Peyrard–Bishop model, the existence of DBs has been demonstrated [7,8], and DBs are thought to be the precursors of the bubbles that appear prior to the transcription processes in which large fluctuations of energy

J. Zhou (Ed.): Complex 2009, Part I, LNICST 4, pp. 411–416, 2009.

have been experimentally observed. Some studies about the existence and properties of MBs in the Peyrard–Bishop model including dipole-dipole dispersive interaction are carried out in [9,10].

In this work, we consider the Peyrard-Bishop DNA model, which Hamiltonian can be written as

$$H = \sum_{n=1}^{N} \left(\frac{1}{2} m \dot{u}_n^2 + D(e^{-bu_n} - 1)^2 + \frac{1}{2}\varepsilon_0(u_{n+1} - u_n)^2 \right),$$ (1)

the term $\frac{1}{2}m\dot{u}_n^2$ represents the kinetic energy of the nucleotide of mass m at the nth site of the chain, and u_n is the variable representing the transverse stretching of the hydrogen bond connecting the base at the nth site. The Morse potential, i.e., $D(e^{-bu_n} - 1)^2$, represents the interaction energy due to the hydrogen bonds within the base pairs, being D the well depth, which corresponds to the dissociation energy of a base pair, and b^{-1} is related to the width of the well. The stacking energy is $\frac{1}{2}\varepsilon_0(u_{n+1} - u_n)^2$, where ε_0 is the stacking coupling constant.

In scaled variables this Hamiltonian can be writing as:

$$H = \sum_n \left[\frac{1}{2}\dot{u}_n^2 + V(u_n) + \frac{1}{2}\varepsilon(u_n - u_{n+1})^2 \right],$$ (2)

where u_n represents the displacement of the nth pair-base from the equilibrium position, ε is the coupling parameter and $V(u_n)$ is:

$$V(u_n) = \frac{1}{2}\left(\exp(-u_n) - 1 \right)^2.$$ (3)

Time-reversible, stationary breathers can be obtained using methods based on the anti-continuous limit [11]. At $t = 0$, $\dot{u}_n = 0, \forall n$, and the displacements of a breather centered at n_0 are denoted by $\{u_{SB,n}\}$. A moving breather $\{u_{t,n}\}$ can be obtained with the following initial displacements and velocities:

$$u^0_{MB,n} = u_{SB,n} \cos(\alpha(n - n_0))$$
$$\dot{u}^0_{MB,n} = \pm u_{SB,n} \sin(\alpha(n - n_0)).$$ (4)

The plus-sign corresponds to a breather moving towards the positive direction and the minus one, the opposite. This procedure works as well as the marginal-mode method [6] and gives good mobility for a large range of ε. The translational velocity and the translational kinetic energy of the MB increase with α. We use Eqs. (4) as initial conditions to integrate the dynamical equations using a symplectic algorithm [13].

The study begins generating two identical stationary breathers, with the same frequency, separated by a fixed number of pair-bases between their centers. We call N_c the number of pair-bases separating initially the centers of the two DBs. Both breathers are in phase, that is, before the perturbation, each breather is always like the mirror image of the other one. The perturbation should be given simultaneously to both breathers and the initial conditions of each breather given by Eqs. (4), with the plus sign for one breather and the minus sign for the other one. In this way the MBs travel with the same modulus of velocity, but opposite directions, and they are in phase.

2   Results and Conclusions

We can analyze collisions with a fixed value of the parameter α and different values of N_c so that the colliding MBs keep unchanged. Also, we can analyze collisions varying the parameter α maintaining fixed the number N_c, thus the colliding MBs change for each value of α. We write

$$N_c = N_o + jj, \tag{5}$$

where N_o is a fixed number to guarantee that the breathers are initially far apart, and jj is a positive even number.

In the first approach we fix the parameter α and perturb the DBs varying their separation N_c, thus the only difference between two collisions is the time passed between the initial perturbation and the initiation of the collision.

We consider collisions where the DBs are in phase and perturbed simultaneously. We have taken $N_o = 40$ and jj varies in the interval $[0,100]$ with step size 2. Then, up to fifty different collisions can be analyzed for a fixed value of α and ε.

We have performed an extensive numerical simulations considering different values of the coupling parameter ε, and MBs with different values of the wave number α. The values of ε have been taken in the interval $[0.13,0.35]$ with step size 0.01. For each value of ε the values of α have been taken in the interval $[0.030,0.200]$ with step size 0.002. We present the results obtained with $\varepsilon = 0.32$ and $\alpha = 0.048$; $\alpha = 0.138$; $\alpha = 0.18$, which correspond to MBs with increasing velocities. These values are representative of the different scenarios that can be found. Fig. 1 represents the trapped energy versus jj for these three cases. The qualitative results are similar for other values of the parameters (ε, α).

Fig. 1(left) corresponds to the case with the smallest velocity, i.e., $\alpha = 0.048$, the distribution of points appears in a narrow band and there are no points with trapped energy close to zero. When the MBs have small enough kinetic energy, most of the energy gets trapped after the collision and two small MBs are generated traveling with opposite directions, they transport the remaining energy except a small part that is lost in the form of phonon radiation. Notice that for jj up to 30, the points oscillate following a repetitive regular pattern, and this regularity begin to change as jj increases.

For intermediate values of α the phenomenon of non-trapping, or breather generation, appears for some values of jj. This can be appreciated in Fig. 1(central), obtained with $\alpha = 0.138$, where some points appear with trapped energy close to zero, this means that after the collision almost all the energy is transported by two emerging MBs with the same velocities that the incoming MBs'. For this

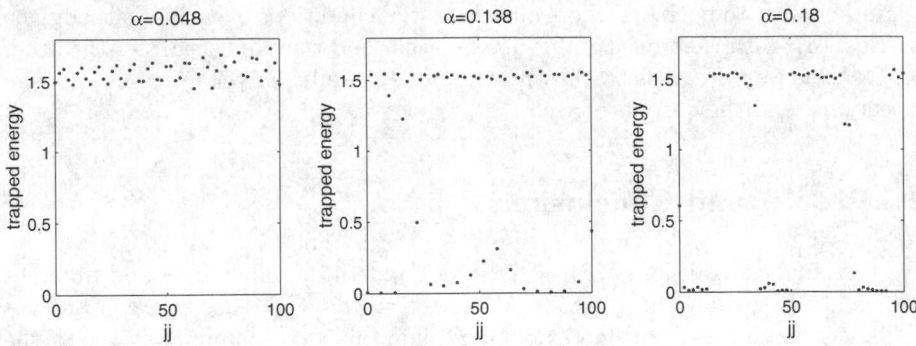

Fig. 1. Three distributions of points representing the trapped energy versus jj, for $\alpha = 0.048$, $\alpha = 0.138$, and $\alpha = 0.18$, respectively. Coupling parameter $\varepsilon = 0.32$ and breather frequency $\omega_b = 0.8$.

value of α two points in the upper band is followed by one point that fall down close to zero.

For $\alpha = 0.18$, the upper band is divided in pieces and another fragmented lower band appears. There are alternating intervals of N_c values corresponding to the upper band and other corresponding to the lower band. This means that there are some successive values of jj associated with trapping, followed by other ones associated with breather generation, see Fig. 1(right).

Fig. 2 shows the displacements versus time for eight collisions of Fig. 1(central), corresponding to $jj = 34, ..., 48$, with the fixed value $\alpha = 0.138$, $\varepsilon = 0.32$ and breather frequency $\omega_b = 0.8$.

Fig. 3 shows the evolution of the trapped energy for the collisions with $jj = 34, ...40$ of Fig. 2. For $jj = 34$ and $jj = 40$ two new breathers are generated. The other cases correspond to breather trapping with breather generation.

It is interesting to study the collisions maintaining fixed the number N_c and varying α for fixed values of ω_b and ε. In real DNA the MBs could be generated at fixed points of the chain by the action of proteins. Obviously, the phenomenology is similar to the previous case and the study has permitted to observe a great sensitivity of the outcomes with respect to the parameter α (Ref.[14]). To see this, let us consider the results for three nearness values of α with $N_c = 40$, $\varepsilon = 0.32$ and $\omega_b = 0.8$:

For $\alpha = 0.1370$, the collision produces three new breathers, a trapped breather containing most of the initial energy and two new MBs.

For $\alpha = 0.1372$, there is a noticeable attenuation of the amplitude of the trapped breather, which anticipates an entirely new outcome. The emerging MBs contain most of the initial energy.

For $\alpha = 0.1374$, there is no trapping and two new MBs emerge with almost the same velocity that the colliding breathers'.

The previous studies let us to conclude that for a given values of ε and ω_b, the relevant parameters to determine the outcomes of the collisions are both α and the number N_c.

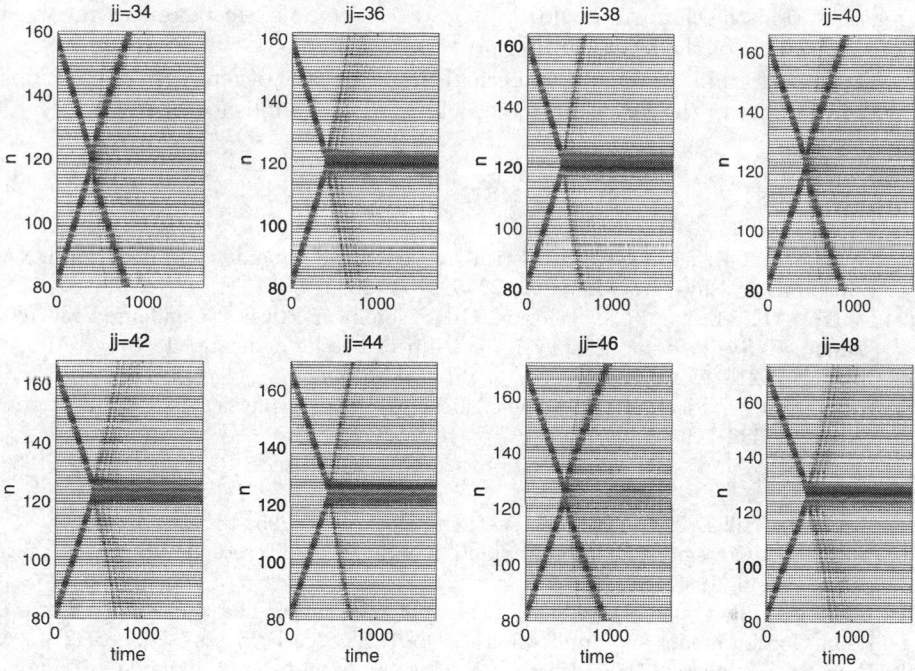

Fig. 2. Displacements versus time for eight collisions corresponding to $jj = 34, ..., 48$, with the fixed value $\alpha = 0.138$. Coupling parameter $\varepsilon = 0.32$ and breather frequency $\omega_b = 0.8$.

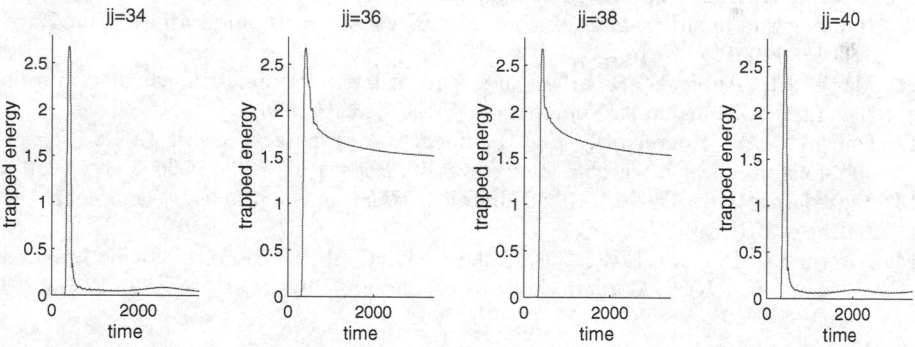

Fig. 3. Trapped energy versus time corresponding to the first four collisions of Fig. 2, respectively

The simulations of MB collisions in the Peyrard-Bishop DNA model show a new mechanism for concentrating energy in DNA. When two MBs collide, it is possible, in some favorable cases, to get stationary trapped breathers with more energy than the colliding breathers. These breathers are also movable and after colliding with other ones, could give rise to even more energetic stationary breathers. This mechanism could be part of the complex mechanisms involved in the initiation of the transcription processes.

We are performing extensive numerical simulations of other types of collisions that can appear in the DNA molecule, which will be published elsewhere.

References

1. Peyrard, M., Bishop, A.R.: Statistical mechanics of a nonlinear model for DNA denaturation. Phys. Rev. Lett. 62, 2755–2758 (1989)
2. Dauxois, T., Mackay, R.S., Tsironis, G.P.: Nonlinear Physics: Condensed Matter, Dynamical Systems and Biophysics - A Special Issue dedicated to Serge Aubry. Physica D 216, 1–246 (2006)
3. Kivshar, Y.S., Flach, S.: Nonlinear localized modes: physics and applications. Chaos 13, 586–799 (2003)
4. Flach, S., Mackay, R.S.: Localization in nonlinear lattices. Physica D 119, 1–238 (1999)
5. Flach, S., Willis, C.R.: Discrete breathers. Phys. Rep. 295, 181–264 (1998)
6. Aubry, S., Cretegny, T.: Mobility and reactivity of discrete breathers. Physica D 119, 34–46 (1998)
7. Dauxois, T., Peyrard, M., Willis, C.R.: Localized breather-like solution in a discrete Klein-Gordon model and application to DNA. Physica D 57, 267–282 (1992)
8. Dauxois, T., Peyrard, M., Bishop, A.R.: Dynamics and thermodynamics of a nonlinear model for DNA denaturation. Phys. Rev. E 47, 684–695 (1993)
9. Cuevas, J., Archilla, J.F.R., Gaididei, Y.B., Romero, F.R.: Moving breathers in a DNA model with competing short- and long-range dispersive interactions. Physica D 163, 106–126 (2002)
10. Alvarez, A., Romero, F.R., Archilla, J.F.R., Cuevas, J., Larsen, P.V.: Breather trapping and breather transmission in a DNA model with an interface. Eur. Phys. J. B 51, 119–130 (2006)
11. Marín, J.L., Aubry, S.: Breathers in nonlinear lattices: Numerical calculation from the anticontinuous limit. Nonlinearity 9, 1501–1528 (1996)
12. Dmitriev, S.V., Kevrekidis, P.G., Malomed, B.A., Frantzeskakis, D.J.: Two-soliton collisions in a near-integrable lattice system. Phys. Rev. E 68, 056603, 1–7 (2003)
13. Sanz-Serna, J.M., Calvo, M.P.: Numerical Hamiltonian problems. Chapman and Hall, Boca Raton (1994)
14. Alvarez, A., Romero, F.R., Cuevas, J., Archilla, J.F.R.: Discrete moving breather collisions in a Klein-Gordon chain of oscillators. Phys. Lett. A 372, 1256–1264 (2008)

Morphological Similarities between DBM and an Economic Geography Model of City Growth

Jean Cavailhès[1], Pierre Frankhauser[2], Geoffrey Caruso[3],
Dominique Peeters[4], Isabelle Thomas[5], and Gilles Vuidel[6]

[1] Research Director, INRA, UMR 1041, CESAER, Dijon,
26 Bd Docteur Petitjean, F-21000, France
Jean.Cavailhes@enesad.inra.fr
[2] Professor, CNRS, ThéMA, University of Franche-Comté, Besançon, France
pierre.frankhauser@univ-fcomte.tr
[3] Professor, University of Luxembourg, Luxembourg
geoffrey.caruso@uni.lu
[4] Professor, Geography and CORE, Catholic University of Louvain, Belgium
dominique.peeters@uclouvain.be
[5] Research Director, FNRS, Geography and CORE, Catholic University of Louvain, Belgium
isabelle.thomasàuclouvain.be
[6] Software engineer, CNRS, ThéMA, University of Franche-Comté, Besançon, France
gilles@vuidel.org

Abstract. An urban microeconomic model of households evolving in a 2D cellular automata allows to simulate the growth of a metropolitan area where land is devoted to housing, road network and agricultural/green areas. This system is self-organised: based on individualistic decisions of economic agents who compete on the land market, the model generates a metropolitan area with houses, roads, and agriculture. Several simulation are performed. The results show strong similarities with physical Dieletric breackdown models (DBM). In particular, phase transitions in the urban morphology occur when a control parameter reaches critical values. Population density in our model and the electric potential in DBM play similar roles, which can explain these resemblances.

1 Introduction

In this paper, we propose a model of self-generated city where households evolve in a cellular automata space. We simulate the growth of a metropolitan area where land is devoted to housing, road network and agricultural/green areas. Such as several previous papers, the results show strong similarities with the *Dieletric breackdown models* (DBM) proposed in physics. Our contribution is to show that this property holds with urban economics micro-foundations, while previous works use diffusion mechanisms close to DBM, which are far removed from household's behaviour in urban economic theory.

J. Zhou (Ed.): Complex 2009, Part I, LNICST 4, pp. 417–428, 2009.

1.1 The Literature

Cellular Automata (CA) have been widely used in urban geography for simulating the development of cities. Various types of modelling strategies can be distinguished. A first group favors abstract approaches for studying the interactions between groups of populations. They are linked to the concept of *self-organisation* (see, e.g., Schelling, 1971; Couclelis, 1985; Phipps, 1989). In a second trend of papers, urban growth is modelled more concretely by referring directly to physical (or biological) models to generate structures that reproduce the morphological evolution of cities (see, e.g., Batty and Longley, 1986; Batty 1991; Frankhauser, 1991; Makse et al., 1995). A third type of models introduce into the previous ones rules that describe in a heuristic way and on an aggregated level the spatial interactions between different land uses see, e.g., White and Engelen,1993; 1994). A common concern of these approaches is the absence of micro-economic fundations. Caruso et al. (2007) and Cavailhès et al. (2004) are attempts to bridge this gap, the first paper dealing with CA, the second with a fractal setting.

In the meantime, other concepts have been proposed to simulate the formation of clusters. One such example in physics is *Diffusion-limited aggregation* (DLA), proposed by Witten and Sander (1981), which is the process whereby particles following a random walk cluster to form aggregates. It was applied to simulate town growth (see, e.g., Benguigui, 1995; 1998; or, in a percolation version: Makse et al., 1998). Combining DLA with electric fields, Niemeyer et al. (1984) proposed the *Dielectric breakdown model* (DBM) to describe the patterns of dielectric breakdown of solids, liquids, and gases, and to explain the formation of the branching, self-similar Lichtenberg figures. DLA and DBM have been applied in physics (e.g., discharges in non-homogeneous material (Peruani et al.), surface thermodynamics (Bogoyavlenskiy et al., 2000)) and chemistry, such as biology (see, e.g., Chikushi and Hirota, 1998; Li et al., 1995) or urban geography (see, e.g., Batty, 1991). DLA and DBM models can only generate connected aggregates. It thus seems difficult to apply them as such at periurban zones where built areas are in patches scattered in the countryside. This lead Benguigui et al. (2001) to propose an approach where they introduce leap-frogging. Another concern of the urban applications of DLA-DBM models is the thin link to urban economics theory. Recently, authors have proposed models with economic foundations that lead to DBM-like diffusion mechanisms (see, e.g., Andersson et al., 2002).

1.2 The Self-organised Growth of a Metropolitan Area

The purpose of this paper is to investigate the development of a metropolis where economic agents living in residences scattered in a mixed residential/agricultural area commute to work to an exogenous Central Business District (CBD) by an endogenous road network. This system is self-organised (except for some elementary urban growth rules): based on individualistic decisions of agents who compete on the land market, the model generates a metropolitan area with houses, agriculture and the road network.

Land belongs to absentee landowners, each renting her parcel to the highest bidder, either a resident or a farmer, hence determining land occupancy. Farmers produce food stuff and a green amenity (open space, landscape, etc), which is a by-product enjoyed by neighbouring residents. Households arrive sequentially in the city and choose their location by maximizing a utility function subject to a budget constraint. Each migrant chooses freely its location, considering the commuting cost to the CBD, the rent of the residential plot determined by the competition on the land market and the surroundings. She enjoys both open space/agricultural amenity and local public goods (respectively negative and positive functions of the neighboring population density).

As households migrate into this growing city, agricultural cells are developed, while the local public authority creates a connected road network so as to provide all new households with an access to the CBD. At each step, a resident can move in another cell of the city to maximise her utility if the migrant changes her surroundings. This competition leads to an adjustment of the residential rent until all the inhabitants obtain the same utility (short run equilibrium, instantaneously reached in our framework). Thus, the city grows according to the individual choice of the economic agents, leading to a self-organization of the residential space, the road network and the open-spaces. The mean commuting cost increases with population and thus households' utility progressively decreases. Migration stops when households' utility equals the utility of the rest of the world(i.e. long run equilibrium).

Our analysis privileges the morphological properties of the emerging road network. Appropriate fractal measures are used to this end. One of the main results is the occurrence of phase transitions between linear and dendritic structures when varying the parameters. This is akin to what is observed in the literature with DBM models. Some interesting analogies between both models are elaborated.

The remainder of the paper is organised as follows. Section 2 presents the microeconomic model and the cellular automata environment where households evolve. Section 3 provides some simulations and their morphological properties. Next, Section 4 compares these findings with similar results obtained in physics and chemistry. Section 5 concludes.

2   An Economic Model for Residential Location

2.1   General Setting

We consider a closed 2D space with a set of cells i. A pointwise *central business district* (CBD) is located exogeneously at the centre of the grid, where two preexisting orthogonal roads intersect. Initially, the rest of the grid is occupied by farmers who produce stuff food under constant returns to scale and sell this output on the world market. Each cell has three possible (mutually exclusive) states: residential, road, or agricultural (or undeveloped), which we respectively denote by j, k and l. The grid is gradually filled in by residences and roads from undeveloped cells. Such conversions are irreversible.

2.2   Residential Growth

Households' preferences are represented by a Cobb-Douglas utility function $U = Z^{\delta} H^{\alpha} E^{\beta} S^{\gamma}$, whose arguments are Z, a non-residential composite good made up of every market good except housing, H, housing, E, open-space externalities (greenness), and S, local public goods externalities. Each household maximizes U under a budget constraint: $Y - \theta d = Z + SR$, where Y is the income, θ the unitary commuting cost, d the distance to the city centre and R the unitary land rent (the composite good is taken as the numeraire, hence $p_z = 1$). The parameters $\delta \in [0,1]$, and $\alpha = 1 - \delta$ indicate respectively the preference for the composite good and for housing, whereas $\beta \geq 0$ and $\gamma \geq 0$ are respectively the preference parameters for open-space and for social externalities. The first-order conditions for a constrained optimum yield the demand functions of Z and S as well as the indirect utility function:

$$V = (Y - \theta d) R^{-\alpha} E^{\beta} S^{\gamma} \tag{1}$$

Households arrive one by one. At each time step t, the migrant picks up the cell l that maximizes its indirect utility function, considering the commuting cost θd_l^t to the city centre, the land rent R_l^t of her residential plot, the neighbourhood open-spaces E_l^t (i.e. the density of agricultural cells within a given radius around l at $t-1$), and the local public goods S_l^t (i.e. the density of residential cells in the same neighbourhood at $t-1$).

A migrant evaluates the indirect utility (1) provided by each agricultural cell. Her bid rent (a reservation level of the land rent) equalises the utility she can obtain in the city with the utility of the rest of the world (open city model of urban economics). Nevertheless, several agricultural cells provide the same utility: the migrant can put the landowners in competition to obtain a decrease of the rent actually paid, until the land rent reaches Φ, the agricultural rent. Her indirect utility function is:

$$V_l^t = (Y - \theta d_l^t) \Phi^{-\alpha} (E_l^t)^{\beta} (S_l^t)^{\gamma}. \tag{2}$$

For more details about the microeconomic program, see Caruso et al. (2007). Let ρ_l be the density of residences within a given neighbourhood around each undeveloped cell l. We define the two neigbourhood externalities as $E_l^t = e^{-\mu \rho_l^{t-1}}$ and $S_l^t = e^{\nu \rho_l^{t-1}}$. (2) can be written (for simplicity, we ommit t and l indexes) $V = (Y - \theta d)\Phi^{-\alpha} e^{-\beta \mu \rho} e^{\gamma \nu \rho}$, where $\beta \mu$ and $\gamma \nu$ are undistinguishables. Thus we fix $\mu = 1$ and $\nu = 0.5$. Let

$$V^t = \max_l V_l^t. \tag{3}$$

The city continues to grow as long as the utility of an entrant as measured by (Uentrant) exceeds a given threshold $\overline{V}$, which is the utility she enjoys in the rest of the world. When $V^t = \overline{V}$, we obtain a so-called *long-run equilibrium*.

2.3 Network

Households access the CBD via a road network gradually built by the local authority, which builds new roads where necessary to provides each new household with an access to the CBD: all the residences must be linked to the existing road network. For deciding which undeveloped cells will be converted to roads, we minimize the number of new road cells to be created at each time step.

3 Morphological Properties of Patterns Generated by the S-Ghost-City Model

A Java-based software (called S-GHOST) was implemented to simulate the model described above. The parameters are the household's income Y, the share of income devoted to housing α, the agricultural rent Φ, the preference for green and social externalities β and γ, the size of a cell's neighborhood $\hat{x}$, and the transportation cost θ. Ex-aequo cells are solved randomly. One of the most striking outcome is the structural change in the urban morphology that occur when β and γ cross some thresholds. In Figure 1, we illustrate the shape of the patterns obtained for fixed $\beta = 0.25$.

We observe that for γ-values lower than 0.11 residents disperse around the CBD since they look for green amenities in their vicinity and are rather indifferent about social amenities (see Figure 1 (a)). Hence, agricultural cells subsit around build-up

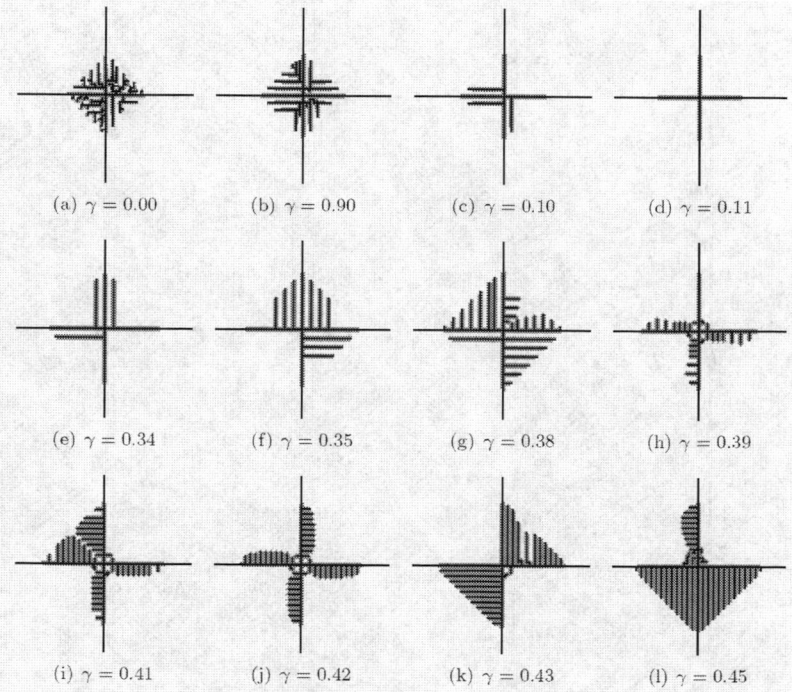

(a) $\gamma = 0.00$ (b) $\gamma = 0.90$ (c) $\gamma = 0.10$ (d) $\gamma = 0.11$

(e) $\gamma = 0.34$ (f) $\gamma = 0.35$ (g) $\gamma = 0.38$ (h) $\gamma = 0.39$

(i) $\gamma = 0.41$ (j) $\gamma = 0.42$ (k) $\gamma = 0.43$ (l) $\gamma = 0.45$

Fig. 1. Long-run equilibria for $\beta = 0.25$ and varying γ

cells. At the threshold $\gamma = 0.11$ and above, preferences for social and environmental amenities compensate. The residents line up on the sides of the two preexisting orthogonal roads in order to minimize the commuting cost, because moving to a lateral location would increase this cost without providing enough social or open-space compensation (Figure 1 (d)). The compensation effect is evident when comparing Figure 1 (d) to Figure 2 (a) where $\beta = 0.0$ and $\gamma = 0.0$, i.e. where residents are indifferent to their neighbourhood.

In Figure 1, beyond a second critical value of $\gamma = 0.34$ the city expands again along lateral streets (surfacic pattern). Migrants accept the higher commuting cost of a lateral cell because this cell provides more local public goods than in the cross-like city, due to a higher density. As γ increases further, residents cluster together more densely as they taste for social contact increases. However, green cells separating the built-up cells still exist up to γ around 0.40. Beyond that value, the preference for social contacts dominates residential choice: compact clusters emerge. Moreover, singular phenomena may occur as shows the example of Figure 1 (i), (j) and (l). Here, the access to green areas is blocked by culs-de-sacs so that migrants have to go further to have still access to the CBD.

We now come to the second series considered, which refers to $\gamma = 0$ and a set of different values of β (Figure 2).

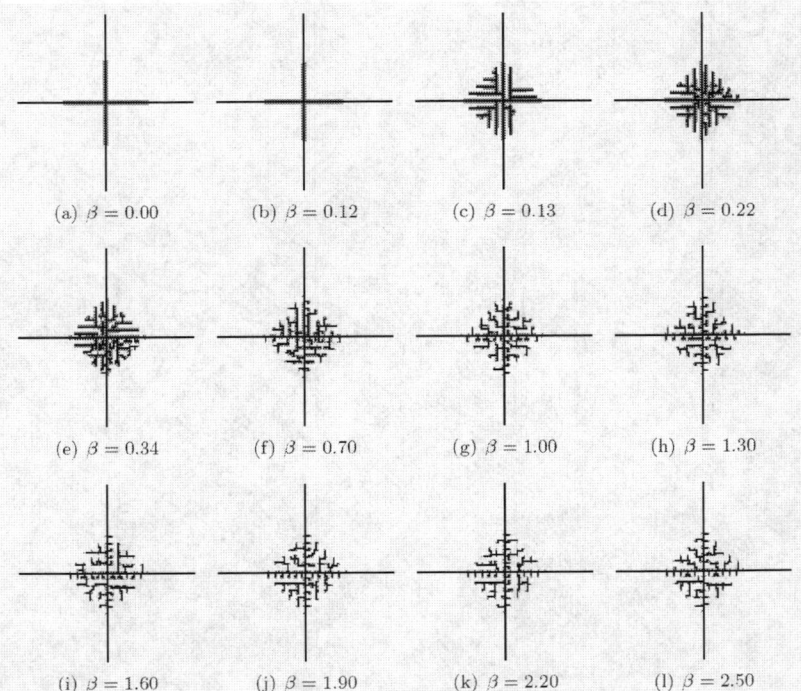

Fig. 2. Long-run equilibria for $\gamma = 0$ and varying β

Hence residents do not have any taste for local public goods, but they are more or less interested by green amenities. For β-values less than 0.12, both preferences are very small, and a cross-like city is again obtained. Beyond this threshold the increasing preference for green amenities dominates and dendritic patterns occur where houses and roads are separated by undeveloped cells (Figures 2 (c) to (l)). This pattern remains stable all over the range considered that runs up to $\beta = 3.0$.

We use fractal analysis to characterize the shape of the road network (see e.g. Benguigui, 1995 or Lu and Tang, 2004). The method used here is described in Frankhauser (19..)   and the results are produced in Figure 3 where the fractal dimension (Y-axis) depends on β ($\gamma = 0.0$) (Display A) or γ ($\beta = 0.25$) (Display B). In Display (A), after the linear city ($\beta < 0.12$) the fractal dimension is $1.7 - 1.8$ or so when β is comprised between 0.3 and 2.5. In Display (B), the fractal dimension decreases from 2 or so when $\gamma = 0$ to 1 in the cross-like city ($\gamma = 0.11$ to $\gamma = 0.34$) and then increases in an uneven pattern, most of the values being comprised between 1.5 and 2.

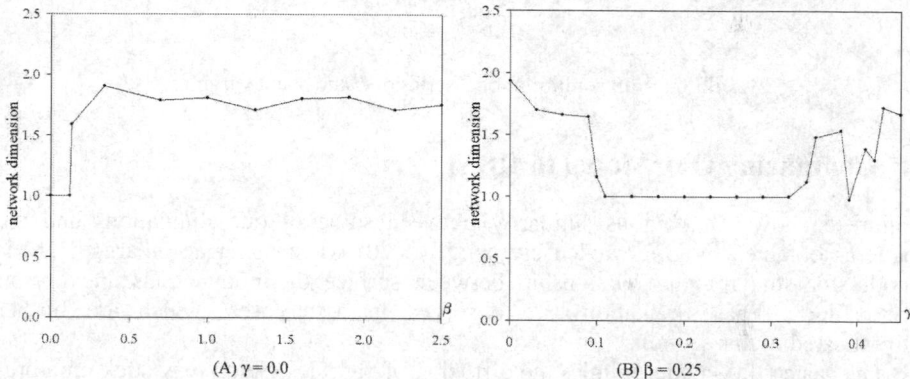

Fig. 3. Fractal dimension for β varying ($\gamma = 0.0$) and for γ varying ($\beta = 0.25$)

Note that similar simulations can be performed with constant values for β and γ but changing the size of the neighbourhood. Results are presented in figure 4 that shows that the length of the neighbourhood is also an important parameter affecting the form of the urban development. We simulate a case where households strongly taste social amenities ($\gamma = 0.5$), without any taste for green externalities ($\beta = 0$). When the length of the neighbourhood is small ($\hat{x} = 4$), agents enjoy a dense environment far from the CBD and they do not care what happen at the city fringe, which is beyond their horizon. When the length increases ($\hat{x} = 7$), agricultural sites enter in the viewshed of peripherical residents who refuse fareway locations: the city shrinks. A phase transition occurs when $\hat{x} = 8$: the pattern becomes regular at this threshold value. Finally, when $\hat{x} = 12$ the pattern collapses again into a cross-like city where few households accept to live in.

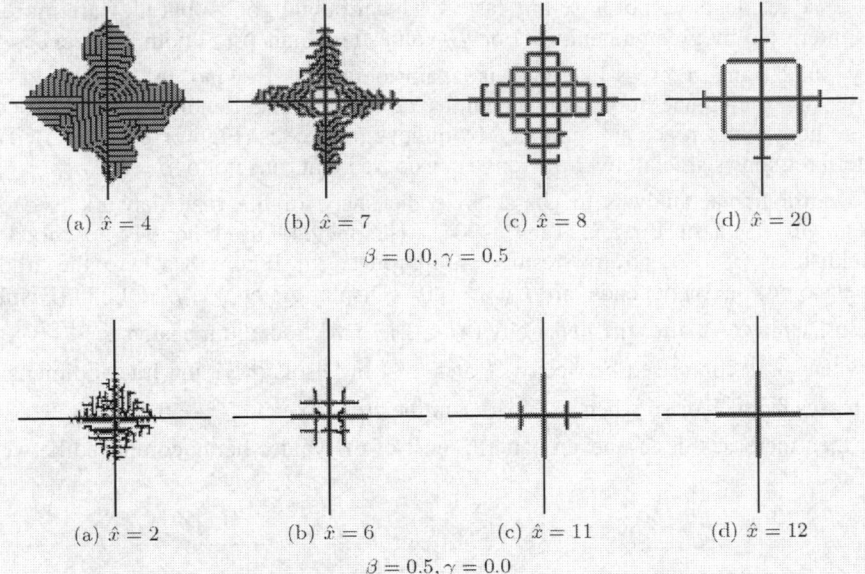

(a) $\hat{x} = 4$ (b) $\hat{x} = 7$ (c) $\hat{x} = 8$ (d) $\hat{x} = 20$

$\beta = 0.0, \gamma = 0.5$

(a) $\hat{x} = 2$ (b) $\hat{x} = 6$ (c) $\hat{x} = 11$ (d) $\hat{x} = 12$

$\beta = 0.5, \gamma = 0.0$

Fig. 4. Long-run equilibria for 2 series of β and γ and varying $\hat{x}$

4 Comparing Our Model to DBM

Figure 5 shows an obvious similarity between some of our simulations and the patterns obtained by Bogoyavlenskiy et al. (2000) when using an enlarged DBM-model for studying the relationship between surface thermodynamics and crystal morphology. Phase transitions are observed in both cases, which are further investigated in this section.

The basic DLA model mimics the diffusion of particles which may stick on a pre-existing seed and added subsequently. They generate a fractal cluster with a fractal dimension of about $D = 1.7$. Niemeyer et al. (1984) introduced the *Dielectric breackdown model* (DBM) to simulate electric discharge patterns. As shown already by Pietronero and Wissman (1984), a formal link can then be established between the DLA-model and the DBM-model. In DBM, the electrodynamic Laplace equation which describes the spatial variation of the electric potential Φ is transcribed into a discrete equation to compute the electric potential for each cell at a given simulation step. Then a site i' located in the immediate neighbourhood of the already generated discharge is chosen randomly, where the probability to choose i' depends on the potential:

$$p_{i \to i'} = \frac{(\Phi_{i'})^\eta}{\sum_{i'} (\Phi_{i'})^\eta}$$ (4)

After having selected the site i', a link is created between i' and the discharge, and the potential in i' falls to zero. Parameter η plays an important role for the

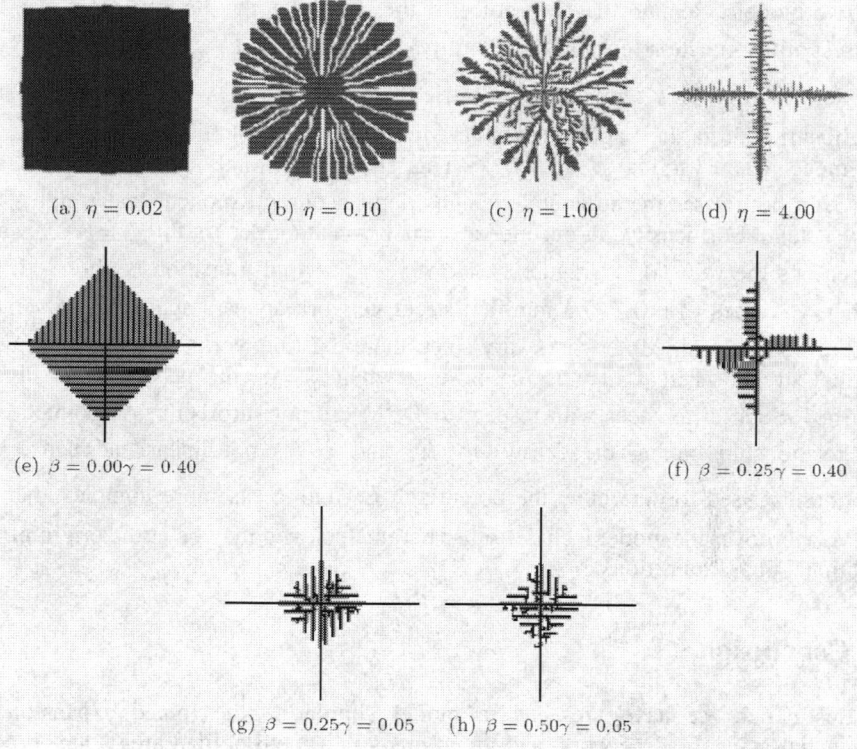

(a) $\eta = 0.02$ (b) $\eta = 0.10$ (c) $\eta = 1.00$ (d) $\eta = 4.00$

(e) $\beta = 0.00 \gamma = 0.40$ (f) $\beta = 0.25 \gamma = 0.40$

(g) $\beta = 0.25 \gamma = 0.05$ (h) $\beta = 0.50 \gamma = 0.05$

Fig. 5. A complex diffusive growth model ((a) to (d)) and our results ((e) to (h))

shape of the generated patterns. As show by Sanchez et al. (1993), for $\eta = 0$ compact clusters appear the fractal dimension of which is two. On the contrary, ribbon-like structures are observed when $\eta > 4$ For higher values the fractal dimension drops even beyond one. Batty (1991) interprets η in the context of urban growth as a parameter describing different types of planning policies.

As saw in Section 1, since his introduction the DBM has been used in many physical domains as well as in biology and urban geography.

Let us remind that in our model an agricultural cell can be converted either into an urbanized or a transportation cell, with a possibility of leap-frogging, but with network cells that remain connected. This allows a comparison with DBM, which is made here on the morphology of the road network.

Let us consider series of pattern where β is fixed (e.g., $\beta = 0.25$; see Figure 1 and Figure 3(B)). For low γ -values ($\gamma < 0.11$) we obtain ramified patterns with a fractal dimension 1.7 or so. When γ exceeds the critical value $\gamma = 0.11$, the road network becomes cross-linear. Even if the morphological changes are here more threshold-like, the resemblance with DBM is clear. Indeed both models tackle with the diffusive growth of a network in space. It is possible to find a formal link between the two models. Remind that the choice of the site for a new house depends on the indirect utility function (eq:fullindirectutility), the variables E_l and S_l depending on the density ρ_l. This reminds the electric potential, which value depends on the neighbourhood via the Laplace-operator. Moreover, morphological changes occur in our model when varying β and/or γ. This situation reminds the role of η in the DBM. In both cases the structural changes depend on an exponent that weights the local potential or density. In our model, a first transition occurs for $\beta = 0.25$ when γ reaches the value 0.11 (Figure 1). However, a second transition is observed: for higher γ-values ($\gamma = 0.34$) a ramified street network appears again. Moreover the features of the ramified networks vary since there exist networks consisting of rather parallel streets whereas in other cases, e.g. when $\gamma = 0$, the networks are highly ramified. Such differences with respect to DBM are not surprising. As pointed out, the relationship linking the density to E_l and S_l is non-linear and hence the relationship used to determine the potentials E_l and S_l is more complex than in DBM. Moreover the model includes other parameters who may be at the origin of the additional phase transitions.

5 Conclusion

In this paper, we have presented a model simulating the joined expansion of residential areas, road network, and green areas in a metropolitan area. Our purpose was to build a simulation model on sound microeconomic foundations. We use a standard urban economics model: a household maximises a utility function, which includes tastes for the residential surroundings (local public goods and green/open space amenity), under a budget constraint including a commuting to work cost and a residential rent. As usual in standard urban economics, equilibrium is reached on the land market. These economic agents arrive sequentially in a 2D cellular automata gird and freely choose their location, considering the cost of each site (commuting, rent) and the enjoyed surroundings. As the city grows, new roads are built by a local authority to provide the migrants with an access to a pre-existing Central Business District (CBD), by a connected road network. Several simulations are made by using a software implemented in Java.

An analogy is observed with results obtained with physical DLA and DBM models. In particular, phase transitions in the urban morphology occur when a control parameter reaches critical values. Such analogy is also obtained by previous works; yet, it occurs here with different theoretical foundations. On the one hand, population density in our model and the electric potential in DBM play similar roles. On the other hand, the spatial model is different: diffusion in DBM (the potential of all the cells of the grill play a part in determining the potential of a cell), and a window vicinity here (only the neighbouring cells play a role in determining the density). Therefore, the resemblance is more surprising. Needles to say, a substantial amount of effort has to be made to decipher the analogy, and to pass from an abstract model to actual urban mophologies (calibration with realistic parameter values, etc.).

References

1. Andersson, C., Lindgren, K., Rasmunssen, S., White, R.: Urban growth simulation from first principles. Physical Review E 66, 026204-1 -9 (2002)
2. Batty, M., Longley, P.: The fractal simulation of urban structure. Environment and Planning A 18, 1143–1179 (1986)
3. Batty, M.: Generating urban forms from diffusive growth. Environment and Planning A 23, 511–544 (1991)
4. Benguigui, L.: A new aggregation model. Application to town growth. Physica A 219, 13–26 (1995)
5. Benguigui, L.: A fractal analysis of the public transportation system of Paris. Environment and Planning A 27(7), 1147–1161 (1995)
6. Benguigui, L.: Aggregation models for town growth. Philosophical Magazine B 77, 1269–1275 (1998)
7. Benguigui, L., Czamanski, D., Marinov, M.: City Growth as a Leap-frogging Process: An Application to the Tel-Aviv Metropolis. Urban Studies 38, 1819–1839 (2001)
8. Bogoyavlenskiy, V.A., Chernova, N.A.: Diffusion-limited aggregation: A relationship between surface thermodynamics and crystal morphology. Physical Review E 61(2), 1629–1633 (2000)
9. Caruso, G., Peeters, D., Cavailhès, J., Rounsevell, M.: Spatial configurations and cellular dynamics in a periurban city. Regional Science and Urban Economics 37, 542–567 (2007)
10. Cavailhès, J., Frankhauser, P., Peeters, D., Thomas, I.: Where Alonso meets Sierpinski: an urban economic model of fractal metropolitan area. Environment and Planning A 36, 1471–1498 (2004)
11. Chikushi, J., Hirota, O.: Simulation of root development based on the dielectric breakdown model. Hydrological Sciences 43(4), 549–559 (1998)
12. Couclelis, H.: Cellular worlds: a framework for modelling micro-macro dynamics. Environment and Planning A 17, 585–596 (1985)
13. Frankhauser, P.: Aspects fractals de structures urbaines. Espace géographique 1, 45–69 (1991)
14. Li, B., Wang, J., Wang, B., Liu, W., Wu, Z.: Computer simulations of bacterial-colony formation. Europhysics Letters 30, 239–243 (1995)
15. Lu, Y., Tang, J.: Fractal dimension of a transportation network and its relationship with urban growth: a study of the Dallas - Fort Worth area. Environment and Planning B 31(6), 895–911 (2004)
16. Makse, H.A., Andrade, J.S., Batty, M., Havlin, S., Stanley, H.E.: Modeling urban growth patterns with correlated percolation. Physical Review E 58, 7054–7062
17. Makse, H.A., Havlin, S., Stanley, H.E.: Modelling Urban Growth Patterns. Nature 377, 608–612 (1995)
18. Mathiesen, J., Jensen, M.H., Bakke, J.O.H.: Dimensions, maximal growth sites, and optimization in the dielectric breakdown model. Phys. Rev. E 77, 066203 (2008)
19. Niemeyer, L., Pietronero, L., Wiesmann, H.J.: Fractal Dimension of Dielectric Breakdown. Phys. Rev. Lett. 52, 1033–1036 (1984)
20. Peruani, F., Solovey, G., Irurzuni, I.M., Mola, E.E., Marzocca, A., Vicente, J.L.: Dielectric breakdown model for composite materials. Phys. Rev. E 67, 066121 (2003)
21. Phipps, M.: Dynamical behavior of cellular automata under the constraint of neighborhood coherence. Geographical Analysis 21(3), 197–216 (1989)

22. Pietronero, L., Wissman, H.J.: Stochastic Model for Dielectric Breakdown. Journal of Statistical Physics 36(5,6), 909–916 (1984)
23. Sánchez, A., Guinea, F., Sander, L.M., Hakim, V., Louis, E.: Growth and forms of Laplacian aggregates. Phys. Rev. E 48, 1296–1304 (1993)
24. Schelling, T.C.: Dynamic Models of Segregation. Journal of Mathematical Sociology 1, 143–186 (1971)
25. White, R., Engelen, G.: Cellular automata and fractal urban form: a cellular modelling approach to the evolution of urban land use patterns. Environment and Planning A 25, 1175–1199 (1993)
26. White, R., Engelen, G.: Cellular dynamics and GIS: modelling spatial complexity. Geographical Systems 1, 237–253 (1994)
27. Witten, T.A., Sander, L.M.: Diffusion-Limited Aggregation, a Kinetic Critical Phenomenon. Phys. Rev. Lett. 47(19), 1400–1403 (1981)

Modular Synchronization in Complex Network with a Gauge Kuramoto Model

C. Choi[1], E. Oh[1,2], B. Kahng[1], and D. Kim[1]

[1] Department of Physics and Astronomy and Center for Theoretical Physics,
Seoul National University, Seoul 151-747, Korea
bkahng@snu.ac.kr
[2] Bioanalysis and Biotransformation Research Center,
Korea Institute of Science and Technology, Seoul 136-791, Korea

Abstract. We modify the Kuramoto equation(KE) by introducing a gauge term which is a function of link betweenness centrality(BC). The gauge term induces the phase difference from 0 to π between two nodes that belong to different modules. Therefore, a synchronization occurs in each module individually even though the whole network is not synchronized globally. By measuring the phase similarity of all pairs of connected nodes, we can detect the modular structure of complex networks. This algorithm requires relatively little computational time $O(NL)$ for network with N nodes and L links.

Keywords: synchronization, module identification, modular complex network, Kuramoto model.

1 Introduction

Every system that has constituents and relationships between themselves can be represented by networks conceptually. Constituents of the system are nodes and their interactions are links which connect a node and another node in networks. Many physical and social systems have been studied via networks.

Among many kinds of dynamics on complex networks, synchronization is one of the most popular subjects. Synchronization is a process of adjusting some properties assigned to each node via interactions between the elements of complex network. This is so useful that it has been investigated in several kinds of fields - physics, biology, sociology, etc.

Module is a set of densely-connected nodes in complex network. Modular complex network has several modules and these modules are connected each other with relatively sparse links.

Synchronization in modular complex network shows some new features compared to synchronization in complex network having no module structure. Links are dense within modules and sparse between modules. So in synchronization process nodes in same module are synchronized first by intra-module links and the whole network becomes synchronized later by inter-module links or not. We reported on this phenomenon in our previous paper [1]. In this lecture note we follow the outline of the paper briefly and all figures in here are also taken from it.

J. Zhou (Ed.): Complex 2009, Part I, LNICST 4, pp. 429–434, 2009.

2   Module Identification

Before we introduce our method to indentify modules of complex network, it is also helpful to review some of the previous methods that are regardless of synchronization.

q-state Potts model [2] is very useful method in statistical mechanics. It concerns about spin-spin interactions. Spins assigned to each node in complex network have q states for spin value and they interact with the nearest neighbors nodes. When the energy of the whole system is minimized, it is likely to that nodes in same module have same value of spin.

Girvan and Newman algorithm(GN) [3] is one of the most famous module-detecting algorithms. Link BC is an important concept for GN algorithm. It is the number of paths between every pair of nodes that pass through the link. One calculates BC for all links and removes the link which has the maximum BC value. Repeating this step until the modularity Q of the network is maximized one can get module structures of the network. Due to the recalculation of BC at each step, the computational time of GN algorithm is relatively high and scales as $O(L^2 N)$, where L is the number of links and N is the number of nodes.

Clauset, Newman and Moore(CNM) [4] introduced a hierarchical agglomeration algorithm for detecting module structure. Its computational time scales to $O(L \log^2 N)$ for sparse networks.

3   Detecting Algorithms in Synchronization

Using synchronization to detect modular structure is another dynamic algorithm as well as q-state Potts model. Arenas et al. [5] showed oscillators of nodes in different modules are synchronized in different time scales and they are ordered hierarchically. But one must choose the characteristic time, t_c at which modular structure is determined.

And Boccaletti et al. [6] introduced another dynamic clustering algorithm called opinion changing rate (OCR) model. The dynamics of node are governed by

$$\frac{d\theta_i}{dt} = \omega_i - \frac{J}{\sum_{j \in \mathrm{nn}(i)} b_{ij}^{\alpha(t)}} \sum_{j \in \mathrm{nn}(i)} b_{ij}^{\alpha(t)} \sin(\theta_i - \theta_j) \beta e^{-\beta|\theta_j - \theta_i|}, \qquad (1)$$

where ω is the natural frequency of node i, σ is the coupling strength and b_{ij} is BC of the link between node i and node j. By adjusting the parameter $\alpha(t)$ and β one can tune the interaction coupling between neighbouring nodes.

4   Kuramoto Model with Gauge Term

We introduce a modified KE [7] which we call gauge KE,

$$\frac{d\phi_i(t)}{dt} = \Omega_i - J \sum_{j=1}^{N} a_{ij} \sin(\phi_i(t) - \phi_j(t) - \eta g(b_{ij})). \qquad (2)$$

Here ϕ_i is the phase of node i, Ω_i is the natural frequency of node i selected from the Gaussian distribution, $e^{-\Omega^2/2}/\sqrt{2\pi}$, J is the overall coupling constant and a_{ij} is the (i,j)-th component of the adjacency matrix, which is one when the node i and j are connected, and zero otherwise. η is a control parameter. When $\eta = 0$ the equation is equal to the original KE and we will set $\eta = 1$ for our algorithm. The gauge term $g(b_{ij})$ is defined as

$$g(b_{ij}) - \frac{b_{ij} - b_{\min}}{b_{\max} - b_{\min}}\pi, \tag{3}$$

where $b_{\min}$ ($b_{\max}$) is the minimum (maximum) value of link BC. Usually BC is relatively large for inter-module links and small for intra-module links. So inter-module links have large gauge term and their couplings become negative in Eq.(2). This means that two different modules which are connected with inter-module links have different average phases and velocities. This property enables us to employ gauge KE as modular structure detecting algorithm.

The followings are 4 steps of our algorithm.

i) With sufficiently large coupling constant J, we obtain the phases $\phi_i(t)$ of each oscillator in the steady state.

ii) We measure the phase similarity defined as $C_{ij} = \langle [1 + \cos(\phi_i(t) - \phi_j(t))]/2 \rangle$ for each links. The brackets are the average over different times, natural frequencies Ω_i and initial random phases $\phi_i(0)$.

iii) From the state in which every link is removed, we connect links one by one in descending order of C_{ij}.

iv) We repeat the step iii) until the modularity of the system becomes maximum. The modularity Q is defined as

$$Q = \sum_{\alpha} e_{\alpha\alpha} - a_{\alpha}^2, \tag{4}$$

where $a_{\alpha} = \sum_{\beta} e_{\alpha\beta}$, and $e_{\alpha\beta}$ is the fraction of edges that connect the nodes belonging to the modules α and β [8].

5   Simulation Results

We know the degree of synchronization of oscillators on the network quantitatively by measuring order parameter M defined as $\mathcal{M}_{\text{tot}} \equiv \langle |\sum_{j=1}^{N} e^{i\phi_j}/N| \rangle$, where $\langle \cdots \rangle$ means time average and ensemble average. We measure the order parameter in the steady state. If $\eta = 0$, the order parameter converges to 1 for sufficiently large J. As η is increased to 1, it has lower values than 1 as shown in Fig. 1(a).

We use ad hoc network for simulation. The network has four equal-sized modules. And the number of nodes is $N = 128$ and the number of links is $L = 1024$. We can adjust z_{out} which is the mean degree of inter-modular links.

We define the local order parameter as $\mathcal{M}_{\alpha} \equiv \langle |\sum_{j=1}^{N_{\alpha}} e^{i\phi_j}/N_{\alpha}| \rangle$, where α is the module index, N_{α} is the number of nodes within the module α and the sum

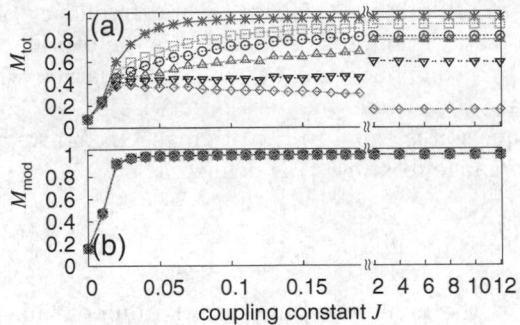

Fig. 1. The order parameter defined over the entire network (a) and within a module (b) versus the coupling constant J for the *ad hoc* network in case of $z_{out}/\langle k \rangle = 0.05$. Data are for $\eta = 0.0$, 0.6, 0.7, 0.8, 0.9 and 1.0 from the top in (a). The same symbols are used for (b), but data for different η collapse onto the single curve.

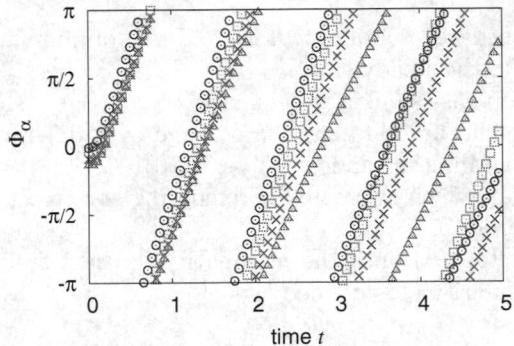

Fig. 2. The time evolution of average phases of the four modules, distinguished by different symbols, for the *ad hoc* network with $z_{out}/\langle k \rangle = 0.05$ when $\eta = 1.0$ and $J = 2.0$

is over nodes within the module. By comparing $\mathcal{M}_{tot}$ and $\mathcal{M}_\alpha$ (Fig. 1(b)) we can verify that the synchronization occurs within each module first.

The average phase of each module as a function of time is shown in Fig. 2. The modules are distinguishable because of the different average phases. The differnce of slopes indicates that the average phase velocities of each module are also different from each other.

To test performance of gauge KE model, we measure the mutual information defined as

$$I(A, B) = \frac{-2 \sum_{i=1}^{M} \sum_{j=1}^{M'} \log(\frac{N_i^j}{N_i N^j})}{\sum_{i=1}^{M} N_i \log(\frac{N_i}{N}) + \sum_{j=1}^{M'} N^j \log(\frac{N^j}{N})} \tag{5}$$

where $M = 4$ is the number of preassigned modules and M' is the number of detected modules. N_i^j is the number of nodes belonging to the i-th preassigned and the j-th detected modules, $N_i = \sum_j N_i^j$ and $N^j = \sum_i N_i^j$. To test performances

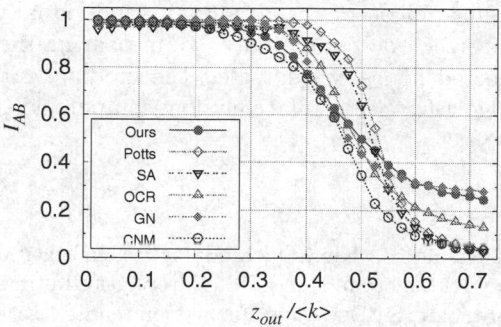

Fig. 3. The mutual information versus $z_{\text{out}}/\langle k \rangle$, the fraction of inter-modular edges per mean degree for the *ad hoc* network

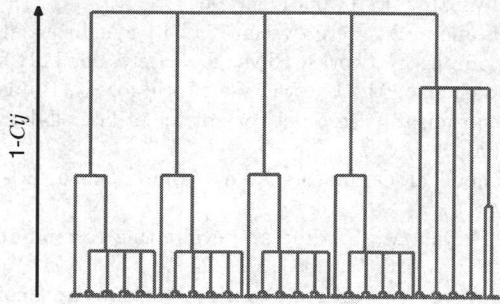

Fig. 4. The dendrogram based on the phase similarity between connected pairs of vertices for the hierarchical network with three levels

of several module-detecting algorithm we measure the mutual information on *ad hoc* networks [9] with several z_{out} values as shown in Fig. 3. The q state Potts model and the simulated annealing(SA) [10] are better than our algorithm in performance. However, if we consider the computational time, then ours($O(NL)$) may be useful for large scale networks in practical view. Our algorithm also does not need to tuning any parameters whereas OCR algorithm requires an extra task of parameter tuning.

Next, we applied our algorithm to the hierarchical network introduced by Ravasz and Barabási [11]. Fig. 4 shows the dendrogram constructed based on the phase similarity C_{ij}. The hub at each level is grouped with one of the four identical modules connected to it in that level.

6   Summary

In summary, we reviewed some module-detecting algorithms. And we introduced a gauge KE in which the gauge term is a function of link BC. The gauge term induces the phase difference between two nodes that belong to different modules.

Therefore, a synchronization occurs in each module individually even though the whole network is not synchronized globally. By measuring the phase similarity of all pairs of connected nodes, we detected the modular structure of *ad hoc* networks and this algorithm needs relatively low computational cost.

References

1. Oh, E., Choi, C., Kahng, B., Kim, D.: Modular synchronization in complex networks with a gauge Kuramoto model. EPL (Europhysics Letters) 83, 68003 (2008)
2. Reichardt, J., Bornholdt, S.: Detecting Fuzzy Community Structures in Complex Networks with a Potts Model. Phys. Rev. Lett. 93, 218701 (2004)
3. Girvan, M., Newman, M.E.J.: Community structure in social and biological networks. Proc. Natl. Acad. Sci. U.S.A. 99, 7821 (2002)
4. Clauset, A., Newman, M.E.J., Moore, C.: Finding community structure in very large networks. Phys. Rev. E 70, 066111 (2004)
5. Arenas, A., Diáz-Guilera, A., Pérez-Vicente, C.J.: Synchronization Reveals Topological Scales in Complex Networks. Phys. Rev. Lett. 96, 114102 (2006)
6. Boccaletti, S., Ivanchenko, M., Latora, V., Pluchino, A., Rapisarda, A.: Opinion dynamics and synchronization in a network of scientific collaborations. Phys. Rev. E 75, 045102(R) (2007)
7. Kuramoto, Y.: Chemical Oscillators, Waves and Turbulence. Springer, Berlin (1984)
8. Newman, M.E.J., Girvan, M.: Finding and evaluating community structure in networks. Phys. Rev. E 69, 026113 (2004)
9. Danon, L., Diaz-Guilera, A., Duch, J., Arenas, A.: Comparing community structure identification. J. Stat. Mech.: Theory Exp. P09008 (2005)
10. Kirkpatrick, S., Gelatt Jr., C.D., Vecchi, M.P.: Optimization by Simulated Annealing. Science 220, 671 (1983)
11. Ravasz, E., Barabasi, A.-L.: Hierarchical organization in complex networks. Phys. Rev. E 67, 026112 (2003)

Modification Propagation in Complex Networks

Mary Luz Mouronte, María Luisa Vargas, Luis Gregorio Moyano,
Francisco Javier García Algarra, and Luis Salvador Del Pozo

Telefónica I+D, Madrid, Spain
mlml@tid.es
http://www.tid.es

Abstract. To keep up with rapidly changing conditions, business systems and their associated networks are growing increasingly intricate as never before. By doing this, network management and operation costs not only rise, but are difficult even to measure. This fact must be regarded as a major constraint to system optimization initiatives, as well as a setback to derived economic benefits. In this work we introduce a simple model in order to estimate the relative cost associated to modification propagation in complex architectures. Our model can be used to anticipate costs caused by network evolution, as well as for planning and evaluating future architecture development while providing benefit optimization.

Keywords: Complex system, complex networks, structures and organization in complex systems.

1   Introduction

In recent years, it has become very clear to large business organizations that their systems and networks are evolving increasingly complex. The amount of different components in a system and the various ways these elements may interconnect each other result in quite intricate network topologies. Moreover, these complex structures are seldom static, but quite the opposite: they grow and change, sometimes in a permanent basis, as different requirements and functionalities develop to satisfy other system or client needs. Meanwhile, the study of complex networks—a growing field of scientific research—is becoming more and more important as a tool to tackle new needs that derive from emergence phenomena in complex structures. There are many such systems which are relevant to illustrate this growing focus, for example, communication networks, transportation networks, computer networks, social networks and several others [1,2,3]. All these are examples of a large number of basic similar units connected in a non-trivial manner and that may be well characterized by the concepts provided by complexity science. This characterization is used to explain certain properties that in turn will lead to new, improved ways to interact with such networks.

An interesting question arises when dealing with complex systems, and refers to how a modification in one point of the network propagates through the rest

J. Zhou (Ed.): Complex 2009, Part I, LNICST 4, pp. 435–440, 2009.

of the interdependent units of the network. In the present (ongoing) work [4,5], we focus in studying the cost associated to this type of modification propagation and the relation with the network's intrinsic complexity. More specifically, we construct a simple model to describe costs associated to required software or hardware modifications (for example, as a consequence of network update or service improvement) in a complex network, as coulbe an operation support system architecture (i.e., a structure composed by software and hardware elements designed to fulfil predetermined tasks or processes, as service provision, requests management, alarm handling, inventory updates, among others). Finally, we analyze the influence of two network parameters (the average shortest distance L and the degree distribution exponent γ) in the cost of such modifications.

The costs we have in mind are those unwanted but inherent costs that arise when introducing changes in interconnected systems. The actual economic cost of implementing the modification is not considered in our model, which only tries to capture the "hidden" cost resulting from the complexity of the architecture substrate, and which is not established *a priori*. These costs can be anything which might be considered a liability, e.g. cost in economical expenditure, cost in time, cost in resources, cost in business management, and others.

2   Model

Our aim is to develop a simple model that successfully captures the observed features and costs of modification propagation due to the topological aspects of the network [6]. There are many reasons for software or hardware modifications in an architecture element, e.g., correction of defects, improvement of performance (or other related properties), adaptation to environment changes (technical updates, strategy or business priority modifications, regulatory changes), novelties in market industry, etc. Any modification of this nature has a certain impact over the rest of the components of the architecture. We aim to quantify this impact globally, particulary indirect effects that in principle are not obvious to detect. Our model will improve initial cost estimates of any modification, allowing for further analysis by implementing other relevant (but possibly secondary) features (i.e., network topology quantities, as clustering coefficient, diameter, etc.) which will improve the accuracy of those initial cost estimates.

In the present work, the impact or influence of modifications is modeled through *impact coefficients*, which to our purposes represent node relationship, i.e., the interaction of interfaces in a given system. In this way, we define a matrix K where an element K_{ij} is the coefficient describing the cost impact over node j of any modification done in node i. In the resulting matrix, links that do not exist have zero impact as well as the principal diagonal, meaning that self-impact is discarded. We consider the values of K_{ij} to be stochastic because they present fluctuations that depend on a very large number of variables. In our model, we consider these coefficients to have a well-defined statistics, in the sense that the stochastic variables involved have a precise origin (e.g. a Poissonian "arrival" of failures) and they can thus be correctly described by a given statistical distribution (more on this below). Moreover, these quantities usually vary in time, as nodes and links dynamically change inside the architecture, but for our initial purposes we consider them as stationary. To choose a suitable distribution, we take into account empirical evidence on how these coefficients should be. We consider three main characteristics: firstly, coefficients should be small (measured in arbitrary units, and compared to unity). Secondly, they must be positive (we assume the probability of a modification propagation to lower costs to be negligible). Finally, we must account also the possibility to have few, unusually high cost events (for example, when a difficult but important request from a client must be introduced) which will be described by an exponentially-tailed distribution.

Regarding the actual causes that generate this impact, we can follow basically two: firstly, costs derived from changes in functionalities or improvements in a particular node, and secondly, costs associated to defects involuntarily introduced in any node which propagate through the architecture. If we suppose these two factors to be uncorrelated, and also consider that they are well described by Gaussian probability distributions, then to account for both simultaneously we can make use of the probability distribution of the modulus, which follows a Rayleigh distribution. The Rayleigh distribution, for sufficiently low values of its variance σ^2, has the three needed characteristics we previously mentioned. Note that this is not the only choice one could make. In particular, there could be more causes of cost propagation, not taken into account here. In this case, we would have to use other (related) distributions: the Maxwell distribution (for an additional third variable), or a Chi distribution (if N variables would be needed). Indeed, we are currently working in determining the influence that a particular distribution has in the results, which in turn may lead to conclusions about the task of providing a useful characterization of the stochastic nature of modification propagation costs. To estimate the magnitude of the variance in such a way to get realistic values of impact coefficients, we choose the standard deviation in such a way that 90% of the coupling coefficients are equal or less than 0.05, i.e., one estimates the appropriate σ for which any coupling constant is less or equal than 0.05 with a 0.9 certainty probability. This is achieved with a standard deviation $\sigma = 0.04$.

To calculate the cost of a propagation of a modification done in node i, we perform the sum of all impact coefficients, taking into account all possible further ramifications of "child" nodes. We model this ramification process through the following (iterative) equation:

$$C_i = \sum_{j=1}^{J_i} K_{ij}(1 + C_j) \tag{1}$$

where J_i stands for the number of child nodes in node i. When computing subsequent C_j terms, and in order to avoid loops, the ith row and column from matrix K are removed, i.e., if one calculates what is the cost on node B due to a modification on node A, afterwards there is no contribution of the impact over node A product of a change in node B. Then, the calculation proceeds on the resulting matrices, until the last nodes are reached.

Our model also takes into account the fact that some nodes suffer modifications, while others do not. This means that at any moment, there is some probability p_i for node i to be modified. To model this situation we consider two different aspects. On one hand, all nodes are taken to have a (small) probability to be modified, so there is a small but positive contribution common to all nodes. In our case, we have assigned this probability to 1% (normalized by the total number of nodes). On the other hand, important nodes (i.e., "hubs") are considered to have a larger probability of requiring some modification, which reflects the need to attend an increasing demand, or to better adapt to the child nodes it already has. Therefore, we take into account another contribution to the probability to be modified that will be proportional to a power of the number of connections k that the node has, which in our case is set to $\frac{1}{2}$. Therefore, the probability to change is

$$p_i = c\sqrt{k_i} + 0.01\,\frac{1}{N} \tag{2}$$

where p_i is the probability that must be applied to node i, k_i the number of links of node i, and N the total number of nodes in the architecture. This probability must satisfy that the sum of all nodes equals unity, so

$$c = 0.09\,(\sum_{i=1}^{N} \sqrt{k_i})^{-1}. \tag{3}$$

The average total cost due to modification propagation will be evaluated as

$$C = \sum_{i=1}^{N} p_i\, C_i. \tag{4}$$

3   Results

To estimate how costs depend on the complexity that a substrate architecture or network exhibits, we generated a set artificial scale-free networks [7] (close to 200 randomly generated networks). These networks have a fixed number of nodes and links ($N = 110, N_l = 270$), but variable topological parameters, such as the (scale-free) degree distribution exponent γ, and average shortest length L (i.e., the average minimum number of links needed to travel between two nodes). Our results were calculated with K_{ij} coefficients taken from a Rayleigh distribution of $\sigma = 0.04$ to describe each node pair. Then, for each artificially generated test network, we calculated the associated cost (applying Eq. 4) and directly observed the relationship between cost and structure. Our preliminary results show that there is a strong influence from the topology of the network on our cost model. In Fig. 1 we show the modification propagation cost as a function of these two complexity parameters, namely, the degree distribution exponent γ (right) and the average shortest length L (left). As we can see from the figure, our results

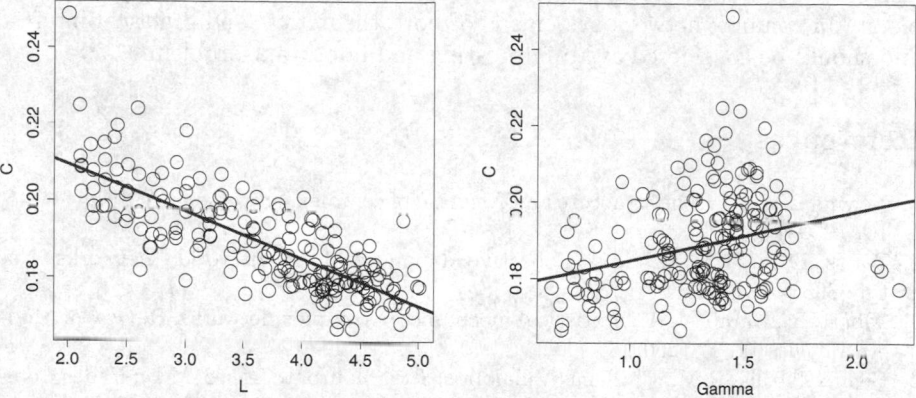

Fig. 1. Left: Propagation cost C as a function of the average shortest length of the network L. Right: Propagation cost C as a function of the degree distribution exponent γ. In solid line, a linear fitting. Both quantities present an overall relative cost change of approximately 20%, with negative slope in the case of the average shortest length L, and with positive slope regarding the γ dependence, in the ranges shown in the figures.

show that, in average, costs decrease as the average shortest length increases. and increases with a higher heterogeneity of the networks. In order to have a general estimative of the relative change, we performed linear fits in our results. In both cases, the absolute change in costs (associated with the chosen range of L and γ) is of about 20%, being the slopes $m_\gamma = 0.012(3)$ and $m_L = 0.0123(7)$. Clearly, a linear estimate is more suit for the lenght dependence than for the γ dependence, where more analysis is needed to determine its influence. Such analysis is currently being addressed.

4 Conclusions

Adapting an organization's business processes to changes in the environment is crucial for maintenance tasks, optimal network evolution, among many other fundamental business activities. The considerable cost of these tasks relates directly to the topology of the operating architecture, i.e., the system designed to carry the organization's processes. In this work, we introduce a simple model to estimate the cost of such modifications in a complex system architecture. Our study is aimed at estimating the unwanted cost of a software modification on an architecture before its actual implementation, by measuring the existing system topology. Furthermore, we plan to explore the possibility to identify the point at which architectures are no longer economically profitable, as a consequence of too high modification costs. Our preliminary results suggest that network complexity (in the sense of scale-free characteristics) implies higher economic expenses when changes need to be made, but larger systems actually decrease this cost. The correct estimation of propagation of costs due to inherent complexity in complex network systems is of both theoretical and practical interest and should be considered as an interesting and important problem.

References

1. Newman, M.E.J.: The structure and function of complex networks. SIAM Review 45, 167 (2003)
2. Watts, D.J., Strogatz, S.H.: Collective dynamics of "small-world" networks. Nature 393, 440 (1998)
3. Albert, R., Barabási, A.: Statistical mechanics of complex networks. Review of Modern Physics 74, 47 (2002)
4. Benito Zafrilla, R.M., Cárdenas Villalobos, J.P., Mouronte López, M.L.: Redes Complejas: El nuevo paradigma. Sociedad de la Información, Tecnología e Información Bulletin (Fundación Telefónica), Madrid (2007)
5. Mouronte, M.L., Armas, A.: Análisis de la Complejidad del Mapa de Sistemas de Telefónica de España, Sociedad de la Información, Tecnología e Información Bulletin (Fundación Telefónica), Madrid (2006)
6. Pastor-Satorras, R., Vespignani, A.: Epidemic dynamics in finite size scale-free networks. Phys. Rev. E. 65, 035108 (2002)
7. Catanzaro, M., Boguña, M., Pastor-Satorras, R.: Generation of uncorrelated random scale-free networks. Phys. Rev. E. 71, 027103 (2005)

Modelling of Population Migration to Reproduce Rank-Size Distribution of Cities in Japan

Hiroto Kuninaka and Mitsugu Matsushita

Dept. of Physics, Chuo University, Kasuga, Bunkyo-ku, 112-8551, Japan
kuninaka@phys.chuo-u.ac.jp
http://www.phys.chuo-u.ac.jp/labs/matusita/

Abstract. We investigate the rank-size distribution of cities in Japan by data analysis and computer simulation. From our previous data analysis of the census data after World War II, it has been clarified that the power exponent of the rank-size distribution of cities changes with time and Zipf's law holds only for a restricted period. We show that Zipf's law broke down owing to the great mergers and recovered by investigating the time evolution of the rank-size distribution of cities without mergers.

Keywords: population, lognormal distribution, power-law distribution, Zipf's law, agent-based modelling.

1 Introduction

Many empirical data that obey the power-law distribution can be observed in both natural and social phenomena. Among them, we often find the special case that the power exponent becomes unity in various phenomena [1]. Generally, we call the empirical law as Zipf's law [2].

In Japan, the rank-size distribution of cities shows the power-law distribution while the rank-size distributions of other municipalities, such as towns and villages, can be approximated much better by the lognormal distribution [4]. Auerbach [3] first reported that the rank-size distribution of the population of cities obeys the power-law distribution which is described by

$$\log R(x) = a - b \log x, \tag{1}$$

where $R(x)$ is the rank of the population x, and both a and b are fitting parameters. Afterward, Zipf proposed that the power exponent b empirically becomes unity in the rank-size distribution of cities [2].

However, the power exponent of the rank-size distribution of cities can easily change due to various factors such as population migration, a change in birth rate, economic situation, etc.[5]. In the case of Japan, after the end of World War II, the rank-size distribution of cities has been changed and strongly affected by the two great mergers of municipalities: the Showa (from 1955 to 1960) and Heisei (from 2000 to now) great mergers [6]. Our previous data analysis and simulation have clarified that the power exponent of the rank-size distribution

J. Zhou (Ed.): Complex 2009, Part I, LNICST 4, pp. 441–445, 2009.

of cities changed under the influence of those great mergers and approached to
unity after the Showa great merger [6].

In this paper, we investigate the time-evolution of the rank-size distribution of
cities without great mergers of municipalities. Our data analysis is based on the
census data from 1950 to 2006, which were obtained from the Statistics Bureau,
Ministry of International Affairs and Communications, Japan [7], and data book
from Japan Statistical Association [8].

2   Data Analysis

Figure 1(a) shows the time evolution of the rank-size distribution of cities in
Japan from 1950 to 2000 [6]. We can find that the head and tail parts of each
distribution can be fitted by discrete power-law distribution functions. In addition, between 1950 and 1960, we can find a remarkable increase in the number
of cities by 307.

Plus points in Fig. 2 show the time evolution of the power exponent of the
rank-size distribution of cities [6]. Between 1950 and 1960, we can find a remarkable increase of the power exponent from the value around unity, which is mainly
due to the increase in the number of cities caused by the great Showa merger. In
addition, during this period, some towns might be promoted to cities when they
fulfilled the promotion condition, one of which is that the population within a
town is larger than 50,000 people. These factors caused the increase of the power
exponent. After the increase, the power exponent shows a monotonic decrease
approaching unity, which means that Zipf's law is recovered. Thus, we can suppose that the power exponent might show monotonic decrease approaching unity
without the increase of the number of cities.

To exclude the effect of the great mergers of municipalities and the promotion
to cities, we trace the rank-size distribution of the cities as of October 1, 2000
back to 1950. Figure 1(b) shows the time evolution of the rank-size distribution
of the 672 cities as of October 1, 2000. The power exponent for each year is

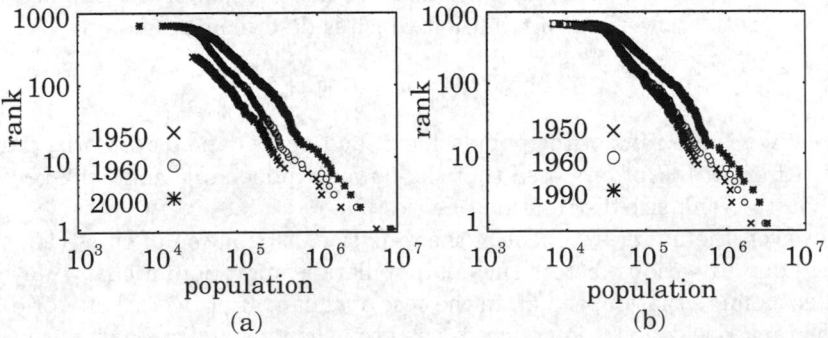

Fig. 1. Time evolution of rank-size distribution of cities in Japan both (a) with and
(b) without mergers of municipalities

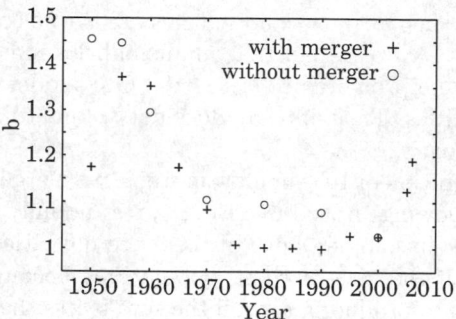

Fig. 2. Time evolution of power exponents of rank-size distribution of cities both with and without mergers

shown as the open circles in Fig.2, where we find the monotonic decrease of the power exponent approaching unity although the values of b are larger than those with mergers in 1980 and 1990. We need to investigate the time evolution of the rank-size distribution of the cities at the instant of 1950 until 2000.

3 Simulation

Here we introduce our model of the population migration to reproduce the increase in b due to the merger of municipalities and its convergence to unity after the merger [6]. Our model is based on an agent-based model that consists of 3500 sites corresponding to all the municipalities. Each site has a uniform random number between 0 and 1 as the initial population. The basic procedure of one simulation steps is summarized as follows:

1. We randomly choose a source site m with the population N_m.
2. We choose a group of sites, $G_{N<N_m}$ or $G_{N>N_m}$, whose populations N are less and more than N_m, respectively. The probability to choose $G_{N<N_m}$ is α (migration parameter), while that to choose $G_{N>N_m}$ is $1 - \alpha$.
3. Among the group of sites chosen in the previous step, we randomly choose the destination site n for migration.
4. The P_{mn} percent of N_m is transferred to the site n, so that the populations of sites m and n vary in quantity as $N_m - P_{mn}N_m$ and $N_n + P_{mn}N_m$, respectively.

In the second step, the migration parameter α is introduced to describe the tendency that people migrate to less populated area from large cities, which was evident after the high economic growth from 1960 to the early 1970s [9]. In addition, P_{mn} is randomly chosen in the range from 0 to 20. We iterate this procedure 10^6 times in our simulation. A sample average is taken over 10 different initial population distributions for all the sites.

When the population of a given site becomes larger than 0.95, we regard the site as a city. Once a site is promoted to a city, the site will not be demoted to a smaller municipality such as towns and villages. This rule corresponds to part of the Local Autonomy Law of Japan that municipalities must have a population of $50,000$ or more to be promoted to cities [10]. Our model does not distinguish between towns and cities. Thus, if a site does not belong to cities, we henceforth call the site as a "town".

After the first migration of 10^6 simulation steps, we merge some municipalities according to the following procedure. First, we randomly choose two sites to merge among all the sites. When both of them are not cities, we merge them to produce a new city if the sum of those populations becomes larger than 0.95, while we merge them to produce a town if the sum is less than 0.95. On the other hand, when at least one site is a city, we merge those two sites with a probability $\beta = 0.5$ to become a new city. The probability β is introduced owing to the fact that the frequency of the merger of towns was much larger than that of cities. We iterate this merging process until the number of cities increases by 77 on average rather than when the first migration stage is finished. In our model, the increase in the number of cities affects the power exponent after the merger. In general, the power exponent increases with an increase in the number of cities generated by the merger.

4   Simulation Result

Figure 3(a) shows the time evolution of the power exponent of rank-size distributions of cities in 10^5, 10^6, and 10^7 simulation steps without mergers. Error bars that are almost invisible on a few data marks are standard deviation obtained by least-squares linear regression. In this simulation, α is fixed at $\alpha = 0.3$. This figure shows that the power exponent converges to the stationary value around $b = 1.094 \pm 0.001$. Thus, our model can reproduce the power-law distribution of cities that converges to Zipf's law although the number of cities keeps increasing after the power exponent becomes $b = 1$, which slightly increases the power exponent.

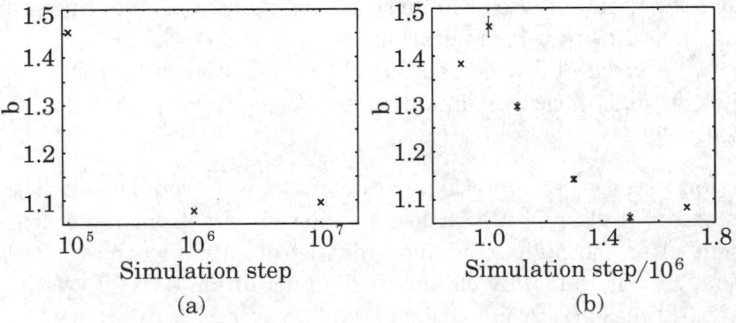

Fig. 3. Time evolution of power exponents of rank-size distribution of cities both (a) without and (b) with mergers

Next, we investigate how the great merger affects the rank-size distributions of cities through the time evolution of the power exponent b. We carry out the first population migration of 10^6 simulation steps. After that, we merge some of those sites, followed by the second population migration of 7×10^5 simulation steps.

Figure 3(b) shows the relation between the power exponent b and the simulation step. Error bars that are almost invisible on a few data marks are standard deviation obtained by least-squares linear regression. Data point at 10^6 steps shows the power exponent b after the merger has finished. We find that b converges to unity after the increase in b due to the merger. Thus, our model can reproduce the time evolution of b qualitatively.

5 Concluding Remarks

We have investigated the time evolution of the rank-size distribution of cities to show how the power exponent might change if there were no merger. The rank-size distribution shows that the power-law behavior and the time evolution of the power exponent markedly change when the great merger of municipalities occurs.

Our future task is, by the use of our model, to reproduce the time evolution of the rank-size distribution of cities without both mergers and the promotion to cities. In addition, we need to investigate the rank-size distribution of all the municipalities.

References

1. Information on Zipf's law, http://www.nslij-genetics.org/wli/zipf/
2. Zipf, G.: Human Behavior and the Principle of Least Effort. Addison-Wesley, Cambridge (1949)
3. Auerbach, F.: Das Gesetz der Bevölkerungskonzentration. Petermanns Geogr. Mitt. LIX, 74–76 (1913)
4. Sasaki, Y., Kuninaka, H., Kobayashi, N., Matsushita, M.: Characteristics of population distribution in municipalities. J. Phys. Soc. Jpn. 76, 074801-1–074801-6 (2007)
5. Soo, K.T.: Zipf's Law for cities: a cross-country investigation. Reg. Sci. Urban. Econ. 35, 239–263 (2005)
6. Kuninaka, H., Matsushita, M.: Why does Zipf's law break down in rank-size distribution of cities? J. Phys. Soc. Jpn. 77, 114801-1-114801-6 (2008)
7. The Statistics Bureau, Ministry of International Affairs and Communications, Japan, http://www.stat.go.jp/
8. Shikuchoson Jinko no Chouki Keiretsu (Long-Term Data for Population of Cities, Wards, Towns and Villages). Nihon Tokei Kyokai, Tokyo (2005)
9. Arai, Y., Kawaguchi, T., Inoue, T.: Nihon no Jinkou Idou (Population Migration in Japan) Kokon Shoin, Tokyo (2002)
10. The Nippon Foundation Library, http://nippon.zaidan.info/seikabutsu/1999/00168/mokuji.htm

Modeling and Robustness Analysis of Biochemical Networks of Glycerol Metabolism by *Klebsiella Pneumoniae*

Jianxiong Ye[1,*], Enmin Feng[1], Lei Wang[1], Zhilong Xiu[2], and Yaqin Sun[2]

[1] Department of Applied Mathematics, Dalian University of Technology, Dalian, 116024, Liaoning, China
yejianxiong128@yahoo.cn

[2] Department of Biotechnology, Dalian University of Technology, Dalian, 116012, Liaoning, China

Abstract. Glycerol bioconversion to 1,3-propanediol (1,3-PD) by *Klebsiella pneumoniae* (*K. pneumoniae*) can be characterized by an intricate network of interactions among biochemical fluxes, metabolic compounds, key enzymes and genetic regulatory. To date, there still exist some uncertain factors in this complex network because of the limitation in biotechniques, especially in measuring techniques for intracellular substances. In this paper, among these uncertain factors, we aim to infer the transport mechanisms of glycerol and 1,3-PD across the cell membrane, which have received intensive interest in recent years. On the basis of different inferences of the transport mechanisms, we reconstruct various metabolic networks correspondingly and subsequently develop their dynamical systems (S-systems). To determine the most reasonable metabolic network from all possible ones, we establish a quantitative definition of biological robustness and undertake parameter identification and robustness analysis for each system. Numerical results show that it is most possible that both glycerol and 1,3-PD pass the cell membrane by active transport and passive diffusion.

Keywords: Metabolic network inference, Biological robustness, Transport across cell membrane, Parameter identification.

1 Introduction

The bioconversion of glycerol to 1,3-PD is particularly attractive to industry because of the increasing glycerol surplus on the market and the potential uses of the product 1,3-PD. The latter is discussed as a bifunctional chemical reagent on a large commercial scale, especially as a monomer for polyesters, polyethers and polyurethanes. Since the 1980s, a number of computational models have been developed to describe the fermentation process of glycerol by *K. pneumoniae*. The latest ones proposed and modified by Zeng et al. [1,2,3] have been widely

---

* Corresponding author.

J. Zhou (Ed.): Complex 2009, Part I, LNICST 4, pp. 446–457, 2009.

used in computer simulation. Nevertheless, some important intracellular intermediate substances (such as 3-hydroxypropionaldehyde) and enzymes (such as glycerol dehydratase and 1,3-PD oxydoreductase), which play important roles in glycerol metabolism, were not taken into consideration in the above models. Therefore, these models are limited in providing a qualitative understanding of the underlying metabolic network, even though they have successful predictions about the fermentation process.

Recently, metabolic network of glycerol by *K. pneumoniae* has received intensive interest [4,5,6,7], since it is helpful for strain selection and genetic modification for increasing 1,3-PD production. The metabolic pathways of glycerol in the intracellular environment consist of oxidative and reductive components. Sun et al. [8] proposed a mathematical model to describe the reductive pathway, in which the interrelationships among substrate, key enzymes, intermediates and target product were considered. By comparing *K. pneumoniae* with *E. coli*, Sun et al. assumed that glycerol passes the cell membrane by both passive diffusion and active transport and 1,3-PD is transported by passive diffusion. The model can successfully simulate experimental results and forecast intracellular metabolites concentration for continuous cultures. But fundamental properties of this metabolic network (such as robustness) were not considered in [8].

Robustness is a property that allows a system to maintain its functions despite external and internal perturbations. It is one of the fundamental characteristics of biological systems. Numerous reports have been published on this topic, especially in the past decades [9,10,11,12,13,14,15]. N. Barkai et al. [11] argued that the key properties of biochemical networks are robust, that is, they are relatively insensitive to the precise values of biochemical parameters. This point of view, which has been observed for a wide variety of experiments [16,17,18], is being gradually accepted by experts in the field of systems biology.

In this paper, we study the fermentations of glycerol covering both extracellular and intracellular environments. The two environments are linked by the transports of substrate (glycerol) and product (1,3-PD) across the cell membrane, which haven't been completely observed in experiments yet. All possible transport mechanisms are under consideration. Different metabolic networks are reconstructed based on distinct inferences of transport mechanisms and correspondingly different dynamic systems are developed. Then parameters in each system are identified to minimize the relative error between experimental data and computational results. Since only extracellular data can be measured in experiments, it is still hard to know which inference of the transport mechanisms is the most reasonable. To cope with this problem, robustness analysis is carried out for each system. We first propose a quantitative definition of biological robustness. After that, robust performance is calculated for each system and the reasonability of these systems are measured by their robust performances.

This paper is organized as follows. In Section 2, all possible dynamical systems of glycerol dissimilation are developed and parameters in each system are identified. In Section 3, a quantitative definition of biological robustness is established and robust performance of each system is calculated to determine the most reasonable metabolic network. Some conclusions are presented at the end of this paper.

2   Modeling and Parameter Identification

In this section, taking the transports of glycerol and 1,3-PD across cell membrane into consideration, we develop three kinetic models to describe the reductive pathway of glycerol dissimilation. Moreover, parameters are identified for each system based on 30 groups of experimental data reported in [1,2,3].

2.1   Kinetic Models

During glycerol metabolism by $K.$ $pneumoniae$ under anaerobic condition, glycerol is first transported across the cell membrane from the extracellular environment to the intracellular environment, and then is further catabolized, reactions catalyzed by enzymes, to generate intermediates and final products, e.g., 3-hydroxypropionaldehyde (3-HPA), 1,3-PD, acetic acid, ethanol, etc. Finally, the products are transported across the cell membrane from the intracellular environment to the extracellular environment. As shown in Fig. 1, the transport mechanisms of glycerol and 1,3-PD across the membrane haven't been observed in experiments yet. Glycerol may pass the membrane by both passive diffusion and active transport or by passive diffusion only. So is the transport of 1,3-PD. In addition, since the molecular weight of glycerol is larger than that of 1,3-PD, it is thought that 1,3-PD will pass the cell membrane only by passive diffusion if glycerol is transported in this pattern [19]. Therefore, three possible cases are discussed, and kinetic systems are developed according to different inferences of the transport mechanisms.

All dynamical models developed in this paper are power-law systems (S-systems), which were first proposed by Savageau et al. [20]. As is known to us, Michaelis-Menten kinetics have been widely used in the study of biological dynamical systems. However, if several substrates or reactions are involved, and if several modulators affect the pathway, the system quickly becomes quite complicated. Another disadvantage of Michaelis-Menten kinetics is that one should clearly know the mechanisms of the constituent enzymes in the pathway before formulating its mathematical model. It is hard to image that complex pathways could be successful described and numerically identified in terms of their detailed enzyme mechanisms by Michaelis-Menten kinetics. Different from Michaelis-Menten kinetics, S-systems are equivalent to linearization in logarithmic coordinates, which are especially convenient for steady-state analysis in large-scale systems.

According to the factual experiments, we make the following assumptions.

(**H1**) The extracellular and intracellular concentrations of substances are uniform in reactor and in cells, respectively.
(**H2**) The effect of the microbial growth on the concentrations of intracellular substances is ignored.

(**H3**) The oxidative pathway can afford the reductive pathway enough reducing power.

Under the above assumptions and according to Fig. 1, three S-systems are developed as follows.

Case 1. Glycerol and 1,3-PD pass cell membrane by both passive diffusion and active transport. Then the dynamical system, denoted by S(1), is formulated as:

$$
\begin{cases}
\dot{x}_1(t) = a_{1,1} x_1^{g_{1,1}} x_2^{g_{1,2}} x_3^{g_{1,3}} x_4^{g_{1,4}} x_5^{g_{1,5}} - b_{1,1} x_1^{h_{1,1}} D^{h_{1,2}} \\
\dot{x}_2(t) = a_{1,2} \left(\frac{c_{s_0}}{x_2}\right)^{g_{1,6}} D^{g_{1,7}} - b_{1,2} x_1^{h_{1,3}} x_2^{h_{1,4}} \left(\frac{x_2}{x_6}\right)^{h_{1,5}} \\
\dot{x}_3(t) = a_{1,3} x_1^{g_{1,8}} x_8^{g_{1,9}} \left(\frac{x_8}{x_3}\right)^{g_{1,10}} - b_{1,3} D^{h_{1,6}} x_3^{h_{1,7}} \\
\dot{x}_4(t) = a_{1,4} x_1^{g_{1,11}} x_2^{g_{1,12}} - b_{1,4} D^{h_{1,8}} x_4^{h_{1,9}} \\
\dot{x}_5(t) = a_{1,5} x_1^{g_{1,13}} x_2^{g_{1,14}} - b_{1,5} D^{h_{1,10}} x_5^{h_{1,11}} \\
\dot{x}_6(t) = a_{1,6} x_2^{g_{1,15}} \left(\frac{x_2}{x_6}\right)^{g_{1,16}} - b_{1,6} x_6^{h_{1,12}} x_7^{h_{1,13}} \\
\dot{x}_7(t) = a_{1,7} x_6^{g_{1,17}} x_7^{g_{1,18}} - b_{1,7} x_7^{h_{1,14}} \\
\dot{x}_8(t) = a_{1,8} x_7^{g_{1,19}} - b_{1,8} x_8^{h_{1,15}} \left(\frac{x_8}{x_3}\right)^{h_{1,16}} \\
x(t_0) = x^0
\end{cases}
\qquad t \in [t_0, t_f^1], \quad (1)
$$

where x_1, x_2, x_3, x_4, x_5 are concentrations of biomass, glycerol, 1,3-PD, acetic acid, ethnol in reactor, respectively. x_6, x_7, x_8 are intracellular concentrations of glycerol, 3-HPA, 1,3-PD, respectively. x^0 is the initial state, which is restricted in

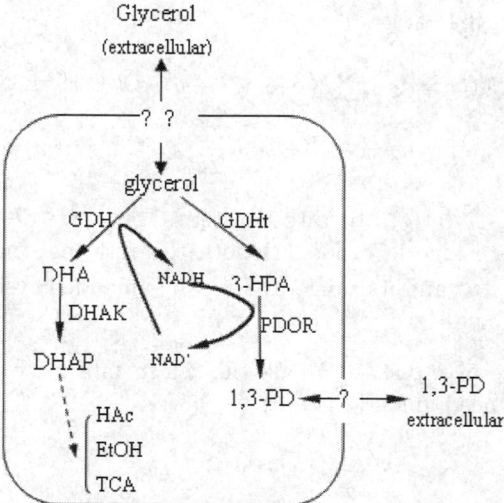

Fig. 1. Anaerobic metabolic pathways of glycerol fermentation. *Abbreviations*: GDHt, glycerol dehydratase; 3-HPA, 3-hydroxypropionaldehyde; 1,3-PD, 1,3-propanediol; PDOR, 1,3-PD oxydoreductase; GDH, glycerol dehydrogenase; DHA, dihydroxyacetone; DHAK, dihydroxyacetone kinase; DHAP, dihydroxyacetonephosphate; HAc, acetic acid; ETOH, ethnol; TCA, TCA cycle.

$S_0 := [0.001, x_1^*] \times [100, x_2^*] \times [0, x_3^*] \times [0, x_4^*] \times [0, x_5^*] \times [0, x_6^*] \times [0, x_7^*] \times [0, x_8^*]$. x_i^*, $i = 1, 2, \cdots, 8$, are the critical values of x_i, respectively. Biomass, glycerol, intermediate substances and products can not exceed their critical concentrations according to the practical production. t_f^1 is the terminal moment of the fermentation, i.e., the moment when the system S(1) reaches its steady state. D is dilution rate and c_{s_0} is substrate concentrate in feed, which are two of the main operating conditions in continuous fermentations. $a_{1,i}$ and $b_{1,i}$ ($i \in I_8 := \{1, 2, \cdots, 8\}$) are rate constants. $g_{1,j}$ ($j \in I_{19}$) and $h_{1,k}$ ($k \in I_{16}$) are kinetic orders.

For convenience, let $x := (x_1, x_2, \cdots, x_8)^T$ and $u^1 := (a_{1,1}, \cdots, a_{1,8}, b_{1,1}, \cdots, b_{1,8}, g_{1,1}, \cdots, g_{1,19}, h_{1,1}, \cdots, h_{1,16})^T$. Generally speaking, kinetic orders representing biochemical reactions or transport steps are very often in the range between 0 and 1, and inhibitory effects typically call for kinetic orders between 0 and -0.5. Rate constants are set in the range between 0 and 1000 in this paper. From the above statements, we set the range of the parameter vector u^1 as $\mathcal{U}^1$. In addition, according to experimental limitation, the range of the operating conditions vector $v = (D, c_{s_0})^T$ and the admissible set of state vector x are denoted by W_{ad} and Λ_{ad}, respectively. Let $f^1 := (f_1^1, f_2^1, \cdots, f_8^1)^T$, $f_i^1 : \Lambda_{ad} \times W_{ad} \times \mathcal{U}^1 \to R$, $i = 1, \cdots, 8$. Then we rewrite S(1) as:

$$\begin{cases} \dot{x}(t) = f^1(x(t), v, u^1), \\ x(t_0) = x^0 . \end{cases} \qquad t \in [t_0, t_f^1], \qquad (2)$$

Case 2. Glycerol passes the cell membrane by both passive diffusion and active transport but 1,3-PD passes the cell membrane by passive diffusion only. Then we just need amend S(1) at:

$$\dot{x}_3(t) = a_{2,3} x_1^{g_{2,8}} \left(\frac{x_8}{x_3}\right)^{g_{2,10}} - b_{2,3} D^{h_{2,6}} x_3^{h_{2,7}}, \qquad (3)$$

$$\dot{x}_8(t) = a_{2,8} x_7^{g_{2,19}} - b_{2,8} \left(\frac{x_8}{x_3}\right)^{h_{2,16}}, \qquad (4)$$

where $a_{2,i}$ and $b_{2,i}$ ($i \in I_8$) are rate constants, $g_{2,j}$ ($j \in I_{19} \setminus \{9\}$) and $h_{2,k}$ ($k \in I_{16} \setminus \{15\}$) are kinetic orders. Denote the terminal moment by t_f^2, the parameter vector by u^2 and its range by $\mathcal{U}^2$. The amended system is denoted by S(2) with its right hand term $f^2(x(t), v, u^2)$.

Case 3. Both glycerol and 1,3-PD pass the cell membrane by passive diffusion only. Then we just need amend S(1) at:

$$\dot{x}_2(t) = a_{3,2} \left(\frac{c_{s_0}}{x_2}\right)^{g_{3,6}} D^{g_{3,7}} - b_{3,2} x_1^{h_{3,3}} \left(\frac{x_2}{x_6}\right)^{h_{3,5}}, \qquad (5)$$

$$\dot{x}_3(t) = a_{3,3} x_1^{g_{3,8}} \left(\frac{x_8}{x_3}\right)^{g_{2,10}} - b_{2,3} D^{h_{2,6}} x_3^{h_{2,7}}, \qquad (6)$$

$$\dot{x}_6(t) = a_{3,6} \left(\frac{x_2}{x_6}\right)^{g_{3,16}} - b_{3,6} x_6^{h_{3,12}} x_7^{h_{3,13}}, \qquad (7)$$

$$\dot{x}_8(t) = a_{3,8} x_7^{g_{3,19}} - b_{3,8} \left(\frac{x_8}{x_3}\right)^{h_{3,16}}, \qquad (8)$$

where $a_{3,i}$ and $b_{3,i}$ $(i \in I_8)$ are rate constants, $g_{3,j}$ $(j \in I_{19} \setminus \{9, 15\})$ and $h_{3,k}$ $(k \in I_{16} \setminus \{4, 15\})$ are kinetic orders. Denote the terminal moment by t_f^3, the parameter vector by u^3 and its range by $\mathcal{U}^3$. The amended system is denoted by S(3) with its right hand term $f^3(x(t), v, u^3)$.

2.2   Parameter Identification

Let $I_l := \{1, \cdots, l\}$ be the serial number set of experiments, where l is the total experiment times. For given $v^s, s \in I_l$, denote the experimental values of extracellular concentrations of reactants at steady stage as $y_1^s, y_2^s, y_3^s, y_4^s, y_5^s$ correspondingly. Let $y^s := (y_1^s, y_2^s, y_3^s, y_4^s, y_5^s)^T \in R^5, s \in I_l$. While reaching the steady state t_f^m, the solution of the system S(m) for given v^s satisfies that

$$f^m(x(t_f^m), v^s, u^m) = 0, \qquad m \in I_3 . \tag{9}$$

Although the concentrations of biomass, glycerol, 1,3-PD, acetic acid and ethanol in reactor are measured in experiments, this paper is only concerned with the relative error between experimental data and computational values of the first three substances for the reason that alkali is intermittently fed into the reactor to maintain its pH value at 7 or so, which has great effect on the extracellular concentrations of acetic acid and ethanol. Therefore, the parameter identification problem can be formulated as follows:

$$P(m): \quad \min \frac{1}{3} \sum_{n=1}^{3} \frac{\sum_{s \in I_l} |x_n^s - y_n^s|}{\sum_{s \in I_l} y_n^s} \tag{10}$$
$$\text{s.t.} \quad f^m(x^s, v^s, u^m) = 0, \quad s \in I_l,$$
$$(x^s, u^m) \in \Lambda_{ad} \times \mathcal{U}^m .$$

After taking logarithms and rearranging, we can rewrite $P(m)$ in the following form:

$$P'(m): \quad \min \frac{1}{3} \sum_{n=1}^{3} \frac{\sum_{s \in I_l} |z_n^s - \ln(y_n^s)|}{\sum_{s \in I_l} \ln(y_n^s)} \tag{11}$$
$$\text{s.t.} \quad A(u^m) \cdot z^s = b(u^m, v^s), \quad s \in I_l,$$
$$(z^s, u^m) \in \Lambda'_{ad} \times \mathcal{U}^m .$$

where $z^s := (z_1^s, \cdots, z_8^s)^T = (\ln(x_1^s), \cdots, \ln(x_8^s))^T$. $A(u^m)$ is an 8×8 matrix determined by u^m and $b(u^m, v^s)$ is a vector in R^8 with its value determined by u^m and v^s. Λ'_{ad} is a set transformed from Λ_{ad} in logarithmic coordinates. The detailed technique can be seen in [20].

The parameter identification problem $P(m)$ is equivalent to $P'(m)$, which is solved by an improved real-coded genetic algorithm introduced in [21]. To

Table 1. Identified parameters for S(1)

$a_{1,1}$	$a_{1,2}$	$a_{1,3}$	$a_{1,4}$	$a_{1,5}$	$a_{1,6}$	$a_{1,7}$	$a_{1,8}$
466.765	364.617	468.720	492.180	138.326	98.737	105.090	435.973
$b_{1,1}$	$b_{1,2}$	$b_{1,3}$	$b_{1,4}$	$b_{1,5}$	$b_{1,6}$	$b_{1,7}$	$b_{1,8}$
345.555	394.919	83.097	414.958	351.909	124.152	171.561	398.340
$g_{1,1}$	$g_{1,2}$	$g_{1,3}$	$g_{1,4}$	$g_{1,5}$	$g_{1,6}$	$g_{1,7}$	$g_{1,8}$
0.0791	-0.0651	-0.0176	-0.0393	-0.0276	0.8128	0.9679	0.02346
$g_{1,9}$	$g_{1,10}$	$g_{1,11}$	$g_{1,12}$	$g_{1,13}$	$g_{1,14}$	$g_{1,15}$	$g_{1,16}$
0.2204	0.6301	0.0596	-0.0630	0.4374	-0.0510	0.4619	0.8843
$g_{1,17}$	$g_{1,18}$	$g_{1,19}$					
0.2815	-0.0702	0.3402					
$h_{1,1}$	$h_{1,2}$	$h_{1,3}$	$h_{1,4}$	$h_{1,5}$	$h_{1,6}$	$h_{1,7}$	$h_{1,8}$
0.4326	0.2242	0.5	0.0005	0.9984	0.7153	0.4697	0.7623
$h_{1,9}$	$h_{1,10}$	$h_{1,11}$	$h_{1,12}$	$h_{1,13}$	$h_{1,14}$	$h_{1,15}$	$h_{1,16}$
0.4936	0.3657	0.4189	0.3807	-0.1195	0.3016	0.4697	0.4985

Table 2. Identified parameters for S(2)

$a_{2,1}$	$a_{2,2}$	$a_{2,3}$	$a_{2,4}$	$a_{2,5}$	$a_{2,6}$	$a_{2,7}$	$a_{2,8}$
50.3511	407.6265	349.4654	356.7966	163.7408	188.1783	177.4258	175.9596
$b_{2,1}$	$b_{2,2}$	$b_{2,3}$	$b_{2,4}$	$b_{2,5}$	$b_{2,6}$	$b_{2,7}$	$b_{2,8}$
64.5248	105.5797	49.3736	407.1377	483.8713	475.0738	104.1135	392.4753
$g_{2,1}$	$g_{2,2}$	$g_{2,3}$	$g_{2,4}$	$g_{2,5}$	$g_{2,6}$	$g_{2,7}$	$g_{2,8}$
0.0010	-0.0875	-0.0278	-0.0451	-0.0948	0.6613	0.8835	0.02102
$g_{2,9}$	$g_{2,10}$	$g_{2,11}$	$g_{2,12}$	$g_{2,13}$	$g_{2,14}$	$g_{2,15}$	$g_{2,16}$
—	0.5042	0.2116	-0.0639	0.4907	-0.0501	0.4971	0.2
$g_{2,17}$	$g_{2,18}$	$g_{2,19}$					
0.2121	-0.0450	0.2146					
$h_{2,1}$	$h_{2,2}$	$h_{2,3}$	$h_{2,4}$	$h_{2,5}$	$h_{2,6}$	$h_{2,7}$	$h_{2,8}$
0.4844	0.6004	0.4585	0.1271	0.5980	0.4166	0.4751	0.5918
$h_{2,9}$	$h_{2,10}$	$h_{2,11}$	$h_{2,12}$	$h_{2,13}$	$h_{2,14}$	$h_{2,15}$	$h_{2,16}$
0.3959	0.3228	0.2512	0.3847	-0.0623	0.4599	—	0.4824

deal with the equality constraint, each parameter vector as an individual is substituted into the linear equations in (11) to compute the corresponding state vector and its performance index. Convex crossover operator is used to ensure the offspring still lie in $\mathcal{U}_{ad}$. In addition, ranking selection and multi-Gaussian mutation operators are adopted in the algorithm.

The relative errors between experimental data and computational values and the identified parameters for each system and are shown in Tables 1-4. Numerical results show that there exists no great difference among the relative errors of the three systems. So it is hardly to determine the best one only by parameter identification.

Table 3. Identified parameters for S(3)

$a_{3,1}$	$a_{3,2}$	$a_{3,3}$	$a_{3,4}$	$a_{3,5}$	$a_{3,6}$	$a_{3,7}$	$a_{3,8}$
123.6634	304.0117	401.7615	39.1099	143.7021	291.7930	31.2899	452.1026

$b_{3,1}$	$b_{3,2}$	$b_{3,3}$	$b_{3,4}$	$b_{3,5}$	$b_{3,6}$	$b_{3,7}$	$b_{3,8}$
358.2629	114.8660	72.8336	417.4015	369.0153	180.3583	244.8731	334.8029

$g_{3,1}$	$g_{3,2}$	$g_{3,3}$	$g_{3,4}$	$g_{3,5}$	$g_{3,6}$	$g_{3,7}$	$g_{3,8}$
0.08264213	-0.0845	-0.022	-0.0378	-0.089	0.7546	0.9945	0.00293

$g_{3,9}$	$g_{3,10}$	$g_{3,11}$	$g_{3,12}$	$g_{3,13}$	$g_{3,14}$	$g_{3,15}$	$g_{3,16}$
—	0.4925	0.4995	-0.0993	0.3783	-0.09589	—	0.8209

$g_{3,17}$	$g_{3,18}$	$g_{3,19}$
0.3113	-0.0141	0.1232

$h_{3,1}$	$h_{0,2}$	$h_{3,3}$	$h_{3,4}$	$h_{3,5}$	$h_{3,6}$	$h_{3,7}$	$h_{3,8}$
0.4985	0.6927	0.4985	—	0.2156	0.3345	0.4614	0.3095

$h_{3,9}$	$h_{3,10}$	$h_{3,11}$	$h_{3,12}$	$h_{3,13}$	$h_{3,14}$	$h_{3,15}$	$h_{3,16}$
0.45797	0.22502	0.1334	0.38563	-0.010	0.4917	—	0.3294

Table 4. Relative errors between experimental data and computational results

$S(m)$	$m = 1$	$m = 2$	$m = 3$
Error	36.9985%	35.032%	37.1262%

3   Robustness Analysis

On the study of biological systems, one always faces a situation that the inter-relationships of the pools (substrates, enzymes, products, etc.) can't be clearly known even though all pools have been detected. In terms of biological networks, we just know the nodes of the networks but have incomplete information of their edges. The true network need to be identified from all possible ones. In this context, some basic features of biological systems (such as robustness) should be taken into consideration.

In this section, the most reasonable dynamical system of those listed in previous section is obtained by means of robustness analysis. We first present a quantitative definition of biological robustness (call it robust performance). Then an algorithm is constructed to calculate the robust performances of the systems discussed in previous section.

3.1   Mathematical Definition of Robustness

Assume that M possible networks with N nodes are taken into consideration, which can be formulated as M dynamical systems with N state variables. $N_{ob} := \{n_1, n_2, \cdots, n_d\} \subset I_N$ is an index set of state variables that can be measured in experiment. For the mth system (denoted by S(m), $m \in I_M$), u^m is a vector of kinetic parameters. S_0 and $\mathcal{U}^m$ still denote the ranges of initial state x^0 and parameter vector u^m, respectively. Λ_{ad} and W_{ad} denote the admissible sets of state vector x and control vector v, respectively. $y^{s,k} := (y_{n_1}^{s,k}, y_{n_2}^{s,k}, \cdots, y_{n_d}^{s,k})$ is experimental result measured at the time point $t_{s,k}$ with control vector v taking value v^s, $s \in I_l$, $k \in I_{l_s}$, where l_s is the total measurement times of the sth experiment. For convenience, we also assume that the mth dynamical system is in the following form:

$$\begin{cases} \dot{x}(t) = f^m(x(t), v, u^m), \\ x(t_0) = x^0, \end{cases} \qquad t \in [t_0, t_f^m], \tag{12}$$

where $f \in C^1(\Lambda_{ad}, W_{ad}, \mathcal{U}^m; R^N)$, t_f^m is the moment when the system $S(m)$ reaches its steady state, i.e.,

$$f^m(x(t_f^m), v, u^m) = 0 . \tag{13}$$

For given $x^0 \in S_0$, $v \in W_{ad}$ and $u^m \in \mathcal{U}^m$, let $x(t; x^0, v, u^m)$ denote the solution of (12) starting from x^0. Then, define

$$p(u^m) = \frac{1}{|N_{ob}|} \sum_{i \in N_{ob}} \frac{\displaystyle\sum_{s \in I_l} \sum_{k \in I_{l_s}} |x_i^s(t_{s,k}; x^0, v^s, u^m) - y_i^{s,k}|}{\displaystyle\sum_{s \in I_l} \sum_{k \in I_{l_s}} y_i^{s,k}}, \tag{14}$$

where $|N_{ob}|$ is the cardinal number of N_{ob}. Let u^{m*} denote an optimal parameter vector for the system $S(m)$ with regard to $\{y^{s,k}\}_{s \in I_l, k \in I_{l_s}}$, i.e., u^{m*} satisfies that

$$u^{m*} \in \text{Argmin}\{p(u^m)|u^m \in \mathcal{U}^m\} . \tag{15}$$

Now, we shall discuss the robustness of $S(m)$ with regard to its optimal parameter vector u^{m*} against a set of perturbations in $\mathcal{U}^m$. In most cases, the robustness can be measured by the variation between the state vector with perturbed parameter vector and that with optimal parameter vector. More generally, assume that only some state variables are under consideration, the index set of which are denoted by N_{rb}. For specified $u^m \in \mathcal{U}^m$, define

$$dp(u^m) = \frac{1}{|N_{rb}|} \sum_{i \in N_{rb}} \frac{\displaystyle\sum_{s \in I_l} \sum_{k \in I_{l_s}} |x_i^s(t_{s,k}; x^0, v^s, u^m) - x_i^s(t_{s,k}; x^0, v^s, u^{m*})|}{\displaystyle\sum_{s \in I_l} \sum_{k \in I_{l_s}} x_i^s(t_{s,k}; x^0, v^s, u^{m*})} \tag{16}$$

and let $\mathcal{U}_{ad}^m := \{u^m \in \mathcal{U}^m| \exists x^0 \in S_0$ such that $x(t; x^0, v, u^m) \in \Lambda_{ad}, \forall v \in W_{ad}, t \in [t_0, t_f^m]\}$ be the feasible set of u^m. Randomly generate q sample points from $\mathcal{U}^m$ by uniform distribution, denoted by $u^{m,1}, \cdots, u^{m,q}$, where q is a sufficiently large positive integer number. Let $\mathcal{U}_{ss}(m) := \{u^{m,i} \in \mathcal{U}_{ad}^m \mid i \in I_q\}$. $|\mathcal{U}_{ss}(m)|$ denotes the cardinal number of $\mathcal{U}_{ss}(m)$. Then the robust performance of the biological dynamical system $S(m)$ can be defined as follows.

Definition 1. *The robust performance (R) of the system $S(m)$ with regard to its optimal parameter vector u^{m*} against a set of perturbations in $\mathcal{U}^m$ is described as:*

$$R_{u^m, \mathcal{U}^m}^m = \frac{1}{|\mathcal{U}_{ss}(m)|} \sum_{u^{m,i} \in \mathcal{U}_{ss}(m)} dp(u^{m,i}) . \tag{17}$$

We can conclude that the larger the value of (17) is, the more robust the system is. This conclusion can be explained in the following two aspects of analysis. On the one hand, observing (16), we know that the value of $dp(u^m)$ is small if $x(t; x^0, v, u^m)$ is insensitive to u^m and is large otherwise; on the other hand, the larger the value of $|\mathcal{U}_{ss}(m)|$ is, the higher the probability of the randomly generated parameter vector being in its feasible set is. Naturally, the comparison of the robustness between two systems can be defined as follows.

Definition 2. *A system $S(i)$ is said to be more robust than a system $S(j)$ when*

$$R^i_{u^i, \mathcal{U}^i} < R^j_{u^j, \mathcal{U}^j} . \tag{18}$$

3.2   Algorithm and Numerical Results

Since only steady states are considered in the study of glycerol continuous fermentations, i.e., $l_s = 1$ for all $s \in I_l$, we have

$$dp(u^m) = \frac{1}{|N_{rb}|} \sum_{i \in N_{rb}} \frac{\sum_{s \in I_l} |x_i^s(t_f^m; x^0, v^s, u^m) - x_i^s(t_f^m; x^0, v^s, u^{m*})|}{\sum_{s \in I_l} x_i^s(t_f^m; x^0, v^s, u^{m*})}, \tag{19}$$

where $N_{rb} = \{1, 2, 3, 6, 7, 8\}$. u^{m*}, $m = 1, 2, 3$, take values from Tables 1-3, respectively. Recall that $x(t_f^m)$ satisfies

$$f^m(x(t_f^m), v^s, u^m) = 0, \qquad s \in I_l . \tag{20}$$

As is mentioned in last section, (20) can be converted into a group of linear equations in logarithmic coordinates.

According to the above analysis, we construct the following algorithm to calculate the robust performances of the system $S(m)$, $m = 1, 2, 3$.

Algorithm 1.

Step 1. Set $w = 0$, $q = 5000$ and $q_1 = 0$. Compute $x^s(t_f^m; x^0, v^s, u^{m*})$ from Eqs.(20) with $u^m = u^{m*}$, $s = 1, 2, \cdots, l$, then goto Step 2.
Step 2. If $w > q$, goto Step 5, else generate a sample point $u^{m,w}$ from $\mathcal{U}^m$ and let $w := w + 1$, then goto Step 3.
Step 3. Compute $x^s(t_f^m; x^0, v^s, u^{m,w})$ from Eqs.(20) with $u^m = u^{m,w}$, $s = 1, 2, \cdots, l$. If $x^s(t_f^m; x^0, v^s, u^{m,w}) \in \Lambda_{ad}$ for all $s \in I_l$, then let $q_1 := q_1 + 1$, goto Step 4, else goto Step 2.
Step 4. Compute $dp(u^{m,w})$ from (19), then goto Step 2.
Step 5. Let $|\mathcal{U}_{ss}(m)| := q_1$ and compute $R^m_{u^m, \mathcal{U}^m}$ from (17), stop.

For the system $S(m)$, according to the above algorithm, we calculate its robust performance 50 times, denoted by $R^m_{u^m, \mathcal{U}^m}(k)$, $k = 1, \cdots, 50$. Then we compute the expectation and variance of $\{R^m_{u^m, \mathcal{U}^m}(k)\}_{k=1}^{50}$, denoted by $\overline{R}^m_{u^m, \mathcal{U}^m}$

Table 5. Expectations and variances of robust performances for the three systems

$S(m)$	$m = 1$	$m = 2$	$m = 3$
$\overline{R^m_{u^m,\mathcal{U}^m}}$	5.9023	22.766	21.625
$VR^m_{u^m,\mathcal{U}^m}$	0.04	0.25	0.574

and $VR^m_{u^m,\mathcal{U}^m}$. As shown in Table 5, on the one hand, $VR^m_{u^m,\mathcal{U}^m}$, $m = 1, 2, 3$, are small enough, which implies that $R^m_{u^m,\mathcal{U}^m}$ is independent of the sample points when q is sufficiently large and therefore the rationality of our definition of robust performance in (17); on the other hand, since the robust performance of the system $S(1)$ is much smaller than that of the other systems, it demonstrates that $S(1)$ is the most reasonable, i.e., it is most possible that both glycerol and 1,3-PD pass the cell membrane by active transport and passive diffusion.

4   Conclusions and Discussions

Metabolic network of glycerol enzyme-catalytic dissimilation by $K.\ pneumoniae$ includes some uncertain factors since the transport mechanisms of glycerol and 1,3-PD across the cell membrane haven't been completely observed, which leads to various inferences of the metabolic network. In this study, different dynamical systems were developed based on distinct inferences of the metabolic network and parameters were identified for each system based on 30 groups of experimental data. To infer the most reasonable metabolic network in the context of lack of intracellular information, we carried out robustness analysis for these systems. A quantitative definition of biological robustness was established and the robust performances of our proposed systems were calculated. Numerical results show that it is most possible that both glycerol and 1,3-PD pass the cell membrane by active transport and passive diffusion.

In the future, we will attempt to search for grounded theoretical basis for the robustness definition proposed in this article and put more numerical examples to verify its rationality and feasibility. It is worthwhile to emphasis that the robustness definition developed in this work can only be applied to the biological systems whose robustness can be reflected in their state vectors, which, of course, is not the only case. So we will be engaged in exploring new feasible theoretical and computational avenues to provide a broad and unified account of robustness of biological systems. In addition, on the basis of our theoretical study of biological systems, we will take further study on the whole metabolic network of glycerol bioconversion to 1,3-PD, including reductive and oxidative pathways and genetic regulation on the metabolic process.

Acknowledgements

This work was supported by 863 Program (Grant No. 2007AA02Z208), 973 Program (Grant No. 2007CB714304), the National Natural Science Foundation of China (Grant Nos. 10671126 and 10871033).

References

1. Zeng, A.P., Rose, A., Biebl, H., Tag, C., Guenzel, B., Deckwer, W.D.: Multiple product inhibition and growth modeling of Clostridium butyricum and Klebsiella pneumoniae in ferentation. Biotechnol. Bioeng. 44, 902–911 (1994)
2. Zeng, A.P., Deckwer, W.D.: A kinetic model for substrate and energy comsumption of microbial growth under substrate-sufficient conditions. Biotechnol. Prog. 11, 71–79 (1995)
3. Zeng, A.P.: A kinetic model for product formation of microbial and mammalian cells. Biotechnol. Bioeng. 46, 314–324 (1995)
4. Biebl, H., Menzel, K., Zeng, A.P., Deckwer, W.D.: Microbial production of 1,3-propanedial. Appl. Microbiol. Biotechnol. 52, 289–297 (1999)
5. Sun, J., van den Heuvel, J., Soucaille, P., Qu, Y., Zeng, A.P.: Comparative genomic analysis of dha regulon and related genes for anaerobic gelycerol metabolism in bacteria. Biotechnol. Prog. 19, 263–272 (2003)
6. Qu, H.J., Wang, F.H., Tian, P.F., Tan, T.W.: Cloning of 1,3-propanediol oxidore-ductase gene from Klebsiella pneumoniae and preliminary study on its expression conditions. Industrial Microbiology 37(1), 25–29 (2008)
7. Tian, P.F., Tan, T.W.: Genetic modification of Klebsiella pneumoniae for increasing 1,3-propanediol production. Chemical Industry and Engineering Progress 27(3), 322–325 (2008)
8. Sun, Y.Q., Qia, W.T., Teng, H., Xiu, Z.L., Zeng, A.P.: Mathematical modeling of glycerol fermentation by Klebsiellapneumoniae: Concerning enzyme-catalytic reductive pathway and transport of glycerol and 1,3-propanediol across cell membrane. Biochem. Eng. J. 38(1), 22–32 (2008)
9. Kitano, H.: Biological robustness. Nat. Rev. Genetic 5(11), 826–837 (2004)
10. Kitano, H.: Towards a theory of biological robustness. Mol. Sys. Biol. 3 (2007)
11. Barkai, N., Leibler, S.: Robustness in simple biochemical networks. Nature 387, 913–917 (1997)
12. Bhalla, U.S., Iyengar, R.: Robustness of the bistable behavior of a biological sig-naling feedback loop. Chaos 11, 221–226 (2001)
13. von Dassow, G., Meir, E., Munro, E.M., Odell, G.M.: The segment polarity network is a robust developmental module. Nature 406, 188–192 (2000)
14. Tian, T.H.: Robustness of mathematical models for biological systems. Austral. Mathematical Soc. 45(E), 565–577 (2004)
15. Chen, B.S., Wang, Y.C., Wu, W.S., Li, W.H.: A new measure of the robustness of biochemical networks. Bioinformatics 21(11), 2698–2705 (2005)
16. Alon, U., Surette, M.G., Barkai, N., Leibler, S.: Robustness in bacterial chemotaxis. Nature 397, 168–171 (1999)
17. Kitano, H.: Cancer as a robust system: implications for anticancer therapy. Nat. Rev. Cancer 4(3), 227–235 (2004)
18. Stelling, J., Sauer, U., Szallasi, Z., Doyle, F.J., Doyle, J.: Robustness of cellular functions. Cell 118(6), 675–685 (2004)
19. Zeng, A.P.: Quantitative Zellphysiologie, Metabolic Engineering und Modellierung der Glycerinfermentation zu 1,3-Propandiol, Habilitationschrift, Technical University of Braunschweig, Germany (2000)
20. Savageau, M.A.: Biochemical systems analysis, II. The steady-state solutions for an n-pool system using a power-law approximation. J. Theor. Biol. 25, 370–379 (1969)
21. Zhang, S.H., Zui, G.: A real-coded adaptive genetic algorithm and its application reseatch in thermal process identification. Chin. Soc. for Elec. Eng. 24(2), 210–214 (2004)

Modeling and Properties of Nonlinear Stochastic Dynamical System of Continuous Culture

Lei Wang[1,*], Enmin Feng[1], Jianxiong Ye[1], and Zhilong Xiu[2]

[1] Department of Applied Mathematics, Dalian University of Technology, Dalian, 116024, Liaoning, China
wanglei@dlut.edu.cn
[2] Department of Biotechnology, Dalian University of Technology, Dalian, 116012, Liaoning, China

Abstract. The stochastic counterpart to the deterministic description of continuous fermentation with ordinary differential equation is investigated in the process of glycerol bio-dissimilation to 1,3-propanediol by *Klebsiella pneumoniae*. We briefly discuss the continuous fermentation process driven by three-dimensional Brownian motion and Lipschitz coefficients, which is suitable for the factual fermentation. Subsequently, we study the existence and uniqueness of solutions for the stochastic system as well as the boundedness of the Two-order Moment and the Markov property of the solution. Finally stochastic simulation is carried out under the Stochastic Euler-Maruyama method.

Keywords: Continuous Culture, Bioconversion, Nonlinear Stochastic System, Stochastic Simulation.

1 Introduction

Stochastic influences play an important role in bioprocess development (engineering). Biotechnical treatment of microorganisms is commonly described by systems of nonlinear ordinary differential equations(ODEs)[1]. It includes an idealization of the technical system component and a qualitative characterization of the biological part.

Over the past several years, 1, 3-propanediol(1, 3-PD) has been paid attention in microbial production throughout the world because of its lower cost, higher production and no pollution [2,3]. Anping Zeng presented the nonlinear dynamical model for substrate consumption and product formation in the course of the bioconversion of glycerol to 1,3-PD by Klebsiella pneumoniae in continuous culture [4,5]. And then, Zhilong Xiu made simulations for the kinetic model [6] and conducted an investigation into the optimal conditions of continuous glycerol fermentation by using the volumetric productivity of 1,3-PD as an optimziation target based on the nonlinear deterministic dynamical system [3].

---

* Corresponding author.

J. Zhou (Ed.): Complex 2009, Part I, LNICST 4, pp. 458–466, 2009.

However, different culture states during long term continuous fermentation of glycerol by *K. pneumoniae* under similar initial fermentation conditions are obtained, resulting in randomness of continuous culture. This random phenomena reveal several different patterns and new features compared with those reported in previous literature. In this paper, since 1,3-PD is a monomer for the production of polycondensates in the bioprocess, we propose a stochastic version of the continuous fermentation process which only considers the inherent stochasticity of microorganism. The process is modeled by a stochastic ordinary differential system driven by three-dimensional Brownian motion, which is time independent and suitable for the factual fermentation. And the global Lipschitz and linear growth conditions of the coefficients of the stochastic system are proved to make sure the existence and uniqueness of solution of the stochastic system. Finally computer simulation is used for the stochastic system behind Monte Carlo and Stochastic Euler-Maruyama method. Compared with the results from the deterministic system, numerical results reveal the peculiar role of randomness in the dynamical responses of the continuous culture.

This paper is organized as follows. In section 2, we present a nonlinear stochastic dynamic system of the continuous fermentation process and prove the properties of the stochastic dynamic system as well as the existence and uniqueness of solutions to the stochastic dynamic system. The stochastic continuity, boundedness and Markov property of solutions to the system are also discussed. Section 3 provides numerical examples to simulate the nonlinear stochastic dynamical system of continuous culture. In section 4, we draw the conclusions and trace the direction for future works.

2   Models and Properties

2.1   Deterministic Model

Mass balances of biomass, substrate and product in contionous microbial cultures are written as follows (see [4]).

$$
\begin{cases}
\frac{dX}{dt} = X(\mu - D), \\
\frac{dC_s}{dt} = (D(C_{s0} - C_s) - Xq_s, \qquad t \in I = (0, T). \\
\frac{dC_p}{dt} = Xq_p - DC_p,
\end{cases}
\tag{1}
$$

where X, C_{s0}, C_s, C_p, and D are biomass, substrate concentration in medium, substrate concentration in reactor, product concentration in reactor, and dilution rate. The specific growth rate μ of biomass, specific consumption rate of substrate q_s and specific formation rate of product q_p are expressed by Eqs. (2)-(4), respectively.

$$
\mu = \mu_{max}(\frac{C_s}{C_s + k_s})(1 - \frac{C_s}{C_s^*})(1 - \frac{C_p}{C_p^*}).
\tag{2}
$$

$$
q_s = m_s + \frac{\mu}{Y_s^m} + \Delta q_s^m \frac{C_s}{C_s + k_s^*}.
\tag{3}
$$

$$q_p = m_p + \mu Y_p^m + \Delta q_p^m \frac{C_s}{C_s + k_p^*} . \tag{4}$$

Under anaerobic conditions at $37°C$ and pH=7.0, the maximum specific growth rate of cells μ_{max} is $0.67h^{-1}$, Monod saturation constant k_s is 0.28 mmol/L. The critical concentrations of biomass, glycerol, 1,3-PD for cell growth are $X^* = 10g/L$, $C_s^* = 2039mmol/L$, $C_p^* = 939.5mmol/L$, respectively. m_s, m_p Y_s^m, Y_p^m, Δq_s^m, Δq_s^m k_s, k_s^*, k_p^* are parameters given in the previous work [3].

As far as the continuous fermentation is concerned, glycerol is added to the reactor continuously, the broth in reactor pours out at the same rate and the volume of the fermentation broth keeps constant in the whole course of bio-conversation. As a result of fact, the following assumptions can be made:

(H1) The concentrations of reactants are uniform in bio-reactor and only as varied as the fermentation time.
(H2) During the process of continuous culture, the substrate added to the reactor only includes glycerol and the fermentation broth is exported by the dilution rate D.

2.2   Stochastic Model

There are many possibilities to express the stochastic behavior of a system. The natural approach is to take the given d-dimensional deterministic model and simply add a stochastic part, given by a map $G : \mathbb{R}^{d+1} \longrightarrow \mathbb{R}^{dm}$ and an R^m-valued white noise process w_t [7]. This leads to the description as a stochastic differential equation(SDE):

$$\dot{x} = F(t, x) + G(t, x)w_t .$$

being an abbreviating notation for the exact expression

$$x_t = \int_{t_0}^t F(\tau, x_\tau)d\tau + \int_{t_0}^t G(\tau, x_\tau)dw_\tau + x_{t_0} .$$

for the stochastic process X_t with values in $\mathbb{R}^d$.

In this paper, we want to study only the influence of the noisy parameter $\mu = \bar{\mu} + \sigma_\mu \dot{w}(t)$, and replace it in Eq.(2), Eq.(3) and Eq.(4). This is a standard technique as far as stochastic population modeling is concerned. It introduces stochasticity into the model.

As a result, we get the SDE model for continuous culture,

$$dx = F(x)dt + G(x)dw, \ t \in I, \ x(0) = x_0 . \tag{5}$$

where

$$F(x) = (X(\mu - D), D(C_{s0} - C_s) - Xq_s, Xq_p - DC_p)^T, \tag{6}$$

$$G(x) = (\sigma_\mu X, \frac{\sigma_\mu X}{Y_s^m}, \sigma_\mu Y_p^m X)^T, \tag{7}$$

$$E(\dot{w}(t)) = 0,$$

$$D(\dot{w}(t)) = 1 .$$

σ_μ is the intensity of the inherent stochasticity disturbance. In Eq. (5), $x = (X, C_s, C_p)^T$ is a stochastic process that reflects the fluctuating trend of the proportion under the inherent stochasticity disturbance.

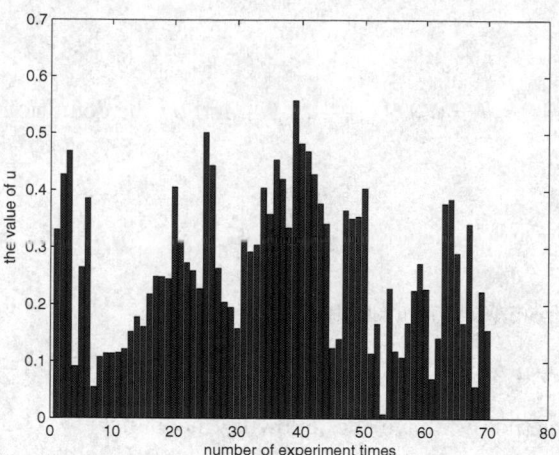

Fig. 1. The variations of the specific growth rate of cells in stable steady state.(data from Ref. [2,3,4])

To determine system noise variances σ_μ additional methods, further experiments are required. Fig. 1 shows the variations of the specific growth rate of cells in stable steady state.

2.3   Existence and Uniqueness of the SDE Solution

Theorem 1. *The vector-valued functions $F(x)$ and $G(x)$ defined by (6) and (7) are measurable for $t \in I$, $x \in R^3$.*

Proof. It is clear from the continuity of the functions $F(x)$ and $G(x)$ on I.

Theorem 2. *For the vector-valued functions $F(x)$ and $G(x)$ defined by (6) and (7), there exist positive constants K and K' such that for $t \in I$ the following conditions hold:*

a) uniform Lipschitz condition

$$\|F(x^1) - F(x^2)\| + \|G(x^1) - G(x^2)\| \leq K\|x^1 - x^2\| ,$$

$$\forall\ x^1\ and\ x^2 \in R^3 .$$

b) growth condition

$$\|F(x)\| + \|G(x)\| \leq K'(1 + \|x\|) .$$

Proof. From [8] We conclude that the function $F(x)$ satisfied the uniform Lipschitz condition and growth condition on R^3 a.e. . Let x^1 and x^2 be in R^3, there exists two constants L and C, we have

$$\|F(x^1) - F(x^2)\| \le L\|x^2 - x^1\| \quad a.e. \tag{8}$$

$$\|F(x)\| \le C\|x\| \le C(1 + \|x\|) \quad a.e. \tag{9}$$

On the other hand, let $a = max\{1, \frac{1}{Y_s^m}, Y_p^m\}$ and by the definition of the function $G(x)$, we have

$$\|G(x^1) - G(x^2)\| \le \sigma_\mu a(\sum_{i=1}^{3}(x_i^1 - x_i^2)^2)^{1/2} \le \sigma_\mu a\|x^2 - x^1\| . \tag{10}$$

Thus, it follows from (6), (7) and (10)

$$\|F(x^1) - F(x^2)\| + \|G(x^1) - G(x^2)\| \le (L + \sigma_\mu a)\|x^2 - x^1\| .$$

Let $K = L + \sigma_\mu a$, then we have the following inequality

$$\|F(x^1) - F(x^2)\| + \|G(x^1) - G(x^2)\| \le K\|x^2 - x^1\| .$$

Next, we will show the growth condition of the function $G(x)$. It is clear from the definition of the function $G(x)$

$$\|G(x)\| \le \sigma_\mu a\|x\| \le \sigma_\mu a(1 + \|x\|) .$$

Therefore, letting $K' = \sigma_\mu a + C$, then we can complete the proof by

$$\|F(x)\| + \|G(x)\| \le K'(1 + \|x\|) . \qquad \square$$

Let the vector-valued functions $F(x)$ and $G(x)$ are defined by (6) and (7). Based on the Theorem 2 and Theorem 5.2.1 [9], we can prove the following theorem.

Theorem 3 (Existence and Uniqueness). *The system* (5) *has a unique solution* $x(t)$ *satisfying for the initial condition* x_0 *on* I.

2.4   Properties of Solutions to the Stochastic Dynamic System

Lemma 1. *[10] Suppose* $G : R^3 \to L(R^3, R^3)$, $\int_0^r E\|G(x(s))\|^2 dt < \infty$ *and* $y(t) = \int_0^t G(x(s))dW(s) < \infty$, *then*

$$E\|y(t)\|^2 \le trQ \cdot \int_0^t E\|G(s)\|^2 ds .$$

Where $L(R^3, R^3)$ *denotes the space of all the bounded linear operators from* R^3 *to* R^3, *and* Q *is the covariance matrix of the Brownian motion* $W(t)$ *and* trQ *denotes the trace of* $Q^T Q$.

Lemma 2. *[11] Let $\varphi(t)$ and $\alpha(t)$ be measurable bounded functions and for some $L > 0$ assume*

$$\varphi(t) \leq \alpha(t) + L \int_0^t \varphi(s)ds .$$

Then

$$\varphi(t) \leq \alpha(t) + L \int_0^t \exp\{L(t-s)\}\alpha(s)ds .$$

According to the proof in Theorem 3, Theorem 5.4 [12] and Theorem 5.2 [9], we can prove the following theorems.

Theorem 4 (Markov Property and Boundedness). *Suppose (H_1) and (H_2) hold. The unique solution $x(t)$ is a Markov process on the interval I whose initial probability distribution at $t = 0$ is the distribution of x_0 and $x(t)$ has continuous paths, moreover*

$$(sup_{0 \leq t \leq T} E\|X(t)\|)^2 < B(1 + E\|x_0\|^2) .$$

where constant B depends only on K and T.

Theorem 5 (Stochastic Continiuty). *Suppose that assumptions H1 and H2 are satisfied. Then, almost all realizations of $x(t)$ are continuous on I.*

Proof. We begin with another version of the linear growth conditions for the coefficients $F(x)$ and $G(x)$.

By (9) and the definition of the function $G(x)$, we can see that

$$\|F(x)\|^2 \leq C^2\|x\|^2.$$

and

$$\|G(x)\|^2 \leq \sigma_\mu^2 a^2\|x\|^2.$$

Hence, let $K_1^2 = C^2 + \sigma_\mu^2 a^2$ it follows that

$$E(\|F(x)\|^2 + \|G(x)\|^2) \leq E[(C^2 + \sigma_\mu^2 a^2)(1 + \|x\|)^2] \leq K_1^2 E\|x\|^2. \qquad (11)$$

Suppose $t \in I$ and $\delta > 0$. Let $s \in I$ be such that $|s - t| < \delta$. From (5), we may have that

$$E\|x(s) - x(t)\|^2 = E\|\int_t^s F(x(\tau))d\tau + \int_t^s G(x(\tau))dw(\tau)\|^2. \qquad (12)$$

Letting $K = \max\{\delta, trQ\}$, it follows from Hölder Inequality, Lemma 1 and Theorem 4 that

$$E\|x(s) - x(t)\|^2 \leq 4E[\int_t^s \|F(x(\tau))\|d\tau]^2 + 4E[\int_t^s G(x(\tau))dw(\tau)]^2$$

$$\leq 4|s - t| \int_t^s E\|F(x(\tau))\|^2 d\tau + 4trQ \int_t^s E\|G(x(\tau))\|^2 d\tau$$

$$\leq 4K \int_t^s E[\|F(x(\tau))\|^2 + \|G(x(\tau))\|^2]d\tau$$

$$\leq 4KK_1^2 \int_t^s [1 + E\|x(\tau)\|^2]d\tau$$

$$\leq 4KK_1^2 B(1 + E\|x_0\|^2)|s - t| .$$

Set $B' = 4KK_1^2 B(1 + E\|x_0\|^2)$, thus the above inequality can be as follows:

$$E\|x(s) - x(t)\|^2 \leq B'|s - t|,$$

or equivalently,

$$\lim_{s \to t} E\|x(s) - x(t)\|^2 = 0 .$$

Thus, for any $\epsilon > 0$, by Tchebycheff inequality, we conclude that

$$\mathcal{P}(\{\omega \in \Omega : \|x(s) - x(t)\| \geq \epsilon\}) \leq E\|x(s) - x(t)\|^2/\epsilon^2 .$$

which completes the proof.   □

3   Numerical Simulation

To elaborate the stochastic nature of continuous fermentation process sufficiently, an numerical example is given. In the example, $D = 0.25/hour$ $C_{s0} = 735mmol/L$ and we use Monte Carlo method to generate five thousand random inputs, which consist of the infinitesimal increment of standard Brownian motion $dw(t)$. Afterwards, we solve the proposed stochastic model using the following Stochastic Euler-Maruyama method and obtain five thousand evaluations of the model. Our numerical approximation to $x(\tau_j)$ will be denoted by X_j.

Stochastic Euler-Maruyama method [12]

$$X_j^k = X_{j-1}^k + F(X_{j-1}^k)\Delta t + a_k X_{j-1}^k(W_t^l(\tau_j) - W_t^l(\tau_{j-1})), \quad j = 1, 2, ..., L .$$

where $\Delta t = T/L$ for some positive integer L. X^k denotes the k'th component of the x. $\tau_j = j\Delta t$, $a_1 = \sigma_\mu = 0.01638$, $a_2 = \frac{\sigma_\mu}{Y^m}$ and $a_3 = \sigma_\mu Y_p^m$. $x_0 = (0.98g/L, 464mmol/L, 184.36mmol/L, 57.83mmol/L, 4.41mmol/L)$, the components of which are the initial concentrations of biomass, substrate, 1,3-PD, acetic acid and Ethanol, respectively. All the parameters of the stochastic system are given in Table 1. Fig. 2 shows the comparison of biomass, substrate and product concentrations between experimental and simulated results, where the points denote the experimental values, written as $y(\tau_i) = (y^1(\tau_i), y^2(\tau_i), y^3(\tau_i))$,

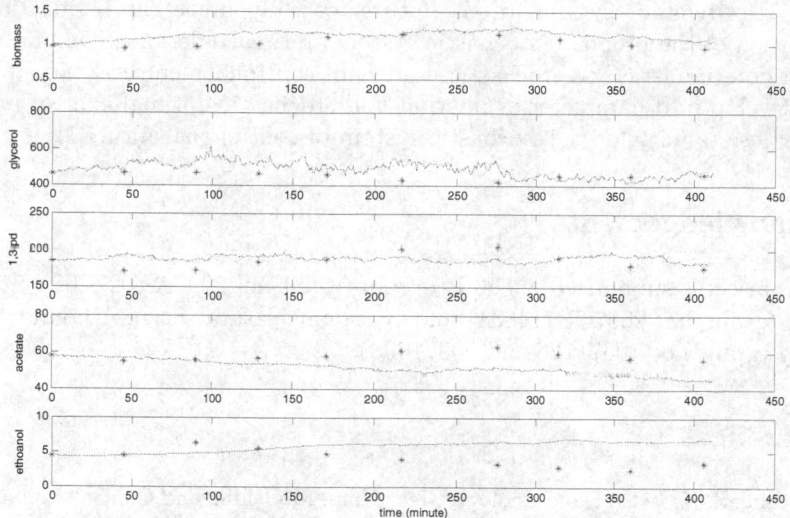

Fig. 2. The comparison of biomass, substrate and product concentrations between experimental and simulated results(acetic acid and ethanol are byproduct)

Table 1. Parameters values of each reactant in the stochastic system

Reactant	μ_m	k_s	m_i^2	Y_i	Δ_i	k_i
Biomass	0.67	0.28	–	–	–	–
Glycerol	–	–	2.20	0.0082	28.58	11.43
$1,3-PD$	–	–	-2.69	67.69	26.59	15.50
Acetic acid	–	–	-0.97	33.07	5.74	85.71
Ethanol	–	–	-0.97	33.07	5.74	85.71

$i = 1, 2..., 10$, and the real lines denote the computational curves $EX^k(t)$, $k \in I_3$. Define errors as follows:

$$e_k = \frac{\sum_{i=1}^{10} |EX^k(\tau_i) - y^k(\tau_i)|}{\sum_{i=1}^{10} y^k(\tau_i)}, \qquad k \in I_3 .$$

We obtain the errors $e_1 = 9.37\%$, $e_2 = 6.37\%$, $e_3 = 19.85\%$. Comparing the errors in this paper with the reported results [8], we conclude that the stochastic system is more fit for modeling actual continuous fermentation under investigation.

4   Conclusions

In this paper, we present a nonlinear stochastic dynamical system of continuous culture and demonstrated the existence and uniqueness of solutions to the stochastic system. Further we proved some properties of the solution to the stochastic dynamic system. In the future we will pursue the verification and validation of the proposed stochastic system and make detailed comparison between deterministic and stochastic models of continuous culture. Moreover, we will develop into parameter estimation and stochastic optimal control problem as well as stability for the stochastic system of continuous culture.

Acknowledgements

This work was supported by 863 Program (grant no. 2007AA02Z208), 973 Program (grant no. 2007CB714304) and National Natural Science Foundation of China (grant nos 10671126 and 10871033).

References

1. Schiigerl, K. (ed.): Biotechnology. Measuring, Modelling and Control, vol. 4. Weinheim (1991)
2. Biebl, H., Menzel, K., Zeng, A.P., Deckwer, W.D.: Microbial production of 1,3-propanedial. Appl. Microbiol. Biotechnol. 52, 289–297 (1999)
3. Xiu, Z.L., Song, B.H., Wang, Z.T., Sun, L.H., Feng, E.M., Zeng, A.P.: Optimization of biodissimilation of glycerol to 1,3-propanediol by klebsiella pneumoniae in one-stage and two-stage anaerobic cultures. Biochemical Engineering Journal 19, 189–197 (2004)
4. Xiu, Z.L., Zeng, A.P., Deckwer, W.: Multiplicity and stability analysis of microorganisms in continuous culture: effects of metabolic overflow and growth inhibition. Biotechnol Bioeng 57(3) February 5 (1998)
5. Zeng, A.P., Rose, A., Biebl, H., Tag, C., Guenzel, B., Deckwer, W.D.: Multiple product inhibition and growth modeling of Clostridium butyricum and Klebsiella pneumo- niae in ferentation. Biotechnol. Bioeng. 44, 902–911 (1994)
6. Xiu, Z.L., Zeng, A.P., An, L.J.: Mathematical modeling of kinetics and research on multiplicity of glycerol bioconversion to 1,3-propanediol. Journal of Dalian University of Technology 40, 428–433 (2000)
7. Kinder, M., Wiechert, W.: Stochastic simulation of biotechnical processes. Mathematics and Computers in Simulation 42, 171–178 (1996)
8. Gao, C.X., Wang, Z.T., Feng, E.M., Xiu, Z.L.: Parameters identification problem of the nonlinear dynamical system in microbial continuous cultures. Applied Mathematics and Computation 169(1), 476–484 (2005)
9. Øksendal, B.: Stochastic differential equations, 6th edn. Springer, Heidelberg (2005)
10. Xu, M.G.: Some properties of the solution of semilinear stochastic evolution equations in Hilbert spaces. Journal of Wuhan University 2, 29–36 (1994)
11. Ying, J.G., Jin, M.W.: Foundation of Stochastic Process. Fudan University Press, Shanghai (2005)
12. Klebaner, I.C.: Intruduction to stochastic calculus with applications, 2nd edn. Imperical College Press (2005)

Modeling a Complex Biological Network with Temporal Heterogeneity: Cardiac Myocyte Plasticity as a Case Study

Amin R. Mazloom[1], Kalyan Basu[2], Subhrangsu S. Mandal[3], and Sajal K. Das[2]

[1] Department of Pharmacology & Systems Therapeutics
Mount Sinai School of Medicine, New York, NY 10029, USA
amin.mazloom@mssm.edu
[2] Department of Computer Science & Engineering
[3] Department of Chemistry & Biochemistry
University of Texas at Arlington, Arlington, Texas, TX 76019, USA

Abstract. Complex biological systems often characterize nonlinear dynamics. Employing traditional deterministic or stochastic approaches to quantify these dynamics either fail to capture their existing deviant effects or lead to combinatorial explosion. In this work we devised a novel approach that projects the biological functions within a pathway to a network of stochastic events that are random in time and space. By applying this approach recursively to the object system we build the event network of the entire system. The dynamics of the system evolves through the execution of the event network by a simulation engine which comprised of a time prioritized event queue. As a case study we utilized the current method and conducted an in-silico experiment on the metabolic plasticity of a cardiac myocyte. We aimed to quantify the down stream effects of insulin signaling that predominantly controls the plasticity in myocardium. Intriguingly, our in-silico results on transcription regulatory effect of insulin showed a good agreement with experimental data. Meanwhile we were able to characterize the flux change across major metabolic pathways over 48 hours of the in-silico experiment. Our simulation performed a remarkable efficiency by conducting 48 hours of simulation-time in less that 2 hours of processor time.

1 Introduction

A complex system is a subset of a world comprised of many components whose interactions with the rest of the world or another subset is properly defined. The behavior of a complex system could properly perceived through the aggregate effects of its components. An organism could be viewed as a system or collection of systems at different hierarchical levels. The boundary of a system is defined by its components which for a bio-system could range from organism to organ, tissue, cell, molecule, and atom. The degree of complexity between levels grows exponentially from top to bottom. In the current study a cell draws the system boundary and molecules are the interacting components of this system. The

J. Zhou (Ed.): Complex 2009, Part I, LNICST 4, pp. 467–486, 2009.

respond of a cell to an exogenous signal (antigens, hormones, pressure or temperature change, etc) shaped from endogenous activities within the cellular networks that attempt to maintain the cell homeostasis. To gain insight on the dynamics of this respond at the system level, interaction of the underlying components must be properly characterized in time and space. In this study we propose a novel stochastic discrete event-based methodology [1] to conduct system level in-silico experiments on a typical eukaryotic cell. We elucidate our method by deploying that on the cardiac myocyte along with insulin signal as a case study to validate the significance of our approach. The idea is to learn the system level effects of an exogenous signal through the changes imposed on the dynamics of Signal Transduction Network, (STN), Transcription Regulatory Network (TRN), and Metabolic Network (MTN) of the cell following the perception of the signal. The rest of the paper is organized as follows: we allotted the rest of this section to provide a background on the quantitative models proposed for the heart cell, on section two the modeling steps in the proposed approach along with the simulation algorithm is described, section three devoted to application of the current schema on the insulin signaling pathway in the heart-cell, section four provides the results for the in-silico experiment, section five discusses the key pros and cons of the approach, and we end the paper by drawing conclusion in section six.

1.1   Background

Since 1960 that Denis Noble proposed the first model of the heart, numerous quantitative models have been proposed for the heart from organ to cell level. These models could broadly be classified into two main families: physiological models and pathway models. The former class includes a combination of mechanical and biochemical models that focuss on capturing the physiological and electrophysiological dynamics of heart and its tissues under different physiological and biochemical conditions. Models in [2,3] are typical examples of this class of models. This top-down modeling approach offers a course grain analysis, therefore is not suitable for detail analysis of intra-cellular networks. The latter class has a more microscopic focus which intends to model one or more pathways from the cellular networks and seeks to quantify their dynamics, discover new pathways, complexes, etc. Instances of such models could be found in [4,5]. In order to predict the dynamics of biological functions latter class subscribes to either one of the following approaches: i) deterministic approach where a sets of ordinary differential equations (ODE) or partial differential equations (PDE) is formed based on biochemical reactions and diffusion to address the rate of change in concentrations of molecular parts. These ODEs/PDEs are then numerically solved to determine the dynamics of the underlaying system. ii) The stochastic approach which comprises strains of Gillespie [6] algorithm to approximate Chemical Master Equation (CME) [7], where the system is mapped into sets of chemical kinetic equations which evolves in Monte Carlo steps. Arkin and Samiolov [8] have shown that non-classical behavior of biological networks cause their dynamics to substantially diverge from their average. Therefore, deterministic approaches based on classical chemical kinetic (CCK) which assumes equilibrium across the entire course of system's evolution would not be an appropriate method to model many of biological systems. This claim remains valid even for the systems with higher molecular abundance. Although Gillespie based family of algorithms are suitable for capturing the behavior of biological functions; however, despite the efficiency enhancements they archived owing to approximation techniques such as *tau-leaping*, they still suffer from high computational complexity. None of these approaches promises a suitable model for a system that comprises network of biological functions (e.g. transcription function, signaling phospo-interaction, metabolic reaction) that manifest several order of magnitude difference in their temporal dynamics. In a system whose components (i.e. network nodes) manifest such temporal heterogeneity sequential evolution of fast processes would exhaust the execution of slower ones. The complex network of cellular processes ($\bigcup\{$STN, TRN, MTN$\}$) in a cardiac myocyte is a typical example of above systems.

To layout a modeling frame-work that contend to such heterogeneity in an in-silico experiment we introduce the concept of myocardial event (*myvent*) that accounts for an individual biological process within a cardiac myocyte. Noting that dynamics of the system is captured through changes in the count of molecular parts in the course of an in-silico experiment. These dynamics evolves through the execution of a network of *myevents*. Each *myevent* is an object from a specific class of *myevents* that has a random execution time with known first and second moments. The ability to bundle one or more *myevents* of a same class grants a *mesoscopic scale* to this modeling approach. This property avoids exhaustive computations for a system that comprises a network of processes that are temporally heterogeneous.

2   Approach

Observations confirm that at the molecular level the cellular behavior arises from the stochastic interaction between molecular parts [8,9]. Such observations is the key motivation in applying stochastic discrete event-based (SDE) method in capturing the dynamics of a cellular function. Hence, identifying molecular functions in a cardiac myocyte and mapping those into sets of *myvents* is fundamental to our approach. Each *myevent* has three attributes: (i) The stochastic physicochemical model that approximates the temporal dynamics for a typical class of myevents (i.e. cytoplasmic reaction, membrane reaction, transcription, etc.), (ii) The molecular resources (input/output) associated with a *myevent*, (iii) The compartment(s) within or across which a *myevent* is executed. In the current study for the first attribute of the a *myevent* we either adopt a physicochemical model from the literature or replace that with a relevant probability distribution that can approximate the experimental data. Also to avoid further complexity we only consider cytoplasm, nucleus, and mitochondria compartments for a cardiac myocyte. In a SDE in-silico experiment, *simulation time* is the representation of the *physical time* of the system being modeled. Each event is associated a *time-stamp* indicating when that event occurs in the physical system being simulated. The event *time-stamp* is computed from the knowledge of the previous event that has triggered the current event, together with the event execution-time which is a realization of the random number that characterizes the event dynamics. The dynamics of resource utilizations with progression in time unveil the complete internal picture of a complex biological system at the molecular level. Applying this doctrine to study the system level dynamics in a cardiac myocyte, demands the following check-list for characterizing the system parameters: (i) Identify the list of discrete *myevents* that can be included in the model based on the available knowledge of the system; (ii) Identify the resources of interest for the execution of the *myevents* function which are being used by the biological process for each discrete event; (iii) Compute the time taken to complete this biological discrete event. For this purpose, it is important to mathematically relate all event parameters which affect the interaction of the resources in a particular biological function; (iv) Identify the next *myevent* or set of *myevents* initiated on the completion of a *myevent*. If multiple discrete *myevents* are possible after completion of a *myevent*, the next *myevent* is chosen probabilistically, based on the biological pathway of the function being modeled. This probability calculation depends on the *myevent-set* and the properties of the *myevents* within the set.

Once the above check list is satisfied the discrete event simulator scheduler which is a time prioritized event queue pops individual *myevents* from the queue and system proceeds. Upon the execution of each *myevent*, molecular resources of the system is updated, system time is moved forward, and new *myevents* are pushed into the event queue from the *next-myevents* list of current *myevent*. The pseudo code for the algorithm that governs a SDE in-silico experiment is given in Fig. 1.(left) and the simulation engine architecture is depicted in Fig. 1.(right).

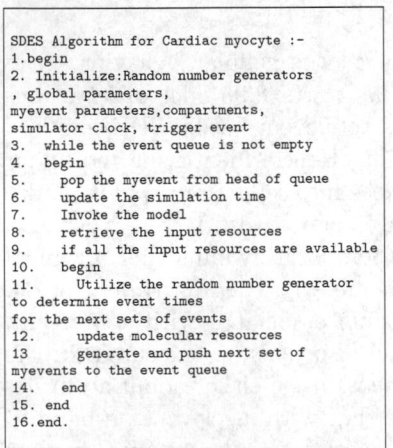

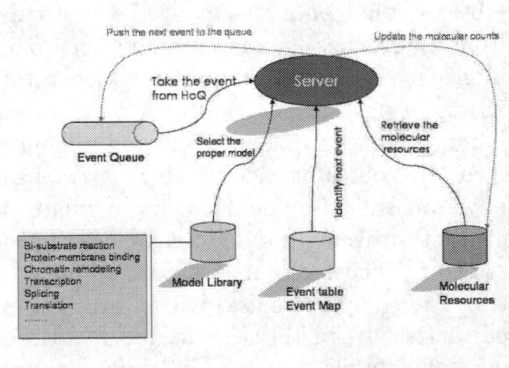

Fig. 1. (Left) The SDES algorithm pseudo code for cardiac myocyte model; (Right) The architecture of simulator engine

In a compartmentalized environment there will be two superclass of *myevents* apart from their types: the *local myevent* (e_{ij}, $i = j$) and *cross compartment myevent* (e_{ij}, $i \neq j$). Subscripts i and j are the source and destination compartments across which the event is executed. Execution of a local *myevent* only effects molecular resources in the compartment local to that *myevent*, where a cross compartment *myevent* potentially changes the resources in both source and destination compartments.

3   Modeling the Cardiac Myocyte Plasticity

Metabolic plasticity is the capacity of a cell to adopt to alternative available metabolic substrates as the source for its energy requirement. In order to model the plasticity of a cardiac myocyte at the system level first we need the to identify the pathways in signal transduction, transcription regulatory, and metabolic networks that pertain to such functionality. Further, identify the *myevents* that comprise each biological function within these pathways. Then associate each *myevent* with the proper stochastic model, input resources, and output resources. Subsequently interconnect those *myevents* in a recursive fashion to form the *myevent* network of that function. The above process that maps a biological functions from its physiological context to an event based context is referred to as *eventology* of that function. By recursively applying the *eventology* to the system we can form the event network of the whole system.

3.1   Eventology of the Signaling Pathways

Cardiac myocytes should have flexibility in their fuel selection in order to be consistent in meeting their energy requirements. Metabolic flux modulation could be regulated at many levels, two of the promising flux modulations in cardiac myocytes are through the control of metabolite uptake and gene expression level [10]. Insulin which is an essential peptide hormone of endocrine system that secretes from β-cells in pancreas is predominantly involved in the fuel selection at both levels. Although the propagation of the insulin signal within the cell influences divers cellular functions such as mitogenic, cell growth, etc.; however, in this work we focus on the signaling information that culminates on the two modulatory effects.

The insulin signal is sensed by binding the insulin to insulin receptors *(INSR)* on the membrane of cardiac myocytes and belong to the family of ligand-activated tyrosine kinase (RTK) receptors [11]. The information of the insulin signal is propagated within the cell through a non-linear signaling network [12]. The inherent robustness is the *de-facto* rule of survival in the evolutionary process of biological systems. Therefore, most of these systems are robust to the large set of stresses and demonstrate the butterfly effect to substantially smaller sets. Setting this fact *vis-a-vis* the complexity of system enables us to reduce the complexity by two strategies: i) by eliminating the components or aggregating their detail to a higher level where it is proven or speculated to have lesser impact on the objective system, ii) exclude a subset of the system from the analysis

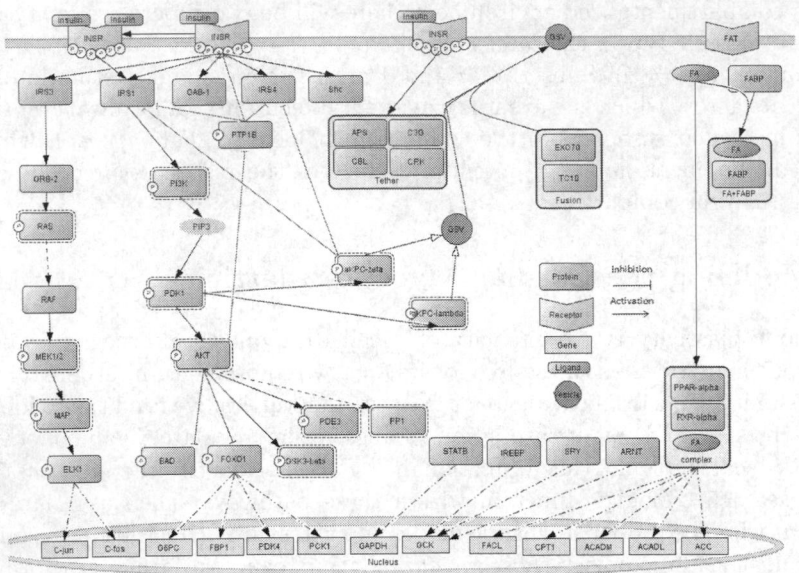

Fig. 2. The insulin (left) and fatty acid (right) signal transduction networks diagram

with the assumption that the rest of the system is in the equilibrium interaction with the current subset. With this strategy we have abstracted the insulin signal transduction hierarchy from excessive details and included those components where a consensus exists on their impact on the cell metabolism [10].

The insulin signal transduction network (STN) that has the above property could be found in *KEGG pathway database* [13]. We imported their STN and modified the original version based on data published elsewhere to include some of missing components that were necessary for our work as well as excluded the excessive details. Fig. 2 shows the signal transduction networks for insulin and fatty acid that we used in our in-silico experiment. The *myevent* diagram for the insulin signaling pathway of Fig. 2 is depicted in Fig. 3. The color code is used to represent the *myevents* with similar physicochemical (e.g phosphorylation, activation, transport) class. The physicochemical class of each *myevent* was explored from literature. A *myevent* whose physicochemical class was unidentified was assigned to a biochemical reaction class. Noting that, since signal transduction and transcription regulatory networks are interrelated we included a subset TRN that is affected by the insulin in the event diagram. Noting that Fig. 3 does not include the events that pertain to the fatty acid signal which is partially depicted in right corner of Fig. 2. In the *myevent* legend the three capital letters following the name of each *myevent* specifies the class of that *myevent* (e.g. TRN: transcription, ACT: activation, PPR: phosphorylation, INA: insulin receptor activation). A self feedback in the event digram induces the signal prorogation by one fold from the feedback point. These loops are added to the map empirically by comparing the in-silico results and experimental data.

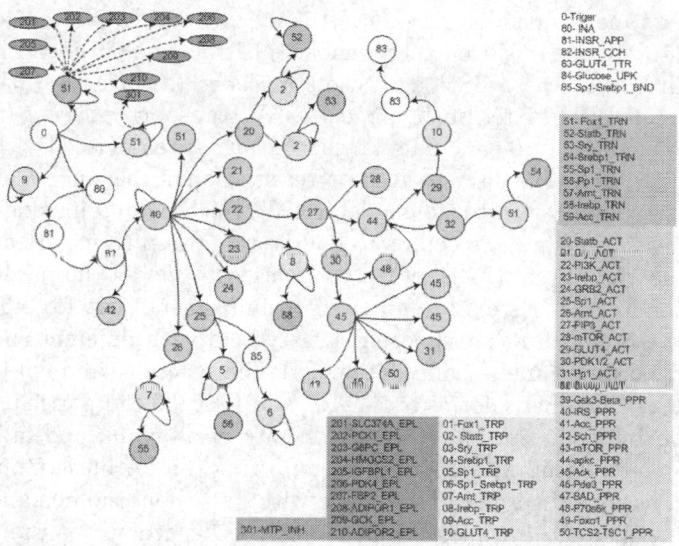

Fig. 3. The *myevent* diagram above depicts selected events for insulin STN in Fig. 2. Events with purple color belong to TRN.

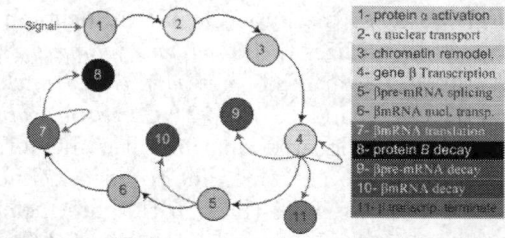

Fig. 4. Event diagram of protein synthesis in eukaryotes

3.2   Models for Glucose and Fatty Acid Uptake

Glucose and Fatty Acids comprise > %90 of the energy resources of cardiac tissues [14]. Hence, in this work we decided to focus on modeling uptake pathways of these two substrates. In adult heart Glucose is taken into the cell mainly by glucose transporter 4 (Glut4) [15]. Insulin promotes the Glut4 membrane transport through two parallel pathways. These two pathways which both originate from the insulin receptor protein (*INSR*) activation complement each others role in mediating the glucose uptake. Phosphorylated *aPKCλ/ξ* is a down stream product of a phosphorylation signaling cascade from the first pathway which enables the Glut4 vesicle transporters (GSV) to move to the vicinity of the membrane [16]. Upon activation *INSR* phospho-activates the *APS* protein that initiates the second pathway. Activation of *APS* initiates a sequence of activations and interactions that involve more than seven proteins [17]. Through a sequence of complex interactions the Glut4 which at the time is present in the vicinity of plasma membrane is first docked then tethered and ultimately fussed into the membrane [16]. In our in-silico experiment we abstract the outlined process into following three *myevents*: GSV activation *myevent* (associated with reaction model in [18] discussed later in the paper), GSV transport *myevent* (associated with diffusion model in [19], and Glut4 tether *myevents* (associated with in model [20]). Fig. 5.(a) shows the discrete even representation of the glucose uptake. We made a subtle modification to the model in [20] which mainly includes adding a capacity to each membrane receptor to handle the group transport activity for GSV. For the Fatty Acid (FA) uptake we focused to model the mechanism for which a strong consensus exists and tried to implement that for a long chain fatty acid (LCFA). Note that choosing a fatty acid with different chain length (e.g. short or medium) mainly affects the oxidation reactions and not the uptake mechanism. The plasma isoforms of FATBP and FAT/36 can participate in passive diffusion by increasing the dissociation rate of albumin, and in facilitated transport by interacting with FATP and importing the FA into cytoplasm [21]. This system adjusts the rate of FA uptake with mitochondrial demand to avoid accumulation of FA in cytoplasm which could be hazardous for the cell. Once the FA entered the cytoplasm, it then binds to cytoplasmic isoform of FATBP and transported to the vicinity of the mitochondrial outer membrane. *Acyl-CoA* Synthase (ACS) converts the Long Chain Fatty Acid (LCFA) to *LC_acyl-CoA*. To participate in the β-Fatty Acid oxidation *LC_acyl-CoA* should be transported into the mitochondria. To cross the impermeable mitochondria membrane the fatty acid transport pathway utilizes the *Carnitine palymitoyltransferase (CPT)* system. CPT composed of *L*−carnitine, *acylcarnitnie translocase* (ACT) and two transfer proteins i.e. *CPT1* and *CPT2* [22]. *Carnitine palymitoyltransferase 1* is a transmembrane protein located on the outer membrane of mitochondria and delivers the *LC_acyl-CoA* to carnitine to form *LC_acylcarnitine*. ACT hands the *LC_acyl-CoA* over to CPT2 through the intermembrane space. The second transfer protein replaces the carnitine group of *LC_acylcarnitine* with *CoA* and releases the *LC_acyl-CoA* in the mitochondria to participate in the β-fatty acid oxidation pathway [22]. *CPT1* is sensitive to *Malonyl_CoA* which is the product of *Acetly_CoA* carboxylation in cytoplasm this reaction is catalyzed by *Acetyl_CoA Carboxylase (ACC)*. Hence *Malonyl_CoA* is a negative regulator of β-fatty acid oxidation. The event-based model of the FA uptake depicted in Fig. 5.(b). The associated model for the designated *myevents* are as follows: FA uptake associated with model in [20], FA transport and FA mitochondrial transport both associated with fast reaction model (described in supplementary materials[1]). In the FA mitochondrial transport we have modeled the process by breaking the transport between the metabolic and signaling networks. More specifically the binding FA to CPT1 is handled by the signaling network as one bimolecular reaction [18]. Shuttling *LC_acylcarnitine* to the CPT2 is handled by a metabolic reaction which is catalyzed by *CPT2*. The reason for breaking the event between metabolic and signaling network originates from the set of

---

[1] Supplementary section not included due to the space limitations and is available upon request from the corresponding author.

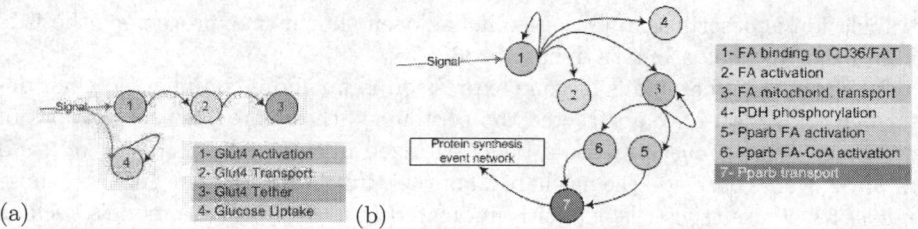

Fig. 5. (a) The *myevent* diagram for glucose uptake process. (b) Fatty acid uptake *myevent* diagram.

metabolic reactions that we used to model the metabolic reaction network in cardiac myocytes.

3.3   Eventology of Protein Synthesis

Down stream effect of a signal might effect the gene expression and consequently the protein synthesis. Protein synthesis is the core process of life which involves a very complex and not completely known regulatory mechanism. Although, we are still far behind from a comprehensive and detailed quantitative model of protein synthesis; however, our knowledge of *central dogma* is just enough to propose an event based abstraction that meets the requirements to fit into the in-silico experiment paradigm.

Protein synthesis in eukaryotes compromises an orchestrated sequence of events including: chromatin remodeling, gene transcription, pre-mRNA splicing, mRNA nuclear transport and mRNA translation. These events involve sophisticated evolution and regulatory mechanisms whose detail discussion is beyond the scope this paper. Hence, we briefly browse through the major concepts that will contribute to our modeling effort. Transcription and translation in eukaryotes is very complex and much of their details yet not properly understood. General mechanism of transcription and translation discussed in [23] and elsewhere. In [24] general concepts involved in mammalian gene transcription is described and a qualitative model for their assembly is proposed. Transcription and RNA II-TFIIB are structurally analyzed in [25] and mechanism of RNA II elongation is discussed in [26]. Binding of TATA Box Proteins (TBP) is essential for gene expression, in [27] regulation of gene expression by TBP is elucidated. Kinetic analysis of gene transcription is provided in [28]. Following the gene expression the pre-mRNA will be spliced to generate messenger mRNA. Each mRNA should be transported to cytoplasm and translated by the ribosomal proteins (tRNA) to give birth to the protein it encodes. The process of transporting the mRNA to cytoplasm is referred to as nuclear transport which it self has divers and complex mechanisms [29]. Also the kinetics of mRNA nuclear transport is studied in [30]. Following the export of mRNA to the cytoplasm ribosomal protein (tRNA) translates the codons in mRNA to the proper amino acids. The mechanism of translation initiation is given in [31] while the molecular mechanism of translation is described in [32]. Also it has been shown that protein synthesis is non-linear and has a bursty dynamic [33].

Nevertheless, the details of gene expression is far more complicated than described above, we have abstracted the protein synthesis process as a network of *myevents*. These *myevents* could be categorized into two classes of *explicit* and *implicit* events based on the mechanism of their initiation. Former, includes those *myevents* whose trigger is explicitly indicated in the qualitative models such as transcription event, splicing event, etc. The latter class includes those *myevents* that will be executed although they are not explicitly included in the qualitative models, examples of those include: protein decay, mRNA decay, transcription termination, etc.

The protein synthesis is the product of collaborative effort between the transcription regulatory and signaling networks. Therefore, suppose an external signal in its downstream activates the transcription factor α, upon activation α is transported into the nucleus. Further assume that as a result gene β is affected. The effect of α on gene expression is interpreted by the gene regulatory network, our abstracted mechanism of gene regulation will be discussed shortly. In the case of positive regulation, $\beta - mRNA$ is produced, transported to cytoplasm, and translated to protein B. Fig. 4 shows the event diagram of this model where the red arrows point to the *implicit events* and black arrows to the *explicit events*. As observed in the diagram a *myevent* could belong to both categories and the only difference is the mechanism for triggering an event with respect to the qualitative model.

For chromatin remodeling we assumed to have SWI/SNF remodeling complex, since it is the most preserved remodeling complex across eukaryotes and has no sequence specificity [34]. Researchers in [35] and elsewhere have reported the data on different aspects of chromatin remodeling. We were able to fit their reported data (result not shown) for the rate of nucleosome *in-cis* translocation (basepair/sec) into a gamma distribution ($\alpha = 2.50 \pm 0.17$, $\beta = 4.67 \pm 0.35$). Therefore, temporal dynamics of the remodeling *myevent* is modeled with gamma distribution. Also we assumed a nucleosome occupancy of 0.3 for all promoter regions. For the transcription event we used the model proposed in [36]. This model uses a birth and death Markov chain to determine the rate of the transcript production. They have modeled the process based on number of RNA PII that binds to the gene and the elongation rate of RNA PII. We have adapted and calibrated the model to become consistent for eukaryotic based on the parameters given in [37,38] (e.g. basal RNA PII elongation rate (40 bases/s), etc.). For the splicing *myevent* we assume to have a *constitutive* splicing [39] where each a pre-mRNA spliced at a rate of 0.25 per minute [30] which is negative exponentially distributed around the mean. For the pre-mRNA and mRNA decay *myevents* we applied the exponential decay processes with a rates according to to half life of these species reported in [40,33]. We used a simple stochastic diffusion model proposed in [19] to estimate temporal dynamics of mRNA nuclear transport based on kinetics reported in [41]. For estimation of translation *myevents* time, we applied the markov model proposed in [36] for translation in prokaryotes and calibrate the parameters based on experimental data in [38,30]. The protein decay *myevent* has an exponential decay process with rate reported in [42]. The transcription termination event has a constant time which we obtained empirically while calibrating the simulator.

3.4 Transcription Regulatory Network and In-Silico Regulatory Model

More than 150 genes have been identified that are positively or negatively regulated by the insulin [43]. Amongst genes affected by insulin, < 50 genes reported as myocardial genes [44]. We sought to collect as many genes that has been reported and is regulated by either insulin or fatty acid signaling pathways in the heart muscle cell [12,43]. To abstract the expression and inhibition of the target gene 'X', we attribute each gene with a status flag and a time stamp. The status flag can hold one of the following three states: being expressed(BE), already inhibited (AI), or no activity (NA). The time stamp indicates the time for the last change in the status flag of the gene. Transition of the gene status from NA to either BE or AI is triggered by the transcription *myevents*. To handle the transition form BE or AI to NA a specific Gene Status Check (GSC) event is predicted that is executed periodically and compare the target gene time stamp with current time. If the difference between the two times is greater that a *GENE_HOLD_STATUS* constant then it shifts the gene status to NA. Based on current model there is no direct shift between BE and AI states.

The input to the a transcription *myevent* is a transcription factor 'T'. Execution of a transcription event indicates that resource for 'T' is available. The non-empty set g includes all the genes that are up/down regulated by transcription factor 'T', upon execution of a transcription event one of these genes is selected for the status change with probability $p = \frac{1}{|g|}$. Based on wether the selected gene 'X' belongs to up-regulated or down-regulated subset of g, its flag is changed accordingly. The set of transcription factors (TFs) that we included in our in-silico along with their target genes is available in supplementary material.

3.5 Metabolic Reaction Model

Glycolysis I, TCA cycle, pyruvate metabolism, and β-fatty acid oxidation pathways are the major pathways dedicated to precursor substrates metabolism in cardiac myocytes. In our experiment we considered the set of metabolic reactions that comprises the above metabolic. This set composed of 109 reactions consistent with human metabolic reactions reported in BiGG database [45]. Each reaction is identified by a unique reaction ID that we borrowed from the original record in the BiGG (list of these reactions is given in supplementary materials).

For a metabolic reaction *myevent* we consider a *lumped metabolic event* whose effects on metabolites is based on the Flux Balance Analysis (FBA) approach [46]. Implementing such strategy requires a metabolic *myevent* to be local. However, keeping metabolic *myevents* local will cause metabolite explosion in some compartment (e.g. mitochondria) and metabolite starvation in the others (e.g. cytosol). To circumvent this issue we define a new cross compartment event called *metabolite squad myevent (MetabSquad)* that executed regularly every τ *squad* unit of time and redistributes the metabolite across pairs of *neighbor compartments*. $\Omega_k(i,j) \leq 1$, $\forall i,j : i \neq j$ is the portion of metabolite k molecules in compartment i to be transported to neighbor compartment j. This ratio is estimated in an iterative fashion. A pair of cellular compartments that can have direct molecular transport between themselves are called *neighbor compartments*.

We employed FBA approach to determine the flux across the metabolic reactions. From a reaction flux we can determine the change in molecular-count of a specific metabolite in the entire set of metabolic reactions in an arbitrary epoch during experiment, given the steady state condition. The essence of the FBA for a metabolic reaction founded on the assumption that the cell tends to maximize the biomass yield in the steady state condition. The emerging problem is then mapped into a linear optimization problem where the solution to this problem are optimum fluxes across sets of metabolic reactions given: the reactants, products, and enzymes concentrations. The method to manipulate the flux for a reaction across each metabolic *myevent* inter-arrival time briefly includes following steps: (i) determine active reactions from the availability of their participant molecular parts, (ii) determine the reaction direction by comparing the equilibrium constant of the reaction to the ratio of $\sum[procuts]$ to $\sum[reactants]$, where brackets in brackets indicate the concentration, (iii) determine the weight of a reaction with respect to all set of reactions (weight of a reaction is inversely proportional to the number of reactions in which its reactants participate), (iv) determine the reaction flux during time t_{mtb} with respect to the enzyme turnover number and metabolite constrains. t_{mtb} is the inter-arrival time between two metabolic *myevetns* which could be set to an arbitrary constant value. The dilemma for setting t_{mtb} value is choosing between efficiency and the precision of the simulation (i.e. large versus short periods).

Noting that any *myevent* that appeared in the event networks and was not discussed individually was modeled using stochastic reaction model proposed in [18]. The model equation for the reaction between reactants A and B is replicated here:

$$p(t_{A \cdot B}) = e^{(-\lambda t_{A \cdot B})}, \quad \lambda^{-1} = \frac{n_A(r_A + r_B)^2}{V} \sqrt{\frac{8\pi k_B T(m_A + m_B)}{m_A \cdot m_B}} \exp(\frac{-E_{Act}}{k_B T})(1)$$

In the above equation λ is reaction rate, n_A is molecular count of reactant A, r_x and m_x are the average molecular radius and molecular mass of reactant x, respectively, k_B is the Boltzmann constant, T is the absolute temperature and E_{Act} is the activation energy of the reaction. Noting that, this model subjects to further approximations given in supplementary materials.

4   In-Silico Results

The traces of plasticity is also observable in the expression profile of those genes contributing to a specific substrate metabolism. On the other hand, metabolism of an abundant metabolite subjects to the promising availability of the transport proteins and metabolic enzymes specific to that metabolite. Hence, a higher gene expression profile is expected for the underlaying genes. Van Bilsen and his colleagues [47] conducted an experiment for the rat heart and identified the expression patterns for some of the genes contributing to the glucose and fatty acid oxidations in the rat heart.

In such experiment the cardiac myocytes were forced to follow a certain pattern in substrate (glucose and fatty acid) metabolism. The pattern imposed by feeding the model animals with glucose rich food for 8 hours (*feeding period*) and then letting them starve for the next 40 hours(*fasting period*). The starvation forces the body to release the fat stored in adipocytes into the blood. This, would let the other cells (e.g. cardiac myocyte) to uptake and oxidize the fatty acids for their functions, which obligates activating the fatty acids-dependent uptake and oxidation pathways. To validate our approach we utilized the proposed methodology to conduct the above experiment in-silico at the molecular level. To date of this paper no in-silico simulation tool or quantitative model has been reported to have the capacity of capturing the system-level dynamics of a cellular network for such a pro-long duration (i.e 48 hours).

To design the experiment we supplied 1.4 nM of each signaling proteins, 1.4 nM of each metabolic enzymes, and a basal level of transcript for each of the genes listed in supplementary materials. Also 11 mM of the Octadecaontate_(n-18:0) which is a saturated stearic fatty acid was supplied as the exogenous fatty acid resource. To mimic the short feeding period followed by a longer fasting period we supplied the initial concentration of the glucose such that it would last for ~ 8 hours, where fatty acid concentration would last for entire course of experiment. Hence, for the 40 hours following the initial 8 hours of experiment only fatty acid would be available as the metabolic substrate. Noting that we suspended the insulin signal once the glucose supply reached %5 of its initial concentration. The choice of the stearic or palmitic fatty acid would not skew the results since the stearic acid is converted to a palmitic acid by metabolic reaction *R_FAOXC180* which is an oxidation-reduction reaction in β-fatty acid oxidation pathway.

The fold change in concentration of the transcripts for those genes whose data could be validated with published data is depicted in Fig. 6. Comparison is shown at two time points for the feeding scenario discussed earlier between the in-silico and empirical results. As observed the *CPT1* which is a member of CPTS increased during the fasting and *ACADL* which is Long-chain specific acyl-CoA dehydrogenase was also induced during that period. The in-silico results shows that HK2 (hexokinase 2) which is a glycolysis pathway metabolic enzyme remained constant during fasting where the empirical data suggested reduction by half fold for the same period. This may suggests a potential inhibitory regulation which is not included in our simulation. Although both results agree on the increase for Fatty acid-binding protein (*FABP*) during the fasting period; however, in-silico results show a significantly higher fold which demands for further regulatory mechanism not implement by our gene regulatory model.

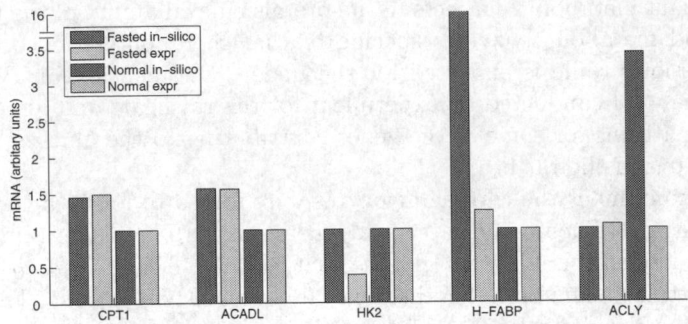

Fig. 6. Change in the expression profile of selected myocardial genes for *normal* feeding period (after 8 hours) and *fasting* period after 48 hours. In-silico results and empirical data are shown in blues and greens, respectively.

This proposition stays valid for ATP-citrate synthase (*ACLY*) too, but this time during the feeding period.

In Fig. 7.(a) we have shown the transcription regulatory effect of current feeding scenario at four time points for the entire set of genes. The genes that induced by *FOXO1* show exponential increase in their expression profile after 5 hours of simulation. This happened because the insulin signal which would negatively regulate those genes gradually diminished. Although the increase in the expression level of these genes was expected; however, the reported quantities in their expression profile subject to further validation with empirical data. Many of the genes involved in fatty acid transport and oxidation pathway show a one to two folds increase which is in agreement with the experimental data reported elsewhere. Since very limited data was available on negative regulatory effects of current transcription factors we were not successful to capture their negative regulatory effects on the gene expression profiles.

To further observe the metabolic plasticity of a myocardial cell we also looked into the metabolic fluxes, ATP synthesis and some substrates concentration profile during the course of experiment. Fig. 7.(b) shows that during the early hours experiment both exogenous substrates were highly utilized in energy production of the cell, as a result *ATP* concentration increased exponentially. The concentration of *D-Glucose-6P* follows an exponential decay which indicates a very high utilization of glucose in cell. After initial raise in the concentration of intermediate metabolites, for the hours between 6 to 20 we observe a decline in the slope of *Stearoyl-CoA(18:0CoA)* decay. The smoother slope is the consequence of negative regulation of *CPT1* by *Malonyl-CoA* as well as marginal inhibitory effect of insulin signal on fatty acid transport system [48], which we incorporated in the event network as a slow reaction *myevent* on *FAT/CD36*. Reduction of *Malonyl-CoA* concentration was followed by increased the activity of CPT1 which further increased the rate of fatty acid oxidation after the first day (24 hours).

Fig. 8 shows the fluxes across all active metabolic reaction in Glycolysis I, TCA cycle, pyruvate metabolism, and β-fatty acid oxidation pathways during the course of in-silico experiment. The radius of circles show *log(flux)* value of

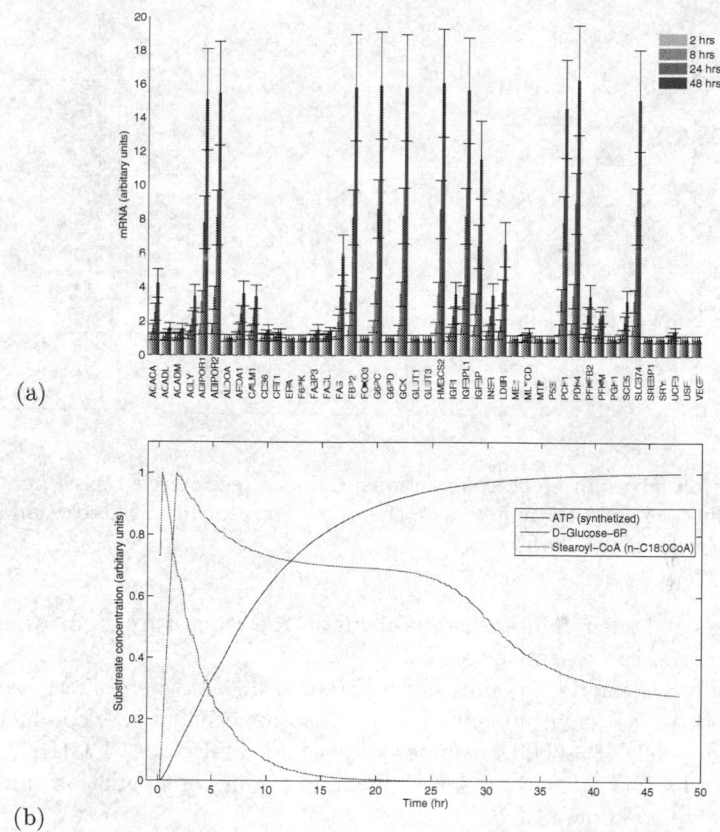

Fig. 7. Effect of 8 hours period of normal feeding followed by 40 hours of fasting on: (a) gene expression profile for all the genes in transcription regulatory network underlaying the current in-silico experiment. (b) concentration of *D-Glucose-6P*(red), *Stearoyl-CoA(18:0CoA)* (gray), and ATP in blue.

the reactions which were measured in 45 minute intervals, x and y axis are the time and reaction index, respectively. From here we reconfirm that roughly there was no flux across glycolysis pathway after 8 hours, where the flux across fatty acid reactions fluxes varied but sustained during the entire course of experiment.

Following is a list of selected parameters along with their values that we used in the simulation: Average myocardial cell volume $= 40 \times 10^{-15}$ m^3 reported in [49], nucleus volume is $\sim \%10$ of the cell volume [23], in myocardial cell mitochondria occupies $\sim \%30$ of the cell volume [50], from data in literature we estimated there are ~ 4660 cardiac myocytes per 1mg wet cells (varies among the samples). To convert any molecular counts in heart muscle to nano-Molar concentration we divided the counts by 240.88×10^2. The weight per amino acid was considered 0.11 KDa and an average weight of a eukaryotic cell $\sim 10^{-9}$ grams. The activation energy $9 < E_a < 21$ k_{bT} was used for the reaction model governed by Eqn. 1 and temperature was set to $T = 300$ K. The complete list

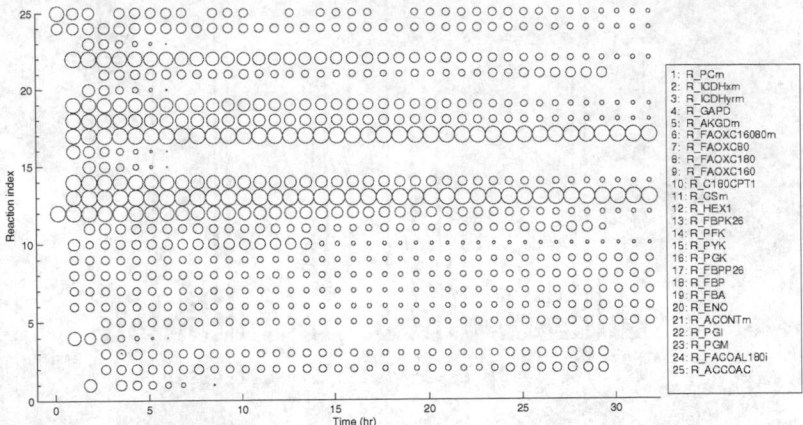

Fig. 8. Reaction fluxes for 25 active reactions: radius of each circle represents the log value of the flux per reaction, numbers on the y axis correspond to the reaction indices on the list to right of the chart

of parameters and their values is available upon the request from corresponding author (amin.mazloom@mssm.edu).

The simulation engine was entirely written in JAVA where JVM was running in Windows XP environment. The in-silico experiment was conducted on a stand alone Dell XPS-3000 machine, which had dual core 2.1 GHz P4 Intel processors and 4 GB of DDR2 RAM. Forty eight hours of simulation time took approximately two hours of CPU time.

5   Discussion

The proposed approach significantly reduced the computational cost of the experiment. Computation complexity is a major factor that challenges most system-level simulation efforts in biological networks. Appropriate design of the insulin and fatty acid event diagrams which forms the road map for evolution of the systems dynamics is essential in the of success of current approach. Although there is no general role to follow for designing the details of an event diagram, for instance where to incorporate a loop or when to aggregate a group of events (e.g GLUT4 tethering and fusion events); nevertheless both experience and practice as well as relevant data from biochemical and biological experiments are particularly important. Including a loop could increase the speed of the signal propagation in the subnetwork originated from that node by one fold. Furthermore on the border of the signal transduction and metabolic networks consistency in selecting of input/output resources is crucial for the evolution of the system. Also most of the laboratory experiments on myocardium are at the tissue and organ-level. Hence, we often have to apply certain approximations or assumptions to re-scale experimental data to be beneficial for our in-silico simulation. Such mappings are not always trivial due to the missing vital data components. The accuracy in capturing the dynamics of a *myevents* is directly proportional to the precision of parametric physicochemical model that is associated with the model. In the current study in several cases we applied a single parameter probability distribution, further study is required to replace those distributions with more realist and accurate formalisms. The proposed regulatory mechanism projects the regulatory effect of a TF to a stochastic binary parameter with constant life time. Undoubtedly, real transcription regulatory mechanisms are far more complex, yet we observed that the current model could be a starting point in designing more complex and yet efficient transcription regulatory models that fit the system level simulation paradigm of complex biological systems. Although the outlined approach demonstrated a high capacity for system level simulation of a complex biological system such cardiac myocytes; however, big knowledge gaps in the structure of the target system could significantly diminish these capacity. Also designing a stochastic phytochemical model that can properly capture the temporal behavior a biological process is a particularly challenging task.

6 Conclusion

We established a novel framework in simulating the dynamics of biological networks with temporal heterogeneity across multiple cellular compartments. Inherent stochasticity that exists in the cell environment is conserved in the current in-silico framework. Furthermore, the proposed approach is scalable, very efficient and fairly accurate compared to available methods for system level modeling of biological systems. The promising capacities of the current approach was demonstrated by utilizing that in conducting an in-silico experiment on the metabolic plasticity of cardiac myocytes. We believe that current method could be very constructive in hypothesis testings experiments, drug target analysis, and cellular level study of diseases . Also the application of this approach is not limited to the heart cell and could potentially be applied in any cell lineage. Although the proposed method is sound in efficiency, yet demands substantial work especially in establishing more promising physicochemical models as well as the gene regulatory model to grant further accuracy to the results.

References

1. Zeigler, B.P., Kim, T.G., Praehofer, H.: Theory of Modeling and Simulation, 2nd edn. (2000)
2. Smith, N.P., Crampin, E.J., Niederer, S.A., Bassingthwaighte, J.B., Beard, D.A.: Computational biology of cardiac myocytes: proposed standards for the physiome. J. Exp. Biol. 210(9), 1576–1583 (2007)
3. Kuijpers, N.H.L., ten Eikelder, H.M.M., Bovendeerd, P.H.M., Verheule, S., Arts, T., Hilbers, P.A.J.: Mechanoelectric Feedback as a Trigger Mechanism for Cardiac Electrical Remodeling: A Model Study. Analys. of Biomedical Eng. 36(11), 1816–1835

4. Sedaghat, A.R., Sherman, A., Quon, M.J.: A mathematical model of metabolic insulin signaling pathways. Physiol. Endocrinal Metab 283, E1084–E1101 (2002)
5. Vinnakota, K., Kemp, M.L., Kushmerick, M.J.: Dynamics of Muscle Glycogenolysis Modeled with pH Time Course Computation and pH-Dependent Reaction Equilibria and Enzyme Kinetics. Biophy. J. 91, 1264–1287 (2006)
6. Gillespie, D.T.: A general method for numerically simulating the stochastic time evolution of coupled chemical reactions. J. Comput. Phys. 22, 403–434 (1976)
7. van Kampen, N.G.: Stochastic Processes in Physics and Chemistry, 2nd edn. North-Holland, Amsterdam (1992)
8. Samoilov, M.S., Arkin, A.P.: Deviant effects in mulecular reaction. Nat. Compu. Biol. 24, 1235–1240 (2006)
9. Gadgil, C., Lee, C.H., Othmer, H.G.: A stochastic analysis of first-order reaction networks. Bull. Math. Biol. 67, 901–946 (2005)
10. Huss, J.M., Kelly, D.P.: Nuclear receptor signaling and cardiac energetics. Circulation Research 95, 568–578 (2004)
11. Nystrom, F.H., Quon, M.J.: Insulin signaling: metabolic pathways and mechanisms for specificity. Cell Signal 11(8), 563–574 (1999)
12. Saltiel, A.R., Kahn, R.: Insulin signaling and the regulation of glucose and lipid metabolism. Nature 414, 799–806 (2001)
13. Kanehisa, M.: KEGG Data base. Novartis found Sym. 247, 91–101 (2001)
14. Kodde, I.F., van der Stok, J., Smolenski, R.T., de Jong, J.W.: Metabolic and genetic regulation of cardiac energy substrate preference. Comparative Biochem. and Physiol. 146, 26–39 (2006)
15. Gould, G.W., Holman, G.D.: The glucose transporter family: structure, function and tissue specific expression. Biochem. 295(2), 329–341 (1993)
16. Watson, R., Pessin, J.E.: Glut4 Translocation: The last 200 nanomters. Cellular Signalling 19, 2209–2217 (2007)
17. Chang, L., Chiang, S., Saltiel, A.R.: Insulin signaling the regulation of glucose transport. Mlecular Medicine 10, 7–12 (2004)
18. Ghosh, P., Ghosh, S., Basu, K., Das, S.K.: Holding time estimation for reactions in stochastic event-based simulation of complex biological systems. Model. Pract. Theory (2007) doi:10.1016/j.simpat.2007.09.002
19. Ghosh, P., Ghosh, S., Basu, K., Das, S.K.: A diffusion model to estimate the inter-arrival time of charged molecules in stochastic event based modeling of complex biological networks. In: Proc. IEEE Comp. Systems Biol. Conf. (2005)
20. Mazloom, A.R., Basu, K., Das, S.K.: A Random Walk Modelling Approach for Passive Metabolic Pathways in Gram-Negative Bacteria. In: Proce. IEEE CIBCB Conf., pp. 1–8 (2006)
21. Koonen, D.P., Glatz, J.F., Bonen, A., Luiken, J.J.: Long chain fatty acid uptake and FAT/CD36 trasnlocation in heart and skeletal muscle. Biochem. Biophys. Acta. 1736(3), 163–180 (2005)
22. Bonnefont, J.P., Demaugre, F., Prip-Buus, C., Saudubray, J.M., Brevit, M., Abadi, N., Thuillier, L.: Carnitine palmitoyltransferase deficiencies. Mol. Genet. Metab. 68, 424–440 (1999)
23. Alberts, B., Bray, D., Lewis, J., Raff, M., Roberts, K., Watson, J.D.: Molecular Biology of the Cell, 4th rev. edn. Garland Publishing Inc. (2002)
24. Flores, O., Lu, H., Reinberg, D.: Factors Involved in Specific Transcription by Mammalian RNA Polymerase II. Biol. Chemistry 267(4), 2786–2793 (1992)
25. Bushnell, D.A., Westover, K.D., Davis, R.E., Kornberg, R.D.: Structural basis of transcription: an RNA Polymerase II-TFIIB Cocrystal at 4.5 Angstroms. Science 303, 983–988 (2004)

26. Sims III, R.J., Belotserkovskaya, R., Reinberg, D.: Elongation by RNA polymerase II: the short and long of it. Genes & Dev. 18, 2437–2468 (2004)
27. Lee, T.I., Young, R.A.: Regulation of gene expression by TBP-associated proteins. Genes & Dev. 12, 1398–1408 (1998)
28. Perez-Ortin, J.E., Alepuz, P.M., Moreno, J.: Genomics and gene transcription kinetics in yeast. TRENDS in Genetics 23(5), 250–257 (2007)
29. Zhang, C., Zobeck, K.L., Burton, Z.F.: Human RNA Polymerase II elongation in slow motion: role of the TFIIF RAP74α1 Helix in nucleoside triphosphate-driven translocation. Molec. and Cell. Biol. 25(9), 3583–3595 (2005)
30. Audibert, A., Weil, D., Dautry, F.: In Vivo Kinetics of mRNA Splicing and Transport in Mammalian Cells. Molc. and Cell. Biol. 22(19), 6706–6718 (2002)
31. Jackson, R.J.: Alternative mechanisms of initiating translation of mammalian mRNAs. Biochem. Society Lectures, pp.1231–1241 (2005)
32. Pestova, T.V., Kolupaeva, V.G., Lomakin, I.B., Pilipenko, E.V., Shatsky, I.N., Agol, V.I., Hellen, C.U.T.: Molecular mechanisms of translation initiation in eukaryotes. Proc. Natl. Acad. Sci. 98(13), 7029–7036 (2001)
33. Lorsch, J.R., Herschlag, D.: Kinetic dissection of fundamental processes of eukaryotic translation initiation in vitro. The EMBO Journal 18(23), 6705–6717 (1999)
34. Saha, A., Wittmeyer, J., Cairns, B.: Chomatin remodeling the industrial revolution of DNA arround histones. Nat. Rew. Mol. Cell Biol. 7, 437–447 (2006)
35. Bustamanate, C., Peterson, C.L., Cairns, B.R., Smith, S.B., Mihardja, S., Grill, S.Q., Saha, A., Smith, C.L., Zhang, Y.: DNA transclocation and loop formation mechanism of charomatin remodeling by SWI/SNF and RSC. Mol. Cell 24, 559–568 (2006)
36. Ghosh, S., Ghosh, P., Basu, K., Das, S.K.: Modeling the stochastic dynamics of gene expression in single cells: a birth and death markov chain analysis. In: IEEE (BIBM) Conf., pp. 308–316 (2007)
37. Singh, S., Yang, H.O., Chena, M., Yu, S.: A kinetic-dynamic model for regulatory RNA processing. J. of Biotechnology 127, 488–495 (2006)
38. Kim, H., Yin, J.: Effects of RNA splicing and post-transcriptional regulation on HIV-1 growth: a quantitative and integrated perspective. IEE Proc. Syst. Biol. 152(3), 138–152 (2005)
39. Sanford, J.R., Caceres, J.F.: Pre-mRNA splicing: life at the centre of the central dogma. J. Cell Science 117, 6261–6263 (2004)
40. Cao, D., Parker, R.: Kinetic dissection of fundamental processes of eukaryotic translation initiation in vitro. RNA Journal 7, 1192–1212 (2002)
41. Kubitscheck, U., Grünwald, D., Hoekstra, A., Rohleder, D., Kues, T., Siebrasse, J.P., Peters, R.: Nuclear transport of single molecules: dwell times at the nuclear pore complex. Cell Biology 168(2), 233–243 (2005)
42. Belle, A., Tanay, A., Bitincka, L., Shamir, R., O'Shea, E.K.: Quantification of protein half-lives in the budding yeast proteome. Proc. Natl. Acad. Sci. 103(35), 13004–13009 (2006)
43. Mounier, C., Poser, B.I.: Transcriptional regulation by insulin: from receptor to the gene. Physiol. Pharmacol 84, 713–721 (2006)
44. Brownsey, R.W., Boone, A.N., Allard, M.F.: Action of insulin on mammalian heart: metabolism, pathology and biochemical mechanism. Cardiovascular Research 34, 3–24 (1997)
45. The BiGG database (2007), http://bigg.ucsd.edu

46. Covert, M.W., Schilling, C.H., Palsson, B.: Regulation of gene expression in Flux balance models of metabolism. Theoritical Biology 213, 73–88 (2001)
47. Van der Lee, K.A.J.M., Willemsen, P.H.M., Samec, S., Seydoux, J., Sulloo, A.G., Pelsers, M.M.A.L., Glatz, J.F.C., Van der Vusse, G.J., Van Bilsen, M.: Fasting-induced changes in the expression of genes controling metabolism in rat heart. J. of Lipid Research 42, 1752–1758 (2001)
48. Saddik, M., Gamble, J., Witter, L.A., Lopaschuk, G.D.: Acetyl-CoA carboxylase resulation of fatty acid oxidation in the heart. J. of Biological Chemistry 268(34), 25836–25845 (1993)
49. Maisch, B.: Enrichment of vital adult cardiac muscle cells by continuous silica sol gradient centrifugation. Basic Res. Cardiol. 76, 622–629 (1981)
50. Drake-Holland, A.J., Noble, M.I.M.: Cardiac metabolism. John Wiley and Sons, New York (1983)

Model and Dynamic Behavior
of Malware Propagation over Wireless Sensor Networks

Yurong Song and Guo-Ping Jiang

Center for Control and Intelligence Technology, Nanjing University of Posts and
Telecommunications, Nanjing, 210003, China
{songyr,jianggp}@njupt.edu.cn

Abstract. Based on the inherent characteristics of wireless sensor networks
(WSN), the dynamic behavior of malware propagation in flat WSN is analyzed
and investigated. A new model is proposed using 2-D cellular automata (CA),
which extends the traditional definition of CA and establishes whole transition
rules for malware propagation in WSN. Meanwhile, the validations of the
model are proved through theoretical analysis and simulations. The theoretical
analysis yields closed-form expressions which show good agreement with the
simulation results of the proposed model. It is shown that the malware propagation in WSN unfolds neighborhood saturation, which dominates the effects of
increasing infectivity and limits the spread of the malware. MAC mechanism of
wireless sensor networks greatly slows down the speed of malware propagation
and reduces the risk of large-scale malware prevalence in these networks. The
proposed model can describe accurately the dynamic behavior of malware
propagation over WSN, which can be applied in developing robust and efficient
defense system on WSN.

Keywords: wireless sensor networks, malware propagation, model, cellular
automata, neighborhood saturation, MAC mechanism, theoretical analysis.

1 Introduction

Wireless sensor networks have been widely used for many interesting and new
applications such as environmental monitoring, patient health care monitoring,
detection of chemical or biological threats, and military surveillance, tracking and
targeting [1]. One key issue is various types of security treats [2, 3] in wireless sensor
networks which are highly distributed and resource constrained environments.
Attacks against wireless sensor networks could be denial of service, worm, Sybil
attack and other malicious codes.

Worm and virus attacks on the Internet have been widely studied [4-8]. Some
studies have greatly contributed to our understandings of various security issues and
threats to wireless Ad hoc networks. However, wireless sensor networks differ from
wireless Ad hoc networks and traditional computer networks in various aspects: First,
wireless sensor networks are highly distributed system and consist of a great number
of distributed nodes (sensor nodes) with the ability to monitor its surroundings.

J. Zhou (Ed.): Complex 2009, Part I, LNICST 4, pp. 487–502, 2009.

Second, sensor nodes are limited in power, computational capacities, and memory[1]. Finally, self-organization is a fundamental feature of wireless sensor networks[9]. Those security mechanisms on Internet or wireless Ad hoc networks could not be applied directly to wireless sensor networks.

Investigation of worms spreading in wireless sensor networks has attracted some researchers [10-12]. Khayam and Radha [10] apply signal processing technique to model space-time propagation dynamics of topologically-aware worms in a sensor network with uniformly distributed nodes. They integrate physical, data link, network and transport protocol characteristics into the proposed model of worm propagation and obtain a closed-form expression of the infected population. De *et al.* [11] model and analyze the node compromise spreading process and identify key factors determining potential outbreaks of such propagations. In particular, they perform their study on random graphs precisely constructed according to the parameters of the network. However, the random graph model is homogeneous, which conflicts with the characteristic that sensor nodes relate closely to the location in WSN. Therefore, the spreading models based on random graph model are not suitable for investigating the propagation over WSN. Afterwards, the authors of Ref.[11], in their another paper [12], point out that the analytical model of [11], based on random graphs, fails to capture the temporal dynamics of the comprise propagation and only succeeds in capturing the outcome of the infection. Furthermore, the authors propose an epidemic theoretic model for evaluating broadcast protocols in wireless sensor networks. However, the model assumes that the number of neighbors that can be infected by any infected node is proportional to all susceptible nodes in the network (see in Eq.(2) and (3) of [12]). The assumption results in an inaccurate evaluation on spreading speed in the process of malware propagation.

Cellular automata (CA) is a mathematical model for complex natural systems [13-16], containing large numbers of simple identical components with local interactions. There is a substantial literature [13, 16-19] on the mathematical model based on cellular automata in epidemiology of theoretical biology. These models based on cellular automata focus on the local characteristics of the disease spreading process influencing the global behavior of the system. However, considering the inherent characteristics of WSN, it is unsuitable that the existing models for epidemiology are applied directly to malware propagation of wireless sensor network.

CA can simulate various uncertain behaviors of complex system, which is difficult for those models based on deterministic equations. Furthermore, CA is easy to realize on computer due to its spatial-temporal discrete property and massively parallel computation. An example of application of cellular automata in wireless sensor networks can be found in [20]. Cunha *et.al.* verify the possibility of using cellular automata to simulate the behavior of a WSN and a simulator has been developed to evaluate an algorithm for a very common problem in sensor networks: the topology control. Particularly, the self–organization and the local interaction are the inherent properties of WSN, which are similar to CA, so CA can be applied to model and simulate the malware propagation in WSN.

In this paper, we focus on the inherent characteristics of WSN and the dynamic process of malware propagation in WSN is modeled and analyzed using cellular automata. The validations of the model are performed through theoretical analysis and simulations. The theoretical analysis yield closed-form expressions which show good agreement with the result of the proposed model. The theoretical analysis and simulation results demonstrate that the proposed model characterizes fully the localization and the spatial-temporal correlation and the model is appropriate to simulate malware propagation in WSN. An evolving pattern in the 2-D cellular space is obtained easily under different time using the model. It is shown that the assumption of homogeneous mixing of nodes causes an inaccurate evaluation on spreading speed in the process of worm propagation. Our model and theoretical analysis show that the initial growth of the malware is significantly slower than the exponential growth observed in malware propagation in [11, 12]. Our research shows malware propagation in WSN unfolds neighborhood saturation, which dominates the effects of increasing infectivity and limits the spread of the malware. In addition, MAC mechanism of wireless sensor networks greatly slows down the speed of malware propagation and reduces the risk of large-scale malware prevalence in these networks. The proposed model is able to describe accurately the dynamic behavior of malware propagation on WSN, and can be used for developing robust and efficient defense system on WSN.

The rest of this paper is organized as follows. In Section 2, a few of related analysis and assumption are described. In Section 3, we propose a new model and establish the transition rules for malware propagation in WSN. Neighborhood saturation in process of malware propagation is pointed out in Section 4. In Section 5, a theoretical analysis is presented. In Section 6, the simulations are presented. Finally, the conclusions are given in Section 7.

2   Related Analysis and Assumption

Considering the inherent characteristics of WSN and the spatial-temporal correlation of malware propagation over WSN, some key factors are given and discussed in this section.

2.1   Routing Mechanism

In general, a more robust mechanism for packets routing in wireless sensor networks is by multi-hop broadcasts[1]. Since the transmission power of a wireless radio is attenuated in a squared or even higher order with the distance, multi-hop routing will consume less energy than direct communication. The attackers take advantage of the broadcast mechanism to propagate malicious codes such that malware spreads quickly to the entire network[21]. We assume that infected nodes adopt multi-hop broadcasting strategy to spread malware to their neighbors. Furthermore, the adopted broadcast protocol ensures that each infected node broadcasts the malware to its neighbors only once for the purpose of preventing broadcast storm.

2.2   Infected Rate

The infected rate is related to many factors, such as authentication mechanism for securing data exchanging, attack characteristic of malware and communication pattern. For simplicity, these factors are integrated into a parameter, namely, the infected rate β, with the value being from 0 to 1.

2.3 Death Rate

It is well known that sensor nodes are severely restricted in terms of computation power and communication capability, especially energy. Malware propagation between nodes results in nodes consuming continuously energy and tending to death. So, the death rate of nodes is defined by:

$$\gamma_{ij}(t) = c\varepsilon_{ij}(t), \tag{1}$$

where c is a constant, $\varepsilon_{ij}(t)$ denotes the cumulative consumption of energy of $cell(i,j)$ until the time t.

2.4 Media Access Control (MAC)

Malwares over wireless sensor networks will face channel collision, which should reduce the spreading rate of malwares. The MAC protocol specifies a set of rules that enable nearby sensor nodes to coordinate their transmissions in a distributed manner[1]. In our model to be given in the next section, a MAC table is designed to solve the problem of channel collision. If a sensor node is transmitting a packet, the states of its neighbors should be set block (denoted by '1') in MAC table, which means neighbors can not transmit packets at the same time. Each sensor node checks its state in the MAC table before starting a data transmission. The sensor nodes transmit packets when the channels are idle (denoted by '0' in MAC table). Therefore, the transmission is restrained if the channels are busy.

3   Modeling Malware Propagation with Cellular Automata

The proposed models focus on the stochastic properties of malware propagation and the intrinsic characteristic of wireless sensor networks. We utilize an 2-D cellular automata to describe the proposed model.

An 2-D cellular automata is a discrete dynamical system formed by a finite number of $l \times r$ identical objects called cells which are arranged uniformly in a two-dimensional cellular space. Each cell is endowed with a state (from a finite state set Q) that changes at every step of time accordingly to a local transition rule.

In this sense, the state of some cell at time t depends on the states of a set of cells, called its neighborhood, at the previous time step t-1. More precisely, a CA is defined by the 4-uplet (C,Q,V,f), where C denotes the cellular space,

$$C = \{(i, j), 1 \le i \le l, 1 \le j \le r\}, \tag{2}$$

and Q is the finite state set whose elements are all possible states of the cells. The neighborhood of each cell can be described by

$$V_{ij} = \{(i+\Delta i, j+\Delta j)\} \subset Z \times Z \tag{3}$$

where $\Delta i, \Delta j$ denote separately the offset of i, j. The local transition function f can be described by

$$s_{ij}^t = f(s_{ij}^{t-1}, \mathbf{s}_{V_{ij}}^{t-1}) \subset Q, \tag{4}$$

where s_{ij}^t denotes the state of *cell(i,j)* and $\mathbf{s}_{V_{ij}}^{t-1}$ denotes the vector of neighborhood of *cell(i,j)* at time t.

3.1 Cellular Space

We consider that a flat WSN (as shown in Fig.1) composed of the maximum N stationary and identical sensors which are randomly placed on rectangular 2-D grid composed of $L \times L$ *units* is exhibited appropriately by a 2-D cellular space. We assume that each cell is occupied by at most one sensor node. Thus, we can make simulations with fewer nodes than the maximum number of cells. Let ρ denote the relative density of sensor nodes, $\rho = N / L^2$. So, the infrastructure of the flat WSN constructs the cellular space and a sensor node denotes a cell in the space. Each sensor node can establish wireless links with only those nodes within a circle of radius R_c due to the limited power. To simplify analysis, we assume that all sensor nodes are equipped with isotropic antennas that have a maximum transmission range R_c. The horizontal and vertical coordinates of a sensor node are represented by i and j in the 2-D grid(cellular space). Namely, *node(i,j)* denotes a node located in the posotion with the coordinate of (i,j).

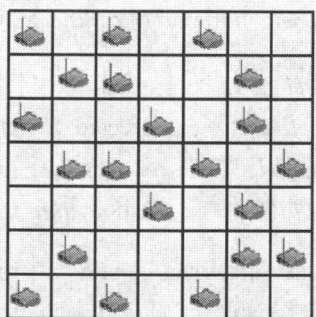

Fig. 1. Distribution of sensor nodes in a flat WSN with L^2 areas and N nodes

3.2 Neighborhood

According to the corresponding transmission range R_c, the neighborhood of each sensor is defined as shown in Fig. 2. Without loss of generality, let the length of a cell of grid be 1 *unit*, if $R_c = 1$ *unit*, each node/cell can have no more than 4 nodes as its neighbors, namely the Von Neumann neighborhood, and if $R_c = 1.5$ *units*, each node/cell can have no more than 8 nodes as neighbors, namely the Moore neighborhood. It is obvious that a node should have more neighbors with the value of R_c increasing, such that the neighborhood of *node(i,j)* is defined by

$$V_{ij} = \{(x,y): \sqrt{(x-i)^2 + (y-j)^2} \leq R_c, (x,y) \subset C\}. \tag{5}$$

Let $N(V_{ij})$ denote the number of neighborhood of *node(i,j)*.

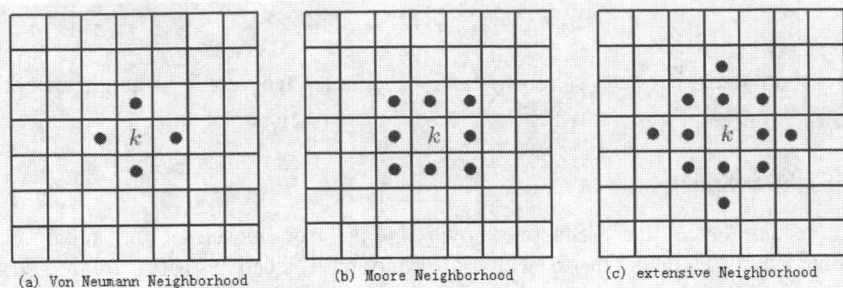

(a) Von Neumann Neighborhood   (b) Moore Neighborhood   (c) extensive Neighborhood

Fig. 2. Cell k and its possible neighborhoods in a 2-D cellular automata

3.3 State Set

At a particular time t, each cell of the cellular space is in a specific state, which depends on a specific application. Considering the MAC mechanism of WSN, we extend the definition of a state variable to that of a state vector. The vector includes two participants that denote epidemic state set Q_1 and channel state set Q_2 separately.

Borrowing the concept of epidemiology, the epidemic state of a sensor node or a cell can be one of following these states: susceptible, infected, recovery or death. Let $Q_1 = \{-1, 0, 1, 2\}$ and denote $s_{ij}(t) \in Q_1$ the epidemic state variable of $cell(i,j)$ and $\mathbf{s}_{V_{ij}}(t) \in Q_1$ the state vector of neighborhood of $cell(i,j)$ at time t, we define

$$s_{ij}(t) = \begin{cases} 0, & cell(i, j) \ \textit{is} \text{ susceptible } \textit{at time } t \\ 1, & cell(i, j) \ \textit{is} \text{ infected } \textit{at time } t \\ 2, & cell(i, j) \ \textit{is} \text{ recovered } \textit{at time } t \\ -1, & cell(i, j) \ \textit{is} \text{ dead } \textit{at time } t \end{cases} \tag{6}$$

The channel state of a sensor node can be idle or busy. Let $Q_2 = \{0, 1\}$ and denote $m_{ij}(t) \in Q_2$ the channel state variable of $cell(i,j)$ and $m_{V_{ij}}(t) \in Q_2$ the state vector of neighborhood of $cell(i,j)$ at time t, we define

$$m_{ij}(t) = \begin{cases} 0, & \text{the channel of } cell(i, j) \ \textit{is} \text{ idle } \textit{at time } t \\ 1, & \text{the channel of } cell(i, j) \ \textit{is} \text{ busy } \textit{at time } t \end{cases} \tag{7}$$

3.4 Transition Function

The transition function can be constructed by

$$s_{ij}(t) = f(s_{ij}(t-1), \mathbf{s}_{V_{ij}}(t-1)) . \tag{8}$$

The detailed rules of (8) are described below and the transition process of states is shown in Fig. 3.

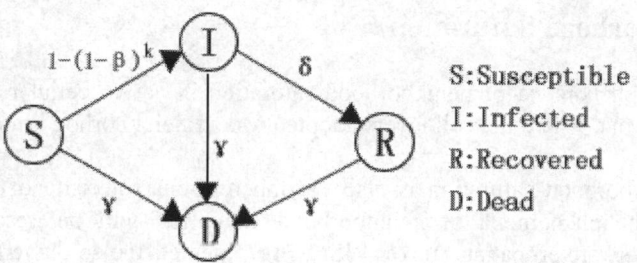

Fig. 3. Process of states transforming of sensor nodes

Susceptible to Infected/Dead. If $s_{ij}(t-1) = 0$, then $s_{ij}(t) = 1$ with probability $1-(1-\beta)^k$, or $s_{ij}(t) = -1$ with probability γ, where β and γ are defined below.

Infected nodes try to spread malware to their neighbors at each time step. A susceptible sensor's node becomes infected with the probability β when the node receives a packet containing a copy of the malware. However, during each time interval, the susceptible node can receive malware from its k neighbors($k \le N(V_{ij})$), so a susceptible node becomes infected node with the probability $1-(1-\beta)^k$. Note that k is the number of neighbors with infected state ($s_{xy}(t-1) = 1$, $(x,y) \subset V_{ij}$) and idle state of channel ($m_{xy}(t-1) = 0$, $(x,y) \subset V_{ij}$) at the time interval between t-1 and t, so,

$$k = \sum_{(x,y) \subset V_{ij}} (s_{xy}(t-1) = 1 \text{ and } m_{xy}(t-1) = 0). \tag{9}$$

Considering the limited power of sensors, some sensor nodes can become dead at the rate γ.

Infected to Recovered/Dead. If $s_{ij}(t-1) = 1$, then $s_{ij}(t) = 2$ with probability δ, or $s_{ij}(t) = -1$ with probability γ;

In particular, infected sensors can get a patch and recover from the infected state with the probability δ .

Recovered to Dead. If $s_{ij}(t) = 2$, then $s_{ij}(t) = -1$ with probability γ.

Let S(t), I(t), R(t) and D(t) denote the population of susceptible, infected, recovered and dead nodes, respectively, the we have

$$\begin{cases} S(t) = \sum_{i,j} (s_{ij}(t)=0), \\ I(t) = \sum_{i,j} (s_{ij}(t)=1), \\ R(t) = \sum_{i,j} (s_{ij}(t)=2), \\ D(t) = \sum_{i,j} (s_{ij}(t)=-1), \\ N = S(t) + I(t) + R(t) + D(t) \end{cases} \tag{10}$$

4 Neighborhood Saturation

In [19], the phenomena of neighborhood saturation in native cellular automata has been pointed out, where the cell layers adopted Moore neighborhood model, shown in Fig. 4.

The neighborhood saturation is also the inherent characteristic of WSN. In the following, the phenomena of neighborhood saturation will be described in the process of malware propagation over WSN. Fig. 5 depicts the cell layers with respect to a central cell in layer1. Layer1 has L_1 neighboring cells in its outer-line layer2. The outer-line neighborhood of $layer_i$ is $layer_{i+1}$ and the inner-line neighborhood is $layer_{i-1}$. L_i is the number of cells in $layer_i$ and is defined in equation (11). It can be visualized as the area enclosed by layer L_{i-1} subtracted from the area enclosed by layer L_i.

$$L_i = \begin{cases} 1 & i = 1 \\ [\rho\pi(R_c i)^2 - \rho\pi R_c^2 (i-1)^2] & i > 1 \end{cases}, \tag{11}$$

where [x] is the nearest integer to x. Furthermore, it is easy to deduce that

$$\frac{L_{i+1}}{L_i} = \frac{2i+1}{2i-1} \to 1, \text{ when } i \to \infty. \tag{12}$$

Equation (12) means that at higher layer, each cell at $layer_i$ is able to infect effectively only one outer-line cell at $layer_{i+1}$. This resulting neighborhood saturation slows effectively down the speed of malware propagation in WSN. The similar result that geographical localization effects on the propagation has also been observed in SARS transmission [22, 23].

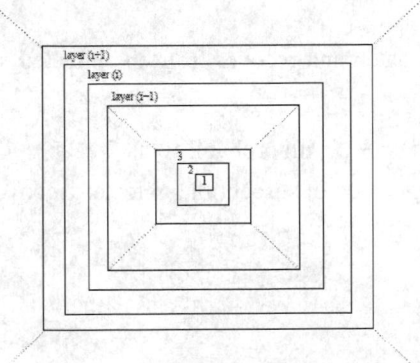

Fig. 4. Cell layers in [19]

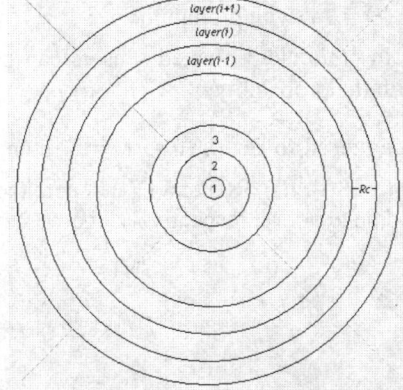

Fig. 5. Cell layers in WSN

5   Theoretical Analysis

Based on SI epidemic theory, a theoretical analysis is presented for analyzing the propagation of worms over wireless sensor networks. The theoretical analysis focuses on capturing the impact of the spatial deployment of sensor nodes to malware propagation. In particular, the sensor nodes have limited communication range R_c. For simplification, we assume that one node located in center of the grid is infected in initial time.

An important characteristic of spreading dynamic in WSN is that there is a circular region of infected nodes centered at the source node which grows outwards. That is the nodes in the infected circular region try to infect their susceptible neighboring nodes lying outside this circle, as shown in Fig. 6.

As shown in Fig. 6, the nodes in region A and B are infected nodes. The difference between region A and B is that the nodes in region A cannot infect any susceptible node because all the susceptible nodes are out of their communication area. The nodes in region B can infect the nodes in region C. The nodes in region D cannot be infected by the nodes in region B due to the limitation of communication distance. We define the width of region B or C to be R_c , the radius of the infected region to be r and the increment of r to be Δr in each time step of malware propagation. $\Delta r \leq R_c$ due to the effect of MAC, density of sensor node, infected rate and security mechanism.

When r<L/2, the area of the potential region C increases with time evolution as shown in Fig. 6(a) and when r>=L/2, the area decreases as shown in Fig. 6(b).

Through above analysis, a function of the infected population is established for considering the two stages separately.

First Stage: $r<L/2$
Under the boundary condition $\rho = 1, \beta = 1$ and without MAC mechanism, we have $\Delta r = R_c$. Namely, in the above ideal case, the radius of the infected region

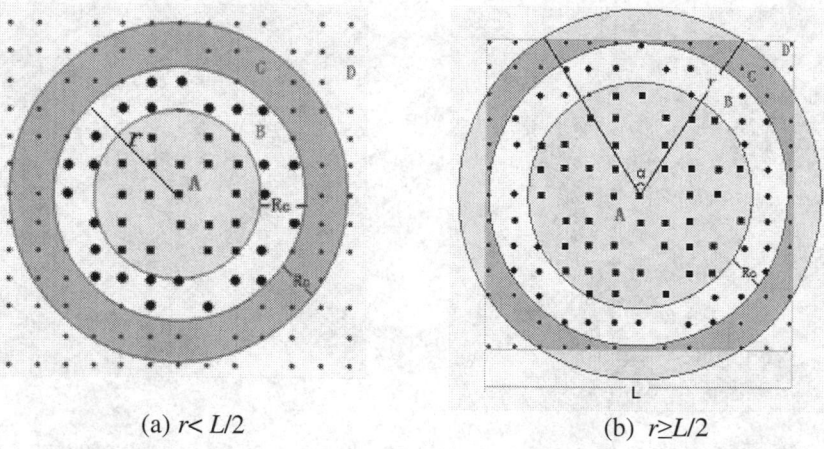

(a) r< L/2   (b) r≥L/2

Fig. 6. Spreading characteristic in flat WSN: (a) first stage: r< L/2; (b) second stage: r≥L/2

extends outside by the increment $\Delta r = R_c$ for each time. After t times, the area $I_{area}(t)$ of infected region should be

$$I_{area}(t) = \pi r^2 = \pi(\Delta rt)^2 = \pi(R_c t)^2 . \tag{13}$$

Note that t is a discrete time.

Considering the real situation, it is necessary that $\Delta r < R_c$, without generality, let $\Delta r = bR_c$, $0 \le b \le 1$. So, the number of infected nodes should be

$$I(t) = \rho \pi r^2 = \rho \pi(bR_c t)^2 . \tag{14}$$

When $bR_c t = L/2$, the first stage ends and $t_{max} = L/2bR_c$.

Second stage: $r \ge L/2$

When $r \ge L/2$, the potential region C will decrease with the time evolution, as shown in Fig. 6 $r \ge L/2$ (b), the population of infected nodes can be expressed by

$$I(t) = \rho \pi r^2 - 4\rho S_{arc} = \rho \pi(bR_c t)^2 - 2\rho(bR_c t)^2(\alpha_t - \sin \alpha_t) , \tag{15}$$

where S_{arc} denotes the area of a camber region and α_t the corner of the camber with the center

$$\sin \alpha_t = \frac{L\sqrt{(bR_c t)^2 - (L/2)^2}}{(bR_c t)^2} . \tag{16}$$

Apparently, b is an important factor affecting the spreading speed. We know that the spreading speed is in proportion to the infected rate and density of nodes. Furthermore, the MAC mechanism constrains the malware propagation. So, we have

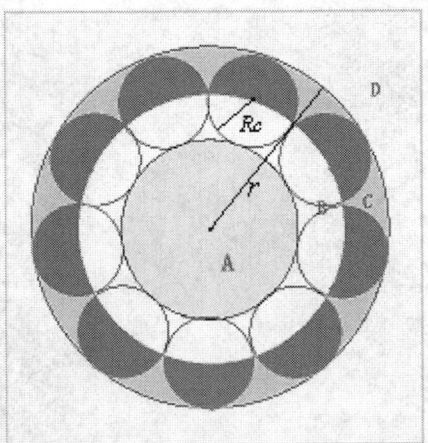

Fig. 7. Maximum coverage area shown in the shadow of region C due to avoiding channel collision for each time step

$b = f(\beta, \rho, p)$, where p is the coverage rate due to MAC mechanism, describing new infected region in the region C. First, the maximum coverage rate cov_{max} is shown in Fig. 7, when $t \to \infty, R_c \ll r$. The infected nodes in region B can spread to the maximum region, i.e., the black area in region C. The closed result of cov_{max} is $\text{cov}_{max} \approx \pi / 4$. It is obvious that $p < \text{cov}_{max}$.

6 Simulations

We simulate the malware propagation in a wireless sensor network consisting of N sensors distributing uniformly and randomly in a cellular space with $L^2 = 200 \times 200 \ unit^2$ cells. $\rho = N / L^2$ denotes the sensor density. The communication range of each sensor is defined by R_c. Each simulation starts by infecting a sensor node located in the center of the WSN. Our goal is to investigate the spreading dynamics of malware in a wireless sensor network and to compare the result of simulations with theoretical analysis and those models in [11,12].

6.1 Simulations for the Proposed Model vs. the Theoretical Analysis

In the simulations, we set $\delta = 0, \gamma = 0$ for the comparison between the proposed model and the theoretical analysis. The other cases can be found in our another paper [24].

There is one initially infected node that locates in the center of the wireless sensor network. We pay more attention to 4 factors impacting on the malware propagation, which include MAC mechanism, node's relative density ρ, infected rate β and the communication range R_c. The time evolutions of the total fraction of infected nodes, $I(t)/N$, under the impact of the 4 factors are exhibited in Fig. 8.(a),(b),(c) and (d) respectively. Solid-line is for our proposed CA model, and Dashed-line denotes the theoretical analysis.

From Fig. 8, we can see that, the time evolution of the proposed CA model agrees perfectly with the theoretical analysis. The malwares show an exponentially increasing transmission from the initial time until reaching approximately 80% infected population, then show a slow transmission until the malwares spread to the whole network. Also we can find that, the MAC mechanism results in a competition between adjacent sensor nodes to access to the shared wireless channel, so MAC mechanism greatly slows down the speed of malware propagation and reduces the risk of large-scale malware prevalence in these networks.

6.2 Simulation for the Proposed Model vs. the Model in [12]

In [12], an ODE model and expression with respect to $I(t)$ has been established. However, the model of the paper does NOT considered the impact of MAC mechanism. For comparison, we add a parameter p concerning MAC mechanism to revise the value of η in the Equation of [12], and let $p=1$ when the network has no

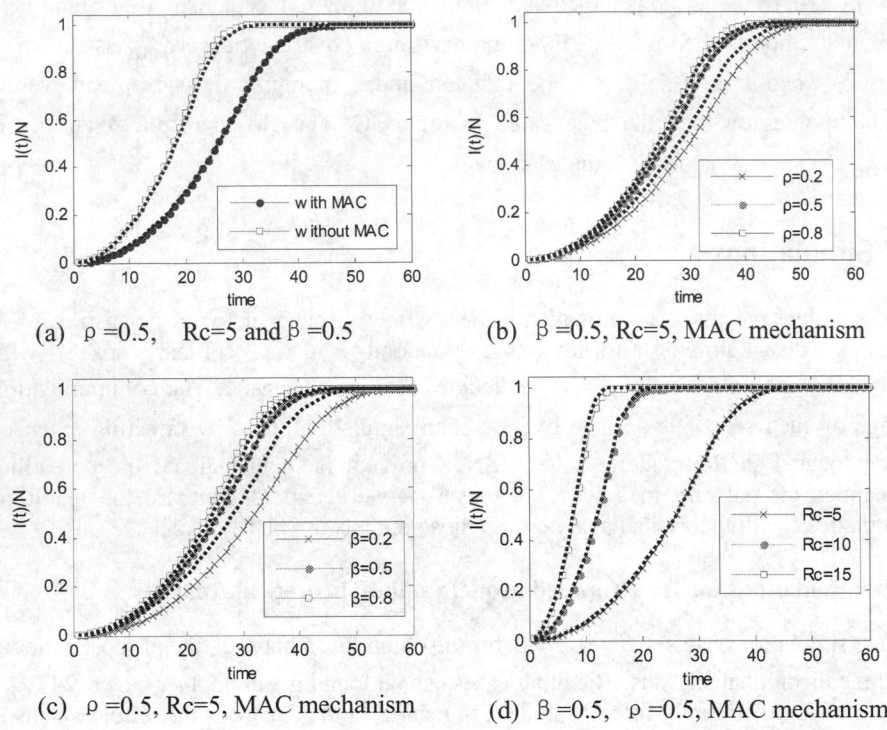

(a) ρ =0.5,   Rc=5 and β =0.5

(b) β =0.5, Rc=5, MAC mechanism

(c) ρ =0.5, Rc=5, MAC mechanism

(d) β =0.5,   ρ =0.5, MAC mechanism

Fig. 8. Time evolution of the fraction of infected nodes on the proposed model(*solid-line*) vs. the theoretical analysis(*dashed-line*): (a) effect of MAC mechanism, (b) effect of relative densityρ, (c) effect of infected rateβand (d) effect of communication range R_c

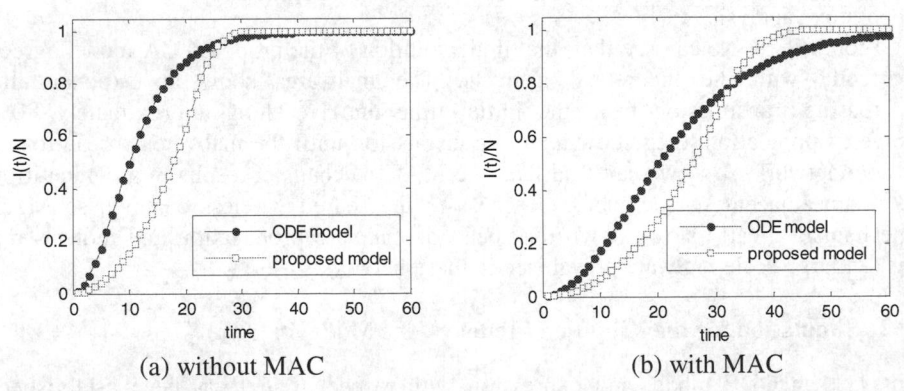

(a) without MAC

(b) with MAC

Fig. 9. Time evolution of the fraction of infected nodes on the proposed model vs. the model in Ref.[12] (a) ρ=0.5,β=0.5, Rc=5, the absence of MAC mechanism (b) ρ=0.5,β=0.5, Rc=5, the presence of MAC mechanism

the MAC mechanism and $p=0.5$ when the MAC mechanism takes action. The revised equation can be described by

$$I(t) = N\left(2/(1+(\frac{\sqrt{N}-1}{\sqrt{N}+1})e^{-\frac{\beta c \eta p_t}{\sqrt{N}}})-1\right)^2, \qquad (17)$$

where η is the average number of neighbors of a node, $\eta = \rho\pi R_c^2$, and $c = 2\sqrt{\rho\pi}R_c$ is a proportional constant. More details can be seen in [12].

By normalizing the time in simulation of Eq.(17), the time evolutions of malware propagation for the two models is shown in Fig. 9.

From Fig. 9, we find the initial speed of propagation is quicker in ODE model than that in our proposed model, but slower in the last stage of propagation. A reasonable explanation is that, ODE model assumes that the number of infected neighbors is in proportion to the fraction of all susceptible nodes in the network, which results in an inaccurate evaluation on spreading speed in the process of malware propagation. In fact, the malware propagation in WSN has the chacteristic of local spatial interaction. And the presence of neighborhood saturation dominates the effects of increasing infectivity and limits the spread of the malware.

6.3 Simulation for Evolution Pattern

Fig. 10 and Fig. 11 show the wireless sensor network evolving patterns in the 2-D cellular space under different time t. The key difference between Fig. 10 and Fig. 11 is that the former neglects the impact of MAC mechanism, whereas the latter is more concerned to the impact of MAC mechanism.

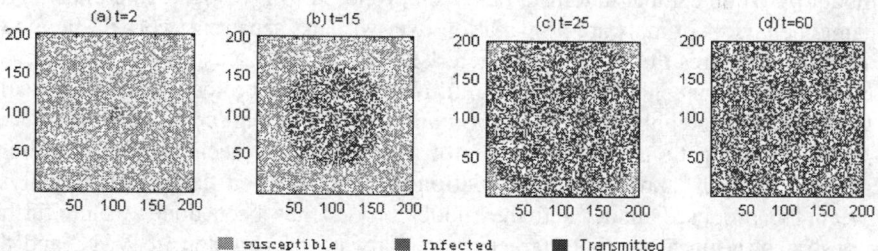

Fig. 10. Evolution pattern without MAC mechanism in the 2-D space with $p=0.5, \beta=0.5, Rc=5$

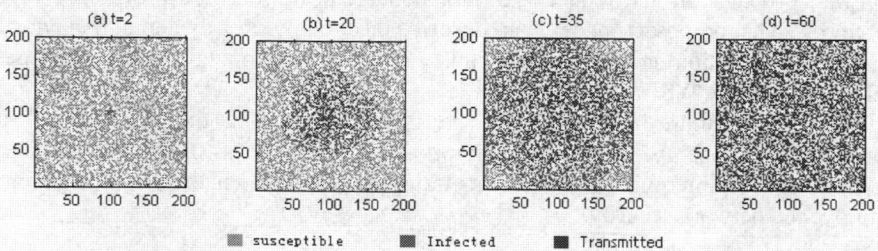

Fig. 11. Evolution pattern with MAC mechanism in the 2-D space with $p=0.5, \beta=0.5, R_c=5$

From the evolution snapshots (Fig. 10 (a)-(d) and Fig. 11(a)-(d)), one can see that the epidemic diffuses continuously from infected source toward outside and the propagation goes along a circular front which is spatially bounded. Specially, in our model, the adopted broadcast protocol ensures that each infected node broadcasts the malware to its neighbors only once for the purpose of preventing broadcast storm. So, those infected nodes that have transmitted packets of malware can not send malware propagation any more. In fact, an infected node inside of the circular has also no chance to infect other susceptible nodes that lie outside of the circular for the reason that interactions among nodes are distance-dependent.

Considering MAC mechanism, Fig. 11 has two key differences with the Fig. 10. First, the diffused speed of malware propagation in Fig. 11 is slower than that in Fig. 10 because the MAC mechanism chokes the malware propagation.

Second, the border between the nodes that have transmitted packet and the nodes that haven't transmitted packets becomes unclear. Due to the constraint of MAC mechanism, only partial nodes win channels to transmit packets in each time step and other nodes must wait for channels to transmit packets, which leads to the wide region occupied by those infected nodes that haven't transmitted packets. In addition, most nodes in the region can never send malware for the limitation of communication range.

The characteristic of propagation with the local spatial interaction between nodes is greatly different from the propagation over Internet and results in slowing the speed of propagation.

7   Conclusions and Future Work

A model based on cellular automata has been proposed to investigate and analyze the dynamic behaviors of malware propagation over wireless sensor networks. The model successfully captures the inherent characteristics of wireless sensor networks, such as limited energy, channel contention and multi-hop broadcast protocols, and reflects the self-organization, neighborhood saturation and spatio-temporal correlation of process of malware propagation. The validations of the model have been performed through theoretical analysis and various simulations. In addition, a comparative analysis between the proposed model and the model in [12] has been done, which further demonstrates the local spatial interaction of malware propagation in WSN, and the presence of neighborhood saturation slows down the spread of the malware. The MAC mechanism of wireless sensor networks greatly slows down the speed of malware propagation and reduces the risk of large-scale malware prevalence in the networks. The proposed model can describe accurately the dynamic behavior of malware propagation on WSN, which can be used for developing robust and efficient defense system on WSN.

In our near future work, we will be further evaluating the influence of the consumption of energy on system behaviors during the process of malware propagation, and developing robust and efficient strategies against the malware propagation to improve the network security.

Acknowledgements

This work was supported in part by the Program for New Century Excellent Talents in University of China under the Contract NCET-06-0510, by the National Natural Science Foundation of China under the Contract 60874091 and by Scientific Innovation Program for University Research Students in Jiangsu Province, China, under the Contract CX08B_081Z.

References

1. Akyildiz, I.F., Su, W., Sankarasubramaniam, Y., Cayirci, E.: Wireless sensor networks: a survey. Computer Networks 38, 393–422 (2002)
2. Pathan, A.S.K., Lee, H.W., Hong, C.S.: Security in Wireless Sensor Networks: Issues and Challenges. Proc. of the 8th IEEE ICACT 2, 1043–1048 (2006)
3. Perrig, A., Stankovic, J., Wagner, D.: Security in wireless sensor networks. Communications of the ACM 47, 53–57 (2004)
4. Staniford, S., Paxson, V., Weaver, N.: How to Own the Internet in Your Spare Time. Usenix Security (2002)
5. Zou, C.C., Towsley, D., Gong, W.B.: Modeling and simulation study of the propagation and defense of internet e-mail worms. IEEE Transactions on Dependable and Secure Computing 4, 105–118 (2007)
6. Zou, C.C., Gong, W., Towsley, D.: Code red worm propagation modeling and analysis. In: Proceedings of the 9th ACM Conference on Computer and Communications Security, p. 10 (2002)
7. Pastor-Satorras, R., Vespignani, A.: Epidemic spreading in scale-free networks. Physical Review Letters 86, 3200–3203 (2001)
8. Newman, M.E.J., Forrest, S., Balthrop, J.: Email networks and the spread of computer viruses. Physical Review E 66, 35101 (2002)
9. Mills, K.L.: A Brief Survey of Self-Organization in Wireless Sensor Networks. Wireless Communications and Mobile Computing 7, 823–834 (2007)
10. Khayam, S.A., Radha, H.: Using signal processing techniques to model worm propagation over wireless sensor networks. Signal Processing Magazine 23, 164–169 (2006)
11. De, P., Liu, Y., Das, S.K.: Modeling Node Compromise Spread in Wireless Sensor Networks Using Epidemic Theory. In: Proceedings of the 2006 International Symposium on on World of Wireless, Mobile and Multimedia Networks, pp. 237–243 (2006)
12. De, P., Liu, Y., Das, S.K.: An Epidemic Theoretic Framework for Evaluating Broadcast Protocols in Wireless Sensor Networks. In: IEEE Internatonal Conference on Mobile Adhoc and Sensor Systems, MASS, Pisa, Italy, pp. 1–9 (2007)
13. White, S.H., Rey, A.M.d., Sanchez, G.R.: Modeling epidemics using cellular automata. Applied Mathematics and Computation 186, 193–202 (2007)
14. Georgoudas, I.G., Sirakoulis, G.C., Andreadis, I.: Modelling earthquake activity features using cellular automata. Mathematical and Computer Modelling 46, 124–137 (2007)
15. Encinas, L.H., Hoya White, S., Del Rey, A.M., Rodriguez Sanchez, G.: Modelling forest fire spread using hexagonal cellular automata. Applied mathematical modelling 31, 1213–1227 (2007)
16. Ahmed, E., Elgazzar, A.S.: Onsome applications of cellular automata. Physica A 296, 529–538 (2001)

17. Liu, Q.-X., Jin, Z.: Cellular automata modelling of SEIRS. Chinese Physics 14, 1370–1377 (2005)
18. Fuentes, M.A., Kuperman, M.N.: Cellular automata and epidemiological models with spatial dependence. Physica A 267, 471–486 (1999)
19. Mikler, A.R., Venkatachalam, S., Abbas, K.: Modeling Infectious Diseases using Global Stochastic Cellular Automata. Journal of Biological Systems 13, 421–439 (2005)
20. Cunha, R.O., Silva, A.P., Loreiro, A.A.F., Ruiz, L.B.: Simulating large wireless sensor networks using cellular automata. In: Proceedings of the 38th annual Symposium on Simulation, pp. 323–330 (2005)
21. Akkaya, K., Younis, M.: A survey on routing protocols for wireless sensor networks. Ad Hoc Networks 3, 325–349 (2005)
22. Small, M., Tse, C.K.: Clustering model for transmission of the SARS virus: application to epidemic control and risk assessment. Physica A: Statistical Mechanics and its Applications 351, 499–511 (2005)
23. Small, M., Tse, C.K., Walker, D.M.: Super-spreaders and the rate of transmission of the SARS virus. Physica D: Nonlinear Phenomena 215, 146–158 (2006)
24. Song, Y., Jiang, G.-P.: Modeling malware propagation in wireless sensor networks using cellular automata. In: IEEE ICNNSP 2008, Zhenjiang, China, pp. 623–627 (2008)

Measuring the Efficiency of Network Designing

Guoqiang Zhang[1] and Guoqing Zhang[2]

[1] Computer Network Information Center, Chinese Academy of Sciences,
Beijing, 100190, China
zhangguoqiang@cnnic.cn
[2] Institute of Computing Technology, Chinese Academy of Sciences,
Beijing, 100190, China

Abstract. Network designing often involves two significant yet contradictive objectives: enhancing the whole network's transmission efficiency while at the same time lowering the whole network's designing cost. Deep study of the interplay between major aspects of network planning–network topology, routing algorithm and node's transmission capability configuration–reveals that good tradeoff can be achieved between these two objectives. By properly combining network topology, routing algorithm and node capability configuration scheme, the network can achieve desirable transmission efficiency at very low cost. This discovery will undoubtedly provide insight into the next generation data network designing.

Keywords: network designing, designing cost, network transmission efficiency, network designing efficiency.

1 Introduction

The Internet becomes more and more complex and susceptible to congestion today. With the emergence of new applications and fast growing population in need of data communication, most experts agree that the existing data network architecture is severely stressed and approaching its capability limits. Thus the move to a brand new next generation data network is in urgent need today.

However, before starting this move, several questions should be properly answered. First, what is the problem with the current data network designing? Second, how to compare between different network designing strategies? Finally, can the network be designed in a cost-effective way and achieve high extensibility.

The answer to the first question requires a close look at the current Internet topology and routing algorithm being used. It has been found that shortest path routing algorithm, which is widely used across Internet literature, has poor performance on BA like networks [1]. Our previous work also found that the most realistic Internet router-level model to date–HOT model–is insensitive to routing algorithm changes, and the only way to improve its transmission efficiency is to upgrade key nodes [2]. In general, when considering network transmission efficiency, network topology, routing algorithm and node capability configuration are closely related to each other. The second question calls for a way to compare and balance the tradeoff between different aspects of network designing

J. Zhou (Ed.): Complex 2009, Part I, LNICST 4, pp. 503–513, 2009.

in a uniform way. For instance, although upgrading critical parts in a network can improve the whole network's transmission efficiency, either economic cost or technical bottleneck will prevent this approach from being used without limitation. That is, either the cost of high-end super computers will exceeds the investment budget or the required processing capability cannot be met due to state-of-art technology constrains. In fact, the network designing can be considered as a multiple objective optimization process. We develop a uniform metric, the network's designing efficiency index(DEI), which reasonably integrates all major designing objectives and simplifies the evaluation process. The answer to the last question can be valuable or insightful to serve as the guidelines for future data network designing. Actually, we find such cost-effective designing schemes do exist.

The rest of this paper is structured as follows: Section 2 provides a brief description of related work. Section 3 introduces the traffic flow model we use. The network designing efficiency index is introduced in section 4 and detailed analysis based on this is given in section 5. Finally, we conclude this paper in section 6.

2   Related Work

Traffic dynamics has been studied extensively on regular networks [3, 4, 5], such as two-dimensional lattices or Cayley trees. However, recent studies on network topologies show that real networks can by no means be characterized by regular networks or random networks [6], but display more complex structural properties such as power-law degree distribution [7, 8]. In Internet router-level topology modeling, two models proposed recently have gained a lot of attentions: BA model [7] and HOT model [9]. BA model is a general model to reproduce the power-law degree distribution by two dynamic mechanisms: growing network and preferential attachment. HOT model more appropriately resembles the real Internet router level topology. It partitions the routers into three hierarchies: the first hierarchy represents the network core, consisting of a number of interconnected low degree nodes with high capabilities, the second hierarchy consists of those high degree access routers connecting to the core routers, and the last hierarchy consists of the low degree peripheral routers connecting to those access routers.

In view of recent evidence that the Internet and many other realistic networks are complex to a significant extent, studies of traffic dynamics on complex network topologies have attracted substantial attentions in recent years [1, 10, 11, 12, 13, 14]. What connects the static network topology and the dynamic traffic behavior is the betweenness centrality. It has been found that if at each time step, R packets with random source and destination addresses are injected into the network and each node can forward only one packet, then the critical packet injection rate R_c can be estimated by

$$R_c = \frac{N(N-1)}{B_{max}} \tag{1}$$

where N is the size of the network and B_{max} is the largest betweenness centrality value of the network [12]. This relationship is intuitive in that the node with the largest betweenness centrality is more susceptible to packet congestion and congestion in this node will quickly spread over the network. Thus, it generally implies that congestion can easily occur in heterogenous networks such as BA network with uniform node transmission capability because in these networks, some nodes have extremely high betweenness centrality values.

With the aim of alleviating traffic congestion and improving the network's transportation capability, three different kind of solutions are proposed. Node heterogeneity is considered in [13], where influences of two node capability models on the traffic dynamics are investigated. A node's capability, i.e, number of packets a node can forward at each time step, is proportionate to its degree and betweenness centrality respectively in these two models. As a conclusion, they suggest a way to alleviate traffic congestion by making nodes with large betweenness as powerful and efficient as possible for processing and transmitting information. The second approach is to change routing algorithms [1]. Instead of following the path with smallest number of links between two nodes as conventionally does, the proposed *efficient routing* algorithm selects a route that minimizes the overall sum of node degrees along it. As the authors have demonstrated, the performance of this routing strategy on BA-like network can be improved over an order of magnitude. The last kind of approach is to modify the network topology. A straightforward way is to improve R_c by creating new edges in the network [15, 16]. A somewhat interesting approach proposed in [14] enhances the BA network's transmission efficiency by kicking out those black sheep edges, i.e, eliminating those edges linking nodes with high betweenness values.

However, from a network designer's perspective, network designing is a multiple objective optimization process. On one hand, it is desirable to achieve high network transmission capability, while on the other hand, it is preferable to lower the cost or the required technology. Unfortunately, these two objectives often contradict with each other. How to design a network in a cost-effective way is thus of significant value. Currently, to the best of our knowledge, all current work strive to optimize the network transmission efficiency, R_c, completely ignoring the lowering of designing cost.

3 Traffic Flow Model

In our traffic flow model, nodes in a network are considered to be capable of generating, forwarding and receiving packets. The traffic dynamics is modeled as follows: at each time step, some packets are injected into the network according to *packet injection mode* and each node forwards some packets according to each node's *transmission capability* towards their destinations based on the particular *routing algorithm*. Each node has a queue for receiving new arriving packets. Packets are processed and transmitted in a first-in-first-out(FIFO) manner so that a packet will be added to the end of the queue when there are other packets waiting for transmission. Once a packet reaches its destination, it is removed from the network.

The above traffic flow model models the traffic dynamics in a network by four constituents: packet injection mode, node transmission capability mode, routing algorithm and queue. Among these four factors, packet injection mode, node transmission capability mode and routing algorithm together determine whether a network will get congested or not. Queue, although affects the network's dynamic behavior, has no influence on whether a network will get congested or not. Therefore, in the following reasoning, we always assume that queues are infinite. And for simplicity, we limit our discussion to the random packet generation mode, i.e, each packet is generated with random source and destination addresses.

4   Measuring a Network Designing Strategy

Since the network designing has two contradictive objectives, we propose a parameterized metric–designing efficiency index(DEI)–to quantitatively measure a designing strategy's efficiency as follows:

$$DEI = \frac{Network\ Transmission\ Capability}{(Designing\ Cost)^\alpha} \tag{2}$$

This metric is intuitively reasonable. If two networks have the same designing cost, then the one with higher network transmission capability outperforms the other. Else, if the network transmission capability is fixed, then the one with lower designing cost should be favored. The α here controls the weight each designing objective places in a particular network designing procedure. For example, if $\alpha=0$, then we can only consider to optimize the network transmission capability, totaly ignoring the designing cost. The exact metrics we choose to represent network transmission capability and designing cost will be discussed in the following two subsections.

4.1   Generalized R_c Computation

Following previous studies, we choose R_c to measure a network's transmission capability, however, to study the network transmission capability for different combinations of network topology, routing algorithm and node capability model, we must extend the computation of R_c to more general context.

In the most generalized case, we have a network topology G, in which each node i can forward $C(i)$ packets at each time step. we no longer assume shortest path routing algorithm, but in stead allow any *topology-based* routing algorithm Γ. By topology-based routing algorithm, we refer to those routing algorithms that only make routing decisions on static topology information, not on dynamic traffic variation.

Since routing algorithm is no longer shortest path routing, betweenness can no longer be used to estimate the possible traffic a node needs to handle at each time step. In this sense, we should replace $B(i)$ with the effective betweenness $B_\Gamma(i)$ to estimate the possible traffic passing through a node, which is formally defined as:

$$B_\Gamma(i) = \sum_{u \neq v} \frac{\delta_\Gamma^{(i)}(u,v)}{\delta_\Gamma(u,v)} \tag{3}$$

where $\delta_\Gamma(u, v)$ is the total number of candidate paths between u and v under routing algorithm Γ and $\delta_\Gamma^{(i)}(u, v)$ is the number of candidate paths under routing algorithm Γ between u and v passing through i.

Based on this definition, at each time step, the expected number of packets arriving at node i in free-flow state is $\frac{RB_\Gamma(i)}{N(N-1)}$. For i not to be congested, it follows that $\frac{RB_\Gamma(i)}{N(N-1)} \leq C(i)$, which leads to $R \leq \frac{C(i)N(N-1)}{B_\Gamma(i)}$. Therefore, for the whole network to be in free-flow state, the critical injection rate is:

$$R_c = min_i \frac{C(i)N(N-1)}{B_\Gamma(i)} \tag{4}$$

4.2   Designing Cost

In order to study the influence of different node capability assignment strategies on the whole network's transmission efficiency, we fix $\sum_i C(i)$ to be a constant. Thus this sum can no longer be chosen as the network's designing cost. In response, we choose the maximum node capability C_{max} to be the designing cost. This is meaningful for two reasons. First, because low-end servers are very cheap today, a network's financial cost is largely determined by high-end servers. Second, designing cost also refers to technology cost. The required maximum node capability defines the boundary of whether the proposed designing strategy is technically feasible according to state-of-art technologies.

5   Experimental Studies

In this section, we will propose several possible network designing strategies by combining different designing components, and study their designing efficiencies to find some cost-effective designs.

5.1   Routing Algorithms

We employ two routing algorithms: one is the traditional shortest path routing, the other is the efficient routing algorithm proposed in [1]. For given source and destination, the efficient routing algorithm chooses a path that minimizes the sum of node degrees along the path. More formally, the efficient routing chooses a path $s = v_0, v_1, v_2, \cdots, v_k = t$ between s and t that minimizes the objective function $\sum_{0 \leq i < k} d(v_i)$, where $d(v_i)$ is the vertex degree of v_i.

5.2   Network Topologies

We apply the above two routing algorithms to six network topologies: ring, lattice, E-R [6], W-S [17], BA and HOT. The ring is constructed by placing all the nodes in a circular ring and connecting each node to its left two nearest nodes and right two nearest nodes. The two-dimensional lattice is constructed in toroidal mode so that the lattice is completely homogeneous. W-S graph is

Table 1. Elementary topological properties of the six networks

Network	Number of Nodes	Number of Edges	Diameter	Average Path Length
Ring	1225	2450	306	153.5
Lattice	1225	2450	34	17.5
WS	1225	2450	14	7.44
ER	1225	2480	11	4.73
BA	1225	2447	7	4.67
HOT	1225	2442	9	6.46

built from the ring by randomly rewiring 20 percent of its edges. BA graph is constructed according to the standard BA model with $m = 2$ [7]. Finally, the HOT model graph contains an 80 nodes random graph with average degree 6 as its network core, 20 high-degree access routers and 1125 periphery nodes. Basic graph properties of the six networks are presented in Table 1.

5.3   Node Transmission Capabilities

Four node capability models are considered: uniform node capability, degree dependent node capability, betweenness dependent node capability and effective betweenness dependent node capability. In uniform node capability mode, each node has the same packet transmission capability. While for the other three node transmission capability modes, a node's transmission capability value is proportionate to its degree, betweenness and effective betweenness respectively.

In order to compare the effects between different node transmission capability modes, we demand that the total capability of a network remains fixed for all the four node capability modes once the network is given, which we set to the sum of all node degrees for simplicity.

Since the assignment will cause a node's capability to be fractional, we treat the fractional capability as follows. Denote $C(i)$ as the transmission capability of node i, then at each time step, i first forwards $\lfloor C(i) \rfloor$ packets towards their destinations, after which a random number $r \in (0, 1)$ is generated and compared against $C(i) - \lfloor C(i) \rfloor$. If $r < C(i) - \lfloor C(i) \rfloor$, i forwards another packet towards its destination. By this means, a node with capability 1.2 will forward 1 packet with probability 0.8 and forward 2 packets with probability 0.2.

5.4   Network Transmission Capability-R_c

R_c can be obtained by applying Equation 4. For shortest path routing, $B_\Gamma(i)$ is equivalent to $B(i)$. However, for efficient routing, an efficient way for calculating $B_\Gamma(i)$ should be first devised. The key to computing $B_\Gamma(i)$ lies in fast detection of paths between any two nodes i and j that minimize the sum of node degrees along the path. We solve this problem as follows. First, we transform each undirected graph into a directed graph by substituting each undirected edge (u, v) with two directed edges (u, v) and (v, u). Second, for each directed edge (u, v), we associate with it a weight $w(u, v) = d(u)$. Fig 1 gives an example of this transformation

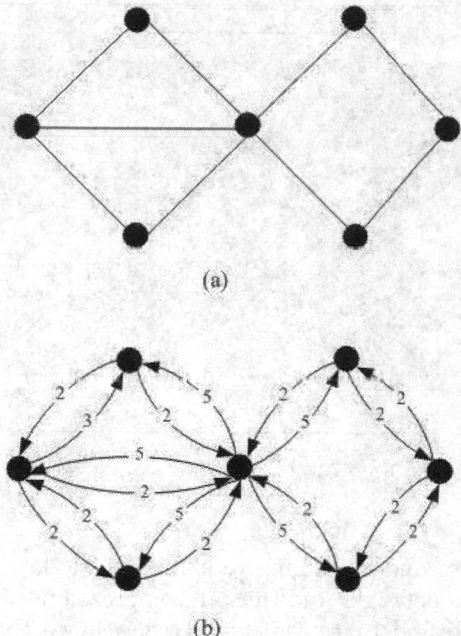

(a)

(b)

Fig. 1. An example showing how to turn the simple undirected graph into an directed weighted graph for effective betweenness calculation

process. In this way, we turn the problem of finding a path between i and j that minimizes the sum of node degrees into the problem of finding the shortest path between i and j in a directed weighted graph, which can be efficiently solved by Dijkstra's algorithm [18].

Table 2 presents the R_c values for different combinations of network topology, node capability assignment and routing algorithm. We further presents this

Table 2. Theoretical results of critical injection rate R_c under different combinations, where UC represents uniform capability mode, DC represents degree dependent capability mode, BC represents betweenness dependent capability mode, EBC represents effective betweenness dependent capability mode, SPR represents shortest path routing and EFR represents efficient routing.

(node capability, routing algorithm)	BA	HOT	ER	WS	Lattice	Ring
(UC, SPR)	8.7	16.7	173.7	132.8	280	32
(UC, EFR)	155.3	17.7	325.8	186.8	280	32
(DC, SPR)	312.9	28.5	414.8	194.8	280	32
(DC, EFR)	238.5	29.3	448.2	213	280	32
(BC, SPR)	975.7	702.0	790.9	591.5	280	32
(BC, EFR)	139.6	197.6	347.7	340.1	280	32
(EBC, EFR)	545.8	540.1	656.9	509.5	280	32

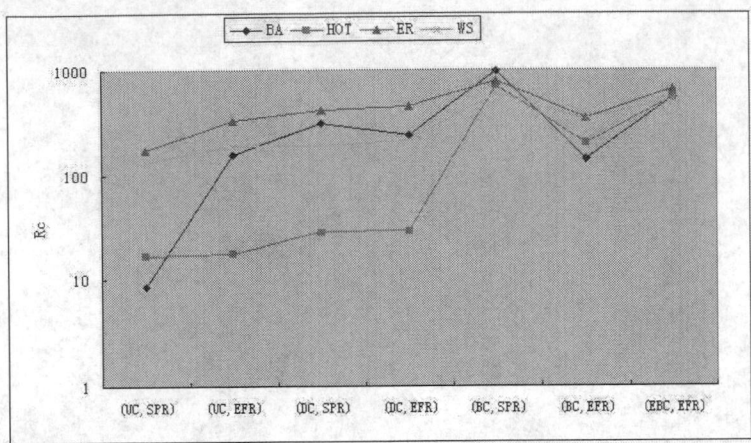

Fig. 2. R_c under different combinations

result in Fig 2 to ease understanding(we eliminate the lattice and ring because they are homogenous networks and thus all combinations have the same effect). From this figure, we find that (a)all networks achieve the highest R_c values when shortest path routing and betweenness based node capability assignment is applied;(b)BA network is sensitive to routing algorithms while HOT network doesn't, the only way to improve HOT network's R_c value is to upgrade key nodes with high betweenness centrality values.

5.5 Designing Cost–C_{max}

Table 3 presents C_{max} values required under different node capability assignment strategies. We also presented C_{max} as the result of different combination strategies in Fig 3. Comparing it with Fig 2, we find that although betweenness based capability assignment enables large R_c values, it also demands the largest C_{max} values, in other words, the largest designing cost. Especially for BA networks, C_{max} spans the widest range. However, BA network also shows good tradeoff property in achieving high R_c and low C_{max}. With efficient routing and effective betweenness based capability model (EBC,EFR), the R_c value is slightly over

Table 3. C_{max} under different combinations of network topology, routing algorithm and node capability mode, where the meaning of short notations are the same as Table 2

(node capability model, routing algorithm)	BA	HOT	ER	WS	Lattice	Ring
(UC, *)	4	4	4	4	4	4
(DC, *)	144.0	160.0	12.0	7.0	4	4
(BC, *)	449.0	169.6	19.5	18.8	4	4
(EBC, EFR)	14.1	122.22	8.1	10.9	4	4

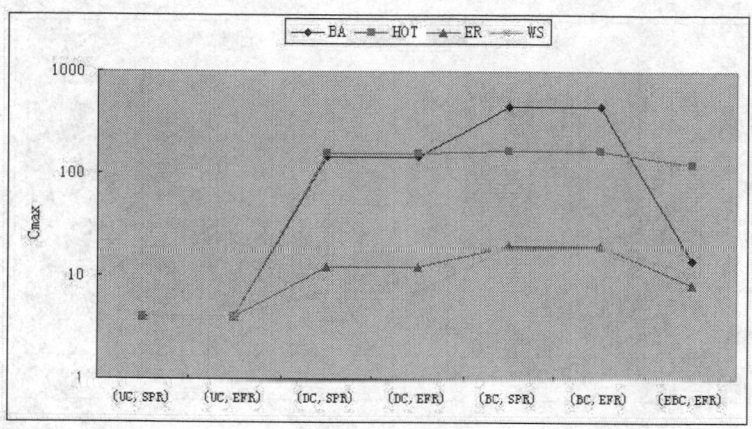

Fig. 3. C_{max} under different combinations

half of the largest R_c value(545.8 vs 975.7), but the C_{max} reduces to less than $\frac{1}{30}$ of the largest C_{max}(14.1 vs 449.0).

5.6   Designing Efficiency

To give a quantitative impression on the efficiency of different network designing strategies, we set $\alpha = \frac{1}{2}$ and presents the DEI indexes for different combinations in Table 4 and further illustrate in Fig 4. In fact, DEI is an indication of whether it can achieve good tradeoff between the two designing objectives, or in other words, whether a cost-effective design exists. We find that BA network shows good property in achieving good tradeoff while HOT network doesn't. On the other hand, for $\alpha = \frac{1}{2}$, we find that the ER random network has the best designing efficiency. Taking all aspects into account, the most efficient designing for $\alpha = \frac{1}{2}$ is the ER network with efficient routing and effective betweenness based node capability model.

Table 4. Network DEI($\alpha = \frac{1}{2}$) for different network designing strategies

(node capability, routing algorithm)	BA	HOT	ER	WS	Lattice	Ring
(UC, SPR)	4.4	8.4	86.9	66.4	140	16
(UC, EFR)	77.6	8.8	162.9	93.4	140	16
(DC, SPR)	26.1	2.2	119.7	73.6	140	16
(DC, EFR)	19.9	2.3	129.4	80.5	140	16
(BC, SPR)	46.0	53.9	179.1	136.5	140	16
(BC, EFR)	6.6	15.2	78.7	78.5	140	16
(EBC, EFR)	145.6	48.9	231.2	154.3	140	16

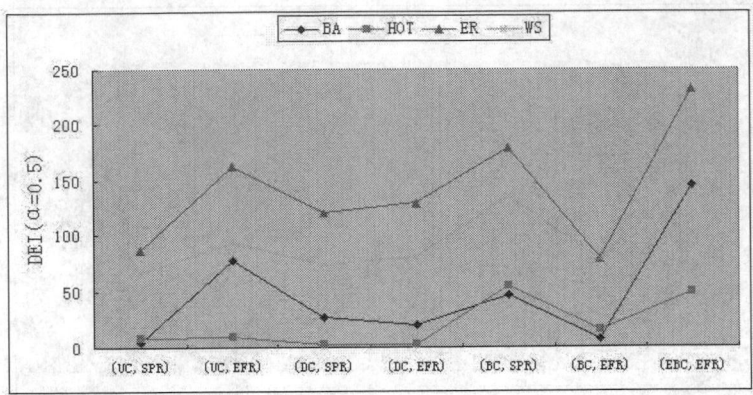

Fig. 4. $DEI(\alpha = \frac{1}{2})$ under different combinations

6   Conclusion

In this paper, we proposed that network designing is a multi-objective designing process and involves several seemingly independent but in fact closely related aspects. We provided a comprehensive view on how to compare different network designing schemes and proposed a quantitative metric, the designing efficiency index, to measure a particular network designing scheme's efficiency. We found that betweenness based capability model combined with shortest path routing can achieve highest network transmission capability, but this scheme also requires the highest cost. By adopting efficient routing and effective betweenness based capability model, BA network can achieve good tradeoff between the two designing goals, thus can be said to have cost-effective designing, while more realistic HOT network doesn't show this property. Taking into account all designing aspects, we found that ER network with efficient routing and effective betweenness based capability model is a good designing choice, possibly the most cost-effective designing among all the designing schemes studied. This may be an interesting finding that shows sometimes random designing is the best.

Regarding these findings, the possible practical significance in reality is listed as follows. First, the difference of traffic dynamics between the networks offers insightful understanding for different network topologies, especially for BA and HOT. Second, these findings provide guidelines for how to upgrade the nodes in a network. We suggest that upgrading should be performed by not only considering a node's structural position in the network, but also the routing algorithm applied. Finally, our findings can help the designing of totally new networks. We pointed out the possible tradeoff between network topology, the routing algorithm applied, the network transmission efficiency achieved and the budgeted payout.

References

1. Yan, G., Zhou, T., Hu, B., Fu, Z.Q., Wang, B.H.: Efficient Routing on Complex Networks. J. Phy. Rev. E, 046108 (2006)
2. Zhang, G.Q., Yuan, B., Zhang, G.Q.: Towards a Comprehensive Understanding of Routing on Complex Networks. In: Proceedings of IEEE Next Generation Internet Networks(NGI 2007), Norway (2007)
3. Leland, W.E., Taqqu, M.S., Willinger, W., Wilson, D.V.: On the Self-similar Nature of Ethernet Traffic. J. IEEE. Trans. Net. 2, 1–15 (1994)
4. Li, H., Maresca, M.: Polymorphic-torus Network. J. IEEE. Trans. Comp. 38, 1345–1351 (1989)
5. Sole, R.V., Valverde, S.: Information Transfer and Phase Transitions in a Model of Internet Traffic. J. Phy. A. 289, 595–605 (2001)
6. Bollobas, B.: Random Grpahs. Cambridge University Press, Cambridge (1985)
7. Barabási, A.L., Albert, R.: Emergence of Scaling in Random Networks. J. Science 286, 509–512 (1999)
8. Faloutsos, M., Faloutsos, P., Faloutsos, C.: On Power-law Relationships of the Internet Topology. J. ACM. SIGCOMM. Comp. Comm. Rev. 29, 251–262 (1999)
9. Li, L., Alderson, D., Willinger, W., Doyle, J.: A First Principles Approach to Understanding the Internet's Router-level Topology. In: SIGCOMM 2004, Oregon (2004)
10. Borgatti, S.P.: Centrality and Network Flow. J. Social. Net. 27, 55–71 (2005)
11. Goh, K.-I., Kahng, B., Kim, D.: Universal Behavior of Load Distribution in Scale-Free Networks. J. Phy. Rev. Lett. 87 (2001)
12. Guimerà, R., Guilera, A.Z., Redondo, F.V., Cabrales, A., Arenas, A.: Optimal Network Topologies for Local Search with Congestion. J. Phy. Rew. Lett. 89, 328170 (2002)
13. Zhao, L., Lai, Y.C., Park, K., Ye, N.: Onset of Traffic Congestion in Complex Networks. J. Phy. Rev. E. 71, 026125 (2005)
14. Zhang, G.Q., Wang, D., Li, G.J.: Enhancing the Transmission Efficiency by Edge-deletion in Scale-free Networks. J. Phy. Rev. E. 71, 017101 (2007)
15. Gupte, N., Singh, B.K.: Role of Connectivity in Congestion and Decongestion in Networks. J. Eur. Phy. B. 50, 227–230 (2006)
16. Gupte, N., Singh, B.K., Janaki, T.M.: Networks:Structure, Function and Optimization. J. Phy. A. 346, 75–81 (2005)
17. Watts, D.J., Strogatz, S.H.: Collective Dynamics of Small World Networks. J. Nature 393 (1998)
18. Cormen, T.H., Leiserson, C.E., Rivest, R.L., Stein, C.: Introduction to Algorithms, 2nd edn. MIT Press, Cambridge (2001)

Gravity Model for Transportation Network Based on Optimal Expected Traffic

Jiang-Hai Qian and Ding-Ding Han*

School of Information Science and Technology, East China Normal University,
Shanghai 200241, China
ddhan@ee.ecnu.edu.cn

Abstract. We propose a spatial network model for transportation system based on the optimal expected traffic. The expected traffic represents the prediction of the flow created by two vertices and is calculated by the improved gravity equation $w_{ij} = K \frac{M_i^\alpha M_j^\alpha}{D_{ij}^\gamma}$. The model maximizes the total expected traffic of the network. By changing the two parameters α and γ which controls the fitness and the geographical constraints, the model can vary its topology from the star-like network to the decentralized road-like network. The simulation for the Chinese city airline network reproduced many properties of the real network. In the end of this paper the relationship of the expected traffic and the real traffic is discussed.

Keywords: spatial network, expected traffic, gravity.

1 Introduction

Since the initial studies on the small-world phenomenon by Watts and Strogatz [1] and the scale-free property by Barabasi and Albert [2], lots of achievements on complex network have been gotten. And our research group have studied some works [3]. Most previous works focus on the topological properties of the network. However many networks are those embedded in the real space whose nodes occupy a precise position in Euclidean space and whose links are constrained by the geographic distance. The typical examples are the transportation systems ranging from river [4] to airport [5,6,7], street [8], railway and subway [9]. To model these spatial networks, geographical ingredient is demonstrated to play an important role on the network's topology. In the previous studies the large cost to establish long-distance link is considered to be the main reason that causes the nodes to connect to their geographical neighbors [10,11,12,13]. In addition to the spatial preference, the topology preferential attachment, namely nodes with larger degree have larger probability to be linked, is also important in the formation of the complex network [14,15,16]. Such mechanism can form hubs, the well-connected nodes, which usually reduce the diameter of the network.

* Corresponding author.

J. Zhou (Ed.): Complex 2009, Part I, LNICST 4, pp. 514–524, 2009.
© ICST Institute for Computer Sciences, Social Informatics and Telecommunications Engineering 2009

Another interesting view of modeling the spatial network is to consider the intrinsic attributes of nodes [18,17]. The intrinsic attributes represent the fitness of nodes to win edges and are interpreted as, for example, capacity, social skills, activity levels, information contents and population of cities, etc. Based on this idea, Naoki Masuda and Hiroyoshi Miwa proposed a non-growing geographical threshold model which generalizes a variety of models such as the Boolean model and the gravity model [19]. R.Xulvi-Brunet and I.M.Sokolov interpret the nodes' intrinsic attributes as their different interaction range [20]. The nodes with large interaction range have larger probability to cover more nodes and gain more links. The novel idea combines the intrinsic attribute and the geographical influence naturally.

These works provide some guidelines in modeling spatial network. However they all concentrate on the cost of constructing networks while the effect of traffic on network's design is paid less attention. Whereas traffic may be an even more important factor because it represents the efficiency of the network. If the traffic between nodes can be predicted in some way, it is likely to construct the network efficiently. Inspired by this idea and its significance, we propose a simple spatial network model. The model is to maximize the whole expected traffic of the network, indicating the highest efficiency that the network may gain. The expected traffic is measured by the gravity equation. In the end of this paper, the relationship between the expected traffic and real traffic is discussed.

2   Expected Traffic and Gravity

Traffic in the real-world network is demonstrated to be strongly correlated to its topology. The empirical evidence coming from the studies on metabolic and airline network has shown that the traffic between nodes has the following form [21,22]:

$$w_{ij} \sim x_{ij}(k_i k_j)^{\theta} . \tag{1}$$

where x_{ij} is a random number and θ is a positive exponent. k_i, k_j are the degree of node i and j respectively and w_{ij} is the weight or traffic between them. This result indicates the traffic can be measured after the topology has been known. However if the aim is to predict the traffic before the network is constructed, it seems useless.

Motivated by the studies on the intrinsic attributes of nodes in modeling complex networks [18,17], we consider the effect of node's fitness on the traffic. It is believed and demonstrated that nodes with better fitness usually gain more links. For example, in airline network, cities with large population are usually hubs. To satisfy this basic fact that node's degree usually has a positive correlation with its fitness, here the correlation of fitness and degree is simply assumed to be following the form:

$$k \sim M^{\beta} . \tag{2}$$

where M and k are respectively the fitness and the degree of node while β is a positive exponent. According to Eq(1) and Eq(2) the relationship of traffic and the fitness of nodes is easily to be obtained:$w_{ij} \sim x_{ij}(M_i M_j)^\alpha$, where M_i, M_j represent the fitness of node i and j and exponent $\alpha = \beta\theta$. On the other hand, for spatial network the traffic decreases with the geographical distance since long distance journey spends more time and cost. So the traffic is assumed to follow $w_{ij} \sim \frac{1}{D_{ij}^\gamma}$, where D_{ij} is the geographical distance between node i and j. In this paper, it is defined as the Euclidean distance. Thus the traffic between two nodes can be described by a gravity equation:

$$w_{ij} = K \frac{M_i^\alpha M_j^\alpha}{D_{ij}^\gamma} . \tag{3}$$

where K is a constant coefficient, α, γ are two tunable parameters which determine the impact of fitness and geographical distance on the traffic.

Eq(3) has been confirmed by a very recent empirical study [23] and the gravity model is considered as a suitable form in describing interaction of particles in geographical space when the physical gravity or similar mass interaction is active [19]. The gravity model provides us a way to predict the traffic of network, because once the nodes are sited and the fitness is assigned, the traffic of any pair of nodes can be calculated even when no link exists. For this reason, we call w_{ij} described by Eq(3) the expected traffic. However we emphasize that Eq(3) only describes the flow created by vertices i and j, that is, the traffic which origins from i(or j) and ends at the other. It does not include the traffic created by other pair of nodes but travels through link (i,j). Thus the expected traffic could be different from the real weight. The relationship of the expected traffic and the real traffic will be discussed in the section 5.

3   Gravity Model for Transportation Network

When people prepare to construct a network, what do they care more? Previous studies concentrate on the cost and expenses. However we argue the traffic that the network can carry is even more important. It is because not only high efficiency will bring much benefit but also an inefficient network will cause even more additional expense or loss. Inefficiency of the transportation network will cause inestimable loss since the infrastructure plays an extremely important role on the development of a country. Now consider n nodes, every two nodes have their demand for some information exchange. What we care is that which of these demands are the most exigent, or in other words, which of the expected traffic among these nodes is the largest. Such information can be obtained in advance by Eq(3). Thus the network can be constructed efficiently by preferentially investing those node pairs with large expected traffic. If there is no other restriction, a fully connected graph is obtained. But the real-world networks are usually sparse because a fully connected network requires too large cost. Thus the number of links is limited. In the present paper the number of edges is defined as the budget of constructing the network. One may argue that the budget should rely more on the spatial distance because the cost of different links might be different and for airline networks very long links might be infeasible. We argue here that this point does not conflict with our model because the spatial ingredient in our work have been considered in the expected traffic(see Eq(3)) and the simplification of the budget only indicates that compared with the expected traffic the cost is a minor ingredient. For a pair of distant nodes, their w_{ij} is not large enough to be connected under the fixed budget. However if the fitness is large enough to eliminate the effect of long geographical distance, the two nodes still have chance to link with each other. This can help us understand the phenomenon observed in the airline networks where small airports usually connect to the nearby hubs while the large airports can connect with each other despite of their long spatial distances.

Now suppose there are n nodes distributed on a two-dimension plane. The fitness and coordinates of each node are known. By Eq(3), the expected traffic w_{ij} of any two nodes can be calculated. We connected preferentially those node pairs with larger w_{ij} and complete such process when the links come to the value we preset. Such process can be described as an optimal model:

$$Max \ W_{exp} = \sum_{i<j} w_{ij} \eta_{ij} = \sum_{i<j} K \frac{M_i^\alpha M_j^\alpha}{D_{ij}^\gamma} \eta_{ij} \ . \tag{4}$$

$$s.t. \ \sum_{i<j} \eta_{ij} = \epsilon \ . \tag{5}$$

where η_{ij} is the adjacency matrix element of the network. ϵ is the number of edges we prescribe and W_{exp} is the whole expectant traffic of network. However, the above-mentioned process may cause some isolated nodes. Two more restrictions are introduced to ensure each node is connected:

$$\sum_i \eta_{ij} \geq 1 (j = 1, 2, 3, ..., n) \ . \ \sum_j \eta_{ij} \geq 1 (i = 1, 2, 3, ..., n) \ . \tag{6}$$

Following this method, a network of thirty nodes is simulated. Set the budget $\epsilon=39$ and the coefficient $K = 1$ (actually K makes no difference to the model). By varying the value of α, γ, we got four networks with different topology as is seen in Fig. 1.

As to Fig. 1(a), the value of $\gamma = 0$ makes the network only rely on the fitness of nodes, which causes the topology to be dominated by two hubs since the nodes of good fitness is easy to magnetize others. Whereas with γ increasing, more effect of the geographic factor makes the node tend to connect to the closer ones and weaken the hub-and-spoke effect. When $\alpha=1,\gamma=2$(Fig. 1(c)),the network exhibits some features similar to the airline network. When $\alpha=0$(Fig. 1(d)), the topology is entirely constrained by the geography, which forms a two-dimensional network strongly reminiscent of roads.

Our model has simple realistic and physical significance. Since high efficiency can bring high profit, from operator's view, the significance of our model is that it provides a possible way for the operators to gain the highest profit. Moreover the idea of the expected traffic indicates that link between nodes may

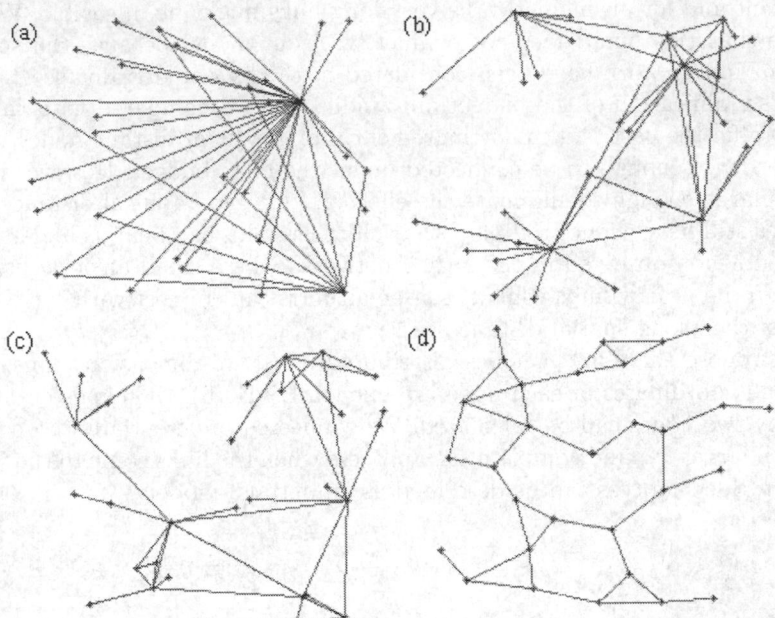

Fig. 1. Four networks with different topology are shown which is controlled by the parameters α,γ. (a)α takes any value,$\gamma=0$. The network is dominated by two large hubs when $\gamma=0$;(b)$\alpha = 1,\gamma = 1$. With γ increasing, the hubs become much smaller than figure (a); (c)$\alpha=1,\gamma=2$;(d)$\alpha=0,\gamma$takes any value. The topologies are strongly reminiscent of airlines and roads respectively.

depend on their potential dynamical strength. If there is no traffic or information exchange demand between two nodes, the link is unwanted even though the cost establishing the link is small. In a circuit, for example, a lead equals to disconnection if its current is zero. On the other hand if the potential traffic is large enough, link will be constructed even though it costs much because large traffic can bring high profit.

4   Simulation for the Chinese City Airline Network

To take the idea of maximizing expected traffic into application, we use our gravity model to simulate the Chinese airline network and make comparison with the real data [24]. In the following simulation, the nodes represent cities and the edges represent the airlines. We set the number of nodes $n = 121$ and number of links $\epsilon = 689$. The fitness M and the distance D_{ij} are respectively defined as the population of the city and the Euclidean distance of city i and j. Selecting $\alpha=1,\gamma \in[1,2]$ [25], they are to describe the interactions of cities. Here we set $\gamma = 1.5$.

   Fig. 2 shows the simulated network. Obviously, the hubs in the real network such as Beijing, Shanghai, Guangzhou, Harbin, Urumchi exhibit the similar

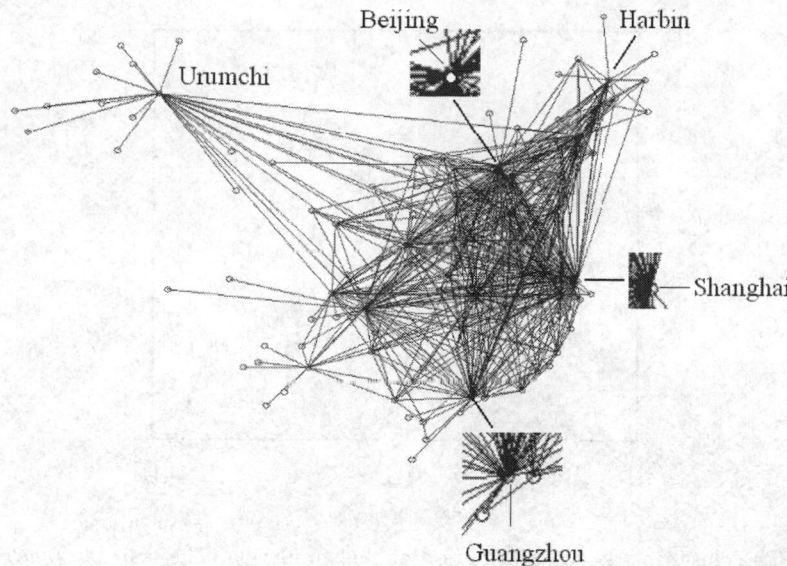

Fig. 2. The simulation for the Chinese airline network. Each link of two nodes reflects an airline between them. There is totally 689 links. The cities, such as Beijing, Shanghai, Guangzhou, Harbin, Urumchi, exhibit hub-and-spoke phenomenon in our simulation just as they do in the real network.

hub-and-spoke phenomenon in our simulation. In real condition, Beijing, Shanghai and Guangzhou are the three cities with the highest degree while in our model they are respectively Shanghai, Beijing, Wuhan (Guangzhou is the fourth). The reason for this difference may be that using the population to denote the fitness of city is intuitive but not exact because the economy and the administration factors are also important indexes for the grade of city. In spite of this difference, we still succeed in reproducing the every hub and their hub-and-spoke phenomenon existing in the real network.

We calculate the average shortest-path length $L[1]$ and the Pearson correlation coefficient $r[26]$ of the model network. The average shortest-path length of a network with N nodes is the average number of edges that has to be crossed on the shortest path from any one node to another. The Pearson correlation coefficient describe that a network is assortative or disassortative. In our simulation $L = 2.302$, $r = -0.401$ while in the real network $L = 2.263$, $r = -0.408$. Fig. 3 shows the clustering-degree distribution. The clustering-degree distribution is the correlation of the degree k and the average clustering coefficient of all nodes with degree k. It meets $C(k) \sim k^{-1}$ which indicates the model network exhibits the same hierarchy [27] as the real network does.

Fig. 4 is the degree distribution of the model network. It satisfies the two-regime power-law distribution. It satisfies the two-regime power-law distribution with the exponent $\gamma_1 = -0.46$ for the first power laws and $\gamma_2 = -2.3$ for the

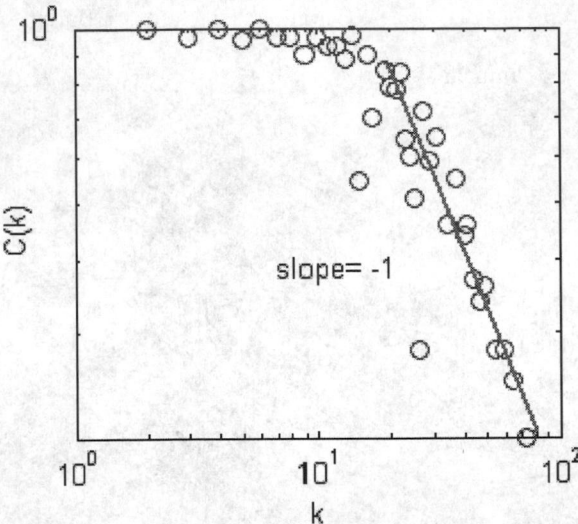

Fig. 3. The clustering-degree distribution of the simulation. The distribution satisfies a linear decreasing feature with slope -1 in log-log coordinate, which indicates that the result of simulation reproduces the hierarchy.

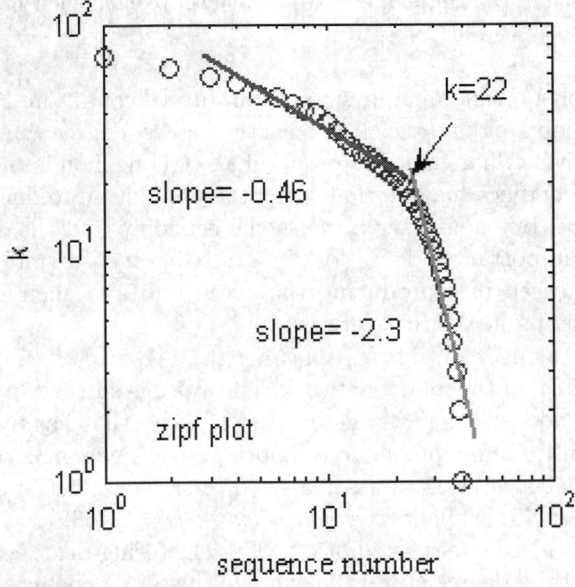

Fig. 4. The degree distribution of the simulation. The rank of the node is the sequence according to its degree. The node with the maximum degree has the rank number one and the second has rank number two, the rest may be deduced by analogy. The result presents a behavior of two-regime power-law degree distribution with the exponent $\gamma_1=-0.46$ for the first power laws and $\gamma_2=-2.3$ for the second. The turning point happens at degree k=22. In comparison, $\gamma_1=-0.53$, $\gamma_2=-2.05$ and the turning point at degree k=20 in the Chinese City Airline network.

second. The turning point happens at degree $k = 22$. Both the exponents and the turning point fit the real network well.

5 Expected Traffic and Real Traffic

Since the expected traffic plays an important role in our network design, it is meaningful to study its relationship with the real traffic. As is defined in section 2, expected traffic w_{ij}, calculated by Eq(3), is a good prediction of the traffic produced by the corresponding nodes. It does well in describing the direct interaction of two nodes. However in the transportation networks a so-called transfer mechanism may causes the difference between expected traffic and real weight. Transfer mechanism widely exists in the technical network. Most commonly the basic role of the routers in Internet is to transfer the packet from the original to the destination. By the transfer mechanism link carries not only the traffic produced by the directly connected nodes but also the traffic of other pairs that travel through it. Thus the real traffic of a link is the sum of the expected traffic of the directly connected nodes and other additional transferred traffic. To make clear this phenomenon the process is visualized in Fig. 5.

Fig. 5 shows that the expected can be different from the real one in a network with transfer mechanism, but it doesn't mean our model based on the prediction of the traffic is uncorrect. Because when a link is to be established, what we care more is the direct interaction between two nodes. As long as the expected traffic is large enough, it is necessary and profitable to construct the link. Besides it is well-known that a self-organized system usually evolves by the local information,

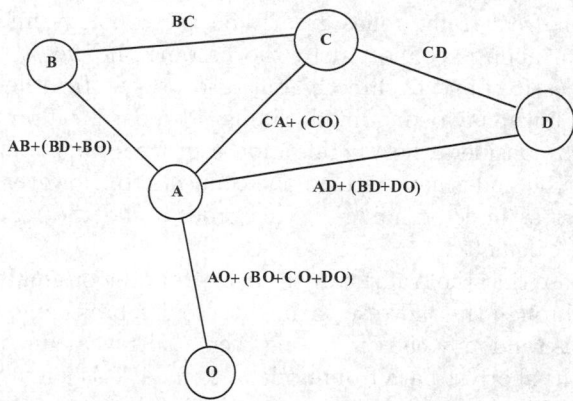

Fig. 5. Relationship of the expectant traffic and the real traffic. As an example, we only focus on the link(A,O). AO,BO,CO,DO in the figure represents the expectant traffic between the node O and the others respectively. However since there is no direct link to node B,C,D, the traffic to these nodes have to be distributed to the link (A,O). In the figure, traffic in the bracket are the transferred ones and out of the bracket are the expectant traffic of the directly connected nodes. The sum of the both represents the real traffic.

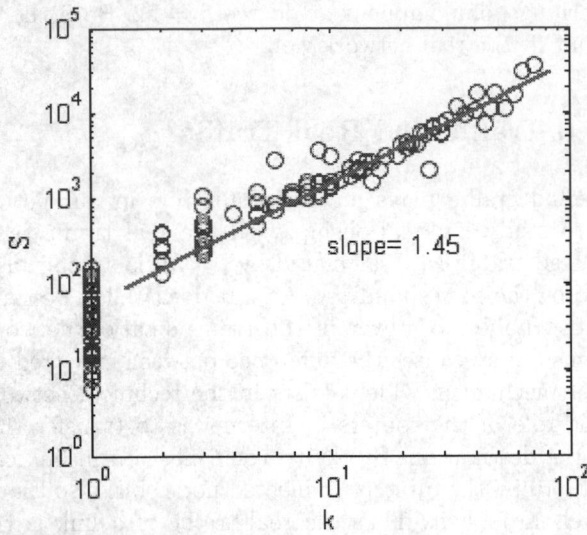

Fig. 6. The correlation of the degree and node strength. The correlation follows $S(k) \sim k^{1.45}$. Compared with $S(k) \sim k^{1.37}$ for real network, our simulation was succeeded in reproducing the non-linearly correlation of degree and node strength.

but the traffic distribution caused by the transfer mechanism depends on the global topology information. So it is difficult and impossible to consider the corresponding effect in the self-organized network design.

The transfer mechanism indicates that although the expected traffic can be predicted in some way and play a role on network construction, once the topology is determined, the real traffic relies greatly on the topology information. And when a new-born node is connected to the network, it affects the weight not only on the nodes it connects directly but also others. In other words traffic within many old links may be modified. Similar idea has been proposed by BBV model [28] which considers the weight update process happening only among the neighbors of the node that the new one connects to. However gravity model provides us a better understanding to this process and the real traffic within each link can be calculated.

First we suppose that the traffic transmitting from the original to the destination always go through the shortest path. If there is more than one such choice, the actual path is randomly chosen among them. Now we define matrix A_N^{mn} to describe one of the shortest path from node m to n. A_N^{mn} is a matrix with $N \times N$ elements, where N is the number of the nodes. The element of A_N^{mn}, denoted by a_{ij}^{mn}, takes one if the shortest path from node m to n goes through the link(i, j), otherwise it takes zero. Then the real traffic of each link can be written as:

$$\Omega_{rcal} = \sum_{(m,n)} w_{mn} A_N^{mn} . \tag{7}$$

where w_{mn} is the expected traffic between node m and n. Ω_{real} is a matrix with $N \times N$ elements and its element denoted by ω_{ij} represents the real traffic within link(i,j). The node strength reads as

$$s_i = \sum_i \omega_{ij} \, . \tag{8}$$

Following the above method, we calculate the strength-degree correlation of the simulated Chinese airline network(see Fig. 6). As is shown in Fig. 6, the node strength increases with the degree, but does quicker than linearly, as 1.45 power, namely satisfies $S(k) \sim k^{1.45}$ while in the real airline network the correlation follows $S(k) \sim k^{1.37}$.

6   Conclusion

In contrast to the previous studied spatial graph, we consider the traffic is the most important factor in network design and the traffic can be predicted by the gravity equation. Based on this idea a simple model for transportation network is proposed whose aim is to maximize the expected traffic of networks. Our model has its realistic significance that it provides a possible way to construct network efficiently. The gravity model can generate different kinds of topology by controlling two parameters such as α and γ . With α decreasing and γ increasing the topology changes from a star-like network to a decentralized road network. The agreement of our simulation with the properties found in real airline network suggest the idea proposed may play a key role in the network topology.

The expected traffic may be different from the real one in networks with transfer mechanism since it only contributes to the direct interaction of two nodes. The relationship between expected and real traffic is obtained by Eq(7) which indicates how the topology information influences the traffic in the network. More studies and demonstration to this question is essential to be done in future. Besides it is interesting to note that transfer mechanism is uncommon in social networks such as citation network and movie actor collaboration network. This essential difference may be a key factor resulting in the difference behavior of weight in all kinds of networks.

This work was partially supported by Shanghai Development Foundation for Science and Technology under Grant Numbers 06JC14082 and 05XD14021.

References

1. Watts, D.J., Strogatz, S.H.: Collective dynamics of "small-world" networks. Nature (London) 393, 440 (1998)
2. Barabsi, A.-L., Albert, R.: Emergence of scaling in random networks. Science 286, 509 (1999); Barabsi, A.-L., Albert, R., Jeong, H.: Mean-field theory for scale-free random networks. Physica A 272, 173 (1999)

3. Han, D.-D., Liu, J.-G., Ma, Y.-G., Cai, X.-Z., Shen, W.-Q.: Scale-Free Download Network for Publication. Chinese Phy. Lett. 21(9), 1855 (2004); Han, D.-D., Liu, J.-G., Ma, Y.-G.: Fluctuation of the Download Network. Chinese Phy. Lett. 25(2), 765 (2008); Han, D.-D., Qian, J.-h., Liu, J.-G.: Network Topology and Correlation Feature Affiliated with the European Airline Companies. Physica A (in press) (2008) doi:10.1016/j.physa.2008.09.021

4. Pitts, F.: The Profess. Geograph. 17, 15 (2004)

5. Guimer, R., Mossa, S., Turtschi, A., Amaral, L.A.N.: The worldwide air rans-portation network: Anomalous centrality, community structure, and cities global roles. Proc. Natl. Acad. Sci. USA 102, 7794 (2005); Guimer, R., Amaral, L.A.N.: Modeling the world-wide airport network. Eur. Phys. J. B 38, 381 (2004)

6. Barrat, A., Barthlemy, M., Pastor-Satorras, R., Vespignani, A.: Proc. Natl. Acad. Sci. USA 101, 3747 (2004)

7. Smith, D.A., Timberlake, M.: Urban Stud. 32, 287 (1995)

8. Crucitti, P., Latora, V., Porta, S.: Preprint physics/0504163

9. Latora, V., Marchiori, M.: Is the Boston subway a small-world network. Physica A 314, 109 (2002)

10. Carlson, J.M., Doyle, J.: Phys. Rev. E 60, 1412 (1999)

11. Gastner, M.T., Newman, M.E.J.: The spatial structure of networks. Eur. Phys. J. B 49, 247–252 (2006)

12. Gastner, M.T., Newman, M.E.J.: Optimal design of spatial distribution networks. Phys. Rev. E 74, 016117 (2006)

13. Kaiser, M., Hilgetag, C.C.: Spatial growth of real-world networks. Phys. Rev. E 69, 036103 (2004)

14. Waxman, B.: Routing of multipoint connections. IEEE J. Selec. Areas Commun. 6, 1617 (1988)

15. Yook, S.-H., Jeong, H., Barabsi, A.-L.: Proc. Natl. Acad. Sci. USA 99, 13382 (2002)

16. Xie, Y.-B., Zhou, T., Bai, W.-j., Chen, G., Xiao, W.-K., Wang, B.-H.: Geographical networks evolving with optimal policy. Phys. Rev. E 75, 36106 (2007)

17. Bianconi, G., Barabási, A.-L.: Competition and multiscaling in evolving networks. Europhys. Lett. 54, 436 (2001)

18. Bianconi, G., Barabási, A.-L.: Bose-Einstein consideration in complex networks. Phy. Rev. Lett. 86, 5632 (2001)

19. Masuda, N., Miwa, H., Konno, N.: Geographical threshold graphs with small-world and scale-free properties. Phys. Rev. E 71, 036108 (2005)

20. Xulvi-Brunet, R., Sokolov, I.M.: Growing networks under geographical constraints. Phys. Rev. E 75, 046117 (2007)

21. Wu, Z., Braunstein, L.A., Colizza, V., Cohen, R., Havlin, S., Eugene Stanley, H.: Phys. Rev. E 74, 056104 (2006)

22. Macdonald, P.J., Almaas, E., Barabsi, A.-L.: Minimum spanning trees of weighted scale-free networks. Europhys. Lett. 72, 308 (2005)

23. Jung, W.-S., Wang, F., Eugene Stanley, H.: Gravity model in the Korean highway. Europhys. Lett. 81, 48005 (2008)

24. Liu, H.-K., Zhou, T.: Empirical study of Chine se city airline network. Acta Physica Sinica 56, 0106 (2007)

25. Liu, J.S., Chen, Y.G.: Scintia Geographica Sinica 20, 0528 (2000)

26. Newman, M.E.J.: Mixing patterns in networks. Phys. Rev. E 67, 026126 (2003)

27. Ravasz, E., Barabási, A.-L.: Hierarchical organization in complex networks. Phys. Rev. E 67, 026112 (2003)

28. Barrat, A., Barthélemy, M., Vespignani, A.: Weighted evolving networks: coupling topology and weight dynamics. Phy. Rev. Lett. 92, 228701 (2004)

A Bipartite Graph Based Model of Protein Domain Networks

J.C. Nacher[1], T. Ochiai[2], M. Hayashida[3], and T. Akutsu[3]

[1] Department of Complex Systems, Future University-Hakodate, Japan
[2] Faculty of Engineering, Toyama Prefectural University, Japan
[3] Bioinformatics Center, Institute for Chemical Research, Kyoto University, Japan

Abstract. Proteins are essential molecules of life in the cell and are involved in multiple and highly specialized tasks encoded in the amino acid sequence. In particular, protein function is closely related to fundamental units of protein structure called *domains*. Here, we investigate the distribution of kinds of domains in human cells. Our findings show that while the number of domain types shared by k proteins follows a scale-free distribution, the number of proteins composed of k types of domains decays as an exponential distribution. In contrast, previous data analyses and mathematical modeling reported a scale-free distribution for the protein domain distribution because the relation between kinds of domains and the number of domains in a protein was not considered. Based on this finding, we have developed an evolutionary model based on (1) growth process and (2) copy mechanism that explains the emergence of this mixing of exponential and scale-free distributions.

Keywords: Growing networks, protein domains, scale-free networks.

1 Introduction

The complexity of a wide variety of systems as the metabolic pathways, protein interaction networks, social relationships or transportation systems, can be investigated in terms of networks where the elementary units of the system are represented by nodes and their interactions as edges. In recent years, empirical analyses and theoretical modeling of networks have rapidly become a highly-active research area, uncovering the existence of unexpected organizing principles and similarities in real systems, with sizes ranging from hundreds to billions of nodes [1,2,3,4]. Whereas at a global level, real complex networks deviate from predictions of random graph theory [5] and display a scale-free and hierarchical organization [6,7], a complementary perspective at a local level reveals a significant prevalence and variety of highly characteristic patterns of interactions, such as motifs, modules, cliques and communities with specific functional tasks [8,9,10].

Recent experimental efforts in proteomics have generated a massive amount of newly sequenced proteins, molecular structures, foldings mechanisms as well as interacting domains data. Using this information, protein interaction maps have

J. Zhou (Ed.): Complex 2009, Part I, LNICST 4, pp. 525–535, 2009.
© ICST Institute for Computer Sciences, Social Informatics and Telecommunications Engineering 2009

been constructed and analyzed. Although these networks are still incomplete, it allows for the first time the study of the large-scale structure of functional interactions within a cell for a variety of organisms. These analyses have shown that cellular networks such as metabolic pathways, protein-protein interaction networks can be classified as scale-free networks [2,11,12].

A protein is a long chain of amino acids encoding important cellular functions. Each protein can be composed of one or more protein *domains* that represent fundamental building blocks with specific structural and functional features. However, a different classification allows the definition of *protein modules* considered as a more compact structural unit in a protein with a length in the range of 20-40 residues [13,14].

In this work, we will focus on proteins composed of domains as fundamental building blocks, in particular we have analyzed the empirical data corresponding to proteins and interacting domains using human proteome information collected from the UniProt[26] (UniProtKB/Swiss-Prot Release 56.0 of 22-Jul-2008) and Integr8[27] (Release 84 constructed from UniProt 14.0) databases. We then investigate the distribution of kinds of domains in human cells. Our findings show that while the number of a domain type shared by k proteins follows a scale-free distribution, the number of proteins composed of k types of domains decays as an exponential distribution. This finding has not been reported before, as previous analyses [15,16,17] did not study the relation between kinds of domains and the number of domains in a protein.

This problem can be investigated using a bipartite graph whose nodes can be classified into two disjoint sets N (proteins) and M (domains) such that each edge connects a node in N and one in M [18]. For example, N_k indicates the number of protein with k edges if the protein is composed of k domains. Similarly, M_k denotes the number of domains with k edges if this domain is shared by k proteins.

Based on our empirical findings on the dissimilar nature of N_k and M_k distributions, we have developed an evolutionary model using *the rate equation approach*, first suggested by Krapivsky et al.[19], that explains the emergence of this mixing of exponential and scale-free distributions. The model requires (1) growth process and (2) copy mechanism. We first use the rate equation approach for constructing the discrete mathematical equations corresponding to bipartite graphs. We then transform them into differential equations and solve them using the continuum limit.

2 Theoretical Model and Experimental Results

2.1 Theoretical Model

Let us consider a bipartite graph, whose nodes are divided into two disjoint sets N (proteins) and M (domains), and only connections between two nodes in different sets N and M are allowed as shown in Fig. 1. In what follows, N_k denotes the number of proteins (square) with k edges (domains). Similarly, M_k denotes the number of domains (circle) shared by k proteins. Furthermore, we consider that each domain represents a specific kind of domain. Therefore, two domains corresponding to the same type of domain are not allowed. This is a crucial point in our analysis. Then, we propose an algorithm that builds a power-law distribution for M_k and an exponential distribution for N_k.

1. The model is initialized with a same small number l of N-nodes and M-nodes. Each node from l_N and from l_M is connected by an edge, then the degree of all N-nodes and M-nodes is only one, where we have assumed $l = l_N = l_M$.

2. At time $t = 1$, with probability α_N, a randomly selected N-node is copied. Otherwise, with probability β_N, a new N-node is added. We then connect this new N-node to n_0 randomly selected M-nodes. In this process, $\alpha_N + \beta_N = 1$.

3. At the same time step, with probability α_M, a randomly selected M-node is copied. Otherwise, with probability β_M, a new M-node is added. We then connect this new M-node to m_0 randomly selected N-nodes. As in the above process, $\alpha_M + \beta_M = 1$.

4. Steps (2) and (3) are iterated t times until a desired number of nodes is generated. At the end, the network will consist of the same number $t+l$ of N-nodes and M-nodes.

Therefore, our model of growing bipartite networks is composed of two main ingredients: (1) growth process and (2) copy mechanism. Fig. 1 illustrates these mechanisms for both sets of nodes. From this algorithm, we construct the rate equation for the bipartite network. The rate equation approach was first introduced in network science by Krapivsky et al., [19] and applied to the study of percolation [20], protein evolution networks [21] and citation networks as well as used in extensive theoretical analyses [22]. Furthermore, it has also been applied to the computation of the node degree correlations [23]. On the other hand, models applied to bipartite graphs are much less numerous and only a very few works have addressed the issue [25]. See also the review on rate equation approach for further information [24]. By following our algorithm, the rate equation for the time evolution of the number of nodes with degree k in both sets of nodes N_k and M_k can be written as:

$$\frac{dN_k}{dt} = \alpha_M \left(\frac{k-1}{M(t)} N_{k-1} - \frac{k}{M(t)} N_k \right) + \beta_M \left(\frac{m_0}{N(t)} N_{k-1} - \frac{m_0}{N(t)} N_k \right)$$

$$+ \alpha_N \frac{N_k}{N(t)} + \beta_N \delta_{kn_0} \quad (1)$$

$$\frac{dM_k}{dt} = \alpha_N \left(\frac{k-1}{N(t)} M_{k-1} - \frac{k}{N(t)} M_k \right) + \beta_N \left(\frac{n_0}{M(t)} M_{k-1} - \frac{n_0}{M(t)} M_k \right)$$

$$+ \alpha_M \frac{M_k}{M(t)} + \beta_M \delta_{km_0} \quad (2)$$

where $N(t) = t + l$ and $M(t) = t + l$ are the total number of N-nodes and M-nodes at time t, respectively. In these equations, δ_{kn_0} and δ_{km_0} indicate the

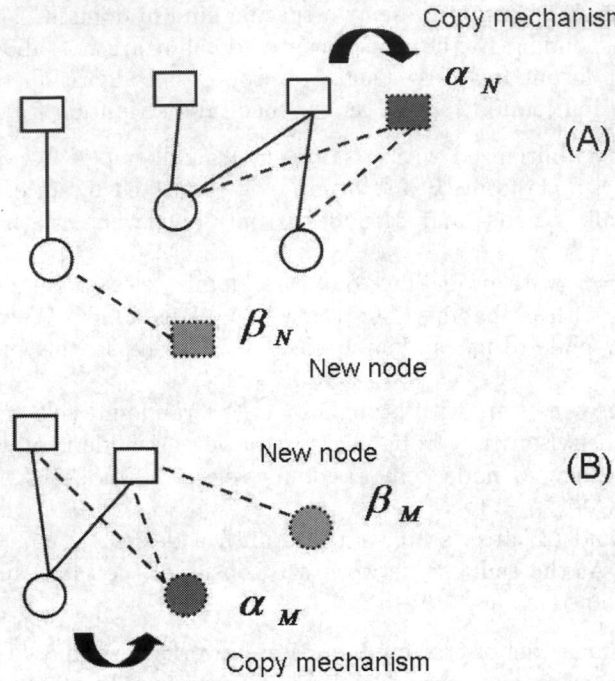

Fig. 1. Description of growth and copy mechanisms in our model for bipartite graphs. Squares (proteins) (A) and circles (kinds of domains) (B) can be added and copied. One protein connected to one (two) kind of domains indicates that this protein consists of one (two) kinds of domains.

contribution of a new node connected to already existing n_0 and m_0 nodes. Next, by introducing the probability distribution $n_k = N_k/N(t)$ and $m_k = M_k/M(t)$, we obtain

$$\frac{d((t+l)n_k)}{dt} = \alpha_M((k-1)n_{k-1} - kn_k) + \beta_M m_0(n_{k-1} - n_k)$$
$$+\alpha_N n_k + \beta_N \delta_{kn_0} \qquad (3)$$

$$\frac{d((t+l)m_k)}{dt} = \alpha_N((k-1)m_{k-1} - m_k) + \beta_N n_0(m_{k-1} - m_k)$$
$$+\alpha_M m_k + \beta_M \delta_{km_0} \qquad (4)$$

In the limit $t \to \infty$, we obtain the equation for the stationary distribution:

$$n_k = \alpha_M((k-1)n_{k-1} - kn_k) + \beta_M m_0(n_{k-1} - n_k)$$
$$+\alpha_N n_k + \beta_N \delta_{kn_0} \qquad (5)$$

$$m_k = \alpha_N((k-1)m_{k-1} - km_k) + \beta_N n_0(m_{k-1} - m_k)$$
$$+\alpha_M m_k + \beta_M \delta_{km_0} \qquad (6)$$

In the continuum k limit, these equations take the following form:

$$n_k = -\frac{d}{dk}\{(\alpha_M k + \beta_M m_0)n_k\} + \alpha_N n_k \tag{7}$$

$$m_k = -\frac{d}{dk}\{(\alpha_N k + \beta_N n_0)m_k\} + \alpha_M m_k \tag{8}$$

Then, from the last equation, we obtain

$$m_k \propto (\alpha_N k + \beta_N n_0)^{-\frac{1-\alpha_M+\alpha_N}{\alpha_N}} \tag{9}$$

In the limit for large k ($k \to \infty$),

$$m_k \propto k^{-\frac{1-\alpha_M+\alpha_N}{\alpha_N}} \tag{10}$$

$$\sim k^{-\frac{1+\alpha_N}{\alpha_N}} \tag{11}$$

where we have used $\alpha_M \sim 0$ in the last equation. Therefore, the degree distribution for M-nodes (number of domains shared by k proteins) obeys a power-law. On the other hand, from Eq. (7), we can write

$$n_k \propto (\alpha_M k + \beta_M m_0)^{-\frac{1-\alpha_N+\alpha_M}{\alpha_M}} \tag{12}$$

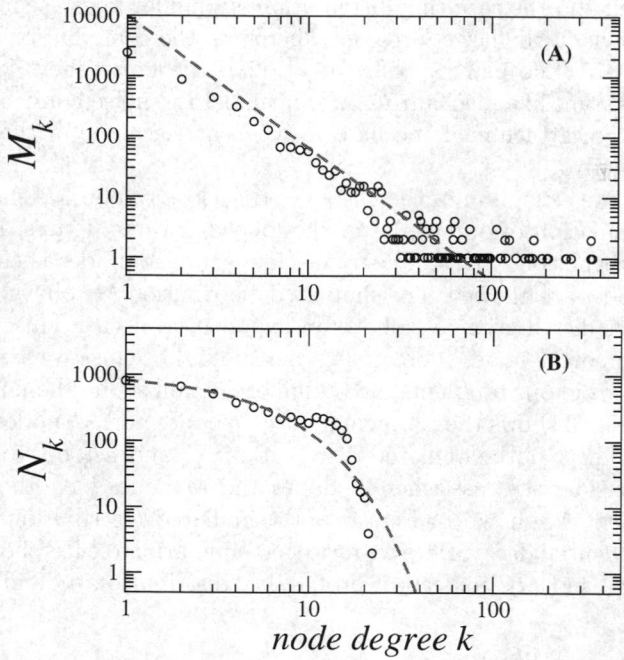

Fig. 2. Theoretical results (red dashed line) of the model and computational simulation (black circles) with $\alpha_N = 0.8$ and $\alpha_M = 0.05$, (A) Power-law distribution with degree exponent 2.18. (B) Exponential decay.

In particular, in the limit $\alpha_M \to 0$,

$$n_k \propto e^{-\frac{\beta_N}{m_0}k}k \qquad (13)$$

Therefore, we obtain that the degree distribution for N-nodes (number of proteins composed of k types of domains) obeys a exponential decay. We highlight main features of the model as follows:

1. By using a bipartite growing network model composed of copy and random attachment processes with supression of copy of M-nodes (types of domains) ($\alpha_M \sim 0$), we reproduce the observed distributions of power-law and exponential decay of several real networks composed of two types of nodes.
2. $\alpha_M \sim 0$ implies that M-nodes (kinds of domains) are unlikely copied, if compared to N-nodes. This is meaningful because kinds of domains are unique and cannot be duplicated by definition. This asymmetry in the growing mechanisms is fundamental to derive the observed mixing distributions.

2.2   Model Simulation

When both parameters α_N, α_M take values close to one simultaneously, a so-called "*giant fluctuation*" occurs [20]. It indicates that a model that only includes the copy mechanism (i.e., a model configuration with α_N, α_M close to one) does not behave well and the resulting distribution is singular and resembles the sum of delta functions in the large k region. Therefore, the contribution of a "*noise*" term is needed. While in Krapivsky et al. [20], the noise effect is introduced through a mutation-like mechanism, in our model the noise contribution comes from the random attachment mechanism when at least one of the parameters β_N, β_M is non zero.

Thus, with the exception of the case α_N, α_M close to one, we show the computational simulation of our model in the following three figures. Fig. 2 shows the degree distribution when the copy mechanism of M-nodes is supressed and copies of N-nodes are allowed. The simulated distribution M_k obeys a power-law, while the other distribution N_k obeys an exponential decay. This copy mechanism supression of M-nodes (domains) is meaningful because we are considering kinds of domains in our problem, and a kind of domain should be unique by definition. Next, Fig. 3 shows the case when both N-nodes and M-nodes are allowed to be copied. Then, both simulated distribution N_k and M_k obey a power-law. Finally, we consider the case when N-nodes and M-nodes have the copy mechanism supressed. As shown in Fig. 4, both simulated distribution N_k and M_k follow an exponential decay. Here we note that simulation results show the degree distribution N_k and M_k, instead of probability distribution n_k and m_k.

2.3   Experimental Results

We have performed an empirical analysis using human proteins collected from the UniProt[26] (UniProtKB/Swiss-Prot Release 56.0 of 22-Jul-2008) and Integr8[27]

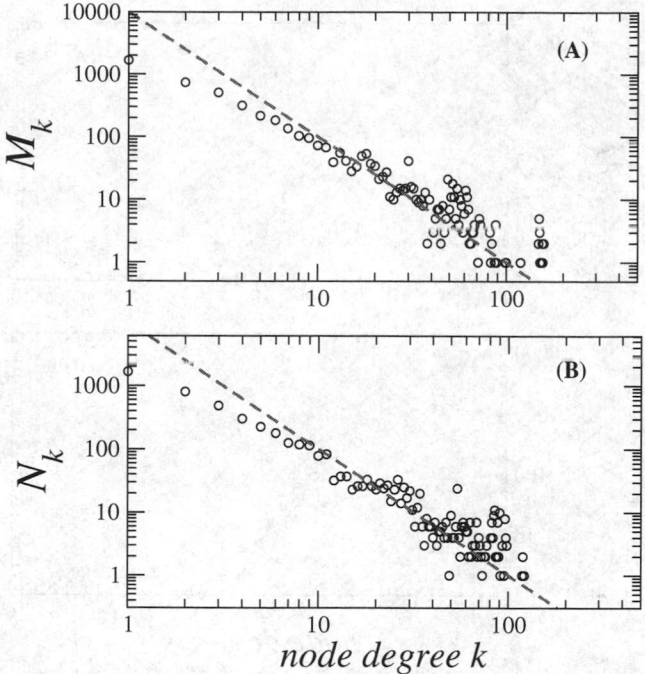

Fig. 3. Theoretical results (red dashed line) of the model and computational simulation (black circles) with $\alpha_N = 0.5, \alpha_M = 0.5$ Both figures (A) and (B) show a power-law distribution with degree exponent 2

(Release 84 constructed from UniProt 14.0) databases. Integr8 database provides non-redundant set of UniProt entries representing each complete proteome. We have obtained Pfam[28] domains for each protein from the DR line of UniProt format.

Fig. 5 shows an exponential distribution for the number of kinds of domains in a protein. Human proteins were downloaded from the UniProt and Integr8 databases. Next, Fig.6 shows the distribution of the number of domain types shared by k human proteins in the UniProt and Integr8 databases. In this case, we can observed that the distribution follows a scale-free distribution. These results are in agreement with the predictions of our evolutionary model shown in Fig. 2. It is worth noticing that our model generates the same number of domains as proteins because the number of M-nodes and N-nodes is the same by construction. However, we have also analyzed and computed this case of asymmetric growth in the number of nodes. Although we omit the main derivation for space reasons, our results show that not only the mixing of scale-free and exponential distributions is conserved but also the exponent degree of power-law is kept invariant under the asymmetric growth. To be precise, only the exponent of the exponential decay distribution depends on the asymmetric growth.

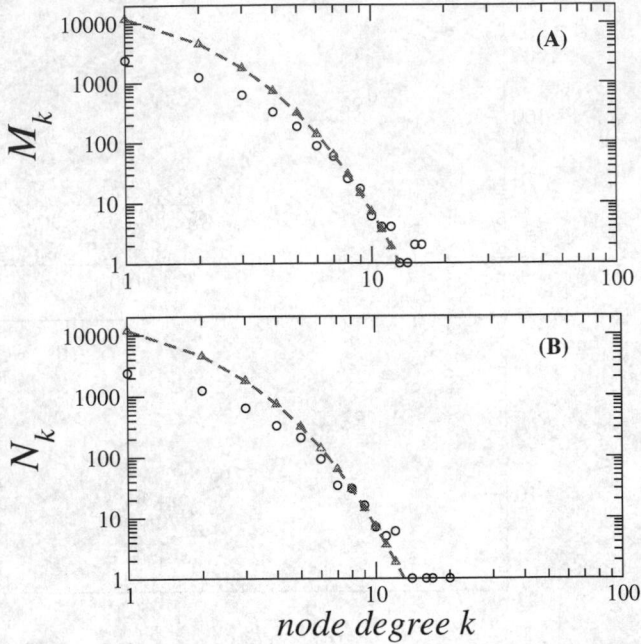

Fig. 4. Theoretical results of the model (red dashed line) and computational simulation (black circles) with $\alpha_N = 0.05, \alpha_M = 0.05$ Both (A) and (B) distributions show an exponential decay

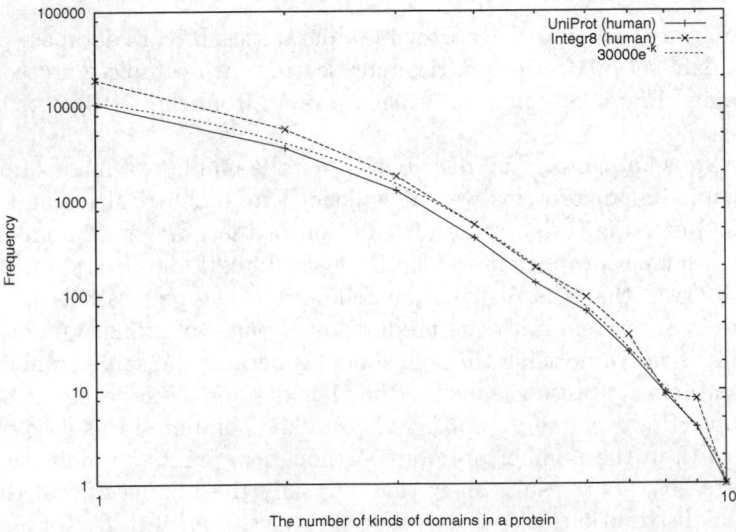

Fig. 5. The distribution of the number of kinds of domains in a protein for human proteome space. Data collected from UniProt and Integr8 databases.

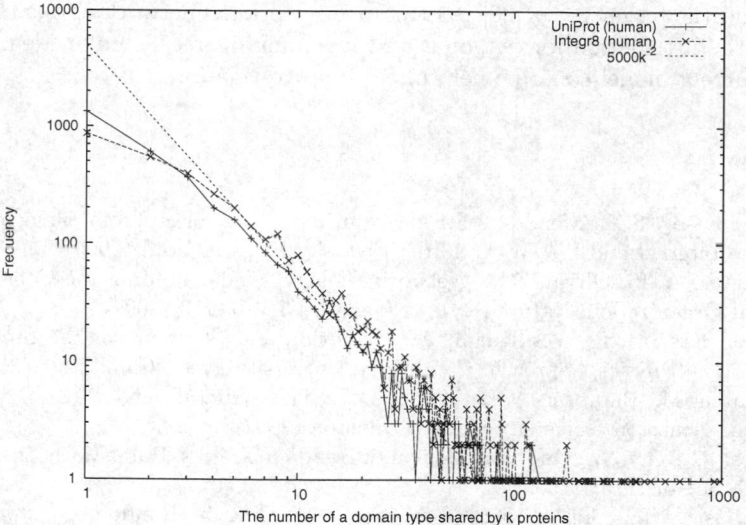

Fig. 6. The distribution of the number of domain types shared by k proteins for human proteome space. Data collected from UniProt and Integr8 databases.

3 Conclusion

In summary, we have investigated the distribution of protein and kinds of domains in human cells. Our results indicate that while the number of a domain type shared by k proteins follows a scale-free distribution, the number of proteins composed of k types of domains decays as an exponential distribution. It is worth noticing that previous data analyses and mathematical modeling reported a scale-free distribution for the protein domain distribution because the relation between kinds of domains and the number of domains in a protein was not considered.

Based on this finding, we have developed a simple evolutionary model based on (1) growth process and (2) copy mechanism. This model based on the rate equation approach for computing bipartite graphs does not only predict the observed asymmetry in the distribution of protein composed of k unique domains and number of domains shared by k proteins but also predicts the degree exponent for the power-law in the vicinity of value 2.

Furthermore, the model elucidates that the supression of copy mechanisms in one set of nodes is enough to create the mixture distribution and the symmetry breaking. This copy mechanism supression of M-nodes (domains) is reasonable because we are considering kinds of domains in this problem, and a kind of domain can be considered unique by definition.

Of particular interest for future work will be to extend the current model of bipartite graphs to be applied to other biological systems like gene regulatory networks where nodes represent operons encoding transcriptional factors

(TFs) and the target genes [29]. As this transcriptional network also exhibits an assymetric distribution for outgoing and incoming degrees, similar ideas shown in the current model could be helpful for its investigation.

References

1. Dorogovtsev, S.N., Mendes, J.F.F.: Evolution of Networks: From Biological Nets to the Internet and WWW. Oxford University Press, Oxford (2003)
2. Barabási, A.-L., Oltvai, Z.N.: Network Biology: Understanding the Cells's Functional Organization. Nature Reviews Genetics 5, 101–113 (2004)
3. Pastor-Satorras, R., Vespignani, A.: Evolution and Structure of the Internet: A Statistical Physics Approach. Cambridge University Press, Cambridge (2004)
4. Newman, M., Barabási, A.-L., Watts, D.J.: The Structure and Dynamics of Networks. Princeton University Press, Princeton (2007)
5. Erdös, P., Rényi, A.: On the Evolution of Random Graphs. Publ. Math. Inst. Hung. Acad. Sci. 5, 17–61 (1960)
6. Barabási, A.-L., Albert, R.: Emergence of Scaling in Random Networks. Science 286, 509–512 (1999)
7. Ravasz, E., Somera, A.L., Mongru, D.A., Oltvai, Z.N., Barabási, A.-L.: Hierarchical Organization of Modularity in Metabolic Networks. Science 297, 1551–1555 (2002)
8. Milo, R., Shen-Orr, S., Itzkovitz, S., Kashtan, N., Chklovskii, D., Alon, U.: Network Motifs: Simple Building Blocks of Complex Networks. Science 298, 824–827 (2002)
9. Shen-Orr, S., Milo, R., Mangan, S., Alon, U.: Network Motifs in the Transcriptional Regulation Network of Escherichia coli. Nat. Genetics 31, 64–68 (2002)
10. Palla, G., Derenyi, I., Farkas, I., Vicsek, T.: Uncovering the Overlapping Community Structure of Complex Networks in Nature and Society. Nature 435, 814–818 (2005)
11. Jeong, H., Tombor, B., Albert, R., Oltvai, Z.N., Barabási, A.-L.: The Large-scale Organization of Metabolic Networks. Nature 407, 651–654 (2000)
12. Jeong, H., Mason, S., Barabási, A.-L., Oltvai, Z.N.: Lethality and Centrality in Protein Networks. Nature 411, 41–42 (2001)
13. Go, M.: Modular Structural Units, Exons, and Function in Chicken Lysozyme. Proc. Natl. Acad. Sci. USA 80, 1964–1968 (1983)
14. Go, M.: Correlation of DNA Exonic Regions with Protein Structural Units in Haemoglobin. Nature 291, 90–92 (1981)
15. Wuchty, S.: Scale-free Behavior in Protein Domain Networks. Mol. Biol. Evo. 18, 1694–1702 (2001)
16. Karev, G.P., Wolf, Y.I., Rzhetsky, A.Y., Berezovskaya, F.S., Koonin, E.V.: Birth and Death of Protein Domains: A Simple Model of Evolution Explains Power Law Behavior. BMC Evo. Biol. 2, 18 (2002)
17. Nacher, J.C., Hayashida, M., Akutsu, T.: Protein Domain Networks: Scale-free Mixing of Positive and Negative Exponents. Physica A 367, 538–552 (2006)
18. Newman, M.E.J., Strogatz, S.H., Watts, D.J.: Random Graphs with Arbitrary Degree Distributions and Their Applications. Phys. Rev. E 64, 026118 (2001)
19. Krapivsky, P.L., Redner, S., Leyvraz, F.: Connectivity of Growing Random Networks. Phys. Rev. Lett. 85, 4629 (2000)
20. Kim, J., Krapivsky, P.L., Kahng, B., Redner, S.: Infinite-order Percolation and Giant Fluctuations in a Protein Interaction Network. Phys. Rev. E. 66, 055101 (2002)

21. Ispolatov, I., Krapivsky, P.L., Yuryev, A.: Duplication-divergence Model of Protein Interaction Network. Phys. Rev. E. 71, 061911 (2005)
22. Krapivsky, P.L., Redner, S.: Organization of Growing Random Networks. Phys. Rev. E. 63, 066123 (2001)
23. Barrat, A., Pastor-Satorras, R.: Rate Equation Approach for Correlations in Growing Network Models. Phys. Rev. E 71, 036127 (2005)
24. Krapivsky, P.L., Redner, S.: Rate Equation Approach for Growing Networks. Lecture Notes in Physics 625, 3–22 (2003)
25. Ergün, G.: Human Sexual Contact Network as a Bipartite Graph. Physica A 308, 483–488 (2002)
26. The UniProt Consortium: The Universal Protein Resource (UniProt). Nucleic Acids Research 36, D190–D195 (2008)
27. Kersey, P., Bower, L., Morris, L., Horne, A., Petryszak, R., Kanz, C., Kanapin, A., Das, U., Michoud, K., Phan, I., Gattiker, A., Kulikova, T., Faruque, N., Duggan, K., Mclaren, P., Reimholz, B., Duret, L., Penel, S., Reuter, I., Apweiler, R.: Integr8 and Genome Reviews: Integrated Views of Complete Genomes and Proteomes. Nucleic Acids Research 33, D297–D302 (2005)
28. Finn, R.D., Tate, J., Mistry, J., Coggill, P.C., Sammut, J.S., Hotz, H.R., Ceric, G., Forslund, K., Eddy, S.R., Sonnhammer, E.L., Bateman, A.: The Pfam protein families database. Nucleic Acids Research 36, D281–D288 (2008)
29. Luscombe, N.M., Babu, M.M., Yu, H., Snyder, M., Teichmann, S.A., Gerstein, M.: Genomic Analysis of Regulatory Network Dynamics Reveals Large Topological Changes. Nature 431, 308–312 (2004)

The Results on the Stability of Glycolytic Metabolic Networks in Different Cells

Qinghua Zhou[1,2,*], Gang Peng[2], Li Jin[2], and Momiao Xiong[2,3]

[1] College of Mathematics & Computer, Hebei University,
Baoding City, 071002, Hebei Province, China
[2] Theoretical Systems Biology Lab, School of Life Science, Fudan University,
Shanghai, 200433, China
[3] Human Genetics Center, University of Texas School of Public Health,
Houston, TX, USA
{qinghua.zhou,gpengfd,ljin007,momiao}@gmail.com

Abstract. Evolutionary forces will affect the structure of metabolic networks and their dynamic behaviors. To examine this hypothesis, in this work we investigate the relationship between the complexity of the metabolic glycolytic networks and the stability of the networks in different cells. By deriving the stoichiometrix from the FBA methods, we develop the models for Sce, Dmgr, Dsmi and Pic in fungi. Based on these models, we analyze the stability of the networks. The results show that the metabolic networks are more complicated with more stable ones.

Keywords: stability, metabolic networks, glycolysis, FBA methods.

1 Introduction

Cellular metabolism and its regulation represent a large scale dynamical system and complex dynamic behavior has been observed for a wide variety of metabolic pathways (Steuer [1]). Generally, the dynamic properties of cellular regulatory systems are considered to be essential for cellular regulation and constitute the conceptual basis for many physiological properties of living cells. As mentioned in Grimbs et al. [2], the dynamic behavior of metabolic networks is governed by numerous regulatory mechanisms, such as reversible phosphorylation, binding of allosteric effectors or temporal gene expression, by which the activity of the participating enzymes can be adjusted to the functional requirements of the different cells.

With the developments in genomics, more and more information on genetic networks has been provided for several micro-organisms. The next logical step is how to use this information to study the integrated behavior of the cellular networks. One of the most concerned areas has been the study of metabolic networks. The analysis of these networks by using mathematical methods can facilitate applying the research results to the real problems. For example, it can guide the metabolic engineering

---

* Correspondent author.

J. Zhou (Ed.): Complex 2009, Part I, LNICST 4, pp. 536–540, 2009.

process. It is well known that the stability and robustness of the biological systems are the most popular rules which all the living creatures live on. The research on the stability of biological systems includes very rich contents: linear and nonlinear systems, variable and invariable, and so on. In this research, we initially study some basic contents on systems' stability. Specifically, we will investigate the stability of glycolytic metabolic networks in some fungi cells and the relationship between the stability and the biological evolution.

There are several approaches which have been proposed to study the metabolic networks, including metabolic control analysis, biochemical systems theory, cybernetic modeling and flux balance analysis (FBA). Except of FBA, these approaches require kinetic information of the cellular reactions (Mahadevan et al. [3]). However, the kinetic information is often unavailable for most of biological cells. Therefore, we choose FBA incorporated with mathematical analysis to study the stability of the metabolic networks.

FBA is the method which assumes that the metabolic networks will reach a steady state constrained by the stoichiometry. This assumption is based on the fact that metabolic transients are typically rapid compared to cellular growth rates and environmental changes. The consequence of this assumption is that all metabolic fluxes on the formation and degradation of any metabolite must balance, which is leading to the flux balance equation (Varma & Palsson [4]):

$$S \cdot v = 0 \tag{1}$$

where S is a matrix containing the stoichiometry of the metabolic reactions, v is a vector of the metabolic reaction rates. In general, the above equation is always underdetermined. To deriving a meaningful result, it must recur to combining other mathematical methods.

In 2007, Grimbs et al. [2] proposed a computational approach based on structural kinetic modeling. The authors applied the approach to the metabolism of human erythrocytes and the results showed that the allosteric enzyme regulation significantly enhances the stability of the network.

In this paper, we investigate the relationship between the complexity of the glycolytic metabolic networks and their stability. In our previous study, we found that the yield of the productions of some important metabolites is higher with more complicated metabolic networks. In this work, we assume the similar results about the stability of the network will be obtained. That is, the metabolic networks are more stable with more complicated ones.

2   The Glycolytic Metabolic Networks of Different Cells in Fungi

In this section, after we introduce the mathematical methods which will be used in analyzing the stability of the networks, we depict the metabolic networks of four different cells which we selected to investigate. All of them belong to the class of Fungi. Specifically, they are: Saccharomyces cerevisiae (sce), Saccharomyces mikatae (dsmi), Magnaporthe grisea (dmgr) and Pichia stipicis (pic).

2.1 Mathematical Methods

A metabolic network is combined by a set of coupled chemical reactions and transport processes. Suppose a network which contains m metabolites and r reactions. Based on the FBA formula, the time-dependent changes of the metabolite concentrations can be described by a set of differential equations of the form $\dot{x} = S \cdot v$, where x denotes the m-dimensional vector of metabolite concentrations, S denotes the $m \times r$-dimensional stoichiometric matrix and v denotes an r-dimensional vector of enzyme kinetic reaction rates. If considering a steady state of the system, then the differential equations system converts to equation (1).

Given a metabolic state characterized by x^0 and v^0, the system of differential equations can be approximated by a Taylor series expansion as follows:

$$\frac{dx}{dt} = S \cdot v(x^0) + S \cdot \frac{\partial v}{\partial x}\Big|_{x^0} (x - x^0) + \dots \tag{2}$$

where the first item describes the steady state properties of the system, as exploited by FBA to constrain the stoichiometrically feasible flux distributions. Let the second item $S \cdot \frac{\partial v}{\partial x}\Big|_{x^0} = J$, then the structure of the Jacobian matrix J constrains the possible dynamics of the system at each metabolic state. Evaluating the eigenvalues of J, then if the largest real part of the eigenvalues is positive, it implies the instability of the metabolic state. And only if all the eigenvalues have a negative real part, the metabolic state is stable. For detailed explanation of the methods, please refer to Grimbs et al. [2].

2.2 The Glycolytic Networks of Sce, Dmgr, Dsmi and Pic

We select four kinds of cells in Fungi to continue our research. Specially, the cells are Sce, Dmgr, Dsmi and Pic.

The metabolic network considered for modeling the glycolysis of Sce is shown in Fig 1. The network consists of 14 metabolites and 12 reactions, in which all the metabolites are:

$$x_1 = G6P, \quad x_2 = F6P, \quad x_3 = FBP, \quad x_4 = DHAP,$$
$$x_5 = GAP, \quad x_6 = 1,3-BPG, \quad x_7 = PEP, \quad x_8 = PYR,$$
$$x_9 = ACA, \quad x_{10} = ETOH, \quad x_{11} = NAD^+, \quad x_{12} = ATP,$$
$$x_{13} = AMP, \quad x_{14} = GLC.$$

Where the substrate is Glucose (GLC), and the important product is Ethanol (ETOH). For the meaning of other abbreviations in Fig 1, please refer to Hynne et al. [5].

Remember that, in our models, the substrate Glc is not included in the metabolic vector x, but treated as a constant input of the network. For Sce, set the concentration of Glc equals to 1mmol per gram (Dry Weight). For the metabolic glycolytic networks of Dmgr, Dsmi and Pic, please refer to the reference [6].

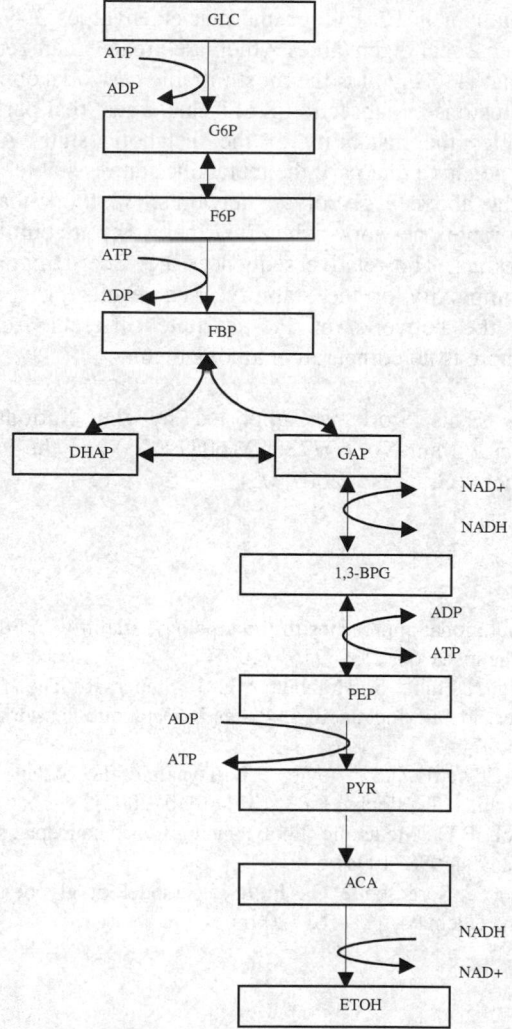

Fig. 1. The glycolytic network of Sce

2.3 Results and Conclusions

According to the methods introduced in subsection 2.1, we compute the eigenvalues of the Jacobian matrix separately. The results are as follows:

For Sce, there are 12 nonzero eigenvalues, in which there are one pair eigenvalues whose real part are greater than zero, say: 3.9257+1.9579i and 3.9527-1.9579i.

For Dmgr, the number is 10 and there are 2 real eigenvalues which are greater than zero: 4.476 and 0.086336. Remember that there is one root which is greater than Sce's.

For Dsmi, the number is 12. Except one pair eigenvalues 3.476+2.62i and 3.476-2.62i, there are other 2 real eigenvalues which are greater than zero, say: 2.2326E-14 and 3.6887E-13. That is to say, it is the most instable networks comparatively.

Follow the conclusions in subsection 2.1, if the largest real part of the eigenvalues is positive, it implies the instability of the metabolic state. And only if all the eigenvalues have a negative real part, the metabolic state is stable.

Therefore, for the above 3 glycolytic networks, we think that it becomes more stable with more complex networks when their networks are similar in producing the same important product. The relative sequences are: Sce>Dmgr>Pic. No matter for considering the complexity or the stability. The results proved the assumption's correctness. Well, the networks of Pic is quite different from the others, our conclusion is that there is no comparative among them.

Acknowledgments. This work is supported by the National Nature Science Foundation of China (Grant No. 60773062, 60772073) and the Foundation of Hebei Education Department (Grant No. 2007105).

References

1. Steuer, R.: Computational approaches to the topology, stability and dynamics of metabolic networks. Phytochemistry 68, 2139–2151 (2007)
2. Grimbs, S., Selbig, J., Bulik, S., Holzhütter, H.G., Steuer, R.: The stability and robustness of metabolic states: identifying stabilizing sites in metabolic networks. Molecular Systems Biology 3, 146 (2007)
3. Mahadevan, M., Edwards, J.S., Doyle, F.J.: Dynamic flux balance analysis of diauxic growth in Escherichia coli. Biophy. J. 83, 1331–1340 (2002)
4. Varma, A., Palsson, B.O.: Metabolic flux balancing: Basic concepts, scientific and practical use. Bio. Tech. 12, 994–998 (1994)
5. Hynne, F., Danø, S., Sørensen, P.G.: Full-scale model of glycolysis in Saccharomyces cerevisiae. Biophy. Chem 94, 121–163 (2001)
6. http://www.ebi.ac.uk

The Probability Distribution of Inter-car Spacings*

Jin Guo Xian and Dong Han

ShangHai Jiao Tong University, ShangHai 200240, China
jgxian@sjtu.edu.cn

Abstract. In this paper, the celluar automation model with Fukui-Ishibashi-type acceleration rule is used to study the inter-car spacing distribution for traffic flow. The method used in complex network analysis is applied to study the spacings distribution. By theoretical analysis, we obtain the result that the distribution of inter-car spacings follows power law when vehicle density is low and spacing is not large, while, when the vehicle density is high or the spacing is large, the distribution can be described by exponential distribution. Moreover, the numerical simulations support the theoretical result.

Keywords: FI model, Power Law, Exponential distribution, Complex network, Spacings distribution.

1 Introduction

Celluar automata(CA) has been considered as a model system in a wide variety of problems and phenomena in statistical physics and other fields[1]. Because it is conceptual simple, easy to use in simulation by computer and can model the complex behavior of the traffic flow, the CA became popular for the microscopic simulation of traffic flow. Two famous CA models are the Nagel-Schreckenberg (NS)model and the Fukui -Ishibashi(FI) model[2]. The CA model share several basic features: the lane is represented as one dimension lattices; the lattices sites are called "cell" which can be either empty or occupied by at most one vehicle at a given instant time, at every time step the speed of every vehicle will be updated following a well defined rule, the main difference between these models is the particular procedure implemented to change the speed of a vehicle. In the FI model, the vehicle can move by v_{max} (v_{max} is the maximum speed of the vehicle) cells at most in one time step if they are not blocked by cars in front. More precisely, if the empty cells $C_n(t)$ in front of a car is larger than v_{max} at time t, then the speed can be update to $v_{max} - 1$ by probability p and v_{max} by $1 - p$. The FI model differs from the NS model in that the increase in speed may not be gradual, and that stochastic delay only applies to high speed cars. In this paper we will consider the FI model due to its simplicity and leads to considerable improvement of the flow for higher velocities.

---

* This research was supported by National Nature Science Foundation of China (10531070).

J. Zhou (Ed.): Complex 2009, Part I, LNICST 4, pp. 541–549, 2009.

Complex network has received much attention from the physics community in recent years[5][6][7]. Employing the theory and method of complex networks, one can hope that it provides some insight into the evolution mechanism of traffic flow. Some work such as [4], study the scale free property in traffic system by numerical simulation. In general, CA model is used to study the traffic system by computer simulation. It is difficult to study the traffic flow by theoretical analysis based on CA model, since in paper [3], author give some result for the traffic character by theoretical analysis and simulation. In this paper we study the inter-car spacings distribution using the methods used in the complex network study. Our result give a clear picture for the spacings distribution which is meaningful for the traffic flow theory.

The paper is organized as follows. In section 2 we give the model definition and theoretical analysis of spacings distribution. In section 3, the simulation results for $p = 0.5$ and all various density are given and compared with the theoretical results. At last, we summary with a discussion of our method.

2   Model and Analytical Solution of Inter-car Spacings Distribution

The modified Nagel-Schrechenberg model working with the Fukui-Ishibashi Acceleration rule is a probabilistic cell automation . In this model, the space, time and velocities are discrete. The one dimensional horizontal road with traffic flowing is divided into L cell. Each cell may either be empty, or may be occupied by a vehicle with speed $0, 1, 2, ...v_{max}$, v_{max} is the maximum speed of the vehicle. If there are N vehicles in the length L road, then the vehicle density is $\rho = N/L$. The speed of nth vehicle at time t is defined as $v_n(t), n \in \{0, 1, 2, v_{max}\}$. If $C_n(t)$ represents the number of empty sites in front of nth vehicle at time t, then the spacings in front of nth vehicle can be calculated by

$$C_n(t+1) = C_n(t) + v_{n+1}(t) - v_n(t) \tag{1}$$

Let $M = v_{max}$, the speed of vehicle v_n is updated through the following rules:

$$v_n(t+1) = \begin{cases} 0 & \text{if } C_n(t) = 0 \\ C_n(t) - 1 & \text{with probability } p, \text{if } 0 < C_n(t) < M \\ C_n(t) & \text{with probability } 1 - p, \text{if } C_n(t) > M \\ M - 1 & \text{with probability } p, \text{if } C_n(t) \geq M \\ M & \text{with probability } 1 - p, \text{if } C_n(t) \geq M \end{cases} \tag{2}$$

If $N_k(t)$ represents the number of inter-car spacings with length k at time t, the probability of finding such a spacing at time t is $P_k(t) = \frac{N_k(t)}{N}$. We assume that $t \to \infty$ is equivalent to $L \to \infty$ and $P_k = \lim_{t \to \infty} P_k(t)$. Next, we will derive the formula for P_k under different condition.

Let $\mathcal{F}_t = \sigma\{N_k(t), 0 \leq k \leq L - N)\}$, and from time step t to $t+1$ the change number of spacing k is $N_k(t+1) - N_k(t)$, then we can calculate the condition expectation $E(N_k(t+1) - N_k(t)|\mathcal{F}_t)$ as follows

$$E(N_k(t+1) - N_k(t)|\mathcal{F}_t) = \sum_{i=-N}^{N} iP(N_{k+1}(t) - N_k(t) = i|\mathcal{F}_t) \qquad (3)$$

Let

$$\begin{aligned}
Q_k^1(t) &= P(N_k(t+1) - N_k = 1|\mathcal{F}_t) \\
Q_k^2(t) &= P(N_k(t+1) - N_k = -1|\mathcal{F}_t)
\end{aligned} \qquad (4)$$

then $Q_k^1(t)$ and $Q_k^2(t)$ can be calculated by the following formulas:

$$Q_0^1(t) = \sum_{i=1}^{M}(1-p)P_i(t)(P_0(t) + pP_1(t))$$

$$Q_0^2(t) = P_0(t)(1 - P_0(t) - pP_1(t))$$

$$Q_1^1(t) = P_0(t)[(1-p)P_1(t) + pP_2(t)] + \sum_{i=2}^{M} P_i(t)(1-p)[(1-p)P_1(t) + pP_2(t)]$$

$$+ \sum_{i=2}^{M} P_i(t)p[P_0(t) + pP_1(t)] + P_{M+1}(t)(1-p)[P_0(t) + pP_1(t)]$$

$$Q_1^2(t) = 1 - P_1(t)((1-p)((1-p)P_1(t) + pP_2(t)) - p(P_0(t) + pP_1(t))$$

$$\qquad (5)$$

when $2 \leq k \leq M$

$$Q_k^1(t) = P_0(t)[(1-p)P_k(t) + pP_k + 1(t)]$$

$$+ \sum_{i=1, i\neq k}^{M} P_i(t)(1-p)[(1-p)P_k(t) + pP_{k+1}(t)]$$

$$+ \sum_{i=1, i\neq k}^{M} P_i(t)p[(1-p)P_{k-1} + pP_k]$$

$$+ \sum_{i=1}^{k-1} P_{M+i}(t)(1-p)[(1-p)P_{k-i}(t) + pP_{k-i+1}(t)]$$

$$+ \sum_{i=1}^{k-2} P_{M+i}(t)p[(1-p)P_{k-i-1}(t) + pP_{k-i}(t)]$$

$$+ [P_{M+k}(t)(1-p) + P_{M+k-1}(t)p][P_0(t) + pP_1(t)]$$

$$Q_k^2(t) = 1 - P_k(t)(1-p)[(1-p)P_k(t)$$

$$+ p(P_{k+1}(t)] - P_k(t)p[(1-p)P_{k-1}(t) + pP_k(t)] \qquad (6)$$

For $k \geq M + 1$

$$Q_k^1(t) = P_{k-1}(t)p(1-p) \sum_{i=M}^{L-N} P_i(t)$$

$$Q_k^2(t) = P_k(t)p(1-p) \sum_{j=M}^{L-N} P_j(t) \tag{7}$$

From (3), one can obtain the condition expectation as follows.

$$E(N_k(t+1) - N_k(t)|\mathcal{F}_t) = Q_k^1(t) - Q_k^2(t) + \delta(k,t) \tag{8}$$

Where the $\delta(k,t), (k = 0, 1, 2, \cdots L - N)$ is a small value, especially when k is large. The reason is that the probability $P(N_k(t+1) - N_k(t) = i|\mathcal{F}_t)$ and $P(N_k(t+1) - N_k(t) = -i|\mathcal{F}_t), i \geq 2$ decrease rapidly with the i increasing. For example, $P(N_k(t+1) - N_k(t) = 2|\mathcal{F}_t) = (Q_k^1)^2$, because $0 \leq Q_k^1 < 1$ is a small value, then $(Q_k^1)^2$ is smaller .When $k > M$, for two cases we derive the formulas for P_k.

Case I: the vehicle density ρ is small and k is not very large
From t step to $t + 1$ step, the $N_k(t)$ updating to $N_k(t+1)$ are decided by the probability $P_k(t) = \frac{kN_k(t)}{L-N}$. From (7)(8), we have

$$E(N_k(t+1) - N_k(t)|\mathcal{F}_t) = \frac{(k-1)N_{k-1}(t)}{L(1-\rho)}p(1-p) \sum_{i=M}^{L-N} \frac{N_i(t)}{L(1-\rho)}$$

$$- \frac{kN_k(t)}{L(1-\rho)}p(1-p) \sum_{j=M}^{L-N} \frac{jN_j(t)}{L(1-\rho)} + \delta_{k,t} \tag{9}$$

Let $I_k(t) = E(N_k(t))$, it follows from (9) that

$$I_k(t+1) = I_k(t)[1 - \frac{kp(1-p)}{L-N} \sum_{j=M}^{L-N} \frac{jN_j(t)}{L-N}]$$

$$+ \frac{(k-1)I_{k-1}(t)p(1-p)}{L-N} \sum_{j=M}^{L-N} \frac{jN_j(t)}{L-N} \tag{10}$$

Because $\sum_{j=M}^{L-N} \frac{jN_j(t)}{L-N} = 1 - \frac{1}{L-N} \sum_{j=1}^{M-1}(jN_j(t)) \to C$ when $t \to \infty (L \to \infty)$ then we have the following equation

$$P_k = \frac{(k-1)P_{k-1}p(1-p)C\frac{1}{\frac{1}{\rho}-1}}{1 + k\frac{Cp(1-p)}{\frac{1}{\rho}-1}}$$

$$= \frac{(k-1)P_{k-1}p(1-p)C}{1/\rho - 1 + kCp(1-p)} \tag{11}$$

Let $a = p(1-p)C$, from (11), we have

$$P_k = \frac{(k-1)P_{k-1}}{k + (1-\rho)/a\rho} \tag{12}$$

Solving this equation we obtain the result

$$P_k \approx \frac{1}{k^{1+\frac{1-\rho)}{a\rho}}} \tag{13}$$

Case II: The vehicle density ρ is large or $k \gg M$
From t step to $t+1$ step, the $N_k(t)$ updating to $N_k(t+1)$ are decided by the probability $P_k(t) = \frac{N_k(t)}{N}$, then from (7)(8), we have

$$E(N_k(t+1) - N_k(t)|\mathcal{F}_t) = \frac{N_{k-1}(t)}{N}p(1-p)\sum_{i=M}^{L-N}\frac{N_i(t)}{N}$$

$$-\frac{N_k(t)}{N}p(1-p)\sum_{i=M}^{L-N}\frac{N_i(t)}{N} + \delta_{k,t} \tag{14}$$

Let $b_M(t) = \sum_{j=M}^{L-N}\frac{N_j(t)}{N}$, by (14) we have following equation

$$I_k(t+1) = I_k(t)[1 - \frac{p(1-p)b_M(t)}{N}] + \frac{I_{k-1}(t)p(1-p)b_M(t)}{N} \tag{15}$$

Because $b_M(t) \to b$ when $t \to \infty$,and $N = L\rho \to \infty$, we have the equation

$$P_k = \frac{bp(1-p)P_{k-1}}{1 + p(1-p)b} \tag{16}$$

Solving the equation, we obtain

$$P_k \approx (\frac{bp(1-p)}{1 + p(1-p)b})^k = e^{-kln(1+\frac{1}{bp(1-p)})} \tag{17}$$

Based on the above induction we can find that, when $k > M$,P_k can be represented as

$$P_k = \begin{cases} C_0\frac{1}{k^\alpha} & \text{if } M < k < k^* \\ C_1e^{-\lambda k} & \text{if } k > k^* \end{cases} \tag{18}$$

Where k^* relates to the density ρ. By the simulation in the next section, we can find that k^* decreases to M gradually while ρ is increasing. That is, at a point of ρ, P_k can be represented by exponential distribution completely.

3   Simulation

We now proceed to present computer simulation and compare the result with the theoretic result. At the beginning, the vehicle number N is chosen as 5000 and the road length L is adjusted so as to the vehicle density ρ can have desired value. The time length is $t^* = 50000$. In order to guarantee $L \to \infty$ increases with t and the density ρ unchanged, the simulations follow the rule that if a car leave the road then a new car will be added into. P_k are calculated by the formula as follows:

$$P_k = \frac{\sum_{t=1}^{t^*} N_k(t)}{Nt^*}, k > M. \qquad (19)$$

Let $x_k = lnP_k$ and $y_k = k$, it is obvious that if (x_k, y_k) are in a line, then the distribution is $P_k = C_0 e^{-\lambda k}$. If $x_k = \ln P_k$ and $y_k = \ln k$, (x_k, y_k) are in a line, then the data fit the power law $P_k = C_1 \frac{1}{k^\alpha}$.

The simulation result are presented to the Fig 1 ~ 5.

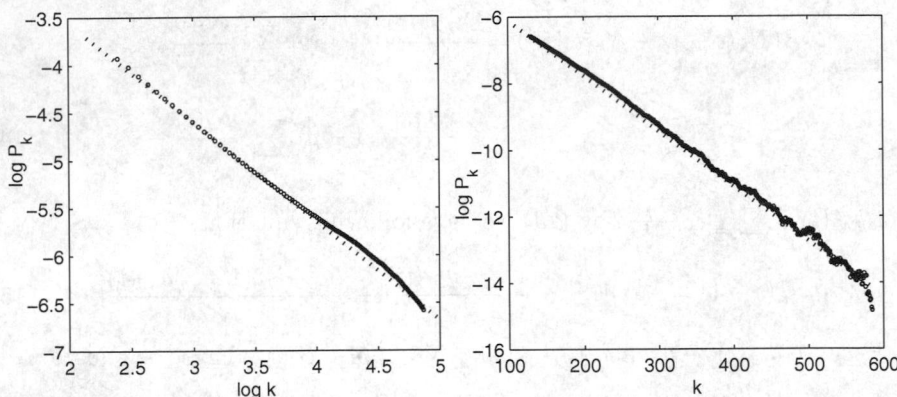

Fig. 1. Show that P_k follows power law when while$10 < k < 130$, while it is exponential distribution when $k \geq 130$. Where $v_{max} = 5, p = 0.5$.

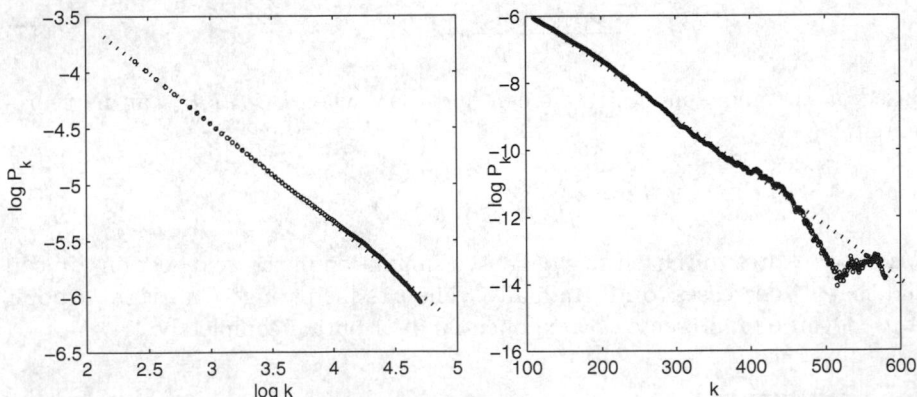

Fig. 2. Show that theP_k follows power law when $10 < k < 100$, while it is exponential distribution when $k \geq 100$. Where $v_{max} = 4, p = 0.5$.

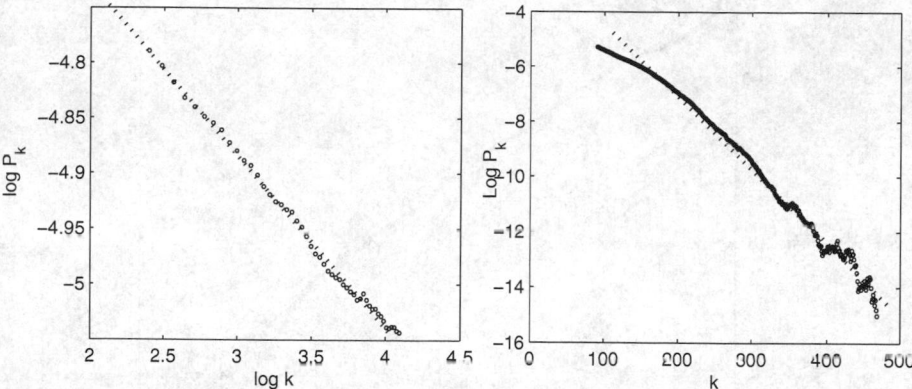

Fig. 3. Show that P_k follows power law when $0 < k < 60$, while it is exponential distribution when $k \geq 60$. Where $v_{max} = 3, p = 0.5$.

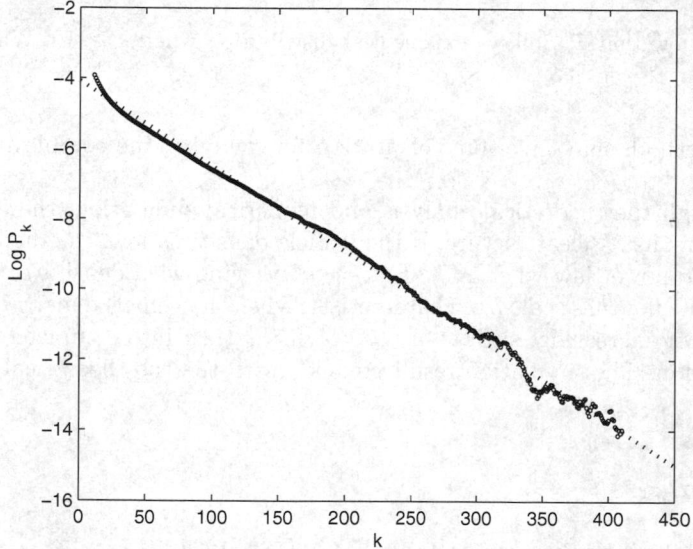

Fig. 4. Show that P_k follows exponential distribution when $k \geq 10$. Where $v_{max} = 2, p = 0.5$.

4   Notes and Comments

CA model have been widely used to model the traffic system, due to its simplicity for computer simulation. In our study, the methods for complex network analysis are applied to study the traffic flow characters. The inter-car spacings distribution is studied by theoretical analysis and numerical simulation. Our

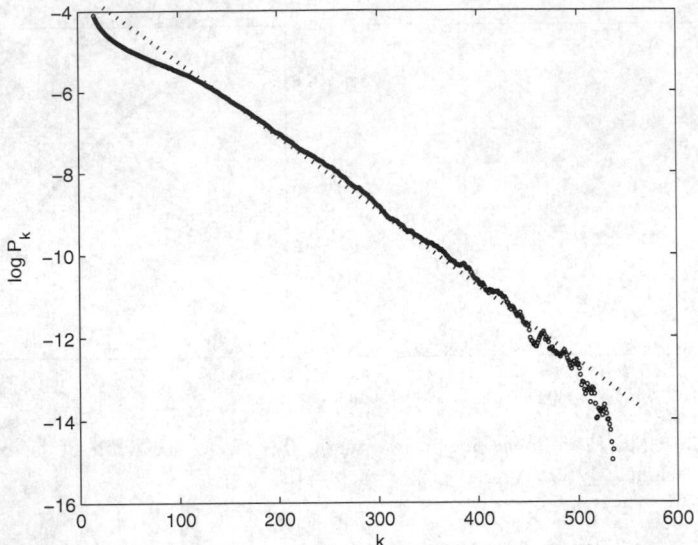

Fig. 5. Show that P_k follows exponential distribution when $k \geq 10$. Where $v_{max} = 1, p = 0.5$.

study methods may give some elicitation for studying the complexity of traffic flow.

Through the theoretical analysis and numerical simulation, the evolution of traffic flow has a clear picture. If the vehicle density is low, the distribution P_k shows the power law when $M < k < k^*$, k^* varying with density ρ. This tell us that traffic flow has scale free characteristic when the vehicle density is low. With the density increasing, such as $v_{max} = 1, 2$, P_k turn into complete exponential distribution. The simulation result coincide with the the theoretical result very well.

References

1. Nagel, k., Schreckenberg, M.J.: Phys. I 2, 2882 (1993)
2. Fukui, M., Ishibashi, Y.: Traffic flow in 1D cellular automaton model including cars moving with high speed. J. Phys. Soc. Jpn 65(6), 1868–1870 (1996)
3. Fu, C.-J., Wang, B.-H., et al.: Analytical studies on a modeified Nagel-Schrechenberg model with the Fukui-Ishibashi Acceleration rule. Elsevier Chaos solitons and Fractals 31, 772–776 (2007)
4. Ke-Ping, L.: Scale-Free Properties in Traffic System Commun. Theor. Phys. (Beijing China) 46(2), 374–380 (2006)
5. Réka, Barabási, A.-L.: Statistical mechanics of complex network. Reviws of modern physics 74 (January 2002)
6. Dorogovtsev, S.N., Mendes, J.F.F., Samukhin, A.N.: Structure of growth networks with preferential linking. Phys. Rev. Lett. 85, 4633–4636 (2000)

7. Krapivsky, P.L., Redner, S., And Leyvraz, F.: Connectivity of growing random networks. Phys. Rev. Lett. 85, 4629–4632 (2000)
8. Fourrate, K., Loulidi, M.: Disordered Cellar automata traffic flow models. M. J. Condensed Matter 5(2), 151–155 (2004)
9. Sasvari, M., Kertesz, J.: Cellular automata models of single lane traffic. Physical Review E 56(4), 4104–4110 (1997)
10. Wolfram, S.: Cellular Automata and Complexity. Addison-Wesley Publishing Company, Baltimore (1994)

The Origin of Evolution in Physical Systems

Jean-Claude Heudin

International Institute of Multimedia
Pôle Universitaire Léonard de Vinci – 92916 Paris La Défense
Jean-Claude.Heudin@devinci.fr

Abstract. A tentative outline for a model for the evolution of physical systems is presented. The universal classes of dynamical behaviors found in Cellular Automata experiments provide the basis for introducing the variation-stabilization principle as a synthetic interpretation of these phenomena. It is suggested that biological evolution takes its root in the evolution of physical systems as a particular case of the variation-stabilization principle that occurs at the transition phase between ordered and chaotic regimes.

Keywords: variation, stabilization, evolution, complex physical system.

1 Introduction

In 1794, Erasmus Darwin, grandfather of the great Charles Darwin, published his book entitled *Zoonomia or the Laws of Organic Life*. In the first lines, he suggested that "the whole of Nature" is governed by physical laws [1]. Reciprocally, many physicists have proposed theories inspired by the Darwinian approach. For example, the path explored by Ludwig Boltzmann was similar in essence to the one of Darwin. Charles Darwin's theory begins with the assumption of the spontaneous fluctuation of species. Then natural selection leads to irreversible biological evolution. Boltzmann's order principle also shows that randomness leads to irreversibility, even if the result is the destruction of initial structures [2]. More recently, Lee Smolin proposed natural selection as a speculative hypothesis for the evolution of cosmological complexity [3]. All these interpretations of "evolution" in physical systems may be more than a coincidence. It seems clear that there is a continuum from the "evolution" of physical systems to biological evolution, but there is no theory that makes a convincing link between them.

In this paper, we try to establish a bridge between the evolution in physical systems and biological evolution by means of the variation-stabilization principle. We begin in section 2 by introducing this principle as a model for the evolution of complex physical phenomena. In section 3, we describe an experiment that illustrates this principle based on a two-dimensional Generalized-Life Cellular Automata [4]. We verify that this system exhibits the four universal complexity classes discovered by Stephen Wolfram [5] and we locate complex dynamics at the transition between ordered and chaotic phases [6]. In section 4, we discuss these results in the framework of the variation-stabilization principle. We argue that this principle represents a valuable synthetic

J. Zhou (Ed.): Complex 2009, Part I, LNICST 4, pp. 550–559, 2009.
© ICST Institute for Computer Sciences, Social Informatics and Telecommunications Engineering 2009

re-conceptualization for the evolution of complex physical systems, replacing biological evolution as a particular case that occurs at the transition phase between ordered and chaotic regimes.

2 The Variation-Stabilization Principle

Any physical system can be represented as a hierarchical model. Evidence have been presented that a hierarchical modeling approach is required to obtain the necessary precision for studying complex natural phenomena [7]. In most cases, this hierarchal model of complexity is pyramidal: as a phenomenon increases in complexity, it decreases in number of elements. Thus, like a pyramid, the model is composed of a finite set of n layers L (vertical complexity) with an order relation:

$$\Delta = \{ L_n \} \text{ with } L_0 \to L_1 \to ... \to L_{n-1} \to L_n.$$

Each layer n is composed of a dynamical network of structural elements S with similar behavioral repertoire (horizontal complexity):

$$L_n = \{ S_i^n \}.$$

Each structural element S_i of a given level L_n details the way in which it reacts to local situation and interactions with other structural elements of this level. The fluctuations of the environment and the structural interactions between elements lead to the formation of transitory structures S_j which are finite set of elements S_i. This stochastic phenomena can be noted as a *variation* function: $V(s)$. Due to their structural properties, some of these S_j elements continue to exist for some arbitrary times. They represent potential structural elements of the higher level L_{n+1}. If there are no more external fluctuations (i.e. closed world hypothesis), the system evolves after some transitory time to a stable state. Note that, the terms "stable state" must be understood as a statistical stability of global dynamics over time. This phenomena can be noted as a *stabilization* function $Z(s)$. The next figure summarizes this principle.

Variations create emergent temporary structures that achieve stability if their structural properties make them "adapted" to the environment. This principle of variation and stabilization applies at the same time in parallel at all levels of the hierarchical model.

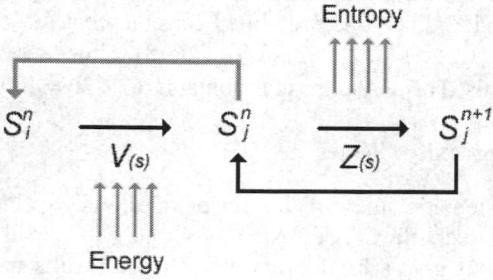

Fig. 1. Global relations in the variation-stabilization principle

The variation-stabilization process potentially leads to different macroscopic states. Throughout the physical world one can observe such different states of matter. The most familiar examples are the solid, liquid and gas phases. Other existing phases include crystals, colloids, glasses, amorphous, and plasma. These different forms can be grouped into four basic classes based on their attractor in the phase space: fixed point, limit cycle, chaotic and complex. It has been shown that Cellular Automata (CA) represent a good approach for modeling these different phases of matter. In particular, Stephen Wolfram proposed a correspondence with four universal complexity classes in one-dimensional Cellular Automata [5]: Class I includes CA whose dynamics reaches a steady state; Class II consists of CA characterized by periodic behaviors ; Class III CA produce structures that seem random; Class IV CA exhibit complex dynamics with both periodic and random patterns. Thus, CA seems a good experimental model for studying the variation-stabilization principle.

3 The Generalized-Life Experiment

3.1 Generalized-Life Cellular Automata

In order to study the variation-stabilization principle using a CA, we used a two-dimensional CA space based on a generalization of John Conway's "Game of Life" [8, 9]. This CA set has shown a wide diversity of dynamics in terms of *growth* (infinite, bounded and fall) and *periodicity* (chaotic, periodic and stable) [4]. In these experiments, a randomized initialization simulates an initial *variation* of the CA environment and its transition rule simulates the *stabilization* function.

Each stabilization rule can be written in the form $E_b E_h F_b F_h$ where E_b is the minimum number of living neighbor cells that must touch a currently "living" cell in order to guarantee that it will remain alive in the next generation. Fb is the minimum number of "living" cells touching a currently "dead" cell in order that it will come to life in the next generation and E_h and F_h are the corresponding upper limits. According to this notation, Conway's Life would be written "Life 2333," that is $E_b = 2$, $E_h = 3$, $F_b = 3$, and $F_h = 3$. More formally, let S_t define the state of a cell and N_t be the number of living cells in the neighborhood at time t. The stabilization function is then:

$$if \ (S_t = 0 \ \& \ N_t \leq F_b \ \& \ N_t \geq F_h) \ | \ (S_t = 1 \ \& \ N_t \leq E_b \ \& \ N_t \geq E_h) \ S_{t+1} = 1, \ else \ S_{t+1} = 0. \quad (1)$$

In a 2D grid, a cell is surrounded at most by 8 neighbors. The rule parameters are bound between 1 and 8, since 0 is prohibited for quiescent reasons. So intervals can be any among (1,1) (1,2) … (1,8) (2,2) (2,3) … (8,8). Thus there are $(8 + 7 + … + 1)$ = 36 possible intervals. This CAspace thus contains $36 \times 36 = 1,296$ rules.

3.2 Universal Complexity Classes

We conducted a systematic study of this set of rules. As expected, we were able to classify the Generalized-Life CA set using Wolfram's universal complexity classes. The following sections give a brief overview of these results with some typical CA examples.

Fixed Dynamics Class. Class I is associated to limit points in the phase space. For almost all initial random configurations, cellular elements have the same value after a relatively short transient period. This stabilization process leads to a homogeneous state of the cell matrix. Figure 2 gives examples of such dynamics.

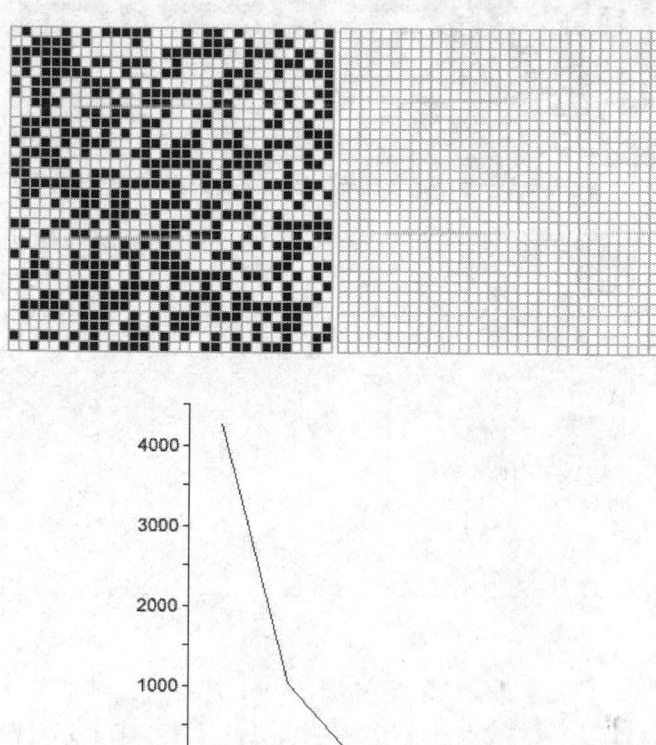

Fig. 2. Two configurations on the upper-left at t = 1 (left) and t = 4 (right) for rule 7788. All initial configurations lead to a homogeneous state of "dead" empty cells. Note that only a part of the CA configurations is shown in these images. The bottom curve is a time series for rule 5566 showing the number of "alive" cells (y-axis) for each generation (x-axis). After only four generations, eall cells are "dead".

Periodic Dynamics Class. Class II is associated with limit cycles in the phase space. Almost all configurations lead to a stable state with nested structures, except for some oscillating patterns. This stabilization process creates a periodic system. Figure 3 illustrates examples of such dynamics.

Chaotic Dynamics Class. Class III is associated with chaotic behaviors which refer to unpredictable space-time behaviors. Each configuration exhibits some rare stable patterns that survive only few generations and then are destroyed by the surrounding chaos. This stabilization process leads to an apparently random sequence, but this sequence is characterized by some statistical stability over time. Figure 4 illustrate examples of such dynamics.

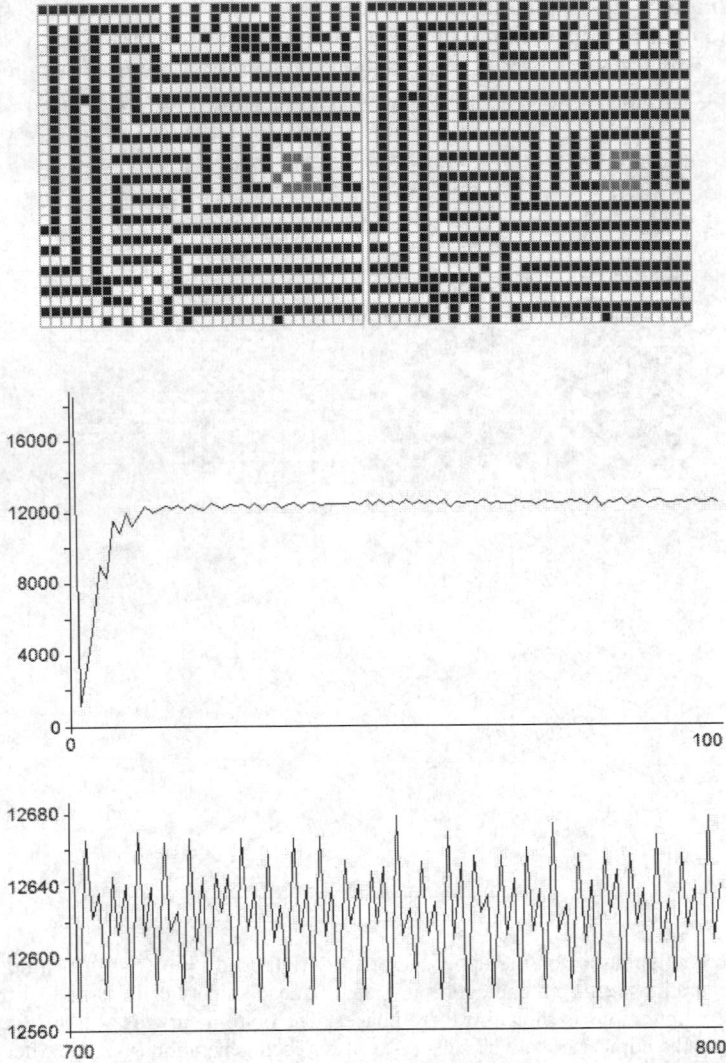

Fig. 3. Two configurations on the top at t = 59 (left) and t = 60 (right) for rule 5613. The cellular automata space becomes fixed after transitory configurations except for some oscillating pattern (in gray). Note that only a part of the CA configurations is shown in these images. The bottom curves are time series for rule 1616 showing the number of "alive" cells (y-axis) for each generation (x-axis). After a short transitory phase (upper curve), the CA oscillates between a small number of states (lower curve).

Complex Dynamics Class. Class IV is associated with complex behaviors characterized by long transients. The stabilization process leads to complex configurations that include both fixed and periodic structures. Some of these periodic patterns can propagate in the cell matrix (i.e. gliders). This emergent dynamical behavior is a sign that potentially indicates the support of universal computation. Figure 5 shows the

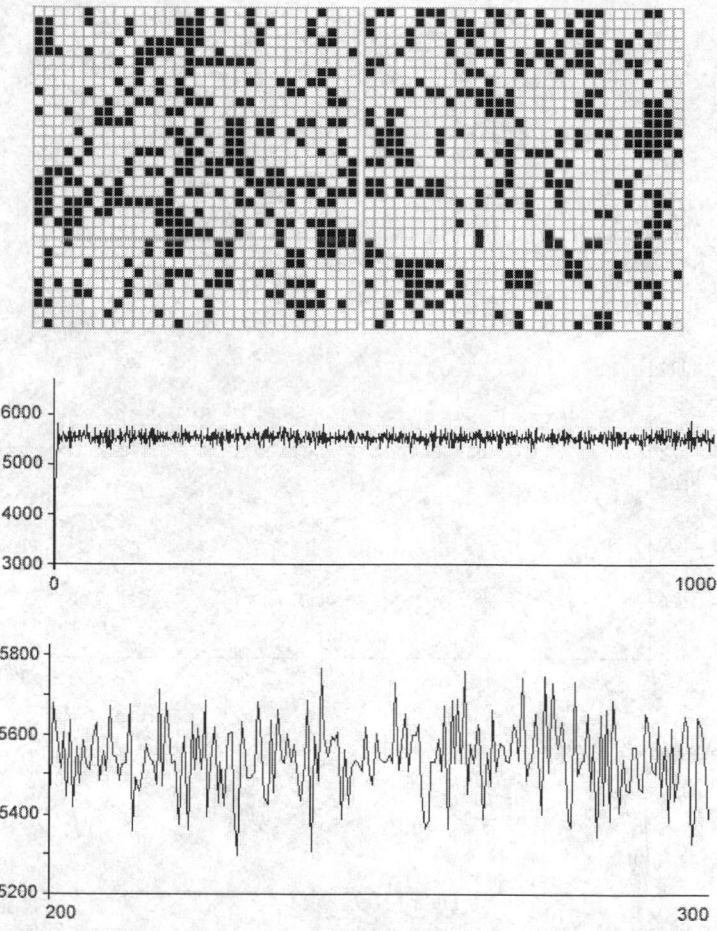

Fig. 4. Two configurations on the top at t = 10 (left) and t = 20 (right) for rule 1122. This rule exhibits a typical example of chaotic dynamics. Any emerging pattern is rapidly destroyed by the fluctuations of the environment. Note that only a part of the CA configurations is shown in these images. The bottom curves are time series for rule 1122 showing the number of "alive" cells (y-axis) for each generation (x-axis). After a short transitory phase (upper curve), the CA shows a typical chaotic behavior (lower curve).

most famous example of such dynamics, that is Conway's rule [8]. This CA has been extensively studied and considered as a metaphor of the emergence of life.

We verified that Complex CA, such as Conway's rule, are most likely to be found at a phase transition between ordered and chaotic CA. However, in contrast with Christopher Langton's continuous progression across the different regimes in his one dimensional CA study [10, 13], the behavioral structure of the Generalized-Life space is complex rather than linear. It is composed of homogeneous areas separated by smooth or sharp phase transitions [4]. The existence of these phase transitions between ordered and chaotic dynamics and between infinite and limited growth, as

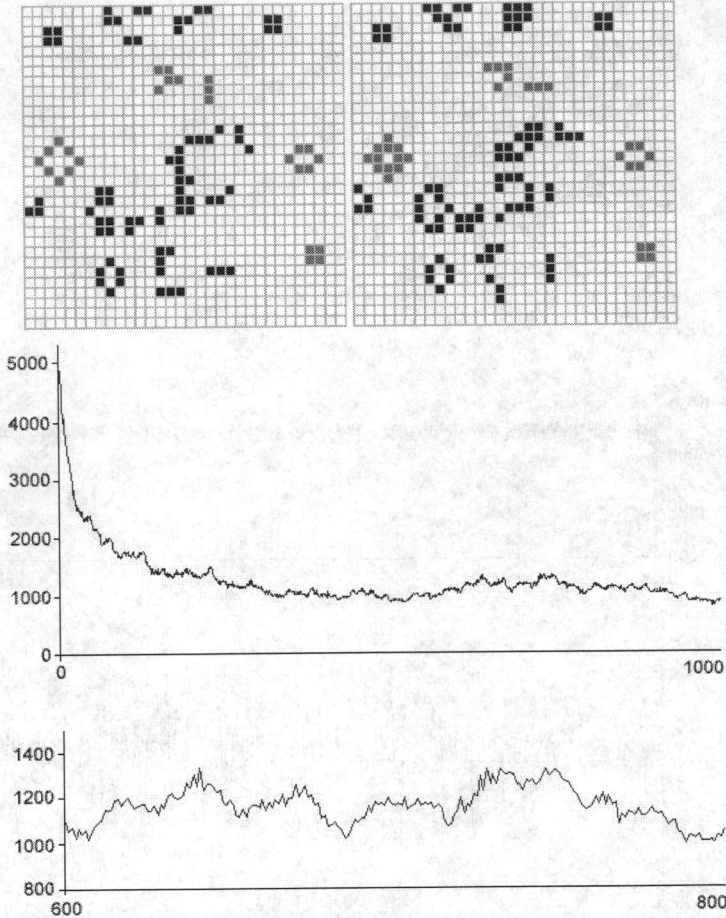

Fig. 5. On the top, two consecutive configurations beginning at t = 302 (left) for rule 2333. In gray, some typical fixed and periodic patterns. A glider is moving in the center at the top of the matrix. Note that only a part of the CA configurations is shown in these images. The two bottom curves are time series for rule 2333 showing the number of "alive" cells (y-axis) for each generation (x-axis). After a long transitory phase (upper curve), typically thousands of generations, the CA stabilizes with a majority of "dead" cells and some rare fixed and periodic patterns. The bottom curve shows a typical time series during the transitory phase. This type of curve seems to be a good signature of complex systems dynamics.

well as the location of complex CA in the vicinity of these transitions globally confirms the "edge of the chaos" hypothesis.

4 Discussion

The previous results show that after an initial variation, the behavior of the CA evolves toward a global statistically-stable state over time. This stabilization process leads to four cases: fixed for Class I, cyclic for Class II and chaotic for Class III. The complex Class IV CA seem to delay this stabilization thanks to the computational properties of the stabilization rule. Norman Packard argued first on the relationship between these dynamical behaviors and computational capability in CA [11]. Since, there has been a continuing effort to study this hypothesis [12–14]. Evidence has been shown that some class IV CA support universal computation in contrast with CA from other classes [15]. One property considered is the amount of memory required for producing these dynamics. The idea is that ordered or cyclic behaviors require memory in proportion to the nature and length of the pattern it repeats, while an ideal random behavior uses no memory to produce its information. In contrast, a theoretical Universal Turing Machine must have its storage space enlarged as needed. In the stabilization rule, the memory capability depends mainly on the E_b and E_h parameters. This rule (1) can be rewritten without any behavioral change:

$$if \ (S_t = 0 \ \& \ N_t \leq F_b \ \& \ N_t \geq F_h) \ S_{t+1} = 1 \tag{2a}$$

$$else \ if \ (S_t = 1 \ \& \ N_t \leq E_b \ \& \ N_t \geq E_h) \ S_{t+1} = S_t \tag{2b}$$

$$else \ S_{t+1} = 0. \tag{2c}$$

This reformulation shows that the (2b) part of the stabilization rule can memorize the previous state of the current cell while parts (2a) and (2c) are necessary for implementing basic logical functions such as AND, OR and NOT.

Another point of discussion is the obvious relationship between the *variation-stabilization* principle and the Darwinian *mutation and natural selection* principle. Our hypothesis is that biological evolution is a particular form of the variation-stabilization principle that occurs at a phase transition between ordered and chaotic regimes. In this framework, the DNA computing capacity seems to be a necessary advantage. More generally, evolution paradigms such as gradualism, punctuated equilibrium and natural drift, rather than being contradictory alternatives, could be interpreted as different forms of the variation-stabilization principle. As with our CA experiment, if the system is within the transition closer to the ordered regime, it leads to gradual evolution. If the system is within the transition closer to the chaotic regime, we have more randomness in the evolution dynamics. More studies must be conducted in order to validate this hypothesis.

The last important question we want to address here is the relationship of our study with the concept of "emergence," which is widely used in the sciences of complexity. Some effort has been made to formalize "emergence", but a consensus on a clear definition is still distant [16]. In our approach, a phenomenon at level L_n is *emergent* with respect to the lower-level L_{n-1} when it arises as a result of the interactions of L_{n-1} structural elements but not deductible from the properties of these elements. A well-known example in our CA experiment is the "glider": a periodic pattern discovered by John Conway. Gliders have been also found in other Class IV CA [4, 17] and there is evidence of their relationship with the universal computation capacity [15]. The fact is that it is practically impossible to deduce their global properties by just looking at their underlying structural elements (i.e. CA cells). The reason is a *causality break* due to the very high number of non-linear interactions between these structural elements. The observation of the two levels can not be made simultaneously. One can observe the global "patterns" level, but cannot observe in details the local behavior of the matrix cells at the same time. We think that this must be related to the *principle of complementarity* originally formulated by Niels Bohr in quantum mechanics [18]. In other words, the emergent properties represent the "wave-like interpretation" of the phenomenon which is not observable in the "particle-like interpretation" at a lower level. The *duality* of this two interpretations is necessary to have a complete description of the phenomenon.

5 Conclusion

We have introduced the variation-stabilization principle as a tentative outline for a general model of evolution in physical systems. We presented an experiment with the Generalized-Life CA that illustrate this principle and locate complex dynamics at the transition between ordered and chaotic regimes. We suggested that biological evolution takes its root in the evolution of physical systems as a particular case of the variation-stabilization principle that occurs at the transition phase between ordered and chaotic regimes. More research must be conducted to validate this theory. First, one could argue that the Generalized-Life CA represents a hierarchical system of only two levels: the cell matrix itself and the observed emergent patterns. This is true, and one research direction is to study the variation-stabilization principle in a more realistic and multi-level model such as the one we have used in cosmological experiments [19, 20]. A second complementary direction is to study the validity of the variation-stabilization principle in species dynamics using a virtual ecosystem such as Lifedrop [21]. Finally, another important direction is to study the relationship between the variation-stabilization principle and the second law of thermodynamics.

References

1. Darwin, E.: Zoonomia or, The Laws of Organic Life, 3rd edn., vol. 1. J. Johnson, London (1801)
2. Prigogine, I., Stengers, I.: Order out of Chaos, Man's New Dialogue with Nature. Bantam Books, New York (1984)
3. Smolin, L.: The Life in the Cosmos. Oxford University Press, New York (1997)
4. Magnier, M., Lattaud, C., Heudin, J.C.: Complexity Classes in the Two-dimensional Life Cellular Automata Subspace. Complex Systems 11, 419–436 (1997)
5. Wolfram, S.: Universality and Complexity in Cellular Automata. Physica D 10, 1–35 (1984)
6. Langton, C.G.: Computation at the edge of chaos: Phase transitions and emergent computation. Physica D 42, 12–37 (1990)
7. Heudin, J.C.: Modeling Complexity using Hierarchical Multi-Agent Systems. In: 6th International Workshop on Data Analysis in Astronomy, Erice (2007)
8. Gardner, M.: The fantastic combinations of John Conway's new solitaire game "Life". Scientific American 223, 120–123 (1970)
9. Bays, C.: Candidates for the game of life in three dimensions. Complex Systems 1, 373–400 (1987)
10. Langton, C.G.: Life at the edge of chaos. In: Artificial Life II, SFI Studies in the Sciences of Complexity, vol. 10, pp. 41–91. Addison Wesley, Reading (1991)

11. Packard, N.H.: Adaptation toward the edge of chaos. In: Kelso, J., Mandell, A., Shlesinger, M. (eds.) Dynamic Patterns in Complex Systems, pp. 239–301. World Scientific, Singapore (1988)
12. Langton, C.G.: Computation at the edge of chaos: Phase transitions and emergent computation. Physica D 42, 12–37 (1990)
13. Mitchell, M., Hraber, P.T., Crutchfield, J.P.: Revisiting the edge of chaos: Evolving cellular automata to perform computations. Complex Systems 7, 89–130 (1993)
14. Crutchfield, J.P., Packard, N.H.: Symbolic dynamics of noisy chaos. Physica D 7, 201 (1983)
15. Berlekamp, E., Conway, J.H., Guy, R.: Winning ways for your mathematical plays. Academic Press, London (1982)
16. Kubik, A.: Toward a Formalization of Emergence. Artificial Life 1, 41–65 (2003)
17. Heudin, J.-C.: A new candidate rule for the game of two dimensional life. Complex systems 10, 367–381 (1996)
18. Bohr, N.: Causality and Complementarity. Philosophy of Science 4, 289–298 (1937)
19. Heudin, J.-C.: Complexity classes in three-dimensional gravitational agents. In: Artificial Life VIII, pp. 9–13. MIT Press, Sydney (2002)
20. Torrel, J.C., Lattaud, C., Heudin, J.C.: Studying complex stellar dynamics using a hierarchical multi-agent model. In: 6th International Workshop on Data Analysis in Astronomy, Erice (2007)
21. Métivier, M., Lattaud, C., Heudin, J.-C.: A Stress-based Speciation Model in LifeDrop. In: 8th International Conference on Artificial Life, pp. 121–126. MIT Press, Sydney (2002)

The Nonlinear Mechanism of Phase Transition in Computer Networks*

Li Yi-Peng[1], Huang Yi-Hua[2], Wang Lei[1], and Ren Yong[1]

[1] Department of Electronic Engineering, Tsinghua University, P.R. China
yp-li05@mails.tsinghua.edu.cn
[2] Department of Electronic and Communication Engineering, SUN YAT-SEN University,
P.R. China

Abstract. In this paper, the nonlinear mechanism of phase transition in computer networks is analyzed, and a distributed proxy approach is introduced to improve network performance based on the two-dimensional coupling model. Theoretical analysis figures out that the nonlinear mechanism of router is the essential reason of network performance phase transition. Simulation results reveal that the extreme clustering characteristic of web access behavior gives arise to left-shift of phase transition critical compared with regular networks; after distributed proxy approach is employed, right-shift of the phase transition critical illustrates performance improvement. Finally, several important issues are mentioned.

Keywords: nonlinear mechanism, phase transition, distributed proxy.

1 Introduction

The rapid development of Internet applications brings more convenience for users collecting information, such as web network and typical P2P file sharing systems, which have become the majority traffic contributors to infrastructure. Millions of computers around the world attach to the Internet through many autonomous regional networks of routers, which interconnect through backbone networks of routers in a distributed, hierarchical fashion. All of these make Internet topology more complex and have negative impacts on infrastructure performance, e.g. poor transmission efficiency and network congestion.

To analyze and solve these problems, relative concepts and theories in statistic physics have already been introduced to characterize the collective dynamics of Internet traffic. Phase transition, for example, has been used to characterize the internet performance fluctuation, distinguishes the network performance status as free-flow and congestion. In the previous relevant research, Willinger et al. [1] provided a simple

---

* This work is supported in part by the National Nature Science Foundation of China (NSFC) grants 60672142, 60772053, the National Basic Research Program of China (973 Program) grants 2007CB307100 / 2007CB307105, and by the China Postdoctoral Science Foundation grants 20080430400.

J. Zhou (Ed.): Complex 2009, Part I, LNICST 4, pp. 560–568, 2009.
© ICST Institute for Computer Sciences, Social Informatics and Telecommunications Engineering 2009

physical explanation for self-organized criticality of Internet traffic, caused by multiple ON/OFF processes. Then Csabai [2] and Takayasu *et al.* [3] characterized Internet traffic statistics property according to spectral characteristics, shown by ping sequences. In 2007, Hu and Wang [4] have extended phases of network performance, they found the fundamental diagram of flow against density, hysteresis inside, and classified the traffic flow with four states: free flow, saturated flow, bistable and jammed. All this kind of researches supply valuable theoretical basis and simulation supports, which is likely to contribute to better network engineering and management.

Internet, as a typical huge complex system, different applications running on it and the corresponding user behavior has shown complicated statistic property. Generally, the resulting performance fluctuation can be dependent on two important factors. First, the existing nonlinear relationship between router input and output, router interactions due to protocol, make Internet infrastructure being dynamic status in processing millions of packets. The critical status, corresponding to the phase transition point, indicates the decline in network performance. Second, clustering characteristic of user access behaviors, shown by traffic analysis, aggravates the nonlinear relationship mentioned and interactions between routers, leading to deterioration of network performance. Unfortunately, the nonlinear mechanism, the essential reason of phase transition, has never been discussed [2-5] to our best knowledge. Hence, it is important to study the nonlinear dynamic mechanism of network phase transition, and introduce a relative improving approach.

In this paper, a novel two-dimensional coupling model is proposed to describe the huge and complex link relations in computer network. Then we provide theoretical analysis to the nonlinear mechanism of phase transition and the distributed proxy approach. Based on the novel model, simulations reveal the negative effect of the nonlinear mechanism and the network performance improvement brought by employing the distributed proxy.

The rest of the paper is organized as follows. In Section 2, we describe the novel two-dimensional couple model in detail. In Section 3, we give a theoretical analysis to the nonlinear mechanism and the distributed proxy approach, which is also introduced into the two-dimensional coupling model. Section 4 shows extensive simulations by using the modified model in two different network sizes. We conclude in Section 5.

2   Two-Dimensional Coupling Model

In 1999, two-dimensional cellular automation [6] is used in network modeling for the first time, which became a useful tool to characterize the internet collective dynamics behavior [7]. For one thing, its non-periodic boundary condition makes the spatial distribution of packets unbalanced, which means some central nodes more congested than the others. For another, this kind of model is always with the default configuration that nodes are homogenous [8], which can not describe the distributed access behavior of users accurately. To describe the huge and complex link relations in computer networks, we proposed a novel two-dimensional coupling model (see Fig.1).

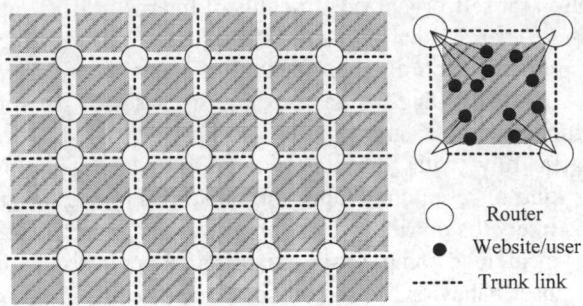

Fig. 1. Two-dimensional cellular automation model

In the two-dimensional coupling model, different websites or source providers and intensive users locate in shadow grid, connecting to the router nodes randomly. Routers, which follow the periodic boundary conditions [9], build up the backbone network, and route packets according to the shortest path routing strategy [10]. L is the size of the backbone network; means there are L routers in the horizontal and vertical direction of the backbone network.

The position of each router is denoted by a discrete space variable $\hat{r}$:

$$\hat{r} = i\hat{c}_x + j\hat{c}_y \tag{1}$$

Let K_r be the number of websites or users connecting to router $\hat{r}$, then the corresponding website or user position can be denoted as follow:

$$\hat{r}_k = i\hat{c}_x + j\hat{c}_y + k\hat{c}_z \tag{2}$$

where $\hat{c}_x$, $\hat{c}_y$, $\hat{c}_z$ are Cartesian unit vectors, and $i, j = 1, \cdots, L$, $k = 1, \cdots, K_r$.

The dynamics behavior of users in the model is governed by the parallel update with discrete time step. During the progress of model evolution, users create new packets with zero life time, which are forwarded by routers to some selected destination. All routers work as FIFO (First Input First Output). During each time step, the processing of packet transmission is shown as follows.

(1) Each user node sends one packet with probability p independently;
(2) According to FIFO rule, each router routes one packet in its buffer to destination by one hop, the next hop router queue length increases by 1, the routed packet lifetime increases by 1;
(3) If the packet is routed to the destination, the relative record ends.

At the time step k, if queue length of router $\hat{r}$ is $q(\hat{r}, k)$, the sum queue length of all routers is

$$Q(k) = \sum_i q(\hat{r}_i, k), \; i = 1, \cdots, L^2 \tag{3}$$

The router's buffer size is set to

$$B_{Router} = 2 \times \sum_i K_{\hat{r}_i}, \; i = 1, \cdots, L^2 \tag{4}$$

In the case of buffer overflowing, routers will drop packets until there is free space in the buffer.

3 The Nonlinear Mechanism of Phase Transition and Improvement

Research in [7] shows that the phase transition of network performance is due to the nonlinear interaction of routers. And our preliminary studies prove that extreme clustering characteristic of web access behaviors gives arise to left-shift of phase transition point [11]. But all these studies are still limited to large-scale or medium-scale.

Consider a simple example as Fig.2 shows.

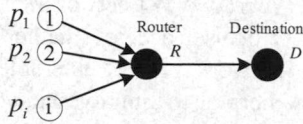

Fig. 2. Simple packet forwarding system

In this simple system, source nodes $1, \cdots, i$ send packets to destination D with independent probability p_i, which will increase from 0 to 1 gradually. Router R, queue length noted by Q_R, can output one packet per time step.

At the time step k, the expectations input and output of R in the sense of probability are:

$$E\left[I_{R,k} \right] = \sum_i p_i \tag{5}$$

$$\begin{cases} E\left[O_{R,k+1} \middle| Q_{R,k} = 0 \right] = 1 - \prod_i (1 - p_i) \\ E\left[O_{R,k+1} \middle| Q_{R,k} \geq 1 \right] = 1 \end{cases} \tag{6}$$

Equ.6 figures out that the current time output of the router only depends on its queue length at the previous time step.

At the beginning, total packets input of the router is less than its maximum output, all the incoming packets will be forwarded real time. So, the system appears linearly and

$Q_R = 0$. However, with the packet sent probability increasing, input packets become greater than the maximum output capacity. Router shows nonlinear characteristics between its input and output, which means packets begin to queue in its buffer and Q_R increasing.

There can be a little interval Δt, during which all the source nodes have the same unchanged packet sent probability $p_{\Delta t}$. The total increase of Q_R is

$$Q_{R,\Delta t} = \max\left\{\sum\left(E\left[O_{R,\Delta t}\right] - E\left[I_{R,\Delta t}\right]\right), 0\right\} \tag{7}$$

When the packet queue length grows beyond the buffer capacity, the router will overflow, resulting in packets dropped. And the lifetimes of the rest packets will increase because of queuing in the buffer.

The nonlinear relation between input and output described above makes the router congested, which further affects packet forwarding of surrounding routers. In other words, the congestion will diffuse as packet transmission over the whole network. Finally, the entire network will be too congested to work efficiently.

In the actual World Wide Web (WWW) network, each webpage has different accessed frequency. The majority of user accesses gather in a few websites with high in-degree, which makes the network more vulnerable to the negative impact of nonlinear. Although it is very normal to improve network congestion by increasing trunk link bandwidth and enhancing router processing capacity in practice, the collapse of web servers still can not be avoid because of the nonlinear existing.

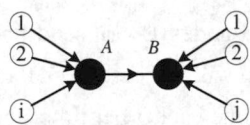

Fig. 3. Reducing Impact of nonlinear using distributed proxy

Fig.3 shows an extreme example without impact of nonlinear using distributed proxy. Nodes $1, \cdots, i, j$ send packets as described before. When A works as a router, $Q_A \neq 0$, B processes the incoming packet, $Q_B = 0$. If A is deployed to be proxy of B, it is obvious that $Q_A = 0$. So, we can see that the network can eliminate the negative impact of nonlinear partly by appropriate distributed proxy.

In the previous novel two-dimensional coupling model, the proxy approach can be explained in Fig.4. Node B in shadow grid 9 processes all the user access packets from the whole network, originally. Then, node A in shadow grid 5 is deployed as a distributed proxy of B. According to the routing strategy described in section 2, all user access packets from shadow grids 1, 2, 4 and part of packets from 3, 5, 7 will be redirected to the proxy node A. Packets from 6, 8, 9 will still be processed by node B. The

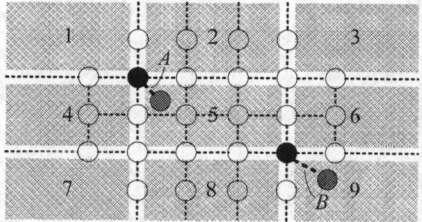

Fig. 4. Modified Two-dimensional coupling network model

deployed proxy node A, not only reduces the queue length of partial routers, but also makes all the user packets, which are original routed to B, spatially balanced in the entire network.

Given the cost constraints and other factors in actual network, it is unrealistic to eliminate the negative impact of nonlinear to the network performance completely. However, it is very feasible to reduce this kind of impact by deploying proxy node appropriately. Based on the novel two-dimensional coupling model, distributed proxy approach will be taken on the simulation verification in section 4.

4   Simulation Results and Analysis

The analysis of actual traffic in Tsinghua CERNET reveals the fact that 10% source nodes attracts 80% user access in WWW network, shown in fig.5.

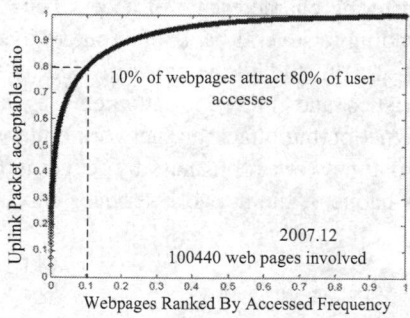

Fig. 5. Clustering Characteristic of User Access Behavior

That is the clustering characteristic of user access behaviors in actual network. The following simulations compare the performance changing of actual network, regular network and network with distributed proxy. In regular network, each node has the same accessed probability. The network model sizes are $L = 10, 20$. In each shadow grid there are 15 users or source providers. After 1000 time steps' evolution, the results are shown in figures 6, 7 and 8. Vertical solid lines indicate the phase transition critical, corresponding to the critical packets injected probability.

Fig. 6 and 7 are statistical results of queue length and packet lifetime, respectively. Clustering characteristic of user access behavior aggravates the negative affects of nonlinear to network, which arise to left-shift of phase transition critical, means the actual network more congested than the regular network.

The performance deterioration makes the network can not accept so many user accesses. Packets are dropped due to overflow of router buffer. In other words, the websites won't be capable of dealing with so many user access demands, and users can not receive requested information from the server. It is very difficult to guarantee the quality of services (QoS).

The power-law distribution of node degree in WWW network means there are a few nodes with high degree, called rich node here. According to the statistical ranked results shown in fig.5, the top 5% rich nodes are selected and deployed $c \lg k$ ($0 \le c \le 1$, k represents the node degree) distributed proxies at random positions in the model respectively. To some extent, these distributed proxy nodes balances the network traffic load, leading to the increasing of network critical injection rate, as shown in fig.6, 7 and 8. Right-shift of the phase transition critical illustrates an obvious performance improvement.

In this case, packets are redirected to the nearest proxy, with routing path shortened. Packets, originally gathered in one backbone trunk, are processed by several different routers connected to proxies. The whole network makes full use of buffer and process capacity of proxies, reducing buffer overflows and packet loss. The faster packets routed to the destination, the fewer life time they end with. Obviously, network performance has been greatly improved.

The routing distribution of data packets influences the balance of overall network traffic. With packet destination selected randomly, the network traffic appears balanced. Unfortunately, clustering characteristic of access behavior breaks up this balance in actual internet, leading to several backbone routers overloaded and inefficient packet transmission. Li et al. [12] studied the network security from the perspective of overall network characteristics, and figured it out when user accesses some important nodes not for useful information but attacking, network collapses rapidly. Distributed proxy approach can improve network robustness by reducing the risk of major nodes being attacked. Even this happens, the network services can be reconstructed and recovered rapidly.

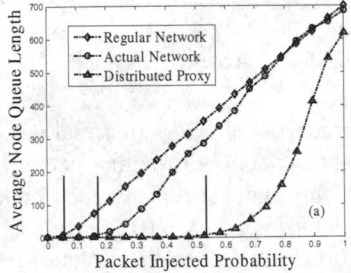 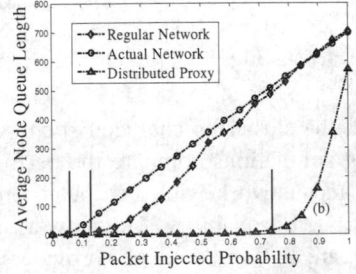

Fig. 6. Average node queue length statistical results in different network size. (a):L=10 (b):L=20.

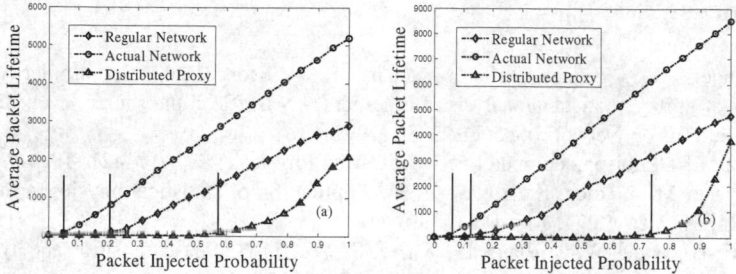

Fig. 7. Average packet lifetime statistical results in different network size. (a):L=10 (b):L=20.

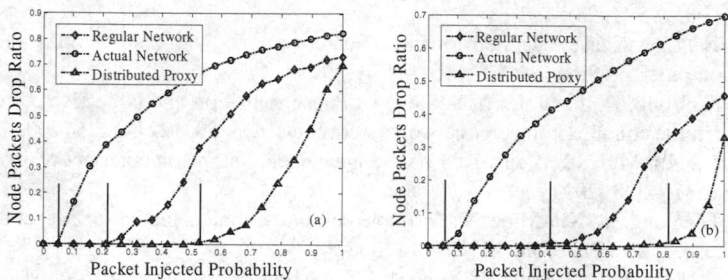

Fig. 8. Node packets dropped ratio statistical results in different network size. (a):L=10 (b):L=20.

5 Conclusions

In this paper, we analyze the nonlinear relationship between input and output of router node in detail, reveal the nonlinear mechanism of network performance phase transition and propose the distributed proxy approach to improve the network performance. Through theoretical analysis and simulation verification based on the novel two-dimensional coupling network model, we conclude as follows: 1) the queue length of router, packet life time and packet dropped ratio all arise phase transition with nonlinear existing; 2) compared to packet destination selected randomly in regular network, the extreme clustering characteristic of access behaviors in WWW networks gives arise to left-shift of phase transition critical, demonstrating deterioration of internet performance; 3) to solve this problem, distributed proxy approach is proposed to deploy several proxies, shortens the packet routing path. As a result, right-shift of the phase transition critical illustrates an improved performance of the network.

This paper is an important part of series studies. Research on various types of network phase transitions and the relationship to measure of proxies, the quantitative characterization of network nonlinear, optimization of network modeling and the bound of network performance, are all challenging research issues.

References

1. Willingers, W., Taqqus, M.S., Shennans, R., Wilson, D.V.: Self-similarity through high-variability: statistical analysis of Ethernet LAN traffic at the source level. IEEE/ACM Transactions on Networking 5(1), 71–86 (1997)
2. Csabai, I.: 1/f noise in computer network traffic. Physica A 27, 417–421 (1994)
3. Takayasu, M., Fukuda, K., Takayasu, H.: Application of statistical physics to the Internet traffics. Physica A 274, 140–148 (1999)
4. Hu, M.-B., Wang, W.-X., Jiang, R., Wu, Q.-S., Wu, Y.-H.: Phase transition and hysteresis in scale-free network traffic. Phys. Rev. E 75, 036102 (2007)
5. Tretyakov, A.Y., Takayasu, M., Takayasu, H.: Phase transition in a computer network model. Physica A 253, 315–322 (1998)
6. Ohira, T., Sawatari, R.: Phase transition in a computer network traffic model. Phys. Rev. E 58, 193–195 (1998)
7. Fuks, H., Lawniczak, A.T.: Performance of data networks with random links. Mathematics and Computers in Simulation 51, 101–117 (1999)
8. Jian, Y., Yong, R., Feng, L., Xiu-Ming, S.: Phase transition and collective correlation behavior in the complex computer network. Acta Phys. Sin. 50, 1221–1225 (2001)
9. Feng, L., Xiu-Ming, S., Yong, R.: Long-range correlation in computer network. Acta Phys. Sin. 53, 373–378 (2004)
10. Feng, L., Yong, R., Xiu-Ming, S.: A simple cellular automata model for packet transport in the Internet. Acta Phys. Sin. 51, 1175–1180 (2002)
11. Lei, W., Shu-Hua, Z., Jian, Y., Yong, R., Xiu-Ming, S.: Influence of virtual networks to internet collective behavior. Acta Phys. Sin. 56, 36–42 (2007)
12. Ying, L., Xiu-Ming, S., Yong, R.: Dynamic properties of epidemic spreading on finite size complex networks. Chinese Physics 14, 2153–2157 (2005)

The Evolution of ICT Markets:
An Agent-Based Model on Complex Networks

Liangjie Zhao[1], Bangtao Wu[1], Zhong Chen[1], and Li Li[1,2]

[1] Antai College of Economics and Management, Shanghai Jiaotong University,
Shanghai, 200052, China
zhaolj1981@gmail.com
[2] School of Education, Shanghai Normal University, Shanghai 200234, China

Abstract. Information and communication technology (ICT) products exhibit positive network effects.The dynamic process of ICT markets evolution has two intrinsic characteristics: (1) customers are influenced by each others' purchasing decision; (2) customers are intelligent agents with bounded rationality.Guided by complex systems theory, we construct an agent-based model and simulate on complex networks to examine how the evolution can arise from the interaction of customers, which occur when they make expectations about the future installed base of a product by the fraction of neighbors who are using the same product in his personal network.We demonstrate that network effects play an important role in the evolution of markets share, which make even an inferior product can dominate the whole market.We also find that the intensity of customers' communication can influence whether the best initial strategy for firms is to improve product quality or expand their installed base.

Keywords: information and communication technology, evolution of market, diffusion of innovations, complex networks, network effects.

1 Introduction

With the development of information and communication technology (ICT), ICT products have become more and more important in our life. The market of ICT is a widely concept. It includes not only PC market, but also the market of telecommunications equipment, customer electronics, and electronic components. According to IDC's estimation, the global ICT market investment has become 2.3 trillion dollars in 2006.

ICT products exhibit positive network effects, which mean the utility of these products will increase with the total number of users or the amount and variety of complementary goods [1]. Economic literature commonly agree that the influence of network effects for the adoption of ICT products, will lead to evolution phenomena including positive feedback, critical mass, compatibility, standardization, lock-in, path dependence and inefficiency [2,3]. Traditional economics approaches of network effects theory, which study on the evolution of ICT markets, can be classified into two kinds of models. On the one hand, the primary

J. Zhou (Ed.): Complex 2009, Part I, LNICST 4, pp. 569–579, 2009.

goal of theoretical models is an analysis of the competition strategies, such as installed base strategy [4], compatibility strategy [5] or pricing strategy [6]. On the other hand, many scholars use empirical approaches (e.g. hazard model, hedonic price model) to estimate the effect of direct or indirect network strength on the of evolution of markets [7,8].

However, by focusing on the effects of supply side policies, traditional theories ignore the impact of demand side on the evolution of ICT markets. The evolution of market is a dynamic process, in which all customers make their collective purchasing decisions. When customers choose between different products, they face a coordination problem [9]. For instance, customers may choose the telecom operator with larger communications network. Such network will bring more value to them, especially when their family members, friends and business partners also join the same network. In ICT markets, customers are influenced by each others' purchasing decision, so there exist interactions of potential adopters within their socio-economical system [10].

Meanwhile, Customers are intelligent agents with bounded rationality, which means they can make expectations about the market share of products. Traditional economics assume customers are perfect rationality. However, the behavior of an agent in reality is "nearly optimal with respect to its goals as its resources will allow" [11]. Because of imperfect information about the market, customers cannot make fulfilled-expectations about the future size of a network which correct with the equilibria of market. In fact, there exist some situations such as local bias-small clusters of users who adopt a product which does not have the largest installed base and not dominate in the whole market [12]. In addition, if the product cannot come up to the expected network benefits of the potential user, it may result in negative feedback. WAP is one such case [13]. The potential users are unwilling to pay for this product, so the real network benefits do not increase and this result in the users begin to abandon the product. Finally, the product will fail because it unable to get enough installed base to overcome the problem of critical mass.

Aiming at better understands of the evolution process of ICT markets, we construct an agent-based simulation model. Our main hypothesis is customer's purchasing decision is sometimes influenced more by the personal network of his or her acquaintances, which also called as "local feedback effects". It may be reasonable to make such assumption, since opinions and choices of family and friends play a significant role in a customer's selection of ICT products, which also called as "opinion leader" and "word-of-mouth". After knowing the early adopters have been satisfied, potential users are less wary of ICT products. Some empirical studies also support our assumption [14,15].

We also use complex networks to model the interactions between customers. [16] points out that complex networks give us a more direct-viewing understanding of the emergence of complex systems, i.e., the behaviors of whole systems. Recent years, theories of complex networks have also been applied into management science. [17] use small-world network and spatial dimension of sales data to predict the success of new product. [12] argue the validity of "Winner-Take-All" hypothesis, which comes from traditional theory of network effects, may depend on the topology of complex networks. We believe that the agent-based simulation models on complex network is a good tool for us to explore the complex social and economic systemsespecially the innovation and opinion dynamics.

The organization of this paper is as follows. The second section provides the agent-based model used for analysis. The third section shows the results of our simulations.Then, we discuss the main findings and their implications.

2   The Model

We model the dynamics of market evolution where two incompatible ICT products compete with each other. Following [18], we assume the total utility of a customer for one single ICT product is constituted by two parts: intrinsic utility and network utility. Intrinsic utility reflects the quality of a product. Meanwhile, network utility comes from the numbers of customers purchasing the same product. However, under the condition of imperfect information, customers do not know about the real market share of each product exactly. So we also assume, with bounded rationality, they make expectations about the market share of each product by the fractions of its neighbors who have already adopt separately.

2.1   Basic Model

Considered a social network size of $N = 1000$ nodes, i.e., there are one thousand customers. The total utility of purchasing product $j\{j = A or B\}$ for individual i at time t is given by

$$U_{it}^j = r_i + q^j + \beta^j N D_{i(t-1)}^j \tag{1}$$

where r_i is customer i 'preference for quality, q^j represents quality, β^j measures the strength of network effects,and $D_{i(t-1)}^j$ is the fractions of customer i'neighbors that have already adopt.

Following assumptions by many prior scholars (e.g., [1,12]), we suppose $r_i < 0$ for most of customers.Meanwhile, it is distributed normally, where mean μ and variance σ^2. Quality q^j reflects theintrinsic utility of products. Further, network utility depends not only on the expected network size $N D_{i(t-1)}^j$,which is equivalent to installed base[1], but also network strength, stemming from some characteristics of customers, such as personal interests, product loyalty[7].

2.2   Adoption and Repurchase Processes

we consider the discrete version of the continuous dynamics, i.e. the purchase decisions of customers are made in a sequential order, so in every period only one agent is chosen to revise his decision.

---

[1] Like [19], we replace installed base by market share.

By obeying dynamic preference rules as follows, Each customer makes a decision whether to buy or repurchase one unit of either product A or B.The dynamics of the adoption process is as follow.

(1) If both $U_{it}^A < 0$ and $U_{it}^B < 0$, customer will choose nothing;
(2) If $U_{it}^A > U_{it}^B > 0$,customer will choose product A ;
(3) If $U_{it}^B > U_{it}^A > 0$, customer will choose product B ;
(4) If both $U_{it}^A > 0$ and $U_{it}^B > 0$, customer will choose a product by probability based on the different total utility of different product:

$$Prob(choice = j) = \frac{U_{it}^j}{U_{it}^A + U_{it}^B} \tag{2}$$

From (2), we can see the larger total utility of a product, the larger probability it will be chosen by customers.

2.3 Simulation Design

Considering the condition when two incompatible ICT products are introduced to the market simultaneously, we use two different average connectivity scale-free networks to conduct simulations. The average connectivity of social networks $< k >$ (approximately $< k >$ neighbors per node) reflects the intensity of communication between customers [20], so we want to explore how the intensity of customers' communication influence the evolution of ICT markets. In order to comparative analysis the different emergence from the diverse average connectivity of network, we keep those two SF networks have the same numbers of nodes. Further, we suppose there have some seeds at the beginning of the simulations, i.e., adopters who have already own one product at first period. Those adopters, similar to "innovators" defined by [21], can be seen as initial market share or installed base of each product.

In particular we set $< k >= 6, 16$, $\mu = -30$ and $\sigma^2 = 5$.The degree distribution of network which $< k >= 6$ follows $P(k) = k^{-2.16}$.The other one with high average node degree $< k >= 16$ follows $P(k) = k^{-2.86}$ .All results are the average value of 100 independent simulation runs. The iteration of each run is 10000 periods.

3 The Results

First of all, we study on the issue whether a firm chooses between investment in initial installed base or in initial quality of its product in the initial periods of market. Because this issue concern about the business strategy of firms whether could dominate in market finally. Traditional views of network effects theory emphasize the importance of installed base products, since there exists a critical mass point. If the installed base of a product exceeds this turning point, the sales of this product will increase quickly by positive network effects. So investment in initial installed base strategy is also called as Get-Big-Fast strategy. However, managers could also use improving quality of its product strategy to compete. Meanwhile, we can see a lot of business cases in which the superior technology win the market finally although it doesn't have bigger installed base than the old inferior technology[2]. Those cases show quality drives the success of ICT products.

So there are two different strategies that firms may face: *quality advantage strategy* (QAS) or *installed base advantage strategy*(IBAS). In order to comparison, we construct a simulation by assuming both initial market shares of product A and B are 5%, i.e., $IMSA = IMSB = 0.05$, both qualities of product A and B are 20, i.e., $q^A = q^B = 20$. Further, we consider two different situations. One is when network strength is weak. i.e., $\beta^A = \beta^B = 0.5 < 1$; another is when network strength is strong. i.e.,$\beta^A = \beta^B = 2 > 1$.

We keep the determinants of product B constant, since we focus on the effects of quality advantage vs. installed base advantage on final market share of product A. Then, we make a paired comparison between initial quality advantage and initial installed base advantage at 20%, 40%, 60%, 80%, 100% level separately. For example, at 20% level, if the firm adopts an initial quality advantage strategy, it improves quality of product A .So the quality become

$$q^A = 20 * (1 + 20\%) = 24 \tag{3}$$

On the contrary, if the firm adopts an initial installed base advantage strategy, it promotes initial market share of product A by some marketing methods such as free sampling.So the initial installed base become

$$IMSA = 0.05 * (1 + 20\%) = 0.06 \tag{4}$$

Figure 1 and Figure 2 show different scenarios. When the intensity of customers' communication is low ($< k >= 6$), QAS is a dominant strategy whether network strength is weak or strong.however, when the intensity of customers' communication is high ($< k >= 16$), which strategy is better depend on the strength of network effects in the market. on the one hand, QAS is better strategy than IBAS when network strength is weak. On the contrary, IBAS strategy is a better choice for firms when network effects is strong.To sum up, we use table 1 to depict the choice of strategy when firm investment in installed base or quality of its product at the initial periods of market.

So our results indicate that whether intrinsic utility (e.g., quality) or network utility (e.g., installed base) drives the success of ICT products may depend on the intensity of communications between customers and the strength of network effects in the market.

Why our findings contradict with the conclusions of traditional economics theory? To answer this question, we examine local bias in four situations. Recall that our basic assumption is that agents are bounded rationality. Without perfect

---

[2] Yet, in fact there also may result in a bad situation in where the inferior technology will still dominate the whole market. The most famous case is QWERTY keyboard against Dvoak [22]. We will discuss this phenomenon next.

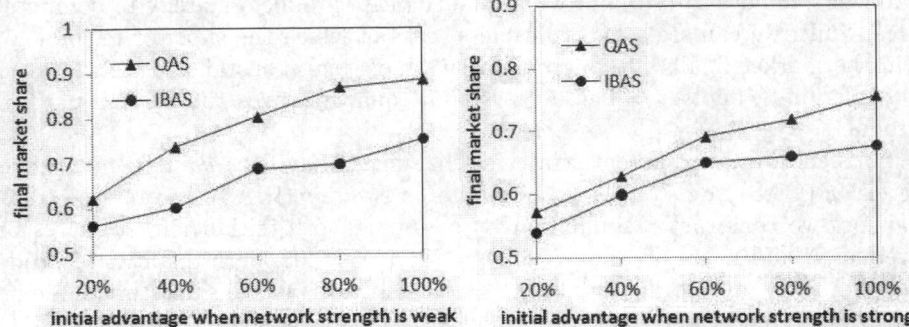

Fig. 1. The effects of initial quality advantage vs. the effects of initial installed base advantage on final market share of product A when the intensity of customers' communication is low

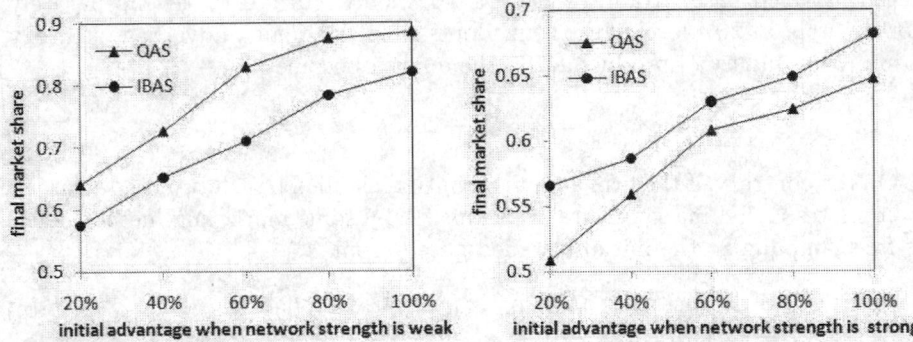

Fig. 2. The effects of initial quality advantage vs. the effects of initial installed base advantage on final market share of product A when the intensity of customers' communication is high

Table 1. Business strategy at the initial periods of market

Network Strength		Intensity of customers' communication	
		Low	High
	Weak	quality strategy	quality strategy
	Strong	quality strategy	Installed base strategy

information, they cannot know the market share of each product completely. So they make expectations about the market by observing the choices of their acquaintances. Like [12] We measure the local bias as follow:

$$Localbias = \sum_{i=i}^{N} \frac{\left| ((s_{it}^A - s_{it}^B) - (s_t^A - s_t^B)) \right|}{N} \tag{5}$$

Where s_{it}^A and s_{it}^B are the shares of products A and B in customer i's neighbor network at t period, respectively, and s_t^A and s_t^B are the market shares of products A and B in the whole market at t period, respectively. From the measurement above, we can know that there is no local bias if it is zero.

Figure 3 shows the different evolution of local bias under four scenarios. Table1 gives the details of parameter values for simulations. The simulation results show the clusters or communities of social networks seem to let customers make locally based choice. As a result, the level of local bias increases at first and reach the steady state finally. When the initial advantage (installed base or quality) is small, i.e. scenario 1 or 2, the level of local bias is much higher. On the contrary, i.e. scenario 3 or 4, the level of local bias will lower because the final market is a dominant market, which means most of market has been occupied by single product.

Then we turn to another question which has been discussed in economies hotly. Under which condition, the market will become an inefficient market.[9] point out that there are two kinds of inefficient markets. One is excess inertia, i.e., customers are reluctant to adopt a new product with incompatible superior technology, although its quality is much better than the old one. Another is insufficient friction, i.e., customers always favor a new product with incompatible superior technology.

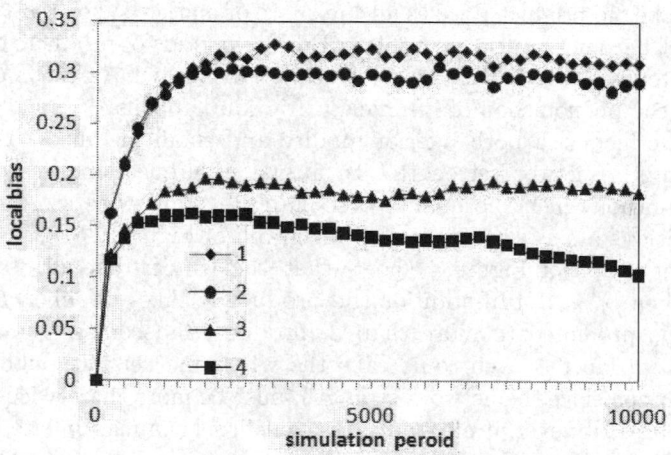

Fig. 3. Local bias over time

Table 2. Simulation parameter values for Local bias

scenario	q^A	q^B	IMSA	IMSB	β^A	β^B	$<k>$
1	20	20	0.06	0.05	2	2	6
2	24	20	0.05	0.05	0.5	0.5	6
3	20	20	0.1	0.05	2	2	16
4	40	20	0.1	0.05	0.5	0.5	16

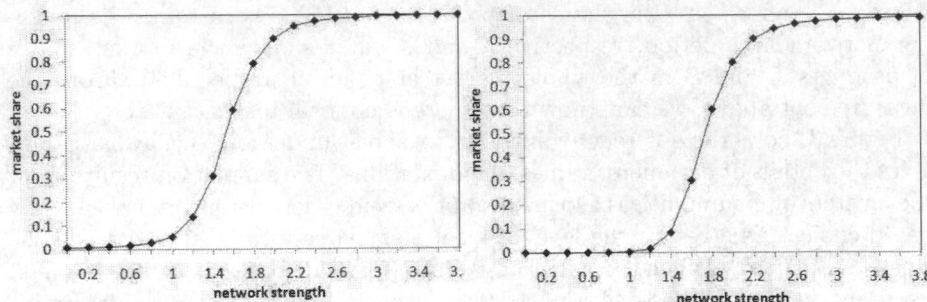

Fig. 4. Left graph:the effects of superior product A's network strength on the evolution of market. Right graph:the effects of inferior product A's network strength on the evolution of market.

In this simulation, we assume the old product B dominates the whole market,so its market share is 95% and network strength $\beta^B = 1.2$.Then a new superior product A is introduced, which has 5% initial market share. The quality of product A is twice as high as the old one. i.e., $q^A = 2 * q^B = 40$. It is to explore whether product A can lead to success by promoting its network strength β^A.The left graph of Figure 4 shows that the market will be excess inertia when $\beta^A < 0.6$. The old product B takes all the share of market.When $0.6 < \beta^A < 2.8$, final market becomes multiple equilibria.Both two kinds of products could coexist at the steady state. When $\beta^A > 2.8$, the new product A always corners the market, so the phenomenon of insufficient friction happens.

Can a product with both inferior quality and small installed base compete by improving its network strength? To answer this question, we consider the extreme situation when the quality of B is more than twice as A. i.e.,.While, other conditions are as same as in the left graph of Figure 4.

The right graph of Figure 4 shows that the winner-take-all of product B happens when $\beta^A < 1.4$.In addition,the product A takes off when $\beta^A > 1.4$.If β^A could be promoted to more than 2, product A becomes dominant.When $\beta^A >= 3.4$, product A even could take the whole market, although this event may not happen since the network strength must be more than twice as much as the competitor. However, it also indicates that firms can manipulate its network strength to win the market.This is similar to empirical study[9].Although we don't report the results when average connectivity is 6, the simulations show there are not qualitatively different.

4   Conclusions

We have developed simulation models to examine the effects of a firm's initial advantage, the intensity of customers' communication, the customer decision-making process and the network strength of products on the evolution of ICT markets. We found that when the intensity of customers' communication is low, quality advantage strategy is always better than installed base advantage strategy whether network strength is weak or strong. However, when the intensity of customers' communication is high, initial installed base advantage strategy is better than initial quality advantage strategy only when network strength is strong. This conclusion gives insight into managerial implications that whether managers choose quality advantage strategy or installed base advantage strategy at the beginning of the market may depend on the intensity of customers' communication and network strength. Traditional theory emphasizes the importance of installed base since firms can get big fast by the positive network effects. Our model suggests that this strategy may be only valid when both network connectivity and network strength are high. So managers should do some marketing research to measure network effects of the market and the intensity of customers' communication.

In addition, we found that whether market evolves into an inefficient market depends on the strength of network effects. When the network strength of superior product is too low, the market will be excess inertia. On the contrary, it may result into insufficient friction. Traditional theories use the heterogeneity of customers' preferences to explain why Apple's Macintosh still survives, although it is incompatible with the dominantly Wintel architecture. Our study suggests that there may be another two reasons. On the one hand, the network strength of PC market is not high. On the other hand, Apple's quality is very good. In fact, according to the American Customer Satisfaction Index (ACSI) published by the University of Michigan, Apple has kept customer satisfaction at the highest level in PC market since 1994, the first year of the index introducing [23].

Further, our simulation also gives implications that firms can beat competitors by increase its network strength. This is due to asymmetries of network effects between different products. In 16-bit home video game market, Nintendo beats Sega by higher network strength, although Sega has larger installed base [7].

However, our model assumes all of customers make expectations about the market by the fractions of adopters of different products in their acquaintances. In fact, there are also many customers know the market clearly through media's reports. So it needs to separate the whole customers. On the other hand, how the topology of complex networks affects the diffusion of innovations, such as degree distribution, connectivity, still needs to examine. We will explore those issues in future.

5   Sensitive Analysis

In order to test the validity of our results, we did a sensitivity analysis. First, we use WS networks[3] ($< k >= 6, 16$) to instead of SF networks for simulation. Then, we change σ^2, the variance of customers' preferences, from 0 to 25. Although the simulation results have some quantitative variance, the qualitative of our conclusions have no change.

---

[3] The probability of rewiring connections between nodes is 0.05.

Acknowledgments

This work was supported by National Natural Science Foundation of China under Grant No.70671070.

References

1. Katz, M.L., Shapiro, C.: Network externalities, competition, and compatibility. The American Economic Review 75(3), 424–440 (1985)
2. Farrell, J., Saloner, G.: Standardization, Compatibility, and Innovation. Rand Journal of Economics 16, 70–83 (1985)
3. Wendt, T.W.O., Westarp, F.v.: Reconsidering Network Effect Theory. In: 8th European Conference on Information Systems (2000)
4. Farrell, J., Saloner, G.: Installed Base and Compatibility: Innovation, Product Preannouncements, and Predation. The American Economic Review 76(5), 940–955 (1986)
5. Gandal, N.: Compatibility, Standardization,Network Effects: Some Policy. Implications, Oxford Review of Economic Policy 18, 80–91 (2002)
6. Economides, N., Viard, V.B.: Pricing of Complementary Goods and Network Effects, working paper, New York University (2002)
7. Shankar, V., Bayus, B.L.: Network Effects and Competition: An Empirical Analysis of the Video Game Industry. Strategic Management Journal 24(4), 375–394 (2003)
8. Stremersch, S., Tellis, G.J., Franses, P.H., Binken, J.L.G.: Indirect network effects in new product growth. Journal of Marketing 71(3), 52–74 (2007)
9. Katz, M.L., Shapiro, C.: Systems Competition and Network Effects. Journal of Economic Perspectives 8, 93–115 (1994)
10. Schoder, D.: Forecasting the success of telecommunication services in the presence of network effects. Information Economics and Policy 12, 181–200 (2002)
11. Simon, H.A.: Models of Man. In: Social and Rational. John Wiley & Sons, Chichester (1957)
12. Lee, E., Lee, J., Lee, J.: Reconsideration of the Winner-Take-All Hypothesis: Complex Networks and Local Bias. Management Science 52(12), 1838–1848 (2006)
13. De Marez Lieven, S.B., Verleye Gino, B.M.: ICT-innovations today: making traditional diffusion patterns obsolete, and preliminary insight of increased importance. Telematics and Informatic 21, 235–260 (2004)
14. Goolsbee, A.: Evidence on Learning and Network Externalities in the Diffusion of Home Computers, Working paper, University of Chicago, GSB (2002)
15. Lim, B.L., Choi, M., Park, M.C.: The late take-off phenomenon in the diffusion of telecommunication services: network effect and the critical mass. Information Economics and Policy 15, 537–557 (2003)
16. Strogatz, S.H.: Exploring complex networks. Nature 401(8), 268–276 (2001)
17. Garber, T., Goldenberg, J., Libai, B., Muller, E.: From density to destiny: Using spatial dimension of sales data for early prediction of new product success. Marketing Science 23(3), 419–428 (2004)
18. Arthur, W.B.: Competing technologies, increasing returns, and lock-in by historical events. The Economic Journal 99, 116–131 (1989)
19. Clark, B.H., Chatterjee, S.: The evolution of dominant market shares: the role of network effects. Journal of Marketing Theory and Practice 7(2), 83–95 (1999)

20. Wendt, O., Westarp, F.V.: Determinants of Diffusion in Network Effect Markets. Working paper. J. W. Goethe-University, Germany (1999)
21. Rogers, E.M.: Diffusion of Innovations, 3rd edn. The Free Press, New York (1985)
22. David, P.A.: Clio and the economics of QWERTY. American Economic Review, Papers and Proceedings 75, 332–337 (1985)
23. Guglielmo, C., Thaw, J.: Dell Falls, Apple Gains in Customer Satisfaction, Survey Says. Bloomberg (retrieved November 14 - August 16) (2007), http://www.bloomberg.com/apps/news?pid=10000103sid=ahi5j9WcL8VMrefer=us

The Effects of Link and Node Capacity on Traffic Dynamics in Weighted Scale-Free Networks

M.B. Hu[1], R. Jiang[1], Y.H. Wu[2], and Q.S. Wu[1]

[1] School of Engineering Science, University of Science and Technology of China,
Hefei 230026, China
[2] Department of Mathematics and Statistics, Curtin University of Technology,
Perth WA6845, Australia

Abstract. The effect of link and node capacity on traffic dynamics are investigated in weighted scale-free networks by adopting a traffic routing model with local node strength information: $P_{l \to i} = \frac{s_i^\alpha}{\sum_j s_j^\alpha}$. The link bandwidth is controlled by: $B_{ij} = \max(\beta w_{ij}, 1)$, and the capacity of nodes is controlled by: $\max(\gamma s_i, 1)$. The phase transition from free flow to congestion is reproduced. The optimal routing strategy is sought out. When β increases from zero, the optimal strategy changes from preferring low-strength nodes to high-strength nodes. When $\beta \approx 1.0$, there will be two optimal routing strategies. When β is low, the system's behavior is controlled by link bandwidth, while it is controlled by node capacity when β is high. Our work may be useful for the design of modern traffic systems and communication networks.

Keywords: Weighted scale-free networks, Routing strategy, Traffic capacity.

1 Introduction

Complex networks have received much attention from physicists, mainly because a wide range of systems in nature and society could be described by complex networks [1-9]. Recently, the traffic dynamics on complex networks has attract much attention from both physical and computational societies in the past decade [1-6]. This is because of the high importance of large communication networks such as the Internet and WWW in modern society. For example, the submarine earthquake near Taiwan in December 2006 broke a few important optical cables, and after that, the information flow on the Internet were significantly delayed over many countries particularly in the Asia-Pacific region. The goal of traffic research work is to enhance traffic flow and to avoid traffic congestion on a growing large communication network. A variety of studies have been focused on developing better routing strategies [10-16]. Other models were dedicated to improve the system efficiency by changing/optimizing the topology of underlying infrastructure of the networked systems.

It is now known that the network structure plays a significant role in the dynamical process taking place on the network. In the current studies, some

J. Zhou (Ed.): Complex 2009, Part I, LNICST 4, pp. 580–588, 2009.

important properties have been discovered, not only in the topology but also in the weights. Empirical evidences have shown that the modern communication and transportation networks have the small-world (SW) and/or scale-free (SF) properties [1,2]. Therefore, it is natural and important to consider traffic dynamics on SW and SF networks in order to better understand and control various traffic-induced problems. Up to now, many traffic models have been proposed and studied [14-21]. The phase transition from free flow to congestion were reproduced and the point of phase transition were often used to characterize the overall capacity of the systems.

However, the network weights have not been taken into consideration in these models while in most real cases, communication networks are often associated with a large heterogeneity in the capacity and intensity of the connections. Moreover, weights have a strong correlation with the network topology [22-27] and the existing weighted features play a significant role in a variety of dynamical processes [28-30]. Therefore, a modeling approach that can capture the effects of weighted characteristics on traffic dynamics is need.

In this paper, a conceptual traffic model in which packets are routed on weighted scale-free networks is proposed and studied. The proposed model is inspired by the local routing model on un-weighted networks [21]. The present model couples the traffic flow and the weighted characteristics of the network. For traffic model on weighted SF network, we found the optimal capacity occurs at a specific value of routing strategy parameter. The overall capacity is quantified by the critical generating rate, at which a phase transition occurs from free flow to congestion. We also found that the optimal routing strategy depends strongly on the network parameters, which can generally reproduce the real observations.

The paper is organized as follows. In the following section, the traffic model is described in detail, in Sec. 3 simulation results of traffic dynamics are provided, and Sec. 4 gives the conclusion.

2   Traffic Model

Now, we briefly describe the traffic model. We adopt the weighted SF model proposed by Wang et al. [27] to generate the underlying network infrastructure. In this model, the power-law distributions of degrees, weights, and strengths are all in good accordance with real observations of weighted technological networks. The network model is generated with a weight-driven preferential attachment with co-evolution of weights and topology. And the weight-topology co-evolution mimics the traffic interactions of vertices. The model rules can be described as follows. Starting from m_0 nodes fully connected by links with assigned weight $w_0 = 1$, the system are driven by two mechanics: (1) the strength dynamics: the weight of each link connecting i and j is updated as $w_{ij} \rightarrow w_{ij} + 1$, with probability $P_{ij} = W \times p_{ij} = W \times \frac{s_i s_j}{\sum_{a<b} s_a s_b}$, where $s_i = \sum_{j \in \Gamma_i} w_{ij}$ is the strength of node i and Γ_i is the neighboring set of node i; (2) the topological growth: a new node n is added with m links that are randomly attached to a node i according

to the strength preferential probability: $\Pi_{n\to i} = \frac{s_i}{\sum_j s_j}$, where j runs over all existing nodes. Analysis of this model [31] shows that the outcome strength distribution follows a power law $P(s) \sim s^{-\Theta}$ with the exponent $\Theta = 2 + m/(m + 2W)$, where the exponent Θ is controlled by both the weight parameter W and the number of newly added links m. The relationship between node strength and degree follows a power law, i.e., $s \sim k^\phi$ with $\phi > 1$. This nonlinear correlation is in good accordance with empirical observations. Then the exponent γ of power-law degree distribution $P(k) \sim k^{-\theta}$ can be expressed as $\theta = \phi(\Theta - 1) + 1$.

The traffic dynamics is modeled on top of the network as follows. At each time step, R packets are generated homogeneously among the nodes in the system. To navigate packets, all the nodes perform a parallel local search among their immediate neighbors. If a packet's destination is found within the searched area of node l, i.e. the immediate neighbors of l, the packet will be delivered from l directly to its target and then removed from the system. Otherwise, the packet will be delivered to a neighboring node i according to the probability:

$$P_{l\to i} = \frac{S_i^\alpha}{\sum_j S_j^\alpha},\qquad(1)$$

where S_i is the strength of the neighboring node, the sum runs over the immediate neighbors of the node l, and α is an introduced tunable parameter characterizing the preferential probability in choosing neighbors to forward packets. Furthermore, the capacity (or bandwidth) of the link connecting nodes l and i is set to $B_{li} = \max(\beta w_{li}, 1)$, i.e., the link can handle at most B_{li} packets from each end per time step. When the link capacity is reached, the delivery of packets will be delayed and wait for the next time step. We treat all the nodes as both hosts and routers and assume that node i can deliver at most $C_i = \max(\gamma s_i, 1)$ packets per time step towards their destinations, where s_i denotes strength of node i. During the evolution of the system, the FIFO (first-in-first-out) rule is applied on the nodes.

3   Simulation Results and Discussions

The network overall capacity is measured by the critical generating rate R_c at which a phase transition occurs from free flow state to congestion. For this purpose, we investigate the order parameter [10]:

$$\eta(R) = \lim_{t\to\infty} \frac{1}{R} \frac{\langle \Delta N_p \rangle}{\Delta t}.\qquad(2)$$

Here $\Delta N_p = N_p(t + \Delta t) - N_p(t)$, $\langle ... \rangle$ denotes taking the average over a time window of width Δt, and $N_p(t)$ is the number of packets in the system at time t. As shown in Fig.1, when $R < R_c$, $\langle \Delta N \rangle = 0$ and $\eta(R) = 0$, the numbers of added and removed packets are balanced, corresponding to the free-flow state. When $R > R_c$, $\eta(R)$ increases suddenly from zero, corresponding to a phase transition from free-flow to congestion, in which packets will accumulate in the

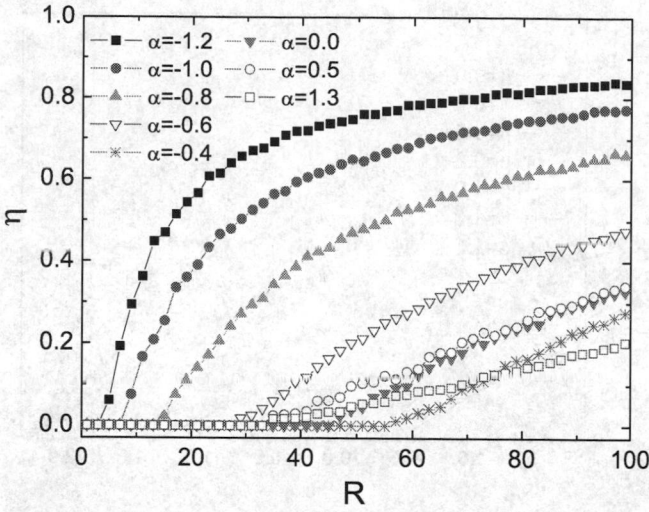

Fig. 1. (Color online) Typical variation of order parameter η versus R for weighted scale-free networks with different routing parameter α. Other parameters are $N = 1000$, $m0 = m = 5$, $W = 2$, $\beta = 1.0$ and $\gamma = 0.5$.

system. Hence, R_c is the maximum packet generating rate under which the system operates effectively. The system's overall handling and delivering capacity can be measured by the critical value of R_c. In Fig.1, one can also see that R_c is different for different α values. When $\alpha = -0.4$, R_c reaches a maximum value of $R_c^{max} \approx 57$. When $\alpha = 0.5$ and 1.3, R_c takes almost the same value. Thus we can investigate the variation of R_c with routing parameters to find an optimal navigation strategy.

Figure 2 and 3 shows the variation of R_c with α for different value of bandwidth parameter β and fix $\gamma = 0.5$, which shows the effect of α and β on the system capacity. One can see that there is a maximum value of R_c at some optimal α_c, which corresponds to the optimal routing strategy. And β has a strong influence on the value of α_c. When β is low, α_c remains at -0.4. With the increment of β, α_c increases from -0.4 to -0.3. When $1.0 \leq \beta \leq 1.5$, there are two peak values of R_c^{max} in the curves. The second peak appears at around $\alpha_c = 1.3$, but with a lower value of R_c^{max}.

When $\beta > 1.5$, the second peak vanishes and the optimal value of α_c increases further from negative to positive. This means that the optimal navigation strategy changes from preferring the low-strength nodes to preferring the high-strength nodes. The system's maximal capacity will increase with β until a threshold is reached at around $\beta = 5$. Then the maximal value of R_c will not increase when $\beta > 5.0$. And α_c remains at 0.3 when $\beta > 5$. This means that the system capacity will not be improved only by increasing the bandwidth of the links. It is because when β is large, all links are operating efficiently under their maximum capacity, so that the network capacity is mainly controlled by node capacity. One should think about some other ways to improve the system

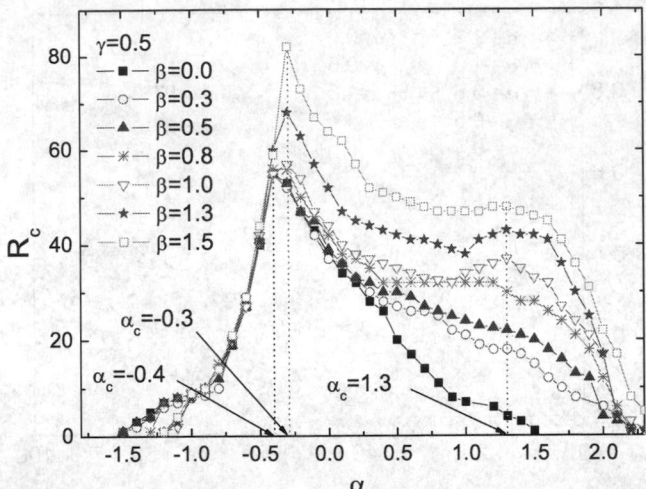

Fig. 2. (Color online) R_c vs α with different value of β. The results are obtained by averaging R_c over ten network realizations. Other network parameters are $N = 1000$, $m0 = m = 5$, $W = 2$.

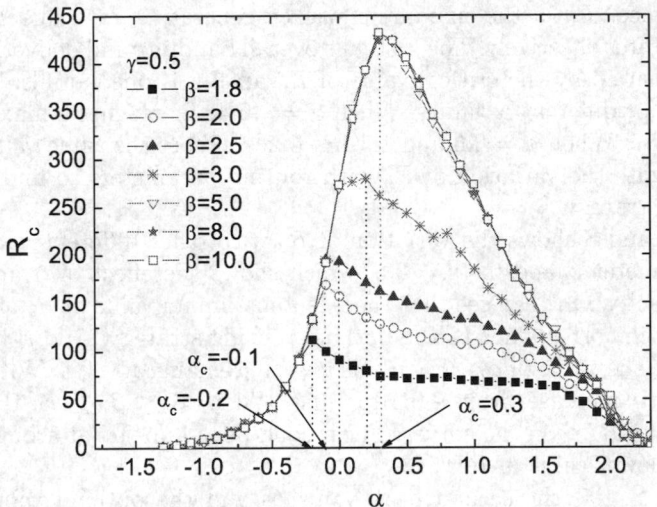

Fig. 3. (Color online) R_c vs α with varying β. Parameters are the same as in Fig.2.

capacity, such as increasing the node capacity, changing the network topology, or developing more efficient routing strategy, and so on.

Further simulations with bigger values of γ show that the system capacity will be greatly improved only when $\beta \geq 5.0$. As shown in Fig.4 and Fig.5, when $\beta < 5.0$, the dynamic behavior of the system remains almost the same. When $\beta \geq 5.0$, R_c^{max} is improved. And the value of R_c^{max} is doubled when $\beta = 8$ and 10.

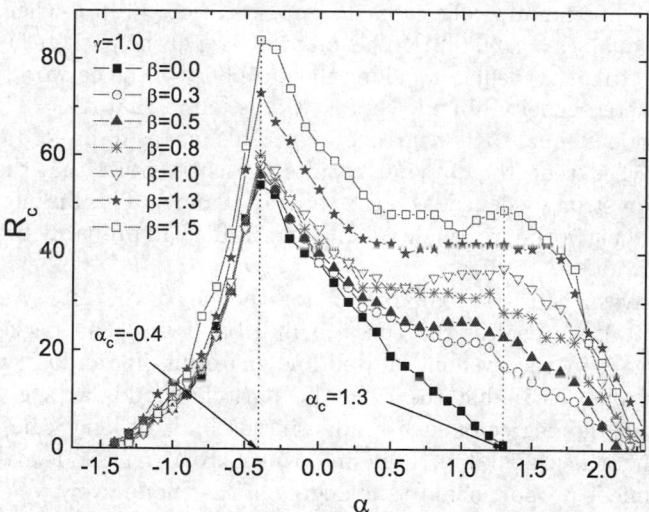

Fig. 4. (Color online) R_c vs α with different value of β and $\gamma = 1.0$. The results are obtained by averaging R_c over ten network realizations. Other network parameters are $N = 1000$, $m0 = m = 5$, $W = 2$.

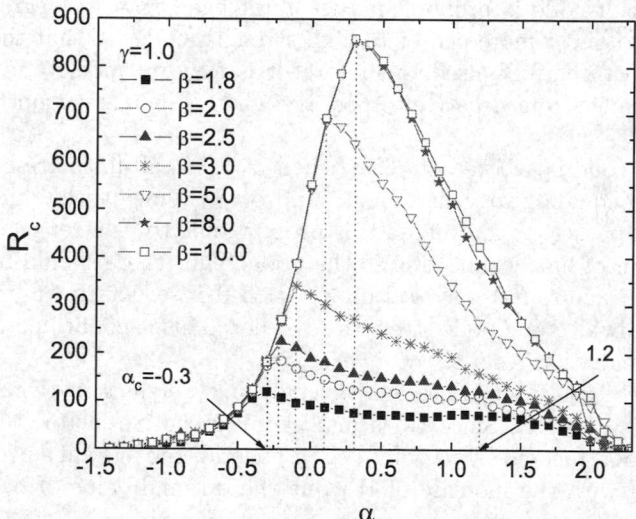

Fig. 5. (Color online) R_c vs α with different value of β and $\gamma = 1.0$. Parameters are the same as in Fig.5.

Thus we can see, when β is low, the system dynamic behavior is mainly controlled by the bandwidth of links. When β is high, it is controlled by the node capacity.

These findings are different from the results on un-weighted scale-free networks. Previous studies with a traffic model routing with node degree information have found that the maximum traffic capacity appears at $\alpha_c = -1.0$ when the nodes' capacity are equal: $C = const$ [21]. This means to repel the packets from the central nodes and to make them move along the periphery of the network. When considering the heterogeneity of node capacity, it is found that $\alpha_c = 0.0$ when the capacity of node is equal to its degree: $C = k$. This means to prefer random walk in the system. Different from previous results, here it is shown that the traffic dynamic behavior strongly depends on the value of β and γ. This finding is valuable since most real networks are weighted, and the traffic will probably be affected by the link bandwidth.

Here we give a heuristic explanation for the peaks and α_c. We note that when $\alpha = 0.0$, if we neglect the effect of link bandwidth, all packets perform randomlike walk in the system. A well-known result in random walk theory valid for our analysis is that the time the particle spends at a given node is proportional to the degree of such a node in the limit of long times [32]. Thus one can see that the probability of sending packets to a given node averaging over a period of time is proportional to the degree of that node. Now we consider the constriction of link bandwidth. When $\beta = 0$, all links have the same bandwidth of 1. The high-strength nodes may cause congestion at the links connecting to them. So that α_c should be negative to avoid sending packets to high-strength nodes. The same effect appears when β is low. On the other hand, when β is very large, the bandwidth of links will not trigger congestion. For the case, because the strength of node is proportional to but bigger than its degree, α_c should be positive to direct more packet to high-strength nodes so that the capacity of these nodes can be fully used. Therefore, it is easy to understand that α_c will gradually change from negative to positive when β increases from zero to some large value.

For the second peak, we note that when $\beta = 1.0$, all links are operated with a bandwidth equaling to their weight. And $\alpha = 1.0$ means that the probability of delivering packets to a given node is proportional to its strength, which is the sum of weight of links connecting to the node. Thus there should be an optimal configuration at around $\beta \approx 1.0$ and $\alpha \approx 1.0$ in order to fully make use of the link bandwidth and node strength together. This is confirmed by previously presented simulation results.

Finally, we briefly introduce the effect of link weight growth rate W on the packet traffic capacity. Since W is just a multiplicative factor, the qualitative behavior is not affected by varying W. In general, the system's overall capacity will increase with the increase of W, but the optimal value of α_c remains the same.

4   Conclusion

In summary, the traffic dynamics on weighted scale-free networks is studied with a local routing strategy. Simulation results show that the link bandwidth, the node capacity and the routing strategy of packets have different effects on the system's dynamic behavior. We found that when the bandwidth is low, it is better to make use of low-strength nodes in the routing strategy. And the optimal routing strategy will change from low-strength nodes to high-strength nodes with the increment of link bandwidth. But the system's capacity will not be improved by merely increasing link bandwidth. This behavior is in agreement with empirical practice. Moreover, we found a second optimal routing strategy when both the link bandwidth and node strength could be fully used.

Our study shows that the traffic dynamics on weighted scale-free network has many new characteristics. It is also different from traffic dynamics on well-organized lattice, on regular or random networks [15,16]. Our work may have practical implications for optimizing some modern traffic and communication systems in the real world.

Acknowledgments

This work is funded by National Basic Research Program of China (No.2006CB705500), the NNSFC under Key Project No.10532060, Project Nos.10872194, 70601026, 10672160. Y.-H. Wu acknowledges the support of Australian Research Council through a Discovery Project Grant.

References

1. Watts, D.J., Strogatz, S.H.: Collective dynamics of 'small-world' networks. Nature 393, 440–442 (1998)
2. Barabási, A.L., Albert, R.: Emergence of scaling in random networks. Science 286, 509–512 (1999)
3. Albert, R., Barabási, A.L.: Statistical mechanics of complex networks. Rev. Mod. Phys. 74, 47–97 (2002)
4. Dorogovtsev, S.N., Mendes, J.F.F.: Evolution of networks. Adv. Phys. 51, 1079–1187 (2002)
5. Newman, M.E.J.: The structure and function of complex networks. SIAM Rev. 45, 167–256 (2003)
6. Boccaletti, S., Latora, V., Moreno, Y., Chavez, M., Hwang, D.U.: Complex networks: Structure and dynamics. Phys. Rep. 424, 175–308 (2006)
7. Latora, V., Marchiori, M.: How the science of complex networks can help developing strategies against terrorism. Chaos, Solitons and Fractals 20(1), 69–75 (2004)
8. Li, C.G., Chen, G.R.: Local stability and Hopf bifurcation in small-world delayed networks. Chaos, Solitons and Fractals 20(2), 353–361 (2004)
9. Zhang, H.F., Wu, R.X., Fu, X.C.: The emergence of chaos in complex dynamical networks. Chaos, Solitons and Fractals 28(2), 472–479 (2006)
10. Arenas, A., Díaz-Guilerà, A., Guimerà, R.: Communication in networks with hierarchical branching. Phys. Rev. Lett. 86, 3196–3199 (2001)
11. Guimerà, R., Arenas, A., Díaz-Guilera, A., Giralt, F.: Dynamical properties of model communication networks. Phys. Rev. E 66, 026704 (2002)
12. Guimerà, R., Díaz-Guilera, A., Vega-Redondo, F., Cabrales, A., Arenas, A.: Optimal network topologies for local search with congestion. Phys. Rev. Lett. 89, 248701 (2002)

13. Guimerà, R., Danon, L., Díaz-Guilerà, A., Giralt, F., Arenas, A.: Self-similar community structure in a network of human interactions. Phys. Rev. E 68, 065103(R) (2003)
14. Tadić, B., Thurner, S., Rodgers, G.J.: Traffic on complex networks: Towards understanding global statistical properties from microscopic density fluctuations. Phys. Rev. E 69, 036102 (2004)
15. Tadić, B., Thurner, S.: Information super-diffusion on structured networks. Physica A 332, 566–584 (2004)
16. Tadić, B., Thurner, S.: Search and topology aspects in transport on scale-free networks. Physica A 346, 183–190 (2005)
17. Noh, J.D., Rieger, H.: Random walks on complex networks. Phys. Rev. Lett. 92, 118701 (2004)
18. Zhao, L., Park, K., Lai, Y.C.: Attack vulnerability of scale-free networks due to cascading breakdown. Phys. Rev. E 70, 035101(R) (2004)
19. Yan, G., Zhou, T., Hu, B., Fu, Z.Q., Wang, B.H.: Efficient routing on complex networks. Phys. Rev. E 73, 046108 (2006)
20. de Moura, A.P.S.: Fermi-Dirac statistics and traffic in complex networks. Phys. Rev. E 71, 066114 (2005)
21. Wang, W.X., Wang, B.H., Yin, C.Y., Xie, Y.B., Zhou, T.: Traffic dynamics based on local routing protocol on a scale-free network. Phys. Rev. E 73, 026111 (2006)
22. Barrat, A., Barthélemy, M., Pastor-Satorras, R., Vespignani, A.: The architecture of complex weighted networks. Proc. Natl. Acad. Sci. U.S.A. 101, 3747–3752 (2004)
23. Macdonald, P.J., Almaas, E., Barabási, A.L.: Minimum spanning trees of weighted scale-free networks. Europhys. Lett. 72, 308–314 (2005)
24. Dorogovtsev, S.N., Mendes, J.F.F.: Scaling behaviour of developing and decaying networks. Europhys. Lett. 52, 33–39 (2000)
25. Barrat, A., Barthélemy, M., Vespignani, A.: Weighted evolving networks: Coupling topology and weight dynamics. Phys. Rev. Lett. 92, 228701 (2004)
26. Abe, S., Thurner, S.: Complex networks emerging from fluctuating random graphs: Analytic formula for the hidden variable distribution. Phys. Rev. E 72, 036102 (2005)
27. Wang, W.X., Wang, B.H., Hu, B., Yan, G., Ou, Q.: General dynamics of topology and traffic on weighted technological networks. Phys. Rev. Lett. 94, 188702 (2005)
28. Motter, A.E., Zhou, C.S., Kurths, J.: Network synchronization, diffusion, and the paradox of heterogeneity. Phys. Rev. E 71, 016116 (2005)
29. Zhou, C.S., Motter, A.E., Kurths, J.: Universality in the synchronization of weighted random networks. Phys. Rev. Lett. 96, 034101 (2006)
30. Zhang, Z.Z., Zhou, S.G., Chen, L.C., Guan, J.H., Fang, L.J., Zhang, Y.C.: Recursive weighted treelike networks. Eur. Phys. J. B 59, 99–107 (2007)
31. Xie, Y.B., Wang, W.X., Wang, B.H.: Modeling the coevolution of topology and traffic on weighted technological networks. Phys. Rev. E 75, 026111 (2007)
32. Bollobás, B.: Modern Graph Theory. Springer, New York (1998)

The Effect of Lane-Changing Time on the Dynamics of Traffic Flow

Xin-Gang Li[1], Bin Jia[1], and Rui Jiang[2]

[1] School of Traffic and Transportation, Beijing Jiaotong University,
Beijing, 100044, P.R. China
lsinban@gmail.com, bjia@bjtu.edu.cn
[2] School of Engineering Science, University of Science and Technology of China,
Hefei, 230026, P.R. China
rjiang@ustc.edu.cn

Abstract. In this paper, the lane-changing time is considered in the cellular automata models for traffic flow. The lower the velocity of a vehicle, the longer the lane-changing time. The simulations are carried out in the two-lane system and the on-ramp system. When the lane-changing time is taken into account, the maximum flux per lane is reduced in the two-lane system compared with the original two-lane model, and it is even lower than that of single-lane road when a lane-changing takes longer time; the capacity drop can be reproduced in the on-ramp system.

Keywords: lane-changing time; cellular automata; traffic flow.

1 Introduction

In recent years, modelling the dynamics of traffic flow has attracted much attention of researchers from the field of physics. Many theoretical models have been proposed to explore the evolution mechanism of traffic flow [1,2,3,4,5]. Among those models, cellular automaton (CA) is an excellent tool for simulating real traffic flow, because its efficient and fast performance when used in computer simulations. In 1992, Nagel and Schreckenberg proposed the well-known Nagel-Schareckenberg (NaSch) model [6]. Although it is very simple, the NaSch model can reproduce some real traffic phenomena, such as the occurrence of phantom traffic jams and the realistic flow-density relation (fundamental diagram). The NaSch model is a minimal model in the sense that any further simplification of the model leads to unrealistic behavior. Later, several extensions of the NaSch model are proposed [7,8,9,10].

Due to the road consists of multi-lane in real traffic, the single lane NaSch model can not simulate realistic traffic, especially when the system is inhomogeneous. As a result, the two-lane models are proposed by introducing the additional lane-changing rules. Lots of lane-changing rules, which consist of symmetric and asymmetric ones, have been implemented to simulate the realistic lane-changing behaviors [11,12,13,14,15,16,17,18,19]. In our previous work, the

J. Zhou (Ed.): Complex 2009, Part I, LNICST 4, pp. 589–598, 2009.

honk effect [17,18] and the aggressive lane-changing behavior [19] were investigated and more realistic results had been obtained.

The traffic dynamics around bottlenecks, where lane-changing behaviors frequently happen, have been widely studied with CA models [20,21,22,23,24,25]. The bottlenecks include on-ramps, off-ramps, lane closings, uphill gradients, narrow road sections, etc. The lane-changing rules are used to model the vehicle entering, preparing to exit or exiting the main road.

A lane change can be described in three parts [26]. First, the head portion is the time and distance required for a vehicle to move from a straight-ahead path to the first intercept of the lane line. The actual lane change is begun when a vehicle first encroaches on the lane line between the original and destination lane. Secondly, the maneuver is ended once the vehicle has completely crossed that line. Finally, the tail portion of the maneuver is the time and distance required for a vehicle to return to a straight-ahead path in the destination lane after crossing the lane line. It is obvious that time is needed during lane-changing and the vehicle takes up two lanes in the range from the point its head reaching the lane line to the point its end leaving the lane line. The lane-changing time for high speed vehicle can be shorter in comparison to the vehicle with low speed. But in most of the present models, a lane-changing usually completes within one time step, no matter how much the speed the vehicle has.

In order to model the realistic driving behavior in lane-changing, Sasoh proposed a model [27], in which a lane-changing needs 2 seconds; Toledo and Zohar presented a model [28], in which the lane-changing time depends on subject speed relative to neighboring vehicles. In those models, the lane-changing can be cancelled in the duration over which the vehicle would complete the lane change maneuver, if the traffic condition on the original lane becomes better. But the influence of the lane-changing vehicle on the following vehicles on the original lane and the destination lane are not considered.

In this paper, the lane-changing time, which depends on the vehicle's current velocity, is introduced into the CA models for traffic flow. We assume that a lane-changing vehicle takes up the two lanes, i.e., the original lane and the destination lane. So the occupancy of the road will become high when a lane-changing happens. The two-lane system and the on-ramp system are taken as two typical examples to study the effect of lane-changing time on the dynamics of traffic flow. More realistic results are obtained.

This paper is organized as follows: In the following Section, the two-lane system and the on-ramp system are introduced. In Section 3, simulation results are analyzed in detail. The conclusion is given in Section 4.

2   Model

Before introducing the two-lane system and the on-ramp system, we briefly recall the definition of the NaSch model [6], which is used to model the forward motion of vehicle. The NaSch model is a discrete model for traffic flow. The road is divided into L cells, which can be either empty or occupied by a vehicle with a velocity $v = 0, 1, ..., v_{\max}$. The vehicles which are numbered $1, 2, 3, ..., N$ move from the left to the right on a lane with periodic boundary conditions. At each discrete time step $t \to t+1$, the system update is performed in parallel according to the following four rules: (i) acceleration: $v_n(t+1/3) \to \min(v_n(t)+1, v_{\max})$; (ii) deceleration: $v_n(t+2/3) \to \min(v_n(t+1/3), d_n)$; (iii) randomization: $v_n(t+1) \to \max(v_n(t+2/3) - 1, 0)$ with probability p; (iv) position update: $x_n(t+1) \to x_n(t) + v_n(t+1)$. Here v_n and x_n denote the velocity and position of the vehicle n respectively; v_{max} is the maximum velocity and $d_n = x_{n+1} - x_n - 1$ denotes the number of empty cells in front of the vehicle n; p is the randomization probability.

2.1 Two-Lane System

The velocity and the position are updated according to the NaSch model. As to the two-lane system, one has to introduce lane-changing rules, which control the parallel motion of vehicles. In two-lane models the update step is usually divided into two sub-steps: In the first sub-step, vehicles may change lanes in parallel according to lane changing rules and in the second sub-step the lanes are considered as independent single-lane NaSch models.

The lane-changing rules can be symmetric or asymmetric with respect to the lanes and to the vehicles. In this paper, we just investigate the symmetric two-lane model. Chowdhury et al. [12] have assumed a symmetric rule set where vehicles change lanes if the following criteria are fulfilled [hereafter referred to as symmetric two-lane cellular automata (STCA) model]:

$$d_n < \min(v_n + 1, v_{max}) \text{ and } d_{n,other} > d_n \text{ and } d_{n,back} > d_{safe}. \tag{1}$$

Here $d_{n,other}, d_{n,back}$ denote the number of free cells between the nth vehicle and its two neighbor vehicles on the other lane at time t, respectively. If there is a vehicle on the destination lane driving side by side with vehicle n, $d_{n,back}$=-1. d_{safe} is a safe distance and equals to the maxspeed of the following vehicle on the destination lane.

2.2 On-Ramp System

The on-ramp system with accelerating lane has been well studied by using CA model in our previous work[22]. The schematic of the on-ramp system is shown in Fig.1. One can see that the road is divided into four sections: region A (main road upstream the point M), B (on-ramp upstream the point M), C (main road downstream the point N) and D (the accelerating section). The NaSch model is used to modelling the movement of vehicle. In section D, the road has two lanes. The vehicle on the right lane must change to the left lane before it reach the end of section D, and the vehicle on the left lane are not allowed to change to the right lane. If the condition

$$d_{n,other} > 1 \text{ and } d_{n,back} > v_{ob} \tag{2}$$

is met, the lane-changing is performed by vehicle n [hereafter referred to as on-ramp with accelerating lane (OAL) model]. Here $d_{n,other}$ and $d_{n,back}$ have the

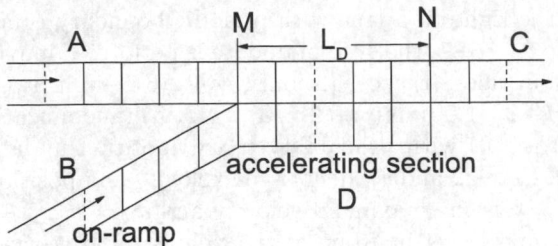

Fig. 1. Schematic illustration of the on-ramp system

same mean as that in the two-lane system. v_{ob} is the velocity of the following vehicle on the destination lane. Condition $d_{n,other} > 1$ means "I can move on the destination lane at next time step"; and condition $d_{n,back} > v_{ob}$ is a safety criterion.

2.3   Lane-Changing Time

The lane-changing time depends on the current velocity. In this paper, $v_{max}=5$ is selected. We assume that the vehicle with velocity 0, 1, 2, 3, 4 and 5 takes $3t_0$, $3t_0$, $2t_0$, $2t_0$, $1t_0$ and $1t_0$ time step(s), respectively, to perform a lane change. During a vehicle changing lane, it takes up the two lanes, i.e., the current lane and the destination lane. Now, d_n of the lane-changing vehicle is calculated by $d_n = \min(d_n^l, d_n^r)$, d_n^l (d_n^r) is the number of empty cells in front of vehicle n on left (right) lane. When the gaps $d_{n,other}$ and $d_{n,back}$ are calculated, one should take the lane-changing vehicle into account. As to the vehicle just driving behind the lane-changing vehicle, it does not change lane for $d_n = d_{n,other}$. When a vehicle is changing lane, it will return to the original lane if the traffic condition on the original lane becomes better. The two-lane system considering lane-changing time is referred as STCA-LT model, and the on-ramp system considering lane-changing time is referred as OAL-LT model.

2.4   Boundary Condition

In the two-lane system, periodic boundary condition is used. While in the on-ramp system, open boundary condition is adopted. We assume that the first cell on section A and B correspond to $x=1$, and the entrance regions of road A and B include v_{max} cells, i.e., the vehicles can enter road A and B from the cells $1, 2, \ldots, v_{max}$. In one time step, when the update of the vehicles on the road is completed, we check the positions of the last vehicles on road A and B and that of the first vehicle on road C, which are denoted as x_A^{last}, x_B^{last} and x_C^{first}, respectively. If $x_A^{last}(x_B^{last}) > v_{max}$, a vehicle with velocity v_{max} is injected with probability $\alpha_A(\alpha_B)$ at the cell $\min[x_A^{last}(x_B^{last}) - v_{max}, v_{max}]$. Near the exit of road C, the leading vehicle is removed if x_C^{first} is larger than the length of lane C and the following vehicle becomes the new leading vehicle and it moves without any hindrance.

3   Simulations and Discussions

In the simulation, the randomization parameter $p = 0.3$ is used. Each cell corresponds to 7.5m and a vehicle has a length of one cell. One time step corresponds to 1 s.

3.1   Two-Lane System

The two-lane road is divided into 2×2000 cells. Here only the homogenous system with one type of vehicle is considered. In Fig.2, we show the fundamental diagrams. One can see that the flux per lane in the intermediate density range is improved in the STCA model compared to that of a single lane road. As to the STCA-LT model, the flux in the intermediate density range is also enhanced but smaller than that of the STCA model when $t_0=1$. However, the flux is depressed when $t_0=2$ compared to that of the single lane road. This indicates that in the homogenous system lane-changing can not improve the flux per lane if a vehicle takes longer time to complete a lane change.

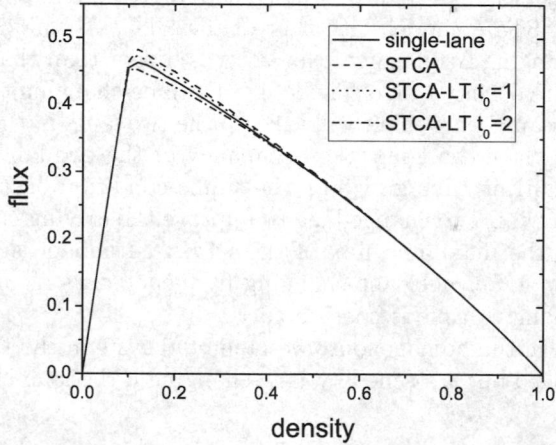

Fig. 2. Fundamental diagrams for the STCA model, the STCA-LT model and the single-lane NaSch model

The lane-changing frequency in the two models are shown in Fig.3. Here the number of lane changes that happen at $v = 0, 1$, $v = 2, 3$ and $v = 4, 5$ are counted seperatedly. The lane-changing frequency is defined as the number of lane changes per time step per vehicle. One can see that as density increasing, the lane-changing frequency first increases then decreases in all cases. In the free flow density range, all the vehicles can drive with free flow speed. There is no vehicle with $v < 4$. So lane-changing frequency with $v < 4$ is 0. As the density increases to the congested flow density range, jams will appear on the road. Vehicles with low speed may change lane to obtain good traffic condition. When the density is

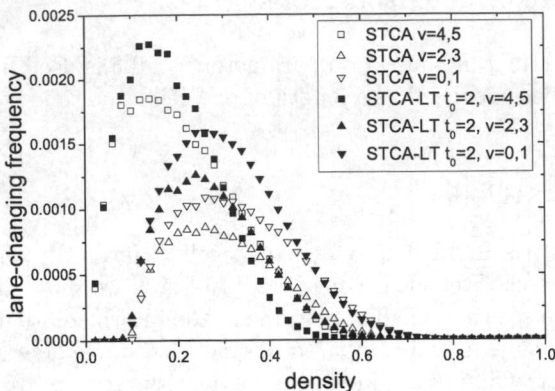

Fig. 3. Lane-changing frequency of the vehicles with different velocity in the two models

not large, the lane-changing with $v=4,5$ takes the highest proportion. But when the density is large, the lane-changing with $v=0,1$ takes the highest proportion. When the density is larger than 0.7, lane-changing rarely happens. Compared the results of the STCA model with that of the STCA-LT model, one can see that, the lane-changing frequency of the latter is higher than that of the former in the intermediate density range. The longer the lane-changing time, the higher the frequency. Because the vehicle will take up the two lanes in the duration over which it performs lane-changing, the occupancy of the two-lane road becomes higher as lane-changing time growing. The traffic condition becomes worse and more vehicles are willing to change lane to improve the driving condition. As we can see in Fig.2, the maximum flux of the STCA-LT model is lower than the of the STCA model. So high lane-changing frequency does not bring high flux when the lane-changing time is considered.

We argue that in the homogenous two-lane road system, the traffic condition can not be improved but wrosened by lane-changing if the lane-changing time is too longer.

3.2   On-Ramp System

In the simulations, section C is divided into 1000 cells; section D into $L_D=20$ cells and sections A and B into 2000 cells. The first 50,000 time steps are discarded to let the transient time die out. The flux is obtained by counting the vehicles that pass a virtual detector in 50,000 time steps. The flux on roads A, B and C are q_A, q_B and q_C, respectively.

In Fig.4, the phase diagram in (α_A, α_B) space of the two models is shown. Similiar to the results in Ref.[22], one can see that the phase diagram is categorized into four regions. In region I, the traffic flows on both roads A and B are free flow; in region II, the traffic is free on road A and it is congested on road B; in region III, the traffic is congested on road A and is free on road B; and in region IV, the traffic flows are congested on both roads A and B.

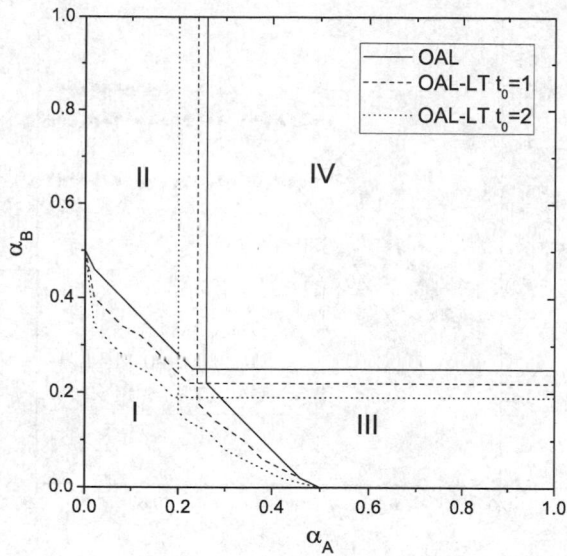

Fig. 4. The phase diagrams in (α_A, α_B) space of the OAL model and the OAL-LT model. Here $L_D = 20$.

Compared with the results of the OAL model, one can see that region I shrinks and region IV enlarges in the OAL-LT model. As t_0 becomes large, regions I and IV also change with the same trend. This indicates that longer lane-changing time deteriorates the traffic condition on both the main road and the on-ramp.

Next, the capacity of the on-ramp system is investigated. Since $q_C = q_A + q_B$, the saturated flux q_C^s on road C is deemed as the capacity of the on-ramp system. Fig.5 shows q_C as a function of α_A with different α_B. One can see that q_C first increases as α_A growing then the saturated flux q_C^s is reached. When there is no vehicles from on-ramp lane $\alpha_B = 0$, the on-ramp system is degenerated into the single-lane system. q_{max} is the maximum flux of the single-lane system. When there are vehicles from on-ramp lane $\alpha_B > 0$, the saturated flux q_C^s equals to the value of q_{max} in the OAL model, while q_C^s is lower than q_{max} in the OAL-LT model. Thus the capacity of the on-ramp system is not reduced in the OAL model, while it drops to lower values in the OAL-LT model. We know that the capacity drop usually happens in real traffic, so we believe that the lane-changing time is a factor for capacity drop. Fig.5 also shows that as t_0 increasing, the capacity drops to smaller values.

Lastly, the impact of the on-ramp on the main road is studied. We choose $\alpha_A = 0.5$, and the flux of road C q_C as a function of α_B is shown in Fig.6. One can see that in the OAL model, q_C does not change as α_B increasing. This is consistent with the above result that the capacity is not reduced in the OAL model. However, q_C will first decrease then become constant in the OAL-LT model. As α_B increasing, the number of vehicles changing from the on-ramp to the main road becomes larger, thus more disturbances are brought to the

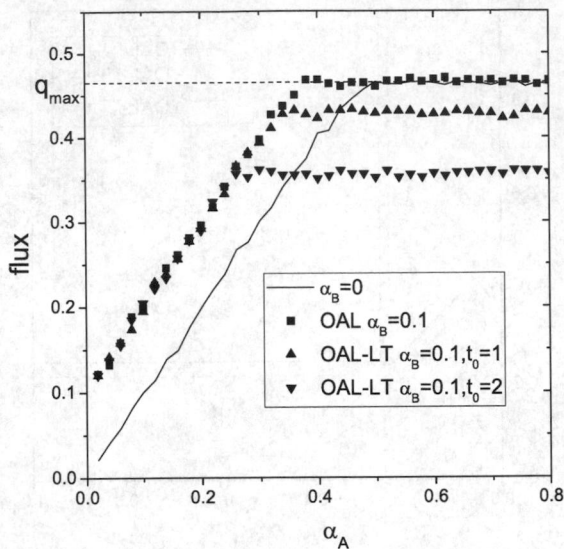

Fig. 5. q_C as a function of α_A. $L_D=20$.

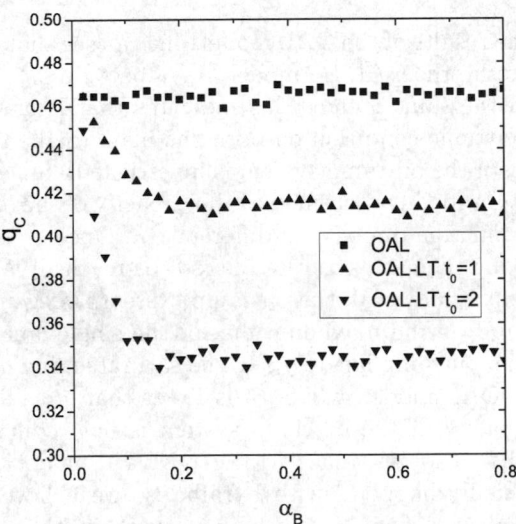

Fig. 6. q_C as a function of α_B. The parameters are $\alpha_A=0.5$ and $L_D=20$.

traffic on the main road and q_C decreases. When α_B is larger than the critical value, the traffic flow on both the main road and the on-ramp are saturated, and the number of vehicles changing from the on-ramp to the main road becomes constant. So q_C does not change when α_B is larger.

4 Conclusion

In this paper, the lane-changing time is taken into account in the CA models for traffic flow. A vehicle with low speed will need more time to complete a lane change. Thus we assume that the lane changing time is related to the velocity of the vehicle.

In the two-lane system, with longer lane changing time the maximum flux per lane is depressed, but the lane-changing frequency is improved.

In the on-ramp system, the traffic conditions on both the main road and the on-ramp are deteriorated and the capacity drop can be simulated when lane-changing time is considered.

Finally, we should mention that only the simulation results are presented here. The empirical data need to be collected to verify our model and this will be done in our future work.

Acknowledgements

This project is financially supported by 973 Program (2006CB705500), the National Natural Science Foundation of China (Nos. 70631001, 70501004 and 70701004), Program for New Century Excellent Talents in University (NCET-07-0057), and the Innovation Foundation of Science and Technology for Excellent Doctorial Candidate of Beijing Jiaotong University (No. 48025).

References

1. Chowdhury, D., Santen, L., Schadschneider, A.: Statistical Physics of Vehicular Traffic and Some Related System. Physics Reports 329, 199–329 (2000)
2. Maerivoet, S., Moor, B.D.: Cellular automata models of road traffic. Phys. Rep. 419, 1–64 (2005)
3. Helbing, D.: Traffic and related self-driven many particle systems. Rev. Mod. Phys. 73, 1067–1141 (2001)
4. Kerner, B.S.: The Physics of Traffic: Empirical Freeway Pattern Features, Engineering Applications, and Theory. Springer, Berlin (2004)
5. Jia, B., Gao, Z.Y., Li, K.P., Li, X.G.: Models and Simulations of traffic System Based on the Theory of Cellular Automaton. Science Press, Beijing (2007)
6. Nagel, K., Schreckenberg, M.: A Cellular automaton model for freeway traffic. J Physique I 2, 2221–2229 (1992)
7. Fukui, M., Ishibashi, Y.: Traffic flow in 1D cellular automaton model including cars moving with high speed. J. Phys. Soc. Jpn 65, 1868–1870 (1996)
8. Barlovic, R., Santen, L., Schadschneider, A., et al.: Metastable states in cellular automata for traffic flow. Eur. Phys. J. B 5, 793–800 (1998)
9. Knospe, W., Santen, L., Schadschneider, A., et al.: Towards a realistic microscopic description of highway traffic. J. Phys. A 33, L477–L485 (2000)
10. Li, X.B., Wu, Q.S., Jiang, R.: Cellular automaton model considering the velocity effect of a car on the successive car. Phys. Rev. E 64, 066128 (2001)
11. Rickert, M., Nagel, K., Schreckenberg, M., Latour, A.: Two lane traffic simulations using cellular automata. Physica A 231, 534–550 (1996)

12. Chowdhury, D., Wolf, D.E., Schreckenberg, M.: Particle hopping models for two-lane traffic with two kinds of vehicles: effects of lane changing rules. Physica A 235, 417–439 (1997)
13. Nagel, K., Wolf, D.E., Wagner, P., Simon, P.: Two-lane traffic rules for cellular automata: A systematic approach. Phys. Rev. E 58, 1425–1437 (1998)
14. Wagner, P., Nagel, K., Wolf, D.E.: Realistic multi-lane traffic rules for cellular automata. Physica A 234, 687–698 (1997)
15. Knospe, W., Santen, L., Schadschneider, A., Schreckenberg, M.: Disorder effects in cellular automata for two-lane traffic. Physica A 265, 614–633 (1999)
16. Knospe, W., Santen, L., Schadschneider, A., Schreckenberg, M.: A realistic two-lane traffic model for highway traffic. J. Phys. A 35, 3369–3388 (2002)
17. Jia, B., Jiang, R., Wu, Q.S., Hu, M.B.: Honk effect in the two-lane cellular automaton model for traffic flow. Physica A 348, 544–552 (2005)
18. Jia, B., Jiang, R., Wu, Q.S.: A realistic two-lane cellular automaton model for traffic flow. Inter. J. Mod. Phys. C 15, 381–392 (2004)
19. Li, X.G., Jia, B., Gao, Z.Y., Jiang, R.: A realistic two-lane cellular automata traffic model considering aggressive lane-changing behavior of fast vehicle. Physica A 367, 479–486 (2006)
20. Jiang, R., Wu, Q.S., Wang, B.H.: Cellular automata model simulating traffic interactions between on-ramp and main road. Phys. Rev. E 66, 036104 (2002)
21. Jiang, R., Jia, B., Wu, Q.S.: The stochastic randomization effect in the on-ramp system: single lane main road and two lane main road situations. J. Phys. A 36, 11713–11723 (2003)
22. Jia, B., Jiang, R., Wu, Q.S.: The effects of accelerating lane in the on-ramp system. Physica A 345, 218–226 (2005)
23. Pederson, M.M., Ruhoff, P.T.: Entry ramps in the Nagel-Schreckenberg model. Phys. Rev. E 65, 056705 (2002)
24. Li, F., Zhang, X.Y., Gao, Z.Y.: The effect of restricted velocity in the two-lane on-ramp system. Physica A 374, 827–834 (2007)
25. Jia, B., Jiang, R., Wu, Q.S.: The traffic behaviors near an off-ramp in the cellular automaton traffic model. Phys. Rev. E 69, 056105 (2004)
26. Worrall, R.D., Bullen, A.G.R.: An empirical analysis of lane changing on multilane highways. Highway Research Board 303, 30–43 (1970)
27. Sasoh, A.: Impact of Unsteady Disturbance on Multi-lane Traffic flow. J. Phys. Soc. Japan 71, 989–996 (2002)
28. Toledo, T., Zohar, D.: Modeling Duration of Lane Changes. Transportation Research Board 1999, 71–78 (2007)

The Difference between Single-Valued and Multi-Valued Cases in the Compact Representation of CPD in Bayesian Networks

Qin Zhang

Automation School of Chongqing University, Chongqing 400044, P.R. China
zhangqin@cqu.edu.cn

Abstract. This paper addresses an important issue about the compact representation of the conditional probability distribution (CPD) applied in the well known Bayesian Networks in uncertain causality representation and probabilistic inference. That is, there is an essential difference between the single-valued cases and the multi-valued cases, while this difference does not exist when the CPD is represented in the conditional probability table (CPT). In other words, the present compact representation and inference methods applicable in the single-valued cases may not be applicable in the multi-valued cases as people usually think. A detailed example is provided to illustrate this problem. The solution is provided in the references by the author.

Keywords: knowledge representation, uncertainty, causality, probabilistic inference.

1 Introduction

It is well known that the typical representation of the conditional probability distribution (CPD) in the well known Bayesian Network (BN) is the conditional probability table (CPT). However, it is also noted that there are too many parameters to be specified in a CPT. For example, suppose a child variable has 5 parent variables and all the 6 variables have 5 states each, the number of conditional probabilities included in the CPT is 5^6=15625. They are too many for the users to specify. To provide the compact representations, many efforts have been made, such as noisy-OR [1], CSI [2], DCD [3], etc. However, it should be noted that many of the compact representations are presented for or illustrated with the binary variables, while actually these cases are single-valued but not multi-valued. The so called single-valued case is such a case in which only the causes of one state (denoted as the true state) of the child variable are specified. In contrast, the case in which the causes of more than one state of the child variable are specified separately is a multi-valued case. Note that the word "valued" indicates the child variable states whose causes are specified directly (not the complement of the other states). Since the binary child variables can be involved in either the single-valued cases or the multi-valued cases, it is not clear whether or not or how the present compact representation models applicable in the single-valued cases are

J. Zhou (Ed.): Complex 2009, Part I, LNICST 4, pp. 599–606, 2009.

also applicable in the multi-valued cases. Note that the binary single-valued cases and the binary multi-valued cases are different. It is explained below.

2 The Detailed Discussion on the Essential Difference

Usually, people need only specify the causes of the true state of a binary variable X_n. Suppose $X_{n,1}$ denotes the true state and $X_{n,2}$ denotes the false state[1]. The reason why I use $X_{n,1}$ instead of X_n to denote the true state and use $X_{n,2}$ instead of $\overline{X}_n$ to denote the false state is for convenience of indicating the difference between the binary multi-valued cases and the binary single-valued cases. In the multi-valued cases, the states of a variable are in the identical positions. For the well known burglary (X_1), earthquake (X_2) and alarm (X_3) example in noisy-OR model [1], $X_{3,1}$ can be caused by either $X_{1,1}$ or $X_{2,1}$ independently. It is easy for the domain experts to give the special conditional probabilities (will be explained later) of $X_{3,1}$ caused by $X_{1,1}$ and $X_{2,1}$ independently, while it is not easy for them to give the CPT directly. This is because the burglary and earthquake are totally different domains. However, it should be noted that in this example, only the causes of $X_{3,1}$ are specified, while the causes of $X_{3,2}$ must *not* be specified, because $X_{3,2}$ has already been given as the complement of $X_{3,1}$. Figure 1 illustrates this binary single-valued case, in which, ⟶ represents the state level causal link.

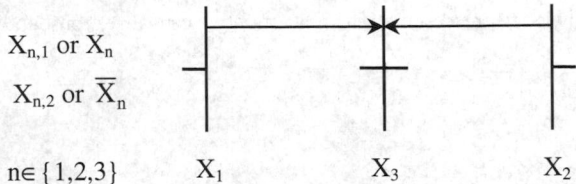

Fig. 1. The illustration for the binary single-valued case

However, the real world is not always so simple. For the example of a simple memory circuit, it has two identical states/outputs: 0 or 1. This is a typical binary variable. Similar to the alarm variable, we may denote this variable as X_3 with $X_{3,1}$ representing state "0" and $X_{3,2}$ representing state "1". But differently, both $X_{3,1}$ and $X_{3,2}$ can be caused by different events. Suppose event $X_{1,1}$ causes $X_{3,1}$ and $X_{2,1}$ causes $X_{3,2}$, with independently given special conditional probabilities $p_{3,1;1,1}$ and $p_{3,2;2,1}$ respectively, where the subscripts nk;ij denotes that event X_{nk} is caused by event X_{ij}.

The reason why the word "special" is put in front of "conditional probabilities" is because usually $p_{nk;ij} \neq \Pr\{X_{nk}|X_{ij}\}$. In fact, $p_{nk;ij}$ is the probability of the linkage event

---

[1] In this paper, X_{nk} denotes either the kth state/value of the variable X_n or the event that X_{nk} is true. The upper case letters denote variables or events. The lower case letters denote the probabilities of events, e.g., $x_{nk} \equiv \Pr\{X_{nk}\}$. The difference between the variable X_n and the event X_{nk} is that X_{nk} has two subscripts, in which the second subscript denotes the state or indexes the specific value of X_n.

in DCD and is the probability of the inhibitor in noisy-OR [1]. Similar notations are also used in [4] in which $p_{3,1;1,1}$ and $p_{3,2;2,1}$ are denoted as $c_{X1,1}(X_{3,1})$ and $c_{X2,1}(X_{3,2})$, so that the two types of conditional probabilities are distinctive.

Another example of multi-valued cases is the sex (X_3) that has two identical states: "male" $(X_{3,1})$ or "female" $(X_{3,2})$. The biological causes of the two states may be different and specified separately.

Abstractly, a simple binary multi-valued case is illustrated in figure 2. The difference of figure 2 from figure 1 is that in figure 2, the causes of both $X_{3,1}$ and $X_{3,2}$ are specified separately.

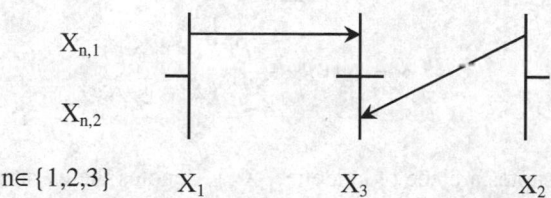

$X_{n,1}$

$X_{n,2}$

$n \in \{1,2,3\}$ X_1 X_3 X_2

Fig. 2. The illustration for the binary multi-valued case

The multi-valued cases are often encountered when the child variable has more than two states. For the example of a temperature variable (X_3), it may have "normal" $(X_{3,1})$, "high" $(X_{3,2})$, "very high" $(X_{3,3})$, "low" $(X_{3,4})$ and "very low" $(X_{3,5})$ five states. The causes of them are usually different and should be specified separately.

It is well known that the probabilities of all states of a variable must sum up to 1 in any case. This can be called the *normalization*. The single-valued cases always satisfy the normalization, because the conditional probability of the true state $\Pr\{\text{true}|\text{condition}\}$ cannot be greater than 1 and the conditional probability of the false state $\Pr\{\text{false}|\text{condition}\}$ is just the complement of the true state, i.e., $\Pr\{\text{false}|\text{condition}\}=1-\Pr\{\text{true}|\text{condition}\}$. However, in the multi-valued cases, the situation is different. To illustrate this, suppose that in figure 2, $X_{1,1}$ causes $X_{3,1}$ with $p_{3,1;1,1}=0.7$ and $X_{2,1}$ causes $X_{3,2}$ with $p_{3,2;2,1}=0.8$. In this situation, if we calculate the conditional probabilities of $X_{3,1}$ and $X_{3,2}$ as in the single-valued case separately, the normalization is not satisfied, because, as two identical states, $\Pr\{X_{3,1}|X_{1,1}X_{2,1}\}=p_{3,1;1,1}$ and $\Pr\{X_{3,2}|X_{1,1}X_{2,1}\}=p_{3,2;2,1}$ separately, while

$$\Pr\{X_{3,1}|X_{1,1}X_{2,1}\}+\Pr\{X_{3,2}|X_{1,1}X_{2,1}\}=p_{3,1;1,1}+p_{3,2;2,1}=0.7+0.8=1.5>1$$

It seems that equation (1) always satisfies the normalization in any multi-valued case:

$$\Pr\{X_{nk}|E\} = \frac{\Pr\{X_{nk}E\}}{\sum_{k}\Pr\{X_{nk}E\}}. \tag{1}$$

In which, E represents the evidence or condition. This equation is generally used to achieve the normalization of the multi-valued cases. For example, in [5], we find the following words:

"In this and subsequent examples, we assume that variables are Boolean (i.e., with domain {*true, false*})"; "The theory and the implementations are not restricted to binary variables".

We also find the following equations in [5]:

$$P(X \mid E_1 = o_1 \wedge \cdots \wedge E_s = o_s) = \frac{P(X \wedge E_1 = o_1 \wedge \cdots \wedge E_s = o_s)}{P(E_1 = o_1 \wedge \cdots \wedge E_s = o_s)};$$

$$P(E_1 = o_1 \wedge \cdots \wedge E_s = o_s) = \sum_{v \in dom(X)} P(X = v \wedge E_1 = o_1 \wedge \cdots \wedge E_s = o_s);$$

$$P(X = v_i \mid e) = \frac{P(X = v_i \wedge e)}{\sum_{v_j} P(X = v_j \wedge e)}.$$

Where $X_i = O_i$ denotes a piece of evidence, $X = v_j$ denotes the event of a hypothesis and $v \in dom(X)$ denotes all the possible states/values that X can have.

These equations are exactly the same as equation (1). It seems that many researchers think that the single-valued cases are generally the same as the multi-valued cases, except that equation (1) should be used to satisfy the normalization. However, this is incorrect, because equation (1) is valid only when $\sum_k \Pr\{X_{nk} \mid E\} = 1$. This can be seen in equation (2):

$$\Pr\{X_{nk} \mid E\} = \frac{\Pr\{X_{nk}E\}}{\Pr\{E\}} = \frac{\Pr\{X_{nk}E\}}{\Pr\{E\} \sum_k \Pr\{X_{nk} \mid E\}} = \frac{\Pr\{X_{nk}E\}}{\sum_k \Pr\{X_{nk}E\}}. \tag{2}$$

In the case of using CPT to represent CPD, $\sum_k \Pr\{X_{nk} \mid E\} = 1$ is always satisfied, because E is a state combination of the conditional variables and the normalization of X_{nk} is always satisfied in the CPT for the given E. However, in the case of the compact representation, the situation is different. If we use the special conditional probabilities to calculate $\Pr\{X_{nk} \mid E\}$ separately as in the single-valued cases, $\sum_k \Pr\{X_{nk} \mid E\} = 1$ is usually not satisfied. The above example has shown this. Therefore, the compact representations and inference algorithms applicable in single-valued cases cannot be automatically applied in multi-valued cases.

To illustrate this in details, let E in equation (1) be $E_j = SCPV_{n;j}$, where $SCPV_{n;j}$ denotes the state combination #j of the parent variables of X_n, i.e., $SCPV_{n;j} = \bigcap_i X_{ik_{ij}}$, k_{ij} indexes the state of X_i included in $SCPV_{n;j}$. Then, equation (1) can be written as

$$\Pr\{X_{nk} \mid E_j\} = \frac{\Pr\{X_{nk}E_j\}}{\sum_k \Pr\{X_{nk}E_j\}} = \frac{\Pr\{X_{nk} \mid E_j\}}{\sum_k \Pr\{X_{nk} \mid E_j\}} = \alpha_{n;j} \Pr\{X_{nk} \mid E_j\}, \tag{3}$$

$$\alpha_{n;j} \equiv 1/\sum_k \Pr\{X_{nk} \mid E_j\}.$$

(4)

Where $\alpha_{n;j}$ is the normalization factor, so that equation (3) sums up to 1 with respect to k. It is obvious that equation (1) is equivalent to equations (3) and (4).

In both equations (3) and (4), the $\Pr\{X_{nk}|E_j\}$ on the right side is the separately calculated conditional probability and the $\Pr\{X_{nk}|E_j\}$ on the left side of equation (3) is the normalized conditional probability. Note that in equation (4), the normalization factor $\alpha_{n;j}$ is not a constant but a variable depending on $E_j=SCPV_{n;j}$.

For the example shown in figure 3, which is the figure 1 in [2], suppose all variables are binary and only the causes of $X_{4,1}$ are specified with CSI. The case is single-valued, because $\Pr\{X_{4,2}|E_j\}=1-\Pr\{X_{4,1}|E_j\}$. Suppose we additionally specify the causes of $X_{4,2}$ separately as shown in figure 4, the case becomes multi-valued. Note that X_1 is not a parent variable of $X_{4,2}$, while X_1 is a parent variable of $X_{4,1}$. This is allowed in the separate specifications in the multi-valued cases.

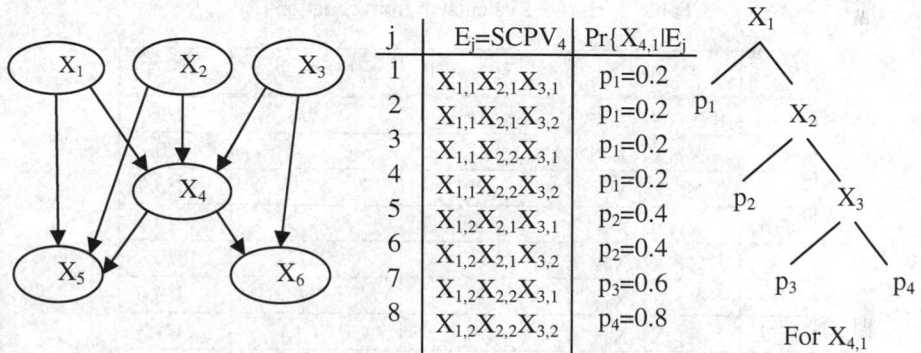

j	$E_j=SCPV_4$	$\Pr\{X_{4,1}|E_j$
1	$X_{1,1}X_{2,1}X_{3,1}$	$p_1=0.2$
2	$X_{1,1}X_{2,1}X_{3,2}$	$p_1=0.2$
3	$X_{1,1}X_{2,2}X_{3,1}$	$p_1=0.2$
4	$X_{1,1}X_{2,2}X_{3,2}$	$p_1=0.2$
5	$X_{1,2}X_{2,1}X_{3,1}$	$p_2=0.4$
6	$X_{1,2}X_{2,1}X_{3,2}$	$p_2=0.4$
7	$X_{1,2}X_{2,2}X_{3,1}$	$p_3=0.6$
8	$X_{1,2}X_{2,2}X_{3,2}$	$p_4=0.8$

Fig. 3. The single-valued case with CSI

As being pointed out earlier, it is obvious that $\Pr\{X_{4,1}|E_j\}$ in figure 3 and $\Pr\{X_{4,2}|E_j\}$ in figure 4 cannot sum up to 1. If we insist on applying equations (3) and (4) to satisfy the normalization, we have to have

$\alpha_{4;1}=1/(\Pr\{X_{4,1}|E_1\}+\Pr\{X_{4,2}|E_1\})=1/(0.2+0.4)=1/0.6$
$\alpha_{4;2}=1/(\Pr\{X_{4,1}|E_2\}+\Pr\{X_{4,2}|E_2\})=1/(0.2+0.2)=1/0.4$
$\alpha_{4;3}=1/(\Pr\{X_{4,1}|E_3\}+\Pr\{X_{4,2}|E_3\})=1/(0.2+0.7)=1/0.9$
$\alpha_{4;4}=1/(\Pr\{X_{4,1}|E_4\}+\Pr\{X_{4,2}|E_4\})=1/(0.2+0.7)=1/0.9$
$\alpha_{4;5}=1/(\Pr\{X_{4,1}|E_5\}+\Pr\{X_{4,2}|E_5\})=1/(0.4+0.4)=1/0.8$
$\alpha_{4;6}=1/(\Pr\{X_{4,1}|E_6\}+\Pr\{X_{4,2}|E_6\})=1/(0.4+0.2)=1/0.6$
$\alpha_{4;7}=1/(\Pr\{X_{4,1}|E_7\}+\Pr\{X_{4,2}|E_7\})=1/(0.6+0.7)=1/1.3$
$\alpha_{4;8}=1/(\Pr\{X_{4,1}|E_8\}+\Pr\{X_{4,2}|E_8\})=1/(0.8+0.7)=1/1.5$

j	$E_j=SCPV_4$	$Pr\{X_{4,2}|E_j\}$
1	$X_{1,1}X_{2,1}X_{3,1}$	$p_5=0.4$
2	$X_{1,1}X_{2,1}X_{3,2}$	$p_6=0.2$
3	$X_{1,1}X_{2,2}X_{3,1}$	$p_7=0.7$
4	$X_{1,1}X_{2,2}X_{3,2}$	$p_7=0.7$
5	$X_{1,2}X_{2,1}X_{3,1}$	$p_5=0.4$
6	$X_{1,2}X_{2,1}X_{3,2}$	$p_6=0.2$
7	$X_{1,2}X_{2,2}X_{3,1}$	$p_7=0.7$
8	$X_{1,2}X_{2,2}X_{3,2}$	$p_7=0.7$

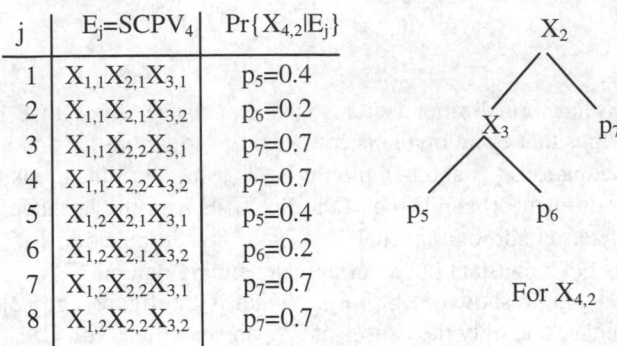

Fig. 4. The specification to $Pr\{X_{4,2}|E_j\}$ with CSI in the multi-valued case

Then, from equation (3), the CPT can be calculated as shown in table 1.

Table 1. The CPT calculated from equation (3)

j	$E_j=SCPV_{4;j}$	$Pr\{X_{4,1}|E_j\}$	$Pr\{X_{4,2}|E_j\}$	$\alpha_{4;j}$
1	$X_{1,1}X_{2,1}X_{3,1}$	1/3	2/3	1/0.6
2	$X_{1,1}X_{2,1}X_{3,2}$	1/2	1/2	1/0.4
3	$X_{1,1}X_{2,2}X_{3,1}$	2/9	7/9	1/0.9
4	$X_{1,1}X_{2,2}X_{3,2}$	2/9	7/9	1/0.9
5	$X_{1,2}X_{2,1}X_{3,1}$	1/2	1/2	1/0.8
6	$X_{1,2}X_{2,1}X_{3,2}$	2/3	1/3	1/0.6
7	$X_{1,2}X_{2,2}X_{3,1}$	6/13	7/13	1/1.3
8	$X_{1,2}X_{2,2}X_{3,2}$	8/15	7/15	1/1.5

It is seen that the normalized CPT in table 1 is based on many different $\alpha_{4;j}$, $j\in\{1,\ldots,8\}$. Theoretically, the number of $\alpha_{n;j}$ equals to the number of $E_j=SCPV_{n;j}$, which means that the number of $\alpha_{n;j}$ can be huge. For the example of a child variable with five parent variables with five states each, the number of $SCPV_{n;j}$ is $5^5=3125$. This is too many for domain experts to realize when they specify the causes and parameters of the states of X_n separately. The questions are: Why do we need so many different normalization factors? Are these different normalization factors realized by the domain experts when they specify the causes and parameters for the different multi-valued states separately? In other words, are these different normalization factors what the domain experts want? I do not think that these questions have always been clearly realized and answered when people apply equation (1).

Moreover, if we change the values of p_i as shown in table 2 (the new set of p_i), the normalized CPT remains same as in table 1. This is another issue that the domain experts may not realize. In fact, although the two sets of p_i work out a same CPT, they have different influences in the probability propagation through the causality chains

Table 2. The comparison between two sets of p_i

		The old set of p_i			The new set of p_i		
j	$E_j=SCPV_4$	$Pr\{X_{4,1}|E_j\}$	$Pr\{X_{4,2}|E_j\}$	$\alpha_{4;j}$	$Pr\{X_{4,1}|E_j\}$	$Pr\{X_{4,2}|E_j\}$	$\alpha_{4;j}$
1	$X_{1,1}X_{2,1}X_{3,1}$	$p_1=0.2$	$p_5=0.4$	1/0.6	$p_1=0.1$	$p_5=0.2$	1/0.3
2	$X_{1,1}X_{2,1}X_{3,2}$	$p_1=0.2$	$p_6=0.2$	1/0.4	$p_1=0.1$	$p_6=0.1$	1/0.2
3	$X_{1,1}X_{2,2}X_{3,1}$	$p_1=0.2$	$p_7=0.7$	1/0.9	$p_1=0.1$	$p_7=0.35$	1/0.45
4	$X_{1,1}X_{2,2}X_{3,2}$	$p_1=0.2$	$p_7=0.7$	1/0.9	$p_1=0.1$	$p_7=0.35$	1/0.45
5	$X_{1,2}X_{2,1}X_{3,1}$	$p_2=0.4$	$p_5=0.4$	1/0.8	$p_2=0.2$	$p_5=0.2$	1/0.4
6	$X_{1,2}X_{2,1}X_{3,2}$	$p_2=0.4$	$p_6=0.2$	1/0.6	$p_2=0.2$	$p_6=0.1$	1/0.3
7	$X_{1,2}X_{2,2}X_{3,1}$	$p_3=0.6$	$p_7=0.7$	1/1.3	$p_3=0.3$	$p_7=0.35$	1/0.65
8	$X_{1,2}X_{2,2}X_{3,2}$	$p_4=0.8$	$p_7=0.7$	1/1.5	$p_4=0.4$	$p_7=0.35$	1/0.75

when we apply the efficient inference algorithms based on the compact representations. To avoid this inconsistence, we may use the CPT in the probabilistic reasoning only. But this means that we give up the efficient algorithms applicable in the single-valued cases.

It should be pointed out that, in a single-valued case, the single-value is associated with only the child variable but *not* the parent variables. The number of *active* states of the parent variables can be more than one. Moreover, the single-valued child variable must be binary. More than two states of a single-valued child variable are meaningless, because the states not specified can be combined as one state. Figure 5 shows a single-valued case in which X_1 has three states and both X_1 and X_2 have two *active* states, while X_3 has only one *valued* state.

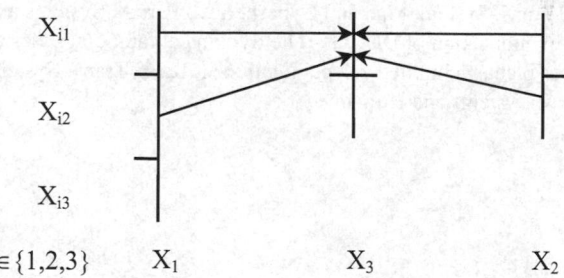

X_{i1}

X_{i2}

X_{i3}

$i\in\{1,2,3\}$ X_1 X_3 X_2

Fig. 5. Example of a single-valued case in which the parent variables have multiple active states

3 Conclusion

In conclusion, we should be careful whether or not equation (1) is what we want. In other words, the representations and algorithms applicable in the single-valued cases may not be applicable in the multi-valued cases when we apply the compact representation model in stead of CPT for representing CPD. The solution for compactly representing CPD and the corresponding inference algorithm has been presented by the author in [6-10].

References

[1] Pearl, J.: Probabilistic reasoning in intelligent systems. Morgan Kaufmann, San Mateo (1988)

[2] Boutilier, C., Friedman, N., Goldszmidt, M., Koller, D.: Context-specific independence in Bayesian Network. In: Proc. of UAI 1996 (1996)

[3] Zhang, Q.: Probabilistic reasoning based on dynamic causality trees/diagrams. Reliability Engineering and System Safety (46), 209–220 (1994)

[4] Poole, D., Zhang, N.L.: Exploiting contextual independence in probabilistic inference. Journal of Artificial Intelligence Research 18, 263–313 (2003)

[5] D'Ambrosio, B.: Local expression languages for probabilistic dependence. Int. J. Approximate Reasoning 139(1), 61–68 (1995)

[6] Zhang, Q.: DUCG - a new approach to deal with dynamical uncertain causality information - part I: the static discrete DAG case. Submitted for publication in IEEE Trans. System, Man and Cybernetics - Part A: Systems and Human

[7] Zhang, Q.: DUCG - a new approach to deal with dynamical uncertain causality information - part II: the dynamical DCG case. Submitted for publication in IEEE Trans. System, Man and Cybernetics - Part A: Systems and Human

[8] Zhang, Q.: DUCG - a new approach to deal with dynamical uncertain causality information - part III: the discrete dynamical case. Submitted for publication in IEEE Trans. System, Man and Cybernetics - Part A: Systems and Human

[9] Zhang, Q.: DUCG - a new approach to deal with dynamical uncertain causality information - part IV: The Continuous Variable and Uncertain Evidence. Submitted for publication in IEEE Trans. System, Man and Cybernetics - Part A: Systems and Human

[10] Zhang, Q.: An application of DUCG – The dynamical fault diagnoses and predictions of a nuclear power plant. Submitted for publication in IEEE Trans. System, Man and Cybernetics - Part A: Systems and Human

The Control Based on Internal Average Kinetic Energy in Complex Environment for Multi-robot System[*]

Mao Yang, Yantao Tian[**], and Xianghua Yin

School of Communication Engineering, Jilin University,
Renmin Street. 5988, 130025 Changchun, China
tianyt@jlu.edu.cn,
Yangmao820@yahoo.com.cn

Abstract. In this paper, reference trajectory is designed according to minimum energy consumed for multi-robot system, which nonlinear programming and cubic spline interpolation are adopted. The control strategy is composed of two levels, which lower-level is simple PD control and the upper-level is based on the internal average kinetic energy for multi-robot system in the complex environment with velocity damping. Simulation tests verify the effectiveness of this control strategy.

Keywords: multi-robot system, velocity damping, trajectory tracking.

1 Introduction

With the constantly expanding of the robot application areas, it is very difficult for the single robot to meet the demand on application. Multiple robots form a system and completing a more complex task through collaboration is increasingly becoming the focus problem in the robotics and intelligent science field [1-2]. The important goal of the development of the multi-robot systems is to design a basic structure of distributed control, so that the robot can implement tasks without the supervision, and it demands strong self-adaptive ability when the robot works in an unknown environment. Trajectory tracking is a comparatively effective control strategy to realize multi-robot foraging mission and map detection, and it is mainly divided into two stages: the generation of reference trajectory and trajectory tracking. Under the condition that the multi-robot model is known, the track meet dynamic stability can be generated offline. The investigation of multi-robot origins from the behavior of swarm organism in nature which can finish a complex work though coordination such as flocks of birds, school of fish. Biologists have been working on understanding and modeling of swarming behavior for a long time. There are two fundamentally different approaches that they have been considering for analysis of swarm dynamics. For one hand, statistics method is adopted for the multi-robot system from the macro aspects; for another hand, the individual-based models from the microeconomic are becoming more and more popular.

---

[*] This work is sponsored by the National Science Foundation of China (Grant No: 60675057).
[**] Corresponding author.

J. Zhou (Ed.): Complex 2009, Part I, LNICST 4, pp. 607–617, 2009.
© ICST Institute for Computer Sciences, Social Informatics and Telecommunications Engineering 2009

In 1986, Reynolds developed a behavior model to animate the coordinated motions of a group of agents named "boids". Vicsek model is a special case of boids. Gazi(2004) established a one-order system for stability analysis of social foraging swarm[3],which is a good survey of previous work in swarm literature. He presented a continuous first-order kinematic model for swarm members, and applied the idea of virtual force to propose a decentralized controller to analyze swarm aggregation in n-dimensional space. Gazi(2004) specified a general class of attraction/repulsion functions that can be used to achieve swarm aggregation. They presented stability analysis for several cases of the functions considered to characterize swarm cohesiveness, size and ultimate motions while in a cohesive group. Above models can be used to handle different control objectives, however, the shortcoming of above models is that it require all the individuals moving with a common velocity which is not suitable for some application. Although second order dynamic model are utilized, most of them are based on kinematic swarm model. Sliding mode controller is developed to force the vehicle motions to obey dynamics of the kinematic swarm in [4]. Pedrami(2007) firstly presents control and analysis of energetic swarms in [5], in particular the internal kinetic energy is investigated in his paper, Pedrami(2007) modify the temperature definition as the internal average kinetic energy in[6].The cohesion of swarm is an important issue for multi-robot system. It is assumed that a swarm internal energy is bounded, then, a relation between the swarm size and internal average kinetic energy studied in his paper. Pedrami(2008) introduce the controller for wheeled mobile robots(WMR) In [7].

In above work, the environment of multi-robot system is ideal, which there not exist any damping. As a matter of fact, the velocity damping is ineluctable which is not investigated in above works. Another point which is worth noting is that control means energy input, e.g. when the temperature in [5] is too large, we have to offer too much energy. This is not we expect. The energy we have is limited, so how to use the limited energy for finishing the work is meaningful for the real project. Therefore nonlinear programming and cubic spline interpolation are adopted in this paper for trajectory tracking which the environment is not ideal i.e. the velocity damping is not zero.

The paper is organized as follows. In section 2 an 2-dimensional second order swarm model is presented. In section 3 the lower-level control is a simple PD control which the generation of reference trajectory is based on nonlinear programming and cubic spline interpolation. In section 4, a complete discussion of the swarm energy is performed and an upper layer internal average kinetic energy controller for multi-robot with the velocity damping is developed. It is also shown how an internal average kinetic energy controller is useful. Simulation results are presented in section 5 to verify the effectiveness of proposed control strategy. The paper ends with conclusions and future research directions in section 6.

2   Swarm Model

Consider a swarm of M members moving in 2-dimensional space. For the multi-robot system we have assumptions as follow

Assumption 1

(1) We ignore each robot's dimension and treat it as a point mass.
(2) All robots move synchronously, i.e., there is no communication delay between each other.
(3) All robots are homogenous, i.e., there is no difference in essence.

Previous theoretical and modeling efforts fall into a number of categories. Many of the original ideas about group spacing were formulated qualitatively, or using simple formulate. For each robot, we can establish the model according to the classical Newtonian mechanics law as follow

$$\dot{x}_i = v_i \tag{1}$$

$$m_i \dot{v}_i = u_i^{ext} + u_i^{in} - b_i v_i \tag{2}$$

where $i = 1,2,3...,M$, $x_i \in \mathfrak{R}^2$ is the position of the i th robot, $v_i \in \mathfrak{R}^2$ is the velocity of the i th robot, $m_i = m$ is the mass of the i th robot, u_i^{ext} is input which we impose. u_i^{in} is the total force on the robot i as a result of inter-robot interaction, obviously, we have $\sum_{i=1}^{M} u_i^{in} = 0$ because the internal interaction offset. The term u_i^{in} is for the cohesion of swarm and is of the form

$$u_i^{in} = -\sum_{j=1, j\neq i}^{M} [g_a(\|x^i - x^i\|) - g_r(\|x^i - x^i\|)](x^i - x^i) \tag{3}$$

where $g_r : \mathfrak{R}^+ \to \mathfrak{R}^+$ and $g_a : \mathfrak{R}^+ \to \mathfrak{R}^+$ represents respectively the magnitude of repulsion force and the attraction force. $-b_i v_i$ represents the velocity damping from the environment. It is worth noting that $b_i = b \neq 0$ in this paper.

Assumption 2. There exist corresponding functions $J_a : \mathfrak{R}^+ \to \mathfrak{R}$ and $J_r : \mathfrak{R}^+ \to \mathfrak{R}$ such that for any $y \in \mathfrak{R}^2$

$$\nabla_y J_a(\|y\|) = yg_a(\|y\|), \nabla_y J_r(\|y\|) = yg_r(\|y\|) \tag{4}$$

Definition 1. The swarm center $\bar{x} \in \mathfrak{R}^2$ is defined by

$$\bar{x} = \frac{1}{M} \sum_{i=1}^{M} x_i \tag{5}$$

Therefore, the velocity of the swarm center $\bar{v} \in \mathfrak{R}^2$ is derived by time differentiation

$$\bar{v} = \frac{1}{M} \sum_{i=1}^{M} v_i \tag{6}$$

The position and velocity can be denoted as follows

$$x = (x^{1^T},...,x^{M^T})^T, v = (v^{1^T},...,v^{M^T})^T$$

Then, the system can be expressed as

$$
\begin{pmatrix} \dot{x}^i \\ \dot{v}^i \end{pmatrix} = \begin{pmatrix} 0 & 1 \\ 0 & -b \end{pmatrix} \begin{pmatrix} x^i \\ v^i \end{pmatrix} + \begin{pmatrix} 0 \\ 1 \end{pmatrix} (u^i_{ext} + u^i_{in}) \tag{7}
$$

We have assumption that $u^{ext}_i = u^{ext}$ because of (4) and $\sum_{i=1}^{M} u^{in}_i = 0$, we have

$$
m\ddot{\overline{x}} + b\dot{\overline{x}} = u^{ext} \tag{8}
$$

3   The Generation of Reference Trajectory Based on Minimum of Energy Consumption

In the field of robotics control, it has always been a goal pursued that multi-robot system can finish a complex mission with minimum of energy consumption. It is of great significance for space exploration which multi-robot system does. The engineering background of this paper can be understand that the multi-robot system moves from an initial position to a position designated in advance in order to achieve the assignment of map detection, foraging and so on. In this paper, the performance index is the minimum of energy consumption by means of nonlinear programming; the optimal position is achieved, then the reference trajectory gained according to cubic spline interpolation. So the reference trajectory can be tracked though PD controller.

3.1   The Generation of Reference Trajectory

For (8), let $\overline{x} = (\overline{s}_x, \overline{d}_x)$, then $\overline{s}_x$ can be expressed as polynomial of degree m over $\overline{d}_x$ as follows

$$
\overline{s}_x = p_0 + p_1 d_x + p_2 d_x^2 + \cdots + p_m d_x^m \tag{9}
$$

Then,

$$
\dot{\overline{s}}_x = p_1 + 2 \cdot p_2 d_x + \cdots + m \cdot p_m d_x^{m-1}
$$
$$
\ddot{\overline{s}}_x = 2 \cdot p_2 + \cdots + m \cdot (m-1) p_m d_x^{m-2}
$$

where $p_0, p_1 \cdots p_m$ are undetermined coefficients; suppose that the initial point is $\overline{x}(0) = (0,0)$; the target point is $\overline{x}(T)$; the tracking time T are unknown. $[\overline{x}(0), x(T)]$ is divided into N parts average according to $\Delta t = x(T)/N$. Take minimum energy as the objective function, the state of initiation point and target point are the constraint conditions. Then according to the definition of definite integral, we have

$$
\min J = \int_0^{x(T)} u(t) \cdot u^T(t) dx = \sum_{k=0}^{N-1} u(k) \cdot u^T(k) \cdot \frac{x(T)}{N} \tag{10}
$$

Furthermore, optimal trajectory is achieved taking the form as follow

$$
\overline{s}_x = p_0 + p_1 d_x + p_2 d_x^2 + \cdots + p_m d_x^m
$$

3.2 Numerical Experiment

The dynamic model for multi-robot system can be expressed in form of (8). $X = (x, y)$ is position of robot in 2-D plane. The structural parameter values in this paper are respectively the quality $m = 1$, the number of robots $M = 5, b = 0.2$; the initial position $\overline{x}(0) = (0,0)$;the target position $\overline{x}(T) = (10,10)$ .The reference trajectory based on minimum of energy consumption is obtained through Matlab numerical experiment:

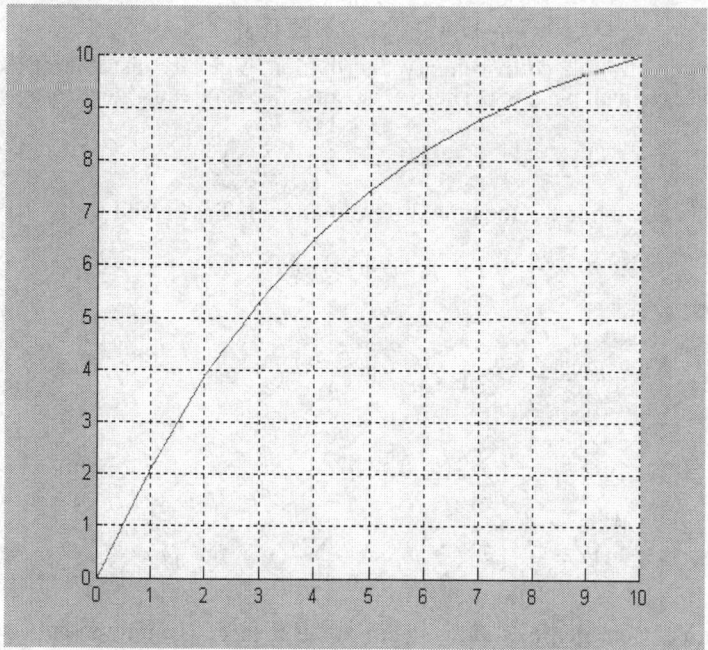

Fig. 1. Reference trajectory

4   Internal Average Kinetic Energy Controller

In this section, a controller based on internal average kinetic energy is introduced. Design procedures include two different steps. The first step is to design the controller such that the swarm center arrives at the specified position as soon as possible. The second step is to modify the control input designed in the first step. Any combination of individual input u_i^{ext} that satisfies (16) guarantees the optimal control for swarm center. However, it does not guarantee the cohesion of swarm. Various types of energy are present firstly. Total potential energy of swarm

$$J(x) = \sum_{i=1}^{M-1} \sum_{j=i+1}^{M} [J_a(\|x^i - x^i\|) - J_r(\|x^i - x^i\|)] \tag{11}$$

Total kinetic energy of swarm

$$E_k(v) = \frac{1}{2}\sum_{i=1}^{M} m_i \left\| v^i \right\|^2 \qquad (12)$$

Average kinetic energy of swarm

$$E_b(v) = \frac{1}{2}(\sum_{i=1}^{M} m_i)\left\| \bar{v} \right\|^2 \qquad (13)$$

Definition 2. Internal average kinetic energy T is defined as follow

$$T = 1/M[E_k(v) - E_b(v)] \qquad (14)$$

where T is directly related to velocity distribution. Velocity distribution is useful to study a coverage path planning. There exist some types of equivalent form

$$T(v) = \frac{m}{2M}(\sum_{i=1}^{M} \left\| v^i \right\|^2 - M \left\| \bar{v} \right\|^2) = \frac{m}{2M^2}\{\sum_{i=1}^{M-1}\sum_{j=i+1}^{M} \left\| v^i - v^j \right\|^2\}$$

According to assumption 2, the time differentiation of T is given by

$$\dot{T}(v) = \phi + \sigma + \psi$$

where ϕ, σ, ψ are

$$\phi = \frac{1}{M}\{\sum_{i=1}^{M}[(u_{ext}^i)^T v^i] - \bar{v}^T(\sum_{i=1}^{M} u_{ext}^i)\}$$

$$\sigma = \frac{1}{M}\{\sum_{i=1}^{M}[(u_{in}^i)^T v^i]\}$$

$$\psi = \frac{1}{M}\{\bar{v}^T(\sum_{i=1}^{M} b_i v^i) - \sum_{i=1}^{M}[(v^i)^T b_i v^i]\}$$

Proposition 1. Consider the following controller in viscous environment

$$u_i^{ext} = u^*(\bar{x}, \bar{v}) + u_i^T$$

where the extra control is given by

$$u_i^T = -\sum_{i=1}^{M} \alpha_{ij}(v^i - v^j) \qquad (15)$$

The α_{ij} is the control parameter and developed as

$$\alpha_{ij} = -\frac{(x^i - x^j)\beta_{ij}(v^i - v^j)}{\left\| v^i - v^j \right\|^2} + (\lambda - b/M)\frac{(\left\| v^i - v^j \right\|^2 - k)}{\left\| v^i - v^j \right\|^2}$$

where λ is a positive constant and k is a parameter for control; β_{ij} is given as

$$\beta_{ij} = g_a(\left\| x^i - x^i \right\|) - g_r(\left\| x^i - x^i \right\|)$$

It can be shown that T is convergent.

Proof. For optimal control, it is required to investigate the input. From (20) the input is calculated as

$$\sum_{i=1}^{M} u^{i}{}_{ext} = Mu^{*}(\overline{x},\overline{v}) + \sum_{i=1}^{M} u^{i}{}_{T}$$

To keep optimal characteristic of above control, the extra input $u^{i}{}_{T}$ should satisfy the following condition $\sum_{i=1}^{M} u_{i}^{T} = 0$. This condition holds since control parameters α_{ij} satisfies the symmetry, e.g., $\alpha_{ij} = \alpha_{ji}$, so

$$\sum_{i=1}^{M} u_{i}^{T} = -\sum_{i=1}^{M-1}\sum_{j=i+1}^{M} [\alpha_{ij}(v^{i}-v^{j}) + \alpha_{ji}(v^{j}-v^{i})] = 0$$

This means that the extra control input u_{i}^{T} can be viewed as an internal interaction. Furthermore, we have

$$\dot{T}(v) = \phi+\sigma+\psi = \sigma+\psi = \frac{1}{M}\{\sum_{i=1}^{M}[(u_{in}^{i}+u_{i}^{T})^{T}v^{i}]\} + \frac{1}{M}\{\overline{v}^{T}(\sum_{i=1}^{M}b_{i}v^{i}) - \sum_{i=1}^{M}[(v^{i})^{T}b_{i}v^{i}]\}$$

$$= 1/M\{\sum_{i=1}^{M}[(-\sum_{j=1,j\neq i}^{M}\beta_{ij}(x^{i}-x^{j}) - \sum_{i=1}^{M}\alpha_{ij}(v^{i}-v^{j}))^{T}v^{i}]\} + 1/M[\overline{v}^{T}bM\overline{v} - b\sum_{i=1}^{M}\|v^{i}\|^{2}]$$

$$= -(1/M)\{\sum_{i=1}^{M-1}\sum_{j=i+1}^{M}\beta_{ij}(x^{i}-x^{j})^{T}(v^{i}-v^{j}) + \sum_{i=1}^{M-1}\sum_{j=i+1}^{M}\alpha_{ij}(x^{i}-x^{j})^{T}(v^{i}-v^{j})\} + 1/M(-2TMb/m)$$

$$= -(1/M)[(\lambda-b/M)\sum_{i=1}^{M-1}\sum_{j=i+1}^{M}(\|v^{i}-v^{j}\|^{2}-k) + (2TM/m)b]$$

$$= -(1/M)[(\lambda-b/M)T - (\lambda-b/M)Mk(M-1)/2 + (2TMb)/m]$$

$$= -\lambda(2M/m)T - k(M-1)(b/M-\lambda)/2$$

In other words, we have

$$\dot{T}(v) + \lambda\frac{2M}{m}T(v) + \frac{k(M-1)(b/M-\lambda)}{2} = 0$$

The differential equation is stable since $\lambda > 0$. This completes the proof.

5   Simulation Results

In this section, simulation will verify the validity of the controller proposed above. The goal of the experimental tasks is that swarm robots system need to arrive at the specified prior location so as to implement rescue as soon as possible, at the same time, the swarm system need to keep cohesion. The swarm system has 5 robots and moving in 2-D space. The attraction and repulsion functions are expressed by

$$g_{a}(y) = A exp(-\frac{y}{a}); g_{r}(y) = R exp(-\frac{y}{r})$$

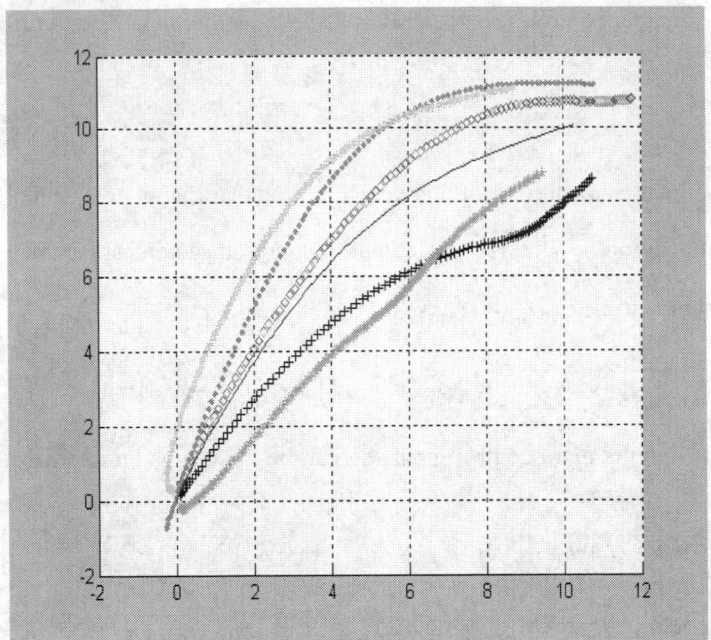

Fig. 2. Each robot trajectory for group 1

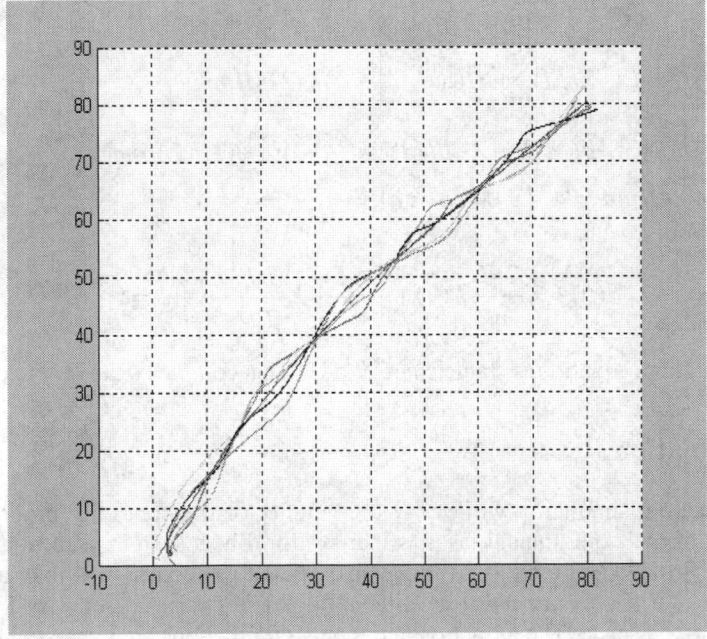

Fig. 3. Each robot trajectory for group 2

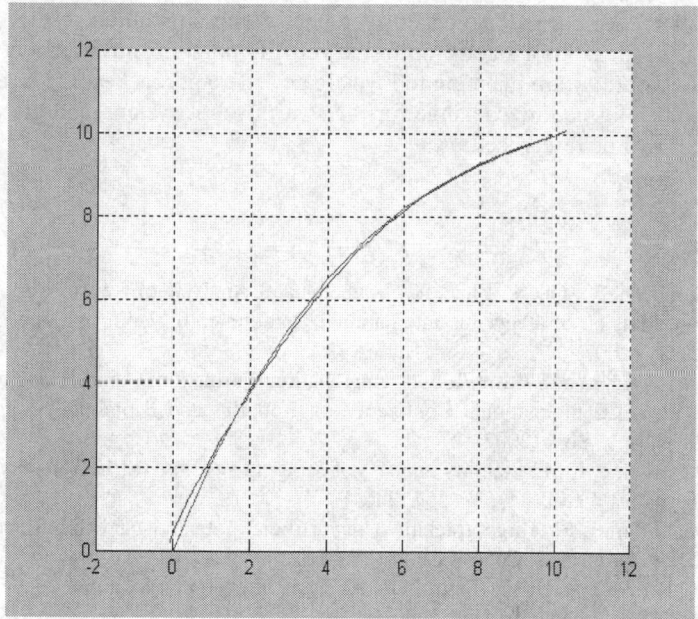

Fig. 4. Reference trajectory and average trajectory

where A is the magnitude of attraction and a is the spatial range of the attraction, while R the magnitude of repulsion and r is the spatial range of the repulsion. In order to keep cohesion, $R > A$ *and* $a > r$ is necessary which is the short-ranged repulsion and long-ranged attraction case. There are 2 groups of experiments for the proposed controller, for group 1, we select the design constants to be $\overline{x}(0) = (0,0)$, $\overline{x}(T) = (10,10)$, $b = 0.2$, $m = 1$, $M = 5$. For group 2, $\overline{x}(T) = (80,80)$.

Fig.2 and Fig.3 show all the trajectories for the robots keep cohesion. And the average trajectory is from $\overline{x}(0) = (0,0)$ to $\overline{x}(T) = (10,10)$ and $\overline{x}(T) = (80,80)$ respectively. Different colors represent different robots and the blue line is the average trajectory of the swarm system. Fig.4 is the effect of PD controller between reference trajectory and average trajectory.

6 Conclusion and Future Work

In this paper, we showed our unique strength in dealing with distributed control for multi-robot in complex environments with velocity damping. Swarm model is built based on the Newton second law. Reference trajectory is designed according to minimum energy consumed for multi-robot system, which nonlinear programming and cubic spline interpolation are adopted. The control strategy is composed of two levels, which lower-level is simple PD control which can track the reference trajectory and the upper-level is based on the internal average kinetic energy for multi-robot system which can control the size of swarm. For the upper-level, convergence proof is presented for the velocity damping environments. The simulation results testify the effectiveness of the control algorithm for the velocity damping environments.

For future work, we are planning to investigate other approaches which are concentrated on local information exchange for the swarm robot system, and the robustness of the system will be researched.

References

1. Yang, Y., Tian, Y., Qi, X., Zhao, X.: Mathematical Analysis of Swarm Robots Foraging Based on Division Strategy. In: International Conference on Computational Intelligence and Security (2007)
2. Yang, Y., Tian, Y.: Swarm Robots Aggregation Formation Control Inspired by Fish School. In: IEEE International Conference on Robotics and Biomimetics, pp. 805–809. IEEE Press, New York (2007)
3. Gazi, V., Passino, K.M.: Stability analysis of foraging swarms. IEEE Trans. Systems, Man and Cybernatics, Part B 34, 539–557 (2004)
4. Gazi, V., Ordonez, R.: Target tracking using artificial potentials and sliding mode control. In: Proc. 2004 American Control Conf. (2004)
5. Pedrami, R., Gordon, B.W.: Control and Analysis of energetic Swarms. Technical Report, CIS LAB, Concordia University (2006)
6. Pedrami, R., Gordon, B.W.: Temperature control of energetic swarms. Technical Report, CIS LAB, Concordia University (2007)
7. Pedrami, R., Gordon, B.W.: Control of energetic robotic swarms systems. In: IEEE International Conference on Robotics and Biomimetics, pp. 547–552. IEEE Press, New York (2007)
8. John, H.R., Hongyan, W.: Social Potential Fields: A Distributed Behavioral Control for Autonomous Robots. Robotics and Autonomous Systems (1999)
9. Brooks, R.A.: A Robust Layered Control System for a Mobile Robot. J. IEEE Transactions on Robotics and Automation (1986)
10. Lerman, K., Galstyan, A.: A general methodology for mathematical analysis of multi-agent systems. USC Information Sciences Technical Report ISI-TR-529 (2001)
11. Gerkey, B.P., Sold, M.M.J.: Auction methods for Multi Robot coordination. IEEE Transactions on Robotics and Automation 18(5), 758–768 (2002)
12. Warburton, K., Lazarus, J.: Tendency-distance models of social cohesion in animal groups. J. Theor. Biol. 150, 473–488 (1991)
13. Lee, C.T., Hoopes, M.F., Diehl, J., Gilliland, W., Huxel, G., Leaverand, E.V., McCann, K., Umbanhowar, J., Mogilner, A.: Non-local concepts and models in biology. J. Theor. Biol. 210, 201–219 (2001)
14. Miller, R.S., Stephen, W.J.D.: Spatial relationships in flocks of sandhill cranes. Ecology 47(2), 323–327 (1966)
15. Misund, O.A.: Sonar observations of schooling herring: school dimensions swimming behaviour and avoidance of vessel and purse seine Rapp. P.-v. Reun. Cons. int. Explor. Mer. 189, 135–146 (1990)
16. Misund, O.A.: Dynamics of moving masses: variability in packing density shape and size among herring sprat and saithe schools. ICES. J. Sci. 50, 145–160 (1993)
17. Mogilner, A., Edelstein-Keshet, L.: A non-local model for a swarm. J. Math. Biol. 38, 534–570 (1999)

18. Newlands, N.: PhD Thesis: Shoaling dynamics and abundance estimation: Atlantic Blue-finTuna (Thunnus thynnus), Fisheries Center UBC, Vancouver BC Canada (2002)
19. Niwa, H.-S.: Self-organizing dynamic model of fish schooling. J. Theor. Biol. 171, 123–136 (1994)
20. Niwa, H.-S.: Newtonian dynamical approach to fish schooling. J. Theor. Biol. 181, 47–63 (1996)
21. Okubo, A.: Diffusion and Ecological Problems: Mathematical Models. Springer, New York (1980)
22. Okubo, A.: Dynamical aspects of animal grouping: swarms, schools, flocks, and herds. Adv. Biophys. 22, 1–94 (1986)
23. Parr, A.E.: A contribution to the theoretical analysis of the schooling behaviour of fishes. Occasional Papers of the Bingham Oceanographic Collection 1, 1–32 (1927)
24. Parrish, J., Edelstein-Keshet, L.: Complexity, pattern, and evolutionary trade offs in animal aggregation. Science 284, 99–101 (1999)
25. Radin, C.: The ground state for soft discs. J. Stat. Phys. 26, 365–373 (1981)
26. Sakai, S.: A model for group structure and its behavior. Biophysics Japan 13, 82–90 (1973)
27. Sinclair, A.R.E.: The African buffalo: a study of resource limitation of populations. University of Chicago Press, Chicago (1977)
28. Suzuki, R., Sakai, S.: Movement of a group of animals. Biophysics Japan 13, 281–282 (1973)
29. Tegeder, R.W., Krause, J.: Density dependence and numerosity in fright stimulated aggregation behaviour in shoaling fish. Phil. Trans. R. Soc. Lond. B 350, 381–390 (1995)
30. Turchin, P.: Beyond simple diffusion: models of not-so-simple movement in animals and cells. Comments on Theor. Biol. 1, 65–83 (1989)
31. Uvarov, B.P.: Locusts and Grasshoppers, Imperial Bureau of Entomology, London (1928)
32. Vabo, R., Nottestad, L.: An individual based model of fish school reactions: predicting antipredator behaviour as observed in nature. Fisheries Oceanography 6(3), 155–171 (1997)
33. Viscido, S.V., Miller, M., Wethey, D.S.: The response of a selfish herd to an attack from outside the group perimeter. J. Theor. Biol. 208, 315–328 (2001)
34. Viscido, S.V., Wethey, D.S.: Quantitative analysis of fiddler crab flock movement: evidence for 'selfish herd' behaviour. Animal Behaviour 63(4), 735–741 (2002)
35. Krause, J., Tegeder, R.W.: The mechanism of aggregation behaviour in fish shoals: individuals minimize approach time to neighbors. Anim. Behav. 48, 353–359 (1994)

The Contrast of Parametric and Nonparametric Volatility Measurement Based on Chinese Stock Market

Xinwu Zhang, Yan Wang, and Handong Li*

School of Management, Beijing Normal University,
Beijing, 100875, People's Republic of China
li_handong@sina.com

Abstract. Most procedures for modeling and forecasting financial asset return volatilities rely on restrictive and complicated parametric GARCH or stochastic volatility models. The method of realized volatility constructed from high-frequency intraday returns is an alternative choice for volatility measurement. In this paper we make an empirical analysis on Chinese stock index data by using the method of nonparametric realized volatility. We find that the realized volatility can describe the Chinese stock index volatility very well. The original Chinese stock index return series show obvious leptokurtic, fat-tailed relative to the Gaussian distribution.We show that the return series standardized instead by the realized volatility are very nearly Gaussian distribution, and we find that the four minutes is a better choice as the best time interval to describe the volatility of Chinese stock market. We also make a contrast with the popular method of GARCH model, but the return series standardized instead by GARCH model don't accord with Gaussian distribution. The result shows that the realized volatility can describe the dynamic behaviors of Chinese stock market well. In a way, it indicates that the Chinese stock market is effective.

Keywords: realized volatility, GARCH, volatility measurement, conditional distribution.

1   Introduction

In econophysics research, much effort has been devoted on both the empirical and the theoretical level to such phenomena like fat-tailed distributions of financial fluctuations, persistent correlations in volatility of financial asset returns.The distributional characteristics of asset returns are the key ingredients for the pricing of financial instruments, portfolio allocation, performance evaluation, and managerial decision making. The most critical feature of the conditional return distribution is arguably its second moment structure, which is empirically the dominant time-varying characteristic of the distribution. Because volatility persistence renders

* Correspondence Author. This research is supported by National Science Foundation of China under grant No. 70601002.

J. Zhou (Ed.): Complex 2009, Part I, LNICST 4, pp. 618–627, 2009.

high-frequency returns temporally dependent (Bollerslev, Chou and Kroner [1]), it is the conditional return distribution, and not the unconditional distribution, that is of relevance for risk management. This is especially true in high-frequency situations, such as monitoring and managing the risk associated with day-to-day operations of a trading desk, where volatility clustering is omnipresent.

It is well known that most of high-frequency financial asset return time series, such as the Shanghai stock market index, are leptokurtic, fat-tailed relative to the Gaussian distribution, and that the fat tails are typically reduced but not eliminated when return series are standardized by volatilities estimated from popular models such as GARCH and SV. A sizable literature explicitly attempts to model the fat-tailed conditional distribution, including, for example, Engle and Gonzalez-Rivera [2], and K. Chen, C. Jayprakash and B. Yuan [3],

Andersen, Bollerslev, Diebold and Labys[4] consider two major dollar exchange rates, and they show that returns standardized instead by the realized volatilities are nearly Gaussian. It indicates that it may be very important to find a suitable volatility measure to depict the conditional distribution of return series.

Assuming that return series dynamics operate only through the conditional variance, a standard decomposition of the time-t return series is:

$$r_t = \sigma_t * \varepsilon_t \tag{1}$$

Where σ_t refers to the time-t conditional standard deviation, and $\varepsilon \sim N(0,1)$. Thus, given σ_t it would be straightforward to back out ε_t and assess its distributional properties. Of course σ_t is not directly observable. When using an estimate of σ_t from GARCH or SV model the distributions of the resulting standardized returns are typically found to be fat-tailed, or leptokurtic.

The research about the volatility of Chinese stock market returns is concentrated in low frequency data by the method of GARCH model in recent years. We analyze high frequency data using the method of realized volatility, and find that the realized volatility method can describe the dynamic volatility behaviors of Chinese stock market return well.

The stock index return series show obvious leptokurtic, fat-tailed distribution, but after we standardize the return series which is processed with realized volatility, it doesn't refuse Gaussian distribution hypothesis. We also make a compare between the realized volatility and GARCH model, the result indicates that return distribution standardized by GARCH model refuse Gaussian distribution hypothesis.

This shows that the realized volatility can describe the dynamic behaviors of Chinese stock market well. Meanwhile the result is according with the hypothesis that the series of stock index obey Semimartingale Stochastic Differential Equations, which suggest that the Chinese stock market is available.

2   Realized Volatility and *GARCH* Model

2.1   Realized Volatility

In 1980, Merton noticed that the variance in the fixed period of independent and identically distributed random variables could be estimated by the square sum of the return realization value ,and as long as the frequency is enough high, the estimation is very exact. French and Schwert etc [5] use the daily income in months to estimate the variance of every month. Andersen and Bollerslev[6], Hesieh [7], and Taylor S, Xu X [8] respectively use the square sum of within-day return to estimate the variance of daily return. In the recent years, Andersen and Bollerslev [9] put forward using the high-frequency to calculate the volatility rate. They put forward the method of measuring the realized volatility, that is to use the square sum of the return for some time as the estimation of the volatility. This estimation method is different from ARCH models and SV models ,and it doesn't depend on the model and doesn't need complex parameter estimation.

We first give some usual symbols.

Logarithm yield: As observed the financial assets price data in the time interval in time length $[0, T]$, we define return:

$$r^*(i, \delta) = lnS_i - lnS_{i-\delta} \tag{2}$$

It denotes the Continuously Compounded Return of financial assets in a certain time interval. Notice it is logarithmic.

In the financial environment of risk-free arbitrage, logarithm yield r_i^* of financial assets obeys Special Half Martingale process. If (Ω, I, P) supposing is a complete probability space, Information Filtration $(I_t)_{t \in [0,T]} \subseteq I$ is a increasing subalgebra series, I_t is p-complete and right continuity $S_t, t \in [0, t]$. This definition represent the price of financial primary assets in this spatial , then S_t is included in the information set I_t in t time.

Logarithm yield in the Δ period: The logarithm yield in the Δ period in the t-day invested in some financial primary assets is

$$r^*(t, \Delta) = X_t - X_{t-\Delta} \tag{3}$$

Where t represent the t-day $\Delta > 0$. As existing market microstructure noise, there exists the deviation in some degree between high-frequency financial data observed and potential real data, so we need proceed data according to following method

$$\tau_i = t - 1 + \frac{i}{n}, (i = 0, 1, \cdots, n) \tag{4}$$

Where τ_i represent observation value at the closing time point of i-observation period in t-day.

Note the observation value of log-price to be

$$Y_{\tau_i} = X_{\tau_i} + \varepsilon_{\tau_i} \tag{5}$$

where ε_{τ_i} is microstructure noise.

For simplification, let

$$E(\varepsilon_{\tau_i}) = E(\varepsilon), E(\varepsilon_{\tau_i}^2) = E(\varepsilon^2), \tag{6}$$

Where ε and log-price process are mutually independent.

$$\text{Let} \qquad r(\tau_i, \frac{1}{n}) = Y_{\tau_i} - Y_{\tau_i-1}, (i = 1, \cdots, n) \qquad (7)$$

It is logarithm yield in $[\tau_{i-1}, \tau_i]$ of financial assets base on observation data. when $\Delta=1$, realized volatility in t-day is

$$RV_t = \sum_{i=1}^{n} [r(\tau_i, \frac{i}{n})]^2 \qquad (8)$$

It is the realized volatility of financial primary assets in t-day in which the observation frequency of financial assets price data in t-day is $n + 1$.

2.2 GARCH Model

GARCH models are very popular for representing the dynamic evolution of the volatility of financial returns and have been extensively analyzed in the literature [see, e.g., Bollerslev, Engle, and Nelson [10], Bera and Higgins [11], Diebold and Lopez [12], and McAleer and Oxley [13], among many others].

The condition variance of GARCH model is represented as follows:

$$y_t = x_t \beta + \varepsilon_t \qquad (9)$$

$$\varepsilon_t \mid \psi_{t-1} \sim N(0, h_t) \qquad (10)$$

$$h_t = \alpha_0 + \sum_{i=1}^{q} \alpha_i \varepsilon_{t-i}^2 + \sum_{j=1}^{p} \beta_j h_{t-j} \qquad (11)$$

$$= \alpha_0 + \alpha(L)\varepsilon_t^2 + \beta(L)h_t \qquad (12)$$

To guarantee the condition variance , demand :

$$\alpha_0 > 0 \qquad (13)$$

$$\alpha_i \geq 0, i = 1, \cdots, q \qquad (14)$$

$$\beta_j \geq 0, j = 1, \cdots, p \qquad (15)$$

$GARCH(p, q)$ is represented as GARCH process in which the order is p and q. Relative to ARCH, the advantages of GARCH model are that the lower orders GARCH model can represent higher order ARCH model, which can reduce the number of lag order of the model and then the recognition and estimation of the model become easier than ARCH. The stationary condition of GARCH model is as following:

$$\sum_{i=1}^{q} \alpha_i + \sum_{j=1}^{p} \beta_j < 1 \qquad (16)$$

The model of $GARCH(1,1)$,which is often used in finance analysis, is the popular and sample model in the GARCH families. It is usually found to be the better choice to describe the volatility of financial assets. It's formulation is

$$h_t = \alpha_0 + \alpha\varepsilon_{t-1}^2 + \beta h_{t-1} \tag{17}$$

Where $\alpha_0 > 0, \alpha \geq 0, \beta \geq 0$.The necessary and sufficient condition of the stationary process of $GARCH(1,1)$ model is $\alpha + \beta < 1$.

3   Return Standardization

The asset return series are naturally decomposed as $r_t = \sigma_t * \varepsilon_t$, where $\varepsilon \sim N(0,1)$, and σ_t is the time conditional standard deviation. On rearranging this decomposition, we get the standardized return series,

$$\frac{r_t}{\sigma_t} = \varepsilon_t \tag{18}$$

Obviously, the result of the distribution is mainly dependent on the σ_t.

In practice, people have made a lot of research about σ_t, and many volatility models have been proposed. However, as formally shown by Andersen, Bollerslev, Diebold and Labys, the ex-post volatility over a day may be estimated to any desired degree of accuracy by summing sufficiently high-frequency returns within the day.

In this paper we use the realized volatility and GARCH model to get the conditional σ_t, and then compare the return distribution standardized instead by realized volatility with the one by GARCH model.

4   Empirical Analysis

4.1   Data

The sample data we used in this paper are the Shanghai securities integrated index from September 11 2006 to March 31 2008, which includes 82560 minutes datum of 344 trading day. There is a lunch break time in the stock market, where has a small fluctuate on the data. Because the fluctuate is no different from the other minute intervals, we connect them directly. For the two methods, we extract two datum groups. One is minute datum for the realized volatility method, the other is daily datum for the GARCH model.

4.2   Original Return Distribution

The Shanghai Securities Integrated Index(SSII) return is gained by following method

$$r_n = lnS_n - lnS_{n-1} \tag{19}$$

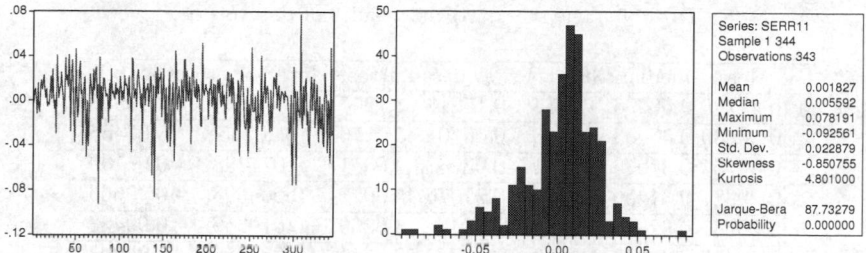

Fig. 1. The Fluctuation Diagram and the Histogram of daily returns

Where S_n represents the closing price of the n-th day in the observable sample series. We can get the original return distribution. The result is described in figure 1.

The value of skewness of original return distribution is -0.850755. The value of kurtosis of one is 4.801. The return distribution is leptokurtic, right deviation, and volatility clustering. The P-value of JB statistic is nearly to zero, so the result refuse the null hypothesis of Gaussian distribution. By testing, we also find that the return series show Heteroscedasticity.

4.3   Realized Volatility Analysis

We can use different time intervals to calculation daily realized volatility. On one hand, smaller the time interval is, less the information lose is, but bigger the error of microstructure is. On the other hand, longer the time interval is, more the information loses is, but the measurement error will become the main error. So the time interval should balance between the two kind errors. Of course, there may be the best time interval about realized volatility. Now we calculate the realized volatility at the time interval from 1 minute to 30 minutes by using the formula.

$$RV_t = \sum_{i=1}^{n} [r(\tau_i, \frac{i}{n})]^2 \tag{20}$$

Next we make the returns distribution standardized by the realized volatilities

$$\frac{r_t}{\sigma_t} = \varepsilon_t \tag{21}$$

The standardized results of day return by different time intervals realized volatility is indicated in table 1.

The JB statistic obeys the $\chi^2(2)$ distribution. The P- value represents the reception level of Gaussian distribution. The null hypothesis of the test is that the standardized return obeys Gaussian distribution. At the test significance level of 0.05, all of the P-values of statistics are not significant, it indicates that

Table 1. The standardized results of day return

	Mean	median	Std.dev	skewness	kurtosis	Jarque-bera	probability
1	0.42766	0.58696	1.9484	-0.15212	3.1275	1.5553	0.45949
2	0.36039	0.43083	1.5168	-0.06704	3.0276	0.26785	0.87465
3	0.31757	0.37075	1.3231	-0.03238	3.0351	0.077498	0.96199
4	0.29917	0.34856	1.2235	-0.01076	3.0022	0.006693	0.99666
5	0.28341	0.32807	1.1775	0.000132	3.0269	0.010369	0.99483
6	0.28425	0.32286	1.1599	0.018464	3.0681	0.085759	0.95803
7	0.27592	0.31516	1.1625	0.044382	3.118	0.31155	0.85575
8	0.28733	0.30697	1.1845	-0.00123	2.9191	0.093557	0.9543
9	0.29608	0.34409	1.2161	0.12277	3.2531	1.7775	0.41117
10	0.29499	0.32648	1.238	0.035407	3.1288	0.30863	0.857
11	0.31367	0.30706	1.2409	0.068909	2.9929	0.27217	0.87277
12	0.29474	0.35038	1.2749	-0.05773	2.9378	0.24584	0.88433
13	0.30299	0.33594	1.2421	-0.02154	2.8266	0.45607	0.7961
14	0.29761	0.33909	1.2292	0.014576	2.9816	0.01698	0.99155
15	0.3166	0.35388	1.2783	0.03567	3.0846	0.17491	0.91626
16	0.29395	0.35401	1.2842	-0.1579	3.2056	2.0294	0.3625
17	0.3009	0.34399	1.2394	0.090443	3.0826	0.56503	0.75388
18	0.3266	0.33246	1.2575	0.11249	3.0649	0.78344	0.67589
19	0.30935	0.36246	1.2972	0.074532	3.1412	0.60254	0.73988
20	0.32698	0.35293	1.3334	0.10732	3.3089	2.0218	0.36389
21	0.30482	0.33763	1.2705	0.083091	3.1353	0.6564	0.72022
22	0.31655	0.32452	1.3127	-0.01417	3.0678	0.077192	0.96214
23	0.30349	0.31914	1.2738	-0.04182	3.0113	0.10179	0.95038
24	0.29482	0.36018	1.3276	-0.21238	3.1978	3.1377	0.20829
25	0.30208	0.3545	1.2686	-0.14879	2.9742	1.275	0.5286
26	0.30632	0.32425	1.2604	-0.0118	2.8626	0.27773	0.87035
27	0.31121	0.32553	1.346	0.081218	3.1796	0.83824	0.65763
28	0.30952	0.3517	1.2952	0.096925	3.1082	0.70443	0.70313
29	0.33009	0.3457	1.2737	0.256	3.3665	5.6659	0.058839
30	0.33415	0.35503	1.4182	0.095115	3.4102	2.9217	0.23204

the test result accepts the null hypothesis of Gaussian distribution. So the return standardized by the realized volatilities is very nearly Gaussian.

There is a big span at different time intervals, we plot the P- values at the figure 2.

From figure 2 we can see that the P-values from 3,4,5,6,8,14,15,22,23 minutes interval realized volatility are larger than 0.9, most of P-values are bigger than 0.6, which show that normality is obvious. Furthermore we can see that the P-values of the time interval from 3 to 6 minutes are bigger than the other sections, so we think that the time section of 3 to 6 minutes is the best time interval of realized volatility about Chinese stock market, and 4 minutes is a proper choice.

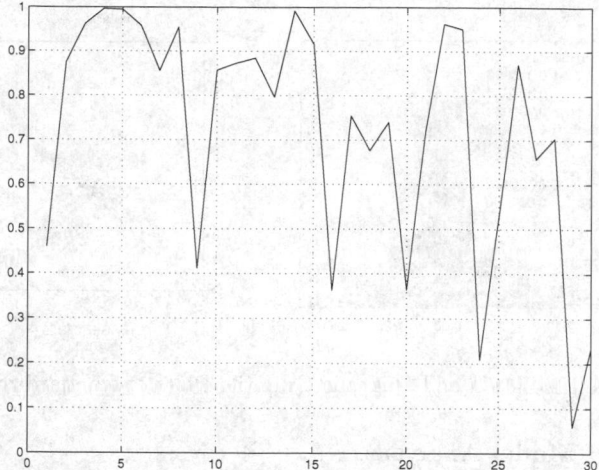

Fig. 2. The P-values of different time intervals

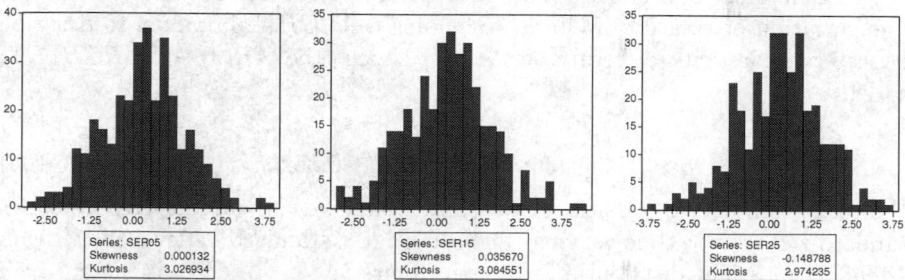

Fig. 3. The Histogram at the time intervals of 5,15,25 min

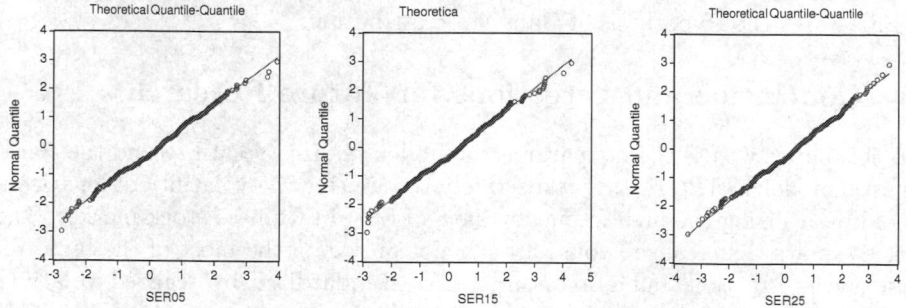

Fig. 4. The Quantile-Quantile Plots at the time intervals of 5,15,25 min

After standardized, the skewness value of the return series is nearly to zero, and the kurtosis value is nearly to 3. For example, the standardized result of 5,15,25 minutes are in figures 3, and 4.

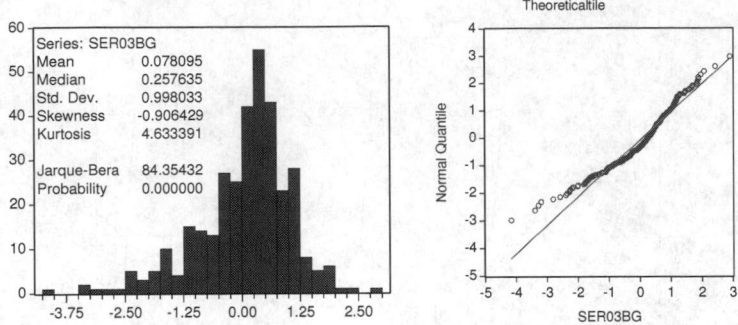

Fig. 5. The Histogram and Quantile-Quantile Plots of standardized return

4.4 GARCH Model Analysis

To understand the differences between the realized volatility and GARCH model, we also make an empirical analysis by utilizing the GARCH model.

A large number of literature show that $GARCH(1,1)$ model can describe the time variation of volatility well. So we choice $GARCH(1,1)$ model to describe the heteroscedesticity of return series. We choose the $AR(3) - GARCH(1,1)$ model.

$$R_n = 0.1138R_{n-3} + \varepsilon_t \tag{22}$$

$$h_t = 0.0000434 + 0.0601\varepsilon_{t-1}^2 + 0.8620h_{t-1} \tag{23}$$

After obtaining the conditional variance series, we can gain day return series standardized by time-varying conditional standard deviation from $GARCH(1,1)$. The distribution plot is in figure 5.

From the figures 5, we can see that $GARCH(1,1)$ model may describe the volatility of SSII, but it isn't accurate. The fitting Residual Error series are leptokurticfat-tailedright deviation. The P-value of JB statistic is nearly to zero, so it refuses the Hypothesis of Gaussian distribution.

5   Conclusion and Directions for Future Research

In this paper we use the nonparametric model, realized volatility and the Parametric model, $GARCH(1,1)$ respectively to describe the volatility of logarithm yield by utilizing the high frequency data of SSII in Chinese stock market. The result shows that realized volatility can not only gain the most of the high frequency information, but also return series standardized by realized volatility is basically accord with Gaussian distribution after taking logarithm and standardizing, which supports the Efficient Market Hypothesis in Chinese stock market. The standardized return series of $GARCH(1,1)$ are leptokurtic, fat-tailed, right deviation, which aren't accord with Gaussian distribution. It indicates that GARCH model can not correctly describe the dynamic of return series in Chinese stock market. The empirical results show that the volatility measurement of realized volatility is more accurate than GARCH model's.

Furthermore, we also study the best time interval of sampling about realized volatility. The result show that 4 minutes is a best time interval, which can make the daily realized volatility more nearly to Gaussian distribution.

Based on our research result, we plan to go on our research about Chinese stock market volatility at the following aspects: The first is to investigate the dynamic distribution of stock index return, and to determine the effects of different sampling frequency realized volatility. The second is to analyze the economics meanings of realized volatility to Chinese stock market, and to investigate the dynamic mechanism of practical volatility. The third is to make a contrast of accuracy and reliability of forecasting by different methods, so as to provide better methods for empirical research in future.

References

1. Bollerslev, T., Chou, R.J., Kroner, K.F.: ARCH Modeling in Finance: A Review of the Theory and Empirical Evidence. Journal of Econometrics 52, 5–59 (1992)
2. Engle, R.F., Gonzalez-Rivera, G.: Semiparametric ARCH Models. Journal of Business and Economic Statistics 9(4), 345–359 (1991)
3. Chen, K., Jayprakash, B.Y.: Conditional Probability as a Measure of Volatility Clustering in Financial Time Series. Europhysics Letters 18, 1–6 (2005)
4. Anderson, T.G., Bollerslev, T.: Exchange rate returns standardized by realized volatility are nearly Gaussian. Multinational Finance Journal 4, 159–179 (2000)
5. Kenneth, F., Schwert, G.W., Stambaugh, R.: Excepted Stock Returns and Volatility. Journal of Financial Economics 19, 3–30 (1987)
6. Anderson, T.G., Bollerslev, T.: Answering the critics: Yes, ARCH models do provide good vohtility forecasts. National Bureau of Economic Research (NBER) Working paper, No. 6023 (1997)
7. Hsieh, D.A.: Chaos and nonlinear dynamics: application to financial markets. The Journal of Finance 46, 1839–1877 (1991)
8. Taylor, S.J., Xu, X.: The incremental volatility information in one million foreign exchange quotations. Journal of Empirical Finance 4, 317–340 (1997)
9. Andersen, T.G., Bollerslev, T., Diebold, F.X.: Parametric and nonparametric volatility measurement. In: Hansen, L.P., AytSahalia, Y. (eds.) Handbook of Financial Econometrics. North Holland, Amsterdam (2002) (forthcoming)
10. Bollerslev, T., Engle, R., Nelson, D.: ARCH Models. In: Handbook of Econometrics, vol. IV, pp. 2959–3038. North-Holland, Amsterdam (1994)
11. Bera, A.K., Higgings, M.L.: A Survey of ARCH Models: Properties, Estimation and Testing. Journal of Economic Surveys 7, 305–366 (1993)
12. Diebold, F.X., Lopez, J.A.: Macroeconomics: Developments, Tensions and Prospects. Blackwell, Oxford (1995)
13. McAleer, M., Oxley, L.: Contributions to Financial Econometrics. Blackwell, Oxford (2003)

The System Dynamics Research on the Private Cars' Amount in Beijing

Jie Fan and Guang-le Yan

Management College, University of Shanghai for Science and Technology,
516 Jungong Road, 200093, Shanghai, China
happyfj2002@163.com, glyan2003@yahoo.com.cn

Abstract. The thesis analyzes the development problem of private cars' amount in Beijing from the perspective of system dynamics. With the flow chart illustrating the relationships of relevant elements, the SD model is established by VENSIM to simulate the growth trend of private autos' amount in the future on the background of "Public Transportation First" policy based on the original data in Beijing. Then the article discusses the forecasting impacts of "Single-and-double license plate number limit" on the number of city vehicles and private cars under the assumption that this policy implemented for long after the 2008 Olympic Games. Finally, some recommendations are put forward for proper control over this problem.

Keywords: Private cars' amount in Beijing, System dynamics, Social economic modeling, Policy simulation and control.

1 Introduction

The 2008 Beijing International Automotive Exhibition, which was the largest one of such kind hitherto in China, had dropped its curtain. Cars purchase had become a fashion since recent years, and the sales volume made the final success of Beijing auto expo. Over the past decade, cars' sales quantity increased by almost 10 times in China. Beijing, as China's political and cultural center, its people's living standards kept ahead compared with others and the amount of individual cars in city was rapidly expanding. On the one hand, along with the entrance into WTO, the automobile industry opened more widely, the prices of cars went in line with equal level of the international market gradually, and the price wars were not uncommon launched in full swing. On the other hand, the potential production capacity in autos manufacture made excess quota and formed a seller's market. In Beijing, with the higher salary and cars' price shrinkage, it's no longer a dream for average residents to possess a small or medium-sized car. However, if everyone owns a personal auto, whether this nice dream would end in a nightmare? The large population in Beijing had resulted in city crowd and housing tension, for the traffic works as the city venation, too many private cars will aggravate the traffic congestion and deteriorate the environment even worse undoubtedly. The theme of Auto China 2008 was impressing as saying "Dream, Harmony, New Vision". Then how to unify this motif with the development of city harmoniously? The research on trend forecast of private cars and finding reasonable ways for good control is significant.

System Dynamics (SD) was created by Forrester, one famous professor of MIT, in the 1960s. This approach emphasizes system internal structure and feedback mechanisms,

J. Zhou (Ed.): Complex 2009, Part I, LNICST 4, pp. 628–639, 2009.
© ICST Institute for Computer Sciences, Social Informatics and Telecommunications Engineering 2009

being good at solving complex system problems featured as long-period, high-order, non-linear and multi-variables (see [1]). This method has been used in various fields universally since it was founded and the problem of urban development belongs to the social complex systems area which suits for SD to make further study. The recent researches were more about city's resources and ecological environment, involving urban water resources, land resources and reserves, waste management etc. On the city traffic aspects, Liu Qing in [2] structured a dynamic model to analyze the effect and structure of urban road transport system, Zhang Lin feng in [3] established SD models about city centers forming and evolvement to make a conclusion that the traffic situation can promote or inhibit city centers development. In [4], Le Ming-bo proposed the implementation of traffic jam fee illustrated by the cause relation chart in urban traffic. Most of these studies focused on macroeconomic structural analysis and model establishment, rarely made simulation runs or forecast with historical data, not to mention expound and adjust on the background of policy. Whereas by using the scientific SD simulation method, this paper analyzed main causes that impact private vehicles' amount in Beijing and its feedback loops, established system dynamic model. Then with the statistical data combined qualitative and quantitative analysis, it made simulation run under the policy that encourage the public transportation career to predict the future trend of private cars in city. Lastly the article renewed the model aiming at the hot issue that whether the "odd-even license traffic restriction" strategy should be applied permanently after the Beijing Olympic Games, compared the simulate results and gave some policy recommendations.

2 System Analysis and Modeling

2.1 System Boundary

The system's behavior is generated by the interactions of system inner elements. SD assumed that the external environment changes which are not controlled by internal factors wouldn't affect the system's behavior essentially. So the conception factors and variables in the internal border of the system which have intrinsic relationships with the dynamic research issue should be brought into the model, while those external variables should be excluded (see [5]). The system of urban dwellers' private cars is complex and socio-economic. It not only includes sections such as circulation, exchange and consumption of motor vehicles, but also contains urban population, economic, environment, control policy, public transportation and other sub-parts. According to the modeling purpose of this paper, the system is defined in a scope involving private cars' amount, price, use expending, per capita disposable income, the number of resident population, the total number of Beijing household, passengers transportation amount by public traffic, the total amount of motor vehicles, the number of buses and taxis, the number of parking spaces, roads areas and other level variables.

2.2 Cause Relationship and Dynamic Model

Causalities determine the system behaviors and functions. Analyzing the factors that affect private cars demand, it can be summed up as follows: urban road construction, traffic management level, parking lot building, public transport, cars' price, use expending, purchase tax, people's income, and control policy.

Some condition changes such as speeding up the road construction in city, increasing the residents' income, declining private cars' price and renovating peoples' consumption conception would help to encourage the purchase of private cars. While others, such as increasing the use expending (included fuel cost, road toll, parking fee, insurance fee etc.), strengthening the policy control power, developing the public transportation will curb private cars' consumption in Beijing. In addition, the increasing amount of private cars raises the per capita ownership, which could stir the comparison psychology of consumption and take a role in promoting the purchase of private cars. Then the volume expansion of the city total motor vehicles decreases per vehicle road area leading in more road traffic pressure and potential environment threats, which should force the government to focus its control intensity. Besides, the supply and demand ratio of parking lot is deflating along with the vehicles growth. The increasing fee of parking would result in use expending rising and effect consumption psychology. Such factors may choke back consumption of the private cars as well.

Based on the analysis above and taken the causalities of elements into account, a SD flow chart of private cars' amount in Beijing is designed in which correlative level variables, rate variables and information flows are combined organically to point out the action mechanism of this system. (Figure 1)

The SD model is built as follow:

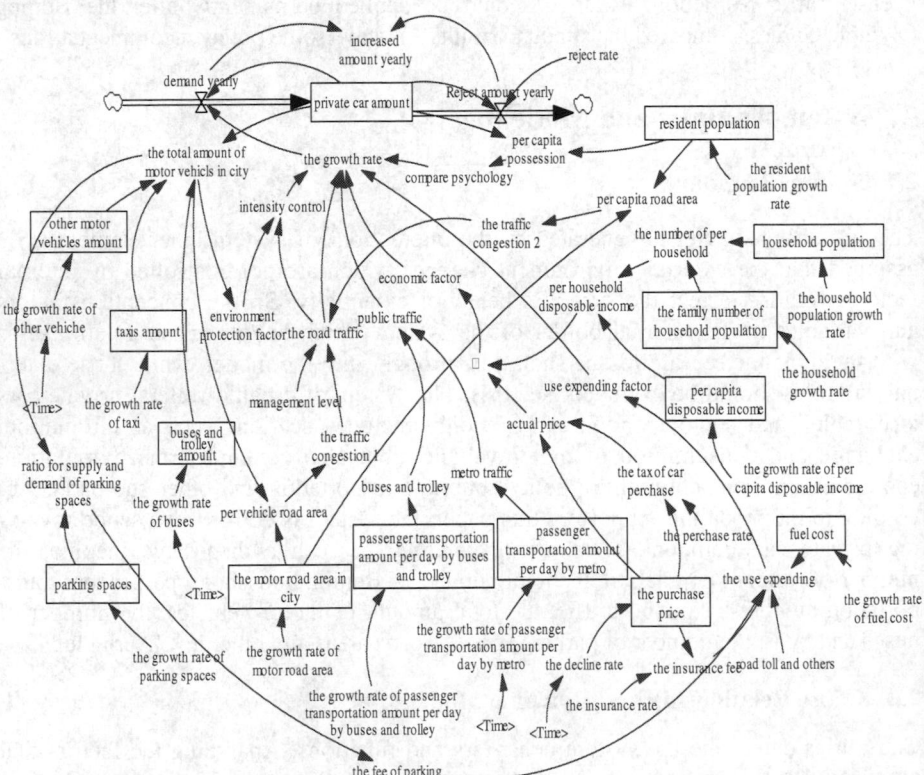

Fig. 1. The SD model for the amount of private cars in Beijing (VENSIM platform)

According to the research finding of Qian Ping-fan, it is the Ⅱ value (the ratio that price of cars is divided by annual income of household) that decides the actual progress of private cars' household in China when choose the best parameter among some reference standards (per capita GDP, R value, Ⅱ value) to judge whether and when private cars enter into families in a region. In his thesis (see [6]), 2001 was the starting year indicating the entry of private cars into resident families in Beijing. Therefore in this paper, it adopts the standard of Ⅱ value to weigh the consumption effect made by residents' income level and adjusts the cars' price to actual price. The statistical data of 2001 are filled as initial values in the follow-up simulation as well.

2.3 Model Equations and Parameters

The System Dynamics model contains level equations, rate equations, auxiliary equations, parameters equations and initial value equations. Make out these model equations, initial parameters and lookup functions and check them as a whole. The parameter values are calculated by statistic samples from 2001 to 2006, and the data of 2007 remained as a test sample.

Referring to "Beijing Statistics Yearbook, 2001", the statistical data collected from Beijing traffic management office and other relevant information gathered, the initial values are set as below.

Table 1. The model initial value

Variable name	Initial value	unit
Private car amount	624000	liang
Taxis amount	65155	liang
Buses and trolley amount	14803	liang
Other motor vehicles amount	995042	liang
The purchase price	200000	yuan
Resident population	1.3851e+003	Ten thousand person
Household population	1.1223e+003	Ten thousand person
The family number of household population	4.053e+002	Ten thousand family
Parking spaces	30	Ten thousand wei
The motor road area in city	5.2087e+003	Ten thousand square meter
passenger transportation amount per day by metro	1.28411e+002	Ten thousand person per day
passenger transportation amount per day by buses and trolley	1.1037e+003	Ten thousand person per day
Fuel cost	4000	yuan
per capita disposable income	11577.8	yuan

Parameters setting:

1. Computer the logarithm of resident population in Beijing from 2001 to 2006. The trend chart is shown as follow by SPSS (Figure 2).

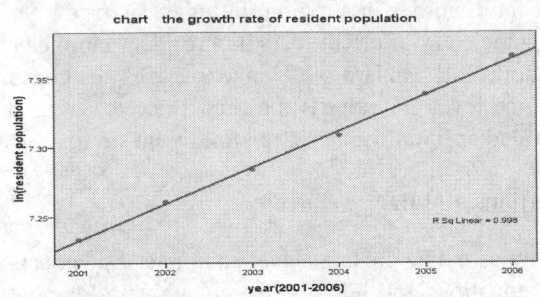

Fig. 2. The growth trend of resident population in Beijing

It's unnecessary here to forecast the resident population growth precisely. As seen from the scatter diagram, the increase of resident population properly fits the linear growth rule by and large. To simplify the model, the resident population growth rate is reckoned as a constant. Fitting the line to get the equation:

$$Y=0.026N-45.295 \tag{1}$$

$R^2=0.998$ Y—the logarithm of resident population   N—year

So the average resident population growth rate is 2.6%. Similarly, based on the historical data collected, figure and fit each level variable into logarithm growth equation and make out the average growth rate. Finally some basic growth parameters of the model are gained: the growth rate of per capita disposable income is 11.7%, the growth rate of household population is 1.31%, the growth rate of the number of urban households is 2.62%, the growth rate of road area is 3.1%, the growth rate of parking spaces is 23% and the growth rate of taxis' amount is 3.78%. Besides, because the linear fitting equation figures the growth rate of buses' amount without a high accuracy, the same as the growth rate of other vehicles' amount counted, these two parameters are given in the form of lookup functions.

According to "Beijing traffic planning--11.5", it's said that till 2010, the public passenger transportation system of city center would assume 40% of total passenger transportation volume, among these, 13~15 million passenger transportation taken by buses and trolley other than 5~6 Million taken by metro. Thus taking the average growth rate computed by historical data into account, the growth rate of passenger transportation amount per day by buses and trolley is set as 0.8%. And in view of the hold of 2008 Olympic Games, along with Metro Transportation Development Policy in Beijing, the construction speed of metro career would be even faster than before, so set the growth rate of passenger transportation amount per day by metro as 6% till 2010 and 15% after then.

2. The initial car purchase price is set as 200,000 Yuan. Since the automotive market is opening more widely, suppose that the price decrease gradually and remain

around 100,000 Yuan till 2015. In addition, the insurance rate of Private car is 4% and the reject rate is 6.67% according to National Automotive Rejection Policy. The parameter of road toll and other costs is set as 6,000 Yuan approximately.
3. The growth rate of private cars' demand is determined by factors combined such as control power, road traffic, public traffic, economic factor, use expending factor and compare psychology. These factors act in the form of lookup functions with a range of 0~1. The formula of lookup function is presented as below.

$$Y = withloopup\,(X,([(x_{min},y_{min})-(x_{max},y_{max})](x_1,y_1)(x_2,y_2)\cdots\cdots(x_n,y_n)))$$

4. Take some equations in the SD model for example:

L private car amount = INTEG (demand yearly - Reject amount yearly, 624000)   (2)

L   resident population = INTEG (the resident population growth rate *

resident population, 1.3851e+007)   (3)

A   the road traffic = the traffic congestion 1^0.6* management level* the

traffic congestion 2^0.4   (4)

C   reject rate = 0.067   (5)

The equation of private cars' growth rate is designed by referred the similar formula in "Impact on Sedan Demands after China Joins the WTO" written by Wang Qi-fan and Jia Jian-guo (see [7]). Then trained by several model adjustment and simulation, it is defined as follow.

R   the growth rate= (1- the road traffic) ^0.05*(1- intensity control)*
economic factor^0.05* compare psychology^0.095* use expending   (6)
factor^0.5*(1 - public traffic)*a   (ratio a=1.5)

The model equations are not listed one by one since the number as a whole is very large. Finally, a complex system dynamics model which includes 64 equations, 14 orders has been established. This model, taking the main factors that influence the development of private cars' amount in Beijing into account, could make a dynamic and long-term demonstration for the object variable.

2.4   Test of the Complex System Model and Simulation

Simulate the model by different running steps as 0.25/0.5/1 and make a comparison for the forecasting trends of private cars' amount in Beijing. It is shown that the system behavior is stable (Fig.3). With non-sensibility to the changes of parameters, the model properly reflects the complexity of the socio-economic system. It has a good robustness character and gains no pathological results.

The SD model simulates the medium and long term development trend of private cars' amount in Beijing based on the statistical data from 2001 to 2006. Selecting some simulating results in 2007 of variables and making comparisons with actual values (Table 2), the contrasted results show that the errors of them are very small and

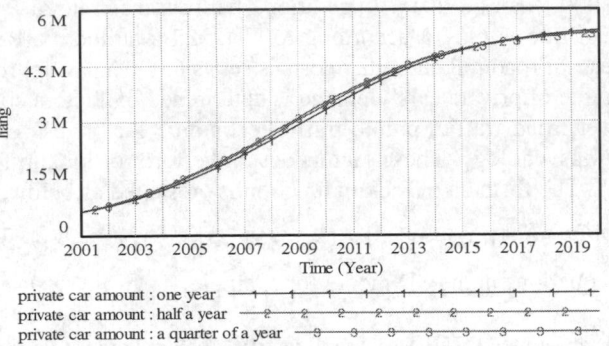

Fig. 3. Model stability analysis

Table 2. Compare the forecasting and actual with values in 2007

Item	Simulate value	real value	error
Private car amount (Ten thousand)	212.97	212.1	0.410%
The total amount of vehicles in city (ten thousand)	313.671	312.8	0.278%
Resident population(ten thousand)	1626.14	1633	-0.420%
Household population(ten thousand)	1213.45	1213.3	0.012%
The family number of household population (ten thousand)	473.335	473	0.071%
The motor road area in city (Ten thousand square meter)	6255.78	6272	-0.259%
per capita disposable income(yuan)	22487.7	21989	2.268%
Buses and trolley amount	19394.9	19395	-0.001%
Taxis amount	666.46	66646	0.000%
passenger transportation amount per day by buses and trolley(Ten thousand person per day)	1157.75	1157.93	-0.016%

the system behavior described by model is consistent with the actual action. The model is constructed effectively.

3 Model Analysis and Policy Adjustment

3.1 Trends Forecast

As shown in the simulating chart (figure 4), the development of private cars' amount in Beijing has an "S-type" growing trend and till 2020, the number will reach about 5 million. This quantity consists with the conclusion cited by "Master Plan of Beijing City (2004-2020)". According to the curve, now the development of Beijing private cars' amount still stays in its fast growth phase and the demand of private cars is expanding largely. The demand potential will peak at 2010, and after then will be decreasing slowly. The amount of private cars will stay stable gradually after 2016.

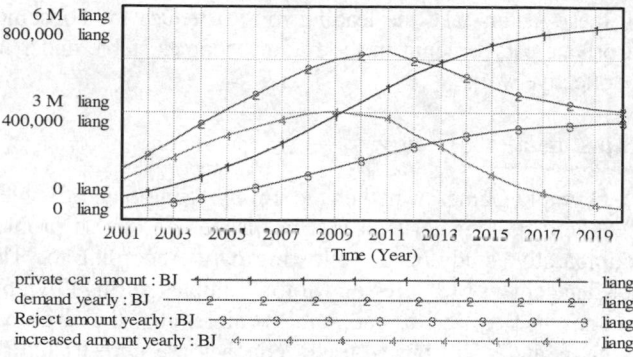

private car amount : BJ	liang
demand yearly : BJ	liang
Reject amount yearly : BJ	liang
increased amount yearly : BJ	liang

Fig. 4. The development trends of private cars in Beijing

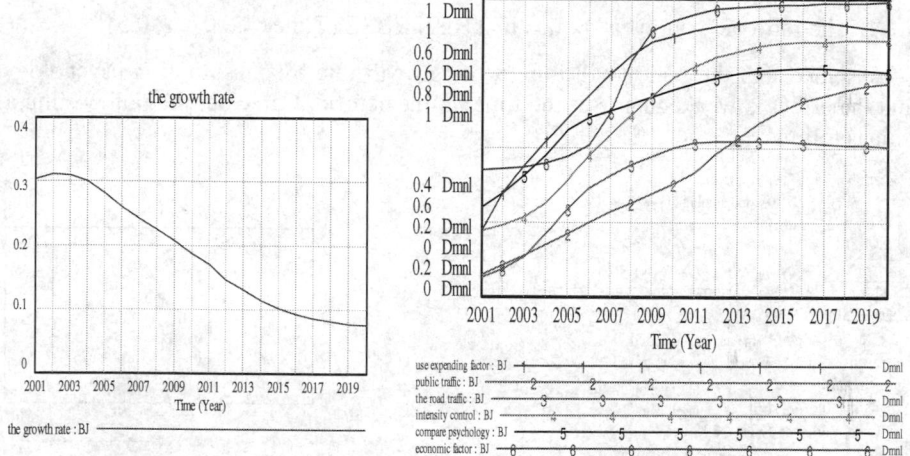

use expending factor : BJ	Dmnl
public traffic : BJ	Dmnl
the road traffic : BJ	Dmnl
intensity control : BJ	Dmnl
compare psychology : BJ	Dmnl
economic factor : BJ	Dmnl

Fig. 5. The growth rate and growth factors of private cars in Beijing

Seen from the charts of growth rate and factors (figure 5) to make some analysis, it can be concluded that the growth rate had experienced modest growth in the years 2001-2002 and then reduced gradually, the forepart of that is also in line with the actual statistical data from 2001 to 2007. As forecasting result shows, the growth rate of private cars in the future will keep stable after 2016. By analyzing the growth factors, at the beginning, some elements, such as the rising of economic level, compare psychology strengthened and the reducing of use expenditure resulted by the decline of cars' price (use expending factor augments and works), promote the consumption of private cars in Beijing. However, with the amount of private cars increasing, the growth rate of urban roads can not keep the pace of motor vehicles' growth, which makes the situation of traffic congestion even worse. Meanwhile, for the increasing of the total motor vehicles' amount, taking the awareness of environment protection and traffic congestion alleviation into account, the government is forcing the control power. Besides, one factor can't be ignored is that the "Public Transportation First" policy gradually presents its advantages. Subsequently and remarkably, all these factors take the leading role to reduce the consumption demand. The growth of private cars' amount in the future becomes stable under impacts by the whole factors' collective work.

3.2 Policy Adjustment

After the 2008 Olympic Games, whether the policy "Single-and-double license plate number limit" should be long insisted is a hot issue. As to this problem, the paper carries out a comparative study in the following part. According to "The 2008 Temporary Traffic Management Measures on Motor Vehicles during Olympic Games and Paralympics Games in Beijing", it said some vehicles that the large passenger cars, buses, trolleys, taxis and small buses, travel coaches are unrestricted by "odd-even-limitation". Then the original model is modified (Figure 7) with simulating step=0.5 for the "odd-even" policy in Beijing implemented on July 20, 2008.

The equation of "odd-even" limit ratio is set as follow:

A   The ratio of "odd-even" limit= IF THEN ELSE (Time<=2007, 1, 0.5)   (7)

Assumed that the policy will impact the growth rate by consumption psychology, and the effect is weakening over the time in linear form. Carry out the policy simulation as follow:

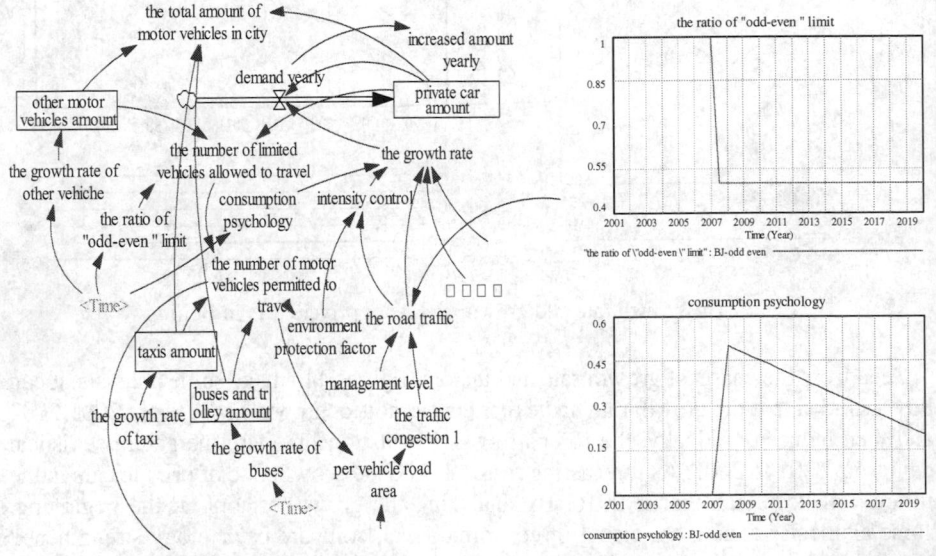

Fig. 6. Renewed part

Fig. 7. The ratio of "odd-even" limit and consumption psychology

Comparing the simulating results, it shows that the growth trends of total amount of motor vehicles and private cars both shrink in recent years under carrying out the "odd-even" policy for a long time because the policy has an impact on consumer psychology, which suppresses the purchase of the private cars. But in the long term, the impact of this policy acts feebly and the stable amount goes close to the original line, because the travel restriction results in less motors traveling on the road, contributing to traffic fluency, environmental pressure alleviation and government control relaxation, all of those would be in favor of the private cars purchase. That is, on this premise the impact of the policy on psychology lessens along with the time, the "odd-even Limit" policy, to some extent, can put out the enthusiasm of consumers to buy cars in early time, but in a long run, under the factors combined, the policy couldn't inhibit the growth of private cars far away. The growth rate increases faster than original level over time, making the amount even larger than that of non-implementation state.

Thus it needs to find a better solution from other way. Make an assumption that the project of Beijing Municipal Rail Transit quicken its construction speed after 2008, and adjust the parameters appropriately to improve the growth rate of passenger transportation amount per day by public traffic after 2008. The new simulating result is shown as follow.

Compared the results with the original forecast, speeding up the development career of public transportation shortens the increasing period of private cars, makes the

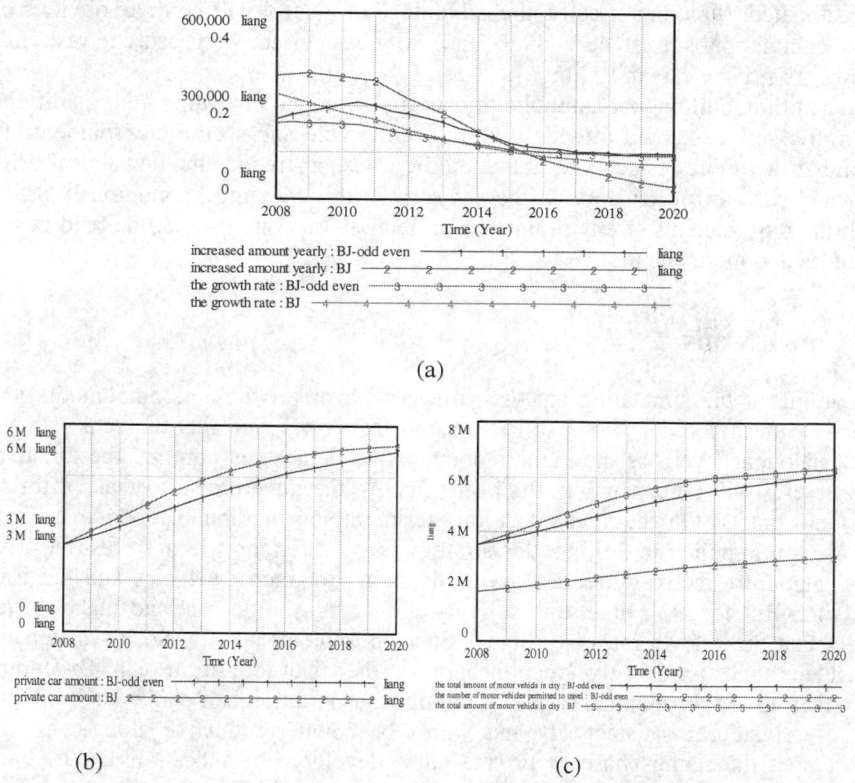

Fig. 8. The comparison chart of simulating results (odd-even license plate limit) (a) Increased amount yearly and the growth rate (b) Private car amount (c) The total amount of motor vehicles in city and the number permitted to travel

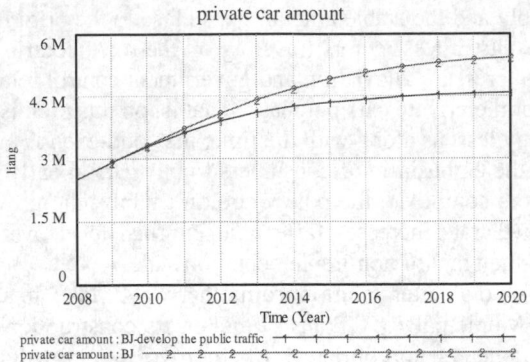

Fig. 9. The comparison chart of simulating results
(Speed up the development career of public transportation)

stable phase achieved much faster and significantly reduces the total amount of original estimated. Most importantly, this advantage of control will be more obvious over time without opposite effects and rebound. So it is an effective measure to restrain the number of private cars in Beijing.

In addition, building urban public transport system and providing high-quality public traffic services could provide a more comfortable and secure environment. That would attract more residents to utilize public transport, reduce the demand of private cars and ease traffic tension. At the same time, this measure harmoniously accords with the requirements of environmental protection, on which should be paid positive attitudes and more attention.

4   Conclusions

According to the simulating analysis, the growth of private cars amount fits an S-shaped pattern. Recently, it is still in a rapid increase period that the peak value has been untouched yet. As the highest point of the demand is coming, the amount of private cars will be expanding. It's necessary for the government to make efforts on amount control. On the study that long-term implementation of "odd–even motor license limit" policy in Beijing, this strategy is good for short term to restrict private cars' amount effectively, reduce the total amount of motor vehicles and the traffic pressure. But for long time, this policy trends to play little role and makes reverse effect that the growth rate rises higher than un-implement state. So it's better to be considered and listed in five-years program rather than long-term plan. By contrast, speeding up the development pace of public transportation can choke back the number of private car satisfactorily and won't be counterproductive along with time. Shortening the rising phase of private cars' development and shrinking the stable quantity, the development project of public transport is a long-term and optimum strategy to control the private cars' amount as well as improve the urban traffic level.

Referring to the "Master Plan of Beijing City (2004-2020)", it is expected that in 2020, urban centers passenger travel by public transport would assume 50% accounting for the total proportion, contrasted 27% in 2000. However, The public transport development level, even if as the ratio planned, is far lower than that of many developed foreign cities, in which the index accounts 75% approximately. Thus, the program of public transport system construction, which is an effective means to control the private cars' quantity, now still, has great potential and large room for development in view of current level. It should be fully supported and encouraged to accelerate its construction speed zealously.

References

1. Wang, Q.F.: System Dynamics. Tsinghua University Press, China (1994)
2. Liu, Q.: On a Model of System Dynamics for Urban Street Transportation. Journal of Wuhan transportation university 01, 89–92 (1999)
3. Zhang, L.F., Fan, B.Q., Yan, G.L., Lu, Z.L.: Dynamic Models About City Centers Evolvement Based on Transportation Effect and Simulation. Systems Engineering 05, 61–65 (2004)
4. Li, M.B.: The System Dynamics Analysis on traffic Congestion policy. Railway Transport and Economy 05, 79–81 (2006)
5. Xu, X.Y., Shi, G.: The application and research of system dynamics to the prediction of automobile market. Journal of Shandong University of Technology 06, 1–4 (2007)
6. Qian, P.F.: The Market status and Forecast prospects of Car's Entering into Families in China. China economic times 02, February 19 (2004)
7. Wang, Q.F., Jia, J.G.: Impact on Sedan Demands after China Joins the WTO. Systems Engineering-theory & Practice 03, 56–62 (2002)
8. May, A.D., Shepherd, S.P., Timms, P.M.: Optimal transport strategies for European cities. Transportation 3, 27–35 (2000)
9. May, A.D., Roberts, M.: The design of integrated transport strategies. Transport Policy 02, 75–83 (1995)
10. Hall, R.I.: A study of policy formation in complex organizations: emulating group decision-making with a simple artificial intelligence and a system model of corporate operations. Journal of Business Research 45, 157–171 (1999)
11. Beijing Statistical bureau: Beijing Statistics Yearbook. China statistics Press, China (2001–2008)

The Topological Characteristics and Community Structure in Consumer-Service Bipartite Graph

Lin Li[1], Bao-Yan Gu[2], and Li Chen[3]

[1] Business School, University of Shanghai for Science and Technology,
Shanghai 200093, China
Lilin@usst.edu.cn
[2] gubaoyan@hotmail.com
[3] lichen0411@163.com

Abstract. We apply network analysis to study bipartite consumer-service graph that represents service transaction to understand consumer demand. Based on real-world computer log files of a library, we found that consumer graph projected from bipartite graph deviates significantly from theoretical predictions based on random bipartite graph. We observed smaller-than-expected average degree, larger-than-expected average path length and stronger-than-expected tendency to cluster. These findings motivated to explore the community structure of the network. As a result, the weighted consumer network showed significant community structure than the unweighted network. Communities picked out by the algorithm revealed that individuals in the same community were due to their common specialties or the overlapping structure of knowledge between their specialties.

Keywords: bipartite graph, consumer demand, topological features, community structure, weighted network.

1 Introduction

Complex systems have been a new paradigm for study of management, physical and technological domains [1]. A great deal of scholars making advances in different areas offers an opportunity to promote the knowledge exchange needed to reap the benefits of this basic research for problems in management, organization, and business. *Management Science* (2007,53(7)) published ten papers that use complexity theory to study the emergence, coordination, efficiency, and innovation in small groups, firms, and markets with an eye to the needs of practicing managers. One analysis tool of complex systems is network analysis which enables one to quantity the components and interactions of any different systems that have actors and relationships. Although network analysis has a long research history in graph theory and developed key concepts in social science, recent advances have shown cross talk among the different disciplines. The central idea of recent studies is to have agents interact with each other according to prescribed rules that may change over time as the agents adapt to their environment and learn from their experiences [2, 3]. Therefore we may understand

J. Zhou (Ed.): Complex 2009, Part I, LNICST 4, pp. 640–650, 2009.

the possible origins of the system and know the key variables leading cause and effect relationships by network analysis methodology.

The underlying hypothesis of marketing literature on consumer purchase behavior is that consumers naturally form cohesive subgroups with consistently correlated preferences and that consumer preferences are adequately expressed in the sales-transaction data. Data on sales-transaction can be obtained relatively easily and they are very popular in data mining. How to transform the sales transaction into a graph is the key to use network analysis tools studying consumer behavior, and the bipartite graph can do this well, which has two types of vertices and edges running only between vertices of unlike types. Many social networks are bipartite, forming what the sociologists call *affiliation networks*, i.e., networks of individuals joined by common membership of groups [4]. Recent studies have adopted the bipartite graph modeling to study sales transaction data; those findings motivated the development of a new recommendation algorithm based on graph partitioning [5].

This paper applies bipartite graph modeling to study reader borrowing behavior in the library of a university. We are interested in whether the real networks deviate from the theoretical predictions based on the generating function method which is introduced by Newman [6]. If these networks exhibit significantly different topological characteristics than expected values, we attempt to identify the underlying mechanism that governs consumer-service behavior. Here we consider the lending book as a kind of service supplied by the library institution. We hope that our research can bring about useful insights to service institutions which try to enhance their service efficiency and service quality by analyzing the interaction between consumer and service.

2   Related Research Work

2.1   Bipartite Graphs

In organizations and events, people gather because they have similar tasks, interests or share a preference for a particular thing. For instance, directors and commissioners on the boards of a corporation are collectively responsible for its financial success and meet regularly to discuss business matters. In such networks there are usually two sets of vertices, which are called *actors* and *events*, and edges connect vertices from different sets only. This type of network is called a *two-mode network* or a *bipartite graph*, which is structurally different from the one-mode network or unipartite graph, in which each vertex can be related to each other vertex. Examples of such networks include the board of directors of companies, co-ownership networks of companies, collaboration of scientists and movie actor collaboration networks. The last two are sometimes called *collaboration networks*. In the case of movie actors, the two types of vertices are movies and actors, and the net work can be represented as a graph with edges running between each movie and the actors that appear in it [6].

In many cases, graphs that are bipartite are actually studied by projecting them down onto one set of vertices or the other-so called "one mode" projections.

In such a projection two actors are considered connected if they have appeared in an event together. That is to say we study relations among one kind of vertices: relation between actors or between events, but not between actors and events. The construction of the one-mode network however involves discarding some of the information contained in the original bipartite network, and for this reason it is more desirable to model networks using the full bipartite structure[6]. But in description of a bipartite network there are more complications: some structural indices must be computed in a different way for bipartite networks, for example, the concept of degree, distance and centrality of vertices. Techniques for analyzing one-mode networks cannot always be applied to two-mode networks without modification or change of meaning. So what can we do? The solution commonly used is to change the two-mode network into a one-mode network, which can be analyzed with standard techniques [7]. We also follow this projection approach in our study. Some studies have done in extracting the hidden information of bipartite network projection, such as the weighting method proposed by Zhou *et al.*[8], which provides a method for compressing bipartite network and highlights a possible way for personal recommendation. Zhang *et al.*[9] proposed a model named a stretched exponential distribution (SED) to explain the topological characteristics of many empirical collaboration networks.

Newman *et al.*[6] derived the theoretical predictions of the topological measures of the one-mode projection based on a given vertex degree distribution of the full bipartite graph using generation function method. These topological measures include average degree, average path length, and clustering coefficient. The generation function $G_0(x)$ of a unipartite undirected graph is defined

$$G_0(x) = \sum_{k=0}^{\infty} p_k x^k \tag{1}$$

where p_k is the probability that a randomly chosen vertex on the graph has degree k. The theoretical predictions of the statistical properties of the unipartite graphs projected from a bipartite graph can be derived from the two generating functions associated with the degree distributions of the two types of vertices.

In our context, we will speak in the language of "readers" and "books" in the bipartite consumer-service graph. Let be the probability distribution of the degree of readers (the number of books which readers have borrowed) and be the distribution of degree of books (the number of readers by which books have been borrowed). Two generating functions can be constructed thus:

$$f_0(x) = \sum_j p_j x^j, \ g_0(x) = \sum_k q_k x^k \tag{2}$$

Newman *et al.*[6] show that the generation function of the unipartite reader graph projected from the he bipartite consumer-service graph is given by

$$G_0(x) = f_0(g_1(x)) = f_0(\frac{g_0'(x)}{g_0'(1)}) \tag{3}$$

The corresponding theoretical predictions of average degree z_1, average path L, and triangle clustering coefficient C are giver by

$$z_1 = G_0'(1), \quad L = 1 + \frac{\log(N/G_0'(1))}{\log((\frac{f_0''(1)}{f_0'(1)})(\frac{g_0''(1)}{g_0'(1)}))}$$

$$C = \frac{M}{N} \frac{g_0'''(1)}{G_0''(1)} \tag{4}$$

where M is the total number of books and N is the total number of readers. All of these results work equally well if "readers" and "books" are interchanged.

2.2 Community Structure

Social networks usually contain dense pockets of people who "stick together." Social interaction is the basis for solidarity, shared norms, identity, and collective behavior, so people who interact intensively are likely to be considered a social group. This phenomenon is called *homophily* [7]. A numbers of recent studies have focused on the statistical properties of networked systems. A few properties seem to be common to many networks: the small-world property, power-law degree distributions, and network transitivity. Another property which is found in many networks is the property of community structure, in which network nodes are joined together in tightly-knit groups between which there are only looser connections [10].

The traditional method for detecting community structure in networks is hierarchical clustering. The networks are represented a nested set of increasing large components (connected subsets of vertices) according to how closely connected the vertices are, which are taken to be the communities. But hierarchical clustering has a tendency to separate single peripheral vertices from the communities to which they should rightly belong [10]. Another deficiency of these methods can't tell us when the communities found by the algorithm are good ones. Algorithms always produce some division of the network into communities, even in completely random networks that have no meaningful community structure [11]. Girvan and Newman [10, 11] proposed an algorithm (GN) based on the iterative removal of edges with high "betweenness" scores that appears to identify community structure with some sensitivity, they also propose a measure called *modularity* for the strength of the community structure, which gives an objective metric for choosing the number of communities into which a network should be divided. As pointed out by Newman and Girvan [11], the principal disadvantage of their algorithm is the high computational demands it makes.

Newman [12] described a new algorithm for extracting community structure from networks, which has a considerable speed advantage over previous algorithms, running to completion in time that scales as the square of the network size. This allows us to study much larger systems than has previously been possible. This algorithm is based on the idea of modularity Q which is defined as follows. Let e_{ij} be the fraction of edges in the network that connect vertices in group i to those in group j, and let $a_i = \sum_j e_{ij}$. Then

$$Q = \sum_i (e_{ii} - a_i^2) = Tre- \parallel e^2 \parallel \tag{5}$$

is the fraction of edges that fall within communities, minus the expected value of the same quantity if edges fall at random without regard for the community structure. If a particular division gives no more within-community edges than would be expected by random chance we will get $Q = 0$. Values other than 0 indicate deviations from randomness, and in practice values greater than about 0.3 appear to indicate significant community structure.

Starting with a state in which each vertex is the sole member of one of n communities, we repeatedly join communities together in pairs, choosing at each step the join that results in the greatest increase (or smallest decrease) in Q. The change in upon joining two communities is given by $\triangle Q = e_{ij} + e_{ji} - 2a_i a_j = 2(e_{ij} - a_i a_j)$, we can select the best cut by looking for the maximal value of Q. The entire algorithm runs in worst-case time $O((m+n)n)$ on a network with m edges and n vertices. It is worth noting that this algorithm can be trivially generalized to weighted networks in which each edge has a numeric strength associated with it, by making the initial values of the matrix elements e_{ij} equal to those strengths rather than just zero or one[12].

3   Empirical Study

3.1   Consumer-Service Network

We constructed consumer-service networks using data sets provided by the library of a university in a seven year period from 2001 to 2006. The raw data for the networks described here is a computer log file containing lists of information, including readers, books they borrow, date, and other information such as readers' department, books' China library classification code, and so forth. Projection of reader-book network is straightforward. In the projected network, two readers are connected if they have borrowed at least one common book. We can also project book network in the similar way, but these results are not in this paper.

For the simplicity of calculation, the network only includes readers who are teachers and graduate students although data for undergraduate students are available in the database. The bipartite reader-book graph has 3205 reader vertices, 44127 book vertices and 135637 edges.

3.2   Topological Characteristics of the Network

The degree of a vertex is the number of links incident with it. The probability of a vertex with degree k (or the degree distribution) p_k is the most important topological property of the network. In bipartite reader-book graph, the degree distributions of both users p_j and books q_k which are calculated from Equation (2) are shown on logarithmic scales in Figure 1, and p_j appears to have power-law tails, and an exponentially truncated power law is a better fit for q_k. The

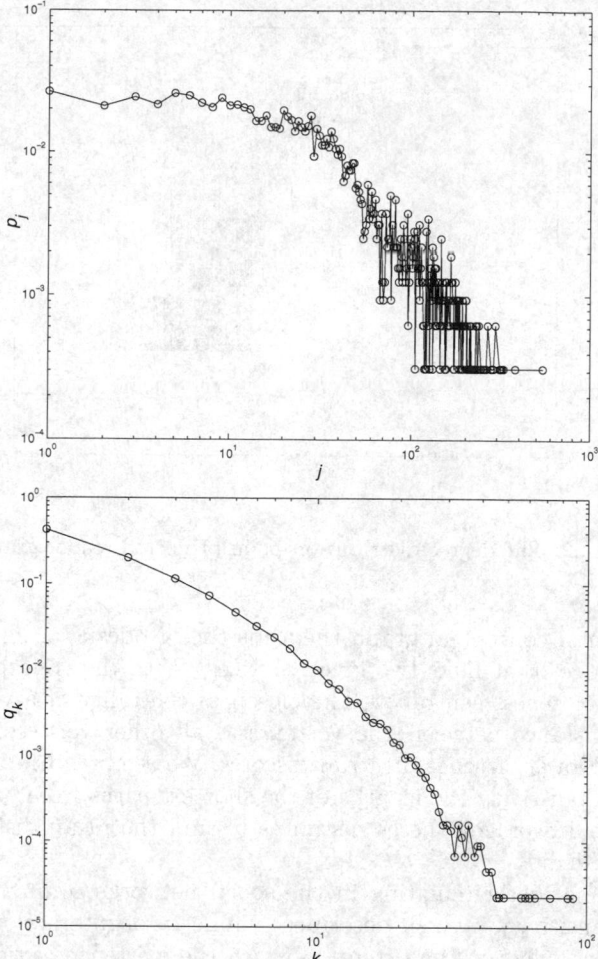

Fig. 1. The degree distributions of both readers p_j and books q_k in bipartite graph

average degree of the network is the average value of all vertices degree. The average degree of readers is 42.30 and that of books is 3.07.

The projected reader network contains only 3175[1] vertices and 269,721 edges. There is only one giant component in the network, i.e., the network is completely connected. In the reader network, degree of vertex means the number of neighbors with which the reader borrowed the one or more same books. The degree distribution ρ_k of the projected reader network is shown in figure 2, and it shows more fluctuation.

---

[1] If the degrees of some readers' adjacent (books) vertices are 1, these reader vertices are lost in the projected reader network [8].

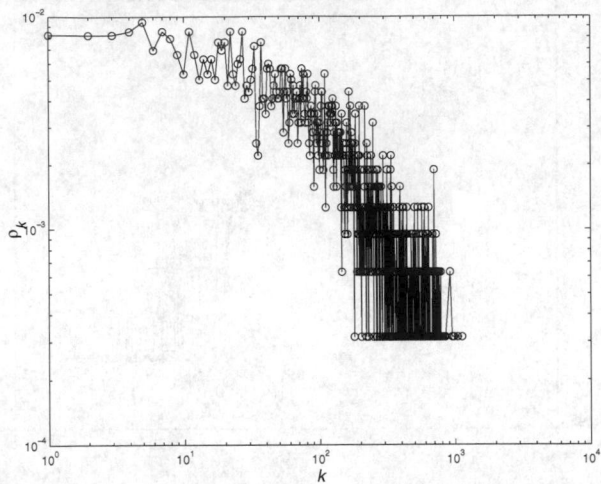

Fig. 2. The degree distribution ρ_k in projected reader graph

A fundamental concept in graph theory is the 'geodesic' or shortest path of vertices and edges that links two given vertices. With the concept of distance, we can define closeness centrality. The closeness centrality of a vertex is based on the total distance between one vertex and all other vertices, where larger distances yield lower closeness centrality scores. We use breadth-first search [12] to calculate exhaustively the lengths of the shortest paths from every vertex on the network and averaged theses distances to find the mean distance between any pair of readers.

An interesting idea circulating in the social networks community currently is that of "transitivity," which describes symmetry of interaction among trios of actors [14]. It refers to the extent to which the existence of ties between actors A and B and between actors B and C implies a tie between A and C. The transitivity is that fraction of connected triples of vertices which also form "triangles" of interaction. Here a connected triple means an actor who is connected to two others. This quantity is usually called the clustering coefficient, and can be written

$$C = \frac{3 \times number\ of\ triangles\ on\ the\ graph}{number\ of\ connected\ triples\ of\ vertices} \tag{6}$$

We calculated the actual topological measures of the projected reader network, such as average degrees, average path lengths, and clustering coefficients. We used equation (4) to calculate the expected average degree, average path length, and clustering coefficients of the projected network with given two-mode network degree distribution. These bipartite consumer-service graph degree distributions were computed directly from the borrowing log file. The deviation of the actual values from the expected values of the three topological measures would indicate that the relationship between readers in projected network is not only determined by the degree distribution of the two-mode network, in other words, there is some underlying mechanism to cause the deviation.

In Table 1 we show values of the three actual topological measures of the reader network and theoretical predictions calculated from Equation (4). The reader network exhibits a substantially larger average path length and higher clustering coefficient than those of random networks, while the average degree measure is smaller than that of its random counterpart. The percentage errors of predictions from actual vales are given in the last row of Table 1. As the table shows, these are all quite large: they vary from 36% for average degree to 90% for cluster coefficient. These findings strongly suggest that the consumers' demand is not random, and we conjecture intuitively that the reader network may show group structure. Communities appear in networks where vertices join together in tight groups that have few connections between them. Dense connections in groups can produce high clustering coefficient and not very high average degree, and looser connections between groups cause larger average path length. One might well imagine that reader network would divide into groups representing particular areas of research interest or specialty. However, this assumption must be empirically confirmed.

Table 1. Summary of the actual and predictive three topological characteristics for the projection reader network studied here

	z_1	L	C
actual	169.903	2.149	0.292
predictive	266.035	1.388	0.153
percentage error of prediction(%)	36.135	54.827	90.850

3.3 Community Structure Analysis

To analyze the community structure we use fast algorithm proposed by Newman [12]. The algorithm is based on the concept of modularity, which is described in section 2.2 of this paper. The algorithm proceeds as follows:

1. Starting with a state in which each vertex is the sole member of n communities.

2. Calculating the change Q in upon joining of a pair of communities between which there are edges. If more than one $\triangle Q$ highest values, then one of them is chosen at random.

3. Incorporating communities resulting in the greatest increase (or smallest decrease) in Q and step 2 is repeated until one community remains.

4. Selecting the best division by looking for the maximal value of Q.

The result derived by feeding the reader network into the algorithm turns beyond expectation. The peak modularity is only $Q = 0.155$ which is too small to indicate significant community structure ($Q > 0.3$). We doubt whether the community structure assumption is correct or there is important details we missed.

Then we bring to mind the fact that in the reader network construction two readers are connected by just one link, although they may borrow more than one same book. The data sets used here present more information than in the simple networks we have constructed from them. In particular, we can count quantity of book each pairs of readers have borrowed during the period of the study. We can use this information to make an estimate of the strength of relations.

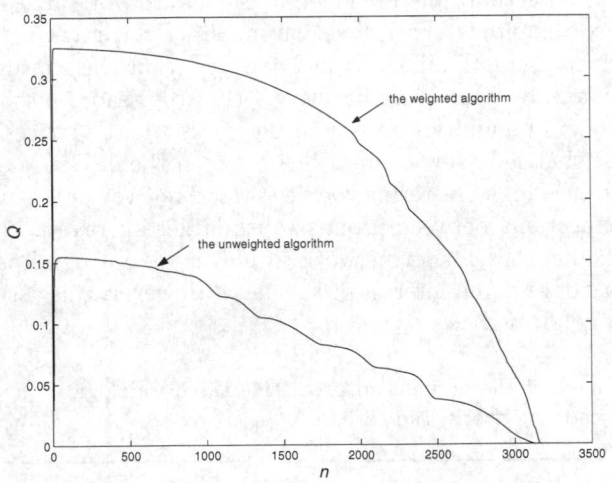

Fig. 3. Plot of the modularity Q versus community number n with unweighted and weighted algorithms in projected reader graph

We introduce weighted reader network, which allows for this by including a measure w_{ij} as the strength of interaction, which is the number of same books they have borrowed. The algorithm of detecting community structure is generalized trivially to weighted networks in which each edge has a numeric strength associated with it, by making the initial values e_{ij} of equation (5) equal to those strengths. The analysis reveals that the network consists of about 26 communities, with a high peak modularity of $Q_w = 0.325$, indicating significant community structure. In figure 3 we show the modularity Q versus community number n with unweighted and weighted algorithms. As we can see, $Q_w = 0.325$ is steeper than Q at the beginning of joining two communities and then the slope of Q_w is gentle, i.e. the weighted algorithm joins the tightly connected communities rapidly.

Eleven of the communities found by weighted algorithm are large, containing between them 89% of all the vertices, while the others are small-see Table 2. There appears to be a strong correlation between the community found by the algorithm and the department division related to the readers. The largest department component of each community is bold font style.

The weighted algorithm seems to find two types of communities: Teachers and students grouped together by similarity research background (community

Table 2. Crosstabulation between the community found by the algorithm and the department divisions related to the readers

Community	Department code									Total
	A	B	C	D	E	F	G	H	+38 smaller departments	
1	**493**	13		6	8		4	36	31	591
2	57	7	42	35	2	**140**	3	5	40	331
3	5	16	**196**	10		7	2	14	39	289
4	1	**234**	1	5	1		16	2	11	280
5		41	10	**168**	1		3	1	14	238
6	26	8	5	8	**154**	3	4	2	24	234
7	**127**	10	6	7	6	5	2	2	31	196
8	**152**	10	2	3	1	5	3	1	16	193
9	18	**116**	8	1	10	3	15		7	178
10	6	26	**60**	46		8	4		18	168
11	11	**46**	8	6	3		26	10	22	132
+15 smaller communities	66	63	38	42	7	14	12	20	83	345
Total	962	599	376	337	193	185	94	93	336	3175

one, four, seven and eight), or by same foundation courses (community two, ten), and farther analysis shows these courses can be concerning foreign language or computer technology. From the view point of department, Readers of a department are mainly distributed a few communities, especially in the large size departments (department A, B). The community structure presents the divisions running along disciplinary lines as well as the mark of interdisciplinary research or knowledge.

4   Conclusions

In this paper we have applied network analysis to study consumer demand in a library setting. We represent service transaction as a bipartite graph, and study the topological characteristics of the reader graph projected from the consumer-service graph. The results show that the topological characteristics of real reader graph deviate significantly from the theoretical predictions based on a random bipartite graph. The graph exhibits smaller-than-expected average degree, larger-than-expected average path length and stronger-than-expected tendency to cluster.

We assume the existence of community structure in the reader network, and use the algorithm for detecting community structure to confirm the assumption. We have found that the unweighted reader network doesn't present community structure, while the network show clear community structure with a simple weighting method. This is consistent with the statement that projection unipartite graph from a bipartite graph causes information loss, and weighting method is proper way to retain the original information [8]. The community structure analysis reveals that this construction captures essential ingredients of disciplinary interactions between readers.

Owing to the fact that knowing the interrelation of consumer's demand is the important thing when any organization wants to improve service quality or enhance service efficiency, we hope that the ideas and methods presented here will prove useful in the analysis of many other types of consumer-service networks.

References

1. Amaral, L.A.N., Uzzi, B.: Complex Systems-A New Paradigm for the Integrative Study of Management, Physical, and Technological Systems. Management Science 53, 1033–1035 (2007)
2. Epstein, J.M., Axtell, R.: Growing Artificial Societies: Social Science from the Bottom Up. The MIT Press, Cambridge (1996)
3. Wolfram, S.: A New Kind of Science. Wolfram Media, Champaign (2002)
4. Newman, M.E.J.: The structure and function of complex networks. SIAM Review 45(2), 167–256 (2003)
5. Huang, Z., Zeng, D.D., Chen, H.: Analyzing Consumer-Product Graphs: Empirical Findings and Applications in Recommender Systems. Management Science 53, 1146–1164 (2007)
6. Newman, M.E.J., Strogatz, S.H., Watts, D.J.: Random graphs with arbitrary degree distributions and their applications. Phys. Rev. E 64, 26118 (2001)
7. Nooy, W., de Mrvar, A., Batagelj, V.: Exploratory Social Network Analysis with Pajek. Cambridge University Press, Cambridge (2005)
8. Zhou, T., Ren, J., Medo, M., Zhang, Y.C.: Bipartite network projection and personal recommendation. Phys. Rev. E 76, 046115 (2007)
9. Zhang, P.P., Chen, K., He, Y., Zhou, T., Su, B.B., Jin, Y.D., Chang, H., Zhou, Y.P., Sun, L.C., Wang, B.H., He, D.R.: Model and empirical study on some collaboration networks. Phys. A 360(2), 599–616 (2006)
10. Girvan, M., Newman, M.E.J.: Community structure in social and biological networks. Proc. Natl. Acad. Sci. USA 99, 7821–7826 (2002)
11. Girvan, M., Newman, M.E.J.: Finding and evaluating community structure in networks. Phys. Rev. E 69, 026113 (2004)
12. Newman, M.E.J.: Fast algorithm for detecting community structure in networks. Phys. Rev. E 69, 066133 (2004)
13. Newman, M.E.J.: Scientific collaboration networks. II. Shortest paths, weighted networks, and centrality. Phys. Rev. E 64, 016132 (2001)
14. Newman, M.E.J.: Scientific collaboration networks. I. Network construction and fundamental results. Phys. Rev. E 64, 016131 (2001)
15. Gleiser, P., Danon, L.: Community structure in jazz. Preprint cond-mat/0307434 (2003)

Time Dependent Virus Replication in Cell Cultures

Juan G. Díaz Ochoa[1,2], Andreas Voigt[1,2], Heiko Briesen[1], and Kai Sundmacher[1,2]

[1] Max Planck Institute for Complex Technical Systems,
Sand Tor Str. 1, D-39106, Magdeburg, Germany
diaz@mpi-magdeburg.mpg.de
[2] Otto-von-Guericke-Universität, Universitätsplatz 2,
D-39106, Magdeburg, Germany

Abstract. We present in this report a stochastic model for the virus replication of influenza A in a cell culture. We consider not only the infection process of individual cells but also the number of intracellular components expressed in virus equivalent. Given that this expression is non constant in time we suggest a variable threshold, related to a viral resistance in the cell population, that could explain the time variation in the viral expression in the cell seen in experiments.

Keywords: Virus Replication, Influenza A, Cellular viral Resistance.

1 Introduction

The inoculation of pathogens has been used for a long time in several cultures around the world as a method to boost the resistance of the population against viral infections. For instance, Voltaire describes how the inoculation of pustules in small children has been used by Circassian womans as protection against small-pox, a method that was later introduced in England in the XVIII century [1]. Since its introduction and further development in Europe, this method has become a fundamental element in modern medicine. A vaccine consists of a weak form of a given pathogen that later is inoculated to an individual that has not been infected. As a consequence the individual is infected in a controlled way, inducing a reaction of the individual's immune system, which not only attacks the virus but also learns to recognize such kind of pathogens. This makes this individual immune against this pathogen.

The use of vaccines requires its efficient production. In order to optimize such production process it is necessary to understand how the infection in a cell culture works. However, there is no enough information about virus-host cell interaction in a cellular level and virus spreading in populations of cells in bioreactors. In this report we describe the replication dynamics of influenza A virus in mammalian cell cultures.

The replication of Influenza A virus has been extensively described in several works [2,3]. In this frame the infection of the cell population, and not the virus

J. Zhou (Ed.): Complex 2009, Part I, LNICST 4, pp. 651–656, 2009.

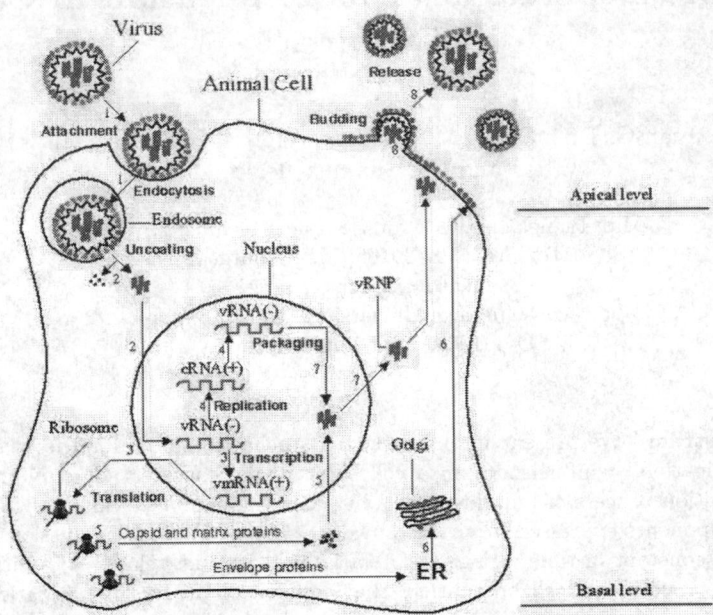

Fig. 1. Replication cycle of Influenza A Virus.Following virus attachment to the surface's receptos of the cell, virions are incorporated to the cell. In diferent steps the genome is transfered to the nucleus, virus protein syntesis and virus genome replication starts and the virus life-cycle ends with the budding and generating of new virus particles (Figure from Sidorenko et al. [6]).

propagation inside a given human population (and the interaction of the virus with the immune system) -see for instance [4,5]-, is the relevant question of this investigation. The studied virus basically consists of a polar DNA molecule encapsulated in a protein membrane forming two basic structures; hemaglutinin and neuramidase. After virus attachment to the cell membrane the genome is transferred to the nucleus. Thereafter virus protein synthesis and virus genome replication starts and the virus life cycle ends with the release of newly generated virus particles (See Fig. 1).

In some works the intracellular process of the infection cycle was considered [6], whereas in other approaches the initial steps of infection and endocytosis were implemented [9]. Other investigations have showed, how the ratio in the production rates of different viral strains is, a relevant problem in order to analyze drug resistance of different viral variants [8,7]. However, the present model describes systems where there is only one class of viral species, focusing more on the specific infection mechanism of the cell rather than on the concurrence with different viral variants. In a nut shell, the virus spreading in populations as well as the differentiation of infected cells is not well understood (those are two important factors for the optimization of vaccine production). In the present work we extend the investigation with the development of different variants of adequate stochastic models for the replication of a virus population in a cell culture based on a model sugested by Sidorenko et. al. [10].

Given that such model is not able to reproduce a time variation in the cells infection grade, observed by means of experimental techniques, we propose in the present approach mechanisms that could explain this variation. In particular we consider the variability of the resistance of the infected cells, represented by a non constant probability of infection, as a plausible mechanism explaining the non monotone infection behavior of the cells. In the next section we introduce the fundamental assumptions of the model and the implementation strategies.

2   Model and Strategies

The present approach considers a distributed balance model, accounting for the stochastic nature of the infection process (See Fig. 1). In previous works a similar approach were adopted, allowing a qualitative and quantitative description of the replication process [10,11]. In such model the interaction between virus and host cells is explicitly represented. It is also assumed that the infection process takes place depending on the concentration of virions. Hence, the fundamental assumption is that the virus infection and replication in the cell culture takes place as a consequence of the adsorption of free virions and not simply as the direct contact among cells. This assumption additionally implies that the spatial distribution of virions and cells is not explicitly considered (from the experimental point of view the spatial distribution is not relevant, because the cell culture is well mixed in a bioreactor). The basic formulation of the cell infection and degradation is similar to the mathematical basis of virus population dynamics introduced by Nowak and May [12], where the population dynamics is represented as a balance of infected and degraded cells. However, here the virus expression is heterogeneous among the cell population. Therefore it is neccesary to introduce an internal coordinate J that corresponds to the intracellular number of viral components expressed in virus equivalents (VE). For the sake of the modeling of the system, this internal coordinate gives the different possible reproduction pathways of the virus inside the cell. Experimentally this number of VE is equivalent to the fluorescence intensity of the expressed cells.

The dynamics is modeled by means of a kinetic Monte Carlo method [13], which requires different transition probabilities for each simulation step. An individual cell can be infected, remain uninfected or suffer for degradation. One process is the degradation of individual cells. If the cell survives then the internal coordinate can adopt either the value $J + 1$, for virus replication, or $J - 1$, for virus release. The change of this internal coordinate represents a change of the class of the cell. Naturally it is assumed that a virus release is only possible if $J > 0$. If the internal state reaches a maximal state J_{max} then the cell does not releases new viruses. The total population of free virions is again described using a population balance equation.

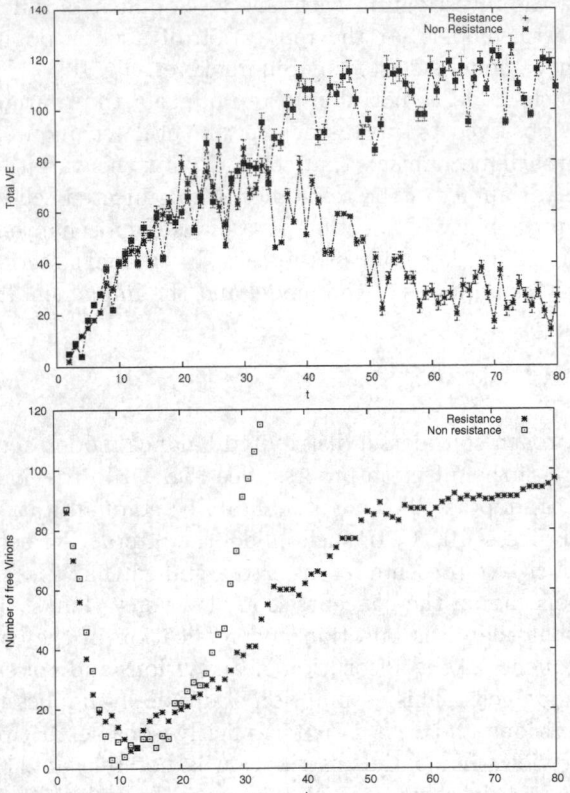

Fig. 2. Frequency of virus replication in the cell population (upper figure) and mean number of free virions (lower figure) as a function of the time for two scenarios: when the cell population develops a kind of resistance agains the virus and when there is no resistance. In the second scenario the frequency of virus replication decrease after some initial time regime (the lines are guides for the eyes).

This basic approach qualitatively (and quantitatively) represents the virus spread in a culture. However, this approach does not account the time variation of the cells infection grade, which has been experimentally obtained using flow calorimetry and fluorescence techniques. Essentially it appears to be that constant infection and release rates do not correctly represent the cellular dynamics. In particular, the initial implementation assumes that the cellular culture is simply infected depending on the concentration and a constant penetration mechanism of the virus into the cell. However, it is also possible to assume that the cells, after some time regime, are able to develop some incipient form of resistance. Naturally, this resistance is a cellular mechanism is not related to the action of an antibody system. In particular, cell signaling mechanism and media mediated transport allow the growth of the concentration of interferons in the cell culture, which have an antiviral function in the cellular host [14].

In order to test this idea we propose that the expression of virus into the cell depends on some threshold related to the whole population of infected cells, assuming that some signaling mechanism advertise the population of non infected cell, increasing their resistance

$$R = \Theta(\Sigma_i V_i - Pr) \qquad (1)$$

where $\Theta(x)$ is the step function, R is the resistance probability, Pr the resistance threshold, where $Pr = (0, V_{Max}]$, with V_{Max} the minimal viral population when there is no cell resistance; V_i are the free virions in the reactor. With this assumption, at some particular time the expression process is stopped, reducing the number of free virions. This simultaneously reduces the number of cells in the population able to express the virus, producing a time variation in the VE. This time variation qualitatively shows that a similar mechanism should be taking place into the cell culture (In [10] experimental results are reported).

The present investigation will be extended to a more detailed description of the role of the host defence system in the dynamics of the virus population; a detailed comparison with experimental results will be shown in future works. Aditionally, a spatial distribution is not considered in this report. However,the present results will help to consider if the the spatial distribution of the virus particles, in particular if the formation of a kind of biofilm, in the microcarriers play also a role in the regulatory process of the virus production.

References

1. Voltaire, F.M.A.: Lettres Philosophiques. The Harvard Classics (1909)
2. Wittaker, G., Bui, M., Helenius, A.: The Role of Nuclear Import and Export in Influenza Virus Infection. Trends in Cell Biology 6, 67 (1996)
3. Ludwig, S., Pleschka, S., Wolff, T.: A fatal Relationship–Influenza Virus Interactions with the Host Cell. Viral Immunol. 12, 175 (1999)
4. Nowak, M.A., McMichael, A.J.: How the HIV defeats the Immune System. Scientific American, Agust (1995)
5. Fergunson, N.M., Cummings, D.A.T., Cauchemez, S., Fraser, C., Riley, S., Meeyai, A., Iamsirithaworn, S., Burke, D.S.: Strategies for Containing an emerging Influenza Pandemic in Southeast Asia. Nature 437, 209 (2005)
6. Sidorenko, Y., Reichl, U.: Structured model of Influenza virus replication in MDCK cells. Biotechnology and Bioengineering 88, 1 (2004)
7. Wu, H., Huang, Y., Dykes, C., Liu, D., Ma, J., Perelson, A.S., Demeter, L.M.: Modeling and Estimation of Replication Fitness of Human Immunodeficiency Virus Type 1 In Vitro Experiments by Using a Growth Competition Assay. J. Virol. 80, 2380 (2006)
8. Marée, A.F.M., Keulen, W., Boucher, C.A.B., De Boer, R.J.: Estimating Relative Fitness in Viral Competition Experiments. J. Virol. 74, 11067 (2000)
9. Mittal, A., Bentz, J.: Comprehensive Kinetic Analysis of Influenza Hemagglutinin-Mediated Membrane Fusion: Role of Sialate Binding. Biophysical Journal 81, 1521 (2001)
10. Sidorenko, Y., Schulze-Horsel, J., Voigt, A., Reichl, U., Kienle, A.: Stochastic Population Balance Modeling of Influenza Virus Replication in Vaccine Production Processes. Chem. Eng. Sci. 63, 157 (2008)

11. Sidorenko, Y., Voigt, A., Schultye-Horsel, J., Reichl, U., Kienle, A.: Stochastic Population Balance Modeling of Influenza Virus Replication in Vaccine Production Processes. Chem. Eng. Sci. (2008) doi: 10.1016-j.ces.2007.12.034
12. Nowak, M.A., May, R.M.: Virus Dynamics. Oxford University Press, New York (2000)
13. Gillespie, D.T.: A General Method for Numerically Simulating the Stochastic Time Evolution of Coupled Chemical Reactions. Jour. Comp. Phys. 22, 403 (1976)
14. Takaoka, A., Yanai, H.: Interferon signalling network in innate defence. Cellular Microbiology 8, 907 (2006)

You Never Walk Alone: Recommending Academic Events Based on Social Network Analysis

Ralf Klamma, Pham Manh Cuong, and Yiwei Cao

Databases & Information Systems
RWTH Aachen University, Ahornstr. 55, D-52056 Aachen, Germany
{klamma,pham,cao}@dbis.rwth-aachen.de

Abstract. Combining Social Network Analysis and recommender systems is a challenging research field. In scientific communities, recommender systems have been applied to provide useful tools for papers, books as well as expert finding. However, academic events (conferences, workshops, international symposiums etc.) are an important driven forces to move forwards cooperation among research communities. We realize a SNA based approach for academic events recommendation problem. Scientific communities analysis and visualization are performed to provide an insight into the communities of event series. A prototype is implemented based on the data from DBLP and EventSeer.net, and the result is observed in order to prove the approach.

Keywords: Recommender systems, Social Network Analysis, community analysis, community of practice, information visualization.

1 Introduction

Academic events play an important role as the major publication and dissemination outlet in scientific communities. In computer science, the number of academic events has increased dramatically in recent years, which is evident in data from DBWorld[1] collected by [6] and data from DBLP and EventSeer.net (see Figure 1). It is challenging, especially for young researchers to find suitable events for submitting papers to and to join in some research communities. There is also the need to identify the research community of a particular researcher.

Until now, tools and methodologies developed for academic events management and documentation still have problems. Event management systems consider event managing process from event announcement, paper submission, paper review to paper acceptance notification. Digital libraries like ACM[2], DBLP[3] or CiteSeer[4] mainly focus on research publications providing tools for papers

---

[1] http://www.cs.wisc.edu/dbworld/
[2] http://portal.acm.org/dl.cfm
[3] http://www.informatik.uni-trier.de/~ley/db/
[4] http://citeseerx.ist.psu.edu/

J. Zhou (Ed.): Complex 2009, Part I, LNICST 4, pp. 657–670, 2009.

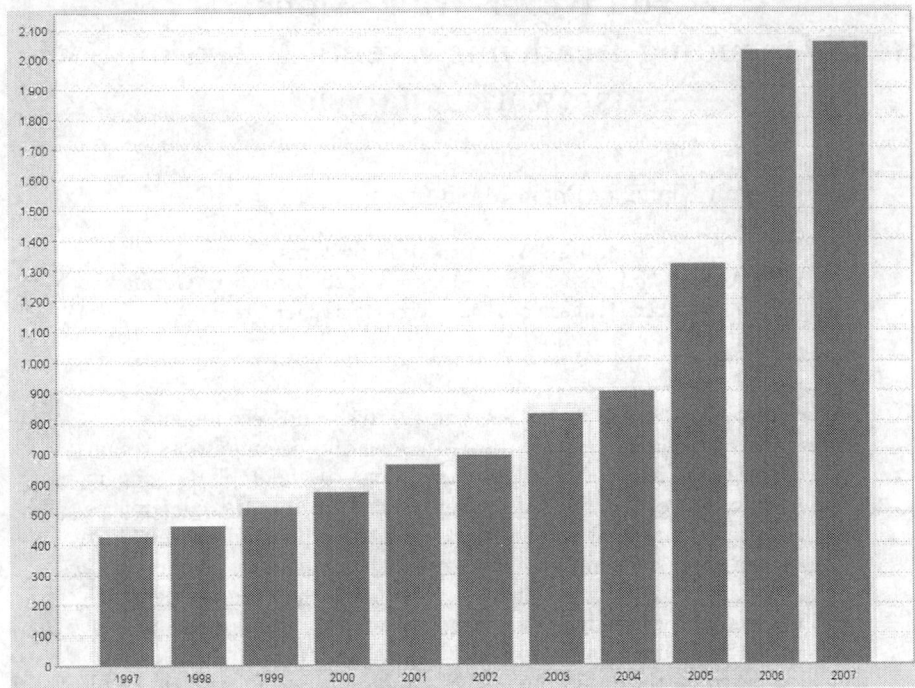

Fig. 1. Number of events in DBLP (by distinct proceedings)

searching. Some other systems like EventSeer.net[5] make a step forward to academic event and community analysis. None of the above mentioned systems provides recommendation tool to help researchers in event finding.

To overcome the aforementioned problem, a model for academic events is required. Event and community data exists but it is unstructured. Past events and their communities are documented by proceedings in digital libraries. Upcoming events are recognized by Call for Papers and detail information can be obtained from their web sites. There is no structured data for academic events. Moreover, with the recent advantages in technical communication as well as the increasing use of digital cooperation mechanism, there is also a requirement to integrate new digital media such as blogs, wikis, mailing-list, images, etc., into one model for events documentation. The model must reflect all aspects of events and their communities as well as be capable to connect and collect data from heterogeneous data sources such as digital libraries and the Web.

In this paper, we propose a model for events and scientific communities. Based on this model, we realize a SNA based approach to recommend the events to researchers. We study how the research communities support individual members in events finding by applying collaborative filtering technique for event

---

[5] http://eventseer.net/

recommendation. The paper is organized as follow. In the next section, we briefly survey the related work on Collaborative Filtering, Actor Network Theory and Social Network Analysis. In Section 3, we present a conceptual model for academic events and communities. In Section 4, the design of recommendation algorithm is discussed. In Section 5, we describe our experimental result with the real dataset from DBLP and EventSeer.net. In Section 6, we conclude our paper with a discussion and an outlook.

2   Related Work

Combining Social Network Analysis and recommender systems have been studied and applied in different application domains. In digital libraries, many approaches have been proposed to provide useful tools to researchers, e.g. citation recommendation [19], book recommendation [20], paper recommendation [21] etc. Generally, recommendation techniques can be categorized into three classes: Collaborative Filtering (CF), Content based and Hybrid approaches. CF is based on users community to generate recommendations, while Content based uses the features of items. Hybrid approaches combine CF and Content based with some other techniques such as demography, utility-based, knowledge-based recommendations to improve the quality of recommendation results. In this paper, we investigate how CF could be applied to event recommendation problem. We leave out hybrid approaches for the future work.

Collaborative Filtering (CF)
CF is widely used in commercial applications. CF provides the recommendations based on previous user's preferences and the opinions of other users who have similar preferences [4]. User's preferences can be expressed explicitly (e.g. rating for an item) or implicitly by interpreting user's behavior like purchase history, browsing data and other types of information access pattern. Collaborative filtering algorithms can be divided into two categories: memory-based collaborative filtering algorithms operate on the entire user-item database to generate the recommendations; model-based collaborative filtering algorithms use the user database to learn a model which is then used for recommending.

In general, a recommender system has three components: background data which is the information that the system has before the recommendation process begins, input data which is the information that user must communicate to the system in order to generate a recommendation, and an algorithm that combines background and input data to arrive at its suggestions [2]. In collaborative filtering, background data is the rating history of users on set of items, input data is rating history of target user. Collaborative filtering works by viewing the above dataset as a rating matrix. Ratings may be binary or real values indicate user's preference on the item. Columns in this matrix are items (called item vectors) and rows represent users (called user vectors). Each entry in the matrix is the user's rating for a particular item.

Actor Network Theory (ANT)

Actor-Network Theory (ANT) was developed by two French scholars, Michel Callon and Bruno Latour [7]. Digital networks are a meeting point for the social and technology. In ANT model, we have a network formulated by actors and relationships [8]. A actor may be a human or an object without any distinction. Any set of actors involved in a certain activity formulates a network. There are three special kinds of actors. The member stands for a person or a community. The medium enables members to do the activities, for example establishing communication links and exchanging the information. Artifacts are objects created by members using some media.

The conceptual model proposed in this paper is based on ANT. As mentioned earlier, digital media need to be integrated into the model for events and communities documentation. ANT tries to explain social order not through the notion of "the social" but through the networks of connections between human agents, technologies and objects [9]. Communities of academic events have been seen as communities of practice in which members exchange the information and communicate with each others using the combination of various communication methods such as face-to-face meeting and technology-enhanced methods, e.g. discussion forums, websites, mailing-list, blogs, wikis etc. Technology-enhanced communication techniques have became more and more important, especially when the international degree of recent conferences increases. Members of the community can live in different country and continents. Sometimes it is hard to organize face-to-face meeting and discussion. Therefore advance communication method is a important mechanism contributing to the successful of a scientific community. All these aspects need to be modeled as a cross-media base for scientific community.

Social Network Analysis

In digital library, it is possible to create the networks that reflects the collaboration between researchers using the references in research papers. In particular, there are many research work have studied the creation of these networks and applied Social Network Analysis for scientific community to understand the structure and pattern of research collaboration [10,11,12]. In the domain of publication and venue ranking, many approaches have been proposed to measure the impact of scientific collection (journals, proceedings) and scholar authors [15,16,17,18], which focuss on citation and co-authorship networks as the professional network between researchers. There are also researches which try to apply Social Network Analysis to evaluate the quality of academic events [6]. We are adding to these work by investigating the role of research community in helping researchers to find academic events and to identify research communities.

3 Model for Academic Events and Scientific Communities

Based on ANT, a model for academic events and communities is proposed as given in Figure 2. In this model, we consider the network of researchers in the relation with academic events. For each event (and event series), we have a network representing research collaboration between members. There are three kinds of network under consideration, including co-authorship network, citation network and co-participation network. *Scientist* entity describes the node of network, *Link* entity represents the connection between nodes and *Subnetwork* entity models the subnetwork extracted from *global network* which is composed of *Scientist* and *Link* entities. *Link* entity has a attribute *type* to differentiate three kinds of network: co-authorship, citation and co-participation networks.

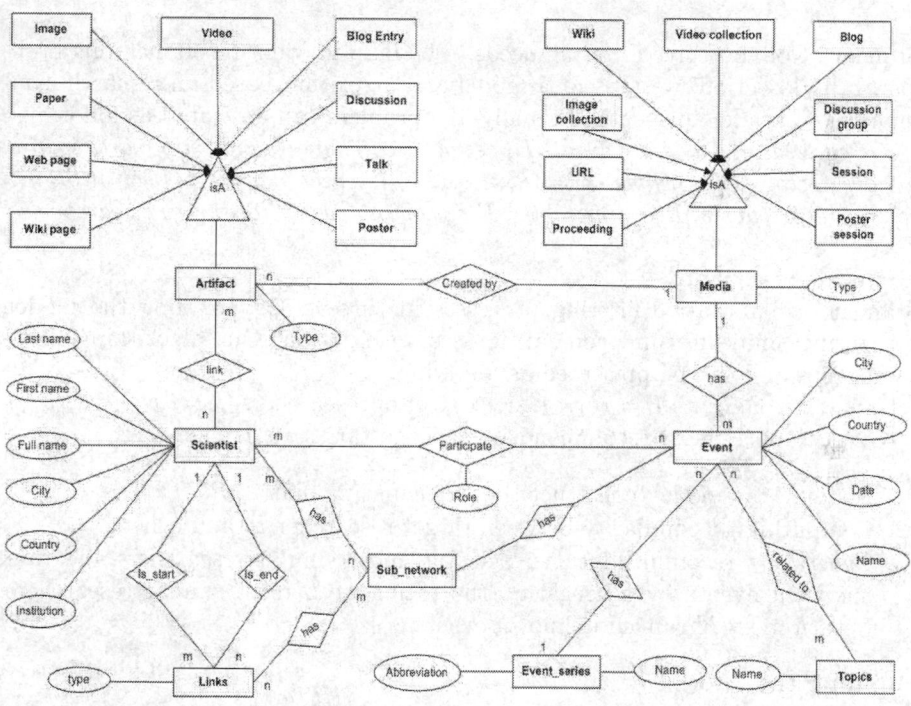

Fig. 2. Model for events and communities

Each *Event* belongs to a *Event series*, e.g. ACM SIGMOD, VLDB series etc. We consider all kinds of academic event, including conferences, workshops, international symposiums, doctoral consortiums as well as winter/summer schools. In general, workshops can be held as independent events (therefore they have their own series) or in combination with conferences, symposiums or consortiums. Each *Event* has a set of *Topics* which presents event's research domains and objectives. In fact, research topics tracking as well as topics classification are complicated problems. Research topics can be categorized in hierarchical structure in which a common topic (e.g. Database, Information Systems) can be divided into sub-classes. To keep it simple, in our model we use a "flat" list of topics used to specify research interests of a event. In mediabase, we integrate all types of digital media, e.g wikis, blogs, web sites, videos, images etc.

This model intends to be the basic on which a recommendation tool is based. It also serves as the foundation for event and community analysis. Mediabase could be extent so that different media management and monitoring tools like BlogWatchers, MailWatcher, WikiWatchers etc. can be applied.

4   Collaborative Filtering and Academic Event Recommendation

Standard Collaborative Filtering needs to be map to event recommending problem. In this section, we present a model and algorithm based on research communities of academic events. Formally, the problem can be stated as following:

Given a set of academic events E, set of researchers U and set of participation history vectors V in which $v_u = (e_1, e_2, ..., e_n)$ represents the participation history of researcher u. Recommend top K upcoming events for target researcher u_t.

General Algorithm
Standard collaborative filtering processes in three steps: building the model, computing similarity and generating recommendation. Our algorithm follows these steps and can be presented as following:

Input: set of events $E = (e_1, e_2, ..., e_N)$, set of researchers $U = (u_1, u_2, ..., u_M)$.
Output: top K most recommended events to target researcher u_t.

1. Building the model: construct the participating matrix $R(MxN)$.
2. Computing the similarity between target researcher u_t and others.
3. Generating recommendations: Select L most similar researchers and rank unknown events by aggregating the rating of L most similar researchers. Return most K ranked unknown events.

Building the Model
As presented in Section 2, collaborative filtering operates on a rating matrix in which each entry is user rating on an item. To map this model to our problem, we use the following approach: we consider academic events as "items" and researchers are users who will get the recommendations. The rating value of a researcher for an event is binary (i.e. 1 and 0), meaning that he participated or he will take part in this event, or not. We use event participation history of researchers as background data and the input data is the participation history of a particular researcher. The rating matrix then can be built using the background data. Formally, given a set of academic events E, set of researchers U then $R(M, N)$ is the rating matrix in which entry $R_{u,e} = 1$ if user u participated in event e and $R_{u,e} = 0$ if user u did not participated in event e. We use U_p to

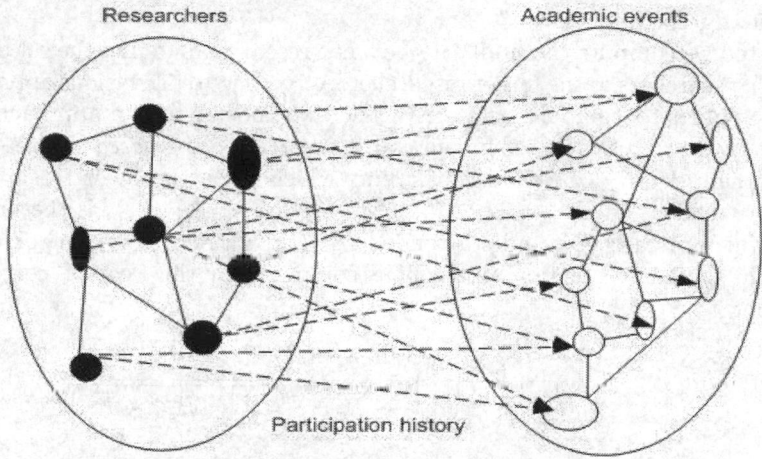

Fig. 3. Collaborative Filtering model mapping

denote the pth row of R which is called the researcher vector of researcher u_p and E_q to denote q^{th} column of R which is called event vector of event e_q.

The above approach suffers from the *start-up* and *generality* problems. For newbie researchers who did not attend any events before nor publish any papers with other researchers, they will not get the recommendations since the system has no information about them. To overcome this problem, we use profile building mechanism as in other recommender systems: users have to rate a sufficient number of items before they can get the recommendations. Under the assumption that normally a newbie researcher starts his researches with the help of his professors or advisors as well as his colleagues. That means implicitly he has a research community. He could also join the communities of events in which he is interested, in order to keep track of what these communities is doing. Overall, by explicitly declaring his own "implicit" community, a newbie researcher can "embed" himself into a scientific community and let that community help him to find events.

Generality problem emerges from the fact that researchers may change their fields as well as work on different fields. For example, a researcher may work on database system and distributed system. Therefore, he attends conferences on database system and distributed system as well. Target researcher attended many conferences with him on database system and then he may be recommended conferences on distributed system. With many researchers like that, it is difficult to find a set of recommended events which satisfy target researcher's preferences. We solve this problem by a subjective classification via profile building mechanism. User's preference on topics is used to filter out events which are not relevant before performing recommendation process. This preprocessing procedure ensures that recommendation algorithm will work on a set of events which satisfies user's needs in general.

Computing Similarity

In this step we compute the similarity between researchers to find the set of most "closed" researchers to the target researcher. According to [5], various approaches can be applied to compute similarity. The two most popular approaches are *correlation* and *cosine-based*. In our work, we use *cosine-based* approach. To present them, let $E_{x,y}$ be the set of events which researcher x or researcher y, or both attended, i.e $E_{x,y} = \{e \in E \mid R_{x,e} = 1 \| R_{y,e} = 1\}$. $E_{x,y}$ is the union of events which researcher x and y attended (E_x and E_y relatively). In correlation approach, similarity function $sim(x,y)$ is computed by the Pearson correlation coefficient:

$$sim(x,y) = \frac{\sum_{e \in E_{x,y}} (R_{x,e} - \overline{R_x})(R_{y,e} - \overline{R_y})}{\sqrt{\sum_{e \in E_{x,y}} (R_{x,e} - \overline{R_x})^2 \sum_{e \in E_{x,y}} (R_{y,e} - \overline{R_y})^2}} \tag{1}$$

in which the average rating of researcher x, $\overline{R_x}$ is:

$$\overline{R_x} = \frac{1}{|E_x|} \sum_{e \in E_x} R_{x,e} \tag{2}$$

which is equals to 1 in our case.

In the cosine-based approach, the two researchers x and y are treated as two vectors $\vec{x}$ and $\vec{y}$ in m-dimensional space, where $m = |E_{x,y}|$. Similarity between two vectors can be measured by computing the cosine of the angle between them:

$$sim(x,y) = cos(\vec{x}, \vec{y}) = \frac{\vec{x} \cdot \vec{y}}{\|\vec{x}\| \times \|\vec{y}\|} = \frac{\sum_{e \in E_{x,y}} R_{x,e} R_{y,e}}{\sqrt{\sum_{e \in E_{x,y}} R_{x,e}^2} \sqrt{\sum_{e \in E_{x,y}} R_{y,e}^2}} \tag{3}$$

where $\vec{x} \cdot \vec{y}$ denotes the dot-product between the vectors $\vec{x}$ and $\vec{y}$.

Generating Recommendations

Recommendation generating is a ranking process in which we compute a ranked values for unknown events. According to [5], ranked value is usually computed as an aggregate of the ratings of L most similar researchers for the same event:

$$R_{c,e} = aggr_{d \in C} R_{d,e} \tag{4}$$

where C denotes the set of L researchers who are most similar to researcher c and have participated in (or will attend) event e. Some of the aggregate functions are:

$$R_{c,e} = \frac{1}{L} \sum_{d \in C} R_{d,e} \tag{5}$$

$$R_{c,e} = k \sum_{d \in C} sim(c,d) \times R_{d,e} \tag{6}$$

$$R_{c,e} = \overline{R_c} + k \sum_{d \in C} sim(c,d) \times (R_{d,e} - \overline{R_d}) \tag{7}$$

where $\overline{R_c}$ is computed as in previous section and multiplier k serves as a normalizing factor and is usually selected as:

$$k = \frac{1}{\sum_{d \in C} sim(c,d)} \qquad (8)$$

We use aggregate as an average (defined in the first case). However, in more complicated cases, the aggregate could be a weighted sum in which the similarity between c and d is used as a weight, i.e the more similar c and d are, the more weight $R_{d,e}$ will carry in the ranked value $R_{c,e}$.

5 Prototype Evaluation

Datasets

To evaluate the approach, a prototype is implemented based on the data from DBLP XML record and EventSeer.net[6]. First, DBLP XML record is parsed to get the list of past events and co-authorship network of each event. Event series are taken by parsing DBLP Website. Events then are bound into series by the unique URL prefixes of event and event series. Location information of events is also taken from DBLP Website. Upcoming events are extracted from EventSeer.net Web site. EventSeer.net contains most of Call for Papers for conferences in Computer Science. From EventSeer.net, we got a list of upcoming (and past) conferences with the information about time, locations, topics, persons and organizations. Overall, we have a dataset as summarized in Table 1.

Table 1. Dataset summary

Data	Quantity
Events	16821
Series	2099
Authors	522938
Topics	4910
Co-authorship of events	1282796 links

Data from DBLP and EventSeer.net is enough for the evaluation, although it is not complete. Ideally, we should have the list of all participants, authors and programm committee members (PC members) of each event. DBLP contains only the authors, while EventSeer.net indexes persons who are mentioned in Call for Papers, so they are mostly PC members. However, using authors and PC members as background data for recommendation algorithm is reasonable. Authors and PC members of each event have a closed relation via papers review process. Authors also have the knowledge about each others since they have worked on the same problems.

---

[6] http://bosch.informatik.rwth-aachen.de:5080/AERCS/

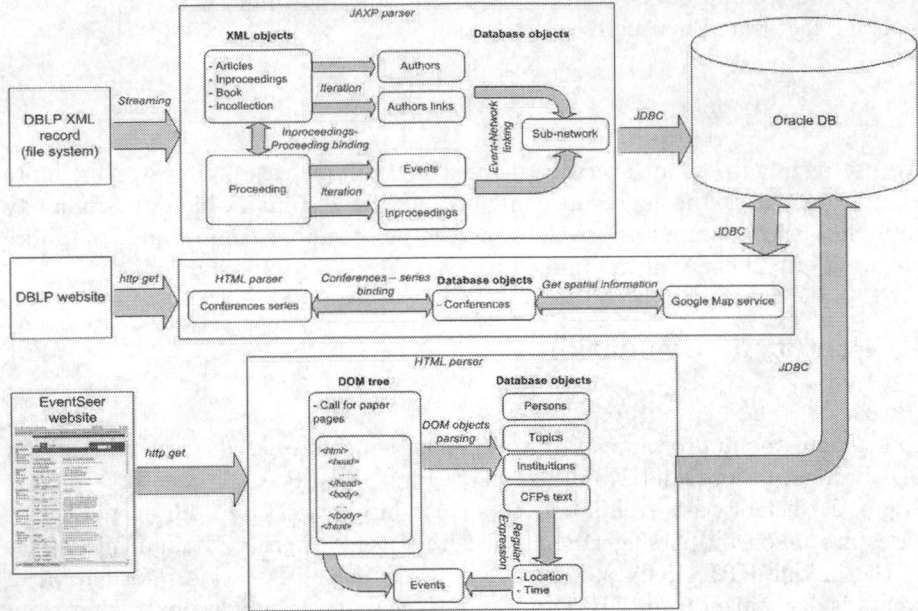

Fig. 4. Data preparation process

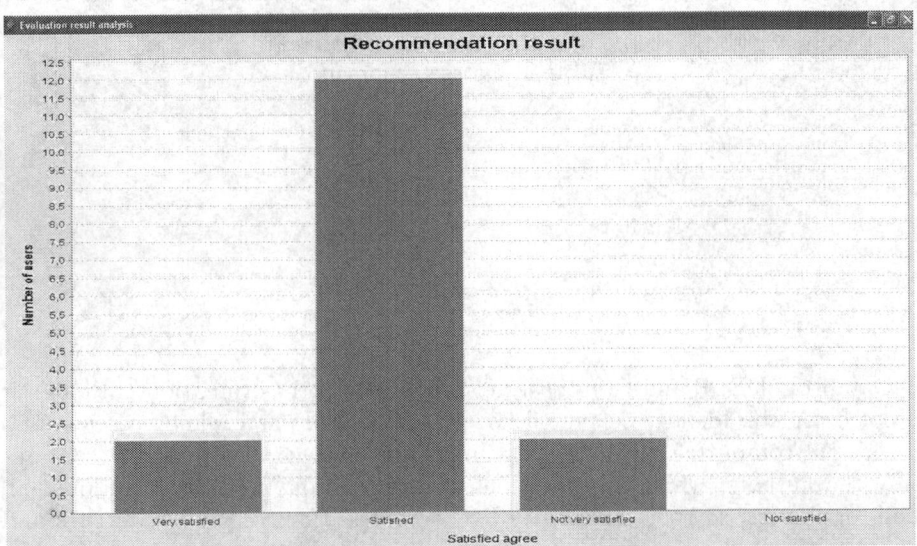

Fig. 5. Users satisfaction with recommendation result

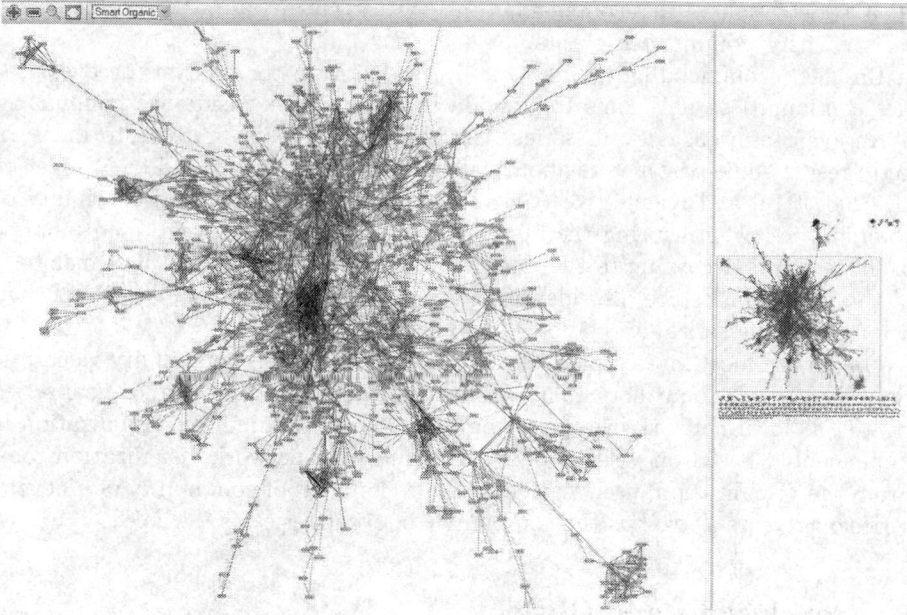

Fig. 6. ACM SIGMOD community visualization

Online Experiment

To evaluate the approach, we conducted a online survey on a set of users to get the opinion about recommendation result and community analysis provided by the system. Users are selected from colleagues and students working and studying at the Chair of Database and Information Systems, RWTH Aachen, Germany. A short tutorial is given to users and a questionaire is put online to let users answer a set of questions. The tutorial guides users through several tasks in order to get to know the concepts of the system, e.g. profile building, getting recommendation, finding events and event series as well as community analysis and visualization.

The system gains over 20 feedbacks in which most of the questions are filled in. First, we analyze the feedbacks to see users experiences in academic events as well as their roles in the events they attended. Most of users participated in 6 to 20 events, others attended 1 to 5 events. Among them, about 11 users took part in the events as participants, 6 users as presenters and a small number (about 4 users) as PC members. This result shows that our users community are young researchers.

In the second step, we assess the feedbacks to know users opinion about recommendation result. Users are asked to build their profiles in which they have to declare the preferences on topics, locations, persons and events. System then generates a list of upcoming events recommended to them. Users can compare the list with events they are interested in as well as discover new events which they do not know. Users express their opinion by answering a question about their satisfaction with recommendations. As shown in Figure 5, most of users satisfy with recommended events.

Besides recommending events to users, we perform the analysis on event series communities. This aims to provide to users a look inside the community of an event as well as event series. With our dataset, we are able to measure and present some parameters about the communities as proposed by Wenger et al. (2002) [14] and Kienle [13]. We analyze the development and continuity of communities by measuring the number of participants over years and number of participants according to the number of events they attended. Key members of the communities are also identified according to the number of events they attended in the series.

One of the most interesting features of the prototype is community visualization. We provide co-authorship network visualization of an event and event series as well as local network of a particular researcher. Community visualization is implemented based on yFile AJAX - a commercial network visualization tool. From the visualization, users can see the development of community of an event series over years as well as the community of event series as a whole.

6   Conclusions and Outlook

Recommender systems for digital libraries and scientific communities is an ongoing research domain. A recommender system could be a great tool for young researchers to find academic events to which they can submit papers. Our experiments show that applying a community based recommendation algorithm supports researchers in events finding. By using event participation history as background data for a Collaborative Filtering based algorithm, we are able to recommend the most relevant academic events to researchers. The algorithm works on the dataset which can be easily extracted from references in papers documented in digital libraries like DBLP, ACM or EventSeer.net.

The dataset of our system should be enhanced with some other data sources. Currently, data from DBLP and EventSeer.net is imported into our database. To have better recommendation results and analysis, we need also data from other digital libraries such as ACM, CiteSeer. The problem here is how to connect these data sources to provide a unique repository for academic events. We are working on this problem by investigating and applying different data and Web mining techniques in order to create a mesh data source network. Based on this, useful services could be further designed and implemented.

In the future, it would be interest to investigate other recommendation techniques as well as algorithms for event recommendation problem. Content-based recommendation and the combination of content-based with CF and other recommendation techniques is a promising direction. It would also be interesting to see these recommendation approaches in other domains. Currently, we are performing the social network analysis of 45.000 schools in Europe.

Another idea is to follow the dynamic behaviour of researchers. The movement of researchers between communities could be captured. The question is that what are important factors affecting this movement and the role of spanners in the communities. By tracking and analysis the dynamic movement of members, we could be able to recommend the future directions in research as well as carrier for researchers.

References

1. Schafer, J.B., Konstan, J.A., Riedl, J.: E-commerce recommendation applications. Data Min. Knowl. Discov. 5(1-2), 115–153 (2001)
2. Burke, R.: Hybrid recommender systems: Survey and experiments. User Modeling and User-Adapted Interaction 12(4), 331–370 (2002)
3. Sarwar, D., Karypis, G., Konstan, J., Reidl, J.: Item-based collaborative filtering recommendation algorithms. In: Proceedings of the 10th international conference on World Wide Web, pp. 285–295. ACM Press, New York (2001)
4. Breese, J.S., Heckerman, D., Kadie, C.M.: Empirical analysis of predictive algorithms for collaborative filtering, pp. 43–52 (1998)
5. Adomavicius, G., Tuzhilin, A.: Toward the next generation of recommender systems: A survey of the stateof-the-art and possible extensions. IEEE Transactions on Knowledge and Data Engineering 17(6), 734–749 (2005)
6. Zhuang, Z., Elmacioglu, E., Lee, D., Giles, C.L.: Measuring conference quality by mining program committee characteristics. In: Proceedings of the 2007 conference on Digital libraries, pp. 225–234. ACM Press, New York (2007)
7. Latour, B.: On recalling ant. In: Law, J., Hassard, J. (eds.) Actor-Network Theory and After, pp. 15–25 (1999)
8. Denev, D.: Multidimensional Patterns of Disturbance in Digital Social Networks. Master's thesis, RWTH Aachen University (2006)
9. Couldry, N.: Actor Network Theory and Media: Do They Connect and On What Terms? In: Hepp, A., et al. (eds.) Cultures of Connectivity. School of Economics and Political Science, London, pp. 1–14 (2004)
10. Newman, M.E.: Scientific collaboration networks. i. network construction and fundamental results. Phys. Rev. E. Stat. Nonlin. Soft. Matter. Phys. 64(1-2) (2001)
11. Newman, M.E.: Coauthorship networks and patterns of scientific collaboration. Proc. Natl. Acad. Sci. USA 101, 5200–5205 (2004)
12. Huang, T.H., Huang, M.L.: Analysis and visualization of co-authorship networks for understanding academic collaboration and knowledge domain of individual researchers. In: CGIV 2006: Proceedings of the International Conference on Computer Graphics, Imaging and Visualisation, pp. 18–23. IEEE Computer Society, Washington (2006)
13. Kienle, A., Wesser, M.: Principles for cultivating scientific communities of practice. In: Proceedings of the 2nd International Conference on Communities and Technologies, pp. 283–299. Springer Netherlands (2005)
14. Wenger, E., McDermott, R., Snyder, W.M.: Cutivating Communities of Practice: A guid to Managing Knowledge. Havard Business School Press, Campridge (2002)
15. Yan, S., Lee, D.: Toward alternative measures for ranking venues: a case of database research community. In: Proceedings of the 2007 conference on Digital libraries, pp. 235–244. ACM Press, New York (2007)
16. Rahm, E., Thor, A.: Citation analysis of database publications. SIGMOD Record 34(4), 48–53 (2005)

17. Sidiropoulos, A., Manolopoulos, Y.: A citation-based system to assist prize awarding. SIGMOD Record 34(4), 54–60 (2005)
18. Sidiropoulos, A., Manolopoulos, Y.: A new perspective to automatically rank scientific conferences using digital libraries. Inf. Process. Manage. 41(2), 289–312 (2005)
19. McNee, S.M., Albert, I., Cosley, D., Gopalkrishnan, P., Lam, S.K., Rashid, A.M., Konstan, J.A., Riedl, J.: On the recommending of citations for research papers. In: Proceedings of the 2002 ACM conference on Computer supported cooperative work, pp. 116–125. ACM Press, New York (2002)
20. Mooney, R.J., Roy, L.: Content-based book recommending using learning for text categorization. In: Proceedings of the Fifth ACM Conference on Digital Libraries, pp. 195–204. ACM Press, New York (2000)
21. Torres, R., McNee, S.M., Abel, M., Konstan, J.A., Riedl, J.: Enhancing digital libraries with TechLens+. In: Proceedings of the 4th ACM/IEEE-CS Joint Conference on Digital Libraries, pp. 228–236. ACM Press, New York (2004)

Visualization of Complex Biological Systems: An Immune Response Model Using OpenGL

John Burns[1], Heather J. Ruskin[2], Dimitri Perrin[2], and John Walsh[1]

[1] Dept of Computing, Institute of Technology Tallaght, Dublin 24, Ireland
john.burns@ittdublin.ie
http://computing.dcu.ie/~jburns
[2] School of Computing, Dublin City University, Dublin 9, Ireland

Abstract. In this paper we present an update on our novel visualization technologies based on cellular immune interaction from both large-scale spatial and temporal perspectives. We do so with a primary motive: to present a visually and behaviourally realistic environment to the community of experimental biologists and physicians such that their knowledge and expertise may be more readily integrated into the model creation and calibration process. Visualization aids understanding as we rely on visual perception to make crucial decisions. For example, with our initial model, we can visualize the dynamics of an idealized lymphatic compartment, with antigen presenting cells (APC) and cytotoxic T lymphocyte (CTL) cells. The visualization technology presented here offers the researcher the ability to start, pause, zoom-in, zoom-out and navigate in 3-dimensions through an *idealised* lymphatic compartment.

Keywords: Visualization, Emergent Behavior, Immune Response.

1 Introduction

Emergent behaviour is the process whereby global features or structures emerge naturally from a local system in which such features are not merely aggregates of microscopic interactions. However, emergent behaviour can be a difficult property to unambiguously identify, and therefore emergent behaviour is often characterised by the process of systemic *self-organisation* where the higher-level components of the model take on non-random spatial structures. Thus, self-organisation becomes apparent through visual representation of system components and patterns of change which occur over time.

Visualization is a burgeoning field of scientific computing which seeks to provide multi-dimensional graphical representation of underlying models or data. Such representations can typically be subject to manipulation such as rotation, transformation, animation and so on. Complex data sets are transformed into visually meaningful 2-D or 3-D realizations in order to improve understanding of the underlying data, and to aid in its interpretation. (See, eg, [1],[2] and [3])

One of the key pathways to successful and speedy drug design and development is creation of accurate and realistic computational models and simulations

J. Zhou (Ed.): Complex 2009, Part I, LNICST 4, pp. 671–679, 2009.

of all levels of human biological response. Such models would then enable researchers to both test their hypotheses as well as to form the basis of an advanced biological knowledge repository. We therefore need to address the question of how to effectively create a software application that will draw the experimental biology community in to the process of model design and development. In the past, this has proved to be extremely difficult, primarily, we feel, because such models rarely, if ever, provided the researcher with mechanisms to visualise the model dynamics. Therefore, it follows that this problem can be most effectively addressed by the development of a feature rich 3D Bio-Visualisation platform.

2   Research Objectives

The two key objectives of this research effort are two-fold:

1. To develop better in-silico disease models of human disease progression, in particular, the pathways and dynamics of inflammatory diseases (such as arthritis and asthma). This type of modelling should be based on a mechanistic understanding of the disease process as a function of time, and not merely on individual potential target molecules, in other words, systems simulation versus target simulation.
2. The development of a 3-D software visualisation platform that will enable us to integrate the domain expertise of experimental biologists. In addition, this platform would enhance learning in the laboratory or classroom, in that students of cellular biology might more quickly comprehend the complex dynamical nature of typical immune system functions.

At first, these two objectives are not necessarily related. However, in order to achieve the first of these objectives, we have concluded that the second objective will be a crucial step.

2.1   Contribution to Knowledge

We are strongly encouraged and motivated in this research by the findings of a recent EU report ("The Innovative Medicines Initiative (IMI) Strategic Research Agenda Creating Biomedical R&D Leadership for Europe to Benefit Patients and Society") (http://www.efpia.org/4_pos/SRA.pdf) which indicates that in-silico modeling and simulation of biological processes is a key requirement in the development of cost-effective and timely new disease therapies. We see this platform as a key enabler for this strategy, in that integration of the computing and bio/medical communities continues to be problematic and inconsistent.

3   Model-View-Controller

The architecture of the application is based on the common design pattern known as the *model-view-controller* (or MVC) pattern, popularised by [4]. In MVC, the model represents the information (the data) of the application and the business rules used to manipulate the data; the view corresponds to elements of the user interface such as text, checkbox items, and so forth; and the controller manages details involving the communication to the model of user actions such as keystrokes and mouse movements. This approach is presented in Fig 1.

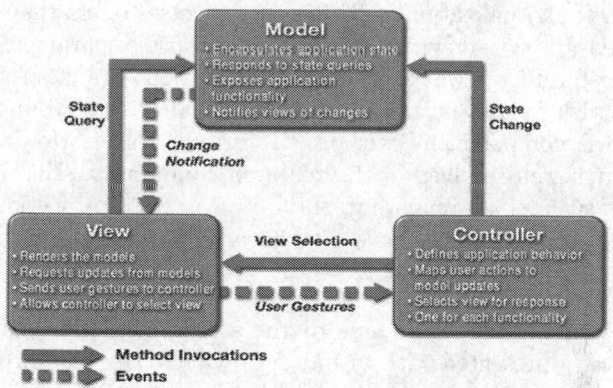

Fig. 1. Model View Controller architecture, (image courtesy of Sun Microsystems)

Thus, in our model, we have the simulation (model) component executing in a thread, with a shared data structure for the view thread to read and render every 10 seconds. A set of keyboard controls allow the user to navigate through the model space in 3 dimensions, moving to sites within the lymphatic compartment that may be of interest, and pausing the simulation to study individual cells, or clusters of cells, that may be of specific interest.

3.1   Model

The model has two classes of nodes, the immune cells (*ctl*) the antigen-presenting cells (*apc*). Although restricting the model to just *apc* and *ctl* cells is a simplification of actual biological systems, it is justified by noting that virus-specific CTL are generally considered to be the principal effectors in mediating recovery in the acute immune response to respiratory viral infection (see references contained in [5]). The model has some key parameters, such as initial cell levels, rates of change in the lifecycle, frequency of infection events and many others. Some viral pathogens are capable of persistent re-infection, in that, although population levels of infected antigen presentation cells may decline in response to clearance pressure by a specific CTL response, over time, the number of infected cells rises to chronic and sometimes acute levels. Examples of such viruses are HIV, HTLV, hepatitis C (HCV), hepatitis B virus, CMV EBV and rubella Such persistent re-infection pathogens have been associated with normal immune function suppression. This means that the model simulates persistent re-infection by randomly scattering a repeat 'dose' of the pathogen, introduced some 300 time-steps in the simulation. This re-infection pattern is a represents a resurgence of infected cells every 6.25 days, in discrete bursts. This model is discussed in more detail in [6], [7] and [8].

3.2 View

As mentioned, the visualization of biological processes offers the researcher the ability to speed-up, slow-down and hypothesize various parameters. Visualization aids understanding as we rely on visual perception to make crucial decisions. For example, with our initial model, we can visualize the dynamics of an idealized lymphatic compartment, with APC and CTL cells. However, the early models, although well received in the research community, still lack important characteristics such as 3D rendering, zoom-in, rotation, projection and instant reply. These initial deficiencies have now been rectified in the results presented in this paper.

OpenGL Texture Mapping. One of the strengths of this platform is that the user is always presented with visual cues which they can directly relate to images of cells that might be encountered in the actual lab setting. With very slight modifications we can convert typical 2D raster images into 3D spherical-based objects.

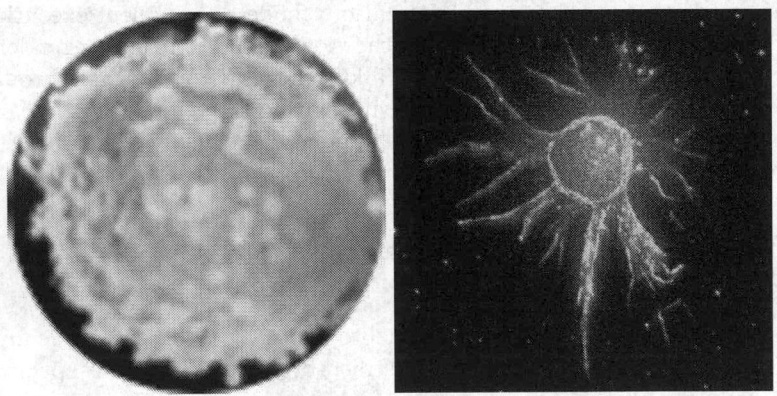

Fig. 2. Left, an actual laboratory photo of a CTL cell and right, an image of an antigen presenting cell. Images courtesy of the La Jolla Institute for Allergy and Immunology.

For this, we take freely available CTL images (see Fig. 2), convert them to .tga format, and then load and bind them into our visualisation engine:

```
//load the naive CTL:
pImage = gltLoadTGA("./images/ctl.tga",
          &iWidth, &iHeight, &iComponents, &eFormat);
glBindTexture(GL_TEXTURE_2D, CTL_IMAGE);
glTexImage2D (GL_TEXTURE_2D, 0, GL_RGBA, iWidth,
```

```
        iHeight, 0, GL_RGBA, GL_BYTE, pImage);
gluBuild2DMipmaps(GL_TEXTURE_2D, GL_RGBA, iWidth,
        iHeight, GL_RGBA, GL_UNSIGNED_BYTE, pImage);
//...
```

Then we use the following OpenGL calls we map these images to `GluSphere` objects:

```
glPushMatrix();
glColor3f(0.0, 0.0, 1.0);
//...
glBindTexture(GL_TEXTURE_2D, CTL_IMAGE);
gluSphere(qobj, 0.01, 20.0, 20.0);
glPopMatrix();
```

in this way, each of our biological entities (or agents) within the model, are visualised in turn, at each update step.

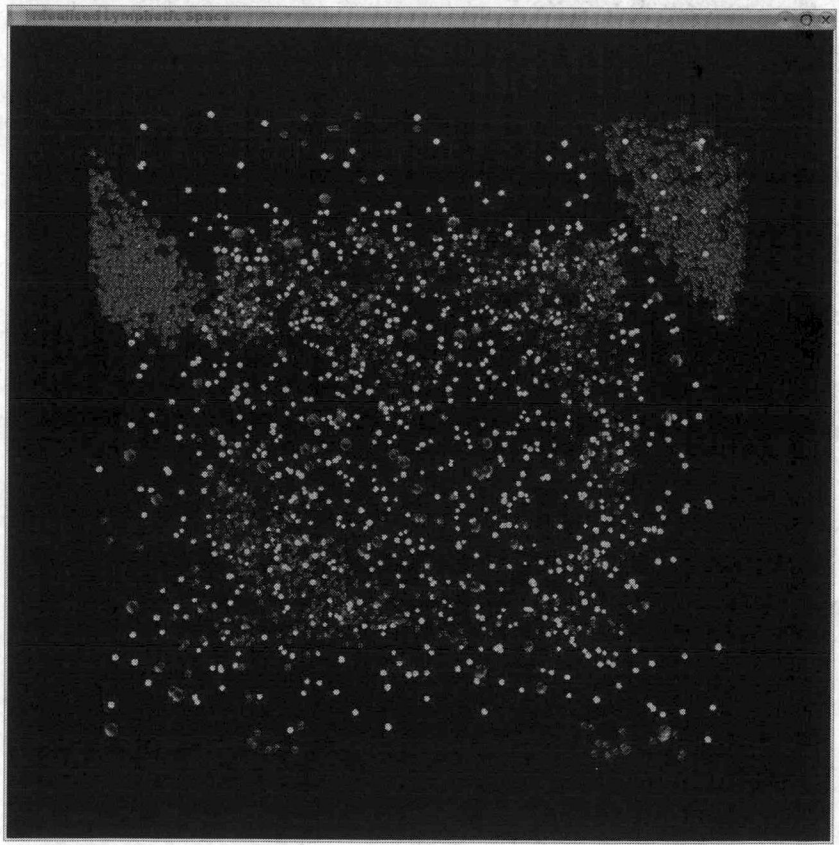

Fig. 3. The initial view of the model lymphatic compartment, with the viewer camera position placed far away along the z axis

4   Visualisation Results

Our results are presented here, in the form of four renderings of the lymphatic compartment outlined in the previous section. Although we model a space of some $x = y = z = 100$ in size, we normalise this space to the OpenGL co-ordinate space, which has each co-ordinate axis run from $[-1, 1]$. When the simulation begins, the viewers perspective always starts out placed $x = y = 0$ with $z = 20$. This has the effect of placing the viewer far outside the space, but looking towards the centre of the space, along the z-axis. This position is shown in the imgage in Fig. 3.

The user can now begin to navigate the space, by using the following key commands:

1. F2 key: move the camera postion along the z-axis towards the origin.
2. F3 key: move the camera postion along the z-axis away from the origin.

Fig. 4. Centre-right - above and behind is the APC, with the CTL visible to the front lower right

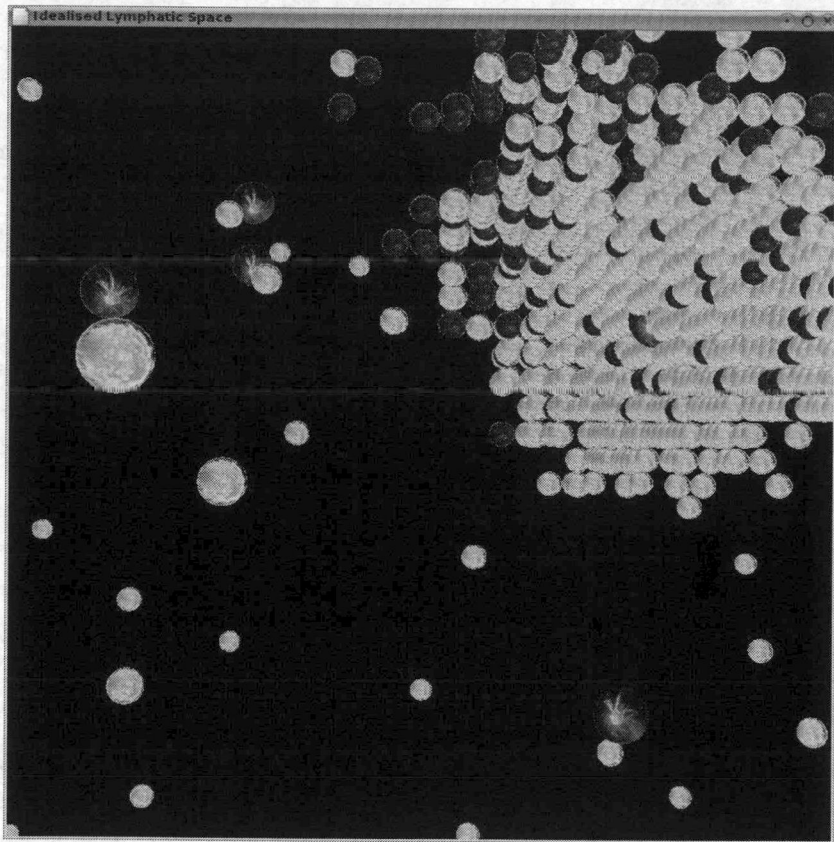

Fig. 5. An APC has been recognised, and the CTL cells begin their clonal expansion phase (the cluster of yellow cells), prior to departing the lymphatic compartment

3. Up arrow: move the camera up the y-plane.
4. Down arrow: move the camera down the y-plane.
5. Left arrow: move left along the x-axis.
6. Right arrow: move right along the x-axis.

In this position, the viewer has moved the camera into the lymphatic space and paused the simulation to examine two cells. Clearly visible in Fig. 4, the results of texture mapping the 2D images from Fig. 2, we can see the antigen presenting cell (with its characteristic centre and extended arms) above and behind a naive CTL. Also visible are more distant APC and CTL cells.

One of the strengths of OpenGL for this kind of visualisation is that completely automates and hides the details for scaling and managing object depth and distance. Using the `gluBuild2DMipmaps()` function call (as shown above), we request the OpenGL engine to build a series of scaled images of our object, which are then called directly by the rendering engine depending on the image postion in the space.

Fig. 5 we can see the camera has been moved to the proximity of a clonal expansion process, triggered by the recognition of an APC by a CTL [1]. This expansion is marked by a steep increase in the number of CTL cells dividing and arming in the search for more APC bearing the genetic material that triggered the recognition. This population increase is carefully controlled in order to prevent the lymphatic compartment becoming congested with cells.

4.1   Simulation Platform

DCU School of Computing has recently upgraded its cluster technology with the purchase of a Linux-based 448-core compute cluster. This advanced platform is crucial for the implementation of our research. We are also in the process of equipping a dedicated research space with large-screen digital display facilities for the projection of 3D graphical applications. In conjunction with ITT Dublin we are jointly developing and funding Bio-visualisation projects in order to promote the important work of our group.

5   Summary and Conclusions

We have developed a *front end* visualization platform, based on the MVC design pattern, to allow student and lecturers in the classroom and lab to experiment with a variety of parameters in order to study a range of possible outcomes from normal to abnormal disease clearance.

In an exciting collaboration between Dublin City University School of Computing, and ITT Dublin, we are developing novel visualization techniques to study cellular interaction from both the large-scale spatial and temporal perspectives. Visualization of processes offers the researcher the ability to speed-up, slow-down and hypothesize various parameters. Visualization aids understanding as we rely on visual perception to make crucial decisions. For example, with our initial model, we can visualize the dynamics of an idealized lymphatic compartment, with APC and CTL cells.

One of the strengths of this platform is that the user is always presented with visual cues which they can directly relate to images of cells that might be encountered in the actual lab setting. With very slight modifications we can convert typical 2D raster images into 3D spherical-based objects. We are strongly motivated in this research by the findings of a recent EU which indicates that in-silico modeling and simulation of biological processes is a key requirement in the development of cost-effective and timely new disease therapies.

References

1. Hagen, H., Ebert, A., van Lengen, R.-H., Scheuermann, G.: Scientific visualization: methods and applications. In: Proceedings of the 19th spring conference on Computer graphics, pp. 23–33. ACM Press, New York (2003)

[1] See [9] for a good general treatment of basic immunology.

2. Efroni, S., Harel, D., Cohen, I.R.: Towards Rigorous Comprehension of Biological Complexity: Modeling, Execution and Visualization of Thymic T cell Maturation. J. Theor. Biol. 197, 507–516 (2003)
3. Everett, P.C., Seldin, E.B., Troulis, M., Kaban, L.B., Kikinis, R.: A 3-D System for Planning and Simulating Minimally-Invasive Distraction Osteogenesis of the Facial Skeleton. In: Hamza, M.H. (ed.) Biomedical Engineering. Proceedings of the IASTED International Conference, pp. 90–95. ACTA Press (2003)
4. Gamma, E.: Design Patterns. Addison-Wesley, Reading (1995)
5. Cole, G.A., Hog, T.L., Coppola, M.A., Woodland, D.L.: Efficient Priming of CD8+ Memory T Cells Specific for a Subdominant Epitope Following Sendai Virus Infection. J. Immunol. 158, 4301–4309 (1997)
6. Burns, J., Ruskin, H.J.: Diversity Emergence and Dynamics During Primary Immune Response: A Shape Space. Physical Space Model. Theor. in Biosci. 123(2), 183–194 (2004)
7. Burns, J., Ruskin, H.J.: Network Topology in Immune System Shape Space. In: Bubak, M., van Albada, G.D., Sloot, P.M.A., Dongarra, J. (eds.) ICCS 2004. LNCS, vol. 3038, pp. 1094–1101. Springer, Heidelberg (2004)
8. Ruskin, H.J., Burns, J.: Weighted networks in immune system shape space. Physica A 365(2), 549–555 (2006)
9. Janeway, C.A., Travers, P., Walport, M., Capra, J.D.: Immunobiology. The Immune System in Health and Disease. Churchill-Livingston (1999)

Using the Weighted Rich-Club Coefficient to Explore Traffic Organization in Mobility Networks

José J. Ramasco[1], Vittoria Colizza[1], and Pietro Panzarasa[2]

[1] Complex Networks and Systems group, ISI Foundation, Turin, Italy
jramasco@isi.it
[2] School of Business and Management, Queen Mary College, University of London,
London, United Kingdom
http://isiosf.isi.it/~jramasco

Abstract. The aim of a transportation system is to enable the movement of goods or persons between any two locations with the highest possible efficiency. This simple principle inspires highly complex structures in a number of real-world mobility networks of different kind that often exhibit a hierarchical organization. In this paper, we rely on a framework that has been recently introduced for the study of the management and distribution of resources in different real-world systems. This framework offers a new method for exploring the tendency of the top elements to form clubs with exclusive control over the system's resources. Such tendency is known as the weighted rich-club effect. We apply the method to three cases of mobility networks at different scales of resolution: the US air transportation network, the US counties daily commuting, and the Italian municipalities commuting datasets. In all cases, a strong weighted rich-club effect is found. We also show that a very simple model can account for part of the intrinsic features of mobility networks, while deviations found between the theoretical predictions and the empirical observations point to the presence of higher levels of organization.

Keywords: complex networks, human mobility, transportation systems.

1 Introduction

The elements of many systems, ranging from technological to economic and social ones, are often organized into hierarchies [1,2,3,4,5,6]. Investigating the nature of the interactions among the highest-ranking elements of a system can offer useful insights into the system's organization and functioning. For example, do the top elements attract and exchange among themselves the vast majority of the resources available in the system, or do they tend to distribute resources homogeneously within the system? By adopting the framework of network theory – where the system is represented in terms of nodes, corresponding to its elements, and links connecting interacting elements [7,8,9,10,11,12,13] – researchers have

J. Zhou (Ed.): Complex 2009, Part I, LNICST 4, pp. 680–692, 2009.

begun to study interactions among top elements by investigating whether the system's structure displays higher interconnectedness among highly connected nodes (also called *rich* nodes) than randomly expected [14]. This feature is known as the rich-club phenomenon [14,15]. By analyzing the topology of a networked system at its top hierarchical level, the rich-club phenomenon helps highlight important organizational principles of the system's structure [14,16,17,18,19].

This approach, however, assumes that the richness of a node is exclusively given by the number of connections departing from the node. In this respect, it is limited by the binary nature of links on which it draws, whereas a crucial piece of information is encoded in the strength of connections that can vary substantially across the network [6]. In infrastructure and information networks, variations in the strength of links correspond to differences in the carrying capacity of connections, measured in terms of the amount of information, energy, people, and goods that can travel along them [5,6,20,21,22]. In social networks, strong links are often found among socially embedded individuals [23,24,25,26,27,28,29,30]. A full understanding of how top nodes are organized, therefore, relies on the study not only of which other nodes they interact with, but also of the strength of their interactions. A recently introduced measure called the weighted rich-club coefficient [31] enables us to study whether and the extent to which the prominent elements of a system attract, control, and share among themselves the vast majority of the system's resources. In this paper, we apply this measure to the case of real-world mobility networks. We introduce systems with different transportation modes and of different scales of resolution, and in these systems we investigate whether and the extent to which transportation hubs manage and distribute traffic flows among themselves. Finally, we compare the empirical results with the predictions obtained from a simple model for transportation fluxes. This comparison allows us to probe which are the basic mechanisms behind the organization of the transportation datasets.

2   Mobility Networks: Air Transportation and Commuting Patterns

Transportation systems and mobility patterns of individuals can be mathematically represented as networks composed of nodes, corresponding to locations, and links describing the movement of individuals from an origin to a destination. In addition, each link connecting a node i to a node j is also characterized by a weight w_{ij} that measures the travel flux, i.e., the amount of travelers moving along that connection. Several examples of mobility networks have been analyzed, and found to exhibit skewed distributions of travel fluxes per connection, as well as large fluctuations in the traffic passing through various locations [6,20,32]. These results have been reported for different mobility types and different geographic scales, from mobility within a city [32], to commuting patterns at regional and country scales [33,34], to the worldwide air transportation network [6,20].

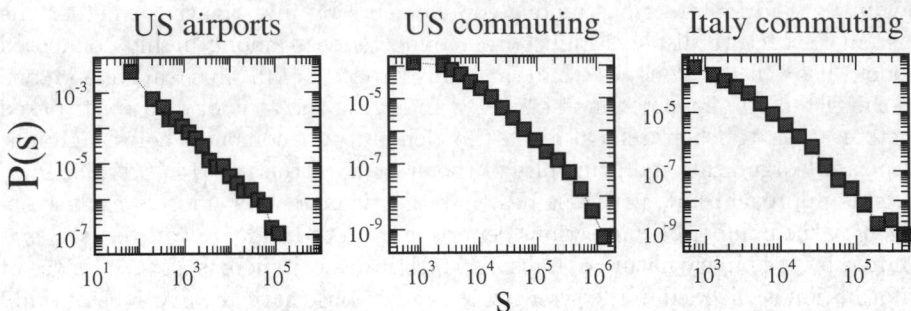

Fig. 1. Probability distribution of the traffic s of each node in the three mobility networks. Left: the US air transportation system; center: the US commuting network; right: the Italian commuting network.

Here we consider three empirical datasets involving various means of transportation and coming from diverse geographic areas: the US air transportation network [35], the commuting patterns among US counties [35], and the commuting patterns among Italian municipalities [36]. The US air transportation network is composed of 676 nodes representing the commercial airports located in the continental United States. The 3,523 links represent direct flights between these airports, while the weights of these links indicate the average number of seats available per day in each connection [6]. The other two networks correspond to commuting patterns of two countries of different size: the US and Italy. In these networks, each node represents a location (i.e., origin or destination of the commuting), and a link corresponds to the existence of a flow of individuals commuting from an origin to a destination. The weight of each link represents the total flux of daily commuters between any two counties in the US, or any two municipalities in Italy. In both cases, the data was collected through national census surveys [35,36]. In the US, there are a total of 3,141 counties connected by 35,340 weighted links, whereas in Italy there are 8,101 municipalities connected by 125,246 weighted links.

Nodes in mobility networks can be characterized in terms of their strength or traffic s, a measure of the number of travelers passing through each node, defined as the sum of the weights of the links departing from a given node [6]:

$$s_i = \sum_{j \in \nu(i)} w_{ij}, \tag{1}$$

where $\nu(i)$ is the set of neighbors of node i. In the three cases under study, the traffic s_i of a node i corresponds to the number of airline travelers and the number of daily commuters passing through i in the airline transportation network and in the two commuting datasets, respectively. Figure 1 reports the probability distribution $P(s)$ of the traffic s, showing in all cases the presence of large fluctuations, and signalling that the nodes are hierarchically organized in terms of their traffic capacities.

In what follows, we consider the traffic of a node as a measure of its prominence in the system, and we measure the weighted rich-club coefficient [31] in order to investigate the extent to which the hubs at the top of the hierarchy control and share among themselves the strongest connections of the system.

3   Weighted Rich-Club Coefficient

The weighted rich-club coefficient is a measure that allows us to study the extent to which the top elements of a given system collude to secure and share resources among themselves [31]. This measure can be applied to any networked system in which ranking relationships can be established among the nodes according to a given property. This property is usually referred to as the *richness parameter* r [31], in analogy with the topological rich-club coefficient [14,15] that measures the tendency of high degree nodes (also called rich nodes) to form tight interconnected subgraphs.

The weighted rich-club coefficient builds on, and extends, the topological one into a new broader framework in which the intensity and capacity of the links are explicitly taken into account. More specifically, if we consider a richness parameter r and aim to determine the relative strength of the links connecting the rich nodes with respect to the system's total capacity, the following weighted rich-club coefficient can be defined [31]:

$$\phi^w(r) = \frac{W_{>r}}{\sum_{l=1}^{E_{>r}} w_l^{\text{rank}}} \quad , \tag{2}$$

where the numerator is the sum of the weights associated to the links connecting rich nodes. Assuming that the total number of links between the rich nodes is $E_{>r}$, the denominator corresponds to the sum of the weights of the $E_{>r}$ strongest links of the graph. The term w_l^{rank} represents an order relationship established among the weights of the links in the network: $w_l^{\text{rank}} \geq w_{l+1}^{\text{rank}}$, with $l = 1, 2, \ldots, E$, and E being the total number of links in the graph. Thus, Eq. (2) measures the fraction of weights shared by the rich nodes compared with the total amount they could share if they were connected through the strongest links available in the network. $\phi^w(r)$ takes values ranging from 0 to 1. It is equal to 0 if there is no link connecting the rich nodes, whereas it reaches the value of 1 when the links connecting the rich nodes are the strongest available ones.

In analogy with the topological rich-club coefficient [14], Eq. (2) in itself is not informative and has to be compared to an appropriate null model [14,37]. In fact, even random graphs can show a non-zero value in Eq. (2). To properly evaluate the weighted rich-club phenomenon, we therefore need to assess it against a null model that is random, but at the same time comparable to the real network. In particular, our choice of an appropriate null model reflects the need to discount for associations between weights and topology. To this end, the null model must meet three main requirements. First, it must have the same number of nodes and links as the original network. Second, it must have the same weight distribution $P(w)$ (i.e., the probability that a given link has weight w) – a crucial constraint since we are looking for non-trivial intensity of interactions among rich nodes. Third, the nodes in the rich club must be the same as in the real network, which also preserves the richness distribution $P(r)$ (i.e., the probability that a given node has richness r) of the real network. A null model that does not fulfill the above three requirements cannot be compared to the real network, and thus does not allow for a proper weighted rich-club assessment (for a full discussion and a comparison with other proposed methods, see Ref. [31]).

In the context of mobility networks, we want to explore the tendency of highly trafficked locations to attract the majority of the passenger fluxes circulating on a system. By defining the richness parameter in terms of the traffic passing through each node, we need a null model that is able to keep the node traffic fixed while destroying all associations between links and weights observed in the real network. In [31], we introduced a procedure to generate null models that keep the value of node strength unchanged. We called this procedure Directed Weight Reshuffle because it is based on the randomization of directed networks that preserves not only the topology and $P(w)$, but also the out-strength distribution $P(s_{out})$ (i.e., the probability that the sum of weights of the outgoing links of a node is s_{out}) of the real network [38]. In the Directed Weight Reshuffle null model, the weights are locally reshuffled for each node across its outgoing links (see Ref. [31] for details). This procedure is applicable to directed graphs but can be easily extended also to the undirected case by duplicating each undirected link into two directed links, one in each direction.

It is now possible to assess the weighted rich-club effect by measuring the ratio:

$$\rho^w(r) = \frac{\phi^w(r)}{\phi^w_{null}(r)}, \tag{3}$$

where the denominator is the weighted rich-club coefficient measured on the null model. When ρ^w is larger than one, the network displays a positive weighted rich-club ordering, with rich nodes concentrating a disproportionately large amount of their efforts towards other rich nodes compared with what happens in the random null model. Conversely, if it is smaller than one, the links among the members of the club are weaker than randomly expected.

4   Results

We measured Eq. (3) in the three empirical datasets. As shown in Fig. 2, we found that all mobility networks under study, while displaying a relatively mild topological ordering, are characterized by a strong weighted rich-club effect. This result clearly points to the presence of large backbones of travel fluxes associated with the connections that link locations with very high traffic. Not only do busy airports direct routes to one another, but they also secure control over travel fluxes by channeling on those routes a larger proportion of passengers than randomly expected [31].

In analogy with what was found in the airport network, also commuting flows between highly trafficked locations are much stronger in terms of number of

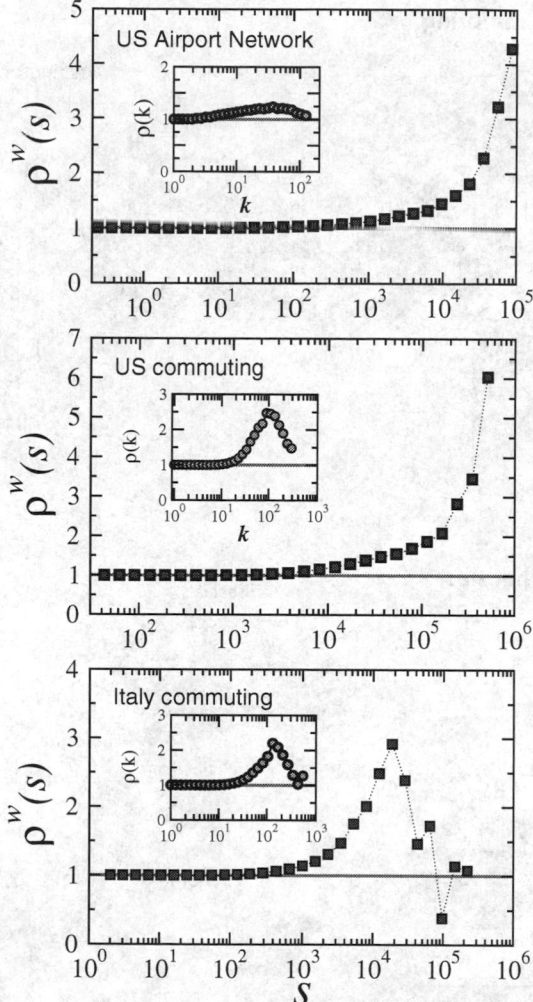

Fig. 2. Weighted rich-club coefficient for the US airport network (top), the US commuting network (center) and the Italian commuting network (bottom), where richness is measured in terms of node traffic. The insets refer to the topological rich-club coefficient.

commuters than would be expected in a fully random model. However, some differences between the commuting patterns in the US and in Italy can be found. If we compare the highest value reached by $\rho^w(s)$ in the two cases, the weighted rich-club effect is indeed twice as strong in the US network as it is in the Italian one. Moreover, the strong fluctuations observed for very large values of s in the Italian network, which bring the value of the ratio $\rho^w(s)$ down to 1, are not observed in the US case, characterized instead by an increasing trend of $\rho^w(s)$ for the whole range of traffic values. This behavior seems to uncover a difference in

US counties commuting

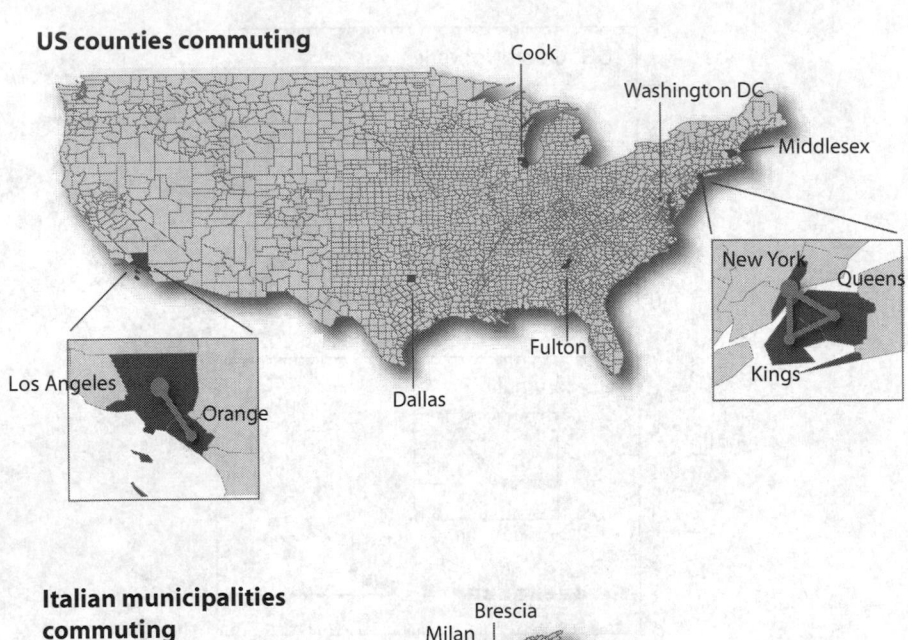

Italian municipalities commuting

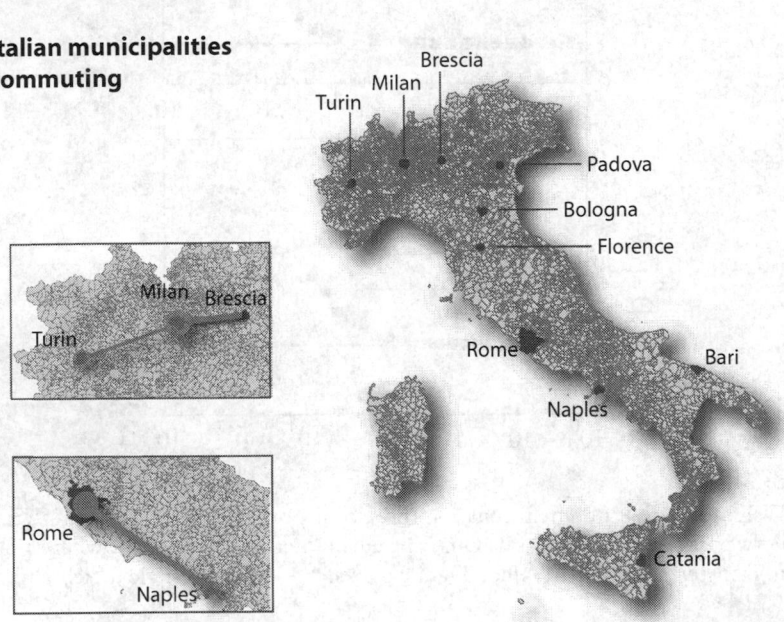

Fig. 3. Map of the ten top locations, counties or municipalities, in the ranking of commuting traffic, top for the US and bottom for Italy. In the plot only the strongest connections in number of passengers are shown.

the commuting patterns between the largest traffic hubs within the two countries under study. In the US, the larger the commuting traffic passing through two locations, the larger the travel flux that connects these two locations. In the Italian case, this is true only up to a certain traffic value, approximately equal to $2 \cdot 10^4$ commuters. Above this threshold, the weighted rich-club ratio decreases, thus showing that the largest commuting hubs in the country are more likely to channel a large proportion of passengers toward less trafficked locations than toward other hubs.

Geographic distances, population and area sizes associated to counties and municipalities, and the costs associated with commuting, can partly explain the variations in the flux of passengers traveling between locations seen in Italy and in the US. Economic considerations come into play also as a result of the likely change of means of transportation as distances between locations increase. While short travels can be faced by means with similar low cost such as train, metro, car or buses, longer travels require the use of planes. This observation finds support especially if we look at the value of $\rho^w(s)$ at the very end of the traffic range in the Italian commuting network. This result shows that the connections between the largest Italian centers of commuting (Milan and Rome) share a travel flow of individuals that does not considerably deviate from the random value.

On the other hand, the commuting fluxes between neighboring counties, such as those forming LA or New York (see Figure 3) strongly enhance the signal observed for $\rho^w(s)$ in Figure 2 at very high values of s. American large cities indeed occupy on average larger surface areas than their European counterparts, therefore typically including several counties, among which there might be a larger commuting flow than among counties which are found at larger distances. This behavior is not observed in Italy where the largest commuting hubs are not found within close distance.

Additional sociological and cultural considerations can help explain the results. Demographic studies have shown that, while people in the US exhibit a pronounced proclivity toward mobility not constrained by distances, not only for enhancing their social and economic status, but also for raising residential satisfaction [39], Italians are typically characterized by a stronger attachment to the places where they started their career [40]. Thus, for cultural reasons, people in Italy might not tend to look for jobs far away from where they live, even when economic opportunities do not abound locally.

In order to gain a better understanding of the mechanisms that are responsible for the observed rich-club effect, in the next section we introduce a simple traffic model based on some statistical laws found in the empirical data, and compare the predictions obtained from that model with the results we found in our datasets.

5   Comparison with a Simple Traffic Model

A very peculiar feature observed in many transportation networks is that there exists a relationship between the weight w_{ij} of a connection from node i to node j and the product of the degrees of the two nodes, $k_i k_j$ [6,41,42,43,44,45,46]:

$$\langle w_{ij} \rangle = (k_i \, k_j)^\theta .$$

(4)

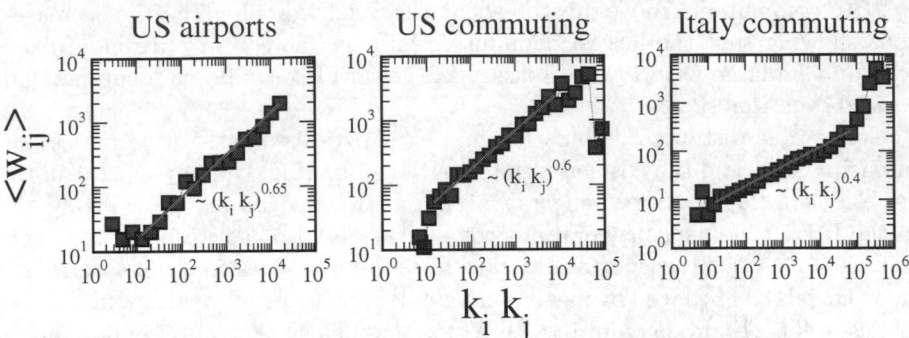

Fig. 4. Relationship between the weight w_{ij} of a link connecting node i to node j and the product of the degrees of the two nodes, $k_i k_j$, in the US airport network (left), the US commuting network (center), and the Italian commuting network (right). The red lines are fits to the empirical data.

This relationship was found in transportation networks of individuals at various scales [6,44] and also in systems of freight and mail transportation [44]. The exponent θ is typically found in the range $0.4 - 0.6$. Figure 4 shows this relationship for the three networks here investigated, supporting previous results observed in other settings.

This finding leads us to compare the real-world networks under study with a simple traffic model for the location of the strong connections in a network. The idea is to maintain the same structure as in the real network, and to generate new weights for the links based on Eq. (4) [41,47,48]. We then check whether the weighted rich-club effect observed in the previous section is just a byproduct of the way weights are created through Eq. (4). It is important to note that, when generating the new weights, we lose some of the variability in the location of the strong connections since the value of the weight of each link only depends on the degree of the two connected nodes. However, this simple traffic model is in itself sufficient to test if the observed weighted rich-club ordering results simply from the *first order* relationship captured by Eq. (4).

Figure 5 reports the rich-club ordering obtained from the model and the one observed in the empirical datasets. In all three cases, the traffic model based on Eq. (4) is able to reproduce approximately the whole range of values of traffic per node s. However, it seems to fail to replicate some of the values for the weighted rich-club effect observed in the empirical networks. In the US airport network and in the US commuting network, the increasing trend of $\rho^w(s)$ is captured by the model but the observed values in the empirical datasets for the most trafficked nodes are approximately $1.5 - 2$ times larger than the ones theoretically predicted. In the Italian commuting network, the model correctly produces the magnitude of the weighted rich-club effect. However, it fails to reproduce the relative strength and capacity of the links connecting the highly trafficked nodes. This result, therefore, highlights a deviation of the trend of commuting flows between the Italian municipalities where the top locations do not share a

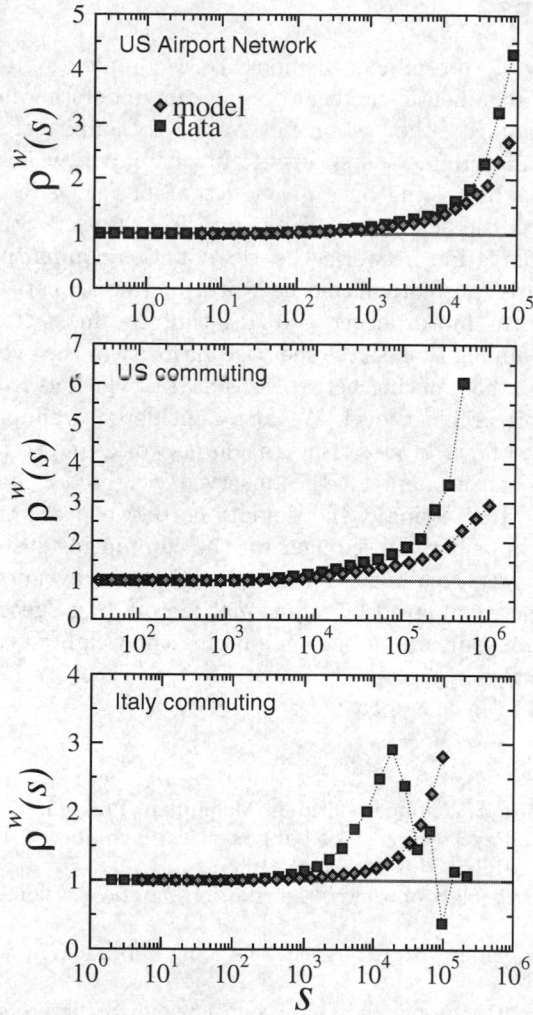

Fig. 5. Comparison of the weighted rich-club effect observed in the mobility networks (squares) and the one obtained by using the traffic model (diamonds). From top to bottom: the US airport network, the US commuting network, and the Italian commuting network.

considerably large amount of commuters, as instead predicted by Eq. (4). In order to explain this deviation and the differences between the two commuting patterns under study, other features ought to be taken into account. Among these are the geographic distance between locations, transportation costs, problems of traffic congestion, geographic concentration and availability of business opportunities, as well as possible cultural discrepancies between the habits, views, and values of the populations on the two sides of the Atlantic.

6 Conclusions

The study of how the prominent elements of a complex system interact with one another has fundamental implications on our understanding of the organization and functioning of the system in a variety of empirical applications and from a number of disciplinary perspectives. In this paper, we investigated transportation networks by relying on a method that has been recently introduced in [31]: the weighted rich-club effect. The empirical datasets considered here are the US air transportation network, the US counties commuting network, and the Italian municipalities commuting network. In the US networks, we found a strong weighted rich-club ordering, showing that the most trafficked locations control and share among themselves the vast majority of the overall traffic in the system. The Italian commuting network displays deviations from this behavior at the very top hierarchical level. We also considered a simple model for the generation of travel flows in a transportation network, and tested the extent to which this model can account for the observed weighted rich-club ordering in the real networks. Interestingly, the deviations between the traffic model and the real networks appear to be stronger for the commuting systems than for the airport network. These results pave the way toward new avenues of investigation concerned with the role played by additional factors (e.g., geographic distance, congestion problems, cultural habits) in shaping the weighted rich-club ordering in mobility networks.

References

1. Pareto, V.: Cours d'èconomie politique. Macmillan, Paris (1897)
2. Zipf, G.K.: The Psycho-Biology of Language: An introduction to dynamic philology. Houghton Mifflin, Massachusetts (1935)
3. Simon, E.H.: On a class of skewed distribution functions. Biometrika 42, 425–440 (1955)
4. Barabási, A.-L., Albert, R.: Emergence of scaling in random networks. Science 286, 509–512 (1999)
5. Pastor-Satorras, R., Vespignani, A.: Evolution and Structure of the Internet: A Statistical Physics Approach. Cambridge Univ. Press, Cambridge (2004)
6. Barrat, A., Barthélemy, M., Pastor-Satorras, R., Vespignani, A.: The architecture of complex weighted networks. Proc. Natl. Acad. Sci (USA) 101, 3747–3752 (2004)
7. Albert, R., Barabási, A.-L.: Statistical mechanics of complex networks. Rev. Mod. Phys. 74, 47–97 (2002)
8. Newman, M.E.J.: The Structure and Function of Complex Networks. SIAM Rev. 45, 167–256 (2003)
9. Dorogotsev, S.N., Mendes, J.F.F.: Evolution of Networks: From Biological Nets to the Internet and WWW. Oxford Univ. Press, Oxford (2003)
10. Wasserman, S., Faust, K.: Social Network Analysis. Cambridge Univ. Press, Cambridge (1994)
11. Boccaletti, S., Latora, V., Moreno, Y., Chavez, M., Hwang, D.-U.: Complex networks: Structure and dynamics. Phys. Rep. 424, 175–308 (2006)
12. Caldarelli, G.: Scale-Free Networks: Complex Webs in Nature and Technology. Oxford Univ. Press, Oxford (2007)

13. Barrat, A., Barthélemy, M., Vespignani, A.: Dynamical Processes on Complex Networks. Cambridge Univ. Press, Cambridge (2008)
14. Colizza, V., Flammini, A., Serrano, M.A., Vespignani, A.: Detecting rich-club ordering in complex networks. Nature Phys. 2, 110–115 (2006)
15. Zhou, S., Mondragon, R.J.: The rich-club phenomenon in the Internet topology. IEEE Commun. Lett. 8, 180–182 (2004)
16. De Masi, G., Iori, G., Caldarelli, G.: Fitness model for the Italian interbank money market. Phys. Rev. E 74, 066112 (2006)
17. Balcan, D., Erzan, A.: Content based networks: A pedagogical overview. Chaos 17, 026108 (2007)
18. Wuchty, S.: Rich-Club Phenomenon in the Interactome of P. falciparum–Artifact or Signature of a Parasitic Life Style? PLoS ONE 2, e355 (2007)
19. Guimerá, R., Sales-Pardo, M., Amaral, L.A.N.: Classes of complex networks defined by role-to-role connectivity profiles. Nature Phys. 3, 63–69 (2007)
20. Guimerá, R., Mossa, S., Turtschi, A., Amaral, L.A.N.: The worldwide air transportation network: Anomalous centrality, community structure, and cities' global roles. Proc. Natl. Acad. Sci. (USA) 102, 7794–7799 (2005)
21. Dorogovtsev, S.N., Mendes, J.F.F.: Evolving Weighted Scale-Free Networks. In: AIP Conference Proceedings, Science of complex networks: < From Biology to the Internet and WWW: CNET 2004, vol. 776, pp. 29–36 (2005)
22. Newman, M.E.J.: Analysis of weighted networks. Phys. Rev. E 70, 056131 (2004)
23. Granovetter, M.: The strength of weak ties. Am. J. Sociology 78, 1360–1380 (1973)
24. Simmel, G.: The Sociology of Georg Simmel. Free Press, New York (1950)
25. Hansen, M.T.: The search-transfer problem: The role of weak ties in sharing knowledge across organization subunits. Administrative Science Quarterly 44, 82–111 (1999)
26. Uzzi, B.: Social structure and competition in interfirm networks: The paradox of embeddedness. Administrative Science Quarterly 42, 35–67 (1997)
27. Panzarasa, P., Opsahl, T.: Growth mechanisms for evolving social networks: Triadic closure, homophily and tie strength. In: Proc. of the XXVII Int. Sunbelt Social Network Conf. (2007)
28. Ramasco, J.J.: Social inertia and diversity in collaboration networks. Eur. J. Phys. ST 143, 47–50 (2007)
29. Ramasco, J.J., Morris, S.: Social inertia in collaboration networks. Phys. Rev. E 73, 016122 (2006)
30. Valverde, S., Solé, R.V.: Self-organization versus hierarchy in open-source social networks. Phys. Rev. E 76, 046118 (2007)
31. Opsahl, T., Colizza, V., Panzarasa, P., Ramasco, J.J.: Prominence and control: The weighted rich-club effect. Phys. Rev. Lett. (2008)
32. Chowell, G., Hyman, J.M., Eubank, S., Castillo-Chavez, C.: Scaling laws for the movement of people between locations in a large city. Phys. Rev. E 68, 066102 (2003)
33. De Montis, A., Barthélemy, M., Chessa, A., Vespignani, A.: The structure of interurban traffic: a weighted network analysis. Environment and Planning B: Planning and Design 34, 905–924 (2007)
34. Patuelli, R., Reggiani, R., Gorman, S.P., Nijkamp, P., Bade, F.-J.: Network analysis of commuting flows: A comparative static approach to German data. Networks and Spatial Economics 7, 315–331 (2007)
35. Bureau of Transportation Statistics (BTS), http://www.bts.gov/
36. Italian National Statistics Institute (ISTAT), http://www.istat.it/

37. Amaral, L.A.N., Guimerá, R.: Complex networks: Lies, damned lies and statistics. Nature Phys. 2, 75–76 (2006)
38. Serrano, M.A., Boguñá, M.: Vespignani: Patterns of dominant flows in the world trade web. J. Econ. Interac. Coor. 2, 111–124 (2007)
39. Florida, R.: The Rise of the Creative Class: And How it's Transforming Work, Leisure, Community and Everyday Life. Basic Books (2003)
40. Nuvolati, G.: Popolazioni in movimento, citta' in trasformazione. Abitanti, pendolari, city users, uomini d'affari e flaneurs. Il Mulino Ricerca (2000)
41. Wu, Z., Braunstein, L.A., Colizza, V., Cohen, R., Havlin, S., Stanley, E.H.: Optimal paths in complex networks with correlated weights: The world-wide airport network. Phys. Rev. E 74, 056104 (2006)
42. Ramasco, J.J., Gonçalves, B.: Transport on weighted networks: When the correlations are independent of the degree. Phys. Rev. E 76, 066106 (2007)
43. Serrano, M.A., Boguñá, M., Pastor-Satorras, R.: Correlations in weighted networks. Phys. Rev. E 74, 055101(R) (2006)
44. Colizza, V., Vespignani, A.: Epidemic modeling in metapopulation systems with heterogeneous coupling pattern: Theory and simulations. J. Theor. Biol. 251, 450–467 (2008)
45. Macdonald, P.-J., Almaas, E., Barabási, A.-L.: Minimum spanning trees of weighted scale-free networks. Europhys. Lett. 72, 308–314 (2005)
46. Almaas, E., Krapivsky, P., Redner, S.: Statistics of weighted treelike networks. Phys. Rev. E 71, 036124 (2004)
47. Barrat, A., Barthélemy, M., Vespignani, A.: Weighted evolving networks: coupling topology and weight dynamics. Phys. Rev. Lett. 92, 228701 (2004)
48. Barrat, A., Barthélemy, M., Vespignani, A.: Modeling the evolution of weighted networks. Phys. Rev. E 70, 066149 (2004)

Tracking the Evolution in Social Network: Methods and Results

Shengqi Yang, Bin Wu, and Bai Wang

Beijing Key Laboratory of Intelligent Telecommunications Software and Multimedia,
School of Computer Science, Beijing University of Posts and Telecommunications,
Beijing 100876, China
sheng_qi.yang@yahoo.com.cn

Abstract. Contrary to previous static knowledge, our dynamic view in social network is so limited. Recent uncovering those hidden dynamic patterns has posed a series of challenging problems in network evolution. To make effective exploration, we present a fundamentally novel framework for uncovering the intricate properties of evolutionary networks. Different from static snapshots methods, we firstly trace the timelines of networks, which could explicitly characterize the network to several evolving segments. Then based on extracted smooth segments from the timeline, a graph approximation algorithm is devised to capture the frequent characteristics of the network and reduce the noise of interactions. Moreover, by employing the relationship between multi-attributes, an innovative community detection algorithm is proposed for detailed analysis on the approximate graphs. Besides the algorithms, to track these dynamic communities, we also introduce a community correlation and evaluation criterion. Finally, applying this framework to several synthetic and real-world datasets, we demonstrate the critical relationship between event and social evolution, and find that close-knit relationship with well-distributed tie strengths among members of large communities will contribute to a longer life span.

Keywords: Social Network, Dynamic patterns, Community Detection.

1 Introduction

Dynamic properties and evolving patterns have currently caught a considerable amount of attention in complex social networks. Many temporal analytic methods have been actually implemented in various social networks [5][6][11]. In these methods, graphs are usually employed to describe the network during a particular snapshot (e.g., one day, one week or one month) rather than the whole lifespan. Based on these sequential graphs, the temporal properties can be extracted [7][9]. However, these seemingly suitable methods neglect the randomness and emergency of the dynamic interactions, which can be helpful to uncover critical social structures and behaviors (such as criminal gangs). In brief, there are mainly two concernful properties neglected in prior *hard snapshot* methods:

J. Zhou (Ed.): Complex 2009, Part I, LNICST 4, pp. 693–706, 2009.

- Interaction Noise: Interactions of individuals may appear alternately. Such noise may be amplified and affect the final results.
- Event: Traditional evolutionary analysis just neglected the function of events. However, in most real-world networks, event is a driving property of social evolution.

Considering both of these problems, the main contributions in this paper can be induced into four parts: (a)Although previous works [3][5] have promoted the concept of *timeline*, it is just a side product or can not be implemented efficiently. In this work, we propose a efficient method that can generate the timeline throughout the lifecycle of networks without any preconditions, which will also guild our further analysis; (b)Different from traditional graph approximation on single sparse graph [8], we propose a novel approximation method to abstract graph sequence of a network. The goal in graph approximation in our framework is to sculpt naturally occurring structure. This makes it an ideal technique for characterizing the graph segments while avoiding noise; (c)To analyze communities and their evolutions in a unified process with our framework, we elaborate a community detection algorithm based on the weighted approximate graphs. Our method, which is well adjusted to suit our framework, can well reveal the hidden group structures of network and yield good results both efficiently and effectively; (d)A variety of recent works have been mining multidimension properties in dynamic networks [1][7][10]. To track the community evolution, we propose a community correlation and evaluation method that can be used practically for uncovering inherent patterns. By applying our method to several real-world networks, we find out that large community with certain compact structure usually survive for a long span.

The rest of this paper is organized as follows: Section 2 presents the notations and definitions used in this paper. Section 3 introduce the datasets that used to verify our framework. In section 4, we describes our framework and algorithms in detail. The corresponding experimental results and statistical analysis are soundly presented. Finally, we conclude our work in Section 5.

2   Notation and Definition

Table 1 lists the basic symbols used throughout this paper. Given a evolving network made up of n snapshots, we describe it as

$$G = \{\mathcal{G}^{(1)}, \mathcal{G}^{(2)}, \ldots, \mathcal{G}^{(n)}\} \tag{1}$$

Definition 1 (GRAPH SEGMENT). *A graph segment consisted of n snapshots is defined as $\mathcal{S}^{(i)} = \{\mathcal{G}^{(t)}, \mathcal{G}^{(t+1)}, \ldots, \mathcal{G}^{(t+n)}\}(n \geq 0)$. i is the index of the segment.*

A network can be represented as a series of sequential and non-overlapping graph segments as $G = \{\mathcal{S}^{(1)}, \mathcal{S}^{(2)}, \ldots\}$.

Table 1. Symbols

Symbol	Definition
G	Graph that describes the whole network
$\mathcal{G}^{(t)}$	Graph snapshot at t
$\mathcal{S}^{(i)}$	Graph segment at index i
$\mathcal{T}^{(i)}$	Approximate Graph based on $\mathcal{S}^{(i)}$
$\mathcal{C}_j^{(i)}$	Community in $\mathcal{T}^{(i)}$ with index j
$V(\mathcal{G}^{(t)})$	The node set of $\mathcal{G}^{(t)}$
$E(\mathcal{G}^{(t)})$	The edge set of $\mathcal{G}^{(t)}$
$adj_t(v)$	The neighbors of node v at time t
$d_t(v)$	The degree of node v in $\mathcal{G}^{(t)}$
$w_t(v,u)$	The weight of edge (v,u) in $\mathcal{G}^{(t)}$
$\delta(\mathcal{G}^{(t)}, \mathcal{G}^{(t+1)})$ or $\delta(t, t+1)$	The distance between $\mathcal{G}^{(t)}$ and $\mathcal{G}^{(t+1)}$
$\tilde{d}_{t,t+1}(v)$	The distance of node v between snapshot t and $t+1$
$Cor_n(\mathcal{C}_j^{(i)}, \mathcal{C}_k^{(i+1)})$	Node correlation rate between community $\mathcal{C}_j^{(i)}$ and $\mathcal{C}_k^{(i+1)}$
$Cor_e(\mathcal{C}_j^{(i)}, \mathcal{C}_k^{(i+1)})$	Edge correlation rate between community $\mathcal{C}_j^{(i)}$ and $\mathcal{C}_k^{(i+1)}$

Definition 2 (APPROXIMATE GRAPH). *An approximate graph $\mathcal{T}^{(i)}$ is an abstract image of a graph segment $\mathcal{S}^{(i)}$ that can characterize the graph sequence of the segment.*

3   Datasets

Here we collect several datasets to demonstrate our algorithm and suggest its generality in solving problems with complex relationships.

Table 2. Datasets

name	N	E	$\overline{N_t}$	$\overline{E_t}$	$\overline{\rho}$	time span
Random	10k	872k	2k	8.7k	4.36	100
BA 1	10k	877k	1.5k	4.6k	3.0	100
BA 2	10k	890k	1.5k	4.6k	3.0	100
BA 3	10k	890k	1.44k	6.6k	4.6	100
VAST [1]	400	9834	373	983	2.64	10(d)
Enron [2]	150	24k	60	219	3.34	111(w)
cond-mat [3]	52k	280k	1k	4k	3.95	117(m)
Cell Calls A	265	113k	167	812	4.83	118(d)
Cell Calls B	352	54k	196	436	2.23	102(d)
Cell Calls C	64k	1,090k	7.4k	10.8k	1.5	101(d)

---

[1] From http://www.cs.umd.edu/hcil/VASTchallenge08/
[2] From http://www.cs.cum.edu/enron/
[3] From http://arxiv.org/archive/cond-mat

Since typical real-world data does not have the ground truth (knowledge of changes in structure) available, we resort to *synthetic datasets* as a proof of concept to show the efficiency of our algorithm. As in table 2, the four synthetic datasets are built up from two models. *Random* comes from random construction (simple ER model) while *BA 1, 2 and 3* are constructed to comply with BA model. During their evolvement, we update 40% nodes between every two snapshots. *BA 2 and 3* also introduce some events with 2% and 5% high degree nodes disposal respectively to simulate structure changes during the process.

Currently increasing concentration on social behavior patterns of human, including from phone calls to e-mails, often offering particular avenues to explore both static and evolving social structures. In this paper, we also elaborate our framework to uncover such undergoing patterns in real world. Here we would like introduce our collected datasets including both public benchmarks and anonymous mobile call records.

The VAST Dataset is a challenge task from *IEEE VAST 2008*. It describes a set of cell calls from a fictitious island over a ten-day period which was narrowed down to about 400 unique cell phones during this period.

Co-authorship Dataset comes from Cornell e-Print cond-mat library spanning 70 months from 03/2001 to 12/2006. In this dataset, each record stands for a co-authorship experience between two authors.

Enron Email Dataset contains data from about 150 users, mostly senior management of Enron and spans 111 weeks from 12/1999 to 03/2002. During this period, several symbolic events happened, including the collapse.

Company Calls A B are the call records from the same company during different periods. Calls A spans 187 days from 10/2005 to 3/2006, while Calls B spans 152 days from 12/2007 to 4/2008. During B period, there was a change in top management of the company.

Cell Calls C is the cell call records of one province of an operator in China from 12/2007 to 04/2008. The detailed information of each call pair, include the duration and the frequency, is also reserved.

The call data used here is obtained from a mobile service operator. Although a single call communication may not carry much information that can reflect all the aspects of the relationship of the two involved individuals, reciprocal calls of long duration can usually explain a profile of them as a signature of some work-, family-, leisure-, ore service-based relationship [4]. Here, we utilize these mobile call datasets as a proxy of real-world communication network.

We would like to mention that for the purpose of keeping the privacy of each customer, all the phone numbers are identified by a surrogate key so that it is not possible to recover the actual customer.

4   Methods, Algorithms and Experiments

As mentioned in *Introduction*, the framework in this paper divides the exploration process into several steps which is applicable to track the network evolution in multi-dimension. Our primary concern in this section in developing the framework that we generate to solve problems is to make sure that the methods and algorithms are more practical and suitable.

4.1   TimeLine Detection

Although previous works [3][5] have promoted the concept of *timeline*, it is just a side product or can not be implemented efficiently (mainly due to their incremental methods based on ready-made community structures). In our framework, however, we propose a efficient method that can generate the timeline throughout the lifecycle of networks without any preconditions, which will also guild our further steps.

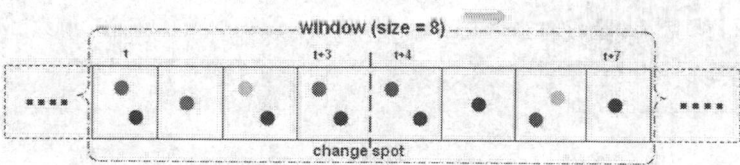

Fig. 1. moving window with size = 8. Green node stands for the nodes dead at change spot. Red node is new born nodes at change spot. Dark blue stands for stable nodes that both appears before and after the change spot while light blue for flush nodes that only appear in one snapshot.

Our algorithm will calculate the value of each change spot illustrated as Fig. 1. Our original method to quantify such change is inspired by the information theories of *relative entropy* and *type method*. The distance $\delta(t, t+1)$ at the change spot is the accumulation of the distance between each corresponding nodes pair $\tilde{d}_{t,t+1}(v)$ in the two graphs, which is defined as:

$$\tilde{d}_{t,t+1}(v) = \begin{cases} |\log \frac{d_t(v)+1}{1}| & v \in \{dead\ nodes\} \\ |\log \frac{1}{d_{t+1}(v)+1}| & v \in \{born\ nodes\} \\ |\log \frac{d_t(v)}{d_{t+1}(v)}| + |\log \frac{adj_t(v) \bigcap adj_{t+1}(v)}{adj_t(v) \bigcup adj_{t+1}(v)}| & v \in \{stable nodes\} \end{cases} \quad (2)$$

In our method, flush nodes are just neglected due to their small proportion[4] and limited effect. For dead or born nodes, we simple focus on their change in degree as Formula 2. While for stable nodes, we depict their change both in degree (their activeness) and neighbors (their environment). The integrated distance is formulated as

$$\delta(t, t+1) = \frac{\sum_{\forall v \in V(dead)} \tilde{d}_{t,t+1}(v) + \sum_{\forall v \in V(born)} \tilde{d}_{t,t+1}(v) + \sum_{\forall v \in V(stable)} \tilde{d}_{t,t+1}(v)}{|V(\mathcal{G}^{(t)}) \cup V(\mathcal{G}^{(t+1)})|}$$

$$(3)$$

To avoid the side-effect brought about by graph size, here the distance is penalized by the node union of graphs. Then given an evolving graph G, we can get the raw timeline by Formula (3).

---

[4] The proportion of flush nodes is dependent on the size of window.

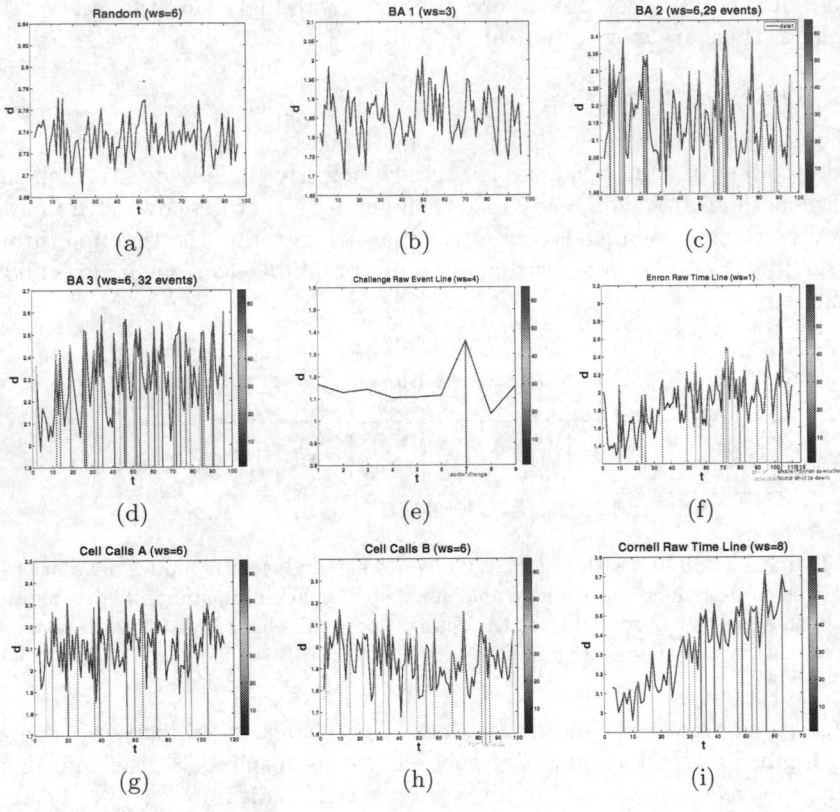

Fig. 2. TimeLines

By applying our algorithm to several synthetic and real-world datasets, the generated timelines are plotted as Fig. 2. Considering the randomness in generation, the timelines of synthetic datasets (a)-(d) are more chaotic than the other 5 real-world[5] datasets (e)-(f). From the perspective of effect, our algorithm locates 24 out of 29 events in BA 2 as marked in (c) and 31 out of 32 events in BA 3 as marked in (d). In real-world, we efficiently find out the critical social emergence in VAST 10 days call records as described in (e) and all the critical events such as company collapse and CEO committed suicide in Enron as marked in (f).

4.2 Graph Segmentation and Approximation

The most recent work which is closely related to mine [5] also segments graph stream in an incremental manner. However, minor deviation could accumulate and form a big distance between graphs, which is also called *butterfly effect*. In

---

[5] Although VAST dataset is also a synthetic one, it is considered to incorporate many social properties in order to simulate such social behaviors. In this article, we take it as a real one.

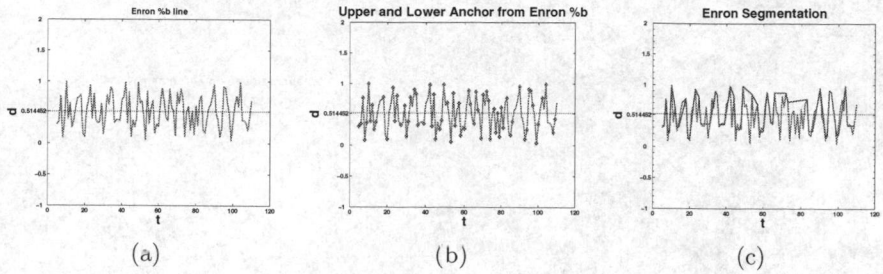

Fig. 3. Segmentation

Table 3. Datasets

	Enron	cond-mat	Cell Calls A	Cell Calls B	Cell Calls C
$|\mathcal{S}^{(i)}|$	13	15	18	18	16
$\overline{\mathcal{S}^{(i)}}$	4	4	5	4	5

general, a network and the membership of groups often evolve gradually [1]. From the social perspective, the evolving network usually alternates between change and smoothness. So the change between two successive smooth segments can be applied to describe the influence of that event. In our framework, we highlight the smooth segments and abstract it into single graphs for farther research.

To obtain the smooth segments, here we propose a two-step auto segmentation algorithm to extract smooth graph segments as follows:

1. We employ *Bolling Bands*[14], %b, to prune the raw event line and generate the first base for linear segmentation, which is normalized between 0 and 1. Fig. 3(a) shows the resulted %b line from Fig. 2(f)
2. The first windowing process is implemented to locate all the upper and lower anchors. The window size is determined by real dataset. Here we apply this process to (a) with window size = 10 and locate all the anchors described as (b). The second windowing process is to smooth the lower anchors and prune upper anchors. The obtained segment line is plotted as black line in (c).

The results by applying the segmentation algorithem to our datasets is summarized as Table 3.

When it comes from segmentation to approximation, we would like to present our objects firstly. (a)The abstraction from $\mathcal{S}^{(i)}$ to a single graph $\mathcal{T}^{(i)}$ should simplify the future evolutionary analysis. (b)$\mathcal{T}^{(i)}$ should well characterize $\mathcal{S}^{(i)}$.

It is intuitively occurred to us that $\mathcal{T}^{(i)}$ should contain only the common structure of a smooth segment. However, it is somewhat over-fit. Our approximation process aims to depict more frequent structures rather then the most.

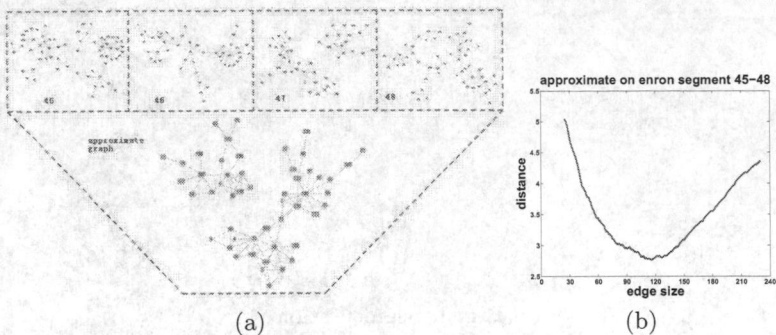

(a)   (b)

Fig. 4. Approximation

Then how to obtain a proper approximate graph from the smooth segment? Here we introduce the *Graph Segment Description Distance* between $\mathcal{T}^{(i)}$ and $\mathcal{S}^{(i)}$ as,

$$d(\mathcal{T}^{(i)}, \mathcal{S}^{(i)}) = \sqrt{\sum_{\mathcal{G}^{(t)} \in \mathcal{S}^{(i)}} w_i \delta(\mathcal{T}^{(i)} \| \mathcal{G}^{(t)})^2} \qquad (4)$$

Even in a smooth segment $\mathcal{S}^{(i)}$, there are also some small events among its snapshots. We find out the lowest point, $\delta(t, t+1)$, in $\mathcal{S}^{(i)}$ and set $w_t = 1$ and $w_{t+1} = 1$. Other weights can be set by $\frac{\delta(j, j+1)}{\delta(t, t+1)}$. Along the adding procedure, we can expect that, at the beginning, when the edges with high weight and frequency are added to $\mathcal{T}^{(i)}$, the value of the $d(\mathcal{T}^{(i)}, \mathcal{S}^{(i)})$ will conspicuously decrease. However when it comes to the low ranked edges, the distance increases reversely. To demonstrate our method, we extract the segment $\mathcal{S}^{(45-48)}$ (include 4 graph snapshots) from Fig. 3(c) and by employing our algorithm, the curve of $d(\mathcal{T}^{(i)}, \mathcal{S}^{(i)})$ which describes the process of approximation on $\mathcal{S}^{(45-48)}$ shows the shape V in Fig. 4(d). Fig 4(a) shows the approximate graph.

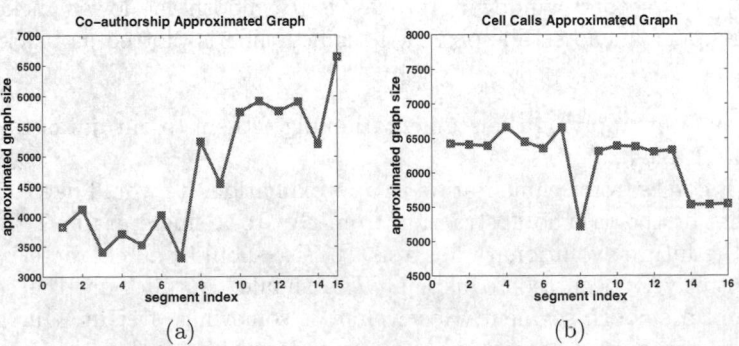

(a)   (b)

Fig. 5. Approximation on cond-mat and cell calls C

The generated graphs are usually tight relevant with their correspond time-lines, which can be observed from Fig. 5. In Fig. 5(a). It is clearly that graph size is almost as the rising shape as the timeline in Fig 2(i). By comparison, the size of approximate graphs in *Cell Calls C* (Fig. 5(d)) is relatively smooth.

Besides precisely depicting segment, as the same time the event or certain great change between two $T^{(i)}$ can also be observed, which can be employed to explore the social transformation. Fig. 6(a) and (b) describe the change effect at the 7th day in Fig. 2(e). It is clear that there is a significant event happened at the high-level leaders. The replacement, such as from node 200 to 300, who are recognized as the leaders, is obvious before and after the event. In Fig. 6(c)

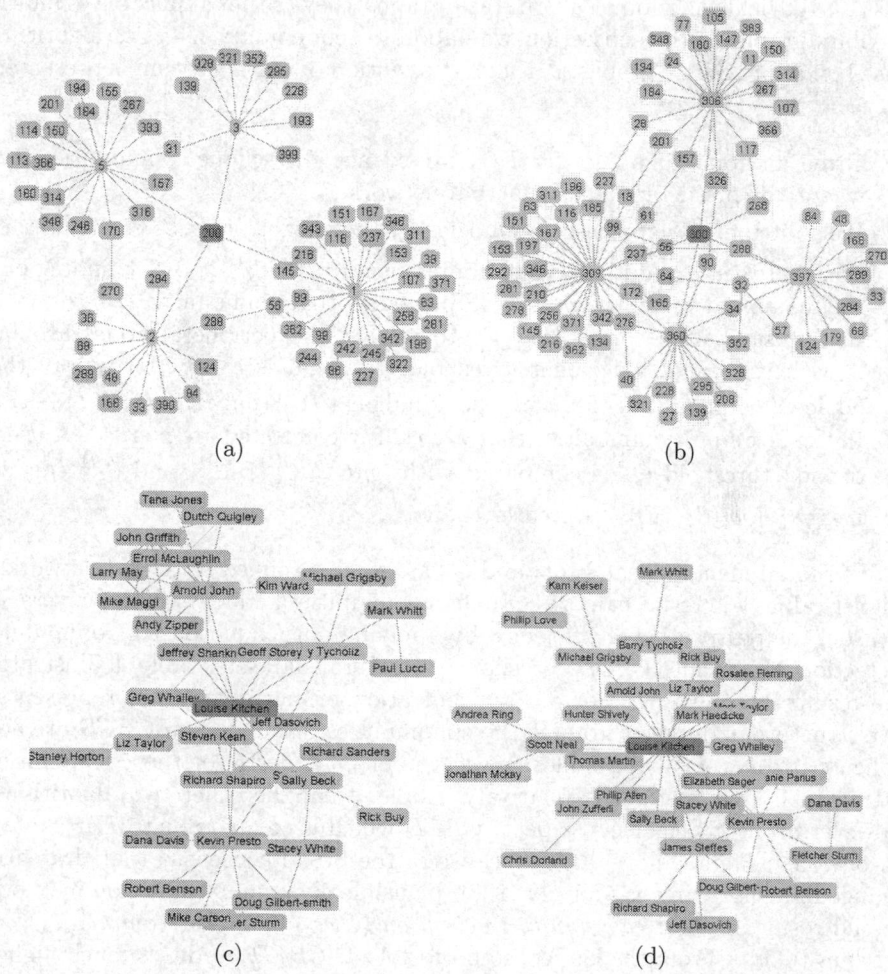

Fig. 6. Graph Tracking: (a) and (b) picture the main structure of VAST before and after the event; (c) and (d) picture the main structure of Enron before and after the company collapse (*Louise Kitchen* labeled in red was the president of Enron)

and (d), we extract the main top-level structures of Enron before and after
the event around time 96 in Fig. 2(f), when the company filed for bankruptcy
protection (2001/12/2). Fig. 6(c) describes the structure before the event. Most
of the people around *Louise Kitchen* are *vise president*. However in Fig. 6(d),
there are also some other people whose position may reflect the event, such as
Rick Buy (the *Chief Risk Management Officer*) and *Mark Haedicke* (a managing
director of *Legal Department*).

4.3   Community Tracking

Community Detection. Although traditional modularity-based optimal methods [13][15] yielded a good result on static graphs, they are not applicable to multi-
attribute graphs. In this subsection, we elaborate a community detection algorithm
based on approximate graphs. A rough description of our algorithm is presented
as:

1. Given an approximate graph $T^{(i)}$, we extract all k-clique communities [12]
   where k is pre-determined by actual networks.
2. We split the overlapping communities by the weight, $w_i(v, \mathcal{C}_j^{(i)})$, of interactions between v and its adjacent communities $\{\mathcal{C}_j^{(i)}, \mathcal{C}_k^{(i)}, \ldots\}$. By finding out
   the maximum value, the node is adjudged to that community.
3. In this step, each community try to attract the peripheral nodes around
   it with the nodes judgement threshold Q_v. That is to say, for any of the
   adjacent communities of v, v can be adjudged to $\mathcal{C}_j^{(i)}$ only if $w_i(v, \mathcal{C}_j^{(i)}) > Q_v$.
4. In the last step, communities that are tightly connected are merged. A user-
   defined threshold Q_c is compared with $w_i(\mathcal{C}_j^{(i)}, \mathcal{C}_k^{(i)})$. $\mathcal{C}_j^{(i)}$ and $\mathcal{C}_k^{(i)}$ can be
   merged if $w_i(\mathcal{C}_j^{(i)}, \mathcal{C}_k^{(i)})$ is greater than Q_c.

We should mention that k, Q_v and Q_c are all determined by actual networks.
Higher values will result in more compact communities and fewer nodes coverage.

From the perspective of efficiency, by applying our Clique-Based Community
Detection algorithm (*CBCD* for short) to several datasets, table 4 illustrates
the results on 4 aspects. For our concentration primarily on more representative structure of dynamic graphs, the communities resulted from *CBCD* are not
obligated to cover all the nodes. So the number and size of the communities
extracted by our method is relatively smaller than the other two algorithms.
However, from the efficiency aspect, *CBCD* which is comparable to *FAST* have
an obvious advantage over *GN*. Because of the broad consensus that dramatic
change in a short time is unlikely [1][3], plausible community partition methods
should result in lower *community fluctuation* (*Fluc*) or higher *community correlation* (*Corc*). From the last column of table 4, *CBCD* produces much higher
Cor_c than the others.

Community Correlation and Evaluation. Community tracking is a continuous topic in evolutionary analysis. Traditional methods in earlier works mainly

Table 4. Results of community detection algorithms on call graphs

name	$\|\{\mathcal{T}^{(i)}\}\|$	Alg.	avg. Comm. N.	avg. Comm. S.	avg. T.[7]	avg. Cor_c
Cell Calls A	18	GN[15]	12.9	12.9	4.3s	0.019
		FAST[13]	10.7	15.7	0.10s	0.015
		CBCD[6]	9.6	12.7	0.46s	0.037
Cell Calls B	19	GN	14	11.1	1.5s	0.016
		FAST	22.6	7.9	0.13s	0.015
		CBCD	6.3	10.1	0.35s	0.017
Cell Calls C	16	GN	356	9.5	59.3s	0.031
		FAST	221.1	12.6	1.8s	0.031
		CBCD	106	8.8	1.9s	0.037

focused on either node overlapping [9] or structure (edge) overlapping [7]. However, these two criterions have apparently weak points. In our framework, we take both these two correlations into account.

To evaluate the evolutionary trend of communities, we propose a *community fluctuation criterion*. For graph $\mathcal{G}^{(i)}$ and $\mathcal{G}^{(i+1)}$, which respectively contain m and n communities, the community correlation is formulated as

$$Cor_c(\mathcal{G}^{(i)}, \mathcal{G}^{(i+1)}) = \sum_{\substack{\mathcal{C}_j^{(i)} \\ 1 \leqslant j \leqslant m}} \left(\frac{|N(\mathcal{C}_j^{(i)})|}{|N(\mathcal{G}^{(i)})|} \sum_{\substack{\mathcal{C}_k^{(i+1)} \\ 1 \leqslant k \leqslant n}} Cor_n(\mathcal{C}_j^{(i)}, \mathcal{C}_k^{(i+1)}) Cor_e(\mathcal{C}_j^{(i)}, \mathcal{C}_k^{(i+1)}) \right) \quad (5)$$

then the community fluctuation, $Fluc(\mathcal{G}^{(i)}, \mathcal{G}^{(i+1)})$, is defined as

$$Fluc(\mathcal{G}^{(i)}, \mathcal{G}^{(i+1)}) = 1 - Cor_c(\mathcal{G}^{(i)}, \mathcal{G}^{(i+1)}) \quad (6)$$

Palla et al. [7] defined the stationarity of a community to evaluate the relationship between its age and the average correlation between subsequent states. However they neglect the effect of inherent structure of communities along the evolvment. To relate the structure of community with its age and evolving correlation, we define the *compactness* of community as the distance between a community and its *standard complete structure* $std(\mathcal{C}_j^{(i)})$, which is not only a complete weighted graph with the same nodes and total edge weight with $\mathcal{C}_j^{(i)}$ but is the most compact structure. The quantity of compactness of $\mathcal{C}_j^{(i)}$ is formulated as

$$Comp(\mathcal{C}_j^{(i)}) = \sum_{\forall (v,u) \in E(std(\mathcal{C}_j^{(i)}))} |\log \frac{w_{\mathcal{C}_j^{(i)}}(v, u)}{w_{std(\mathcal{C}_j^{(i)})}(v, u)}| / |E(\mathcal{C}_j^{(i)})| \quad (7)$$

This representation takes into account both link compactness and weight distribution of the structure. We would like to mention here that a small value of $Comp(\mathcal{C}_j^{(i)})$ stands for a more compact community structure.

---

[6] With parameters $k = 3, Q_v = 0.6, Q_c = 0.8$.
[7] Intel Xeon CPU 2.60GHz×2, 2G memory.

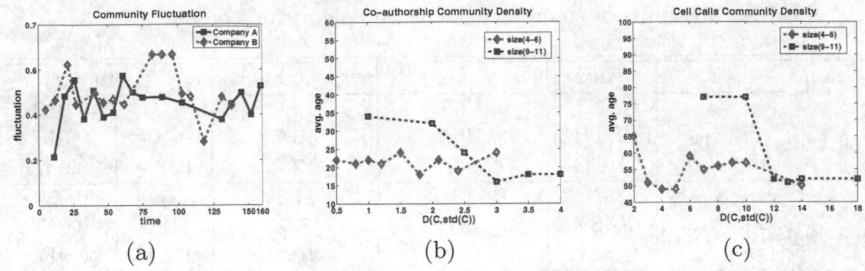

(a)   (b)   (c)

Fig. 7. Community Tracking

Discussion. In reality in social networks, timelines are usually well-distributed. That is to say, sometimes, changes only stand for the randomness of interactions but not the events in infrastructure. A negative effect brought about in this situation is that some significant events happened locally may be submerged by the noise of peripheral random interactions. To find the needles in the hay stack, here we employ our community fluctuation method to community tracking. It is obvious that in Fig. 7(a) the line depicting the community fluctuation along the evolution of Cell Calls A (solid blue line) has a relative smooth line. But for Calls B (dashed green line), there is a quite high valued segment between 80 and 100, which implies there is a significant fluctuation in the community structure. Actually, during this period, this company has a series of changes in personnel which include many high-level replacements.

From the evolving aspect, it is recently a central issue in analyzing evolving communities. Previous works [2][7][10][11] have already promoted a broad range of seminal properties in community evolution mainly in macroscopic view. In order to uncovering hidden structure properties, we employ two massive datasets, *cond-mat* and *Cell Calls* to explore their community structure and evolution trend. By imposing our *community evaluation* method to *cond-mat* and *Cell Calls C*, Fig. 7(b) and (c) represent the relations between the average age span at a given community size and the pre-defined compactness $Comp(\mathcal{C}^{(i)})$. In these two statistical plots, we get similar results. To small communities as the red lines describe, the average age span does not vary greatly along the decreasing of the compactness of their structure. By contrast, to larger communities as the blue lines describing, the average age span firstly maintains a relative high value, then decreases sharply within a limited range and at last tends to be stable.

Our insight into the evolving community structure shows that a close-knit relationship with well-distributed link weight among members of large communities will contribute to a longer life span while the phenomenon is not marked in small communities.

5   Conclusion

In this paper, we aim to uncover the evolving patterns and temporal behaviors related with social changes. In contrast with hard snapshot extraction of the evolving network, we employ timeline to segment networks. By approximating the smooth segments, the generated graphs characterize the graph sequence of smoothly evolving segments. This two-step method effectively avoids noise and is more reasonable to reflect the infrastructure of these segments. The following experiments present the event effect on the structure of those social networks. For community exploration, we find that, in Co-authorship and Cell Calls, the community structure is closely relevant to its age span.

In the future, we will improve our timeline detection method to be more effective. Our exclusive community detection is somewhat arbitrary in reality. A next step of our work is to analyze the activities of core nodes or overlapping nodes along the evolution. There are more factors, which could affect the temporal behaviors of the community or even the whole network, to be explored.

References

1. Tang, L., Liu, H., Zhang, J.P., Nazeri, Z.: Community Evolution in Dyanmic Multi-Mode Networks. In: Proceeding of the 14th ACM SIGKDD international conference on Knowledge discovery and data mining, pp. 677–685 (2008)
2. Leskovec, J., Backstrom, L., Kumar, R., Tomkins, A.: Microscopic Evolution of Social Networks. In: Proceeding of the 14th ACM SIGKDD international conference on Knowledge discovery and data mining, pp. 462–470 (2008)
3. Lin, Y.R., Chi, Y., Zhu, S.H., Sundaram, H., Tseng, B.L.: Facetnet: a framework for analyzing communities and their evolutions in dynamic networks. In: Proceeding of the 17th international conference on World Wide Web, pp. 685–694 (2008)
4. Onnela, J.P., Saramaki, J., Hyvonen, J., Szabo, G., Lazer, D., Kaski, K., Kertesz, J., Barabasi, A.L.: Structure and tie strengths in mobile communication networks. In: Proceedings of the National Academy of Sciences, vol. 104, pp. 7332–7336 (2007)
5. Sun, J., Faloutsos, C., Papadimitriou, S., Yu, P.S.: GraphScope: Parameter-free Mining of Large Time-evolving Graphs. In: Proceedings of the 13th ACM SIGKDD international conference on Knowledge discovery and data mining, pp. 687–696 (2007)
6. Asur, S., Parthasarathy, S., Ucar, D.: An event-based framework for characterizing the evolutionary behavior of interaction graphs. In: Proceedings of the 13th ACM SIGKDD international conference on Knowledge discovery and data mining, pp. 913–921 (2007)
7. Palla, G., Barabasi, A.-L., Vicsek, T.: Quantifying social group evolution. Nature 446, 664–667 (2007)
8. Long, B., Xu, X.Y., Zhang, Z.F., Yu, P.S.: Community Learning by Graph Approximation. In: Proceedings of the 2007 Seventh IEEE International Conference on Data Mining, pp. 232–241 (2007)
9. Tanya, Y., Berger-Wolf, Jared, Saia.: A Framework for Analysis of Dynamic Social Networks. In: Proceedings of the 12th ACM SIGKDD international conference on Knowledge discovery and data mining, pp. 523–528 (2006)
10. Backstrom, L., Huttenlocher, D., Kleinberg, J., Lan, X.Y.: Group Formation in Large Social Networks: Membership, Growth and Evolution. In: Proceedings of the 12th ACM SIGKDD international conference on Knowledge discovery and data mining, pp. 44–54 (2006)

11. Kumar, R., Novak, J., Tomkins, A.: Structure and evolution of online social networks. In: Proceedings of the 12th ACM SIGKDD international conference on Knowledge discovery and data mining, pp. 611–617 (2006)
12. Derényi, I., Palla, G., Vicsek, T.: Clique percolation in random networks. Phys. Rev. Lett. 29, 94(16), 160–202 (2005)
13. Newman, M.E.J.: Fast algorithm for detecting community structure in networks. Phys. Rev. E 69(6), 066133 (2004)
14. Bollinger, J.A.: Bollinger on Bollinger Bands, 1st edn. McGraw-Hill, New York (2001)
15. Girvan, M., Newman, M.E.J.: Community structure in social and biological networks. Proceedings of the National Academy of Sciences 99(12), 7821–7826 (2002)

Towards Network Complexity

Matthias Dehmer[1] and Frank Emmert-Streib[2]

[1] Institute for Bioinformatics and Translational Research, UMIT,
Eduard Wallnoefer Zentrum 1, 6060, Hall in Tyrol, Austria
Matthias.Dehmer@umit.at
[2] Queen's University Belfast, Computational Biology and Machine Learning,
Center for Cancer Research and Cell Biology, School of Medicine,
Dentistry and Biomedical Sciences, 97 Lisburn Road, Belfast BT9 7BL, UK
v@bio-complexity.com

Abstract. In this paper, we briefly present a classification scheme of information-based network complexity measures. We will see that existing as well as novel measures can be divided into four major categories: (i) partition-based measures, (ii) non partition-based measures, (iii) non-parametric local measures and (iv) parametric local measures. In particular, it turns out that (ii)-(iv) can be obtained in polynomial time complexity because we use simple graph invariants, e.g., metrical properties of graphs. Finally, we present a generalization of existing local graph complexity measures to obtain parametric complexity measures.

Keywords: networks, network complexity, information measures.

1 Introduction

To find quantitative measures for detecting the complexity of graph-based systems is a research topic with ongoing interest. Here, we are interested in such systems which can be described as complex networks. For example, quantitative approaches to measure network complexity have been developed by [5,13]. In [5], the complexity of a network was defined to be the number of its containing spanning trees. MINOLI [13] defined the so-called combinatorial complexity of a network. The key feature of this complexity measure is that it increases with the number of each factor which contributes to the complexity of a network structure. Further approaches to measure complexity of networks can be found in [15]. In this paper, we deal with information-based complexity measures [2,3,14,16]. Classical measures are based on inducing vertex partitions of a network to infer a finite probability distribution. Then, by using SHANNON's entropy [17], information-based complexity measures for networks are obtained [1,2,14,16].

The contribution of the paper is twofold: First, we present a classification scheme for information-based network complexity measures. By briefly reviewing existing measures and defining a class of parameterized local complexity measures, we divide the measures into four major categories (see Figure (1)). In

J. Zhou (Ed.): Complex 2009, Part I, LNICST 4, pp. 707–714, 2009.
© ICST Institute for Computer Sciences, Social Informatics and Telecommunications Engineering 2009

particular, we present general definitions to obtain so-called information functionals which are based on using efficiently computable graph invariants. We will see that those information functionals [6] can be used for defining both global and local network complexity measures. Further, by applying this principle presented in [6], we obtain straightforward a generalization of existing local graph complexity measures. Local complexity measures are here understood as measures for assigning a complexity score to each vertex of a network. As a final remark, we emphasize that for each category (see Figure (1)) we only give examples for such measures, especially in terms of the partition-based measures [2,3,16,14].

The paper is organized as follows: In Section (2), we begin with introducing basic mathematical preliminaries. Then, we present the classification scheme by starting with Section (3) that deals with classical partition-based network complexity measures. In contrast, Section (4) outlines complexity measures for networks which are not based on inducing vertex partitions. A generalization of existing local complexity measures is introduced in Section (5). The paper finishes with a summary and conclusion in Section (6).

2   Mathematical Preliminaries

We first state some mathematical preliminaries. Especially, we repeat the definitions of some known metrical properties of graphs [9,10,18]. We define an undirected, finite and connected network represented by $G = (V, E), |V| < \infty$, $E \subseteq \binom{V}{2}$. G is called connected if for arbitrary vertices v_i and v_j there exists an undirected path from v_i to v_j. Otherwise, we call G unconnected. $\mathcal{G}_{UC}$ denotes the set of finite, undirected and connected graphs. The degree of a vertex $v \in V$ is denoted by $\delta(v)$ and equals the number of edges $e \in E$ which are incident with v. In order to measure distances between vertices in a graph, we denote $d(u, v)$ as distance between $u \in V$ and $v \in V$ expressed as the minimum length of a path between u, v. $d(u, v)$ is a metric. We call the quantity $\sigma(v) = \max_{u \in V} d(u, v)$ the eccentricity of $v \in V$. Further, $\rho(G) = \max_{v \in V} \sigma(v)$ is called the diameter of G. The j-sphere of a vertex v_i regarding $G \in \mathcal{G}_{UC}$ is defined as $S_j(v_i, G) := \{v \in V \mid d(v_i, v) = j, j \geq 1\}$. Further metrical properties of graphs can be found in [18].

3   Partition-Based Complexity Measures

In this section, we give a short overview on classical partition-based information measures for determining complexity of networks [2,3,16,14]. These measures which are based on SHANNON's entropy formulas are graph entropy measures which can be interpreted as so-called structural information contents. To understand how to apply SHANNON's entropy to networks, we start with a network $G = (V, E)$, an equivalence criterion α, and X denotes an arbitrary graph invariant. By applying α, we get the following scheme:

$$\begin{pmatrix} 1 & 2 & \cdots & k \\ |X_1| & |X_1| & \cdots & |X_k| \\ p_1 & p_2 & \cdots & p_k \end{pmatrix}. \tag{1}$$

The first row represents the obtained equivalence classes whereas the second row represents the cardinalities of the induced vertex partitions. Now, by defining the quantities $p_i = \frac{|X_i|}{|X|}$, one obtains directly a finite probability distribution $\mathcal{P}_G = (p_1, \ldots, p_k)$ indicated by the third row. If we now apply the well know formulas for expressing the total and mean information content [3,4], we yield [3]

$$I(G, \alpha) = |X| \log(|X|) - \sum_{i=1}^{k} |X_i| \log(|X_i|), \tag{2}$$

$$\bar{I}(G, \alpha) = -\sum_{i=1}^{k} p_i \log(p_i) = -\sum_{i=1}^{k} \frac{|X_i|}{|X|} \log\left(\frac{|X_i|}{|X|}\right). \tag{3}$$

According to α, Equation (2) denotes the total structural information content of $G = (V, E)$ and Equation (3) denotes the mean structural information content of G, respectively. As examples for such measures, we express, e.g.,

$$I_V(G) = |V| \log(|V|) - \sum_{i=1}^{k} \frac{|N_i|}{|V|} \log\left(\frac{|N_i|}{|V|}\right), \tag{4}$$

$$\bar{I}_V(G) = -\sum_{i=1}^{k} \frac{|N_i|}{|V|} \log\left(\frac{|N_i|}{|V|}\right), \tag{5}$$

$$I_c(G) = \min_{\hat{V}} \left\{ -\sum_{i=1}^{h} \frac{n_i(\hat{V})}{|V|} \log\left(\frac{n_i(\hat{V})}{|V|}\right) \right\};$$
$$\hat{V} = \{V_i | 1 \le i \le h\}; |V_i| = n_i(\hat{V}); h = \chi(G). \tag{6}$$

The first equation (see Equation (4)) is often called the total topological information content [16] of G. Here, the equivalence criterion α corresponds to determine the automorphism group (vertex orbits) of G. The second equation (see Equation (5)) represents the mean topological information content [16] of G by using the same equivalence criterion. $|N_i|$ stands for the cardinality of the i-th vertex orbit. The third equation (see Equation (6)) representing the so-called chromatic information content was originally developed by MOWSHOWITZ [14]. $h = \chi(G)$ denotes the chromatic number whereas $n_i(\hat{V})$ are the cardinalities of the vertex partitions induced by the underlying chromatic decomposition. As an important remark, we note that the time complexity of an algorithm to compute those vertex partitions can be very costly. For example, the problem to calculate the automorphism group of a graph is equivalent to decide wether to graphs are isomorphic. But for arbitrary graphs, there is no efficient algorithm to check if there exists an isomorphism for two given graphs [12]. Similarly, the problem of determining the chromatic number of an undirected graph is NP-complete [10].

4   Non Partition-Based Complexity Measures

To avoid applying algorithms for inducing vertex partitions which are based on algebraic principles (see Section (3)), we outline a method that results in non partition-based complexity measures for networks [6]. We start with arbitrary $G = (V, E)$. If S represents a certain set, e.g., a set of vertices or paths etc., the monotonous and positive mapping $f : S \longrightarrow \mathbb{R}_+$ is called an information functional of G. f captures structural information of a network. If we now define the quantities $p^f(v_i) := \frac{f(v_i)}{\sum_{j=1}^{|V|} f(v_j)}$, we infer a probability distribution $\mathcal{P}_G^f = (p^f(v_1), \ldots, p^f(v_{|V|}))$. Instead of determining a probability value for induced vertex partitions, we now assign a probability to each vertex $v \in V$. As a result, we obtain the complexity measure

$$\bar{I}_f(G) := -\sum_{i=1}^{|V|} \frac{f(v_i)}{\sum_{j=1}^{|V|} f(v_j)} \log \left(\frac{f(v_i)}{\sum_{j=1}^{|V|} f(v_j)} \right). \tag{7}$$

Equation (7) represents a family of information-based network complexity measures. We now state a definition for expressing two possible types of novel information functionals.

Definition 1. *We define two types of information functionals as*

$$f = f(v_i, X_1, \ldots, X_\mu, c_1, \ldots, c_\mu), \tag{8}$$
$$f = f(v_i, X_1, \ldots, X_\mu). \tag{9}$$

X_i represents a graph-theoretical quantity (generally speaking a graph invariant) and c_k positive coefficients, respectively.

Example 1. We exemplary consider the information functionals

$$f_1(v_i) := c_i \delta(v_i), \; c_i > 0, 1 \le i \le |V|, \tag{10}$$
$$f_2(v_i) := \alpha^{c_i \delta(v_i)}, \; c_i > 0, 1 \le i \le |V|, \alpha > 0, \tag{11}$$
$$f_3(v_i) := \alpha^{c_1 |S_1(v_i, G)| + c_2 |S_2(v_i, G)| + \cdots + c_{\rho(G)} |S_{\rho(G)}(v_i, G)|}, $$
$$c_k > 0, \; 1 \le i \le |V|, \alpha > 0, \tag{12}$$
$$f_4(v_i) := d(v_i, v_1) + d(v_i, v_2) + \cdots + d(v_i, v_{|V|}), \tag{13}$$
$$f_5(v_i) := \sigma(v_i). \tag{14}$$

The shown information functionals are defined for $G \in \mathcal{G}_{UC}$. To simplify the notation, we write, e.g., $f_1(v_i)$ instead of $f_1(v_i, \delta(v_i), c_i)$. Especially, it holds $\mu = 1$. Equation (10), Equation (11) and Equation (12) are parametric information functionals which lead to parametric graph complexity measures (using Equation (7)). $f_1(v_i)$ is a linear information functional whereas $f_2(v_i), f_3(v_i)$ are exponential functionals. $f_3(v_i)$ was originally defined in [8] where here the functional can be derived as special case (by setting certain parameters in Definition (1)). Equation (13) and Equation (14) are non-parametric information

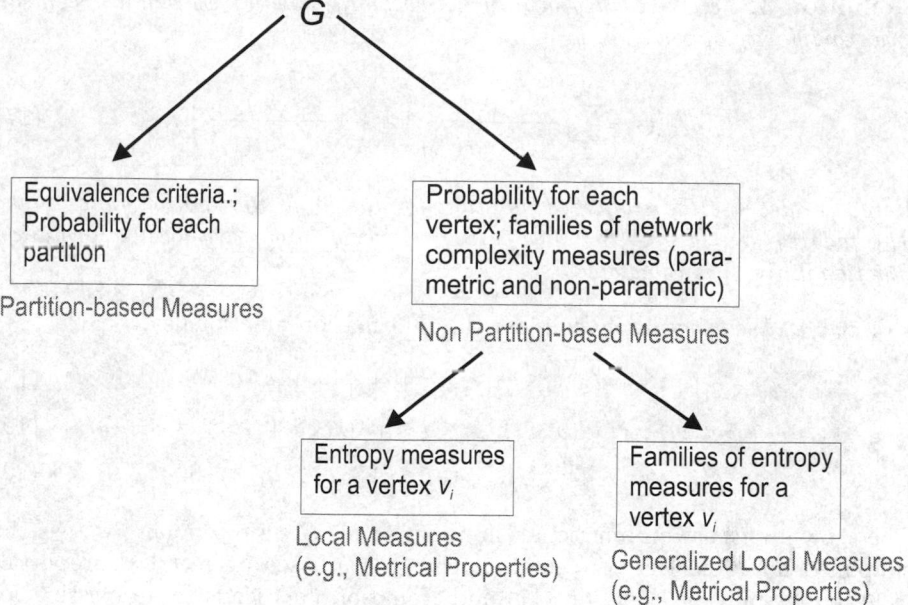

Fig. 1. Classification scheme for information-based network complexity measures

functionals based on metrical properties of graphs. As an important result, we note that especially by using Equation (12) as information functional, the calculation of the resulting graph complexity measure requires polynomial time complexity [7]. For the remaining information functionals, this result can be similarly proven.

5   Local Complexity Measures

In Section (3) and Section (4), we discussed (global) information-based complexity measures for networks. These measures can be used for determining the structural complexity of networks globally. In the following, we sketch a method to obtain local complexity measures. This relates to measure the complexity on local features or substructures of a graph. In particular, we now state a definition to determine information-based complexity for the vertices in a network.

Definition 2. *We define two types of information functionals as*

$$g = g^j(v_i, X_j, c_j), \tag{15}$$
$$g = g^j(v_i, X_j). \tag{16}$$

X_j *represents a graph-theoretical quantity (generally speaking a graph invariant) and* c_j *positive coefficients, respectively.*

Definition 3. *Let G be an arbitrary network. In general, we define the mean local complexity of vertex v_i by*

$$\bar{I}_g(v_i) := -\sum_{j=1}^{|V|} \frac{g^j(v_i)}{\sum_{j=1}^{|V|} g^j(v_i)} \log\left(\frac{g^j(v_i)}{\sum_{j=1}^{|V|} g^j(v_i)}\right). \qquad (17)$$

Here, we also use the simplified notation for the local information functionals. The index j $(1 \leq j \leq |V|)$ indicates that starting from v_i, a local information functional $g^j(v_i)$ is calculated.

Example 2. We exemplarily consider the information functionals

$$g_1^j(v_i) := d(v_i, v_j), \ 1 \leq i \leq |V|, \qquad (18)$$

$$g_2^j(v_i) := c_j d(v_i, v_j), \ 1 \leq i \leq |V|, \ c_i > 0, \qquad (19)$$

$$g_3^j(v_i) := \alpha^{c_j d(v_i, v_j)}, \ 1 \leq i \leq |V|, \ c_i > 0, \alpha > 0. \qquad (20)$$

The shown information functionals are defined for $G \in \mathcal{G}_{UC}$. Again, we use the simplified notation. Here, we see that the g^j are based on metrical properties of graphs. Applying Equation (17) and Equation (18) leads to an existing local information complexity measure [11]. Equation (19) and Equation (20) are (possible) generalized versions which finally lead to parametric local complexity measures. The proof to show that the final local graph complexity measure can be computed in polynomial time is very similar to the proof presented in [7].

The overall classification scheme for the presented approaches is depicted in Figure (1).

6   Summary and Conclusion

In this paper, we presented a classification scheme of information-based network complexity measures. We began with classical partition-based complexity measures which are mostly based on inducing vertex partitions of the graph in question. By using pure algebraic principles, this can be a difficult problem. Then, for obtaining measures with better time complexity, we outlined a recently contributed method that leads to families of information-based network complexity measures. Based on Definition (1), we saw that novel information functionals can be easily inferred (see, e.g., Equation (10), (11), (14)). Finally, we expressed Definition (2) for obtaining local complexity measures (for vertices). By stating this definition, we generalized existing local complexity measures to measure the information distance for vertices of a graph. As result, local parametric measures can be easily obtained. In general, these information measures can be used to detect structural complexity of graph-based systems, e.g., in biology and chemistry. Especially in QSPR (quantitative structure property relationship) that is branch of mathematical chemistry, a main problem is to characterize molecules by using information-theoretic complexity measures. The complexity measures presented in Section (3) and Section (4) are suitable to characterize graphs (e.g., chemical structures) globally. That means starting from a graph, inferred structural features and using SHANNON's entropy, we obtain a value for its information content. In contrast, the measures given in Section (5) address the problem of calculating entropies of local graph elements, e.g., vertices. In principle, this gives us the possibility to study the importance of such graph elements among each other or between different graphs.

As future work, we will apply these measures (local and global) for analyzing biological networks and combine them with statistical techniques. Moreover, we are interested in comparing the measures presented in this paper numerically (measures of the four major categories). From this, one can gain novel insights regarding the problem of studying the interplay between the measures.

Acknowledgments

We would like to thank Danail Bonchev, Alexandru T. Balaban and Abbe Mowshowitz for fruitful discussions. Support from the 2006 Ciência fund (Portugal) is gratefully acknowledged.

References

1. Bonchev, D.: Information indices for atoms and molecules. MATCH 7, 65–113 (1979)
2. Bonchev, D.: Information Theoretic Indices for Characterization of Chemical Structures. Research Studies Press, Chichester (1983)
3. Bonchev, D., Rouvray, D.H.: Complexity in Chemistry, Biology, and Ecology. In: Mathematical and Computational Chemistry. Springer, Heidelberg (2005)
4. Brillouin, L.: Science and Information Theory. Academic Press, New York (1956)
5. Constantine, G.: Graph complexity and the laplacian matrix in blocked experiments. Linear and Multilinear Algebra 28, 49–56 (1990)
6. Dehmer, M.: Information processing in complex networks: Graph entropy and information functionals. Applied Mathematics and Computation 201, 82–94 (2008)
7. Dehmer, M.: Information-theoretic concepts for the analysis of complex networks. Applied Artificial Intelligence (in press) (2008)
8. Dehmer, M.: A novel method for measuring the structural information content of networks. In: Cybernetics and Systems (in press) (2008)
9. Halin, R.: Graphentheorie. Akademie Verlag (1989)
10. Harary, F.: Graph Theory. Addison-Wesley, Reading (1969)
11. Konstantinova, E.V., Skorobogatov, V.A., Vidyuk, M.V.: Applications of information theory in chemical graph theory. Indian Journal of Chemistry 42, 1227–1240 (2002)
12. McKay, B.D.: Graph isomorphisms. Congressus Numerantium 730, 45–87 (1981)
13. Minoli, D.: Combinatorial graph complexity. Atti. Accad. Naz. Lincei, VIII. Ser., Rend., Cl. Sci. Fis. Mat. Nat. 59, 651–661 (1975)

14. Mowshowitz, A.: Entropy and the complexity of the graphs I: An index of the relative complexity of a graph. Bull. Math. Biophys. 30, 175–204 (1968)
15. Neel, D.L., Orrison, M.E.: The linear complexity of a graph. The Electronic Journal of Combinatorics 13 (2006)
16. Rashevsky, N.: Life, information theory, and topology. Bull. Math. Biophys. 17, 229–235 (1955)
17. Shannon, C.E., Weaver, W.: The Mathematical Theory of Communication. University of Illinois Press (1997)
18. Skorobogatov, V.A., Dobrynin, A.A.: Metrical analysis of graphs. MATCH 23, 105–155 (1988)

Towards a Partitioning of the Input Space of Boolean Networks: Variable Selection Using Bagging

Frank Emmert-Streib[1] and Matthias Dehmer[2]

[1] Queen's University Belfast, Computational Biology and Machine Learning,
Center for Cancer Research and Cell Biology, School of Medicine,
Dentistry and Biomedical Sciences, 97 Lisburn Road, Belfast BT9 7BL, UK
v@bio-complexity.com
[2] Institute for Bioinformatics and Translational Research, UMIT,
Eduard Wallnoefer Zentrum 1, 6060, Hall in Tyrol, Austria
Matthias.Dehmer@umit.at

Abstract. In this paper we present an algorithm that allows to select the input variables of Boolean networks from incomplete data. More precisely, sets of input variables, instead of single variables, are evaluated using mutual information to find the combination that maximizes the mutual information of input and output variables. To account for the incompleteness of the data bootstrap aggregation is used to find a stable solution that is numerically demonstrated to be superior in many cases to the solution found by using the complete data set all at once.

Keywords: Bootstrap aggregation, Mutual Information, Boolean networks, Causality.

1 Introduction

The analysis of networks and their inference has gained much attention during the last years. This interest is at least twofold. First, networks are very interesting objects from a mathematical point of view that possess a multitude of properties that are still to be investigated [1,5,6,16,22,21]. Second, networks can serve as representation of phenomena, e.g., from physics, chemistry or biology [11] to allow their systematic investigation. Especially, in molecular biology networks are nowadays found omnipresently representing, e.g., signaling, metabolic or protein networks [2,3,12,17]. It is important to emphasize that in many of the cases mentioned above networks represent some form of 'interaction' occurring within the system. That means the network structure represents causal dependencies or independencies among the variables in the system [9,19]. For this reason, the inference of, e.g., gene networks from experimental data represents one of the major goals in molecular biology because the inferred networks allow to gain insights in the causal working mechanism of living cells.

In this paper, we present an information-theoretic method that allows to identify a set of input variables of a Boolean network. That means, we are aiming to

J. Zhou (Ed.): Complex 2009, Part I, LNICST 4, pp. 715–723, 2009.

identify a set of variables that effects, potentially causally, the outcome of another variable. We use Boolean networks because gene networks are frequently modeled as Boolean networks [13] and the inference of gene networks is an application we have in mind when designing our method. The algorithm we suggest is based on a recent method by LIANG et al. [14] which they called REVEAL (Reverse Engineering Algorithm). We extend this algorithm regarding two important points. First, we modify the algorithm that it can also deal with incomplete data. Second, we use bagging (bootstrap aggregation) to find the optimal input set that is more robust against noise or outliers in the data [7]. We want to emphasize that such a selection mechanism is useful with respect to the partitioning of the input space of Boolean networks because it may allows to reduce the complexity of post-processing steps. Further, the obtained sets of variables can be seen as larger components (larger than single variables) that might be used to construct the overall network implying approaches that are beyond node-to-node based tests like d-separation [10,18,20] or ARACNE [4,15].

This paper is organized as follows. In the next two sections, we present our method and in section 3 we present numerical results. This paper finishes in section 4 with conclusions.

2   Methods

For our study we assume that we have a given Boolean network that is defined via its lookup table (LUT). The LUT provide a mapping from the binary input variables to the binary output variables. For a Boolean gate with n input variables there is a total of 2^n different combinations that can be realized by n binary variables. Hence, a complete LUT consists of 2^n entries. The method we propose is intended to be used for an incomplete LUT. That means, only a certain fraction of all possible input combinations is observed and used as training set to identify the set of input variables.

Our method consists of a modified version of the REVEAL algorithm. The principle idea of the REVEAL algorithm [14] is that under ideal conditions the entropy of an output is completely determined by the mutual information between the output and its input I_s, i.e.,

$$H(O) = I(O; I_s). \tag{1}$$

Because the mutual information can be written as

$$I(O; I_s) = H(O) - H(O|I_s) = H(O) + H(I_s) - H(O, I_s) \tag{2}$$

this implies that under ideal conditions

$$H(I_s) = H(O, I_s) \tag{3}$$

holds. The REVEAL algorithm is an iterative algorithm starting with an input I_s consisting of just one variable. If condition 1 does not hold combinations of

---

Algorithm 1. Maximization of mutual information

1: $d^* = 0$
2: $I_s^* = \{\}$
3: **for all** allowed input sets **do**
4:    update I_s
5:    calculate $H(O)$
6:    calculate $I(O; I_s)$
7:    $d - I(O, I_s)/H(O)$
8:    **if** $d > d^*$ **then**
9:       $d = d^*$
10:      $I_s^* = I_s$
11:   **end if**
12: **end for**

---

variables are used as input I_s until the perfect configuration is found. If no perfect solution is found the algorithm does not make any suggestion for a candidate set.

In this paper we extend the algorithm above by allowing data that are not sufficient for a perfect recovery of input variables, e.g., due to noise. This corresponds to more realistic situations because, e.g., experimental data will always contain noise to some extend that counteracts our goal to identify the perfect input set of an output variable. For this reason we need to modify the optimality criteria in Eq. 1 to account for experimental data in general. This can be accomplished by two modifications. First, note that entropies are always non-negative [8]. This implies that (from Eq. 2)

$$I(O; I_s) \leq H(O), \tag{4}$$

holds. Hence, we are searching an input set that maximizes the fraction, i.e.,

$$I_s^* = \underset{I_s}{\operatorname{argmax}} \left\{ \frac{I(O; I_s)}{H(O)} \right\}. \tag{5}$$

Here I_s^* corresponds to the optimal input set that can be found for given data from which all entropies are estimated. The second modification consists in the stopping criteria. Because we do no longer expect to find a perfect solution we calculate $I(O; I_s)$ for all input sets we want to consider. It is clear that the number of possible input sets I_s increases rapidly with the number of available inputs for this reason we restrict this complexity by selecting only a subset thereof. More precisely, in this study we allowed only input sets of size up to $3 = |I_s|$. Here $|.|$ measures the cardinality of the set I_s. Algorithm 1 gives pseudo code of the principle mechanism of our approach.

In addition to these two modification we use bagging (bootstrap aggregation) [7] to produce a more stable result. Briefly, we sample B bootstrap samples from the original data of the same size (with replacement) and obtain this way B solution sets I_s^b, $b \in \{1, \ldots, B\}$. Hereby we consider each possible input set I_s as a model m. For these models we calculate the probability p_m that model m has

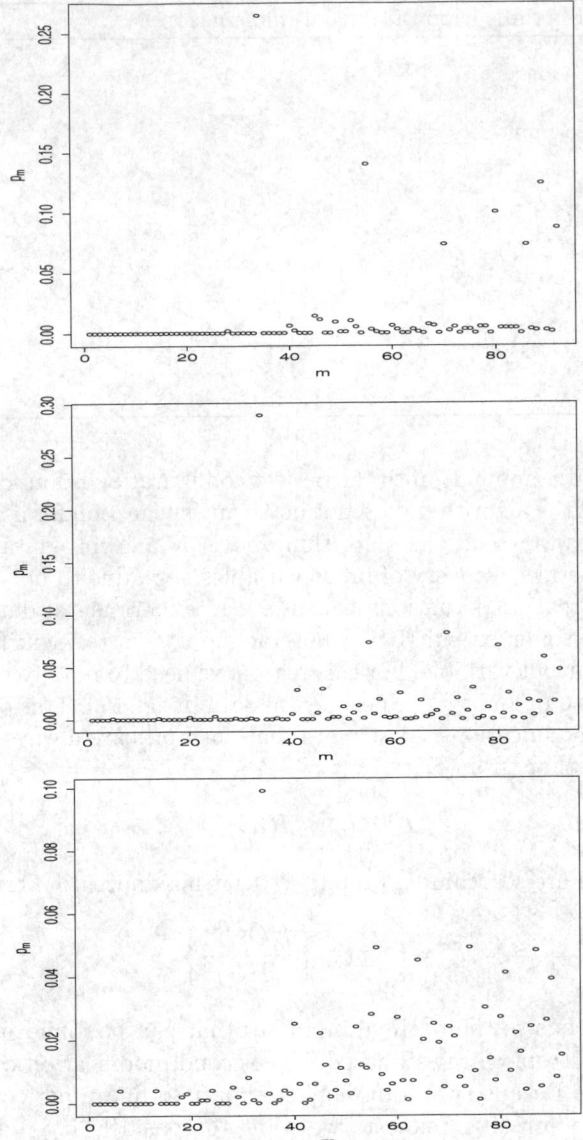

Fig. 1. Results for node 8. Top: $N_s = 77$ (30%) with $V_B = \{6, 7\}$, $V_T = \{1, 5\}$. Middle: $N_s = 51$ (20%) with $V_B = \{6, 7\}$, $V_T = \{5\}$. Bottom: $N_s = 25$ (10%) with $V_B = \{6, 7\}$, $V_T = \{3\}$.

been chosen by all B bootstrap samples. This give us finally the model, input set, that gives the most stable result,

$$V_B = m^* = \operatorname*{argmax}_{m} \{p_m\}. \tag{6}$$

We apply bagging to obtain a probability distribution over all possible models. This allows us in addition to obtain an optimal solution for given data to see how probable other models are for the same data.

3   Results

In this paper we study a Boolean network consisting of 8 variables. The network is defined by the following equations corresponding to a synchronous updating (as in [14])

$$O_1 = I_1 \tag{7}$$
$$O_2 = I_2 \tag{8}$$
$$O_3 = I_3 \tag{9}$$
$$O_4 = I_4 \tag{10}$$
$$O_5 = I_5 \tag{11}$$

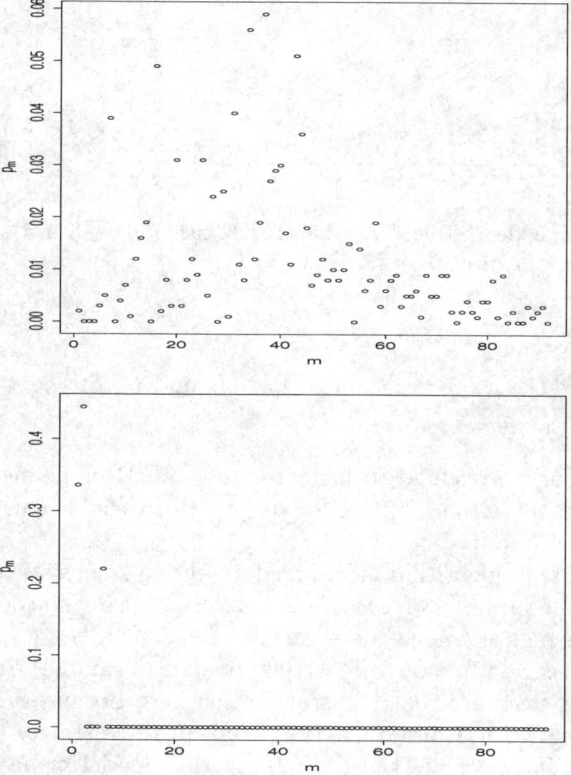

Fig. 2. Results for node 8. Top: $N_s = 13$ (5%) with $V_B = \{1,2,3\}$, $V_T = \{1,3,8\}$. Bottom: $N_s = 3$ (1%) with $V_B = \{2\}$, $V_T = \{1\}$.

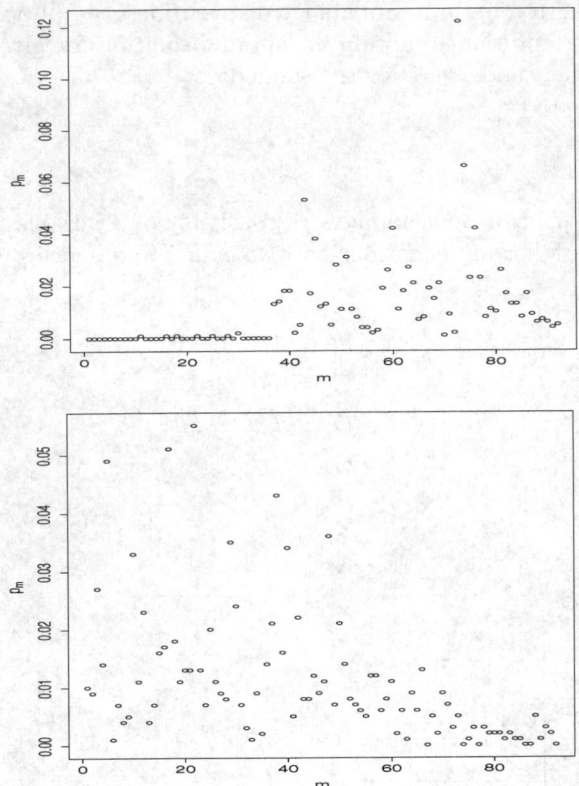

Fig. 3. Results for node 7. Top: $N_s = 25$ (10%) with $V_B = \{3,4,5\}$, $V_T = \{3,7,8\}$. Bottom: $N_s = 13$ (5%) with $V_B = \{3,4\}$, $V_T = \{4,6\}$.

$$O_6 = I_1 \text{ or } I_2 \tag{12}$$
$$O_7 = D(I_3, I_4, I_5) = (I_3 \text{ and } I_4) \text{ or } (I_3 \text{ and } I_5) \text{ or } (I_4 \text{ and } I_5) \tag{13}$$
$$O_8 = I_6 \text{ or } I_7 \tag{14}$$

Here 'and' and 'or' correspond to logical gates and 'D' is defined by the right hand side of Eqn. 13. Each input (I) or output (O) variable can assume values in $\{0, 1\}$.

The purpose of our algorithm introduced in section 2 consists in finding input sets for the output variables of a Boolean network that form at least candidates for a causal dependence among these variables.

For our analysis we use always $B = 1000$ bootstrap samples from given data. This implies that there are $256 = 2^8$ state transitions. Because we are aiming to realistic situations we use just a fraction of all possible states. In Figure 1 we show results for O_8. p_m gives the probability that model m has been selected by the bootstrap samples. For example, for the top figure $N_s = 77$ state transitions (randomly sampled from all 256) have been used as data. All of these results

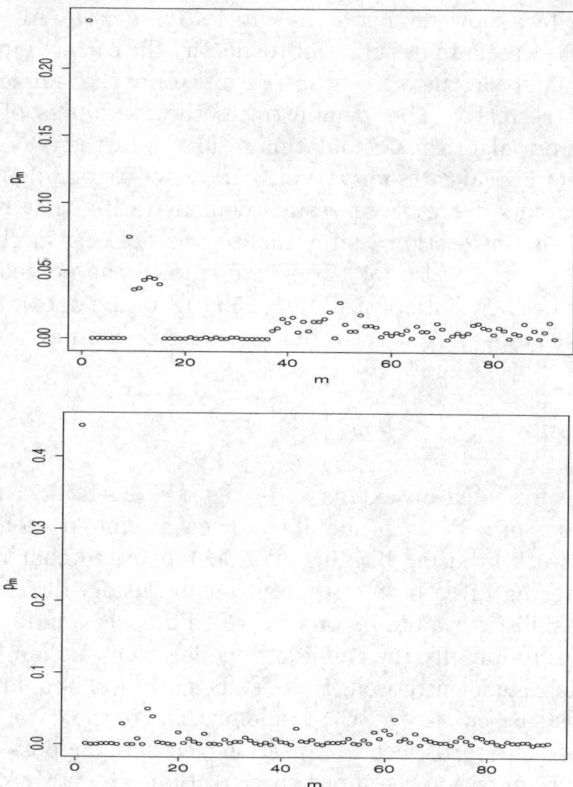

Fig. 4. Results for node 1. Top: $N_s = 25$ (10%) with $V_B = \{1\}$, $V_T = \{1\}$. Bottom: $N_s = 13$ (5%) with $V_B = \{1\}$, $V_T = \{1\}$.

V_B find the correct input set $\{I_6, I_7\}$. Even if we use only $N_s = 25$ samples (bottom figure), corresponding to 10% of all state transitions, gives the correct result. Instead, when we use all data, not applying bagging, we find in all three cases suboptimal input sets V_T. Further reducing the amount of data finally results also in suboptimal input sets as can be seen in Fig. 2. An interesting observation from Fig. 1 and 2 is that the probability distribution over all possible models involving three or less variables is shifted from the right to the left [1]. The interpretation for this shift is that the less data are available (low N_s values) the less complex (lower number of variables) is the input set our algorithm suggests. This behavior is reasonable because less data allow only simpler models without risking over-fitting the data. Figure 3 and 4 also show this behavior. We repeated our analysis 100 times drawing new test data of size $N_s = 25$ (10% of the data) to study the behavior of the population. We found that in 70% of the cases our algorithm identifies the correct input set.

---

[1] The models are enumerated from simple (left) to more complex sets (right) comprising more variables.

In figure 3 and 4 we show two further results for O_7 and O_1. Also for these two cases 10% of the data seem to be sufficient to identify the correct input set. Again, using all data without bagging gives worse results, only the simple case $O_1 = I_1$ can be identified correctly. This demonstrates the usefulness of bagging and justifies its use in our algorithm. Simulation results of further Boolean networks containing different logical gates for varying parameters of our algorithm and the number of data points confirm our results demonstrating that the exemplary results presented in this section that visualize the working mechanism of our algorithm hold in general. Also for node 7 we studied the population behavior by drawing 100 times new data of size $N_s = 25$ (10% of the data). Here we found that in 62% of the cases the correct input set could be found.

4   Conclusions

In this paper we presented an extension of the REVEAL algorithm [14]. The extended algorithm provides a probability for each allowed input set (model) of a Boolean network by using bagging. We demonstrated that for incomplete lookup tables bagging gives better results than by using all data at once. In general, the probabilistic evaluation of all allowed models could be exploited by ranking all models to identify the candidate models that explain the entropy of the output variable sufficiently well. Here 'sufficiently well' could be quantified by significance tests based on, e.g., the randomization of the data. Our approach represents a first step towards such a realization containing important ingredients that would allow to partition the input space of Boolean networks according to the significance of input sets rather than single variables. In this article we just focused on the most probable set. In future work we will study the more general problem.

Acknowledgements

M. D. thanks the Ciência 2006 fund (Portugal) for financial support.

References

1. Albert, R., Barabasi, A.: Statistical mechanics of complex networks. Rev. of Modern Physics 74, 47 (2002)
2. Alon, U.: An Introduction to Systems Biology: Design Principles of Biological Circuits. Chapman & Hall/CRC, Boca Raton (2006)
3. Barabasi, A.L., Oltvai, Z.N.: Network biology: Understanding the cell's functional organization. Nature Reviews 5, 101–113 (2004)
4. Basso, K., Margolin, A.A., Stolovitzky, G., Klein, U., Dalla-Favera, R., Califano, A.: Reverse engineering of regulatory networks in human b cells. Nature Genetics 37(4), 382–390 (2005)
5. Bollobas, B.: Modern Graph Theory. Springer, Heidelberg (1998)

6. Bornholdt, S., Schuster, H. (eds.): Handbook of Graphs and Networks: From the Genome to the Internet. Wiley, Chichester (2003)
7. Breiman, L.: Bagging predictors. Machine Learning 24(2), 123–140 (1996)
8. Cover, T., Thomas, J.: Information Theory. John Wiley & Sons, Inc., Chichester (1991)
9. Cox, D., Wermuth, N.: Multivariate dependencies: Models, analysis and interpretation. Chapman & Hall/CRC, Boca Raton (1996)
10. de la Fuente, A., Bing, N., Hoeschele, I., Mendes, P.: Discovery of meaningful associations in genomic data using partial correlation coefficients. Bioinformatics 20(18), 3565–3574 (2004)
11. Dehmer, M., Emmert-Streib, F. (eds.): Analysis of Complex Networks: From Biology to Linguistics. Wiley-VCH, Chichester (in press, 2009)
12. Jeong, H., Mason, S.P., Barabasi, A.L., Oltvai, Z.N.: Lethality and centrality in protein networks. Nature 411, 41–42 (2001)
13. Kauffman, S.: Metabolic stability and epigenesis in randomly constructed genetic nets. Journal of Theoretical Biology 22, 37–467 (1969)
14. Liang, S., Fuhrman, S., Somogyi, R.: Reveal, a general reverse engineering algorithm for inference of genetic network architectures. In: Pac. Symp. Biocomput., pp. 18–29 (1998)
15. Margolin, A., Nemenman, I., Basso, K., Wiggins, C., Stolovitzky, G., Dalla Favera, R., Califano, A.: Aracne: an algorithm for the reconstruction of gene regulatory networks in a mammalian cellular context. BMC Bioinformatics 7, S7 (2006)
16. Newman, M.E.J.: The structure and function of complex networks. SIAM Review 45, 167–256 (2003)
17. Palsson, B.: Systems Biology. Cambridge University Press, Cambridge (2006)
18. Pearl, J.: Probabilistic Reasoning in Intelligent Systems. Morgan Kaufmann, San Francisco (1988)
19. Pearl, J.: Causality: Models, Reasoning, and Inference, Cambridge (2000)
20. Shipley, B.: Cause and Correlation in Biology. Cambridge University Press, Cambridge (2000)
21. Watts, D.: Small Worlds: The Dynamics of Networks between Order and Randomness. Princeton University Press, Princeton (1999)
22. Watts, D., Strogatz, S.: Collective dynamics of 'small-world' networks. Nature 393, 440–442 (1998)

Toward Automatic Discovery of Malware Signature for Anti-Virus Cloud Computing

Wei Yan and Erik Wu

Advanced Threats Research
Trend Micro, Inc.
USA

Abstract. Security vendors are facing a serious problem of defeating the complexity of malwares. With the popularity and the variety of zero-day malware over the Internet, generating their signatures for detecting via anti-virus (AV) scan engines becomes an important reactive security function. However, AV security products consume much of the PC memory and resources due to their large signature files. AV cloud computing becomes a popular solution for this problem. In this paper, a novel Automatic Malware Signature Discovery System for AV cloud (AMSDS) is proposed to generate malware signatures from both static and dynamic aspects. Our experiments on millions-scale samples suggest that AMSDS outperforms most state-of-the-art automatic signature generation techniques of both industry and academia.

Keywords: anti-virus, network security, malware, cloud computing.

1 Introduction

Malwares are used to compromise computers and to steal the users private data by exploiting software vulnerabilities[1,2]. In cases which malwares are the zero-day threats, generating their signatures for detecting via anti-virus (AV) scan engine becomes an important reactive security function. However, modern malwares can easily bypass AV scanners by using code obfuscation, which can prevent malicious file contents from being detected. Current malware signature generation technique always involves in heavy manual work by studying emulation traces with hours or even days delay. Therefore, security researchers are facing great challenges in overcoming the complexity of malwares, and fighting against the malware backlog is nothing new.

To effectively handle the scale and magnitude of new malware variants, anti-virus functionality is moved into the cloud. In this paper, Automatic Malware Signature Discovery System (AMSDS), a novel and lightweight desktop agent for AV cloud is described. AMSDS keeps a good workload balance between the desktop and cloud services. It can automatically generate a lightweight signature database with the size hundreds times smaller than traditional signature ones. In the AV cloud model, users do not need to install a large virus signature file, but a lightweight set of "cloud signatures". The benefits include easy deployment,

J. Zhou (Ed.): Complex 2009, Part I, LNICST 4, pp. 724–728, 2009.

low costs of operation, and fast signature updating. Further, AMSDS signatures can be easily integrated into existing AV products.

We begin in Section 2 by presenting a brief introduction of virus executable file format. In Section 3, we expound how AMSDS generates cloud desktop patterns. We present experimental results in Section 4, and close in Section 5.

2   Virus Executable File Format

Executable files are special-formatted file objects that can be understood and executed by operating systems. Examples of modern executable formats include Portable Executable format (PE) for Windows, Executable and Linkable Format (ELF) for Linux and Mach Object (Mach-O) for Mac OS. This paper is focused on the PE format[3] as it is the most popular format for executables, libraries, and drivers in Windows.

A PE file comprises various sections and headers which describe the section data, import table, export table, resources, etc. A PE file starts with the DOS executable header, which is followed by the PE header. The PE header begins with the signature bits "PE". The PE header also includes some general file properties, such as the number of sections, machine type, and time stamp. Another type of header is called optional header, which contains an array of important information segments. The optional header is followed by the section table headers, summarizing each section's raw size, virtual size, section name, etc. Finally, at the end of the PE file is the section data, which contains the file's Original Entry Point (OEP). OEP refers to the execution entry point of a PE file, where the file execution begins. To search a PE file for malwares, a scanner typically scans the segments at certain offsets from OEP for the known signatures. PE tools facilitate the ease to view, analyze and edit WIN32 PE files.

Existing commercial security applications search the binary files for pre-defined signatures to identify known malwares. Unfortunately, this technique can be easily fooled by obfuscated viruses, which use software packers [2](programs that compress and encrypt executable files in disk and restore the original executable image, when loaded into memory), to protect the viruses' internal code and data structures from being detected by security software.

3   AMSDS

AMSDS aims to generate intelligent malware signatures for AV cloud desktop. Each client has a lightweight AMSDS signature file, whose size is hundreds times smaller than traditional signature databases. Only when a suspicious file cannot be detected by AMSDS patterns, clients will send a request to a cloud server, where exists the full traditional pattern database.

In this paper, our assumption is based on the fact that the samples of the same malware family must have some identical binary raw sequences. Among those sequences, there exists a set of identical binary strings. Our AMSDS signatures are generated from those strings. AMSDS firstly parses a suspicious PE sample

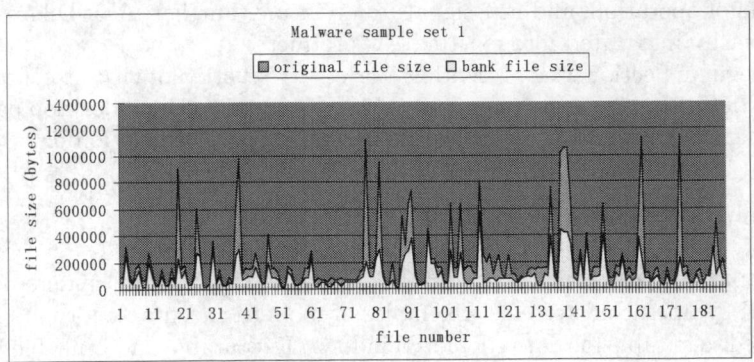

Fig. 1. AMSDS de-noise PE sample files

and list its internal structures, such as PE header, optional header, section table, import table, export table, and the resources. To speed up the signature generation, AMSDS will discard some raw data of this sample, and only reserves the segments where hackers may insert their malicious codes. This process is called "de-noise". After the denoising, AMSDS can make the testing environment safer by destroying malware's PE formats. Therefore, the malicious sample can not be executed. Fig. 1 shows the de-noise performance for some malware samples. Owing to the de-noise stage, AMSDS can shrink the original file sizes almost by half.

AMSDS is able to generate malware signatures from both static and dynamic aspects. With the incoming of malware samples, the intelligent converter in AMSDS parses malwares static information, and uses machine learning technique to find multiple disjoint binary sequences. Based on those invariant substrings, the static compound signatures are generated for inline matching. AMSDS can be also used to automatically generate behavior signaures. Nowadays, the emulator or sandbox is used to capture malware dynamic traces. However, the overwhelming static and dynamic "noise" generated by using code obfuscation make it hard to analyze. Therefore, current malware signature generation techniques always involve in heavy manual work by studying emulation traces with hours or even days delay. A malware emulator and its tailored malware behavior ontology are used to collect the dynamic execution token traces. Those tokens are abstract representations by traversing the malware behavior ontology. Afterwards, the malware behavior signatures are generated from those token streams.

4   Simulation

We describe our experimental results on on millions-scale samples in this section. By using the disjoint invariant segments, AMSDS achieves low false positive rate; By using "back up" signatures, AMSDS can defeat code obfuscation efficiently. Our testing sets include more than 10 millions benign samples and almost 35k malware samples from 40 families. We choose small training set from 30 to 50 samples for each family. The average malware detection rate is around 80%. (For larger training set, the detection rate will be higher.) As shown in Fig. 4, the false positive rate for benign samples is 0.001%. In the aspect of speed, AMSDS is comparable with current AV products. For example, AMSDS takes 1m26s seconds to scan a folder, which includes 20,000 files.

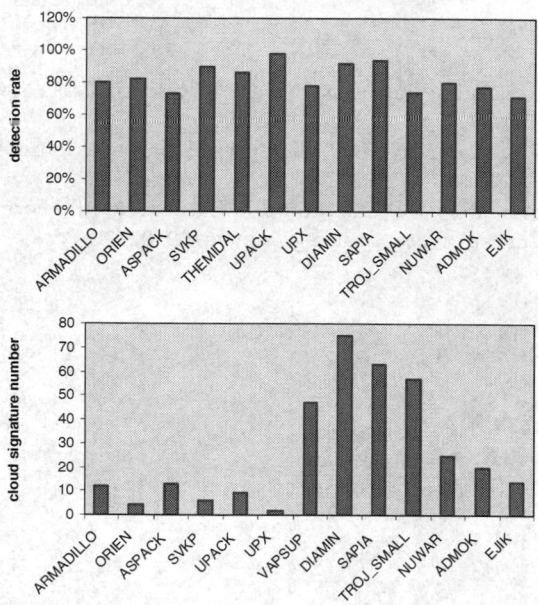

Fig. 2. Detection rate and pattern number

We also measured the malware detection rate for 30 families, which includes 10 Windows PE packer and 20 malware families. Fig.2 presents the detection rates and pattern number for each family. The total pattern file size is less than 10k. Our testing suggests that the average malware detection rate is around 80%. We also found an interesting result: the pattern number for each packer family is much less than that of malware family. The only reason we think is that few patterns are enough to capture the packers' unpacking semantics. Normally a packer's unpacking process involves four consecutive steps: decompression or decryption, anti-debugging checks, import table rebuilding, and jumping to OEP[2]. For packers, each above step always involves similar working flow. For example, in the import table building stage, a packer extracts the DLL names, followed by the trunk table addresses and APIs. For files with relocation tables, the packer stores the Relative Virtual Address (RVA) of the relocatable data blocks, which will be relocated when the relocation table needs rebuilding.

5   Conclusion

In this paper, we propose AMSDS, a ontology-based automatic signature generation system for zero-day malwares, which generates both static and dynamic tokens for AV cloud computing. Our approach is generic and flexible. Different otologies can be plugged in for various detecting purposes. The experiments show it outperforms state-of-the-art automatic signature generation techniques.

References

1. Grace, C.: Understanding intrusion detection systems. PC Network Advisor 122, 11–15 (2000)
2. Yan, W., Zhang, Z., Ansari, N.: Revealing packed malware. Journal of IEEE Security and Privacy 6(5), 65–69 (2008)
3. An In-Depth Look into the Win32 Portable Executable File Format, http://msdn.microsoft.com/msdnmag/issues/02/02/PE/

Topological Structure and Interest Spectrum of the Group Interest Network

Ning Zhang

Business School, University of Shanghai for Science and Technology,
Shanghai 200093, P. R. China
zhangning@usst.edu.cn

Abstract. In this paper, the behavior characteristics that the specifical campus group users accessing world wide web has been studied, the dynamic group interest network has been constructed, which was a para-bipartite graph and the topological structure had been discussed. Although the users' visiting time is random and the web pages they visited are different but the interests of a majority of the campus group are accordant. The results indicate that the incoming degree distribution of the group interest network follows power law. And the group interest spectrum was basically steady. The visiting behavior of the campus group had their special disciplinarian.

Keywords: Complex network, para-bipartite graph, group interest spectrum, behavior characteristics, in-degree.

1 Introduction

The rapid development of information technology has brought the great challenges in theoretical study and practical application, and its significant social and economic values attracted significant coverage from all disciplines. The research of information systems' complexity has become one of the important problems for international academic community, especially for the cross-frontier scientific research. Recently, with the research and development of the complex networks, the information system regard as a complex system, has become a cross-research focal point [1-4]. Complex networks are available for studying a great deal of practical system, such as the World Wide Web, the Internet, the electrical power-grid networks, the biological nets and social networks [5-9]. Many empirical evidences indicate that the topological characteristics of practical networks are neither regular nor random [10], they belong to both small world [11] and scale-free [12, 13]. The findings of complex network reflect the basic characteristics of many complex systems, bringing material breakthrough to these systems' research. For instance, scientist collaboration networks [14-16] can make us clear about the relationships among the scientists in different fields, which have short average path length but big clustering coefficient. That is to say, the scientist collaboration networks have the characters of good connectivity and strong clustering. The power law degree distribution of World Wide Web let it has dual-characteristics, robust and frangible [17].

J. Zhou (Ed.): Complex 2009, Part I, LNICST 4, pp. 729–736, 2009.

These universal characteristics have significant theoretical meanings and engineering application values. As for the engineering application, the values are obviously. If we can identify all kinds of groups of 137 million Internet users in China and the popular resources of 0.843 million webs [18], then we can pick the most popular resources according to their sort order to store. In this way, the power law distribution can ensure that the mainstream resources are able to meet the needs of the majority.

Information in our life become more and more important. How to conquer 'figure gulf' to let everyone share information resources fairly is always being the global issue and also a full concerned issue to government administrator. Different groups of people need different web resources. In the limited resources, we should to consider which kind of information resources can meet the needs of the majority, and use the most proper way to let the individuals far from cities sharing sunshine information [19,20]. That means we should to study different groups' interests. With the help of clustering analytical method, linearity regressive analytical method, we often dig web users' interests and constitute interest model, according to word, text structure characters, paragraph and sorts expression ability [21]. For user's preference, we can use self-adaptive theory to study user's preference [22] and constitute user's preference model base one data cube [23]. For the data digging method, we can use Markov model to searching for user's behavior characters [24] and find out user interest profile according to the implicit feedback [25], and then combining web content and behavior analysis [26] or base on its searching history [27] to invest interesting model. There are plenty of searching engines and information filtration methods. But none of those researches deal with group users' interest structure characters, or explain group interest spectrum's structure and stability mechanism, or reveal the clustering phenomena and rules of group interests.

Our researches are aimed at studying group users' behavior characters and topological characters of group interesting network, finding out interest spectrum of group interest network and it self's evolvement rules. This essay is only involved in introduction researching results. The further results will be published in other papers as the researches go deeper.

2 Data Analysis

The group users in this paper refer to the faculty and the students in our campus. They visit the Internet by local area network in their offices or dormitories, there are one fixed IP address for one room, but one room may have several computers and more than one users, so we call these users as group users.

There are 3 sets of records were collected during the continuous period (see table 1). For the dataset 1, the time period is during the second semester of our school year. In this semester, the senior students begin their graduation project, they have more free time to visite the Internet. For the dataset 2, the time period is during the first semester of our new school year. In this semester the new senior sdudents had their courses and the freshers just came into the school, began their military Training. For the dataset 3, the time period is during our ordinary school activity and all of the students and faculties began their normal life. These data exist clearly periods. From fig.1, we can see that the lower traffic volume appeared during 23:00 Pm to 7:00 Am in every day, and the average traffic volume per hour in dataset 1 is more than that in dataset 2 and 3. From fig.2, we can see that the web traffic volume reduced during legal holidays, and the traffic volume in holydays was less than the working days, lower traffic volume appeared during the weekends in every week, the lowest traffic volume appeared at October golden week. From fig.3, we can see common fluctuating cycles and some times of the day the variations are busier than others.

Table 1. The data statistics

Data set	Begin time	End time
1	4:00 on Mar. 14, 2006	3:59 on Mar. 20, 2006
2	4:00 on Aug. 30, 2006	3:59 on Oct. 8, 2006
3	4:00 on Nov. 19, 2006	3:59 on Nov. 22, 2006

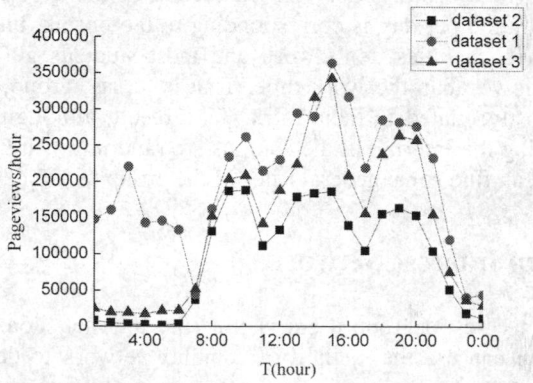

Fig. 1. The average hourly activity per hour

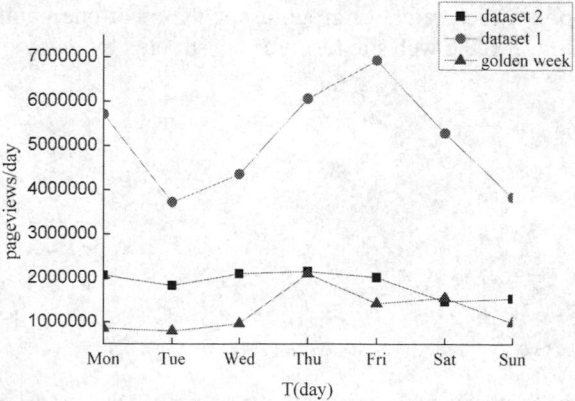

Fig. 2. The average week activity in Data Set 1 and 2

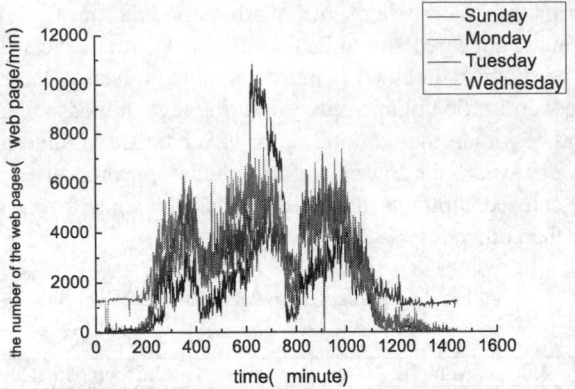

Fig. 3. The total traffic volume per minute in Data Set 3

All these analysis results reflect the characters of the group user' activities, the lowest traffic volume per day is corresponding to the dorms' black out time. During the legal holidays, faculties don't work and most students go home, so the traffic volume is also lower than the work time. That is to say, group users access campus network are mostly related to their work. As a result, although the visiting time of each users visiting the Internet in the campus are random, with the helpp of data analyzing, we still can find some general rules of the group users' behavior.

3   The Group Interest Network

Actually, the activities of group users visiting the Internet should be a dynamic random process, we can use the method of complex network to describe it. The group users' web visiting at each moment can construct a group interest network. The group users' web visiting during a period can constrcte dynamic group interest network. The network have the format of para-bipartite graph and it contain two kinds of vertices, one is user vertex, which refers to group users, the other one is information resource vertex, which refers to the web site resource constructed by many web pages and need

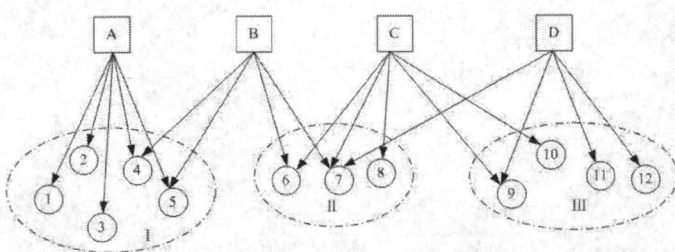

Fig. 4. Schematic illustration of the group interest network. □Nodes A-D represent group users, ○ nodes 1-12 represent web pages, Dash dot circles I -III represent resources (web sites)

to transfer. Since it's not like the bipartite graph with explicit two different kinds of vertices like the definition in the graph theory, we call it as para-bipartite graph. Group users' visiting activities set up the relationship between group users and information resources, which can be expressed by directed connection. And the complex relationship of many users corresponding to many information resources construct the group interest network (see fig.4). According to the group interest network, we can get users relation network and resources relation network by the projection of users and resources respectively.

We construct group interest network by classifying the web pages to different web site according to group users accessing habit per day (see table 2), which is a directed dynamic complex network. Basing on dataset 1, the Tuesday's network contains 658 users and 16078 web sites. The Wednesday's network contains 676 users and 17491 web sites. The Thursday's network contains 674 users and 18233 web sites and etc. The network contains 1023 users and 60079 web sites in total during one week. We calculate this group interst networks' in-degree. The in-degree here means the numbers of group users' visiting a web site. Such as in fig.4, the group user A visited 5 pages of web site 1 and B visited 2 pages of web site 1, then the visiting number of web site 1 is 7. In the same way, the visiting number of website 2 is 6 and number of website 3 is 5. Even though the number of user are different and the web sites they visited are different in every day, the in-degree distribution of group interest network have power law character (see fig. 5). Which means that lots of web sites are with a few links (visiting amount), a few web sites are with a medium number of links and a very few noteworthy web sites are with a large number of links in this network. The in-degree frequencies and their per centum of the group interest network can be seen in table 3. The in-degree frequencies refer to the numbers of a in-degree. Such as in fig.4, the in-degree frequencies of in-degree 1 are 7, in-degree frequencies of in-degree 2 are 4, and 3 are 1.

Table 2. The statistics of the group user visiting the Internet

Week	User	Web page	Web site	Degree exponent
3.14 (Tuesday)	658	3727905	16078	1.530477
3.15 (Wednesday)	676	4361211	17491	1.533105
3.16 (Thursday)	674	6068099	18233	1520289
3.17 (Friday)	647	6931486	18390	1.528237
3.18 (Saturday)	381	5291921	12215	1.520854
3.19 (Sunday)	393	3844392	14488	1.532267
3.20 (Monday)	663	5710836	17584	1.521534

Table 3. The in-degree frequencies and their per centum of the group interest network

In-degree	in-degree frequencies	per centum
1-100	51218	85.25%
101-800	6845	11.39%
801-1681974	2016	3.36%

The in-degree distribution of group interest network follows power law, $P_{in}(k) \propto k^{-\gamma}$, $\gamma = 1.52$ (see fig.5), so this network was scale free.

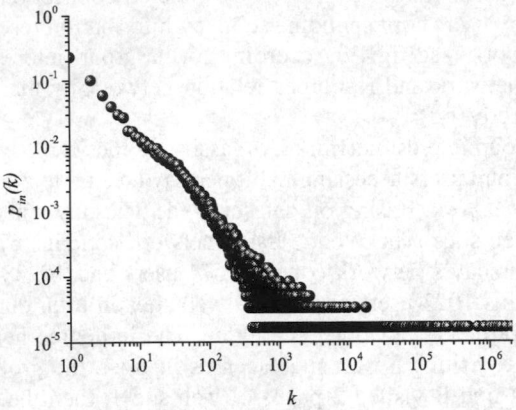

Fig. 5. The in-degree distribution of the group interest network

4   The Interest Spectrum

By ordering the web resources of data set 1, in the web sites with the top 20 visiting volumes, users'number of 7 web sites is 1, users' number, which is no more than 10, web sites are 3. Only 10 popular web sites have more group users. So we got the top ten popular web sites and page numbers of the group users visited (see table 4). These results reveal that which kind of web sites can attract much more our campus' group user, and these users' a part of interest spectrum. We find that the group interest spectrum have a good stability. For instance, although the top 10 web sites of the group users accessed have different orders each day, they are still in each day's top 10 list. We can find out the number of the group users which visited the top ten web sites (see table 5). In the further research, we can get every group users' interst spectrum.

Table 4. The top 10 web sites of the group user surfing during one week

Web site	Web pages
sina. com. cn	1681974
163. com	1451668
sohu. com	834362
usst. edu. cn	662827
online. sh. cn	444066
msn. com	424835
allyes. com	390524
pconline. com. cn	385017
taobao. com	321813
chinaren. com	251568

Table 5. The users' number of visiting the top 10th web sites during one week

Web site / Date	Mar.14	Mar.15	Mar.16	Mar.17	Mar.18	Mar.19	Mar.20
sina.com.cn	341	461	351	339	202	216	351
163.com	338	360	343	331	199	224	349
sohu.com	279	305	293	282	161	179	293
usst.edu.cn	252	268	271	260	119	128	295
online.sh.cn	128	129	130	114	63	61	123
msn.com	283	287	301	283	243	190	318
allyes.com	435	439	432	422	251	260	467
pconline.com.cn	74	79	84	67	53	59	74
taobao.com	111	126	124	128	80	85	147
chinaren.com	112	134	127	109	75	78	123

Researches indicate that the major users have the similar interests, and if we can conform the information resources interested by most individuals and use the abroad storage technique to maintain most peoples request, then we can let people to obtain sunshine information in the most economical way.

5 Conclusions

According to the relationship between group users and information resources, the special group users' web visiting behaviour has been studyed, the time features that the group user visited world-wide-web has been observed, the group interst network has been set up, the topological structure has been discussed in this paper. With the help of complex network's method, the study indicates that the group interest network's in-degree distribution belongs to power law distribution. The given group's interest spectrum is basically stable and the visiting behavior of the campus group had their special disciplinarian, and the interests of a majority of the campus' group users are accordant.

Acknowledgements. This work was supported by Shanghai Leading Academic Discipline Project (No. S30501) and the Natural Science Foundation of Shanghai (06ZR14144).

References

1. Simkin, M.V., Roychowdhury, V.P.: A theory of Web traffic. EuroPhys. Lett. 82, 28006 (2007)
2. Goncalves, B., Ramasco, J.J.: Human dynamics revealed through Web analytics. Phys. Rev. E 78, 26123 (2008)
3. Golder, S., Wilkinson, D., Huberman, B.A.: Rythms of social interaction: messaging within a massive online network, e-print ArXiv cs/0611137 (2006)

4. Meiss, M.R., Menczer, F., Fortunato, S., Flammini, A., Vespignani, A.: A, Ranking Web sites with real user traffic. In: Proc. WSDM (2008)
5. Albert, R., Barabási, A.-L.: Statistical mechanics of complex networks. Rev. Mod. Phys. 74, 47–97 (2002)
6. Dorogovtsev, S.N., Mendes, J.F.F.: Evolution of Networks: From Biological Nets to the Internet and the WWW. Oxford Univ. Press, Oxford (2003)
7. Pastor-Satorras, R., Vespignani, A.: Evolution and Structure of the Internet: a Statistical Physics Approach. Cambridge Univ. Press, Cambridge (2004)
8. Newman, M.E.J.: The structure and function of complex networks. SIAM Rev. 45, 167–256 (2003)
9. Amaral, L.A.N., Ottino, J.M.: Complex networks—augmenting the framework for the study of complex systems. Eur. Phys. J. B 38, 147–162 (2004)
10. Erdös, P., Rényi, A.: On the evolution of random graphs. Publ. Math. Inst. Hung. Acad. Sci. 5, 17–61 (1960)
11. Watts, D.J., Strogatz, S.H.: Collective dynamics of 'small-world' networks. Nature 393, 440–442 (1998)
12. Albert, R., Jeong, H., Barabási, A.-L.: Diameter of the World Wide Web. Nature 401, 130–131 (1999)
13. Faloutsos, M., Faloutsos, P., Faloutsos, C.: On power-law relationships of the Internet topology. Comput. Commun. Rev. 29, 251–262 (1999)
14. Newman, M.E.J.: Scientific collaboration networks: I.Network construction and fundamental results. Phys. Rev. E 64, 16131 (2001)
15. Newman, M.E.J.: Scientific collaboration networks: II.Shortest paths, weighted networks, and centrality. Phys. Rev. E 64, 16132 (2001)
16. Newman, M.E.J.: The structure of scientific collaboration networks. Proc. Natl. Acad. Sci. USA 98, 404–409 (2001)
17. Albert, R., Jeong, H., Barabási, A.-L.: Error and attack tolerance of complex networks. Nature (London) 406, 378 (2000)
18. China Internet Network Information Center, Statistical Reports on the Internet Development in China (in Chinese),
http://www.cnnic.cn/html/Dir/2007/01/22/4395.htm
19. Li, Y.-P.: Sunshine information——conflict-free share structure. China engineering science 2(1), 24–27 (2000) (in Chinese)
20. Li, Y.-P.: Construct Broad-Storage grid. Computer world 37 (2005) (in Chinese)
21. Lin, H.-F., Yang, Y.-S.: The representation and update mechanism for user profile. Journal of computer research and development 39(7), 843–847 (2002) (in Chinese)
22. Lei, Y.-S., Gan, R.-C., Du, D.: A framework of adaptive vertical website based on user preferences. Computer Engineering 31(24), 18–20 (2005) (in Chinese)
23. Chen, J.-L.: User's interest model based on data cube. Journal of Gulin university of technology 25(1), 84–88 (2005) (in Chinese)
24. Hu, Y.-H., Zhao, H.-J., Lu, H.-R., Wang, H.-J.: Research on extracting patterns from web user behavior. Computer Engineering and Design 27(18), 3416–3418 (2006) (in Chinese)
25. Sun, T.-L., Yang, F.-Q.: An approach of building and updating user interest profile according to the implicit feedback. Journal of northeast normal university 35(3), 99–104 (2003) (in Chinese)
26. Zhao, Y.-C., Fu, G.-Y., Zhu, Z.-Y.: User interest mining of combining web content and behavior analysis. Computer Engineering 31(12), 93–94 (2005) (in Chinese)
27. Xu, K., Cui, Z.-M.: User profile model based on user search histories. Computer technology and development 16(5), 18–20 (2006) (in Chinese)

Topological Analysis and Measurements of an Online Chinese Student Social Network

Duoyong Sun[1], Jiang Wu[2], Shenghua Zheng[3], Bin Hu[2], and Kathleen M. Carley[4]

[1] College of Information System and Management, National University of Defense Technology, Changsha, Hunan, China, 410073
duoyongsun@gmail.com
[2] Huazhong University of Science and Technology
1037, Luoyu Road, Wuhan, China, 430074
jiangwu.john@gmail.com, bin_hu@mail.hust.edu.cn
[3] Zhejiang University of Technology
18 Chaowang Road, Hangzhou, China, 310032
zheng0210@163.com
[4] CASOS, ISRI, Carnegie Mellon University
5000, Forbes Avenue, Pittsburgh, USA, 15213
kathleen.carley@cs.cmu.edu

Abstract. Online social network attracts more researchers now. In this paper, we topologically analyze an online Chinese student social network--Xiaonei.com. We use Python language to crawl two datasets of Xiaonei in January and February, 2008. The degree distribution and small world phenomena are testified. We also use a social network analysis tool to analyze these two datasets from the viewpoint of social network structure. Seventeen measurements such as Fragmentation, Component Count, Strong/Weak are summarized to identify the exogenous attributes of Xiaonei.com. Additionally, two latent applications of online social network service are proposed in the discussion section.

Keywords: Online Social Network, Topological Analysis, SNA, Complex Network Analysis.

1 Introduction

Nowadays, online social networks (OSN) such as MySpace, Facebook, Friendster, LinkedIn and Orkut have attracted millions of users, many of whom have integrated these sites into their daily lives. These sites have various objectives including connecting those with shared interests such as music or politics (e.g., MySpace.com), focusing on the college student population (e.g., Facebook.com), dating through one's own friends to create a romantic relationship (e.g., Friendster.com), creating networks of co-workers and business associates (e.g., LinkedIn.com) and linking with Google site developers (e.g., Orkut). Besides these pure online social networks, there are also some online community sites such as the Flickr photo-sharing site, the Youtube

J. Zhou (Ed.): Complex 2009, Part I, LNICST 4, pp. 737–748, 2009.

video-sharing site and the SinaBlog blog-sharing site, all of which inside maintain a latent social network.

Out of all these social network services, Facebook is different because one must provide a real campus email address or a valid student identification if one wants to attend a certain network such as the CMU network. Thus the users in Facebook usually use their real names, real photos and have made highly identifiable profiles [1]. Facebook is a social networking site that reinforces and expands real-world social connections. In Facebook, there are more than 65 million active users and over 6 million active user groups. Facebook has 85% market share of U.S. university students over half of whom return daily and spend an average of 20mins per day on it. In the view of the huge potential of Facebook, some cloned sites are emerging now. Xiaonei (www.xiaonei.com) is the Chinese version of Facebook.

A social network represents relationships among friends and its structure has attracted a lot of interest from scholars. The evolution of the structure within the larger online social network of Flickr and Yahoo! 360 has been studied, characterizing users as either passive members of the network; inviters who encourage offline friends and acquaintances to migrate online; and linkers who fully participate in the social evolution of the network [2]. From the viewpoint of complex network theory, Alan Mislove et.al analyze Flickr, YouTube, LiveJournal and Orkut at the same time to confirm the power-law, small-world and scale-free properties of online social networks [3]. In this paper, we focus on researching college student-oriented network services because they more closely approximate a social network in the reality. In fact, these online users like to search to create links with friends with whom they have offline relationships [4]. About 396,836 nodes and 7,097,144 edges [5] in the Xiaonei network and about 4,200,000 users [6] in Facebook network have all manifested their small-world and power-law phenomena. Besides, personal privacy and visualization [7] in these types of network services have attracted more and more scholarly in-depth studies [8].

In this study, in addition to analyzing online social networks from the viewpoint of complex system theory, we analyze them also from the viewpoint of topological structure using a social network analysis method. Social network analysis methods include centrality measures, subgroup identification, role analysis, elementary graph theory, and permutation-based statistical analysis. From this study, we answer sociological questions such as who is the most important in the network when information diffuses, what is the impact on the entire network when one is isolated from the network and how do informal groups form to influence the information diffusion. Our contributions are listed as follows: 1) we analyze the degree distribution and small world phenomena of an entire Chinese student online social network; 2) we use a social network analysis tool—ORA[9] to determine shortest paths and to analyze connected components for online social network; 3) we introduce the new measurements from the viewpoint of social network analysis.

This paper is organized as follows. In section 2, Data sets and crawling method are introduced. In Section 3, we use topological analysis to analyze the structure of online social networks and give some measurements for SNA. In Section 4, we discuss the practical applications of this study. Finally, conclusions and suggestions for further research are presented.

2 Data Sets

Because of the privacy policy of Xiaonei.com, we use the network of Huazhong University of Science and Technology (HUST). In the HUST network, there were 44,419 nodes and 803,987 links in the beginning of January, 2008. In the end of February, 2008, there were 47,546 nodes and 876,983 links. We used these two datasets to analyze the topological structure and evolution of social networks among friends online. Compared to an ego-network, HUST network is a complete social network within a famous Chinese college online social network. It is entirely complete and closed, thus it is suitable for analysis as a sample of online social networks [10].

2.1 An Overview of Xiaonei.com

Xiaonei.com was created in December of 2005 and was an absolutely dominant college social network service in China. Currently, more than 1000 colleges outside China, 3000 Chinese colleges, 8000 Chinese high schools and 7000 companies have opened their network service in Xiaonei.com. Users can find their old friends, make new friends, share photos, music, movies, and share their blogs and personal news. College-oriented social networking sites such as Facebook and Xiaonei provide opportunities to combine online and face-to-face interactions within an ostensibly bounded domain. This makes them different from traditional networking sites: they are communities based "on a shared real space"[11]. Since the majority of these sites require a college's email account for a participant to be admitted to the online social network of that college, expectations of the validity of certain personal information provided by others on the network may increase. Together with the apparent sharing of a physical environment with other members of the network, that expectation may increase the sense of trust and intimacy across the online community.

2.2 Crawling Methods

We used Python to design a script program to analyze the webpage in order to find user profiles and relationships with friends, and then to automatically download user profiles to store into the Mysql database. Due to personal privacy, we were only able to access the HUST network. Furthermore, there were also 5% users who didn't open their profiles for non-friends in the HUST network. First, we used "browse users" function to obtain a fraction of the user ids randomly till the number of selected nodes didn't increase dominantly. Second, we used these user ids to do a snowball sampling by breadth-first search (BFS) in the HUST network. It is known that the power-law nature in the degree distribution is well conserved under snowball sampling since the snowball sampling method easily picks up hubs. This property reduces the degree exponent and produces a heavier tail[12]. It was also annoyed that the verification code was required to input every time when we downloaded 100 profiles. Therefore, we use multi accounts and multi threads to download the profiles.

2.3 Demographics

The majority of users of the Xiaonei at HUST are undergraduate students (83.3% of all profiles). Furthermore, the majority of users are male (65.4% vs. 34.6%). The average age is 22.14 years. Figure 1 shows the distribution for different users according to their entrance year that represent the year of using Xiaonei because Xiaonei only permits IP addresses from college campuses to register on the HUST network. Every September is the entrance month for students, thus students with entrance year of 2006 have used Xiaonei for one and a half years at most, and the students who entered into college in 2007 have just used Xiaonei for half a year. From their online times and last online dates of two sample data sets, we performed statistical analysis,, finding 71.4% of users login to Xiaonei more than 2 times one week. These users are active users who contribute more to the creation of a virtual community than non-active users.

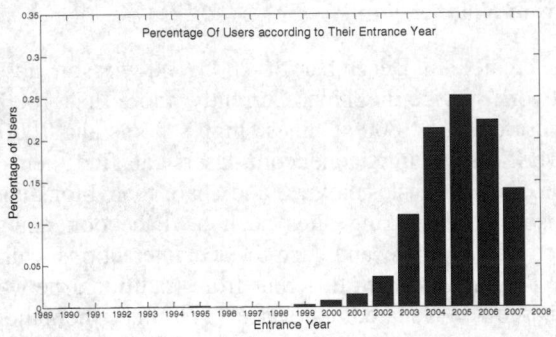

Fig. 1. Percentage of Users according to Their Entrance year

3   Topological Analysis

A social network is a social structure made of nodes (which are generally individuals or organizations) that are tied together by one or more specific types of interdependency, such as values, visions, ideas, financial exchange, friends, kinship, dislikes, conflicts, trades, web links, sexual relations, disease transmission (epidemiology), or airline routes. The resulting structures are often very complex. The Social network analysis method originated in the sociology field and has been widely used in the analysis of online social networks [10]. We pay more attention to the topological structure of online social networks and use ORA [9] plus Matlab to analyze them.

In the Xiaonei network, relationships between friends are reciprocal. The fact that one is directed to certain links can be useful for locating content in information networks. The links in the social networks we studied are regarded as undirected and users may link to each other. This property of symmetry is consistent with that of offline social networks [13].

3.1   Topological Analysis Tool

UCINET is a very popular tool for social network analysis (SNS). It can only handle a maximum of 32,767 nodes (with some exceptions) although practically speaking many procedures get too slow at around 5,000 - 10,000 nodes. We tried to convert our data into the UCINET data type and input it into UCINET, however the data file was very huge (2.3G) and the process speed was very slow. Therefore, UCINET is not suitable to analyze our data set because the Xiaonei network has 47,546 nodes that exceed the process ability of UCINET.

In this paper, we used ORA [9] that was developed by the Center for Computational Analysis of Social and Organizational Systems (CASOS) of Carnegie Mellon University. Beside being a SNS tool, ORA is also a risk assessment tool for locating individuals or groups that are potential risks given social, knowledge and task network information. ORA is based on JAVA and uses XML file to store data, thus it is compatible with other tools in various operation systems. The converted XML file is just 79.6M large and importing it into ORA is very fast.

3.2 Degrees Distribution

For the college students' online network, the entire Facebook and Xiaonei networks have been proven to satisfy the Power-law [14] rule in degree distribution [5, 6]. The partial samples of the virtual online networks Orkut and MySpace also satisfy Power-law. Here, we also observed the degree distribution of Xiaonei HUST network. In Figure 2, we report the cumulative distribution of degree P (>k), which indicates the probability that a randomly selected node has more than k links. It can satisfy the Power-law (see the right log-log sub-figure of Figure 2), and it also can fit exponential function f(x) = a*exp(b*x) where a=4.247e+004 and b = -0.05131 in data1 and a=4.363e+004 and b = -0.04898 in data2. Here, data1 and data2 were obtained at the beginning of January, 2008 and at the end of February, 2008 respectively.

Furthermore, we observed the distribution of "online times" in the Xiaonei HUST network. The ratio of users according to their online times is shown in Figure 3. We used SPSS to compute the correlation of Degree and Online Times, and a strong correlation (Correlation Coefficient is 0.883) between them was found. That is to say, if one person had more online times, he/she would have more time to search for and make friends. Also he/she could spend more time to construct his/her own webpage to attract more users.

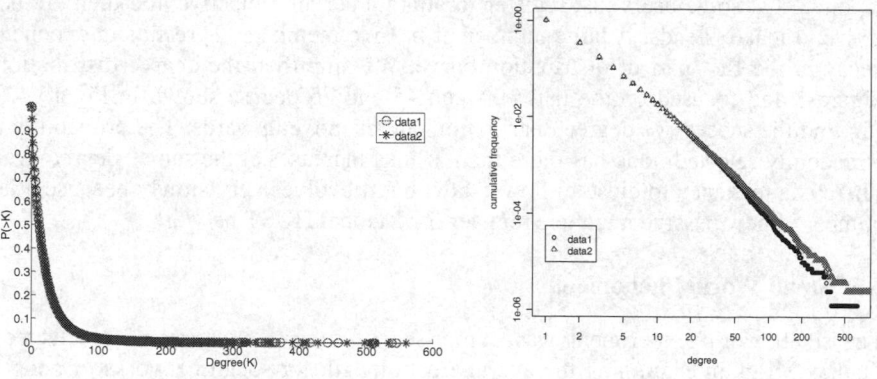

Fig. 2. Degree Distribution of Xiaonei HUST network

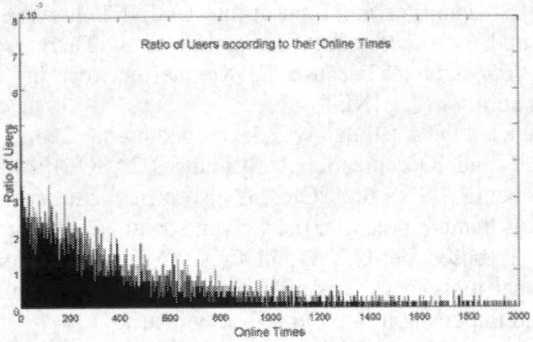

Fig. 3. Ratio of Users according to their Online Times

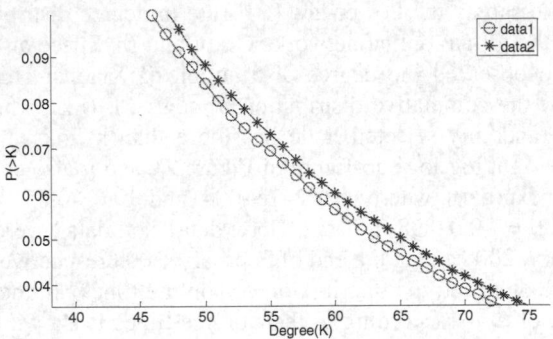

Fig. 4. Degree (45~75) Distribution of HUST network

The users in Xiaonei increased very quickly, from the beginning of January to the end of February 2008. More than 3000 users register Xiaonei account in two month. These users need time to develop their friendships. Also, the CTO of Xiaonei Inc. reported that some users just register to attain a certain objective and then are never active to make friends. A large number of passive members [2] results in exponential decay in the first part of distribution curve. We amplified the degree distribution in Figure 2 and focused on the part between 45 and 75 degree shown in Figure 4. We can find the successive degree distribution curves move upwards. The probability that a randomly selected node has more than k links increases at the same "degree" value. There is a tendency to close to Power-law, but it evolves very slowly because a large number of new passive members register for Xiaonei HUST network.

3.3 Small World Phenomena

The small world experiment was comprised of several experiments conducted by Stanley Milgram examining the average path length for social networks of people in the United States. The research was groundbreaking in that it revealed that human society is a small world type network characterized by shorter-than-expected path

Table 1. Topological Measurements of Online Social Network

MEASURE	TYPE	Data1	Data2
Average Distance[16]	Graph	4.15207	4.16321
Clustering Coefficient, Watts-Strogatz[16]	Graph	0.198627	0.187391
Component Count, Strong[10]	Graph	*NA*	*NA*
Component Count, Weak[10]	Graph	173	764
Connectedness, Krackhardt[17]	Graph	0.991732	1.03602
Density[10]	Graph	0.000412741	0.000426163
Diameter[10]	Graph	44419	47546
Edge Count Ratio, Lateral[18]	Graph	0.563342	0.556119
Efficiency, Global[19]	Graph	0.253951	0.249533
Efficiency, Local[19]	Graph	0.303109	0.304126
Fragmentation[20]	Graph	0.0082677	0.0360216
Span of Control[21]	Graph	*NA*	*NA*
Speed, Average[18]	Graph	0.240843	0.2402
Upper Boundedness, Krackhardt[17]	Graph	0.999871	0.997303
Network Centralization/Total Degree[22]	Graph	0.0102928	0.00994435
Hierarchy, Krackhardt[17]	Graph	0.0082217	0.0080065
Centrality, Closeness[22]	Graph	8.88937e-05	8.81563e-05

lengths. The experiments are often associated with the term six degrees of separation [14], which was a small world phenomenon. We check the average shortest path length between entities. As shown in Table 1, Average Distance on two data sets is 4.15207 and 4.16321 respectively. Values less than six represent small world phenomena. Therefore, it can be proven that there exist small world phenomena in the Xiaonei HUST network.

3.4 Shortest Path Finder

Given a pair of two selected entities, we computed the shortest path between two entities according to Newman's algorithm[15]. We used Java to implement Newman's algorithm and compared it with ORA shortest path finder. The average consuming time of Newman's algorithm is 25 second; however ORA just take 20 second. For example, we computed the shortest path from agent 0-100013893 to agent 17868-223752718 and determined the number of shortest paths. There were 12 shortest paths and the shortest path length was 5, as shown in Figure 5. The Sphere of Influence of each entity represents a series of relationships with friends who have a direct relationship with this entity. It can also be computed using ORA. As shown in Figure 6 and 7, where the degree of entity 0-100013893 is 3 and the degree of entity 17868-223752718 is 12, the sphere of influence of entity 0-10001389 is obviously larger than that of entity 17868-223752718. The computation of the shortest path and

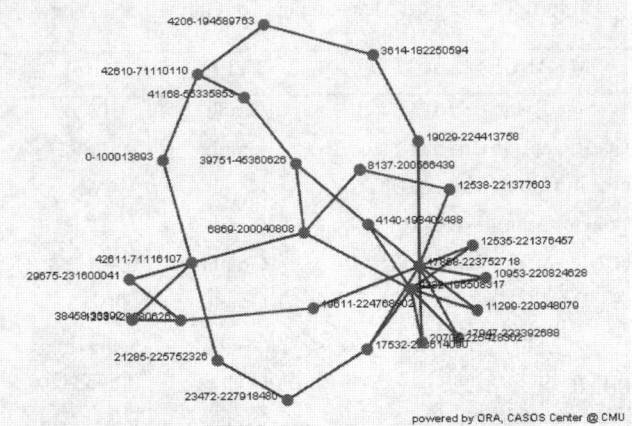

Fig. 5. Shortest paths from agent 0-100013893 to entity 17868-223752718

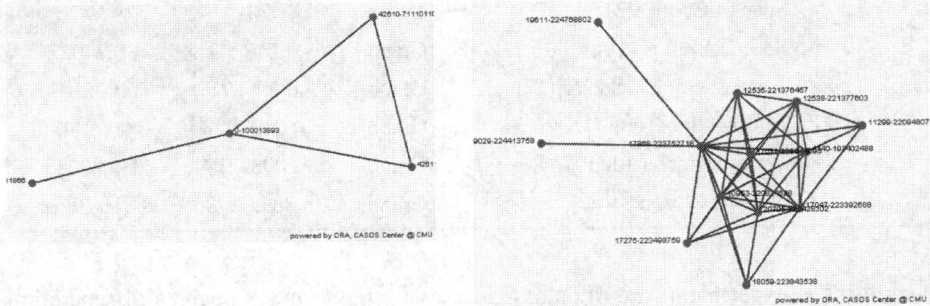

Fig. 6. Sphere of influence for 0-100013893 and 17868-223752718

the sphere of influence can provide a latent application for online social network, which will be discussed in Section 4.

3.5 Connected Component Analysis

The nodes in a disconnected graph may be partitioned into two or more subsets in which there are no paths between the nodes in different subsets. The connected sub-graphs in a graph are called components [10]. A component in an online social network is a sub-network in which there are links between all pairs of users in the sub-network, and no link between a user in the network and any user not in the network. These disconnected components represent the informal groups in a network organization. Although online network services provide interest groups for users, it cannot be ignored that informal groups also take an important role in the construction of online social networks, which will be discussed more in Section 4. There are two types of connected components: strongly connected components and weakly connected components [9]. Because our Xiaonei HUST network is undirected, we chose the weakly connected component method to compute the number of connected components. As shown in Table 1, through two months, the number of components increased from 173 to 764. More informal groups were born with the interaction of users in online social networks.

3.6 Measurements on SNA

For the purpose of analyzing online social networks, we had to filter some measures from more than 100 measurements in ORA. We paid more attention to the entire topological structure of online social networks, thus we filtered on the base of the Graph type of the measurements. In Table 1, we choose 17 measurements. The relative references are also listed in the Table 1. We briefly analyzed the results according to these measurements as follows.

1) Density: The number of edges divided by the number of possible edges including self-reference.

2) Diameter: The maximum shortest path length between any two nodes in a network.

3) Connectedness: Measures the degree to which a square network's underlying (undirected) network is connected.

4) Edge Count Ratio Lateral: Fixing a root entity x, a lateral edge (i,j) is one in which the distance from x to i is the same as the distance from x to j.

5) Efficiency, Global: Measures the closeness of the entities in the network.

6) Efficiency, Local: Measures the closeness of the entities in each ego network in the network.

7) *Fragmentation:* The proportion of entities in a network that are disconnected. It is related to Component Count. In two month, the number of fragmentations in Xiaonei HUST network increases.

8) Span of Control: The average number of out edges per node with non-zero out degrees. It is only used in the directed network. Therefore, for Xiaonei HUST undirected network, we can not obtain this result.

9) Speed, Average: The average inverse geodesic distance between all entity pairs. The highest score is achieved for a clique, and the lowest for all isolates.

10) Upper Boundedness: The degree to which pairs of agents have a common ancestor.

11) Network Centralization, Total Degree: A centralization of a square network based on total degree centrality of each entity.

12) Hierarchy: The degree to which a square network N exhibits a pure hierarchical structure.

13) Clustering Coefficient, Watts-Strogatz: Measures the degree of clustering in a network by averaging the clustering coefficient of each entity. The clustering coefficient of a entity is the density of its ego network which is the sub graph induced by its immediate neighbors.

14) *Component Count, Strong/Weak:* The number of strongly/ weekly connected components in a network.

15) Centrality, Closeness: The average closeness of an entity to the other entities in a network. Loosely, Closeness is the inverse of the average distance in the network between the entity and all other entities. This is defined for directed networks.

4 Discussions

Facebook has opened its API so that anyone can develop his/her own application, and then upload the application for using it online. Unfortunately, Xiaonei has not opened its API yet. Therefore, the latent applications demonstrated here don't specify the application development of Xiaonei network. Furthermore, offline social networks also benefit from online social network services.

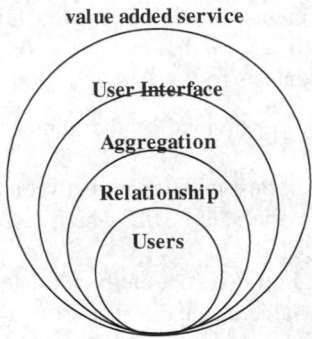

Fig. 7. Applications of Online Social Network

In Figure 7, we classify applications of online social networks five categories. The basic application allows users to edit their own information freely. Second, for the purpose of creating friendships, we need to provide user-friendly application to display and maintain friend lists. Third, aggregation lets online social network services push all the information to users and lets users efficiently use that information. Fourth, the user interface on the website is very important for attracting new users and keeping old users. Finally, a value added service such as mobile SNA can allow users to enjoy a full virtual life. From the study on which paper is based, two latent applications emerge for online social networks at the level of relationships and aggregation.

Shortest path finder. The ultimate objective of using online social networks is for people to extend their circle of friends. Since users even want to meet new people online, it is very useful to find the shortest path to do so. Even though, users can find this certain people according to his/her name, work place or other information and send an invitation to him/her. Generally, users do not like to add strangers as new friends[4]. Rather they prefer to add users whom they have known offline. Also, some people would like to become initially acquainted with each other online but actually interact in the real world. SNA is just a tool to make offline friends by first meeting them online. Therefore, shortest path finder can help users find efficient ways to make offline friends by search shortest path online.

Informal Online Group Mining. The objective of this application is to help users to jump out of his/her own social group. By judging if the ego-network is disconnected components or not or by finding shortest path for jumping out of this disconnected component, people can expend their social circles. Also, by mining informal groups that are disconnected components, online social network services can provide a lot of references to help users take part in certain informal groups.

5 Conclusions

In this paper, we used a topological analysis tool to analyze an online Chinese student social network—Xiaonei.com. We used Python to crawl two datasets of Xiaonei from January to February 2008. By these two datasets, degree distribution and small world phenomena were analyzed using Matlab and SPSS. It was found that degree distribution has a tendency to close to Power-law, however it evolves very slowly because a large number of new passive members routinely register for the Xiaonei HUST network. Xiaonei HUST network also has small world attribute. Then we used a social network tool, ORA, to analyze our datasets from the viewpoint of social network analysis. We summarized 17 measurements such as Fragmentation, Component Count, Strong/Weak. Also, 2 latent applications of online social network services were proposed in the discussion section. This paper also shows that ORA[1] have the advantages in finding shortest path and analyzing connected component.

This is only a preliminary report for our work-in-progress research. In the future, we will contact Xiaonei Inc. to obtain their cooperation in order to analyze the entire Xiaonei network dataset. Further study for applying classical social network analysis to the analysis of online social networks will be carried out.

Acknowledgments. The authors gratefully acknowledge China Scholarship Council to provide scholarship for my visiting to Carnegie Mellon University and CASOS-the center for Computational Analysis of Social and Organizational Systems. This work was funded by the China National Natural Science Fund (No. 70671048).

References

1. Ralph Gross, A.A.: Information Revelation and Privacy in Online Social Networks (The Facebook case). In: ACM Workshop on Privacy in the Electronic Society (WPES) (2005)
2. Ravi Kumar, J.N., Tomkins, A.: Structure and Evolution of Online Social Networks. In: Proceedings of 12th International Conference on Knowledge Discovery in Data Mining, ACM Press, New York (2006)
3. Alan Mislove, M.M., Gummadi, K.P., Druschel, P., Bhattacharjee, B.: Measurement and analysis of online social networks. In: Proceedings of the 7th ACM SIGCOMM conference on Internet measurement. 2007, San Diego, California, USA. ACM, New York (2007)
4. Lampe, C., Ellison, N., Steinfeld, C.: A familiar Face (book): Profile elements as signals in an online social network. In: Proceedings of Conference on Human Factors in Computing Systems. ACM Press, New York (2007)
5. Fu, F., Long Wang, L.L.: Empirical analysis of online social networks in the age of Web 2.0. Physica A 387, 675–684 (2008)

---

[1] Please visit http://www.casos.cs.cmu.edu/projects/ora/

6. Scott, A., Golder, D.W., Huberman, B.A.: Rhythms of social interaction: messaging within a massive online network. In: 3rd International Conference on Communities and Technologies (CT 2007), East Lansing, MI, June 28-30 (2007)
7. Jeffrey Heer, D.B.: Vizster: Visualizing Online Social Networks. In: Proceedings of the 2005 IEEE Symposium on Information Visualization (INFOVIS 2005) (2005)
8. Acquisti, A., Gross, R.: Imagined communities: Awareness, information sharing, and privacy on the Facebook. In: Proceedings of 6th Workshop on Privacy Enhancing Technologies. Robinson College, Cambridge (2006)
9. Carley Kathleen, C.D., Matt, D., Jeffrey, R., Moon, I.-C.: ORA User's Guide, 2007, Carnegie Mellon University, School of Computer Science, Institute for Software Research, p. 388 (2007)
10. Wasserman Stanley, K.F.: Social Network Analysis: Methods and Applications. Cambridge University Press, Cambridge (1994)
11. Sege, I.: Where everybody knows your name (April 27, 2005) Boston.com
12. YongYeol Ahn, S.H., Kwak, H.: Analysis of Topological Characteristics of Huge Online Social Networking Services. In: International World Wide Web Conference, Banff, Alberta, Canada (2007)
13. Thomas Lento, H.T.W., Gu, L.: The Ties that Blog: Examining the Relationship Between Social Ties and Continued Participation in the Wallop Weblogging System (2007)
14. Newman, M.: The Structure and Function of Complex Networks. SIAM Review 56, 167–256 (2003)
15. Newman, M.E.J.: Scientific collaboration networks. II. Shortest paths, weighted networks, and centrality. Physical Review E 64 (016132) (2001)
16. Watts, D.J., Strogatz, S.: Collective dynamics of 'small-world' networks. Nature 393(6684), 440–442 (1998)
17. David, K.: Graph Theoretical Dimensions of Informal Organizations. In: Carley, M.J.P.H., Kathleen, M. (eds.) Computational Organization Theory. Lawrence Erlbaum Associates, NJ (1994)
18. Carley, K.: Summary of Key Network Measures for Characterizing Organizational Architectures, Carnegie Mellon University (2002)
19. Latora, V., Marchiori, M.: Efficient Behavior of Small-World Networks. Phys. Rev. Lett. 87(19) (2001)
20. Borgatti, S.P.: The Key Player Problem. Dynamic Social Network Modeling and Analysis. In: Committee on Human Factors, National Research Council, Workshop Summary and Papers (2003)
21. David, K., Tie, S.: Super Strong and Sticky. Power and Influence in Organizations. In: Roderick Kramer, M.N. (ed.), pp. 21–38. Sage, Thousand Oaks (1998)
22. Freeman, L.C.: Centrality in Social Networks I: Conceptual Clarification. Social Networks 1, 215–239 (1979)

Time, Incompleteness and Singularity in Quantum Cosmology

Philip V. Fellman[1], Jonathan Vos Post[2], Christine Carmichael[3], Alexandru Manus[1],
and Dawna Lee Attig[1]

[1] Southern New Hampshire University, Manchester, NH, USA
Shirogitsune99@yahoo.com, alexandru.manus@snhu.edu,
adawnalee@msn.com
[2] Computer Futures, Inc., Altadena, CA, USA
jvospost3@gmail.com
[3] Woodbury University, Burbank, CA USA
cmcarmichael@yahoo.com

Abstract. In this paper we extend our 2007 paper, "Comparative Quantum Cosmology: Causality, Singularity, and Boundary Conditions", http://arxiv.org/ftp/arxiv/papers/0710/0710.5046.pdf, to include consideration of universal expansion, various implications of extendibility and incompleteness in spacetime metrics and, absent the treatment of Feynman diagrams, the use of Penning trap dynamics to describe the Hamiltonians of space-times with no characteristic upper or lower bound.

Keywords: Quantum Cosmology, Incomplete Geodesic, Cosmic Inflation, Time, Singularity.

1 Introduction

Cosmology, and more particularly, quantum cosmology, has generated a number of different, often radically different viewpoints about both the beginning and the ending of the universe as well as a variety of possibilities with respect to cyclical theories of expansion and contraction. Despite the manifold differences between these theories, many of them, in particular, the theories expounded by Hawking (1999), Carroll and Chen (2004, 2005) and Peter Lynds (2003, 2006) often have more elements in common than those among which they differ. In particular, the commonalities tend to cluster around the concepts of "weak singularity", "no boundary condition", and "the problem of specialness" with respect to scale, entropy and initial conditions.

In the present paper we examine these concepts with particular attention to the nature and role of singularities relying in large part on two particular sources for our analysis. The first is the 1994 work of C.J.S. Clark on the analysis of space-time singularities, a work of general relativity largely exclusive of quantum mechanical effects and the second is a very recent paper by David J. Fernandez C. and Mercedes

J. Zhou (Ed.): Complex 2009, Part I, LNICST 4, pp. 749–762, 2009.
© ICST Institute for Computer Sciences, Social Informatics and Telecommunications Engineering 2009

Velazquez (2008) whose mathematics we employ to explore at a quantum mechanical level the turning points of cosmic expansion (i.e. maximal spaces with no further possible extensions) and the "big bang/big crunch" particularly in the context of the limitations imposed on the order of events suggested by Lynds (2006).

1.1 The Mechanics of Contraction – The Standard Approach

"Big Crunch" theories all depend on some sort of universal contraction although not nearly enough of these theories offer a plausible mechanism for why the universe should suddenly contract at a particular point. That is, "by what mechanism should the universe cease expanding, at which point gravity initiates an apparently unstoppable contraction?" The problem is sufficiently deep that for many years most physicists assumed a "steady state" model of the universe. In his treatment of incomplete geodesics and maximal spaces, C.J.S. Clarke offers at least one possibility in his treatment of the maximality assumption (p. 8):

> The foregoing result has shown that if M is inextendible then there is some timelike curve (actually a geodesic) - i.e. a possible worldline of a particle - which could continue in some extension of M but which in M itself simply stops. This seems unreasonable: why should M be cut short this way? It seems natural to demand that "if a space-time can continue then it will"; in other words, to demand that any reasonable space-time should not be inextendible. This is an assumption imposed upon space-time in addition to the field equations of Einstein.
> It can easily be shown that any space-time can in fact be extended until no further extension is possible. At this point the space-time is called maximal, and so we are lead to the idea that we need only consider maximal space-times. But this idea is really not as innocuous as it might seem because of the problem that the extension of a space-time, when it exists, cannot usually be determined uniquely. In special cases there are unique extensions: an analytic space-time has (subject to some conditions) a unique maximal analytic extension; similarly, a global hyperbolic solution of the field equations (with a specified level of differentiability) is contained in a unique maximal solution. In both these cases a "principle of sufficient reason" demands that the maximal solution be taken. But suppose one has a non-analytic space-time where Einstein's field equations fail to predict a unique extension (either because there is a Cauchy horizon or because there is some sort of failure of the differentiability needed for the existence of unique solutions). Or suppose a situation arises in which there is a set of incomplete curves, each one of which can be extended in some extension of the space-time, but where there is no extension in which they can all be extended. (There exists admittedly artificial examples of this (Misner, 1967). In cases such as these, the same principle of sufficient reason would not allow one extension to

exist at the expense of another. Perhaps the space-time, like Buridan's ass between two bales of hay, unable to decide which way to go, brings the whole of history to a halt.

2 The Quantum Mechanical Approach to Maximal States

In at least one sense we must "jump the gun" a bit in order to discuss the quantum mechanical argument for the maximal extension of space-time never being achieved because this result is merely a byproduct of a quantum mechanical argument against the existence of strong singularity. The gist of Fernandez and Velazques proof is that (8.0) "in identifying the appropriate displacement operator as well as an adequate 'extremal' state for the Penning trap cavity...the corresponding Hamiltonian has neither a ground nor a top energy eigenvalue. The proof of this claim, which we will return to in our discussion of singularities is as follows and is drawn directly from the previously reference paper by Fernandez and Velazquez. In order to understand the proof, we must begin with their demonstration of the extremal state wave function:

This existence of the extremal state $|0,0,0\rangle$ is guaranteed by a theorem which is proven elsewhere [5]. It ensures that, if the operators

$$B_j = i\vec{P} \cdot \vec{\alpha} + \vec{R} \cdot \beta_j, \quad B_j^\dagger = -i\vec{\alpha}_j^\dagger \cdot \vec{P} + \vec{\beta}_j^\dagger \cdot \vec{R}, \quad j = 1,2,3, \tag{1}$$

Obey the commutation relations (17), then the system of partial differential equations $\langle\vec{r}|B_j|0,0,0\rangle$ $0, j = 1,2,3$, for the extremal state wave function $\phi_0(\vec{r}) \equiv \langle\vec{r}|0,0,0\rangle$ has a square integrable solution given by

$$\phi_0(\vec{r}) = c\exp\left(-\frac{1}{2}a_{ij}x_ix_j\right) = c\exp\left(-\frac{1}{2}\vec{r}^{\mathrm{T}}\mathbf{a}\vec{r}\right), \tag{2}$$

Where $\mathbf{a} = (a_{ij})$ is a complex symmetric matrix satisfying

$$\mathbf{a}\vec{\alpha}_j = \vec{\beta}_j, \quad j = 1,2,3, \tag{3}$$

According to (22), through equations (12.16) we identify the vectors

$$\vec{\alpha}_1 = \frac{1}{2(b^2+v)^{1/4}}(1,-i,\ 0)^{\mathrm{T}}, \qquad \beta_1 = (b^2+v)^{1/2}\vec{\alpha}_1,$$

$$\vec{\alpha}_2 = -\frac{1}{2(b^2+v)^{1/4}}(1,i,\ 0)^{\mathrm{T}}, \qquad \beta_2 = (b^2+v)^{1/2}\vec{\alpha}_2, \tag{4}$$

$$\vec{\alpha}_3 = -\frac{1}{\sqrt{2}(-2v)^{1/4}}(0,0,\ 1)^{\mathrm{T}}, \qquad \beta_3 = (-2v)^{1/2}\vec{\alpha}_3,$$

Thus, $\mathbf{a} = \mathrm{diag}\left[\sqrt{b^2+v}, \sqrt{b^2+v}, \sqrt{-2v},\right]$, and from (23) we finally get the extremal state wave function we were looking for:

$$\phi_0(\vec{r}) = c\exp\left(-\frac{\sqrt{b^2+v}}{2}(x^2+y^2) - \sqrt{\frac{-v}{2}}z^2\right). \tag{5}$$

What is particularly interesting about this extremal state is that the Hamiltonian has neither a ground nor a top energy eigenvalue (p. 9) This is given a further exposition in the section on mean values of physical quantities. In particular, this calculation again demonstrates that the values of the extremal state as well as the CS wave function coincide with the previous calculation. Mathematically this is shown by the following exposition (pp. 7-8):

Let us evaluate next the mean values $\langle X_j \rangle_z \equiv \langle z|X_j|z \rangle, \langle P_j \rangle_z \equiv \langle z|P_j|z \rangle, j = 1,2,3$, and the corresponding mean square deviations in a given $CS|z \rangle$. To do that, we analyze first how the operators X_j, X_j^2, P_j, P_j^2 are transformed under $D(z)$. By using equation (35) it is straightforward to show that:

$$D^\dagger(z)X_j^\eta D(z) = (X_j + \Gamma_j)^\eta. \quad D^\dagger(z)P_j^\eta D(z) = (P_j + \Sigma_j)^\eta, \eta = 1,2 \dots \quad (6)$$

Therefore:

$$\langle X_j \rangle_z = \langle X_j \rangle_0 + \Gamma_j, \langle X_j^2 \rangle_z = \langle X_j^2 \rangle_0 + 2\Gamma_j\langle X_j \rangle_0 + \Gamma_j^2, (\Delta X_j)_z^2 = (\Delta X_j)_0^2 \quad (7)$$

$$\langle P_j \rangle_z = \langle P_j \rangle_0 + \Sigma_j, \langle P_j^2 \rangle_z = \langle P_j^2 \rangle_0 + 2\Sigma_j\langle P_j \rangle_0 + \Sigma_j^2, (\Delta P_j)_z^2 = (\Delta P)_0^2 \quad (8)$$

Notice that the mean square deviations of X_j and P_j are independent of z_1, z_2, z_3 but depend on $\langle X_j \rangle_0, \langle P_j \rangle_0, \langle X_j^2 \rangle_0, \langle P_j^2 \rangle_0, j = 1,2,3$, which need to be evaluated. The first six quantities are obtained from the homogeneous equations $\langle B_k \rangle_0 = i(\vec{\alpha}_k)_j\langle P_j \rangle_0 + (\vec{\beta}_k)_j\langle X_j \rangle_0 = 0, \langle B_k^\dagger \rangle_0 = -i(\vec{\alpha}_k^*)_j\langle P_j \rangle_0 + (\vec{\beta}_k^*)_j\langle X_j \rangle_0 = 0, k = 1,2,3$ (see (22)) and use that $B_k|0,0,0 \rangle = \langle 0,0,0|B_k^\dagger = 0$. By using (25), the system to be solved becomes:

$$-i\sqrt{-2v} \langle Z \rangle_0 + \langle P_z \rangle_0 = 0$$

$$\sqrt{b^2 + v}(\langle X \rangle_0 - iY_0) + i(\langle P_x \rangle_0 - i\langle P_y \rangle_0) = 0,$$

$$-\sqrt{b^2 + v}(\langle X \rangle_0 + iY_0) - i(\langle P_x \rangle_0 + i\langle P_y \rangle_0) = 0,$$

and the complex conjugate equations. Its solution is given by

$$\langle X_j \rangle_0 = \langle P_j \rangle_0 = 0, \qquad j = 1,2,3. \quad (9)$$

In order to obtain $\langle X_j^2 \rangle_0, \langle P_j^2 \rangle_0$, we calculate the mean values for the several products of pairs involving $B_j, B_k^\dagger$. From these thirty six equations just twenty one are linearly independent: $\langle B_j B_k \rangle_0 = 0, j = 1,2,3, k \leq j$ (six equations); $\langle B_j^\dagger B_k^\dagger \rangle_0 = 0$, $j = 1,2,3, k \leq j$ (six equations); $\langle B_k^\dagger B_j \rangle_0 = 0, j,k = 1,2,3$, (nine equations). By solving this linear system, the non-null results for the mean values of the twenty one independent products of X_i and P_j are now:

$$\langle X^2 \rangle_0 = \langle Y^2 \rangle_0 = [4(b^2 + v)]^{-\frac{1}{2}}, \qquad \langle Z^2 \rangle_0 = (-8v)^{-\frac{1}{2}},$$

$$\langle P_x^2 \rangle_0 = \langle P_y^2 \rangle_0 = [(b^2 + v)/4]^{\frac{1}{2}}, \qquad \langle P_z^2 \rangle_0 = (-v/2)^{\frac{1}{2}},$$

$$\langle XP_x \rangle_0 = \langle YP_y \rangle_0 = \langle ZP_z \rangle_0 = i/2.$$

The previous formulas imply that equations (39, 40) become

$$(\Delta X)_z^2 = (\Delta Y)_z^2 = [4(b^2 + v)]^{-\frac{1}{2}}, \quad (\Delta Z)_z^2 = (-8v)^{-\frac{1}{2}},$$

$$(\Delta P_x)_z^2 = (\Delta P_y)_z^2 = [(b^2 + v)/4]^{\frac{1}{2}}, \quad (\Delta P_z)_z^2 = (-v/2)^{\frac{1}{2}},$$

and therefore

$$(\Delta X)_z(\Delta P_x)_z = (\Delta Y)_z(\Delta P_y)_z = (\Delta Z)_z(\Delta P_z)_z = 1/2.$$

This means that our CS have minimum Heisenberg uncertainty relations.

Finally, by using equations (15, 21) we calculate the mean value of the Hamiltonian H in a given CS $|z\rangle$:

$$\langle H \rangle_z = \omega_1 |z_1|^2 - \omega_2 |z_2|^2 + \omega_3 |z_3|^2 + E_{0,0,0}. \tag{10}$$

A similar calculation for $(H^2)_z$ can be done, leading to:

$$(\Delta H)_z^2 = \left(b + \sqrt{b^2 + v}\right)^2 |z_1|^2 + \left(b - \sqrt{b^2 + v}\right)^2 |z_2|^2 - 2v|z_3|^2. \tag{11}$$

Once again, the fact that H is not positive definite is clearly reflected in (42).

Along this work we have assumed that $b = -\dfrac{eB}{2c} > 0$. For $b < 0$, small differences concerning the identification of the appropriate annihilation and creation operators arise. However, the extremal state and CS wave functions $\phi_0(\vec{r}), \phi_z(\vec{r})$ as well as the corresponding mean values, will coincide with those previously calculated. In particular, the Heisenberg uncertainty relation will achieve once again its minimum value [14].

3 Singularity – Early Arguments

One of the simplest arguments against strong singularity is simply that of a choice of improper metric. In this case, an inappropriate choice of the Schwarzschild metric as a representation of strong or true singularity. As Clarke explains:

> "In 1924 Eddington showed there was an isometry between the space-time M defined by the region r>2m in the Schwarzschild metric and part of a larger space-time M'. Incomplete curves in M on which r -> 2m were mapped by this isometry into curves which were extensible in M': The singularity at 2m was no longer present. So if we identify the Schwarzschild space-time with the part of the Eddington space-time M' with which it is isometric, we see that it is not just incomplete in the formal sense defined above: it actually had a piece missing from it, a piece that is restored at M'. The singularity at r=2m is thus a mathematical artifact, a consequence of the fact that the procedure used to solve the field equations had fortuitously produced only a part of the complete space.
>
> We note that, despite this, there are still some authors who regard the Schwarzschild 'singularity' at r=2m as genuine; but this is only

justified if (as done by Rosen, 1974) one uses a non-standard physical theory in which there is some additional structure (such as a background metric) which itself becomes singular under the under the isometry of the metric into M', so that one structure, the metric, or the background is always singular at r = 2m.

The situation in Schwarzschild clearly contrasts with that of the Friedmann metrics. For these, on any of the incomplete curves the Ricci scalar tends to infinity. For the smooth space-times that we are considering at the moment this is impossible on a curve which has an endpoint in space-time, and so there can in this case, be no isometric M' in which these curves have an endpoint."

Clarke's subsequent arguments rest on the properties of extensibility and global hyperbolicity[1] and ends up with various cosmic censorship models (both strong and weak forms) which either prevent the existence of strong singularities or which make them inaccessible to any particular observer in an inertial frame with bounded acceleration. Much of the distinction between these theories is a function of globally hyperbolic space-time and past-simplicity/past hyperbolicity such that for all timelike geodesics k in $U_0(s)$, $\mathscr{I}(k) < 2$; also geodesics in $U_0(s)$ have no conjugate points (p. 127) From here he proceeds to eliminate what her refers to "primal singularities" and "dragging geodesics". In this context, cosmic censorship provides a number of mechanisms for explaining why strong singularities are not accessible. However, Peter Lynds offers a rather different approach, based on the second law of thermodynamics suggesting that the very nature of the big bang-big crunch makes any strong singularity inaccessible, even if it exists within the light cone and that this inaccessible history also explains the problems of scale and naturalness raised by Carroll and Chen (2004, 2005).

4 Time

In 2003, Peter Lynds published a controversial paper, "Time and Classical and Quantum Mechanics: Indeterminacy vs. Discontinuity" in Foundations of Physics Letters. Lynds' theory does away with the notion of "instants" of time , relegates the "flow of time" to the psychological domain. While we have commented elsewhere on the implications of Lynds' theory for mathematical modeling it might save a bit of time simply to borrow Wikipedia's summary of this paper:

> Lynds' work involves the subject of time. The main conclusion
> of his paper is that there is a necessary trade off of all precise

---

[1] For every pair of points, $p,q \in U$, $I^-(p) \cap I^+(q)$ is compact. Here $I^\pm$ is the future(past of a set S in space-time. "Causality" holds on U (no closed timelike curves exist). Classically, a more restrictive and technical assumption is required, namely, "strong causality" – that no "almost closed" timelike curves exist, but the recent work of Hawking and Penrose (1996) shows that causality suffices. Global hyperbolicity implies that there is a family of Cauchy surfaces for U. Essentially, it means that everything that happens on U is determined by the equations of motion, together with initial data specified on a surface. (Wikipedia).

physical magnitudes at a time, for their continuity over time. More specifically, that there is not an instant in time underlying an object's motion, and as its position is constantly changing over time, and as such, never determined, it also does not have a determined relative position. Lynds posits that this is also the correct resolution of Zeno's paradoxes, with the paradoxes arising because people have wrongly assumed that an object in motion has a determined relative position at any given instant in time, thus rendering the body's motion static and frozen at that instant and enabling the impossible situation of the paradoxes to be derived. A further implication of this conclusion is that if there is no such thing as determined relative position, velocity, acceleration, momentum, mass, energy and all other physical magnitudes, cannot be precisely determined at any time either. Other implications of Lynds' work are that time does not flow, that in relation to indeterminacy in precise physical magnitude, the micro and macroscopic are inextricably linked and both a part of the same parcel, rather than just a case of the former underlying and contributing to the latter, that Chronons, proposed atoms of time, cannot exist, that it does not appear necessary for time to emerge or congeal from the big bang, and that Stephen Hawking's theory of Imaginary time would appear to be meaningless, as it is the relative order of events that is relevant, not the direction of time itself, because time does not go in any direction. Consequently, it is meaningless for the order of a sequence of events to be imaginary, or at right angles, relative to another order of events.

One can see from the above summary that this radical reformulation of the concept of time is bound to have significant cosmological implications. We discussed some of these implications in a brief paper in 2004, "Time and Classical and Quantum Mechanics and the Arrow of Time".[2] We began with a discussion of John Gribbin's analysis, "Quantum Time Waits for No Cosmos", and Gribbin's argument against the mechanics of time reversibility as an explanation for the origin of the universe, where he cites Raymond LaFlamme:[3]

> The intriguing notion that time might run backwards when the Universe collapses has run into difficulties. Raymond LaFlamme, of the Los Alamos National Laboratory in New Mexico, has carried out a new calculation which suggests that the Universe cannot start out uniform, go through a cycle of expansion and collapse, and end up in a uniform state. It could start out disordered, expand, and then collapse back into disorder. But, since the COBE data show that our Universe was born in a

---

[2] "Time and Classical and Quantum Mechanics and the Arrow of Time", paper presented at the annual meeting of the North American Association for Computation in the Social and Organizational Sciences, Carnegie Mellon University, June, 2004.
[3] http://www.lifesci.sussex.ac.uk/home/John_Gribbin/timetrav.htm\

smooth and uniform state, this symmetric possibility cannot be
applied to the real Universe.

The concept of time reversibility, while apparently quite straightforward in many
cases, seems never to be without considerable difficulty in cosmology, and indeed,
explaining the mechanics of time reversibility and its relationship to Einstein's
cosmological constant is one of the major enterprises of quantum cosmology (Sorkin,
2007). The terns of the debate expressed by Gribbin above, have been extended by
Wald (2005) who acknowledges Carroll and Chen's work, discussed earlier in this
paper, but who argues that at some point an anthropic principle introduces a
circularity into the causal logic. Sorkin covers much of the distance needed for a
Carroll-Chen type counterargument in his 2007 paper, "Is the cosmological
"constant" a nonlocal quantum residue of discreteness of the causal set type?",
however a complete discussion of Sorkin's model is beyond the scope of our present
exploration.[4]

4.1 The Arrow of Time

One set of problems with the thermodynamic arrow of time for the very early
universe (and this is partially addressed by Sorkin) is that in addition to the problem
of possible inhomogeneities in the early universe there is still a lack of consensus or
unequivocal evidence on the invariance of fundamental physical constants during the
early history of the universe. Various authors have recently suggested that the speed
of light may have been greater during the earliest period of the universe's formation
(Murphy, Webb and Flambaum, 2002).

In our 2004 review of Lynds work on how we might better understand the
thermodynamic arrow of time we also raised two moderately troubling complexity
issues which at present remain largely unanswered. The first is the problem of
heteroskedastic time behavior in the early universe. This question is not unconnected
to the question of changes in the values of fundamental physical constants in the early
universe. In most models of early universe formation a smooth or linear flow of time is

---

[4] In his conclusion, Sorkin argues "Heuristic reasoning rooted in the basic hypotheses of causal
set theory predicted $\Lambda \sim \pm 1/\sqrt{V}$, in agreement with current data. But a fuller understanding of
this prediction awaits the "new QCD" ("quantum causet dynamics"). Meanwhile, a reasonably
coherent phenomenological model exists, based on simple general arguments. It is broadly
consistent with observations but a fuller comparison is needed. It solves the "why now"
problem: Λ is "ever-present". It predicts further that $p_\Lambda \neq - p_\Lambda$ ($w \neq -1$) and that Λ has
probably changed its sign many times in the past. The model contains a single free parameter
of order unity that must be neither too big nor too small. In principle the value of this
parameter is calculable, but for now it can only be set by hand. In this connection, it's
intriguing that there exists an analog condensed matter system the "fluid membrane", whose
analogous parameter is not only calculable in principle from known physics, but might also be
measurable in the laboratory! That our model so far presupposes spatial homogeneity and
isotropy is no doubt its weakest feature. Indeed, the ansatz on which it is based strongly
suggests a generalization such that Λ -fluctuations in "causally disconnected" regions would
be independent of each other; and in such a generalization, spatial inhomogeneities would
inevitably arise.

assumed. However, it is possible to imagine inflationary models where the expansion of time dimension or the time-like dimensions of a higher order manifold inflate in a heteroskedastic fashion. To the extent that the thermodynamic arrow of time is invoked as an element of cosmological explanation, It would need to be able to explain the dynamical evolution of the universe, not just as we know it today, but at those particularly difficult to characterize beginning and end points of the system. The difficulty with heteroskedastic time distributions is that they may or may not allow recovery of the standard Boltzmann expression. At a deeper level, it is likely that in characterizing the development of the early universe, one may have to incorporate a significant number of non-commuting quantum operators. Further, in this context, our knowledge of the early universe is both substantively incomplete, because we lack any system of measurement for the first three hundred thousand years of time evolution of the system (i.e. the period prior to the decoupling of baryons and photons) and very likely theoretically incomplete as well. We can compare the problem to one of discrete time series evolutions with low dimensionality and discrete combinatorics. For example, the random order of a shuffled deck of cards can eventually be repeated because the dimensionality of the system is low. As dynamical systems take on higher orders of dimensionality their asymptotes become ill-defined (in at least one sense, this is the objection raised by Gribbin and LaFlamme).[5]

Another problem is what Freeman Dyson characterizes as the struggle between order and entropy in the big crunch. As the universe approaches infinity and the average density approaches zero, temperature does not approach zero, and thus the nature of the struggle between order and entropy may actually be characterized by very different time evolutions that those with which we are familiar. In addition, there is the "Maxwell's Demon" family of arguments. This is a systems dynamic which is particularly relevant to complexity science. The problem here is that there may be emergent phenomena at the end of the life of the universe which causes the system's time evolution to then behave in unexpected ways. In some sense this is logical trap lurking behind statistical reasoning. Under normal conditions, the descriptive and inferential statistical conjecture that the near future will look like the recent past (or more boldly that fundamental physical constants are perfect invariants) is entirely reasonable. However, in the face of emergent phenomena, this assumption may not hold. Indeed, this problem is at the center of much of the debate over "relic" radiation and arguments over the age of the universe.

Yet another problem, which may also encompass emergent behavior, has to do with symmetry breaking. The universe has undergone several phase transitions by symmetry breaking. As a result, additional forces have emerged at each of these transitions. First gravity separated out of other forces, and it is for that reason that we can expect gravity wave detectors to probe more deeply into the early history of the universe than any other technology. To return to the emergent properties argument, we cannot definitively rule out (by means of present theory and observations) the possibility that at some future time (presumably near the end of the system's time

---

[5] A substantial amount of work on non-extensive statistical mechanics has been done by Tsallis, et al. most recently (2007) "Nonergodicity and Central Limit Behavior for Long-range Hamiltonians" http://arxiv.org/PS_cache/arxiv/pdf/0706/0706.4021v3.pdf

evolution) that some fifth force will separate out from the known four forces.[6] At the classical level, time reversibility and a thermodynamic arrow of time is no longer problematical, but at the quantum level, and at the cosmological level, the concept remains murky at best.

5 Lynds' Conjecture

Peter Lynds has developed an alternative cosmology, or an alternative foundation for cosmology which flows in part from his treatment of time. He introduces his approach by stating:[7]

> Based on the conjecture that rather than the second law of thermodynamics inevitably be breached as matter approaches a big crunch or a black hole singularity, the order of events should reverse, a model of the universe that resolves a number of longstanding problems and paradoxes in cosmology is presented. A universe that has no beginning (and no need for one), no ending, but yet is finite, is without singularities, precludes time travel, in which events are neither determined by initial or final conditions, and problems such as why the universe has a low entropy past, or conditions at the big bang appear to be so special, require no causal explanation, is the result. This model also has some profound philosophical implications.

The model arises in part as a consequence of Lynds' unique treatment of time, and his ability to present a scientific framework which dispenses with the conventional notion of "instants" and a concomitant "flow" of these instants of time. He develops his cosmology based on the conjecture that in a "big crunch", at precisely the moment where the second law of thermodynamics would necessarily be breached in order to preserve symmetrical event structure, and just before the universe reaches a singularity, instead of breaching the second law of thermodynamics, the order of events would be reversed and universal expansion would begin without a singularity actually having been reached. In Lynds words:[8]

> The natural question then became, what would happen if the second law of thermodynamics were breached? People such as Hawking (1996, 1999) and Gold had assumed that all physical processes would go into reverse. In other words, they had assumed

[6] Admittedly, this is a significant part of the epistemological argument put forth by Carroll and Carroll and Chen in the "natural" universe conjecture. Implicit in their theory is the idea that if cosmos formation is a "natural" phenomena, rather than the "unnatural" situation suggested by the differences in scale values for fundamental constants, then an additional, emergent force would be "unnatural".
[7] Lynds, P. (2006) "On a finite universe with no beginning or end", http://arxiv.org/ftp/physics/papers/0612/0612053.pdf
[8] Ibid.

that events would take place in the direction in which entropy was decreasing, rather than increasing as we observe today. Furthermore, they had assumed that entropy would decrease in the direction in which the universe contracted towards a big crunch (in their case, towards what we call the big bang). But if the second law correctly holds, on a large scale, entropy should still always increase. Indeed, what marks it out so much from the other laws of physics in the first place, is that it is asymmetric – it is not reversible. If all of the laws of physics, with the exception of the second law of thermodynamics, are time symmetric and can equally be reversed, it became apparent that if faced with a situation where entropy might be forced to decrease rather than increase, rather than actually doing so, the order of events should simply reverse, so that the order in which they took place would still be in the direction in which entropy was increasing. The second law would continue to hold, events would remain continuous, and no other law of physics would be contravened.

Hence, in Lynds' model, any events which would have taken place in a situation where entropy was decreasing would experience a reversal of the time ordering of events, and in the subsequent expansion of the universe, "events would immediately take up at where the big crunch singularity would have been had events not reversed, and in this direction, no singularity would be encountered. The universe would then expand from where the big crunch singularity would have been had events not reversed (i.e. the big crunch reversed), and with events going in this direction, entropy would still be increasing, no singularity would be encountered, and no laws of physics would be contravened. They would all still hold." (p. 6)

Lynds' argument necessarily bounds this reversal in the ordering of events to a very small region, and quite shortly thereafter, normal processes of inflation, including increasing entropy resume. Both the physical and the philosophical implications of this position are profound. On the philosophical side, Lynds has introduced a new concept, not only of the ultimate origin of the universe, but also a complex redefinition of "past" and "future":

At this point, it becomes apparent that this would not only lead things back to the big bang, but it would actually cause it. The universe would then expand, cool, and eventually our solar system would take shape. It would also mean that this would be the exact repeat of the universe we live in now. Something further becomes evident, however, and it is perhaps the most important (and will probably be the most misunderstood and puzzled over) feature of this model. If one asks the question, what caused the big bang? The answer here is the big crunch. This is strange enough. But is the big crunch in the past or the future of the big bang? It could equally be said to be either. Likewise, is the big bang in the past or future of the big crunch? Again, it could equally be said to be either. The differentiation between past and future becomes completely meaningless. Moreover, one is now faced with a universe that has neither a beginning nor end in time, but yet is also finite and needs no beginning. The finite vs. infinite paradox of Kant completely disappears.

Although if viewed from our normal conception of past and future (where we make a differentiation), the universe would repeat over and over an infinite number of times, and could also be said to have done so in the past. Crucially, however, if one thinks about what is actually happening in respect to time, no universe is in the future or past of another one. It is exactly the same version, once, and it is non-cyclic. If so desired, one might also picture the situation as an infinite number of the same universe repeating at exactly the same time. But again, if properly taking into account what is happening in respect to time, in actuality, there is no infinite number of universes. It is one and the same.

As previously indicated, this conjecture represents another radical and novel interpretation of time. However, one of the most interesting features of Lynds' conjecture is that it actually meets the two primary criteria of Hawking's M-Theory, (a) weak singularity (in Lynds' case the singularity is there, but it is, in some sense, outside the light cone and outside the observable event horizon) and (b) no boundary condition (albeit, not in the precise fashion that Hawking interprets the no boundary condition restriction).[9] Admittedly, the model is in some ways profoundly counter-intuitive, but that is largely because in a very curious way, even when treating subjects in both relativity and quantum mechanics, we have a tendency to either overtly or covertly introduce Newtonian notions of time. Some of this is addressed in Smolin's critiques of general relativity and quantum mechanics.[10] Lynds himself addresses a potential source of difficulty in the section of his article entitled "Potential Criticisms": (p.9)

---

[9] It is important to note that when the clock restarts at the big bang, the universe is not in the future or past of another one. In a sense, it is time itself that restarts (although, again, nothing in fact actually "restarts"), so there is no past or future universe. Because of this, no conservation laws are violated. It is also important to note that it is simply just the order of events that reverse - something that would be immediate. Time does not begin "flowing" backwards to the big bang, nor does anything travel into the future or past of anything, including time and some imagined "present moment". Indeed, this model contains another interesting consequence. As there is no differentiation between past and future in it, and, strictly speaking, no event could ever be said to be in the future or past of another one, it would appear to provide a clear reason as to why time travel is not possible. In relation to future and past, there is clearly nothing there to travel into. Physically speaking, the same can be said for travel through an interval of time, a flow of it, as well as space-time. (p. 8)

[10] In "Three Roads to Quantum Gravity", Smolin argues that the fundamental flaw in relativity is that it fails to incorporate the effect of the observer on observed phenomena and that quantum mechanics, while achieving the former has a tendency to treat quantum-mechanical events as occurring in traditional, Newtonian spacetime. He then argues that the unification of the partial completeness of these two new physical paradigms will be required to develop an adequate theory of quantum gravity. A quantum cosmology is likewise implicit in such a unification. Lynds provides some interesting clues to this unification insofar as puts time on all scales on a firm Einsteinian footing. In fact, one might answer Smolin's provocative essay title "Where are the Einsteinians?" with the retort "In New Zealand".

An obvious criticism for the model seems to raise itself. It implies that the universe can somehow anticipate future or past events in exact detail, and then play them over at will. At first glance, this just seems too far-fetched. How could it possibly *know*? With a little more thought, however, one recognizes that such a contention would assume that there actually was a differentiation between past and future events in the universe. With this model, it is clear there would not be. Events could neither be said to be in the future or past of one another; they would just be. Moreover, as there is nothing to make one time (as indicated by a clock) any more special than another, there is no present moment gradually unfolding; all different events and times share equal reality (in respect to time, none except for the interval used as the reference). Although physical continuity (i.e. the capability for events to be continuous), and as such, the capability for motion and of clocks and rulers to represent intervals, would stop them from all happening at once (and to happen at all), all events and times in the universe would already be mapped out. As such, as long as it still obeyed all of its own physical laws, the universe would be free and able to play any order of events it wished. Please note that this timeless picture of reality is actually the same as that provided by relativity and the "block" universe model, the formalized view of space-time resulting from the lack of a "preferred" present moment in Einstein's relativity theories, in which all times and events in the universe – past, present and future – are all mapped out together, fixed, and share equal status.

Lynds model contains a number of additional features, including some novel treatments of Kaon decay, black holes and "white holes", all of which are successfully incorporated in his model. While it is beyond the scope of the present paper to discuss these details, they deserve mention as indicators of the level of sophistication in what some might initially imagine to be a naïve interpretation of quantum cosmology.

6 Conclusion

In the foregoing paper we have examined a number of cosmological dynamics, particularly in light of the recent theories of time and cosmology put forward by Peter Lynds. Further, we have noted how the connection between the necessary conditions of extendibility and incompleteness with respect to Einstein's field equations leads under a variety of conditions to "unobservable singularity". The novelty of Lynds' solution is that it suggests that while the primordial singularity, including the problematic initial conditions of "specialness" explained by Carroll and Chen, exists within the light cone, it is nonetheless an inaccessible geodesic. Further, Lynds argument offers the novelty of a closed causal loop between the Big Bang and Big Crunch, no longer requiring an explanation for the special or natural entropic and scale conditions of the observable universe.

References

1. Carroll, S.: Is Our Universe Natural? http://www.arXIv:hepth/0512148v1
2. Carroll, S., Chen, J.: Does Inflation Provide Natural Initial Conditions for the Universe? Gen. Rel. Grav. 37, 1671–1674 (2005); Int. J. Mod. Phys. D14, 2335–2340 (2005)
3. Carroll, S., Chen, J.: Spontaneous Inflation and the Origin of the Arrow of Time, http://arxiv.org/PS_cache/hep-th/pdf/0410/0410270v1.pdf
4. Clarke, C.J.S.: The Analysis of Space-Time Singularities. Cambridge Lecture Notes in Physics. Cambridge University Press, Cambridge (1994)
5. Fellman, P., Post, J.V., Carmichael, C., Post, A.C.: Comparative Quantum Cosmology: Causality, Singularity, and Boundary Conditions, http://arxiv.org/ftp/arxiv/papers/0710/0710.5046.pdf
6. Fernandez, D.J., Velazquez, C.M.: Coherent states approach to Penning trap, http://arxiv.org/PS_cache/arxiv/pdf/0809/0809.1684v1.pdf
7. Gribbin, J.: Quantum Time Waits for No Cosmos, http://www.lifesci.sussex.ac.uk/home/John_Gribbin/timetrav.htm
8. Hawking, S., Penrose, R.: The Nature of Space and Time. Princeton University Press, Princeton (1996)
9. Hawking, S.: Physics Colloquiums - Quantum Cosmology, M-theory and the Anthropic Principle (January 1999), http://www.hawking.org.uk/text/physics/quantum.html
10. Kauffman, S., Smolin, L.: A Possible Solution For The Problem Of Time In Quantum Cosmology. The Third Culture, http://www.edge.org/3rd_culture/smolin/smolin_p1.html
11. Lynds, P.: Time and Classical and Quantum Mechanics: Indeterminacy vs. Discontinuity. Foundations of Physics Letters 16(4) (2003)
12. Lynds, P.: On a Finite Universe with no Beginning or End, http://arxiv.org/ftp/physics/papers/0612/0612053.pdf
13. Lynds, P.: Time for a Change – The Instantaneous, Present and Existence of Time, http://www.fqxi.org/community/forum/category/10
14. Misner, C.W.: Taub-NUT space as a counter-example to almost anything. In: Ehlers, J. (ed.) Lectures in Applied Mathematics, vol. 8. American Mathematical Society (1967)
15. Murphy, M.T., Webb, J.K., Flambaum, V.V., Curran, S.J.: Does the fine structure constant vary? A detailed investigation into systematic effects (2002) arXiv:astro-ph/0210532
16. Platt, J.R.: Science, Strong Inference – Proper Scientific Method. Science Magazine 146(3642) (October 1964)
17. Pluchino, A., Rapisarda, A., Tsallis, C.: Nonergodicity and Central Limit Behavior for Long-range Hamiltonians, http://arxiv.org/PS_cache/arxiv/pdf/0706/0706.4021v3.pdf
18. Smolin, L.: Three Roads to Quantum Gravity. Perseus Press (2002)
19. Smolin, L., Wan, Y.: Propagation and interaction of chiral states in quantum gravity, http://arxiv.org/PS_cache/arxiv/pdf/0710/0710.1548v1.pdf
20. Sorkin, R.D.: Is the cosmological constant a nonlocal quantum residue of discreteness of the causal set type? http://arxiv.org/PS_cache/arxiv/pdf/0710/0710.1675v1.pdf
21. Wald, R.M.: The Arrow of Time and the Initial Conditions of the Universe (2005), http://arxiv.org/PS_cache/gr-qc/df/0507/0507094v1.pdf

The Complex Economic System of
Supply Chain Financing

Lili Zhang and Guangle Yan

Business School, University of Shanghai for Science and Technology,
516 Jungong Road, 200093 Shanghai, China
radfahrer@163.com, glyan2003@yahoo.com.cn

Abstract. Supply Chain Financing (SCF) refers to a series of innovative and complicated financial services based on supply chain. The SCF set-up is a complex system, where the supply chain management and Small and Medium Enterprises (SMEs) financing services interpenetrate systematically. This paper establishes the organization structure of SCF System, and presents two financing models respectively, with or without the participation of the third-party logistic provider (3PL). Using Information Economics and Game Theory, the interrelationship among diverse economic sectors is analyzed, and the economic mechanism of development and existent for SCF system is demonstrated. New thoughts and approaches to solve SMEs financing problem are given.

Keywords: supply chain financing, small and medium enterprises financing, information economics, optimization.

1 Introduction

Around the world, the source of external financing for small and medium enterprises (SMEs) is an interesting topic to academics, and it's also an issue of great importance to policy makers. The practice of supply chain financing has expanded greatly in recent years, which brought new fields of study on the solutions to SMEs financing and financial innovation in the Banking Industry.

As a new topic, limited systematic research results of SCF have been revealed. Some of the extant literature just put forward the idea or perception of SCF, described its consequence or worth [1-4]; others mostly focused on logistics to analyze the conception, members, operational process etc of SCF models [5-8]. They have paid scant attentions to quantitative, systematic, and intensive study on SCF.

In this paper, the economic organization of SCF is regard as a complex system composed of many interest parties. After establishing the organization structure of SCF System, we analyze the basic financing models of SCF systematically, and demonstrate the economic mechanism of its development and existence by game theory. Then the advantage, properties and financing mechanism come out clearly.

J. Zhou (Ed.): Complex 2009, Part I, LNICST 4, pp. 763–772, 2009.

2 SCF's Properties and Financing Models

As a new concept, there has not a vulgate definition for SCF. In this paper, the SCF system is defined as a complex financial system, which means a bank measures the quality of company's supply chain, and then provides finance to one or several suppliers and distributors (especially the SMEs) to ensure the integrated chain running stably and smoothly. The commercial bank, the big enterprises and SMEs on the chain, the 3LP, the government and other institutions co-exist in a mutualism environment. The organization structure of SCF system is given in Figure 1.

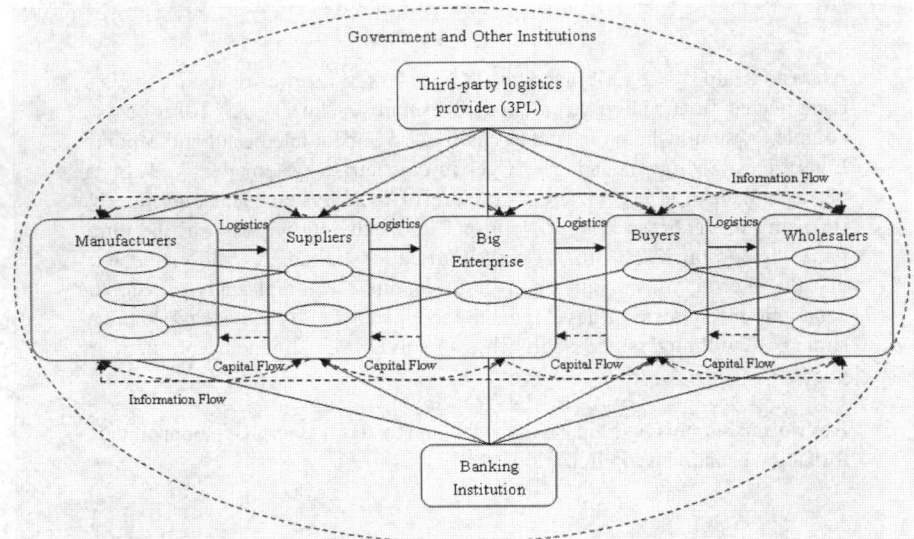

Fig. 1. The organization structure of SCF system

2.1 SCF's Conception and Properties

According to the definition above, we can get three properties of the SCF models as follows.

(1) An organ system. The SCF system is composed of series of interdependent subsystems with specific functions. It consists of three units, inputting, processing and outputting. Its ultimate aim is to expand the supply chain value and upgrade the overall competitiveness.

(2) Collective rationality. In SCF system, the enterprises' (including the banking institution) decisions react directly with other's, which means chain parties' present interests and long-term interests should be considered in decision making. And the collective rationality comes to effect because of the good competitive and cooperative relation and enhanced visibility among the subsystems.

(3) Complexity. The SCF system's participants are excessive, where include capital flow, logistics flow and information flow. They are all complicated and uncertain.

2.2 SCF's Financing Models

The cash gap [9] is the number of days between a business's payment of cash for goods and services bought and the receipt of cash from its customers for goods or services sold. As defined above, the cash gap model is made up of three parts: receivables, payables and inventory. Correspondingly, the models of SCF consist of three ones which are reflected in table 1. In details, the financial models can be divided into two main types, one is debt financing model without the participation of 3PL, the other is personalty financing model with the participation of 3PL.

Table 1. The types of SCF models

	Financial Model	3PL	Stage	Hypothecation	Position on the Supply Chain
Debt Financing Model	Receivables	Non-participation	Inventory Shipped to Cash Received	Creditor Right	Supplier
Personalty Financing Model	Payables	Participation	Inventory Arrives to Cash paid	Real Right (Goods will be bought)	Distributor
	Inventory	Participation	All times with stable inventory	Real Right (Inventory)	Any Enterprises

3 Game Theory Analysis of Debt Financing without the Participation of 3PL

First, let's analyze the debt financing model without the participation of 3PL, whose fundamental idea is: a SME, the creditor of a big enterprise, can apply for loan to bank through the creditor right. The availability of financing for a SME increases while the bank credit risk decreases.

3.1 Repeated Game with Complete Information between Bank and Enterprise

We assume that a SME needs loan L for an investment project, whose success probability is α ($0 < \alpha < 1$). Let β be the rate of investment if successfully, then the SME's expected revenue from the project loan is $\alpha\beta L$. To the bank, let R be interest income and C be monitoring costs, then the return to the bank can be written as

$$\alpha\beta L - R(\alpha\beta L - R > 0, 0 < C < R).$$

For simplicity, we shall assume that the interest rate is fully liberalized on the market and transaction cost between the bank and SME is zero. Their discount rate is $\delta(0 < \delta < 1)$.

With the information presented above, it is possible to specify the pay-off matrix of the game in Table 2. [10]

Table 2. Pay-off matrix of repeated game with complete information between bank and enterprise

		SME Strategies	
		Repay	Don't repay
Commercial bank Strategies	Loan	$\dfrac{R-C}{1-\delta}, \dfrac{\alpha\beta L - R}{1-\delta}$	$-L-C, \alpha\beta L + L$
	Don't loan	0,0	0,0

Whether he bank provide the loan or not is given on the success probability of the project α.

Then

$$\frac{\alpha\beta L - R}{1-\delta} > \alpha\beta L + L \ ,$$

that is

$$\alpha \geq \frac{R+L-\delta L}{\delta\beta L} \ ,$$

the Nash Equilibrium in perfect information games sets. As a matter of fact, it cannot attain in practical because of "information asymmetry" and "credit grudging". Then SCF emerges and provides new thoughts and approaches to solve the problem.

3.2 Signalling Game between Bank and Enterprise in SCF System

We assume that there are two types of SMEs. One is good SME whose success probability is

$$\alpha_1 \geq \frac{R+L-\delta L}{\delta\beta L} \ .$$

It will repay the loan under complete information to keep good terms with the bank. The other one is bad SME whose success probability is lower,

$$\alpha_2 < \frac{R+L-\delta L}{\delta\beta L} \ .$$

It will not repay the loan without restrained mechanism.

Suppose the SME has two strategies to select.

The first one is to "be in" SCF system, that is, the SME has real and stable trade contacts with the big enterprise. If the SME fails to repay, the big enterprise will also make up for the lost of bank. To SME, we resume the reverse guarantee costs provided by big enterprise are S (As debtor-creditor relationship between SME and big enterprise exists originally, the costs S is too small and can be neglected.) and default costs are F. In SCF system, the information set the bank can observe is f.

The second one is to "be out of" SCF system, which information set is i. With the information presented above it is possible to specify the extensive game in Figure 2.

According to the strategy set of the SME, four aspects can be discussed. First of all, we analyze the situation (Be in, Be in).

Although the bank can't judge a SME whether a good enterprise or not just by the signal sent by SME, the probability of a SME to be considered as a good one, $p(0.5 < p < 1)$, is high provided that the SME has closer economic partnership with the big enterprise.

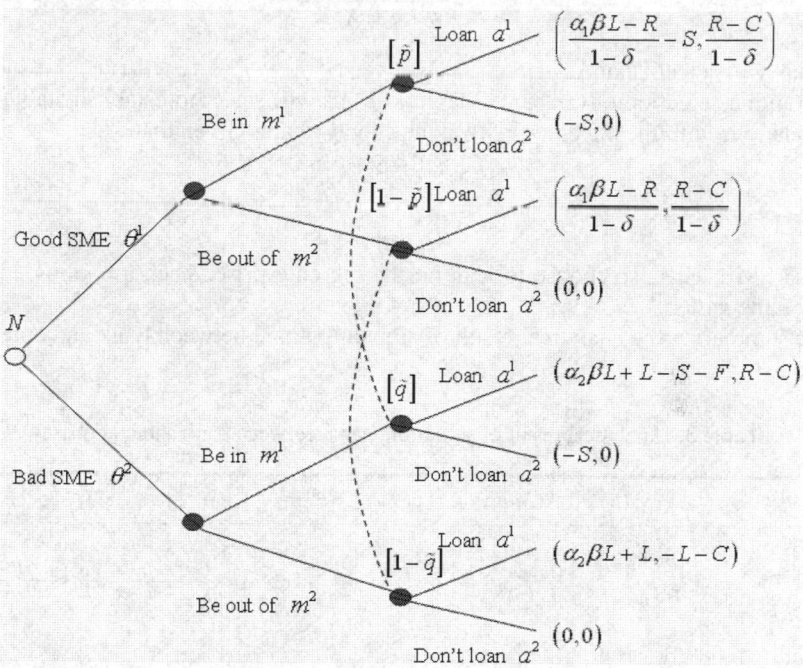

Fig. 2. Signaling game between the bank and enterprise

Within the information set f, the return to the bank if it selects "loan" is

$$\Pi_{B_1}^f = p\frac{R-C}{1-\delta} + (1-p)(R-C) ,$$ (1)

and the return to the bank if it select "don't loan" is

$$\Pi_{B_2}^f = 0 .$$ (2)

It was patently obvious that $\Pi_{B_1}^f > \Pi_{B_2}^f$, and the bank will select "loan".

Within the information set i, the return to the bank if it select "loan" is

$$\Pi_{B_1}^i = q\frac{R-C}{1-\delta} + (1-q)(-L-C) ,$$ (3)

and the return to the bank if it select "Don't loan" is

$$\Pi_{B_2}^i = 0 .$$ (4)

If

$$\Pi^i_{B_1} > \Pi^i_{B_2} \; ,$$

then

$$q > \frac{L+C-\delta L-\delta C}{R+L-\delta L-\delta C} \; ,$$

the bank will select "loan".

Given the selection of the bank, the good SME will select to establish stable business relations with the big enterprise on the supply chain. Given that

$$\alpha_2 \beta L + L - S - F > 0 \; ,$$

that is

$$F < \alpha_2 \beta L + L - S \; ,$$

the bad SME will also choose to lean on the big enterprise. Then a pooling equilibrium can be gotten.

We summarize the analytical result of signaling game between bank and enterprise in Table 3.

Table 3. Analytical result of signalling game between bank and enterprise

	S(SME)	S(Bank)	$\left[\tilde{p}(\theta^1 \vert m^1), \tilde{p}(\theta^2 \vert m^1)\right]$	$\left[\tilde{p}(\theta^1 \vert m^2), \tilde{p}(\theta^2 \vert m^2)\right]$
Pooling Equilibrium $F < \alpha_2 \beta L + L - S$	(Be in, Be in)	(Loan, Loan)	$(0.5 < p < 1, 0 < 1-p < 0.5)$	$\left(q > \dfrac{L+C-\delta L-\delta C}{R+L-\delta L-\delta C}, 1-q < \dfrac{R-C}{R+L-\delta L-\delta C}\right)$
Out-of Equilibrium	(Be out of, Be out of)	/	/	/
Separating Equilibrium $F > \alpha_2 \beta L + L - S$	(Be in, Be out of)	(Loan, Don't loan)	$(1,0)$	$(0,1)$
Out-of Equilibrium	(Be out of, Be in)		/	/	/

3.3 Research Revelation of Debt Financing Model

The above discussion thus shows that a pooling equilibrium can be gotten, if

$$\Pi^f_{C_2} > \Pi^i_{C_2} \; ,$$

that is

$$F < \alpha_2 \beta L + L - S \approx \alpha_2 \beta L + L \; .$$

In such a case, the return of "be in" is greater than that of "be out of", and so both the SMEs, good or bad, will choose to apply to the bank for a loan under the SCF system. Under this model, the bank cannot discriminate good SEM from bad, it will provide the loan to the "Be out of" SME given its prior belief

$$\tilde{p}\left(\theta^2 \middle| m^2\right) < \frac{R-C}{R+L-\delta L-\delta C}.$$

Then the bank can transfers the credit risk of the loan to the suppliers' high-quality buyers as a joint-liability guarantor and will provide the loan to both good and bad SMEs, which is bound to cause loss to the bad SME's counterpart. The big enterprise will "raise the threshold" accordingly to avoid this pooling equilibrium.

On the other hand, a separating equilibrium can be gotten,
if

$$\Pi_{C_2}^f < \Pi_{C_2}^i,$$

that is

$$F > \alpha_2 \beta L + L - S \approx \alpha_2 \beta L + L.$$

In such a case, the good SME chooses to "be in" and bad one chooses to "be out of". Consequently, the good one can secure the loan and the bad not. Then a "virtuous cycle" is starting between the bank and SME. Furthermore, the whole industrial chain can develop more stably and coordinately.

4   Optimal Analysis of Personalty Financing with the Participation of 3PL

"Payables" and "Inventory" belong to "Personalty Financing" model with the participation of 3PL, whose fundamental idea is: The bank B holds the eminent domain of finance; the logistics provider L holds supervision and control right over the inventory. They make an alliance A to provide financing services to SME. In such a case, the bank and logistics provider are principal, and the SME is agent. [11]

4.1   Optimal Decision Analysis of Personalty Financing Model

We assume that ultimate demand for products depends on the price, i.e., $q = \alpha - \beta p$ (α, β is nonnegative parameter). Supervision cost per unit product of the bank and logistics provider is c_B and c_L respectively. Product unit revenue defines as u_i, $i = B, L, A$. The SME's production profit gained by demand can be measured by the bank's profit simply, if the production cost is standardized to 0 and the interest is left out of account.

Let a random variable X_i be the posterior belief of $c_i (i = B, L)$, its cumulative distribution function is $\psi_i(x)$ and density function is $\varphi_i(x)$. Meanwhile let a random variable $\hat{X}_i$ signs the indirect information, its cumulative distribution function is $\hat{\psi}_i(x)$ and density function is $\hat{\varphi}_i(x)$. All the distribution functions are assumed strictly positive and have increasing failure rate.

For further analyses, let's give a useful property. [12]

Property 1. Suppose a random variable X_i is a log-concave function, whose cumulative distribution function is $F_i(x)$ and density function is $f_i(x), i = 1,2$. Then,

(1) The increasing failure rate $\dfrac{F_i(x)}{f_i(x)}$ is monotone decreasing.

(2) If $t(x)$ is twice differentiable and strictly log-concave, $\dfrac{F_i(x)}{f_i(x)}$ is a log-concave function.

(3) Integral equation $F(x) = \int_0^x G_1(x - x_2) G_2(x_2)$ is a log-concave function.

When $u_A \geq c_B + c_L$, the alliance will accept agreement offered by SME. The supervision cost of the alliance observed by SME can be denoted by $\hat{X}_B + X_L$, so that the probability of signing the contract agreement at the first stage is given by

$$P\{\hat{X}_B + X_L \leq u_A\} = \int_0^{u_A} \psi_L(u_A - x_B) d\hat{\psi}_B(x_B) . \tag{5}$$

The SME wishes to maximize its expected profit at the first stage

$$\max \Pi(p, u_A) = (p - u_A)(\alpha - \beta p) \int_0^{u_A} \psi_L(u_A - x_B) d\hat{\psi}_B(x_B) . \tag{6}$$

Since $\Pi(p, u_A)$, u_A is concave function with respect to p, find the critical numbers of p, then have

$$\begin{cases} p = \dfrac{\alpha + \beta u_A}{2\beta} \\ q = \dfrac{\alpha - \beta u_A}{2} \end{cases} . \tag{7}$$

By (6) and (7) we may consider the optimization problem of solving the following equation

$$\max \Pi = \dfrac{(\alpha - \beta u_A)^2}{4\beta} \int_0^{u_A} \psi_L(u_A - x_B) d\hat{\psi}_B(x_B) . \tag{8}$$

Calculate the first derivative and get

$$\dfrac{\alpha - \beta u_A}{2\beta} = \dfrac{\int_0^{u_A} \psi_L(u_A - x_B) d\hat{\psi}_B(x_B)}{\int_0^{u_A} \varphi_L(u_A - x_B) d\hat{\psi}_B(x_B)} . \tag{9}$$

Since $\int_0^{u_A} \psi_L(u_A - x_B) d\hat{\psi}_B(x_B)$ is the integral of ψ_L and $\hat{\psi}_B$, it is also a log-concave function according to Property 1. The right side of (9) is the increasing function of u_A, while the left side of (9) is decreasing function of u_A. Hence, we get the unique optimal solution u_L^* of (9), which means we attain the unique optimal solution $\left(p^*, u_A^*\right)$ of (6).

4.2 Research Revelation of Personalty Financing Model

Based on the optimism analysis above, we can state that the three participants, namely, the bank, 3PL and SME, can achieve the optimum and realize multi-win in the personalty financing model. Firstly, for banking institution, SCF represents an opportunity to generate new revenues, deepen relationships with regular clients, and reduce credit risk. Secondly, for logistics provider, SCF helps 3PL to upgrade service quality standards and enhance the value-added groups in marketing integration. Thirdly, for SME, SCF allows SME to increase its cash stock, strengthen its financial profile, and take advantage of its raw materials or finished goods on market as pledge, all of which contribute greatly to develop a more stable supply chain.

5 Conclusion

As an efficient and multi-win complex financial system, all participants in Supply Chain Financing system seek to maximize their profits by playing various roles respectively, inter-coordinating and sharing the risks or profits together. In this paper, by using information economics and game theory, we try to analyze the two SCF models with and without the participation of third-party logistic providers respectively. Then, by proving the existence of unique optimum solution, this paper explains the economic mechanism why the SCF model can solve the problem of SMEs financing.

Acknowledgments. This work was supported by Shanghai Leading Academic Discipline Project under Grant T0502 and Scientific Research Planned Projects of Science and Technology Commission of Shanghai Municipality under Grant 06JC14057.

References

1. Berger, A.N., Udell, G.F.: A More Complete Conceptual Framework for SME Finance. Journal of Banking and Finance 30, 2945–2966 (2006)
2. Yang, S.H.: Research on Supply Chain Financing Service. Logistics Technology 10, 179–182 (2005) (in Chinese)
3. Gonzalo, G., Badell, M., Puigjaner, L.: A Holistic Framework for Short-term Supply Chain Management Integrating Production and Corporate Financial Planning. International Journal of Production Economics 106, 288–306 (2007)
4. Feng, Y.: Supply Chain Financing: Financial Innovation Services to Realize Muilt-win. New Finance 2, 60–63 (2008) (in Chinese)
5. Richard, G.: Longer Chains, Lower Costs: to Create A Seamless Supply Chain, Every Link Must Feel Like It's Winning. Treasury and Risk Management 14, 40–46 (2004)
6. Klapper, L.F.: The Roal of Factoring for Financing Small and Medium Enterprises. Journal of Banking and Finance 30, 3111–3130 (2006)

7. Tang, S.Y.: Study of interflow of commodities service in finance. Inner Mongolia Coal Economics 05, 13–16 (2005) (in Chinese)
8. Yan, J.H., Xu, X.Q.: SME Financing model based on Supply Chain Financing. Shanghai Finance 2, 14–16 (2007) (in Chinese)
9. Boer, G.: Managing the Cash Gap. Journal of Accountancy 10, 27–32 (1999)
10. Chen, Q.A., Chen, L.: Study on Debt Financing Game Model of SMEs based on Credit Guarantee. On Economic Problems 7, 21–24 (2008) (in Chinese)
11. Li, J., Xu, Y., Feng, G.Z., Li, Y.X.: Research on Decision of Bank's Outsourcing in Warehouse Financing. Operations Research and Management Science 2, 84–87 (2007) (in Chinese)
12. Bergstrom, T., Bagnoli, M.: Log-concave Probability and Its Applications. Economic Theory 26, 445–469 (2005)

The Bipartite Network Study of the Library Book Lending System

Nan-nan Li and Ning Zhang

Business School, University of Shanghai for Science and Technology
Shanghai, China 200093
lnn19850625@126.com, zhangning@usst.edu.cn

Abstract. Through collecting the library lending information of the University of Shanghai for Science and Technology during one year, we build the database between the books and readers, and then construct a bipartite network to describe the relationships. We respectively establish the corresponding un-weighted and weighted bipartite network through the borrowing relationship and the reading days, thereout obtain the statistical properties via the theory and methods of complex network. We find all the properties follow exponential distribution and there is a positive correlation between the relevant properties in un-weighted and weighted networks. The un-weighted properties can describe the cooperation situation and configuration, but the properties with node weight may describe the competition results. Besides, we discuss the practical significance for the double relationship and the statistical properties. Further more, we propose a library personal recommendation system for developing the library humanity design resumptively.

Keywords: complex network, bipartite network, node weight, personal recommendation.

1 Introduction

The analytic methods of complex network have been widely used in various fields [1-7] to describe the relationship between the individual and the collective behavior of the system. Thereinto, each individual corresponds to the different node, and individual interactions correspond to the edge of the two nodes. In the un-weighted network, the edge only gives the qualitative description that whether there is an edge between the nodes. However, in most cases, the distinctness of the interactional strength between nodes plays a vital role. In this way, the edge-weight is introduced to describe the inter-action difference, thus form the weighted network [8, 9]. As a major expressive form of complex network, the bipartite network can commendably represent the original information, which has been received more and more attention by researchers. A series of cooperation network in nature and society can be described as the bipartite network constituted with act and actor [10-12].

Library is a treasury trove of spiritual wealth of mankind, an important part of human spiritual civilization and an inexhaustible knowledge of human resources. Books lending is one of the ways to provide services of the library. The quantities of lending

J. Zhou (Ed.): Complex 2009, Part I, LNICST 4, pp. 773–782, 2009.

books directly reflect the readers' demand, also is used to measure the book using effectiveness and be regarded as an important factor toward the purchase of books. From the view of system, library lending network is a typical complex network, as well as a typical cooperation and competition network. Through lending process, we can establish certain links between books and readers, thus constitute the bipartite network. However, the prior empirical researches of the library network [13, 14] are un-weighted and lack definite rationality due to the unclassified books. Compared with the prior research, this paper improves as follows:

(i) In the previous study, book is expressed by book barcode. However, in the library every book has different barcodes, and the number of the book with a same edition is usually more than one. Therefore, if a reader borrows some books with the same edition, it will be disposed as different books. Here, based on the disposal about the bibliography storeroom and library collection, we use call number instead of barcode, and then obtain the statistic of the books with same edition.

(ii) Suppose a reader read a book with one or twenty days, the significance of the book is clearly different for this reader. Hence, if only using the times that the reader borrow a same edition book as the weight, the significance is not great. Dealing with the borrow-return record time based on the library computer system, we get the duration that the reader borrow and return the same edition book with days, and regard them as the weight between this reader and book. If the reader borrows the same edition book several times, the weight will be the sum of each time.

Therefore, through the reasonable classification towards all books, we study the un-weighted and weighted bipartite network of the library lending relations, then obtain several accurate results and analyze the collaboration-competition relationship, hope to provide some new empirical foundation for the library research and help the procurement staffs to better understand the needs situation of readers. Finally, using a proper weighting method [15], we propose a library recommendation system.

2 Constructing the Library Lending Bipartite Network

In this paper, the data are from the library lending situation of the University of Shanghai for Science and Technology (USST) during one year, table 1 shows the specific data format. During this period, the total number of the borrowed books is 83,959 (different book barcodes), which with different editions is about 51,084 kinds (different call numbers), and the total number of readers is 12,610. First, we introduce the statistical situation of original data in the library, as table 1, each reader corresponds to a reader barcode, each book corresponds to a book barcode, and all the barcodes are different with each other. If one reader borrows a book, there will be a connection between the reader barcode and the book barcode. In the format data of the library bibliographic collection, a book call number corresponds to several different book barcodes in table 2. This is because the library will provide several books of the same edition (the title, author, publishing company, publishing time are the same) to lend for readers, these books of a same edition is indicated with the same call number (convenience for readers to lookup), each book has a different bar code for distinction (convenience for the library staffs to take notes). In the actual situation, if a

Table 1. The data format of the library book

Book barcode	Reader barcode	Reader grade	Department	Operation	Dealing time
842851	691	graduate student	Business School	borrow	2007-10-26
801569	983	graduate student	Business School	return	2007-10-26
689855	46	student	English School	return	2007-10-26
848798	7938	student	Business School	return	2007-10-26
850723	246	teacher	Physics School	return	2007-10-26
734438	1658	student	Business School	borrow	2007-10-26

Table 2. The data format of the library bibliography collection

Call number	Book barcode	Superscription
O551/Z53	E031208	Heat and trermodynamics
O551/Z53	E038805	Heat and trermodynamics
I565.44/A933	E010449	Pride and prejudice
I565.44/A933	E028330	Pride and prejudice
I565.44/A933	E031156	Pride and prejudice
I565.44/A933	E028331	Pride and prejudice

reader borrows two books with the same edition, we should regard them as the same book. Thus, in order to exhibit the lending situation more reasonably, we use call number instead of book bar code to establish the library network by SQL.

In the library bipartite network, the nodes can be divided into two types. One type expresses the books, named "acts"; the other expresses the readers who participate in the acts, called "actors". If there is a borrowing relationship between them, the two will be connected by a line forming an edge. In the entire bipartite network, there is a kind of collaboration-competition relationship among the acts, that is, the lending quantity of all the books present the level of library service , and the quality and popularity of the books form a kind of borrowing competition for readers. On the other hand, the books borrowed by the same reader form a type of collaboration-competition relationship, namely, these books together constitute the reader's knowledge systems, and the books compete against each other to provide service owing to the reader's limited energy. In each act, the relations between the actors also represent both collaboration and competition, the number of readers that borrow the same book constitutes the reading value of this book, and because the number of readers is relatively larger than the books, the processes form the competition to borrow the same book for the readers.

3   The Properties of the Bipartite Network for the Library Lending System

For un-weighted bipartite network, we study two properties named act size and act degree which merely consider the borrowed times. In the study of the weighted bipartite network, we regard the reading days as the edge-weight between the reader and the book, and obtain the node strength of books and readers respectively.

3.1 Act Size and Act Degree

Act size is the number of actors connected with an act, that is, the number of readers who borrow this book during the year, roughly indicate the competitiveness of the book. As shown in Fig. 1, except for the small impact on the tail, the distribution can be well approximated by an exponential form

$$y = 0.9519e^{-0.258x}$$

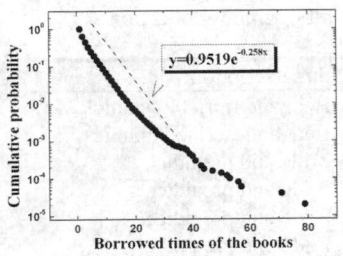 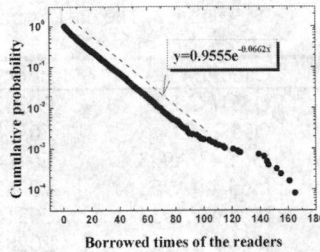

Fig. 1. The cumulative distribution of act size   **Fig. 2.** The cumulative distribution of act degree

Table 3 shows the distribution of the number of borrowed times, two books are borrowed more than 70 times at best, 54.62 percent of the books are borrowed 2-9 times, and 36.70 percent are one time. The average size of the act is 9.44, that is, the average

Table 3. The distribution of the book borrowed times

Borrowed times	79, 71	50-59	40-49	30-39	20-29	10-19	2-9	1
Books' number	2	5	13	42	268	3115	27903	18750
Percent	0.00	0.01	0.03	0.08	0.53	6.10	54.62	36.70

Table 4. The category distribution of the former hundred books

Book category	Chinese literature	English language	Computer	Anglo-American Literature	Mathematics	physics
Books' number	29	20	16	11	4	4

Table 5. The borrowed times distribution of the readers

Borrowed times	1	2-9	10-19	20-29	30-39	40-49	50-59	60-69	70-79	79-165
Amount	1040	5452	3010	1491	750	434	202	111	57	62
Percent	8.25	43.24	23.9	11.8	5.95	3.44	1.60	0.88	0.45	0.49

borrowed times of all books in the year are 9.44. Surveying the category of the first 100 books with the highest borrowed times as shown in table 4, the most frequent types are Chinese literature, English language and computer science. In the computer category, the number of books on Matlab and C++ is seven and four respectively accounting for the most. Similarly, among the four mathematic books, there are three about probability theory and mathematical statistics. Therefore, the library should be increase relevant procurement, as far as possible to meet the needs of readers.

Act degree of an actor node is defined by the total number of borrowed books for a reader during one year, which expresses the competitive size of the reader. Act degree distribution describes the situation of the readers in the library, Fig. 2 shows the cumulative distribution is approximated by an exponential function $y = 0.9555e^{-0.0662x}$ on the whole. Known from table 5, 43.24 percent of readers borrow 2-9 books in the year. The average of actor degree for all readers is 14.16, that is, the average number of the borrowed books for every reader is 14.16 during the year. Through the statistic about the first 100 readers with the highest times, there are 4 teachers, 71 graduate students and 24 other students. With practice, teachers have abundant knowledge and more engage in single areas, so they borrow books more specifically, also can purchase books continually owing to better economic status, while graduate students have fewer knowledge relatively, thirst for knowledge strongly, and need to gain a large number of relevant literature in study and research, also have better advantage of library privileges (borrowing 10 books one time) than other students (5 books), so their times are higher than other students.

3.2 Node Strength Distribution

Owing to the detailed records that readers borrow-return each book during one year, we can obtain the reading time using SQL and ACCESS. If a reader has been borrowed a book, then we establish an edge between them, and regard the reading time as the weight of this edge. In this way, the weighted bipartite network is established. The node strength is a natural extend corresponding to node degree, and the role of the node strength distribution $p(s)$ is similar to degree distribution $p(k)$, which denote the probability of one node having node strength S. It is defined as

$$s_i = \sum_{j \in N_i} w_{ij} \tag{1}$$

Where N_i represents the collection of neighbor nodes of node i. Node strength considers not only the close neighbor of a node, but also the linked weights between this node and the neighbors, which represent the integrated information of this node.

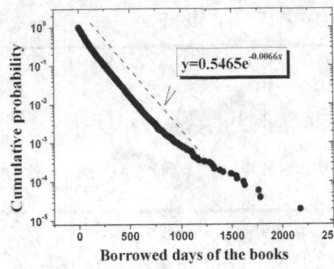 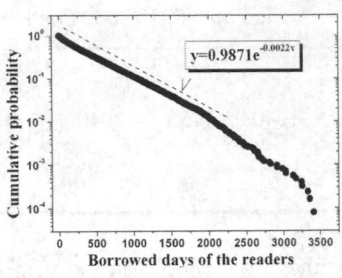

Fig. 3. The cumulative distribution of the node strength, the left is about book and right for reader

Table 6. The distribution of the book node strength

Node strength	0-30	31-60	61-100	101-200	201-300	301-400	401-500	501-600	601-2176
Amount	14431	9740	8635	10204	4139	1955	960	425	415
Percent	28.25	19.07	16.90	19.97	8.10	3.82	1.88	0.83	0.81

Table 7. The former hundred books of the highest node strength

Specialty	English language	Computer	Math	Kinetics	Mechanics	Physics	Anglo-American literature	Telecom
Number	29	25	9	8	6	5	5	4

The node strength of a book denotes the sum of the reading days by every reader for this book, which indicate the competitive size more accurately. Fig. 3 shows the cumulative distribution can well fit with an exponential function $y = 0.5465e^{-0.0066x}$. The average of the node strength is 106.74, namely, every book is kept for 106.74 days by readers during the year. Table 6 lists the specific situation distribution of the days, 92.29 percent is within 300 days, less than a year's time. This shows that the book in the library can meet the reader's demand basically. As the books with larger node strength represent the longer time of this book between borrow and return, namely more competitiveness, which indicate the reader's interest and trend, so we analyze them as focus. Table 7 shows that English language and computer books are still on the top, but the literature books decrease significantly. This indicates the concept of readers is to place study first, entertainment second. At the same time, the readers pay great attention to English and computer books which have become an indispensable tool. In addition, the number of books in table 7 is uniform on dynamic engineering, mechanics, physics, radio and telecommunications technologies, which indicate that the readers of different specialties all will borrow the relatively professional books during study and research, which further explain the accuracy of the empirical work.

Fig. 3 shows the cumulative distribution of the reader node strength, the main part of the distribution follows an exponential function $y = 0.9871e^{-0.0022x}$. This node strength denotes the total number of the days the reader reads every book in one year, a more accurate competitive expression. The average of the node strength for all readers is 423.44 days, and the node with the greatest strength is a teacher in kinetic college. Through the analysis of prior 100 readers with the greatest strength, these readers are 29 teachers, 70 graduate students and one other student. Because the stated reading time of teacher (90 days) is longer than graduate student (Doctor for 90 days, 30 days for master) and other students (30 days), the teacher total day increase accordingly.

3.3 The Contrast of Un-weighted and Weighted Network

In the bipartite network, a book act size in un-weighted network is the total borrowed times of the book by all readers, and book node strength in weighted network is the sum of the reading days by every reader, both express the book competitiveness. Similarly, actor degree is the total borrowed times of a reader in un-weighted network, reader node strength in weighted network is the total reading days of the reader, and also both show the reader competitiveness. Then, is there a certain relation between the two pairs of distribution in the form of un-weight and weight network? First of all, they follow the same distribution on the whole, the same conclusion in previous study [16], the reasons of such mechanism are explained deep in a network evolution model [17, 18]. Secondly, as shown in Figure 4, there is a positive correlation between them. More exactly, the node strength follows a power law with respective to act size, also to actor degree. This means that, the more times a book is borrowed, the longer time it can be kept. In the same way, the more times a reader borrow, the longer time he can keep. Through studying the statistical relationship between the act degree and node strength, we know that there is larger competitiveness when the actor participates more acts. In practice, as the library restricts the number of the borrowed books and the preserving-book time for the readers, the readers usually faster return the less useful books in order to borrow the needed books for themselves. Then borrowing a book does not represent that the reading value of this book is existent, but the length of the reading

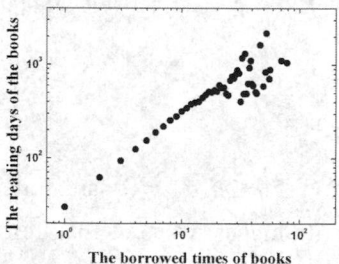 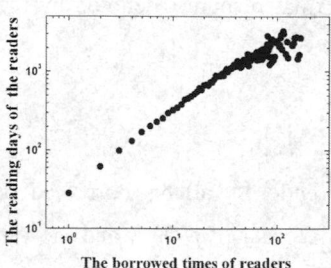

Fig. 4. The empirical results of relationship between book node strength and act size (*the left*), also reader node strength and act degree (*the right*)

days reflect the different value of the book for each readers. Therefore, weighted bipartite network can reflect the actual situation better. All these show that both of them can describe this collaboration situation and configuration, but only the latter can describe the competition results and other competition properties accurately.

4 The Library Personal Recommendation System

There are large numbers of books in library, which maybe have hundreds of kinds in one direction, and most of the same direction books also have different emphasis. Therefore, it is troublesome to find the most correlative book in them for readers. And using the historical books which a reader has been borrowed, we can discover his habits and consider them for him in the future. Here, we propose the library recommendation system which make use of the reader's historical information, and hope to provide several suggestions for the library humanity design.

A reasonable assumption is that the books which a reader have borrowed are what he like, and a recommendation algorithm aims at predicting his personal opinions on those books he have not yet collected. Based on our database, the recommendation system consists of readers and books, and denotes the book-set as $B = \{b_1, b_2, \cdots b_n,\}$ and reader-set as $R = \{r_1, r_2, \cdots, r_m\}$. If readers have borrowed some books, the recommendation system can be described by an $n \times m$ adjacent matrix $\{a_{ij}\}$, where $a_{ij} = 1$ if r_j has already borrowed b_i and $a_{ij} = 0$ otherwise. Through the application of the weighting method for bipartite networks presented in [15], we propose a recommendation algorithm. First, we construct the bipartite reader-book network by book-projection which is named G. Then, for a given reader r_i, we set the initial resource located on each node of G as $f(b_j) = a_{ji}$. That is to say, if the book b_j has been borrowed by r_j, then its initial resource is unit, otherwise it is zero. Apparently, the initial configuration which captures personal preferences is different for every reader. The initial resource can be understood as giving a unit recommending capacity to each collected book. According to the weighted resource allocation process discussed in [15], the final resource, denoted by the vector $\overrightarrow{f'}$, is $\overrightarrow{f'} = W \overrightarrow{f}$. Thus components of f' are

$$f'(b_j) = \sum_{l=1}^{n} w_{jl} f(b_l) = \sum_{l=1}^{n} w_{jl} a_{li} \tag{2}$$

For any reader r_i, all his borrowed books $b_j (1 \leq j \leq n, a_{ji} = 0)$ are sorted in the descending order of $f'(b_j)$, and those books with highest value of final resource are recommended. Note that, the calculation of $f'(b_j)$ should be repeated m times, since the initial configurations are different for every reader.

5   Conclusions

Through collecting the detailed information of the library of USST during one year, we obtain the characteristics of the library lending network, including act size distribution, act degree distribution, node strength distribution, and find they all follow exponential which imply the connecting numbers of most nodes are similar by nature, namely the nodes that is much higher or lower than average rarely exist. In this paper, it shows that the borrowed books in the library are balanced, the lending system is perfect basically, and the library can meet the readers' needs on the whole. In addition, there is a positive correlation between the distribution of act size and book node strength, the same as act degree and reader node strength. Thus, compared with the un-weighted network, the weighted network can not only describe the cooperation situation and configuration, but also the competition results.

From the analysis of the relevant statistics, we can see that the readers adhere to the concept of learning-oriented, while still read the literary book to enrich life. English and computer books are the favor, therefore the library should increase the corresponding books. In the groups of readers, the node strength of teachers are far greater than students, which fully shows that they always stand at the forefront of the discipline, and enhance academic standards actively. At the same time, graduate student have a strong thirst for knowledge and is the backbone of reader's groups.

Furthermore, we proposed a library personal recommendation system based on a proper weighting method. Through the historical borrowed books for a reader, we will give some suggestions to him when he needs some books. But now, the presented recommendation algorithm is just a rough framework whose details have not been exhaustively explored yet. Next, we will study further and model this mechanism in library.

Acknowledgements. We would like to thank the sublibrarian Ling Han and teacher Lan Chen for supplying the data of the USST library. Thanks also to teacher Da-ren He of Yang Zhou University, Ji-ming Li, Dan-rong Zhang, Geng-sheng Zhao of business School in USST for the helpful discussion. This work is supported by Shanghai Leading Academic Discipline Project (S30501).

References

1. Albert, R., Albert, I., Nakarado, G.L.: Structural vulnerability of the North American power grid. J. Physical Review E 69, 25103 (2004)
2. Albert, R., Jeong, H., Barabási, A.-L.: Diameter of the World Wide Web. J. Nature 401, 130–131 (1999)
3. Motter, A.E., Moura, A.P.S., Lai, Y.C., et al.: Topology of the conceptual network of language. J. Physical Review E 65, 065102(R)(2002)
4. Redner, S.: How popular is your paper? An empirical study of the citation distribution. J. The European Physical Journal B 4, 131–134 (1998)
5. Jeong, H., Mason, S., Barabási, A.-L., et al.: Lethality and centrality in protein networks. J. Nature 411, 41–42 (2001)

6. Jeong, H., Tombor, B., Albert, R., et al.: The large-scale organization of metabolic networks. J. Nature 407, 651–654 (2000)
7. Battiston, S., Catanzaro, M.: Statistical properties of corporate board and director networks. J. The European Physical Journal B 438, 345–352 (2004)
8. Boccaletti, S., Latora, V., Moreno, Y., et al.: Complex Networks: Structure and Dynamics. J. Physics Reports 424, 175–308 (2006)
9. Wu, J.S., Di, Z.R.: COmplex networks in statistical physics. J. Progress in Physics 24(1), 18–46 (2004) (in Chinese)
10. Lambiotte, R., Ausloos, M.: Uncovering collective listening habits and music genres in bipartite Networks. J. Physical Review E 72, 066107 (2005)
11. Garey, M.R., Johnson, D.S.: A Guide to the Theory of NP Completeness. In: Computers and Intractability. Freeman Publishers, San Francisco (1979)
12. Scott, J.: Social Network Analysis. In: A Handbook, 2nd edn. Sage Publishers, Thousand Oaks (2002)
13. Guo, M.R.: Statistics and analysis of books lent to West Normal University's Library(2001~2004). J. The journal of the library science in Jiangxi 36(4), 27–29 (2006) (In Chinese)
14. Zhang, D.R., Zhang, N.: The study of the library lending network in complex network. Shanghai: Business School in University of Shanghai for Science and Technology, 17–20 (2007) (in Chinese)
15. Zhou, T., Ren, J., Matúš, M., Zhang, Y.C.: Bipartite network projection and personal recommendation. J. Physical Review E 76, 046115 (2007)
16. Fu, C.H., Zhang, Z.P., Chang, H., et al.: A kind of collaboration–competition networks. J. Physica A 387, 1411–1420 (2008)
17. Zhang, P., Chen, K., He, Y., et al.: Model and empirical study on some collaboration networks. J. Physica A 360, 599–616 (2006)
18. Chang, H., Su, B.B., Zhou, Y.P., et al.: Statistical Mechanics and its Applications. J. Physica A 383, 687–702 (2007)

Temperature-Induced Domain Shrinking in Ising Ferromagnets Frustrated by a Long-Range Interaction

Alessandro Vindigni[1], Oliver Portmann[1], Niculin Saratz[1], Fabio Cinti[2], Paolo Politi[3], and Danilo Pescia[1]

[1] Laboratorium für Festkörperphysik, ETH Zürich, 8093 Zürich, Switzerland
[2] Dipartimento di Fisica, Università di Firenze I-50019 Sesto Fiorentino, Italy
[3] Istituto dei Sistemi Complessi, CNR, I-50019 Sesto Fiorentino, Italy

Abstract. We investigate a spin model in which a ferromagnetic short-range interaction competes with a long-range antiferromagnetic interaction decaying spatially as $\frac{1}{r^{d+\sigma}}$, d being the dimensionality of the lattice. For σ smaller than a certain threshold $\hat{\sigma}$ (with $\hat{\sigma} > 1$), the long-range interaction is able to prevent global phase separation, the uniformly magnetized state favored by the exchange interaction for spin systems. The ground state then consists of a mono-dimensional modulation of the order parameter resulting in a superlattice of domains with positive and negative magnetization. We find that the period of modulation shrinks with increasing temperature T and suggest that this is a universal property of the considered model. For $d = 2$ and $\sigma = 1$ (dipolar interaction) Mean-Field (MF) calculations find a striking agreement with experiments performed on atomically-thin Fe/Cu(001) films. Monte Carlo (MC) results for $d = 1$ also support the generality of our arguments beyond the MF approach.

Keywords: frustrated systems, modulated systems, long-range interactions, competing interactions, Ising model.

The competition between a short-ranged interaction favoring a uniformly *charged* state and a long-range interaction preventing its realization on larger spatial scales is often assumed to be the mechanism underlying pattern formation in chemistry, biology and physics as well as opinion cluster emergence in social networks. A minimal spin model in which the ferromagnetic nearest-neighbor exchange interaction, J, competes with a long-range antiferromagnetic interaction of strength g may hopefully contain enough complexity to be paradigmatic for a variety of realistic systems. For $\sigma \leq \hat{\sigma}$ (see next Sect.), the lowest energy configuration – which is indeed realized at $T = 0$ – is given by a succession of domains with saturated positive and negative magnetization, which alternate in a sharp mono-dimensional modulation of period $2h_{gs}$. At finite T, the spins located at the interface between two oppositely magnetized domains are significantly more susceptible to thermal fluctuations than spins in the interior of the domains. As a result, the balance between the ferromagnetic exchange and the antiferromagnetic long-range interaction is biased in favor of the latter, which finally makes the modulation period shrink as T is increased.

J. Zhou (Ed.): Complex 2009, Part I, LNICST 4, pp. 783–786, 2009.

Ground-State Properties. The spin Hamiltonian we consider is

$$\mathcal{H} = -J \sum_{\langle i,j \rangle} \sigma_i \sigma_j + \frac{g}{2} \sum_{\{i \neq j\}} \frac{\sigma_i \sigma_j}{|r_{ij}|^{d+\sigma}}, \tag{1}$$

where $\sigma_i = \pm 1$ (Ising variables), J and g are positive constants, $\langle i,j \rangle$ and $\{i \neq j\}$ indicate that the sum is extended either to nearest neighbors only or to all the couples respectively. The site indices have to be thought of as integer coordinates locating a spin in a lattice of any dimension, e.g. $i \equiv (i_x, i_y, i_z)$ if $d = 3$; equivalently $r_{ij} = i - j$. When $g = 0$, the uniform state has the lowest energy. If $g \neq 0$, the creation of one domain wall in the uniform state causes an increase of the exchange energy of $2J$ and a net decrease of the long-range interaction energy ΔE_{LR}. The scaling of ΔE_{LR} with the number of spins in the lattice N can be easily estimated in the continuum limit. For $d = 1$, integrating over the sites on the left- (dx) and right-hand (dx') side of the domain wall yields

$$\Delta E_{LR} \sim \frac{1}{r^{\sigma+d}} \overset{\int dx}{\Rightarrow} \frac{1}{r^{\sigma+d-1}} \overset{\int dx'}{\Rightarrow} \frac{1}{r^{\sigma+d-2}}.$$

This estimation can be generalized to any lattice dimension to get

$$\Delta E_{LR} \sim \left[\frac{1}{r^{\sigma-1}} \right]_1^N \underset{N \to \infty}{\sim} \begin{cases} \infty & \text{for } \sigma \leq 1 \Rightarrow \text{domain ground state} \\ < \infty & \text{for } \sigma > 1 \Rightarrow \text{domain/uniform ground state.} \end{cases}$$

When $\Delta E_{LR} \to \infty$ (in the thermodynamic limit $N \to \infty$), the system prefers to split into domains and the ground state turns out to have a mono-dimensionally modulated structure [1,2]. The half-period of modulation h_{gs} depends on the ratio $\frac{J}{g}$. For $\sigma > 1$ a more detailed analysis is required to define the threshold $\hat{\sigma}$ which separates the uniform- from the patterned-ground-state phase[1]. We represent [6] the ground-state configuration in the regime $\sigma \leq \hat{\sigma}$ with a square-wave profile of period $2h$ modulated along the x direction $\sigma_j = \text{Sq}(k_0 j_x) = \sum_{m \geq 0} a_m \sin(k_m j_x)$ (with $k_0 = \frac{\pi}{h}$, $k_m = (2m+1)k_0$ and $a_m = \frac{4}{\pi} \frac{1}{2m+1}$). Exploiting the orthogonality relation $\sum_{j_x=1}^{N_x} e^{-i(k-k')j_x} = N_x \delta_{k,k'}$ (with $j_x + N_x = j_x$), two-point correlations at $T = 0$ can be computed by averaging over the site variables j ($\langle \ldots \rangle_j$ henceforth)

$$\langle \sigma_{j+r} \sigma_j \rangle_j = \frac{1}{N_x} \sum_{j_x} \text{Sq}(k_0(j_x + r_x)) \text{Sq}(k_0 j_x) = \frac{1}{2} \sum_{m \geq 0} a_m^2 \cos(k_m r_x). \tag{2}$$

The form of Eq. (2) suggests the observation of a peak in the structure factor located at every odd higher-harmonic of $k_0 = \frac{\pi}{h}$ at sufficiently low temperatures. This is actually observed in MC simulations performed for $d = 1$ [6], in spite of the fact that no long-range order is expected to occur at any $T \neq 0$ [1]. Eq. (2) allows writing the energy per spin for a square-wave profile

---

[1] For $\sigma > 1$, $\hat{\sigma}$ depends on the ratio $\frac{J}{g}$. In $d = 1$, a straightforward discrete-lattice calculation gives a closed equation involving the Riemann zeta function: $\zeta(\hat{\sigma}) = \frac{J}{g}$ [6].

$$\mathcal{E}_h = 2J\frac{1}{h} + \frac{g}{2}\sum_{m\geq 0} a_m^2 \sum_{r_\nu}\sum_{r_x\geq 1} \frac{\cos{(k_m r_x)}}{|\underline{r}|^{d+\sigma}} = 2J\frac{1}{h} + \sum_{m\geq 0} a_m^2 f_\sigma(k_m); \qquad (3)$$

the exchange contribution comes just from counting the number of domain walls, $\frac{N_x}{h}$, while the sum $\sum_{r_\nu}$ is performed over $d-1$ integer variables according to the lattice dimension. The whole energy (3) depends parametrically on the half-period of modulation, h, so that the ground-state is found by minimizing it with respect to this variable to obtain h_{gs} [1,2,6].

MF Approach and Experiments. An experimental counterpart of our model (1) with $d = 2$ and $\sigma = 1$ is represented by ultrathin Fe films grown epitaxially on Cu(001) [3]. For these specific d and σ, the ground state is expected to be a striped pattern [2]. In the experimental system such a magnetic-domain pattern is indeed encountered together with a variety of different ones [3]. Throughout all these patterns a significant domain width reduction is observed as T is increased [4]. Fig. 1a shows how this experimental fact is well reproduced by a MF treatment of Hamiltonian (1) [5]. In the shadowed region slow-dynamics effects become important so that the equilibrium-thermodynamic description does not apply anymore. In Fig. 1b a typical MF magnetization profile inside a single stripe domain is reported for different T. Apart from $T \sim 0$, the average magnetization of the domain-wall spins (full triangles), is systematically lower than that of inside-domain spins. Consequently, the creation of new domain walls "costs" less and less as T is increased: the balance between exchange and long-range interaction is thus biased with respect to what happens at $T = 0$ and domains with smaller equilibrium size are ultimately favored.

Elastic Model. To the aim of recovering the same phenomenology as with the MF approach but with an alternative treatment of thermal fluctuations, let us consider first the effect of a perturbative displacement field along x, u_{j_x}, of the whole square-wave profile:

$$\sigma_j = \mathrm{Sq}\,(k_0(j + u_j)) = \sum_{m\geq 0} a_m \sin{(k_m(j + u_j))}. \qquad (4)$$

After some algebra [6] and recalling the definition of $f_\sigma(k_m)$ (3), the perturbed energy can be written as

$$\Delta\mathcal{E}_h = \frac{1}{N}\sum_q\sum_{m\geq 0}\left\{a_m^2 k_m^2\left[\frac{1}{2}f_\sigma(k_m - q) + \frac{1}{2}f_\sigma(k_m + q) - f_\sigma(k_m)\right]|\tilde{u}_q|^2\right\}, (5)$$

where $\tilde{u}_q$ is the Fourier transform of the displacement field. As far as the large-distance behavior is concerned, Eq. (5) can be expanded for $q \ll k_0$:

$$\Delta\mathcal{E}_h = \frac{1}{N}\sum_q\left[\frac{1}{2}k_0^2\frac{\partial^2\mathcal{E}_h}{\partial k_0^2}q^2|\tilde{u}_q|^2\right]. \qquad (6)$$

Eq. (6) is formally equivalent to a Planar Degenerate System Hamiltonian. For such systems and $d = 1$ the correlation length is expected to behave like $\xi \sim \frac{1}{T}$, which is in good agreement with our MC results [6]. Even in the absence of

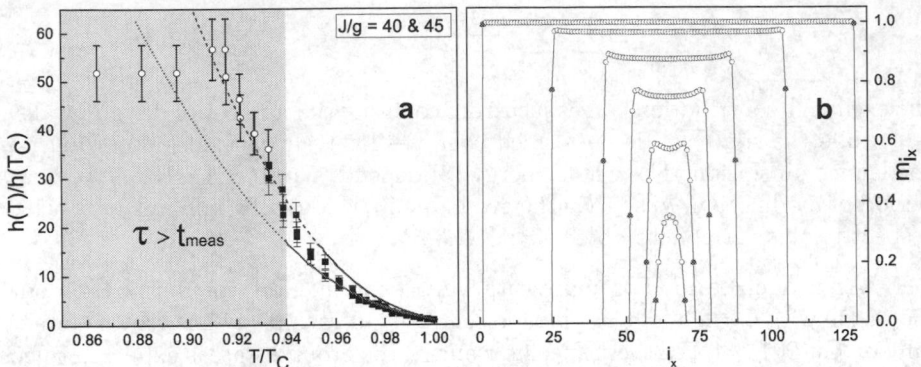

Fig. 1. a) Relative domain width variation as a function of the reduced temperature, T_C being the Curie temperature. Experimental points correspond to labyrinthine (open circles) and striped (full square) patterns. Lines correspond to MF calculations with $d = 2$ and $\sigma = 1$ performed for $\frac{J}{g} = 40$ (violet) and $\frac{J}{g} = 45$ (blue): exact (solid), parabolic extrapolation (dotted). In the shadowed region the relaxation time – estimated independently – becomes larger than the characteristic time of the measurements. **b)** MF magnetization profile inside a striped domain. The average magnetization on each site m_{i_x} is plotted *versus* the site index i_x itself.

long-range order, a characteristic length scale is preserved at finite temperatures in the form of the modulation period of the two-point correlation function. In the limit $T \to 0$, this period approaches h_{gs} continuously from below so that it can be considered the $d = 1$ counterpart of the temperature-dependent domain width in ultrathin Fe/Cu(001) films [4,5].

Justifying – in the framework of the elastic model – the decrease of the modulation period with the increase of T independently of the lattice dimensionality d and of the occurrence of long-range order is a goal for future work.

References

1. Giuliani, A., Lebowitz, J.L., Lieb, E.H.: Ising models with long-range dipolar and short range ferromagnetic interactions. Phys. Rev. B 74, 064420 (2006)
2. MacIsaac, A.B., Whitehead, J.P., Robinson, M.C., De'Bell, K.: Striped phases in two-dimensional dipolar ferromagnets. Phys. Rev. B 51, 16033 (1995)
3. Portmann, O., Vaterlaus, A., Pescia, D.: An inverse transition of magnetic domain patterns in ultrathin films. Nature 422, 701–704 (2003)
4. Portmann, O., Vaterlaus, A., Pescia, D.: Observation of Stripe Mobility in a Dipolar Frustrated Ferromagnet. Phys. Rev. Lett. 96, 047212 (2006)
5. Vindigni, A., Saratz, N., Portmann, O., Pescia, D., Politi, P.: Stripe width and nonlocal domain walls in the two-dimensional dipolar frustrated Ising ferromagnet. Phys. Rev. B 77, 092414 (2008)
6. Cinti, F., Portmann, O., Pescia, D., Vindigni, A.: One-dimensional Ising ferromagnet frustrated by long-range interactions at finite temperatures. ArXiv: 0812.0907v1 [cond-mat]

Slowdown in the Annihilation of Two Species Diffusion-Limited Reaction on Fractal Scale-Free Networks

Chang-Keun Yun, Byungnam Kahng, and Doochul Kim

Department of Physics and Astronomy and Center for Theoretical Physics,
Seoul National University, Seoul 151-747, Korea
bkahng@snu.ac.kr

Abstract. In the diffusion-limited reaction process $A + B \to \emptyset$ on random scale-free networks, particle density decays as $\rho(t) \sim t^{-\alpha}$ when $\rho_A(0) = \rho_B(0)$, where $\alpha > 1$ for the degree exponent $2 < \gamma < 3$ and $\alpha = 1$ for $\gamma \leq 3$. We investigate the reaction on fractal scale-free networks numerically, finding $\rho(t)$ decays slowly with the exponent $\alpha \approx d_s/4 < 1$, where d_s is the spectral dimension of the network.

Keywords: diffusion-limited reaction, fractal scale-free network, segregation.

1 Introduction

Diffusion-limited reaction kinetics has been studied for long time. It can be used for modeling chemical reactions, epidemic spreading, and so on. Here, we limit our interest to the two-species annihilation process $A + B \to \emptyset$.

It is known that if the initial densities of A and B particles are equal, the density decays as $\rho(t) \sim t^{-\alpha}$. In a mean-field approximation, α is 1. This approximation is valid for the processes in Euclidean space with $d > d_c = 4$. For $d < 4$, however, the mean-field approach is invalid, and α is d/d_c, which is less than 1. This behavior is caused by the segregation effect: A-rich or B-rich domains form, and reactions can take place only at the limited area, that is, the boundary between those domains [1,2,3]. And this argument can be extended to the fractal case, giving $\alpha = d_s/4$ when $d_s < 4$, where d_s is the spectral dimension. Although this value is in good agreement with many numerical results [15,16], this extension has been questioned [3] in some cases.

Recent studies show that in complex networks, particle density can decay faster, and α can be larger than 1 [4]. In random scale-free networks, α can be obtained analytically and is $1/(\gamma - 2)$ for $2 < \gamma < 3$ and 1 for $\gamma > 3$ where γ is the exponent of the degree distribution $P_d(k) \sim k^{-\gamma}$ [8]. This fast decay can be explained in terms of the existence of hubs and extremely small diameter. Particles tends to move towards hubs which are closely located. Distance between particles of different species are close because of small diameter. These two factors cause A and B particles to mix, accelerating the reactions.

J. Zhou (Ed.): Complex 2009, Part I, LNICST 4, pp. 787–791, 2009.

2   Two-Species Annihilation on Fractal Scale-Free Networks

We study two-species annihilation reaction on fractal scale-free networks. When a network is fractal, this network satisfies the scaling $N_B(l_B) \sim l_B^{-d_f}$, where l_B is the size of boxes and N_B is the number of boxes needed to cover the network. In a fractal scale-free network, hubs are located repulsively. Particles move towards the local hubs nearby, and reactions occur. As a result, it is likely to remain particles of the same species in the vicinity of the local hubs, forming domains. After forming domains, reactions take place at the boundaries between domains, causing particle density to decay slowly.

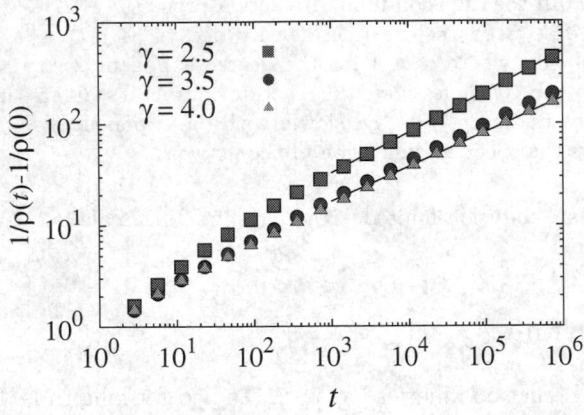

Fig. 1. The particle density as a function of time on the critical branching trees. Guidelines have slopes 0.40 (top) and 0.34 (bottom).

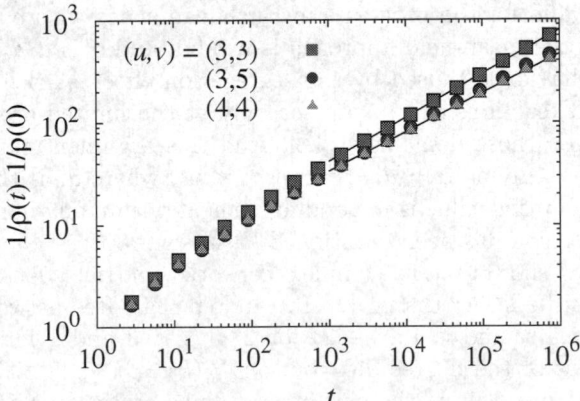

Fig. 2. The particle density as a function of time on the (u, v)-flower networks. Guidelines have slopes 0.43 (top) and 0.38 (bottom).

Table 1. Comparison of the numerically obtained exponent α_{num} with $d_s/4$

	CBT			(u,v)-flower				(u,v)-tree		
γ	α_{num}	$d_s/4$	(u,v)	γ	α_{num}	$d_s/4$	(u,v)	γ	α_{num}	$d_s/4$
2.5	0.40	0.38	(2,2)	3	0.53	0.5	(2,2)	3	0.34	0.33
2.7	0.37	0.35	(2,4)	3.58	0.45	0.43	(2,4)	3.58	0.37	0.36
3.5	0.36	0.33	(3,3)	3.58	0.43	0.41	(3,3)	3.58	0.31	0.31
4.0	0.34	0.33	(2,6)	4	0.43	0.42	(2,6)	4	0.40	0.38
4.5	0.34	0.33	(4,4)	4	0.38	0.38	(4,4)	4	0.31	0.30

We use the critical branching tree [10], (u,v)-flower and (u,v)-tree [19] as substrates. Simulation results show that the particle density decays slowly as shown Figs. 1, 2. Numerical values of α are less than 1 and close to $d_s/4$ as can be seen in Table 1 for the critical branching tree, (u,v)-flower, and (u,v)-tree.

3   Role of Local Hubs in Fractal SF Networks

To confirm the role of local hubs, we measure the particle density on the networks generated by rewiring the links of $(3,3)$-flower under the conservation of the degree distribution. Fig. 3 shows the results. As we rewire the links, the density of particle decays faster and α increases from 0.43 to 1, which is the value on scale-free network with the same degree exponent γ.

Next, we measure a quantity introduced in Ref. [20],

$$Q_{AB} = \frac{N_{AB}}{N_{AA} + N_{BB}}, \tag{1}$$

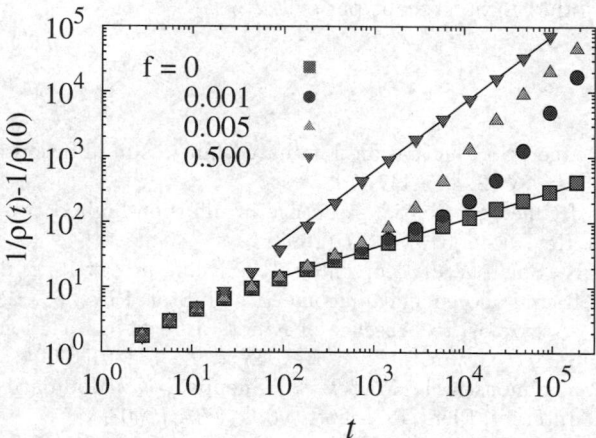

Fig. 3. The particle density on a $(3,3)$-flower($f = 0$) and its rewired networks. f is the fraction of rewired links. Guidelines have slopes 0.43 (bottom) and 1.0 (top).

where N_{AB} is the the number of contacts between A and B particles, N_{AA} and N_{BB} are defined similarly. Q_{AB} close to 0 indicates that segregation occurs, and the value close to 1 means A and B particles are mixed completely. Fig. 4 is the plot of Q_{AB} as a function of time. From the figure, we confirm strong segregation on a fractal network. And as more links are rewired and the distance between hubs are closer, the segregation effect decreases.

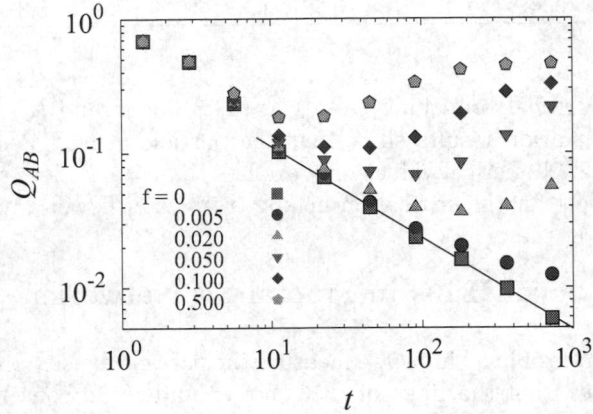

Fig. 4. Plot of Q_{AB} for the networks in Fig. 3

4   Summary

To sum up, in the two-species annihilation process $A + B \rightarrow \emptyset$ on fractal SF networks, segregation forms originated from the existence of local hubs, causing the particle density to decay slowly with the exponent $\alpha < 1$. More detailed results can be found in our recent paper [22].

References

1. Kang, K., Redner, S.: Fluctuation-dominated kinetics in diffusion-controlled reactions. Phys. Rev. A 32, 435–447 (1985)
2. Leyvraz, F., Redner, S.: Spatial structure in diffusion-limited two-species annihilation. Phys. Rev. A 46, 3132–3147 (1992)
3. Lindenberg, K., Sheu, W.-S., Kopelman, R.: Scaling properties of diffusion-limited reactions on fractal and euclidean geometries. J. Stat. Phys. 65, 1269–1283 (1991)
4. Gallos, L.K., Argyrakis, P.: Absence of Kinetic Effects in Reaction-Diffusion Processes in Scale-Free Networks. Phys. Rev. Lett. 92, 138301 (2004)
5. Eriksen, K.A., Simonsen, I., Maslov, S., Sneppen, K.: Modularity and Extreme Edges of the Internet. Phys. Rev. Lett. 90, 148701 (2003)
6. Noh, J.D., Rieger, H.: Random Walks on Complex Networks. Phys. Rev. Lett. 92, 118701 (2004)
7. Catanzaro, M., Boguñá, M., Pastor-Satorras, R.: Diffusion-annihilation processes in complex networks. Phys. Rev. E 71, 056104 (2005)

8. Weber, S., Porto, M.: Multicomponent reaction-diffusion processes on complex networks. Phys. Rev. E 74, 046108 (2006)
9. Song, C., Havlin, S., Makse, H.A.: Self-similarity of complex networks. Nature (London) 433, 392–395 (2005)
10. Goh, K.-I., Salvi, G., Kahng, B., Kim, D.: Skeleton and Fractal Scaling in Complex Networks. Phys. Rev. Lett. 96, 018701 (2006)
11. Yook, S.-H., Radicchi, F., Meyer-Ortmanns, H.: Self-similar scale-free networks and disassortativity. Phys. Rev. E 72, 045105(R) (2005)
12. Song, C., Havlin, S., Makse, H.A.: Origins of fractality in the growth of complex networks. Nat. Phys. 2, 275–281 (2006)
13. Barabási, A.-L., Albert, R.: Emergence of Scaling in Random Networks. Science 286, 509–512 (1999)
14. Kim, J.S., Goh, K.-I., Salvi, G., Oh, E., Kahng, B., Kim, D.: Fractality in complex networks: Critical and supercritical skeletons. Phys. Rev. E 75, 016110 (2007)
15. Meakin, P., Stanley, H.E.: Novel dimension-independent behaviour for diffusive annihilation on percolation fractals. J. Phys. A 17, L173 (1984)
16. Zumofen, G., Klafter, J., Blumen, A.: Scaling properties of diffusion-limited reactions: Simulation results. Phys. Rev. A 43, 7068–7069 (1991)
17. Burda, Z., Correia, J.D., Krzywicki, A.: Statistical ensemble of scale-free random graphs. Phys. Rev. E 64, 046118 (2001)
18. Gallos, L.K., Song, C., Havlin, S., Makse, H.A.: Scaling theory of transport in complex biological networks. Proc. Nat. Aca. Sci. USA 104, 7746–7751 (2007)
19. Rozenfeld, H.D., Havlin, S., ben-Avraham, D.: Fractal and transfractal recursive scale-free nets. New J. Phys. 9, 175 (2007)
20. Gallos, L.K., Argyrakis, P.: Influence of a complex network substrate on reaction-diffusion processes. J. Phys., Condens. Matter 19, 065123 (2007)
21. Lee, S., Yook, S.-H., Kim, Y.: Diffusive capture processes for information search. Physica A 385, 743–749 (2007)
22. Yun, C.-K., Kahng, B., Kim, D.: Segregation in the annihilation of two-species reaction-diffusion processes on fractal scale-free networks. arXiv:0811.2293v1 [cond-mat.dis-nn] (2008)

SIRS Dynamics on Random Networks: Simulations and Analytical Models

Ganna Rozhnova and Ana Nunes

Centro de Física Teórica e Computacional and Departamento de Física,
Faculdade de Ciências da Universidade de Lisboa,
P-1649-003 Lisboa Codex, Portugal
a_rozhnova@cii.fc.ul.pt

Abstract. The standard pair approximation equations (PA) for the Susceptible-Infective-Recovered-Susceptible (SIRS) model of infection spread on a network of homogeneous degree k predict a thin phase of sustained oscillations for parameter values that correspond to diseases that confer long lasting immunity. Here we present a study of the dependence of this oscillatory phase on the parameter k and of its relevance to understand the behaviour of simulations on networks. For $k = 4$, we compare the phase diagram of the PA model with the results of simulations on regular random graphs (RRG) of the same degree. We show that for parameter values in the oscillatory phase, and even for large system sizes, the simulations either die out or exhibit damped oscillations, depending on the initial conditions. This failure of the standard PA model to capture the qualitative behaviour of the simulations on large RRGs is currently being investigated.

Keywords: stochastic epidemic models, oscillations, pair approximations, random regular graphs.

A number of approaches has been used to study the spreading dynamics of an infectious disease. A common paradigm, emerging from a simple deterministic framework, is to assume that populations are not spatially distributed so that individuals mix perfectly and contact each other with equal probability. Thus in the limit of infinite populations, the time evolution of the disease is described in terms of the densities of infectives and susceptibles as a function of time, and governed by a system of coupled ordinary differential equations which can be deduced from the law of mass action [1]. Another approach is to use stochastic dynamics on a lattice (or more general graphs) where the variables at each node represent the state of an individual. The effects of spatial correlations that mass action models disregard play an important role in the behaviour of infection dynamics on graphs, and therefore also in real populations [2]. The ordinary pair approximation (PA) as well as various improvements to include higher order correlations have been proposed in the context of ecological and epidemiological deterministic models [3]. In [4], the performance of the PA in the description of the steady states and the dynamics of the Susceptible-Infective-Recovered-Susceptible (SIRS) model on the hypercubic lattice was analyzed in detail.

J. Zhou (Ed.): Complex 2009, Part I, LNICST 4, pp. 792–797, 2009.

In this study, we consider the dynamics of the same epidemic model on a random network of homogeneous degree k and N nodes, a regular random graph of degree k (RRG-k). Each node can be occupied by an individual in susceptible (S), infected (I), or recovered (R) state. Infected individuals recover at rate δ, recovered individuals lose immunity at rate γ, and infection of the susceptible node occurs at infection rate λ multiplied by the number of its infected nearest neighbours n, $n \in \{0, 1, \ldots, k\}$:

$$I \xrightarrow{\delta} R \,,$$
$$R \xrightarrow{\gamma} S \,, \tag{1}$$
$$S \xrightarrow{\lambda n} I \,.$$

In the infinite population limit, with the assumptions of spatial homogeneity and uncorrelated pairs, the system is described by the deterministic equations of the standard or uncorrelated PA [4]:

$$\frac{d\,s}{d\,t} = \gamma\,(1 - i - s) - k\lambda\,si \,,$$

$$\frac{d\,i}{d\,t} = k\lambda\,si - \delta\,i \,,$$

$$\frac{d\,si}{d\,t} = \gamma\,ri - (\lambda + \delta)\,si + \frac{(k-1)\lambda\,si}{s}(s - sr - 2si) \,, \tag{2}$$

$$\frac{d\,sr}{d\,t} = \delta\,si + \gamma\,(1 - s - i - ri - 2sr) - \frac{(k-1)\lambda\,si\,sr}{s} \,,$$

$$\frac{d\,ri}{d\,t} = \delta\,(i - si) - (\gamma + 2\delta)\,ri + \frac{(k-1)\lambda\,si\,sr}{s} \,.$$

In the above equations the variables s, i stand for the probability that a randomly chosen node is in state S, I, and the variables si, sr, ri stand for the probability that a randomly chosen pair of nearest neighbour nodes is an SI, SR, RI pair. As expected, neglecting the pair correlations and setting the pair state probabilities equal to the product of the node state probabilities these equations reduce to the classic equations of the randomly mixed SIRS model.

The phase diagram of the PA SIRS model (2) for $k = 4$ is plotted in Fig. 1a). We have set the time scale so that $\delta = 1$. Region I represents susceptible-absorbing states and region II corresponds to active states that can be asymptotically stable nodes or asymptotically stable foci. The critical line separating regions I and II corresponds to the transcritical bifurcation curve that is given by $\lambda_c(\gamma) = (\gamma + 1)/(3\gamma + 2)$ (black dashed line). In addition, for small values of γ we find a new phase boundary (black solid line), that corresponds to a supercritical Hopf bifurcation of the nontrivial equilibrium and has been missed in previous studies of this model [4]. This boundary separates the active phase with constant densities from an active phase with oscillatory behavior, that is stable at low γ.

In the thin phase of region III, the PA model predicts sustained oscillations in the thermodynamic limit. We have performed a systematic study of the

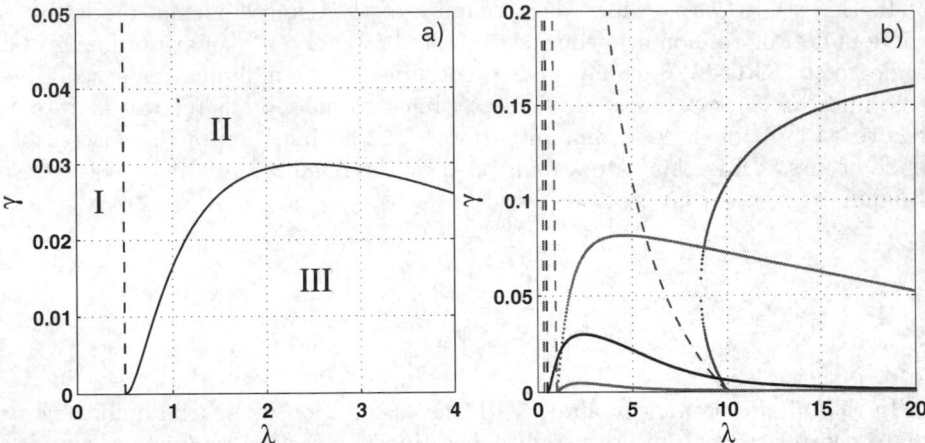

Fig. 1. a) Phase diagram in the (λ, γ) plane for the PA SIRS model, where region I corresponds to susceptible-absorbing states and region II corresponds to active states with nonzero infective densities. The critical line between regions I and II is the dashed line. The second critical curve (solid line) bounds a region with limit cycle solutions (region III). Parameters: $\delta = 1$, $k = 4$. b) Phase diagram of the PA SIRS model for $\delta = 1$ and $k = 2.1$, $k = 3$, $k = 4$, $k = 5$. Dashed (dotted) lines correspond to the transcritical (supercritical Hopf) bifurcation curves. The oscillatory region III becomes smaller as k increases.

dependence of this oscillatory phase on the parameter k and of its relevance to understand the behaviour of simulations on networks.

The phase diagram of the PA model for $\delta = 1$ and several values of k in the range $k > 2$ is shown in Fig. 1b). The critical lines separating the absorbing and the active phases (dashed lines) are given by $\lambda_c(\gamma) = (\gamma + 1)/((k - 1)\gamma + k - 2)$. Within the active phase, the dotted lines are numerical plots of Hopf bifurcation curves. The oscillatory phase is large for $k \gtrsim 2$ and it gets thinner as k increases, but it persists for the whole range of $2 < k \lesssim 6$. A similar phase diagram, with a Hopf bifurcation critical line bounding an oscillatory phase, was reported in other studies of related models [5], where SIR dynamics with different mechanisms of replenishment of susceptibles is modelled at the level of pairs with the standard or another closure approximation. These different models all exhibit an oscillatory phase in the regime of slow driving through introduction of new susceptible individuals (small γ in the present case). This suggests that this oscillatory phase may be related with the phenomenon of recurrent epidemics in infectious diseases that confer permanent or long lasting immunity.

We have compared the behaviour of the PA SIRS model (2) for $k = 4$ with the results of stochastic simulations on RRG-4 for several system sizes. In the stochastic simulations, the system was set in a random initial condition with given node and pair densities and an efficient algorithm for stochastic processes in spatially structured systems based on Gillespie's method [6] was used to update

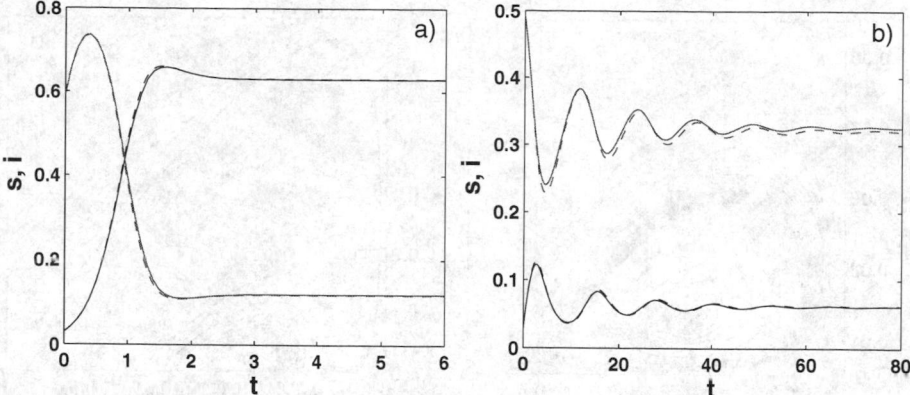

Fig. 2. For $\delta = 1$, $k = 4$, comparison of the solutions of the PA deterministic model (dashed lines) with the results of stochastic simulations (solid lines) on a RRG-4 with $N = 10^6$ for parameter values in region II. Susceptible and infective densities are plotted starting from an initial condition with a small density of infectives. Parameters: a) $\gamma = 2.5$, $\lambda = 2.5$; b) $\gamma = 0.1$, $\lambda = 2.5$.

the states of the nodes according to the processes of infection, recovery and immunity waning (1). For each set of parameter values and initial conditions, the simulations were averaged over 10^3 realizations of the RRG-4 graph.

The results of stochastic simulations for $N = 10^6$ and solutions of the PA SIRS equations (2) are shown in Figs. 2 and 3. The susceptible (blue lines) and the infective (black lines) densities are shown in Fig. 2 for two sets of parameter values: $\gamma = 2.5$, $\lambda = 2.5$ (Fig. 2a)) and $\gamma = 0.1$, $\lambda = 2.5$ (Fig. 2b)). The numerical solutions of the PA SIRS equations are plotted in dashed lines, and the results of the simulations in solid lines. For parameter values well within region II of the phase diagram as in Fig. 2a) there is excellent agreement between the solutions of the PA SIRS model for the same initial densities and the results of the stochastic simulations, both for the transient behaviour and for the steady states. This agreement deteriorates as γ decreases and the boundary of the oscillatory region is approached as can be seen in Fig. 2b). For parameter values in the oscillatory region III most simulations (black solid line) die out after a short transient (Fig. 3b)) while the corresponding solutions of the PA SIRS deterministic model (blue solid line) converge to the stable limit cycle for all initial conditions (a typical set is denoted by B in the plot). By choosing initial conditions not far from the stable cycle predicted by the PA SIRS model to avoid extreme susceptible depletion during the transient, damped oscillations towards a non trivial equilibrium may also be observed in region III. In Fig. 3a) a plot is shown of one of these surviving simulations (black solid line), together with the solution of the PA equations (blue solid line) for the same parameter values and initial conditions (in the plot denoted by A). Thus, instead of an oscillatory phase, the stochastic model on RRGs exhibits in region III a bistability phase, even for large system sizes.

This failure of the PA model to capture the qualitative behaviour of the simulations on large RRGs is currently being investigated. Extinctions due to finite

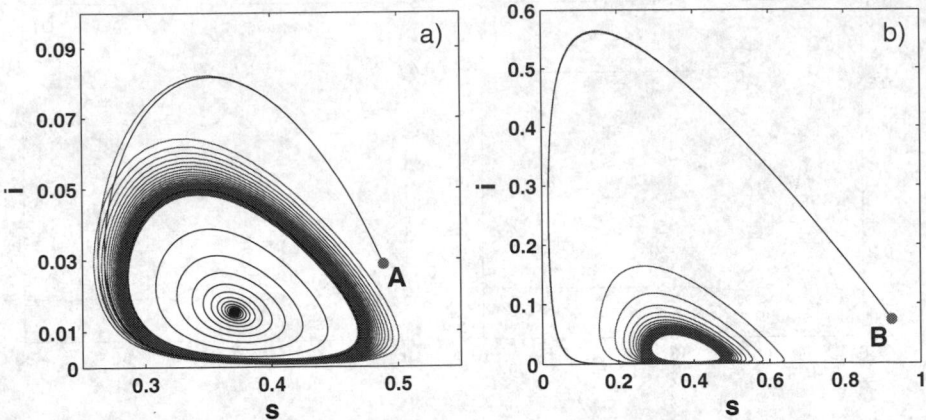

Fig. 3. For $\delta = 1$, $k = 4$, comparison of the solutions of the PA deterministic model with the results of stochastic simulations on a RRG-4 with $N = 10^6$ for parameter values in region III, $\gamma = 0.025$, $\lambda = 2.5$, and two sets of initial conditions A and B marked by dots in the (s, i) plane. The solutions of the PA equations spiral towards a limit cycle. In the simulations, the disease either goes extinct (right panel) or a non-oscillatory steady state is approached (left panel). Initial conditions: a) $s \approx 0.4889$, $i \approx 0.0287$, $si \approx 0.0104$, $sr \approx 0.2369$, $ri \approx 0.0129$; b) $s \approx 0.9240$, $i \approx 0.0731$, $si \approx 0.0558$, $sr \approx 0.0024$, $ri \approx 0.0005$.

size are one of the reasons why the oscillatory phase is seen as an absorbing phase in the stochastic simulations. Indeed, as can be seen in Fig. 3, the oscillations predicted by the PA SIRS deterministic model attain very small densities of infectives during a significant fraction of the period ($i < 10^{-5}$ for initial conditions B in the transient regime). It would be interesting to check whether the more regular oscillations that have been observed in the standard PA for some predator-prey models [7] persist in stochastic simulations of these models on RRGs.

However, the breakdown of the PA SIRS model as the boundary of the oscillatory phase is approached from above and the bistability regime found in region III show that there are other effects at play. The standard pair approximation is only valid for tree-like structures where each node has exactly the same number of contacts and there are no loops, the Bethe lattices. These infinite structures cannot be simulated on a computer. On the other hand, classic results of graph theory show that a particular realization of a RRG-k will contain a large number of loops, of which the large majority are long (with respect to the average path length), so that locally the graph is essentially tree-like. One would expect then the PA to perform well on RRGs, provided they are large enough.

Increasing system size up to $N = 10^7$ we still find suppression of oscillations in region III and significant discrepancies between the transient and steady states of the PA SIRS solutions and the results of the simulations in region II close to the boundary with region III. A similar problem of oscillation emergence and suppression and quantitative differences in Monte Carlo simulations versus mean-field approximation and PA of an evolutionary Rock-Scissors-Paper game on different structures was carefully investigated in [8]. For this problem, a more accurate multi-site approximation instead of the PA was shown to solve the qualitative and quantitative discrepancies with the simulations. The study of improved models beyond the PA for SIRS dynamics on RRGs will be the subject of future work.

Acknowledgments. Financial support from the Foundation of the University of Lisbon and the Portuguese Foundation for Science and Technology (FCT) under contracts POCI/FIS/55592/2004 and POCTI/ISFL/2/618 is gratefully acknowledged. The first author (GR) was also supported by FCT under grant SFRH/BD/32164/2006 and by Calouste Gulbenkian Foundation under its Program 'Stimulus for Research'.

References

1. Murray, J.D.: Mathematical Biology I: An Introduction. Springer, New York (2002)
2. Keeling, M.J., Eames, K.T.D.: Networks and Epidemic Models. J. R. Soc. Interface 2, 295–307 (2005)
3. Matsuda, H., Ogita, N., Sasaki, A., Sato, K.: Statistical Mechanics of Population: The Lattice Lotka-Volterra Model. Prog. Theor. Phys. 88, 1035–1049 (1992); Keeling, M.J., Rand, D.A., Morris, A.J.: Correlation Models for Childhood Epidemics. Proc. R. Soc. Lond. B 264, 1149–1156 (1997); van Baalen, M.: Pair Approximations for Different Spatial Geometries. In: Dieckmann, U., Law, R., Metz, J.A.J. (eds.) The Geometry of Ecological Interactions: Simplifying Spatial Complexity, pp. 359–387. Cambridge University Press, Cambridge (2000)
4. Joo, J., Lebowitz, J.L.: Pair Approximation of the Stochastic Susceptible-Infected-Recovered-Susceptible Epidemic Model on the Hypercubic Lattice. Phys. Rev. E 70, 036114 (2004)
5. Rand, D.A.: Correlation Equations and Pair Approximations for Spatial Ecologies. In: McGlade, J. (ed.) Advanced Ecological Theory: Principles and Applications, pp. 100–142. Blackwell Science, Oxford (1999); Morris, A.J.: Representing Spatial Interactions in Simple Ecological Models. PhD dissertation, University of Warwick, Coventry, UK (1997); Benoit, J., Nunes, A., Telo da Gama, M.M.: Pair Approximation Models for Disease Spread. Eur. Phys. J. B 50, 177–181 (2006)
6. Bortz, A.B., Kalos, M.H., Lebowitz, J.L.: A New Algorithm for Monte Carlo Simulation of Ising Spin Systems. J. Comput. Phys. 17, 10–18 (1975); Gillespie, D.T.: A General Method for Numerically Simulating the Stochastic Time Evolution of Coupled Chemical Reactions. J. Comput. Phys. 22, 403–434 (1976)
7. Satulovsky, J.E., Tomé, T.: Stochastic Lattice Gas Model for a Predator-Prey Sytem. Phys. Rev. E 49, 5073–5079 (1994); Tomé, T., de Carvalho, K.C.: Stable Oscillations of a Predator-Prey Probabilistic Cellular Automaton: a Mean-Field Approach. J. Phys. A: Math. Theor. 40, 12901–12915 (2007)
8. Szabó, G., Szolnoki, A., Izsák, R.: Rock-Scissors-Paper Game on Regular Small-World Networks. J. Phys. A: Math. Gen. 37, 2599–2609 (2004); Szolnoki, A., Szabó, G.: Phase Transitions for Rock-Scissors-Paper Game on Different Networks. Phys. Rev. E 70, 037102 (2004)

Self-organized Collaboration Network Model Based on Module Emerging

Hongyong Yang, Lan Lu, and Qiming Liu

School of Computer Science and Technology, Ludong University, Yantai 264025, China
hyyang_ld@yahoo.com.cn

Abstract. Recently, the studies of the complex network have gone deep into many scientific fields, such as computer science, physics, mathematics, sociology, etc. These researches enrich the realization for complex network, and increase understands for the new characteristic of complex network. Based on the evolvement characteristic of the author collaboration in the scientific thesis, a self-organized network model of the scientific cooperation network is presented by module emerging. By applying the theoretical analysis, it is shown that this network model is a scale-free network, and the strength degree distribution and the module degree distribution of the network nodes have the same power law. In order to make sure the validity of the theoretical analysis for the network model, we create the computer simulation and demonstration collaboration network. By analyzing the data of the network, the results of the demonstration network and the computer simulation are consistent with that of the theoretical analysis of the model.

Keywords: Module emerging, Scientific Collaboration, Self-organized, Network model, Complexity.

1 Introduction

The studies of the complex network have opened out the internal essence of many phenomena in real world. Erdos and Renyi presented the random network model (called ER model) [1] since 1960, ER random graph has been used as a basic model to study complex network. In 1998, a paper, written by Watts and Strogatz on small-world network, was appeared in Nature [2] to depict the feature that there exist the shorter paths among nodes in a large network. Another seminal paper was written by Barabasi and Albert on scale-free networks in Science in 1999[3]. The scale-free network (also called BA model) is a network model inspired to the formation of the World Wide Web and is based on two basic ingredients: growth and preferential attachment. The degree distribution of BA model produces the form $p(k) \propto k^{-\gamma}$. Because the power distribution absents distinct characteristic length, this class network is called scale-free network. It has been shown that many real complex networks are small-world and scale-free network, these include transportation networks, phone call networks, Internet, WWW, the actors collaboration networks, scientific coauthorship and citation networks so on [4,5,6].

J. Zhou (Ed.): Complex 2009, Part I, LNICST 4, pp. 798–805, 2009.

The presentation of these two original results affords new theory to realize the complex network, and the upsurge of reaching complex network is raised in many science fields. The statistic of the static state parameter is analyzed for Internet, WWW, cinema and television actors' collaboration network, scientists' collaboration network, people relation network, and linguistics network so on. Based on the network model, the dynamics features are studied to infectious disease, percolation model, network searching, and network navigation. Especially, for the research of the coauthorship network in scientific articles, many new results have been obtained recently [7-9]. These researches enrich the realization for complex network, and increase understands for the new characteristic of complex network.

The collaboration network of scientific researcher is a network describing the collaboration relation of scientific researcher. In this network, each node is denoted as a scientific researcher of the network, and each link between two nodes is created if two scientific researchers have published an article together. In this paper, based on the complex network, we establish an evolution network model to study the collaboration relation of scientific researchers.

The remainder of this paper is organized as follows. In section 2, the definition of the module (also called motif) is presented, and a scientific collaboration evolving model is taken on based on the motif emerging. The complex characteristics of the network model are calculated in the section 3, and a computer simulation is created to validate the theoretic results. In section 4, a demonstration network is constructed to show the efficiency of the network model. Finally, some conclusions are made in section 5.

2 Self-organized Network Evolution Model

The scientific cooperation network is used to describe the collaboration relation among the scientific thesis authors. Generally, every scientific researcher is depicted as a node of the network, and the collaboration relation of the people (publishing a paper together) is pictured as the link between the nodes. When analyzing the collaboration network, we found there are a lot of authors in a paper and these authors compose of a full-connected sub-network. Here we call this sub-network as a module (also called motif).

Since there are a lot of the same authors in the different papers, a module is embedding into the scientific collaboration network by the same author. Then a self-organized collaboration network model based on motif emerging is created. Following the analysis, we present a new scientific collaboration network evolution model:

Initially, suppose m_0 articles and k_0 authors in the network. The authors of a paper will compose of a motif; let the sum of the initialized motif be n_0. With the time t changing, we increase an article every time, and author collaboration network evolves as following (Here, we suppose K is the most author number in a paper):

(1) There are i authors in a new paper with a probability q_i, and these authors will group a full-connected sub-network as a motif, where $\sum_{i=1}^{K} q_i = 1$. When $i=1$, there is only one node in the motif, i.e., this paper is written by one author.

(2) There are j authors in the motif who are identical as that in the primary network with a probability p_j, where $\sum_{j=0}^{K} p_j = 1$. The new motif will be inset the primary network with the same authors. When $j=0$, the nodes in the motif will have none of the identical nodes, and the motif will set into the network without connected. When $j=K$, all node in the motif have been included in the network, and the nodes will be invariable after the motif inset.

Rule 1: The motif degree of the node is defined as 1 if an author publishes a paper; and the motif degree will add 1 if this author publishes a new paper again, and so on.

Rule 2: The link strength between the nodes is defined as the strength is 1 if two nodes are linked firstly; and the strength adds 1 if these two nodes are linked again, and so on.

Rule 3: The node is selected to inset the network, following the method of motif degree preference, the selecting probability

$$\Pi(i) = n_i / \sum_j n_j \tag{1}$$

where the numerator n_i as the motif degree of node i, which is the summation of the paper of node i, the denominator is the summation of the motif degree of all nodes in the network.

3 Degree Distribution of Network Model

3.1 Analysis on the Characteristics of the Network Model

In this section, we will research the complex feature of the evolution model. In the scientific collaboration, one author can publish several papers, i.e., the module degree of one node can be very large. One author can cooperate with many authors, i.e., the link strength of the node can be very large. In the description of the network, the degree distribution $p(k)$ is used to denote the probability of a node with degree k. Next, we calculate the module degree distribution and the strength distribution of the node in the new evolution model.

The module degree of a node is the number of the paper published by the author. Following the algorithm description, the new module insets the network connected with j nodes that are decided by the probability p_j. The diversification of the module degree of the node i

$$\frac{\partial n_i}{\partial t} = \sum_{l=1}^{K} q_l \sum_{j=0}^{l} p_j j \Pi(i). \tag{2}$$

Suppose $l=1, \ldots\ldots, K$, and let $\sum_{j=0}^{l} p_j j = J_l$, we have

$$\sum_{l=1}^{K} q_l J_l = \bar{J}.$$

Since the sum of the module degree satisfying

$$\sum_{j} n_j = n_0 + t(\sum_{l=1}^{K} q_l l),$$ (3)

Let $\sum_{l=1}^{K} q_l l = L$, when t is large enough, we get

$$\frac{\partial n_i}{\partial t} = \bar{J}\Pi(i) = \frac{\bar{J}}{L} \frac{n_i}{t}.$$ (4)

From $n_i(t_i) = 1$ at $t=t$, we obtain

$$n_i(t) = (\frac{t}{t_i})^{\beta},$$ (5)

where $\beta = \bar{J}/L$. We can calculate the module degree distribution using the method in [4,5]

$$p(n) = \frac{t}{(n_0 + t)\beta} n^{-(\frac{1}{\beta}+1)} \propto n^{-(\frac{1}{\beta}+1)}.$$ (6)

Now we will calculate the strength distribution of the node. In the scientific collaboration network, if two scientific researchers publish one paper together, the link strength of the nodes is 1 with a line between two authors. If these two researchers collaborate once again, the link strength will increase 1. The strength of a node is equal to the sum of the link strengths of its neighbors. Following the evolving process of the model, the diversification of the strength of the node will satisfy

$$\frac{\partial s_i}{\partial t} = \sum_{l=1}^{K} q_l \sum_{j=0}^{l} p_j j(l-1)\Pi(i)$$

$$= \sum_{l=1}^{K} q_l J_l (l-1)\Pi(i)$$

$$= (M - \bar{J})\Pi(i),$$

where $M = \sum_{l=1}^{K} q_l J_l l$, since

$$\frac{\partial n_i}{\partial t} = \bar{J}\Pi(i),$$

we have

$$\frac{\partial s_i}{\partial t} = \frac{M - \bar{J}}{\bar{J}} \frac{\partial n_i}{\partial t}.$$

From Eq.(5), we obtain

$$s_i = \frac{M - \bar{J}}{\bar{J}} n_i = \frac{M - \bar{J}}{\bar{J}} \left(\frac{t}{t_i}\right)^{\beta}, \tag{7}$$

where $\beta = \bar{J} / L$. Let $A = \dfrac{M - \bar{J}}{\bar{J}}$, We can calculate the strength distribution

$$p(s) = \frac{t A^{\frac{1}{\beta}}}{(n_0 + t)\beta} s^{-(\frac{1}{\beta}+1)} \propto s^{-(\frac{1}{\beta}+1)}. \tag{8}$$

From the theories analysis, it can be obtained that the module degree distribution of the node has the same power law as the strength degree distribution, and the scale-free feature is put up by the module degree and strength degree.

3.2 Simulation

Next, we apply the computer simulation to study the complex characteristic of the network model. Suppose $K=4$, $q_1 = 0.2$, $q_2 = 0.3$, $q_3 = 0.3$, $q_4 = 0.2$;

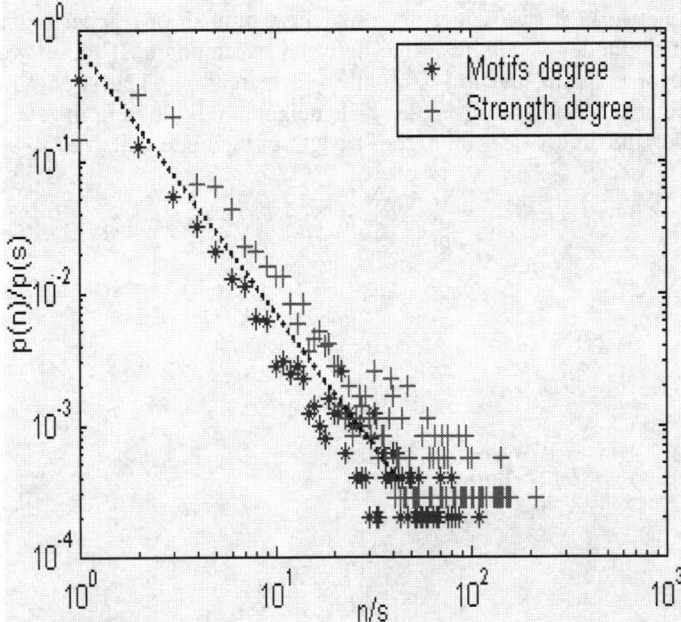

Fig. 1. The distribution plots (log-log plots) of the model

and $p_0 = 0.2$, $p_1 = 0.15$, $p_2 = 0.25$, $p_3 = 0.25$, $p_4 = 0.15$. In the initialized network, there are 10 authors with 3 paper published. Based on the network evolving model, a scientific collaboration network with 500 papers is built by the computer simulation, and the motif degree distribution and the strength degree distribution are calculated. The distribution plots (Log-Log) are shown Fig. 1 whose data are obtained from the average of 20 times simulations. These two distributions have the same power law and take on the feature of the scale-free network, which is accord with the theoretical results.

4 Analysis on Demonstration Network

We have been working a statistic of the papers in Journal of Information from January 2001 to December 2006 in database of China National Knowledge Infrastructure (CNKI). A coauthorship database is created on the articles and their authors; there are 801 articles and 1078 authors in the database. A scientific collaboration network is made from the coauthorship database, where a node denotes au author, link of two nodes denotes the cooperation between two authors (they have vended an article together). The module is defined as the collaboration authors in an article, which is full connected sub-network. It is easy to know that the cooperation network is a sort of self-organized network. The information of the database is shown in Table 1.

Table 1. The information of the database of the Journal of Information

Author Number	1	2	3	≥ 4
Paper number	223	325	212	41
Percent	27.8%	40.6%	26.5%	5.1%

Then, we analyze the characteristic of this network by calculating the module degree of the node in the coauthorship network, which is the paper number of every author. The statistic result is shown in Table 2, where a lots of nodes have small module degree (there are 1018 authors with module degree less 3) and few nodes with large module degree (only 6 authors with module degree greater than 10).

Table 2. The module degree of the authors

Module degree	1	2	3	4	5	6	7	8	9	10	11	12	15	22
Number of the authors	805	157	56	23	9	8	4	5	5	1	1	2	1	1

In the demonstration network, we calculate the strength degree of the node, i.e., the cooperation times with other authors (Table 3). There are many nodes with little strength degree (792 authors with module degree less than 3), however, there exists a larger node with strength degree 37 who has a very large workgroup and has more cooperation researchers.

Table 3. The strength degree of the authors

Strength Degree	0	1	2	3	4	5	6	7	8	9	10	11
Authors Number	135	340	317	107	79	31	16	11	10	4	4	4

Strength Degree	12	13	15	16	17	20	30	37
Authors Number	5	3	4	2	3	1	1	1

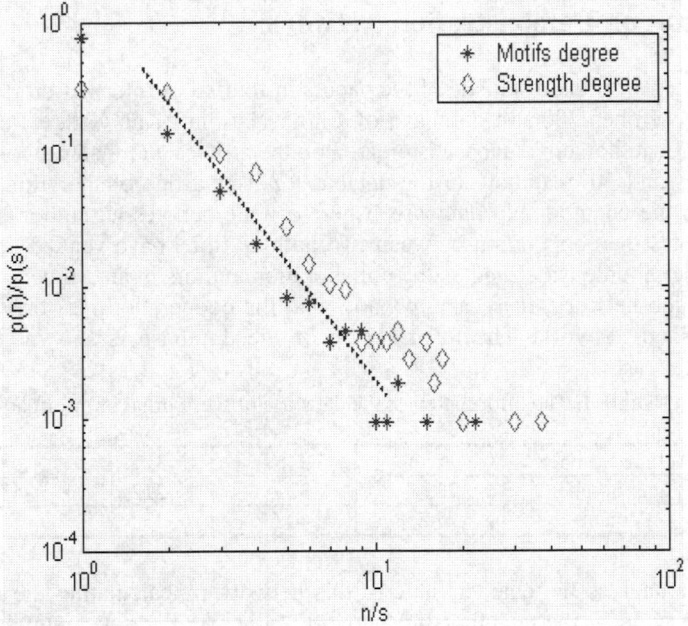

Fig. 2. The plot of the distributions in the demonstration network (Log-Log)

Based on the results of statistic, a plot of the distributions is drawn by using the data of Table 2 and Table 3 with log-log figure, where "*" defines the motif degree distribution of node and "◇" defines the strength degree of the node in the demonstration network. In Fig. 2, the power law of the motifs degree is the same as that of the strength degree, which is consensus with the results in section 3.

5   Conclusions

In this paper, based on the collaboration ways of the authors, a self-organized collaboration network model is presented. By defining the module with the authors subnetwork full-connected in a paper, the network is evolved by inset the module. It is validated by the theoretical analysis and computer simulation that the distributions of the module degree and strength degree satisfy the same power law and this network is a scale-free network. By analyzing the data of the demonstration network in CNKI, the demonstration results are consistent with that of the theoretical analysis. Therefore, this model can be applied to study the evolvement of the author collaboration network.

Acknowledgements

This work is supported by Chinese National Natural Science Foundation (under the grant 60774016, 60875039), the Science Foundation of Office of Education of Shandong Province (under the grant J08LJ01) and the Natural Science Foundation of Ludong University (under the grant L20074102, Z0704) of China.

References

1. Erdos, P., Renyi, A.: On the evolution of random graphs. Publ. Math. Inst. Hung. Acad. Sci. 5, 17–60 (1960)
2. Watts, D., Strogatz, S.: Collective dynamics of "small world" networks. Nature 393(4), 440–442 (1998)
3. Barabási, A., Albert, R.: Emergence of scaling in random networks. Science 286(5439), 509–512 (1999)
4. Albert, R., Barabási, A.: Statistical mechanics of complex networks. Rev. Modern Phys. 74(1), 47–97 (2002)
5. Newman, M.: The structure and function of complex network. SIAM Review 45(2), 167–256 (2003)
6. Boccaletti, S., Latora, V., Moreno, Y., et al.: Complex Networks: Structure and dynamics. Physics Reports 424(4-5), 175–308 (2006)
7. Newman, M.: Scientific collaboration. I. Network construction and fundamental results. Phys. Rev. E 64(1), 016131 (2001)
8. Newman, M.: Scientific collaboration. II. Shortest paths, weighted networks, and centrality. Phys. Rev. E 64(1), 016132 (2001)
9. Wang, F.-S., Yang, H.-Y.: Analysis on Scientific Collaboration Network of Author. Journal of the China Society for Scientific and Technical Information 26(5), 659–663 (2007)

Self-organized Balanced Resources in Random Networks with Transportation Bandwidths

Chi Ho Yeung and K.Y. Michael Wong

Department of Physics, The Hong Kong University of Science and Technology, Hong Kong, China

Abstract. We apply statistical physics to study the task of resource allocation in random networks with limited bandwidths for the transportation of resources along the links. We derive algorithms which searches the optimal solution without the need of a global optimizer. For networks with uniformly high connectivity, the resource shortage of a node becomes a well-defined function of its capacity. An efficient profile of the allocated resources is found, with clusters of node interconnected by an extensive fraction of unsaturated links, enabling the resource shortages among the nodes to remain balanced. The capacity-shortage relation exhibits features similar to the Maxwell's construction. For scale-free networks, such an efficient profile is observed even for nodes of low connectivity.

Keywords: resource allocation, bandwidth, Maxwell's construction, scale-free networks, Bethe approximation, message-passing.

1 Introduction

Analytical techniques developed in statistical physics have been widely employed in the analysis of complex systems in a wide variety of fields, such as neural networks [1,2], econophysical models [3], and error-correcting codes [2,5]. Recently, a statistical physics perspective was successfully applied to the problem of resource allocation on sparse random networks [6,7]. Resource allocation is a well known network problem in the areas of computer science and operations management [8,9]. It is relevant to applications such as load balancing in computer networks, reducing Internet traffic congestion, and streamlining network flow of commodities [10,11].

In this paper, we analyze resource allocation on networks with finite bandwidths. We derive algorithms which enable us to find the optimal solutions without the need of a global optimizer. Compared with conventional techniques such as linear or quadratic programming [14], the adopted approach in this paper reduces the computational complexity. Furthermore, the analysis allows us to understand the underlying mechanisms during resource redistribution, on both scale-free and regular networks (i.e. networks with uniform connectivity). An efficient profile of the allocated resource in found wth features similar to the Maxwell's construction.

J. Zhou (Ed.): Complex 2009, Part I, LNICST 4, pp. 806–818, 2009.

2   The Model

We consider a network with N nodes, labelled $i = 1, \ldots, N$. Each node i is randomly connected to c other nodes. The connectivity matrix is given by $\mathcal{A}_{ij} = 1, 0$ for connected and unconnected node pairs respectively. We first develop a theory for sparse networks, namely, those of intensive connectivity $c \sim O(1) \ll N$, and subsequently consider its validity in networks of general connectivity, such as scale-free networks.

Each node i has a capacity Λ_i randomly drawn from a distribution $\rho(\Lambda_i)$. Positive and negative values of Λ_i correspond to supply and demand of resources respectively. The task of resource allocation involves transporting resources between nodes such that the demands of the nodes can be satisfied to the largest extent. Hence we assign $y_{ij} \equiv -y_{ji}$ to be the *current* drawn from node j to i, aiming at reducing the *shortage* ξ_i of node i defined by

$$\xi_i = \max\left(-\Lambda_i - \sum_{(ij)} \mathcal{A}_{ij} y_{ij}, 0\right). \tag{1}$$

The magnitudes of the currents are bounded by the *bandwidth* W, i.e., $|y_{ij}| \leq W$.

To minimize the shortage of resources after their allocation, we include in the total cost both the shortage cost and the transportation cost. Hence, the general cost function of the system can be written as

$$E = R \sum_{(ij)} \mathcal{A}_{ij} \phi(y_{ij}) + \sum_i \psi(\Lambda_i, \{y_{ij} | \mathcal{A}_{ij} = 1\}). \tag{2}$$

The summation (ij) corresponds to summation over all node pairs, and Λ_i is a quenched variable defined on node i.

In the present model of resource allocation, the first and second terms correspond to the transportation and shortage costs respectively. The parameter R corresponds to the *resistance* on the currents, and Λ_i is the capacity of node i. The transportation cost $\phi(y_{ij})$ can be a general even function of y_{ij}. In this paper, we consider ϕ and ψ to be concave functions of their arguments, that is, $\phi'(y)$ and $\psi'(\xi)$ are non-decreasing functions. Specifically, we have the quadratic transportation cost $\phi(y) = y^2/2$, and the quadratic shortage cost $\psi(\Lambda_i, \{y_{ij} | \mathcal{A}_{ij} = 1\}) = \xi_i^2/2$.

3   Analysis

The analysis of the model is made convenient by the introduction of the variables ξ_i. It can be written as the minimization of Eq. (2) in the space of y_{ij} and ξ_i, subject to the constraints

$$\Lambda_i + \sum_{(ij)} \mathcal{A}_{ij} y_{ij} + \xi_i \geq 0, \qquad \xi_i \geq 0, \tag{3}$$

and the constraints on the bandwidths of the links $|y_{ij}| \leq W$.

Introducing Lagrange multipliers to the above inequality constraints with the Kuhn-Tucker condition, the function to be minimized becomes

$$L = \sum_i \left[\psi(\xi_i) + \mu_i \left(\Lambda_i + \sum_{(ij)} \mathcal{A}_{ij} y_{ij} + \xi_i \right) + \alpha_i \xi_i \right]$$
$$+ \sum_{(ij)} \mathcal{A}_{ij} \left[R\phi(y_{ij}) + \gamma_{ij}^+ (W - y_{ij}) + \gamma_{ij}^- (W + y_{ij}) \right], \quad (4)$$

where $\mu_i \leq 0$, $\alpha_i \leq 0$, $\gamma_{ij}^+ \leq 0$ and $\gamma_{ij}^- \leq 0$. Optimizing L with respect to y_{ij}, one obtains

$$y_{ij} = Y(\mu_j - \mu_i) \quad \text{with} \quad Y(x) = \max\left\{ -W, \min\left[W, [\phi']^{-1}\left(\frac{x}{R}\right) \right] \right\}. \quad (5)$$

The Lagrange multiplier μ_i is referred to as the *chemical potential* of node i, and ϕ' is the derivative of ϕ with respect to its argument. The function $Y(\mu_j - \mu_i)$ relates the potential difference between nodes i and j to the current driven from node j to i. For the quadratic cost, it consists of a linear segment between $\mu_j - \mu_i = \pm WR$ reminiscent of Ohm's law in electric circuits. Beyond this range, y is bounded above and below by $\pm W$ respectively. Thus, obtaining the optimized configuration of currents y_{ij} among the nodes is equivalent to finding the corresponding set of chemical potentials μ_i, from which the optimized y_{ij}'s are then derived from $Y(\mu_j - \mu_i)$. This implies that we can consider the original optimization problem in the space of chemical potentials.

We introduce the free energy at a temperature $T \equiv \beta^{-1}$,

$$F = -T \ln Z, \quad (6)$$

where Z is the partition function

$$Z = \prod_{(ij)} \left(\int_{-W}^{W} dy_{ij} \right) \exp\left[-\beta R \sum_{(ij)} \mathcal{A}_{ij} \phi(y_{ij}) - \beta \sum_i \psi(\Lambda_i, \{y_{ij} | \mathcal{A}_{ij} = 1\}) \right]. \quad (7)$$

The statistical mechanical analysis of the free energy can be carried out using the Bethe approximation, which is valid in the limit of low connectivity. In this approximation, a node is connected to c branches of the tree, and the correlations among the branches are neglected. In each branch, nodes are arranged in generations, A node is connected to an ancestor node of the previous generation, and another $c - 1$ descendent nodes of the next generation.

We consider the vertex $V(\mathbf{T})$ of a tree $\mathbf{T}$. We let $F(y|\mathbf{T})$ be the free energy of the tree when a current y is drawn from the vertex by its ancestor node. One can express $F(y|\mathbf{T})$ in terms of the free energies $F(y_k|\mathbf{T}_k)$ of its descendents $k = 1, \ldots, c - 1$,

$$F(y|\mathbf{T}) = -T\ln\left\{\prod_{k=1}^{c-1}\left(\int_{-W}^{W}dy_k\right)\exp\left[-\beta\sum_{k=1}^{c-1}F(y_k|\mathbf{T}_k) - \beta R\sum_{k=1}^{c-1}\phi(y_k)\right.\right.$$
$$\left.\left. -\beta\psi\left(\max\left(-\Lambda_{V(\mathbf{T})} - \sum_{k=1}^{c-1}y_k + y, 0\right)\right)\right]\right\},\tag{8}$$

where $\mathbf{T}_k$ represents the tree terminated at the kth descendent of the vertex, and $\Lambda_{V(\mathbf{T})}$ is the capacity of $V(\mathbf{T})$. We then consider the free energy as,

$$F(y|\mathbf{T}) = N_{\mathbf{T}}F_{\text{av}} + F_V(y|\mathbf{T}),\tag{9}$$

where $N_{\mathbf{T}}$ is the number of nodes in the tree $\mathbf{T}$, and F_{av} is the vertex free energy per node. $F_V(y|\mathbf{T})$ is referred to as the *vertex free energy*. Note that when a vertex is added to a tree, there is a change in the free energy due to the added vertex. In the language of the cavity method [4], $F_V(y|\mathbf{T})$ are equivalent to the *cavity fields*, since they describe the state of the system when the ancestor node is absent. In the zero temperature limit, we obtain a recursion relation,

$$F_V(y|\mathbf{T}) = \min_{\{y_k||y_k|\leq W\}}\left[\sum_{k=1}^{c-1}\left(F_V(y_k|\mathbf{T}_k) + R\phi(y_k)\right)\right.$$
$$\left. +\psi\left(\max\left(-\Lambda_V(\mathbf{T}) - \sum_{k=1}^{c-1}y_k + y, 0\right)\right)\right] - F_{\text{av}}.\tag{10}$$

$$F_{\text{av}}(y|\mathbf{T}) = \left\langle\min_{\{y_k||y_k|\leq W\}}\left[\sum_{k=1}^{c}\left(F_V(y_k|\mathbf{T}_k) + R\phi(y_k)\right)\right.\right.$$
$$\left.\left. +\psi\left(\max\left(-\Lambda_V(\mathbf{T}) - \sum_{k=1}^{c}y_k, 0\right)\right)\right]\right\rangle_{\Lambda}.\tag{11}$$

4 Distributed Algorithms

A distributed algorithm can be obtained by iterating the chemical potentials of the nodes. The optimal currents are given by Eq. (5) in terms of the chemical potentials μ_i which, from Eqs. (1) and (4), are related to their neighbors via

$$\mu_i = \begin{cases} 0 & \text{for } h_i^{-1}(0) > 0, \\ h_i^{-1}(0) & \text{for } -\psi'(0) \leq h_i^{-1}(0) \leq 0, \\ g_i^{-1}(0) & \text{for } h_i^{-1}(0) < -\psi'(0), \end{cases}\tag{12}$$

where $h_i(x)$ and $g_i(x)$ are given by

$$h_i(x) = -\Lambda_i - \sum_j A_{ij}Y(\mu_j - x), \qquad g_i(x) = \psi' \circ h_i(x) + x,\tag{13}$$

with function Y again given Eq. (5). $h_i(x)$ is the shortage of resource at node i when μ_i takes the value x. $\psi' \circ h_i(x)$ is then the corresponding dissatisfaction cost per unit resource of node j. This provides a simple local iteration method for the optimization problem in which the optimal currents can be evaluated from the potential differences of neighboring nodes.

An alternative algorithm can be obtained by adopting message-passing approaches, which have been successful in problems such as error-correcting codes [12] and probabilistic inference [13]. However, in contrast to other message-passing algorithms which pass conditional probability estimates of discrete variables to neighboring nodes, the messages in the present context are more complex, since they are free energy functions $F_V(y|\mathbf{T})$ of the continuous variable y. Inspired by the success of replacing the function messages by their first and second derivatives in [7], we follow the same route and form two-parameter messages Let $(A_{ij}, B_{ij}) \equiv (\partial F_V(y_{ij}|\mathbf{T}_j)/\partial y_{ij}.\partial^2 F_V(y_{ij}|\mathbf{T}_j)/\partial y_{ij}^2)$. These are the messages passed from node j to its ancestor node i, based on the messages received from its descendents in the tree $\mathbf{T}_j$. To obtain recursion relation of the messages, we minimize in the space of the current adjustments ϵ_{jk} the vertex free energy

$$F_{ij} = \sum_{k \neq i} A_{jk} \left[A_{jk}\varepsilon_{jk} + \frac{1}{2}B_{jk}\varepsilon_{jk}^2 + R\phi'_{jk}\varepsilon_{jk} + \frac{R}{2}\phi''_{jk}\varepsilon_{jk}^2 \right] + \psi(\xi_j), \quad (14)$$

subject to the constraints

$$\sum_{k \neq i} A_{jk}(y_{jk} + \varepsilon_{jk}) - y_{ij} + \Lambda_j + \xi_j \geq 0, \qquad \xi_j \geq 0, \quad (15)$$

together with the constraints on bandwidths $|y_{jk} + \epsilon_{jk}| \leq W$. ϕ'_{jk} and ϕ''_{jk} represent the first and second derivatives of $\phi(y)$ at $y = y_{jk}$ respectively. We introduce Lagrange multiplier μ_{ij} for constraints (15). After optimizing the energy function of node j, the messages from node j to i are given by

$$A_{ij} \leftarrow -\mu_{ij}, \quad (16)$$

$$B_{ij} \leftarrow \begin{cases} 0 & \text{for } h_{ij}^{-1}(0) > 0, \\[2em] \left\{ \sum_{k \neq i} A_{jk}(R\phi''_{jk} + B_{jk})^{-1} \right. \\ \left. \times \Theta\left[W - \left| y_{jk} - \frac{R\phi'_{jk} + A_{jk} + \mu_{ij}}{R\phi''_{jk} + B_{jk}} \right| \right] \right\}^{-1} \\ \text{for } -\psi'(0) \leq h_{ij}^{-1}(0) \leq 0, \\[2em] \left\{ \psi''(\xi)^{-1} + \sum_{k \neq i} A_{jk}(R\phi''_{jk} + B_{jk})^{-1} \right. \\ \left. \times \Theta\left[W - \left| y_{jk} - \frac{R\phi'_{jk} + A_{jk} + \mu_{ij}}{R\phi''_{jk} + B_{jk}} \right| \right] \right\}^{-1} \\ \text{for } h_{ij}^{-1}(0) < -\psi'(0), \end{cases}$$

$$(17)$$

where

$$g_{ij}(x) = [\psi' \circ h_{ij}](x) + x, \tag{18}$$

$$\mu_{ij} = \begin{cases} 0 & \text{for } h_{ij}^{-1}(0) > 0, \\ h_{ij}^{-1}(0) & \text{for } -\psi'(0) \le h_{ij}^{-1}(0) \le 0, \\ g_{ij}^{-1}(0) & \text{for } h_{ij}^{-1}(0) < -\psi'(0), \end{cases} \tag{19}$$

and $h_{ij}(x)$ is defined by

$$h_{ij}(x) = y_{ij} - \Lambda_j - \sum_{k \neq i}' \max\left\{-W, \min\left[W, y_{jk} - \frac{R\phi'_{jk} + A_{jk} + x}{R\phi''_{jk} + B_{jk}}\right]\right\}. \tag{20}$$

Since the messages are simplified to be the first two derivatives of the vertex free energies, it is essential for the nodes to determine the *working points* at which the derivatives are taken. Optimal currents y_{jk} are thus computed and sent backward from node j to the descendent nodes $k \neq i$. These backward messages serve as a key in information provision to descendents, so that the derivatives in the subsequent messages are to be taken at the updated working points. Minimizing the free energy (14) with respect to y_{jk}, the backward message is found to be

$$y_{jk} \leftarrow \max\left\{-W, \min\left[W, y_{jk} - \frac{R\phi'_{jk} + A_{jk} + \mu_{ij}}{R\phi''_{jk} + B_{jk}}\right]\right\}. \tag{21}$$

An important result of our study is that for the frictionless case with $\psi'(0) = 0$, the message-passing algorithm, in the two-parameter approximation, yield solutions *identical to* the previous algorithm, *which is exact* for all connectivities, as long as the algorithms converges. This is a remarkable result since the message-passing algorithm is originally derived for dilute networks only.

5   The High Connectivity Limit

We consider the case that the bandwidth of individual links scales as $\tilde{W}/c$ when the connectivity increases, where $\tilde{W}$ is a constant. Thus the total bandwidth $\tilde{W}$ available to an individual node remains a constant.

We start by writing the chemical potentials using Eq. (12),

$$\mu_i = \min\left[\Lambda_i + \sum_{j=1}^{N} \mathcal{A}_{ij} Y(\mu_j - \mu_i), 0\right]. \tag{22}$$

In the high connectivity limit, the interaction of a node with all its connected neighbors become self-averaging, making it a function which is singly dependent on its own chemical potential, namely,

$$\sum_{j=1}^{N} \mathcal{A}_{ij} Y(\mu_j - \mu_i) \approx cM(\mu_i). \tag{23}$$

Physically, the function $M(\mu)$ corresponds to the average interaction of a node with its neighbors when its chemical potential is μ. Thus, we can write Eq. (22) as

$$\mu = \min[\Lambda + cM(\mu), 0], \tag{24}$$

where μ is now a function of Λ, and we have

$$M(\mu_i) = \int_{-\infty}^{\infty} d\Lambda \rho(\Lambda) Y(\mu(\Lambda) - \mu_i) \tag{25}$$

where we have written the chemical potential of the neighbors as $\mu(\Lambda)$, assuming that they are well-defined functions of their capacities Λ.

To explicitly derive $M(\mu)$, we take advantage of the fact that the rescaled bandwidth, $\tilde{W}/c$ vanishes in the high connectivity limit, so that the current function $Y(\mu_j - \mu_i)$ is effectively a sign function, which implies that the current on a link is always saturated. (This approximation is not fully valid if c is large but finite and will be further refined in subsequent discussions) Thus, we approximate

$$M(\mu_i) = \frac{\tilde{W}}{c} \int_{-\infty}^{\infty} d\Lambda \rho(\Lambda) \operatorname{sgn}[\mu(\Lambda) - \mu_i]. \tag{26}$$

Assuming that $\mu(\Lambda)$ is a monotonic function of Λ, and for Gaussian distribution of capacities, $\mu(\Lambda)$ is explicitly given by

$$\mu = \min\left[\Lambda - \tilde{W}\operatorname{erf}\left(\frac{\Lambda - \langle\Lambda\rangle}{\sqrt{2}}\right), 0\right]. \tag{27}$$

This equation relates the chemical potential of a node, i.e. the shortage after resource allocation, to its initial resource before. Resource allocation through a large number of links results in a well-defined function relating the two quantities.

Eq. (27) gives a well-defined function $\mu(\Lambda)$ as long as $\tilde{W} \leq \sqrt{\pi/2}$. However, when $\tilde{W} > \sqrt{\pi/2}$, turning points exists in $\mu(\Lambda)$ as shown in Fig. 1(a). This creates a thermodynamically unstable scenario, since in the region of $\mu(\Lambda)$ with negative slope, nodes with lower capacities have higher chemical potentials than their neighbors with higher capacities. Mathematically, the non-monotonicity of $\mu(\Lambda)$ means that $\operatorname{sgn}[\mu(\Lambda) - \mu_i]$ and $\operatorname{sgn}(\Lambda - \Lambda_i)$ are no longer necessarily equal, and Eq. (27) is no longer valid.

Nevertheless, Eq. (22) permits another solution of constant μ in a range of Λ. Hence, we propose that the unstable region of $\mu(\Lambda)$ should be replaced by a range of constant μ as shown in Fig. 1(b) analogous to Maxwell's construction in thermodynamics.

In the high connectivity limit, resources are so efficiently allocated that the resources of the rich nodes are maximally allocated to the poor nodes. By

(a)   (b)

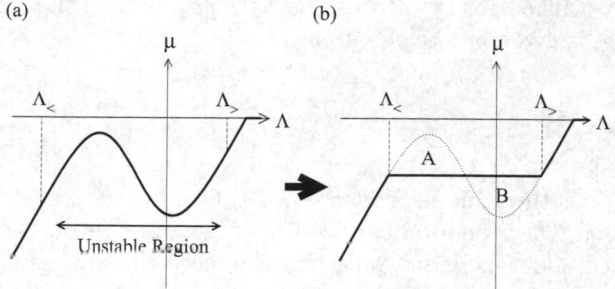

Fig. 1. Maxwell's construction on $\mu(\Lambda)$

considering the conservation of resources, and letting $(\Lambda_<, \mu_o)$ and $(\Lambda_>, \mu_o)$ be the end points of the Maxwell's construction as shown in Fig. 1(b). we arrive at

$$-\int_{\Lambda_<}^{\Lambda_>} d\Lambda \rho(\Lambda) \mu_o - \left(\int_{\Lambda_>}^{\Lambda_o} + \int_{-\infty}^{\Lambda_<}\right) d\Lambda \rho(\Lambda) \mu(\Lambda)$$

$$= -\int_{-\infty}^{\Lambda_o} d\Lambda \rho(\Lambda) \Lambda - \Lambda_o \int_{\Lambda_o}^{\infty} d\Lambda \rho(\Lambda). \tag{28}$$

where Λ_o is given by $\Lambda_o = \tilde{W} \int_{-\infty}^{\Lambda_o} d\Lambda \rho(\Lambda)$. Nodes with $\Lambda \geq \Lambda_o$ send out their resources without drawing inward currents from their neighbors, and can be regarded as *donors*. Substituting Eqs. (24), (25) in the range $\Lambda < \Lambda_<$ and $\Lambda > \Lambda_>$, we arrive at the condition

$$\mu_o \int_{\Lambda_<}^{\Lambda_>} d\Lambda \rho(\Lambda) = \int_{\Lambda_<}^{\Lambda_>} d\Lambda \rho(\Lambda) \mu(\Lambda), \tag{29}$$

which implies that the value of μ_o should be chosen such that the areas A and B in Fig. 1(b), weighted by the distribution $\rho(\Lambda)$, should be equal.

For capacity distributions $\rho(\Lambda)$ symmetric with respect to $\langle \Lambda \rangle$, we have $\mu_o = \langle \Lambda \rangle = (\Lambda_< + \Lambda_>)/2$. As a result, the function $\mu(\Lambda)$ is given by

$$\mu(\Lambda) = \begin{cases} \langle \Lambda \rangle & \text{for } \mu_< < \mu < \mu_>, \\ \min\left[\Lambda - \tilde{W}\text{erf}\left(\frac{\Lambda - \langle \Lambda \rangle}{\sqrt{2}}\right), 0\right] & \text{otherwise,} \end{cases} \tag{30}$$

where as $\Lambda_<$ and $\Lambda_>$ are respectively given by the lesser and greater roots of the equation $x = \langle \Lambda \rangle + \tilde{W}\text{erf}[(x - \langle \Lambda \rangle)/\sqrt{2}]$.

Nodes i with chemical potentials $\mu_i = \langle \Lambda \rangle$ represent clusters of nodes interconnected by an extensive fraction of unsaturated links, which provides the freedom to fine tune their currents so that the shortages among the nodes are uniform. They will be referred to as the *balanced* nodes. The fraction f_{bal} of balanced nodes is given by the equation

$$f_{\text{bal}} = \text{erf}\left(\frac{\tilde{W} f_{\text{bal}}}{\sqrt{2}}\right). \tag{31}$$

Note that f_{bal} has the same dependence on $\tilde{W}$ for all negative $\langle \Lambda \rangle$. The inset of Fig. 4 shows that when the total bandwidth $\tilde{W}$ increases beyond $\sqrt{\pi/2}$, the fraction of balanced nodes increases, reflecting the more efficient resource allocation brought by the convenience of increased bandwidths. When $\tilde{W}$ becomes very large, a uniform chemical potential of $\langle \Lambda \rangle$ networkwide is recovered, converging to the case of non-vanishing bandwidths.

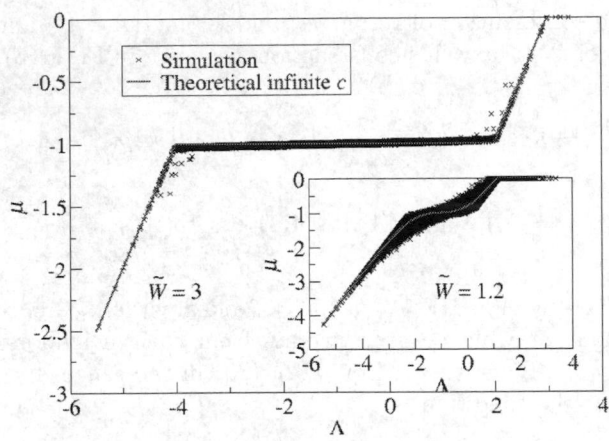

Fig. 2. The simulation results of $\mu(\Lambda)$ for $N = 10000$, $c = 15$, $R = 0.1$, $\langle \Lambda \rangle = -1$ and $\tilde{W} = 3$ with 70000 data points, compared with theoretical prediction. Inset: The corresponding results for $\tilde{W} = 1.2$.

We compare the analytical result of $\mu(\Lambda)$ in Eq. (30) with simulations in Fig. 2. For $\tilde{W} > \sqrt{\pi/2}$, data points (Λ, μ) of individual nodes from network simulations follow the analytical result of $\mu(\Lambda)$, giving an almost perfect overlap of data. The presence of the balanced nodes with effectively constant chemical potentials is obvious and essential to explain the behavior of the majority of data points from simulations. On the other hand, for $\tilde{W} < \sqrt{\pi/2}$, the analytical $\mu(\Lambda)$ shows no turning point as shown in the inset of Fig. 2. Despite the scattering of data points, they generally follow the trend of the theoretical $\mu(\Lambda)$.

Our analysis can be generalized to the case of large but finite connectivity, where the approximation in Eq. (26) is not fully valid. This modifies the chemical potentials of the balanced nodes, for which Eq. (26) has to be replaced by

$$M(\mu) = \frac{\tilde{W}}{c}\left[\int_{\Lambda_>}^{\infty} d\Lambda \rho(\Lambda) - \int_{-\infty}^{\Lambda_<} d\Lambda \rho(\Lambda)\right] + \int_{\Lambda_<}^{\Lambda_>} d\Lambda \rho(\Lambda)\left(\frac{\mu(\Lambda) - \mu}{R}\right). \tag{32}$$

We introduce an ansatz of a linear relationship between μ and Λ for the balanced nodes, namely,

$$\mu = m\Lambda + b. \tag{33}$$

After direct substitution of Eq. (33) into $M(\mu)$ given by Eq. (32), we get the self-consistent equations for m and b,

$$m - \frac{R}{R + c\,\mathrm{erf}\left(\frac{\Lambda_> - \langle\Lambda\rangle}{\sqrt{2}}\right)}, \quad b = \frac{c\,\mathrm{erf}\left(\frac{\Lambda_> - \langle\Lambda\rangle}{\sqrt{2}}\right)}{R + c\,\mathrm{erf}\left(\frac{\Lambda_> - \langle\Lambda\rangle}{\sqrt{2}}\right)}\langle\Lambda\rangle. \tag{34}$$

Thus, the Maxwell's construction has a non-zero slope when the connectivity is finite.

We remark that the approximation in Eq. (32) assumes that the potential differences of the balanced nodes lie in the range of $2R\tilde{W}/c$, so that their connecting links remain unsaturated. Note that the end points of the Maxwell's construction have chemical potentials $\langle\Lambda\rangle \pm R\tilde{W}/c$ respectively, rendering the approximation in Eq. (32) *exact* at one special point, namely, the central point of the Maxwell's construction. Hence, this approximation works well in the central region of the Maxwell's construction, while deviations are expected near the end points.

In the simulation data shown in Fig. 3, the data points of (Λ, μ) from different ratios of R/c follow the trend of the corresponding analytical results, both within and outside the linear region, with increasing scattering within the linear region as R/c increases. As expected, there are derivations between the analytical and simulational results at the two ends of the linear region, with smoothened corners appearing in the simulation data, especially in the case of $R/c = 2/20$.

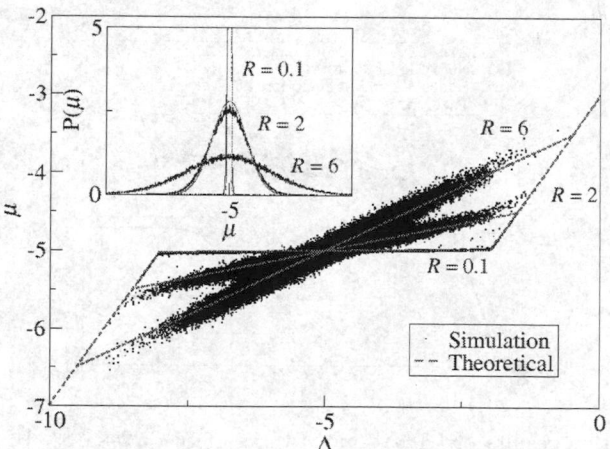

Fig. 3. Simulation results of (Λ, μ) for $N = 2 \times 10^5$, $\tilde{W} = 3$, $c = 12$ and $\langle\Lambda\rangle = -5$ at different values of R, each with 65000 data points. as compared to the theoretical predictions. Inset: the corresponding chemical potential distribution $P(\mu)$ of the 3 cases.

We note that when R/c increases, the gradient of the linear region increases, corresponding to a less uniform allocation of resources.

Remarkably, as evident from Eq. (34), even with constant available bandwidth $\tilde{W}$, increasing connectivity causes m to decrease, and hence sharpens the chemical potential distribution. The narrower distributions correspond to higher efficiency in resource allocation. It leads us to realize the potential benefits of increasing connectivity in network optimization even for a given constant total bandwidth connecting a node.

6   Scale-Free Networks

We have considered the allocation of resources in regular networks in the high connectivity limit. However, recent studies of complex networks show that many realistic communication networks have highly heterogeneous structure, and the connectivity distribution obeys a power law [15]. These networks, commonly known as scale-free networks, are characterized by the presence of hubs, which are nodes with very high connectivities, and are found to modify the network behavior significantly. Hence, it is interesting to study the allocation of resources in scale-free networks.

The simulation results are presented in Fig. 4, where we plot the data points of (Λ, μ) from nodes of $c = 3$ in scale-free networks. Despite their low connectivity, their capacity-shortage relation exhibit the flat distribution characteristic of the Maxwell's construction, coinciding with the analytical results of the high connectivity limit. This shows that the presence of hubs in scale-free networks increases the global efficiency of resource allocation, leading to a more uniform distribution of resources.

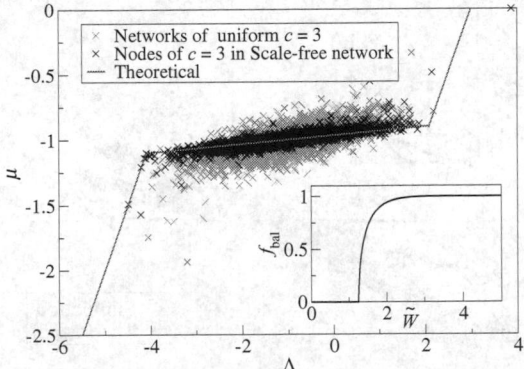

Fig. 4. Simulation results of (Λ, μ) for networks of $N = 2 \times 10^5$, $\tilde{W} = 3$, $R = 0.1$ and $\langle \Lambda \rangle = -1$ with (a) uniform connectivity of $c = 3$ and (b) scale-free network of $P(c) \sim c^{-3}$ with $c \geq 3$. each with 2500 data points. as compared to the theoretical predictions Eq. (34). Inset: the dependence of the f_{bal} in Eq. (31) on the bandwidth $\tilde{W}$.

To confirm this advantage of the scale-free topology, we also plot in the figure the data points obtained from networks of uniform connectivity $c = 3$. Evidently, the data points are much more scattered away from the Maxwell's construction.

7   Conclusion

We have applied statistical mechanics to study an optimization task of resource allocation on a network, in which nodes with different capacities are connected by links of finite bandwidths. By adopting suitable cost functions, such as quadratic transportation and shortage costs, the model can be applied to the study of realistic networks. We employ the Bethe approximation to derive recursive relations of the vertex free energies, which are useful in both algorithmic and analytic aspects.

In particular, the study reveals interesting effects due to finite bandwidths. A remarkable phenomenon is found in networks with fixed total bandwidths per node, where bandwidths per link vanish in the high connectivity limit. For sufficiently large total bandwidths, clusters of balanced nodes self-organized to have a uniform shortage reminiscent of the Maxwell's construction in thermodynamics. In scale-free networks, such clusters even include nodes with low connectivity, implying a more efficient resource allocation compared to networks with uniform connectivity. We believe that the techniques presented in this paper are useful in many different network optimization problems and will lead to a large variety of potential applications.

Acknowledgements

We thank David Saad for very meaningful discussions. This work is supported by the Research Grant Council of Hong Kong (grant numbers HKUST 603606, HKUST 603607 and HKUST 604008).

References

1. Hertz, J., Krogh, A., Palmer, R.G.: Introduction to the Theory of Neural Computation. Addison-Wesley, Redwood City (1991)
2. Nishimori, H.: Statistical Physics of Spin Glasses and Information Processing. Oxford University Press, Oxford (2001)
3. Challet, D., Marsili, M., Zhang, Y.-C.: Minority Games. Oxford University Press, Oxford (2005)
4. Mézard, M., Parissi, G., Virasoro, M.A.: Spin Glass Theory and Beyond. World Scientific, Singapore (1987)
5. Kabashima, Y., Saad, D.: J. Phys. A 37, R1 (2004)
6. Wong, K.Y.M., Saad, D.: Phys. Rev. E 74, 010104 (2006)
7. Wong, K.Y.M., Saad, D.: Phys. Rev. E 76, 011115 (2007)

8. Peterson, L., Davie, B.S.: Computer Networks: A Systems Approach. Academic Press, San Diego (2000)
9. Ho, Y.C., Servi, L., Suri, R.: Large Scale Syst. 1, 51 (1980)
10. Shenker, S., Clark, D., Estrin, D., Herzog, S.: Comput. Commun. Rev. 26, 19 (1996)
11. Rardin, R.L.: Optimization in Operations Research. Prentice Hall, Englewood Cliffs (1998)
12. Opper, M., Saad, D. (eds.): Advanced Mean Field Methods. MIT Press, Cambridge (1999)
13. Mackey, D.J.C.: Information Theory, Inference and Learning Algorithms. Cambridge University Press, Cambridge (2003)
14. Bertsekas, D.: Linear Network Optimization. MIT Press, Cambridge (1991)
15. Barabási, A.L., Albert, R.: Science 286, 509 (1999)

Selection of Imitation Strategies in Populations When to Learn or When to Replicate?

Juan G. Díaz Ochoa

Institute for Theoretical Physics, Fachbereich 1, Bremen University,
Otto Hahn Allee, D-28334 Bremen, Germany
diazochoa@itp.uni-bremen.de

Abstract. A question in the modeling of populations of imitators is if simple imitation or imitation based on learning rules can improve the fitness of the individuals. In this investigation this problem is analyzed for two kinds of imitators involved in a cooperative dilemma: One kind of imitators has a replicator heuristics, i.e. individuals which decide its new action based on actions of their neighbors, whereas a second type has a learning heuristics, i.e. individuals which use a learning rule (for short learner) in order to determine their new action. The probability that a population of learners penetrates in a population of replicators depends on a training error parameter assigned to the replicators. I show that this penetration is similar to a site percolation process which is robust to changes in the individual learning rule.

Keywords: Learning, Population Dynamics, Game theory, Percolation.

1 Introduction

Intuitively we know that being part of a community or a group makes us less vulnerable to external influences (or technically speaking, increases our fitness). The best strategy to being part of a group is behaving like its members, an attitude that requires good imitator abilities. Imitation is a paradigm that not only belongs to the study of social and biological sciences [1,2,3], but also to the study of artificial intelligence [4]. In populations the problem consists to find the rule that an individual uses to select an action from a finite menu. Every individual is conditioning its decision on his endowed private signal about the state of the world and observed predecessor's decisions, without knowing the private signals of these predecessor's. The problem is, with which criteria is possible to select the best heuristics to make imitation?

Several works analyze social dynamics that an imitator heuristics generates, assuming that a group of individuals behave according to the majority choice [5,6,7]. However, there is no criterion that dictates which specific heuristics is the right imitation rule. The diversity of imitation heuristics imposes a difficult problem in searching a criterion that matches to the desired rule [9]. One approach to solve this problem is to use an evolutionary perspective, assuming that imitation heuristics depends on natural selection [8,10]. In general imitation in

J. Zhou (Ed.): Complex 2009, Part I, LNICST 4, pp. 819–831, 2009.

social dilemmas can be explored defining individuals without memory. However, the reason to include individuals with memory in this investigation is based on an argument introduced by Bull, Holland and Blackmore [16], namely the larger the memory, the easier the imitation process is (where memory represents the individual's endowment of the private signals coming from its environment.). Hence, the selection of better imitators will favor individuals with larger brains, i.e. the human brain expanded in size by memetic and not genetic reasons.

In particular, the effect of memory in a social dilemma has been explored in previous works, showing that memory (in particular larger memories) could induce individuals to act in a more cooperative way [17,18]. The combined effect of this mechanism together with spatial extension allow individuals with a prospect cooperative character to disseminate, form clusters and minimize interactions with non-cooperative individuals [19,20,21,14]. In several of such models individuals are modeled as a kind of automata that give an answer depending of a mechanism that uses past information from the game. Two prominent examples from the literature are Pavlov, i.e. the individual's action switch whenever the individuals was punished in the last time step (Nowak and Sigmund, 1993 [22]), and TFT, i.e. the actual individual's action is just the opposite as its neighbor's action [1]. These examples consider individuals with memory size one, i.e. these individuals are able to store the information (either fitness or individual's action) from the previous time steep. However, there are other models considering individuals with longer memory size. In such case the individuals store several actions from its opponent in a vector; the new action is the result of the projection of this vector on the individual's strategy [21,17]. Recently an alternative learning schema combining past actions and past payoffs were introduced, allowing the individual to adapt to the character of its opponent as well as the must promising strategy behavior [24].

The present work is also based on the selection by evolution of different behavioral characteristics of a population. However, the present contribution is not aimed to select different learning rules, but to find how evolution selects two representative imitation heuristics in the context of a social dilemma (represented by a prisoner's dilemma game). As the author is aware, in the literature there is no analysis of the behavior and concurrence of different learning rules in large populations and spatial games. The present is the selection of two representative mimetic heuristics that goes beyond the simple individual imitation rules usually defined in these kind of problems.

The two basic rules to be considered are: mimetic rules based on replication, i.e. where the individual imitate the behavior of its neighbors, and mimetic rules based on learning, i.e. where the individual attempt to 'learn' and reproduce the behavior of its neighbors. The first kind of individual is called **replicator**, which tries to minimize the generalization error by utilizing the available information in a statistical optimal way [7,11,12], whereas the second kind of individual is called *individual with learning* (for short **learner**), which is based on the idea of coincidence training [13,12]. The behavior of both individuals is similar to the behavior of conformists with memory [1,14,15], i.e. individuals that basically makes an imitation of the action of its opponent. The aim of the present work is to consider if in a given population of learners can or cannot invade a population of replicators. The question is: is an individual with a pay-off based learning rule more effective than a simple imitator?

The remaining of this work is divided as follows. In the next section the model is explained, in particular imitation heuristics and the co-evolving dynamics, based on a game, are introduced as two coupled processes that determine the selection of individuals with a given imitation rule. In section three the simulation method and the main results are presented. In section four I present a discussion of the principal implications of this investigation. The last section is devoted to the main conclussions.

2   Model

Two learning heuristics are considered in this model. The first schema assumes that an individual takes decisions under the influence of an informational band-wagon effect. In this heuristics the individual tries to minimize the error in the implementation of a new action by utilizing the available information in a sta-tistical optimal way [12,7]. The second schema assumes that the individual's decisions are made as the result of an individual learning process based on coin-cidence training [12,25]. Actions have uncertain payoffs. Additionally, individuals receive signals from other individuals and not from the environment.

Before choosing an action, the decision maker i observes the history of his private signal, $\sigma^i(t-l)$, and the history of the actions of the opponent j, $\sigma^j(t-l)$, where $1 \leq l \leq M$ is the step in the time history , where M is the number of signals that it is able to keep in its memory. This history is stored in a compact space Ω_i assigned to each individual i and is used in order to compute the probability distribution of the individual's new action $\sigma^i(t)$, which is an element of the set of the population's actions A.

Once an action is defined the individual becomes a payoff in the frame of a prisoner's dilemma game. Hence the individual's action can be either cooperate, C, or defect, D (Both actions will be defined in the next subsections). If the agent has low payoff respect its neighbors, then its learning schema is replaced by the complementary heuristics. Otherwise, the individual conserves its actual heuristics. In a nutshell, the population of agents owning a learning heuristics depends on the actions adopted by this population, and the population of actions depends on the population of learning heuristics. Furthermore remember that relative large memories are defined in order to facilitate the imitation process [16]. The agents are placed in a square lattice with periodic boundary conditions.

In the next subsections a detailed description of both the learning heuristics and the population dynamics is done.

2.1   Replicators: Bayes Rule

The input signal σ^i_I and σ^j_I can be either 1 for C or -1 for D. The individual i can follow a history Ω_i composed by the signals of $1 \leq l \leq M_i$ actions ($\sigma^j(t-l)$)

of its opponent j and M_i own signals $(\sigma i(t - l))$, where m_i is the memory size of individual i. The decision maker i uses the observed action history and its own private belief in order to compute the probability of the new action WR^{ij} against individual j. The conditional probability that individual i cooperates, i.e. $\sigma^i = 1$, is given by

$$WR^{ij} = min[1, e^{-\beta \Delta S^{ij}}], \tag{1}$$

where $\Delta S^{ij} = S^{ij}(t) - S^{ij}(t-1)$. Here, $S^{ij}(t)$ is the history of the signals $\sigma^i(t-M)$ and $\sigma^j(t - M)$ in the memory of individual i (this memory has a M_i size) given by $S^{ij}(t) = \sum_{l=1}^{M_i} J^{ij} \sigma^j(t - l)\sigma^i(t - l)$, where J_{ij} is the connectivity between both signals. For simplicity, $J_{ij} \sim M_i$. Observe that j also makes simultaneously a new decision against i with a probability WR^{ji}. Hence $S^{ij}(t) \neq S^{ji}(t)$ and $WR^{ij} \neq WR^{ji}$. The parameter $\beta = 1/\Lambda$ is a control parameter (valid for the whole population) related to the training error of the replicator [12].

Hence, for large β the individual tends to adopt C because the whole population reduces its fluctuations and a bandwagon effect appears. If $\beta \to 0$ then the individual must choose in a set of mixed strategies. Indeed, in this heuristics the imitation of actions depends on the fluctuations of actions of the whole population.

2.2 Minimal Neuronal Network

According to this schema the learner try to find some patron in the behavior of their opponents. Hence, this heuristics requires a kind of individual's mind, which is modeled using a simple neuronal network. Imitation heuristics is based on the pay-off of the individual, with a rule based on a linkage function that depends on the individual's utility.

The function $\tilde{S}^{ij}(t)$ for the learner is defined as

$$\tilde{S}^{ij}(t) = \sum_{l=1}^{m_i} \tilde{J}^{ij} \sigma^j(t - l), \tag{2}$$

where $\tilde{J}_{ij}$ is the linkage parameter defined according to the following learning rule: if $\sigma^i(t - l) = 1$ and $\sigma^j(t - l) = 1$ then $\tilde{J}_{ij} = 1$; if $\sigma^i(t - l) = 1$ and $\sigma^j(t - l) = -1$ then $\tilde{J}_{ij} = 1$ (i.e. it is better to imitate the opponents actions); if $\sigma^i(t-l) = -1$ and $\sigma^j(t-l) = 1$ then $\tilde{J}_{ij} = 1$ (i.e. the individual transforms a non-cooperative attitude into a cooperative attitude); if $\sigma^i(t-l) = -1$ and $\sigma^j(t-l) = -1$ then $\tilde{J}_{ij} = 1$ (i.e. it is better to imitate the opponents actions). This definition corresponds to a strategy of the kind $((1, 1), (-1, 1))$. Nevertheless an additional analysis with a strategy of the form $((1, 1), (-1, -1))$ (when $\sigma^j(t - l) = -1$ then $\tilde{J}_{ij} = 1$) will also be made in this investigation.

A simple example of the typical moves for a learner with $M = 6$ is presented in table 1. In this example the opponent present to the learner two different patrons that it should recognize. This response strongly depends on the initial conditions.

Table 1. An example of the learning heuristcs of a learner with $M = 6$. The elements in the discrete table represent the first component of the output σ^i, where i is for the reference individual and j its opponent. This table presents the learning process (negative time t) and the actions in the first six time steps.

t	$\hat{\sigma}^j$	$\hat{\sigma}^i$	$\tilde{J}_{ij}$	$\tilde{S}^{ij}$	t	$\hat{\sigma}^j$	$\hat{\sigma}^i$	$\tilde{J}_{ij}$	$\tilde{S}^{ij}$	t	$\hat{\sigma}^j$	$\hat{\sigma}^i$	$\tilde{J}_{ij}$	$\tilde{S}^{ij}$	t	$\hat{\sigma}^j$	$\hat{\sigma}^i$	$\tilde{J}_{ij}$	$\tilde{S}^{ij}$
-6	-1	1	1	-	-6	-1	-1	1	-	-6	-1	1	1	-	-6	-1	-1	1	-
-5	1	1	1	-	-5	1	-1	-1	-	5	1	1	1	-	-5	1	-1	-1	-
-4	1	1	1	-	-4	1	-1	-1	-	-4	-1	1	1	-	-4	-1	-1	1	-
-3	1	1	1	-	-3	1	-1	-1	-	-3	1	1	1	-	-3	1	-1	-1	-
-2	-1	1	1	-	-2	-1	-1	1	-	-2	-1	1	1	-	-2	-1	-1	1	-
-1	-1	1	1	-	-1	-1	-1	1	-	-1	1	1	1	-	-1	1	-1	-1	-
0	-1	(-1,1)	-	0	0	-1	-1	-	-6	0	-1	(0,1)	-	0	0	-1	-1	-	-6
1	1	(-1,1)	-	0	1	1	-1	-	-2	1	1	(0,1)	-	0	1	1	1	-	6
2	1	(-1,1)	-	0	2	1	1	-	2	2	-1	(0,1)	-	0	2	-1	-1	-	-6
3	1	(-1,1)	-	0	3	1	1	-	6	3	1	(0,1)	-	0	3	1	1	-	6
4	-1	(-1,1)	-	0	4	-1	1	-	2	4	-1	(0,1)	-	0	4	-1	-1	-	-6
5	-1	(-1,1)	-	0	5	-1	-1	-	-2	5	1	(0,1)	-	0	5	1	1	-	6

In the present simulation this learning process does not only take place at the beginning of the process but also could take again place after some random period of time. The transition probability for σ^i is in this case given by the following function [25]

$$WP^{ij} = \begin{cases} C & \text{if } \tilde{S}^{ij} > 0; \\ (C,D) & \text{if } \tilde{S}^{ij} = 0; \\ D & \text{if } \tilde{S}^{ij} < 0 \end{cases} \qquad (3)$$

This definition does not depends on any noise source, but on changes of $\tilde{J}_{ij}$ according to the individual's learning rules. The mathematical meaning of C and D will be given in the next section.

This model for a learner is equivalent to similar definitions of individuals owning a memory and that react according to the stored information and a fixed strategy [21]. However, instead of a sigmoid function, the transition probability is given by a delta function $\delta(\tilde{S}^{ij} - S_0)$, where S_0 is the restriction imposed by the strategy. Hence, such kind of individuals does not adapt to their environment (as the learner), but own a strategy that is more or less fit to the given environment.

2.3   Co-evolutive Dynamics

The action of individual i is represented by the action σ^i, that can be $\sigma^i = 1$ for C (Cooperate) or $\sigma^i = -1$ for D (Defect). The pay-off of individual i relative to individual j, given by U^{ij}, and individual j relative to i, given by U^{ij}, are defined according to the matrix defined in table 2 [26], with $W > R > P > S$, where W is for temptation (or winner), R for reward, P for punishment and S

Table 2. Definition of the pay off of individuals i and j in the prisoner's dilemma game. The first horizontal line corresponds to individual i; the first column to individual j. D represents defect and C represents cooperate. W is for winner, R for reward, S for sucker and P for punishment, such that $W > R > P > S$ [23,17].

-	C	D
C	R	W
D	S	P

for sucker (please do not confuse S with $\mathcal{S}^{ij}$, the history of the signals for the Bayes rule). The values of the matrix are defined as $R > \frac{1}{2}(W + S)$, such that for an individual is better to switch from cooperate to non-cooperate, but all the other individuals will profit if a single individual switches from defection to cooperation. Setting $W = 5$, $R = 3$, $P = 2$ and $S = 0$ this rule approaches the game to the pay-off matrix introduced by Axelrod (1984) [23,17].

The total outcome for the individual i respect to its opponents is $f^i = \sum_{j=1}^{K} U^{ji}$ and the total outcome of these opponents respect to i is given by $f'^i = \sum_{j=1}^{K} U^{ij}$, where K is the number of neighbors. The utility is a fundamental quantity that should determine the co-evolutionary process. The structure of the pay-off matrix is in general non-commutative; therefore, $f'^i \neq f^i$.

Suppose an individual in the lattice, that has a given heuristics, is selected and plays simultaneously with its four neighbors. Also suppose that i is a replicator. In this example the following actions were obtained: C^{i1} vs. C^{1i}, D^{i2} vs. C^{2i}, D^{i3} vs. D^{3i} and D^{i4} vs. D^{4i}. Hence $f^i = 3+5+2+2 = 12$, whereas $f'^i = 3+0+2+2 = 7$; hence, in this case $f^i > f'^i$ and, therefore, the agent can preserve its learning heuristics. Otherwise this heuristics is replaced by its complementary heuristics.

2.4   Simulation

The results were obtained from computer simulations of this model. The population has 2500 individuals placed on places of a two dimensional square lattice with periodic boundary conditions. Each individual has a memory size $M = 18$ and interacts with its four neighbors randomly selected inside a circular radius $R = 2$. The simulations were started introducing an initial random distribution of actions in an initial period of time, equivalent to the memory size of the individuals, in order to compute the initial learning heuristics of the learner and the initial action of the replicator. The final results do not contain the information of the initial period of time.

3   Results

In a first analysis the reaction of two learners is shown. If both individuals have a small memory size, the willingness to cooperate is relative high. But when one of both individuals increases its memory, the frequency of cooperation decreases. One suspect that the frequency of cooperation depends on the memory size,

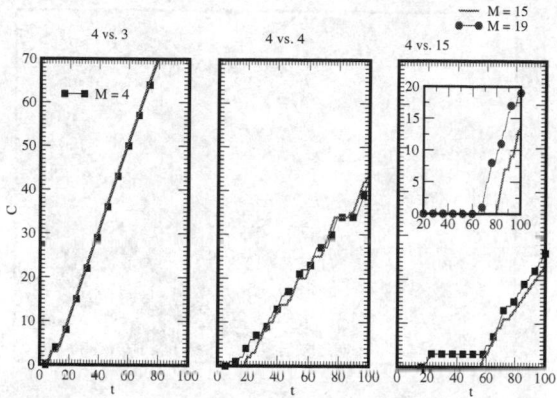

Fig. 1. Cumulative cooperation frequency of two learners with different memory (information storage) size as a function of the time. The strategy in this example is $((1,1),(-1,1))$.

in particular those larger memories have a smaller cooperation rate. However, the interaction of two learners with $M > 10$ show that the individual with small memory size has a lower cooperation frequency. In general is possible to conclude that individuals with $M > 10$ are more 'cautious'.

The analysis of the behavior of the individuals is extended to learners vs. replicators. Given that the behavior of the replicators depends on the fluctuations of the environment, the cooperation frequency also depends on the fluctuation sizes, given by the parameter β. Low values of β imply large fluctuations in the system. For low β the learner cannot imitate the replicator, because the fluctuations in the behavior of the replicator does not help the individual in its learning process. On the other side, the larger the value of β, the larger is the cooperation frequency, because the learner can imitate the behavior of its opponent. In this sense we suspect that the fluctuations of the environment affects the form that a class of individuals can invade a given population.

The previous results were obtained for fixed memory sizes. However, there is no clear criteria for the selection of the memory size of the individuals of the population. One plausible starting point is to assume that the memory of the individuals also co-evolve with the population. This option implies not only an ensemble of imitators but also, additionally, an ensemble of memories assigned to this population of imitators. The combination of both ensembles could represent a more realistic system, but this assumption do not helps to understand the specific role that the imitation rules have in the co-evolution of the population. For this reason the memory size of the individuals will remain fixed in order to restrict the dynamics of the population to the effects induced by the imitation rules.

In order to analyze such a system it is necessary to find a criterion to fix the memory size (storage size) of individuals. For the replicators, previous investigations shown that relative memory sizes are associated to larger individual fitness.

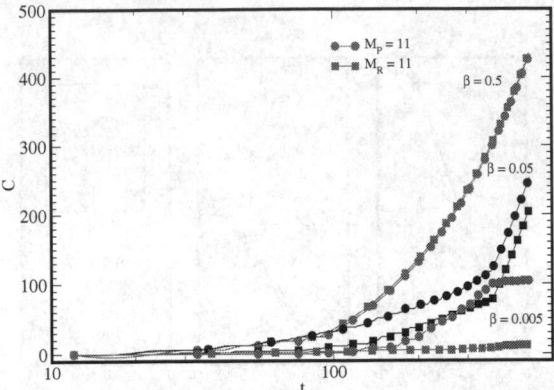

Fig. 2. Cumulative cooperation frequency C (i.e. the number of times that anindividual selects C) of a learner vs. a replicator as a function of the time. Different cooperation frequencies were computed as a function of the fluctuations of the environment. The learner's strategy is $((1,1),(-1,1))$.

Nevertheless, this result depends on β [27]; but we can assume in general that this kind of individual's large memory size represents an advantage in evolution. In general it is reasonable to assume that both imitators share the same relative large memory size. In the present investigation the individual memory size were fixed to $M = 18$ (i.e. the individual can store 18 individual bits).

The initial configuration considers more replicators in order to analyze if the learners can invade the initial population. If the control parameter β is small, then the chance that a population of learners invades the population of replicators increases. However, this final population of learners does not extinguish the population of replicators, i.e. there is a coexistence of both groups of individuals. If β increases then the initial population of learners is almost extinguished; this result is logic because the learner has a very poor performance while the replicator respnses to the noise of the environment (fig 3). For this analysis the population of learners grows as a function of the time until it reaches an equilibrium state.

The dynamics of the population depends on the probability that a given imitator is replaced by another imitator. This probability depends on the imitator's fitness. This transition has not only effect in the population of imitators, but also in the number of individuals with a cooperative (or non-cooperative) behavior. This process is similar to remove a given individual with a probability given by W_p. Of course, this removal is preferential, depending on the individual's fitness, which also depends on β. Hence, the parameter β determines the probability that a given class of individual will be replaced or not. Therefore, this is the order parameter to be used in the subsequent analysis.

The strategy matrix for the learner allows the representation of different individual behaviors. In particular, if the strategy is $((1,1),(1,1))$ the individual simply repeats the same actions as its neighbors, i.e. the individual behaves using

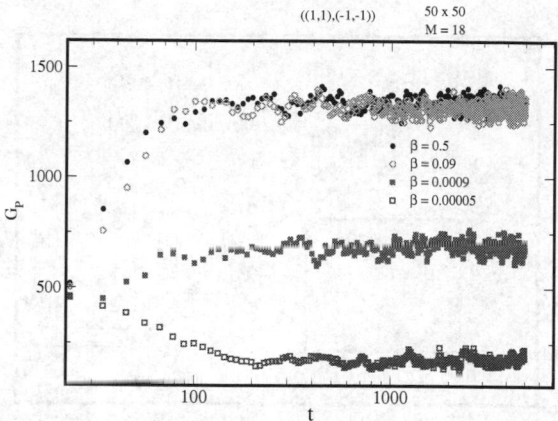

Fig. 3. Average number of learners G_P as a function of the time for different parameters β

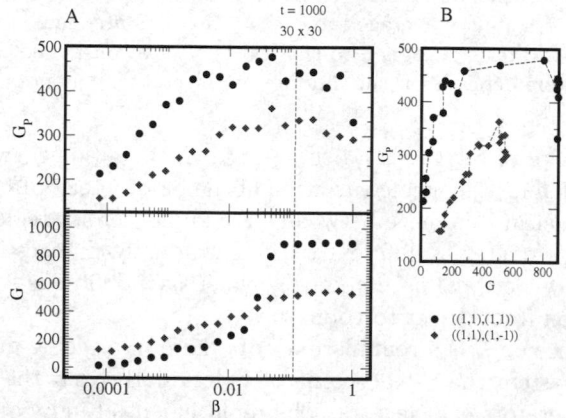

Fig. 4. Comparison of Tit for Tat $((1,1),(1,1))$ against a $((1,1),(-1,1))$ strategy. The frequency of coopeators (mean number of cooperators represented by G) as well as the population of learners (mean number of learers, represented by G_P) as a function of β is shown (part **A**). In **B** the Population of learners as a function of the frequency of cooperators (G_P vs. G) is shown.

a kind of Tit for Tat strategy (TFT). In several experiments and computer tournaments it has been shown that this strategy is very successful [1,23]. In the present work we also found that individuals with a large information endowment are also successful if it adopts a TFT strategy against the replicators (See Fig. 4, part **A** and **B**). In this simulation the frequency of cooperators G with TFT shows a behavior similar to a ferromagnetic material, i.e. close to the critical parameter β_C there is a spontaneous change of non-cooperative to cooperative actions; this behavior is logic because the learners simply reproduce the behavior of the replicators, which essentially are a kind of Ising model. In contrast to this

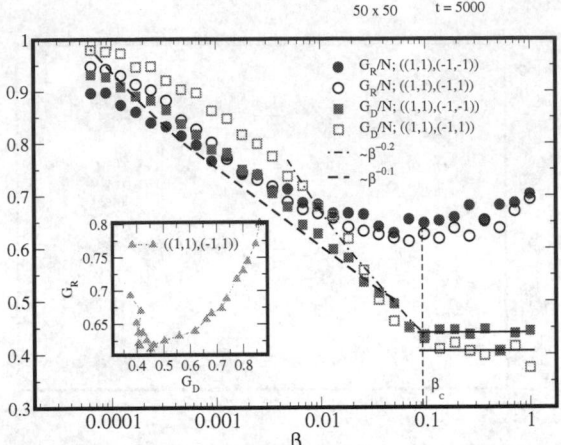

Fig. 5. Averaged number of defectors G_D (i.e. individuals with actions of type D) and replicators G_R (i.e. individuals with a replicate heuristics) as a function of the parameter β. The results are averaged and the error bars are smaller than the size of the symbols. Two strategies assigned to the learners were considered. In the inset, the number of replicators depends on the number of defectors in the population.

result, the behavior of $((1,1),(-1,1))$ shows a kind of cross over, with a constant exponent $\gamma << 0.5$ for the whole variation of the parameter β. In resume, in this system there is a clear difference between TFT and other strategies, namely that a population with strategies different to TFT gradually changes its frequency of cooperation below a critical parameter β_C, whereas individuals with TFT shows a phase transition from defect to cooperate.

The dependence on β is a reminiscence of a kind of an Ising model. However, in the present system the replacement of individuals avoid the appearance of domains of cooperators or defectors. The replacement of individuals may imply that this process is similar to a percolation process. In the fig. 5, the average of the number of defectors and the average of the number of replicators is shown.

In the first case the increase of the β parameter suggest a kind of inverse percolation process in the number of defectors. This result is equivalent to a kind of net magnetization of the system. However, the critical parameter β_c (equivalent to a critical temperature) is $\beta \sim 0.09$, which is much smaller than the critical temperature of a common two-dimensional Ising model. Above this parameter the population of defectors does not get totally extinguished. The critical exponent is $\gamma \sim 0.1$. This result shows a slight similitude to the random node removal of scale-free networks [28], where β is equivalent to the fraction of nodes to be removed.

In the second case there is also a decrease in the number of replicators. However, above the critical parameter β_C the number of replicators increases again. Contrary to the actions, the population does not clearly show a kind of reverse percolation process. The phase diagram of number of defectors against the number of replicators shows that there is a critical number of defectors where the population is almost invaded by learners, at $G_D \sim 0.45$; below this value there is again an increase in the number of replicators.

The present was not only restricted to a single strategy for the learners. A test with a second strategy was also performed. For $((1,1)(-1,-1))$ the critical exponent is $\gamma \sim 0.1$, which is much lower than the critical exponent for the original strategy. Additionally, the number of learners above the critical value is much slower than the original strategy. Hence, a change in the strategy introduces changes in the whole population and in the critical exponents.

4   Discussion

The learners can reproduce some behavioral patron when the number of errors in the training of the replicators is low. From an intuitive point of view this result is reasonable because an efficient learning rule (as the rule used by learners) requires low corruption of information by noise. However, this fact does not exclude the chance that a set of learners cannot invade a population of replicators when the number of errors in the imitation process is large. Even with $\beta \sim 0.0009$ (i.e. the replicators change frequently of actions) a set of learners can also invade a population of replicators. Only for very low values of β there is an extinction process of learners. Hence, a learning heuristics can invade a simple replicator heuristics if the individuals can make use of the informational stability available from its environment. Naturally, trying to reproduce a patron is less effective than doing a replication of the opponents actions; the present computation shows that a population of learners with a TFT strategy (i.e. the heuristics for a learner but without learning) is larger than a population of infividuals with learning.

This result is based on two representative heuristics, which are at the same time a representative example within other several specific imitation heuristics. The reader could have the opinion that this model is defined in an ad-hoc way restricted to only two behavioral rules. Furthermore, the system is a population constrained in a 2-Dimensional lattice with periodic boundary conditions. However, the construction and the results computed with this model could have potential implications in the use of tools coming from the statistical mechanics in artificial intelligence to describe populations of individuals trying to define some trend in the memetic heuristics, when errors in the imitation process (for instance induced by the environment) are present.

5   Conclusion

In a population with two kinds of heuristics the control parameter β, related to the training error of the replicator, has not only influence in the individual's actions, but also in the size of the population of individuals owning a particular heuristics. At the same time this behavior has influence in the frequency of actions within the population, i.e. the willingness to cooperate or not with its neighbors. Above a critical parameter β_C the cooperative attitude of the individuals do not depends in this control parameter. Above this critical value β_C

the population of learners reaches also an optimal value. In general it has been shown that the penetration process of learners in a population of replicators is similar to a percolation process, and that this processis robust against changes in the individual's learning rule.

In this frame the following question could be made: is preferable to restrict the actions to the information available after some decision process in the neighbor or is better to try to find some reaction patron? In this case it is not clear if learning or imitation is the basis in the formation of attitudes, in particular in the consideration of social dilemmas and collective behavior of heterogeneous individuals [29,30]. In the formulation of individual actions there is not only imitation but also learning rules that influence individual attitudes: the combined effect of both (and not simply imitation, as is usually assumed) is relevant when an individual takes some decision.

References

1. Axelrod, R., Hamilton, W.D.: The Evolution of Cooperation. Science 211, 1390 (1981)
2. Dawkins, R.: The Selfish Gene. Oxford University Press, Oxford (1990)
3. Selten, R., Ostmann, A.: Imitation Equilibrium. Homo Oeconomicus 16, 114 (2001)
4. Breazeal, C., Buchsbaum, D., Gray, J., Gatenby, D., Blumberg, B.: Learning from and About Others: Towards Using Imitation to Bootstrap the Social Understanding of Others by Robots. Artificial Life 11, 31 (2005)
5. Kirman, A.: Ants, Rationality, and Recruitment. The Quartery Journal of Economics 108, 137 (1993)
6. Nadeau, R., Cloutier, E., Guay, J.-H.: New Evidence About the Existence of a Bandwagon Effect in the Opinion Formation Process. International Political Science Review 14, 203 (1993)
7. Smith, L., Sörensen, P.N.: Cowles foundation discussion paper NO. 1552. Yale University (2005)
8. Henrich, J., Boyd, R.: The Evolution of Conformist Transmission and the Emergence of Between-Group Differences. Evolution and Human Behavior 19, 215 (1998)
9. Chavalarias, D.: Metamimetic Games: Modeling Metadynamics in Social Cognition (2005), jass.soc.surrey.ac.uk/9/2/5.html
10. Guzmán, R.A., Rodríguez-Sickert, C., Rowthorn, R.: When in Rome, do as the Romans do: the coevolution of altruistic punishment, conformist learning, and cooperation. Evolution and Human Behavior 28, 112 (2007)
11. Lasner, A., Ekenberg, O.: A One-Layer Feedback Artificial Neural Network With Bayesian Learning Rule. Int. J. Neural Systems 1, 77 (1989)
12. Engel, A., van den Broeck, C.: Statististical Mechanics of Learning. Cambridge University Press, Cambridge (2001)
13. Hebb, D.O.: The organization of Behavior. Wiley, New York (1949)
14. Nowak, M., May, R.M.: Evolutionary Games and Spatial Chaos. Nature 359, 826 (1992)
15. Galam S.: Behind the Shirts, is Symmetry (1999), physics/9901012 v1 11
16. Bull, L., Holland, O., Blackmore, S.: On Meme-Gene Coevolution. Artificial Life 6, 227 (2000)

17. Hauert, C.H., Schuster, H.G.: Effects of increasing the Number of Players and Memory Size in the Iterated Prisoner's Dilemma: a numerical approach. Proc. R. Soc. Lond. B 264, 513 (1997)
18. Sanz, A., Martin, M.: Memory boots Cooperation. Int. Jour. Mod. Phys. C 17, 841 (2006)
19. Hauert, C.: Fundamental Clusters in Spatial 2–2 games. Proc. R. Soc. Lond. B 268, 761 (2001)
20. Herz, A.V.M.: Collective Phenomena in Spatially Extended Evolutionary Games. J. Theor. Biol. 169, 65 (1994)
21. Lindgren, K., Nordhal, M.G.: Evolutionary dynamics of spatial games. Physica D 75, 292 (1994)
22. Nowak, M.A., Sigmund, K.: A Strategy of Win-Stay, Lose-Shift that outperforms Tit-for-Tat in the Prisoner's Dilemma Game. Nature 364, 56 (1993)
23. Axelrod, R.: The evolution of cooperation. Basic Books, New York (1984)
24. Hauert, C., Stenull, O.: Simple adaptive Strategy wins the Prisoners Dilemma. J. Theor. Biol. 218, 261 (2002)
25. Schneider, J.J., Kirkpatrick, S.: Stochastic optimizazion. Springer, Heidelberg (2007)
26. Hofbauer, J., Sigmund, K.: Evolutionary Games and Population Dynamics. Cambridge University Press, Cambridge (1998)
27. Díaz Ochoa, J.G. (ed.): Diversity under Variability and extreme Variability of Environments. arXiv:0804.2898v1 (2008)
28. Albert, R., Barabasi, A.L.: Statistical mechanics of complex networks. Reviews of Modern Physics 74, 47 (2002)
29. Granovetter, M.: Threshold Models of Collective Behavior. The American Journal of Sociology 83, 1420 (1978)
30. Huberman, B.: The Dynamics of Social Dilemmas. Scientific American 76 (1994)

Sediment Transport Dynamics in River Networks: A Model for Higher-Water Seasons

Jie Huo[1], Xu-Ming Wang[1], Rui Hao[1], and Jin-Feng Zhang[2]

[1] School of Physics and Electric Information Sciences,
Ningxia University, Yinchuan 750021, China
wxmwang@nxu.edu.cn
[2] School of Physics and Electronic Information,
Huaibei Coal Industry Teachers' College, Huaibei 235000, China

Abstract. A dynamical model is proposed to study sediment transport in river networks in higher-water seasons. The model emphasizes the difference between the sediment-carrying capability of the stream in higher-water seasons and that in lower-water seasons. The dynamics of sediment transport shows some complexities such as the complex dependence of the sediment-carrying capability on sediment concentration, the response of the channel(via erosion or sedimentation) to the changes of discharge.

Keywords: sediment transport dynamics, fast transience, complexity, self-organization, self-adaptation.

In decades past, natural river networks have attracted a good deal of attention in physics and geophysics communities [1,2,3,4,5,6]. Three progresses can generalize the main scientific achievements. The first one is the empirical observations on natural river networks that revealed many power-law relationships between some parameters of river networks [1,2]. The second one is the modeling studies based on the local erosion rules to get deeper understandings of how these natural events occur in the evolutive processes [3,4,5,6]. The main spirit of the models embodies the nature of the water: flows downhill and changes the local height of landscape due to the erosion. Some parallel achievements are the theoretical studies based on the minimum energy dissipation [7,8], which introduce the assumption that an open system with fixed energy input tends to form a structure that minimizes the total energy dissipation in the process of the precipitated water flowing downhill. The third one, the insight into the mechanisms of what creates the scaling laws, was carried into execution by Dodds and Rothman via recurring to statistic physics method [9].

It is known that river patterns bear striking fractal forms and self-organized characteristics [2]. As stated by Dhar [8], this self-organized structure is the result of the feedback between the water flow and the landscape itself via the variation of the erosion. This implies that sedimentation can not be neglected, at least, for the purpose of investigating the evolution of the river patterns [10,11].

J. Zhou (Ed.): Complex 2009, Part I, LNICST 4, pp. 832–840, 2009.

Recently, we proposed a dynamical model to simulate sediment transport in matured river networks in lower-water season by consideration of both functions of erosion and sedimentation [12,13]. The core spirit of this model embodies the feedback mechanism between erosion and sedimentation via the adjustment of sediment-carrying capability(SCC)of runoff. A steady state predicated by the model shows scaling laws that the quantity of erosion or sedimentation(QES) distributes exponentially along the channel in the downriver direction. The result under one of the combinations of the two free model parameters implies a catastrophe due to a great deal of sedimentation or erosion in the downriver reach of the channel [13]. However, the way into the steady state will be interrupted by frequent fluctuations in runoff, caused by random precipitations. The response of the river to the abrupt change of the input shows self-organized and self-adaptive behaviors. The self-organized characteristic is represented by the observation that the variation of the SCC is opposite to that of the QES as the discharge changes, which shows that the response of the river trends to depress the increase of erosion as water flow increases and that of sedimentation as water flow decreases. This is actually incarnated by the rapid change of QES, the preceding response to the change of water flow. After this, the dynamic shows self-adaptive characteristics: the channel is forced to change its formation by sediment erosion or sedimentation via adjusting SCC of water: erosion may be enhanced(or sedimentation may be mitigated) as water flow goes up, while sedimentation may be strengthened(or erosion may be mitigated) as water flow goes down [see Ref. [13]].

This letter is aimed at a further investigation in a higher-water season to test whether a river is commonly dominated by the aforementioned dynamics. The model should give emphasis to the difference between the sediment transport in higher-water season and in lower-water season. Sediment concentration usually sharply increases in a higher-water season in comparison with a lower-water season due to the erosion of the relative more precipitation. Therefore, the SCC of water is not only adjusted by the undergone erosion or sedimentation state, but also directly driven by the sediment concentration. This means that the adjustments of SCC can be fallen into two types. The former is positive and the latter is passive.

For the convenience of discussions, the passive adjustment of SCC is incarnated in the relation between the transported sediment S and the discharge Q as well as the sediment concentration Q_s(which is actually sediment concentration of the nearest neighboring upriver stream for the discussed segment). Then we have

$$S = kQ^\alpha Q_s^\beta. \tag{1}$$

Where k denotes a part of SCC, which corresponds to the positive adjustment of SCC based on the undergone erosion or sedimentation. α and β are indices, which indicate, without losing generality, that the transported sediment S nonlinearly depends on both of Q and Q_s.

In our discussions, the ordering scheme of river is designed by differing the lateral and not lateral between two injecting streams as the confluence takes place. If one goes against a stream, say rank h, at each of nodes the laterally injecting stream belongs to the branches of rank $h + 1$, but the non laterally injecting stream is actually one segment of this stream. So the rank of a stream can be uniquely determined by the times of its laterally injecting confluences as it drains to the sea. And the segments of a stream can be also numbered with the times of its non-laterally injecting confluences as it drains to the sea. The first segment is at the upper end, which is actually taken as the source. And then the segment number increases one by one as branches of rank $i+1$ inject into the stream in turn. For convenience, we denote the stream flowing on a channel segment by the segment itself.

Let's firstly consider the confluence on the segments in a time interval, $t \rightarrow t + \delta t$, which can be simply denoted by t(one of the discrete time variables, i.e., $t = 1, 2, \cdots$). So the confluence, in the tth time step, can be directly expressed as

$$Q_h^l(i+1, t) = Q_h^l(i, t) + Q_{h+1}^i(i_{last}, t). \tag{2}$$

Where $h = 0, 1, 2, \cdots$, $i, l = 1, 2, \cdots$, and $i = 1$ denotes the source of the stream. The character Q denotes the stream flow. The left side of Eq. (1) represents the outflow of the $i+1$th segment of lth branch of rank h, the right side is the inflows respectively coming from the ith segment of this branch and the ith branch of rank $h+1$(which is in fact denoted by its last, i_{last}, segment). The outflow balances the inflows. It implies that the change of water can only be attributed to the confluence on condition that the rainfall in the basin is denoted by the change of water from headstreams.

Generally, the sediment carried by the outflow may not be equal to that by the inflows. So the erosion or sedimentation will occur in these segments. Based on the confluence expressed by Eq. (1) and the definition presented above, the QES can be determined by equation,

$$\Delta S_h^l(i+1, t) = S_h^l(i+1, t) - S_h^l(i, t) - S_{h+1}^i(i_{last}, t). \tag{3}$$

On the right side of the equation, the first term denotes the sediment carried by the outflow, the second and third denote that carried by the inflows, respectively from the upper segment and the injecting branch of higher rank. $\Delta S_h^l(i+1, t) > 0$ indicates that the $i+1$th segment is in a scouring state in the tth time interval, while $\Delta S_h^l(i + 1, t) < 0$ implies that it is in a deposition state.

According to the above discussions, the positive adjustment of SCC via the self-variation of k can be conducted by the QES on a segment and that on its neighboring segments. This is similar to that of lower-water season.

(1) For a natural river, the sedimentation implies that both the deposited and the suspended sediment particles are relatively smaller, and the viscosity intensity of fluid is enhanced, so that the friction of river-bed against the flow will increase and the stream can carry more sediment. On the contrary, the erosion implies the deposited material over the river-bed and the suspended particles become coarser, the friction will decrease, thus the stream can transport lesser sediment. It is the adjustment of k of a segment conducted by QES on itself.

(2) Whereas, the QES on the upriver neighboring segment will influence k of the discussed segment, in comparison with what has been presented above,

in the reversed way. If a segment is in scouring state, the increasing sediment may cause k of its downriver neighboring segment to increase so that the stream can discharge more sediment, while if the segment is in deposited state, the decreasing sediment may let k of the downriver neighboring segment to decrease. Naturally, the QES on a higher rank branch may affect k in the same way.

In fact, k of a segment should be determined by the integration of aforementioned factors. This integration may imply competition between the two opposite effects, increasing and decreasing k of a segment. So the expression that presents the above-mentioned self-adjustment mechanism can be given

$$
\begin{aligned}
k_h^l(i+1,t+1) = {} & k_h^l(i+1,t) - A_1 \Delta S_h^l(i+1,t)/(Q_h^l(i+1,t+1))^\alpha (Q_{sh}^l(i,t+1) \\
& + Q_{sh+1}^i(i_{last},t+1))^\beta + A_2(\wedge S_h^l(i,t) \\
& + \Delta S_{h+1}^i(i_{last},t))/(Q_h^l(i+1,t+1))^\alpha (Q_{sh}^l(i,t+1) \\
& + Q_{sh+1}^i(i_{last},t+1))^\beta.
\end{aligned}
\tag{4}
$$

Where A_1 and A_2 are free parameters of the model. They denote the adjustment strength of k executed by the QES on a segment and its nearest upriver neighbors, respectively.

In our model, any one upriver segment can be chosen as the most top one, and also treated as the source of the stream. The regular or irregular changes, $\xi(t)$, of the discharge and that of sediment concentration, $\zeta(t)$, may represent the rainfalls. Then we have

$$
Q_h^l(1,t) = Q_{h_0}^l + \xi(t),
\tag{5}
$$

$$
Q_{sh}^l(1,t) = Q_{sh_0}^l + \zeta(t),
\tag{6}
$$

and

$$
\Delta S_h^l(1,t) = \Delta S_{h_0}^l
\tag{7}
$$

for all possible h, l and t. Where $\Delta S_{h_0}^l$ may have a fixed value for the simpleness.

The core spirit of the model consists in the description of self-adjustability of a river in the sediment transport process, which is manifested by the positive adjustment of SCC that is expressed by Eq.(4). Similar self-adjustability often dominates the dynamics of a completely open system. The simulation work may not only help us to understand the nature of sediment transport on river networks but also provide us with some illumination for general dynamics of self-adaptive systems else.

Before carrying out the simulation with the model, it is necessary to test it. The database about the Yellow River can provide us with the observed data to perform this test [14]. However, a problem may occur: how to conduct the test? Two objects to compare are not math in time scale: the observed data is the average over a month, but the simulating results are instantaneous if we follow the real scene in the river network(random rainfalls). So we have to introduce random fluctuation of discharge to mimic the rainfalls confined by the condition that the average of the discharge is equal to the average observed. The results calculated by this scheme are presented in Fig. 1 and show that they are in good

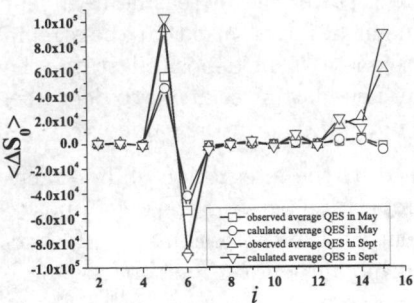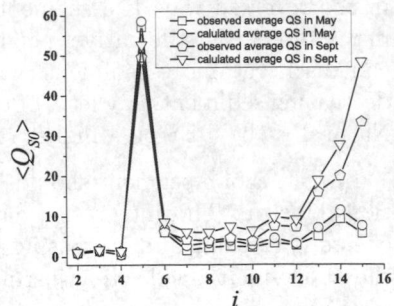

Fig. 1. The comparisons of (a) the calculated average QES($< S_0 >$) and (b) sediment concentration $< Q_{s0} >$, with that of the observed on 15 segments from Tangnaihai to Sanmenxia reach of the Yellow River in 1982. The formers are obtained by introducing the random fluctuation each other 50 time steps, and taking average over 500 time steps.

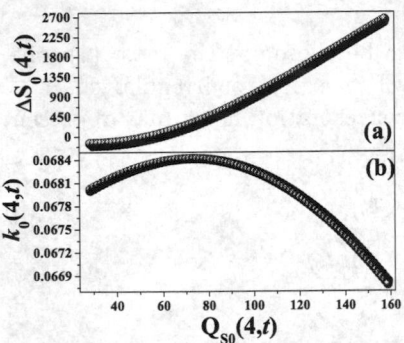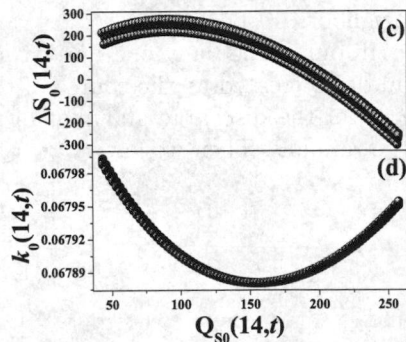

Fig. 2. The calculated results to show the dependence of (a) ΔS_0 and (b) k_0 on Q_{s0} of segment 4 and that of segment 14 showed respectively by (c) and (d). The details are presented in the text.

agreement with the average observed QES and sediment concentration. These indicate that our model can describe the sediment transport in river networks in higher-water seasons. The model can be employed to mimic the process in a natural river network. It is necessary to point out that data in the database are only about the main stream and its first rank branches, we have to change our model by neglecting the adjustment of k made by the first rank branches. So $\Delta S_{h+1}^i(i_{last}, t)$ in Eq. (4) can be deleted, and the superscripts of the terms can be canceled.

Let's firstly see about the influence of sediment concentration on the positive self-adjustable part of SCC, k, in the simplest case, that is, only the injections of the first rank branches into the main stream are considered. The result as sediment concentration of the first rank branches, Q_{s1}, increases is presented by Fig. 2, which is obtained as the related variables take the following values: $\Delta S_0(1,t) = 50$, $Q_{s0}(1,t) = 100$, $k_0(i,1) = 5.7 \times 10^{-2}$, $Q_0(i,1) = RN(180, 350, N_1)(RN(180, 350, N_1)$ denotes N_1 random numbers between 180 and 350), $Q_{s0}(i,1) = RN(30, 108, N_1)$, $Q_1(i,t) = 40$, $A_1 = 0.6 \times 10^{-4}$, $A_2 = 0.9 \times 10^{-5}$, $\alpha = 1.3$, $\beta = 0.98$, $Q_{s1}(i,t) = 50 + \zeta(t)(\zeta(t) = 0.1\frac{t}{260}$ for $mod(t, 260) = 0$, which implies that Q_{s1} will increase 0.1 at each regular time-step interval, 260). The number of the first rank branches, $N_1 = 30$. As shown in Fig. 2 (evolves for 23400 time steps), the influence of Q_{s1} on k_0 or ΔS_0 bears complex behaviors.

Two strong impressions of the figures on one may be the varying trends of k_0 and ΔS_0 as Q_{s1} increases. The first one is that the variation of k_0 is almost opposite to that of ΔS_0 on the same segment, which implies that the self-adjustment of the SCC tends, as a whole, to depress large increase of the erosion or sedimentation. The second one consists in that the variation of k_0 on a upriver segment, say 4, are opposite to that of a downriver segment, say, 14. This show a complex response mechanism of k_0 to Q_{s0}. Firstly, it demonstrates a complex dependence of k_0 on Q_{s0}, and basically agrees with the fact that the SCC will weaken incipiently and then strengthen when sediment concentration monotonously increases from a relative smaller initial value. The experimental results [15] indicated that the suspension of sediment is mainly attributed to the eddy motion of a stream with smaller value of Q_s. However, the eddy motion will be damped as Q_s gets larger. Therefore, as Q_s increases to a threshold the SCC of the stream will strengthen since the suspension of sediment particles is governed now by the buoyancy force of water instead of the eddy motion. Secondly, the complexity of the dynamics is also indicated by the dependence of the varying trend of SCC on the initial state, erosion or sedimentation. As the core idea of our model, the SCC of stream is apt to go down when the segment is in a scoring state, while SCC of stream is inclined to go up when the segment is in a deposited state. Obviously, Fig. 2 represents the above-mentioned complexities. Fig. 2(a) shows that segment 4 is initially in deposited state the SCC should go up(Fig. 2(b)), meanwhile the stream has lower sediment concentration, the SCC should go down. The competition of these two varying trends makes the SCC to transform from increase to decrease at threshold, about 70, of Q_{s0}. Fig. 2(c) and (d) indicates that both of the initial erosion of segment 14 and the increasing sediment concentration of the stream can lead the SCC to decrease. At threshold, about 160, of Q_{s0}, the curve has an inflexion, which means that adjustment mechanism of SCC dominated by the eddy motion is replaced by that governed by the increasing buoyancy force due to the increase of sediment concentration.

Now let's discuss the transient characteristic of sediment transport dynamics in the scouring or deposited process. It can be presented by the dependence of QES on discharge. To show more related details, we calculate the QES by considering the influences of the second order branches on the dynamics of the mainstream in the case of the parameters and variables are set with the following values: for all possible i, i' and t, $\Delta S_0(1,t) = 50$, $Q_0(i,1) = RN(60, 120, N_1 + 1)$, $Q_{s0}(i,1) = RN(25, 90, N_1 + 1)$, $k_0(i,1) = 5.7 \times 10^{-2}$, $Q_1(i',1) = RN(25, 90,$

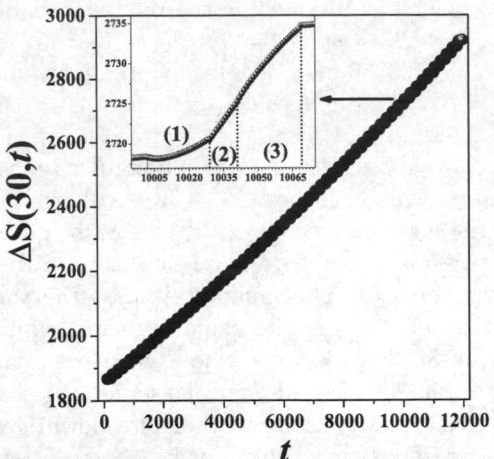

Fig. 3. The transient QES on segment 30 as the confluence of the second rank branches is considered. The inset is its partial magnification for showing the three SFTs. The details are presented in the text.

$N_2 + 1$), $Q_{s1}(i', 1) = RN(50, 180, N_2 + 1)$, $k_1(i', 1) = 2.0 \times 10^{-1}$, A_1, A_2, α and β are the same of Fig. 2. $Q_2(i'', t) = RN(0.8, 1.6, N_2) + \xi(t)(\xi(t) = 0.005\frac{t}{200}$ for $mod(t, 200) = 0$). The number of the first and the second rank branches is $N_1 = 30$ and $N_2 = 40$, respectively.

The calculated results shown in Fig. 3 is similar to the characteristics of transient dynamics, in a lower-water season, of which the response of QES to the abrupt changes of discharge can actually be fallen into the faster and the slower phases. The former is the rapid response of river to the change and will last for the stream flow getting a new stationary value, the latter can lead the dynamics to the steady state if the discharge keeps changeless. The partial magnification, the insets, of Fig. 3, show that the faster transience can be divided into three sub-phases, three so-called sub-faster-transiencies(SFTs). Ref.[13] has reported the mechanism that SFT (1) is caused by the changing branches(in the water flux) of first rank injecting, one by one, into this segment. It will last for all of the changing stream from its upriver branches of the first rank reach the segment; SFT (2) is induced by the more and more changed streams from the branches of rank 2 reach the segment. This SFT will go on up to its nearest upriver stream of rank 1 becoming stationary; then the transience enters into SFT (3), and comes to the end time step at which the uppermost stream of rank 1 also get stationary. The situation is opposite when the discharge of rank 2 stream decreases.

Obviously, SFT(1) embodies self-organized characteristics of the sediment transport process: a river trends to depress a mass of erosion when the stream flow increases, while control a vast sum of sedimentation when the stream flow decreases. However, when the input of a river changes, it may be forced to adjust its state to acclimatize itself to the new conditions. SFT(2) and SFT(3) are just the case. They indicate that the QES has to increase as the discharge increases, but decrease as the discharge decreases. This may be a common characteristic of self-adaptive dynamics for a kind of completely open systems. These dynamic characteristics may have engineering significance. If the majority of random fluctuations of discharge takes place within the faster transient phase, especially within SFT(1), the gross of erosion trends to balance that of sedimentation, so the channel can hold relative steadier. If we can control the discharge increasing(decreasing) at regular time-step intervals within SFT(2) or SFT(3)(the latter may be better), eroding(accumulating) can go on to improve the conditions of river channels for transporting sediment.

Although the natural fluctuation of water flux is attributed to all of the upriver branches of higher rank with random precipitation. This may weaken or erase the divisions of the three SFTs. However, according to the above discussions, engineers can take use some man-made-flood-peaks to control sediment transport processes for different engineering purposes. The detailed descriptions of this research and the other understandings to sediment transport in river networks are in process.

This study is supported by the National Natural Science Foundation of China under Grant No. 10565002 and the Program for New Century Excellent Talents in University of China under Grant No. NCET-06-0914.

References

1. Leopold, L.B.: Downstream Change of Velocity in Rivers. Am. J. Sci. 251, 606 (1953)
2. Rodríguez-Iturbe, I., Rinaldo, A.: Fractal River Basins: Chance and Self-Organization. Cambridge University Press, Cambridge (1997)
3. Banavar, J.R., Colaiori, F., Flammini, A., Giacometti, A., Maritan, A., Rinaldo, A.: Sculpting of a Fractal River Basin. Phys. Rev. Lett. 78, 4522–4525 (1997)
4. Manna, S.S.: Branched Tree Structures: from Polymers to River Networks. Physica A 254, 190 (1998)
5. Maritan, A., Colaiori, F., Flammini, A., Cieplak, M., Banavar, J.R.: Universality Classes of Optimal Channel Networks. Science 272, 984 (1996)
6. Giacometti, A.: Local Minimal Energy Landscapes in River Networks. Phys. Rev. E 62, 6042–6051 (2000)
7. Sun, T., Meakin, P., Jøssang, T.: A Minimum Energy Dissipation Model for Drainage Basins that Explicitly Differentiates Between Channel Networks and Hillslopes. Physica A 210, 24 (1994)
8. Dhar, D.: Theoretical Studies of Self-organized Criticality. Physica A 369, 29–70 (2006)
9. Dodds, P.S., Rothman, D.H.: Geometry of River Networks. I. Scaling, Fluctuations, and Deviations. Phys. Rev. E 63, 016115–016117 (2000)
10. Murray, A.B., Paola, C.: A Cellular Model of Braided Rivers. Nature 371, 54–57 (1994)
11. Meakin, P., Sun, T., Jossang, T., Schwarz, K.: A Simulation Model for Meandering Rivers and Their Associated Sedimentary Environments. Physica A 233, 606–618 (1996)

12. Hao, R., Wang, X.-M., Huo, J., Zhang, J.-F.: Simulating Sediment Transport on River Networks. Mod. Phys. Lett. B 22, 127–137 (2008)
13. Wang, X.-M., Hao, R., Huo, J., Zhang, J.-F.: Modeling Sediment Transport in River Networks. Physica A 387, 6421–6430 (2008)
14. http://www.loess.csdb.cn/hyd/user/index.jsp
15. Cao, Z., Egashira, S., Carling, P.A.: Role of Suspended-Sediment Particle Size in Modifying Velocity Profiles in Open Channel Flows. Water Resour. Res. 39, 1029 (2003)

Scaling Relations in Absorbing Phase Transitions with a Conserved Field in One Dimension

Sang-Gui Lee and Sang Bub Lee[*]

Department of Physics, Kyungpook National University,
Daegu, 702-701, Republic of Korea
sblee@knu.ac.kr

Abstract. Validity of two scaling relations $\beta = \nu_\parallel \theta$ and $z = \nu_\parallel/\nu_\perp$ widely known in absorbing phase transitions is studied for the conserved lattice gas (CLG) model and the conserved threshold transfer process CTTP) both in one dimension. For the CLG model, it is found that both relations hold when the critical exponents calculated from the all-sample average density of active particles are considered. For the CTTP model, various exponents are calculated via Monte Carlo simulations and they are confirmed by the off-critical scaling and the finite-size scaling analyses. The exponents estimated from the all-sample averages again satisfy both relations. These observations are in strict disagreement with earlier observations in two dimensions [Phys. Rev. Lett. **85**, 1803 (2000); Phys Rev. E **68**, 056102 (2003)] but support the more recent observation for the CLG model [Phys. Rev. E **78**, 040103(R) (2008)].

Keywords: absorbing phase transition, conserved lattice gas, conserved threshold transfer process, critical exponents, scaling relations.

1 Introduction

Nonequilibrium, continuous phase transition from a fluctuating active phase to single or multiple absorbing states has attracted great attention during last several decades [1,2,3,4]. So far, only few universality classes were identified; i.e., the directed percolation (DP) class [5,6,7,8] and the parity conserving (PC) class [9,10,11,12] were firmly established, but the pair-contact process with diffusion (PCPD) class is still under controversy [13,14,15,16]. The triplet and quadruplet reaction-diffusion models were also claimed to belong to the new universality class [17,18]. Besides these classes, a new universality class was proposed by Rossi *et al.* for the models with a conserved field, generated from the symmetry that the order parameter is locally coupled to a short-range nondiffusive conserved field [19]. The conserved lattice gas (CLG) model, the conserved threshold transfer process (CTTP), and the stochastic sandpile model were found to belong to this universality class [20,21,22].

In usual critical phenomena, many physical quantities near criticality are known to be described by the power-law behaviors [23,24]. For example, for the

---

[*] Corresponding author.

J. Zhou (Ed.): Complex 2009, Part I, LNICST 4, pp. 841–852, 2009.
© ICST Institute for Computer Sciences, Social Informatics and Telecommunications Engineering 2009

models with a conserved field, the density of active particles exhibits a power-law decay against the evolution time, and the steady-state density in the supercritical region also yields a power-law behavior against the distance from criticality. The universality classes are classified by the powers of such behaviors, known as critical exponents. The critical exponents are not independent but they are linked according to the scaling relationships derived from the scaling hypothesis. The two scaling relations widely known in absorbing phase transitions are

$$\beta = \theta \nu_{\parallel} \tag{1}$$

and

$$z = \nu_{\parallel}/\nu_{\perp}, \tag{2}$$

where β and θ are the exponents associated with, respectively, the order parameter against the distance from criticality and the density of active particles in time, $\nu_{\parallel}$ and $\nu_{\perp}$ are the exponents characterizing temporal and spatial correlation lengths, and z is the dynamic exponent.

Rossi $et\ al.$ [19] claimed that the former relation broke the "simple" scaling for the CLG model in two dimensions. Failure of the scaling was first reported on a sandpile model by Vespignani $et\ al.$ [25] and was also discussed in other works [22]. Lübeck and his collaborators also reported supported results [26,27]. Very recently, however, the present authors studied the CLG model in one dimension and found that the former scaling relation held precisely but the latter relation was questioned [28]. Further extensive works in two dimensions revealed that both relations held but the exponent $\nu_{\perp}$ calculated from the surviving-sample averages were found to be inconsistent with the value in the thermodynamic limit of an infinite size system. Similar results were also found for the CTTP model [29].

In the CLG model, each lattice site may be occupied by at most one particle, and a particle is defined to be active if it has at least one particle in the nearest-neighbor sites; otherwise, it is inactive. The dynamics proceeds with the hopping of active particles to one of the nearest-neighbor empty sites. In the CTTP model, on the other hand, each lattice site may be occupied by up to two particles, and the doubly occupied sites are assumed to be active sites. In each process, particles on each active site attempt to hop to randomly selected nearest-neighbor inactive sites. If such inactive site is not available, the particle on an active site does not move. The density of active particles (sites) are considered to be the order parameter for the CLG (CTTP) model. The critical behaviors of the CLG model and the CTTP model in one dimension are particularly important by two reasons. Firstly, because of the simplicity of the models by dimensional reduction, the critical density and some of the critical exponents for the CLG model are known exactly, thus, enabling one to examine the scaling relations. Indeed, de Oliviera calculated the exponents θ, β, and $\nu_{\perp}$ analytically [30], and the present authors estimated the rests of the exponents very accurately by Monte Carlo simulations [28]. Secondly, the universality split between the CLG model and the CTTP model is known to occur in one dimension, whereas in higher dimensions the two models are known to belong to the same universality class [31]. The cause of the universality split is known to be the different hopping mechanisms for the two models. For the CLG model, since only one of the two nearest-neighbor sites is empty, the hopping is deterministic in one dimension, while for the CTTP model the hopping is stochastic because the direction of hopping may be selected randomly when both neighboring sites arc inactive and at least one of them is empty.

In the previous work of the present authors [28], the scaling relations in Eqs. (1) and (2) were examined with the known exponents for the CLG model in one dimension, and the former relation was found to be satisfied while the latter was not. This observation is in disagreement with the more recent work in two dimensions, where both relations were found to hold precisely [29]. In this paper, the scaling relations will be reexamined for the CLG model and the CTTP model both in one dimension. For the CLG model, the known, presumably exact, results will be used in the scaling analyses. For the CTTP model, various critical exponents will be calculated and crosschecked via the off-critical and finite-size scaling analyses. Since the scaling relations were examined in the previous work only for the CLG model and the CLG model and the CTTP model are known to exhibit different behaviors in one dimension, the present work completes examinations of the validity of the scaling relations for the models with a conserved field. (Note that in two and higher dimensions all variant models with a conserved field are known to belong to the same universality class.) It is found that both scaling relations appear to hold precisely for both models, but the critical exponent $\nu_\perp$ obtained from the surviving-sample averages appears to be invalid in the thermodynamic limit, in agreement with the earlier observation [29].

In Sec. 2, the known scaling theory will be reviewed briefly, introducing the off-critical and finite-size scaling functions. In Secs. 3 and 4, the results for the CLG model and the CTTP model will be presented, with appropriate discussions. The concluding remarked will be made in the last section.

2 Scaling Theory

In both the CLG model and the CTTP model, initially ρL particles are distributed randomly, following the rules for each model as described in Sec. 1, in a given system of size L. If the selected site is already occupied by the allowed number of particles, the particle is not added and a new site is selected. As the dynamics proceeds, the density of active particles, ρ_a, decreases in time due to the repulsive contribution of hopping of active particles. If the density of particles, ρ, is too small, all particles eventually become inactive and, if ρ is sufficiently large, ρ_a saturates. Therefore, there exists a critical density ρ_c at which ρ_a decreases, following the power law

$$\rho_a \sim t^{-\theta}, \tag{3}$$

θ being the decay exponent of the density of active particles. For $\rho > \rho_c$, ρ_a converges to the steady-state density ρ_{sat}, which exhibits the power-law behavior against the distance from criticality, i.e.,

$$\rho_{\text{sat}} \sim (\rho - \rho_c)^{\beta} \tag{4}$$

for $\rho > \rho_c$, where β is the order-parameter exponent. Therefore, the off-critical values of ρ_a depend on the evolution time and the distance from criticality via the correlation time $\tau \sim |\rho - \rho_c|^{-\nu_{\parallel}}$. Thus,

$$\rho_a(t) = t^{-\theta}\mathcal{F}(t/\tau) \equiv t^{-\theta}\mathcal{F}(t|\rho - \rho_c|^{\nu_{\parallel}}), \tag{5}$$

where $\mathcal{F}(x)$ is the universal off-critical scaling function. Since $\rho_a \to \rho_{\text{sat}}$ in the $t \gg \tau$ limit, the scaling relation in Eq. (1) follows. On the other hand, for a finite system, since the correlation length cannot exceed the size of system L near criticality, i.e., $\xi \sim |\rho - \rho_c|^{-\nu_{\perp}} \sim L$, it is obtained that $|\rho - \rho_c| \sim L^{-1/\nu_{\perp}}$. Therefore, Eq. (5) becomes

$$\rho_a(t) \sim t^{-\theta}\mathcal{G}(t/L^z) \tag{6}$$

with the aid of Eq. (2), where $\mathcal{G}(x)$ is the finite-size scaling function. If, on the other hand, one focuses on the late time, Eq. (6) can be rewritten as

$$\rho_a(t) \sim L^{-\beta/\nu_{\perp}}\mathcal{H}(t/L^z), \tag{7}$$

where $z\theta = \beta/\nu_{\perp}$ has been used.

The scaling in Eq. (7) is the simple scaling claimed to be broken by an anomalous exponent θ. This scaling is particularly useful when ρ_a saturates in the $t \gg L^z$ region; thus, at ρ_c, Eq. (7) yields $\rho_a(t) \to \rho_{\text{sat}} \propto L^{-\beta/\nu_{\perp}}$ as $t \to \infty$, which enables one to estimate the value of $\beta/\nu_{\perp}$.

3   Conserved Lattice Gas Model

For the CLG model in one dimension, the absorbing phase is one of the two states $010101\cdots$ and $101010\cdots$ at the critical density of $\rho_c = \frac{1}{2}$. Using the simplicity of dynamics, de Oliveira obtained $\beta = 1$ by considering the density of nearest-neighbor pairs of occupied sites close to ρ_c. He also obtained $\nu_{\perp}$ using the transfer matrix approach; although he proposed the classical result of $\nu_{\perp} = \frac{1}{2}$, the present authors pointed out that his result was erroneous and the correct one was $\nu_{\perp} = 1$ (see Ref. [25] of Ref. [28]). The rests of the exponents were obtained by the present authors via the numerical simulations. Summarizing the results, they are

$$\theta = \frac{1}{4}, \quad \beta = 1, \quad \nu_{\parallel} = 4, \quad \nu_{\perp} = 1, \quad z = 2.$$

It should be noted that the value of θ was obtained by direct simulations of ρ_a against the evolution time at ρ_c, and the value of $\nu_{\parallel}$ was obtained from the best data collapse for the scaling function in Eq. (5) and was also confirmed by the scaling of the persistence distribution, i.e., the distribution of average time that the system persists in one of the phases, e.g., in the phase that the density of active particles is larger than the mean density of active particles. With these results, it is clear that the relation in Eq. (1) holds.

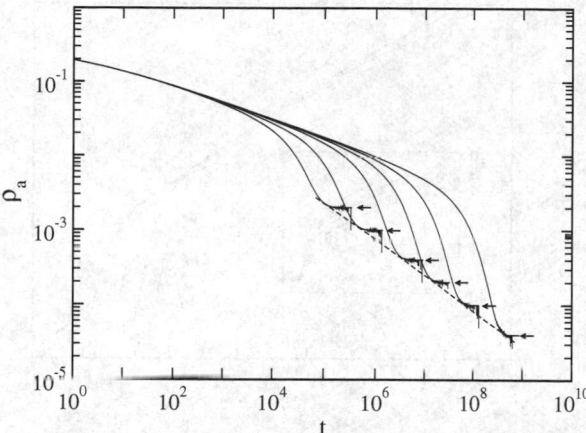

Fig. 1. The surviving-sample averages of $\rho_a(t)$ against the evolution time at the critical density $\rho_c = 0.5$ for the CLG model in one dimension. Data are for, from left to right, $L = 1000, 2000, 5000, 10000, 20000,$ and 50000.

The value of z was obtained from the best collapse of the data for the finite-size scaling analysis of Eq. (6) using the all-sample average data. [Note that the all-sample average is the average over all samples attempted, whereas the surviving-sample average is the average over the samples remaining in an active phase up to the time of interest.] The same result was also conjectured from the spreading of the active particles in the close vicinity of the absorbing state [28]. On the other hand, the exponent $\nu_\perp = 1$ was obtained by transfer matrix approach by de Oliveira [30] and was also obtained from the data for ρ_{sat} for systems of various sizes L at ρ_c [29]. Since Eq. (7) focuses the steady-state density of active particles in the late-time limit, the surviving-sample average should be used to estimate the value $\beta/\nu_\perp$ because the all-sample average densities decay in the long-time limit. Figure 1 shows the data for $\rho_a(t)$ for L ranging from $L = 1000$ to $L = 50000$ at ρ_c. The density of active particles for a given size system yields the power-law behavior in the early time, decays sharply, then saturates in the long time region, and eventually falls into the absorbing state. In the close vicinity of an absorbing state, the active particles consist of a single dimer, i.e., $\rho_{\mathrm{sat}} = \frac{2}{L}$, which implies $\beta/\nu_\perp = 1$ or, equivalently, $\nu_\perp = 1$. The saturation values for various L (marked as arrows in Fig. 1) are precisely $\frac{2}{L}$. With the estimate of $\nu_\perp$ and the values of $\nu_\parallel$ and z, the scaling relation in Eq. (2) appears to be violated. On the contrary, if one assumes that the scaling relation in Eq. (2) is correct, then, $z = \frac{\nu_\parallel}{\nu_\perp} = 4$ would be obtained.

In the previous works in which the simple scaling was claimed to be broken by an anomalous value of θ, the value of z was obtained from the finite-size scaling analysis of Eq. (6), the value of $\nu_\perp$ was measured from the data for ρ_{sat} at ρ_c against the size of system L, and the scaling in Eq. (7) was examined with the surviving-sample average data. We here test the scaling in Eq. (7) using both

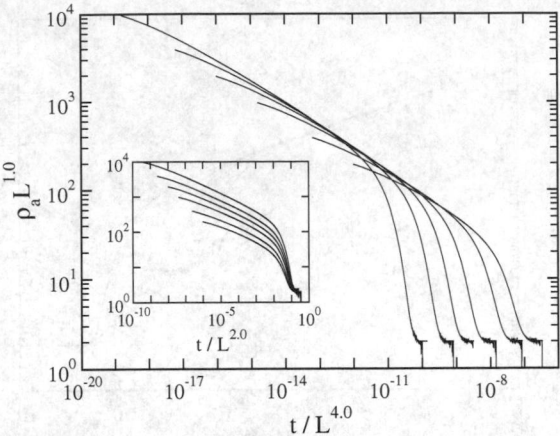

Fig. 2. The scaled density of active particles $\rho_a L^{\beta/\nu_\perp}$ for the surviving-sample averages against the scaled time t/L^z, using $\beta/\nu_\perp = 1$ and $z = 4$. The inset is the same data scaled with $\beta/\nu_\perp = 1$ and $z = 2$.

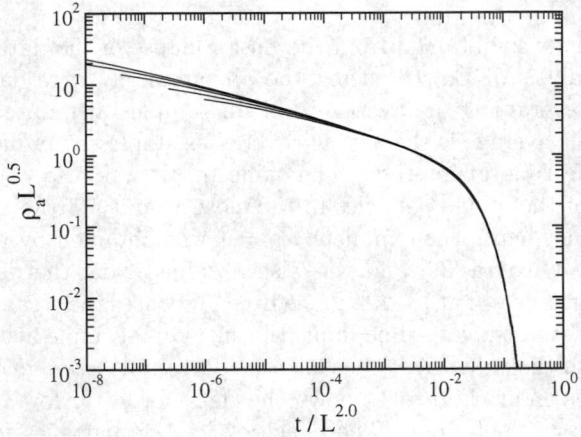

Fig. 3. The scaled density of active particles $\rho_a L^{\beta/\nu_\perp}$ for the all-sample averages against the scaled time t/L^z, using $\beta/\nu_\perp = \frac{1}{2}$ and $z = 2$

$z = 2$ and $z = 4$. In Fig. 2, the main plot is the scaling function $\rho_a(t)L^{\beta/\nu_\perp}$ against the evolution time scaled with $z = 4$ and the inset is the same data against the time scaled with $z = 2$. It is clear that both plots do not show data collapsing, implying that at least one or more critical exponents used in the scaling analysis might be incorrect. In the previous work of the present authors in two dimensions, it was found that the scaling was failed due to the two time scales for the surviving-sample averages, one at the first inflection point (which appears on both the all-sample data and the surviving-sample data) at $t = L^z$

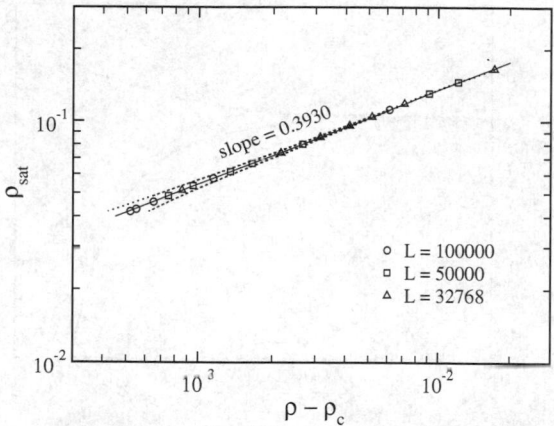

Fig. 4. The double logarithmic plot of the steady-state density ρ_{sat} against the distance from criticality, with $\rho_c = 0.98285$. Two dotted lines which veer up and veer down are the trial plots for $\rho_c^t = 0.98295$ (above) and for 0.98275 (below), and the solid line is the regression fit over the data for $\rho_c = 0.98285$.

with $z = 2$ and the other at the time when saturation sets in, i.e., at $t = L^{z'}$ with $z' \neq z$. The value of z' may be estimated from the point that the data of ρ_a touches the dashed line in Fig. 1 and is found to be $z' \approx 2.15$.

When the same scaling is examined with the all-sample data, the value $\nu_\perp = 1$ should not be used because it was measured with the surviving-sample data. Instead, it should be obtained from Eq. (2), i.e., $\nu_\perp = \nu_\parallel/z = 2$. (Note that the value $z = 2$ was obtained from the finite size scaling of the all-sample data.) Figure 3 shows the all-sample averages of $\rho_a(t)$ scaled by $L^{-\beta/\nu}$ against the scaled time, using $\beta/\nu_\perp = \frac{1}{2}$ and $z = 2$. It is clear that scaling holds perfectly, indicating that the exponents used in this analysis are correct. This assures us that the exponent $\nu_\perp$ obtained from the surviving-samples is not valid.

4   Conserved Threshold Transfer Process

For the CTTP model in one dimension, the critical exponents β, θ, $\nu_\parallel$, $\nu_\perp$, and z were calculated by Lübeck and Heger [27] and, with the values, the scaling relation in Eq. (1) was found to be violated. (In fact, Lübeck and Heger assumed that the relation in Eq. (1) was invalid and they obtained $\nu_\parallel$ from Eq. (2) using the estimates of $\nu_\perp$ and z.) However, since most of the estimates were not crosschecked by alternative methods, such as the scaling analyses in Eqs. (5), (6), and (7), it is necessary to recalculate the exponents in order for the close examination of the scaling relations.

The critical density was estimated by Lübeck and Heger [27] from the best power-law fit of ρ_{sat} against $\rho - \rho_c$ using the surviving-sample average data up to predetermined time steps. Since for any finite size systems there remain some

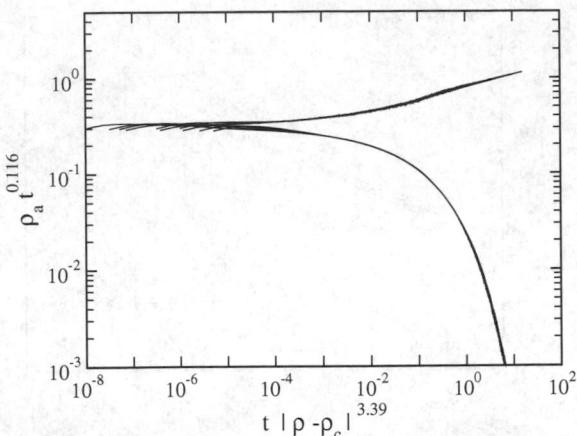

Fig. 5. The off-critical scaling function of the active site density $\rho_a t^\theta$ against the scaled time $t|\rho - \rho_c|^{\nu_\parallel}$ for $\rho = 0.955$, 0.96, 0.965, 0.97, and 0.975 (below) and for $\rho = 0.987$, 0.989, 0.99, 0.995, and 1.0 (above) with $\theta = 0.116$ and $\nu_\parallel = 3.39$, and $\rho_c = 0.98285$, for the CTTP model in one dimension.

surviving samples even below criticality, the value of ρ_c estimated in this way with surviving samples would be underestimated, because the densities for which samples remain in an active phase would be assumed supercritical. In this work, ρ_c is predetermined from the power-law behavior of ρ_a and, with this value, the power law of ρ_{sat} in the supercritical region is analyzed. Since we do not know an accurate value of ρ_c, we use only the data which are the same for all-sample averages and surviving-sample averages. [Note that two samples are identical if no sample falls into absorbing states.] The predetermined value which yields the best results for the test is determined as true ρ_c.

The critical density is obtained as $\rho_c = 0.98285$, with the regression slope $\beta = 0.393(4)$, as shown in Fig. 4. The two dotted curves which veer up and veer down are for the trial densities $\rho_c^t = 0.98295$ (above) and 0.98275 (below), and the solid line is the power-law fit over the data using $\rho_c = 0.98285$. It is thus sufficient to estimate ρ_c up to four significant digits. The critical density obtained is larger than that by Lübeck and Heger, but the exponent β is close to the known values for the stochastic fixed-energy sandpile model, $\beta = 0.42$ [32], and for CTTP model by Lübeck and Heger, $\beta = 0.382$ [27]. The regression slope of $\rho_a(t)$ yields the exponent $\theta = 0.116(3)$, which is slightly smaller than that by Lübeck and Heger, the difference being apparently attributed to the larger value of ρ_c. With the estimates of β and θ, it follows $\nu_\parallel = \beta/\theta \simeq 3.39(4)$.

In order to crosscheck the value of $\nu_\parallel$, the off-critical scaling in Eq. (5) is examined for the data of $\rho_a(t)$. Plotted in Fig. 5 are the scaled densities $\rho_a(t)t^\theta$ against the scaled time $t|\rho - \rho_c|^{\nu_\parallel}$ using $\theta = 0.116$ and $\nu_\parallel = 3.39$. It is clear that data for various densities fall on the two separate curves, one for $\rho > \rho_c$ (above) and the other for $\rho < \rho_c$ (below), indicating that the scaling indeed holds. This

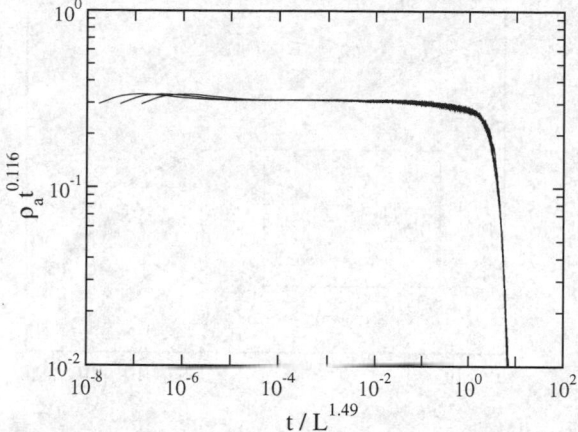

Fig. 6. The finite-size scaling function of the active site density $\rho_a t^\theta$ against the scaled time t/L^z for systems of sizes $L = 40000, 80000, 160000$ for the CTTP model in one dimension

confirms that the value of $\nu_\parallel$ is correct and, accordingly, the scaling relation in Eq. (1) holds.

The finite size scaling in Eq. (6) is also tested with the trial value of z. Plotted in Fig. 6 is the scaled density of active sites $\rho_a(t)t^\theta$ against the scaled time t/L^z for three selected sizes of systems, i.e., for $L = 4 \times 10^4, 8 \times 10^4$, and 1.6×10^5. Data for different size systems exhibit the best collapse for $z = 1.49$, but the quality of scaling is not as good as for the off-critical scaling. If we, however, choose the size of system twice as large as the largest size selected, i.e., $L = 3.2 \times 10^5$, data near the inflection point deviate slightly. Similar behavior was also observed previously for the CTTP model on a checkerboard fractal substrate [33]. We believe this to be that the scaling region is relatively narrow for the CTTP model. Accepting $z = 1.49$, it is obtained that $\nu_\perp = \nu_\parallel/z = 2.28$. In Table 1, the estimates are summarized and compared with the estimates by Lübeck and Heger [27]. It should be noted that, with the estimates in the present work, we did not find any precursor of the violation of scaling relations and, thus, we believe that our estimates are correct and both relations in Eqs. (1) and (2)

Table 1. Summary of the critical exponents for the CLG model and the CTTP model both in one dimension, in comparison with the reported results. The values for the CLG model are conjecture to be exact.

Exponents	θ	β	$\nu_\parallel$	$\nu_\perp$	z
CLG model	$\frac{1}{4}$	1	4	2	2
CTTP model	0.116	0.393	3.39	2.28	1.49
Ref. [27]	0.141	0.382	2.452	1.760	1.393

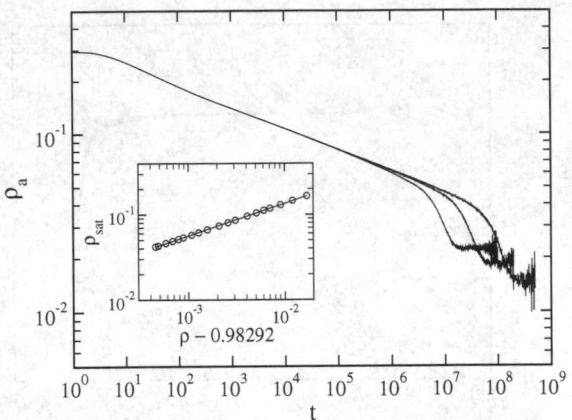

Fig. 7. The surviving sample averages of $\rho_a(t)$ at the trial value $\rho_c^t = 0.98292$ for $L = 25000, 50000$ for the CTTP model in one dimension

hold. It should also be noted that the scaling relation in Eq. (1) does not hold with the estimates by Lübeck and Heger.

In order to test the scaling in Eq. (7), the surviving-sample average data for ρ_a are also calculated for the CTTP model. Selecting the size of system as a multiple of 2 of the base size, i.e., $L_0, 2L_0, 2^2L_0, \cdots$, with L_0 ranging from 10^3 to as large as 2×10^4, the steady-state values at ρ_a should be placed with an equal vertical spacing of $(\beta/\nu_\perp)\ln 2$ on a double logarithmic plot if the scaling $\rho_{\text{sat}} \propto L^{-\beta/\nu_\perp}$ holds. For our estimate of $\rho_c = 0.98285$, the steady-state value does not exhibit such behaviors. We also calculate ρ_{sat} for different trial values of ρ_c. It is found that the steady-state densities are displayed with nearly equal spacing for $\rho_c^t = 0.98292$, as shown in Fig. 7 for $L = 2.5 \times 10^4$, 5×10^4, and 10^5. However, choosing the larger size $L = 2 \times 10^5$ or smaller size $L = 1.25 \times 10^4$, the steady-state value is no longer equally spaced and, with this value of ρ_c, the power-law fit of ρ_{sat} against $\rho - \rho_c$ becomes worse, though the exponent β is found to be similar. Therefore, it is not possible to estimate $\nu_\perp$ with the surviving-sample averages of ρ_{sat} for the CTTP model in one dimension. Similar behavior was also observed for the CTTP model on a checkerboard fractal substrate. Since the dynamics of the CTTP model on a checkerboard fractal is more likely one-dimensional as was claimed in Ref. [33], it is plausible that the critical behaviors of the CTTP model on a one-dimensional lattice and on a checkerboard fractal are similar. Indeed, some of the exponents on a checkerboard fractal are similar to those estimated in the present work.

5   Concluding Remarks

The two widely known scaling relations in absorbing phase transitions are examined carefully for the CLG model and the CTTP model both in one dimension. For the CLG model, it was found that both relations appeared to hold when the exponents obtained using the all-sample average data of ρ_a are considered. It was also found that the scaling in Eq. (7) was broken when the value of $\nu_\perp$ obtained from the surviving-sample averages was used, as was observed in the earlier works in two dimensions. Such a failure of the scaling appeared to be due to that the two time scales existing in the surviving-sample average ρ_a, i.e., one at the first inflection point where the finite-size effect first comes into the system and the other at the time when the saturation sets in, are different. The latter time exists only on the surviving-sample averages and appears to have influenced the value of $\nu_\perp$. In the thermodynamic limit of $L \to \infty$, since the second time scale associated with the saturation diminishes and a single value of $\nu_\perp$ would be obtained. It is believed that this value of $\nu_\perp$ is identical to that obtained from the finite-size scaling analysis with all-sample average data. Therefore, with the exponents valid in the thermodynamic limit, both scaling relations appear to hold.

For the CTTP model, we estimated the critical exponents β, θ, $\nu_\parallel$ and z from direct simulations and off-critical scaling and finite size scaling analyses. During the analyses, there was no precursor of violation of the scaling relations when all-sample average data were used. For the surviving sample averages for the CTTP model, it appeared that the scaling in Eq. (7) did not hold and, as a consequence, the value of $\nu_\perp$ was not obtained.

Acknowledgments

This work was supported in part by the Korea Science and Engineering Foundation Grant (R01-2008-000-10886-0) and by the Korea Research Foundation Grant funded by the Korean Government (KRF-2008-313-C00329). The authors are gratefully for the supports.

References

1. Marro, J., Dickman, R.: Nonequilibrium Phase Transitions in Lattice Models. Cambridge University Press, Cambridge (1999)
2. Hinrichsen, H.: Adv. Phys. 49, 815 (2000); Hinrichsen, H.: Braz. J. Phys. 30, 69 (2000)
3. Ben-Avraham, D., Havlin, S.: Diffusion and Reaction in Fractals and Disordered Systems. Cambridge University Press, Cambridge (2000)
4. Ódor, G.: Rev. Mod. Phys. 76, 663 (2004)
5. Janssen, H.K.: Z. Phys. B: Condens. Matt. 42, 151 (1981)
6. Grassberger, P.: Z. Phys. B Condens. Matt. 47, 365 (1982)
7. Jensen, I., Dickman, R.: Phys. Rev. E 48, 1710 (1993); Jensen, I.: J. Phys. A 27, L61 (1994)
8. Muñoz, M.A., Grinstein, G., Dickamn, R., Livi, R.: Phys. Rev. Lett. 76, 451 (1996)
9. Takayasu, H., Tretyakov, A.Y.: Phys. Rev. Lett. 68, 3060 (1992)
10. Jensen, I.: Phys. Rev. E 50, 3623 (1994)
11. Kwon, S., Park, H.: Phys. Rev. E 52, 5955 (1995)
12. Cardy, J., Tauber, U.C.: Phys. Rev. Lett. 77, 4780 (1996)

13. For the bosonic mversion of the PCPD model, Howard, M.J., Täuber, J.C.: J. Phys. A 30, 7721 (1997); for the fermionic version, Carton, E., Henkel, M., Schollwöck, U.: Phys. Rev. E 63, 36101 (2001)
14. Kochelkoren, J., Chate, H.: Phys. Rev. Lett. 90, 125701 (2002)
15. Noh, J.D., Park, H.: Phys. Rev. E 69, 016122 (2004)
16. Hinrichsen, H.: Physica A 361, 457 (2006)
17. Park, K., Hinrichsen, H., Kim, I.-M.: Phys. Rev. E 66, 025101 (2002)
18. Ódor, G.: Phys. Rev. E 67, 056114 (2003)
19. Rossi, M., Pastor-Satorras, R., Vespignani, A.: Phys. Rev. Lett. 85, 1803 (2000)
20. Manna, S.S.: J. Phys. A 24, L363 (1991)
21. Dhar, D.: Physica 263A, 4 (1999)
22. Vespignani, A., Dickman, R., Muñoz, M.A., Zapperi, S.: Phys. Rev. E 62, 4564 (2000)
23. Stanley, H.E.: Introduction to Phase Transitions and Critical Phenomena. Oxford University Press, London (1971)
24. Amit, D.J.: Field Theory, the Renormalization Group, and Critical Phenomena. World Scientific, Singapore (1984)
25. Vespignani, A., Dickman, R., Muñoz, M.A., Zapperi, S.: Phys. Rev. Lett. 81, 5676 (1998); Dickman, R., Vespignani, A., Zapperi, S.: Phys. Rev. E 57, 5095 (1998)
26. Lübeck, S., Misra, A.: Eur. Phys. J. B 26, 75 (2002)
27. Lübeck, S., Heger, P.C.: Phys. Rev. E 68, 56102 (2003); Lübeck, S.: ibid 66, 046114 (2002)
28. Lee, S.-G., Lee, S.B.: Phys. Rev. E 77, 021113 (2008)
29. Lee, S.B., Lee, S.-G.: Phys. Rev. E 78, 040103(R) (2008)
30. de Oliveira, M.J.: Phys. Rev. E 71, 016112 (2005)
31. Lübeck, S.: Phys. Rev. E 64, 16123 (2001); Lübeck, S.: ibid 66, 46114 (2002)
32. Dickman, R., Alava, M., Muñoz, M.A., Peltola, J., Vespignani, A., Zapperi, S.: Phys. Rev. E 64, 056104 (2001)
33. Lee, S.-G., Lee, S.B.: Phys. Rev. E 77, 041122 (2008)

Scaling Law between Urban Electrical Consumption and Population in China

Xiaowu Zhu[1,2], Aimin Xiong[3], Liangsheng Li[1], Maoxin Liu[1], and X.S. Chen[1]

[1] Institute of Theoretical Physics, Chinese Academy of Sciences,
P.O. Box 2735, Beijing 100080, P.R. of China
[2] Business School, China University of Political Science And Law,
No.27 FuXue Road, Beijing 102249, P.R. of China
[3] Department of Systems Science, School of Management,
Beijing Normal University, Beijing 100875, P.R. of China
{zxw,xiong,liliangsheng,mxliu,chenxs}@itp.ac.cn

Abstract. The relation between the household electrical consumption Y and population N for Chinese cities in 2006 has been investigated with the power law scaling form $Y = A_0 N^\beta$. It is found that the Chinese cities should be divided into three categories characterized by different scaling exponent β. The first category, which includes the biggest and coastal cities of China, has the scaling exponent $\beta > 1$. The second category, which includes mostly the cities in central China, has the scaling exponent $\beta \approx 1$. The third category, which consists of the cities in northwestern China, has the scaling exponent $\beta < 1$. Using a urban growth equation, different ways of city population evolution can be obtained for different β. For $\beta < 1$, population evolutes always to a fixed point population N_f from below or above depending on the initial population. For $\beta > 1$, there is also a fixed point population N_f. If the initial population $N(0) > N_f$, the population increases very fast with time and diverges within a finite time. If the initial population $N(0) < N_f$, the population decreases with time and collapse finally. The pattern of population evolution in a city is determined by its scaling exponent and initial population.

Keywords: population, electrical consumption, scaling law, urban growth.

1 Introduction

Cities can be considered as complex systems which organize in a decentralized manner via interactions between agents, variables and the system itself [1]. An urban system is also a manifestation of human adaptation to the natural environment [2]. In recent research, cities are suggested to be complex systems that mainly grow from the bottom up and follow scaling laws [3]. With extensive body of data, L.M.A. Bettencourt et al. have shown that important demographic, socio-economic, and behavioral urban indicators are, on average,

J. Zhou (Ed.): Complex 2009, Part I, LNICST 4, pp. 853–864, 2009.

scaling functions of city size [4]. Taking population $N(t)$ as the measure of city size at time t, the power law scaling of material resources or measures of social activity in a city $Y(t)$ take the form [4]

$$Y(t) = A_0 N(t)^\beta , \tag{1}$$

where A_0 is a normalization constant. The scaling exponent β reflects the scaling behavior of urban indicator. Scaling, as a manifestation of the underlying general dynamics and geometry, exists throughout physics. It has been applied in understanding problems across the entire spectrum of science. Critical phenomena are significant examples in which scaling has illuminated important universal principles and provided responses to practical problems [8]. In the Bettencourt et al.'s researches, the data of U.S. for many decades mostly and the data of China and European countries also were used. They concluded that the scaling relation Eq.(1) exist across different nations and times.

Among the indicators of a city, we take here the household electrical consumption of cities for a more careful investigation. Electricity is the most important energy in a modern society. It is related closely with the living standard and the living style of cities. For household electrical consumption of Germany in 2002, Bettencourt et al. have found that the data follow quite well the power law scaling form with exponent $\beta = 1.00$ [4,6]. The scaling exponent of the household electrical consumption of Chinese cities in 2002 has been calculated by them also and $\beta = 1.05$ [4]. At $\beta = 1.00$, the household electrical consumption per person is equal to the constant A_0 in Eq.(1) actually. For $\beta \approx 1.00$, the household electrical consumption per person deviates a little from the constant A_0, but the deviation is small. In developed countries, like Germany, the difference of living standard between different cities is small and there is no large difference in the household electrical consumption per person. With single constant A_0, we can relate the household electrical consumption with population of cities by the power law scaling form Eq.(1). The situation for Chinese cities is different. Over the last 30 years, China has experienced rapid development both in economics and urbanization. But the development is quite heterogeneous in the different parts of China. The development speeds of province-level and coastal cities are much larger than the development speeds of the cities in inland China. In inland China, the development speed in central China is larger than the development speed in northwestern and southwestern China. Correspondingly the living standard and urbanization in different parts of China are also quite different. We give two examples to demonstrate this heterogeneity. In 2006, the population of prefecture-level and province-level cities in China is about 28 percent of the national population, but they consume about 55 percent of the national household electricity [7]. For the cities above the prefecture-level, the highest household electrical consumption per person is about 71 times of the lowest one. In many aspects, China is a multi-class society with rapid development and urbanization. In the analysis of Bettencourt et al.[4], they describe the data of Chinese cities in 2002 by a power law scaling form with the scaling exponent $\beta = 1.05$ and single coefficient constant A_0. This is possible because of the usage of logarithmic binning method, which has erased the heterogeneity of Chinese cities.

It is to be expected that China will experience further rapid urbanization in the next 20 years. During this period, the urban population in China will increase from about 40 percent at moment to 80 percent. To know the situation and development of China, it is extremely essential for us to investigate the global properties of cities quantitatively. In the investigation of Chinese cities, it is necessary to consider heterogeneity of Chinese cities. It needs to be checked if Chinese cities satisfy still the power law scaling form after considering their heterogeneity.

In this paper, we choose the province-level and prefecture-level cities of China to explore the relation between household electrical consumption and population. In section 2, we discuss the data of urban population and household electrical consumption of the cities in 2006. In section 3, we analyze the data of the cities with the power law scaling form. It will be found that the cities should be divided into three categories, which follow the power law scaling with different scaling exponent β and coefficient constant A_0. In Section 4, we discuss the population evolution of the cities in different categories with a urban growth equation. In section 5 we make some conclusions.

2   Data of Urban Population and Household Electrical Consumption

The data of household electricity for Chinese cities in 2006 are from Chinese Urban Statistic Year Book 2007 [7]. The prefecture-level and province-level cities contribute the main part of Chinese economics, even though their population share is only about 30% of the whole nation. In 2006, there are 268 prefecture-level cities , 4 provincial cities and 15 sub-provincial cities. Here sub-provincial city is the city which has a little bit higher position in administration than normal prefecture-level city. A provincial city is equal to a province in administration and has more autonomy than a prefecture-level city. In the Chinese Urban Statistic Year Book 2007, the data of cities Tibet and Wuzhou have been missed. So there are only data of 285 cities for the discussion in the following. Provincial-level and prefecture-level cities in China have not only urban area, but also rural area. We will use only population and household electrical consumption of urban area for discussion.

Table 1. Descriptive statistics of the data

	N	Min	Max	Sum	Mean	Max/Min
Population of the city(ten thousand)	285	14.93	1510.99	36652.79	128.1566	101.2
Household Electrical Consumption(ten thousand kWh)	285	2010	1223700	17784259	62400.91	608.8
Household Electrical Consumption per person(kWh/person)	285	42	2964	– –	395.593	70.6

The statistics of data is given in Table 1. The largest city has a population of 15109.9 thousand. The smallest city has a population of 149.3 thousand. The population of the largest city is about 101 times of the smallest city's population. Among 285 cities, the minimum of household electrical consumption is 20100 thousand kWh and the maximum is 12237000 thousand kWh, which is 608 times of the minimum. The household electrical consumption per person has the minimum 42 kWh/person and the maximum 2964 kWh/person, which is about 71 times of the minimum.

3   Data Analysis

3.1   Power Law Scaling Form

We use the power law scaling form Eq.(1), introduced by Bettencourt et al. [4], to investigate the relation between urban population and household electrical consumption of 285 Chinese cities in 2006. With a double-logarithmic representation, the power law scaling form can be rewritten as

$$\ln Y(t) = \ln A_0 + \beta \ln N(t). \qquad (2)$$

Using this representation, the data of Chinese cities are shown in Fig. 1. A fitting line with slope β has been plotted in Fig.1. The scaling exponent β is equal to 1.182 with lower limit 1.086 and upper limit 1.278 at 95% confidence. The constant $\ln A_0$ is equal to 4.961 with lower limit 4.522 and upper limit 5.4 at 95% confidence. The R square of the linear fitting is equal to 0.674, which means that the fitting has not caught the essential points of the data and is not a good description. It can be seen in Fig. 1 that the data have been scattered in a broad range, which is related to the very large difference of the household electrical consumption per person in China.

Considering the large heterogeneity of cities in China, it is natural for us to divide the 285 cities into different categories. For the cities in each category, we can analyze the relation between household electrical consumption and population with the power law scaling form.

3.2   Categorization of Cities in Term of Power Law Scaling

To improve the linear fitting, One of the methods is the logarithmic binning [9]. The logarithmic binning can reduce the stochastic noise when dealing with experimental data. In the fitting, the data are averaged in the bins. The value of R square for "binned" data can be remarkably better than original data in the linear fitting.

Here we do not use the logarithmic binning because of the features of Chinese cities. The strong scattering of the data in Fig. 1 is not only because of the stochastic noise, but mostly because of the large difference of the household electrical consumption per person for cities with similar sizes. If we use logarithmic binning, we will not only reduce the stochastic noises of data, but lost also

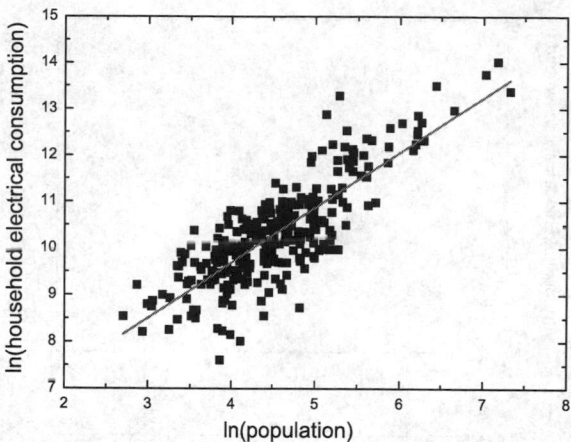

Fig. 1. Double-logarithmic representation of household electrical consumption as a function of the population size of cities in China

the detailed information about the heterogeneity of Chinese cities. For cities in a developed country, like Germany, the living standard is similar and the scattering of data are related mostly to the stochastic noises of data. The logarithmic binning can increase the accuracy of scaling exponent β substantially.

As mentioned before, the household electrical consumptions per person in different cities of China are quite different. Even for cities above the prefecture-level, the difference is almost of two order of magnitude. So we will divide the 285 cities into different category. For each category "j", we introduce the power law scaling form

$$Y(t) = A_j N(t)^{\beta_j}, \tag{3}$$

where the scaling exponent β_j and the constant A_j can be taken as the characterization of category "j".

There are many methods for defining category, such as cluster analysis. Clustering is the classification of objects into different groups, or more precisely, the partitioning of a data set into subsets (clusters). In this way, the data in each subset share some common trait - often proximity according to some defined distance measure. In our study here, it is hard for us to define a distance when using cluster analysis. Here we need to find a different way.

We will define a category by requiring the R square of cities belonging to this category is larger than a value. The R square is the criterion which characterizes the quality of a fitting. For population of cities et al., there is always stochastic inaccuracy in their data. So the value of R square is always smaller than 1 and limited. At the same time, the value of R square should not be too small. Otherwise the fitting cannot describe the data correctly.

At first, we choose the R square to be at least 0.90 for cities in a category. For the first category, we begin from the city with the highest household electrical

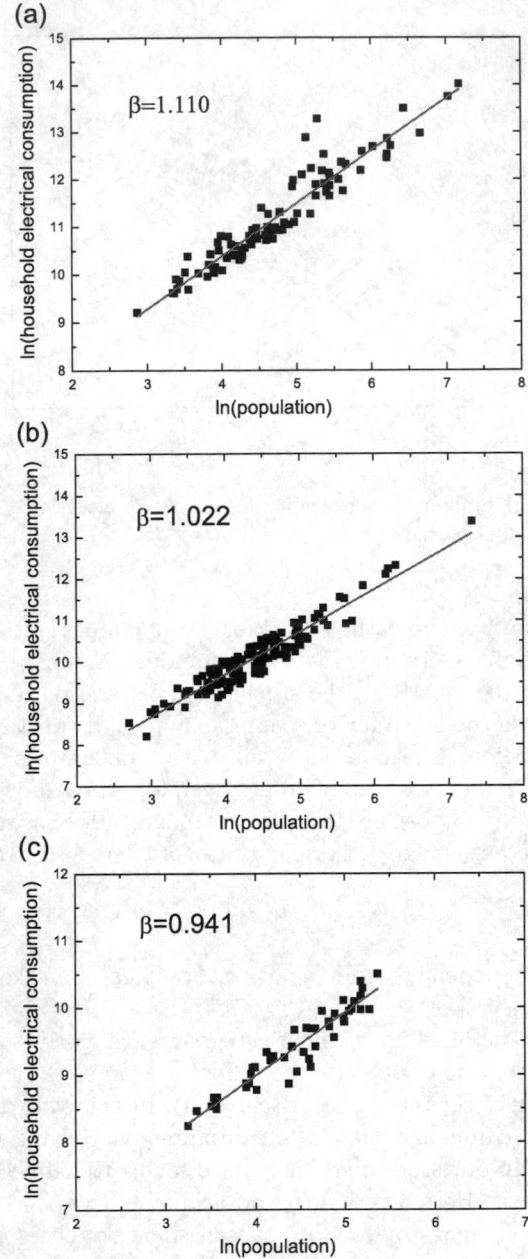

Fig. 2. Double-logarithmic representation of household electrical consumption as a function of the population size of cities in China. (a) the first category with household electrical consumption per person ≥ 430kWh/person; (b) the second category with household electrical consumption per person between 189kWh/person and 430kWh/person; (c) the third category with household electrical consumption per person between 89kWh/person and 189kWh/person.

Table 2. Categories of household electrical consumption per person of Chinese cities

	Observations	Max(kWh)	Min(kWh)
Whole	285	2964(Shenzhen)	42(Yulin)
Category 1	90	2964(Shenzhen)	430(Huzhou)
Category 2	140	428(Chongqing)	189(Chaohu)
Category 3	48	188(Xuancheng)	91(Zhaotong)
Left	7	85(Dingxi)	42(Yulin)

Table 3. Scaling exponents for urban electrical consumption vs. population with $R^2 = 0.90$

	Observations	β	$\ln A_j$	95% CI	R^2
Whole samples	285	1.18	4.961	[1.086, 1.278]	0.674
Category 1	90	1.110	5.943	[1.031,1.188]	0.90
Category 2	140	1.022	5.611	[0.964,1.079]	0.90
Category 3	48	0.941	5.227	[0.849,1.032]	0.90

consumption per person, which is the city Shengzhen with 2962 kWh/person in 2006. Then we add other cities with lower household electrical consumption per person, such as Dongguan, Changsha, Guangzhou, etc. For the cities included, we make a fitting to their data using the power law scaling form Eq.(3) with a suitable scaling exponent β_1 and a suitable constant A_1. The R square decreases when more cities are included. It reaches the value 0.90 after the city Huzhou is taken into account. There are totally 90 cities in the first category. The data of the cities in the first category are shown in Fig. 2a with a linear fitting, which has the scaling exponent $\beta_1 = 1.110$ and $\ln A_1 = 5.943$.

In this way, we can determine further the cities in the second and third category. There are 140 cities in the second category whose household electrical consumption is between 428 kWh/person and 189 kWh/person. The scaling exponent in the second category is very near 1 with $\beta_2 = 1.022$ and the corresponding coefficient constant is $\ln A_2 = 5.611$. The data and the fitting line are shown in Fig. 2b. In the third category, there are 48 cities whose household electrical consumption is between 188 kWh/person and 89 kWh/person. Its scaling exponent is less than 1 with $\beta_3 = 0.941$ and its coefficient constant is $\ln A_3 = 5.227$. We show the data and the fitting line of the third category in Fig. 2c. There are still 7 cities left which cannot be described by a power law scaling form.

The scaling exponents of different categories are summarize in Table 3, where the lower and upper limit of the scaling exponents at 95% confidence interval are given also.

In general, the category with higher household electrical consumption has larger scaling exponent. The scaling exponent of the first category is definitely larger than 1. The exponent of the second category is almost equal to 1. The exponent β of the third category is obviously smaller than 1.

Table 4. Scaling exponents of different categories divided by $R^2 = 0.90, 0.91, 0.89$

R^2	Category	Observations	β
0.674	all cities	285	1.18
0.90	Category 1	90	1.110
	Category 2	140	1.022
	Category 3	48	0.941
	Left	7	$--$
0.91	Category 1	78	1.108
	Category 2	140	1.033
	Category 3	58	0.960
	Left	9	$--$
0.89	Category 1	103	1.083
	Category 2	125	0.983
	Category 3	50	0.942
	Left	7	$--$

To understand our results, we identify the locations of the cities in different categories. It is found that the first category includes the province-level and coastal cities. The cities of the second category are in central China and the cities of the third category are in northwestern China.

The province-level and coastal cities has $\beta_1 > 1$. This means that the household electrical consumption of these cities increases faster than their population. The larger cities in this category consume super-proportionally more household electricity. The cities in northwestern China has $\beta_3 < 1$. The larger cities in the third category consume sub-proportionally less household electricity. For cities in central China, $\beta_2 \simeq 1$ and the cities in this category consume household electricity in proportion with their population.

The classification of categories above is obtained by requiring that the R square of each category is at least 0.90. To verify the robustness of our classification, we need to use different values of R square to classify the category of the cities. As we have discussed before, the R square should has a reasonable value which cannot be too large or too small. So we have chosen the R square to be 0.89 and 0.91. The results of different R square are given in Table 3. Of course, the city numbers of different categories and their scaling exponents will be modified with the change of R square. But their dependence is weak. For the three values of R square, we have always three categories of cities which satisfy the power law scaling form and a few cities left, which do not satisfy. The scaling exponents of different categories are kept to be almost the same. So our classification of the categories of cities is robust and reliable.

4   Growth of City in Different Categories

As we have demonstrated in the last section, the cities of China can be divided into three categories which follow the power law scaling form with different scaling exponents. We are interested in the urban growth of the cities of different categories. Related to household electrical consumption and population, Bettencourt et al. have introduced a simple resource balance equation [4]

$$Y = r_0 N + e_0(dN/dt), \qquad (4)$$

where r_0 is the quantity per unit time to maintain an individual of the population and e_0 is the quantity to add a new individual to the population. This equation can be rewritten into a general growth equation

$$\frac{dN(t)}{dt} = \frac{A_0}{e_0} N(t)^\beta - \frac{r_0}{e_0} N(t). \qquad (5)$$

The solution of this equation [4] is

$$N(t) = \left[A_0/r_0 + \left(N(0)^{(1-\beta)} - A_0/r_0 \right) \exp\left[-(1-\beta)\frac{r_0}{e_0}t \right] \right]^{1/(1-\beta)}. \qquad (6)$$

In the following, we will discuss the solutions in for different β in detail.

(1) $\beta = 1$:
The solution reduces to an exponential form

$$N(t) = N(0)e^{(A_0-r_0)t/e_0}. \qquad (7)$$

Because r_0 is the electrical consumption per unit time to maintain an individual, therefore

$$A_0 = \frac{Y}{N^\beta} = r_0 + \frac{e_0}{N}(dN/dt) > r_0. \qquad (8)$$

In Fig. 3a, the population growth in this case is shown. From this result, we can conclude that the cities in the second category with $\beta \simeq 1$ can have a rapid population increase.

For scaling exponent $\beta \neq 1$, the solution Eq. (6) has a fixed point $N_f \equiv (A_0/r_0)^{1/(1-\beta)}$. If the initial population $N(0) = N_f$, the population keeps to be constant.

With the fixed point population N_f we can rewrite Eq.(6) as

$$N(t) = \left[N_f^{1-\beta} + \left(N(0)^{(1-\beta)} - N_f^{1-\beta} \right) \exp\left[-(1-\beta)\frac{r_0}{e_0}t \right] \right]^{1/(1-\beta)}. \qquad (9)$$

If the initial population $N(0) \neq N_f$, the population evolutes then with time. The evolution at $\beta < 1$ is quite different from the evolution at $\beta > 1$.

(2) $\beta > 1$:
When $N(0) > N_f$, the population increases with time very fast and becomes divergent in a finite time t_c given by

$$t_c = -\frac{e_0}{(\beta - 1)r_0} \ln\left[1 - (N(0)/N_f)^{(1-\beta)} \right]. \qquad (10)$$

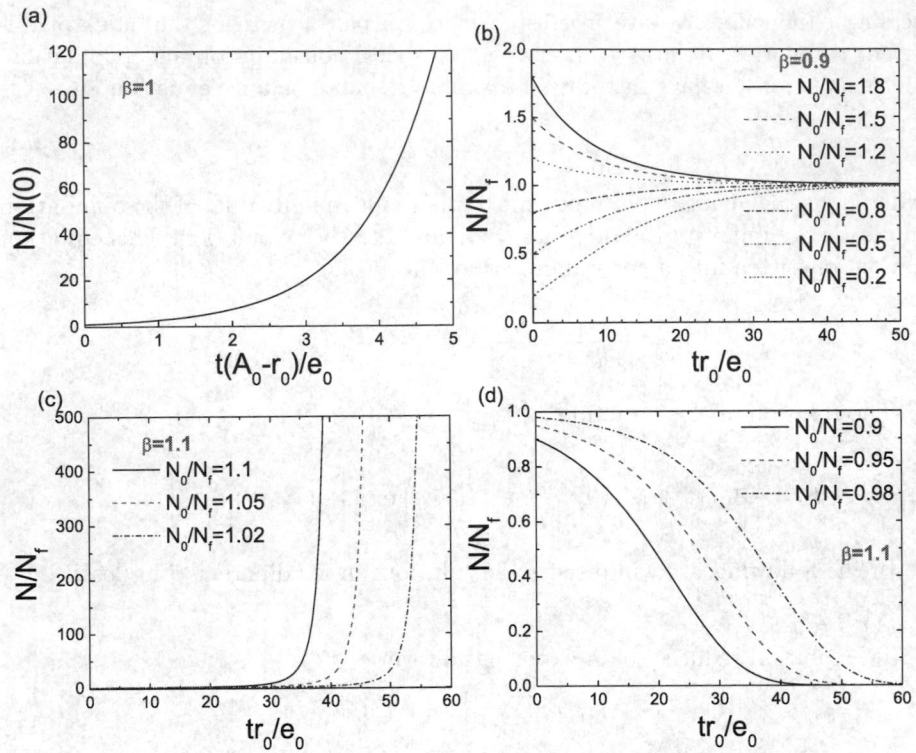

Fig. 3. The growth of population: (a) $\beta = 1$; (b) $\beta < 1$; (c) $\beta > 1$ with $N(0) > N_f$; (d) $\beta > 1$ with $N(0) < N_f$

In Fig. 3c, we show the evolution of population for different initial populations $N(0) = 1.1N_f, 1.05N_f, 1.02N_f$. When $N(0) < N_f$, the population decreases with time and collapse finally. The evolutions of population for different initial populations $N(0) = 0.98N_f, 0.95N_f, 0.9N_f$ are shown in Fig. 3d.

(3) $\beta < 1$:

If $N(0) > N_f$, the population decreases with time and reaches the fixed point N_f finally. If $N(0) < N_f$, the population increases with time and approaches N_f. The evolutions of population with $\beta = 0.9$ for initial population $N(0) = 1.8N_f, 1.5N_f, 1.2N_f, 0.8N_f, 0.5N_f, 0.2N_f$ have been demonstrated in Fig. 3b.

In the analysis of Bettencourt et al. [4], the dependence of population evolution on initial population in cases 2) and 3) has been ignored.

5 Conclusions

Using the power law scaling form introduced by Bettencourt et al. [4] for material resources and population of cities, we have investigated the relation of household electrical consumption and population of Chinese province-level and prefecture-level cities in 2006. In our discussion, the heterogeneity of Chinese cities has been taken into account, which has been ignored in the investigation of Chinese cities by Bettencourt et al. with the logarithmic binning [4]. From the data of 285 cities of China without the logarithmic binning, we find that the Chinese cities can be divided into three categories which follow the power law scaling form with different scaling exponents. The first category includes the province-level and coastal cities and has the scaling exponent larger than 1. The second category, which includes the cities of central China, has the scaling exponent $\beta_2 \approx 1$. In the third category, the scaling exponent is less than 1 and the cities of this category are in northwestern China.

With a urban growth equation, we have studied the population evolution of cities in different categories with different scaling exponents and initial population. We find that the population evolution depends essentially on the scaling exponent and the initial population. For $\beta < 1$, the population will increase or decrease with time to a fixed value N_f. When $\beta > 1$, the population will be divergent in a finite time if initial population $N(0) > N_f$ and collapse if $N(0) < N_f$.

Electricity is just one kind of material resources of city. To get more understanding about Chinese cities, we need to explore the relation between other kinds of material resources and population. Further it is very interesting to take also individual needs and social activity of Chinese cities into account. In these further investigations about Chinese cities, our results here have demonstrated that the heterogeneity of Chinese cities must be considered. The Chinese cities should be divided into different categories for discussion. In this way, we will work on a quantitative description of the global properties of Chinese cities.

Acknowledgement

The project was supported by the National Natural Science Foundation of China under Grant 10325418.

References

1. Bonabeau, E.: Social Insect Colonies as Complex Adaptive Systems. Ecosystems 1, 437–443 (1998)
2. Bessey, K.M.: Structure and Dynamics in an Urban Landscape: Toward a Multiscale View. Ecosystems 5, 360–375 (2002)
3. Batty, M.: The Size, Scale, and Shape of Cities. Science 319, 769–771 (2008)
4. Bettencourt, L.M.A., Lobo, J., Helbing, D., Kuhnert, C., West, G.B.: Growth, innovation, scaling, and the pace of life in cities. Proceedings of the National Academy of Sciences of the United States of America 104, 7301–7306 (2007)
5. Bettencourt, L.M.A., Lobo, J., West, G.B.: Why are large cities faster? Universal scaling and self-similarity in urban organization and dynamics. European Physical Journal B 63, 285–293 (2008)

6. Kühnert, C., Helbing, D., West, G.B.: Scaling laws in urban supply networks. Physica A: Statistical Mechanics and its Applications 363, 96–103 (2006)
7. NBS: Urban Statistical Yearbook of China. China Statistical Press, Beijing (2007)
8. Isalgue, A., Coch, H., Serra, R.: Scaling laws and the modern city. Physica A: Statistical Mechanics and its Applications 382, 643–649 (2007)
9. McManus, O.B., Blatz, A.L., Magleby, K.L.: Sampling, log binning, fitting, and plotting durations of open and shut intervals from single channels and the effects of noise. Pflügers Archiv European Journal of Physiology 410, 530–553 (1987)

Scaling in Modulated Systems

Oliver Portmann, Alessandro Vindigni, and Danilo Pescia

Laboratorium für Festkörperphysik, ETH Zürich, 8093 Zürich, Switzerland

Abstract. How can we understand a system that is too complicated to be simulated and in which some important quantities cannot be determined from observation? This is a question we were confronted with when we analyzed experimentally observed magnetic patterns in ultrathin ferromagnetic films. We have now, for this specific case, found a method that gives a good qualitative understanding of a surprising reentrance of order observed experimentally. This method is based on scaling arguments and may prove useful in the study of other complex systems.

Keywords: frustrated systems, modulated systems, long-range interactions, competing interactions, scaling, pattern formation.

Pattern formation is found in a wide range of physical and chemical systems and often the result of competing interactions. Such a competition leads the basic units to form modulated structures. This self-organization expresses itself in geometrical configurations such as stripes, labyrinths or bubbles that can extend over length scales much larger than the basic units [1]. In such a situation, the basic interactions cannot be completely satisfied individually and the system is in a frustrated state. The difference in (free) energy of geometrically distinct configurations can be very small. Nevertheless, the occurrence of geometrically distinct configurations is deterministic and transitions between such configurations can be triggered by varying an external parameter such as the temperature. Normally, order tends to decrease with the increased excitation energy at higher temperatures [2]. This, however odd it may seem, is not universal [3]. There are rare instances where a system displays a lower symmetry (i.e. higher geometrical order) at a higher temperature even though the overall entropy has to increase. It is a challenge to understand the mechanisms that are responsible for such a peculiar and unexpected behaviour.

We use a scaling analysis to investigate the experimentally observed reentrant order in ultrathin ferromagnetic films [4,5]. Reentrance of order means that a less symmetric pattern (here: stripes) that is present at lower temperatures reoccurs at higher temperatures after a more symmetric intermediate state (labyrinth), see Fig. 1. The basic units in this system are electrons that interact with each other via the quantum-mechanical exchange interaction that favours a parallel alignment of their magnetic moments and via the magnetostatic interaction between their magnetic moments that favours an antiparallel alignment. In principle, it is possible to study this system with Monte-Carlo simulations and test the appropriateness of various models, but so far no single simulation has succeeded

J. Zhou (Ed.): Complex 2009, Part I, LNICST 4, pp. 865–867, 2009.

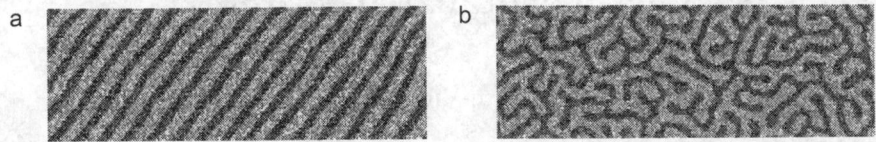

Fig. 1. Images of magnetic patterns in ultrathin iron films on Cu(001) taken with SEMPA (Scanning Electron Microscope with Polarization Analysis). The magnetic moments in this system are perpendicular to the plane of the film. Areas where the magnetic moments point downward are represented by dark grey levels, areas where the magnetic moments point upward are marked by light grey levels. a Stripe pattern. b Labyrinthine pattern. The width of the images is 92 μm.

in reproducing the experimentally observed behaviour [6,7,8]. We find that the system sizes required to observe the phenomenon of reentrance in simulations are beyond present-day computational capabilities, the main problem being the long-range character of the magnetostatic interaction. Nonetheless, we have been able to obtain a good qualitative understanding of the system by analytically reducing this problem in two spatial dimensions to an effectively one-dimensional problem that retains important properties of the original system. In a mean-field approach [9], we find a highly anomalous temperature dependence of an elastic constant. By means of a scaling analysis, we can relate this experimentally inaccessible elastic constant to experimentally measurable quantities [10]. Comparison with experiment suggests that the driving force for the reentrance of order is indeed the strongly anomalous behaviour of this elastic constant.

Our scaling analysis is based on one basic assumption for the effect of small deviations from the equilibrium state and correctly predicts anomalous behaviour. We already have indications that our scaling approach adequately describes further phenomena in ultrathin magnetic films and anticipate that a similar approach may prove fruitful in the analysis of other systems where system-size requirements prohibit numerical simulations and interesting quantities are not accessible experimentally, as in the observation of phenomena that cannot be influenced at will.

References

1. Seul, M., Andelman, D.: Domain Shapes and Patterns: The Phenomenology of Modulated Patterns. Science 267, 476–483 (1995)
2. Strandburg, K.J.: Two-dimensional melting. Rev. Mod. Phys. 60, 161–207 (1988)
3. Greer, A.L.: Too hot to melt. Nature 404, 134–135 (2000)
4. Portmann, O., Vaterlaus, A., Pescia, D.: An inverse transition of magnetic domain patterns in ultrathin films. Nature 422, 701–704 (2003)
5. Portmann, O., Vaterlaus, A., Pescia, D.: Observation of Stripe Mobility in a Dipolar Frustrated Ferromagnet. Phys. Rev. Lett. 96, 047212 (2006)

6. De'Bell, K., MacIsaac, A.B., Whitehead, J.P.: Dipolar effects in magnetic thin films and quasi-two-dimensional systems. Rev. Mod. Phys. 72, 225 (2000)
7. Stoycheva, A.D., Singer, S.J.: Stripe melting in a two-dimensional system with competing interactions. Phys. Rev. Lett. 84, 4657–4660 (2000)
8. Cannas, S.A., Stariolo, D.A., Tamarit, F.A.: Stripe-tetragonal first-order phase transition in ultrathin magnetic films. Phys. Rev. B 69, 092409 (2004)
9. Vindigni, A., Saratz, N., Portmann, O., Pescia, D., Politi, P.: Stripe width and nonlocal domain walls in the two-dimensional dipolar frustrated Ising ferromagnet. Phys. Rev. B. 77, 002414 (2008)
10. Portmann, O., Vindigni, A., Pescia, D.: A Scaling Hypothesis and Re-entrance of Order in Modulated Systems. ArXiv: 0812.3811v1 [cond-mat.soft]

Scaling Behavior of Chinese City Size Distribution

Xiaowu Zhu[1,2], Aimin Xiong[3], Liangsheng Li[1], Maoxin Liu[1], and Xiaosong Chen[1]

[1] Institute of Theoretical Physics, Chinese Academy of Sciences,
P.O. Box 2735, Beijing 100080, P.R. of China
[2] Business School, China University of Political Science And Law,
No.27 FuXue Road, Beijing 102249, P.R. of China
[3] Department of Systems Science, School of Management,
Beijing Normal University, Beijing 100875, P.R. of China
{zxw,xiong,liliangsheng,mxliu,chenxs}@itp.ac.cn

Abstract. We have investigated the population distribution of Chinese cities from 1997 to 2006. The rank-size distributions of Chinese cities deviate from the Pareto distribution. For city size distribution of each year we can find a population threshold P_c that characterizes the boundary of the deviation. The cities with population more than P_c follow the Pareto distribution, while the smaller cities deviate from the Pareto distribution. Using P_c for every year, the rank-size distribution from 1997 to 2006 can be written into a scaling form $R(P,T) = C(T)P^{-\alpha(T)}f(P/P_c(T))$, where the Pareto exponent $\alpha(T)$ is not equal to the value of Zipf's law and evolutes with time. According this scaling form, the data of the city size distributions of Chinese cities from 1997 to 2006 can collapses to a single curve, which is the scaling function of the city size distribution.

Keywords: city size distribution, Pareto distribution, scaling, Zipf's law.

1   Introduction

Over the last 30 years, China has experienced rapid urbanization. From 1978 to 2006 the Chinese national population increased from 962.59 million in 1978 to 1.314 billion in 2006, a 37 percent increase. The non-agricultural population in urban areas had a 144 percent increase, which is much greater than the national population growth rate. As a result, China has become more urbanized, with an urban population share that has increased from 17 percent in 1978 to 44 percent in 2006 [1]. The urban population share of China now is still much lower than the 70 percent of developed nations, but the Chinese urbanization speed is almost the three times of the world average. With the large number of cities and very active dynamics of urban organization, China's urban system and urbanization provide an important area for research.

The most economic activities of human being are happened in cities. One of the most striking regularities in the location of economic activity is the size

J. Zhou (Ed.): Complex 2009, Part I, LNICST 4, pp. 868–875, 2009.

distribution of cities in a country [2]. The empirical study of city size distribution has engaged scientists and economists since the beginning of last century [3]. In 1913 the German geographer Felix Auerbach found an interesting empirical regularity that the product of the population size of a city and its rank in the city size distribution appears to be roughly constant for a country. Since Auerbach proposed this basic proposition, it has been widely accepted by scholars in a variety of disciplines and refined by others. In 1941 Zipf [4] provided an empirical analysis which suggested that the city size distribution can be represented by a Pareto distribution with an exponent equal to 1.

Recently, the city size distribution has attracted renewed attention. The researches of Rosen&Resnick [5] in 1980 and others recently [2] have shown, using empirical analysis of the data from many countries, that the Zipf's law is not always tenable. The city size distribution as a Pareto distribution has an exponent seldom equal to one. Even some doubts were raised regarding the validity of the Pareto distribution. In 1990 Husing [6] argued that not only Zipf's law, but also the Pareto distribution as well were not supported by the empirical data. In 1998 Laherrere and Sornette [7] suggested the use of a stretched exponential distribution. Later it was thought that the Zipf's law is spurious in explaining city size distribution [8]. In 2007 Benguigui and Blumenfeld-Lieberthal [9] went beyond the power law and proposed a new approach to analyze city size distributions. More recently, it was suggested that cities are complex systems that mainly grow from the bottom up [10]. Like a physical system in nature, the size et al of cities follow scaling laws. As a manifestation of the underlying dynamics and structure, scaling is well known in physics. It has been instrumental in understanding problems across the entire spectrum of science [11]. Typically scaling laws reflect generic features of the systems which consist of a large number of interacting particles [12]. They are universal and independent of the microscopic details of systems. The scaling hypothesis has been well verified by a wealth of experimental data on diverse systems [13]. One prediction of the scaling hypothesis is the scaling law which relates the various critical point exponents characterizing the singular behavior of functions such as thermodynamic functions. Another prediction of the scaling hypothesis is a sort of data collapse, where diverse data collapse onto a single curve called a scaling function under appropriate axis normalization.

In this paper, we analyze the city size distributions of Chinese cities from 1997 to 2006. The large number of cities in China make us be able to check if the sizes of Chinese cities follow the Pareto distribution. The rapid changes of Chinese cities from 1997 to 2006 make us be able to investigate the scaling behavior of the city size distributions. The rest of our paper is organized as follows. In Section 2, we describe the date of Chinese cities. In section 3, we analyze the data of Chinese cities and propose a scaling for the city size distributions of China. With the scaling form the city size distributions from 1997 to 2006 can collapse onto a single curve, which is the scaling function of city size distributions. In Section 4 we give some conclusions.

2   Data of Chinese Cities

The definition of city in China is not straightforward. According to administrative level, Chinese cities are classified into three different groups: county-level cities (XianJiShi), prefecture-level cities (DiJiShi) and province-level cities (ShengJiShi). The small settlements with townships or lower administrative levels are not called as city. The administrative criteria distinguishing cities are the scale of urban population, the economic and political importance of an urban agglomeration [14]. The counties or the county-level cities are administrated by prefecture-level cities. The economic situations play the principal roles in the dynamics of city population [15]. The dynamic movement of population in China has resulted in that many counties had been reclassified as cities. The total number of county-level cities increased sharply from 223 in 1980 to 656 in 2006. The prefecture-level cities increase from 78 in 1978 to 283 in 2006. Even though prefecture-level and province-level cities share only about 30 percent of the national population, but they contribute the main part of economics in the whole nation.

When we look up the Chinese Urban Statistic Year Book, only the data of prefecture-level and province-level cities are provided year by year from 1997 to 2006. There are three categories of population in the Chinese Urban Statistic Year Book: the population of whole city including all rural and urban population, the urban population who live in the urban region, the non-agriculture population of the urban population who do non-agricultural work. In this article, we study the city size distribution of the urban population in prefecture-level and province-level cities from 1997 to 2006. We discuss the exponent of Pareto distribution and the scaling behavior of the city size distributions. The urban population can characterize really the size of a city. Other scholars have taken the data of Chinese cities for discussions similarly [14,16].

3   Data Analysis

3.1   City Size Distribution Function

Auerbach(1913), Singer(1936) and Zipf(1949) demonstrated that the city size distributions could be represented by a Pareto distribution

$$R = CP^{-\alpha}, \tag{1}$$

where P is the population of a city and R is its rank which is ordered from the largest to the smallest. α is the exponent of Pareto distribution and C is a constant.

In Fig.1a, the rank-size distribution of Chinese cities in 2006 is shown. There is a threshold of population P_c in the city size distribution. The cities with population $P > P_c$ follow the Pareto distribution, but the cities with population $P < P_c$ deviate from the Pareto distribution. The rank-size distributions of Chinese cities in other years are similar to the year 2006. We can conclude that

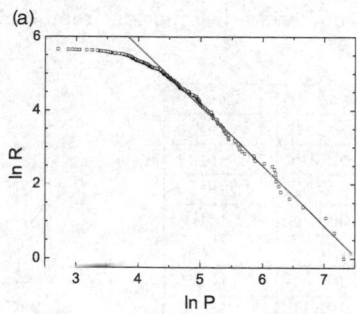
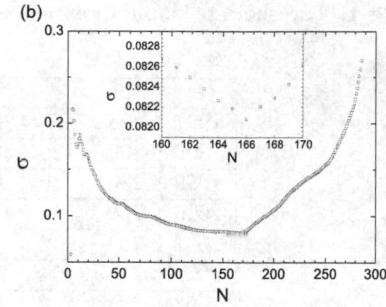

Fig. 1. (a) The rank-size distribution of 2006. (b) The mean square deviation versus number of sample N

the city size distributions of Chinese cities do not follow the Pareto distribution as a whole. The small cities have obvious deviation. To describe the city size distributions of most Chinese cities in different years, a modified form of the usual Pareto distribution (1) is introduced as

$$R(P,T) = C(T)P^{-\alpha(T)}f(P,T) \tag{2}$$

Where T denotes the year. The function $f(P,T)$ of population and year characterizes the modification to the Pareto distribution. There is a Pareto distribution when $f(P,T) = 1$ and there is deviation when $f(P,T) \neq 1$.

We can rewrite Eq.(2) in a double-logarithmic representation as

$$\ln R(P,T) = -\alpha(T)\ln P + M(P,T) + \ln C(T) , \tag{3}$$

where $M(P,T)$ is defined as

$$M(P,T) \equiv \ln f(P,T) = \ln R(P,T) + \alpha(T)\ln P - \ln C(T). \tag{4}$$

The function $M(P,T)$ characterizes precisely the deviation of city size distribution $R(P,T)$ from the Pareto distribution.

3.2 Scaling Behavior of Cities Size Distribution

To calculate the function $M(P,T)$, we need to determine the exponent $\alpha(T)$ from the rank-size distribution of the cities with population $P > P_c$ at first. We estimate P_c by using OLS method. We sort the cities with the order from larger to smaller city. With suitable number of cities that $N > 10$, we can fit the rank-size distribution by a Pareto distribution. Of course, the real data have deviations from a Pareto distribution. Using the OLS method we can calculate the mean square deviation σ between the data and the Pareto distribution with an exponent α which fits the data at best. In Fig. 1b the mean square deviation σ is shown as a function of N. σ has a minimum of σ at the number of cities

Table 1. Threshold of population and exponent of Pareto distribution from 1997 to 2006

year	N	N_c	P_c	α	$\ln C$
1997	223	99	91.15	1.79148	12.60805
1998	229	125	68.33	1.62408	11.80417
1999	236	130	67.6	1.65552	11.96679
2000	262	144	67.14	1.68128	12.1731
2001	266	152	70	1.66078	12.16288
2002	278	154	74.53	1.64056	12.170393
2003	284	162	72.15	1.62331	12.1327
2004	285	166	74.53	1.63023	12.20675
2005	286	167	75.03	1.5887	12.04478
2006	286	166	77.36	1.61828	12.21067

$N = N_c$. When $N > N_c$, σ increases with the number of cities. So we define the population of N_c-th city as the threshold of population P_c. For the city size distribution of 2006, we find that $N_c = 166$ (see the sub-figure in Fig. 1b). In the same way, we can obtain N_c, P_c, α and $\ln C$ of each year from 1997 to 2005. All results are given in Table 1.

With the parameters $\alpha(T)$ and $C(T)$ given in the Table 1, we can calculate the deviation function $M(P,T)$ by using Eq.4 for each year. For $P < P_c$ of each year, $M(P,T)$ is approximately equal to zero. $M(P,T)$ at $P < P_c$ from 1997 to 2006 is shown in Fig.2.

In physics, there is scaling when a system is near its critical point. As a function of temperature and external field, the correlation length and thermo-dynamic quantities can be written in a scaling form which is the product of a power law term and an one-variable scaling function. The variable of the scaling function is the ratio of a physical quantity to its characteristic quantity. Inspired by the scaling in physics, we introduce a scaling form for the evolution of city size distribution in China as

$$R(P,T) = C(T)P^{-\alpha(T)}f_s\left(P/P_c(T)\right). \tag{5}$$

Here we define the ratio of population to the threshold P_c as the scaling variable. Correspondingly we can define the scaling form of the deviation function $M(P,T)$ as

$$M_s(P/P_c) = \ln f_s(P/P_c). \tag{6}$$

The validity of our scaling form Eq.(5) for the evolution of city size distribution should be tested by the data of cities. In Fig.2a, it has been shown that the deviation function $M(P,T)$ of Chinese cities depends really on population P and time T. If $M(P,T)$ has a scaling form, the different curves in Fig.2a could be collapsed onto a single curve after using the scaling variable $P/P_c(T)$. So the two-variable function $M(P,T)$ becomes a single-variable function $M_s\left(P/P_c(T)\right)$. In Fig.2b, it is shown that the data of Chinese cities from 1997 to 2006 collapse

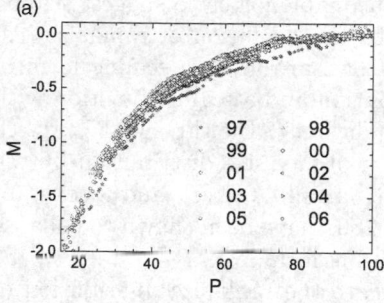
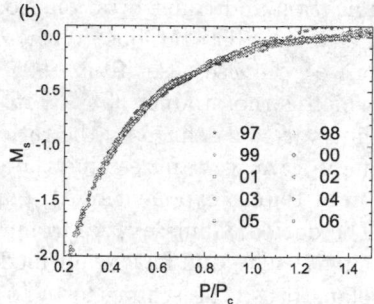

Fig. 2. The deviation function $M(P,T)$ at $P < P_c$ from 1997 to 2006. (a)with population P as variable,(b) with the ratio $P/P_c(T)$ as variable

quite well onto a single curve. We can conclude that the evolution of city size distributions in China from 1997 to 2006 satisfies the scaling form Eq.(5). In this case, the evolution of city size distribution can be described by the evolution of the exponent $\alpha(T)$ and the population threshold $P_c(T)$. The scaling functions in the critical phenomena are universal. Here we could suggest that the scaling functions $f_s(x)$ and $M_s(x)$ here have also universality which depends on the macroscopic properties of a country, but not the microscopic details.

4   Conclusions

Nowadays cities play more and more important role in the world. It is estimated that sometime in 2007 more people are living in cities than outside them, for the first time in history. Cities are undeniably centers of politics, economics and innovation. At the same time, they are also centers for the production of waste, pollution and heat. It is of essential interests for mankind to find out the regularities of organization and evolution of cities. Among them, the regularity of city size distribution is a relatively simple and important one. Even this problem has been investigated for nearly one century and many progresses have been obtained, there are still many fundamental questions which are open.

In this paper, we investigate the rank-size distribution of Chinese cities from 1997 to 2006. In these rank-size distributions cities can be divided into two parts. The cities with population more than a threshold P_c follow the Pareto distribution. The exponent of the Pareto distribution α varies with time from the smallest value 1.58 in 2005 to the largest value 1.79 in 1997. They are obviously larger than $\alpha = 1$ of the Zipf's law. The cities with population less than the threshold P_c deviate from the Pareto distribution. So the rank-size distributions of Chinese cities cannot be described by a Pareto distribution as a whole.

In physics, thermodynamic quantities can follow scaling laws when the correlation length of a system becomes very large. In the societies today, the correlations between cities and people are also very large due to modern technologies in communication and transportation. So it is reasonable for us to suggest scaling law for city size distributions. With respect to the evolution of rank-size distributions of Chinese cities from 1997 to 2006, we introduce a scaling form in Eq. 5, which is the product of a Pareto distribution and a scaling function with the scaling variable defined by the ratio of population to the threshold P_c. With this scaling form we can describe the evolution of city size distribution by the evolution of Pareto exponent $\alpha(T)$, population threshold $P_c(T)$ and a constant $C(T)$. The data of Chinese cities from 1997 to 2007 have been shown to collapse onto a single curve and have confirmed the scaling form.

We plan to use the scaling form to investigate the rank-size distribution of other countries in the future. We will check if there is also such scaling form for cities of other countries and investigate the relationship between the scaling functions of different countries if they exist. Taking cities as complex systems that grow from bottom up, many ideas and methods in physics, especially statistical physics, can be learned for the researches of phenomena in societies.

Acknowledgement

The project was supported by the National Natural Science Foundation of China under Grant 10325418.

References

1. National Bureau of Statistics: China Urban Statistical Yearbook. China Statistics Press, Beijing (1998-2007)
2. Soo, K.T.: Zipf's Law for cities: a cross-country investigation. Regional Science and Urban Economics 35(3), 239–263 (2005)
3. Dobkins, L.H., Ioannides, Y.M.: Dynamic Evolution of the U.S. City Size Distribution. Discussion Papers Series, Department of Economics, Tufts University (1999)
4. Zipf, G.K.: National Unity and Disunity. The Principia Press, Bloomington Indiana (1941)
5. Rosen, K.T., Resnick, M.: The Size Distribution of Cities: An Examination of the Pareto Law and Primacy. Journal of Urban Economics 8(2), 165–186 (1980)
6. Husing, Y.: A note on functional forms and the urban size distribution. Journal of Urban Economics 27, 70–73 (1990)
7. Laherrere, J., Sornette, D.: Stretched exponential distributions in nature and economy: "fat tails" with characteristic scales. The European Physical Journal B 2, 525–539 (1998)
8. Gan, L., Li, D., Song, S.: Is the Zipf law spurious in explaining city-size distributions? Economics Letters 92(2), 256–262 (2006)
9. Benguigui, L., Blumenfeld-Lieberthal, E.: Beyond the power law - a new approach to analyze city size distributions. Computers, Environment and Urban Systems 31, 648–666 (2007)
10. Batty, M.: The Size, Scale, and Shape of Cities. Science 319, 769–771 (2008)
11. Bettencourt, L.M.A., Lobo, J., Helbing, D., Kuhnert, C., West, G.B.: Growth, innovation, scaling, and the pace of life in cities. Proceedings of the National Academy of Sciences of the United States of America 104(17), 7301–7306 (2007)

12. Isalgue, A., Coch, H., Serra, R.: Scaling laws and the modern city. Physica A: Statistical Mechanics and its Applications 382(2), 643–649 (2007)
13. Privman, V., Hohencerg, P.C., Aharony, A.: Universal Critical-Point Amplitude Relations. In: Domb, C., Lebowitz, J.L. (eds.) Phase Transitions and Critical Phenomena, vol. 14, p. 1. Academic Press, New York (1991)
14. Song, S., Zhang, K.H.: Urbanisation and City Size Distribution in China. Urban Studies 39(12), 2317–2327 (2002)
15. Eeckhout, J.: Gibrat's Law for (All) Cities. The American Economic Review 94(5), 1429–1451 (2004)
16. Anderson, G., Ge, Y.: The size distribution of Chinese cities. Regional Science and Urban Economics 35(6), 756–776 (2005)

Social Network as Double-Edged Sword to Exchange: Frictions and the Emerging of Intellectual Intermediary Service

Li Li[1,2], Bangtao Wu[1], Zhong Chen[1], and Liangjie Zhao[1]

[1] Antai College of Economics and Management, Shanghai Jiaotong University,
Shanghai, 200052, China
lilycb163@163.com
[2] School of Education, Shanghai Normal University, Shanghai 200234, China

Abstract. The value of complex social network and the optimization of it are determined by the structure and nodes' characteristics. Direct friction and indirect friction are defined to describe the possible exchange difficulty each node meets with its neighbors in exchange network. Exogenous intermediary and endogenous intermediary can decrease these frictions by adding links. Agent-based Simulating results show that both frictions and the optimization of them are influenced by demander and supplier rate, the exchange network structure as well as the environment constrains and exogenous intermediation acts better than endogenous intermediation in decreasing both frictions. While assists exchange, the results of this paper also implies social network as origin of impefect market.

Keywords: Complex Social network, Exchange network, Direct friction, Indirect friction, Intellectual Intermediary service.

1 Introduction

The relationship between structure and dynamics such as robustness, fragibility, diffusion and spreading of complex network has been discussed by many literatures of complex networks. It is essential for a deeper understanding of the development and self-optimization of the society as a whole [1]. There are four kinds of optimization at the leading edge of the current research on network optimization [2]. The optimization of social network is also an aspect of this field. Intermediation plays a key role as an optimizer of our society. To understand the role and function of the intermediation along the developing process of social network is crucial to our comprehension of the formation and optimization of complex social network.

Exchange network which possesses two kinds of nodes is common in complex social network. In marrige, labor market as well as business dealings, a node realizes its value no other than he meet other side and exchange successfully. These relationship could be abstracted to two complementary services exists in

J. Zhou (Ed.): Complex 2009, Part I, LNICST 4, pp. 876–888, 2009.

exchange network, two sides yield value only by cooperation. The value of holistic exchange network is to facilitate trades among the nodes. Analogizing with friction in mechanics, frictions to exchange refer to the forces blocking the realization of nodes' value. The preference of demander [3], prices two sides willing to accept [4] and other factors can be the source of frictions. Besides information, time [5] and price concession [4][6] are identifiers to describe the frictions to exchange. Frictions to exchange network should involve network structure as substrate of these factors, the relation between nodes as well as the attributs of nodes. Typically, there exists two kinds of intermediary to coordinate trade of exchange network. One is exogenous intermediary such as government. The other is endogenous intermediary emerging from the nodes of exchange network, such as banks, brokers and so on. There are two ways for intermediary to optimize exchange network. One is to link nodes and get the optimal outcome [7][8]. The other is to link nodes and decrease the frictions to exchange. Traditional economics literatures discussed the exist of intermediation at different situation [9]–[11]. Coase's traditional analysis model [12] is used to explain the presence of intermediaries between demand and supply sides. Information is a main topic in the function of intermediation [13]–[17]. Watanabe [18] has promoted a uniform framework to analyze the function of intermediation.

Related with intermediation in complex social networks, key nodes and crucial links such as structure holes, weak ties and strong ties prompt concern of sociologists [19][20][21]. Efficiency and stability can coexist in a network with intermediation by defining critical link and intermediary position [22]. How brokers gain competitive advantage in a certain network [23], endogenous and exogenous intermediations in complex social networks are discussed. For example, Based on Banknet developed by Askenazi [24], banking activity emerges from the interaction of a continuum series of financial transactions between heterogeneous economics agents [25]. Total payoff contributed by intermediation as a whole with different match mechanisms is discussed [7]. Goyal et al. [8] studies a model of network formation where agents provided ability to block bilateral interaction between two players and to be intermediations.

Based on general exchange network generated from a social network with certain structure, exchange network frictions are considered to measure the difficulty nodes trade to each other. The influences that nodes' attributes and the exchange network structure bring to these frictions are discussed in section 2. In section 3 and section 4, the optimization of exogenous and endogenous intermediary are investigated respectively. Section 5 is conclusions and section 6 is acknowledgement.

2   Frictions of Exchange Network

In social network G with $|G|$ nodes such as figure 1(a), nodes refer to individuals or organizations and links present their relationship. With trust $b(b \in [0,1])$ between two nodes and time limit $T(T > 0)$, exchange network K^T is generated from G by the communication between linked nodes with time T, see figure 1(b).

Denotes average degree of G and K^T as $\langle G \rangle$ and $\langle K^T \rangle$. With probability ρ and $1 - \rho$, nodes in K^T are divided into two parts: suppliers and demanders who provide complementary services. Either side supplies homogenous service and achieves his goals only by coorperating with complementary side. The value of exchange network is to assist all nodes achieve their objectives. But the structure of exchange network may block exchange while assist it. For example, a man only knows men and his goal is to get married on one hand. A demander's objective can't obtained if he meet none supplier(woman) within any T and b. On the other hand, if a man meet some women but these women know more other men at the same time, then the chance to success of his merriage will be small than these women only know him. Phenomenons like this are far-ranging in the social and natural world. Merriage, kindney exchange, labor market, risk investment and trade are involved here. Like fiction which block objects moving in classical mechanics, frictions of exchange network are defined as the strenth or probability to obstruct exchange.

2.1   Direct and Indirect Friction of Exchange Network

Just like the difficulty the man who want to get married can meets, frictions of exchange network are rooted in two aspects: one is how many agents an agent meet can't be partners and the other is how many corrivals it has. The first illustrates the probability agents can't meet trade partner and the second accounts for the probability they meet corrivals. Average frictions present the holistic friction level of exchange network and the standard deviations indicate the discrepancy or heterogeneity among nodes.

Definition 1. Direct friction (DF) at time t defined as

$$F_i^t = \begin{cases} 1 & |D_i^t| = 0 \\ \frac{|S_i^t|}{|D_i^t|} & |D_i^t| > 0 \end{cases} \tag{1}$$

Definition 2. Indirect friction (IDF) at time t defined as

$$I_i^t = \begin{cases} 1 & F_i^t = 1 \\ 0 & \exists j \in N_i^t \quad and \quad |D_j^t| = 1 \\ \dfrac{\sum\limits_{j \in N_i^t} (|N_j^t| - 1)/|D_j^t|}{|N_i^t|} & Otherwise \end{cases} \tag{2}$$

Definition 3. Average direct friction(ADF) of K at time t defined as

$$A_F^t = \frac{\sum\limits_{i=1}^{|K^t|} F_i^t}{|K^t|} \tag{3}$$

Definition 4. Average indirected friction$(AIDF)$ of K at time t defined as

$$A_I^t = \frac{\sum\limits_{i=1}^{|K^t|} I_i^t}{|K^t|} \tag{4}$$

Where S_i^t denoted the set of i's neighbors in K^t which possess same attribute–supply or demand to i and N_i^t denoted the set of i's neighbors in K^t which possess different attribute to i. $D_i^t = S_i^t \cup N_i^t$ is the set of i's neighbors. $|D_i^t|$ denoted the degree of node i and $|D_i^t| = |S_i^t| + |N_i^t|$. Define $A_F^t = 0$ and $A_I^t = 0$ where $|K^t| = 0$ or $|K^t| = 1$. The standard deviations of F_i^t and I_i^t which shows the equality or discrepancy of nodes are S_F and S_I. As shown in figure 1(b), frictions of nodes and network can be computed according to formula (1)-(4)($\langle G \rangle = 1.25$, $\langle K^t \rangle = 0.875$, $\rho = 3/8$, each pair of numbers present F_i^t and I_i^t of a node, $A_F^t = 0.29$, $A_I^t = 0.38$, $S_F^t = 0.341$, $S_I^t = 0.324$).

Two nodes linked by an link with probability $\rho^2 + (1 - \rho)^2$ possess same attribute. Assume the probability of node i with degree d is $f(d) = Prob(|D_i^t| = d)$, $E[F_i^t] = f(0) + [1 - f(0)](2\rho^2 - 2\rho + 1) = 1 + 2(\rho^2 - \rho)[1 - f(0)]$ where there are some isolated nodes and $f(0) \neq 0$. $\frac{\partial E[F_i^t]}{\partial \rho} = 2[1 - f(0)](2\rho - 1)$ and the minimum A_F is $A_F^{min} = f(0) + \frac{1}{2}[1 - f(0)]$ at $\rho = \frac{1}{2}$. If there is no isolated node in K^t and $f(0) = 0$, then expected F_i of K^t is $E[F_i^t] = 2\rho^2 - 2\rho + 1$ and the minimum A_F is $A_F^{min}(\frac{1}{2}) = \frac{1}{2}$ at $\rho = \frac{1}{2}$. A_F is only related to the probability of isolated nodes instead of the distribution of degree of K^t. If K^t is random network with connected probability p, $E[F_i^t] = (1-p)^{|K^t|-1} + [1 - (1-p)^{|K^t|-1}](2\rho^2 - 2\rho + 1)$, $A_F^{min} = (1-p)^{|K^t|-1} + \frac{1}{2}[1 - (1-p)^{|K^t|-1}]$ at $\rho = \frac{1}{2}$.

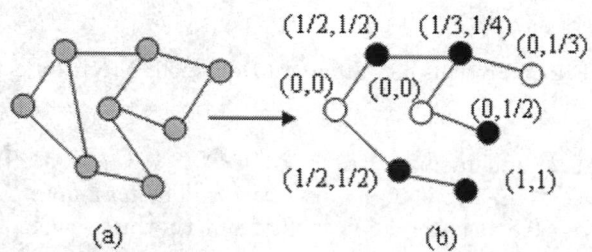

Fig. 1. Exchange network Generated from Social Network

2.2   Features of Average Frictions in Exchange Network

Frictions are influenced by the features of the nodes ρ, the enviornment constrain T and the relationship of pairs of nodes b. T, b and G determine the average degree $\langle K^T \rangle$. Lower $\langle K^T \rangle$ means more probability with isolated nodes and higher A_F^T and A_I^T with same ρ.

Simulated with 500 networks with $|G| = 2 - 500$($b = 0 - 1$, $T = 1 - 30$, $\rho \in (0, 1)$) in order to investigate how frictions influenced by ρ and $\langle K^T \rangle$. Respectively, figure 2(a) and figure 2(b) show A_F^T, A_I^T, S_F^T and S_I^T changed with ρ. A_F^T and A_I^T are almost same at lower connectivity($\langle K^T \rangle = 0.2$). But A_F is higher than A_I at medium connectivity($\langle K^T \rangle = 2.0$) and the disparity shrinked at higher connectivity($\langle K^T \rangle = 29.1$). The relations between S_F^T, S_I^T and $\rho(\rho < 0.5)$ are concave. S_F^T and S_I^T are increased with $\rho(\rho < 0.5)$ at

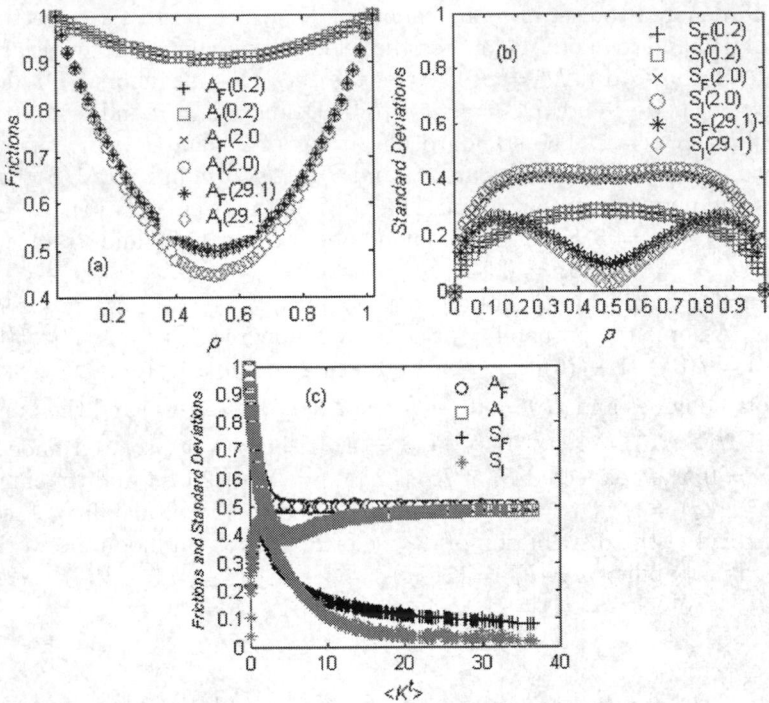

Fig. 2. Frctions and Standard Deviations of Networks

lower($\langle K^T \rangle = 0.2$) and medium($\langle K^T \rangle < \langle K^{T*} \rangle$, $\langle K^{T*} \rangle \approx 1.98$) connectivity. S_F^T is more gentle than S_I^T at high connectivity. The maximum S_F^T and S_I^T are reached where $\rho = 0.5$ at lower connectivity. But they are reached where $\rho \neq 0.5$ at medium($\langle K^T \rangle > \langle K^{T*} \rangle$) and higher connectivity($\langle K^T \rangle = 29.1$). Figure 2(c) shows A_F^T, A_I^T, S_F^T and S_D^T influenced by $\langle K^T \rangle$ at $\rho = 0.5$. $A_F^T(0.5)$ converged to 0.5 where $\langle K^T \rangle > 5$ for there is not any isolated nodes in K^T. $A_I^T(0.5)$ declines to 0.4 where $\langle K^T \rangle > 5$ then goes up to 0.5 where go up $\langle K^T \rangle > 30$. The maximum S_F^T and S_I^T are obtained about $\langle K^T \rangle \approx 2$.

Microscopically, frictions a node meet come from its neighbors and its neighbor's neighbors. Macroscopically, A_F^T are influenced mainly by the proportion of demanders and suppliers as well as the probability of isolated nodes in the network. A_I^T and A_F^T varied in different pattern where $\langle K^T \rangle = 5 - 15$, A_I converged to A_F^T while K^T increased.

3   The Optimization of Exogenous Intermediary Service

3.1   Optimization Algorithm

Assume there is an exogenous intermediary such as labor broker, government, e-commerce web site who know the globe information of K^T and his function is to decrease the frictions of K^T. The intermediary may not be located in the network and add links to the nodes possessing different attributes where there still has no link in K^T. The new link may not exists in G.

Added link to node i and j can influence both direct friction and indirect friction of them and indirect friction of there neighbors which possess different attribute to them. There are two alternatives for exogenous intermediation to decreasing the A_F^t or A_I^t of $K^t (t \geq T)$ by adding links. Firstly, both frictions will be reduced if the objective of exogenous intermediation is to reduce A_I^t by adding links to two nodes with different attributes. Adding a link to two arbitrary nodes i and j with different attributes can reduce A_F^t. An optimization algorithm(IRA) for exogenous intermediation is developed to decrease frictions as follows. The added link must reduce the maximum average friction of K^t.

Indirect friction reducing algorithm(IRA):

Step 0: t=T;

Step 1: Calculate friction of network K^t.

Step 2: Find each pair of nodes i and j with opposite attributes have no link in K^t. If there are no such nodes, step 5.

Step 3: To all pair of nodes not linked in K^t, calculate ΔI_i^t, ΔI_j^t, ΔI_k^t, ΔI_l^t), and ΔA_I^{ijt}. Where $k \in D_i^t$ and $l \in D_j^t$.

Step 4: Find $\{i^*, j^*\} = min_{\Delta A_I^t}\{i, j | \Delta A_I^{ijt} < 0\}$ and add a link between i^* and j^*. $t = t + 1$. If there is no such $\{i^*, j^*\}$, step 5.

Step 5: End optimization.

Secondly, if the objective of exogenous intermediary is to reduce A_F^t, he can simply adds links to all nodes with opposite attributes. To reduce A_I^{ijt} can decrease of both $F_i(t)$ and $I_i(t)$ but to reduce A_F^t is not always true. If $|D_i^t| = 0$ or $|D_j^t| = 0$ before the link added, I_i^t or I_j^t can be reduced by the added link between them. But if $N_i^t > 0$ or $N_j^t > 0$, the indirect friction of nodes in N_i^t or N_j^t can be increased. So if the objective of exogenous intermediary is just adding links to reduce direct friction of nodes, the link-adding process will be terminated once additional added link may make A_I^t increased. Another algorithm called direct friction reducing algorithm(DRA) which changed $min\{\Delta A_I^t\}$ to $min\{\Delta A_F^t\}$ is different to IRA at step 3 and 4. The links will be added to the nodes with $D_i^t = 0$ or $N_i^t = 0$ where $F_i^t = 1$ and $I_i^t = 1$firstly to make $\Delta F_i^t < 0$. If there is no such nodes, the network can not be optimized.

Step 3(a): To all pair of nodes not linked in K^t, calculate ΔI_i^t, ΔI_j^t, ΔI_k^t, ΔI_l^t, and ΔA_I^{ijt}. Calculate ΔF_i^t, ΔF_j^t, ΔF_k^t, ΔF_l^t, and ΔA_F^{ijt} at the same time. Where $k \in D_i^t$ and $l \in D_j^t$.

Step 4(a): Find $\{i^*, j^*\} = min_{\Delta A_F^t}\{i, j | \Delta A_F^{ijt} < 0\}$, if $\Delta A_I^{i^* j^* t} < 0$ then add a link between i^* and j^*, $t = t + 1$; else step 5.

The least links can be added to K^T to get a bipartite graph if $\langle K^T \rangle = 0$. Figure 3(a) and (b) shows the results of DRA and IRA where $\langle K^T \rangle = 0$. Figure 3(a) shows there are 4 suppliers and 4 demanders. With DRA or IRA, exogenous intermediary links them to pairs to make $A_F^{(T+4)} = 0$ and $A_I^{(T+4)} = 0$. Another

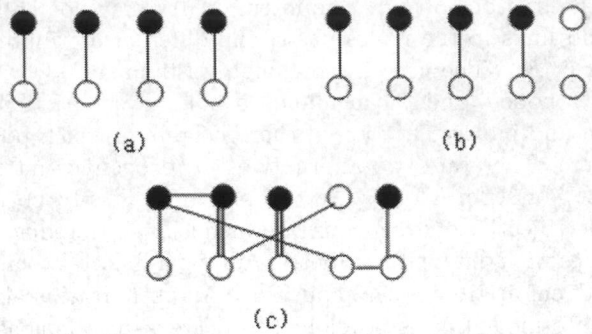

(a)   (b)

(c)

Fig. 3. Optimized network where $\langle K^T \rangle = 0$

Table 1. Different networks between DRA and IRA of 1000 samples

	Result network	Amount of links added	A_F	A_I
max	74	41	57	63
min	28	10	21	23
mean	49.054	24.158	37.271	40.146
SD	7.1305	4.7922	6.0713	6.3824

circumstance is that their are 4 suppliers and 6 demanders. Optimized result with DRA or IRA form 4 links and 2 isolated nodes and $A_F^{(T+4)} = 0.1$ and $A_I^{(T+4)} = 0.1$, as shown in figure 3(b). Figure 3(a) may match along with perfect market and figure 3(b) implies one kind of imperfect market [29]. Figure 3(c) shows the optimization result of $\langle K^T \rangle = 1.2$. Two double lines are links added by IRA and DRA. Here $A_F^{(T)} = 0.6333$ and $A_I^{(T)} = 0.5667$, $A_F^{(T+2)} = 0.3333$ and $A_I^{(T+2)} = 0.1667$. The result network is more complex than figure 3(a) and (b). That means more negotiation and decision will be taken to realize the exchange. It is interesting that monogamy as evolving institute practiced by many nations, will be explained to reduce frictions of family. But to find partner in exchange network is a frictional task.

DRA and IRA play same roles where K^T is empty and complete graph. To compare the effects of these 2 optimization mechanics where K^T is arbitrary network. Simulated with 1000 networks with 1000 times($|K^T| = 2{-}500, b = 0{-}1$, $T = 1 - 30$, $\rho \in (0,1)$), the number of different result networks are shown in table 1. There is positive probability that DRA and IRA create same result network as well as different result networks to any network. The probability to same result is obviously higher than that of different results according to the simulation results. Most networks can be optimized by DRA and IRA without difference. But it is remarkable that even the numbers of links added are same by DRA and IRA, these links added to different pairs of nodes and result to different result networks. Some pairs of different result networks have same A_F^T and A_I^T. Result networks of a certain original network with different A_I^T may not possess different A_F^T at the same time. A_F^T and A_I^T of DRA and IRA are statistically indifference with t-test($\alpha = 0.01$, $p^F = 0.0578$, $p^I = 0.4367$).

3.2 Optimization Results Analysis

Figure 4 shows the optimization rate influenced by the connectivity of K^T. In figure 4(a), average degree of K^T influenced by the number of the nodes in K^T where T is bigger($T-5$). It is obvious the optimization rate(rate of optimized networks from 500 networks) decreased with the increase of T and decrease of ρ, see figure 4(b). In figure 4(c) the optimization rate decreased with the increasing of network density $\langle K^T \rangle/(|K^T| - 1)$. Even if at same density, the optimization rate is much higher at small $T(T-1)$ than bigger $T(T=5)$ because bigger T leads to higher $\langle K^T \rangle$.

Figure 5 indicates the optimization efficiency decreasing with the increasing of $\langle K^T \rangle$. There are several measures be defined to describe the efficiency of the optimization on figure 5. The first is the proportion of networks can be optimized(network rate) and the second is the decreasing rate of $A_F^T(\Delta A_F/A_F)$

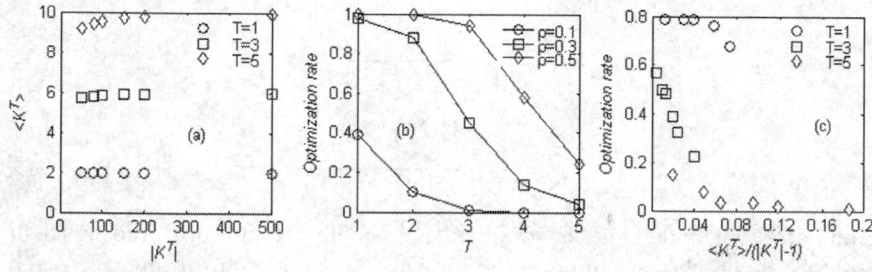

Fig. 4. Optimization Effect of Exogenous Intermediary Services

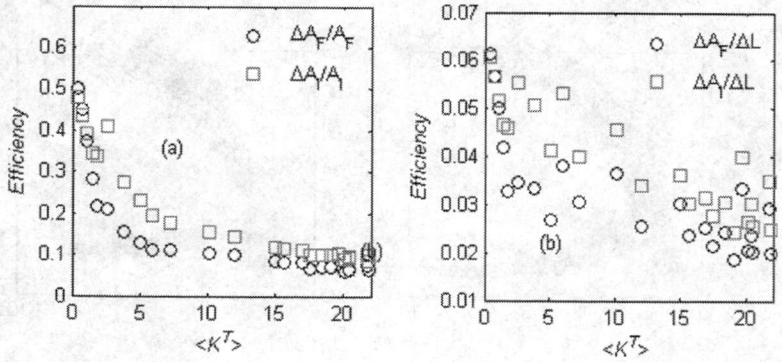

Fig. 5. Optimization Efficiency of Exogenous Intermediary Services

and $A_I^T(\Delta A_I/A_I)$. The second is the decreasing of A_F^T and A_I^T by each added link and indicates the contribution of intermediation($\Delta A_F/\Delta L$ and $\Delta A_I/\Delta L$). Both indexes are decreased with the increasing of $\langle K^T \rangle$.

4   The Optimization of Endogenous Intermediary Service

4.1   Probability of Endogenous Intermediary Service

A node may be an exogenous intermediary while considering his all neighbors as G and he will act on the sub-network consisted of his neighbors. But he may be endogenous intermediary while considering a more extensive network including him. If there exists nodes i in K^T that $S_i^T > 0$ and $N_i^T > 0$, he will be potential intermediary. Compared with exogenous intermediary, endogenous intermediary is embedded into K^T and only has local information of his neighbors. So what can he do is just link nodes remained unkown in K^T and reduce directed friction of K^T. A_F^T and A_I^T will be influenced if a node acts as endogenous intermediary and linked his neighbors with different attributes. In view of the direct friction and indirect friction, define two kinds of intermediary probability of node i as follows:

$$P_i^{Ft} = \frac{\sum\limits_{j \in D_i^t} F_j}{D_i^t - ||N_i^t| - |S_i^t||} \tag{5}$$

$$P_i^{It} = \frac{\sum\limits_{j \in D_i^t} I_j}{D_i^t - ||N_i^t| - |S_i^t||} \tag{6}$$

Simulated with 500 random network($|K^T| = 500$), figure 6(a) and figure 6(b) separately shows the rate of potential intermediary nodes according to P_F(R_F^B is the rate before optimization by DRA and R_F^A means the rate after optimization) and P_I(R_I^B is the rate before optimization by DRA and R_I^A means the rate after

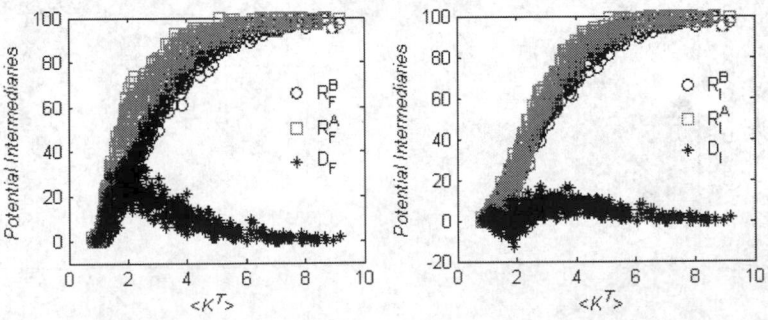

Fig. 6. Probability of Endogenous Intermediary Service

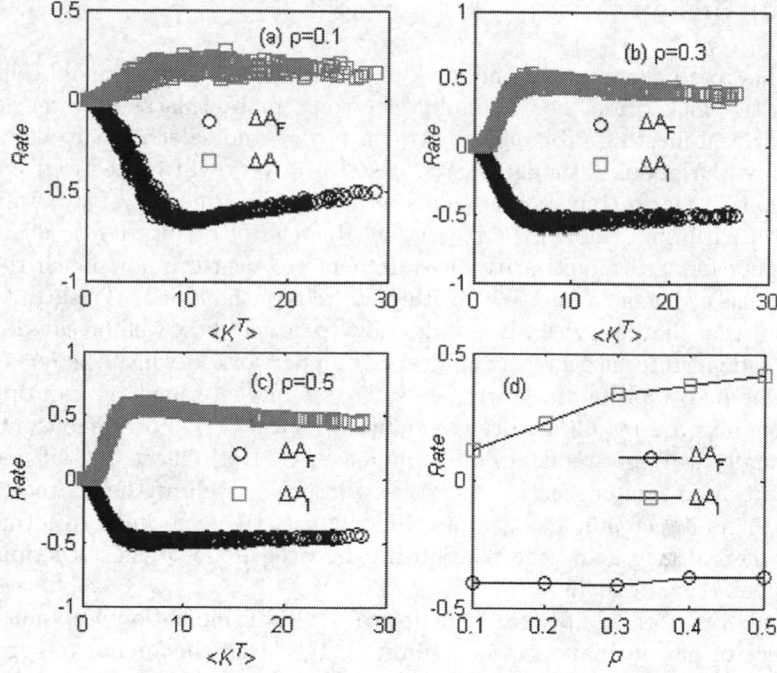

Fig. 7. Optimization Effects of Endogenous Intermediary Service

optimization) varies with the increase of $\langle K^T \rangle$. Both are increased with $\langle K^T \rangle$ increasing. But that of the optimized network is increased faster than the original network. D_F and D_I mean the rate balances between optimized networks and oringinal networks. The maximum difference between optimized network and original network can be obtained at $\langle K^T \rangle = (2-3)$. There also exist some networks that their rates reduced after optimization.

4.2 Optimization Results Analysis

Can K^T be optimized if node i can link his neighbors with different attributes? It is obvious that linked nodes with degree 0 or 1 can reduce both frictions. But the information of i's neighbors may be private to i. He links nodes as possible as he can to ensure his intermediary income. Simulated with 1000 networks of different ρ, Figure 7 shows the optimization efficiency influenced by ρ and $\langle K^T \rangle$. In figure 7(a)($\rho = 0.1$), the maximum decreased direct friction(ΔA_F) and the maximum increased indirect friction(ΔA_I) reached their peak value at $\langle K^T \rangle = 8$. ΔA_F is at about $\langle K^T \rangle = 5$ and ΔA_I is at about $\langle K^t \rangle = 8$ while ρ increased. Figure 7(b) shows the average level of 1000 networks. The average ΔA_F is practically not changed with ρ increased.

5   Conclusion

The value to all agents on the network is propitious to trade. The opposite side of it is to block trade. For example, electronic marketplaces as intermediary can improved meeting probability between buyers and salers, but it also raised the indirect friction. Although search costs of buyers will be reduced by electronic marketplaces, buyers and salers as well as intermediary may "wait and see" to learn from others' experiences [26]. Instead of qualitative analysis, this paper puts forward quantitative measurements to measure how much the network blocks exchange and give algorithms to reduce the blocks. This can help to decide intermediary service what to do, how to do and how well he can do. Frictions are defined to measure the hindrance of a network in which the nodes have two different complementary attributes. Exogenous intermediary can optimize exchange network by adding links to reduced A_F and A_I. Endogenous intermediay usually can only reduce A_F but increase A_I. Both effect and efficiency of optimization are influenced by network structure as well as the characteristics of nodes. As direct influence factors to structure, the time limit to form K as environmental factor and the relationship between nodes b effect frictions and the optimization of them.

The results of this paper can explain we need intermediation very much, but the effect of intermediary service is limited. How well that market as invisible hand [27] and government as the visible hand [28] coordinate with complex social network at least depend on the network structure and the characteristics of agents in it. That means bounded rationality which leads to imperfect market [29] is rooted the embededness of social and exchange networks as well as it is regarded as the origin of the social and exchange networks.

It is reflects embededness of social networks [30][31] and ecology networks that exchange network K is generated from existing network G. To map, diagnose and improve the network consisted of individuals, brokers of social network are crucial to the performance of the network [32]. The nodes of the network mentioned here is not only individuals but also organizations or subnetworks can be regarded as systems. The links could be explained as the supply and demand of both material product and intellectual products. Cooperation for ecological and organization networks [33][34] are the examples of such bipartite relation discussed in this paper. By abstracting general attributes of two side exchange, our results implies that as fundus of nodes' dynamics occuring on, network structure is a double-edged sword to agents to achieve their utilities. To close to real social and ecology systems, homogenous services discussed here is not enough. Future research will be penetrated deeply in heterogeneous services and the dynamics of nodes besides more dynamics of ndogenous intermediary.

Acknowledgement

This work was supported by National Natural Science Foundation of China under Grant No.70671070.

References

1. Palla, G., Barabási, A.-L., Vicsek, T.: Quantifying social group evolution. Nature 446, 664–667 (2007)
2. Motter, A.E., Toroczkai, Z.: Introduction: Optimization in networks. CHAOS 17, 026101 (2007)
3. Gurley, J.G., Shaw, E.S.: Money in a theory of finance. The Brooking Institution Press, Washington (1960)
4. Stoll, H.R.: Friction. In: Papers and Proceedings of the Sixtieth Annual Meeting of the American Finance Association; Boston, Massachusetts, January 7-9, The Journal of Finance 55, 1479–1514 (2000)
5. Lippman, S., McCall, J.J.: An Operational Measure of Liquidity. American Economic Review 76, 43–55 (1086)
6. Demsetz, H.: The Cost of Transacting. Quarterly Journal of Economics 82, 33–53 (1968)
7. Li, L., Chen, Z., Chen, B., Zhao, Z.L.: Research on the Payoff of Intermediary Network with Different Match Mechanism. DCDIS B-Applications & Algorithms 14(S7), 205–209 (2007)
8. Goyal, S., Vega-Redondo, F.: Structural Holes in Scial Networks, working paper, University of Essex (2007)
9. Rubinstein, A., Wolinsky, A.: Middlemen. Quarterly Journal of Economics 102, 581–593 (1987)
10. Biglaiser, G.: Middlemen as Experts. The RAND Journal of Economics 24, 212–223 (1993)
11. Li, Y.T.: Middlemen and Private Information. Journal of Monetary Economics 1, 131–159 (1998)
12. Coase, R.H.: The Nature of the Firm. Economics 4, 386–405 (1937)
13. Heffernan, S.: Modern banking in Theory and in Practice. John Wiley and Sons, London (1996)
14. Diamond, D.W.: Financial Intermediation and Delegated Monitoring. Review of Economic Studies 51, 393–414 (1984)
15. Fama, E.: What's Different about Banks? Journal of Economics 15, 29–39 (1985)
16. Stiglitz, J.E., Weiss, A.: Credit Rating in Market with Imperfect Information. American Economic Review 71, 393–410 (1981)
17. Stiglitz, J.E., Weiss, A.: Banks as Social Accounts and Screening Devices for the Allocation of Credit. National Bureau of Economic Research Working Paper, 2710 (1988)
18. Watanabe, M.: Middlemen: The visible market makers, working paper (2006)
19. Burt, R.: Structural Holes: The Social Structure of Competition. Harvard University Press, Cambridge (1995) (reprint edn.)
20. Granovetter, M.: The Strength of Weak Ties. American Journal of Sociology 78, 1360–1380 (1973)
21. Bian, Y.J.: Bring Strong Ties Back. In: Indirect Ties, Network Bridges, and Job Searches in China; American Sociological Review 62, 266–285 (1997)
22. Gilles, R.P., Chakrabarti, S., Sarangi, S., Badasyan, N.: Middlemen in Peer-to-Peer Networks: Stability and Efficiency, working paper (2004)
23. Ryall, D.R., Sorenson, O.: Brokers and Competitive Advantage. Management Science 53(4), 566–583 (2007)
24. Askenazi, M.: Some notes on the BankNet model, Santa Fe Institute (1996)

25. Sapienza, M.D.: An Experimental Approach to the Study of Banking: The BankNet Simulator. In: Luna, F., Stefansson, B. (eds.) Economic Simulation with Swarm. Kluwer Academic Press, Dordrecht (2000)
26. Bakos, J.Y.: Reducing Buyer Search Costs: Implications for Electronic Marketplaces. Management Science 43(12), 1613–1630 (1997)
27. Smith, A.: An Inquiry into the Nature and Causes of the Wealth of Nations. Chigago University Press, Chicago (1976)
28. Chandler, A.: The Visible Hand: The Managerial Revolution in American Business. Harvard University Press, Cambridge (1977)
29. Samuelson, P.: Proof that Properly Anticipated Prices Fluctuate Randomly. Industrial Management Review 6(2), 41–49 (1965)
30. Granovetter, M.: Economic Action and Social Structure: The Problem of Embeddedness. American Journal of Sociology 91, 481–510 (1985)
31. Uzzi, B.: The Sources and Consequences of Embeddedness for the Economic Performance of Organizations: The Network Effect. American Sociological Review 61, 674–698 (1996)
32. Uzzi, B., Shannon, D.: How To Build a Better Network. Harvard Business Review 83, 53–60 (2005)
33. Cowan, R., Jonard, N.: Zimmermann: Bilateral Collaboration and the Emergence of Innovation Networks Management Science 53(7) 1051–1067 (2007)
34. Saavedra, S., Reed-Tsochas, F., Uzzi, F.: A simple model of bipartite cooperation for ecological and organizational networks. Nature Fall (2008)

Spam Source Clustering by Constructing Spammer Network with Correlation Measure

Jeongkyu Shin and Seunghwan Kim

Asia Pacific Center for Theoretical Physics
Nonlinear and Complex System Laboratory,
Pohang University of Science and Technology, Pohang 790-784, Korea
jkshin@physics.postech.ac.kr, swan@postech.ac.kr

Abstract. Spam filtering is one of the most challenging problems in electric message systems. In general, recent studies on specifying real spam source are based on content filtering because spammers usually falsify their origin. We propose a method to specify spam source based on structural analysis with complex network. We assume that each spam sources either has the same victim list or uses the same spam-hosting program. We treat spam source - target relationship as a bipartite network and construct weighted spam source network by network projection using correlation measure. We find that community clustering methods are inappropriate with spammer network. We group spammers with gradient-based grouping, which uses correlations between nodes as gradient between nodes. We convert them into local minima, which helps to cluster spammers into a few spam source groups. We investigate the weblog spam data with the proposed method and validate it. The method that we propose can be applied to diverse categorization problems, such as multiple text categorization and network subunit clustering.

Keywords: Electronic spam, complex network, clustering method.

Undesired electronic messages on World-Wide Web (for instance, E-Mail, comments, trackbacks, and etc.) are becoming a serious problem of electric communication. These 'spam' messages are usually sent for advertising services and products, and some of them contain malwares that is made for cracking receivers' computers.

Spam and spam-blocker development is a good example of fast evolving system with competition. Spam usually does not contain helpful information for receivers. At the beginning of online advertisement market, spam could be treated as noise in electronic communication because the portion of spam was relatively small. Now the situation changed: most electronic messages are spams. Spam consumes traffic, which is strongly related with communication resource. As many studies are performed to avoid spam, spams also start to evolve: it is a stiff fight between a spear and a shield.

Blacklist is one of well-known traditional filtering methods. If a message is marked as a spam, spam filter adds the signature (usually URL or IP address)

J. Zhou (Ed.): Complex 2009, Part I, LNICST 4, pp. 889–893, 2009.

to blacklist. Blacklist method is easiest and strongest filtering method. However IP spoofing disturbs finding the spam sources [1]. Spam senders (called *Spammers*) usually deceive spam referrer IP using zombie computers which are hacked by spammers, thus the number of spammer looks quite big even though it is small [2].

A modern approach, *Bayesian spam filtering*[3], based on Bayes theorem has become a common filtering method due to its flexibility and effectiveness. Bayesian filtering is one of *content pattern matching methods* which requires more computing resource than blacklist. In spite of the power of Bayesian filtering, Spammers use some tricks to evade spam filtering. For instance, *Bayesian poisoning* is a well-known method to spoil Bayesian filtering. Basically, modern spam filtering consists of combining blacklist and Bayesian filtering which are common spam filtering methods.

IP spoofing exaggerates the number of spammers, and IP blacklist can be too complex as a side effect. By the reason that spammer has content or URL pattern for their spam, grouping zombie computers and specifying spammer can help content-based filtering filter to extract spam pattern. Thus specifying exact spam sources helps developing anti-spam methods.

In this paper, we propose a method to group diverse spammer IPs based on complex network analysis with *spam source - target relation* data. Specifying spam source is usually based on content (including URL, IP address) analogy. We suggest a hypothesis that even though spammers use zombie computers to spread spam messages, same spammer uses the same target list. If targets are almost the same, those attacks can be originated from the same spam source (or same spamming program). If this hypothesis is true, spammers might be grouped by reconstructing spammer-spammer relation network even if spammers falsify their IP address. We treat *spam source - target relationships* as bipartite network [4], and project spam-target vector into spam sources relation matrix using correlation measures.

Bipartite network structure can be converted into unipartite one [5]. Let $\bar{\sigma}^{\mu}$ as signature vector of spammer IP address μ, where

$$\bar{\sigma}_i^{\mu} = \begin{cases} 1 \text{ if spam attacks victim } i \\ 0 \text{ otherwise} \end{cases} \quad (1)$$

Note that every spammer and victims are labeled. $\bar{\sigma}^{\mu}$ denotes the record which victims are attacked by spammer μ. Thus the information of victim network layer is treated as the connection information of spammer network. After construction of every spammer's signature vector, relations between spammer IP addresses can be calculated with similarity measure which is a kind of symmetric correlation measure [6]

$$C^{\mu\lambda} = \frac{\bar{\sigma}^{\mu} \cdot \bar{\sigma}^{\lambda}}{|\bar{\sigma}^{\mu}||\bar{\sigma}^{\lambda}|} \equiv \cos\theta_{\mu\lambda} \quad (2)$$

where $\bar{\sigma}^{\mu} \cdot \bar{\sigma}^{\lambda}$ is the scalar product between signature vectors. We can calculate every correlation between all pairs of spammers and construct spammer-spammer weighted adjacency network C.

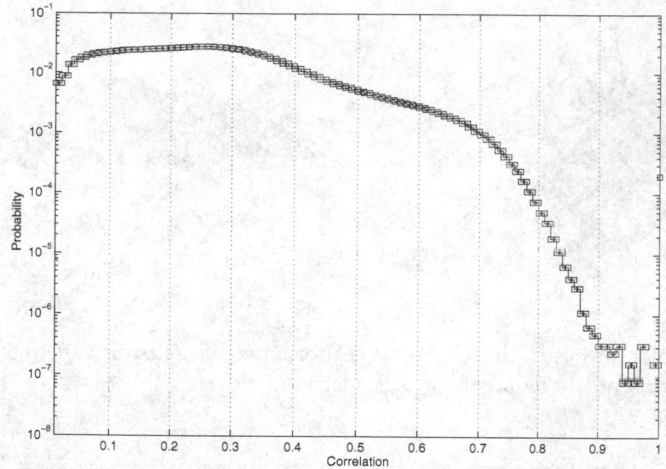

Fig. 1. Probability distribution of spammer network link weight. Result indicates that the percolation-based filtering does not work well with weblog spammer network.

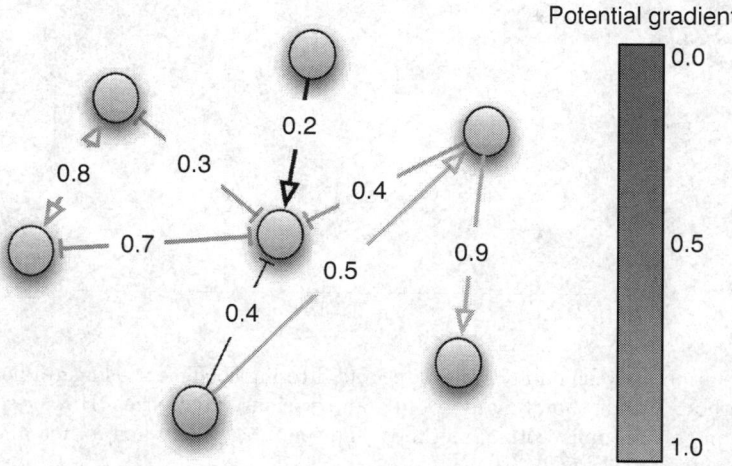

Fig. 2. Gradient-based grouping method example. Each node makes group with another node with highest link weight. Note that this graph is not an directed graph; arrow indicates the belongings, not link direction.

Once the weighted network is determined, various group methods are studied. Percolation-based filtering[4], which is the commonly used grouping method by disconnecting nodes lower than specific filtering coefficient value, has a problem with weblog spam network. Fig.1 shows that there is no transition point as correlation varies. Thus it is hard to determine the proper filtering coefficient.

We propose *gradient-based grouping method*, inspired from energy potential landscape network[7]. Gradient-based grouping method works as follows: first,

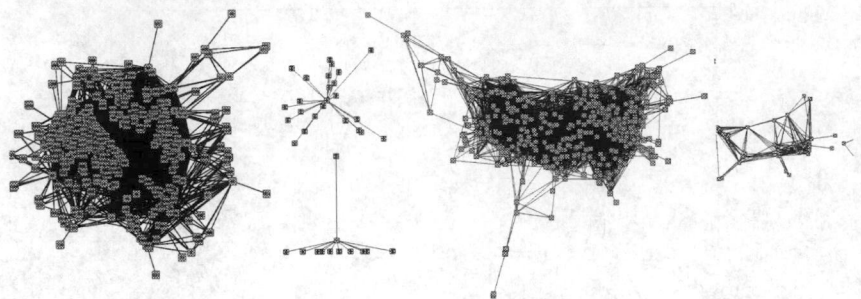

Fig. 3. Percolation-based filtering result. Filtering coefficients are *left*) 0.68 and *right*) 0.75. Groups are made by breaking biggest island.

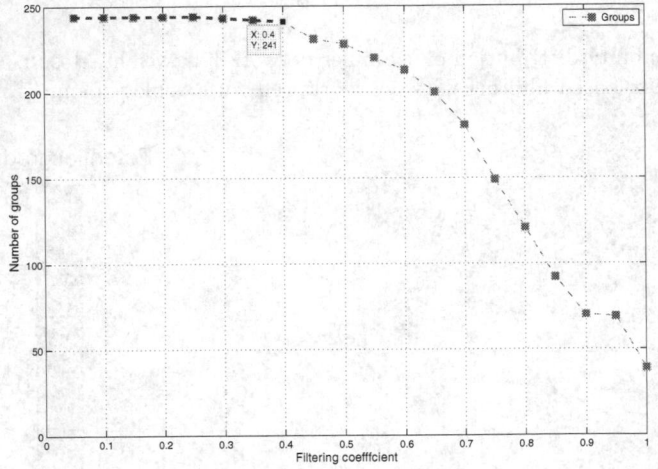

Fig. 4. Relation between number of groups and filtering coefficient using gradient-based group method. Result shows many groups and various group sizes. However, there is no dominant group unlike other methods. The number of groups is not affected by filtering coefficient.

check the weight of links of node i. Find link and connected node j with maximum weight. Finally, mark node i as a same group with j. Repeat these process for every nodes. Time complexity is $\sim O(n)$ for finding link with maximum weight, and $O(n)$ for loop. Thus the time complexity is around $\sim O(n^2)$.

We investigate the weblog (commonly called as *blog*) spam data from Eolin anti-spam service (EAS) [8]. We use spam data between July. 2006 and Nov. 2006. We extract 3118667 unique trackback spam pairs from spammer-victim IP and prune 500th to 5500th spammers (861328 pairs) to reduce side-effect from data irregularity. We find that the number of weblog spammers is usually bigger than number of victims similar to the E-mail spam, where spammers are known to counterfeit the sender information.

We extract *spam source IP - spam target* relationship as a vector sets, and construct the spam source adjacency matrix. Percolation-based grouping shows the linear fragmentations as filtering coefficient varies. Groups are made by breaking biggest island, thus the number of island heavily depends on the filtering coefficient (Fig. 3). In contrast, result from gradient-based grouping method shows uniformly sized, same number of group even if the filtering coefficients varies (See Fig. 4).

To validate our grouping method, we compare the result with those from other common methods. We perform content-based grouping by extracting representative words from spam content. By comparing the content-based grouping with the gradient-structure-based grouping, result shows that 81% of groups are same. We emphasize that, even content-based grouping and our network structure based IP grouping are totally different methods, test results show meaningful similarities; this result supports out hypothesis about spam target lists.

In conclusion, we address a grouping method in spam source specifying problem. We propose a methodology to group spam sources, which helps analyzing spam patterns and characterizing the properties of them, based on complex network analysis. We construct spammer IP address network from a spam source-target bipartite network using correlation measures, and classify spammer groups with gradient-based grouping method. We validate our method by comparing with other methods.

Our method can be applied to diverse categorization problems with sparse data, such as multiple text categorization, time-series analyses and network subunit clustering.

Acknowledgments. Raw data and some tests for this study has been supported by *Tatter and Company (TNC)*, acquired by Google inc. at Sep. 2008.

References

1. Spamlinks.net, http://spamlinks.net/filter-bl.htm
2. Song, S., Manikopoulos, C.N.: IP Spoofing Detection Approach(ISDA) for Network Intrusion Detection System. In: Sarnoff Symposium. IEEE, Los Alamitos (2006)
3. Sahami, M., Dumais, S., Heckerman, D., Horvitz, E.: A Bayesian Approach to Filtering Junk E-Mail. In: AAAI 1998 Workshop on Learning for Text Categorization (1998)
4. Newman, M.E.J., Strogatz, S.H., Watts, D.J.: Random graphs with arbitrary degree distributions and their applications. Phys. Rev. E. 64, 026118 (2001)
5. Newman, M.E.J.: The structure of scientific collaboration networks. Proc. Natl. Acad. Sci. U.S.A. 98, 404–409 (2001)
6. Lambiotte, R., Ausloos, M.: Uncovering collective listening habits and music genres in bipartite networks. Phys. Rev. E. 72, 066107 (2005)
7. Doye, J.P.K.: The network topology of a potential energy landscape: A static scale-free network. Phys. Rev. Lett. 88, 238701 (2002)
8. Eolin Antispam Service, http://antispam.eolin.com

Spiral Waves Emergence in a Cyclic Predator-Prey Model

Luo-Luo Jiang[1], Wen-Xu Wang[2], Xin Huang[3], and Bing-Hong Wang[1]

[1] Department of Modern Physics, University of Science and Technology of China,
Hefei 230026 P.R. China
jiangluo@mail.ustc.edu.cn, bhwang@ustc.edu.cn
[2] Department of Electronic Engineering, Arizona State University,
Tempe, Arizona 85287-5706, USA
[3] Department Physics, University of Science and Technology of China,
Hefei 230026 P.R. China

Abstract. Based on a cyclic predator-prey model of three species, spiral waves on global level of the system are obtained. It is found that the predation intensity greatly affects on the behaviors of spiral waves. The wavelength of spiral waves alter with the mobility in the form of $\lambda \sim D^\theta$. Values of θ are determined by predation rates between species. It indicates the behaviors of spiral waves varying with mobility are universal at the same predation rate which reveals competition of resources among species.

Keywords: cyclic predator-prey model, pattern formation, spiral waves.

1 Introduction

Spatial distribution of individuals is a common feature of ecological systems. Recently, spatial heterogeneity of species has attracted much attention because it is closely related with the stability and coexistence of species in ecological and evolutionary systems [1,2,3]. Two factors concern spatial heterogeneity as well as spatial patterns in which populations distribute spatially and individuals interact locally. The first is internal noise which induce spatio-temporal pattern of species in concerning range [4,5]. The second is predation intensity of species [6]. In this paper, we systemically investigate spiral waves emergence on global level of the system concerning the two factors.

Cyclical interactions of species which dominate each other in a cyclic manner emerge widely in nature such as rodents in the hight-Arctic tundra in Greenland [7], lizards in the inner Coast Range of California [8], and microbial populations of E.coli [9,10]. Recent experiment reveals that cyclical interactions promote biodiversity of three strains of E.coli [10]. Reichenbach *et al.* have found that noise induced by mobility of individuals greatly affects the biodiversity and spatial heterogeneity [4,5]. However, the intensity of interaction is seldom concerned.

J. Zhou (Ed.): Complex 2009, Part I, LNICST 4, pp. 894–899, 2009.

In community food webs, each species have different predation intensity which exhibits as the intensity of interaction among species. Predation intensity has been confirmed to be strong coupled with diversity in ecological systems [6] and invoked as an evolutionary force [11,12]. However, the effect of predation intensity on spatial heterogeneity is not very clear when diversity is promoted.

In this paper, we investigate the effect of predation intensity on spiral waves induced by mobility of individuals. It is found that predation rates affect the local interactions of species which display global effect via mobility of individuals. Spiral waves emerge when global oscillations are achieved. And the wavelength of spiral waves is satisfied power law with the mobility of individuals, $\lambda \sim D^\theta$. The value of θ is determined by the predation rate. It is confirmed that the behaviors of wavelength altering with mobility of individuals is universal at the same value of predation rate. In addition, preying rates are related with vacancy resources in the systems. Our work provides basic understanding of effects of predation intensity on the spatial heterogeneity of species as well as pattern formation.

2 Model

Based on the previous work of Reichenbach *et al.* [4,5], we introduce a cyclic predator-prey model: nodes of spatial lattice present mobile individuals of three species (marked by 1, 2, 3)in the microcosmic bacteria system. Each node can be located at most one individual of a species or a vacancy (denoted by V) which presents resource. There are three interactions, namely predation, reproduction, and exchange which only occur between neighboring nodes. *Predation.—* 1 beats 2 at a selection rate α and 2 becomes a vacancy, in the same way, 2 beats 3, and 3 beats 1. *Reproduction.—* An individual can reproduce an offspring to a neighboring V node at a rate of β. *Exchange.—* An individual could exchange positions with one of its neighbors at a rate γ due to its mobility.

Unlike the deterministic approach which regards the time evolution as a continuous process, here, the applied stochastic approach regards the time evolution as a kind of random-walk process. A standard algorithm for stochastic approach simulation was developed by Gillespie [13,14]. In this model, reactions occur in a random manner: preying happens with probability of $\alpha/(\alpha+\beta+\gamma)$, reproducing with probability of $\beta/(\alpha+\beta+\gamma)$, and moving with probability of $\gamma/(\alpha+\beta+\gamma)$.

3 Results

The initial condition are shown in Fig. 1. It is worth noting that nodes out of these three area are copied by vacancies which present spatial resources to reproduce new individuals. A node in a regular lattice can be occupied at most by an individual or a vacancy. By using an efficient algorithm of Gillespie introduced in section 2, we simulate the evolving process with Monte Carlo (MC) method. At each simulation step, a randomly chosen individual interact with one of its four nearest neighbors which is randomly determined. One step of Mote Carlo time is defined as all the individuals having been chosen once on average. We set

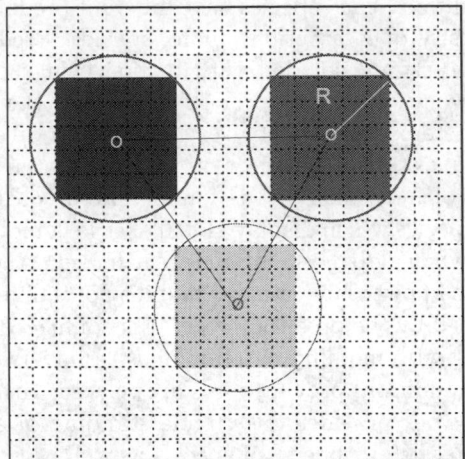

Fig. 1. Illustration of the initial condition. Here, R is set as 3.5, and the lattice completely inside the circle with radius R consist of the species' initially locating area, and three colors correspond three species. In this paper, the injected radius R is fixed at 10.5.

reproducing rate $\beta=1$, and the size of system $N = 1024 \times 1024$. The mobility of individuals D is defined as follow[4,5]:

$$D = 2\gamma/N, \tag{1}$$

As shown in Fig. 2, pattern formations depend on predation rate. All patterns in this paper are obtained after the system reach a stationary state of non-equilibrium. Low predation rate seems to promote spiral waves of global level. In the top of Fig. 2 with $D = 5.0 \times 10^{-6}$, spiral waves format at $\alpha = 0.01$, while edges of spiral waves break up at $\alpha = 0.1$, and spiral waves break into fragmentation at $\alpha = 1.0$ and $\alpha = 5.0$. The α becoming larger makes internal noise larger, which induce spiral waves breaking into fragmentation. The wavelength of spiral waves is defines as $\lambda = X/L$, where X is length of a spiral waves as shown in Fig. 2(a). In the bottom of Fig. 2, the wavelength of spiral waves decreases as increasing of α, and the arm of spiral waves becomes rough, which falls to pieces in the end, as shown in Fig. 2(h). In conditions of the same ability of individuals, predation rate determines sizes of spirals' arm which change into small one at higher predation rate.

Fig. 3 shows oscillation of species' percentage evolving with time for the same parameter of Fig. 1. When the global oscillations of species' percentage are achieved, spiral waves emerge in the systems. It is interesting to find that amplitudes of oscillation increase with the values of D, which is quite different from cases in target waves. In conditions of the same mobility of individuals, the average percentage of species decrease with the increasing of predation rate. It is confirmed that with the increasing of predation rate make the increasing of vacancies the same mobility, which induces strong internal noise to destroy

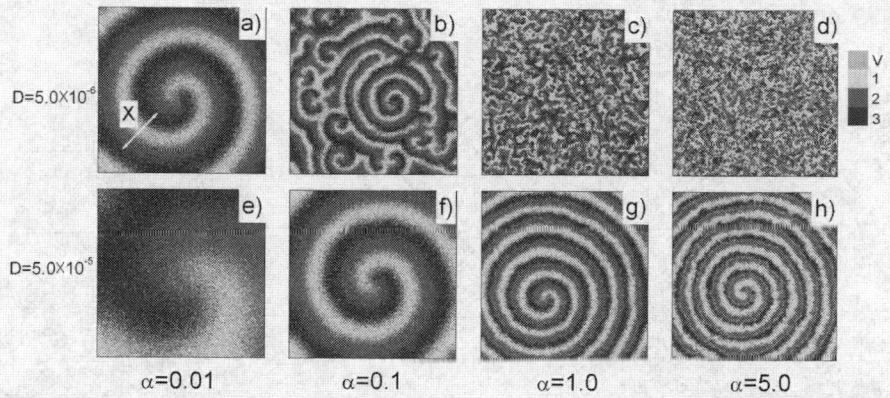

Fig. 2. Pattern formation for different value of α and D. (a) and (e) for $\alpha = 0.01$,(b) and (f) for $\alpha = 0.1$, (c) and (g) for $\alpha = 1.0$, (d) and (h) for $\alpha = 5.0$, at the same time (a),(b),(c),(d) for $D = 5.0 \times 10^{-6}$ and (e),(f),(g),(h) for $D = 5.0 \times 10^{-5}$.

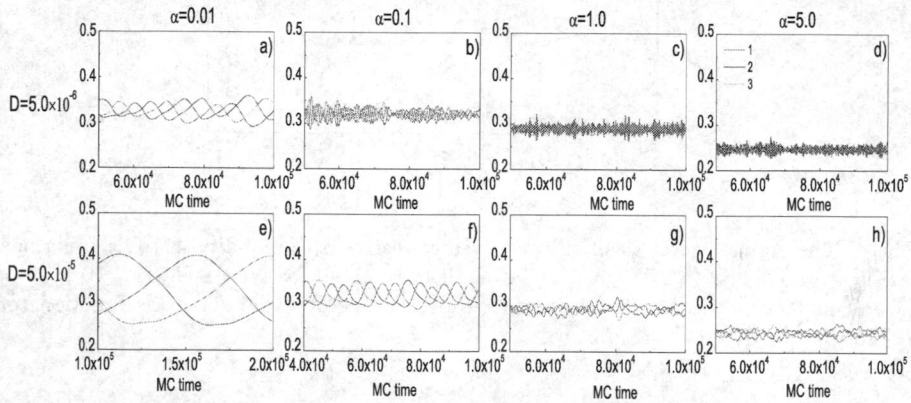

Fig. 3. Oscillation of species' percentage evolves with time for the same parameters of Fig. 2

global oscillation. Therefore, global oscillation can not be promoted at high predation rate, and the system comes into spatiotemporal chaos, as shown in Fig. 3(c) and Fig. 3(d) corresponds patterns in Fig. 2(c) and Fig. 2(d).

We systemically study the effect of predation rate on spiral waves. As shown in Fig. 4(a), for the same value of α, wavelengths of spiral waves vary with the mobility D in the form of $\lambda \sim D^\theta$. The exponent θ decrease with the increasing of α. It means that the behaviors of wavelengths varying with the mobility are universal at the same predation rate. $\lambda \sim \sqrt{D}$ at $\alpha = 1.0$, which confirms the results of Reichenbach at el [4]. When α deviates from 1.0, the exponent θ depart from $\frac{1}{2}$. At $\alpha = 1.0$ the system can be described by stochastic partial equation in which the mobility of individuals can be seen as diffusion. While

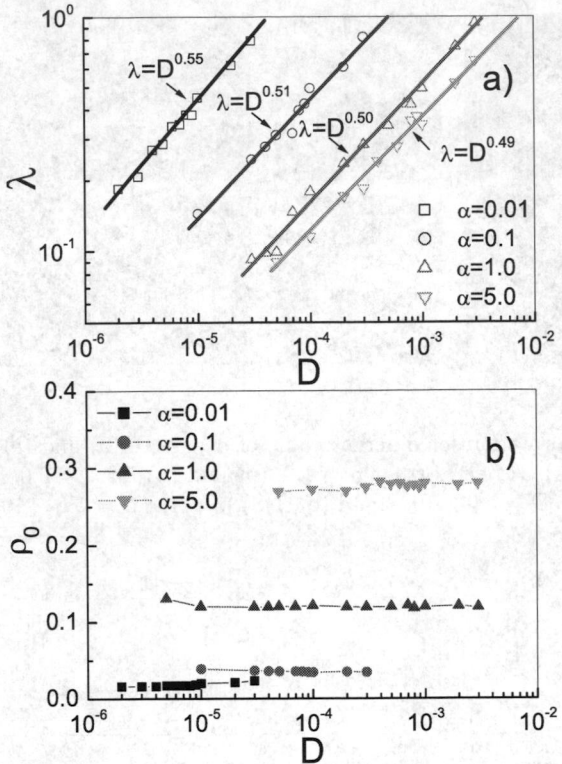

Fig. 4. The top: wavelength of spiral waves as function of mobility D. The functions satisfy $\lambda = D^{0.55}, \lambda = D^{0.51}, \lambda = D^{1/2}$, and $\lambda = D^{0.49}$ for $\alpha = 0.01$, $\alpha = 0.1$, $\alpha = 1.0$, $\alpha = 5.0$ respectively. The bottom: the percentage of vacancies ρ_0 as functions of mobility D for different value of α.

the system can be described by stochastic partial equation when α deviates from 1.0. To study why behaviors of wavelengths varying with the mobility are determined by predation rates, we define ρ_0 as percentage of vacancies which shows variation of resources. In the Fig. 4(b), one can find that ρ_0 keeps the same value ranging different mobility of individuals at a certain predation rate. It is confirmed that percentage of vacancies control kinds of university in spiral waves' formation. Because all individuals reproduce via vacancies, the number of vacancies determines properties of spiral waves' propagation.

4   Conclusion and Discussion

To conclusion, we study emergence of spiral waves affected by predation intensity. Spiral waves emerge when global oscillations are achieved. And the wavelength of spiral waves is satisfied power law with the mobility of individuals, $\lambda \sim D^{\theta}$. The value of θ is determined by the predation rate. It is confirmed that the behaviors of wavelength altering with mobility of individuals is universal at the same value of predation rate. In addition, predation density is related with vacancy resources in the systems. Our work provides basic understanding of effects of predation intensity on the spatial heterogeneity of species as well as pattern formation [15].

Acknowledgments. The authors would like to thank Tao Zhou, Xiao-Pu Han for their assistances in their preparing this paper. This work is supported by the National Basic Research Program of China (973 Program No. 2006CB705500), the National Natural Science Foundation of China (Grant Nos. 10635040, 10532060 and 10472116).

References

1. Lai, Y.-C., Liu, Y.-R.: Noise Promotes Species Diversity in Nature. Phys. Rev. Lett. 94, 038102 (2005)
2. Hanski, I.: Metapopulation Ecology. Oxford University Press, New York (1999)
3. Sayama, H., Kaufman, L., Bar-Yam, Y.: Symmetry Breaking and Coarsening in Spatially Distributed Evolutionary Processes including Sexual Reproduction and Disruptive Selection. Phys. Rev. E 62, 7065 (2000)
4. Reichenbach, T., Mobilia, M., Frey, E.: Mobility Promotes and Jeopardizes Biodiversity in Rock-Paper-Scissor Games. Nature 448, 1046–1049 (2007)
5. Reichenbach, T., Mobilia, M., Frey, E.: Noise and Correlations in a Spatial Population Model with Cyclic Competition. Phys. Rev. Lett. 99, 238105 (2007)
6. Huntley, J.W., Kowalewski, M.: Strong Coupling of Predation Intensity and Diversity in the Phanerozoic Fossil Record. Proc. Natl. Acad. Sci. U. S. A. 104, 15006–15010 (2007)
7. Gilg, O., Hanski, I., Sittler, B.: Cyclic Dynamics in a Simple Vertebrate Predator-Prey Community. Science 302, 866–868 (2001)
8. Sinervo, B., Lively, C.M.: The Rock-Paper-Scissors Game and the Evolution of Alternative Male Strategies. Nature 380, 240–243 (1996)
9. Kirkup, B.C., Riley, M.A.: Antibiotic-mediated Antagonism Leads to a Bacterial Game of Rock-Paper-Scissors in Vivo. Nature 428, 412–414 (2004)
10. Kerr, B., Riley, M.A., Feldman, M.W., Bohannan, B.J.M.: Local Dispersion Promotes Biodiversity in a Real Game of Rock-Paper-Scissors. Nature 418, 171–174 (2002)
11. Stanley, S.M.: An Ecological Theory for the Sudden Origin of Multicellular Life in the Late Precambrian. Proc. Natl. Acad. Sci. U. S. A. 70, 1486–1489 (1973)
12. Vermeij, G.J.: Evolution and Escalation: An Ecological History of Life. Princeton Univ. Press, Princeton (1987)
13. Gillespie, D.T.: A General Method for Numerically Simulating the Stochastic Time Evolution of Coupled Chemical Reactions. J. Comput. Phys. 22, 403–434 (1976)
14. Gillespie, D.T.: Exact Stochastic Simulation of Coupled Chemical Reactions. J. Phys. Chem. 81, 2340–2361 (1977)
15. Jiang, L.-L., Zhou, T., Huang, X., Wang, B.-H.: How Target Waves Emerge in Population Dynamics. arXiv:0807.4390 (2008)

Synchronization Stability of Coupled Near-Identical Oscillator Network

Jie Sun*, Erik M. Bollt, and Takashi Nishikawa

Department of Mathematics and Computer Science
Clarkson University
Potsdam, NY 13699-5815, USA
{sunj,bolltem,tnishikawa}@clarkson.edu

Abstract. To study the effect of parameter mismatch on the stability in a general fashion, we derive variational equations to analyze the stability of synchronization for coupled near-identical oscillators. We define master stability equations and associated master stability functions, which are independent of the network structure. In particular, we present several examples of coupled near-identical Lorenz systems configured in small networks (a ring graph and sequence networks) with a fixed parameter mismatch and a large Barabasi-Albert scale-free network with random parameter mismatch. We find that several different network architectures permit similar results despite various mismatch patterns. *abstract* environment.

Keywords: Synchronization Stability, Coupled Network Dynamics, Near-Identical Oscillators, Master Stability Functions.

1 Introduction

The phenomena of synchronization has been found in various aspects of nature and science [13]. Its applications have ranged widely, including from biology [4,10] to mathematical epidemiology [14], and chaotic oscillators [2], to communicational devices in engineering [3]. With the development of theory and application in complex networks [12], the study of synchronization between a large number of coupled dynamically driven oscillators has become a popular and exciting developing topic, see for example [11,15,16,18,19,20].

To model the coupled dynamics on a network (assumed to be unweighted and undirected and connected throughout this paper), we consider, for $i = 1, 2, ..., N$:

$$\dot{w}_i = f(w_i, \mu_i) - g \sum_{j=1}^{N} L_{ij} H(w_j) \tag{1}$$

where $w_i \in \Re^m$ is used to represent the dynamical variable on the ith unit; $f : \Re^m \times \Re^p \to \Re^m$ is the individual node dynamics (usually chaotic dynamics

* J.S. and E.M.B gratefully acknowledges the support of the ARO grant 51950-MA.

J. Zhou (Ed.): Complex 2009, Part I, LNICST 4, pp. 900–911, 2009.

for most interesting problems) and $\mu_i \in \Re^p$ is the corresponding parameter; $L \in \Re^{N \times N}$ is the graph Laplacian defined by $L_{ij} \equiv -1$ if there is an edge connecting node i and j and the diagonal element L_{ii} is defined to be the total number of edges incident to node i in the network; $H : \Re^m \to \Re^m$ is a uniform coupling function on the net; and $g \in \Re$ is the uniform coupling strength (usually > 0 for diffusive coupling). The whole system can be represented compactly with the use of Kronecker product:

$$\dot{w} - f(w, \mu) - y \cdot L \otimes H(w) \tag{2}$$

where $w = (w_1^T, w_2^T, ... w_N^T)^T$ is a column vector of all the dynamic variables, and likewise for μ and f; and $\otimes$ is the usual Kronecker product [1].

The majority of the theoretical work has been focused on *identical synchronization* where $\max_{i,j} \|w_i(t) - w_j(t)\| \to 0$ as $t \to \infty$, since it is in this situation the stability analysis can be carried forward by using the master stability functions proposed in the seminal work [8]. However, realistically it is impossible to find or construct a coupled dynamical system made up of exactly identical units, in which case identical synchronization rarely happens, but instead, a *nearly synchronous state* often takes place instead, where $\max_{i,j} \|w_i(t) - w_j(t)\| \le C$ for some small constant $C > 0$ as $t \to \infty$. We emphasize that Eq. (1) allows for nonidentical components due to parameter mismatch.

It is therefore important to analyze how systems such as Eq. (1) evolve, when parameter mismatch appears. In [17], similar variational equations were used to study the impact of parameter mismatch on the possible de-synchronization. To study the effect of parameter mismatch on the stability of synchronization, and more specifically, to find the distance bound C in terms of the given parameters in Eq. (1), we derive variational equations of system such as Eq. (1) and extend the master stability function approach to this case, which decomposes the problem into two parts that depend on the individual dynamics and network structure respectively.

2 Theory: Master Stability Equations and Functions

2.1 Derivation of Variational Equations

When the parameters μ_i of individual units in Eq. (1) are close to each other, centered around their mean $\bar{\mu}$, the coupled units w_i are found empirically to satisfy $\max_{i,j} \|w_i(t) - w_j(t)\| \le C$ for some $C > 0$ as $t \to \infty$, referred to as *near synchronization*[17]. When such near synchronization state exists, the *average trajectory* well represents the collective behavior of all the units. The average trajectory $\bar{w} \equiv \frac{1}{N} \sum_{i=1}^N w_i$ of Eq. (1) satisfies:

$$\dot{\bar{w}} = \frac{1}{N} \sum_{i=1}^N f(w_i, \mu_i) - g \sum_{i=1}^N \sum_{j=1}^N L_{ij} H(w_j)$$

$$= \frac{1}{N} \sum_{i=1}^N f(w_i, \mu_i), \tag{3}$$

since $\sum_{i=1}^{N} L_{ij} = 0$ by the definition of L. The variation $\eta_i \equiv w_i - \bar{w}$ of each individual unit is found to satisfy the following *variational equation*:

$$\dot{\eta}_i = D_w f(\bar{w}, \bar{\mu})\eta_i - g \sum_{j=1}^{N} L_{ij} DH(\bar{w})\eta_j + D_\mu f(\bar{w}, \bar{\mu})\delta\mu_i, \tag{4}$$

where $\bar{\mu} \equiv \sum_{i=1}^{N} \mu_i$ and $\delta\mu_i \equiv \mu_i - \bar{\mu}$; and D_w represents the derivative matrix with respect to w and likewise for D_μ and DH. The above variational equations can be represented in Kronecker product form as:

$$\dot{\eta} = \left[I_N \otimes D_w f - g \cdot L \otimes DH \right] \eta + \left[I_N \otimes D_\mu f \right] \delta\mu, \tag{5}$$

where η_i are stacked into a column vector η and likewise for $\delta\mu$.

2.2 Decomposition of Variational Equations

Since we are dealing with undirected graph, the associated L is symmetric and positive semi-definite, and thus L is diagonalizable: $L = P\Lambda P^T$[1], where Λ is the diagonal matrix whose ith diagonal entry λ_i is the ith eigenvalue of L (arranged in the order $\lambda_1 \le \lambda_2 \le ... \le \lambda_N$); and P is the orthogonal matrix whose ith column $v_i = (v_{1,i}, ..., v_{N,i})^T$ is the normalized eigenvector associated with λ_i, and all these v_i form an orthonormal basis of $\Re^N$. Note that because of $\sum_{i=1}^{N} L_{ij} = 0$, we always have $\lambda_1 = 0$ with $v_1 = \frac{1}{\sqrt{N}}(1, ..., 1)^T$; and since we have assumed that the graph is connected, the following holds: $\lambda_1 \equiv 0 < \lambda_2 \le ... \le \lambda_N$.

We may uncouple the variational equation Eq. (5) by making the change of variables

$$\zeta \equiv (P^T \otimes I_m)\eta, \tag{6}$$

or more explicitly, for each i,

$$\zeta_i \equiv v_{1,i}\eta_1 + v_{2,i}\eta_2 + ... + v_{N,i}\eta_N, \tag{7}$$

to yield:

$$\dot{\zeta} = \left[I_N \otimes D_w f - g \cdot \Lambda \otimes DH \right] \zeta + \left[P^T \otimes D_\mu f \right] \delta\mu. \tag{8}$$

where $\zeta \equiv (\zeta_1^T, ..., \zeta_N^T)^T$. Note that since $\sum_{i=1}^{N} \eta_i \equiv \sum_{i=1}^{N} (w_i - \bar{w}) = 0$, and $v_1 = \frac{1}{\sqrt{N}}(1, ..., 1)^T$, the following holds: $\zeta_1 \equiv 0$, by Eq. (7).

Note that since the transformation $\zeta \equiv (P^T \otimes I_m)\eta$ is an orthogonal transformation, $||\zeta|| \equiv ||\eta||$ with the choice of Euclidean norm. In other words, for $||.||$ being the usual Euclidean distance, we have:

$$\sum_{i=1}^{N} ||\zeta_i||^2 \equiv \sum_{i=1}^{N} ||\eta_i||^2. \tag{9}$$

The homogeneous part in Eq. (8) has block diagonal structure and we may write for each eigenmode ($i = 2, 3, ..., N$):

$$\dot{\zeta}_i = \Big[D_w f - g\lambda_i DH\Big]\zeta_i + D_\mu f \cdot \Big(\sum_{j=1}^{N} v_{j,i}\delta\mu_j\Big). \tag{10}$$

The vector $\sum_{j=1}^{N} v_{j,i}\delta\mu_j$ is the weighted average of parameter mismatch vectors, weighted by the eigenvector components associated with λ_i, and may be thought of as the length of projection of the parameter mismatch vector onto the eigenvector v_i.

2.3   Extended Master Stability Equations and Functions

The variational equation in the new coordinate system Eq. (10) suggests a generic approach [8] to study the stability of synchronization for a given network coupled dynamical system investigating on the effect of λ_i and $\sum_{j=1}^{N} v_{j,i}\delta\mu_j$ on the solution of Eq. (10). We define an *extended master stability equation* [1] for near identical coupled dynamical systems:

$$\dot{\xi} = \Big[D_w f - \alpha \cdot DH\Big] + D_\mu f \cdot \psi \tag{11}$$

where we have introduced two auxiliary parameters, $\alpha \in \Re$ and $\psi \in \Re^p$. This generic equation decomposes the stability problem into two separate parts: one that depends only on the individual dynamics and the coupling function, and one that depends only on the graph Laplacian and parameter mismatch. Note that the latter not only depends on the spectrum of L as in [8], but also on the combination of the eigenvectors and parameter mismatch vector.

Once the stability of Eq. (11) is determined as a function of α and ψ, the stability of any coupled network oscillators as described by Eq. (1), for the given f and H used in Eq. (11), can be found by simply setting

$$\alpha = g\lambda_i \tag{12}$$

and

$$\psi = \sum_{j=1}^{N} v_{j,i}\delta\mu_j \tag{13}$$

where $\lambda_i, v_{j,i}, \delta\mu_j$ can be obtained by the knowledge of the underlying network structure L and parameter mismatch pattern. Thus, we have reduced the stability analysis of the original mN-dimensional problem to that of an m-dimensional problem with one additional parameter, combined with an eigen-problem.

---

[1] Note here that to obtain the MSF based on Eq. (11), we need the actual average trajectory $\bar{w}(t)$, which can only be obtained by solving the whole system Eq. (1). However, we found that the trajectory solved from a single system $\dot{s} = f(s, \bar{\mu})$ could be used instead, resulting in good approximation of Ω. The supporting work for proving the shadowability of $\bar{w}$ by s will be reported elsewhere.

The associated *master stability function* (MSF) $\Omega(\alpha, \psi)$ of Eq. (11) is defined as:

$$\Omega(\alpha, \psi) \equiv \lim_{T \to \infty} \sqrt{\frac{1}{T} \int_0^T ||\xi(t)||^2 dt} \tag{14}$$

when the limit exists, where ξ is a solution of Eq. (11) for the given (α, ψ) pair.

For a given coupled oscillator network described by Eq. (1), we have the following equation, based on the generic MSF Ω:

$$\lim_{T \to \infty} \sqrt{\frac{1}{T} \int_0^T \sum_{i=1}^N ||w_i(t) - \bar{w}_i(t)||^2 dt}$$

$$\equiv \lim_{T \to \infty} \sqrt{\frac{1}{T} \int_0^T \sum_{i=1}^N ||\eta_i(t)||^2 dt} \equiv \lim_{T \to \infty} \sqrt{\frac{1}{T} \int_0^T \sum_{i=2}^N ||\zeta_i(t)||^2 dt}$$

$$= \sqrt{\sum_{i=2}^N \Omega^2(g\lambda_i, \psi_i)} \tag{15}$$

where λ_i are the eigenvalues of the graph Laplacian and ψ_i is obtained through Eq. (13). Thus, once the MSF for the dynamics f and coupling function H has been computed, it can be used to compute the asymptotic total distance from single units to the average trajectory: $\langle \sum_{i=1}^N ||w_i(t) - \bar{w}_i(t)||^2 \rangle$ [2] for any coupled oscillator network by summing up the corresponding $\Omega^2(g\lambda_i, \psi_i)$ and take the square root.

In Fig. 1 we plot the MSF for f being Lorenz equations:

$$\dot{x} = \sigma(y - x)$$
$$\dot{y} = x(r - z) - y$$
$$\dot{z} = xy - \beta z \tag{16}$$

as individual dynamics in Eq. (1) ($w = [x, y, z]^T$). The parameters are chosen as: $\sigma = 10, \beta = \frac{8}{3}$, and r is allowed to be adjustable. Here r serves as the μ in Eq. (1). The coupling function H is taken as: $H(w) = w$, i.e., an identity matrix operator.

2.4   Conditions for Stable Near Synchronization

For near synchronization to appear in the presence of parameter mismatch, it is required that the system described by Eq. (1) in the absence of parameter mismatch undergoes stable identical synchronization, which can be checked by

---

[2] Notation $\langle a(t) \rangle$ is introduced and used throughout, to represent the *asymptotic root mean square*: $\sqrt{\lim_{T \to \infty} \frac{1}{T} \int_0^T a(t) dt}$ for the trajectory $a(t)$.

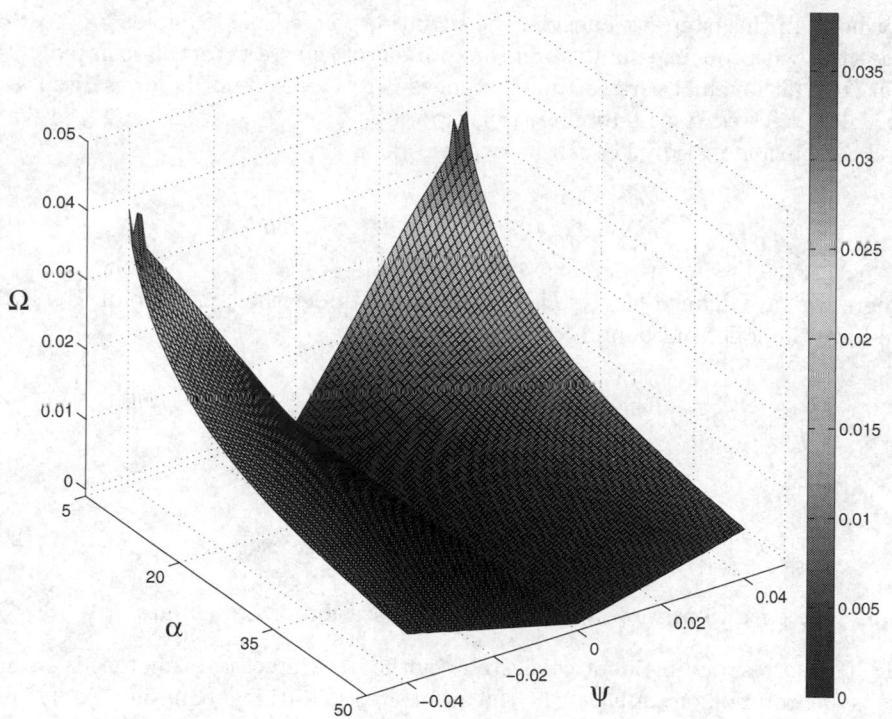

Fig. 1. (Color online) MSF for Lorenz system Eq. (16), with $\sigma = 10, \beta = \frac{8}{3}$, adjustable parameter being r, and coupling function H being an identity matrix operator. The domain shown here is for α from 5 to 50 and ψ from -0.04 to 0.04, while the actual valid domain of MSF could be as large as the stability region of the identical Lorenz system (in this case is the region $\alpha > \lambda_0$ where λ_0 is the largest lyapunov exponent of the original Lorenz system (≈ 1)).

using MSF [8]. In this case, the largest Lyapunov exponent of the synchronous trajectory associated with the homogeneous variational equation:

$$\dot{\xi} = \Big[D_w f - \alpha \cdot DH \Big] \xi \tag{17}$$

is negative, and its solution can be written as $\xi^*(t) = \Phi(t, 0)\xi(0)$ where $\Phi(t, \tau)$ is the fundamental transition matrix [3], satisfying

$$\|\Phi(t, \tau)\| \leq \gamma e^{-\lambda(t-\tau)} \tag{18}$$

for $t \geq \tau$ and some finite positive constants γ and λ.

Note that the transition to loss of stability at certain time instances can occur due to the embedded periodic orbits [7,17], in which case the above inequality will

---

[3] This transition matrix, as a function of two time variables t and τ, can be obtained by the *Peano-Baker series* as long as $D_w f - \alpha \cdot DH$ is continuous. See [5], Ch.3.

not hold. In this paper we consider the situation where Eq. (18) holds for most of the time, with λ being the Lyapunov exponent of the trajectory associated with Eq. (17), although at certain time instances Eq. (18) need not hold, as discussed in [7,17], referred to as bubbling transition [6].

The solution to Eq. (11) can be expressed as:

$$\xi(t) = \Phi(t,0)\xi(0) + \int_0^t \Phi(t,\tau)b(\tau)d\tau, \tag{19}$$

where we have defined $b(\tau) \equiv D_\mu f(s(\tau),\bar{\mu}) \cdot \psi$. Under the condition of Eq. (18), we have the following bound for $\xi(t)$:

$$\|\xi(t)\| \leq \|\Phi(t,0)\| \cdot \|\xi(0)\| + \int_0^t \|\Phi(t,\tau)\|d\tau \cdot \sup_t \|b(t)\|$$

$$\leq \gamma e^{-\lambda t}\|\xi(0)\| + \frac{\gamma}{\lambda}(1 - e^{-\lambda t})\sup_t \|b(t)\|$$

$$\rightarrow \frac{\gamma}{\lambda}\sup_t \|b(t)\| \quad \text{as } t \rightarrow \infty. \tag{20}$$

Thus, the conditions for stable near synchronization of Eq. (1) are:

1. The corresponding identical system (without parameter mismatch) is stably synchronized, or equivalently, the associated variational equation Eq. (17) is exponentially stable;
2. The inhomogeneous part $b(\tau) \equiv D_\mu f(s(\tau),\bar{\mu}) \cdot \psi$ in Eq. (11) is bounded.

These conditions are sufficient to guarantee the boundness of pairwise distance between any two units, so that near synchronous state is stable.

Eq. (18) and Eq. (19) also allow us to analyze quantitatively the magnitude of asymptotic error of a near-identical system such as Eq. (1). For all other variables being the same, if the magnitude of parameter mismatch is scaled by a factor k, then the corresponding variation will become:

$$\widetilde{\xi}(t) = \Phi(t,0)\xi(0) + k \cdot \int_0^t \Phi(t,\tau)b(\tau)d\tau \tag{21}$$

where $\xi(t)$ denotes the variation of the original unscaled near-identical system, which follows Eq. (19). The first term of both Eq. (19) and Eq. (21) goes to zero according to Eq. (18), so that asymptotically the following holds: $\widetilde{\xi}(t) = k \cdot \xi(t)$, i.e., the variation is scaled by the same factor correspondingly.

3   Examples of Application

3.1   Methodology

When the units coupling through the network are known exactly, meaning that the parameter of each unit is known, then from Eq. (12) and Eq. (13) we may use the Ω obtained from MSF at the corresponding (α, ψ) pairs. In Sec. 3.2 and Sec. 3.3 we illustrate this with examples of small networks.

On the other hand, for large networks, in the case that parameters of individual units are not known exactly, but follow a Gaussian distribution: $\delta\mu_i \sim N(\bar{\mu}, \epsilon^2)$, then in Eq. (10) we have:

$$\sum_{j=1}^{N} v_{j,i}\delta\mu_j \sim N(\bar{\mu}, \sum_{j=1}^{N} v_{j,i}^2\epsilon^2)$$
$$\sim N(\bar{\mu}, \epsilon^2) \tag{22}$$

assuming the $\delta\mu_i$ are identical and independent. The standard deviation ϵ may be used, as an expected bound for ψ in Eq. (13), to compute an *expected MSF* to predict the possible variation of individual units to the average trajectory. In Sec. 3.3 a scale-free network with $N = 500$ vertices is used to illustrate.

In all the examples, the individual dynamics is the Lorenz equation Eq. (16), with parameters $\sigma = 10, \beta = \frac{8}{3}$, and $r_i = 28 + \delta r_i$ where δr_i is the parameter mismatch on unit i. The coupling function is chosen as $H(w) = w$ with coupling strength g specified differently in each example. The variation of individual units to the average trajectory $\langle \sum_{i=1}^{N} ||\eta_i(t)||^2 \rangle$ is approximated by $T = 200$ with equally time spacing $\tau = 0.01$.

3.2 Example: Ring Graph

We consider a small and simple graph to illustrate. The graph as well as three different patterns of parameter mismatch are shown in Fig. 2. In Fig. 3 we show the actual variation on individual units and that by MSF.

The MSF predicts well the actual variations found in this near-identical oscillator network, in all three cases. Furthermore, the way parameter mismatch are distributed in the graph is relevant, as a consequence of Eq. (10). From left to right in Fig. (2), the parameter mismatch is distributed more heterogeneously, resulting in larger variation along the near synchronous trajectory.

3.3 Example: Sequence Networks

Sequence networks [21] are a special class of networks that can be encoded by the so called *creation sequence*. In Fig. 4 three different sequence networks of the creation sequence (A, A, A, B, B, B) under different connection rules are shown. Interestingly, despite the fact that the structure of these networks are different, the variation of individual units to the average trajectory are the same, under the mismatch pattern $[-\epsilon, -\epsilon, -\epsilon, +\epsilon, +\epsilon, +\epsilon]$, see Fig. 5.

Study on the eigenvector structure on these networks shows that this comes from the fact that the eigenvectors of all these three networks are the same, and more importantly, the parameter mismatch vector $[-\epsilon, -\epsilon, -\epsilon, +\epsilon, +\epsilon, +\epsilon]$ is parallel to one of the eigenvectors, corresponding to the same eigenvalue $\lambda = 6$ in all three cases. Thus, the only active error mode in the eigenvector basis are the same for all three networks, resulting in the same variations.

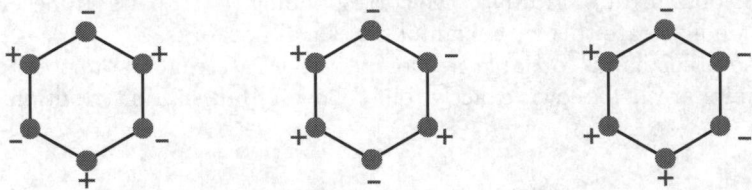

Fig. 2. (Color online) Ring graph (red circles represent vertices and black lines represent edges) with specific parameter mismatch on each unit. The magnitude of parameter mismatch on each unit is assumed to be the same, ϵ. The plus/minus sign on a vertex represents the corresponding sign of mismatch on that unit, $+$ for $+\epsilon$ and $-$ for $-\epsilon$. So the left graph has the parameter mismatch pattern (starting from the top unit): $[-\epsilon, +\epsilon, -\epsilon, +\epsilon, -\epsilon, +\epsilon]$, the middle graph has the pattern $[-\epsilon, -\epsilon, +\epsilon, -\epsilon, +\epsilon, +\epsilon]$, and the right graph has the pattern $[-\epsilon, -\epsilon, -\epsilon, +\epsilon, +\epsilon, +\epsilon]$.

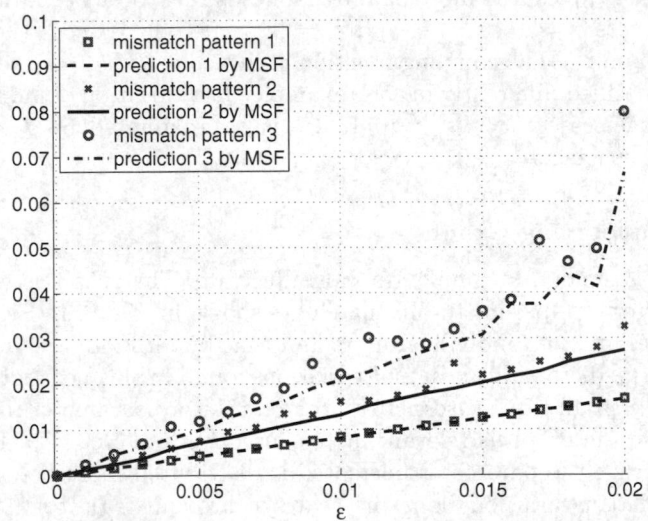

Fig. 3. (Color online) Validating MSF on a ring graph. Here the coupling strength is $g = 5$. The units are coupled through a ring graph, with specific parameter mismatch patterns as shown in Fig. 2. The vertical axis represents the average variation at each given ϵ. Blue squares, crosses, and circles are obtained from actual time series, computed through $\langle \sum_{i=1}^{N} \|\eta_i(t)\|^2 \rangle$ where $\eta_i(t)$ is the distance from unit i to the average trajectory at time t. Black lines (dashed, solid, and dotted) are theoretical prediction $\sqrt{\sum_{i=1}^{N} \Omega^2(\alpha_i, \psi_i)}$ from MSF at (α_i, ψ_i) paris, where (α_i, ψ) are computed according to Eq. (12) and Eq. (13).

3.4   Example: Scale-Free Networks

The synchronization stability of a large network, with the knowledge of the probability distribution of parameters, is another interesting problem. To show how an expected MSF will apply, we use a scale-free network as an example. The

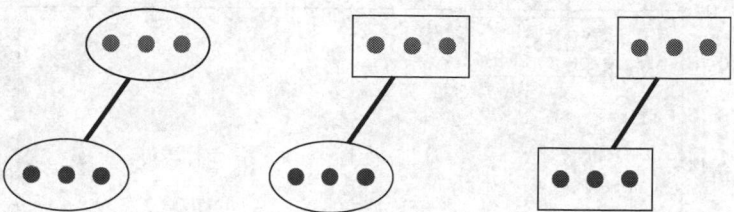

Fig. 4. (Color online) 2-letter sequence nets [21] consisting of 6 vertices and 2 layers (red and blue circles represent vertices and black lines represent edges) obtained from the same creation sequence (A, A, A, B, B, B) under three different rules, on the left the connection rule is $B \to A$, meaning that whenever a vertex of type B is added into the current net, it connects to all previous vertices of type A, thus a bipartite complete graph is created based on this sequence; on the middle the connection rule is $B \to A, B$, resulting in a threshold graph; while on the right the rule $A \to A, B; B \to A, B$ is applied to yield a complete graph. Ovals and boxes are used to highlight the layer structure: vertices within an oval do not have connections, while vertices within a box connect to each other; a thick edge going from one group to the other connects every vertex in one group to all the vertices in the other group. The parameter mismatch pattern here is prescribed to coincide with the type of vertices, which is, for the given ϵ: $[-\epsilon, -\epsilon, -\epsilon, +\epsilon, +\epsilon, +\epsilon]$.

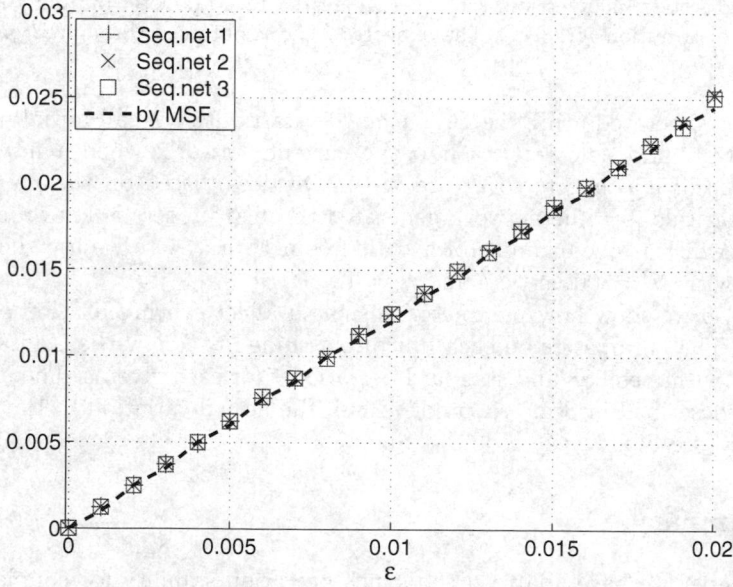

Fig. 5. (Color online) Validating MSF on sequence networks. Here the coupling strength is $g = 2$. The parameter mismatch pattern is shown in Fig. 4. The vertical axis represents the average variation at each given ϵ. (Blue) markers represent variations obtained from actual time series, and (black) dashed lines is the prediction obtained by MSF. Here the MSF line for all three networks are the same.

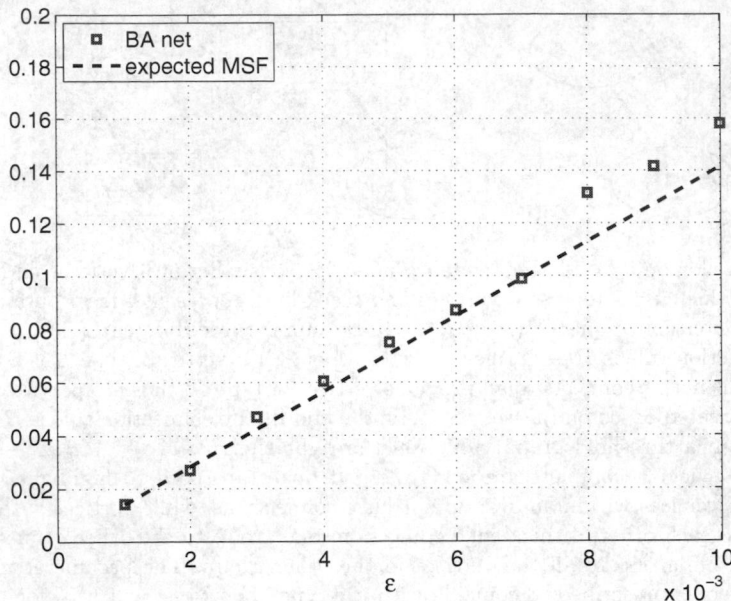

Fig. 6. (Color online) Validating MSF prediction for a BA scale-free network. The network consists of 500 vertices with average degree 12. Parameter on unit i follows a Gaussian distribution $N(28, \epsilon^2)$. The expected MSF is obtained through $\Omega(g\lambda_i, \epsilon)$.

network is generated using the *BA model* [9]: start with a small initial network, consecutively add new vertices into the current network; when a new vertex is introduced, it connects to m preexisting vertices, based on the *preferential attachment* rule [9]. The network generated through process is known as a *BA network*, which is one example of a scale-free network. Here we use such a BA network with $N = 500$ vertices and $m = 12$.

In Fig. 6 we show how parameter mismatch affect synchronization on a BA network. The parameters on each unit are assumed to follow the Gaussian distribution with mean 28 and standard deviation ϵ for each given ϵ. The expected MSF, as described in Sec. 3.1, predicts well the actual variation to the average trajectory, see Fig. 6.

4   Summary

In this paper we have analyzed the synchronization stability for coupled near-identical oscillator networks such as Eq. 1. We show that the master stability equations and functions can be extended to this general case as to analyze the synchronization stability. The variational equations in the near-identical oscillator case highlight the relevance of eigenvectors as well as eigenvalues on the effect of parameter mismatch, which indicates the importance of knowledge of the detailed network structure in designing dynamical systems that are more reliable.

References

1. Lancaster, P., Tismenetsky, M.: The Theory of Matrices with Applications, 2nd edn. Academic Press, London (1985)
2. Pecora, L.M., Carroll, T.L.: Synchronization in Chaotic Systems. Phys. Rev. Lett. 64, 821 (1990)
3. Cuomo, K.M., Oppenheim, A.V.O.: Circuit implementation of synchronized chaos with applications to communications. Phys. Rev. Lett. 71, 65 (1993)
4. Strogatz, S.H., Stewart, I.: Coupled oscillators and biological synchronization. Scientific American 269, 102 (1993)
5. Rugh, W.J.: Linear System Theory, 2nd edn. Prentice Hall, New Jersey (1996)
6. Venkataramani, S.C., Hunt, B.R., Ott, E.: Bubbling Transition. Phys. Rev. E 54, 1346 (1996)
7. Venkataramani, S.C., Hunt, B.R., Ott, E., Gauthier, D.J., Bienfang, J.C.: Transition to Bubbling of Chaotic Systems. Phys. Rev. Lett. 77, 5361 (1996)
8. Peroca, L.M., Carroll, T.L.: Master Stability Functions for Synchronized Coupled Systems. Phys. Rev. Lett. 80, 2109 (1998)
9. Barabasi, A.-L., Albert, R.: Emergence of Scaling in Random Networks. Science 286, 509 (1999)
10. Buono, P.L., Golubitsky, M.: Models of central pattern generators for quadruped locomotion: I. primary gaits. J. Math. Biol. 42, 291 (2001)
11. Bocccaletti, S., Kurths, J., Osipov, G., Valladares, D.L., Zhou, C.S.: The synchronization of chaotic systems. Phys. Rep. 366, 1 (2002)
12. Albert, R., Barabasi, A.-L.: Statistical mechanics of complex networks. Rev. Mod. Phys. 74, 47–97 (2002)
13. Strogatz, S.H.: Sync: The Emerging Science of Spontaneous Order. Hyperion, New York (2003)
14. He, D., Stone, L.: Spatio-temporal synchronization of recurrent epidemics. Proc. R. Soc. Lond. B 270, 1519 (2003)
15. Nishikawa, T., Motter, A.E., Lai, Y.-C., Hoppensteadt, F.C.: Heterogeneity in Oscillator Networks: Are Smaller Worlds Easier to Synchronize? Phys. Rev. Lett. 91, 014101 (2003)
16. Li, X., Chen, G.: Synchronization and desynchronization of complex dynamical networks: an engineering viewpoint. IEEE Trans. on Circ. Syst. 50, 1381 (2003)
17. Restrepo, J.G., Ott, E., Hunt, B.R.: Spatial patterns of desynchronization bursts in networks. Phys. Rev. E 69, 066215 (2004)
18. Skufca, J.D., Bollt, E.M.: Communication and Synchronization in Disconnected Networks with Dynamic Topology Moving Neighborhood Networks. Mathematical Biosciences and Engineering 1, 347 (2004)
19. Stilwell, D.J., Bollt, E.M., Roberson, D.G.: Sufficient Conditions for Fast Switching Synchronization in Time-Varying Network Topologies. SIAM J. Applied Dynamical Systems 5, 140 (2006)
20. Arenas, A., Diaz-Guilera, A., Perez-Vicente, C.J.: Synchronization Reveals Topological Scales in Complex Networks. Phys. Rev. Lett. 96, 114102 (2006)
21. Sun, J., Nishikawa, T., ben-Avraham, D.: Sequence Nets. Phys. Rev. E 78, 026104 (2008)

Synchronization of Complex Networks with Time-Varying Coupling Delay via Impulsive Control

Yang Dai[1], Yunze Cai[1], and Xiaoming Xu[1,2,3]

[1] Department of Automation, Shanghai Jiao Tong University, P.R. China
daiyang1980@gmail.com
[2] University of Shanghai For Science and Technology, P.R. China
[3] Shanghai Academy of Systems Science, P.R. China

Abstract. Impulsive control and exponential synchronization analysis of a class of complex networks with time-varying coupling delay is investigated in this paper. Our aim is to enhance the synchronizability of the complex networks by applying impulsive control. By introducing a comparison system and estimating the corresponding Cauchy matrix sufficient conditions on global exponential synchronization are derived. An impulsive controller is explicitly designed not only to achieve synchronized dynamics for the complex networks, but simultaneously to ensure the states of synchronous error converging with a given decay rate. A numerical example is presented to illustrate the theoretical results and proposed controller design procedure.

Keywords: Complex Networks, Impulsive Control, Synchronization, Delay.

1 Introduction

In recent years, complex networks have been extensively studied by researchers from disciplines as diverse as physics, chemistry, biology, social science, telecommunication and engineering [1,2,3,4] since the remarkable small-world networks [5] and scale-free networks [6] were put forward. With the dynamical units regarded as nodes and the interplay between them expressed by the links of nodes, network topology structure provides a powerful metaphor for describing sophisticated collaborative dynamics of many practical systems in essence.

As a collective dynamical behavior, synchronization of complex networks has been actively investigated in the literature [7,8,9,10,11,12,13]. There are different types of definition for synchronization, including complete or identical synchronization, phase synchronization, lag synchronization, generalized synchronization and anticipating synchronization. The emergence of synchronous phenomenon has a close relation with network attributes, such as network topology structure, average path length, degree distribution, betweenness and coupling delays [11,12,13,14]; hence much effort has been made to assess and compare synchronization propensity for different complex networks. On the other

J. Zhou (Ed.): Complex 2009, Part I, LNICST 4, pp. 912–923, 2009.
© ICST Institute for Computer Sciences, Social Informatics and Telecommunications Engineering 2009

hand, some control strategies have been developed to achieve the synchronized dynamics for complex networks. In the pinning control technique [15], a small fraction of network nodes are chosen to carry out local feedback control strategy and ultimately the dynamics of the whole network was pinned to its equilibrium state. Some other control algorithms, such as adaptive control [16], state feedback control [17] and hybrid control [18], are also proposed as strategies to synchronize the interacting dynamical nodes toward an identical orbit.

As is well known, impulsive control is characterized by the abrupt changes in the system dynamics at certain instants, which is an advantage in reducing the amount of information transmission and improving the security and robustness against disturbances especially in telecommunication network and power grid [19]. In some cases, the scheme of impulsive control cannot be substituted by continuous control. For example in a financial system, with the amount of money in a market and saving rates of a central bank serving as two state variables, a monetary control policy is usually implemented at some particular instants. Considering these traits, impulsive control has been introduced into complex networks to achieve the synchronous dynamics [20,21,22]. Time delays are ubiquitous in biological and physical networks and often their variation is too significant to be ignored owing to the finite speeds of transmission [23]. However, most available literature on impulsive synchronization only take account of constant delay case for simplicity.

In this note, we investigate the impulsive exponential synchronization of complex networks with time-varying coupling delays. A dynamical network model is given and reformulated into the direct product form. By some transformations, the impulsive synchronization problem of the complex network is converted equally into the stability problem of the impulsive control system with time delays. Further, some sufficient conditions of impulsive synchronization are derived based on the comparison method and Cauchy matrix estimating. The impulsive controller is delicately designed, which can ensure the network dynamics exponentially synchronizing with a given decay rate. It is noteworthy that the purpose of introducing impulsive control is not to compete with the other control schemes, but to provide a new viewpoint to deal with the specific synchronization problem.

The rest of the paper is organized as follows. In section 2, complex network model is presented, and some relevant definitions and lemmas are presented. Section 3 deals with the synchronization analysis and gives the controller design procedure in detail. In Sectionx 4, theoretical results are verified by a numerical example to illustrate their effectiveness. Conclusions are drawn in Section 5.

Notation: The notation used throughout the paper is fairly standard. Let $\mathbb{N}$ be the set of natural number; $\mathbb{R}^n$ denotes the n-dimensional Euclidean space; Let $A \otimes B$ be the direct product of the matrices A and B. For a matrix A, the largest eigenvalue and the smallest one are denoted by $\lambda_{max}(A)$ and $\lambda_{min}(A)$, respectively; the induced matrix norm and matrix measure are

$$\|A\| = \sqrt{\lambda_{max}(A^T A)}, \quad \mu(A) = \frac{1}{2}[\lambda_{max}(A^T + A)].$$

2   Problem Formulation and Preliminaries

Consider the ensemble of N identical diffusively coupled nodes, with each one being an n-dimensional dynamical system. The proposed weighted dynamical network is described by

$$\dot{x}_i(t) = Ax_i(t) + f(x_i(t)) + \sum_{j=1}^{N} G_{ij}\Gamma x_j(t - \tau(t)), \qquad i = 1, \ldots, N \qquad (1)$$

and the initial condition function

$$x_i(\theta) = \varphi_i(\theta), \ \theta \in [-\bar{\tau}, 0].$$

where $t \in \mathbb{R}$ is the continuous time variable, $x_i \in \mathbb{R}^n$ is the state variable of node i, f is continuously differentiable map, $\tau(t)$ is bounded time-varying delay with $0 < \tau \leq \bar{\tau}$, A is the constant matrix, and Γ is a constant inner coupling matrix of the nodes of network at time t, $G = (G_{ij}) \in \mathbb{R}^{N \times N}$ is the outer-coupling matrix combining both configuration and weights of the entire networks. Accordingly, this is a general weighted complex network [9].

We design an impulsive control law as below to globally exponentially synchronize the complex dynamical network (1)

$$U_i(k, x_i(t_k)) = D_{ik}x_i(t_k), \qquad i = 1, \ldots, N, \quad k \in \mathbb{N} \qquad (2)$$

The complex networks under impulsive control is obtained

$$\begin{cases} \dot{x}_i(t) = Ax_i(t) + f(x_i(t)) + \sum_{j=1}^{N} G_{ij}\Gamma x_j(t - \tau(t)), & t \in (t_k, t_{k+1}] \\ \triangle x_i(t) = x_i(t_k^+) - x_i(t_k^-) = D_{ik}x_i(t_k^-), & k \in \mathbb{N} \\ x_i(\theta) = \varphi_i(\theta), & \theta \in [-\bar{\tau}, 0] \end{cases} \qquad (3)$$

It is assumed that $x_i(t)$ is left continuous at $t = t_k$, that is, $x_i(t_k) = x_i(t_k^-)$. Therefore, the solutions to (3) are piecewise left-hand continuous functions with discontinuities only at t_k.

Let $\mathbf{x}(t) = [x_1^T(t), \ldots, x_N^T(t)]^T$, $\mathbf{F}(\mathbf{x}(t)) = [f^T(x_1(t)), \ldots, f^T(x_N(t))]^T$, $\mathbf{D}_k = \text{diag}\{D_{1k}, \ldots, D_{Nk}\}$, then the coupled dynamical network (3) can be rewritten in the compact form

$$\begin{cases} \dot{\mathbf{x}}(t) = (I_N \otimes A)\mathbf{x}(t) + \mathbf{F}(\mathbf{x}(t)) + (G \otimes \Gamma)\mathbf{x}(t - \tau(t)), & t \in (t_k, t_{k+1}] \\ \triangle \mathbf{x}(t) = \mathbf{x}(t_k^+) - \mathbf{x}(t_k) = \mathbf{D}_k\mathbf{x}(t_k), & k \in \mathbb{N} \\ \mathbf{x}(\theta) = \varphi(\theta), & \theta \in [-\bar{\tau}, 0]. \end{cases} \qquad (4)$$

Our aim is to find a set of sufficient conditions on the impulsive control time sequence t_k and the constant control gain matrices $\mathbf{D}_k$, such that the closed loop complex network (4) is exponentially synchronized with a given decay rate.

In the sequel, we present some definitions and useful lemmas required throughout this paper.

Definition 1 ([13]). The hyperplane

$$\mathcal{S} = \{(x_1^T, \ldots, x_N^T)^T \in \mathbb{R}^{n \times N}; x_i = x_j; i, j = 1, \ldots, N.\} \tag{5}$$

is said to be synchronization manifold of dynamical complex network (1).

Let $S(t) = (s^T(t), \ldots, s^T(t))^T \in \mathcal{S}$ be the synchronized state, where $s(t)$ is the solution to $\dot{s}(t) = As(t) + f(s(t))$, which may be an equilibrium, aperiodic trajectory or a chaotic attractor of the uncoupled dynamical behavior of each node.

Definition 2. The complex network (1) is said to be globally exponentially synchronized with respect to the state $S(t)$ via impulsive controller $\mathbf{U}(k, \mathbf{x}(t_k)) = \mathbf{D}_k \mathbf{x}(t_k)$ $(k \in \mathbb{N})$, if for any initial function $\varphi(\theta)$, there exist the constant $\lambda > 0$ and $M > 0$ such that the following exponential estimates hold:

$$\|\mathbf{x}(t) - S(t)\| \leq M e^{-\lambda t} \|\varphi(\theta)\|, \tag{6}$$

where λ is called the decay rate of exponential synchronization.

Lemma 1 ([24]). Let $P \in \mathbb{R}^{n \times n}$ be a symmetric positive definite matrix and $P = Q^T Q$. For any $x, y \in \mathbb{R}^n$ and $A \in \mathbb{R}^{n \times n}$, then

1. $x^T A^T P A x \leq \|QAQ^{-1}\|^2 x^T P x$
2. $x^T (A^T P + P A) x \leq 2\mu(QAQ^{-1}) x^T P x$
3. $|x^T P y| \leq \sqrt{x^T P x} \sqrt{y^T P y}$

Lemma 2 ([25]). For the matrices A, B and the scalar α, by the definition of direct product, the following properties are satisfied:

1. $(\alpha A) \otimes B = A \otimes (\alpha B)$
2. $(A + B) \otimes C = (A \otimes C) + (B \otimes C)$
3. $(A \otimes B)(C \otimes D) = (AC) \otimes (BD)$

3   Main Results

In this section, the global exponential synchronization conditions of complex dynamical networks are presented. Moreover, the controller design procedure is given to ensure the network dynamics converging with a given decay rate.

3.1   Synchronization Analysis

Define the network synchronization error of node i with respect to $s(t)$ as $e_i(t) = x_i(t) - s(t)$ and $\mathbf{e}(t) = [e_1^T(t), \ldots, e_N^T(t)]^T$, then the error dynamical network is denoted by:

$$\begin{cases} \dot{\mathbf{e}}(t) = (I_N \otimes A)\mathbf{e}(t) + \mathbf{F}(\mathbf{e}(t)) + (G \otimes \Gamma)\mathbf{e}(t - \tau(t)), & t \in (t_{k-1}, t_k] \\ \mathbf{e}(t_k^+) = \mathbf{x}(t_k^+) - S(t_k^+) = (E + \mathbf{D}_k)\mathbf{e}(t_k), & k \in \mathbb{N} \end{cases} \tag{7}$$

where $\mathbf{F}(e(t)) = [f^T(x_1) - f^T(s), \ldots, f^T(x_1) - f^T(s)]^T$, I_N is a $N \times N$ dimension identity matrix and E is a $Nn \times Nn$ dimension one. It may be seen that the exponential stability of the impulsive time delay system (7) is equivalent to the impulsive synchronization of the complex network (1) with respect to S(t).

Assumption 1 (A1). Suppose that there exists a nonnegative constant L_i ($i = 1, \ldots, N$) such that

$$\|f(x_i, t) - f(s, t)\| \leq L_i \|x_i(t) - s(t)\| = L_i \|e_i(t)\|, \tag{8}$$

We may obtain that

$$\|\mathbf{F}(e(t))\| \leq \max\{L_1, \ldots, L_N\} \|e(t)\| = \mathbf{L}\|e(t)\|, \tag{9}$$

where $\mathbf{L} = \max\{L_1, \ldots, L_N\}$ is a positive constant.

Theorem 1. Under A1, if there exist $0 < \rho = \sup_{k \in \mathbb{N}}\{t_k - t_{k-1}\} < \infty$ and a nonsingular matrix $Q \in \mathbb{R}^{Nn \times Nn}$ such that

$$-p > q \tag{10}$$

$$\|E + Q\mathbf{D}_k Q^{-1}\| \leq \beta, \ 0 < \beta < 1 \tag{11}$$

where

$$p = \frac{2 \ln \beta}{\rho} + 2\mu[Q(I_N \otimes A)Q^{-1}] + 2\mathbf{L}\frac{\lambda_{max}(Q^T Q)}{\lambda_{min}(Q^T Q)} + 1,$$
$$q = \beta^{-2}\|Q(G \otimes \Gamma)Q^{-1}\|^2,$$

then the complex dynamical network (1) is globally synchronized via impulsive control (2) in the following sense:

$$\|\mathbf{x}(t) - S(t)\| \leq Me^{-\frac{\lambda}{2}t}, \quad t \geq 0 \tag{12}$$

where $M = \frac{1}{\beta}\sqrt{\frac{\lambda_{max}(Q^T Q)}{\lambda_{min}(Q^T Q)}} \sup_{-\bar{\tau} \leq \theta \leq 0}\{\|\phi(\theta)\|\}$, $\lambda > 0$ is the solution of

$$\lambda + p + qe^{\lambda \bar{\tau}} = 0. \tag{13}$$

Proof. Choose a Lyapunov function as follows:

$$V(t) = \mathbf{e}^T(t)P\mathbf{e}(t), \tag{14}$$

where P be a symmetric positive matrix and $P = Q^T Q$.

The derivative of V along the trajectories of the error dynamical network (7) is given by

$$\dot{V}(t) = \mathbf{e}^T(t)\big[P(I_N \otimes A) + (I_N \otimes A)^T P\big]\mathbf{e}(t) + 2\mathbf{e}^T(t)P\mathbf{F}(\mathbf{e}(t))$$
$$+ 2\mathbf{e}^T(t)P(G \otimes \Gamma)\mathbf{e}(t - \tau(t))$$
$$\leq 2\mu\big[Q(I_N \otimes A)Q^{-1}\big]\mathbf{e}^T(t)P\mathbf{e}(t) + 2\mathbf{L}\lambda_{max}P\|\mathbf{e}(t)\|^2 + 2\sqrt{e^T(t)Pe(t)}$$
$$\times \sqrt{e^T(t - \tau(t))(G \otimes \Gamma)^T P(G \otimes \Gamma)e(t - \tau(t))}$$
$$\leq 2\mu\big[Q(I_N \otimes A)Q^{-1}\big]\mathbf{e}^T(t)P\mathbf{e}(t) + 2\mathbf{L}\frac{\lambda_{max}(P)}{\lambda_{min}(P)}\mathbf{e}^T(t)P\mathbf{e}(t) + \mathbf{e}^T(t)P\mathbf{e}(t)$$
$$+ \|Q(G \otimes \Gamma)Q^{-1}\|^2\mathbf{e}^T(t - \tau(t))P\mathbf{e}(t - \tau(t))$$
$$\leq \Big(2\mu\big[Q(I_N \otimes A)Q^{-1}\big] + 2\mathbf{L}\frac{\lambda_{max}(P)}{\lambda_{min}(P)} + 1\Big)V(t)$$
$$+ \|Q(G \otimes \Gamma)Q^{-1}\|^2 V(t - \tau(t)), \qquad t \in (t_{k-1}, t_k]$$

And also
$$V(t_k^+) = \mathbf{e}^T(t_k)(E + \mathbf{D}_k)^T P(E + \mathbf{D}_k)\mathbf{e}(t_k)$$
$$\leq \|E + Q\mathbf{D}_kQ^{-1}\|^2 V(t_k)$$
$$\leq \beta^2 V(t_k), \qquad k \in \mathbb{N}$$

Let $\epsilon > 0$ be an arbitrary constant. Construct the comparison system as follows:

$$\begin{cases} \dot{\nu}(t) = \Big(2\mu\big[Q(I_N \otimes A)Q^{-1}\big] + 2\mathbf{L}\frac{\lambda_{max}(P)}{\lambda_{min}(P)} + 1\Big)\nu(t) \\ \qquad + \|Q(G \otimes \Gamma)Q^{-1}\|^2\nu(t - \tau(t)) + \epsilon, \quad t \in (t_{k-1}, t_k], \\ \nu(t_k^+) = \beta^2\nu(t_k), \qquad k \in \mathbb{N} \\ \nu(\theta) = \lambda_{max}(P)\|\phi(\theta)\|^2, \qquad -\bar{\tau} \leq \theta \leq 0 \end{cases}$$

It may be seen that $V(\theta) \leq \nu(\theta)$, for $-\bar{\tau} \leq \theta \leq 0$. According to [24], it leads to

$$V(t) \leq \nu(t), \qquad \text{for} \quad t \geq 0$$

The trivial solution of the comparison system is

$$\nu(t) = W(t,0)\nu(0) + \int_0^t W(t,s)\Big(\|Q(G \otimes \Gamma) \times Q^{-1}\|\nu(s - \tau(s)) + \epsilon\Big)ds, \quad t \geq 0$$

where $W(t,s)$ is the Cauchy matrix and estimated by

$$W(t,s) = \beta^{2\eta(t,s)}\exp\Big\{1 + 2\mathbf{L}\frac{\lambda_{max}(P)}{\lambda_{min}(P)} + 2\mu[Q(I_N \otimes A)Q^{-1}](t-s)\Big\}$$
$$\leq \beta^{2(\frac{t-s}{\rho}-1)}e^{(p - \frac{2\ln\beta}{\rho})(t-s)}$$
$$= \beta^{-2}e^{p(t-s)}$$

in which $\eta(t, s)$ is the number of the control impulses in the interval $(s, t]$. Accordingly, for $t > 0$, we have

$$\nu(t) \leq \beta^{-2}\lambda_{max}(P)\|\phi(0)\|^2 + \int_0^t \beta^{-2}e^{p(t-s)}\left(\|Q(G \otimes \Gamma)Q^{-1}\|^2\nu(s - \tau(s)) + \epsilon\right)ds$$

$$\leq \gamma e^{pt} + \int_0^t e^{p(t-s)}\left(q\nu(s - \tau(s)) + \epsilon\right)ds \tag{15}$$

where $\gamma = \beta^{-2}\lambda_{max}(P)\sup_{-\bar\tau \leq s \leq 0}\|\phi(s)\|^2$.

In the following, we utilize the method of reduction to absurdity to verify that

$$\nu(t) \leq \gamma e^{-\lambda t} - \frac{\epsilon}{\beta^2(p+q)}, \qquad t \geq 0 \tag{16}$$

Since $\epsilon > 0$ and $-p > q$, then we have $\frac{\epsilon}{\beta^2(p+q)} < 0$. Firstly, it is assumed that there exists a $t^* > 0$ such that

$$\nu(t^*) \geq \gamma e^{-\lambda t^*} - \frac{\epsilon}{\beta^2(p+q)}, \tag{17}$$

$$\nu(t) < \gamma e^{-\lambda t} - \frac{\epsilon}{\beta^2(p+q)}, \qquad t < t^* \tag{18}$$

According to (13), (15) and (18), it holds

$$\nu(t^*) \leq \gamma e^{pt^*} + \int_0^{t^*} e^{p(t^*-s)}[q\nu(s - \tau(s)) + \epsilon]ds$$

$$< e^{pt^*}\{\gamma - \frac{\epsilon}{\beta^2(p+q)} + \int_0^{t^*} e^{-ps}[\gamma q e^{-\lambda(s-\tau(s))} - \frac{\epsilon q}{\beta^2(p+q)} + \frac{\epsilon}{\beta^2}]ds\}$$

$$\leq e^{pt^*}\{\gamma - \frac{\epsilon}{\beta^2(p+q)} + \gamma q e^{\lambda\bar\tau}\int_0^{t^*} e^{-(p+\lambda)s}ds + \frac{\epsilon p}{\beta^2(p+q)}\int_0^{t^*} e^{-ps}ds\}$$

$$\leq e^{pt^*}\{\gamma - \frac{\epsilon}{\beta^2(p+q)} + \gamma[e^{-(p+\lambda)t^*} - 1] - \frac{\epsilon}{\beta^2(p+q)}(e^{-pt^*} - 1)\}$$

$$= \gamma e^{-\lambda t^*} - \frac{\epsilon}{\beta^2(p+q)}$$

This contradicts (17), thus the assumption is not tenable and the estimate (16) holds. Let $\epsilon \to 0$, then

$$V(t) \leq \nu(t) \leq \gamma e^{-\lambda t}, \qquad t \geq 0 \tag{19}$$

Moreover,

$$V(t) \geq \lambda_{min}(I_N \otimes P)\|e(t)\|^2, \qquad t \geq 0 \tag{20}$$

Combining the inequality of Eq. (19) and Eq. (20),

$$\|e(t)\| \leq \frac{1}{\beta}\sqrt{\frac{\lambda_{max}(P)}{\lambda_{min}(P)}}\sup_{-\bar\tau \leq \theta \leq 0}\{\|\phi(\theta)\|\}e^{-\frac{\lambda}{2}t}$$

which implies the conclusion (12) and this completes the proof. □

Due to the limited space, similar results for the case $\beta \geq 1$ is omitted.

3.2   Impulsive Controller Design

In the sequel, a design procedure of impulsive controller is provided based on Theorem 1. For a given scalar $\lambda_0 > 0$, we shall provide a set of steps such that the complex network (1) may be synchronized under the impulsive control (2) with an exponential decay rate $\lambda \geq \lambda_0$.

Design procedure:

1. Calculate the parameters $\mathbf{L}, \tau$;
2. Choose a symmetric positive definite matrix P, which is factorized as $P = Q^T Q$. Select the matrices series $\{\mathbf{D}_k\}$ such that $\|E + Q\mathbf{D}_k Q^{-1}\| \leq \beta$;
3. For a given λ_0, determine the set of the impulsive control instants $\{t_k\}$, $t \in \mathbb{N}$ as below: let $\Theta := 2\mu\left[Q(I_N \otimes A)Q^{-1}\right] + 2\mathbf{L}\frac{\lambda_{max}(\Gamma)}{\lambda_{min}(P)} + 1 + \beta^{-2}\|Q(G \otimes \Gamma)Q^{-1}\|^2(1 + e^{\lambda_0 \bar{\tau}}) > 0$, then the upper bound of control intervals can be taken as $\rho = \sup_{k \in \mathbb{N}}\{t_k - t_{k-1}\} = -(ln\beta)/\Theta$.

4   Application to the Network of Coupled Lorenz Oscillators

In this section, a numerical example is presented to illustrate the effectiveness of derived results.

For the sake of simplicity, consider a dynamical network consisting of 6 identical Chen systems. A single Chen system is described by

$$\begin{pmatrix} \dot{x}_{i1} \\ \dot{x}_{i2} \\ \dot{x}_{i3} \end{pmatrix} = \begin{pmatrix} a(x_{i2} - x_{i1}) \\ (c - a)x_{i1} - x_{i1}x_{i3} + cx_{i2} \\ x_{i1}x_{i2} - bx_{i3} \end{pmatrix}, \tag{21}$$

with $a = 35, b = 3, c = 28$. For these parameter settings the dynamics of the system has a chaotic attractor as shown in Fig.1.

Rewrite (21) in the Lur'e form as

$$x_i = Ax_i + f(x_i),$$

where

$$A = \begin{pmatrix} -a & a & 0 \\ c-a & c & 0 \\ 0 & 0 & -b \end{pmatrix}, \quad f(x_i) = \begin{pmatrix} 0 \\ -x_{i1}x_{i3} \\ x_{i1}x_{i2} \end{pmatrix},$$

As shown in Fig.1, the trajectory of Lorenz attractor is restricted in a bounded region $\Omega \in \mathbb{R}^3$. An estimation of the upper bound for a chaotic system is $R = 30.23$, such that $|x_{ij}| < R, s_j < R$ $(i = 1, \ldots, N; j = 1, 2, 3)$. Thus

$$\|f(x_i) - f(s)\| < 2R\|e_i\|.$$

Since each node has the same dynamics and the identical bound of trajectory, then we have $\mathbf{L} = 2R = 60.46$.

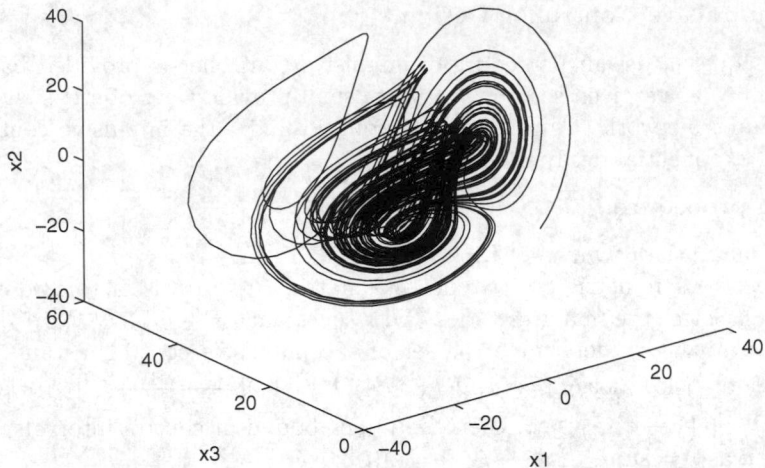

Fig. 1. Dynamical behavior of Chen chaotic system.($a = 35, b = 3, c = 28$)

Without loss of generality, we assume that the asymmetric outer-coupling matrix is

$$G = \begin{pmatrix} -3 & 1 & 1 & 0 & 0 & 1 \\ 3 & -4 & 0 & 1 & 0 & 0 \\ 1 & 0 & -5 & 2 & 0 & 2 \\ 1 & 1 & 2 & -5 & 1 & 0 \\ 0 & 0 & 0 & 1 & -3 & 2 \\ 2 & 0 & 2 & 0 & 2 & -6 \end{pmatrix},$$

and the inner-coupling matrix is

$$\Gamma = \begin{pmatrix} 1 & 1 & 0 \\ 0 & 1 & 1 \\ 1 & 0 & 1 \end{pmatrix}.$$

Let $\tau(t) = 0.05 \sin t$, which is bounded by $\bar{\tau} = 0.05$. Take $Q = 0.1E$ and $P = Q^T Q$. Select the impulsive feedback controller gain as $D_{ik} = \text{diag}\{-0.8, -0.8, -0.8\}$, then $\beta = \|E + \mathbf{D}_k\| = 0.2 < 1$. Let the decay rate $\lambda_0 = 50$, then $\Theta = 224.4997$. Accordingly, the upper bound of the impulsive interval is $\rho = \sup_{k \in \mathbb{N}}\{t_k - t_{k-1}\} = -\ln \beta / \Theta = 7.2 \times 10^{-3}$.

We implement the equidistant impulsive control and denote $\Delta T = t_{k+1} - t_k = 5 \times 10^{-3}$, $k \in \mathbb{N}$. Define the i-th element of the synchronous state error between node 1 and node i as $e_{ji} = x_{j1} - x_{ji}(j = 1, 2, 3$ and $i = 2, \ldots, 6)$, then the results of the dynamical network under impulsive control law $\{t_k, \mathbf{D}_k \mathbf{x}(t_k)\}$ is shown in Fig. 2. For comparison, state error of original complex networks without controller is visualized in Fig.3.

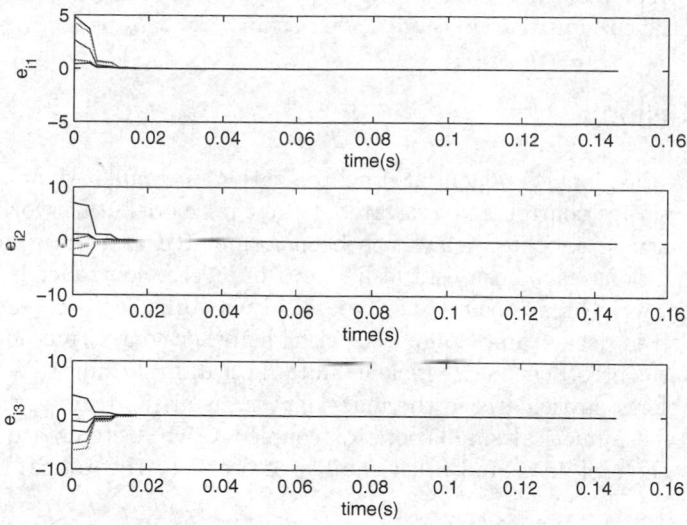

Fig. 2. Synchronous state error for the complex networks via impulsive control. ($\lambda_0 = 50$, $\Delta T = t_{k+1} - t_k = 5 \times 10^{-3}$, $D_{ik} = \text{diag}\{-0.8, -0.8, -0.8\}$).

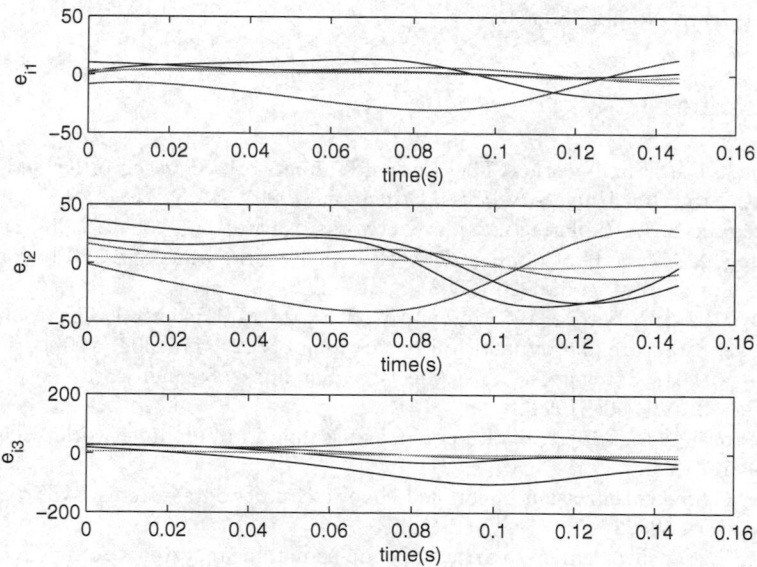

Fig. 3. Synchronous state error for the complex networks without control

From the Fig. 2 and Fig. 3, we can see that synchronizability of the complex networks has been significantly improved by introducing impulsive controller strategy. On the other hand, control actions are performed for the dynamical network at every 0.005 interval, which is prominent in reducing information exchange in the circumstance of large scale network.

5   Conclusion

In this paper, the global exponential synchronization of complex dynamical networks via impulsive control is investigated. A general model of network consisting of time-varying coupling delays has been formulated and the synchronous sufficient conditions have been established. An impulsive controller is designed and analyzed, which may ensure the dynamical networks achieve synchronization with a given decay rate. Compared with the other control methods in the literature, our control scheme is efficient and practical in dealing with synchronization problems particularly in the mass data transmission circumstances. As application, the numerical simulations of coupled Chen systems are given to demonstrate the usefulness and practicability of proposed theoretical results.

Acknowledgments

This work was supported by the National Key Fundamental Research Program (Grant No.2002CB312201-03) and the National Natural Science Foundation of China (Grant No.60575036).

References

1. Watts, D.J.: Small-worlds: The dynamics of networks between order and randomness. Princeton University Press, Princeton (1999)
2. Strogatz, S.H.: Exploring complex networks. Nature 410, 268–276 (2001)
3. Wang, X., Chen, G.: Complex networks: small-world, scale-free and beyond. IEEE Circuits Syst. Mag. 3, 6–20 (2003)
4. Boccaletti, S., Latora, V., Moreno, Y., Chavez, M., Hwang, D.U.: Complex networks: Structure and dynamics. Physics Reports 424, 175–308 (2006)
5. Watts, D.J., Strogatz, S.H.: Collective dynamics of small-world networks. Nature 393, 440–442 (1998)
6. Barabási, A.L., Albert, R.: Emergence of scaling in random networks. Science 286, 509–512 (1991)
7. Wu, C.: Synchronization in coupled chaotic circuits and systems. World Scientific, Singapore (2002)
8. Wu, C.: Synchronization in arrays of coupled nonlinear systems with delay and nonreciprocal time-varying coupling. IEEE Trans. Circuit Syst. II 52, 282–286 (2005)
9. Zhou, C., Kurths, J.: Dynamical weights and enhanced synchronization in adaptive complex networks. Phys. Rev. Lett. 96, 164102 (2006)
10. Li, Z., Chen, G.: Global synchronization and asymptotic stability of complex dynamical networks. IEEE Trans. Circuit Syst. II 53, 28–33 (2006)
11. Lü, J., Yu, X., Chen, G., Cheng, D.: Characterizing the synchronizability of small-world dynamical networks. IEEE Trans. Circuit Syst. I 51, 787–796 (2004)

12. Wang, X., Chen, G.: Synchronization in scale-free dynamical networks: Robustness and fragility. IEEE Trans. Circuit Syst. I 49, 54–62 (2002)
13. Zhou, J., Chen, T.: Synchronization in general complex delayed dynamical networks. IEEE Trans. Circuit Syst. I 53, 733–744 (2006)
14. Dai, Y., Cai, Y., Xu, X.: Synchronization criteria for complex dynamical networks with neutral-type coupling delay. Physica A 387, 4673–4682 (2008)
15. Li, X., Wang, X., Chen, G.: Pinning a complex dynamical network to its equilibrium. IEEE Trans. Circuit Syst. I 51, 2074–2087 (2004)
16. Zhou, J., Lu, J., Lü, J.: Adaptive synchronization of an uncertain complex dynamical network. IEEE Trans. Autom. Control 51, 652–656 (2006)
17. Wu, J., Jiao, L.: Synchronization in dynamic networks with nonsymmetrical time-delay coupling based on linear feedback controllers. Physica A 8-9, 2111–2119 (2008)
18. Li, P., Cao, J.: Stabilisation and synchronisation of chaotic systems via hybrid control. IET Control Theory Appl. 1, 795–801 (2007)
19. Yang, T.: Impulsive control theory. Springer, Berlin (2001)
20. Liu, B., Liu, X., Chen, G., Wang, H.: Robust impulsive synchronization of uncertain dynamical networks. IEEE Trans. Circuit Syst. I 52, 1431–1441 (2005)
21. Zhang, G., Liu, Z., Ma, Z.: Synchronization of complex dynamical networks via impulsive control. Chaos 17, 043126 (2007)
22. Li, K., Lai, C.: Adaptive impulsive synchronization of uncertain complex dynamical networks. Physics Letters A 372, 1601–1606 (2008)
23. Niculescu, S.I., Gu, K.: Advances in time-delay systems. Springer, Berlin (2004)
24. Yang, Z., Xu, D.: Stability analysis and design of impulsive control systems with time delay. IEEE Trans. Autom. Control 52, 1448–1454 (2007)
25. Horn, R.A., Johnson, C.R.: Topics in matrix analysis. Cambridge University Press, Cambridge (1991)

Synchronization in Complex Networks with Different Sort of Communities

Ming Zhao[1], Tao Zhou[1], Hui-Jie Yang[1], Gang Yan[2], and Bing-Hong Wang[1]

[1] Department of Modern Physics, University of Science and Technology of China,
Hefei 230026 P.R. China
zhaom17@mail.ustc.edu.cn
[2] Department of Electronic Science and Technology,
University of Science and Technology of China, Hefei Anhui, 230026, P.R. China

Abstract. In this paper, inspired by the idea that many real networks are composed by sorts of communities, we investigate the synchronization property of oscillators on such community networks. We identify the communities by two ways, one is by the structure of individual community and the other by the intrinsic frequencies probability density $g(\omega)$ of Kuramoto oscillators on different communities. For the two sorts of community networks, when the community structure is strong, only the oscillators on the same community synchronize. With the weakening of the community strength, an interesting phenomenon appears: although the global synchronization is not achieved, oscillators on the same sort of communities will synchronize independently. Global synchronization will appear with the further weakening of community structure.

Keywords: complex network, community structure, synchronization.

1 Introduction

Many social, physical and biological systems have the structure of networks. It is found recently that these networks are not simply regular lattices or random networks but bear some common and important characters, such as short average distance, large clustering coefficient and the power-law degree distribution [1]. Networks possessing these characters are often called complex networks. Besides these characters, many real-world networks have the so-called community structure [2,3]. Community network can be divided into several subsets of nodes, where the edges within a subset are much denser than those between them. It is worth notice that communities in networks may not be the same sort: they may be different in the structure or in the characters of nodes.

Collective synchronization phenomena have been observed for hundreds of years and also exist in a variety of field, including natural, physical, chemical and biological systems [4]. Because of the limitation of knowledge of networks, the studies of collective synchronization are restricted to either on the regular lattices or the random networks for a long time. Recently, with the development of physical networks, the focus of the studies of collective synchronization shifts

J. Zhou (Ed.): Complex 2009, Part I, LNICST 4, pp. 924–933, 2009.

to dynamical networks with complex structures. Without surprise, it is found that the property of collective synchronization on complex networks is much different from that in regular lattices or random networks. To date, scientists have investigated the relationship between some topological coefficients and the network synchronizability, and found that only the combination of short average distance and homogeneous degree distribution that ensures better network synchronizability [5,6,7,8]. Furthermore, the synchronization on community networks is studied [9,10,11], and community structure is proved to inhibit the global synchronization of oscillators. However, the synchronization property of dynamical networks having different sorts of communities is still unclear.

In this paper, with the help of Kuramoto model [12,13,14,15,16], we investigate the synchronization property of complex networks with different sorts of communities from two aspects: we take the famous Watts-Strogatz (WS) small-world [17] model with different rewiring probability p to represent different community structure, and use the natural frequencies probability density $g(\omega)$ of Kuramoto oscillators to identify the communities.

This paper is organized as follows. In section 2, the Kuramoto model and the order parameter are introduced. In section 3 and section 4, we will give the simulation results of synchronization properties of Kuramoto oscillators on complex networks with the communities different in the structure and the characters of nodes separately. The conclusion remarks are drawn in section 5.

2   Kuramoto Model and the Order Parameter

In this paper, we use the coupled phase oscillators, Kuramoto model, to analyze the collective synchronization on complex networks. A modified Kuramoto model is as follow,

$$\frac{d\phi_i}{dt} = \omega_i - \frac{\sigma}{k_i} \sum_{j \in \Lambda_i} \sin(\phi_i - \phi_j), \tag{1}$$

where ϕ_i, ω_i and Λ_i are the phase, the intrinsic frequency and the neighbor set of node i, respectively, and σ is the coupling strength. ω_i is chosen from the probability density $g(\omega)$.

The order parameter M is defined as

$$M \equiv \left[\left\langle \left| \frac{1}{N'} \sum_{j=1}^{N'} e^{i\phi_j} \right| \right\rangle \right], \tag{2}$$

where $\langle \cdots \rangle$ and $[\cdots]$ denote the average over time and over different configurations, respectively. N' is the number of nodes. In this paper, we not only consider the order parameter of all the nodes in the networks, but also take into account the order parameter of oscillators on individual community and that of the group of nodes composed by the same sort of communities. For different situation, the sum goes over the group of nodes we interested in, and N' is taken accordingly.

3   Communities Identified by Structure

WS small-wold model is built on a low-dimensional regular lattice and then rewire one end of each edge with probability p to create some "shortcuts". that join remote parts of the lattice to one another. In this procedure, the number of edges in the network keeps fixed, and it has been found that with the rewiring probability p's increasing, the network synchronizability gets stronger [18,19,20]. Thus, take some WS networks with different rewiring probabilities as different sorts of communities, and then add a few edges randomly among them, a community network is composed.

In this paper, we take a one-dimensional lattice of 500 nodes with periodic boundary conditions, and join each node to its neighbors 3 lattice spacings away, rewire one end of all the 1500 edges with probability 0.15, then a small-world model with about 225 short cuts is created. With the same procedure, create another small-wold model with rewiring probability 0.5. These two networks we created have the same size (the same number of nodes and edges), but different structures, and the latter is much easier to synchronize than the former. Create five WS small-world networks of each kind, and take the ten networks as community units, then add some edges among them, a network of 5000 nodes with two sort of communities are composed. The *community strength* C is defined as the ratio of the number of edges between communities (external edges) and in the communities (internal edges), it is used to measure the strength of the community structure. Clearly, a smaller C corresponds to spars external edges thus a stronger community structure.

In the simulation, the phase ϕ_i and the intrinsic frequency ω_i are randomly and uniformly distributed in the intervals $[0, 2\pi]$ and $[0, 1]$ initially. The numerical results are obtained by integrating Eqs. (1) using the Runge-Kutta method with step size 0.01. After 2000 time steps to allow for relaxation to a steady state, the order parameter M are obtained from the average over 2000 time steps. All the presented data are the average over 100 realizations of configurations.

Figure 1 displays the relationship between the order parameter m and the coupling strength σ for different community strength C. From the figure one can see that no matter how strong the community structure is, individual community synchronize soon with the coupling strength's increasing, but the community with larger rewiring probability shows stronger synchronizability than with the smaller one. However, when the community structure is very strong (C is small), although some communities of the network are composed by similar structure units (WS small-world networks with the same p), because of the external edges is very sparse, the coherent phenomenon is not appear among them, and the network part composed by the easier to synchronize community (WS small-world networks with smaller p) show better synchronizability than the other part. When the community strength C is large, the above-mentioned phenomenon is not evident.

When the oscillators are coupled, after some times' iteration, they will rotate at some stable frequencies, which are not equal to their intrinsic frequencies usually. We define them as stable frequencies Ω_i. Figure 2 shows the stable

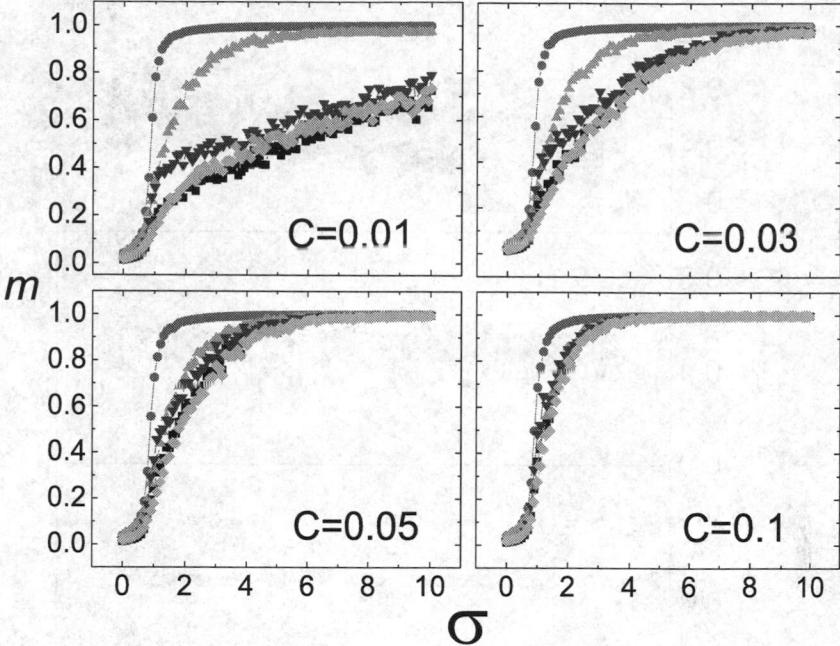

Fig. 1. (color online) The relationship between order parameter m and the coupling strength σ for the community network identified by the community structure. Red circles and green triangles represents the order parameter of the individual community with rewiring probability $p = 0.5$ and $p = 0.15$ separately, blue up triangles and bright blue diamonds represents that of the group of communities that composed by the units of similar structure, and the black squares represent the order parameter of the whole network. There are two sort of communities in the network and the number of communities of each sort are equal.

frequencies Ω_i vs the intrinsic frequencies ω_i at community strength $C = 0.03$. It can be seen that with the coupling strength's increasing, individual community synchronize first, and then the communities composed by the same structure unit and then the whole network, but the communities composed by easier synchronized units also synchronize easier than composed by harder synchronized units, which is consistent with the former results.

Community networks with different size and community number and community sort number have been verified, and there is no essential difference.

4 Communities Identified by the Intrinsic Frequencies Probability Density

Communities can also be identified by the sort of oscillators on the communities. In this paper, intrinsic frequencies probability density $g(\omega)$ are used to brand the communities, i.e., the intrinsic frequencies of the oscillators in different

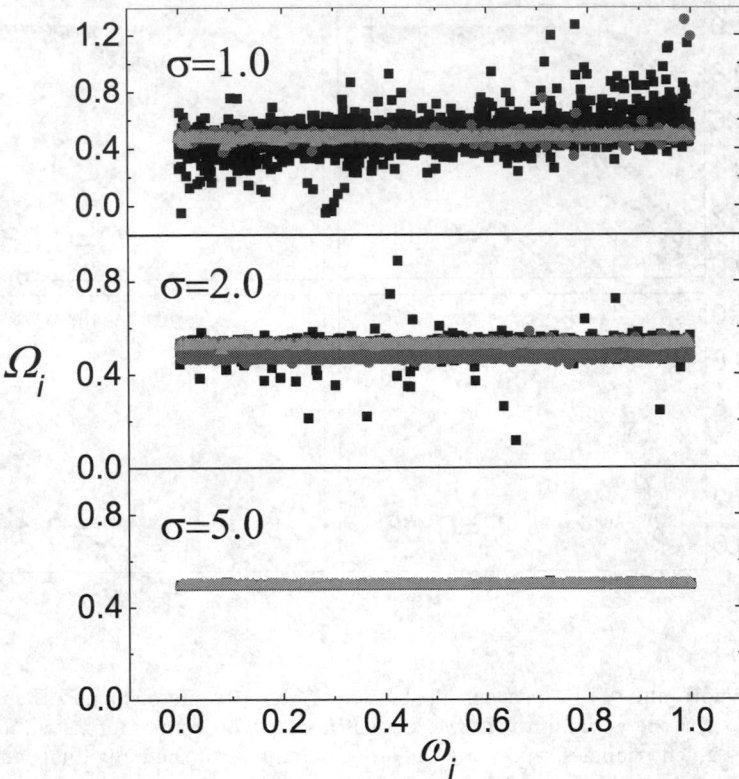

Fig. 2. (color online) The relationship between stable frequencies Ω_i and the intrinsic frequencies ω_i for the community network identified by the community structure. Green triangles, red circles and black squares represent a community that composed by a WS small-world network with $p = 0.5$, the communities that composed by similar structure units with $p = 0.5$ and the communities that composed by similar structure units with $p = 0.15$, respectively. There are two sort of communities in the network and the number of communities of each sort are equal.

communities are distributed with different probability density $g(\omega)$. In this section, we use a toy model that has power-law degree distribution and tunable community strength to simulate the synchronization process.

The model is created as follows. Start from n community cores, each core contains m_0 fully connected nodes. Initially, there are no connections among different community cores. Thus, there is in total n new nodes being added in one time step. Each node will attach m edges to existing nodes within the same community core, and simultaneously m' edges to existing nodes outside this community core. The former are internal edges, and the latter are external edges. Then the network's community strength $C = \frac{m'}{m}$. Similar to the evolutionary mechanism of Barabási-Albert (BA) networks [21], we assume the probability of choosing an existing node i to connect to is proportional to is degree k_i.

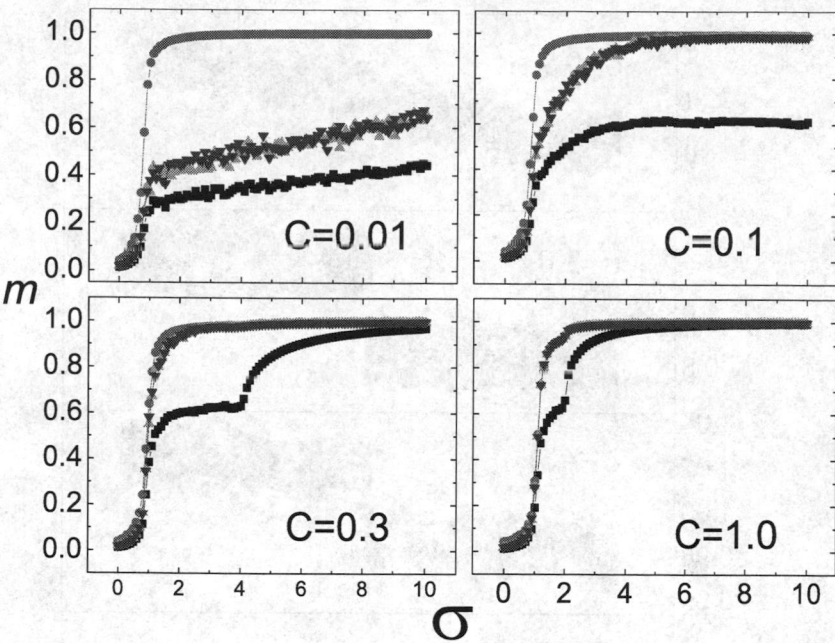

Fig. 3. (color online) (color online) The relationship between order parameter m and the coupling strength σ for the community network identified by the intrinsic frequency distribution probability $g(\omega)$. Red circles represent the order parameters that a community with the oscillators' intrinsic frequencies distributed in $[0,1]$. Green triangles and blue up triangles represent that of communities that the oscillators intrinsic frequencies distributed in $[0,1]$ and $[1,2]$, respectively. And the black squares represent the order parameter of the whole network. There are two sort of communities in the network and the number of each sort are equal.

Each community core will finally become a single community of size N_c, and the network size $N = nN_c$. By using the rate-equation approach [22], one can easily obtain the degree distribution of the whole network, $p(k) \propto k^3$.

From 10 community cores, and with $m + m' = 3$, network of 5000 nodes are created and the average degree is $\bar{k} = 6$. Five communities are located Kuromoto oscillators with intrinsic frequencies distributed randomly and uniformly distributed in the interval $[0,1]$, and the other five in $[1,2]$. Figure 3 shows the simulation results. It is easily drawn that for stronger community structure, the oscillators on each community are phase synchronized but the phase synchronization dose not emerge between different communities. With the community structure's weakening, phase synchronization appear among the same sort of communities but not the whole network and then all the oscillators are synchronized.

Figure 4 shows the relationships between stable frequencies and intrinsic frequencies of each oscillator at different coupling strength for $C = 0.3$. From the figure it can be seen that for a weaker coupling strength, the stable frequencies Ω_i

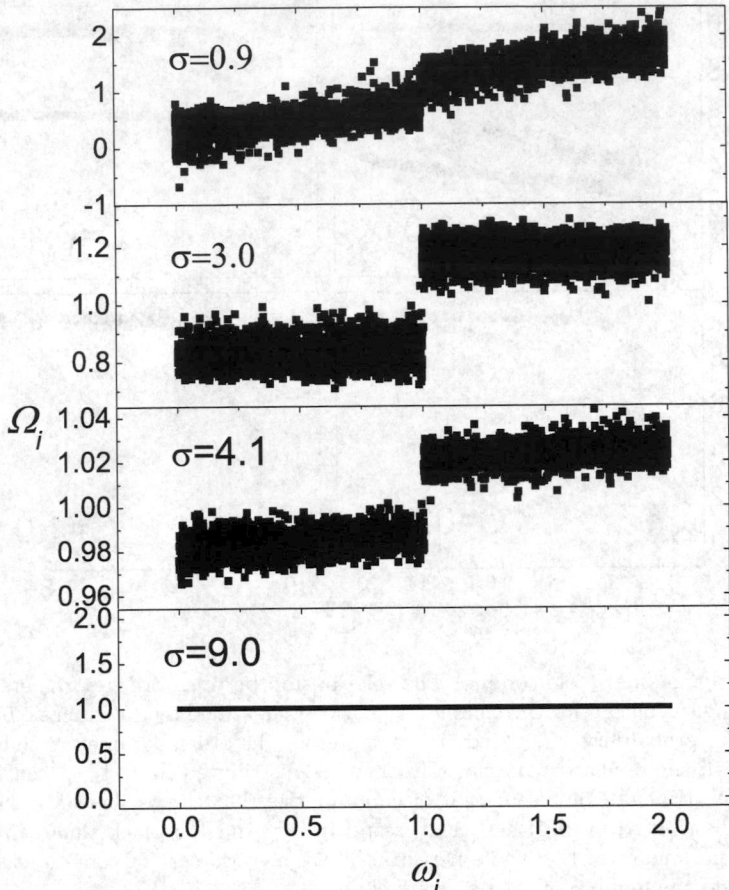

Fig. 4. The relationship between stable frequencies Ω_i and the intrinsic frequencies ω_i for the community network identified by the community structure. There are two sort of communities in the network and the number of each sort are equal.

are almost equal to the intrinsic frequencies ω_i, increasing the coupling strength σ will make oscillators on communities with the same intrinsic frequency distribution rotate at about their own average frequencies separately. Increasing the coupling strength furthermore will make all the oscillators rotate at almost the same frequency and then the synchronization is realized.

We also test the community networks with different size and composed by communities with more structure sorts but with each sort the same number of communities, and no essential difference is found. However, we notice that when the number of communities are not equal between different sort of communities, the synchronization property is not the same as the conclusion drawn above. Figure 5 shows the order parameter m vs coupling strength σ in network with two sort of communities. The meaning of geometry configurations in the figure

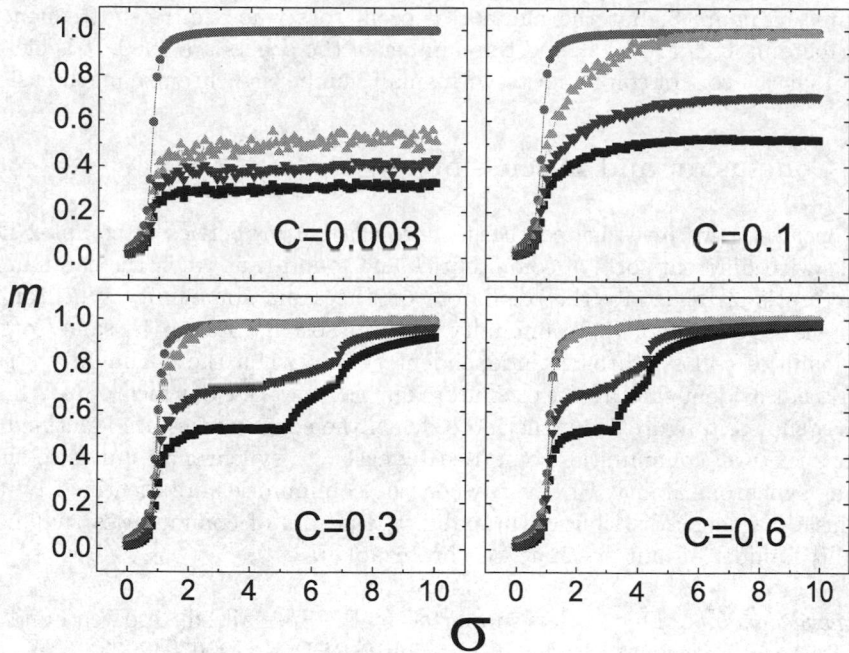

Fig. 5. (color online) (color online) The relationship between order parameter m and the coupling strength σ for the community network identified by the intrinsic frequency distribution probability $g(\omega)$. There are two sort of communities in the network and the ratio of the number of each sort are 2:3.

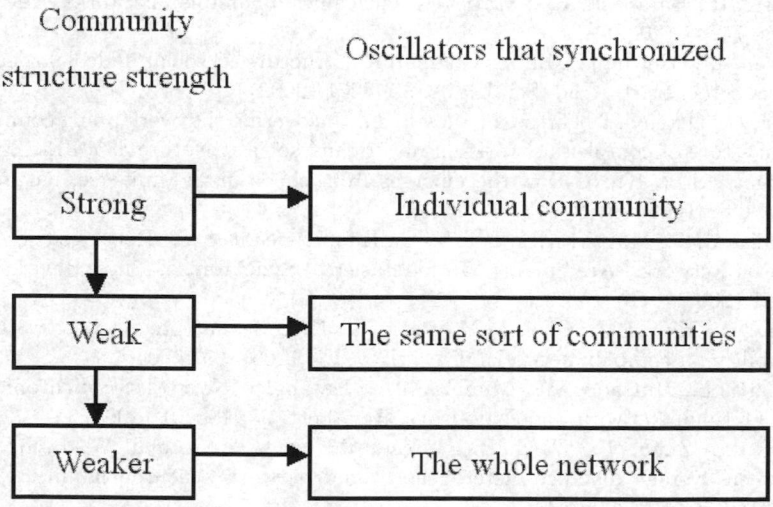

Fig. 6. With the community structure strength changes from strong to weaker, the synchronization phenomenon appears on the individual community, the same sort of communities and then the whole network.

is same to figure 3, but the number of oscillators whose intrinsic frequencies distribute in $[0, 1]$ are 2000, and the number of the others are 3000. It is clearly seen that the sort of communities with small number synchronize much easier.

5   Conclusion and Discussion

In conclusion, we investigated the synchronization properties in complex networks with different sorts of communities and found that when the community structure is strong, only the oscillators on the same community synchronize, with the weakening of the community strength, oscillators on the same sort of communities will synchronize independently, only when the community structure is not evident that all the oscillators on the network can synchronize, this is clearly shown in figure 6. Furthermore, when the communities are identified by their structure, communities composed by easier to synchronize unit also show strong synchronizability. However, when the communities are identified by the intrinsic frequency distribution probability, the sort of communities that have smaller number of unit are more synchronizable.

Acknowledgments. This work is supported by the Specialized Research Fund for the Doctoral Program of Higher Education (SRFD No. 20070420734), and the National Natural Science Foundation of China (Grant No. 10805045).

References

1. Albert, R., Barabási, A.L.: Statistical mechanics of complex networks. Rev. Mod. Phys. 74, 47 (2002)
2. Girvan, M., Newnam, M.E.J.: Community structure in social and biological networks. Proc. Natl. Acad. Sci. U.S.A. 99, 8271 (2002)
3. Palla, G., Derényi, I., Farkas, I., Vicsek, T.: Uncovering the overlapping community struc e ture of complex networks in nature and society. Nature 435, 814 (2005)
4. Strogatz, S.H.: SYNC-How the emerges from chaos in the universe, nature, and daily life. Hyperion, New York (2003)
5. Nishikawa, T., Motter, A.E., Lai, Y.-C., Hoppensteadt, F.C.: Heterogeneity in Oscillator Networks: Are Smaller Worlds Easier to Synchronize? Phys. Rev. Lett. 91, 14101 (2003)
6. Hong, H., Kim, B.J., Choi, M.Y., Park, H.: Factors that predict better synchronizability on complex networks. Phys, Rev. E 69, 067105 (2004)
7. Donetti, L., Hurtado, P.I., Muñoz, M.A.: Entangled Networks, Synchronization, and Optimal Network Topology. Phys. Rev. Lett. 95, 188701 (2005)
8. Zhao, M., Zhou, T., Wang, B.-H., Yan, G., Yang, H.-J., Bai, W.-J.: Relations between average distance, heterogeneity and network synchronizability. Physica A 371, 773–780 (2006)
9. Oh, E., Rho, K., Hong, H., Kahng, B.: Modular synchronization in complex networks. Phys. Rev. E 72, 047101 (2005)
10. Zhou, T., Zhao, M., Chen, G., Yan, G., Wang, B.-H.: Phase synchronization on scale-free networks with community structure. Phys. Lett. A 368, 431 (2007)

11. Park, K., Lai, Y.-C., Gupte, S., Kim, J.-W.: Synchronization in complex networks with a modular structure. Chaos 16, 015105 (2006)
12. Kuramoto, Y.: Araki, H. (ed.) Proceedings of the International Symposium on Mathematical Problems in Theoretical Physics. Springer, New York (1975)
13. Kuramoto, Y.: Chemical Oscillations, Waves, and Turbulence. Springer, Berlin (1984)
14. Kuramoto, Y., Nishikawa, I.: Statistical Macrodynamics of Large Dynamical Systems. Case of a Phase Transition in Oscillator Communities. J. Stat. Phys. 49, 569 (1987)
15. Daido, H.: Population Dynamics of Randomly Interacting Self-Oscillators. I. Prog. Theor. Phys. 77, 622 (1987)
16. Daido, H.: Quasientrainment and slow relaxation in a population of oscillators with random and frustrated interactions. Phys. Rev. Lett. 68, 1073 (1992)
17. Watts, D.J., Strogatz, S.H.: Collective dynamics of small-world networks. Nature 393, 440 (1998)
18. Lago-Fernández, L.F., Huerta, R., Corbacho, F., Sigüenza, J.A.: Fast Response and Temporal Coherent Oscillations in Small-world Networks. Phys. Rev. Lett. 84, 2758 (2000)
19. Wang, X.F., Chen, G.: Synchronization in small-world dynamical networks. Int. J. Bifurcation Chaos Appl. Sci. Eng. 12, 187 (2002)
20. Barahona, M., Pecora, L.M.: Synchronization in Small-World Systems. Phys. Rev. Lett. 89, 054101 (2002)
21. Barabási, A.-L., Albert, R.: Emergence of scaling in random networks. Science 286, 509 (1999)
22. Krapivsky, P.L., Render, S., Leyvraz, F.: Connectivity of Growing Random Networks. Phys. Rev. Lett. 85, 4629 (2000)

Symmetry Breaking in the
Evolution of World Economic Structure

Hui Wang and Guangle Yan

Business School, University of ShangHai for Science and Technology,
JunGong Road. 516, 200093 ShangHai, China
wanghuilele@gmail.com

Abstract. Over the centuries, world economic system and the corresponding economic structure have been in a state of continuous evolution. In this paper, through the empirical analysis on the evolution history of world economic structure, we show that the underlying driving force for the evolution of world economic structure is Technology Innovation. Specifically, we find that symmetry breakings not only emerge in the whole economic structure, but also take place in the local economic relation and economic status of inner countries along the long evolution history of world economic structure. We also elaborate the detailed mechanism of symmetry breaking of world economic structure. That is, in the evolution of world economic structure, all those countries participating in world economic open market are affected to varying degrees by symmetry breaking that is caused by technology innovation, which eventually determines current world economic structure with competitive countries evolving into economic centers and countries completely marginalized evolving into 'singularities' of world economic network.

Keywords: world economic structure, symmetry breaking, technology innovation, economic long wave, international division of labor.

1 Introduction

World economic system is an open and complex system. One of defining features of such systems is its continuous evolution. It is theoretically and practically meaningful to find out the mechanism of evolution of the system and characterize the evolution orbit of the system by exploring the structure and function of the economic system. Thus, people can continuously optimize resource allocation of the economic system, improve the economic structure and strengthen its functions.

'World system' theories have attracted considerable attentions since their birth [1-3]. One of important branches following the theory is the Immauel Wallerstein and Braudel research center, which studies world economy and its transformation from the perspective of structuralism. In light of the underlying theory, their notions are not too much different with that of Andre Frank's [4]. That is, they also believe that the origin of world economic system can be traced back to Age of Discovery. However, their theory emphasize on world economic structure that is made up of economic

J. Zhou (Ed.): Complex 2009, Part I, LNICST 4, pp. 934–943, 2009.

center, half periphery market and periphery market, namely the center-half periphery-periphery structure. Hereafter, different opinions on the formation of the world economic system are proposed in the subsequent papers [5-10]. Kasja Ekholm and Jonathan Friedman think the center-periphery structure lie in world economic system since ancient times, which is also an important organization form in the ancient economy [11]. However, center-periphery means different things in different times. In ancient times, it is the center and periphery position in the sense of economic geography, while at present it means center periphery structure in the international division of labor and is codetermined by the economic geography and industrial structure. Besides these efforts to understand the structure of world economic system, Robert Cox further employed "History structure" concept to analyze the world economy and its transformation, which make a significant contribution to the innovation of methodology on International Political Economy [12-14].

Although we can easily describe the evolutional history of world economic system, it is not trivial to explain the mechanism of the evolution of world economic system. In recent years, symmetry and symmetry breaking has attracted certain interest in the study of structure of complex systems [15-18]. It is widely believed that the evolution of the system is caused by symmetry breaking [16]. Hence, it is meaningful to investigate the evolution of world economic system from the perspective of symmetry and symmetry breaking of world economic structure.

Symmetry in physics has been generalized to characterize invariance-that is, lack of any visible change-under any kind of transformations. It is a significant property to describe the state of the system. It is believed that symmetry dominates modern natural science [19-20]. Hence, it's necessary to analyze symmetry of the system when exploring static or dynamic properties of systems [21]. Symmetry is also shown to be one fundamental property of world economic structure [18]. Symmetry breaking can be described as a phenomenon where small fluctuations acting on a system crossing a critical point decide a system's fate, by determining which branch of a bifurcation is taken [21]. It is symmetry breaking motivate the system to evolve. The evolution of the cosmos, life and society all experience the process from complete symmetry to local symmetry, and eventually asymmetry [21].

Similar to concepts of other symmetries in physics, symmetry of network structure characterizes the invariance under certain transformations. Symmetry in network structure characterizes the invariance of adjacency of vertices under the permutations on vertex set, which implies that 'invariance' of the symmetry in network structure is the relation among vertices and the 'transformation' is permutations on vertex set [18].

With the increase of the openness of the world economy and high degree of economic interdependence, the world economic system evolved from early bilateral structure to present multilateral network. National countries are not independent but coexistent in the world economic network which is increasingly complicated. Once the economic behaviors and trade start, and subsequently markets come into being, the original homogeneous and symmetric nodes will evolve into economic center or periphery market spontaneously. Due to different position in the geographical and international division of labor, countries are located in different hierarchy of the economic network. However, national countries in the same hierarchy are symmetric in terms of their economic spatial position, their status and function in the world economic network are also similar.

2 Technology Innovation – The Driving Force for Symmetry Breakings Emerging in World Economic Structure

In 'Innovation Theory', Schumpeter elaborated great effect of technology innovation on the economic development for the first time, which pioneered the studies about the correlation between technology innovation and the evolution of the economic system [22]. Hereafter, a number of economists including R.Solow, T.W.Schultz, K. Arrow, G.Grossman, E. Helpman, R.Barro, P.Aghion, P.Krugman, A.Young, G.Becker, demonstrated that technology progress is a decisive factor to promote economic continuous growth from the perspective of human capital accumulation, product variety increase, product quality improvement, technology imitation, specialized division of labor, respectively.

The function of technology innovation in the economic history is similar to that of gene mutation in the biological evolution. Enormous energy conserved in the process of economic development burst out, consequently, the equilibrium and symmetry of the system is broken. In the long development history of world economic structure, technology innovation motivates the economic structural transformation, leading to continuous symmetry breaking of the economic structure, and accordingly continuously evolution of the economic system.

The economic history has periods of dramatic transformation, which is the same as human evolution history. Multi-factors will affect dramatic transformation of economic system. The most important factor is always technology innovation, which is "the kernel engine to drive the capitalistic economic growth" and "the driving force for the transformation of economic pattern". During the last 100 years, the world economic system has experienced five dramatic transformation periods. Each transformation period is characterized by certain fundamental technology innovation.

In the first economic long wave, the use of steam engine leads to fundamentally improvement of instruments for production, which give rise to the fist leap of industrial productivity. In this period, the central country of the first industrial revolution is UK.

In the second economic long wave, iron and steel industry and coal mining industry stand out from the various industries, growing to be leading industries with rapid growth. In this period, the center is still UK; while French, US, Germany, Russia and other countries in the half periphery region achieved economic 'take-off'.

The third economic long wave is marked by the electric power in the 1880s. The developments of electric power result in the renovation of vehicles and communication tools. Automobiles, airplanes, telegraph, telephone, fax and radio appeared in succession. With the wide usage of new technique, electric industry, chemical industry, machinery manufacturing, and other heavy industries have replaced the light industry represented by textile industry, becoming the new leading industries. In this period, Germany and US become economic leaders. Japan and Canada cut a figure at this time.

In the 1950s, with the advent of the fourth economic long wave, new technologies, including computer, atomic energy and space technology, are emerging. In this period, instruments of production, transportation and communication system have reached a new level. As the initiator of new industries, US further consolidate its role as world economic center.

In the 1970s, information, network technology suddenly come to the force, and result in a new economic long wave coming unconsciously.

From the past five economic long waves, we can see that each of the major technological innovations arising from fluctuations in the economic cycle change world economic structure at least in two aspects: first, changing industry structure, which is caused by technology innovation and adjustment of supply and demand; second, changing of relative economic strength and corresponding status among countries, which is caused by the unbalanced innovation and development [23]. With the ups and downs of economic long wave, organization form of the world economy, the patterns of cooperation and competition will lead to some fundamental changes in the economic structure, and consequently, symmetry breaking emerges in the economic structure.

3 Continuous Symmetry Breakings in World Economic Structure

During the past five economic long waves, for world economic system, both its whole structure and internal structure have been in a state of constant evolution and continuous symmetry breakings.

3.1 Symmetry Breakings Emerging in the Whole World Economic Structure

After the former four economic long waves, world economic system take on a traditional center-half periphery- periphery structure (which is illustrated in Fig 1) with US, Japan and EU as the economic centers.

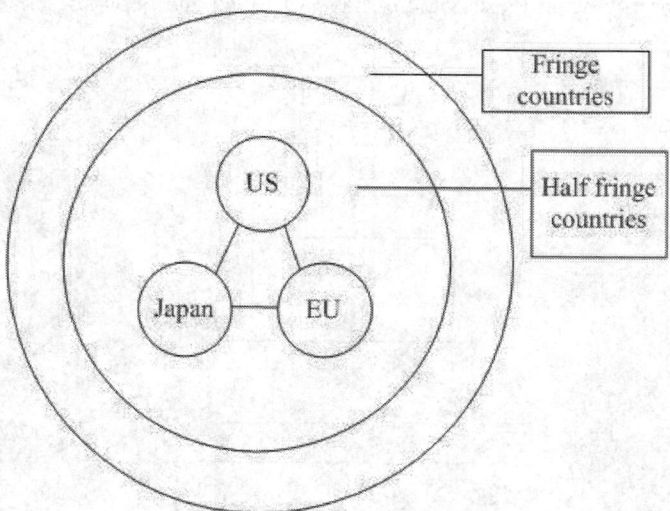

Fig. 1. Illustration of old world economic structure

In the 1990s, with the advent of another technological revolution-information revolution, as well as ensuing changes of production and organization pattern, symmetry breakings of the whole world economic structure emerge, new economic structure comes into being.

At first, owing to the difference in the competitive capacity and adjustment cost of labor market, symmetry breaking took place in the traditional economic centers in the process of economic globalization. US is distinctively outstanding in the upsurge of new economy in the 90s and became a large country inventing and producing information product, consequently, it surpasses other countries in the international division of labor. In light of EU, European countries led by Germany, also have outstanding performance in the process of globalization. They participate in part of the international division of information products. However, for Japan, 90s is the lost ten years. Due to its unsuccessful economic restructuring, Japan's economic still mainly relies on automobile manufacturing industries, which decrease its status in the international division of labor. But in other high-end consumer product manufacturing, Japan still takes priority in the world. It is the formation of vertical structure of international division of labor that results in the ten years of economic boom of US and economic recession of Japan, the richest country in Asia.

Secondly, countries near the three traditional economic centers, including North America, west Europe and east Asia have benefited from economic diffusion to the fullest extent. They get more opportunities than other regions through open economy, which increase their production efficiency and promote their economic growth. These three regional economic groups are the main beneficiary in the process of international division of labor, also become main battlefield of international markets.

Finally, from both geographical and industrial perspectives, periphery participants of international division of labor have become the periphery of the world economic structure. In Europe, Asia and America, represented by Russia, India, Brazil and Argentina, respectively, all these counties have vast land and population. However, due

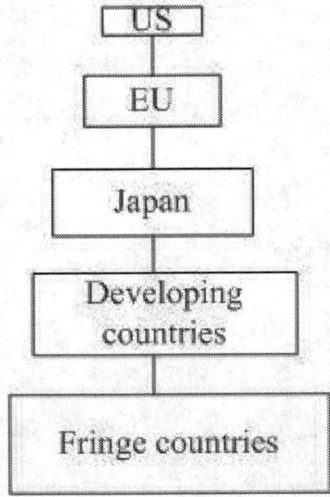

Fig. 2. Illustration of new world economic structure

to lack of enough consuming power or steady political and economical environment, they do not participate in the globalization process completely but also become periphery beneficial groups. Some industries in these periphery countries, such as software industry in India and high technology military industry in Russia, are outstanding and internationalized, however, their globalization index are very low or they participate in international division of labor just as raw material export countries. Even worse cases are irrelevant countries that are completely excluded from globalization. Most of countries in Africa fall into such a predicament.

World economic structure in the current world (illustrated in Fig 2) is similar to the structure of pyramid, which can be described as follows: On the top of the pyramid is US, which creates new economy and steers human being into information society. On the second ladder of the pyramid is EU represented by UK and Germany. UK is a leading financial services provider in international division of labor, while Germany specializes in the capital goods production. On the third ladder of the pyramid are some developed countries represented by Japan. They specialize in final consumer products with intensive capital and high technology and added value. On the fourth ladder of the pyramid are a large number of developing countries. They specialize in final consumer products with intensive labor. On the last ladder of the pyramid are periphery participants in the process of international division of labor.

3.2 Symmetry Breakings Taking Place in Economic Relations and Economic Status between Inner Countries

US, EU and Japan are three traditional economic centers, their status in world economic structure are symmetric. However, symmetry breaking emerges in their relative economic status with the advent of the information age. US are expanding its economic predominance, and still maintain the economic center in the world economic structure. As for EU, its inherent rigidity of systems (labor market, capital market) result in innovation deficiency, even worse, the "brain drain" is intensified. However, However, EU is carrying out positive eastward expansion and integration strategy. Although in the short run, there are still great obstacles such as instability and structural adjustment cost, but from the source of economic growth, long-term effects driven by markets expansion are obvious. Hence, EU can be classified into the second ladder of international division of labor. Japan, on the one hand is the same as EU in structural defects in the labor and capital markets, which takes her a decade to adjust the economic structure. Even worse, neither its currency policies nor financial policies, turn out to be effective. Hence, it is difficult for Japan to go through a renaissance. On the other hand Japan is different from EU in its physical space which is relatively narrow, and consequently the absolute space for economic enlargement is limited. Besides these negative factors, East Asian countries are traditionally dubitable in foreign policies to Japan. All these factors together are great obstacles for Japan to develop its economy rapidly in the East Asian markets, which caused Japan to drop from the economic center into the third ladder. In the 90s, the economic growth (annual percent change) of three traditional economic centers is illustrated in fig.3.

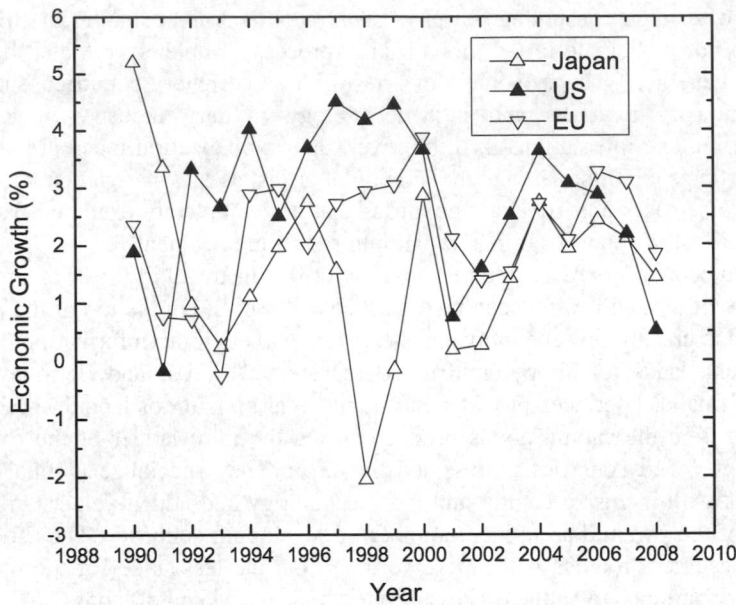

Fig. 3. The economic growth (annual percent change) of US, EU and Japan (Data source: www.imf.org. and www.oecd.org)

In the 1990s, the status of developing countries is advancing in world economic system. The eye-catching change in the world economy is the development of developing countries. Taking advantage of their cost advantages, developing countries get more and more opportunities to participate in international division of labor, consequently, they get more benefits from spill-over effects. For example, they are acquiring more advanced technologies and improving supply and demand of labor force. They enter the export market of technology-intensive production which they can not produce previously, and become part of international production system. In 2001, China, Mexico, Singapore and Turkey became big countries with high foreign direct investment (FDI), while FDI in US, UK, Canada and Germany were slowing down.

The international competition of some country can be evaluated in the following aspects: 1) the ability of economic restructuring and technical capacity under conditions of globalization; 2) the ability of entering big competitive open markets, such as markets of developed countries; 3) the price gap between production cost of original place and the price of target markets, which determines the potential profits in the future; 4) the political prospects and development policies. It is very difficult to measure the above-mentioned aspects directly. Therefore, we choose two comprehensive indicators, the total amount of FDI and total imports and exports, which can be quantified to evaluate the international competition of a country.

China is the biggest winner among developing countries benefiting from international production system. It plays increasing role in the world economic system. China used to be an economically-backward country in the Third World. But now the Chinese market has a pivotal position in the East Asian region. As is evident from

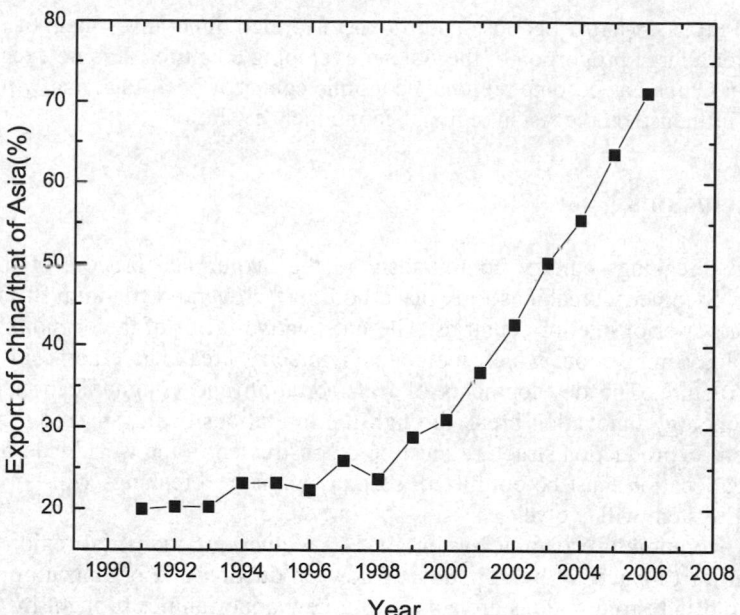

Fig. 4. Exports of China as a share of newly industrialized Asian economies (Data source: www.imf.org. and www.oecd.org)

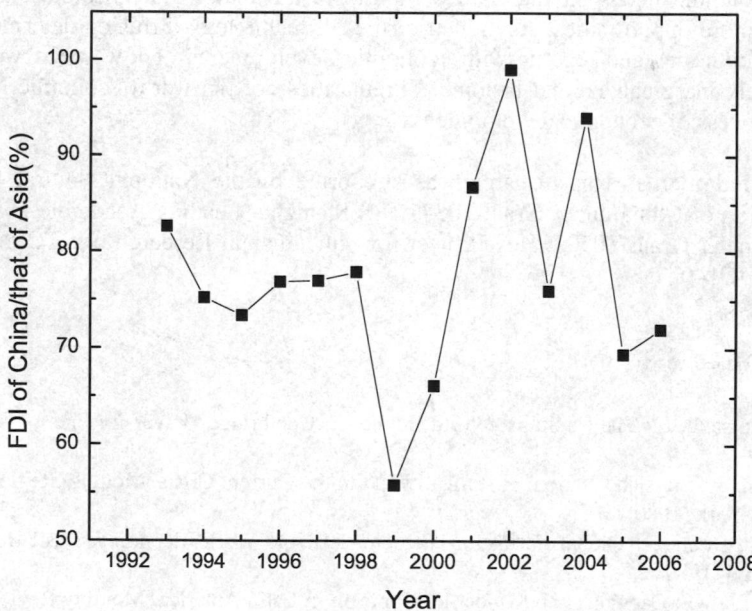

Fig. 5. Inflows of foreign direct investment of China as a share of developing Asia (Data source: www.imf.org. and www.oecd.org)

Fig.4 and Fig.5, exports of goods and inflows of foreign direct investment of China all account for a large proportion of the Asian developing countries. It is no exaggeration to say that China has become regional economic center of East Asia, and will has far-reaching influence on the Asian regional economical evolution.

4 Conclusions

Symmetry breakings emerge continuously in the evolutional process of the world economy. Symmetry breakings took place both in the overall economic structure and economic power of interior countries. The most active factor of the economic system is technology innovation, which motivates symmetry breakings emerge in the economic structure. The developments of transportation and communication resulting from technology innovation break through the limitations of the space distance, and consequently, production structure must be reconstructed in the worldwide range and resource allocation must be optimized, so that symmetry breakings will emerge, and economic system will evolve.

In the evolution of economic system, not only emerging industrial regions can get opportunities to catch up by adoption of new production and organization patterns, but also old industrial regions are also facing new opportunities brought by 'symmetry breaking'. Once the transformation of production and organization patterns is successful, the economy of the whole old industrial regions may get recovery.

In this economic network, countries that can take advantage of economy of scale and economic externalities will be outstanding in the competition, In other words, those economic nodes having both comparative advantage and absolute advantage will get more opportunities. Asymmetric rules of technology diffusion determine that a lot of countries and regions with economic developments of low speed would be completely marginalized and become "singularities" of the world economic network in the process of economic development.

Acknowledgments. This research was supported by the National Natural Science Foundation of China under Grant 70371070; Shanghai Leading Academic Discipline Project under Grant T0502; Key Project for Fundamental Research of STCSM under Grant 06JC14057.

References

1. Wallerstein, L.: The Capitalist-World Economy. Cambridge University Press, Cambridge (1979)
2. Amin, S.: Class and Nation: Historically and in the Current Crisis. Monthly Review Press, New York (1980)
3. Chase-Dunn, C.: Global Formation: Structures of the World Economy. Basil Blackwell, Oxford (1989)
4. Frank, A.G.: Capitalism and Underdevelopment in Latin America. Monthly Review Press, New York (1967)
5. Hirschman, A.O.: National power and the structure of foreign trade. University of California Press, Berkeley (1980)

6. Hollist, W.L.: Conclusion: anticipating world system theory synthesis. International Studies Quarterly 25, 149–160 (1981)
7. Chase-Dunn, C.: Global formation: structures of the world-economy. Basil Blackwell Inc., New York (1989)
8. Rubinson, R.: Dynamics of World Development. Sage Publications, Beverly Hills (1981)
9. Arrighi, G.: Transforming the Revolution: Social Movements and the World System. Monthly Review Press, New York (1990)
10. Arrighi, G., Drangel, J.: The stratification of the world-economy: an exploration of the semiperipheral zone. Review 10(1), 9–74 (1986)
11. Ekholm, K., Friedman, J.: Toward a Global Anthropology. Critique of Anthropology 5(1), 97–119 (1985)
12. Cox, R.: Social Forces, States and World Orders: Beyond International Relations Theory. Millennium: Journal of International Studies 10(2), 126–155 (1981)
13. Cox, R.: Multinationalism and World Order. Review of International Studies 18, 161–180 (1992)
14. Cox, R.: Production, Power, and World Order: social forces in the making of history. Columbia University Press, New York (1987)
15. Xiao, Y.H., MacArthur, B.D., Wang, H., Xiong, M.M., Wang, W.: Network Quotient: Structural Skeletons of Complex Systems. Physical Review E 78, 046102 (2008)
16. Xiao, Y.H., Wu, W.T., Wang, H., Xiong, M.M., Wang, W.: Symmetry-based Structure Entropy of Complex Networks. Physica A 387, 2611–2619 (2008)
17. Xiao, Y.H., Xiong, M.M., Wang, W., Wang, H.: Emergence of Symmetry in Complex Networks. Physical Review E 77(6), 066108(2008)
18. Wang, H., Yan, G.l., Xiao, Y.H.: Symmetry in World Trade Network. Journal of Systems Science and Complexity (in press) (2008)
19. Wigner, E.P.: Symmetries and reflections: Scientific essays. MIT Press, Cambridge (1970)
20. Zee, A.: Fearful symmetry: The search for beauty in modern physics. Princeton Univ. Press, Princeton (1999)
21. Xu, G.Z., Gu, J., Che, H.A.: System Science (in Chinese). Shanghai Scientific and Technological Education Publishing House, Shanghai (2000)
22. Schumpeter, J.A.: Theoretical Problems: Theoretical Problems of Economic Growth. Journal of Economic History 7 (1947)
23. Reuveny, R., Thompson, W.: Leading sectors, lead economies and economic growth. Review of International Political Economy 8(4), 689–719 (winter 2001)
24. Rennstich, J.K.: The new economy, the leadership long cycle and the nineteenth K-wave. Review of International Political Economy, Match, 150–182 (2002)
25. Mendez, R.: Creative Destruction and the Rise of Inequality. Journal of Economic Crowth 7, 259–281 (2002)
26. Pohjola, M.: The New Economy: facts, impacts and policies. Information Economics and Policy, 133–144 (2002)

Studies on Interpretive Structural Model
for Forest Ecosystem Management Decision-Making

Suqing Liu[1,*], Xiumei Gao[1], Qunying Zen[1], Yuanman Zhou[1], Yuequn Huang[1],
Weidong Han[1], Linfeng Li[1], Jiping Li[2], and Yingshan Pu[1,**]

[1] Department of Forestry Science, Faculty of Agriculture Science,
Guangdong Ocean University, Zhanjiang, Guangdong, 524088, P.R. China
liusuqing2001@yahoo.com.cn, Yshp6666@yahoo.com.cn
[2] Central South University of Forest Science and Technology,
Changsha, Hunan 310004, P.R China

Abstract. Characterized by their openness, complexity and large scale, forest ecosystems interweave themselves with social system, economic system and other natural ecosystems, thus complicating both their researches and management decision-making. According to the theories of sustainable development, hierarchy-competence levels, cybernetics and feedback, 25 factors have been chosen from human society, economy and nature that affect forest ecosystem management so that they are systematically analyzed via developing an interpretive structural model (ISM) to reveal their relationships and positions in the forest ecosystem management. The ISM consists of 7 layers with the 3 objectives for ecosystem management being the top layer (the seventh layer). The ratio between agricultural production value and industrial production value as the bases of management decision-making in forest ecosystems becomes the first layer at the bottom because it has great impacts on the values of society and the development trends of forestry, while the factors of climatic environments, intensive management extent, management measures, input-output ratio as well as landscape and productivity are arranged from the second to sixth layers respectively.

Keywords: Forest ecosystem management, Interpretive structural model, Factors, Decision-making.

1 Introduction

Although they have played certain positive roles in the history of forest management, traditional forest management modes have been seriously challenged because of their relatively single-purpose and irrational pursuit of timber production, and neglect of forest ecosystem functions in most cases. Since the concept of New Forestry [1, 2] was developed in 1980s, things have been gradually bettered via establishing and recognizing

---

* The first author: Suqing Liu, Professor & PhD. Tel.: +86-759-2382294; Fax: +86-759-2383398.
** Corresponding author: Yingshan Pu, Professor at Guangdong Ocean University,
Tel.: +86-759-2396038; Fax: +86-759-2383398.

J. Zhou (Ed.): Complex 2009, Part I, LNICST 4, pp. 944–953, 2009.
© ICST Institute for Computer Sciences, Social Informatics and Telecommunications Engineering 2009

the new forestry theories focused on forest ecosystem functions. Consequently an ecological approach to forest ecosystem management, namely, a method for forest ecosystem management has been formed, and perfected step by step [3, 4]. Hereafter, forest ecosystem management has become a vital development trend of modern forestry [5]. Because it is very complicated and characterized by its diversity, complexity, openness and large scale [6], a forest ecosystem unavoidably interweaves with social system, economic system and other physical ecosystems [7, 8], thus forming a more complicated macro-system. In order to manage such a complicated forest ecosystem effectively, it is, first and foremost, necessary to understand fully the constituents that compose the system, the factors that fundamentally affect the system's formation and development, and the correlations that exist among the constituents. Interpretative Structural Modeling (ISM) is a process, with which groups can structure complex issues to form interpretable patterns. Developed in the period from 1971 to 1973 by John N. Warfield, an American scholar, as an initial process for ISM of a complex system at the Battelle Memorial Institute, the process was first described in Battelle Monograph Number 4, titled Structuring Complex Systems [9, 10], and it proved to be an effective means of studying complicated systems as well. Later, a variety of papers were published to further the studies of this field. Report by Warfield [11] showed how to order the elements of a system in such a way that much of the data required could be computed from supplied data. Another study described computer operations which assisted in the interpretation of complex structural models, thus a weighting matrix applied to the elements of a maximal cycle set permitted a set of digraphs to be developed [12]. Also in his another study, the problem of interconnecting two multilevel subsystems models defined by binary matrices A and B and a common, transitive, contextual relation to form a system model defined by matrix M is solved [13]. As a result, the overall structure in complex systems have been analyzed by means of relational matrix principles in graph theory, the correlations and both direct and indirect restraint conditions among the system constituents are interpreted via the relational matrix [14], and the results can be directly elaborated by graphics. From what has been described above, therefore, the correlations between the forest ecosystem constituents and factors can be systematically analyzed to lay reliable theoretical bases for the decision-making in forest ecosystem management by applying ISM principles and methods in spite of the extreme complexity of the forest ecosystem.

2 Determination of Decision-Making Constituents for Forest Ecosystem Management

As the most complicated and largest ecosystem on land, the forest ecosystem maintains its existence and development via circulation of materials, flow of energy and exchange of information. Owing to its openness, it is incessantly in exchange with exterior environments [15, 16]. Included in the exterior environments are not only the factors of climates and other living things beyond forest ecosystem, but also the more complicated factors of humans, society and economy, none of which have no great impacts on forest ecosystem. Only by taking all the factors above into full consideration according to the characteristics of forest ecosystems and modeling analyses on the ecosystems based on the ISM theory, can solid and reliable theoretical bases be laid for forest ecosystem management decision-making.

2.1   The Theoretical Base in Determining Decision-Making Constituents

The factors affecting forest ecosystem management decision-making are very complicated since their complexity lies chiefly in the multitude of the factors, and the nonlinear relations between system and factors as well as among the factors. Although there might be some other different approaches to dealing with these anfractuous relations, the authors of this present paper have selected and determined the major factors for forest ecosystem management decision-making according to the theories of sustainable development, hierarchy-competence levels, cybernetics and feedback.

The Theory of Sustainable Development

Sustainable development is the development that meets the needs of the present without compromising the ability of future generations to meet their own needs, as defined by United Nations Environment Program. The theory focuses that a healthy economy should be developed on the decision-making bases of the ecological sustainability, social equity and people's active participation [17]. Since the sustainability of development lies on the sustainability of environments and resources, sustained yield can only be realized via rational exploitation and wise preservation of natural resources for present and future needs respectively. The index system for forest ecosystem management decision-making must, therefore, be determined in accordance with the above principles as an indispensable restriction of human exploitation activities, for the purpose of good circulation of ecological environments and optimal management for sustained yield of natural resources [18, 19].

The Theory of Hierarchy-Competence Levels

The theory of hierarchy- competence levels, or the theory of hierarchy-capacity levels, means that different hierarchies are functionally compartmentalized according to the status of various managed objects, a competence level of a hierarchy indicates the division of its function and the weight of its power, and the problems of various hierarchies should be treated in their relevant hierarchy- competence levels [20]. Involving the different hierarchies of economic, social and environmental macro-systems, forest ecosystem per se is composed of various subsystems connected with one another at various levels. The application of the theory of hierarchy-competence levels in forest ecosystem management, therefore, can not only reflect the exact features of forest ecosystem, but also make the decision-making more scientific and rational.

The Theory of Feedback and Cybernetics

Information transfer is one of the three major functions in ecosystem [21]. It is found that information transfers not only from input to output in the system, but also represents its feedback from output to input. In the light of cybernetics, it is the feedback of information that makes ecosystem adjustable and controllable. In the process of ecosystem management, information of constant feedback is collected from the dynamic ecosystem by means of monitoring and administration to control the activities and production modes in the controlled areas so as to maintain the optimum status of forest ecosystem management decision-making when the management objectives are fixed on the basis of the theory.

2.2 The Factors of Forest Ecosystem Management

Based on the theory of sustainable development, it is required that the objectives of forest ecosystem management be unified with economic objectives, social objectives and ecological objectives. The attainment of all these objectives, however, is affected by social factors, economic factors, physical environment factors, management factors, landscape factors, system productivity factors, etc. [22]. According to the theory of hierarchy-competence levels, feedback theory and cybernetics, it is revealed that the management objectives are the results integrated with these factors having their relevant impacts on the one hand, and in turn the objectives have effects on the selection of these factors on the other.

In this present study, the factors are categorized into seven groups as follows: Groups of management objective factors, socioeconomic factors, physical environment factors, management factors, landscape factors and system productivity factors. Group of management objective factors includes the factors: (1) economic objective, (2) social objective and (3) ecological objective. In group of socioeconomic factors, there are factors of: (4) the ratio between agricultural production value and industrial production value, (5) gross domestic product, and (6) policy, law and regulation. In group of physical environment factors, there are ten factors, i.e., the factors of: (7) annual average air temperature, (8) frostless period, (9) annual accumulated temperature, (10) annual precipitation, (11) altitude, (12) land feature, (13) aspect, (14) slope position and (15) vegetation type. Group of management factors consists of the factors of: (16) forest type, (17) investment per unit area, (18) amount of work per unit area, and (19) management measures. Group of landscape factors contains the factors of (20) landscape type and (21) landscape fragmentation degree. Included in group of system productivity factors are those of: (22) stocking volume per unit area, (23) growth increment per unit area, (24) input-output ratio and (25) employees' quality.

3 The Establishment of Interpretive Structural Model

A table of relationships among factors for forest ecosystem management (as shown in Tab. 1) is first established in the light of the factors for forest ecosystem management described in 2.2. In this table, if the element x_{ij} is 1, then this means Factor i has effect on Factor j, otherwise x_{ij} is 0. In step 2, an accessible matrix of all the forest ecosystem management factors as shown in Table 2 is obtained by means of computer operations. In step 3, the accessible matrix is divided into hierarchies, eliciting a distribution table of the layers and their factors in the ISM as presented in Table 3. In step 4, a figure of the ISM for forest ecosystem management decision-making (See Figure 1) has finally been developed on the basis of both Table 2 and Table 3 [23]. The results have systematically revealed the correlations among factors, thus laying solid theoretical bases and offering an excellent decision-making support for forest ecosystem management.

Table 1. The relationships among factors for forest ecosystem management

No. of Factors	1	2	3	4	5	6	7	8	9	10	11	12	13	14	15	16	17	18	19	20	21	22	23	24	25
1	0	0	0	0	0	0	0	0	0	0	0	0	0	0	0	0	0	0	0	0	0	0	0	0	0
2	0	0	0	0	0	0	0	0	0	0	0	0	0	0	0	0	0	0	0	0	0	0	0	0	0
3	0	0	0	0	0	0	0	0	0	0	0	0	0	0	0	0	1	1	0	0	0	0	0	0	0
4	0	0	0	0	1	1	0	0	0	0	0	0	0	0	0	0	1	1	0	0	0	0	0	0	0
5	0	0	0	0	0	0	0	0	0	0	0	0	0	0	0	0	0	0	1	0	0	0	0	0	0
6	0	0	0	0	0	0	0	0	0	0	0	0	0	0	0	0	0	0	0	0	0	0	0	0	0
7	0	0	0	0	0	0	0	0	0	0	0	0	0	0	1	0	0	0	0	0	0	0	0	0	0
8	0	0	0	0	0	0	0	0	0	0	0	0	0	0	1	0	0	0	0	1	0	0	0	0	0
9	0	0	0	0	0	0	0	0	0	0	0	0	0	0	1	0	0	0	0	1	0	0	0	0	0
10	0	0	0	0	0	0	0	0	0	0	0	0	0	0	1	0	0	0	0	0	0	0	0	0	0
11	0	0	0	0	0	0	0	0	0	0	0	0	0	0	1	0	0	0	0	1	0	0	0	0	0
12	0	0	0	0	0	0	0	0	0	0	0	0	0	0	0	0	0	0	0	1	0	0	0	0	0
13	1	1	1	0	0	0	0	0	0	0	0	0	0	0	1	0	0	0	0	1	1	1	1	0	0
14	0	0	1	0	0	0	0	0	0	0	0	0	0	0	0	0	0	0	0	0	0	0	0	0	0
15	0	1	0	0	0	0	0	0	0	0	0	0	0	0	0	0	0	0	0	1	0	0	0	0	0
16	1	0	1	0	0	0	0	0	0	0	0	0	0	0	0	0	0	0	0	1	0	0	0	0	0
17	0	1	1	0	0	0	0	0	0	0	0	0	0	0	0	0	0	0	0	0	0	1	1	0	0
18	1	1	1	0	0	0	0	0	0	0	0	0	0	0	0	0	0	0	0	1	0	1	0	1	0
19	0	1	0	0	0	0	0	0	0	0	0	0	0	0	0	0	0	0	0	0	0	1	0	1	0
20	0	0	1	0	0	0	0	0	0	0	0	0	0	0	0	0	0	0	0	0	0	0	1	1	0
21	1	0	0	0	0	0	0	0	0	0	0	0	0	0	0	0	0	0	0	0	0	0	0	0	0
22	1	0	0	0	0	0	0	0	0	0	0	0	0	0	0	0	0	0	0	0	0	0	0	0	0
23	1	1	1	0	0	0	0	0	0	0	0	0	0	0	0	0	0	0	0	0	0	0	0	0	0
24	1	1	0	0	0	0	0	0	0	0	0	0	0	0	0	0	0	0	0	0	1	1	0	0	0
25	0	1	1	0	0	0	0	0	0	0	0	0	0	0	0	0	0	1	1	1	1	1	1	0	0

Table 2. The accessible matrix of forest ecosystem management factors

No. of Factors	1	2	3	4	5	6	7	8	9	10	11	12	13	14	15	16	17	18	19	20	21	22	23	24	25
1	1	0	0	0	0	0	0	0	0	0	0	0	0	0	0	0	0	0	0	0	1	0	0	0	0
2	0	1	0	0	0	0	0	0	0	0	0	0	0	0	0	0	0	0	0	0	1	0	0	0	0
3	0	0	1	0	0	0	0	0	0	0	0	0	0	0	0	0	0	0	0	0	1	0	0	0	0
4	1	1	1	1	1	1	1	0	0	0	0	0	0	0	1	0	0	1	1	1	1	1	1	1	0
5	1	1	1	0	1	0	0	0	0	0	0	0	0	0	1	0	0	1	0	0	1	1	1	1	0
6	1	1	1	0	0	1	0	0	0	0	0	0	0	0	0	0	0	0	1	0	0	1	1	1	0
7	0	1	1	1	0	0	1	1	1	0	0	0	0	0	0	0	0	0	0	1	0	0	0	0	0
8	0	0	0	0	0	0	0	1	0	0	0	0	0	0	0	0	0	0	0	0	0	0	0	0	0
9	0	0	1	1	0	0	1	0	1	0	0	0	0	0	0	0	0	0	0	0	0	0	0	0	0
10	0	0	0	0	0	0	0	0	0	1	0	0	0	0	0	0	1	0	0	0	0	0	0	0	0
11	0	0	1	0	0	0	0	0	0	0	1	0	0	0	0	0	0	0	0	1	0	0	0	0	0
12	0	0	1	0	0	0	0	0	0	0	0	1	0	0	0	0	0	0	0	0	0	0	0	0	0
13	1	0	1	0	0	0	0	0	0	0	0	0	1	0	0	0	0	0	0	1	0	0	0	0	0
14	0	0	1	0	0	0	0	0	0	0	0	0	0	1	0	0	0	0	0	0	0	0	0	0	0
15	0	1	1	0	0	0	0	0	0	0	0	0	0	0	1	0	0	0	1	1	1	1	1	0	0
16	1	1	1	0	0	0	0	0	0	0	0	0	0	0	0	1	0	0	1	0	0	0	0	0	0
17	1	1	1	0	0	0	0	0	0	0	0	0	0	0	0	0	1	0	1	0	1	0	0	0	0
18	1	1	1	0	0	0	0	0	0	0	0	0	0	0	1	0	0	1	1	1	1	1	1	1	0
19	0	0	1	0	0	0	0	0	0	0	0	0	0	0	0	0	0	0	1	0	0	0	0	0	0
20	1	0	1	0	0	0	0	0	0	0	0	0	0	0	0	0	0	0	0	1	1	0	0	0	0
21	1	0	1	0	0	0	0	0	0	0	0	0	0	0	0	0	0	0	0	0	1	0	0	0	0
22	1	0	0	0	0	0	0	0	0	0	0	0	0	0	0	0	0	0	0	0	1	1	0	0	0
23	1	1	1	0	0	0	0	0	0	0	0	0	0	0	0	0	0	0	0	0	1	0	1	0	0
24	1	1	0	0	0	0	0	0	0	0	0	0	0	0	0	0	0	0	0	0	1	1	1	1	0
25	1	1	1	0	0	0	0	0	0	0	0	0	0	0	0	0	0	0	1	1	1	1	1	1	1

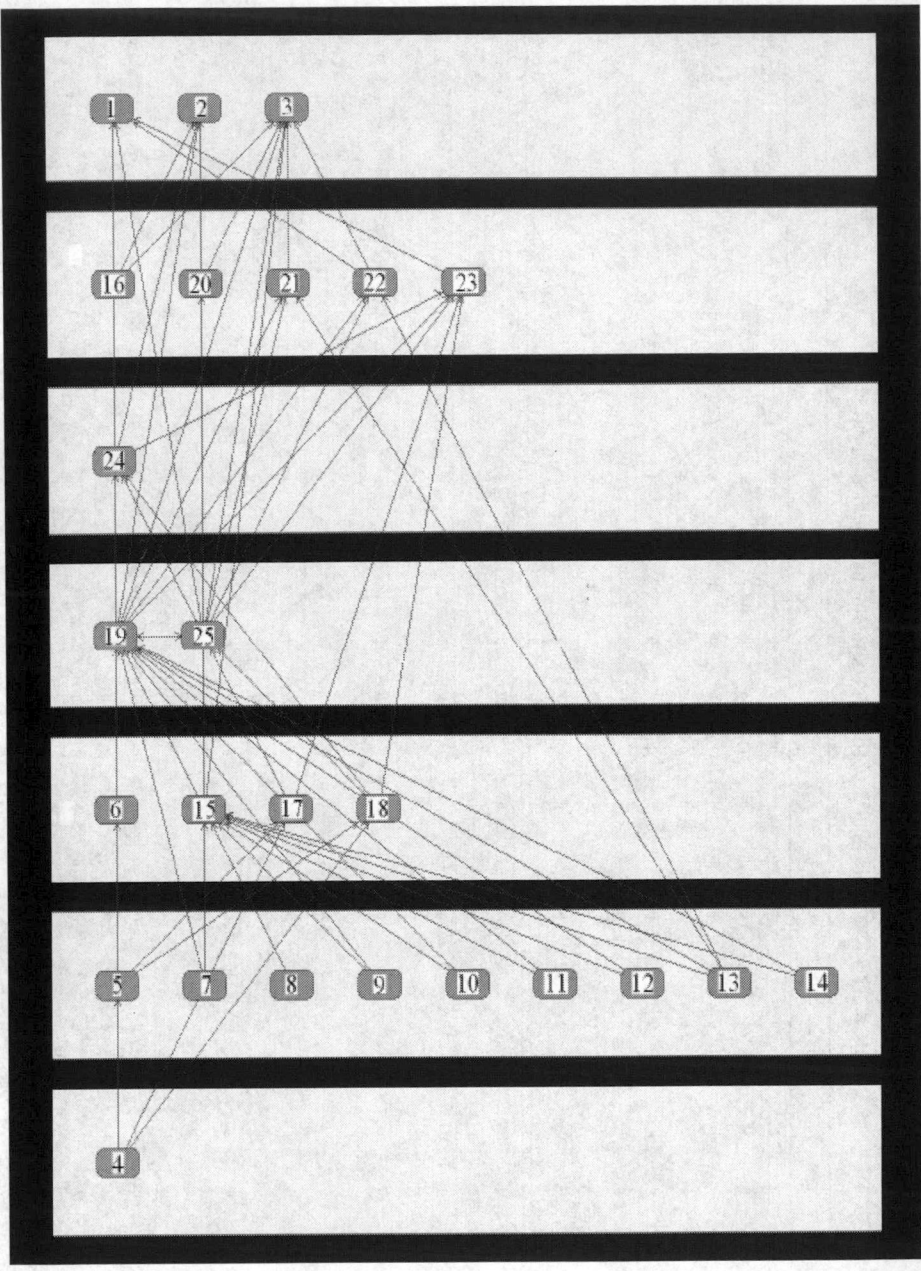

Fig. 1. The interpretive structural model for forest ecosystem management decision-making

No 1 to 25 represents the selected factors as follows: economical objective, social objective, and ecological objective as No 1, 2, 3; the ratio between agricultural production value and industrial production value, gross domestic product, and policy, law and regulation as No 4, 5, 6; annual average air temperature, frostless period, annual accumulated temperature, annual precipitation, altitude, land feature, aspect, slope position, and vegetation type as No 7, 8, 9, 10, 11, 12, 13, 14, 15; forest type, investment per unit area, amount of work per unit area, and management measures as No 16, 17, 18, 19; landscape type, and landscape fragmentation degree as No 20, 21; stocking volume per unit area, growth increment per unit area, input-output ratio, and employees' quality as No 22, 23, 24, 25 respectively.

Table 3. The layers and their factors in the interpretive structural model

Layers	Factors
VII (Top)	1, 2, 3
VI	16, 20, 21, 22, 23
V	24
IV	19, 25
III	6, 15, 17, 18
II	5, 7, 8, 9, 10, 11, 12, 13, 14
I (Bottom)	4

4 Discussion

(i) Forest ecosystem management decision-making is very complicated system engineering with multitudinous factors, including the environmental and socioeconomic factors outside of the system as well as the ones of its own. In stead of being isolated among them, all the factors are interwoven, interacted on and affected one another, they have been proved to be in various layers, and the factors in the lower layers dominate those in the higher layers, thus the factors in the lowest layer at the bottom are the most important and influential ones [24, 25].

(ii) Forest ecosystem is a multiple-objective management system [26], and its objectives include economic, social and ecological ones that compose the top layer of the ISM for forest ecosystem management decision-making.

(iii) The first layer at the bottom of the ISM consists of socioeconomic factor indicated by the ratio between agricultural production value and industrial production value. Since the ratio is one of important indices in socioeconomic development, it influences the values of a society greatly and the development trend of forestry as well. With the development of economy, people tend to require multiple demands including ecological ones from forestry rather than the traditional single need for timber. No longer should forestry pursue just economic benefits, therefore, it should pay more attention to social and ecological benefits instead [27].

(iv) The second layer from the bottom of the ISM consists chiefly of the factors of climatic environments that affect the distribution of forests [28] and produce the ecotypes of forests which, in turn, are the objects and bases of forest ecosystem management since different management measures should be taken according to different forest types. The factors are also the important factors for the division for forest ecosystem management while the division is a base of the ecosystem management. Consequently, the factors are selected in this layer.

(v) The third and the fourth layers are mainly composed of the factors of intensive management extent and management measures. Forestry production is implemented according to different types of the forest ecosystem management division, and relevant intensive management extents and measures are adopted to insure the realization of the objectives of forest ecosystem management. The factors as such are arranged in these two layers respectively.

(vi) The fifth layer is made up of the factor of input-output ratio that is also regarded as special one. Considering that an input-output ratio not only reveals how well a forest ecosystem is managed, but also affects the investment for the ecosystem management, the factor itself covers a layer. As the input-output ratio is proved to be an important factor that deals with whether forest ecosystem management is intensive or not, thus showing close effects on the factors in the sixth layer, the factor of input-output ratio is arranged in the fifth layer.

(vii) The sixth layer consists of both the factors of landscapes and productivity of a forest ecosystem. The types of a forest ecosystem are determined by the landscape factors, and every forest ecosystem type, in turn, has its own ecological service functions, whereas the productivity factors determine the degrees of realizing the system functions. As a result, the factors of landscapes and productivity are located at the top of the ISM as the seventh layer.

Acknowledgements. The authors would like to thank the scientists working at the foundations supported by the grant from the National Natural Science Foundation of China (No.30771724), by the grant from the Guangdong Province Science Foundation (8152408801000007), by the grant from the Cooperative Fund for Production, Education & Research of Education Ministry and Guangdong Province of PR China(2008A030203007) and by the grant from the Key Project Foundation by Educational Department of Hunan Province of China (No.05A027).

References

1. Franklin, J.: The New Forestry. Journal of Soil and Water Conservation 44, 549 (1989a)
2. Franklin, J.: Toward a New Forestry. American Forests 6, 69–74 (1989b)
3. Franklin, J.: Thoughts on Applications of Silvicultural Systems under New Forestry. Forest Watch 10(7), 8–11 (1990)
4. Thomas, J.W.: FEMAT: Objective, Process, and Options. Journal of Forestry 92(4), 66–70 (1994)
5. Xu, D.Y., Zhang, X.Q.: Forest Ecosystem Management — the Focus of the 21st Century (in Chinese, with English abstract). World Forestry Research 2, 1–7 (1998)
6. Xu, G.Z.: Promote the Development of Forest Management by Attach More Importance to the Study on the Complexity of System and Management (in Chinese, with English abstract). Forest Inventory and Planning 4, 1–4 (2002)
7. Chen, X.L.: Structure Exploration of Forest Ecosystem Management (in Chinese, with English abstract). J. of Central South Forestry University 19(2), 43–47 (1999)

8. Xu, G.Z.: Ecological Problems and Forest Ecosystem Management (in Chinese, with English abstract). Central South Forest Inventory and Planning 23(1), 1–5 (2004)
9. Warfield, J.N.: Binary Matrices in System Modeling. IEEE Transactions on Systems, Man, and Cybernetics 3(5), 441–449 (1973)
10. Warfield., J.N.: Structuring Complex Systems. Battelle Memorial Institute Monograph, Columbus, Ohio, vol. 4 (1974a)
11. Warfield, J.N.: Developing Subsystem Matrices in Structural Modeling. IEEE Transactions on Systems, Man, and Cybernetics 4(1), 74–80 (1974b)
12. Warfield, J.N.: Toward Interpretation of Complex Structural Models. IEEE Transactions on Systems, Man, and Cybernetics 4(5), 405–417 (1974c)
13. Warfield, J.N.: Implication Structures for Systems Interconnection Matrices. IEEE Transactions on Systems, Man, and Cybernetics 6(1), 18–24 (1976)
14. Kenneth, L.R.: Structural Examination of Identity in an Individual with Severe Physical Disabilities. Journal of Developmental and Physical Disabilities 9(2), 91–100 (1997)
15. Aber, J.D., Melillo, J.M., Mcclaugherty, C.A.: Predicting Long-Term Patterns of Mass Loss, Nitrogen Dynamics and Soil Organic Matter Formation from Initial Fine Litter Chemistry in Temperate Forest Ecosystems. Canadian Journal of Botany 68, 2201–2208 (1990)
16. Frangi, J.L., Lugu, A.E.: Biomass and Nutrient Accumulation in Ten Year Old Bryophyte Communities Inside a Flood Plain in the Luquillo Experimental Forest, Puerto Rica. Biotropica 24, 106–112 (1992)
17. Leone, A., Marini, R.: Assessment and Mitigation of the Effects of Land Use in a Lake Basin (Lake Vicoin in Central Italy). Journal of Environmental Management 39, 39–50 (1993)
18. Pu, Y.S., Zhang, Z.Y., Pu, L.N., Hui, C.M.: Biodiversity and Its Fragility in Yunnan, China. Journal of Forestry Research 18(1), 39–47 (2007a)
19. Pu, Y.S., Zhang, Z.Y., Pu, L.N.: Strategic Studies on the Biodiversity Sustainability in Yunnan Province, Southwest China. Forestry Studies in China 9(3), 225–237 (2007b)
20. Laudon, K.C., Laudon, J.P.: Management Information Systems: New Approaches to Organization and Technology. Prentice Hall, Upper Saddle River (1998)
21. Wilson, J.B., Agnew, A.D.Q.: Positive Feedback Switches in Plant Communities. Advances in Ecological Research 23, 263–336 (1992)
22. Christine, F., Harald, V., Carsten, L., Franz, M., Vilem, P., Vladimir, J.: Meeting the Challenges of Process-Oriented Forest Management. Forest Ecology and Management 248(1-2), 1–5 (2007)
23. Andrew, P.S.: System Engineering: Methodology & Application. IEEE Press, Los Alamitos (1977)
24. Richard, H.W.: Interpretive Structural Modeling — a Useful Tool for Technology Assessment. Techn. Fcst. Soc. Chg. (2), 165–185 (1978)
25. Watson, R.H.: Interpretative Structural Modeling — a Useful Tool for Technology Assessment. Technological Forecasting and Social Change 11(2), 165–185 (1978)
26. Costanza, R.: The Value of the World's Ecosystem Services and Natural Capital. Nature 387, 253–256 (1997)
27. Frederick, C., Patrice, H., Erin, S.: Policy Instruments to Enhance Multi-functional Forest Management. Forest Policy and Economics 9(7), 833–851 (2007)
28. Daniel, W.M., John, H.P.: Spatial Models of Site Index Based on Climate and Soil Properties for Two Boreal Tree Species in Ontario. Forest Ecology and Management 175(1-3), 497–507 (2003)

Structure of Mutualistic Complex Networks

Jun Kyung Hwang, Seong Eun Maeng, Moon Yong Cha, and Jae Woo Lee

Department of Physics, Inha University, Incheon 402-751 Korea
jaewlee@inha.ac.kr
http://statmech.inha.ac.kr

Abstract. We consider the structures of six plant-pollinator mutualistic networks. The plants and pollinators are linked by the plant-pollinating relation. We assigned the visiting frequency of pollinators to a plant as a weight of each link. We calculated the cumulative distribution functions of the degree and strength for the networks. We observed a power-law, linear, and stretched exponential dependence of the cumulative distribution function. We also calculated the disparity and the strength of the nodes $s(k)$ with degree k. We observed that the plant-pollinator networks exhibit an disassortative behaviors and nonlinear dependence of the strength on the nodes. In mutualistic networks links with large weight are connected to the neighbors with small degrees.

Keywords: complex network, food web, mutualistic network, disparity, disassortative.

1 Introduction

Complex networks appear in many areas such as social, economic, and biological systems[1-20]. Food webs are the well known networks in ecology. Recently the complex structures of the food web are an interesting topic in the complex systems. Ecological networks are very interesting networks with weights of links. The species strongly interact with some species, but another species interact weakly with some species. We can characterize the interacting strength as a weight of link in ecological networks. Therefore, ecological network is a typical example of weighting complex networks.

In ecological network the total number of nodes is smaller than abiotic networks. The total number of nodes is less than a thousand in the reported ecological networks. These small numbers of nodes make difficulty in analysing the ecological network. Network properties of the food webs were reported for many prey-predator and mutualistic networks[21-27]. In the mutualistic network such as plant-pollinator network, the species interact with each other and the network is described by bipartite graph of species interaction[28-37]. The total number of visits by pollinators to plants consider as the strength of interaction. The pollinators have some favoring plants. They visit a plant frequently, but rarely another plants. The weight of link is assigned by the strength of the interaction. In the plant-pollinator network, pollinators prefer some plants. The

J. Zhou (Ed.): Complex 2009, Part I, LNICST 4, pp. 954–959, 2009.
© ICST Institute for Computer Sciences, Social Informatics and Telecommunications Engineering 2009

network properties of the mutualistic networks are important to understand the dynamics and diversity of the ecological system. In this work we consider three plant-pollinator mutualistic networks. We calculate the cumulative distribution function of the degree and strength for the networks. We observed very rich functional dependence of the CDF. This article is organized as followings: In section 2 we introduce the information for data. In section 3 we present the results of the CDF for networks and the properties of the mutualistic network. In section 4 we give the concluding remarks.

2 Data of Food Web

We considered six plant-pollinator food webs. We selected these food web because they had enough nodes to analyze the network properties. We presented basic informations for networks in Table 1. Bar food web (Barrett and Helenurm 1987) was a breal forest in Canada with $N_{po} = 102$ pollinators and $N_{pl} = 12$ plants[32]. Ino food web (Inouye and Pkye 1998) belonged to montane forest which consists of $N_{po} = 91$ pollinators and $N_{pl} = 42$ plants[33]. Mem food web (Mommett 1999) had $N_{po} = 79$ pollinators and $N_{pl} = 25$ plants and it belongs to meadow food web[34]. Kat food web (Kato et al. 1990) was a beech forest food web and contained $N_{po} = 679$ pollinators and $N_{pl} = 93$ plants[35]. Oll food web (Ollerton et al. 2003) was an upland grassland which consists of $N_{po} = 56$ pollinator and $N_{pl} = 9$ plants[36]. Sma food web (Small 1976) was a peat bog food web and it contained $N_{po} = 34$ pollinators and $N_{pl} = 13$ plants[37]. We summarized the total number of links L and connectance $C = L/N^2$ where $N = N_{po} + N_{pl}$ is the total number of species in the food web.

Table 1. Structural properties of six plant-pollinator food web

food web	type	pollinator	plant	link	weight	connectance
Bar	boreal forest	102	12	167	550	0.0129
Ino	montane forest	91	42	281	1459	0.0159
Kat	beech forest	679	93	1202	2384	0.0020
Mem	meadow	79	25	299	2183	0.0276
Oll	upland grassland	56	9	103	594	0.0244
Sma	peat bog	34	13	141	992	0.0638

3 Structure of Mutualistic Networks

We calculated the network properties from the mutualistic networks such as the cumulative probability distribution function(CDF) of the degree and strength, the average disparity $Y(k)$, and the average strength $s(k)$ of the node with the degree k. When we calculated the probability distribution function, we considered all species in the food web including the pollinators and plants. We assigned

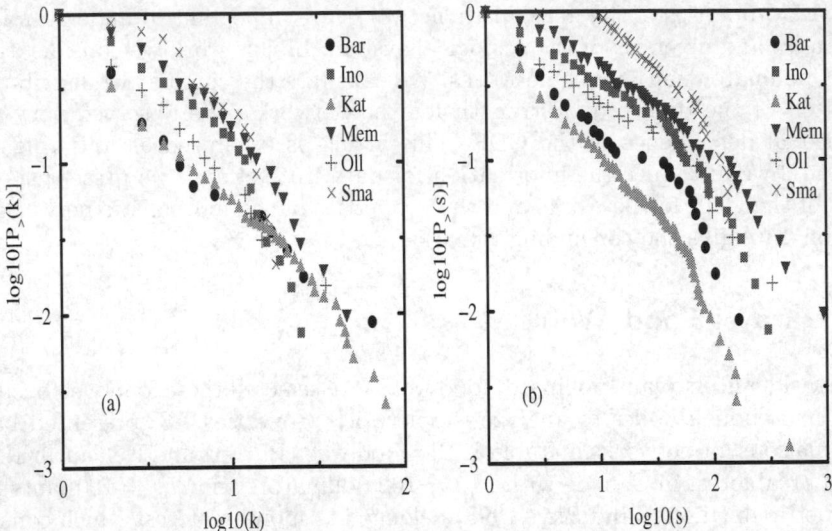

Fig. 1. The log-log plot of the cumulative probability distribution function of (a) the degree and (b) strength for the six mutualistic pollinator-plant networks. When we calculate the distribution function, we include all species in the network.

the weighting w_{ij} of the link as the visiting frequencies of the pollinators to the plants. Then, the network constructed was a weighted mutualistic network.

In Fig. 1 we present the CDF of the degree amd strength for all species in the six food webs. The degree distribution is not universal and it depends on the food webs. The CDF of the degree for all species also follows the power law except Mem and Sma food web. The stretched exponential functions are the best fitting functions in Ino, Mem and Sma food webs. Similarly, we also consider the CDF of the strength for all species. We observe the CDF of the strength for all species follows the power law. The CDF of the strength in the Mem food web,we observe two scaling regions. We summerize the functional forms of the CDF.

Table 2. The functional forms of the cumulative probability distribution function for the degree and strength in six plant-pollinator food webs. PL means the power-law and PL(2) means two regions of the power-law. SE means the stretched exponential dependence.

food web	degree	strength
Bar	PL	PL
Ino	SE	PL
Kat	PL	PL
Mem	SE	PL(2)
Oll	PL	PL
Sma	SE	PL

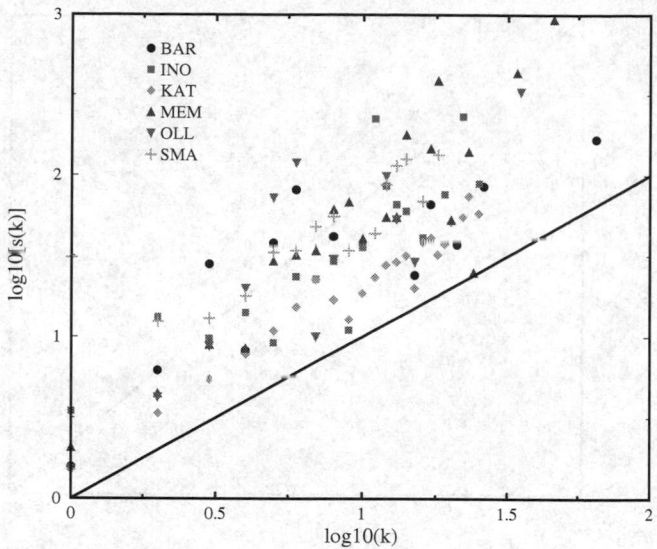

Fig. 2. The log-log plot of the average strength as a function of the degree. The solid line is a linear function $s(k) = k$.

The strength s_i of the node is defined by

$$s_i = \sum_{j=N_i} w_{ij} \tag{1}$$

where N_i is the nearest neighbors of the species i. In the homogeneous linear weighted network, the average strength $s(k)$ is proportional to the degree k. In Fig. 2 we present the average strength as a function of the degree. All mutualistic network shows the nonlinear dependence on the degree. The average strength has the functional form like $s(k) \sim k^\beta$. We calculate the exponents β by the least square fits. We obtain the exponents, $\beta =0.9(2)$(Bar), $1.1(2)$(Ino), $1.11(5)$(Kat), $1.43(7)$(Mem), $1.3(2)$(Oll), and $0.74(16)$(Sma).

The disparity of a node i is defined by

$$Y_i = \sum_{j=N_i} \left(\frac{w_{ij}}{s_i} \right)^2 \tag{2}$$

If the links have the comparatively same weights, then the average disparity is inversely proportional to the degree, $Y(k) \sim 1/k$. If the weight of a link or a few link are dominated, the average disparity is constant, $Y(k) \sim 1$. In general the average disparity shows the power law, $Y(k) \sim k^{-\delta}$. We observe that the mutualistic network exhibits an disassortative properties. Larger weight links are connected to the neighbor with smaller degree, or vice verse. In Fig. 3 we show the average disparity as a function of the degree. The data points are above the

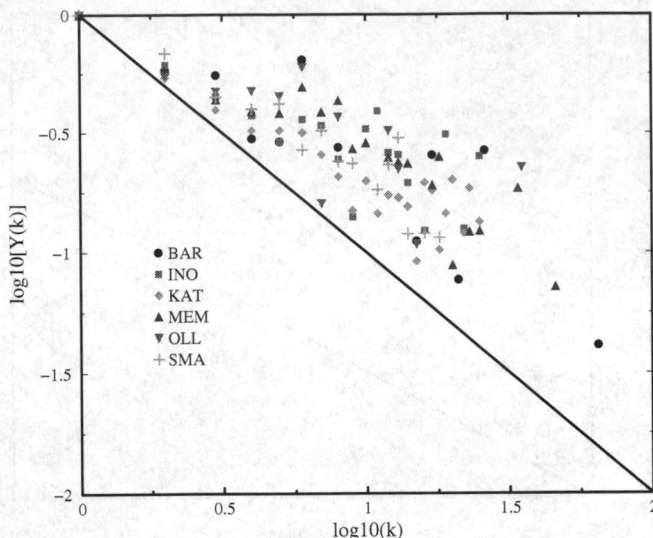

Fig. 3. The log-log plot of the disparity $Y(k)$ as a function of the node. The solid line is a inverse function $Y(k) = 1/k$.

$1/k$ function. We obtain the exponents δ =0.7(1) (Bar), 0.49(9) (Ino), 0.58(6) (Kat), 0.62(7) (Mem), 0.5(1) (Olll), and 0.71(7) (Sma).

In the mutualistic network, the network is strongly influenced by the competitions between the pollinators to get the pollens and the conditions of the environment. The plant prefers to connect to many pollinators to disperse their pollens. However, the pollinators favor plants to collect their foods easily. These competing interactions determine the structure of the networks.

4   Conclusions

We consider six mutualistic pollinators-plant food webs. The CDF of the degree is classified into the power-law and stretched exponential functions. The CDF of the strength follows the power law in all food webs. We calculate the average strength as a function of the degree and the average disparity. The average strength shows a nonlinear dependence on the degree. From the average disparity we observed that the pollinator-plant mutualistic network exhibits a disassortative behavior.

Acknowledgments. This work has been supported by Research Fund of Inha University. We use the data of the food web in the interaction web database.

References

1. Watts, D.J., Strogatz, S.H.: Nature 393, 440 (1998)
2. Barabási, A.-L., Albert, R.: Science 286, 509 (1999)

3. Strogatz, S.H.: Nature 410, 268 (2001)
4. Albert, R., Barabási, A.-L.: Rev. Mod. Phys. 74, 47 (2002)
5. Newman, M.E.J.: SIAM Review 45, 167 (2003)
6. Dorogovtsev, S.N., Mendes, J.F.F.: Adv. Phys. 51, 1079 (2002)
7. Albert, R., Jeong, H., Barabási, A.-L.: Nature 401, 130 (1999)
8. Adamic, L.A., Huberman, B.A.: Science 287, 2115 (2000)
9. Pastor-Satorras, R., Vespignani, A.: Phys. Rev. Lett. 86, 3200 (2000)
10. Stanley, M.H.R., Amaral, L.N., Buldyrev, S.V., Havlin, S., Leschorn, H., Maass, P., Salinger, M.A, Stanley, H.E.: Nature 379, 804 (1996)
11. Lee, K.E., Lee, J.W.: J. Korean Phys. Soc. 44, 668 (2004)
12. Lee, K.E., Lee, J.W.: J. Korean Phys. Soc. 47, 185 (2005)
13. Axtell, R.L.: Science 293, 1818 (2001)
14. Lee, J.W., Lee, K.E., Rikvold, P.A.: Physica A 364, 355 (2006)
15. Hong, B.H., Lee, K.E., Lee, J.W.: Phys. Lett. A 361, 6 (2007)
16. Mantegna, R.N.: Eur. Phys. J. B 11, 193 (1999)
17. Bonanno, G., Caldarelli, G., Lillo, F., Mantegna, R.N.: Phys. Rev. E 68, 046130 (2003)
18. Onnela, J.-P., Chakraborti, A., Kaski, K., Kertesz, J., Kanto, A.: Phys. Rev. E 68, 056110 (2003)
19. Jeong, H., Tombor, B., Albert, R., Oltvai, Z.N., Barabási, A.-L.: Nature 407, 651 (2000)
20. Garlaschelli, D., Caldarelli, G., Pietronero, L.: Nature 423, 165 (2003)
21. Scotti, M., Podani, J., Jordan, F.: Ecol. Lett. 4, 148 (2007)
22. Jordano, P., Bascompte, J., Olsen, J.M.: Ecol. Lett. 6, 69 (2003)
23. Cohen, J.E.: Food web and niche space. Princeton University Press, Princeton (1978)
24. Pimm, S.L., Lawton, J.H., Cohen, J.E.: Nature 350, 669 (1991)
25. Beiser, L.F., Sugihara, G.: Proc. Natl. Acad. Sci. USA 94, 1247 (1997)
26. Dunne, J.A., Williams, R.J., Martinez, N.D.: Ecol. Lett. 5, 558 (2002)
27. Williams, R.J., Berlow, E.L., Dunne, J.A., Barabasi, A.L., Martinez, N.D.: Proc. Natl. Acad. Sci. USA 99, 12913 (2002)
28. Jordano, P.: Am. Nat. 129, 657 (1987)
29. Olesen, J.M., Jordano, P.: Ecology 83, 2416 (2002)
30. Thompson, J.N.: Oikos 84, 5 (1999)
31. Waser, N.M., Chittka, L., Price, M.V., Williams, N.M., Ollerton, J.: Ecology 77, 1043 (1996)
32. Barret, S.C.H., Helenurm, K.: Canadian J. Botany 65, 2036 (1987)
33. Inouye, D.W., Pyke, G.H.: Aust. J. Ecol. 13, 191 (1988)
34. Memmott, J.: Ecol. Lett. 2, 276 (1999)
35. Kato, M., Kakutani, T., Inouye, T., Itino, T.: Contrib. Biol. Lab. Kyto Univ. 27, 309 (1990)
36. Ollerton, J., Johnson, S.D., Cranmer, L., Kellie, S.: South Africa Annals of Botany 92, 807 (2003)
37. Small, E.: Canadian Field Naturalist 90, 22 (1976)

Strong Dependence of Infection Profiles on Grouping Dynamics during Epidemiological Spreading

Zhenyuan Zhao[1], Guannan Zhao[1], Chen Xu[2],
Pak Ming Hui[3], and Neil F. Johnson[1,*]

[1] Physics Department, University of Miami, Coral Gables, FL 33146, USA
njohnson@physics.miami.edu
[2] Department of Physics, Suzhou University,
Suzhou 215006, People's Republic of China
[3] Department of Physics, Chinese University of Hong Kong,
Shatin, Hong Kong, China

Abstract. The spreading of an epidemic depends on the connectivity of the underlying host population. Because of the inherent difficulties in addressing such a problem, research to date on epidemics in networks has focused either on static networks, or networks with relatively few rewirings per timestep. Here we employ a simple, yet highly non-trivial, model of dynamical grouping to investigate the extent to which the underlying dynamics of tightly-knit communities can affect the resulting infection profile. Individual realizations of the spreading tend to be dominated by large peaks corresponding to infection resurgence, and a generally slow decay of the outbreak. In addition to our simulation results, we provide an analytical analysis of the run-averaged behaviour in the regime of fast grouping dynamics. We show that the true run-averaged infection profile can be closely mimicked by employing a suitably weighted static network, thereby dramatically simplifying the level of difficulty.

Keywords: complex systems, networks, epidemics, group dynamics.

1 Introduction

The way in which a given virus, idea, rumour or activity spreads throughout a population, will depend on the underlying connectivity structure of the network describing the population. If the population is sufficiently well connected that it can be considered well-mixed, then the standard approximations of mass-action epidemiology should apply [1]. Much research in the complex systems literature has focused on understanding spreading phenomena in static networks with limited connectivity [1,2,3]. More recently, researchers have begun to relax the assumptions of a static network, by allowing rewirings from one timestep to the next [4,5,6,7,8].

The implicit complexity of competing time and length scales, means that there is still an outstanding challenge to describe spreading phenomena in networks in which there may be an arbitrary number of rewirings per timestep, and where communities

---

* Corresponding author.

J. Zhou (Ed.): Complex 2009, Part I, LNICST 4, pp. 960–970, 2009.

can abruptly change their identity and number in time. This is particularly important since dynamical grouping processes, in which groups coalesce and fragment over time, are thought to occur widely in biological and social systems [9,10,11,12,13].

In this paper, we focus on this problem by considering a variety of spreading processes in a population which is characterized by a particularly simple, but highly non-trivial, dynamical grouping mechanism [11]. Our focus is to understand how this underlying group dynamics (GD) affects the spreading process for a number of standard epidemiological models. The shift from grouping models to dynamical networks, can be achieved simply by allowing the network nodes to represent individuals within the population – and then specifying that any two individuals have a strong link whenever they are within the same group, but have a negligibly weak link when they are in different groups. Since the groupings change in time, so too will the links between them. Most importantly, as groups break up (fragment) or join together (coalesce), a variable number of links will be broken or formed. At the two extremes, a very large group of size $\sim N$ breaking up will represent the net loss of $\sim N(N-1)/2$ links in a given timestep, while two groups of size $\sim N/2$ joining together will represent an increase in the number of links from $\sim 2.\frac{N}{2}(\frac{N}{2}-1)/2$ to $\sim N(N-1)/2$. This shift in descriptive emphasis from groups to networks also means that the group size distribution $n(k)$, which is proportional to the probability that a randomly chosen node is in a group of size k, is simply related to the network connectivity distribution $P(k)$, which is the probability that a randomly chosen node has k contacts.

Our interest in this paper is confined to problems in which there is no obvious spatial component. In other words, instead of dealing with the well-studied patch setups typified by the spread of a real virus between geographically distinct regions [1], we are interested in populations within which the connections have a more general and variable topology – such as communities on the Internet, or in a financial market – and hence are arguably better described by general coalescence and fragmentation probabilities with no direct spatial interpretation. With these applications in mind, we employ a generalization of a simple but highly non-trivial model of grouping dynamics that exhibits a power-law distribution of group sizes with an exponential cutoff [11,13]. In addition to being qualitatively consistent with a variety of real-world social phenomena, including financial market behaviour [11,13], the model has the beauty of just featuring two parameters: the probability that a chosen group fragments in a given timestep ν_{frag}, and the probability that it instead coalesces with another group ν_{coal} [11,13]. We consider the dynamics of several different types of epidemic model: SIR, SIS, SIRS. Using standard terminology [1], individual nodes are either **S**usceptible, **I**nfected or **R**ecovered. There are two particular forms of infection process $S \rightarrow I$ which one might reasonably consider using. The linear form states that the susceptible node gets infected with rate $i.p$ if it is connected to i infected nodes; the more correct nonlinear form says that if it is connected to i infected nodes, the probability will be $1 - (1 - p)^i$ [3]. We use this more correct latter version in our simulations, noting that for $p \ll 1$ they tend to converge as can be seen from a simple binomial expansion of the nonlinear form.

In the limit of a well-mixed population, the mass-action differential equations for all these standard epidemic processes can be written in the following compact form:

$$\frac{dS}{dt} = \mu - (1 - (1-p)^I)S - \mu S + \gamma I + \omega R$$

$$\frac{dI}{dt} = (1 - (1-p)^I)S - qI - \mu I - \gamma I \qquad (1)$$

$$\frac{dR}{dt} = qI - \mu R - \omega R.$$

where p is the infection rate, q is the recovery rate, μ is birth and death rate, γ is the rate of transmission from I to S, and ω is the rate of transmission from R to S.

We now discuss the initial setup for the simulations. Prior to the initial infection, we suppose that the grouping dynamics have been underway for a sufficiently long time that the system has reached its steady state. At some arbitrary time which we call $t = 0$, the largest group is selected and an arbitrary individual in this group becomes infected. This mimics a situation of real-world relevance, such as a financial market [11,13], in which there is a power-law steady-state distribution of group sizes or communities, and then one member of the population gets infected with a virus. Comparing the time scales for epidemic spreading and grouping dynamics, our investigations show that there are three broad regimes of subsequent spreading behaviour:

(1) The grouping dynamics are much slower than the epidemic spreading: The virus then tends to remain within the initial group. The spreading dynamics therefore become equivalent to mass-action spreading in a population whose size is equal to the initial group size. The infection is oblivious to the larger population of susceptibles. This large population of susceptibles is granted an accidental but effective immunity, by happening to be members of different groups at the right time.

(2) The grouping dynamics are of comparable timescale to the epidemic spreading: The grouping dynamics can now play a significant role in suppressing or amplifying the spreading, because they will determine the contact condition for the infected nodes at any given timestep. Due to the existence of relatively large groups which can change their membership on a timescale similar to the spreading, a rather generic spiky infection profile $I(t)$ tends to emerge.

(3) The grouping dynamics are much faster than the epidemic spreading: The virus now sees a time-average of the network's connectivity, where this average is taken over some suitably large time-window. In the limit of very fast grouping dynamics, this static, time-averaged network will correspond to a fully-connected weighted network, in stark contrast to the sparse, disconnected network which is characteristic of individual timesteps. Section 3 explains how we have managed to develop an analytical analysis in this regime, based on a suitably averaged measure of the grouping dynamics.

2 Epidemics within a Population Undergoing Dynamical Grouping

Since the dynamics describing the spreading of an infectious disease will depend on the contact structure which underlies the population, our simulations and analysis are chosen to highlight this interplay between epidemic spreading and the grouping dynamics – in particular, the virus is able to spread within groups but not between groups. As a result of the grouping dynamics, the infectious nodes can themselves move between groups as a result of the fragmentation-coalescence process, and hence propagate the virus. As mentioned earlier, we wish to mimic aspects of modern-day human interaction, e.g. long distance travel and communication, as well as rapidly evolving social networks in cyberspace, hence the grouping dynamics we use is independent of any spatial length scale. In addition, the model allows a group's size and membership to change over time – moreover, groups may disappear or form in time and hence the total number of groups may change in time. We consider a fixed population N of individual objects (e.g. people). At each timestep, a group or cluster is chosen by random-picking a node among the N possible nodes, and then looking to see to which group this node belongs at that timestep. In other words, groups are picked proportional to their group size, reflecting the fact that any one individual is equally likely to initiate a group formation or group-breakup event. With probability ν_{frag}, the group fragments, while with probability ν_{coal} a second group is picked in a similar way, to join with the first one [11,13]. Naturally, $\nu_{frag} + \nu_{coal} \leq 1$.

2.1   SIR Process within the Dynamical Grouping Model

Here we consider the results for the basic SIR process passing through the dynamical grouping model. In the language of the well-mixed mass-action equations of Eq. 1, this process is described by setting μ, γ, and ω equal to zero (see Fig. 1(a)).

Figures 1(b) and (c) show the fraction of the population N who are infected at time t, obtained in a single run of the simulation. It is remarkable that even though the profile looks reasonably standard in Fig. 1 (b), the profile in Fig. 1(c) is completely different in character. Most noticeably, Fig. 1(c) features multiple peaks and a far slower decay than that of Fig. 1(b). Indeed, the overall decay of infection in Fig. 1(b) is of order 20 timesteps while that in Fig. 1(c) is 2000, which is two orders of magnitude larger. Despite the fact that there are only four parameters – the p and q values of the SIR process together with ν_{frag} and ν_{coal} for the dynamical grouping – our exploration of the parameter space shows that it is easy to produce a broad range of qualitatively different behaviors [14]. Common to many of the runs, are the features of multiple, large peaks and a very slow, possibly fat-tailed decay. These $I(t)$ profiles cannot be reproduced in general by the well-mixed SIR model, either in its deterministic form (Eq. 1) or in stochastic simulation. Instead, the changing contact structure acts to drive these large fluctuations and slow decay. Even if we employ a weighted network (Fig. 1(d)) as described in Section 3, the individual run profiles $I(t)$ *and* the run-averaged behaviour $I(t)_{\text{RunAvge}}$ differ significantly from it.

2.2   SIS Process within the Dynamical Grouping Model

We now turn to the standard SIS process (Fig. 2(a)) in which any infected individual recovers to the susceptible state – in other words, there is no immunity conferred by having had the disease [1]. To analyze this case in the well-mixed limit corresponding to the mass-action equations of Eq. 1, we set q, μ and ω equal to zero.

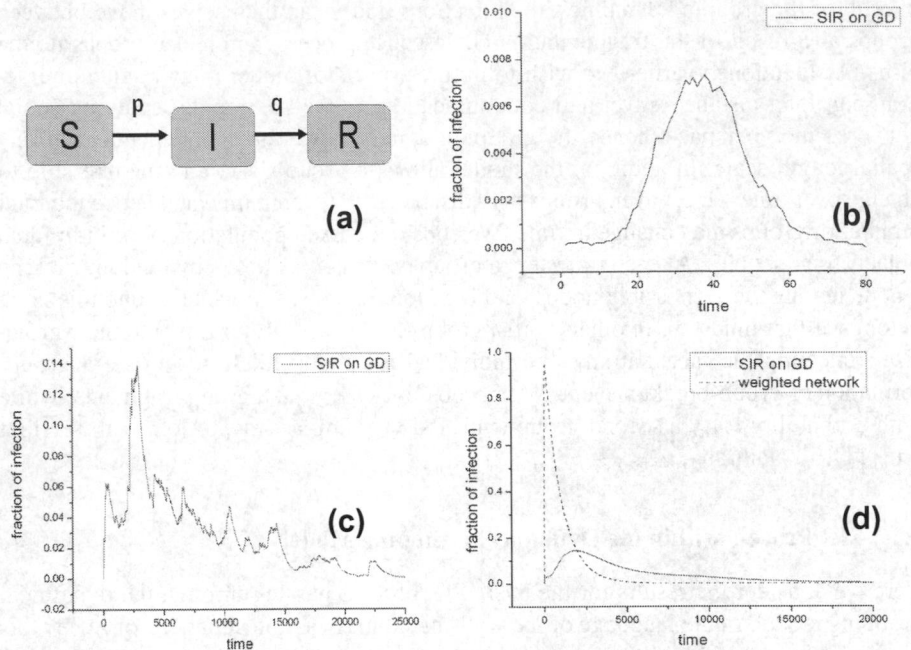

Fig. 1. SIR process within the dynamical grouping model. (a) Schematic of the SIR process. (b) An individual run showing the infection profile $I(t)$ under GD, which corresponds to the fraction of the entire population N who are infected at time t. Here $N = 10^4$, $\nu_{frag} = 0.001$, $\nu_{coal} = 0.99$, $p = 0.001$ and $q = 0.1$. (c) An individual run showing $I(t)$ for $N = 10^4$, $\nu_{frag} = 0.01$, $\nu_{coal} = 0.9$, $p = 0.001$ and $q = 0.001$. (d) Solid red curve shows the run-averaged profile $I_{\mathrm{RunAvge}}(t)$, obtained by averaging $I(t)$ over many runs. Dashed blue curve shows the run-averaged $I(t)$ for a *static* weighted network, where nodes i and j are connected by a time-independent link of strength $P_{i,j}$. The same parameters are used as in panel (c).

The finite value of $I(t)$ in Fig. 2(b) corresponds to an endemic equilibrium, and is reached after a slow growth. Both these features arise as a direct result of the suppressing effect of the grouping dynamics. Specifically, the grouping dynamics only allows a small fraction of the entire population N to be in contact with an infected individual at a given timestep. The remainder of the population are isolated from the infection since they are located in groups comprising entirely susceptibles. As a result, before any infected node can contact the rest of the population and infect them, it gets turned back to susceptible.

Comparing Figs. 2(b) and (c) shows how changing the probabilities γ and ν_{frag}, affects the stability of the $I(t)$ curve in the steady state. Increasing γ aids in producing large jumps because it replenishes the pool of susceptibles. Likewise reducing the fragmentation probability ν_{frag} increases the effective timescale over which groups break up – this in turn allows larger spikes to appear by allowing time for significant decay between jumps. Once again, the weighted network cannot reproduce the run-averaged behavior, as evidenced in Fig. 2(d).

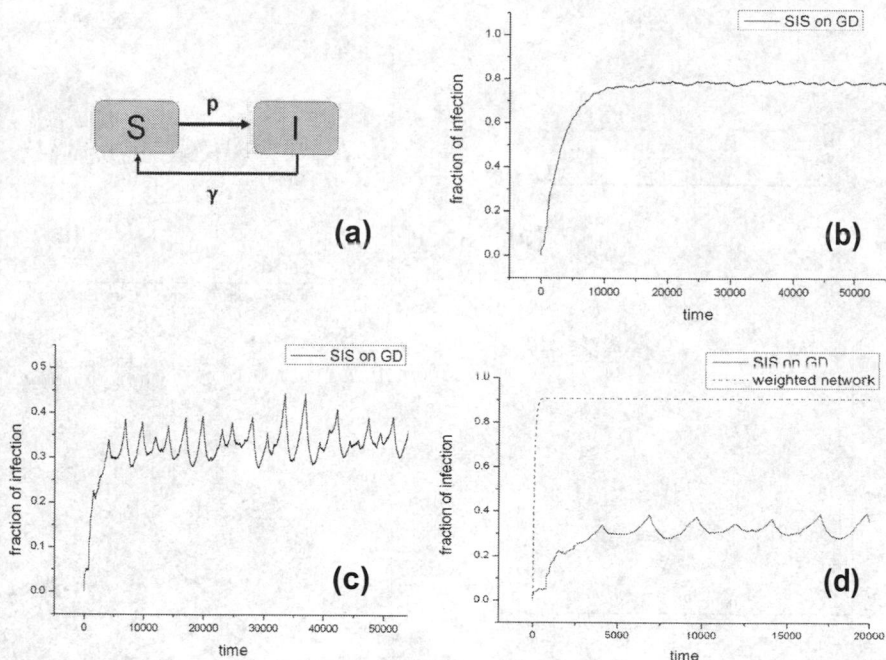

Fig. 2. SIS process within the dynamical grouping model. (a) Schematic of the SIS process. (b) An individual run showing the infection profile $I(t)$ with $N = 10^4$, $\nu_{frag} = 0.01$, $\nu_{coal} = 0.99$, $p = 0.01$ and $\gamma = 0.0001$. (c) An individual run showing $I(t)$ with $N = 10^4$, $\nu_{frag} = 0.001$, $\nu_{coal} = 0.99$, $p = 0.01$ and $\gamma = 0.001$. (d) Solid red curve shows the run-averaged profile $I_{\mathrm{RunAvge}}(t)$, obtained by averaging $I(t)$ over many runs. Dashed blue curve shows the run-averaged $I(t)$ for a *static* weighted network, where nodes i and j are connected by a time-independent link of strength $P_{i,j}$. The same parameters are used as in panel (c).

2.3   SIRS Process within the Dynamical Grouping Model

The SIRS process is summarized in Fig. 3(a), and corresponds to setting μ and γ equal to zero in Eq. 1. As can be seen in Figs. 3(b) and (c) it is qualitatively similar to SIS. However, the presence of a temporary, intermediate R-state (i.e. waning immunity) allows the dynamics to build up even stronger decays – hence the observable peaks appear even stronger as compared to the SIS case.

2.4   Demographic SIR Process within the Dynamical Grouping Model

Finally we discuss the introduction of demography into the SIR model, which we call SIRD. This epidemiological process corresponds to setting γ and ω equal to zero in Eq. 1 – in addition, we assume that the natural mortality is μ. In other words, each individual has a lifespan given by $1/\mu$ [1]. In order to keep the total population constant $S + I + R = N$, μ is also taken to represent the birth rate within the population (Fig. 4(a)). Figure 4(b) shows that when the birth-death processes are much slower than the infection and recovery processes, it is effectively the same as the SIR model

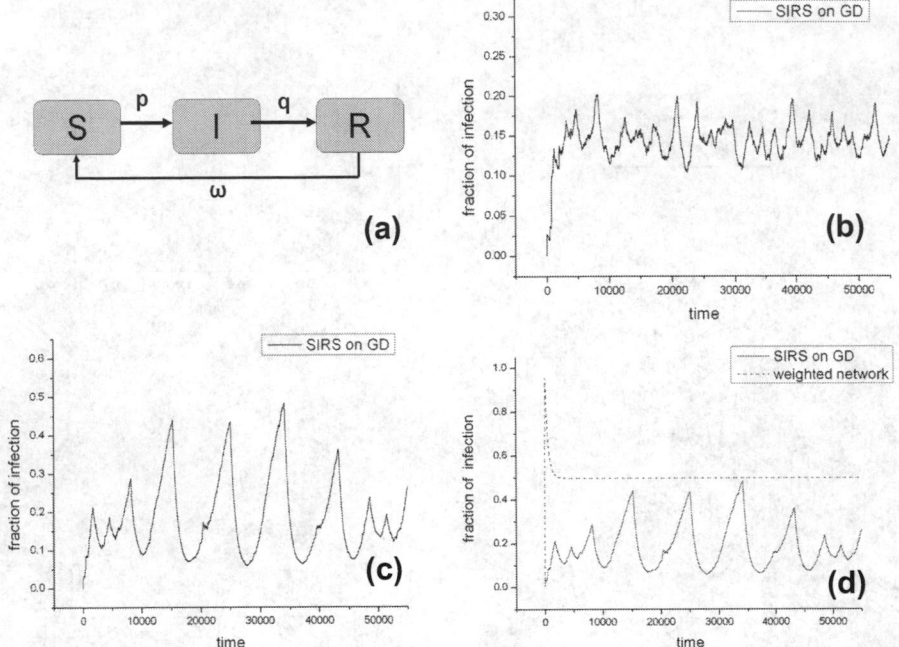

Fig. 3. SIRS process within the dynamical grouping model. (a) Schematic of the SIRS process. (b) An individual run showing the infection profile $I(t)$ with $N = 10^4$, $\nu_{frag} = 0.01$, $\nu_{coal} = 0.99$, $p = 0.01$, $q = 0.001$ and $\omega = 0.001$. (c) An individual run showing $I(t)$ with $N = 10^4$, $\nu_{frag} = 0.001$, $\nu_{coal} = 0.99$, $p = 0.01$, $q = 0.001$ and $\omega = 0.001$. (d) Solid red curve shows the run-averaged profile $I_{\mathrm{RunAvge}}(t)$, obtained by averaging $I(t)$ over many runs. Dashed blue curve shows the run-averaged $I(t)$ for a *static* weighted network, where nodes i and j are connected by a time-independent link of strength $P_{i,j}$. The same parameters are used as in panel (c).

within the dynamical grouping model, as discussed earlier. The associated endemic equilibrium is shown not to be stable in Fig. 4(b), however this stability returns by increasing μ, as shown by comparing Fig. 4(b) to Fig. 4(c). Again, the results for the run-averaged case cannot be reproduced by the static weighted network, as evidenced by Fig. 4(d).

3   Regime of Fast Grouping Dynamics

So far, we have established that new features occur in the $I(t)$ profile both for individual runs, and when run-averaged. In particular, the $I(t)$ curve typically has more peaks and has a slower general decay – in short, it appears more bursty or 'noisy', however we emphasize that this 'noise' is actually meaningful (and interpretable) dynamics within our model, resulting from the intrinsic group formation and breakup processes of coalescence and fragmentation. Even though other group fusion-fission mechanisms could be introduced and the same procedure followed to investigate the various epidemiological processes, we expect the features that we have found to remain qualitatively similar.

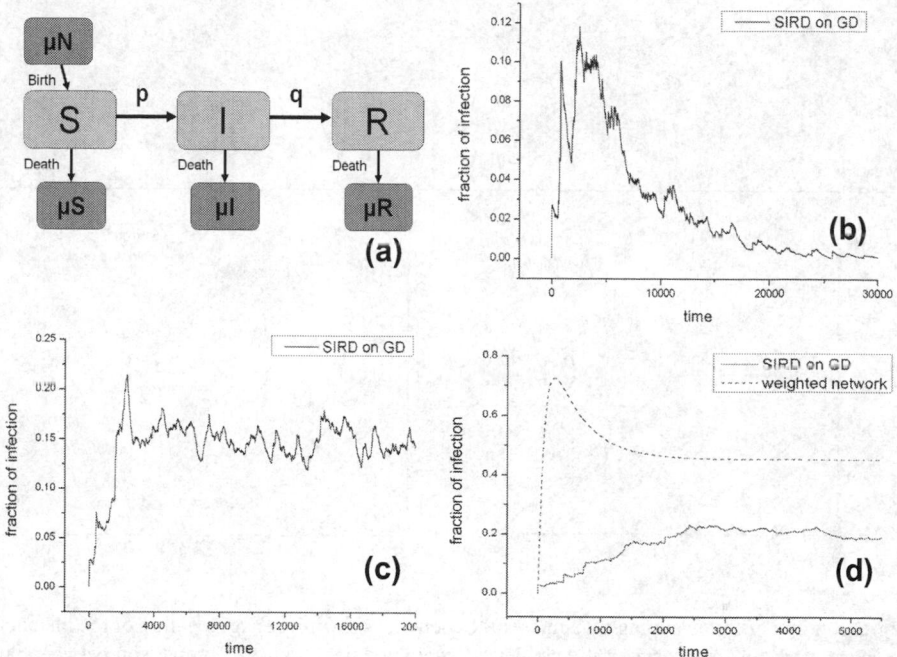

Fig. 4. SIR process with demography, within the dynamical grouping model. (a) Schematic of the SIRS process. (b) An individual run showing the infection profile $I(t)$ with $N = 10^4$, $\nu_{frag} = 0.01$, $\nu_{coal} = 0.99$, $p = 0.01$, $q = 0.001$ and $\mu = 0.000001$. (c) An individual run showing $I(t)$ with $N = 10^4$, $\nu_{frag} = 0.01$, $\nu_{coal} = 0.99$, $p = 0.01$, $q = 0.001$ and $\mu = 0.001$. (d) Solid red curve shows the run-averaged profile $I_{RunAvge}(t)$, obtained by averaging $I(t)$ over many runs. Dashed blue curve shows the run-averaged $I(t)$ for a *static* weighted network, where nodes i and j are connected by a time-independent link of strength $P_{i,j}$. The same parameters are used as in panel (b), but with $\mu = 0.001$.

This is indeed exactly what we find when, for example, we consider another popular dynamical grouping mechanism such as that introduce by Levin [9].

Looking back at the first case that we discussed, of the basic SIR process within the dynamical grouping model, there are essentially four timescales: the group fragmentation timescale $\tau_{frag} \sim \nu_{frag}^{-1}$, the group coalescence timescale $\tau_{coal} \sim \nu_{coal}^{-1}$, the infection timescale $\tau_p \sim p^{-1}$, and an individual's recovery timescale $\tau_q \sim q^{-1}$. The first two timescales correspond to the timescale of the grouping dynamics, while the latter two are instead related to the epidemic process itself. Although there are only four parameters, and they are all physically meaningful, this is still a very large parameter space to analyze – hence for brevity, we will here focus on the regime of fast grouping dynamics, i.e. the first two timescales τ_{frag} and τ_{coal} are much smaller than $\tau_p \sim p^{-1}$ and $\tau_q \sim q^{-1}$.

With fast group breaking and merging processes, one can imagine that the grouping dynamics would manage to refresh or reshuffle the whole population within an infection/recovery period. In other words, the heterogenous population would become

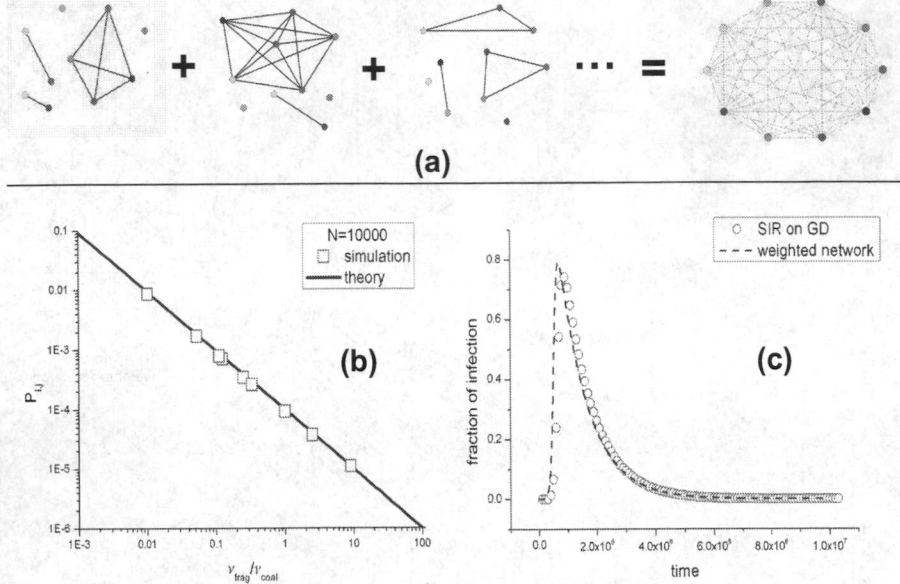

Fig. 5. Regime of fast grouping dynamics. (a) Schematic showing the aggregation of the instantaneous network links, to form a fully connected, weighted network. (b) Comparison between our analytic result for $P_{i,j}$ and numerical results obtained directly from the simulation. (c) Results for the run-averaged infection profile I_{RunAvge} obtained by simulation (red circles) and using the generalized mass-action partial differential equations of Eq.5. Here $\nu_{frag} = 0.05$, $\nu_{coal} = 0.95$, $p = 10^{-6}$, and $q = 10^{-6}$.

effectively homogenous as far as the spread of epidemics is concerned. In this limit, we find that we can capture the features of the run-averaged dynamics using a suitably averaged quantity to capture the average network properties. With this purpose in mind, we focus on a probability $P_{i,j}$ corresponding to the averaged probability that two random nodes i and j are connected. Figure 5(a) shows the effect of aggregating the network dynamics over some sufficiently large time-window that it eventually becomes a fully connected, weighted network. $P_{i,j}$ is the resulting weight of the links, which for our grouping dynamics then satisfies the following master equation:

$$\frac{dP_{i,j}}{dt} = -P_{i,j}^3 \frac{1}{N} \nu_{frag} + (1 - P_{i,j}) \frac{1}{N^2} P_{i,j}^2 \nu_{coal}, \tag{2}$$

where we have assumed that all $P_{i,j}$'s are equal. The first term on the right-hand side of the master equation corresponds to the fragmentation of a group, while the second term corresponds to the coalescence of two groups into one larger group. Solving Eq. 2 by setting the left-hand side equal to zero, yields:

$$P_{i,j} = \frac{1}{1 + N \frac{\nu_{frag}}{\nu_{coal}}}. \tag{3}$$

This expression for $P_{i,j}$ fits remarkably well with simulation results, as evidenced by the excellent agreement in Fig. 5(b). In the limit $N\nu_{frag} \gg \nu_{coal}$, Eq. 3 reduces to the more approximate form

$$P_{i,j} = \frac{\nu_{coal}}{N\nu_{frag}} \quad . \tag{4}$$

Using this result, the effective infection rate becomes $p \cdot P_{i,j}$ instead of p. In this particular regime, the mass-action partial differential equations from Eq. (1) can be modified to the following approximate form for the SIR process in the presence of a dynamically evolving network:

$$\frac{dS}{dt} = -pP_{i,j}SI$$
$$\frac{dI}{dt} = pP_{i,j}SI - qI \tag{5}$$
$$\frac{dR}{dt} = qI \quad .$$

Because of the small values of the epidemic parameters (i.e. small p, q, μ, γ, and ω) we can use the linear form for the infection process. The situation therefore becomes a virus spreading on top of a static weighted network, where all the nodes are fully connected with each other and where the links all have a uniform strength, $P_{i,j}$. Figure 5(c) shows that a good fit can be achieved for the SIR process. As a result, the basic reproductive ratio R_0 should instead be written as R_0^*, with $p \cdot P_{i,j}$ replacing p. For a large population $N \to \infty$, we therefore obtain $R_0^* = \frac{p\nu_{coal}}{qN\nu_{frag}}$.

4 Discussion

We have addressed the fascinating theoretical, yet practically relevant, question of how a population's underlying grouping dynamics might affect epidemiological spreading processes. We have considered the specific case of a coalescence-fragmentation grouping model subject to various standard epidemiological processes. This led to the appearance of multiple peaks and a slower overall decay, as compared to the usual mass-action limit for well-mixed populations. Such features are not uncommon in empirical infection data both in real viral infections, and the profile of activity in online communities [15]. For the specific regime of fast group dynamics, we were able to provide a modified mathematical model to explain the run-averaged infection profile.

We hope that our work provides useful insight and additional motivation to other researchers in the field of complex networks and complex systems, in the common quest to develop a general theory of epidemic spreading in the presence of arbitrarily complex population dynamics – in particular, in the presence of internal group formation and breakup processes.

P.M.H. acknowledges the support of a grant CUHK-401005 from the Research Grants Council of the Hong Kong SAR Government.

References

1. Keeling, M.J., Rohani, P.: Modeling Infectious Diseases in Humans and Animals. Princeton University Press, New York (2007)
2. Pastor-Satorras, R., Vespignani, A.: Epidemic dynamics in finite size scale-free networks. Phys. Rev. E 65, 035108–035112 (2002)

3. Petermann, T., Rios, P.D.L.: The role of clustering and gridlike ordering in epidemic spreading. Phys. Rev. E 69, 066116 (2004)
4. Watts, D.J., Muhamad, R., Medina, D.C., Dodds, P.S.: Multiscale, resurgent epidemics in a hierarchical metapopulation model. Proc. Natl. Acad. of Sci. 102, 11157–11162 (2005)
5. Gross, T., Dommar, C., Blasius, B.: Epidemic dynamics on an adaptive network. Phys. Rev. Lett. 96, 20–23 (2006)
6. Gross, T., Blasius, B.: Adaptive Coevolutionary Networks: A Review. J. R. Soc. Interface 5, 259–271 (2008)
7. Colizza, V., Vespignani, A.: Invasion threshold in heterogenous metapopulation networks. Phys. Rev. Lett. 99, 148701–148705 (2007)
8. Shaw, L.B., Schwartz, I.B.: Noise induced dynamics in adaptive networks with applications to epidemiology. E-print arXiv:0807.3455 on xxx.lanl.gov
9. Gueron, S., Levin, S.A.: The dynamics of group formation. Mathematical Biosciences 128, 243–246 (1995)
10. Gonzalez, M.C., Hidalgo, C.A., Barabási, A.-L.: Understanding individual human mobility pattens. Nature 453, 779–782 (2008)
11. Eguíluz, V.M., Zimmermann, M.G.: Transmission of information and herd behaviour: an application to financial markets. Phys. Rev. Lett. 85, 5659–5662 (2000)
12. McDonald, M., Suleman, O., Williams, S., Howison, S., Johnson, N.F.: Impact of unexpected events, shocking news, and rumors on foregin exchange market dynamics. Phys. Rev. E 77, 046110–046122 (2008)
13. Johnson, N.F., Jefferies, P., Hui, P.M.: Financial Market Complexity. Oxford University Press, Oxford (2003)
14. Zhao, Z., Calderon, J.P., Xu, C., Hui, P.M., Johnson, N.F.: (in preparation)
15. Sornette, D., Deschâtres, F., Gilbert, T., Ageon, Y.: Endogenous Versus Exogenous Shocks in Complex Networks: An Empirical Test Using Book Sale Rankings. Phys. Rev. Lett. 93, 228701 (2004)

Statistical Properties of Cell Topology and Geometry in a Tissue-Growth Model

Patrik Sahlin[1], Olivier Hamant[2], and Henrik Jönsson[1]

[1] Computational Biology & Biological Physics, Department of Theoretical Physics,
Lund University, Sölvegatan 14A, SE-223 62 Lund, Sweden
{sahlin,henrik}@thep.lu.se
[2] INRA, CNRS, ENS, Université de Lyon, 46 Allée d'Italie, 69364 Lyon Cedex 07,
France
Olivier.Hamant@ens-lyon.fr

Abstract. Statistical properties of cell topologies in two-dimensional tissues have recently been suggested to be a consequence of cell divisions. Different rules for the positioning of new walls in plants have been proposed, where e.g. Errara's rule state that new walls are added with the shortest possible path dividing the mother cell's volume into two equal parts. Here, we show that for an isotropically growing tissue Errara's rule results in the correct distributions of number of cell neighbors as well as cellular geometries, in contrast to a random division rule. Further we show that wall mechanics constrain the isotropic growth such that the resulting cell shape distributions more closely agree with experimental data extracted from the shoot apex of *Arabidopsis thaliana*.

1 Introduction

Cell division in plants has been studied by plant biologists for over one hundred years (see review in [1]). From simple microscope observations biologists have formulated rules for cell division. During mitosis plant cells are divided into two daughter cells by introducing a dividing cell wall. Hofmeister suggested a rule where new cell walls are formed perpendicular to the main axis of growth, i.e. perpendicular to the main axis of the cell [2]. Sachs noted that new walls form almost perpendicular to old walls [3]. Similarly to Hofmeister's rule, Errara's rule state that the division is along the shortest path dividing the mother cell into two parts of equal volume [4]. More recently, experiments where spherical cells have been compressed into oval shapes agrees with these rules [5,6]. It has also been seen that the arrangement of cytoskeletal structures reveal the placement of new cell walls [1].

Many biological tissues develop in two-dimensional sheets. The epidermal layer in plants is an example, where anticlinal divisions and the lack of cell migration assure the two-dimensional structure of the layer. The epidermal layer can then be described by a network of connected polygons (cells), edges (walls) and vertices, where the connections are updated at divisions only. The predominant existance of three-vertices leads to the average number of cell neighbors (walls)

J. Zhou (Ed.): Complex 2009, Part I, LNICST 4, pp. 971–979, 2009.

to be six following Euler's rule. But the average can be fulfilled by many neighbor distributions, and already in the 1920's F. T. Lewis studied this in growing and proliferating cucumber epithelia [7,8]. He found that although most cells had six neighbors (47%), the distribution was not symmetric, with more five-sided cells (25%) compared to seven-sided (22%). He also noted the non-existance of triangular cells as well as cells with more than nine neighbors. Recently, Gibson *et. al.* found similar asymmetric distributions in epithelial cell layers in several animal tissues including epithelia from *Drosophila melanogaster* wing primordium [9]. Interestingly, they also introduced a probabilistic model where a discrete Markov chain was used to describe topological updates due to cell divisions and the model was able to predict the experimental distribution of the number of cell neighbors.

The approach by Gibson *et al.* focused on the topology of the cells in the tissue, but disregarded details of cellular geometry, growth and proliferation. Here, we use a two-dimensional cell-based tissue growth model to analyze how explicit division rules and wall mechanics lead to different topological as well as cellular shape distributions. We assume isotropic growth, and mainly study tissues with quite homogeneous cell sizes. This resembles the situation in the plant shoot apex, and we compare our models with novel data from the *Arabidopsis thaliana* shoot apical meristem.

Our model allow us to compare Errara's classical division rule (new walls are placed such that the cell is divided into two equally sized daughters along the shortest path) with a random-direction division rule. We also investigate how wall mechanics constraining the purely isotropic growth affect the topology and cell geometry. We compare the resulting tissues with the distributions of the number of neighbors (topology) as well as cell shapes (geometry) from experimental data.

2   Materials and Methods

2.1   Experimental Data

The model results are compared with the experimental data presented in Gibson *et al.* [9], as well as new data from the shoot apex in *Arabidopsis thaliana* (Fig. 1). The shoot data was extracted from a confocal projection using the merryproj software [10]. It is interesting to note that although the statistics is sparse, the overall topological distribution in the shoot data is very similar to the *Drosophila* case as well as the Lewis data [7,8].

2.2   Tissue Model

The model is a two-dimensional model where the spatial degrees of freedom are for vertices, which are connected via edges that represent cell walls. Each cell is described by a polygon, i.e a number of vertices together with corresponding edges. The vertex positions are updated viscously, where we assume that velocities are proportional to the forces acting upon them. The cell walls are treated

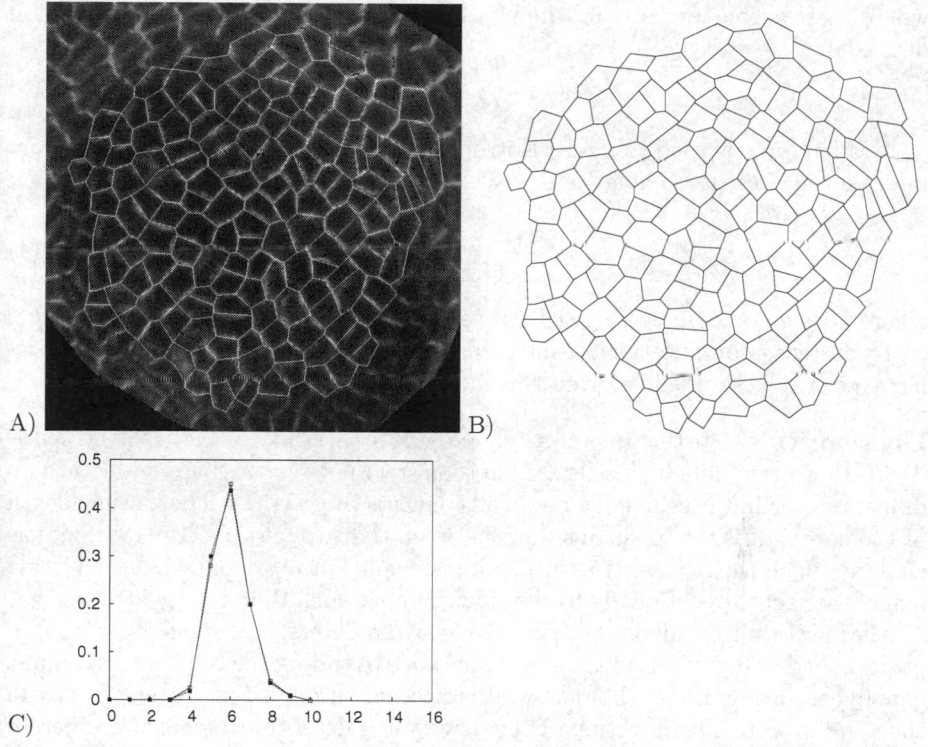

Fig. 1. Data from an image of the meristem of *Arabidopsis* is compared with data from *Drosophila* presented by Gibson *et al.*. A) The original image taken with confocal microscopy. Cell walls are marked manually. B) The template extracted from the image. C) The distribution of the number of neighbors for *Arabidopsis thaliana* marked with filled black squares (■) (110 cells) and *Drosophila melanogaster* marked with empty white squares (□) (2,172 cells) [9].

as mechanical springs. We describe the dynamics with a system of ordinary differential equations originating from an isotropic growth term, wall mechanics, and plastic wall growth. The contribution from wall springs is described by

$$\frac{d\mathbf{v}_i}{dt} = k_w \sum_{j \in \mathcal{C}(i)} \frac{\mathbf{u}_{ij}}{|\mathbf{u}_{ij}|} \left(\frac{|\mathbf{u}_{ij}| - L_{ij}}{L_{ij}} \right) , \tag{1}$$

where $\mathbf{v}_i$ is the position of vertex i, k_w is a material constant setting the strength of the wall springs, L_{ij} is the resting length of the wall spring between vertex i and j, and $\mathbf{u}_{ij} = \mathbf{v}_j - \mathbf{v}_i$. The summation is over vertices connected via edges to vertex i.

Cell walls under tension grow plastically. The change in resting length of a wall spring is

$$\frac{dL_{ij}}{dt} = k_g \Theta \left(\frac{|\mathbf{u}_{ij}| - L_{ij}}{L_{ij}} \right) , \tag{2}$$

where k_g is a constant setting the rate of growth and Θ is the ramp function defined as

$$\Theta(x) = \begin{cases} x \text{ if } x \geq 0 \\ 0 \text{ if } x < 0 \end{cases}. \tag{3}$$

A radial force is used to model isotropic growth of the tissue originating from internal cell pressure. The force on a vertex is described by

$$\frac{d\mathbf{v}_i}{dt} = k_r \mathbf{v}_i, \tag{4}$$

where k_r is a constant setting the strength of the radial force.

To decrease computational time, cells on the boundary of the tissue are removed if a cell is outside a given threshold radius, R_t.

Division Rules. If the area of a cell exceeds a threshold value, D_t, the cell is divided into two daughter cells. At division, two new vertices are added at two different walls in the cell, and a new wall connects the vertices. The resting length of the new wall is set to the distance between the two vertices. In addition, the walls at which the new vertices are added are split into two, and the new resting lengths are set proportionally to the split distance such that $L_1^{\text{new}} + L_2^{\text{new}} = L^{\text{old}}$.

The new wall dividing the cell into two daughters is defined by a spatial position and a direction. As an approximation of dividing the cell into two almost equally sized daughters, the new wall is passing through the center of mass of the dividing cell. The direction of the new wall is determined from two different rules. In the first rule, called *random direction*, a random (uniform) direction is chosen. In the second rule, called *shortest path*, the direction of the dividing wall is chosen such that the path through the cell is the shortest possible. This is the model definition of the Errara rule. To avoid four-vertices, walls that are closer to a vertex than a threshold, $w_t L_{ij}$ is moved away from the vertex to the threshold position.

2.3   Simulations

The system of ordinary differential equations is solved numerically with a 5th order Runge-Kutta ODE solver using adaptive stepsize. Data is sampled at ten different time points. As we remove cells at the boundary a new generation of cells is present at each time point. The number of cell neighbors are collected for all cells, excluding cells at the boundary. Five different intial states are used for each model to gather statistics. The initial states are all one single cell represented by a regular polygon with three, five, seven, nine, or eleven vertices. Data from the 50 different time points is averaged to give final distributions for each model.

We use in house developed software allowing for discrete updates between each time step taken by the numerical ODE solver. In these updates we check cells for division and removal. Parameter values used in the simulations are presented in Table 1.

Table 1. Parameter values used during simulations

Parameter	With mechanics	Without mechanics
k_w	0.05	-
k_g	0.01	-
k_r	0.05	0.05
D_t	1	1
R_t	10	10
w_θ	0.1	0.1

3   Results

3.1   Comparing Different Division Rules

First we compared the topology distributions from the two different cell division rules. The result is presented in Figs. 2A and C, where the data from the simulations are presented together with the experimental distributions. For both division rules the distributions have their maximum at six cell neighbors, but while the distribution for the shortest path division rule match the experimental data well, the distribution for the random direction division rule is broader compared to the experimental data.

3.2   Removal of Spring Wall Mechanics

To study how removal of wall mechanics affects the distribution of cell neighbors, we kept the radial force that drives the isotropic growth, but removed the cell wall springs.

The result is presented in Figs. 2B and D. What might be found surprising is that the differences between the distributions with and without wall spring mechanics are very small. This is true for both the shortest path division rule and the random direction division rule.

3.3   A Quantative Measurement of Cell Shapes

A striking result from our simulations is that even if the mechanics has no or little effect on the distributions, there is an obvious visual difference of cell shapes between simulations with and without mechanics. Examples of cell shapes from simulations with the two different division rules are presented in Fig. 3. Clearly, the cell shapes emerging using the shortest path division rule is more plant-like, and also the simulations with mechanics look more like cellular tissue in comparison with the non-mechanical simulations.

To quantify differences in cell shape, we measure the ratio between the length of the boundary of a cell squared and the area of the cell. In Fig. 4 this measure is presented for different simulations and compared with the *in vivo* data. First, it can be seen that the shortest path division rule generally has a closer match to the experimental values than the random direction division rule. The random

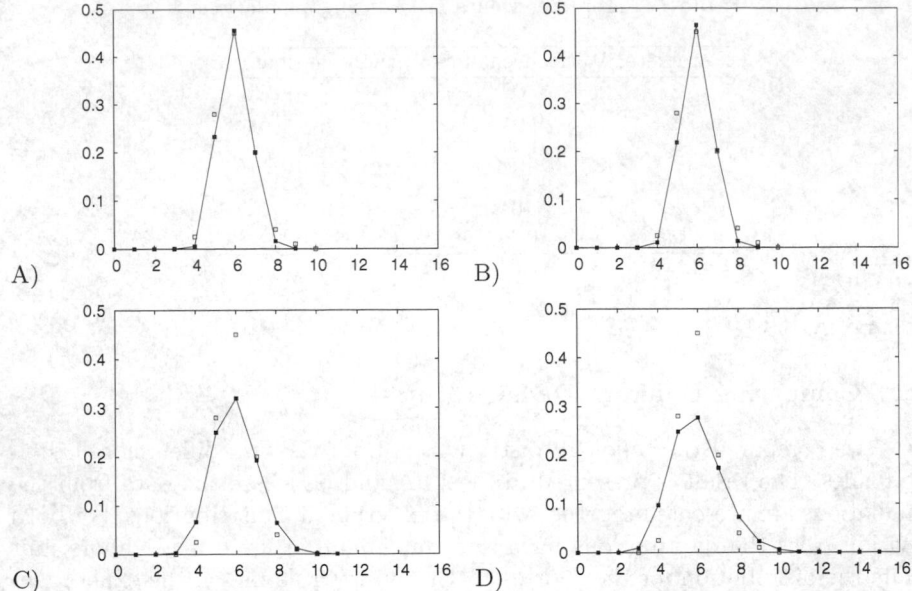

Fig. 2. Distributions of number of neighbors for simulations with different division rules with and without mechanics. Filled black squares (■) marks results from simulations while empty white squares (□) marks experimental data. A) Shortest path division rule with mechanics. B) Shortest path division rule without mechanics. C) Random direction division rule with mechanics. D) Random direction division rule without mechanics.

direction division rule not only differ more in the average value, but it also has a much larger spread among cells. A closer inspection also reveals that simulations with mechanics is closer to the experimental values than simulations without mechanics.

4   Discussion

We have used a simple two-dimensional cell-based tissue growth and proliferation model to investigate the dependence on cell division rules on statistical properties of cell topology and geometry. We used an isotropically growing tissue and showed that one of the classical rules for plant division (Errara's rule), where new plant walls appear at the shortest path that divides the cell in two equally sized dughter cells, indeed do produce a skewed topology distribution seen *in vivo* with an average of six neighbors but with more five than seven neighbor cells (Fig. 2A). On the contrary, the 'control' model with new walls placed in a random direction did not follow this distribution (Fig. 2C).

For each division rule we performed two sets of simulations, one with wall mechanics and one without. While the non-mechanics simulations follow pure isotropic growth, wall mechanics constrains the growth via a wall growth model.

Fig. 3. Examples of simulations with different division rules. Simulations with the shortest path division rule with (A) and without (B) mechanics. Simulations with the random division rule with (C) and without (D) mechanics.

When comparing distributions of number of neighbors there was no difference between the sets with mechanics and the simulations without mechanics, although a visual difference could be seen, where the simulations with mechanics produced more plant-like cells (Fig. 3).

To investigate this further we quantified cell shapes and compared with novel data from the *Arabidopsis* shoot apex. We could again see that Errara's division rule produced a statistical distribution very similar to our measured data, while the random division rule produced far more asymmetric cell shapes. In this case, we could also see a small difference between simulations with or without mechanics, where including wall mechanics generated shapes more similar to the *in vivo* data (Fig 4).

In conclusion, we have showed that statistical properties of cell topology and shape indeed can be used to discern among different model hypotheses for cell proliferation. Interestingly, divisions at random directions do not lead to correct

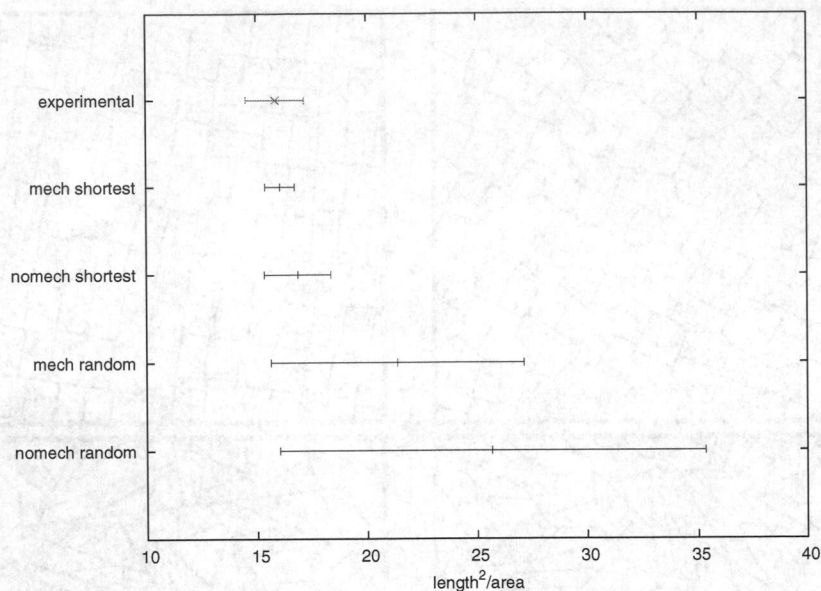

Fig. 4. Mean and standard deviation for the ratio between length of cell boundary squared and cell area. The data point marked with a cross (×) shows the experimental data. Other data points marked with a vertical dash (|) show results from simulations. Simulations have been done with the two different division rules, with and without mechanics.

topology or cell geometries, while Errara's classical cell division rule do agree with the statistical properties from experimental data. Of course, statistical agreement is only a first test which many hypotheses may pass, and ultimately the hypotheses must be compared with statistics of single cell data. Still, we have presented a useful methodology, where explicit and mechanistic hypotheses that combine into a cellular plant growth and proliferation model can be compared on merits based on experimental data.

Acknowledgement. We thank Pawel Krupinski and Pontus Melke for discussions. This work was in part supported by the Swedish Research Council and the Human Frontier Science Program.

References

1. Smith, L.G.: Plant Cell Division: Building Walls in the Right Places. Nat. Rev. Mol. Cell. Biol. 2, 33–39 (2001)
2. Hofmeister, W.: Zusatze und Berichtigungen zu den 1851 Veröffentlichen Untersuchungengen der Entwicklung Höherer Kryptogamen. Jahrbucher für Wissenschaft und Botanik 3, 259–293 (1863)
3. Sachs, J.: Über die Anordnung der Zellen in Jüngsten Pflanzentheilen. Arb. bot. Inst. Wurzburg 2, 46 (1878)

4. Errera, L.: Über Zellformen und Siefenblasen. Botanisches Centralblatt 34, 395–399 (1888)
5. Lintilhac, P.M., Vesecky, T.B.: Stress-induced Alignment of Division Plane in Plant Tissues Grown in Vitro. Nature 307, 363–364 (1984)
6. Lynch, T.M., Lintilhac, P.M.: Mechanical Signals in Plant Development: A New Method for Single Cell Studies. Dev. Biol. 181, 246–256 (1997)
7. Lewis, F.T.: The effect of cell division on the shape and size of hexagonal cells. Anatomical Records 33, 331–355 (1926)
8. Lewis, F.T.: The correlation between cell division and the shapes and sizes of prismatic cells in the epidermis of cucumis. Anatomical Records 38, 341–376 (1928)
9. Gibson, M.C., Patel, A.B., Nagpal, R., Perrimon, N.: The Emergence of Geometric Order in Proliferating Metazoan Epithelia. Nature 442, 1038–1041 (2006)
10. de Reuille, P.B., Bohn-Courseau, I., Godin, C., Traas, J.: A Protocol to Analyse Cellular Dynamics During Plant Development. Plant J. 44, 1045–1053 (2005)

Stability of Non-diagonalizable Networks: Eigenvalue Analysis

Linying Xiang[1,3], Zengqiang Chen[1], and Jonathan J.H. Zhu[2,3]

[1] Department of Automation, Nankai University, Tianjin, 300071, China
xlyzhjl1980@gmail.com
[2] School of Journalism, Renmin University of China, Beijing, China
[3] Department of Media and Communication, City University of Hong Kong, Hong Kong

Abstract. The stability of non-diagonalizable networks of dynamical systems are investigated in detail based on eigenvalue analysis. Pinning control is suggested to stabilize the synchronization state of the whole coupled network. The complicated coupled problem is reduced to two independent problems: clarifying the stable region of the modified system and specifying the eigenvalue distribution of the coupling and control matrix. The dependence of the stability on both pinning density and pinning strength is studied.

Keywords: non-diagonalizable network, pinning control, stability, eigenvalue analysis.

1 Introduction

The last decade has eyewitnessed the birth of a new movement of research interest in the study of complex networks [1-7], which is pervading all disciplines of sciences today, ranging from physics to chemistry, biology, information science, mathematics, and even social sciences. Recently, the interplay between the complexity of the overall topology and the collective dynamics of complex networks gives rise to a host of interesting effects. Especially, there are attempts to control the dynamics of complex networks and guide it to a desired state [8-16]. Previous work on this problem [8-16] has typically focused on diagonalizable networks, where the corresponding Laplacian matrix is assumed to be diagonalizable. However, most optimal networks are non-diagonalizable [17], in particular when the networks are directed.

In this paper, the pinning control problem is further visited for nondiagonalizable networks with identical node dynamics. The main contribution of this paper is to developing a new stability analysis scheme, named eigenvalue analysis, for non-diagonalizable networks of dynamical systems. Briefly, from this approach, the network stability problem is operationalized in two independent tasks: one is to characterize the stable region of the modified system and the other to analyze the eigenvalue distribution of the coupling and control matrix. The former is determined by the dynamical rules governing the isolated node whereas the latter is determined by both the topology of the network and the control scheme.

J. Zhou (Ed.): Complex 2009, Part I, LNICST 4, pp. 980–990, 2009.

An outline of this paper is as follows. The design of pinning controllers of a non-diagonalizable network and the stability of the pinned network are discussed in Sec. 2, in which stable and unstable regions are identified. Various essentially different structures of stable regions are shown and eigenvalue distributions of different coupling matrices are investigated. The dependence of the stability on both pinning density and pinning strength is discussed based on eigenvalue analysis and numerical simulation. Finally, Sec. 3 gives a few concluding remarks.

2   Stability of Synchronization State by Pinning Control

Consider a network of N coupled dynamical systems whose state equations are written in the following form:

$$\dot{x}_i(t) = f(x_i(t)) - \sigma \sum_{j=1}^{N} L_{ij}\Gamma x_j(t), \quad i = 1, 2, \cdots, N, \tag{1}$$

where $x_i \in R^m$ represents the state vector of the i-th node, and the nonlinear function $f(\cdot)$, describing the local dynamics of the nodes, is continuously differentiable and capable of producing various rich dynamical behaviors, including periodic orbits and chaotic states. The parameter σ is positive ruling the overall coupling strength. Also, $\Gamma \in R^{m \times m}$ is a constant matrix linking coupled variables, while the real matrix $L = (L_{ij})$ is called the Laplacian matrix of the non-diagonalizable network, satisfying zero row-sum. The topological information on the network in terms of the connections and the weights is contained in the Laplacian matrix L, whose entries L_{ij} are zero if node i is not connected to node j ($j \neq i$), but are negative if there is a directed influence from node j to node i. In addition, L is not necessarily diagonalizable and symmetric because the network is not constrained to be undirected and unweighted.

Supposing the isolated node accepts a chaotic solution, our central task is to synchronize network (1) onto a prescribed state $\bar{x}$, which is the solution of the individual system $\dot{x}(t) = f(x(t))$ and satisfies $f(\bar{x}) = 0$. To do so, feedback pinning control is acted on the network (1) and the controlled network can be described as

$$\dot{x}_i(t) = f(x_i(t)) - \sigma(\sum_{j=1}^{N} L_{ij}\Gamma x_j(t) + d_i\Gamma(x_i(t) - \bar{x})), \quad i = 1, 2, \cdots, N, \tag{2}$$

where $d_i = d > 0$ if control is applied to the i-th node and $d_i = 0$ otherwise. Without loss of generality, we rearrange the order of nodes in the network such that the pinned nodes $i = 1, \cdots, l$ are the first l nodes in the rearranged network. Note that, l corresponds to the "pinning density", while d is the feedback gain to be designed and furthermore determines the "pinning strength".

We suggest that the pinned system (2) is stable at $\bar{x}$, if $\lim_{t \to \infty} \|x_i(t) - \bar{x}\| = 0$ for all $i = 1, 2, \cdots, N$.

The stability of the system (2) can be analyzed exactly by setting $e_i(t) = x_i(t) - \bar{x}$ and linearizing it at state $\bar{x}$. This leads to

$$\dot{E} = EJ^T(\bar{x}) - \sigma CE\Gamma^T, \tag{3}$$

where $J(\bar{x})$ is the Jacobian matrix of f evaluated at $\bar{x}$. $E^T = [e_1, e_2, \cdots, e_N] \in R^{m \times N}$ and $C = L + D$ with feedback gain matrix $D = diag(d_1, d_2, \cdots, d_N)$.

For convenience, we denote the matrix C as the "coupling and control matrix". It is easy to prove that the matrix C is non-diagonalizable while all the real parts of its eigenvalues are strictly positive.

Consider C written in Frobenius normal form [18], i.e.,

$$C = PBP^T = P \begin{bmatrix} B_1 & & & \\ & B_2 & & \\ & & \ddots & \\ & & & B_n \end{bmatrix} P^T, \tag{4}$$

where P is a permutation matrix and B_k are blocks of the form

$$B_k = \begin{bmatrix} \lambda_k & & & \\ 1 & \lambda_k & & \\ & \ddots & \ddots & \\ & & 1 & \lambda_k \end{bmatrix}, \tag{5}$$

where λ_k is one of the eigenvalues of C.

Introducing a transformation

$$E = P\eta, \tag{6}$$

along with (3), leads to

$$\dot{\eta} = \eta J^T(\bar{x}) - \sigma B\eta\Gamma^T. \tag{7}$$

Each block of the Jordan canonical form corresponds to a subset of these columns in η, which obeys a subset of equations in (7). If block B_k is $n_k \times n_k$, then the equations take the form

$$\dot{\eta}_1 = (J(\bar{x}) - \sigma\lambda_k\Gamma)\eta_1, \tag{8}$$

$$\dot{\eta}_2 = (J(\bar{x}) - \sigma\lambda_k\Gamma)\eta_2 - \sigma\Gamma\eta_1, \tag{9}$$

$$\cdots$$

$$\dot{\eta}_{n_k} = (J(\bar{x}) - \sigma\lambda_k\Gamma)\eta_{n_k} - \sigma\Gamma\eta_{n_k-1}. \tag{10}$$

Here $\eta_1, \eta_2, \cdots, \eta_{n_k}$ represent the modes of perturbation in the generalized eigenspace associated with eigenvalue λ_k. Clearly, η_1 converges exponentially to zero as $t \longrightarrow \infty$, if and only if all the real parts of the eigenvalues of the matrix $(J(\bar{x}) - \sigma\lambda_k\Gamma)$ are less than 0. If this condition holds and the norm of Γ is bounded, then the second term in Eq.(9) is exponentially small as well, which results in exponential convergence of η_2 to zero as $t \longrightarrow \infty$. The same argument when applied repeatedly shows that $\eta_3, \cdots, \eta_{n_k}$ must also converge to zero if all the real parts of the eigenvalues of the matrix $(J(\bar{x}) - \sigma\lambda_k\Gamma)$ are less than 0. Therefore, when all the Jordan blocks are taken into account, we see that the stability condition for the synchronous solution in the general non-diagonalizable case is

$$Re(J(\bar{x}) - \sigma\lambda_k\Gamma) < 0, \quad k = 1, 2, \cdots, n, \tag{11}$$

where $Re(\cdot)$ denotes the real part of an eigenvalue.

The significance of (11) is that the stability problem of the controlled network (2) can be separated into two independent tasks: one is to analyze the stable regions of the modified systems (8)-(10), which depends on the dynamics of the isolated node such as the Jacobian matrix $J(\bar{x})$ and the inner linking structure Γ; the other task is to analyze the eigenvalue distribution of σC, which is independent of the inner dynamics including $J(\bar{x})$ and Γ.

This criterion as shown in (11) provides the stability boundary in the complex plane. Of course, this stability boundary depends on the dynamics of the isolated node and the inner linking matrix Γ. In the following subsection we clarify various essentially different structures of stable regions.

2.1   Stable and Unstable Regions

Now we specify the well-known Rössler model as an example. A single Rössler oscillator [19] is described by

$$\begin{bmatrix} \dot{x}_1 \\ \dot{x}_2 \\ \dot{x}_3 \end{bmatrix} = \begin{bmatrix} -(x_2 + x_3) \\ x_1 + \alpha x_2 \\ x_3(x_1 - \gamma) + \beta \end{bmatrix}, \tag{12}$$

which has a chaotic attractor when $\alpha = \beta = 0.2$ and $\gamma = 5.7$. With this set of system parameters, one unstable equilibrium point is $\bar{x} = [0.007, -0.0351, 0.0351]^T$.

Here we consider the full diagonal coupling $\Gamma = diag(1, 1, 1)$ and the partial diagonal couplings $\Gamma = diag(1, 0, 0)$, $\Gamma = diag(0, 1, 0)$, and $\Gamma = diag(1, 1, 0)$ respectively. In Fig. 1 we plot the stable regions of the synchronization state $\bar{x}$ of Rössler model in the complex plane for different coupling links. The curves represent the critical condition at which the largest real part of the eigenvalues of the matrix $(J(\bar{x}) - \sigma\lambda_k\Gamma)$ is equal to zero. In the region marked by "S" (stable), the largest real part of the eigenvalues of the matrix $(J(\bar{x}) - \sigma\lambda_k\Gamma)$ is negative, while it is positive in the region marked by "U" (unstable). It is interesting to notice that the structure of the stable regions in Fig. 1 can be classified into three groups. Class (i), shown in Fig. 1(a): the critical curve is a straight line, and then larger $Re(\lambda)$ is favorable for stable synchronization of the homogenous states. Class (ii), shown in Fig. 1(b): the critical curve forms a closed circle, and then the stable region is localized in a certain finite $Re(\lambda) - Im(\lambda)$ region. Too large and too small $Re(\lambda)$ and too large $|Im(\lambda)|$ can definitely destroy the stability of the

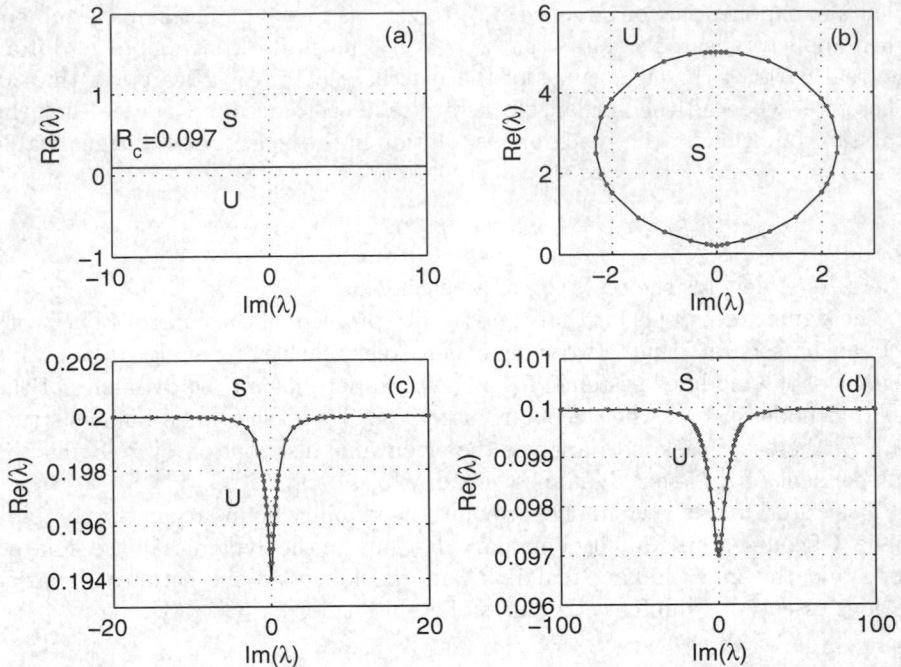

Fig. 1. Distributions of stable ("S") and unstable ("U") regions of the synchronization state $\bar{x}$ for the Rössler model. (a) $\Gamma = diag(1,1,1)$. (b) $\Gamma = diag(1,0,0)$. (c) $\Gamma = diag(0,1,0)$. (d) $\Gamma = diag(1,1,0)$. The black solid lines represent the zero maximum real part of the eigenvalues.

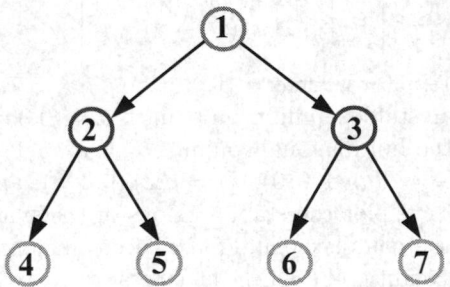

Fig. 2. A non-diagonalizable network with 7 nodes

synchronization state. Class (iii), shown in Figs. 1(c) and 1(d): the critical curve is V-shaped, then larger $Re(\lambda)$ and smaller $|Im(\lambda)|$ are favorable for stabilizing the homogenous states. Note that for the case of the full diagonal coupling, the stability is controlled by the value of $Re(\lambda)$ only [not $Im(\lambda)$], and the threshold value of the stable-unstable boundary at the imaginary axis $R_c = 0.097$.

2.2   Eigenvalue Distribution

To stabilize the synchronization state, the key point is to move all the unstable eigenvalues of L to the stable region by adding suitable control signal. In the following we consider a non-diagonalizable network with 7 nodes as shown in Fig. 2. The network involves three levels. The top level contains the unique node without input, which is the root node. The Laplacian matrix L is

$$
L = \begin{bmatrix}
0 & 0 & 0 & 0 & 0 & 0 & 0 \\
-1 & 1 & 0 & 0 & 0 & 0 & 0 \\
-1 & 0 & 1 & 0 & 0 & 0 & 0 \\
0 & -1 & 0 & 1 & 0 & 0 & 0 \\
0 & -1 & 0 & 0 & 1 & 0 & 0 \\
0 & 0 & -1 & 0 & 0 & 1 & 0 \\
0 & 0 & -1 & 0 & 0 & 0 & 1
\end{bmatrix},
$$

whose eigenvalues are $0, 1, 1, 1, 1, 1, 1$. The root node 1 must be pinned in order to synchronize the network.

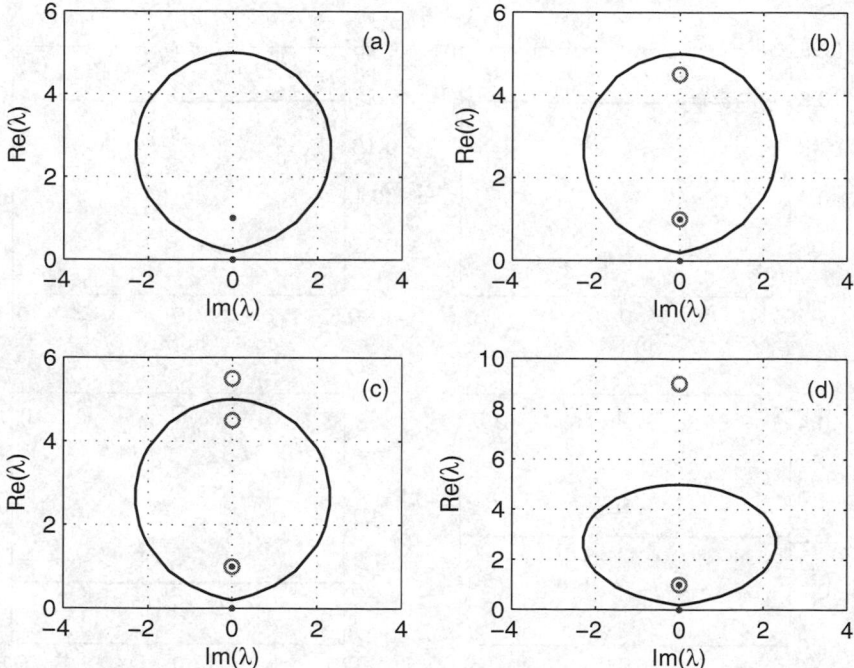

Fig. 3. Eigenvalue distribution of the matrix C: $\sigma = 1$. (a) $l = 0, d = 0$. (b) $l = 1, d = 4.5$. (c) $l = 2, d = 4.5$. (d) $l = 1, d = 9$. The blue dots represent the eigenvalues of C with $l = d = 0$, while the red empty circles represent the eigenvalues of C with $l \neq 0$ and $d \neq 0$. These notations are valid for Fig.4. The black lines denote the critical curves of the stable region in the case of Fig. 1(b).

In Fig. 3 we plot various eigenvalue distributions at $\sigma = 1$ for different d and l. It is observed from Fig. 3(a) that without control (i.e., $l = 0$ and $d = 0$), there is one unstable eigenvalue, 0, located in the unstable region. Keeping all parameters unchanged except setting $l = 1$ and $d = 4.5$ ($l = 1$ means the root node 1 is pinned), the unstable zero eigenvalue moves up and crosses the critical line and finally enters the stable region, as shown in Fig. 3(b). Continuously increasing the pinning density until $l = 2$, an interesting phenomenon occurs: one of the bottom eigenvalues first crosses the upper critical curve and then enters the unstable region, which leads to desynchronization. This feature is still observed in Fig. 3(d) when the feedback gain is increased to $d = 9$ from Fig. 3(b). It is concluded that too large d and/or l can definitely destroy the stabilization of the synchronization state.

In Fig. 4, we do the same as in Fig. 3 except that $\sigma = 0.25$ and the stable region distribution Fig. 1(c) are considered. Consider $l = 1$ and $d = 0.65$, the origin zero eigenvalue moves up as shown in Fig. 4(b). Continuously increasing pinning density until $l = 3$, there is still one nonzero eigenvalue sitting in the unstable region. Increasing pinning strength until $d = 1.95$ from Fig. 4(c), the unstable eigenvalue crosses the critical line and enters the stable region [see Fig. 4(d)].

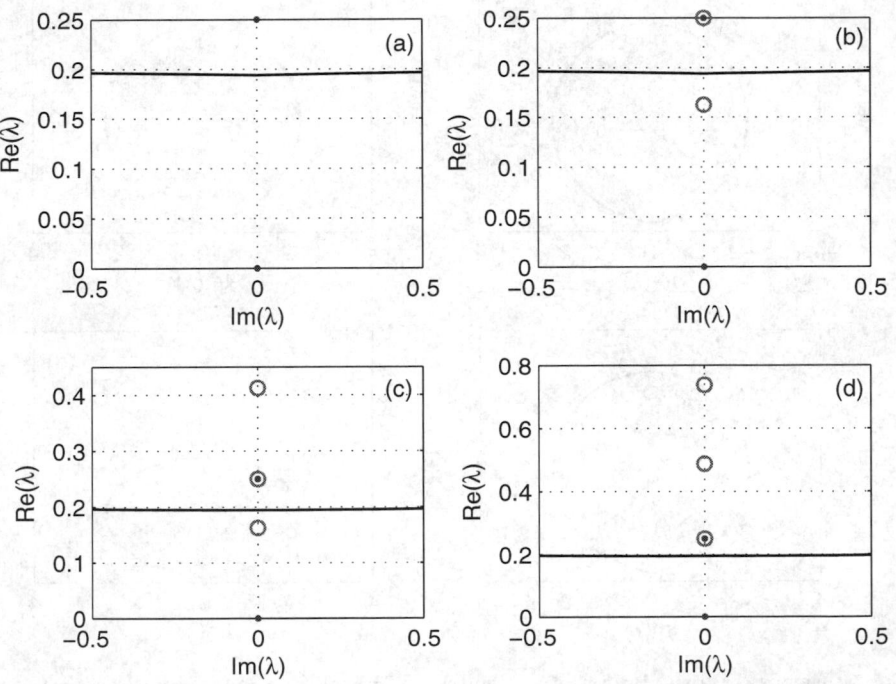

Fig. 4. Eigenvalue distribution of the matrix C: $\sigma = 0.25$. (a) $l = 0, d = 0$. (b) $l = 1, d = 0.65$. (c) $l = 3, d = 0.65$. (d) $l = 3, d = 1.95$. The black lines denote the critical curves of the stable region in the case of Fig. 1(c).

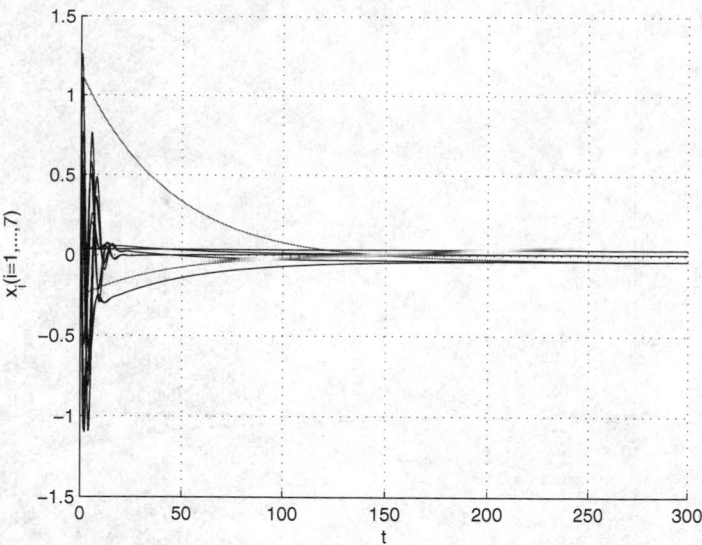

Fig. 5. The evolution of network states corresponding to Fig.3(b). The red lines denote the states of the pinned nodes. These notations are valid for Figs. 6-8.

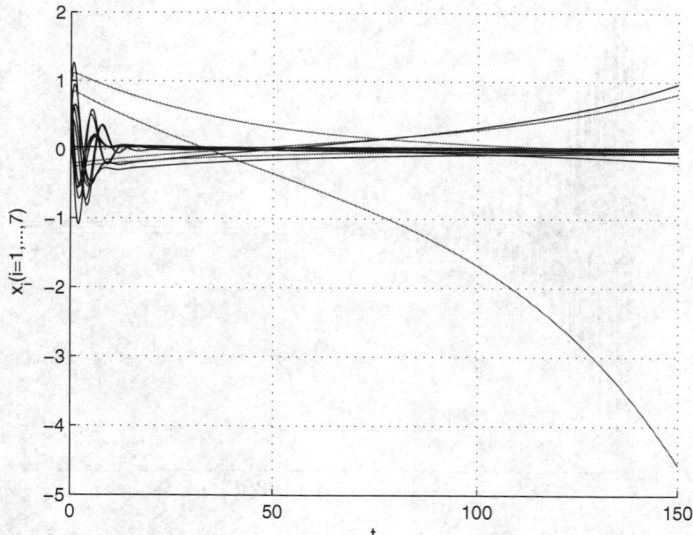

Fig. 6. The evolution of network states corresponding to Fig. 3(c)

It is concluded that increasing the pinning density and/or pinning strength can considerably enhance the controlling efficiency in the case of V-shaped region.

We should emphasize that the stabilization effect depends sensitively on the structure of stable region. For the case of the full diagonal coupling, only if the condition $\sigma\lambda > R_c$ holds, the pinned system can be stable. A different case

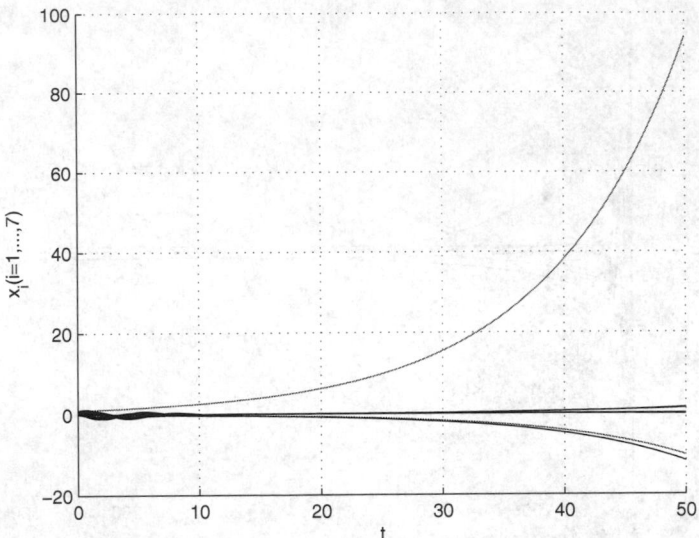

Fig. 7. The evolution of network states corresponding to Fig. 3(d)

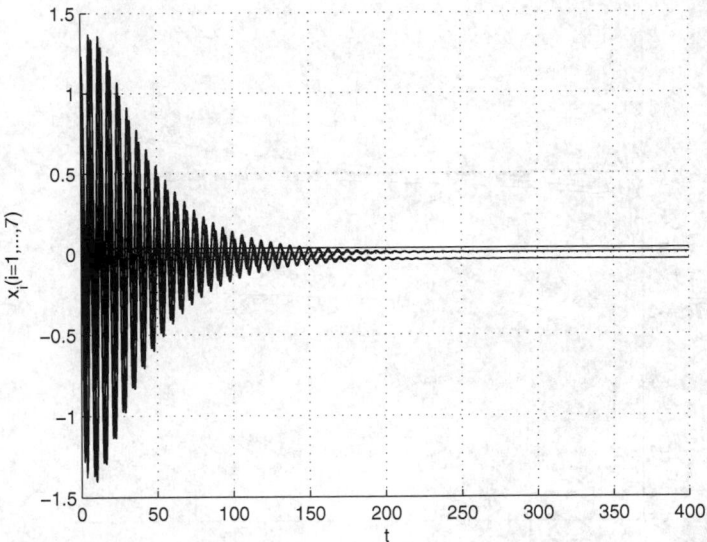

Fig. 8. The evolution of network states corresponding to Fig. 4(d)

is class (ii) structure, in which increasing sufficiently the pinning density or pinning strength definitely spoils stabilization, because some eigenvalues will be pushed upward by the control to the unstable region. This behavior is essentially different from the one displayed in Fig. 4.

Figures 5, 6, 7 and 8 show the process of controlling the 7-node network corresponding to Figs. 3(b), 3(c), 3(d) and Fig. 4(d), respectively. It is clear that the aim state is stabilized well after control, which is consistent with Figs. 3(b) and 4(d). Also, the stability is destroyed by larger l or larger d as shown in Figs. 6 and 7 when the stable region is a closed circle.

3 Conclusions

In this paper, the stability of non-diagonalizable networks under pinning control is examined by applying eigenvalue analysis. The effects of pinning density and pinning strength are investigated in detail.

Based on eigenvalue analysis, the stabilization problem of complicated high-dimensional systems can be divided into two independent problems: One is the description of stable and unstable regions of the isolated node modified by an eigenvalue forcing $\sigma \lambda_k \Gamma$ [see (8)-(10)]; the other is the eigenvalue analysis of the node coupling and control matrix C. The former is independent of the node interaction scheme and the control mechanism, whereas the latter is independent of the inner dynamics, the synchronization state and the inner linking matrix. Both problems have been solved easily. They, together, provide definite answers to the problems of stability of non-diagonalizable coupled networks. For instance, one can easily reveal and classify the stability of the synchronization state by examining whether and how some unstable eigenvalues enter into the stable region. Moreover, one can apply control matrix D to stabilize a synchronization state by moving all the unstable eigenvalues into the stable region. Therefore, the investigation of the distribution of stable and unstable regions of the isolated system becomes extremely important and widely significant for the stability problems of coupled networks with large size. The ideas in this paper can be applied to general coupled extended networks: by changing $f(x), \bar{x}$ and Γ, we can obtain different distributions of stable and unstable regions; by adjusting the control matrix D, we can flexibly change the distribution of the matrix C; and by combining all these manipulations we can stabilize the homogenous stationary state of a general coupled network.

Acknowledgments

This work was supported by the CNSF grant nos 60774088 and 60574036, the Program for New Century Excellent Talents of China (NCET-05-229), and HKRGC CERG CityU1456/06H.

References

1. Watts, D.J., Strogatz, S.H.: Collective Dynamics of 'Small World' Networks. Nature 393, 440–442 (1998)
2. Barabási, A.-L., Albert, R.: Emergence of Scaling in Random Networks. Science 289, 509–512 (1999)

3. Li, C.G., Chen, G.: Synchronization in General Complex Networks with Coupling Delays. Physica A 343, 263–278 (2004)
4. Lü, J., Chen, G.: A Time-varying Complex Dynamical Network: Model and Its Controlled Synchronization Criteria. IEEE. Trans. Autom. Control. 50, 841–846 (2005)
5. Motter, A.E., Zhou, C., Kurths, J.: Network Synchronization, Diffusion, and the Paradox of Heterogeneity. Phys. Rev. E. 71, 016116 (2005)
6. Wu, C.W.: Synchronization and Convergence of Linear Dynamics in Random Directed Networks. IEEE. Trans. Autom. Control. 51, 1207–1210 (2006)
7. Zhu, J.J.H., Meng, T., Xie, Z.M., Li, G., Li, X.M.: A Teapot Graph and Its Hierarchical Structure of the Chinese Web. In: Proceedings of the 17th International Conference on World Wide Web, pp. 1133–1134 (2008)
8. Wang, X.F., Chen, G.: Pinning Control of Scale-Free Dynamical Networks. Physica A 310, 521–531 (2002)
9. Li, X., Wang, X.F., Chen, G.: Pinning a Complex Dynamical Network to Its Equilibrium. IEEE. Trans. Circuits. Syst-I 51, 2074–2086 (2004)
10. Xiang, L.Y., Liu, Z.X., Chen, Z.Q., Yuan, Z.Z.: Pinning Control of Complex Dynamical Networks with General Topology. Physica A 379, 298–306 (2007)
11. Xiang, L.Y., Chen, Z.Q., Liu, Z.X., Chen, F., Yuan, Z.Z.: Stabilizing Weighted Complex Networks. J. Phys. A: Math. Theor. 40, 14369–14382 (2007)
12. Sorrentino, F., di Bernardo, M., Garofalo, F., Chen, G.: Controllability of Complex Networks via Pinning. Phys. Rev. E. 75, 046103 (2007)
13. Sorrentino, F.: Effects of the Network Structural Properties on Its Controllability. Chaos 17, 033101 (2007)
14. Chen, T.P., Liu, X.W., Lu, W.L.: Pinning Complex Networks by a Single Controller. IEEE. Trans. Circuits. Syst-I 54, 1317–1326 (2007)
15. Xiang, J., Chen, G.: On the V-stability of Complex Dynamical Networks. Automatica 43, 1049–1057 (2007)
16. Duan, Z.S., Chen, G., Huang, L.: Complex Network Synchronizability: Analysis and Control. Phys. Rev. E. 76, 056103 (2007)
17. Nishikawa, T., Motter, A.E.: Synchronization is Optimal in Nondiagonalizable Networks. Phys. Rev. E. 73, 065106 (2006)
18. Brualdi, R.A., Ryser, H.J.: Combinatorial Matrix Theory. Cambridge University Press, Cambridge (1991)
19. Rössler, O.E.: An Equation for Continuous Chaos. Phys. Lett. A. 57, 397–398 (1976)

Scale-Free Networks with Different Types of Nodes

Juan Zhang and Wenfeng Wu

Department of Applied Mathematics, College of Science,
Donghua University, Shanghai, P.R. China
zhangjuan@dhu.edu.cn

Abstract. In many natural and social networks, nodes may play different roles or have different functions. In this paper, we propose a simple model with different types of nodes and deterministic selective linking rule. We investigate the structural properties by theoretical predictions. It is found that the given model exhibits a power-law distribution. In addition, we make the model become the weighted network by giving the links the weight and analyze the probability distribution of the node strength.

Keywords: power-law distribution, degree distribution, clustering coefficient, weighted network, mean-field method.

1 Introduction

In recent years, complex network systems [1-4] have received remarkable attention, and many scale-free networks have been created because of the pioneer work finished by Barabási and Albert [2], and most of these models are homogeneous since they are composed of the same type of nodes. However, in many real networks, nodes can be divided into different types according to their importance or other properties. For example, in the transports network (the cities and roads being nodes and edges, respectively), the province seats are more important than county ones [5]. For another example, a researcher may work in two or more fields and cooperate with different authors. He writes papers in one field with collaborators who are always engaged in this field while he may also contribute to the other field by participating in another group. He may also have an independent friendship network [6].

Recently, Shi-Jie Yang and Hu Zhao [7] have developed a heterogenous network in which the nodes are catalogued into two types according to the interactions between them, and they found that the node degree exhibits a multi-scaling law distribution with the scaling exponent of each type of nodes. However, they did not consider the clustering coefficient and the average path length of the network. Shou-liang Bu, Bing-Hong Wang and Tao Zhou [5] have gained a scale-free and high clustering complex networks with three types of nodes by large simulations, however, they did not give the theoretic compute.

In this paper, we develop the network with s ($s \geq 1$) types of nodes, and we consider the relations among not only the same types of nodes but also the different types of nodes. That is to say, every type of nodes can be linked with their own type or linked

J. Zhou (Ed.): Complex 2009, Part I, LNICST 4, pp. 991–1000, 2009.

with their different types. An evident real system obeying this rule is that people of the different sex can have relations in the network, while people of the same sex can also establish relations. Noticeably, the theoretic calculations of two important topological properties (the degree distribution and the clustering coefficient) are developed in this paper. Lastly, we make the model become the weighted network and compute the probability distribution of node strength.

2 The Model

We consider a growing network as follows:

(i) There are s types of nodes in the network, with fractions $\left(p_1, p_2, p_3, \cdots, p_s \right)$, respectively, here $\sum_{i=1}^{s} p_i = 1$.

(ii) Selective rule: for every step, a new node of the type j with probability $p_j (1 \le j \le s)$ and m new edges are added to the network. The new node is preferentially attached to the same type of existing nodes with probability q and to the different types of existing nodes with probability $1-q$.

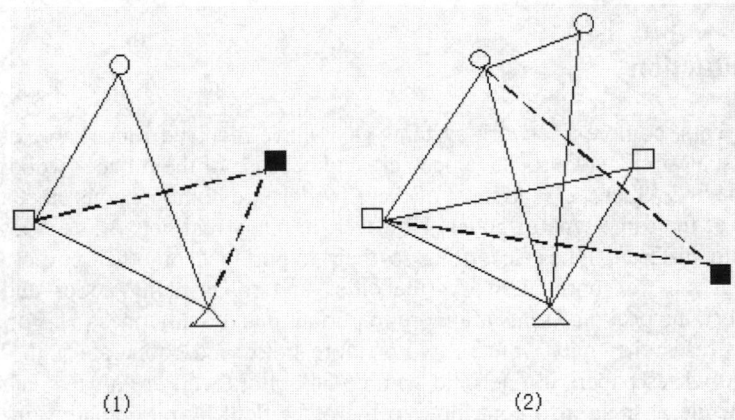

(1)   (2)

Fig. 1. (1) the network consists of three nodes, denoting as one circle, one triangle and one square, respectively, here; (2) each instant a new node and m new links (dashed line) are added (here, $m = 2, q = 0.5$)

(iii) At each instant, when a new node is generated, we assume that the probability $\Pi(k_i)$ with which the new node will be connected to an existing node i depends on the degree k_i of node i:

$$\Pi(k_i) = \frac{k_i}{\sum_j k_j}$$

In addition, we assume that the growing network starts with $N_0 (N_0 \geq m)$ nodes consisting of different types of nodes, and they link with each other. It is known that after t steps the model leads to a network with $N_t = N_0 + t$ nodes and $E_t = mt + N_0(N_0 - 1)/2$ edges. The average node degree is $\langle k \rangle_t = 2E_t / N_t$ which is approximately equal to $2m$ for infinite network size. Fig 1 is an example when $s = 3, m = 2, q = 0.5$.

3 Analytical Calculation of Relevant Network Parameters

Topology properties are very important to understand the complex dynamics of real-life systems. Here we focus on the important characteristics: degree distribution and clustering coefficient.

3.1 Degree Distribution

The degree distribution is one of the most important statistical characteristics of a network. In this section, we use the mean-field method [8] to analyze the property of our network model, we assume that the degree k_i of node i is continuous, and node i is the existing node in the network.

In every step, when node i is a node of the type j, it increases its degree with the rate becomes:

$$\frac{\partial k_i}{\partial t} = m[qp_j + (1-q)(p_1 + p_2 + \cdots + p_{j-1} + p_{j+1} + \cdots + p_s)]\frac{k_i}{\sum_l k_l}$$
$$= m[(2q-1)p_j + 1 - q]\frac{k_i}{\sum_l k_l} \quad (1)$$

Then when node i is an arbitrary node, we have

$$\frac{\partial k_i}{\partial t} = p_1 m[(2q-1)p_1 + 1 - q]\frac{k_i}{\sum_l k_l} + p_2 m[(2q-1)p_2 + 1 - q]\frac{k_i}{\sum_l k_l}$$
$$+ p_3 m[(2q-1)p_3 + 1 - q]\frac{k_i}{\sum_l k_l} + \cdots + p_s m[(2q-1)p_s + 1 - q]\frac{k_i}{\sum_l k_l} \quad (2)$$
$$= m[(2q-1)(p_1^2 + p_2^2 + p_3^2 + \cdots + p_s^2) + 1 - q]\frac{k_i}{\sum_l k_l}$$

Where $\sum_l k_l = 2mt + N_0(N_0-1)$, so Eq. (2) becomes

$$\frac{\partial k_i}{\partial t} = m[(2q-1)(p_1^2 + p_2^2 + p_3^2 + \cdots + p_s^2)+1-q]\frac{k_i}{2mt + N_0(N_0-1)}$$

$$\approx [(2q-1)(p_1^2 + p_2^2 + p_3^2 + \cdots + p_s^2)+1-q]\frac{k_i}{2t}$$

(3)

Denoting $g = (2q-1)(p_1^2 + p_2^2 + p_3^2 + \cdots + p_s^2)+1-q$, there is the initial condition that node i was added to the system at time t_i with the expected value of connectivity $k_i(t_i) = m$, so the solution of Eq. (3) is:

$$k_i(t) = m(\frac{t}{t_i})^{\frac{g}{2}} .$$

(4)

Therefore, the probability with a degree $k_i(t)$ smaller than k, $P(k_i(t) < k)$, can be written as

$$P(k_i(t) < k) = P(t_i > (\frac{m}{k})^{\frac{2}{g}} t) = 1 - P(t_i \le (\frac{m}{k})^{\frac{2}{g}} t) .$$

(5)

Assuming that we add the nodes at equal time intervals to the system, the probability density of t_i is [8]:

$$P_i(t_i) = \frac{1}{N_0 + t} \approx \frac{1}{t} .$$

(6)

Thus, Eq. (5) can be rewritten as

$$P(k_i(t) < k) = 1 - (\frac{m}{k})^{\frac{2}{g}} .$$

(7)

Then the degree distribution $P(k)$ is obtained:

$$P(k) = \frac{\partial P(k_i(t) < k)}{\partial k} = \frac{\frac{2}{g} m^{\frac{2}{g}}}{k^{\frac{2}{g}+1}} .$$

(8)

Obviously, Eq. (8) exhibits the extended power-law form as

$$P(k) \sim k^{-\gamma} .$$

(9)

where $\gamma = \frac{2}{g} + 1 = \frac{2}{(2q-1)(p_1^2 + p_2^2 + p_3^2 + \cdots + p_s^2)+1-q} + 1 .$

3.2 Clustering Coefficient

Clustering coefficient is another important statistic characteristic for a network because it can reflect the clustering extent of the network. Most real-life networks show a cluster structure quantified by it. By definition, clustering coefficient C_i of a node i is $C_i = \dfrac{2e_i}{k_i(k_i-1)}$, where e_i is the total number of existing edges between all its nearest neighbors and $\dfrac{k_i(k_i-1)}{2}$ is the number of all possible edges between them.

Using the mean-field rate-equation theory, we can calculate C_i analytically.

When the new node is a node of the type u $(1 \le u \le s)$,

(i) when node i is a node of the type u,

$$
\begin{aligned}
\frac{\partial e_i}{\partial t} &= \frac{mqk_i}{\sum_l k_l} \sum_{n \in \Omega} [p_u \frac{(m-1)qk_n}{\sum_l k_l} + (1-p_u) \frac{(m-1)(1-q)k_n}{\sum_l k_l}] \\
&= \frac{m(m-1)qk_i}{(\sum_l k_l)^2} [p_u q + (1-p_u)(1-q)] \sum_{n \in \Omega} k_n \\
&= \frac{m(m-1)qk_i}{(\sum_l k_l)^2} [(2q-1)p_u + 1 - q] \sum_{n \in \Omega} k_n
\end{aligned}
\tag{10}
$$

Where k_n denotes the degree of a neighbor of node i, $\sum_{n \in \Omega} k_n$ is the sum of the degrees of all neighbors of node i.

(ii) when node i is a node of the type $j(j = 1, 2, \cdots, u-1, u+1, \cdots, s)$,

$$
\frac{\partial e_i}{\partial t} = \frac{m(m-1)(1-q)k_i}{(\sum_l k_l)^2} [(2q-1)p_u + 1 - q] \sum_{n \in \Omega} k_n .
\tag{11}
$$

Then when node i is an arbitrary node, we have

$$
\begin{aligned}
\frac{\partial e_i}{\partial t} &= \frac{m(m-1)k_i}{(\sum_l k_l)^2} \{ p_u q[(2q-1)p_u + 1 - q] \\
&\quad + (1-p_u)(1-q)[(2q-1)p_u + 1 - q] \} \sum_{n \in \Omega} k_n . \\
&= \frac{m(m-1)k_i}{(\sum_l k_l)^2} [(2q-1)p_u + 1 - q]^2 \sum_{n \in \Omega} k_n
\end{aligned}
\tag{12}
$$

Therefore, when the new node is an arbitrary node, we have

$$
\frac{\partial e_i}{\partial t} = \sum_{u=1}^{s} \{ p_u \frac{m(m-1)k_i}{(\sum_l k_l)^2} [(2q-1)p_u + 1 - q]^2 \sum_{n \in \Omega} k_n \}
$$

$$
= \frac{m(m-1)k_i}{(\sum_l k_l)^2} [(2q-1)^2 \sum_{u=1}^{s} p_u^3 + 2(3q - 2q^2 - 1)\sum_{u=1}^{s} p_u^2 + s(1-q)^2]\sum_{n \in \Omega} k_n
$$

(13)

Denoting $a = (2q-1)^2 \sum_{u=1}^{s} p_u^3 + 2(3q - 2q^2 - 1)\sum_{u=1}^{s} p_u^2 + s(1-q)^2$.

In addition, for uncorrelated random networks we have [2]

$$
\sum_{n \in \Omega} k_n = k_i \frac{\langle k \rangle}{4} \ln t = k_i \frac{m}{2} \ln t .
$$

(14)

So Eq. (13) becomes,

$$
\frac{\partial e_i}{\partial t} = \frac{m(m-1)ak_i}{(\sum_l k_l)^2} \sum_{n \in \Omega} k_n = \frac{m(m-1)ak_i}{(\sum_l k_l)^2} k_i \frac{m}{2} \ln t = \frac{m^2(m-1)a}{2} \frac{k_i^2}{(\sum_l k_l)^2} \ln t
$$

(15)

Integrating both sides of Eq. (15),

$$
e_i = e_{i,0} + \int_1^N \frac{m^2(m-1)a}{2} \frac{k_i^2}{(\sum_l k_l)^2} \ln t\, dt = e_{i,0} + \frac{m^2(m-1)a}{2} \int_1^N (\frac{k_i}{\sum_l k_l})^2 \ln t\, dt
$$

$$
= e_{i,0} + \frac{m^2(m-1)a}{2} \int_1^N \frac{1}{m^2 g^2} (\frac{dk_i}{dt})^2 \ln t\, dt
$$

(16)

$$
= e_{i,0} + \frac{(m-1)a}{2g^2} \int_1^N (\frac{dk_i}{dt})^2 \ln t\, dt
$$

$$
= e_{i,0} + \frac{(m-1)a}{8g^2} \frac{(\ln N)^2}{N} k_i^2(N)
$$

After neglecting $e_{i,0}$, the clustering coefficient becomes:

$$
C_i(k_i) = \frac{e_i}{k_i(k_i - 1)/2} \approx \frac{(m-1)a}{4g^2} \frac{(\ln N)^2}{N} .
$$

(17)

Therefore, the clustering coefficient of this network can be obtained:

$$
C = \frac{1}{N} \sum_{i=1}^{N} C_i \approx \frac{(m-1)a}{4g^2} \frac{(\ln N)^2}{N}
$$

$$
= \frac{(m-1)[(2q-1)^2 \sum_{u=1}^{s} p_u^3 + 2(3q - 2q^2 - 1)\sum_{u=1}^{s} p_u^2 + s(1-q)^2]}{4[(2q-1)\sum_{u=1}^{s} p_u^2 + 1 - q]^2} \frac{(\ln N)^2}{N}
$$

(18)

From Eq. (18), C decreases with N according to $\ln^2 N/N$, this result is same to Ref [9].

4 Weighted Network

We propose a weighted network by imposing the essential features of strength preferential attachment on the above model. The rule of weight evolution is based on the notion that "the rich always gets richer". In other words, high-weighted link has higher probability to evolve. The weight of the links from the same types of nodes are set to w_1, while The weight of the links from the different types of nodes are set to w_2. At each instant, when a new node is generated, these links are randomly connected to the existing nodes according to the strength preferential probability Λ_i, which is defined as

$$\Lambda_i = \frac{s_i}{\sum_j s_j}. \tag{19}$$

By mean field approximation and by treating all discrete variables as continuous [10-11], we assume that node i is a node, which is added at time t_i.

(1) when the new node and node i have the same type, the strength s_i of the node added at time t_i satisfies at time t,

$$\frac{ds_i(t)}{dt} = mqw_1 \frac{s_i(t)}{\sum_l s_l(t)}. \tag{20}$$

(2) when the new node and node i have the different types, the strength s_i of the node added at time t_i satisfies at time t,

$$\frac{ds_i(t)}{dt} = m(1-q)w_2 \frac{s_i(t)}{\sum_l s_l(t)}. \tag{21}$$

When the new node is a node of the type j $(1 \le j \le s)$,

$$\frac{ds_i(t)}{dt} = p_j mqw_1 \frac{s_i(t)}{\sum_l s_l(t)} + (1-p_j)m(1-q)w_2 \frac{s_i(t)}{\sum_l s_l(t)}. \tag{22}$$

Then when node i is an arbitrary node, we have

$$\frac{ds_i(t)}{dt} = \sum_{j=1}^{s}[p_j{}^2mqw_1\frac{s_i(t)}{\sum_l s_l(t)} + p_j(1-p_j)m(1-q)w_2\frac{s_i(t)}{\sum_l s_l(t)}]$$

$$= \frac{s_i(t)}{\sum_l s_l(t)}\sum_{j=1}^{s}[p_j{}^2mqw_1 + p_j(1-p_j)m(1-q)w_2] \tag{23}$$

Since one node is added to the network in each time step, the network size $N_t \approx t$. Then we have

$$\sum_l s_l(t) \approx 2mqw_1t + 2m(1-q)w_2t \quad . \tag{24}$$

Putting Eq. (24) into Eq. (23) and using the initial condition $s_i(t=t_i) = mqw_1 + m(1-q)w_2$, we conclude that

$$\frac{ds_i(t)}{dt} = \frac{s_i(t)}{t}\frac{\sum_{j=1}^{s}[p_j{}^2mqw_1 + p_j(1-p_j)m(1-q)w_2]}{2mqw_1 + 2m(1-q)w_2} \quad . \tag{25}$$

We assume that $n = \dfrac{\sum_{j=1}^{s}[p_j{}^2mqw_1 + p_j(1-p_j)m(1-q)w_2]}{2mqw_1 + 2m(1-q)w_2}$, then

$$s_i(t) \approx [mqw_1 + m(1-q)w_2](\frac{t}{t_i})^n \quad . \tag{26}$$

The probability distribution of node strength can be computed by

$$P(s,t) = -\frac{1}{t}\frac{\partial t_i}{\partial s_i(t)}\Big|_{s_i(t)=s(t)} \quad . \tag{27}$$

From Eq. (26) we obtain:

$$P(s,t) = s^{-\frac{n+1}{n}}\frac{[mqw_1 + m(1-q)w_2]^{\frac{1}{n}}}{n} \quad . \tag{28}$$

Then $P(s,t) \sim s^{-r_s}$

$$r_s = \frac{n+1}{n} = \frac{\frac{\sum_{j=1}^{s}[p_j^2 mqw_1 + p_j(1-p_j)m(1-q)w_2]}{2mqw_1 + 2m(1-q)w_2} + 1}{\frac{\sum_{j=1}^{s}[p_j^2 mqw_1 + p_j(1-p_j)m(1-q)w_2]}{2mqw_1 + 2m(1-q)w_2}}$$

$$= \frac{2qw_1 + 2(1-q)w_2 + \sum_{j=1}^{s}[p_j^2 qw_1 + p_j(1-p_j)(1-q)w_2]}{\sum_{j=1}^{s}[p_j^2 qw_1 + p_j(1-p_j)(1-q)w_2]}$$

$$(29)$$

Obviously this weighted model successfully reproduces the scale-free behavior of the probability distributions of strength with a tunable exponent.

5 Conclusions

In conclusion, we have presented a simple model that consists of different types of nodes. One finds that this model is a scale-free network, and the analytical solution of clustering coefficient is obtained. In addition, we have proposed a simple weighted model with different types of nodes. The weighted model successfully reproduces the scale-free behavior of the probability distributions of strength with a tunable exponent that depends on the microscopic mechanism ruling the weight evolution.

References

1. Stoats, S.H., Tao Zhou, M.S.: Exploring Complex Networks. J. Nature (London) 410, 268 (2001)
2. Albert, R., Barabási, A.L., Tao Zhou, M.S.: Statistical mechanics of complex networks. J. Rev. Mod. Phys. 74, 47 (2002)
3. Dorogovtsev, S.N., Mendes, J.F.F., Tao Zhou, M.S.: Evolution of networks. J. Adv. Phys. 51, 1079 (2002)
4. Newman, M.E.J., Tao Zhou, M.S.: The structure and function of complex networks. J. SIAM Rev. 45, 167 (2003)
5. Bu, S.-l., Wang, B.-H., Tao Zhou, M.S.: Gaining scale-free and high clustering complex networks. J. Physica A 374, 864–868 (2007)
6. Palla, G., Derenyi, I., Farakas, I., Vicsek, T., Tao Zhou, M.S.: Uncovering the Overlapping Community Structure of Complex Networks in Nature and Society. J. Nature 435, 814 (2005)

7. Yang, S.-J., Zhao, H., Tao Zhou, M.S.: Generating multi-scaling networks with two types of vertices. J. Physica A. 370, 863–868 (2006)
8. Albert, R., Barabási, A.L., Jeong, H., Tao Zhou, M.S.: Mean-field theory for scale-free random networks. J. Physica A 272, 173–187 (1999)
9. Ben-Naim, E., Frauenfelder, H., Toroczkai, Z., Tao Zhou, M.S.: Complex Networks. Lecture Notes in Physics, vol. 650. Springer, Heidelberg (2004)
10. Leung, C.C., Chau, H.F., Tao Zhou, M.S.: Weighted assortative and disassortative networks model. J. Physica A 378, 591–602 (2007)
11. Masucci, A.P., Rodgers, G.J., Tao Zhou, M.S.: Multi-directed Eulerian growing networks. J. Physica A 386, 557–563 (2007)

Global Synchronization of Generalized Complex Networks with Mixed Coupling Delays

Yang Dai[1], Yunze Cai[1], and Xiaoming Xu[1,2,3]

[1] Department of Automation, Shanghai Jiao Tong University, P.R. China
daiyang1980@gmail.com
[2] University of Shanghai For Science and Technology, P.R. China
[3] Shanghai Academy of Systems Science, P.R. China

Abstract. In this paper we propose a generalized complex networks model, which concerns asymmetric network configuration including both neutral-type coupling delay and retarded-type one. The synchronization problem of this generalized complex networks is reformulated into the asymptotical stability problem of neutral delay functional differential equations. By introducing descriptor system transformation strategy, the less conservative sufficient condition of delay-independent and independent-of-delay global synchronization criteria are derived in terms of linear matrix inequalities. A numerical example is given to support the theoretical results.

Keywords: Complex Networks, Synchronization, Retarded Delay, Neutral Delay, Linear Matrix Inequalities (LMI).

1 Introduction

During the past decade, complex networks have attracted a lot of interests in the fields of biology, physics, chemistry, engineering and human society [1,2,3,4,5,6]. With each unit regarded as a node, many nonlinear coupling nodes form the complex networks, which describe the sophisticated properties of many systems in nature. As a significant collective behavior, synchronized dynamical propensity in large networks of coupled units has been widely investigated by many researchers [7,8,9,10,11,12,13,14,15,16], since synchronization has potential applications in such places as semiconductor lasers, secure communication and electronic circuits.

Among the methods of synchronization analysis, a common technique is to linearize the dynamical nodes with respect to certain synchronized state and then some local synchronization criteria are derived based on this linearized model [8,9,10,11]. In some situations, if the synchronized state is unavailable in advance, one cannot implement linearization technique. It should be further noticeable that this approach does not solve the inherent nonlinear difficult problem of complex networks. In order to overcome this disadvantage, the global synchronization of complex networks was studied [12,13,14,15,16]. For simplicity, most of these papers considered that the network has symmetric topology structure.

J. Zhou (Ed.): Complex 2009, Part I, LNICST 4, pp. 1001–1010, 2009.

This paper focuses on a more generalized model which may be weighted network, directed network or any other configurations if only coupling matrix can be explicitly represented.

On the other hand, time delays occur commonly in the process of synchronization due to the limited transmitting capability and possible network traffic congestions. In the previous work, complex networks with coupling delay were modeled as interacting retarded delay functional differential equations [9,17,18]. However, neutral delay functional differential equations may better describe the realistic networks in the case of population ecology, distributed networks containing lossless transmission lines, heat exchangers and financial market [19,20,21]. For instance, when the complex dynamical networks are used to model a stock transaction system, each node's state is defined as an agent's behavior such as buying, selling and holding, which is driven dynamically by judging the situation at the time and the historical fluctuating rate records. Hence a complex dynamical networks model with neutral coupling delay is established [22]. In this note, we address the question of how to choose the value of coupling delay such that a given network may achieve synchronous behavior. In order to obtain the less conservative sufficient conditions, descriptor system transformation strategy is utilized during the derivative of the criteria.

The rest of paper is organized as follows. In section 2, a general complex dynamical networks with mixed coupling delays and some useful lemmas are given. In Section 3, descriptor system transformation strategy is introduced to reformulate the network model, then delay-dependent and independent-of-delay synchronization criteria are presented based on this novel model. In Section 4, a numerical example is given to illustrate the main results of this paper. Finally, conclusions are given in Section 5.

Notation: The notation used throughout the paper is fairly standard. The superscripts $'T'$ stands for matrix transposition; $\mathbb{R}^n$ denotes the n-dimensional Euclidean space; $P > 0$ (≥ 0) means P is real symmetric and positive definite (semi-definite). In symmetric block matrices, we use an asterisk $(*)$ to represent a term that is induced by symmetry and $\mathrm{diag}(\ldots)$ stands for a block-diagonal matrix. The norm of a vector or matrix is denoted by $\| \cdot \|$, and $\rho(\cdot)$ denotes the spectral radius of a matrix.

2   Complex Dynamical Networks Model and Preliminaries

Consider the ensemble of N identical diffusively coupled nodes, with each one being an n-dimensional dynamical system. The proposed complex dynamical networks model is described by

$$\dot{x}_i(t) = Ax_i(t) + f(x_i(t)) + \sum_{j=1}^{N} G_{ij}(Bx_j(t - \tau) + C\dot{x}_j(t - \tau)), \quad i = 1, \ldots, N \quad (1)$$

where $x_i = (x_{i1}, \ldots, x_{in})^T \in \mathbb{R}^n$ is the state variable of node i, $t \in \mathbb{R}$ is the continuous time variable, f is continuously differentiable map, τ is the positive constant coupling delay between nodes (it is assumed that the network has identical retarded delay and neutral delay), $B = (a_{ij}) \in \mathbb{R}^{n \times n}$ is a constant inner coupling matrix of the nodes about the retarded delay and $C = (b_{ij}) \in \mathbb{R}^{n \times n}$ regarding to the neutral one, $G = (G_{ij}) \in \mathbb{R}^{N \times N}$ is the outer-coupling matrix combining both configuration and weights of the entire networks.

When $C = 0$, the system model (1) becomes the retarded-type delay dynamical networks as [9]

$$\dot{x}_i(t) = Ax_i(t) + f(x_i(t), t) + \sum_{j=1}^{N} G_{ij} B x_j(t - \tau) \qquad i = 1, \ldots, N \qquad (2)$$

Further when $B = 0$ and $\tau = 0$, the system model (1) turns into the simple uniform dynamical networks proposed by Wang and Chen [23]

$$\dot{x}_i(t) = Ax_i(t) + f(x_i(t), t) + \sum_{j=1}^{N} G_{ij} B x_j(t) \qquad i = 1, \ldots, N \qquad (3)$$

Therefore, (1) is a general complex networks model, with (2) and (3) as the special case.

Let $s(t)$ be the synchronized state of the generalized networks satisfying $\dot{s}(t) = As(t) + f(s(t))$, which may be an equilibrium, aperiodic trajectory or a chaotic attractor of the uncoupled dynamical behavior of each node.

In the sequel, we present some definitions and useful lemmas required throughout the paper.

Definition 1. *The complex dynamical networks (1) is said to be globally asymptotically synchronized if for any initial function $\varphi_i(\theta)$ and $\dot{\varphi}_i(\theta)$, it holds:*

$$\lim_{t \to \infty} \|x_i(t) - s(t)\| = 0, \qquad i = 1, \ldots, N. \qquad (4)$$

Lemma 1. *[24](Schur Complement) Suppose A_1, A_2, A_3 are respectively $p \times p$, $p \times q$ and $q \times q$ matrices, and A_1 is invertible, then the inequality*

$$\begin{pmatrix} A_1 & A_2 \\ A_2^T & A_3 \end{pmatrix} > 0$$

is equivalent to the following two inequalities

$$A_1 > 0, A_3 - A_2^T A_1^{-1} A_2 > 0$$

Lemma 2. *([25]) For the matrices $A \in \mathbb{R}^{m \times n}$, $B \in \mathbb{R}^{p \times q}$, the Kronecker product of A and B is a $mp \times nq$ matrix defined as*

$$A \otimes B \triangleq \begin{pmatrix} a_{11}B & a_{12}B & \ldots & a_{1n}B \\ a_{21}B & a_{22}B & \ldots & a_{2n}B \\ \vdots & \vdots & \ddots & \vdots \\ a_{m1}B & a_{n2}B & \ldots & a_{mn}B \end{pmatrix}$$

3   Main Results

First, the global synchronization of general complex networks is transferred into the stability of $N \times n$ dimensional neutral-type time delay systems. Then delay-dependent and independent-of-delay synchronization criteria are derived.

3.1   Model Transformation

Let $\mathbf{x}(t) = [x_1^T(t), \ldots, x_N^T(t)]^T$, $\mathbf{F}(\mathbf{x}(t)) = [f^T(x_1(t)), \ldots, f^T(x_N(t))]^T$, then the coupled dynamical network (1) can be rewritten as the compact form

$$\dot{\mathbf{x}}(t) = \bar{A}\mathbf{x}(t) + \mathbf{F}(\mathbf{x}(t)) + \bar{B}\mathbf{x}(t - \tau) + \bar{C}\dot{\mathbf{x}}(t - \tau), \tag{5}$$

where $\bar{A} = (I_N \otimes A)$, $\bar{B} = (G \otimes B)$, $\bar{C} = (G \otimes C)$, I_N is a $N \times N$ dimension identity matrix.

Define the synchronous error state of node i with respect to $s(t)$ as $e_i(t) = x_i(t) - s(t)$ and $\mathbf{e}(t) = [e_1^T(t), \ldots, e_N^T(t)]^T$, then the error dynamical network is denoted by:

$$\dot{\mathbf{e}}(t) = \bar{A}\mathbf{e}(t) + \mathbf{F}(\mathbf{e}(t)) + \bar{B}\mathbf{e}(t - \tau) + \bar{C}\dot{\mathbf{e}}(t - \tau), \tag{6}$$

where $\mathbf{F}(\mathbf{e}(t)) = [f^T(x_1) - f^T(s), \ldots, f^T(x_1) - f^T(s)]^T$. It may be seen that the asymptotical stability of neutral time delay system (6) is equivalent to the global synchronization of the complex networks (1) with respect to $s(t)$.

Many efforts have been made to obtain less conservative conditions when the Lyapunov-Krasovskii theory is employed [26]. There are four basic fixed transformation methods, among which the descriptor system transformation method introduced by Fridman [27] is much better because this method may attain less conservative criteria than the others. Thus descriptor system transformation strategy is introduced in the derivation of the global synchronization criterion for generalized complex networks.

Eq. (5) may be represented in the equivalent descriptor form as follows:

$$\dot{\mathbf{e}}(t) = \mathbf{y}(t),$$
$$\mathbf{y}(t) = \bar{A}\mathbf{e}(t) + \mathbf{F}(\mathbf{e}(t)) + \bar{B}\mathbf{e}(t - \tau) + \bar{C}\dot{\mathbf{e}}(t - \tau), \tag{7}$$

According to the Leibniz-Newton formula $\mathbf{e}(t) - \mathbf{e}(t - \tau) = \int_{t-\tau}^{t} \mathbf{y}(s)ds$, then we have:

$$\dot{\mathbf{e}}(t) = \mathbf{y}(t),$$
$$0 = -\mathbf{y}(t) + (\bar{A} + \bar{B})\mathbf{e}(t) + \mathbf{F}(\mathbf{e}(t)) + \bar{C}\dot{\mathbf{e}}(t - \tau) + \bar{B}\int_{t-\tau}^{t} \mathbf{y}(s)ds \tag{8}$$

It is well known that for a neutral delay system to be stable, it is necessary that its neutral part must be stable. This requirement concerns the following hypothesis.

Assumption 1 (A1). The operator is defined as

$$\mathcal{D}\mathbf{e}(t) = \mathbf{e}(t) - \bar{C}\mathbf{e}(t - \tau)$$

Suppose that $\rho(\bar{C}) \leq 1$, then the operator $\mathcal{D}\mathbf{e}(t)$ is stable.

Assumption 2 (A2). Suppose that there exist a set of constant $\sigma_k \geq 0$ ($k = 1, \ldots, n$) such that the nonlinear part of uncoupled node $f(\cdot)$ satisfies local Lipschitz condition:

$$|f_k(a) - f_k(b)| \leq \sigma_k|a - b|, \quad a, b \in \mathbb{R} \tag{9}$$

Under **A2**, the existence and uniqueness of the solution to Eq.(5) is guaranteed.

Let $\Sigma = \text{diag}\{\sigma_k, \ldots, \sigma_n\}$ and $L = I_N \otimes \Sigma$. For a non-negative definite matrix S, there exists a decomposition $S = U^T \Lambda U$ where $U = [\mathbf{u}_1, \ldots, \mathbf{u}_N]^T$, $\mathbf{u}_i = [u_{i1}, \ldots, u_{in}]^T$ and $\Lambda = \text{diag}\{\lambda_1, \ldots, \lambda_{N \times n}\}$ with $\lambda_i \geq 0$ ($i = 1, \ldots, N$). According to Eq.(9), it may be readily deduced that:

$$\begin{aligned}
M(t) &= \mathbf{e}^T(t)LSL\mathbf{e}(t) - F^T(\mathbf{e})SF(\mathbf{e}) \\
&= \mathbf{e}^T(t)LU^T \Lambda U L\mathbf{e}(t) - F^T(\mathbf{e})U^T \Lambda U F(\mathbf{e}) \\
&= \sum_{i=1}^{N} \sum_{j=1}^{n} \lambda_{i \times j} \left[e_{ij}^2 \sigma_j^2 - f_j^2(e_{ij}) \right] \\
&< 0
\end{aligned} \tag{10}$$

3.2 Synchronization Criteria

For the given dynamical network, we can establish the following delay-dependent synchronization criterion in terms of linear matrix inequality.

Proposition 1. *Under A1 and A2, the states of complex networks (1) is globally asymptotically synchronized for a given scalar $\tau > 0$, if there exist positive definite matrices $P_1 = P_1^T > 0$, $Q = Q^T > 0$, $R = R^T > 0$, non-negative definite matrix $S \geq 0$, and any matrices P_2, P_3 such that the following LMI is feasible:*

$$\begin{pmatrix}
(\bar{A} + \bar{B})^T P_2 + P_2^T(\bar{A} + \bar{B}) + LSL & * & * & * & * \\
P_1 - P_2 + P_3^T(\bar{A} + \bar{B}) & -P_3 - P_3^T + Q + \tau R & * & * & * \\
\tau P_2^T \bar{B} & \tau P_3^T \bar{B} & -\tau R & * & * \\
P_2^T \bar{C} & P_3^T \bar{C} & 0 & -Q & * \\
0 & 0 & 0 & 0 & -S
\end{pmatrix} < 0 \tag{11}$$

Proof. Choose the Lyapunov-Krasovskii functional as follows:

$$V(\mathbf{e}_t, t) = V_1 + V_2 + V_3 \tag{12}$$

where

$$V_1 = \left[\mathbf{e}^T(t) \ \mathbf{y}^T(t) \right] EP \begin{bmatrix} \mathbf{e}(t) \\ \mathbf{y}(t) \end{bmatrix} = \mathbf{e}^T(t)P_1\mathbf{e}(t) \tag{13}$$

in which $E = \begin{pmatrix} I & 0 \\ 0 & 0 \end{pmatrix}$, $P = \begin{pmatrix} P_1 & 0 \\ P_2 & P_3 \end{pmatrix}$, $P_1 = P_1^T > 0$

$$V_2 = \int_{t-\tau}^{t} \mathbf{y}^T(s)Q\mathbf{y}(s)ds, \quad Q > 0 \tag{14}$$

$$V_3 = \int_{-\tau}^{0} \int_{t+\theta}^{t} \mathbf{y}^T(s)R\mathbf{y}(s)ds, \qquad R > 0 \tag{15}$$

The time derivative of $V(\mathbf{e}_t, t)$ along the trajectories of (8) is given by

$$\dot{V}(\mathbf{e}_t, t) = \dot{V}_1 + \dot{V}_2 + \dot{V}_3 \tag{16}$$

From (13) to (15), we have

$$\dot{V}_1 = 2\mathbf{e}(t)P_1\dot{\mathbf{e}}(t) = 2\left[\mathbf{e}^T(t)\ \mathbf{y}^T(t)\right] P^T \begin{bmatrix} \mathbf{e}(t) \\ 0 \end{bmatrix}$$

$$= 2\left[\mathbf{e}^T(t)\ \mathbf{y}^T(t)\right] P^T$$

$$\times \begin{bmatrix} \mathbf{y}(t) \\ -\mathbf{y}(t) + (\bar{A} + \bar{B})\mathbf{e}(t) + F(\mathbf{e}) + \bar{C}\mathbf{y}(t-\tau) - \bar{B}\int_{t-\tau}^{t}\mathbf{y}(s)ds \end{bmatrix} \tag{17}$$

$$\dot{V}_2 = \mathbf{y}^T(t)Q\mathbf{y}(t) - \mathbf{y}^T(t-\tau)Q\mathbf{y}(t-\tau) \tag{18}$$

$$\dot{V}_3 = \tau\mathbf{y}^T(t)R\mathbf{y}(t) - \int_{t-\tau}^{t}\mathbf{y}^T(s)R\mathbf{y}(s)ds \tag{19}$$

Let $\xi(t) = [\mathbf{e}^T(t), \mathbf{y}^T(t), \mathbf{y}^T(t-\tau), F^T(\mathbf{e})]^T$. Consider the fact that $M(t) \geq 0$, then $\dot{V}(\mathbf{e}_t, t)$ satisfies that

$$\dot{V}(\mathbf{e}_t, t) \leq \dot{V}(\mathbf{e}_t, t) + M(t) = \xi^T(t)\Psi\xi(t) + \nu - \int_{t-\tau}^{t}\mathbf{y}^T(s)R\mathbf{y}(s)ds \tag{20}$$

where

$$\Psi = \begin{bmatrix} \Psi_{11} & P^T\begin{bmatrix} 0 \\ \bar{C} \end{bmatrix} & 0 \\ \begin{bmatrix} 0\ \bar{C} \end{bmatrix}P & -Q & 0 \\ 0 & 0 & -S \end{bmatrix} \tag{21}$$

$$\Psi_{11} = P^T\begin{bmatrix} 0 & I \\ \bar{A}+\bar{B} & -I \end{bmatrix} + \begin{bmatrix} 0 & \bar{A}^T+\bar{B}^T \\ I & -I \end{bmatrix}P + \begin{bmatrix} 0 & 0 \\ 0 & \tau R+Q \end{bmatrix} + \begin{bmatrix} LSL & 0 \\ 0 & 0 \end{bmatrix} \tag{22}$$

$$\nu = -2\int_{t-\tau}^{t}\left[\mathbf{e}^T(t)\ \mathbf{y}^T(t)\right] P^T \begin{bmatrix} 0 \\ \bar{B} \end{bmatrix}\mathbf{y}(s)ds \tag{23}$$

For any $Nn \times Nn$ dimension positive definite matrix $R > 0$, it holds

$$\nu \leq \tau\left[\mathbf{e}^T\ \mathbf{y}^T\right] P^T \begin{bmatrix} 0 \\ \bar{B} \end{bmatrix} R^{-1} \begin{bmatrix} 0\ \bar{B}^T \end{bmatrix} P \begin{bmatrix} \mathbf{e}^T \\ \mathbf{y}^T \end{bmatrix} + \int_{t-\tau}^{t}\mathbf{y}^T(s)R\mathbf{y}(s)ds \tag{24}$$

Combining (20) and (24), then apply Schur complements to the first term of Ψ_{11}. It yields that $\dot{V}(\mathbf{e}_t, t) < 0$ if the LMI shown in (25) are feasible.

$$\begin{bmatrix} \Psi & \tau P^T\begin{bmatrix} 0 \\ \bar{B} \end{bmatrix} & P^T\begin{bmatrix} 0 \\ \bar{C} \end{bmatrix} & 0 \\ \tau\begin{bmatrix} 0\ \bar{B}^T \end{bmatrix}P & -\tau R & 0 & 0 \\ \begin{bmatrix} 0\ C^T \end{bmatrix}P & 0 & -Q & 0 \\ 0 & 0 & 0 & -S \end{bmatrix} < 0 \tag{25}$$

It can been easily seen that (25) is equivalent to (11). According to the Lyapunov theory, the synchronous error states of the network are asymptotically stable, which implies Proposition 1. This completes the proof.   □

Remark 1. For a certain network, all the parameters are given explicitly. Applying Proposition 1, the upper bound of coupling delay τ_{max} can be obtained by solving (11) with the help of MATLAB LMI toolbox [24]. It is worth noting that $\tau \leq \tau_{max}$ is only sufficient condition to guarantee the global synchronization for the complex networks. That is, there may exist a scalar $\tau' > \tau_{max}$ such that the network can be synchronized with this coupling delay. It is an intrinsically limitation of Lyapunov theory, but there is no better substitution analysis tool for this sophisticated problem.

Remark 2. As it is seen above, the delay-dependent criterion of Proposition 1 is so powerful that the independent-of-delay case is also involved. Let $\tau = 0$, then the independent-of-delay synchronization criterion may be readily derived.

Corollary 1. *Under A1 and A2, the states of complex networks (1) is globally asymptotically synchronized for all $0 \leq \tau < \infty$, if there exist positive definite matrices $P_1 = P_1^T > 0$, $Q = Q^T > 0$, $R = R^T > 0$, non-negative definite matrix $S \geq 0$, and any matrices P_2, P_3 such that the following LMI is feasible:*

$$\begin{pmatrix} (\bar{A} + \bar{B})^T P_2 + P_2^T(\bar{A} + \bar{B}) + LSL & * & * & * \\ P_1 - P_2 + P_3^T(\bar{A} + \bar{B}) & -P_3 - P_3^T + Q & * & * \\ P_2^T \bar{C} & P_3^T \bar{C} & -Q & * \\ 0 & 0 & 0 & -S \end{pmatrix} < 0 \qquad (26)$$

4   Numerical Examples

In this section, a numerical example is presented to illustrate the usefulness of developed theoretical results.

Consider a 2-order nonlinear system as the dynamical node of the complex networks (1) which is described by

$$\dot{x}_i = Ax_i + f(x_i, t) \qquad (27)$$

where $A = \begin{pmatrix} -0.9 & 0.2 \\ 0.1 & -0.9 \end{pmatrix}$, $f(x_i, t) = [f_1(x_i), f_2(x_i)]^T$, $f_1(x_i) = 0.4(|x_{i1} + 1| - |x_{i1} - 1|)$, $f_2(x_i) = 0.1sin(x_{i2})$. It may be easily calculated that the Lipschiz constant matrix of function $f(\cdot)$ is $\Sigma = \text{diag}\{0.8, 0.1\}$.

The dynamical networks consist of 8 identical nodes, which interconnect in the nearest-neighbor topology structure. The coupling configuration are given as follows: the inner-coupling matrices are

$$B = \begin{pmatrix} -1.1 & -0.2 \\ -0.1 & -1.1 \end{pmatrix}, \qquad C = \begin{pmatrix} -0.2 & 0 \\ 0.2 & -0.1 \end{pmatrix},$$

and the outer-coupling matrix is

$$G = 0.5 \times \begin{bmatrix} 2 & 1 & 0 & 0 & 0 & 0 & 0 & 1 \\ 1 & 2 & 1 & 0 & 0 & 0 & 0 & 0 \\ 0 & 1 & 2 & 1 & 0 & 0 & 0 & 0 \\ 0 & 0 & 1 & 2 & 1 & 0 & 0 & 0 \\ 0 & 0 & 1 & 2 & 1 & 0 & 0 & 0 \\ 0 & 0 & 0 & 1 & 2 & 1 & 0 & 0 \\ 0 & 0 & 0 & 0 & 0 & 1 & 2 & 1 \\ 1 & 0 & 0 & 0 & 0 & 0 & 1 & 2 \end{bmatrix}$$

The spectral radius of neutral part matrix is $\rho(G \otimes C) = 0.4 < 1$, which satisfies the precondition of Proposition 1. In sequence, the delay-dependent synchronization criterion of the network is validated. Applying Proposition 1, we can obtain the upper bound of delay $\tau_{max} = 0.22$. It guarantees that the complex networks is globally synchronized for $\tau \leq \tau_{max}$.

Define the k-th element of the state errors between node 1 and node i as $e_{ki} = x_{k1} - x_{ki}$ ($k = 1, 2$ and $i = 2, \ldots, 8$). Take the initial values of the network as random number in $[0, 1]$, then the curves of synchronous state errors for the case of $\tau = 0.1$ and $\tau = 0.8$ are shown in Fig. 1 and 2, respectively.

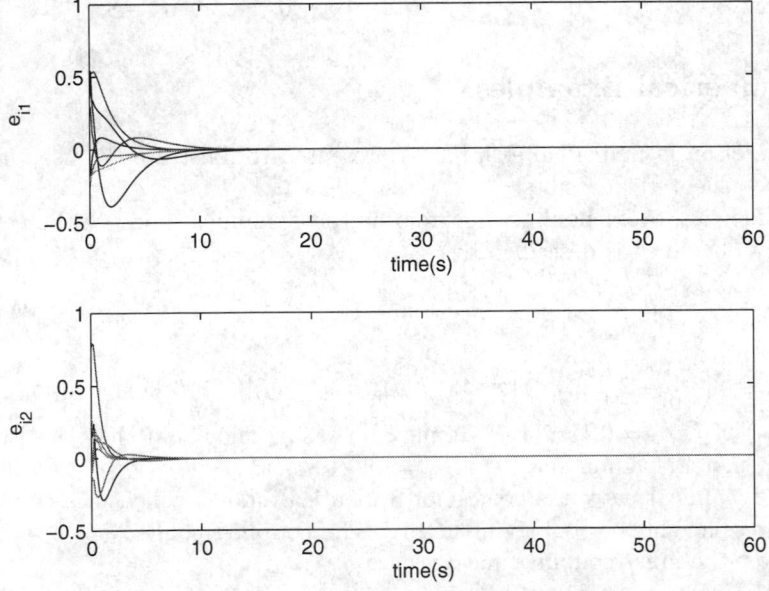

Fig. 1. Synchronous state errors of complex networks (27) for the case of $\tau = 0.1$

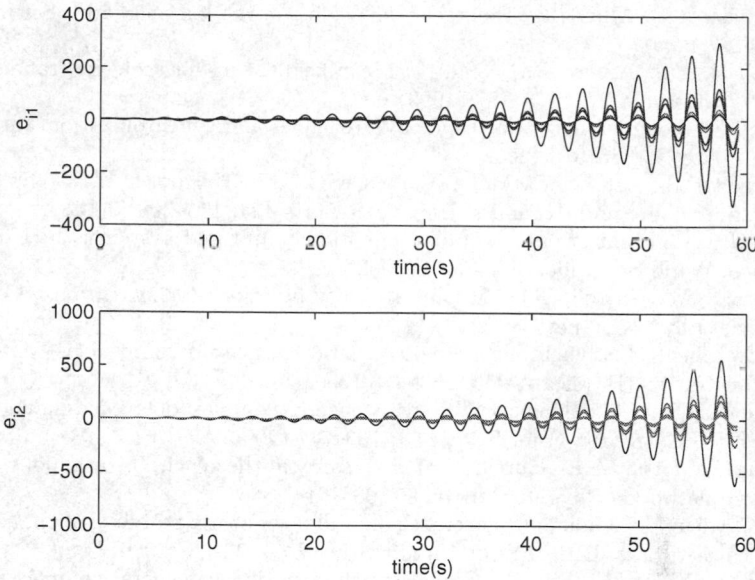

Fig. 2. Synchronous state error of complex networks (27) for the case of $\tau = 0.8$

5   Conclusions

In this paper, we discuss the global asymptotical synchronization problem for the generalized complex dynamical networks with mixed coupling delay. Compared with the previous work, the proposed model provides additionally mathematical description including the derivative of past states in the present system equation. Under some assumptions, the synchronization problem is reformulated into the asymptotically stability of nonlinear neutral delay functional differential equation. Furthermore the delay-dependent and delay-independent synchronization criterion are derived. Both of them can be readily verified by MATLAB LMI toolbox.

Acknowledgments

This work was supported by the National Key Fundamental Research Program (Grant No.2002CB312201-03) and the National Natural Science Foundation of China (Grant No.60575036).

References

1. Watts, D.J., Strogatz, S.H.: Collective dynamics of small-world networks. Nature 393, 440–442 (1998)
2. Watts, D.J.: Small-Worlds: The dynamics of networks between order and randomness. Princeton University Press, Princeton (1999)

3. Barabási, A.L., Albert, R.: Emergence of scaling in random networks. Science 286, 509–512 (1999)
4. Barabási, A.L., Albert, R., Jeong, H.: Mean-field theory for scale-free random networks. Physcia A 272, 173–187 (1999)
5. Wang, X.: Complex networks: topology, dynamics and synchronization. Int. J. Bifurc. Chaos 12, 885–916 (2002)
6. Boccaletti, S., Latora, V., Moreno, Y., Chavez, M., Hwang, D.U.: Complex networks: Structure and dynamics. Physics Reports 424, 175–308 (2006)
7. Wu, C.: Synchronization in coupled chaotic circuits and systems. In: Nonlinear science. World Scientific, Singapore (2002)
8. Pecora, L.M., Carroll, T.L.: Master stability functions for synchronized coupled systems. Phys. Rev. Lett. 80, 2109 (1998)
9. Li, C., Chen, G.: Synchronization in general complex dynamical networks with coupling delays. Physica A 343, 263–278 (2004)
10. Zhou, J., Chen, T.: Synchronization in general complex delayed dynamical networks. IEEE Trans. Circuit Syst. I 53, 733–744 (2006)
11. Zhou, C., Motter, A.E., Kurths, J.: Universality in the synchronization of weighted random networks. Phys. Rev. Lett. 96, 034101 (2006)
12. Li, Z., Chen, G.: Global synchronization and asymptotic stability of complex dynamical networks. IEEE Trans. Circuit Syst. II 53, 28–33 (2006)
13. Zhou, J., Xiang, L., Liu, Z.: Global synchronization in general complex delayed dynamical networks and its applications. Physica A 385, 729–742 (2007)
14. Liu, B., Teoc, K.L., Liu, X.: Global synchronization of dynamical networks with coupling time delays. Physics Letters A 368, 53–63 (2007)
15. Liu, X., Wang, J., Huang, L.: Global synchronization for a class of dynamical complex networks. Physica A 386, 543–556 (2007)
16. Hung, Y.: Paths to globally generalized synchronization in scale-free networks. Physical Review E 77, 016202 (2008)
17. Lu, W., Chen, T., Chen, G.: Synchronization analysis of linearly coupled systems described by differential equations with a coupling delay. Physica D 221, 118–134 (2006)
18. Wang, L., Dai, H., Sun, Y.: Synchronization criteria for a generalized complex delayed dynamical network model. Physica A 383, 703–713 (2007)
19. Hale, J.K., Verduyn Lunel, S.M.: Introduction to functional differential equations. Springer, New York (1993)
20. Kuang, Y.: Delay differential equations with applications in population dynamics. Academic Press, Boston (1993)
21. Niculescu, S.I.: Delay effects on stability: a robust control approach. Lecture Notes in Control and Information Sciences, vol. 269. Springer, London (2001)
22. Dai, Y., Cai, Y., Xu, X.: Synchronization criteria for complex dynamical networks with neutral-type coupling delay. Physica A 387, 4673–4682 (2008)
23. Wang, X., Chen, G.: Synchronization in scale-free dynamical networks: robustness and fragility. IEEE Trans. Circuit Syst. I 49, 54–62 (2002)
24. Boyd, S., Ghaoui, L., Feron, E., Balakrishnan, V.: Linear matrix inequalities in system and control theory. SIAM, Philadelphia (1994)
25. Horn, P.A., Johnson, C.R.: Matrix analysis. Cambridge University Press, New York (1985)
26. Kolmanovskii, V.B., Myshkis, A.D.: Introduction to the theory and applications of functional differential equations. Kluwer Academic Publishers, Dordrecht (1999)
27. Fridman, E.: New Lyapunov-Krasovskii functionals for stability of linear retarded and neutral type systems. Syst. Control Lett. 43, 309–319 (2001)

Community Division of Heterogeneous Networks

Tsuyoshi Murata

Tokyo Institute of Technology
2-12-1 W8-59 Ookayama, Meguro, Tokyo, 152-8552 Japan
murata@cs.titech.ac.jp

Abstract. Many real world data can be represented as heterogeneous networks that are composed of more than one types of nodes, such as paper-author networks (two types) and user-resource-tag networks (three types) of social tagging systems. Discovering communities from such heterogeneous networks is important for finding similar nodes, which are useful for information recommendation and visualization. Although modularity is a famous criterion for evaluating division of given networks, it is not applicable to heterogeneous networks. This paper proposes new modularity for bipartite networks, as the first step for heterogeneous networks. Experimental results using artificial networks and real networks are shown.

Keywords: modularity, community, bipartite networks.

1   Introduction

There are various relations among entities that are represented as networks, such as citations of papers and friendships of SNS (Social Network Service) users. In order to recognize their overall structures and to recommend similar entities, discovering communities from networks attracts many researchers from physics, computer science, and sociology. Quality of the divisions of networks is often measured by modularity, which is a scalar value that measure the density of edges inside communities as compared to edges between communities. As the strategy for finding divisions of given networks, modularity optimization is often employed.

Modularity is, however, appropriate for homogeneous networks that are composed of only one type of vertices (such as papers and SNS users in the above examples). In real-world situations, there are many heterogeneous networks that are composed of more than one types of vertices, such as paper-author networks and movie-actor networks (Figure 1). Modularity is not appropriate for community division of such heterogeneous networks since the density of edges inside communities of same type of vertices is sparser than that of edges between communities of different types of vertices.

As the first step for generalizing the definition of modularity, this paper proposes a new definition of modularity for bipartite networks, which we call bipartite modularity. As far as the author knows, this is the first attempt for

J. Zhou (Ed.): Complex 2009, Part I, LNICST 4, pp. 1011–1022, 2009.

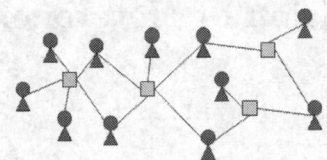

Fig. 1. Example of Bipartite Network

generalizing the definition of modularity for bipartite networks. Experimental results using artificial network data show that our bipartite modularity can clearly detect the existence of community structures of bipartite networks rather than original modularity. Another experiments using real network data show that bipartite modularity is useful also for characterizing each community.

The structure of this paper is as follows: related works about modularity and bipartite networks are reviewed in Section 2. Definition of our new modularity for bipartite networks is shown in Section 3. Experiments using artificial and real network data are shown in Section 4, followed by conclusion in Section 5.

2   Related Work

In this section, the definition of Newman's modularity [9] is reviewed as the basis of the following discussion. Research on bipartite networks are also described.

2.1   Modularity

Modularity is a quantitative measurement for the quality of a particular division of a network. Let us consider a particular division of a network into k communities. Let us suppose M is the number of edges in a network, V is a set of all vertices in the network, and V_l and V_m are the communities. $A(i, j)$ is an adjacency matrix of the network whose (i, j) element is equal to 1 if there is an edge between vertices i and j, and is equal to 0 otherwise. Then we can define e_{lm}, the fraction of all edges in the network that connect vertices in community l to vertices in community m:

$$e_{lm} = \frac{1}{2M} \sum_{i \in V_l} \sum_{j \in V_m} A(i, j)$$

We further define a $k \times k$ symmetric matrix E composed of e_{ij} as its (i, j) element, and its row sums a_i:

$$a_i = \sum_j e_{ij} = \frac{1}{2M} \sum_{i \in V_l} \sum_{j \in V} A(i, j)$$

In a network in which edges fall between vertices without regard for the communities they belong to, we would have $e_{ij} = a_i a_j$. Therefore modularity is defined as follows:

$$Q = \sum_i (e_{ii} - a_i^2)$$

Modularity measures the fraction of the edges in the network that connect vertices within the same community minus the expected value of the same quantity in a network with the same community divisions but random connection between vertices. If the number of edges inside communities is no better than random, we will get $Q = 0$. Values approaching the maximum ($Q = 1$) indicate strong community structures.

There are many related work regarding modularity. Clauset [3] proposes fast modularity algorithm for efficient search for division of high modularity. Newman [10] propose a spectral algorithm for improving the quality of community division. Wakita [13] and Blondel [2] attempt the division of large-scale networks. Danon [4] performs comparison of several network division methods. Fortunato [6] clarifies resolution limits of modularity-based network division methods.

2.2   Research on Bipartite Networks

Although most of the research for social networks focus on homogeneous networks, the following research focus on bipartite networks.

Neighborhood Formation. Sun [11] proposes algorithms for computing the neighborhood of the nodes of bipartite networks using random walk with restarts and network partitioning. Algorithms for identifying abnormal nodes are also proposed, and their effectiveness and efficiency are confirmed by the experiments on several real datasets.

Co-ranking. Zhou [15] proposes a framework for co-ranking authors and documents in heterogeneous networks. The framework is based on coupling two random walks that separately rank authors and documents. As the result of the coupling, both document ranking and author ranking are improved since both ranking depend on each other in a mutually reinforcing way.

Projection. Zhou [16] proposes a method for projecting bipartite networks to weighted homogeneous networks. Bipartite networks are regarded as resource allocation processes between X-vertices and Y-vertices. Initially assigned weights on X-vertices are propagated to Y-vertices and then back to X-vertices in order to obtain weighted homogeneous networks.

Community variance. Murata [8] proposes a criterion for evaluating correspondence between communities of two types of vertices. Although the criterion (community variance) is useful for clarifying structural properties of communities, it has no relation with modularity.

Although the goals of these research are different from ours, these research put stress on the importance of processing bipartite networks appropriately. Our goal in this paper is to propose new criterion for evaluating division of bipartite networks.

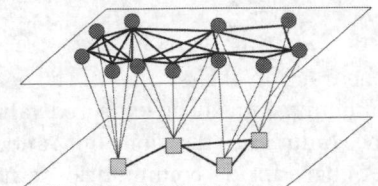

Fig. 2. Projection of Bipartite Network

3   Bipartite Modularity

3.1   Bipartite Networks

In general, social networks can be divided into the following categories: 1) direct connection between persons (such as MySpace or Twitter) and 2) indirect connection through different types of entities (such as film co-starring or paper co-authoring). We call the former "homogeneous networks", and the latter "heterogeneous networks". There are bipartite, tripartite and n-partite networks as the examples of heterogeneous networks. As the first step for processing heterogeneous networks, we focus on bipartite networks composed of two types of vertices. Zhou [16] claims that there are two types of bipartite networks: collaboration network and opinion network. The former is generally defined as a network of users connected by common collaboration acts. The latter is defined as a network of users connected by common objects.

As a naive approach for transforming bipartite networks into homogeneous networks, projection is often used. Suppose a bipartite network is composed of X-vertices $\{x_0, x_1, ...\}$ and Y-vertices $\{y_0, y_1, ...\}$, and y_i is connected to both x_j and x_k. Projection is a transformation of such $x_j - y_i - x_k$ connection into $x_j - x_k$ connection so that a network composed of only X-vertices is obtained. However, projection loses information about the correspondence between X-vertices and Y-vertices, which is often quite valuable.

3.2   Bipartite Modularity

In the case of community division of bipartite networks, finding the correspondence between communities of different types of vertices is often important. Let us suppose that communities of papers and communities of authors are discovered from a paper-author network. If there is one-to-one correspondence between a paper community and an author community, it shows that the topics of the papers attract only limited authors (Figure 3). On the other hand, if there is one-to-many correspondence between a paper community and author communities, it shows that the topics of the papers attract several communities of authors (Figure 4).

In order to evaluate the division of such bipartite networks, we define modularity for bipartite network, which we call bipartite modularity. Bipartite modularity is for measuring the degree of correspondence between communities of different types of vertices.

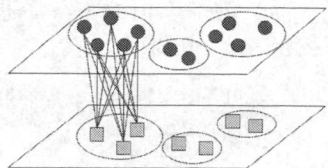

Fig. 3. One-to-one Correspondence between Communities

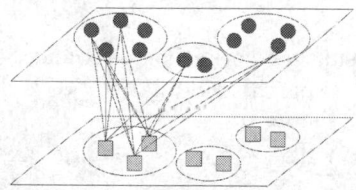

Fig. 4. One-to-many Correspondence between Communities

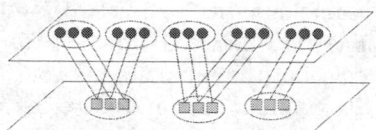

Fig. 5. Communities in a Bipartite Network

Newman's modularity is not appropriate for evaluating community divisions of bipartite networks. Let us suppose that a bipartite network composed of X-vertices and Y-vertices is given, and both X-vertex communities and Y-vertex communities are specified. Since bipartite network does not have any direct edge between X-vertices (and between Y-vertices), $e_{ii} = 0$ for all X-vertex (Y-vertex) communities V_i, and modularity for community division is quite low. For example, modularity of the community division of the bipartite network shown in Figure 5 is -0.14.

Our definition of bipartite modularity is as follows. Let us suppose that M is the number of edges in a bipartite network, and V is a set of all vertices in the bipartite network. Consider a particular division of the bipartite network into X-vertex communities and Y-vertex communities, and the numbers of the communities are L^+ and L^-, respectively. V^+ and V^- are the sets of the communities of X-vertices and Y-vertices, and V_l^+ and V_m^- are the individual communities that belong to the sets ($V^+ = \{V_1^+, ..., V_{L^+}^+\}$, $V^- = \{V_1^-, ..., V_{L^-}^-\}$). $A(i,j)$ is an adjacency matrix of the network whose (i,j) element is equal to 1 if vertices i and j are connected, and is equal to 0 otherwise.

Under the condition that the vertices of V_l and V_m are different types (which means $(V_l \in V^+ \wedge V_m \in V^-) \vee (V_l \in V^- \wedge V_m \in V^+)$), we can define e_{lm} (the fraction of all edges that connect vertices in V_l to vertices in V_m) and a_i (its row sums) just the same as those in section 2.1.

$$e_{lm} = \frac{1}{2M} \sum_{i \in V_l} \sum_{j \in V_m} A(i,j)$$

$$a_i = \sum_j e_{ij} = \frac{1}{2M} \sum_{i \in V_l} \sum_{j \in V} A(i,j)$$

As in the case of homogeneous networks, if edge connections are made at random, we would have $e_{ij} = a_i a_j$. Bipartite modularity Q_B is defined as follows:

$$Q_B = \sum_i (e_{ij} - a_i a_j), \quad j = \underset{k}{\mathrm{argmax}}(e_{ik})$$

As shown in section 2.1, original modularity measures the fraction of the edges in the network that connect vertices within the same community minus the expected value of the same quantity in a network with the same community divisions but random connection between vertices. Bipartite modularity measures the fraction of the edges in the bipartite network that connect vertices of the corresponding X-vertex communities and Y-vertex communities minus the expected value of the same quantity with random connections between X-vertices and Y-vertices. If given network is not bipartite, you can see that $Q_B = Q$, which means that bipartite modularity is a straightforward generalization of original modularity.

If the connection between X-vertices and Y-vertices is no better than random, we will get $Q_B = 0$. High Q_B value indicates strong community structure in a bipartite network. Bipartite modularity of the network shown in Figure 5 is 0.66. If you take a closer look at the expression of Q_B, you will find that the value is the sum of bipartite modularities of different directions ($V^+ \rightarrow V^-$ and $V^- \rightarrow V^+$). Q_B can be divided as follows:

$$Q_{B\pm} = \sum_{i \in V^+} (e_{ij} - a_i a_j), \quad j = \underset{k \in V^-}{\mathrm{argmax}}(e_{ik})$$

$$Q_{B\mp} = \sum_{i \in V^-} (e_{ij} - a_i a_j), \quad j = \underset{k \in V^+}{\mathrm{argmax}}(e_{ik})$$

$$Q_B = Q_{B\pm} + Q_{B\mp}$$

$Q_{B\pm}$ is the bipartite modularity for $V^+ \rightarrow V^-$ direction, and $Q_{B\mp}$ is the bipartite modularity for $V^- \rightarrow V^+$ direction. In the example shown in Figure 5, $Q_{B\pm} = 0.41$ and $Q_{B\mp} = 0.25$, which means that downward connections are relatively focused rather than upward connections in the figure.

The matrix E composed of e_{ij} as its (i,j) element is represented as follows if rows and columns are reordered appropriately.

$$\begin{pmatrix} 0 & \cdots & 0 & e_{1,L^++1} & \cdots & e_{1,L^++L^-} \\ \vdots & \ddots & \vdots & \vdots & \ddots & \vdots \\ 0 & \cdots & 0 & e_{L^+,L^++1} & \cdots & e_{L^+,L^++L^-} \\ \hline e_{L^++1,1} & \cdots & e_{L^++1,L^+} & 0 & \cdots & 0 \\ \vdots & \ddots & \vdots & \vdots & \ddots & \vdots \\ e_{L^++L^-,1} & \cdots & e_{L^++L^-,L^+} & 0 & \cdots & 0 \end{pmatrix}$$

The upper right quarter of the matrix (E_{UR}) corresponds to $Q_{B\pm}$, and the lower left quarter of the matrix (E_{LL}) corresponds to $Q_{B\mp}$. Since E is a symmetric matrix, it is clear that $E_{UR}^T = E_{LL}$. But $Q_{B\pm} \neq Q_{B\mp}$ in general. This is because a set of (i,j) under the condition that $i \in V^+, j = \operatorname*{argmax}\limits_{k \in V^-}(e_{ik})$ is different from a set of (i,j) under the condition that $i \in V^-, j = \operatorname*{argmax}\limits_{k \in V^+}(e_{ik})$.

When two upper-left communities in Figure 5 are merged, Q_B increases to 0.67. But if all upper communities are merged into one community, Q_B decrease to 0.35. By maximizing bipartite modularity, unobvious community structure will be obtained from bipartite networks.

4   Experiments

4.1   Artificial Four-Community Networks

In order to clarify the properties of our bipartite modularity, modularity and bipartite modularity are compared in the following experiments. Networks with known community structure are used to see whether our bipartite modularity has abilities of detecting the structure.

We have generated many networks with 128 vertices, divided into four communities of 32 vertices each. Edges are placed independently at random with probability p_{in} for an edge to fall between vertices in the same community and p_{out} to fall between vertices in different communities. Such artificial network data are used by Newman [9] and Danon [4]. Figure 6 illustrates an example of the networks. Figure 7 shows the average values of modularity and bipartite modularity of 100 artificial networks.

You can see from the figure that modularity and bipartite modularity are the same for networks of high p_{in}. This is obvious from the definition of bipartite modularity. For the networks with high p_{in}, diagonal elements of matrix E are the biggest among the all elements in the same row ($\forall j\ e_{ii} \geq e_{ij}$). Therefore $j = \operatorname*{argmax}\limits_{k}(e_{ik}) = i$ and $Q_B = Q$.

For networks of smaller p_{in} ($p_{in} < p_{out}$), diagonal elements of matrix E are *not* the biggest ($\exists j\ e_{ii} \leq e_{ij}$) and their modularities are below zero. On the other hand, bipartite modularities of the networks are positive because $j = \operatorname*{argmax}\limits_{k}(e_{ik})$ is set to the community that is densely connected with community i.

The above networks are not bipartite because four communities are connected to each other. For the next experiment, we have generated bipartite networks

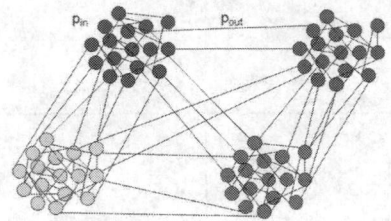

Fig. 6. Network with Four Communities

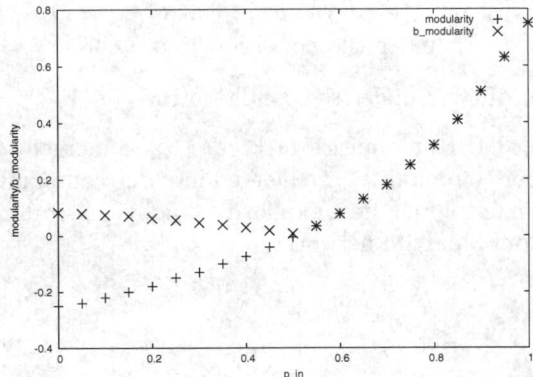

Fig. 7. Modularity and Bipartite Modularity of Four-community Networks

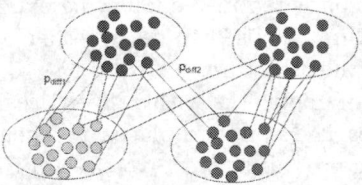

Fig. 8. Bipartite Network with Four Communities

with 128 vertices, divided into four communities of 32 vertices each. Edges are placed independently at random with probability p_{in} for an edge to fall between vertices in the same community, p_{same} to fall between vertices in the communities of same type of vertices, and p_{diff1} p_{diff2} to fall between vertices in the communities of different types of vertices. Suppose there are two communities for each type of vertices. p_{in} and p_{same} are set to zero because there are no edges between vertices of the same type in bipartite networks. Figure 8 illustrates an example of such networks. Networks with various p_{diff1} and p_{diff2} are generated and their modularity and bipartite modularity are calculated. Figure 9 shows the average values of modularity and bipartite modularity of 100 artificial bipartite networks.

Figure 9 shows that original modularity is not appropriate for bipartite networks because there is no edge between vertices of the same type. Bipartite modularity is effective for detecting the existence of community structures for

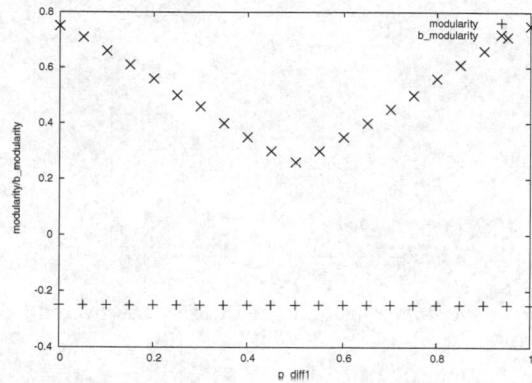

Fig. 9. Modularity and Bipartite Modularity for Bipartite Networks

bipartite networks, and it also shows the degree of correspondence between communities of different types of vertices. In the case of networks with $p_{diff1} = 1.0$ or $p_{diff2} = 1.0$, there are complete one-to-one correspondence between communities of different types of vertices, and the values of bipartite modularity are the highest.

4.2   Real Online Social Networks

Bipartite modularity described above is applied also for real-world networks. We have generated heterogeneous social networks composed of users and boards from the data of Yahoo! Chiebukuro (Japanese Yahoo! Answers, http://chiebukuro. yahoo.co.jp). The site is one of the most popular question-answering forums in Japan. The network of Yahoo! Chiebukuro is summarized in Table 1. From the heterogeneous networks, user communities and board communities are discovered by 1) projecting the heterogeneous networks into homogeneous ones (user networks and board networks), and 2) applying Clauset's fast modularity algorithm [3] for finding community divisions of high modularities. Details of the discovery method is described in [8]. Both user communities and board communities are extracted from the network data. Bipartite modularity and original modularity of the division are as follows. Modularity (Q) and bipartite modularity (Q_B) of the network are -0.1021 and 0.2919, respectively. This shows that the

Table 1. Statistics of the Network of Yahoo! Chiebukuro

number of vertices	6,309,737
number of edges	357,834
average degree	0.1134
clustering coefficient	0
average path length	7.7587

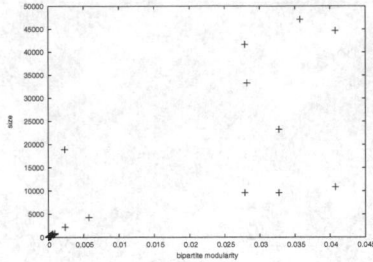

Fig. 10. Bipartite Modularities and the Sizes of Discovered Communities

community division is not good in the sense of homogeneous network division, but it is not bad in the sense of bipartite network division.

Bipartite modularity is for evaluating the whole division of a bipartite network into communities. In addition to that, bipartite modularity of each community (Q_{B_i}) can be used for measuring the degree of "close-knitness" to the communities of the other type of vertices. Figure 10 shows a distribution of bipartite modularity Q_{B_i} (X-axis) and the sizes (Y-axis) of discovered communities. Communities of upper half of the figure (more than 15,000 vertices) are board communities. Their bipartite modularities are high except the one located at middle left position. This community is like the one in Figure 4: the main topics of the community (such as "entertainment and hobby" and "health and fashion") attract many users and thus its bipartite modularity is low. On the other hand, other communities of high bipartite modularity are relatively focused (such as "child care", "mental health", and "cars"), like the one in Figure 3.

This paper focus on the criterion for evaluating given community division for bipartite networks. Finding the best community division is NP-complete [7]. In the case of homogeneous networks, many approaches are proposed for finding appropriate division with the smallest computational cost possible by modularity optimization. Bipartite modularity can be used in the same manner for finding and evaluating division of bipartite networks.

Biclustering algorithms [7][12][5] also aim at finding division of incident matrices. These algorithms are mainly for the purpose of bioinformatics and document clustering, and the size of the incident matrices are at most thousands times tens of thousands. One of the weakness of these algorithms is that most of these algorithms do not scale to large networks. Finding community division of high bipartite modularity from given large-scale networks is another challenging research topic, which is left for our future work.

5   Conclusion

This paper proposes a new criterion for community division of bipartite networks. As far as the author knows, this is the first attempt for generalizing the definition of modularity for bipartite networks. Bipartite modularity is a straightforward generalization of Newman's modularity. Experimental results show that our bipartite modularity is appropriate for detecting the existence of community structures from bipartite networks. In addition to that, bipartite modularity for each community is the degree of correspondence to the communities of the other type of vertices, which can be used for analyzing the characteristics of the community.

Bipartite modularity proposed in this paper is the first step for intelligent processing of heterogeneous networks in the Web. There are several bipartite, tripartite, and n-partite networks in the Web. Social tagging systems can be represented as tripartite networks composed of three types of vertices (users, URLs and tags). Discovering and evaluating communities of such heterogeneous networks is one of the important and challenging topics of Web mining.

References

1. Adamic, L.A., Zhang, J., Bakshy, E., Ackerman, M.S.: Knowledge Sharing and Yahoo Answers: Everyone Knows Something. In: Proceedings of the 17th International World Wide Web Conference, pp. 665–674 (2008)
2. Blondel, V.D., Guillaume, J.-L., Lambiotte, R., Lefebvre, E.: Fast Unfolding of Community Hierarchies in Large Networks, 1–6 (2008) arXiv:0803.0476v1
3. Clauset, A., Newman, M.E.J., Moore, C.: Finding Community Structure in Very Large Networks. Physical Review E 70, 066111, 1–6 (2004)
4. Danon, L., Diaz-Guiler, A., Duch, J., Arenas, A.: Comparing Community Structure Identification. Journal of Statistical Mechanics, P09008, 1–10 (2005)
5. Dhillon, I.S.: Co-clustering Documents and Words using Bipartite Spectral Graph Partitioning. In: Proceedings of the Seventh ACM SIGKDD International Conference on Knowledge Discovery and Data Mining, pp. 269–274 (2001)
6. Fortunato, S., Barthelemy, M.: Resolution Limit in Community Detection. Proceedings of the National Academy of Sciences of the United States of America 104(1), 36–41 (2007)
7. Madeira, S.C., Oliveira, A.L.: Biclustering Algorithms for Biological Data Analysis: A Survey. IEEE/ACM Transactions on Computational Biology and Bioinformatics 1(1), 24–45 (2004)
8. Murata, T., Ikeya, T.: Analysis of Online Question-Answering Forums as Heterogeneous Networks. In: Proceedings of the Second International Conference on Weblogs and Social Media, pp. 210–211 (2008)
9. Newman, M.E.J., Girvan, M.: Finding and Evaluating Community Structure in Networks. Physical Review E 69, 026113, 1–15 (2004)
10. Newman, M.E.J.: Modularity and Community Structure in Networks. Proceedings of the National Academy of Sciences of the United States of America 103, 8577–8582 (2006)
11. Sun, J., Qu, H., Chakrabarti, D., Faloutsos, C.: Neighborhood Formation and Anomaly Detection in Bipartite Graphs. In: Proceedings of the Fifth IEEE International Conference on Data Mining, pp. 418–425 (2005)
12. Tanay, A., Sharan, R., Shamir, R.: Discovering Statistically Significant Biclusters in Gene Expression Data. Bioinformatics 18 (suppl.1), S136–S144 (2002)
13. Wakita, K., Tsurumi, T.: Finding Community Structure in Mega-scale Social Networks. In: Proceedings of the 16th International World Wide Web Conference, pp. 1275–1276 (2007)

14. Xi, W., Zhang, B., Chen, Z., Lu, Y., Yan, S., Ma, W.-Y., Fox, E.A.: Link Fusion: A Unified Link Analysis Framework for Multi-Type Interrelated Data Objects. In: Proceedings of the 13th World Wide Web Conference, pp. 319–327 (2004)
15. Zhou, D., Orchanskiy, S.A., Zha, H., Giles, C.L.: Co-Ranking Authors and Documents in a Heterogeneous Network. In: Proceedings of the Seventh IEEE International Conference on Data Mining, pp. 739–744 (2007)
16. Zhou, T., Ren, J., Medo, M., Zhang, Y.-C.: Bipartite network projection and personal recommendation. Physical Review E 76, 046115, 1–7 (2007)

Autonomous Co-operation and Control in Complex Adaptive Logistic Systems – Contributions and Limitations for the Innovation Capability of International Supply Networks

Michael Hülsmann and Philip Cordes

University of Bremen, Management of Sustainable System Development,
Wilhelm-Herbst-Str. 12, 28359 Bremen
{michael.huelsmann,pcordes}@uni-bremen.de
http://www.wiwi.uni-bremen.de/mh/

Abstract. This paper aims to analyze the potential contributions of the organization principle autonomous co-operation and control to the innovation capabilities of logistics systems and their sub-systems like single organizations. Therefore, the concept of Complex Adaptive Logistics Systems (CALS) will be introduced and the essentiality of the heterogeneity of the elements within logistics systems for their innovation capabilities will be emphasized. One possible driver for homogeneity is the so-called dominant logic.

Keywords: Complex Adaptive Logistics Systems, Logistics, Complexity Science, Autonomous Co-operation, Innovations, Dominant Logic.

1 Introduction

Innovations are – among others – one important factor for the creation of corporate value and competitive advantage [e.g. 1,2,3,4]. Thereby, innovations can be understood as ideas, concepts, and practises in order to improve characteristics and features of products as well as of processes, which are perceived as new and valuable by any stakeholder of the respective organization [e.g. 5,6]. Innovations promise a certain benefit for which the relevant stakeholder groups might be willing to pay an extra premium [7]. Thus, the company's capability to be innovative can have a significant effect on its options for positioning on the markets and, in consequence, on the company's value [e.g. 8,9,10]. Therefore, the corporate innovation capability is crucial for the company's sustainable wealth.

But in times of real-time economies [11] and globalization [e.g. 12] companies are not isolated planning and acting organizations. Furthermore, companies are embedded in the diverse organizational structures of different international supply networks (ISN) and compete with other companies in the same or in other global production and distribution structures to a certain degree [13,14]. Companies, and therewith as ISN, do depend on the plans, performances, strategies, structures, and resources of other companies as well as on those of other networks. That is, why processes of

J. Zhou (Ed.): Complex 2009, Part I, LNICST 4, pp. 1023–1032, 2009.

co-evolution can be observed on both levels: the company level as well as the network level. Therefore, ISN can be understood and described as complex adaptive logistic systems (CALS) [15,16,17,18]. This understanding establishes – on the basis of a complexity science perspective – an analogy between supply networks and complex adaptive systems as they have been described in the literature e.g. [19,20,21,22]. In consequence, the question occurs, how value and competitive advantage can be achieved in ISN as CALS. More precisely, how can innovations be generated in the complex structures of value creating networks due to establish, maintain, and develop an innovation capability for the whole network.

An innovation aims on a successful market penetration. It is based on an efficient and effective innovation management [23], which comprises – at least - the planning of portfolios of technologies, resources, and innovations, the creation of innovations in cross-functional processes across the organization and beyond the organization's boundaries with its partners, like suppliers, customers, etc. [24]. Finally, a successful innovation management needs the intro-duction, integration, and implementation of the new features of products and processes into the markets in order to satisfy stakeholders' demands [25]. Therefore, one precondition of innovation capability in ISN is the existence of an efficient and effective innovation management in such a CALS, especially its before mentioned ability to initiate and maintain cross-functional processes for the creation of new ideas, concepts, and practises throughout the network organization with its different agents. From a complexity science perspective this leads to the question, what characteristics of CALS do promote or hinder the innovation capability of ISN.

To answer this question the paper intends to focus on the phenomenon of heterogeneity, as a precondition for any creation process in complex organizations as well as systems. In order to identify fields of activities needed for an innovation management impediments and facilitators of heterogeneity in CALS shall be identified. Due to the adaptivity feature of CALS the question of dominant logic becomes important for the innovation-aimed design of heterogeneity in such complex networks. Therefore, the paper would like to discuss autonomous co-operation and control as a learning-based approach to avoid problems of dominant logic in collective decision-making in CALS and, in order to establish and maintain a certain degree of heterogeneity, as a facilitating determinant of the ISN's innovation capability.

2   Heterogeneity as a Pre-condition for the Innovation-Capability of Complex Adaptive Logistics Systems (CALS)

The perspective, with which logistics systems are regarded in associated recent research, changed from the analysis of linear supply chains to complex and cross-national supply networks [16,17,26]. Due to that, several authors applied the concept of complex adaptive systems in the field of logistics (15,16,17,18), which leaded to the term of Complex Adaptive Logistics Systems (CALS).

According to McKelvey et al. (2008) [27] the main characteristics of CALS can be classified into an individual, an intra-systemic and an inter-systemic level. To begin with the perspective of the individual, in which the characteristics of the system's single entities are regarded, CALS consist of a large number of agents, which are interacting with each other simultaneously. Holland [21] called this parallelism, which means that the agents send signals to, and receive signals (e.g. resources, information, finances [16]) from other agents at the same time. Their incentives to interact with each other originate in their varying endowment with information [15], in other words, the agents are heterogeneous [20]. This is essential for the interaction, because if the agents would equal each other, they would be equipped with the same resources and therefore, they would not have any reason to interact and to exchange them [15]. Furthermore, interaction implies, that the agents react to the other agents' actions. Therefore, the agents' actions are dependent on the signals they receive, which means, that an agent's action can be a signal as well, that is, in turn, received by another agent [21]. This intra-systemic co-evolution highlights the complexity of CALS that can evolve by interaction between their elements. The agents' ability to react requires the existence of rules [20], which are, in turn, based on local information [16]. These rules need to be changeable over time to assure the system's capability to adapt to environmental changes [21]. What Holland [21] calls modularity means, that every agent develops routines of action sequences (combinations of single actions, which are called building blocks) by reacting to signals, which were send from the environment or by other agents. In new situations agents can revert to this further developed building blocks and test their appropriateness concerning their defined goals. The building blocks that experience positive feedbacks will be applied again [20,21]. Wycisk et al. (2008) [15] states this as the agent's ability to learn.

This leads to the intra-systemic perspective in which the system's organization is regarded. In order to develop own rules how to act and to react, the agents have to be able to render decisions without having to ask a super-ordinate controlling entity. Hence, the agents are autonomous, which means, they initiate their actions by themselves [19,20,22]. They are responsible for their own design, direction and development. In consequence, the system's developing structure is as well not controlled by any super-ordinate entity [28]. Instead, the system's structure evolves by itself with the autonomous decision making capabilities of its agents, it is self-organizing [29]. An essential requirement for a self-organizing system, in turn, is, that it does neither pass over a threshold, from which on it is totally uncontrolled and chaotic (the edge of chaos) [30,31], nor pass over a threshold, from which on its order is pre-configured (the edge of order) [32,33]. Kauffman [22] called this the melting zone.

Finally, and with regard to the mentioned co-evolution on the intra-systemic level, that results from the reciprocal reactions between the agents, co-evolution also takes place on the inter-systemic level. Agents do not only receive signals from other agents in the system, but also signals from the environment, to which they also send signals by their actions and interactions [22]. Therefore, CALS co-evolve with their environment and hence, possibly with other non-logistics systems, like the financial systems on which they are as well dependent to a certain degree.

With recourse to the above-mentioned definition of innovations, it becomes obvious that the existence and the degree of the characteristics of the CALS' individual levels, and therefore, the characteristics of their single elements, are essential for the whole systems to be innovative. The improvement of characteristics and features of products and processes in logistics systems is only possible, when their single elements are able to change their rules, in other words, if they are able to learn. By reacting to other element's respectively agent's actions, the others get feedbacks on their own actions and change, if necessary or meaningful, their own behavioral rules, to improve their own accomplishments and therewith the whole system's performance [34]. This reflects the co-evolution within a CALS as well as with its environment. The learning abilities of the systems' agents, in turn, are, as shown above, dependent on the degrees of interaction between them, and interaction can, as shown above, only take place if the elements are heterogeneous, at least to a certain degree. In consequence, the heterogeneity of a logistics system's agents can be regarded as a driver and homogeneity as a barrier for its innovation capability. Hence, the question arises, which aspects determine a logistics system's heterogeneity and what kinds of phenomena intensify their homogeneity.

3 Dominant Logic of Collective Decision-Making as a Barrier for Innovations in CALS

One phenomena that could possibly homogenate the behavioral rules of a logistics system's agents has been described by Prahalad and Bettis [35] as the so-called dominant logic. This concept was originally developed to explain problems that occur in firms that try to diversify their activities from areas in which they are successful to new areas and describes cognitive barriers of managers. They define the "(…) dominant general management logic (…) as the way in which managers conceptualize the business and make critical resource allocation decisions (...)" [35, p. 490]. Figure 1 illustrates the concept.

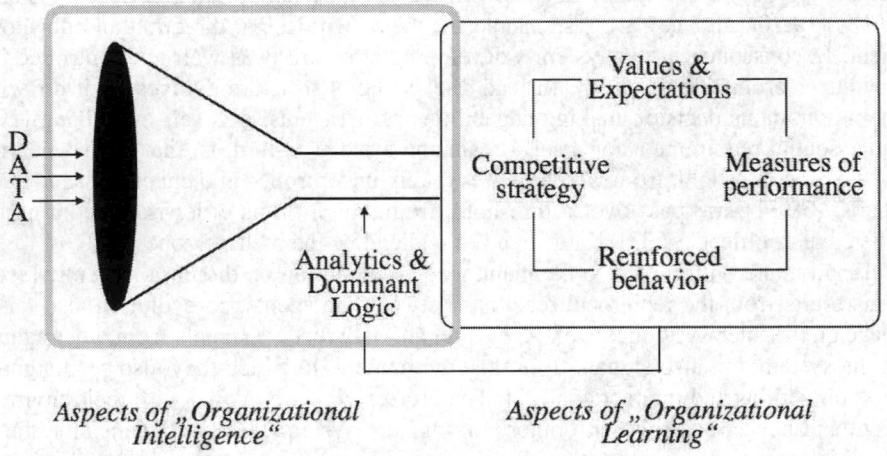

Fig. 1. The dominant logic [36, p. 7]

The ability to learn, as it was mentioned above as an inherent characteristic of CALS, plays an important role in the concept of the dominant logic [36]. The agents within a system repeat decision rules, that leaded to a good performance in the past and to them they got positive feedbacks from other agents via interaction [34]. This reinforced behavior of the single system's elements influences the behavioral rules of the agents within a logistics system (e.g. managers) and therefore the competitive strategies of the logistics system's sub-systems (e.g. logistics service providers, warehouses). These strategies as well as the resulting performances and how they are evaluated, in turn, influence the values and expectations of the single agents (e.g. managers), whereas these values and expectations reinforce, in turn, the single agents behaviors and therewith influence the strategies of the logistics systems' sub-systems [36]. This does not only apply to human elements in logistics systems, but as well to certain non-human ones, the so-called smart parts. Their programming include behavioral rules, which might, in turn, include values and expectations that are influenced by previous actions (in analogy to competitive strategy), certain ways to measure the actions' performances, and in the result possibly a reinforced behavior. Prahalad and Bettis [36] stated this circle as an organization's, or in this context a logistics system's, ability to learn. This ability shapes the system's modality, how to analyze information that flows into the system and is called a system's dominant logic [36].

The adaptivity of a system reflects its ability to, on the one hand, open its boundaries [37] to enable information inflow [38], and on the other hand to close the system's boundaries [39,40], to keep this inflow at a manageable level and therewith to prevent the system from information overload. Whereas the former is needed to react flexible to changes in the relevant environment, the latter maintains the system's stability [41]. Hence, the data that arrives at the system's boundaries has to be selected respectively filtered from irrelevant or less relevant data [36]. This selection, in turn, underlies a certain principle, which evolves over time and was called by Bettis and Prahalad [36] the organizational intelligence. The degree of filtering the incoming data is dependent on the analytics and the dominant logic, that is in turn dependent on the agent's and the sub-system's abilities to learn [36]. In logistics systems, that consist of a large number of organizations (sub-systems) and their agents [18], this intelligence might be cumulated and can therefore be called the intelligence of whole systems, in this case of CALS.

Subsuming these observations, the dominant logic of a system contributes on the one hand to select the incoming data and therewith to prevent the system from an information overload. On the other hand, it limits the amount of information that flows into the system and can therewith lead to a system's under-supply of relevant information. Latter occurs, if the dominant logic filters to much data or if it simply filters the wrong data, which means that the system's information processing capacity is burdened with irrelevant information. Therewith, the dominant logic decreases the adaptivity of CALS by negatively affecting its flexibility by decreasing the inflow of relevant information. This, in turn, leads to a decrease of the amount of information the system's sub-systems as well as their single elements are supplied with. Hence, the elements are equipped with a decreasing amount of information, which they can exchange between each other, which means, that their incentives to interact are decreasing as well. In other words, the dominant logic of CALS diminishes the heterogeneity of their elements and therewith decreases their innovation capabilities.

4 Autonomous Co-operation and Control – A Driver for Heterogeneity and Innovation Capability

In connection with the adaptivity and learning features of CALS the organizational principle of autonomous co-operation and control has been discussed [e.g. 15] and might therefore be a fruitful approach to countersteer against the described effects of a dominant logic in logistics systems as well as to decrease the intensity of a system's dominant logic itself. Because more and more centralized decision-making configurations have shown a lack of capacities and capabilities to cope with increasing complexity and dynamics in ISN respectively in CALS, the approach of autonomous co-operation and control gains more relevance for the design of global logistics network structures [42]. It becomes more feasible and realistic, because modern technologies (e.g. RFID), methods (e.g. collaborative route-planning), and instruments (e.g. multi-agent-modelling) deliver the necessary fundamentals [43]. The organizational principle of autonomous co-operation and control is based on the idea of self-organization, which has its origins in different fields of science, like cybernetics [44], chemistry [45], physics [46], and biology [47]. It aims at an explanation of the autonomous creation of ordered structures in complex systems and can be understood as "(...) processes of decentralized decision-making in heterarchical structures. It presumes interacting elements in non-deterministic systems, which possess the capability and possibility to render decisions. The objective of Autonomous Control is the achievement of increased robustness and positive emergence of the total system due to distributed and flexible coping with dynamics and complexity." [42, p. 9]. Besides all specific descriptions some common characteristics can be identified, which comprise decentralized decision-making, autonomy, interaction, heterarchy and non-determinism [42].

Decentralized decision-making refers to the delegation of decision-making power from a centralized entity (e.g. the disposition department or another planning unit) to the single elements of the system (e.g. employer, packages, industrial trucks) [42]. Therefore, not only one specific unit is capable and responsible for a certain number and quality of decisions, which are made for all the other elements in the system, but much more elements have the rights and capabilities to make decisions for their own area of responsibility. This is also described by the term of heterarchy. That means for an ISN that more elements that are equally enabled to undertake decision-making and that there is a significantly increasing decision-making capacity in comparison to a single-but-central-planning-unit. For the management of ISN this implies that the overall learning capabilities are increasing, too.

Another characteristic item, which is closely connected with the decentralized decision-making and which might be a pre-condition, is autonomy [48], which means, that every element in a system is responsible for itself [28]. That might include the responsibility for its own goals, strategies, means, organizational design, resources, etc. Autonomy needs a situation of a high-degree of non-determinism, otherwise the degrees of freedom were low and there were no chance for a free choice. Therefore, autonomy has a positive effect on the heterogeneity, if it leads to a situation, in which all elements of a system define their own aims, explore their own ways to achieve those aims, and learn about their environmental demands and the effects of their adaptivity and behavior. In this case, autonomy might be a facilitator for learning.

Interaction in ISN describes the fact that the system's elements are able to communicate directly with each other. Therefore, the elements (e.g. smart parts) are able to exchange more information for a certain problem-solving quicker, more precisely and demand-driven – in comparison to a centralized decision-making and order creation [42]. This implies a more target-oriented, effective, and efficient exchange of information. Because the single elements (e.g. RFID tags, intelligent sensors as well as personal executives) might be more capable to collect and process information they need for an upcoming decision, because only the needed portion of information is exchanged, not all the data for a whole ISN. Therefore, the overall amount of information an organization has to acquire and to compute might be increased, because more processes of the acquisition and procession of information can be executed parallel. This leads as well to higher and quicker learning processes, because feedbacks from the environment regarding a certain system's behavior can be recognized, interpreted, and exchanged more directly. The higher the degree of interaction is, the higher the learning capabilities will be, up to the point, which was already introduced as the edge of chaos [30,31].

These examples highlight the potentials of the implementation of autonomous co-operation and control and its associated technologies, in order to increase a logistics system's ability to learn. With recourse to the single aspects of organizational learning, mentioned by Bettis and Prahalad [36], as the determinants for the intensity of an organization's respectively a system's dominant logic, it can be stated, that the implementation of autonomous co-operation into logistics systems as well as the increase of its degree, might decrease the intensity of their dominant logics. Hence, logistics systems respectively CALS, that are organized by principles of autonomous cooperation have a higher degree of heterogeneity than systems that are organized by external control, which contributes therewith to their abilities to be innovative.

5 Conclusions

It has been shown, that the innovation capabilities of organizations within logistics systems respectively sub-systems of CALS are dependent on the degrees of heterogeneity of the systems' elements. One driver of homogeneity and therewith a barrier for innovations is the dominant logic, which in turn is dependent on the systems' as well as the systems' sub-systems' and elements' learning abilities. The adaptivity of a CALS as well as the phenomenon of dominant logic depend on the learning capabilities of such a system: The dominant logic diminishes the heterogeneity in a system; the learning capabilities increase the adaptivity of the system, but might be limited by an existing or evolving dominant logic; too much heterogeneity might lead to an excessive demand of the existing learning capabilities, which results in a dominant logic that reduces the heterogeneity, which might effect the innovation capability negatively. As one can see, there is a complex and ambiguous intertwining between learning capabilities, heterogeneity, and dominant logic. The implementation of the organization principle autonomous co-operation and control is a possible approach to countersteer against the intensity of the dominant logic as well as against the negative effects emanating from dominant logic by increasing the learning capabilities of CALS. Hence, autonomous co-operation facilitates the heterogeneity of the elements within CALS such as ISN and increases therewith their innovation capabilities.

Acknowledgement. This research was supported by the German Research Foundation (DFG) as part of the Collaborative Research Centre 637 »Autonomous Cooperating Logistic Processes – A Paradigm Shift and its Limitations«.

References

1. Henard, D.H., Szymanski, D.M.: Why Some New Products Are More Successful Than Others. Journal of Marketing Research 38(3), 362–375 (2001)
2. Ulijn, J., O'Hair, D., Weggeman, M., Ledlow, G., Hail, H.T.: Innovation, Corporate Strategy, and Cultural Context: What Is the Mission for International Business Communication? The Journal of Business Communication 37(3), 293–317 (2000)
3. Johne, A.: Successful market innovation Successful market innovation. European Journal of Innovation Management 2(1), 6–11 (1999)
4. Johnson, J.D.: Effects of Communication Factors on Participation in Innovations. The Journal of Business Communication 27(1), 7–23 (1990)
5. Rogers, E.M.: Diffusion of Innovations, 5th edn. The Free Press, New York (2003)
6. Westphal, J.D., Gulati, R., Shortell, S.M.: Customization or Conformity? An Institutional and Network Perspective on the Content and Consequences of TQM Adoption. Administrative Science Quarterly 42(2), 366–394 (1997)
7. McDermott, C.M., O'Connor, G.C.: Managing radical innovation: an overview of emergent strategy issues. The Journal of Product Innovation Management 19, 424–438 (2002)
8. Song, M., Thieme, J.: A cross-national investigation of the R&D-marketing interface in the product innovation process. Industrial Marketing Management 35, 308–322 (2006)
9. Chapman, R., Hyland, P.: Complexity and learning behaviors in product innovation. Technovation 24, 553–561 (2004)
10. Gjerde, K.A.P., Slotnick, S.A., Sobel, M.J.: New Product Innovation with Multiple Features and Technology Constraints. Management Science 48(10), 1268–1284 (2002)
11. Siegele, L.: How about now? A survey of the real-time economy. The Economist 362, 18–24 (2002)
12. Welge, M., Holtbrügge, D.: Internationales Management: Theorien, Funktionen, Fallstudien. Schäffer-Poeschel, Stuttgart (2003)
13. Lambert, D.M., Cooper, M.C., Pagh, J.D.: Supply Chain Management: Implementation Issues and Research Opportunities. The International Journal of Logistics Management 9(2), 1–19 (1998)
14. Hülsmann, M., Grapp, J.: Autonomous Cooperation in International-Supply-Networks – The Need for a Shift from Centralized Planning to Decentralized Decision Making in Logistic Processes. In: Pawar, K.S., et al. (eds.) Proceedings of the 10th International Symposium on Logistics (10th ISL), Loughborough, United Kingdom, pp. 243–249 (2005)
15. Wycisk, C., McKelvey, B., Hülsmann, M.: 'Smart parts' logistics systems as complex adaptive systems. International Journal of Physical Distribution and Logistics Management 38(2), 108–125 (2008)
16. Surana, A., Kumara, S., Greaves, M., Raghavan, U.N.: Supply-chain networks: a complex adaptive systems perspective. International Journal of Production Research 43(20), 4235–4265 (2005)
17. Choi, T.Y., Dooley, K.J., Rungtusanatham, M.: Supply networks and complex adaptive systems: control versus emergence. Journal of Operations Management 19(3), 351–366 (2001)
18. Pathak, S.D., Day, J., Nair, A., Sawaya, W.J., Kristal, M.: Complexity and adaptivity in supply networks: building supply network theory using a complex adaptive systems perspective. Decision Science Journal 38(4), 547–580 (2007)
19. Holland, J.H.: The global economy as an adaptive system. In: Anderson, P.W., Arrow, K.J., Pines, D. (eds.) The Economy as an Evolving Complex System, vol. V, pp. 117–124. Addison-Wesley, Reading (1988)

20. Holland, J.H.: Complex Adaptive Systems and Spontaneous Emergence. In: Quadrio Curzio, A., Fortis, M. (eds.) Complexity and Industrial Clusters, pp. 25–34. Physica-Verl., Heidelberg (2002)
21. Holland, J.H.: Studying Complex Adaptive Systems. Journal of Systems Science and Complexity 19(1), 1–8 (2006)
22. Kauffman, S.A.: The Origins of Order: Self-organization and Selection in Evolution. Oxford University Press, New York (1993)
23. Cooper, R.G.: From Experience: The Invisible Success Factors in Product Innovation. The Journal of Product Innovation Management 16, 115–133 (1999)
24. Chesbrough, H.W.: The Era of Open Innovation. MIT Sloan management review 44(3), 35–41 (2003)
25. Drejer, A.: Situations for innovation management: towards a contingency model. European Journal of Innovation Management 5(1), 4–17 (2002)
26. Hülsmann, M., Scholz-Reiter, B., Austerschulte, L., de Beer, C., Grapp, J.: Autonomous Cooperation – A Capable Way to Cope with External Risiks in International Supply Networks? In: Pawar, K.S., Lalwani, C.S., de Carvalho, J.C., Muffatto, M. (eds.) Proceedings of the 12th International Symposium on Logistics (12th ISL), Loughborough, United Kingdom, pp. 172–178 (2007)
27. McKelvey, B., Wycisk, C., Hülsmann, M.: Designing Learning Capabilities of Complex 'Smart Parts' Logistics Markets: Lessons from LeBaron's Stock Market Computational Model. International Journal of Production Economics (submitted) (2008)
28. Probst, G.J.B.: Selbst-Organisation: Ordnungsprozesse in sozialen Systemen aus ganzheitlicher Sicht. Parey, Berlin (1987)
29. Mainzer, K.: Thinking in Complexity: The Complex Dynamics of Matter, Mind, and Mankind. Springer, New York (1994)
30. Langton, C.G.: Artificial Life. Addison-Wesley, Reading (1989)
31. Lewin, R.: Complexity. University of Chicago Press, Chicago (1992)
32. Bérnard, H.: Les Tourbillons Cellulaires dans une Nappe Liquide Transportant de la Chaleur par Convection en Régime Permanent. Annales de Chimie et de Physique 23, 62–144 (1901)
33. Prigogine, I.: An Introduction to Thermodynamics of Irreversible Processes. Thomas, Springfield (1955)
34. Heylighen, F.: The Science of Self-organization and Adaptivity. In: Knowledge Management, Organizational Intelligence and Learning, and Complexity. The Encyclopedia of Life Support Systems, Oxford (2003)
35. Prahalad, C.K., Bettis, R.A.: The Dominant Logic: A New Linkage between Diversity and Performance. Strategic Management Journal 7(6), 485–501 (1986)
36. Bettis, R.A., Prahalad, C.K.: The Dominant Logic: Retrospective and Extension. Strategic Management Journal 16(1), 5–14 (1995)
37. Garavelli, A.C.: Flexibility configurations for the supply chain management. International Journal of Production Economics 8(2), 141–153 (2003)
38. Hicks, H.G., Gullett, C.R.: Organizations: theory and behavior. McGraw-Hill, New York (1975)
39. Luhmann, N.: Zweckbegriff und Systemrationalität. Suhrkamp, Frankfurt am Main (1973)
40. Luhmann, N.: Soziale Systeme: Grundriss einer allgemeinen Theorie. Suhrkamp, Frankfurt am Main (1994)
41. Hülsmann, M., Grapp, J., Li, Y.: Strategic Adaptivity in Global Supply Chains – Competitive Advantage by Autonomous Cooperation. Special Edition of the International Journal of Production Economics (forthcoming) (2008)

42. Windt, K., Hülsmann, M.: Changing Paradigms in Logistics. In: Hülsmann, M., Windt, K. (eds.) Understanding Autonomous Cooperation & Control: The Impact of Autonomy on Management, Information, Communication, and Material Flow, pp. 1–12. Springer, Berlin (2007)

43. Scholz-Reiter, B., Windt, K., Freitag, M.: Autonomous logistic processes: new demands and first approaches. In: Monostori, L. (ed.) Proceedings of the 37th CIRP International Seminar on Manufacturing Systems, Budapest, pp. 357–362 (2004)

44. von Foerster, H.: Cybernetics of Cybernetics. In: Krippendorff, K. (ed.) Communication and Control in Society, pp. 5–8. Gordon and Breach Science Publishers, New York (1979)

45. Glansdorff, P., Prigogine, I.: Thermodynamic theory of structure, stability and fluctuations. Wiley, New York (1971)

46. Haken, H.: Synergetics: cooperative phenomena in multi-component systems. In: Symposium on Synergetics, April 30, May 6, 1972. Schloß Elmau, Stuttgart (1973)

47. Maturana, H.R., Varela, F.: Autopoiesis and cognition: the realization of living. Reidel, Dordrecht (1980)

48. Kappler, E.: Autonomie. In: Frese, E. (ed.) Handwörterbuch der Organisation, 3rd edn., pp. 272–280. Poeschel, Stuttgart (1992)

Asymptotic Behavior of Ruin Probability in Insurance Risk Model with Large Claims

Tao Jiang

Zhejiang Gongshang University, Hangzhou 310018, P.R. China
jtao@263.net

Abstract. For the renewal risk model with subexponential claim sizes, we established for the finite time ruin probability a lower asymptotic estimate as initial surplus increases, subject to the demand that it should hold uniformly over all time horizons in an infinite interval. In the case of Poisson model, we also obtained the upper asymptotic formula so that an equivalent formula was derived. These extended a recent work partly on the topic from the case of Pareto-type claim sizes to the case of subexponential claim sizes and, simplified the proof of lower bound in Leipus and Siaulys ([9]).

Keywords: asymptotic formula, finite time ruin probability, strongly subexponential distributions, the compound Poisson/renewal model, uniform convergence.

1 Introduction

Consider the renewal risk model, in which the claim sizes Z_i, $i = 1, 2, \ldots$, form a sequence of independent, identically distributed (i.i.d.), nonnegative random variables with common distribution B, while the inter-occurrence times θ_i, $i = 1, 2, \ldots$, form another sequence of i.i.d. positive random variables with common finite mean $1/\lambda$. Two sequences $\{Z_i, i = 1, 2, \ldots\}$ and $\{\theta_i, i = 1, 2, \ldots\}$ are assumed to be mutually independent. The locations of claims $\tau_k = \sum_{i=1}^{k} \theta_i$, $k = 1, 2, \ldots$, constitute a renewal counting process

$$N(t) = \max\{k = 1, 2, \ldots : \tau_k \in (0, t]\}, \qquad t \geq 0, \tag{1}$$

with a mean function $\lambda(t) = \mathbb{E}N(t) \sim \lambda t$ as $t \to \infty$. The meaning of $\sim$ will be given in the following. The surplus process is then defined as

$$R(t) = x + ct - \sum_{i=1}^{N(t)} Z_i, \qquad t \geq 0, \tag{2}$$

where $R(0) = x \geq 0$ denotes the initial surplus, $c > 0$ denotes the constant premium rate, and a summation over an empty set of index is 0 by convention.

Ruin probability is one of the most important concept in modern actuarial risk theory.

J. Zhou (Ed.): Complex 2009, Part I, LNICST 4, pp. 1033–1043, 2009.
© ICST Institute for Computer Sciences, Social Informatics and Telecommunications Engineering 2009

Definition 1. *Finite time ruin probability within time t is defined as*

$$\psi(x;t) = \Pr\left(\inf_{0 \leq s \leq t} R(s) < 0 \;\middle|\; R(0) = x \right), \qquad t \geq 0. \tag{3}$$

If $t = \infty$,

$$\psi(x;\infty) = \lim_{t \to \infty} \psi(x;t) = \Pr\left(\inf_{0 \leq s < \infty} R(s) < 0 \;\middle|\; R(0) = x \right) \tag{4}$$

is called ultimate ruin probability.

In order for the ultimate ruin not to be certain, it is natural to assume the safety loading condition

$$\mu = \frac{c}{\lambda} - EZ_1 > 0. \tag{5}$$

We refer readers to Asmussen ([1]) and ([2]) for a nice reviews on the study of the finite time ruin probability and to Tang ([12]) for a list of references devoted to this study. Our goal in the current paper is to derive an asymptotic estimate as the initial surplus x increases for the finite time ruin probability $\psi(x;t)$, subject to the requirement that the asymptotic result should hold uniformly over all time horizons t in an infinite interval.

Hereafter, all limit relationships are for $x \to \infty$ unless stated otherwise. For two positive functions $a(\cdot)$ and $b(\cdot)$, we write $a(x) \overset{\scriptstyle <}{\sim} b(x)$ if $\limsup a(x)/b(x) \leq 1$, write $a(x) \overset{\scriptstyle >}{\sim} b(x)$ if $\liminf a(x)/b(x) \geq 1$, and write $a(x) \sim b(x)$ if both. As done in the main result of this paper, we shall assign a certain uniformity property to some asymptotic relations under discussion. Let us take an example to clarify the meaning of uniformity. For two positive bivariate functions $a(\cdot;\cdot)$ and $b(\cdot;\cdot)$, we say that the asymptotic relation $a(x;t) \sim b(x;t)$ holds uniformly over all t in a nonempty set Δ if

$$\lim_{x \to \infty} \sup_{t \in \Delta} \left| \frac{a(x;t)}{b(x;t)} - 1 \right| = 0. \tag{6}$$

That is, for each fixed $\varepsilon > 0$, there exists some $x_0 > 0$ irrespective to t such that the two-sided inequality

$$(1 - \varepsilon)b(x;t) \leq a(x;t) \leq (1 + \varepsilon)b(x;t) \tag{7}$$

holds for all $x \geq x_0$ and $t \in \Delta$. This is further equivalent to that both $a(x) \overset{\scriptstyle <}{\sim} b(x)$ and $a(x) \overset{\scriptstyle >}{\sim} b(x)$ hold uniformly over all $t \in \Delta$. Admittedly, results that hold with such a uniformity property are of higher theoretical and practical interest.

Heavy-tailed risk has played an important role in insurance and finance because it can describe large claims; see Embrechts et al. ([3]). We shall mainly discuss heavy-tailed claims in this paper. The most important class of heavy-tailed distributions is the subexponential class.

Definition 2. *a distribution F on $[0, \infty)$ is said to be subexponential, written as $F \in \mathcal{S}$, if its right tail $\overline{F} = 1 - F$ satisfies $\overline{F}(x) > 0$ for all x and the relation*

$$\overline{F^{*2}}(x) \sim 2\overline{F}(x) \tag{8}$$

holds, where F^{*2} denotes the convolution of F with itself.

More generally, a distribution F on $(-\infty, \infty)$ is still said to be subexponential if the distribution $F^+(x) = F(x)1_{(0 \le x < \infty)}$ is subexponential, where 1_A denotes the indicator function of A. It is well known that every subexponential distribution F is long tailed, written as $F \in \mathcal{L}$, in the sense that the relation

$$\overline{F}(x+y) \sim \overline{F}(x) \tag{9}$$

holds for each fixed real number y; see, for example, Embrechts et al. ([3], Lemma 1.3.5).

Very often the class $\mathcal{S}$ appears to be too wide to possess desirable probabilistic properties. For this reason, researchers in applied probability have introduced many subclasses of $\mathcal{S}$ to meet certain special requirements. In this regard, Korshunov ([8]) introduced the class of strongly subexponential distributions. For a distribution F on $(-\infty, \infty)$ with $0 < m = \int_0^\infty \overline{F}(u)du < \infty$ and for each fixed $l \in (0, \infty]$, we write

$$\overline{F_l}(x) = \begin{cases} \min\left\{1, \int_x^{x+l} \overline{F}(u)du\right\}, & x \ge 0, \\ 1, & x < 0. \end{cases} \tag{10}$$

Clearly, for each $l \in (0, \infty]$ the function F_l defines a standard distribution on $[0, \infty)$. In the terminology of Korshunov ([8]), the distribution F is said to be strongly subexponential, denoted by $F \in \mathcal{S}_*$, if the relation

$$\lim_{x \to \infty} \frac{\overline{F_l^{*2}}(x)}{\overline{F_l}(x)} = 2 \tag{11}$$

holds uniformly over all $l \in [1, \infty]$. It is easy to check that relation (11) with an arbitrarily fixed number $l \in [1, \infty)$ implies $F \in \mathcal{S}$; see Kaas and Tang ([6]). Hence, $\mathcal{S}_*$ is a subclass of $\mathcal{S}$. From the discussions of Korshunov ([8]), we see that the class $\mathcal{S}_*$ covers almost all useful subexponential distributions with $m < \infty$. Specifically, the class $\mathcal{S}_*$ contains all Pareto-like distributions with $m < \infty$, all lognormal-like distributions, and all heavy-tailed Weibull-like distributions. We should point out that, Pareto-like function class with index $-\alpha$ is usually denoted by $\mathcal{R}_{-\alpha}$: if

$$\overline{F}(x) = x^{-\alpha}L(x), \quad x > 0,$$

where $L(x)$ is a slowly varying function as $x \to \infty$ and index $-\alpha < 0$. $\mathcal{R}_{-\alpha}$ is also called regularly varying function class.

2   Main Result and Insurance Significance

The main results of this paper are as follows:

Theorem 1. *Consider the renewal model with the safety loading condition (5), which is introduced at the very beginning of this paper. If $B \in \mathcal{L}$, then for every positive function $f(\cdot)$ with $f(x) \to \infty$, it holds uniformly over all $t \in [f(x), \infty]$ that*

$$\psi(x;t) \gtrsim \frac{1}{\mu} \int_x^{x+\mu\lambda t} \overline{B}(u)du. \tag{12}$$

When $t = \infty$, formula (12) is reduced to

$$\psi(x;\infty) \gtrsim \frac{1}{\mu} \int_x^\infty \overline{B}(u)du. \tag{13}$$

Specially, in the case of Poisson risk model, we can get the following asymptotic formula:

Theorem 2. *Consider the Compound Poisson model with the safety loading condition (5). If $B \in \mathcal{L}$, then for every positive function $f(\cdot)$ with $f(x) \to \infty$, it holds uniformly over all $t \in [f(x), \infty]$ that*

$$\psi(x;t) \lesssim \frac{1}{\mu} \int_x^{x+\mu\lambda t} \overline{B}(u)du. \tag{14}$$

When $t = \infty$, formula (16) is reduced to

$$\psi(x;\infty) \lesssim \frac{1}{\mu} \int_x^\infty \overline{B}(u)du, \tag{15}$$

which is well known, first established by Veraverbeke ([13]) and Embrechts and Veraverbeke ([4]). which is well known, first established by Veraverbeke ([13]) and Embrechts and Veraverbeke ([4]).

An corollary of These two Theorems can be obtained directly.

Corollary 1. *Consider the Compound Poisson model with the safety loading condition (5). If $B \in \mathcal{L}$, then for every positive function $f(\cdot)$ with $f(x) \to \infty$, it holds uniformly over all $t \in [f(x), \infty]$ that*

$$\psi(x;t) \sim \frac{1}{\mu} \int_x^{x+\mu\lambda t} \overline{B}(u)du. \tag{16}$$

When $t = \infty$, formula (16) is reduced to

$$\psi(x;\infty) \sim \frac{1}{\mu} \int_x^\infty \overline{B}(u)du, \tag{17}$$

Tang ([12]) established (16) in the form of equivalence in the renewal model under the assumption, among others, that the distribution B is consistently varying tailed in the sense that

$$\lim_{l \nearrow 1} \limsup_{x \to \infty} \frac{\overline{B}(lx)}{\overline{B}(x)} = 1. \tag{18}$$

Hence, his result works essentially only for the case of Pareto-like claim sizes. Recently, under the three assumptions as following:

(1) There exists a nonnegative function q: $R_+ \to R_+$ such that

$$Q(u) = \int_0^u q(v)dv, \quad u \in R_+ \quad and \quad \limsup_{u \to \infty} \frac{uq(u)}{Q(u)} =: r \quad is \quad finite;$$

(2) The hazard rate $q(u)$ satisfies $\liminf_{u \to \infty} uq(u) \geq \max\left\{1, \frac{1}{1-r}\right\}$;

(3) The random variable θ is such that $P(0 \leq \theta < \epsilon)$ and $P(\theta = 0) = 1$ for every positive $\epsilon > 0$,

Leipus and Siaulys ([9]) obtained that, in the renewal risk model with the safety loading condition (5), if $B \in \mathcal{S}_*$, then for every positive function $f(\cdot)$ with $f(x) \to \infty$, (16) holds uniformly over all $t \in [f(x), \gamma x]$.

By analyzing this result carefully, we could see that, firstly, the assumptions it demands seem to be too strong. Hence, it is very difficult to be suitable for more general case. Secondly, the proof of their results is too complicated. It isn't pretty mathematically. Finally, that fact that class $\mathcal{L}$ is much bigger than class $\mathcal{S}_*$ illustrates that, Theorem 1 has much wider usage.

In modelling extremal events, heavy-tailed risk has played an important role in insurance and finance because they can describe large claims efficiently; see Embrechts et al. ([3]) and Goldie & Klüppelberg ([5]) for a nice review. We give here several important classes of heavy-tailed distributions for further references.

The insurance significance of Theorem 1 is as the following: first, it provides a lower bound of ruin probability of an insurance company. It is useful in risk management of the insurance company. Second, asymptotic formula when $x \to \infty$ often means that large claim is concerns. In other words, even very large initial capital is paid out by claims. This is just the case of extremal event! In fact, this is one of the reasons that we study heavy-tailed claims.

Uniformity is an important concept in mathematics. The main advantage of uniformity is, changing order of limit and integral is permissable. Thus, some results, say, about finite time ruin probability, can be extended easily to the ultimate time ruin probability.

The following three lemmas play the important roles in obtaining these two Theorems.

Lemma 1. *Let $\{X_i, i = 1, 2, \ldots\}$ be a sequence of i.i.d. random variables with common distribution F and finite mean $\mathrm{E}X_1 = -\mu < 0$. If $F \in \mathcal{L}$, then it holds uniformly over all $n = 1, 2, \ldots$ that*

$$\Pr\left(\max_{1 \leq k \leq n} \sum_{i=1}^{k} X_i > x\right) \gtrsim \frac{1}{\mu} \int_x^{x+\mu n} \overline{F}(u)\mathrm{d}u. \tag{19}$$

This lemma is the extension to the Theorem of Korshunov ([8]); See also Tang (2004a). Lemma 2 reflects the basic property of homogenous Poisson process and it can be found, for example, in Theorem 2.3.1 of Ross ([10]):

Lemma 2. *Let $\{N(t), t \geq 0\}$ be a homogenous Poisson process with arrival times τ_k, $k = 1, 2, \ldots$. Given $N(t) = n$ for arbitrarily fixed $t > 0$ and $n = 1, 2, \ldots$, the random vector $(\tau_1, \cdots, \tau_n)$ is equal in distribution to the random vector $(tU_{(1,n)}, \cdots, tU_{(n,n)})$ with $U_{(1,n)}, \ldots, U_{(n,n)}$ being the order statistics of n independent and uniformly distributed random variables $U_1, \ldots, U_n$ in $(0,1)$.*

Lemma 3 is from Lemma 2.1 of Klüppelberg and Mikosch ([7]):

Lemma 3. *Let $\{N(t), t \geq 0\}$ be a homogenous Poisson process with intensity $\lambda > 0$. Then, it holds for every $\varepsilon > 0$ and $\delta > 0$ that*

$$\lim_{t \to \infty} \sum_{n > (1+\delta)\lambda t} (1+\varepsilon)^n \Pr\left(N(t) = n\right) = 0. \tag{20}$$

3   The proof of the Main Results

3.1   Proof of Theorem 1

Since, by definition, $B \in \mathcal{L}$ implies $B_I \in \mathcal{S}$, by virtue of relation (15), it suffices to prove the uniformity of (16) over all $t \in [f(x), \infty)$. For arbitrarily fixed $\delta > 0$, we write

$$M_-(\delta) = \min_{0 \leq k < \infty} \left(\frac{(1+\delta)k}{\lambda} - \tau_k \right), \tag{21}$$

which is nonpositive and finite almost surely. From the equivalent definition of finite time ruin probability

$$\psi(x; t) = \Pr\left(\max_{0 \leq k \leq N(t)} \left(\sum_{i=1}^{k} Z_i - c\tau_k \right) > x \right), \qquad t > 0, \tag{22}$$

for each fixed $L > 0$, we have

$$\psi(x; t)$$
$$= \Pr\left(\max_{0 \leq k \leq N(t)} \left(\sum_{i=1}^{k} \left(Z_i - \frac{c(1+\delta)}{\lambda} \right) + c\left(\frac{(1+\delta)k}{\lambda} - \tau_k \right) \right) > x \right)$$
$$\geq \Pr\left(\max_{0 \leq k \leq N(t)} \sum_{i=1}^{k} \left(Z_i - \frac{c(1+\delta)}{\lambda} \right) > x + cL, \, M_-(\delta) > -L \right)$$
$$= \sum_{n=1}^{\infty} \Pr\left(\max_{0 \leq k \leq n} \sum_{i=1}^{k} \left(Z_i - \frac{c(1+\delta)}{\lambda} \right) > x + cL \right) \Pr\left(N(t) = n, M_-(\delta) > -L \right) \tag{23}$$

We write

$$\mu_2(\delta) = \frac{c(1+\delta)}{\lambda} - \mathbb{E}Z_1 > 0. \tag{24}$$

Applying Lemma 1, it holds uniformly over all $n = 1, 2, \ldots$ that

$$\Pr\left(\max_{0 \leq k \leq n} \sum_{i=1}^{k} \left(Z_i - \frac{c(1+\delta)}{\lambda} \right) > x + cL \right) \gtrsim \frac{1}{\mu_2(\delta)} \int_x^{x+\mu_2(\delta)n} \overline{B}(u + cL) \, du. \tag{25}$$

Substituting this into (23) and considering an arbitrarily fixed number $0 < l < 1$, we have that, uniformly over all $t \in [f(x), \infty)$,

$$\psi(x; t)$$

$$\gtrsim \frac{1}{\mu_2(\delta)} \sum_{n=1}^{\infty} \int_x^{x+\mu_2(\delta)n} \overline{B}(u + cL) \, du \cdot \Pr(N(t) = n, \, M_-(\delta) > -L)$$

$$\geq \frac{1}{\mu_2(\delta)} \sum_{n \geq (1-l)\lambda t} \int_x^{x+\mu n} \overline{B}(u + cL) \, du \cdot \Pr(N(t) - n, \, M_-(\delta) > -L)$$

$$\geq \frac{1}{\mu_2(\delta)} \int_x^{x+(1-l)\mu\lambda t} \overline{B}(u + cL) \, du \cdot \Pr\left(\frac{N(t)}{\lambda t} \geq 1 - l, \, M_-(\delta) > -L\right). \quad (26)$$

We apply an elementary inequality, $\Pr(AB) \geq \Pr(A) + \Pr(B) - 1$, to obtain that

$$\Pr\left(\frac{N(t)}{\lambda t} \geq 1 - l, \, M_-(\delta) > -L\right) \geq \Pr\left(\frac{N(t)}{\lambda t} \geq 1 - l\right) + \Pr(M_-(\delta) > -L) - 1.$$

$$(27)$$

As $t \to \infty$, it is well known that $N(t)/\lambda t \to 1$ holds almost surely; see, for example, Section 2.5 of Embrechts et al. ([3]). Hence for each $\varepsilon > 0$, we may find some $x_0 > 0$ and $L_0 > 0$ such that the inequality

$$\Pr\left(\frac{N(t)}{\lambda t} \geq 1 - l, M_-(\delta) > -L_0\right) \geq 1 - \varepsilon \quad (28)$$

holds for all $t \in [f(x_0), \infty)$. Substitution of this into (26) with $L = L_0$ gives that, uniformly over all $t \in [f(x), \infty)$,

$$\psi(x; t) \gtrsim \frac{1 - \varepsilon}{\mu_2(\delta)} \int_x^{x+(1-l)\mu\lambda t} \overline{B}(u + cL_0) \, du$$

$$\sim \frac{1 - \varepsilon}{\mu_2(\delta)} \left(\int_x^{x+\mu\lambda t} - \int_{x+(1-l)\mu\lambda t}^{x+\mu\lambda t} \right) \overline{B}(u) \, du$$

$$\geq \frac{1 - \varepsilon}{\mu_2(\delta)} \int_x^{x+\mu\lambda t} \overline{B}(u) \, du \left(1 - \frac{\int_{x+(1-l)\mu\lambda t}^{x+\mu\lambda t} \overline{B}(u) \, du}{\int_x^{x+\mu\lambda t} \overline{B}(u) \, du} \right)$$

$$\geq \frac{1 - \varepsilon}{\mu_2(\delta)} \int_x^{x+\mu\lambda t} \overline{B}(u) \, du \left(1 - \frac{l\mu\lambda t \overline{B}(x + (1 - l)\mu\lambda t)}{(1 - l)\mu\lambda t \overline{B}(x + (1 - l)\mu\lambda t)} \right)$$

$$= \frac{1 - \varepsilon}{\mu_2(\delta)} \frac{1 - 2l}{1 - l} \int_x^{x+\mu\lambda t} \overline{B}(u) \, du.$$

Since the constants $\delta > 0$, $0 < l < 1$, and $\varepsilon > 0$ can be arbitrarily small, we finally obtain the desired relation (16) with the indicated uniformity property.   □

3.2   Proof of Theorem 2

We complete the proof by proving that, when claim-arrival follows Poisson process, the right hand of (12) is also the upper bound, uniformly over all $n = 1, 2, \ldots$.

For arbitrarily fixed $0 < \delta < 1$ such that $(1 - \delta)v > EZ_1$ and for each $n = 1, 2, \ldots$, we have

$$\Pr\left(\max_{1 \leq k \leq n}\left(\sum_{i=1}^{k} Z_i - vnU_{(k,n)}\right) > x\right) \leq \Pr\left(\xi_n + \eta_n > x\right), \qquad (29)$$

where ξ_n and η_n are given by

$$\xi_n = \max_{1 \leq k \leq n}\left(\sum_{i=1}^{k} Z_i - (1 - \delta)vk\right) \qquad (30)$$

and

$$\eta_n = \max_{1 \leq k \leq n}\left((1 - \delta)vk - vnU_{(k,n)}\right). \qquad (31)$$

For the random variables ξ_n, $n = 1, 2, \ldots$, by Korshunov (19), it holds uniformly over all $n = 1, 2, \ldots$ that

$$\Pr\left(\xi_n > x\right) \stackrel{<}{\sim} \frac{1}{(1 - \delta)v - EZ_1}\int_x^{x + (v - EZ_1)n} \overline{B}(u)\mathrm{d}u.$$

Hence for each $\varepsilon > 0$, we may choose some $M > 0$ such that for all $n = 1, 2, \ldots$ and $x \geq M$,

$$\Pr\left(\xi_n > x\right) \leq \frac{1 + \varepsilon}{(1 - \delta)v - EZ_1}\int_x^{x + (v - EZ_1)n} \overline{B}(u)\mathrm{d}u. \qquad (32)$$

For the random variables η_n, $n = 1, 2, \ldots$, we aim to prove that there exists some nonnegative random variable η independent of $\{\xi_n, n = 1, 2, \ldots\}$ and satisfying

$$\overline{G}(x) = \Pr\left(\eta > x\right) \leq C_1 e^{-C_2 x} \qquad (33)$$

with some $C_i = C_i(\delta) > 0$, $i = 1, 2$, such that for all $n = 1, 2, \ldots$ and $x \geq 0$,

$$\Pr\left(\eta_n > x\right) \leq \Pr\left(\eta > x\right). \qquad (34)$$

In view that the identity $\left(U_{(k,n)} \leq u\right) = \left(\sum_{i=1}^{n} 1_{(U_i \leq u)} \geq k\right)$ holds for all $k = 1, 2, \ldots, n$ and $u \in [0, 1]$, we have

$$\Pr\left(\eta_n > x\right) = \Pr\left(\bigcup_{k=1}^{n}\left(U_{(k,n)} < \frac{(1 - \delta)vk - x}{vn}\right)\right)$$

$$\leq \sum_{\frac{x}{(1-\delta)v} \leq k \leq n} \Pr\left(\sum_{i=1}^{n} 1_{\left(U_i \leq \frac{(1-\delta)vk-x}{vn}\right)} \geq k\right).$$

For arbitrarily fixed $h > 0$, an application of Chebyshev's inequality gives that

$$\Pr\left(\eta_n > x\right) \leq \sum_{\frac{x}{(1-\delta)v} \leq k \leq n} e^{-hk}E\left(\exp\left\{h\sum_{i=1}^{n} 1_{\left(U_i \leq \frac{(1-\delta)vk-x}{vn}\right)}\right\}\right)$$

$$\leq \sum_{\frac{x}{(1-\delta)v}\leq k\leq n} e^{-hk}\exp\left\{(e^h-1)\frac{(1-\delta)vk-x}{v}\right\}$$

$$= \exp\left\{-\frac{(e^h-1)\,x}{v}\right\}\sum_{\frac{x}{(1-\delta)v}\leq k\leq n}\exp\left\{[(e^h-1)(1-\delta)-h]\,k\right\}.$$

Choose $h_0 > 0$ so that $(e^{h_0}-1)(1-\delta)-h_0 < 0$. It follows that for all $n = 1, 2, \ldots$ and $x \geq 0$,

$$\Pr(\eta_n > x)$$

$$\leq \exp\left\{-\frac{(e^{h_0}-1)\,x}{v}\right\}\frac{\exp\left\{[(e^{h_0}-1)(1-\delta)-h_0]\frac{x}{(1-\delta)v}\right\}}{1-\exp\left\{(e^{h_0}-1)(1-\delta)-h_0\right\}}$$

$$= \frac{\exp\left\{-\frac{xh_0}{(1-\delta)v}\right\}}{1-\exp\left\{(e^{h_0}-1)(1-\delta)-h_0\right\}}.$$

The last inequality illustrates that for some positive numbers C_1 and C_2, the tail probability $\Pr(\eta_n > x)$ is bounded by $\min\left\{1, C_1 e^{-C_2 x}\right\}$ for all $n = 1, 2, \ldots$ and $x \geq 0$. We introduce a random variable η independent of the sequence $\{\xi_n, n = 1, 2, \ldots\}$ and with an exact tail probability $\min\left\{1, C_1 e^{-C_2 x}\right\}$. In this way, as announced, relations (33) and (34) are fulfilled immediately. Starting from (29) and using (34), (32), and Fubini's theorem in turn, we obtain that for all $n = 1, 2, \ldots$ and $x \geq M$,

$$\Pr\left(\max_{1\leq k\leq n}\left(\sum_{i=1}^{k}Z_i-vnU_{(k,n)}\right)>x\right)$$

$$\leq \int_{0-}^{x-M}\Pr(\xi_n+y>x)\,G(\mathrm{d}y)+\overline{G}(x-M)$$

$$\leq \frac{1+\varepsilon}{(1-\delta)v-\mathrm{E}Z_1}\int_{0-}^{x}\int_{x}^{x+(v-\mathrm{E}Z_1)n}\overline{B}(u-y)\mathrm{d}uG(\mathrm{d}y)+\overline{G}(x-M)$$

$$\leq \frac{1+\varepsilon}{(1-\delta)v-\mathrm{E}Z_1}\int_{x}^{x+(v-\mathrm{E}Z_1)n}\overline{B*G}(u)\mathrm{d}u+\overline{G}(x-M). \tag{35}$$

Since $\overline{G}(x)$ decreases exponentially, as described in (33), and $B \in \mathcal{S}_* \subset \mathcal{S} \subset \mathcal{L}$, we know that $\overline{G}(x) = o\left(\overline{B}(x)\right)$ by part (2) of Lemma 2.1, hence that $\overline{B*G}(x) \sim \overline{B}(x)$ by part (3) of Lemma 2.1. Moreover, for all $n = 1, 2, \ldots$ and $x \geq M$,

$$\frac{\overline{G}(x-M)}{\int_{x}^{x+(v-\mathrm{E}Z_1)n}\overline{B}(u)\mathrm{d}u}$$

$$\leq \frac{\overline{G}(x-M)}{\overline{B}(x-M)}\frac{\overline{B}(x-M)}{(v-\mathrm{E}Z_1)\overline{B}(x+(v-\mathrm{E}Z_1))} \to 0.$$

It follows from (35) that uniformly over all $n = 1, 2, \ldots,$

$$\Pr\left(\max_{1 \le k \le n}\left(\sum_{i=1}^{k} Z_i - vnU_{(k,n)}\right) > x\right) \lesssim \frac{1+\varepsilon}{(1-\delta)v - EZ_1}\int_x^{x+(v-EZ_1)n}\overline{B}(u)\mathrm{d}u.$$

By the arbitrariness of ε and δ, we obtain the upper bound of Theorem 2.   □

3.3   Discussions

Example 1 (Special case). In the case of Poisson process. We take $t = \infty$. Then it holds that

$$\psi(x;\infty) \sim \frac{1}{\mu}\int_x^{\infty}\overline{B}(u)\mathrm{d}u, \tag{36}$$

which is consistence with the result of Embrechts and Veraverbeke ([4]). If we take the distribution as $\mathcal{R}_{-\alpha}$ and $\alpha > 1$ specially. Then

$$\psi(x;\infty) \sim \frac{1}{\mu}x^{1-\alpha}. \tag{37}$$

In other words, we can calculate the ruin probability when the claim follows Pareto distribution. Similarly, when the claim follows Weiull distribution, it holds that

$$\psi(x;\infty) \sim \frac{1}{\mu}\int_x^{\infty}e^{-cu^{\tau}}\mathrm{d}u. \tag{38}$$

Thus, by calculating the integral in the right hand of (38), we could easily estimate the approximation probability.

Acknowledgement. This work was supported by the National Natural Science Foundation of China (Grant No. 70471071 and Grant No. 70871104) and the planning project of National Educational Bureau(Grant No. 08JA630078).

References

1. Asmussen, S.: Approximations for the probability of ruin within finite time. Scand. Actuar. J. (1), 31–57 (1984)
2. Asmussen, S.: Ruin probabilities. World Scientific Publishing Co., Inc., River Edge (2000)
3. Embrechts, P., Klüppelberg, C., Mikosch, T.: Modelling extremal events for insurance and finance. Springer, Berlin (1997)
4. Embrechts, P., Veraverbeke, N.: Estimates for the probability of ruin with special emphasis on the possibility of large claims. Insurance Math. Econom. 1(1), 55–72 (1982)
5. Goldie, C.M., Klüppelberg, C.: Subexponential distributions. In: Adler, R., Feldman, R., Taqqu, M.S. (eds.) A practical Guide to Heavy-Tails: Statistical Techniques and Applications. Birkhäuser, Boston (1998)
6. Kaas, R., Tang, Q.: Note on the tail behavior of random walk maxima with heavy tails and negative drift. N. Am. Actuar. J. 7(3), 57–61 (2003)

7. Klüppelberg, C., Mikosch, T.: Large deviations of heavy-tailed random sums with applications in insurance and finance. J. Appl. Probab. 34(2), 293–308 (1997)
8. Korshunov, D.A.: Large deviation probabilities for the maxima of sums of independent summands with a negative mean and a subexponential distribution. Theory Probab. Appl. 46(2), 355–365 (2003)
9. Leipus, R., Siaulys, J.: Asymptotic behaviour of the finite-time ruin probability under subexponential claim sizes. Insurance Math. Econom. 7(7), 1016–1027 (2006)
10. Ross, S.M.: Stochastic processes. John Wiley & Sons, Inc., New York (1983)
11. Tang, Q.: Uniform estimates for the tail probability of maxima over finite horizons with subexponential tails. Probab. Engrg. Inform. Sci. 18(1), 71–86 (2004a)
12. Tang, Q.: Asymptotics for the finite time ruin probability in the renewal model with consistent variation. Stoch. Models. 20(3), 281–297 (2004b)
13. Veraverbeke, N.: Asymptotic behaviour of Wiener-Hopf factors of a random walk. Stochastic Processes Appl. 5(1), 27–37 (1977)

Approaching the Linguistic Complexity

Stanisław Drożdż[1,2], Jarosław Kwapień[1], and Adam Orczyk[1]

[1] Polish Academy of Science, Institute of Nuclear Physics, 31-342 Kraków, Poland
Stanislaw.Drozdz@ifj.edu.pl
[2] University of Rzeszów, Faculty of Mathematics and Natural Sciences,
35-310 Rzeszów, Poland

Abstract. We analyze the rank-frequency distributions of words in selected English and Polish texts. We compare scaling properties of these
distributions in both languages. We also study a few small corpora of
Polish literary texts and find that for a corpus consisting of texts written by different authors the basic scaling regime is broken more strongly
than in the case of comparable corpus consisting of texts written by the
same author. Similarly, for a corpus consisting of texts translated into
Polish from other languages the scaling regime is broken more strongly
than for a comparable corpus of native Polish texts. Moreover, based on
the British National Corpus, we consider the rank-frequency distributions of the grammatically basic forms of words (lemmas) tagged with
their proper part of speech. We find that these distributions do not scale
if each part of speech is analyzed separately. The only part of speech
that independently develops a trace of scaling is verbs.

Keywords: Complexity, natural language, Zipf law, word classes.

1   Introduction

Even though central to the contemporary science the concept of complexity -
by its very nature - still leaves its precise definition a somewhat open issue.
In intuitive and qualitative terms this concept refers to diversity of forms, to
emergence of coherent and orderly patterns out of randomness, but also to a
significant flexibility that allows switching among such patterns on a way towards
searching for the ones that are optimal in relation to environment. Physics offers
the methodology and concepts that seem promising for formalizing complexity.
One of those concepts is criticality implying a lack of characteristic scale which
indeed finds evidence in abundance of power laws and fractals in Nature.

Whatever definition of complexity one however adopts, the human language
deserves a special status in the related investigations. It not only led humans
to develop civilization but it also constitutes - from a scientific perspective - an
extremely fascinating and complex dynamical structure [1]. Like many natural
systems, language during its evolution developed remarkable complex patterns
of behaviour such as hierarchical structure, syntactic organization, long-range
correlations, and - what is particularly relevant here - a lack of characteristic
scale. This latter phenomenon - in quantitative linguistics commonly referred

J. Zhou (Ed.): Complex 2009, Part I, LNICST 4, pp. 1044–1050, 2009.

to as the Zipf law - describes the rank-frequency distribution of words in a (sufficiently large) piece of text. This well-known, quantitatively formulated in 1949 by G.K. Zipf observation [2], based originally on "Ulysses" and latter on confirmed for many other literary texts, states that frequency of the rank-ordered words is inversely proportional to the words' rank. It needs to be added here that this law constitutes a principal reference in quantitative linguistics and inspiration for ideas and development in many different areas of science.

Zipf suggested interpretation of this law in terms of the so-called principle of least-effort [2,3]. This interpretation was however soon questioned after it had been shown that the Zipfian relation applies also to a "typewriting monkey" example [4], i.e. an essentially purely random process. This pointed to the Zipf law as too indiscriminate to reflect the complex organization of languages.

2   Results and Discussion

Our recent studies based on English as well as on Polish texts open a new perspective to comprehend the linguistic complexity and sheds a new light on an involved message encoded in the Zipf law. First, referring to the classic analysis of "Ulysses" by James Joyce, carried out by G.K. Zipf [2], we compare properties of the rank-frequency distributions of words in the English and the Polish text [5] of this novel. Results are shown in Figure 1(a) and Figure 1(b). Both versions of the text show scaling behaviour over three decades between the ranks 10 and 10,000. However, the scaling exponent for the Polish text ($\alpha \simeq 0.90$) is significantly smaller than for the English original ($\alpha \simeq 1.05$). This difference might originate from a far more inflectable character of the Polish language, which demands a larger set of words (understood as particular sequences of letters) to reproduce the course of narration in "Ulysses". This observation may be considered a regularity, since typical Polish texts have smaller α than typical English texts. There are, however, Polish texts which have a value of α that is similar to English standards as it is documented in Figure 1(c). It is noteworthy that each of the three texts are well approximated by power-law distributions almost over the whole range of ranks.

Any text, as a piece of human syntactic communication, is not a series of grammatically equivalent words, but rather a convoluted mixture of words belonging to different parts of speech (classes). Tagging all the words with their proper parts of speech allows us to compare statistical properties of words within each class separately. Figure 2 shows the rank-frequency distributions of the most frequent nouns, verbs, adjectives, and adverbs in the British National Corpus [6] which is a representative sample of the contemporary written and spoken English. The data comes from [7]. As it can be seen, despite the fact that the whole corpus exhibits a Zipf-type scaling for ranks up to several thousands [8], the corresponding distributions may not necessarily show any scaling if only words representing a single part of speech are considered. Verbs are the only class of words that develop some trace of scaling behaviour with the scaling index $\alpha \approx 1$. Looking from this perspective at the global distribution of all the words belonging

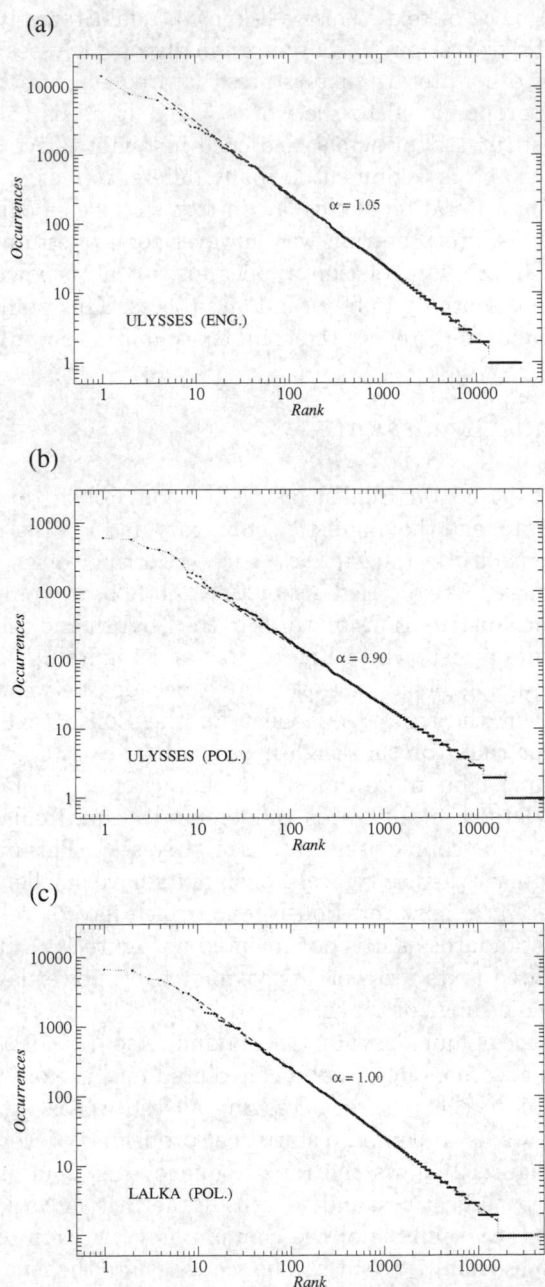

Fig. 1. Rank-frequency distributions of words in the English (a) and Polish (b) text of "Ulysses" by James Joyce, as well as in the native Polish text of "Lalka" by Bolesław Prus. All the three texts have comparable lengths of 240,000-260,000 words.

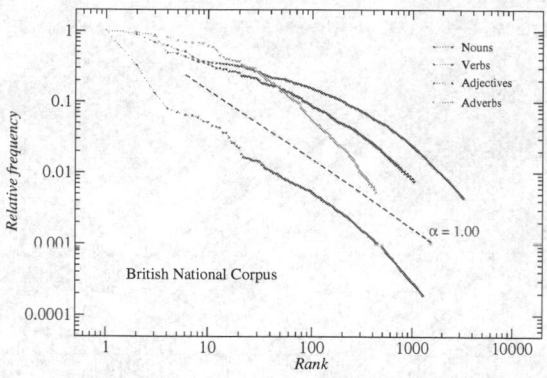

Fig. 2. Rank-frequency distributions of the most frequent lemmatized nouns (red), verbs (green), adjectives (violet), and adverbs (turquoise) taken from the British National Corpus. A slope with exponent $\alpha = 1.00$ is shown as a benchmark.

to all the classes together it is extremely interesting to reiterate that the entire rank-frequency distribution is Zipf-like.

This result can reflect a highly convoluted syntactic organization of human language, which may be considered as a complex system. From this angle, the linguistic complexity primarily manifests itself through the logic of mutual dressing among words belonging to different parts of speech that the entire proportions emerge scale-free even though in majority of these parts separately the proportions do not respect such a kind of organization. Another interesting issue open for speculation is whether the above results actually reflect the fact that the Zipf's principle of least action is more applicable to verbs - a part of speech related to action - than to nouns that are linked to objects. Worth considering is also a possibility that this reflects mapping of the well-established physical principle of least action onto the frequency distribution of verbs in a text.

As it was already mentioned above, passing from a single English text to a larger literary corpora is associated with reaching the limits of the applicability of the Zipf law in its classical form with the scaling index $\alpha \simeq 1$. Typically, after a short transient, for ranks larger than a few thousands another scaling regime is observed with $\alpha > 1$ [9,8]. Presence of two distinct scaling exponents in the rank-frequency distribution of English words can be explained by the existence of two sets of words: the first one comprises common words which are frequently used by all the authors (thus forming the language core), and the second one comprises the remaining words among which are technical words, words typical for a specific author or words which are otherwise rarely used. However, we propose another complimentary explanation of the breaking of the Zipf law for higher word ranks. Based on books of a few different authors we observe that the Zipf law is better realized for single texts than it is for corpora, even if we consider a corpus to be a collection of works of the same author. Figure 3 shows the rank-frequency distributions of words for two small corpora of Polish texts: the first one (black symbols) comprises 26 novels and stories by

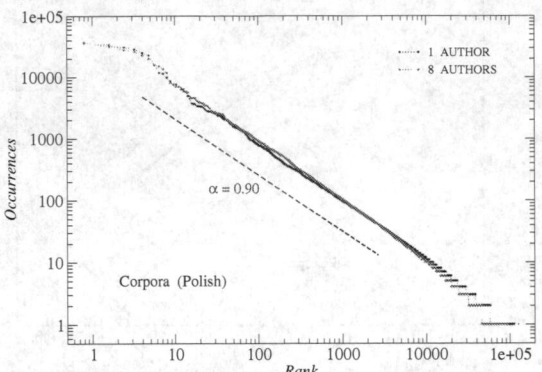

Fig. 3. Rank-frequency distributions of words for two Polish corpora consisting of works of the same author (black) and of different authors (green). A power-law with best-fitted exponent $\alpha \simeq 0.90$ is denoted by blue dashed line.

Polish fantasy writer Andrzej Sapkowski, while the second one (green symbols) is formed of 41 novels and stories written by 8 different authors. The texts in the second corpus were selected in such a way that the total length of each corpus is comparable (1.3 million words). It can easily be noted that for ranks larger that 6,000-8,000 the distribution for the second corpus shows a slightly faster decay than the distribution for the first corpus. In turn, the distribution for the first corpus seems to deviate from the unique scaling behaviour more than any single member text of this corpus (not shown here, but compare this with the result for a single novel in Figure 1(c)). This conclusion, however, must be treated with care since single texts have much smaller vocabulary than larger corpora.

The above evidence suggests that the long-range correlations originating from a given book's continuous narration can be a strong source of scaling behaviour. These correlations are distorted if we form a corpus consisting of different works, in the same manner as the correlations which are allowed to exist in each particular realization of a system are suppressed if one forms a statistical ensemble from a number of different realizations of this system. This possibility opens space for contesting the traditional model of analysis in quantitative linguistics, according to which the corpora, due to their larger size, are more useful subject of analysis than single works. In our opinion this leads to losing a significant amount of information. The above outcome is another argument in favour of the concept that the words extracted from their context are rather different objects even from a purely statistical point of view than the same words embedded in a contextual environment.

Finally, let us look once again at Figure 1(a) and Figure 1(b), where the Zipf plots for an English text and its Polish translation are presented. Both distributions show the undistorted power-law slope for the whole range of ranks, which means that the scale-free character of "Ulysses" was preserved by the translator. Actually, if one takes into account the peculiar character of this novel, especially the unequally rich vocabulary, this result has to be considered remarkable.

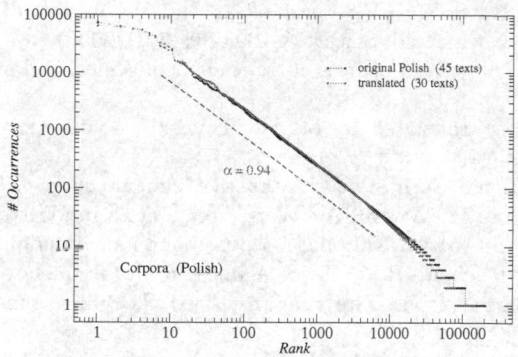

Fig. 4. Rank-frequency distributions of words for two Polish corpora consisting of 45 original texts written in Polish (black) and of 30 foreign texts translated into Polish (red). A power law with exponent $\alpha \simeq 0.94$ well-fitting both distributions within the ranks 10-2000 is denoted by green dahed line.

Motivated by this observation, we more systematically look at the rank-frequency distributions of words from texts which were translated into Polish from other languages. We find that although such texts show scale-free behaviour for the smallest ranks, for larger ones a breakdown of scaling occurs and we see a deflection towards smaller frequencies. In order to compare scaling properties of the translated and the native Polish texts, we constructed the following two small corpora: the first one consisting of 45 texts written originally in Polish, and the second one consisting of 30 translated texts. Both corpora have similar size of 2.3 million words. The results are presented in Figure 4. For the ranks < 2000 both corpora develop roughly the same distributions with $\alpha \simeq 0.94$. However, for higher ranks the translated corpus is represented by less frequent words than the native corpus. The observed difference is sufficiently significant to consider it as an actual property of both considered groups of texts. This result does not seem to be unexpected. Generally, the higher is α for a sample, the poorer is the vocabulary of a corresponding author. It seems natural to expect that vocabulary of a writer is richer than a vocabulary of a translator. The first one works under no lexical constraints, while the second one has to concentrate principally on preserving the sense and the style of the original, what can lead to impoverishment of the lexicon. Moreover, differences in grammar also may play a role here.

The famous statement of P.W. Anderson [10], "More is different", seems particularly adequate in relation to the human language, indeed.

References

1. Nowak, M.A., Plotkin, J.B., Jansen, V.A.A.: The evolution of syntactic communication. Nature 404, 495–498 (2000)
2. Zipf, G.K.: Human behavior and the principle of least effort. Addison-Wesley, Cambridge (1949)

3. Ferrer Cancho, R., Solé, R.V.: Least effort and the origins of scaling in human language. Proc. Natl. Acad. Sci. USA 100, 788–791 (2003)
4. Miller, G.A.: Some effects of intermittent silence. Amer. J. Psychol. 70, 311–314 (1957)
5. Joyce, J.: Ulisses, translated by M. Słomczyński, Wydawnictwo Pomorze (Bydgoszcz) (1992)
6. The British National Corpus website: http://www.natcorp.ox.ac.uk/
7. Leech, G., Rayson, P., Wilson, A.: Word Frequencies in Written and Spoken English: based on the British National Corpus. Longman, London (2001)
8. Ferrer Cancho, R., Solé, R.V.: Two regimes in the frequency of words and the origins of complex lexicons: Zipf's law revisited. J. Quant. Linguistics 8, 165–173 (2001)
9. Montemurro, M.A.: Beyond the Zipf-Mandelbrot law in quantitative linguistics. Physica A 300, 567–578 (2001)
10. Anderson, P.W.: More is different. Science 177, 393–396 (1972)

Application of the Kelly Criterion to Ornstein-Uhlenbeck Processes

Yingdong Lv and Bernhard K. Meister

Department of Physics, Renmin University of China
lyd08250@163.com, b_meister@ruc.edu.cn

Abstract. In this paper, we study the Kelly criterion in the continuous time framework building on the work of E.O. Thorp and others. The existence of an optimal strategy is proven in a general setting and the corresponding optimal wealth process is found. A simple formula is provided for calculating the optimal portfolio for a set of price processes satisfying some simple conditions. Properties of the optimal investment strategy for assets governed by multiple Ornstein-Uhlenbeck processes are studied. The paper ends with a short discussion of the implications of these ideas for financial markets.

Keywords: utility function, optimal investment strategy, self-financing, complete market, risk-neutral measure, Brownian motion, Ornstein-Uhlenbeck.

1 Introduction

The Kelly Criterion [1], [2] was initially introduced in 1956 to find the optimal betting amount in games with fixed known odds, and was later extended to the field of financial investments by E. O. Thorp and others. The strategy maximizes the entropy and with probability one outperforms any other strategy asymptotically [3]. This approach was recently further developed by Kargin [4], who applied the criterion to a mean-reverting asset process under liquidity and credit constraints.

The Kelly Criterion tells us that the optimal betting fraction is given by p-q, if a gambler is faced with a bet, where the probability to double the money is p and to lose the initial stake is q ($p>q$). The optimal betting fraction maximizes the expected log wealth. The question, why investors should choose to maximize the log wealth, has a simple answer: according to Breiman's theorem [3], it gives the asymptotically optimal pay-out and dominates any other strategy.

In this paper, we start by extending the original idea to the general continuous time framework with n correlated assets. Our task is to find the optimal self-financing trading strategy. We will prove that if the market is complete, this optimal self-financing trading strategy always exists. A limited number of applications are discussed in the context of Ornstein-Uhlenbeck processes.

The paper is organized as follows. In section 2 we review the standard assumptions and prove the optimization theorem. The theorem covers both the existence of the optimal trading strategy and the explicit form of the associated optimal investment fraction. In section 3 we apply the theory to a market of n correlated assets given by

J. Zhou (Ed.): Complex 2009, Part I, LNICST 4, pp. 1051–1062, 2009.

Ornstein-Uhlenbeck mean-reverting processes. The optimal investment strategy is calculated for some representative examples. In the last section we put the results into the financial context and describe some open problems.

2   General Theory

In the first section, we will cover the assumptions and the theoretical framework. In 2.2 we will provide the main result, which is contained in Theorem 1.

2.1   Basic Assumptions and Other Preliminaries

We will use the standard notation and conventions of financial mathematics. In section 2 the basic assumption is that the market is complete and frictionless [5]. Let us further assume that we consider all the processes in the finite time interval 0 to the terminal time T. There exists a probability space $(\Omega,\ P_T,\mathscr{F}_T)$, on which all of the random variables are constructed, where Ω is the sample space, $\mathscr{F}_T$ is a σ-algebra which denotes the information accumulated up to time T and P_T is the spot probability measure [5]. The filtration $\mathscr{F}_t$, $t \in [0,T]$, represents the information accumulated up to time t. The sub-probability space $(\Omega,\ P_t,\mathscr{F}_t)$ is introduced at time t, where P_t is the restriction of P_T on the filtration $\mathscr{F}_t$.

We assume there are $n+1$ investable assets in the market including the wealth process B_t, representing a saving account with value 1 at the initial time 0. We assume B_t follows

$$dB_t = B_t r_t dt ,\tag{1}$$

where r_t is the short term rate at time t.

The other n assets in the market are denoted by $S_i(t)$, $t \in [0,T]$, $i = 1,2,...,n$, and we define a n-dimensional vector by $\mathbf{S}_t = (S_1(t),S_2(t),...,S_n(t))^T$, where '$T$' represents the transposition of a matrix. Let us define the relative assets price process by $\tilde{\mathbf{S}}_t = \mathbf{S}_t B_t^{-1}$. Let $\phi_0(t)$ denotes the number of units of B_t an investor holds at time t and $\phi_i(t)$, $t \in [0,T]$, $i = 1,2,...,n$, denotes the number of units of the ith asset an investor holds at time t. In addition, the n-dimensional vector ϕ_t is defined as $\phi_t = (\phi_1(t),\phi_2(t),...,\phi_n(t))^T$. $V_t(\psi)$ is the total value of the portfolio $\psi_t = (\phi_0(t),\phi_t)$. So we have

$$V_t(\psi) = \phi_0(t)B_t + \phi_t \cdot \mathbf{S}_t ,\tag{2}$$

where $\phi_t \cdot S_t$ is the inner product of two vectors.

Definition 1. *A self-financing trading strategy* $\psi_t = (\phi_0(t), \phi_t)$ *is a strategy that satisfies:*

$$dV_t(\psi) = \phi_0(t) dB_t + \phi_t \cdot dS_t, \quad \forall t \in [0,T] . \tag{3}$$

We assume $V_0(\psi) = 1$.

Definition 2. *A self-financing trading strategy* $\psi_t = (\phi_0(t), \phi_t)$ *is said to be admissible if and only if*

$$V_t(\psi) \geq 0, \; P_T \text{ a.s.} \quad \forall t \in [0,T] . \tag{4}$$

$U(x), x \geq 0$, is defined to be a concave function representing the utility of wealth. Here concaveness means

$$U\big((1-p)x_1 + px_2\big) \geq (1-p)U(x_1) + pU(x_2), \forall x_2 \geq x_1 \geq 0 \text{ and } 0 \leq p \leq 1 . \tag{5}$$

Further it is assumed that $U(x)$ has a first order derivative for $\forall x \in (0, +\infty)$. The first order derivative at $x = 0$ can be either finite or infinite, and the first order derivative of $U(x)$, $x \geq 0$, is a strictly decreasing function of x with $\lim_{x \to +\infty} U'(x) = 0$. If $U'(0) = +\infty$, let $I(x), x \geq 0$, be the inverse function of $U'(x)$ with $I(0) = +\infty$ and $I(+\infty) = 0$. For $U'(0) = b > 0$, we denote by $I_b(x), x \in [0,b]$, the inverse function of $U'(x)$, with $I(0) = +\infty$. In this case, we define $I(x)$ as

$$I(x) = \begin{cases} I_b(x) , & x \in [0,b] \\ 0 , & x \in (b, +\infty) \end{cases} . \tag{6}$$

Let us denote by $\mathscr{D}$ the class of all of the admissible self-financing trading strategies. We say a self-financing trading strategy $\psi^* \in \mathscr{D}$ is the optimal trading strategy, if and only if

$$E_{P_T}\big[U(V_T(\psi^*))\big] \geq E_{P_T}\big[U(V_T(\psi))\big], \quad \forall \psi \in \mathscr{D} . \tag{7}$$

Our task is to find an optimal $\psi^* \in \mathscr{D}$, which satisfies eq.7.

2.2 The Optimal Strategy

To find the optimal strategy, we will first need to introduce the following lemma..

Lemma 1. *The function* $I(x), x \in [0, +\infty)$ *satisfies the following inequality:*

$$U(I(y)) - yI(y) \geq U(c) - yc, \quad \forall y, c \in [0, +\infty) . \tag{8}$$

Proof. If $I(y) = c$, then eq.8 is obviously satisfied. If $I(y) > c$, then the average growth rate of $U(x)$ from c to $I(y)$ should be larger than the first order derivative of $U(x)$ at $I(y)$, which is given by $U'(I(y))$, since the first order derivative of $U(x)$ is a strictly decreasing function of x.

The average growth rate of $U(x)$ from c to $I(y)$ is

$$\frac{U(I(y)) - U(c)}{I(y) - c}.$$

This yields the following inequality

$$\frac{U(I(y)) - U(c)}{I(y) - c} \geq U'(I(y)) = y$$

$$\Rightarrow U(I(y)) - yI(y) \geq U(c) - yc.$$

An almost identical argument can be applied in the case $I(y) < c$.   □

Let us define $\tilde{P}_T$ as the martingale of the market and $Z_t = \dfrac{dP_t}{d\tilde{P}_t}$. Then $(Z_t, \mathscr{F}_t)$ is a $\tilde{P}_T$ martingale [5]. Define $\eta_t^* = y_t Z_t^{-1} B_t^{-1}$, $V_t^* = I(\eta_t^*)$, where we assume y_t is a deterministic function of t and is defined in such a way that $\tilde{V}_t^* = V_t^* B_t^{-1}$ is a $\tilde{P}_T$ martingale. So y_t solves the equation

$$\tilde{E}\left[B_t^{-1} I\left(y_t Z_t^{-1} B_t^{-1} \right) \right] = 1 . \tag{9}$$

The deterministic property of y_t seems contrived, but is necessary for the proof of Proposition 1. As we shall see in section 3, in the case of the log utility function the deterministic function y_t indeed exists and is a constant.

Proposition 1. $V_T^* = I(\eta_T^*)$ *satisfies the inequality given by eq.7.*

Proof. Let $V_T(\psi)$, $\forall \psi \in \mathscr{D}$, be the wealth process corresponding to a special trading strategy ψ, then

$$E_{P_T}\left[U\left(V_T^* \right) \right] - E_{P_T}\left[U\left(V_T(\psi) \right) \right]$$

$$= E_{P_T}\left[\left(U\left(I(\eta_T^*) \right) - \eta_T^* I(\eta_T^*) \right) - \left(U\left(V_T(\psi) \right) - \eta_T^* V_T(\psi) \right) \right] + E_{P_T}\left[\eta_T^* \left(V_T^* - V_T(\psi) \right) \right] \tag{10}$$

According to lemma 1, the first term of the right hand side of eq.10 is positive. The second term is equal to zero,

$$E_{P_T}\left[\eta_T^*\left(V_T^*-V_T(\psi)\right)\right]=\tilde{E}\left[Z_T\eta_T^*\left(V_T^*-V_T(\psi)\right)\right]=y_T\tilde{E}\left[\tilde{V}_T^*-\tilde{V}_T(\psi)\right]=0 . \quad (11)$$

The last equality of the above equation is deduced from the fact that both $\tilde{V}_t^*$ and $\tilde{V}_t(\psi)$ are martingales under the martingale measure.

Combining eq.10 and eq.11, we directly get

$$E_{P_T}\left[U\left(V_T^*\right)\right]\ge E_{P_T}\left[U\left(V_T(\psi)\right)\right] \qquad\qquad \square$$

Proposition 1 only state the fact that $I\left(\eta_t^*\right)$ satisfies eq.7. It doesn't necessarily mean that $I\left(\eta_t^*\right)$ is the optimal wealth process. We will prove in the following theorem that $I\left(\eta_t^*\right)$ is in fact the optimal wealth process.

Theorem 1. *Given a concave utility function $U(x)$, there exists an optimal self-financing trading strategy ψ^*, such that for each time $t\in[0,T]$, the wealth process $V_t\left(\psi^*\right)$ of this strategy satisfies*

$$E_{P_T}\left[U\left(V_t\left(\psi^*\right)\right)\right]\ge E_{P_T}\left[U\left(V_t(\psi)\right)\right], \quad \forall\psi\in\mathscr{D} .$$

And the optimal wealth process is given by: $V_t\left(\psi^*\right)=I\left(\eta_t^*\right),\ t\in[0,T]$.

Proof. Define $\tilde{V}_t^*$ to be $B_t^{-1}I\left(\eta_t^*\right)$. For $t=T$, we have $V_T^*=B_T\tilde{V}^*=I\left(\eta_T^*\right)$, which represents a general contingent claim in the market. Since the market is complete, the contingent claim V_T^* is attainable. This means there exists a self-financing trading strategy ψ^* such that $\tilde{V}_T^*=V_T\left(\psi^*\right)B_T^{-1}$, where $V_t\left(\psi^*\right)$ is the wealth process of this self-financing trading strategy. So the relative wealth process $\tilde{V}_t\left(\psi^*\right)=V_t\left(\psi^*\right)B_t^{-1}$ is a martingale under the martingale measure $\tilde{P}_T$. $\tilde{V}_t^*$ is also a martingale under the martingale measure $\tilde{P}_T$, and we have

$$V_t^*=B_t\tilde{E}\left[\tilde{V}_T^*\mid\mathscr{F}_t\right]=B_t\tilde{E}\left[\tilde{V}_T\left(\psi^*\right)\mid\mathscr{F}_t\right]=V_t\left(\psi^*\right), \quad \forall t\in[0,T] . \quad (12)$$

Eq.12 shows that ψ^* is a self-financing trading strategy and replicates the optimal wealth process $\tilde{V}_t^*=B_t^{-1}I\left(\eta_t^*\right)$. From the combination of V_t^*, satisfying eq.7 for any time t before T, and eq.12, we can see that $V_t\left(\psi^*\right)$ also satisfies eq.7. This proves that the strategy ψ^* is both self-financing and optimal. $\square$

It follows from Proposition 1 and Theorem 1 that the existence of a self-financing trading strategy $\psi^* \in \mathcal{D}$, where the total wealth at a fixed time T is consistent with eq.7, implies that the wealth process $V_t(\psi^*)$ satisfies

$$E_{P_T}\left[U\left(V_t(\psi^*)\right)\right] \geq E_{P_T}\left[U\left(V_t(\psi)\right)\right], \quad \forall \psi \in \mathcal{D} .$$

Therefore, an optimal trading strategy for a fixed time T will be optimal for any time before T. It follows further that an optimal trading strategy is only based on the information up to time t. The optimal trading strategy ψ_t^* is an adapted process with respect to the filtration $\{\mathcal{F}_t, t \in [0, +\infty)\}$. In the next section, we will apply the theorem to the case of a financial market containing a saving account and n correlated assets, whose price processes follow Ornstein- Uhlenbeck mean-reverting processes.

3   Implications for Ornstein-Uhlenbeck Processes

In this section we set $r_t = r$ and $S_i(t) = \exp(x_i(t))$, $t \in [0, T]$, $i = 1, 2, ..., n$. Each $x_i(t)$ is governed by

$$dx_i(t) = \left[a_i - b_i x_i(t)\right]dt + \sum_{j=1}^{n} \sigma_{i,j} dW_t^j, \quad i, j = 1, 2, ..., n .$$

where a_i is some fixed real number, $b_i > 0$ is some nonnegative real number and $\sigma_{i,j}$ are constants. $\mathbf{W}_t = \left(W_t^1, W_t^2, ..., W_t^n\right)^T$ is a standard n-dimensional Brownian motion.

Define $\mathbf{a}$ to be the vector $(a_1, a_2, ..., a_n)^T$ (' τ ' is the transposed of a matrix), $\mathbf{b}$ to be the $n \times n$ matrix of the form $\mathbf{b} = \begin{cases} \mathbf{b}_{i,j} = b_i, & i = j \\ \mathbf{b}_{i,j} = 0, & i \neq j \end{cases}$, and σ to be the matrix $\sigma_{i,j} = \sigma_{i,j}$, for $1 \leq i, j \leq n$. The matrix σ has a non-zero determinant. Then the dynamic equation of $\mathbf{x}_t = \left(x_1(t), x_2(t), ..., x_n(t)\right)^T$ can be expressed as

$$d\mathbf{x}_t = \left[\mathbf{a} - \mathbf{b}\mathbf{x}_t\right]dt + \sigma d\mathbf{W}_t . \tag{13}$$

Let $\mathbf{S}_t = \left(S_1(t), S_2(t), ..., S_n(t)\right)^T$, $\tilde{\mathbf{S}}_t = B_t^{-1}\mathbf{S}_t$.

According to Ito's lemma, the dynamic of $\mathbf{S}_t$ is

$$dS_i(t) = S_i(t)\mu_i(t)dt + S_i(t)\sum_{j=1}^{n} \sigma_{i,j} dW_t^j, \quad i = 1, 2, ..., n .$$

where $\mu_i(t) = a_i - b_i \log\left(S_i(t)\right) + \frac{1}{2}\|\sigma_i\|^2$ and $\sigma_i = \left(\sigma_{i,1}, \sigma_{i,2}, ..., \sigma_{i,n}\right)^T$.

Define

$$\frac{dP_T}{d\tilde{P}_T} = \exp\left(\int_0^T \boldsymbol{\theta}_u \cdot d\tilde{\mathbf{W}}_u - \frac{1}{2}\int_0^T \|\boldsymbol{\theta}_u\|^2 \, du \right), \tag{14}$$

where $\boldsymbol{\theta}_u = \left(\theta_1(u), \theta_2(u), ..., \theta_n(u)\right)^T$ is a n-dimensional adapted stochastic process and $\|\boldsymbol{\theta}_u\| = \sqrt{\theta_1^2(u) + ... + \theta_n^2(u)}$ is the Euclidean vector norm, and $\tilde{\mathbf{W}}_t$ a Girsanov transformed Brownian motion, i.e. $\mathbf{W}_t = \tilde{\mathbf{W}}_t - \int_0^t \boldsymbol{\theta}_u \, du$.

Then under $\tilde{P}_T$, $\tilde{\mathbf{S}}_t$ it follows

$$d\tilde{S}_i(t) = \tilde{S}_i(t)\left[\mu_i(t) - r - \sum_{j=1}^n \sigma_{i,j}\theta_j(t) \right]dt + \tilde{S}_i(t)\sum_{j=1}^n \sigma_{i,j}d\tilde{W}_t^j \ .$$

Define $c_i(t) = \mu_i(t) - r$ and in vector form $\mathbf{c}_t = \left(c_1(t), c_2(t), ..., c_n(t)\right)^T$. If $\boldsymbol{\theta}_t$ solves

$$\boldsymbol{\sigma}\boldsymbol{\theta}_t = \mathbf{c}_t \Rightarrow \boldsymbol{\theta}_t = \boldsymbol{\sigma}^{-1}\mathbf{c}_t \ , \tag{15}$$

then under $\tilde{P}_T$ it follows that

$$d\tilde{\mathbf{S}}_t = \tilde{\mathscr{S}}_t\boldsymbol{\sigma}d\tilde{\mathbf{W}}_t \ , \tag{16}$$

where the matrix $\tilde{\mathscr{S}}_t$ is defined as: $\tilde{\mathscr{S}}_t = \begin{cases} \tilde{\mathscr{S}}_{i,j}(t) = \tilde{S}_i(t) \ , \ i = j \\ \tilde{\mathscr{S}}_{i,j}(t) = 0 \quad , i \neq j \end{cases}$. $\tilde{\mathbf{S}}_t$ is a martingale under $\tilde{P}_T$.

To apply theorem 1, we need first to prove the completeness of the market price processes under consideration. The next lemma tells us that indeed the market is complete. The proof is given in an earlier presentation [7].

Lemma 2. *The mean-reverting market given above is complete.*

In case $U(x) = \log(x)$, we find $I(x) = 1/x$. Using eq.9, we can show $y_t = 1$. Then the optimal discounted wealth process is

$$\tilde{V}_t^* = Z_t = \exp\left(\int_0^t \boldsymbol{\theta}_u \cdot d\tilde{\mathbf{W}}_u - \frac{1}{2}\int_0^t \|\boldsymbol{\theta}_u\|^2 \, du \right).$$

Now we are in a position to derive a general result for Ornstein-Uhlenbeck processes.

Theorem 2. *The optimal trading strategy* $\psi_t^* = \left(\phi_0^*(t), \phi_t^*\right)$ *is given by:*

$$\phi_0^*(t) = B_t^{-1}V_t^*\left(1 - B_t^{-1}\boldsymbol{\theta}_t^T\boldsymbol{\lambda}_t\mathbf{S}_t\right), \phi_i^*(t) = B_t^{-1}V_t^*\sum_{j=1}^n \theta_j(t)\lambda_{j,i}(t), i = 1, 2, ..., n \ . \tag{17}$$

where $\boldsymbol{\lambda}_t = \left(\tilde{\mathscr{S}}_t\boldsymbol{\sigma}\right)^{-1}$.

Proof. First, we can show immediately

$$V_t^* = \phi_t^* \cdot \mathbf{S}_t + \phi_0^*(t) B_t ,$$

where V_t^* is the optimal wealth process given by

$$V_t^* = B_t \exp\left(\int_0^t \boldsymbol{\theta}_u \cdot d\tilde{\mathbf{W}}_u - \frac{1}{2} \int_0^t \|\boldsymbol{\theta}_u\|^2 \, du \right) .$$

Using Ito's lemma for $\tilde{V}_t^*$, we get

$$d\tilde{V}_t^* = dZ_t = \tilde{V}_t^* \boldsymbol{\theta}_t^T d\tilde{\mathbf{W}}_t .$$

From eq.16 we know that

$$d\tilde{\mathbf{W}}_t = \lambda_t d\tilde{\mathbf{S}}_t .$$

Combining the above two equations, we have

$$d\tilde{V}_t^* = \tilde{V}_t^* \boldsymbol{\theta}_t^T \lambda_t d\tilde{\mathbf{S}}_t = B_t^{-1} \tilde{V}_t^* \left(\boldsymbol{\theta}_t^T \lambda_t d\mathbf{S}_t - r\boldsymbol{\theta}_t^T \lambda_t \mathbf{S}_t dt \right) , \tag{18}$$

and

$$d\tilde{V}_t^* = B_t^{-1} dV_t^* - rB_t^{-1} V_t^* dt . \tag{19}$$

Combining eq.18 and eq.19, we arrive at

$$dV_t^* = V_t^* \left[B_t^{-1} \boldsymbol{\theta}_t^T \lambda_t d\mathbf{S}_t + B_t^{-1} \left(1 - B_t^{-1} \boldsymbol{\theta}_t^T \lambda_t \mathbf{S}_t \right) dB_t \right] = \phi_t^* \cdot d\mathbf{S}_t + \phi_0^*(t) dB_t . \tag{20}$$

Eq.20 shows directly that $\psi_t^* = \left(\phi_0^*(t), \phi_t^* \right)$ given by eq.17 is the optimal self-financing trading strategy.   □

The optimal fraction vector $\mathbf{f}_t^* = \left(f_1^*(t), f_2^*(t), ..., f_n^*(t) \right)^T$ is composed of the individual $f_i^*(t)$, e.g. $\phi_i^*(t) S_i^*(t)/V_t^*$. By simple calculations based on Theorem 2, we have

$$\mathbf{f}_t^* = \mathbf{R}^{-1} \mathbf{c}_t , \tag{21}$$

where $\mathbf{R} = \boldsymbol{\sigma}\boldsymbol{\sigma}^T$ is a symmetric matrix and called correlation matrix. We will show in a separate paper that the matrix $\mathbf{R}$ denotes the correlations of the yield rates, e.g. the correlation of the ith and j^{th} assets is a deterministic function of $\boldsymbol{\sigma}_i \cdot \boldsymbol{\sigma}_j$. If the standard inverse of the volatility matrix does not exist, then one can resort to the generalized Moore-Penrose inverse to obtain a related result for the optimal investment fractions in markets without arbitrage.

Another derivation of the optimal fraction can be based on the function

$$F(\mathbf{x}) = \mathbf{c}_t^T \mathbf{x} - \frac{1}{2}\mathbf{x}^T \mathbf{R}\mathbf{x}, \quad \forall \mathbf{x} \in \mathbb{R}^n, \tag{22}$$

linked to the mean-variance approach, since the optimal fraction given by eq.21 is the maximum of the function F. This indicates the close relationship between the utility maximization and the mean-variance method.

In the special case where the market is composed of only one stock, by eq.21, the optimal fraction is $f_t^* = (\mu_t - r)/\sigma^2$, where $\mu_t = a - b\log(S(t)) + 0.5\sigma^2$. Fig. 1 shows a sample path for the stock process and the associated optimal investment fraction and wealth process. As an aside, if one assumes zero interest rates, than the sensitivity of the optimal fraction to a percentage estimation error in the drift μ is twice the negative of a similar error in σ, e.g. a *1%* overestimation in volatility has approximately the same impact as an underestimation of the drift by *2%*.

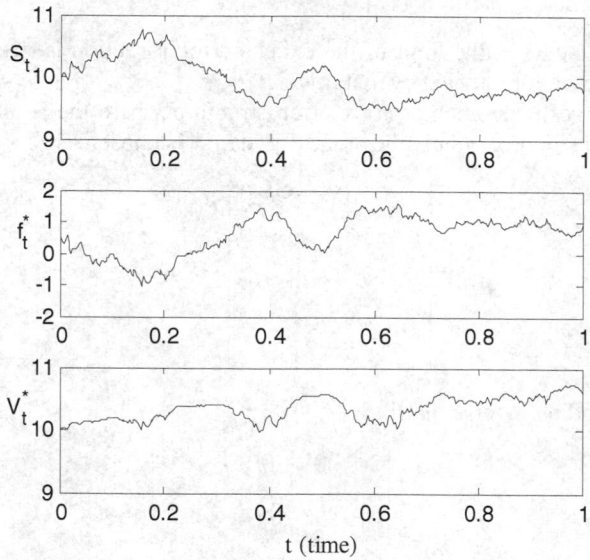

Fig. 1. Simulation of the stock price process, the corresponding optimal strategy f_t^*, and the wealth process V_t^* with parameters a=0.5, b=0.2, σ =0.1, r=0.03, S_0 =10 and V_0 =10

Besides the sensitivity to estimation errors, it would be interesting to analyze the impact of the correlation matrix on the optimal trading strategy. Positive correlation has a tendency to reduce the number of 'independent' assets and forces investors to reduce leverage, e.g. the sum of the absolute values of the investment fractions is smaller.

Next, we study a special case where the assets have local correlations. The different assets only correlate to the neighboring assets but have no correlation to the rest assets. Let's set the risk-free rate to zero and the volatility matrix to be

$$\sigma = \begin{bmatrix} \sigma & \sigma & 0 & \cdots & 0 \\ 0 & \sigma & \sigma & \ddots & \vdots \\ 0 & 0 & \ddots & \ddots & 0 \\ \vdots & \ddots & \ddots & \ddots & \sigma \\ 0 & 0 & 0 & \cdots & \sigma \end{bmatrix}_{n \times n} , n \geq 2 .$$

In this case, we will only study the large time limit. Let us denote by $\bar{f}_S^*(\infty)$

$$\bar{f}_S^*(\infty) := \lim_{t \to +\infty} E_{P_r}\left[\sum_{i=1}^{n} f_i^*(t)\right]. \tag{23}$$

After some simple manipulations we get

$$\bar{f}_S^*(\infty) = \begin{cases} \dfrac{n+1}{4} & \text{for n odd} \\ \dfrac{n}{4} & \text{for n even} \end{cases}. \tag{24}$$

For a fixed odd integer n, the limit of the expected total fraction is $(n+1)/4$, which is identical to the value for the next even number.

Now, let us investigate another correlation structure where the assets have global correlations. As a simple example the volatility matrix is chosen as

$$\sigma = \sigma \begin{bmatrix} 1 & 0 & 0 & \cdots & 0 \\ 1 & 1 & 0 & \cdots & 0 \\ 1 & 1 & \ddots & \ddots & \vdots \\ \vdots & \vdots & \ddots & 1 & 0 \\ 1 & 1 & \cdots & 1 & 1 \end{bmatrix}_{n \times n} , n \geq 2 ,$$

and the corresponding inverse matrix is

$$\sigma^{-1} = \sigma^{-1} \begin{bmatrix} 1 & 0 & 0 & \cdots & 0 \\ -1 & 1 & 0 & \cdots & 0 \\ 0 & -1 & \ddots & \ddots & \vdots \\ \vdots & \vdots & \ddots & 1 & 0 \\ 0 & 0 & \cdots & -1 & 1 \end{bmatrix}_{n \times n} .$$

Thus the optimal fraction $f_i^*(t)$ is given by

$$f_i^*(t) = \begin{cases} c_1(t) - (c_2(t) - c_1(t)) & i=1 \\ (c_i(t) - c_{i-1}(t)) - (c_{i+1}(t) - c_i(t)) & i=2,3,\dots,(n-1) \\ c_n(t) - c_{n-1}(t) & i=n \end{cases}. \tag{25}$$

The total optimal fraction is

$$f^*(t) = \sum_{i=1}^{n} f_i^*(t) = c_1(t) = \frac{\mu_1(t) - r}{\sigma^2} . \tag{26}$$

The total fraction here is equal to the optimal fraction in another market containing only the first asset. This surprising result is partly due to the fact that in the multi-dimensional case the investment fractions are likely to have both positive and negative signs. The expected total optimal fraction is

$$\overline{f}^*(t) = E\left[f^*(t)\right] = \overline{c}_1(t) = \frac{\left[a_1 - b_1 \log\left(S_1(0)\right)\right] e^{-b_1 t} + \frac{1}{2}\sigma^2 - r}{\sigma^2} . \tag{27}$$

As time approaches infinity, the following limit is reached:

$$\overline{f}^*(\infty) = \lim_{t \to +\infty} \overline{f}^*(t) = \begin{cases} \dfrac{a_1 + \dfrac{1}{2}\sigma^2 - r}{\sigma^2} & \text{for } b_1 = 0 \\[3mm] \dfrac{\dfrac{1}{2}\sigma^2 - r}{\sigma^2} & \text{for } b_1 \neq 0 \end{cases} . \tag{28}$$

For additional examples, in particular in higher dimensions, we refer the reader to [7].

4 Conclusions

In the earlier sections we presented a discussion of the Kelly criterion in the continuous-time framework. The main theorem shows that in a complete market there exists an optimal self-financing trading strategy that maximizes the logarithmic utility function. The optimal investment fractions were explicitly calculated.

One general implication of the Kelly's criterion is maybe worth mentioning. It follows from Breiman's theorem [2], which shows that a logarithmic utility maximizer outperforms with probability one in the long run any substantially different trading strategy. This theorem has surprising consequences, for example it has spanned a smallish field called 'evolutionary finance' [8]. According to evolutionary finance 'natural selection' should favor agents with log utility. Such agents maximize the growth rate of their wealth with probability one, and thus dominate eventually the market. The stark claim is that either the investor maximizes utility or is marginalized. The authors are doubtful, if such a strong claim is justified, since only in the long time limit does the utility maximizer almost surely outperform. In the real world, where one has multiple independent agents and frequent paradigm shifts, maybe an even more aggressive strategy is warranted. Being 'overinvested' can be 'superior' (lower utility, but higher winning probability) in the short term. Even in the medium term the log maximizer has difficulties to outperform, if many independent agents exist. This could be a partial explanation for the regular crisis in financial markets, e.g. investors, who seek a short-term competitive advantage, invest over-aggressively. This is on top of the significant inherent volatility of utility maximization. A proper understanding of the impact of the Kelly criterion on the optimal behavior of individual agents is the precursor to consistent multi-agent modeling.

As a speculative aside, maybe utility maximization has a role in the study of the punctuated equilibrium observed in the evolutionary history of the earth, since utility maximization could provide a potential explanation without necessarily having to resort to external causes like asteroid impacts or volcanic eruptions for rare widespread extinction events.

Due the space limitations, we are not able to provide even a brief description of the application of the result in the area of statistical arbitrage. The present discussions are based on the continuous time framework, but realistic markets have an inherent discreteness. Furthermore there are different types of frictions, e.g. transaction cost, bid-offer spreads and liquidity constraints, which impose portfolio readjustment frequency restrictions. Not all of those influences are small and can be neglected. In an earlier presentation [7] correction terms for reducing the investment fractions were explicitly calculated. It would be of interest to give a comprehensive analysis of the impact of the different types of frictions for statistical arbitrage strategies. This will be done by the authors in a separate paper.

In conclusion this article gives a quantitative insight into the trade-off between risk and return as diversification opportunities are added, correlation structure changed, and other constraints modified.

Acknowledgements. YL expresses his thanks to one of his dearest friends, Huiqing Chen, for her kind encouragement and support. BKM acknowledges support from the NSF of China and multiple informative discussions with DC Brody and O Peters.

References

1. Kelly, J.L.: A New Interpretation of Information Rate. Bell System Technical Journal 35, 917–926 (1956)
2. Thorp, E.O.: The Kelly Criterion in Blackjack, Sports Betting, and the Stock Market. In: The 10th International Conference on Gambling and Risk Taking (1997)
3. Breiman, L.: Optimal Gambling Systems for Favorable Games. In: Neyman, J. (ed.) Proceedings of the Berkeley Symposium on Mathematical Statistics and Probability, vol. 1, pp. 65–78 (1961)
4. Kargin, V.: Optimal Convergence Trading, arXiv: math.OC/0302104 (2003)
5. Musiela, M., Rutkowski, M.: Martingale Methods in Financial Modelling. Springer, New York (1997)
6. Karatzas, I., Shreve, S.: Brownian Motion and Stochastic Calculus. Springer, Heidelberg (1998)
7. Lv, Y., Meister, B.K., Peters, O.: Implications of the Kelly Criterion for Multiple Ornstein-Uhlenbeck Processes. In: Bachelier Finance Society Fifth World Congress (2008)
8. Hakansson, N.H.: On Optimal Myopic Portfolio Policies, with and without Serial Correlation of Yields. Journal of Business 44, 324–334 (1971)

Application of SRM to Diverse Populations

Sahin Delipinar and Haluk Bingol

Department of Computer Engineering
Bogazici University, Istanbul, Turkey
sahin_dp@yahoo.com

Abstract. In today modern industrial cities we see that many people having different cultures share the same settlement and form a typical social complex system. People come from other cities or even foreign countries Newcomers bring their cultural values such as clothing, meals, likes and dislikes. As a result of interacting with other people some cultural values change, some completely forgotten while others become popular and known by the majority of people. There should be a mechanism helping some cultural values being more popular and causing other people being assimilated by majorities. Different cultures' interactions with each other and consequences of their interactions will be investigated by the principle rules of Simple Recommendation Model which is proposed by Bingol in 2006. The agents will be grouped according to their national origin and remember and forget the choices instead of agents. Also selections of interacted agents will be made according to people's choices.

Keywords: Emergence of fame, cultural choices, assimilation.

1 Introduction

Humans are social creatures and exchange ideas by interacting with each other. By doing so they learn new people, habituates or cultural values from their parents or from the people they interact.

In Simple Recommendation Model (SRM) each agent has a limited memory capacity and keeps other agents in his memory [1]. The agents interact with each other by exchanging agents in their memories. Since the memory capacities are limited, an agent is known in price of forgetting the other. There are giver, taker, recommended and forgotten agents in a recommendation process. Selections of the agents are random. Hence the model is called the Simple Recommended Model (SRM). As a result of simple recommendations, some agents become extremely known. This observation is interpreted as emergence of fame.

We will extend SRM by applying the model into the real life scenarios of today's world. In the SRM, an agent interacted with others completely randomly but in real life they interact within groups they belong to. Groups can be formed from friends, work, occupation or clubs. In our work we take ethno-national groups.

We will try to make predictions about the result of interactions of different ethno-national cultures by composing the SRM and work of Wimmer. We will propose a

J. Zhou (Ed.): Complex 2009, Part I, LNICST 4, pp. 1063–1071, 2009.

model in which agents interact with each other according to their choice of interactions according to their choice of interactions and impose their choices to others. We will extend the recommendation model based on Wimmer's work on a Swiss society with Italian and Turkish immigrants [2].

There are theoretical paradigms in Sociology, Order Theory of Durkheim that emphasizes the ways in which different groups progressively become more unified and indistinct and Conflict Theory of Marx and Weber that emphasizes the inequality among ethnic groups [3]. Our work will show whether the groups in our model will go to unify or distinguish in having common choices.

The rest of the paper is outlined as follows. Section 2 explains the Wimmer's work. Section 3 briefly explains Bingol's SRM since our model will extend the mechanisms of SRM. In Section 4 we will introduce our choice recommendation model. Section 5 will give the simulation results and finally we will give a related work and conclude the subject.

2   Wimmer's Work of Swiss Population

Wimmer conducted a series of researches in three Swiss Towns, namely St. Johann, the Breitenrain and the Hard neighborhood. All of the residential areas are highly populated with immigrants and suitable to group formation. In his research, he has worked on the relations between the native Swiss and immigrant populations [2].

Althouh there are many social groups, Wimmer has focused on the three largest groups, namely Swiss, Italians and Turks. The percentages of relations are as given in Table 1.

There is a sharp distinction between relations of Swiss Italians and Turks. We will run our simulations in the light of those interaction ratios given in Table 1.

As expected, a group prefers itself to interact. In this respect Swiss is the closest community and prefere Italians when they interact. Italians are slightly more closer than Turks. Note that this is a highly asymmetric system. For example Swiss prefer Turks with 0.8%, while Turks prefer Swiss with 20%.

Table 1. Ethno-national background of the people according to their choice of interactions

		National	Background	of Alteri		
		Swiss	Italian	Turkish	Others	Total
National	Swiss	85.5%	5.0%	0.8%	8.7%	100 %
Background	Italian	17.8%	68.9%	0.7%	12.6%	100 %
of Respondent	Turkish	20.8%	3.9%	66.6%	8.7%	100 %

3   Bingol's Recommendation Model

We will briefly mention the SRM of *Fame* [1] since our model will be based on the same principles. Then, we will give the variations of our model. Here is a brief description of SRM:

There are n agents. Each agent has a limited memory capacity m and initially randomly filled with other agents. If an agent a_i resides in the memory of another agent a_j then a_j knows a_i, if not then a_j does not know a_i. The *knownness* or let us say the *fame* of a_i is the percentage of agents that know a_i. An agent can know only m agents which is $m<<n$.

The memory contents of the agents change as the recommendation takes place. In any simulation cycle there are recommender, taker, recommended and forgotten agents. The recommendation operation happens as follows; the recommender agent selects the recommended agent from its memory and recommends it to the taker agent. If the recommended agent is not in the taker's memory yet, the taker agent replaces the recommended agent with the forgotten agent in its memory slot and learns the recommended agent in the price of forgetting the forgotten agent. If it is already in his memory nothing is done. This is a simple recommendation operation of the model. All the selections are made randomly.

4   Choice Recommendation from Different Populations

Our model basically differentiates from SRM in that the population is divided into sub-groups, namely A_S, A_I and A_t, that represents Swiss, Italian and Turkish populations respectively. The other important point is that our agents will keep their choices, let's say their cultural values, in their m memories instead of keeping other agents. Those choices will be represented by consecutive non-overlapping numbers. Let C_S, C_I and C_t be the sets of choices of Swiss, Italians and Turks respectively. Every agent will have the same m. Assume that M_S, M_I and M_t are the memory contents of a randomly selected agent of Swiss, Italians and Turkish people respectively and, $M_s \subseteq C_s$, $M_i \subseteq C_i$ and $M_t \subseteq C_t$. The Number of choices will be proportional to the group's size, the bigger population will have the more choices. Initial popularity of a choice is calculated by the number of agents who keeps that choice in his memory. In other words, let $a_i \in A_s$ and $c_i \in C_s$ be a choice of a_i. Then initial popularity of b_i is $P_i = |\{c_i \in M_j \ |c_i \in A_s\}|$. If a choice has zero fame at the end of the simulations, it will be completely forgotten and if it is known by all the agents then it will be completely known.

There is an important differences in the recommendation operation in our model, that is selection of taker agent. Taker agent will be selected according to the preference of the giver agent. Let a_i is a giver agent in A_S and a_j is a taker agent in A_I. Then, P_{AsAi} is the probability of A_S interacting to A_I and the taker will be in A_I with P_{AsAi}. Giver agent is selected randomly. The rest of the recommendation process occurs as follows:

Giver agent selects the recommended choice from it's memory content randomly and recommends it to the taker agent. Taker agent is selected according to the ratios given in Table 1. by a random number generator. We care with the memory contents of Swiss, Italian and Turks choices. If the random number generator selects *others*, we just skip to the next simulation cycle. If the recommended choice is already in the memory of the taker agent nothing is done. If not, a choice selected randomly from the taker's memory and replaced by the recommended choice.

In our model $m=5$. The simulations will be made over $n=1000$ agents. The population ratios of Swiss, Italians and Turks living in three Swiss towns are given in Wimmer's work [2]. The number and percentages of empirical and simulation populations as well as the number of choices are given in Table 2.

Table 2. Etho-national background of the people according to their choice of interactions

	Emprical	Data		Simulation	Data	
	Populations	% of Populations	Sim. Populations	% of Sim. Populations	Memory Capacity of Each Agent	Number of Choices
Swiss	23000	58.97	916	91.58	5	153
Italian	1360	3.48	54	5.40	5	9
Turkish	760	1.94	30	3.02	5	5
Others	21000	35.61	-	-	-	-
Total	46120	100.00	1000	100.00	-	167

Total number of choices for a group is selected to make initial popularity of choices be equal. For example, Italians are 54 agents. Each agent has a memory capacity of $m=5$. Then, total memory capacity is 270. When we distribute 9 choices to memory slots by a regular memory initialization scheme like in SRM model [1], the popularity of each choice will be 30. The other groups' choices are selected so as to make initial popularities 30. This popularity size is not strict and may be any number as long as initial popularity of every choice is equal.

5 Simulation Results

Simulations were held for 10^{11} cycles. 10 different simulations were held and their averages were taken as the result. We have inspected the results for maximum, minimum, average popularities and forgotten ratios of choices. The figures of the results are given in below figures.

5.1 Maximum Popularity

Maximum popularity is the total number of agents who know that choice. In our model, maximum popularity can be at most 1000 which means to be known by everybody.

Maximum popularity always belongs to Swiss population's choices but is not higher than 321 even after 10^{11}. simulation cycles. It is far beyond to reach to be completely known. Although Turks' choices are almost half of the Italians' choices their average maximum popularities are almost the same and even gets higher than Italians. The result is given in Figure 1.

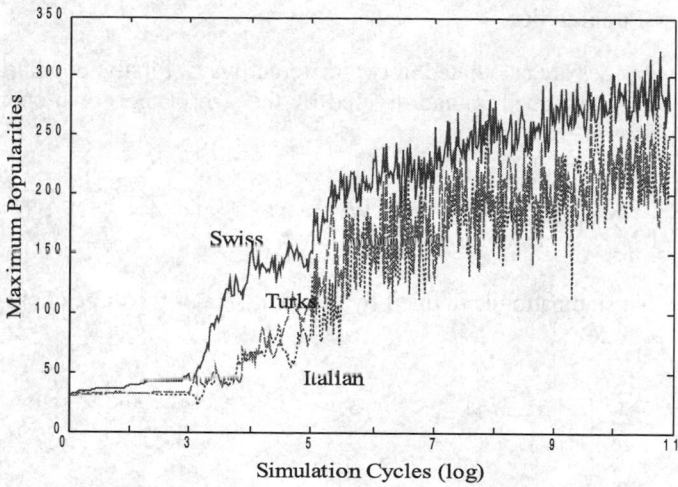

Fig. 1. Maximum popularities of choices

5.2 Minimum Popularity

Minimum popularity means to be least known by the whole agents. It is the smallest total number of agents who know that choice. Minimum popularity can be at least zero which means to be completely forgotten. Once a choice is completely forgotten there is no way to be known again. There is a sharp increase in the minimum popularities of all three societies after 10^3 simulation cycles. Turks choices have less minimum popularities among others. Although their average popularities are less than Swiss, some of Swiss choices' popularities drops faster than Turks. The result is given in Figure 2.

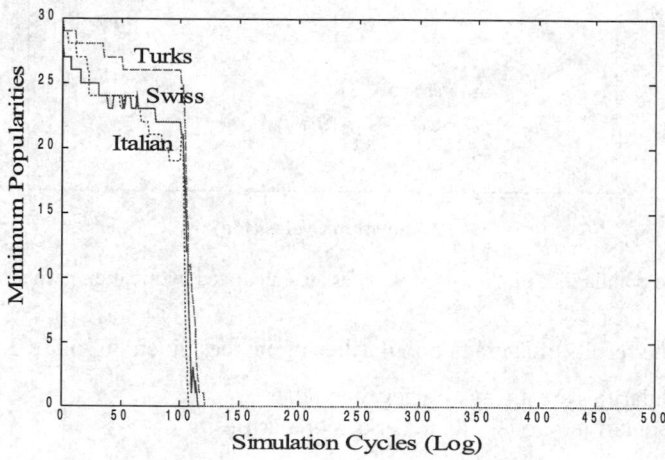

Fig. 2. Minimum popularities of choices

5.3 Average Popularities

Average popularities are calculated in two different ways. Firstly, popularities of each choice of a set are summed up and divided by the size of their own choice set. It is defined as follows;

$$p_{own}^{avg} = \frac{1}{|C_k|} \sum_{a_j \in A} |\{a_j \mid c_i \in M_j\}| \qquad \text{Where } c_i \in C_k \text{ and } A = A_S \cup A_i \cup A_T \qquad (1)$$

Secondly, the summation is divided by the total size of the three choice sets (167). It is defined as follows;

$$p_{all}^{avg} = \frac{1}{|C|} \sum_{a_j \in A} |\{a_j \mid c_i \in M_j\}| \qquad \text{Where } C = C_S \cup C_I \cup C_T \text{ and } A = A_S \cup A_I \cup A \qquad (2)$$

The figure of the first way of calculation is given in figure 3. It is trivial that Swiss' averages will be higher than others when calculated with the second way since Swiss are outnumbered Italians and Turks. Italians' popularities are also higher than Turks' popularities but at the end of 10^{11} simulation cycles, Turks' popularities are around 0.69 while Italians' popularities are around 0.38, even though Italians' choices almost double Turks' choices.

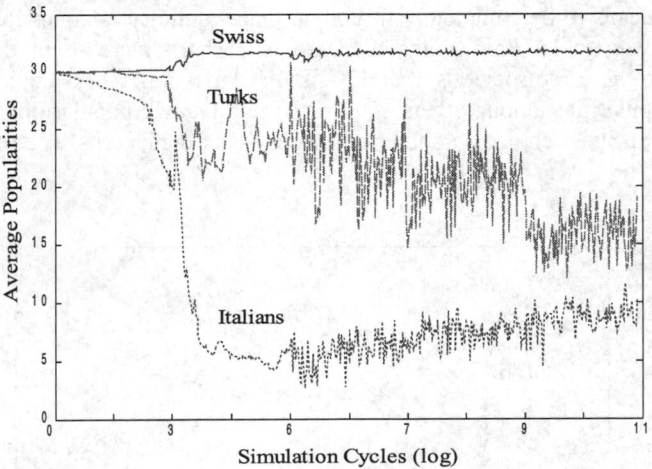

Fig. 3. Average popularities of choices. Averages are calculated within their own set of choices.

Here are the results of average popularities of choices given in Figure 3.

1. Swiss' popularities are always higher than others.
2. Italians' popularities are less than Turks' popularities.

There are two reasons. Firstly, Swiss prefer Italians five times more than Turks. So Swiss may impose their choices to Italians five times more than they do to Turks.

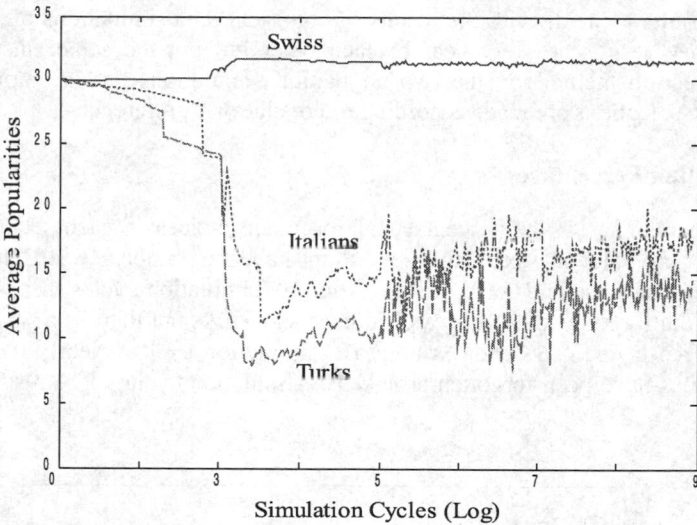

Fig. 4. Average popularities of choices when interactions of Swiss and Italians have been modified

Secondly, Italians prefer to communicate with other smaller minorities like Spanish and other Eastern European groups more than Turks do. Their preference of those minorities instead of Swiss adversely affect their choices' popularities.

We adjusted the relation percentages given in Table 1 to prove that hypothesis. We equalized Swiss preference of Italians and Turks. We also decreased Italians' preference of other groups to equalize it with those Swiss' and Turks' preferences. The result is given in Figure 4. As expected, Italians' average popularities got higher than Turks' averages.

We saw that although Turks have the fewest population they managed to overcome over Italians whose population are almost double of Turks. Swiss managed to impose their choices, let us say assimilate, other groups. We also conducted a series of simulations representing the interactions of two populations having the same memory capacities and population sizes. Population sizes were at the same ratios such as $n_1=32$ and $n_2=968$ representing Turks and Swiss respectively. Both of Swiss and Turks populations have $m=5$. Simulations were held for 10^9 cycles.

Table 3 gives the results of above simulations. The ratios of maximum and average popularities and forgotten choices of two populations are given.

Table 3. Etho-national background of the people according to their choice of interactions

Turks / Swiss	Two population simulations	Emprical data simulations
Maximum popularities	0.81	0.80
Forgotten ratios	1.25	1.23
Average popularities	0.83	0.85

Those results coincide with the results of empirical data simulations of Turks and Swiss. Above results are very near to each other but not the same since there is random selection methods in the two population simulations but in empirical data simulations, selections are made according to populations' preferences.

5.4 Forgotten Percentages

Forgotten percentage is the percentage of forgotten choices in each set of choices. Swiss have zero value at 16×10^3 th step, Italians have zero value at 8×10^3 th step. and Turks have zero value at 21×10^3 th step. After 10^4 simulation cycles there is a linear increase in the forgotten percentages of Italians and Turks but there is a sharp (almost double) increase for Swiss choices after 10^6 simulation cycles. Nearly two third of Swiss' choices have been forgotten around 10^6 simulation cycles. The result is given in Figure 5.

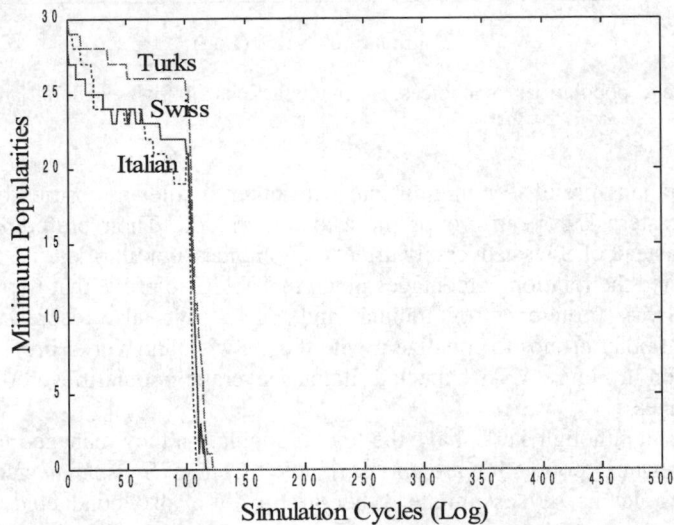

Fig. 5. Minimum popularities of choices

6 Related Work and Coclusions

Robert Axelrod and Ross A. Hammond have introduced *ethnocentrism* syndrome which can be described as in-group favoritism and out-group hostility of different groups living together. They have made simulations of agent based models in a square lattice [4]. They found that groups tend to show ethnocentric strategy with 76%, compared to 25% by chance. The results claims that groups will tend to diversify by choosing their own groups members for interactions rather than unify. Our researches show that they will unify after many simulations later.

In this research we have extended the SRM by using large populations of agents having small memory sizes. Agents have interacted by exchanging their choices instead of exchanging agents. Here are some foresights about the consequences of their future interactions in the light of our simulations:

1. A group which is larger in population, has more chance to increase the popularities of its choices.
2. Close communities which interacts mostly among themselves have more chance to preserve their cultural values like Swiss population. Since social networks are scale-free, higher interactions rates between the agents of the same group makes topology more robust to outer influences and assimilations [5].
3. If an immigrant group is more popular for the dominant population than other immigrant groups, then it is more probable to be assimilated.
4. Assimilations take very long time. At the end of 10^{11} simulation cycles 95.5% of Italians' choices and 92% of Turks' choices have been forgotten. No complete assimilation is detected.
5. Although Swiss population's choices are more popular, none of them is known more than 32% of all populations even after 10^{11} simulations later.

We have found that communities tent to unify rather than diversify. All the interacting communities may show an ethnocentric behavior at first but sooner or later they will began to unify as they interact with each other. This will cause either integration or assimilation of minorities The model can be extended by applying other factors such as population's choices of religions and languages. Then some choices will not be easily changed by a simple recommendation model. It can be used some socio-economic researches of nations and cultural assimilation processes.

References

1. Bingol, H.: Emergence of Fame. ArXiv:nlin. AO/0609033 vl (2006)
2. Wimmer, A.: Does ethnicity matter? Everyday group formation in three Swiss immigrant neighborhoods. Ethnic and Racial Studies 27(1), 1–36 (2004)
3. Smith, A.: Civic and Citizenship Education in Contested and Divided Societies. UNESCO Chair, University of Ulster, Northern Ireland (2003)
4. Axelrod, R., Hammond, R.A.: The Evolution of Ethnocentric Behavior. Revised version of a paper delivery at Midwest Political Science Convention, Chicago IL (2003)
5. Ghanea-Hercock, R.: Assimilation and Survival in Cyberspace. BTexact Future Technologies Group Adastral Park, UK (2002)

Antisynchronization of Two Complex Dynamical Networks

Ranjib Banerjee[1], Ioan Grosu[2], and Syamal K. Dana[1]

[1] Central Instrumentation, Indian Institute of Chemical Biology,
(Council of Scientific and Industrial Research), Kolkata 700032, India
{ranjib_rs,skdana}@iicb.res.in
[2] Faculty of Bioengineering, University of Medicine and Pharmacy, "Gr.T.Popa", Iasi, Romania
ioan.grosu@instbi.umfiasi.ro

Abstract. A nonlinear type open-plus-closed-loop (OPCL) coupling is investigated for antisynchronization of two complex networks under unidirectional and bidirectional interactions where each node of the networks is considered as a continuous dynamical system. We present analytical results for antisynchronization in identical networks. A numerical example is given for unidirectional coupling with each node represented by a spiking-bursting type Hindmarsh-Rose neuron model. Antisynchronization for mutual interaction is allowed only to inversion symmetric dynamical systems as chosen nodes.

Keywords: Antisynchronization, dynamical networks, OPCL coupling.

1 Introduction

In recent years, studies on collective behavior of nonlinear dynamical systems has inclined more to dynamical processes in complex networks [1-7] since many real-life systems, living and nonliving, show complex network topology instead of regular links like nearest-neighbor or all-to-all global coupling. A complex network consists of a large number of nodes connected by links or edges where their connectivity, instead of being random as proposed earlier [8], shows statistical properties like small-world [3-4] or scale-free effect [2, 6, 7] in real world. In a complex dynamical network, each node is considered as a dynamical system, either continuous-time or discrete time. Understanding the process of collective behavior or synchronization in a crowd of dynamical nodes within a complex topology then becomes interesting and important [9, 10] to explain many real world phenomena in engineering networks [11] like Internet, World Wide Web, World Trade Web and in biological networks like neurons in brain, pacemaker cells in heart and genetic networks [9]. Particularly, the nodes of the complex networks are assumed as dynamical, as for example in biological systems, which evolve with time. In this context, synchronization in complex networks called as *inner synchronization* has been investigated [9, 10-13] recently to understand the interplay between dynamics of nodes and the topology of a

J. Zhou (Ed.): Complex 2009, Part I, LNICST 4, pp. 1072–1082, 2009.

complex network. In *inner synchronization*, all the nodes have a common dynamics both in amplitude and phase. Establishing conditions of synchronization and desynchronization between two or more networks is an important task of practical relevance [14]. Inducing desynchronization in networks is important from a viewpoint of overcrowding or jamming in networks like Internet, which may be avoided by breaking a state of synchrony within a network. Here, we address a process of antisynchronization in two complex dynamical networks, which may work as an alternative to desynchronization. In a state of antisynchronization in two dynamical networks, the evolution of each of the nodes of a network is locked with the corresponding node of another network in alternate time.

A complex dynamical network is described by

$$\dot{x}_i = f(x_i) + \varepsilon \sum_{j=1}^{N} a_{ij} \Gamma x_j; \quad i = 1, 2, 3...., N \tag{1}$$

where $f : R^n \rightarrow R^n$ is a continuous dynamical flow that governs the local dynamics of each uncoupled node i and $x_i = (x_{i1}, x_{i2}, ..., x_{in})^T \in R^n$ is the state variable of a node; N is the number of nodes. The matrix $A = (a_{ij}) \in R^{N \times N}$ defines the connectivity of nodes in a network whose entries follow a rule: if there is a connection between the nodes i and j ($j \neq i$), then $a_{ij} = 1$; otherwise $a_{ij} = 0$ ($j \neq i$); the diagonal elements of A are

defined as $a_{ii} = - \sum_{j=1, j \neq i}^{N} a_{ij} = - \sum_{j=1, j \neq i}^{N} a_{ji}$, and clearly if the degree of ith node is k_i then

$a_{ii} = -k_i$, $i = 1, 2, 3, ..., N$. $\varepsilon > 0$ is the coupling strength between the nodes of individual networks. $\Gamma \in R^{n \times n}$ is a constant diagonal matrix whose elements are 0 or 1 and it defines the links between the state variables of any two nodes. In a Γ matrix, if all the elements are 1, then any pair of nodes is connected by all state variables, otherwise they are partially connected if any of the elements is zero. A synchronous state of all nodes of the network is defined by $\dot{x} = f(x)$. We are concerned here with the process of synchronization, particularly, antisynchronization and how to implement them in two complex dynamical networks.

Synchronization of two complex dynamical networks was reported earlier [15] using a master-slave type unidirectional open-plus-closed-loop (OPCL) coupling [16]. In a recent Letter [17], we extended the OPCL method to establish antisynchronization and amplification or attenuation in two chaotic oscillators. Here, we extend the results further to achieve antisynchronization and/or attenuation in two dynamical networks. Once the node dynamics is known, one can design an appropriate coupling using the OPCL scheme to realize antisynchronization or to attenuate any undesired effect in one dynamical network from being transmitted to another response dynamical network.

The relevance of synchronization between two dynamical networks was explained in [15] by citing an interesting example of two economic worlds: one developing and another developed. It explained how a developed economy influences the developing world economy and considered unidirectional influence while studying synchronization in two such networks. Although this unidirectional effect is strongly felt in a recent economic crisis in the United States that is followed by immediate crash in the share market network of many countries, the reality is more complex. It is natural that both the economies (developed or developing) influence each other to evolve a new world economic order. It is obviously more realistic to consider mutual interactions between the networks either economic networks or social networks to derive a true picture. Accordingly, in addition to the unidirectional effect, we address the mutual OPCL coupling issue to realize antisynchronization in two complex dynamical networks.

The paper is structured as follows. In the next section, antisynchronization using unidirectional OPCL coupling in two oscillators is described. The theory is then extended to complex dynamical network in section 3. Mutual interaction in two dynamical networks is described in section 4. Results are summarized in section 5.

2 Antisynchronization in Two Oscillators

We briefly introduce the general scheme [16, 17] of unidirectional OPCL coupling in two chaotic oscillators: a chaotic driver is defined by $\dot{y} = f(y)$, $y \in R^n$. The model of the chaotic oscillator with parameters is assumed known *a priori*. It drives another chaotic oscillator $\dot{x} = f(x), x \in R^n$ to achieve a goal dynamics $g(t)=\alpha y(t)$ as a desired response, where α is a constant. The response system after coupling is given by

$$\dot{x} = f(x) + D(x, g).\tag{2}$$

where the coupling term $D(x, g)$ is defined by

$$D(x, g) = \dot{g} - f(g) + \left(H - \frac{\partial f(g)}{\partial g} \right)(x - g).\tag{3}$$

$\dfrac{\partial f(g)}{\partial g}$ is the *Jacobian* and H is an arbitrary constant Hurwitz matrix $(n \times n)$, whose eigenvalues have all negative real parts. The error signal of the coupled system is defined by $e = x - g$ when $f(x)$ can be written, using the Taylor series expansion,

$$f(x) = f(g) + \frac{\partial f(g)}{\partial g}(x - g) + \dots\tag{4}$$

Keeping the first order terms in (4) and substituting in (3), the error dynamics is obtained as $\dot{e} = He$ from (2) and this ensures that $e \to 0$ as $t \to \infty$ and the synchronization is asymptotically stable. The Hurwitz matrix can be easily constructed from the *Jacobian* of the known model of the interacting oscillators. The elements of the Hurwitz matrix, H_{ij}, are then chosen such that $\left(H - \dfrac{\partial f(g)}{\partial g} \right)_{ij}$ is zero when $\left(\dfrac{\partial f(g)}{\partial g}\right)_{ij}$ is a constant in (3). If $\left(\dfrac{\partial f(g)}{\partial g}\right)_{ij}$ involves a state variable, we

define $H_{ij} = p_{ij}$ where p_{ij} is a constant. The parameter values, p_{ij}, are so selected as to satisfy the Routh-Hurwitz (RH) criterion. For a 3D dynamical system, the characteristic equation of the H matrix is

$$\lambda^3 + a_1 \lambda^2 + a_2 \lambda + u_3 = 0 \qquad (5)$$

where a_i (i=1, 2, 3) are coefficients.

The corresponding RH criterion [16] is given by

$$a_1 > 0, \quad a_1 a_2 > u_3, \quad u_3 > 0 \qquad (6)$$

The selection of the parameters p_{ij} is so appropriately made that the RH criterion is fulfilled and thereby ensures synchronization that is asymptotically stable even in presence of any parameter mismatch [18]. The multiplying constant α in the goal dynamics can be used as a control parameter to realize CS (α=1), AS (α=-1), attenuation ($|\alpha| < 1$) or amplification ($|\alpha| > 1$).

3 Complex Dynamical Network: Unidirectional Coupling

We extend the unidirectional coupling scheme to complex dynamical networks to realize antisynchronization and attenuation. As described earlier [15], an analytical approach is possible to establish synchronization using OPCL coupling between two dynamical networks for identical connectivity matrix and it is found unchanged for the proposed generalization here. The driving network may be expressed by (1) and the response network is defined by

$$\dot{y}_i = f(y_i) + \alpha \dot{x}_i - f(\alpha x_i) + \left(H - \frac{\partial f(\alpha x_i)}{\partial (\alpha x_i)} \right)[y_i(t) - \alpha x_i(t)] + \varepsilon \sum_{j=1}^{N} b_{ij} \Gamma y_j \qquad (7)$$

$y_i = [y_{i1}, y_{i2}, y_{i3}]^T$ is the state variable of the ith node of the response network; other notations have similar meaning as above, and A=$(a_{ij}) \in R^{n \times n}$, B=$(b_{ij}) \in R^{n \times n}$ are symmetric or asymmetric matrices; each row sum of A and B equal to zero. Networks (1) and (7) achieve synchronization if

$$\lim_{t \to +\infty} \| y_i(t) - \alpha x_i(t) \| = 0, \; i = 1, 2, 3..., N \qquad (8)$$

For simplification, we assume two networks having identical topology (A=B). Then linearizing the error system, $e_i(t) = y_i(t) - \alpha x_i(t)$, around x_i, we obtain

$$\dot{e}_i = H e_i + \varepsilon \sum_{j=1}^{N} a_{ij} \Gamma e_j; \; i = 1, 2, 3...., N \qquad (9)$$

which can be simplified as

$$\dot{e} = He + \varepsilon \Gamma e A^T \tag{10}$$

T stands for transpose and $e = [e_1, e_2, \ldots, e_N]$ denotes $n \times N$ matrix. The coupling matrix may be decomposed by taking $A^T = SJS^{-1}$ where J is a Jordan canonical form with complex eigenvalues $\lambda \in C$ and S contains the corresponding eigenvectors. If we define, $\eta = eS$, using eq.(10), we can easily derive

$$\dot{\eta} = H\eta + \varepsilon \Gamma \eta J \tag{11}$$

where $J = [J_1, J_2, \ldots, J_h]^T$ is a block diagonal matrix and J_k is a block corresponding to the m_k multiple eigenvalues λ_k of A.

$$J_k = \begin{bmatrix} \lambda_k & 1 & 0 & \ldots & & 0 \\ 0 & \lambda_k & 1 & \ldots & & 0 \\ . & . & . & . & . & . \\ 0 & 0 & . & & \lambda_k & 1 \\ 0 & 0 & \ldots & & 0 & \lambda_k \end{bmatrix} \tag{12}$$

Assuming $\eta = [\eta_1, \eta_2, \ldots, \eta_h]^T$, $\eta_k = [\eta_{k,1}, \eta_{k,2}, \ldots, \eta_{h,m_k}]^T$ and, since the sum of every row of the matrix A is zero and J_1 is a 1×1 matrix, we can assume $\lambda_1 = 0$. Now if $\lambda_1 = 0$, it satisfies $\dot{\eta}_1 = H\eta_1$ and hence the zero solution of $\dot{\eta}_1 = H\eta_1$ is asymptotically stable if H is a Hurwitz matrix. In this way one can easily establish asymptotic stability of all zero solutions for k>1; details may be found in [15] that confirms synchronization of the dynamical networks (1) and (7) once A=B and H is a Hurwitz matrix. The analysis presented in ref.15 remains unaffected by the introduction of the parameter α, where we set a goal dynamics at each node of the networks as $y(t) = \alpha x(t)$. Hence we can realize synchronization ($\alpha = 1$), antisynchronization ($\alpha = -1$) or attenuation ($|\alpha| < 1$) simply by a choice of the α-value.

We present a numerical example where each ith node of both the networks is described by spiking-bursting type Hindmarsh-Rose neuron model [19],

$$\dot{x}_{i1} = x_{i2} - a x_{i1}^3 + b x_{i1}^2 - x_{i3} + I, \quad \dot{x}_{i2} = c - d x_{i1}^2 - x_{i2}, \\ \dot{x}_{i3} = r\{s(x_{i1} + 1.6) - x_{i3}\}. \tag{13}$$

where $a=1$, $b=3$, $c=1.0$, $d=5.0$ and $s=5.0$. The state variables x_{i1} and x_{i2} correspond to fast oscillation and x_{i3} represents the slow dynamics as decided by a choice of $r=0.003$. The bias current I=4.1 sets the oscillatory mode in a chaotic regime. The H matrix of the model (13) is given by

$$H = \begin{bmatrix} p_1 & 1 & -1 \\ p_2 & -1 & 0 \\ rs & 0 & -r \end{bmatrix} \tag{14}$$

where p_1 and p_2 are parameters. It can be analytically established [16, 17] that if $p_2=0$ and $p_1<1+r$, H is a Hurwitz matrix with eigenvalues all with negative real parts. We set the $\Gamma=\mathrm{diag}(1, 0, 0)$ to establish a scalar coupling between the nodes within the individual networks. In [15], the authors assumed that all state variables of each node of the individual networks were coupled. It is found, in numerical simulations, that scalar coupling or fewer coupling as set by $\Gamma=\mathrm{diag}(1, 0, 0)$ suffices to realize synchronization or antisynchronization. We choose two undirected networks each having N=10 nodes, where A is given by

$$A=\begin{bmatrix} -4 & 1 & 1 & 0 & 0 & 0 & 1 & 0 & 0 & 1 \\ 1 & -5 & 1 & 1 & 0 & 0 & 0 & 0 & 1 & 1 \\ 1 & 1 & -5 & 1 & 1 & 1 & 0 & 0 & 0 & 0 \\ 0 & 1 & 1 & -4 & 1 & 1 & 0 & 0 & 0 & 0 \\ 0 & 0 & 1 & 1 & -5 & 1 & 1 & 1 & 0 & 0 \\ 0 & 0 & 1 & 1 & 1 & -5 & 1 & 1 & 0 & 0 \\ 1 & 0 & 0 & 0 & 1 & 1 & -6 & 1 & 1 & 1 \\ 0 & 0 & 0 & 0 & 1 & 1 & 1 & -5 & 1 & 1 \\ 0 & 1 & 0 & 0 & 0 & 0 & 1 & 1 & -4 & 1 \\ 1 & 1 & 0 & 0 & 0 & 0 & 1 & 1 & 1 & -5 \end{bmatrix} \tag{15}$$

Since the coupling matrix A is symmetric, its first eigenvalue is zero and the rest are negative. Once the parameter $p_1<1+r$ ($p_2=0$) is ensured, eq.(10) confirms that the real parts of the eignenvalues of $He+\varepsilon\lambda_k\Gamma$ (λ_k is the set of eigenvalue of A) are negative for arbitrary value of $p_1<1+r$. Networks (1) and (7) will develop synchronization for $\alpha=1$ when each node of the network (1) develops an identical dynamics with each corresponding node of the network (7). This result is already reported earlier [15], however, we introduced a general framework here to choose any desired value of α. As a result, antisynchronization can also be established by a choice of $\alpha=-1$, when corresponding nodes of the networks develop identical dynamics but in opposite phase as shown in Fig.1. Attenuating the amplitude of the dynamics in a driver network is also possible at a response network by simply choosing ($|\alpha|<1$), details of which are redundant. For numerical simulations, the initial conditions are randomly chosen and the synchronization error is measured by

$$\|e(t)\| = \max\{\max_{1\leq i\leq 10}|x_{i1}(t)\mp y_{i1}(t)|,$$
$$\max_{1\leq i\leq 10}|x_{i2}(t)\mp y_{i2}(t)|, \tag{16}$$
$$\max_{1\leq i\leq 10}|x_{i3}(t)\mp y_{i3}(t)|\}, \text{for } t\in[0,+\infty).$$

Minus (-) sign denotes synchronization and plus (+) sign for antisynchronization.

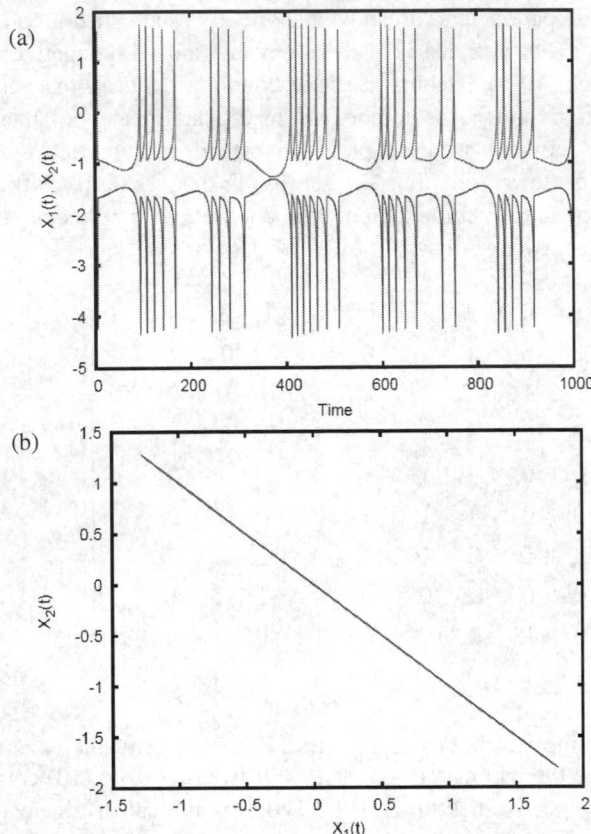

Fig. 1. Antisynchronization in complex networks, (a) time series [$x_{i1}(t)$ in red and $y_{i1}(t)$ in blue] in any two corresponding nodes of the driver and response networks, (b) state variable $x_{i1}(t)$ of a node of driver network is plotted against state variable $y_{i1}(t)$ of a corresponding node in the response. $p_1=-1.5$, $\varepsilon=10^{-6}$.

Synchronization between the networks is independent of *inner synchronization* of individual networks. It is obtained even for very low value of $\varepsilon=10^{-6}$ when there is no *inner synchronization*. The synchronization between the two networks is fastest when there is no *inner synchronization*. The speed of synchronization is shown in Fig.2 for different ε -values. For larger coupling $\varepsilon>0.4$, the speed is not much changing with increase in coupling (ε). Similarly, antisynchronization can also be achieved for nonsymmetric A, i.e., when the inner connectivity of the network is directed. We obtained antisynchronization for A≠B in similar vain as described in [15], however, no analytical treatment is possible; numerical results is only done, for which details are redundant since it is almost a repetition of the results in ref.15. We rather prefer to extend the results to mutual interactions in two complex dynamical networks. Note that the results are checked with larger number of nodes in the networks (N=100).

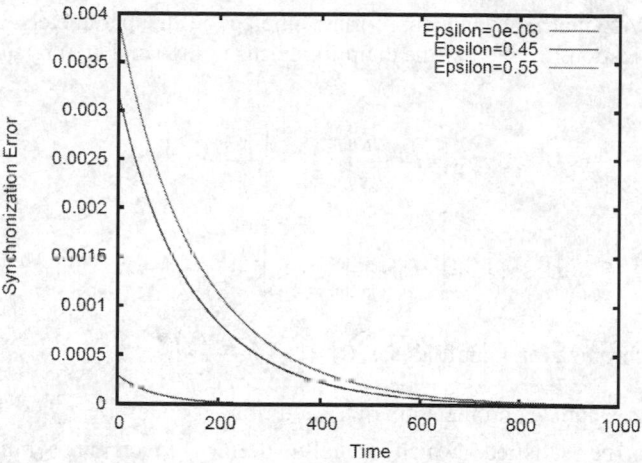

Fig. 2. Dependence of synchronization with coupling. $\varepsilon=10^{-6}$, 0.45 and 0.55, $p_1=-1.5$, $p_2=0$

4   Complex Networks: Mutual Coupling

Synchronization for mutual or bidirectional interactions in two complex networks was not investigated in the previous study [15]. We develop the theory of synchronization in dynamical networks for mutual interaction using the OPCL coupling scheme and then extend it to antisynchronization. Note that the mutual OPCL coupling in two chaotic oscillators was reported earlier [20, 21] for synchronization, but antisynchronization was never investigated. A modification in the theory is needed to realize mutual antisynchronization in two chaotic oscillators, however, it is found limited to inversion symmetric dynamical systems only. Details of antisynchronization using mutual OPCL coupling are reported elsewhere [22]. Two oscillators under mutual OPCL coupling are given by

$$\dot{x} = f(x) + D_x(x, y) ; \quad x \in R^n ,\tag{17}$$

$$\dot{y} = f(y) + D_y(x, y) ; \quad y \in R^n ,$$

where

$$D_x(x, y) = \left(H - \frac{df}{dx}\bigg|_{x = s_+} \right)\left(\frac{x - y}{2}\right) ,\tag{18}$$

and

$$D_y(x, y) = \left(H - \frac{df}{dy}\bigg|_{y = s_+} \right)\left(\frac{y - x}{2}\right) ,\tag{19}$$

$$s_+(t) = \left(\frac{x(t) + y(t)}{2}\right) \text{ is the synchronization manifold.}$$

It can be easily established [21] that the error dynamics $e = (x-y)$ is now governed by $\dot{e} = He$ and its zero error solution or the synchronization is asymptotically stable once

H is a Hurwitz matrix by an appropriate choice of the parameters. For realizing antisynchronization, we modify the coupling terms in (18) and (19) by

$$D_x(x, y) = \left(H - \frac{df}{dx} \Big|_{x = s_-} \right) (\frac{x + y}{2}), \tag{20}$$

$$D_y(x, y) = \left(H - \frac{df}{dy} \Big|_{x = s_-} \right) (\frac{x + y}{2}), \tag{21}$$

and antisynchronization manifold is $s_-(t) = \left(\frac{x(t) - y(t)}{2} \right)$.

To realize antisynchronization, an additional condition $f(y) = -f(-y)$ is necessary to be satisfied, which actually defines inversion symmetry of any dynamical flow. The asymptotically stable antisynchronization is then ensured once H is a Hurwitz matrix. The error dynamics is again governed by $\dot{e} = He$ where the error state is $e = (x + y)$ for antisynchronization. The antisynchronization in two dynamical networks is thus restricted by the inversion symmetry property of a dynamical node as also reported earlier [23] for two chaotic oscillators. We define two mutually coupled complex dynamical networks by

$$\dot{x}_i = f(x_i) + D_x(x_i, y_i) + c \sum_{j=1}^{N} a_{ij} \Gamma x_j; i = 1, 2, \ldots, N \tag{22}$$

$$\dot{y}_i = f(y_i) + D_y(x_i, y_i) + c \sum_{j=1}^{N} b_{ij} \Gamma y_j; j = 1, 2, \ldots, N \tag{23}$$

where

$$D_x(x_i, y_i) = \left(H - \frac{df}{dx_i} \Big|_{x_i = s_i} \right) (\frac{x_i \mp y_i}{2}), \tag{24}$$

$$D_y(x_i, y_i) = \left(H - \frac{df}{dy_i} \Big|_{y_i = s_i} \right) (\frac{y_i \mp x_i}{2}), \tag{25}$$

and

$$s_i = (\frac{x_i \mp y_i}{2})$$

The mathematical structure of the error dynamics remains similar to (10) and (11). All notations and their meanings remain same as earlier. Once the connectivity matrix is again assumed symmetric (A=B), the analytical approach in (10)-(11) for unidirectional coupling remains same for implementing synchronization and antisynchronization in complex networks under mutual coupling and hence we do not repeat them here. However, in numerical examples, we take a model system [24, 25] that is inversion symmetric. The model of the dynamics of ith node of a network is

$$\dot{x}_{i1} = 0.49x_{i2} - x_{i3}, \quad \dot{x}_{i2} = x_{i1} - x_{i2}, \quad \dot{x}_{i3} = x_{i1}^{3} - x_{i2}. \tag{26}$$

We confirm antisynchronization in two mutually coupled complex dynamical networks in Fig.3 using each node as represented by the model (26). Synchronization in two mutually interacting complex networks using the Hindmarsh-Rose model representing unit node dynamics is achieved in numerical simulations but details are not presented here.

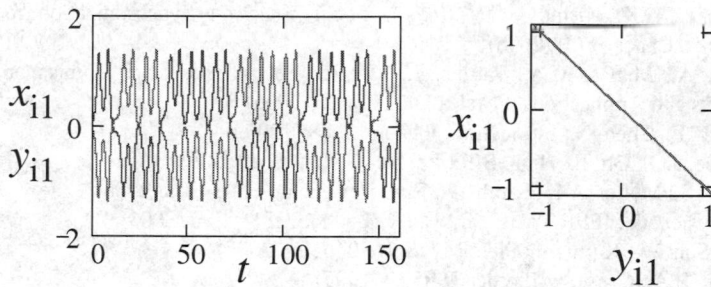

Fig. 3. Antisynchronization of complex dynamical networks for mutual coupling. x_{i1} and y_{i1} are the similar state variables of corresponding ith nodes of the driver and response networks, (a) time series of x_{i1} and y_{i1} in red and in blue, (b) $x_{i1}(t)$ is plotted against $y_{i1}(t)$.

5 Summary

We focused on antisynchronization in two complex dynamical networks for unidirectional as well as bidirectional interactions using OPCL coupling. We mainly extended the previous results [15] on synchronization in two complex dynamical networks under unidirectional interaction. However, we encounter one limitation in realizing antisynchronization in mutually coupled networks. The dynamical flow at each node of the networks must have the inversion symmetry property. While the unidirectional OPCL coupling has no such restriction in inducing antisynchronization, but its bidirectional version fails to overcome this restriction. The synchronization between the networks is independent of inner synchrony of the individual network. It is interesting to note that the speed of synchrony is faster when there is no *inner synchronization* in the individual networks.

Acknowledgments. I.G. and S.K.D. acknowledge support by the Ministry of Education and Research, Romania and the Ministry of Science & Technology, India. I.G. acknowledges support from FUNCDYN programme of ESF and PI-II-CDI grant from CNCSIS, Romania. S.K.D. also acknowledges support by the DST, India (grant # SR/S2/HEP-03/2005). R.B. is a Junior Research Fellow of the CSIR, India.

References

1. Boccaletti, S., Laora, V., Moreno, Y., Chavez, M., Hwang, D.U.: Phy. Rep. 424, 175, (2006)
2. Albert, R., Barabási, A.L.: Rev. Mod. Phys. 74, 47 (2002)
3. Newman, M.E.J., Watts, D.J.: Phys. Lett. A 263, 341 (1999)
4. Watts, D.J., Srogatz, S.H.: Nature (London) 393, 440 (1998)
5. Newman, M.E.J.: SIAM Rev. 45, 167 (2003)
6. Barabási, A.L., Albert, R., Jeong, H.: Phys. Lett. A 272, 173 (1999)
7. Albert, R., Barabási, A.L.: Science 286, 509 (1999)
8. Erdos, P., Renyi, A.: Publ. Math. Inst. Hung. Acad. Sci. 5, 7 (1960)
9. Suykens, J.A.K., Osipov, G.V. (eds.): Focus issue on Synchronization in Complex Networks. Chaos 18(3) (2008)
10. Motter, A., Matías, M.A., Kurths, J., Ott, E.: Focus issue on Dynamics on Complex networks and Applications. Physica D 224 (1-2) (2006)
11. Wang, X.F., Chen, G.: Int. J. Bifur. Chaos 12, 885 (2002)
12. Restrepo, J.G., Ott, E., Hunt, B.R.: Physica D 224, 114 (2006)
13. Zhou, C.S., Motter, A.E., Kurths, J.: Phy. Rev. Lett. 96, 034101 (2006)
14. Li, X., Chen, G.: IEEE Trans. Cir. Systs. 50 (11), 1381–1390 (2003)
15. Li, C., Sun, W., Kurths, J.: Phy. Rev. E 76, 046204 (2007)
16. Jackson, E.A., Grosu, I.: Physica D 85, 1 (1995)
17. Grosu, I.: Phy. Rev. E 56, 3709 (1997)
18. Grosu, I., Padmanaban, E., Roy, P.K., Dana, S.K.: Phy. Rev. Lett. 100, 0234102 (2008)
19. Hindmarsh, J.L., Rose, R.M.: Proc. R. Soc. London, Ser. B 221, 87 (1984)
20. Lerescu, A.I., Oancea, S., Grosu, I.: Phy. Lett. A 352, 222 (2006)
21. Grosu, I.: Int. J. Bifur. Chaos 17 (10), 3519 (2007)
22. Banerjee, R., Grosu, I., Roy, P.K., Dana, S.K. (submitted)
23. Belykh, V.N., Belykh, I.V., Hasler, M.: Phy. Rev. E 62, 6332 (2000)
24. Thomas, R., Kaufman, M.: Chaos 11, 170 (2001)
25. Roy, P.K., Mukherjee, T.K., Dana, S.K.: Proc. Nat. Conf. Nonlin. Syst. Dynamics, India (2008)

Analysis and Modeling on the Government's Co-agglomeration in Industrial Clustering

Ying-Chao Zhang, Chao Chen, Xin-Yi Huang, Xiao-Ling Ye, and Yi-Lu Cai

Nanjing University of Information Science and Technology,
Nanjing 210044, China
yc.nim@163.com

Abstract. Industry clusters have been the focus of scholars and governments since the second half of the 20th century. During a cluster's growing process, the government plays an important role. In order to show the growing law of the clusters and how government did for co-agglomeration, we proposed two kinds of models: Logistic model and BA model with parameter α to describe the single and mass clusters separately, and we choose the gross industrial output value of the 13 cities in Jiangsu province as numerical verification, showing that the government is part and parcel of the industrial clusters.

Keywords: Industrial clustering, co-agglomeration, BA model, Logistic model.

1 Introduction

The last few years have witnessed tremendous development devote to the effect of industrial clusters. Today, the industrial cluster has become a global economic phenomenon. Industrial clusters are existed in many developed and developing countries both in high-tech industries and the labor-intensive industries. For example, the electronics and information industry cluster in U. S. Silicon Valley, auto industry cluster in Detroit, machine tool industry in Stuttgart, surgical instruments industry cluster in Tuttlingen, and so on. Italy, which we called "Kingdom of SMEs", has had the number of 199 clusters by the year 2002, of which there are 69 clusters of textile, 27 shoes, 39 furniture, 32 mechanical, 17 food, and a cluster of metal products, 4 clusters of chemical products and 6 clusters of paper and printing. Clusters in China have also led to Rapid economic growth.

More and more clusters are coming into being, such as clusters in Guangdong, Zhejiang and Jiangsu Province. Being interested in industrial cluster, the government is also taking action to promulgate mass of cluster policies for assistance.

As the reason for the forming of industrial cluster is not unique, government's behavior is also an important factor, especially in China. Whether government support or not is rather key to any sorts of industrial clusters. Recently, much attention has been paid to analyzing and modeling industrial clusters. An industrial cluster, however, is more similar to a population, so we choose Logistic model and complex network to describe the industrial cluster, and also we can see the importance of government during its development.

J. Zhou (Ed.): Complex 2009, Part I, LNICST 4, pp. 1083–1092, 2009.

2 Methods

2.1 Logistic Model for Single Cluster

Clusters' life cycle plays an important role on the competitiveness of industrial clusters. Tichy(1998) divided it into four phases: formative phase, growth phase, maturity phase and petrify phase.

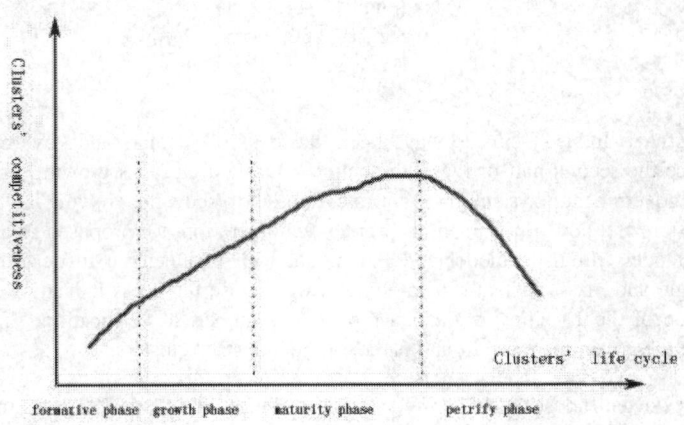

Fig. 1. The relationship between cluster's life cycle and its competitiveness

Population Ecology of the Logistic model described that in certain circumstances, at the force of self copy ability as well as limited resources and other populations' restrictions, the population growth process. Similarly, enterprises entering the cluster can be understood as a certain type of an economic population increase as well. As the industrial cluster has the characteristics of ecological, we try to use the Logistic model to describe clusters' life cycle.

The Logistic model of single industry cluster can be defined as

$$\begin{cases} \dfrac{dn}{dt} = r(1 - \dfrac{n}{K}) \\ N(0) = n_0 \end{cases} \tag{1}$$

where n is the number of enterprises in the cluster which depends on time t, $N(0)$ referrers to the enterprises number in year 0 (the original number) . In a certain period of time and space conditions, the elements of endowment unchanged. K is the maximum scale of the cluster, as $K = \lim_{t \to \infty} n(t)$. At the same time, r represents initial (or maximum) increasing rate. While the cluster's ability to use resource doesn't change with the cluster's scale changes, r can be considered as const. As the industrial cluster theory expounds, the cluster's increasing rate is the remainder of enterprise that enter and exit rate at a certain time frame.

According to Eq. (1), we can get

$$n = \frac{Kn_0 e^{rt}}{K + n_0(e^{rt} - 1)} \tag{2}$$

Eq. (1)(2)show that an industrial cluster's Logistic increasing process from formative phase to maturity phase is associated with balanced scale and increasing rate.

With the help of government, in Logistic model, K and r are controllable parameters. The maximum number of enterprises K is mainly depends on nature resources, infrastructure, institutional environment, labor resources, technology and market demand. The increasing rate r is related with the ability a cluster uses resources. Except those nature resources which naturally existed, the government's co-agglomeration embodies the role of improving infrastructure such as electricity, traffic and communication equipment, completing the legal system to make a suitable environment, and also high level of education can improve the quality of labor force at the same time. The appropriate government policies and measures can promote cluster innovation, cluster learning and cluster brand building, so that its core competitiveness will be greatly enhanced (as the dotted line in Fig. 2.).

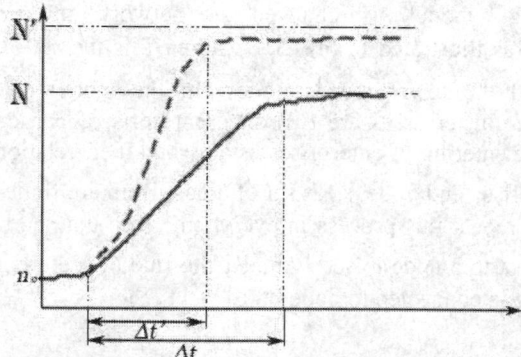

Fig. 2. Logistic growth process of an industry cluster from its formative to maturity phase is related to its balance scale N and growth rate r. When the value of K is higher, the upper limit N' of the curve in Fig.4. is higher, that is to say, the maximal number of enterprises in this industry cluster is larger; when the value of r is higher, the curve is steeper, that is to say, the time Δt from its growth to maturity phase is smaller and the growth rate is faster.

2.2 BA Model for Multiple Clusters

Self agglomeration is an important section during the evolution of complex network. Watts and Strogatz [8] brought up the concept of small-world networks and Clustering coefficient, and attracted much attention to the phenomenon of clustering, such as the analysis on the world wide web [9] and cell network [7]. At the end of the 20th century, Barabasi, Albert and Jeong first brought up the concept of scale free network (BA model) [1]. Their analysis from the dynamic, growth perspectives showed the complex network has the characteristic of power-law distribution. BA model provides the concept of "priority connections" which greatly influenced Zhang Siying's [6] self agglomeration model. However, in industrial clustering, enterprises entering a certain area rely on government's co-agglomeration as well.

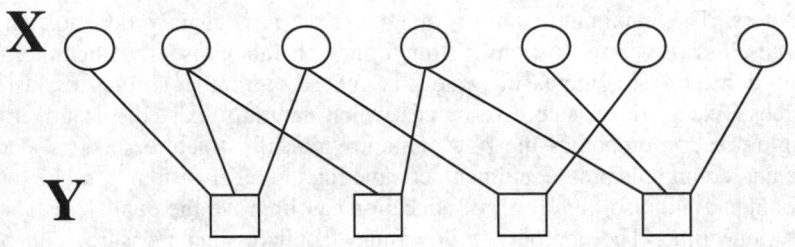

Fig. 3. A basic bipartite network

Basic on Zhang's model, we changed the network into a bipartite network $G =< X, E, Y >$ [as illustrated in Fig. 3.], where E is the set of edges. The upper are X nodes, and the lower are Y nodes. In the model of industrial cluster, enterprises and the cluster areas are two different sorts of participants in the dual-mode network. The entering of enterprises established their relationship. Define node X as set of enterprises and node Y as set of areas. Both are finite sets. According to BA model [3], we repeat this process in every time step that, add a new node to set X and then connect to a node in set Y under the rule of preferential attachment. So the rate of a new node connected to node $y_i \in Y$ is

$$\prod(k_i) = \frac{k_i}{\sum_j k_j} \tag{3}$$

where k_i is the degree of the old node y_i. However, as the influence of government, the co-agglomeration model cannot be exactly linear. In other words, the connection rate is not fully proportional to the degree of the node. More reasonable we use

$$\prod(k_i) = \frac{\alpha_i k_i}{\sum_j \alpha_j k_j} \tag{4}$$

where α_i is a parameter mainly decided by the government's intention of area i. Cluster's scale usually grows slower while it's getting close to maturity phase, at this time α_i will be smaller or even minus sometime, so that the government must try to take some active measure to raise the value of α_i.

Besides, at the beginning of industrial clustering, we cannot ignore the original attraction [4], because each area has its nature resources more or less. That is $k_i \neq 0$ or $\prod(0) \neq 0$. So Eq. 4. will be amended as

$$\prod(k_i) = \frac{(k_i + \beta_i)}{\sum_j k_j + \beta_j} \tag{5}$$

where β_i is original attraction depended on the nature resources of area i. β_i managed to make the distribution of industries more balanced to a certain extent. It conforms to the policy that narrowing the gap between east and west China as well as southern and northern Jiangsu, inspiring weak regional governments actively creating a proper environment for industrial clusters.

Based on the method above, finally we get the equality as this

$$\prod(k_i) = \frac{f(k_i)}{\sum_j f(k_j)} \tag{6}$$

where $f(k_i) = \alpha_i k_i + \beta_i$ is pre-set function.

3 Numerical Results

We downloaded the data-set from The Statistics Information Network of Jiangsu (www.jssb.gov.cn), range from 1999 to 2006.

Fig. 4 and 5 are the numbers of industrial enterprises in Wuxi and Suzhou. The curve in Fig.4 is more approach to the Logistic model. It has experienced the former three phases of cluster's lifecycle: formative phase (1999-2000), growth phase (2000-2004) and maturity phase (2004-2006). Clearly we can get $n_0 = 677$, $K \approx 3000$. In the phase of growth (2000-2004), the average growth rate is $r = 554.75$. As a result, the government of Wuxi may pay more attention to find ways to break through the ceiling of 3000 from now on. The experience of Suzhou can be divided into two different periods. The former period is from the year 1999 to 2004. It is a classic Logistic model, and its $n_0 = 610$, $K \approx 1500$, $r = 672$, but it spent only one year to reach the first maturity phase. However, after three years, Suzhou experienced another growth phase. From 2004 to 2005, its growth rate, we express it as r', also reached 627, and the maximum number of enterprises, we express it as K', get to a new high of 2100. It was the effect of Suzhou government's co-agglomeration.

During the year 2002 to 2004, the government was busy in completing the building of traffic, communication, and other basic equipment. Its favorable conditions and desirable policy attracted lots of large enterprises. In these years, Suzhou government did a good job.

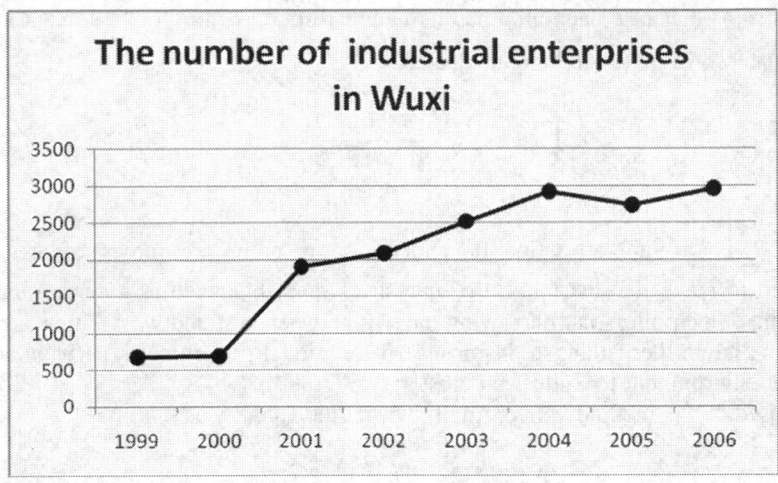

Fig. 4. The number of industrial enterprises in Wuxi

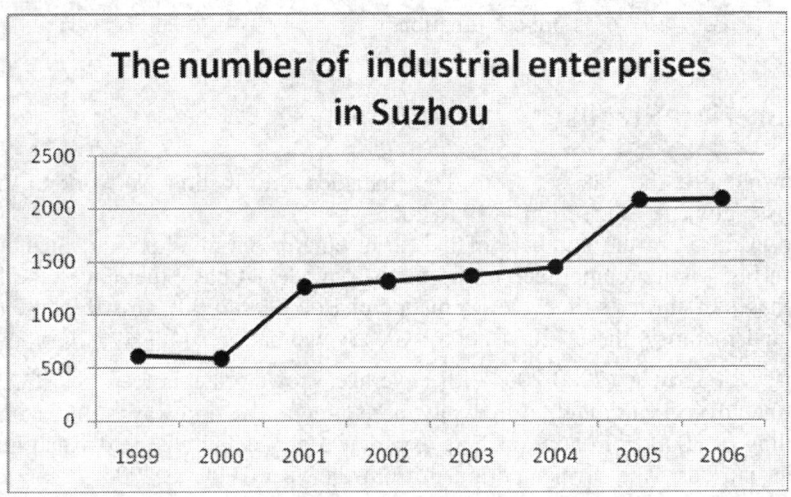

Fig. 5. The number of industrial enterprises in Suzhou

The following table shows the number of industrial enterprises in the 13 cities in Jiangsu province from 2000 to 2006.

Table 1. The number of industrial enterprises in the main 13 cities in Jiangsu province from 2000 to 2006

AREA \ YEAR	2000	2001	2002	2003	2004	2005	2006
Nanjing	878	1408	1739	1818	1862	2024	1605
Wuxi	697	1903	2090	2517	2916	2732	2963
Xuzhou	268	257	282	329	375	486	576
Changzhou	727	804	2003	2290	2560	2885	3421
Suzhou	584	1256	1311	1366	1444	2071	2084
Nantong	375	476	521	583	674	829	939
Lianyungang	309	270	260	248	261	269	270
Huaian	132	344	379	417	498	584	716
Yancheng	140	144	149	172	409	462	573
Yangzhou	236	517	535	586	663	750	793
Zhenjiang	230	234	408	459	456	518	595
Taizhou	178	188	207	232	302	378	414
Suqian	19	21	26	28	152	182	279

From 2000 to 2006, the total number of industrial enterprises in Jiangsu province was increasing. Several industrial clusters have been in forming. In general, from the aspect of the number of enterprises, Changzhou, Wuxi, Suzhou and Nanjing were the top 4 cities. [as illustrated in Fig. 6.] Nantong, Huaian, Yangzhou and Zhenjiang were in the second group, and Xuzhou, Lianyungang, Yancheng and Suqian were the third. This is the same as the three area of Jiangsu province.

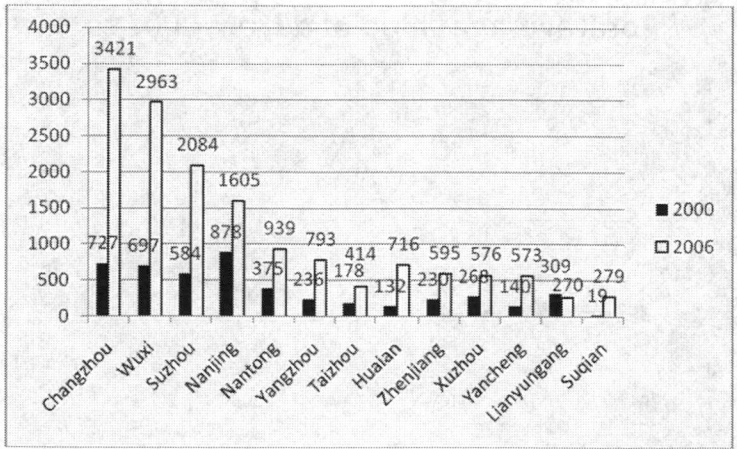

Fig. 6. The total number of industrial enterprises of the 13 cities in Jiangsu province

As the original attraction β_i does little to the co-aggregation model, we can ignore the parameter. Thus we get the average value of α in Fig. 7. We can get the conclution that in most cities, the average value of α is between 1 and 2, which means during these years, most clusters are still growing. Yancheng, especially Suqian, the value of α is larger than other cities, for the two clusters are in the beginning of growth phase. Although the αs of Nanjing and Lianyungang are both very close to 0, they are different. Nanjing is more possibly in the phase of maturity while Lianyungang is formative. Generally, the growing pace of the industrial clusters in Jiangsu province are complied with preferential attachment.

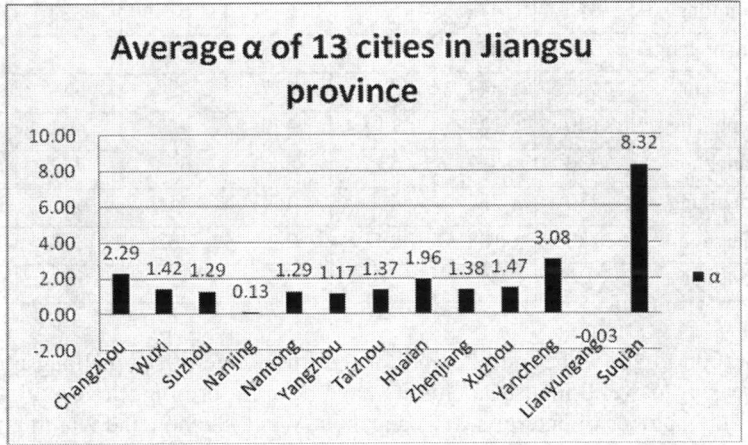

Fig. 7. Average α of the 13 cities in Jiangsu province

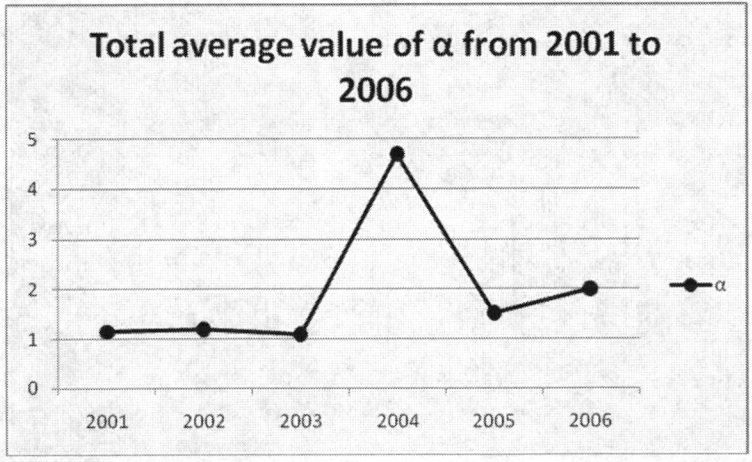

Fig. 8. This figure shows the total average value of α of the 13 cities from 2001 to 2006. Except in 2004, affected by Yancheng (11.77) and Suqian (37.82), the tendency of α is keeping on a rise.

Table 2. From the different average value of α, we can see that with the improvement of industrialization, the growth rate get slower. Pay attention to the total average. We removed the value of the year 2004, because of its erratic behavior.

Comparison of several average value of α	
Total average	1.3837
Average of 1st group	1.2821
Average of 2nd group	1.4512
Average of 3rd group	2.4203

4 Conclusion and Discussion

Bipartite network perfectly divided the nodes into two groups: the areas and the enterprises. The number of former is stable and the latter changes. As an enterprise enters a certain area, their relationship is established. Enlightened by BA mode and reference [5], we adopt the concept of preferential attachment. Parameter of co-agglomeration is the key problem in the co-agglomeration model. In this article we proposed the parameter α as a multiplicator, because the cluster may get smaller in some years, so that the parameter shall be minus which k^α cannot express. We also proved a single industrial cluster's increasing is similar to the Logistic model, and it may reveal another trend under the control of government, such as Suzhou is going through the second time of cluster's life cycle and Wuxi has a long growth phase.

As in most city, the value of α is normal, and the clusters in those cities are healthily and steadily growing under the control of government. If we have a more accurate data set, such as monthly data, or a longer time range data set, we can probably get a more accuracy result and can also analysize its distribution. For the total value of average α, it is better to set a weight to reduce the influence of less developed clusters.

Finally, as Michael Porter said, cluster policy must be an important part of state and local economic policy [2], so we cannot ignore the affect of government in the process of industrial clustering, from the angle of either qualitative or quantitative analysis. With the help of government's co-agglomeration, the industrial clusters can grow steady and healthy continuously.

Acknowledgments. This work is supported in part by the National Natural Science Funds (Grant No. 60874111), the National Scientific and Technological Special Program of Commonweal Industry (for Meteorology) (Grant No. 200806050), Jiangsu Province Philosophy and Social Sciences (Grant No. 08EYB018) and NUIST(Grant No. 20070084).

References

1. Vianconi, G., Barabasi, A.-L.: Competition and multiscaling in evolving networks. Europhys. Lett. 54, 436–442 (2001)
2. Porter, M.E.: Location, Competition and Economic Development: Local Clusters in a Global Economy. Economic Development Quarterly 14 (2000)

3. Barabasi, A.-L., Albert, R., Jeong, H.: Mean-field theory for scale-free random networks. Physica A 272, 173–187 (1999)
4. Jeong, H., Nda, Z., Barabasi, A.-L.: Measuring preferential attachment in evolving networks. Europhys. Lett. 61, 567–572 (2003)
5. Zhang, S.-y.: Self-Clustering, Attraction Kernel and Quantity of Clustering 自聚集、吸引核与聚集量. Complex Systems and Complexity Science 2(4), 84–92 (2005)
6. Mukkala, K.: Agglomeration Economies in the Finnish Manufacturing Sector. In: 43rd Congress of the European Regional Science Association, pp. 27–30. Jyväskylä, Finland (2003)
7. Gladarellia, G., Catanzaro, M.: The corporate boards networks. Physica A 338, 98 (2004)
8. Watts, D.J., Strogatz, S.H.: Collective dynamics of small-world networks. Nature 393, 440–442 (1998)
9. Barabasi, A.-L., Albert, R., Jeong, H.: The diameter of the world wide web. Nature (1999)

Analysing Weighted Networks: An Approach via Maximum Flows

Markus Brede[1] and Fabio Boschetti[2]

[1] Centre for Complex Systems Science, CSIRO Marine and Athmospheric Research,
GPO Box 284, Grace Canberra, ACT 2601, Australia
Markus.Brede@Csiro.au
[2] CSIRO Marine and Athmospheric Research, PB 5, Floreat, WA 6913

Abstract. We present an approach for analysing weighted networks based on maximum flows between nodes and generalize to weighted networks 'global' measures that are well-established for binary networks, such as pathlengths, component size or betweenness centrality. This leads to a generalization of the algorithm of Girvan and Newman for community identification. The application of the weighted network measures to two real-world example networks, the international trade network and the passenger flow network between EU member countries, demonstrates that further insights about the systems' architectures can be gained this way.

Keywords: Complex Networks, Weighted Networks.

1 Introduction

Describing systems of many interacting elements as networks has lead to considerable advances in the understanding of complex systems. Remarkably, such an analysis seems to indicate that systems from many different contexts, be it biology, ecology, sociology or even human-designed systems, share a plethora of common characteristics. Thus they can be roughly classified by the structure of their interaction topology. Common features are scale-free degree distributions, large degrees of clustering and modularity, small sizes and particular degree mixing patterns [1,2,3].

So far most of this analysis has treated networks as binary. However, in many applications such as transport networks [4,5] financial networks [6,7,8] collaboration networks [9], networks of metabolic fluxes [10] or gene regulatory and protein interaction networks [11,12] networks are weighted. Moreover, in some cases the distribution of link weights has been found to be very skewed, ranging over several orders of magnitude [4,7,10]. As an extreme example, imagine two pairs of cities, one coupled by a highway with a flow capacity of a thousand cars a day and the other one connected by a gravel road with a flow capacity of 10 cars a day. Should both connections be treated in the same way? Other studies indicate that strong and weak links have different roles in the systems' organization [13,14]. Naturally, this raises a number of questions: first, whether

J. Zhou (Ed.): Complex 2009, Part I, LNICST 4, pp. 1093–1104, 2009.

a binary representation of such systems is justified; second, whether more accurate information could be obtained by analyzing the weighted networks in its natural form; and third whether the general properties discussed above (degree distribution, clustering, etc.) are also found in weighted networks. In order to investigate these issues means of translating binary graph theoretical characteristics to weighted networks are needed.

The recent literature shows several efforts in this direction [4,5,11,15,16,19,20] which can broadly be grouped into three lines of research. In one approach, the 'Ensemble Approach' [5] link weights w_{ij} are mapped to probabilities $p_{ij} = F(w_{ij})$ for the presence of binary links between nodes and weighted network characteristics are defined as averages of binary quantities over the ensemble of networks defined by the connection probabilities p_{ij}. While this idea provides a nice general principle, there are two shortcomings. First, the definition of the map F is arbitrary; in [5] its definition requires the introduction of a lower cutoff probability whose setting is problematic due to the fact that weight distributions are often found to be very skewed with many very weak and few strong links. Small changes in this cutoff then determine the strength of the impact of all weak links —and hence the impact of a major fraction of the total link weights in the network— and can thus strongly influence the network measures. Second, network measures will typically have to be calculated by Monte Carlo simulations, such that the analysis of large networks can be very time-consuming and error-prone, if weak links are located in critical positions.

In the other line of research, differently motivated local weighted measures, mainly vertex strength, clustering coefficients or weighted nearest neighbour degrees, have been introduced [4,11,15,16]. From this approach there does not appear to be an obvious generalization of non-local measures, such as the size of connected components or distances.

Third, Ref. [17] introduces a mapping of weighted networks to multigraphs. On the basis that link weights represent capacities this allows for the generalization of global measures, notably the betweenness centrality in [17]. In this work we follow the spirit of [17] and elaborate global network measures based on the notion that weights in the network represent capacities or maximum flows. Focusing on 'global' network characteristics, we elaborate a number of measures, particularly distances, betweenness centralities and the definition of component sizes. Also, the introduction of a weighted betweenness centrality allows for a refinement of the cluster partitioning algorithm of Girvan and Newman [22].

2   Weighted Global Network Measures from Flow Principles

In this section we develop a set of weighted network characteristics and demonstrate their usefulness to understand the structure of two example networks, the trade flow network in 2000 and the passenger flow network between EU countries in 2004. For this, our guiding idea is the analysis of maximum flows along links. For both examples flows have a natural meaning in terms of the underlying system, i.e. bilateral trade flows for the trade network and passenger flows in the passenger network. Both are undirected networks, but the generalization of the network measures we propose to directed networks is straightforward.

Let us consider a weighted network of N nodes as given by a matrix $\{w_{ij}\}_{i,j=1}^{N}$, where an entry w_{ij} gives the weight of the link between the nodes i and j. Since we interpret link weights as capacities it is reasonable to assume $w_{ij} > 0 \forall i, j$. The binary representation of the matrix W is given by the adjacency matrix B, where $b_{ij} = H(w_{ij})$, $i, j = 1, ..., N$, where $H(x) = 1$ if $x > 0$ and $H(x) = 0$ otherwise In the following, measures for standard binary networks refer to the matrix $\{b_{ij}\}$, whereas measures for weighted network refer to the matrix $\{w_{ij}\}$. In this terminology the number of links is given by $L = \sum_{i<j} b_{ij}$ and the (total) link weight by $S = \sum_{i<j} w_{ij}$. The average weight of a connection is obtained as $\overline{w} = S/L$.

Often a better understanding of the pecularities of a given network becomes possible by comparing it to a suitable null model. For the case of weighted networks, such a model is given by the ensemble of weight-randomized networks, i.e. networks with the same binary links and the same link weights, but an uncorrelated arrangement of the latter. In the following, results will be compared to averages over the ensemble of weight-randomized networks.

2.1 Component Size

In binary networks, cluster or component sizes give the sizes of maximum sets of connected nodes. Since in a densely connected weighted network all nodes can typically be reached from each other, a definition equivalent to the one for binary networks would not provide much information. However, even though all nodes can be reached from each other, the flow that can pass through the paths connecting them may be different. Thus, a sensible definition for the strength of the connection from a node i to a node j is the amount of flow $F_{\max}(i, j)$ that can simultaneously be passed along all links from i to j. Consequently, the average pairwise flow

$$F = \frac{1}{S} \sum_{i<j} F_{\max}(i, j) \tag{1}$$

provides a measure for the overall transport capacity between all nodes in the network, a measure roughly corresponding to component sizes in binary networks. Measuring F in units of the average link weight in (1) then allows for a better comparison between networks of different total link weight and generates a dimensionless measure. One should note that, despite (1) is not the exact equivalent of the cluster size for binary networks, the general concept of cluster size is respected, since (1) corresponds to the average number of independent path between nodes.

2.2 Pathlength

In binary networks pathlengths $l(i, j)$ are measured as the minimum number of edges that need to be traversed to establish a path between nodes

$$l(i,j) = \min_{P(i,j)} L(P(i,j)), \tag{2}$$

where $P(i,j) = (p_1 = i, p_2, ..., p_l = j)$ denotes a path from i to j and $L((p_1 = i, p_2, ..., p_l = j)) = l - 1$ its length. As in the previous section, such a quantity does not carry much information about the system structure in almost fully connected weighted systems such as the ITN or PFN, since one has $L = 1$ for almost every pair of nodes. A more sensible measure for pathlength must take into account the differences between strong and weak links.

From the flow perspective, we obtain a natural generalization of the discrete pathlength via the introduction of transport rates. To elaborate this concept, we need to introduce weighted paths as sequences of nodes and link weights traversed, i.e. $P_w(i,j) = ((p_1 = i, p_2, ..., p_l = j), (v_{p_1,p_2}, ..., v_{p_{l-1},p_l}))$ with $v_{p_i,p_{i+1}} \leq w_{p_i,p_{i+1}}$ is a weighted path composed of all or parts of the edges $w_{p_1,p_2}, w_{p_2,p_3}, ...,$ etc. Given such a weighted path $P_w(i,j)$ from i to j, we define its capacity t as $t(P_w(i,j)) = \min_{1 \leq k < l} v_{p_k,p_{k+1}}$ and, assuming that the transport along each link takes one unit of time, we define the transport rate as $r(P_w(i,j)) = t(P_w(i,j))/L(P_w(i,j))$, where $L(P_w(i,j)) = l - 1$ is the path's discrete length. A direct translation of the discrete pathlength for binary networks to weighted networks is thus obtained from the maximum transport rate $t_{\max}(i,j) = \max_{P_w(i,j)} r(P_w(i,j))$ via

$$l'_w(i,j) = \overline{w}/t_{\max}(i,j). \tag{3}$$

The definition (3) reduces to (2) for unweighted binary networks. However, a major caveat of (3) is that the computation of $t_{\max}$ requires the evaluation of all paths between two nodes, which is computationally demanding and impracticable for large networks.

A way around this difficulty is to consider the maximum average simultaneous transport rate between nodes, i.e. the average transport rates when the maximum possible amount of flow is carried between the nodes in the optimal way. To elaborate this concept, let us define that two weighted paths $P_w(i,j) = ((p_1 = i, p_2, ..., p_l = j), (x_{p_1,p_2}, ..., x_{p_{l-1}p_l}))$ and $Q_w(i,j) = ((q_1 = i, q_2, ..., q_l = j), (y_{q_1,q_2}, ..., y_{q_{l-1}q_l}))$ are independant, if any (allowed) flow through a part of one does not impede any flow through any part of the other. This is the case if $x_{mn} + y_{mn} \leq w_{mn}$ for all links m, n that occur in P_w and Q_w. A set of independant paths is a set of paths that are pairwise independant. Again, this definition is the natural equivalent of the definition of path independance in binary networks.

Then, the optimum maximum transport from i to j can be constructed in the following fashion:

1. The set of all considered paths $S(i,j)$ is empty.
2. Find the shortest (in the discrete sense) path $P_w(i,j)$ from i to j that is independent from all paths already in $S(i,j)$.
3. If no path fulfilling the requirements in 1. is found the algorithm terminates. Otherwise, add P to S and continue with 1.

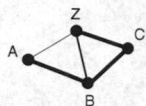

Fig. 1. This example illustrates the concept of weighted distances introduced in (5). Thin lines have weight 1, intermediate lines weight 2 and thick lines weight 3; thus the average link weight is $\overline{w} = 12/5$. Consider the distance between nodes A and Z; the maximum flow is $F_{\max}(A, Z) = 4$. The optimum maximum transport is obtained by transporting one unit directly via AZ, two units via ABZ and one unit via $ABCZ$. The corresponding independant paths are $P_1 = ((A, Z), (1))$, $P_2 = ((A, B, Z), (2, 2))$, and $P_3 = ((A, B, C, Z), (1, 1, 1))$, with the respective transport rates $1/1$, $2/2$ and $1/3$, i.e. one has $r_{\mathrm{avg.}}(A, Z) = 5/6$ and $l_w(A, Z) = 72/25$.

The sum $\sum_{P_w \in S(i,j)} t(P_w)$ gives the maximum simultaneous flow $F_{\max}(i, j)$ between i and j introduced in subsection 2.1. Thus,

$$r_{\mathrm{avg}}(i, j) = 1/F_{\max}(i, j) \sum_{P_w \in S(i,j)} t(P_w) r(P_w) \tag{4}$$

is the weighted optimal average simultaneous transport rate between i and j. A measure for weighted pathlengths can be obtained from Eq. (4) by taking the inverse of the average transport rate measured in units of the average link weight $\overline{w}$

$$l_w(i, j) = \overline{w}/r_{\mathrm{avg}}(i, j). \tag{5}$$

The concept is illustrated with an example in Fig. 1. An overall measure for distances between nodes in the weighted network is obtained by taking the average of (5) over all pairs of nodes.

Our definition of weighted pathlengths takes into account the discrete lengths of paths between nodes as well as their capacity. It is a measure of how much can be transported between nodes and how fast. However, similar to the definition of a component size in subsection 2.1, this also does not directly correspond to the unweighted measure for a pathlength in binary networks. Instead, definition (5) measures the average length of paths in the set of shortest independant paths between two nodes.

Eq. (4) can also be used as the basis for a definition of a weighted diameter. A sensible measure for the longest weighted path is $d_w = \max_{i \neq j} l_w(i, j)$.

To illustrate the concept we have measured the average weighted distances for the ITN and the PFN. In the ITN one finds $l_w = 1.47$ and in the PFN $l_w = 1.58$, compared to $l_w = 1.35$ and $l_w = 1.44$ for the null models, i.e. both networks are slightly larger than expected in a random link arrangement. For diameters one finds $d_w = 27920$ (between Andorra and Nassau) and $d_w = 45.3$ (Slovenia and Slovakia) for the ITN and PFN, respectively. Average diameters for the randomized networks are $d_w = 42400$ and $d_w = 12.3$, respectively. Thus, even though the average distance in the ITN is larger than expected, the diameter is considerably smaller than expected, i.e. for the ITN average pairwise

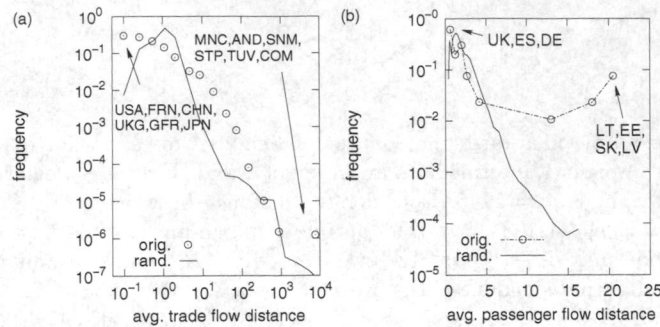

Fig. 2. Distribution of averages distances in the (a) ITN and (b) PFN. In the ITN the major industrialized countries and China have small average distances while small countries have large average distances. Andorra (AND) with $l_w = 11700$ has the largest average distance in the network. In the PFN the United Kingdom (UK), Germany (DE) and Spain (ES) have the smallest average distances while Listhuania (LT), Estonia (EE), Latvia (LV) and Slovakia (SK) have the largest.

distances are more 'concentrated' around the mean distance than in the respective randomized reference case. The opposite is found for the PFN, for which the distribution of distances is bimodal, see below.

Further information about both networks can be gained by analysing the distribution of average weighted pathlengths $\overline{l_w(i)} = 1/(N-1) \sum_j l_w(i,j)$ which are displayed in Figure 2. For both networks these distributions are skewed, exhibiting pronounced differences between countries in the centre (short average weighted distances) and at the periphery (large average distances). For instance, the centre of the ITN is composed of a core comprising the US, Canada, Mexico, Germany, Japan, the UK, France and China while very small countries appear on the right tail. The distribution of average distances is more skewed than expected for random link arrangements, for which a peaked distribution of distances with most countries having similar average distances is expected, cf. Fig. 2a.

Even more in the PFN, a bimodal distribution of distances, clearly separating a 'core' from a 'periphery' is found (Fig. 2b). The UK, Germany and Spain form this clearly distinct core, while Slovakia and some Baltic countries are found at the periphery of the network. In contrast, in a random link arrangement a strongly peaked distribution of distances with only a few countries at the periphery would be expected.

2.3   Betweenness Centrality

In binary networks the betweenness centrality is a measure for the traffic through nodes (or links). It is usually defined as the number of shortest paths passing through a node (or link), i.e.

$$b(i) = \sum_{k,l} \delta(i,k,l) \tag{6}$$

with $\delta(i, k, l) = 1$ if a shortest path between the nodes k and l passes through the node i and $\delta(i, k, l) = 0$ otherwise. Again, as in the previous cases of discrete cluster size and pathlength, $b(i) \approx$ const. in almost fully connected weighted networks. That is, the discrete definition of betweenness centrality does not carry much information about densely connected weighted networks.

However, using the definition of a weighted pathlength in the previous subsection, a possible generalization of the discrete definition (6) to the weighted case becomes obvious. Let

$$b_w(i) = \sum_{k,l} \delta_w(i, k, l), \tag{7}$$

where $\delta_w(i, k, l)$ is the sum of the capacities of all paths through i used in constructing the optimum average simultaneous transport rate as in subsection 2.2. The definition (7) is thus an approximate measure for how much flow passes through nodes. The equivalent definition for a link centrality is obvious.

2.4   Vulnerability

One question about many distributed systems that is of considerable importance is that of vulnerability to random failure or targetted attack. The issue has been extensively studied for binary networks and recently also for weighted networks [18]. In Ref. [18] the vulnerability of the airtraffic network to random and targetted node removal is investigated. The authors investigate measures for structural damage to the network when nodes are removed according to different attack strategies. Different measures of structural system integrity are introduced, of which the most relevant one for the present study is the measure that computes the total link weight in the giant component. A major finding is that the topological structure of the binary network is a poor indicator of damage — the network can still be largely connected while node removals have already significantly impaired the system structure as captured by indicators that take account of the weighted architecture.

Though important, this result is not surprising: removing nodes based on a weighted centrality ranking will typically remove nodes of largest strength first, thus also removing larger amounts of link weight than when removing randomly chosen nodes. Thus naturally the amount of link weight remaining in the system decays faster than for random node removal. On the other hand, weak links tend to remain in the network and guarantee connectedness as long as many enough of them remain.

In the following, we analyse the integrity of the ITN subject to weighted link removal. This is equivalent to measuring system function when the overall trade volume is systematically reduced. For the analysis we use the measure of average pairwise maximum flows F of subsection 2.1 as an indicator of overall system integrity. Similar to [18] the fraction of nodes in the giant component N captures the system's topological integrity. More precisely, the ratios F_g/F_0 and N_g/N_0 are recorded. The indices indicate the measure when a fraction g of the total link weight has been removed.

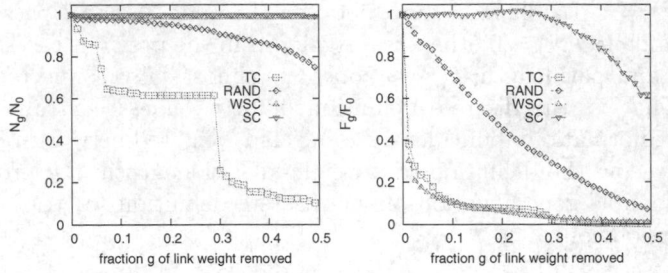

Fig. 3. Vulnerability of the ITN to random and targetted link removal. (a) fraction N_g/N_0 of the giant component remaining and (b) fraction of the maximum average pairwise flow (1) F_g/F remaining after the removal of a fraction g of the total link weight in the system. The labels indicate different removal strategies, TC removal based on ranking according to Eq. (7), RAND random link removal, WSC removal based on ranking according to the link centrality defined in [20] and SC removal according to standard unweighted centrality based ranking.

We consider the following strategies for gradual link removal, which, for practical reasons, is conducted in steps of $w_{\text{step}} = S/500$. For every step the removal procedure is repeated till an amount of w_{step} of total link weight has been removed. The ranking measures are recalculated after every removal step.

1. Random link removal (RAND).
2. Removal based on the centrality defined in Eq. (7) (TC).
3. Removal based on the weighted centrality defined in [20] (WSC).
4. Removal based on the standard link centrality (SC), that is purely based on the networks' topological structure.

Figures 3a and 3b compare simulation results for the topological and overall integrity of the system for various link removal strategies. One notes that for random and weight-based removal strategies the topological system integrity is largely unimpaired over a large parameter range whereas weighted indicators of system function already record major damage.

The data of figure 3b show a clear ranking of the attack strategies. Targetted removal according to an unweighted system indicator proves even less effective than random link removal. This observation underlines the importance of taking into account the arrangements of link weights to analyse the system. Most effective among the considered strategies are link removals based on the weighted centralities, where the weighted centrality measure introduced in [20] is slightly more effective for small amounts of removed weight, whereas a strategy based on the measure introduces in subsection 2.3 has larger impacts for large amounts of removed link weight. It is interesting to note that both strategies inflict comparable amounts of damage to the system in completely different ways. Note that for SC the topological integrity is completely unimpaired, wheras for TC considerable parts of the system are split off even when only relatively little link weight is removed, cf.Fig. 3a,b. Thus link removal according to SC impairs system function by thinning out one all-comprising giant component, whereas TC operates by gradually removing strategically placed links towards disconnecting the network.

Another observation appears particularly important: targetted removal of less than 3% of total trade volume can reduce the overall integrity of the system by around two thirds! Thus, even though the system is almost completely connected in a binary sense it can be classified as extremely fragile for targetted link removal.

2.5   Community Analysis and Modularity

The definition of centrality in the previous section can be used to generalize the algorithm of [22] to detect community structures in weighted networks. The algorithm of [22] dissects the network by sequentially removing the links with the largest centrality, thus gradually disconnecting the network. The basic idea behind the procedure is that more traffic flows along links connecting seperate communities, i.e. bridging links, than via those between nodes of the same community. In Ref. [20] a generalization of this procedure to weighted networks has already been suggested, thereby the traffic through a link (which is estimated from shortest paths on the binary network) is divided by the link weight. Then, links with the largest traffic per link weight ratios are removed. This procedure, however, does not account that weak links can become congested, consequently diverting a considerable fraction of the flow to strong links.

Our definition of centrality in (7) overcomes this problem. In more detail, to dissect weighted networks into communities we proceed as follows:

1. As long as the number of components is not equal to network size, calculate the weighted betweenness centrality of every link.
2. Remove the most 'overused' link, that is, the one with the largest weighted centrality over link weight ratio. Proceed with 1.

Figures 4 and 5 show the full dendrograms from the dissection of the ITN and the PFN into communities. The significance of a community division can

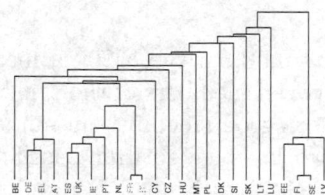

Fig. 4. Dendrogram from the dissection of the PFN into communities. Distances from the top represent the link weight removed to disconnect the network, i.e. highest branches represent early dissections and consequently most strongly disconnected communities. The most significant community division is the following split (1) 'central Europe': DE, EL, ES, FR, IT, AT, IE, NL, UK, PT (red), (2) Scandinavia and the Baltics: EE, FI, LV, SE (blue) and (3) the rest comprising 9 countries, each country as an isolated community (black). For this division the modularity is $Q = .57$.

Fig. 5. Dendrogram from the dissection of the ITN into communities. The most significant community split distinguishes (1) a large clique of 50 mostly South and North American countries, but also some Asian countries and the UK, which is centred around the US, (2) a group of 29 (mostly) European countries, (3) a group of former Soviet Republics, (4) a multitude of very small groups (mostly with just one member) of other countries. The modularity for this split is $Q = .629$.

be classified according to its modularity Q [20], a measure for the ratio of the difference of link weight between members of the same groups and the expected fraction of link weight between group members in random arrangements and the total link weight in the network. The algorithm based on the traffic centrality of Eq. (7) yields $Q = .63$ whereas algorithms based on the standard centrality or weighted centrality of [20] only yield poorer community divisions with $Q = .182$ and $Q = .273$, respectively.

In the case of the PFN one notes that the most significant group division distinguishes three main sets of countries: (1) a 'core Europe' including 10 countries, (2) a block of four Scandinavian and Baltic countries, and (3) a set of nine countries, each defining a group of its own. Whereas the 10 countries of (1) form a clearly defined core containing almost all link weight, the set of countries summarized in (3) and to some extent the Scandinavian-Baltic cluster (2) form a clearly marked periphery. Nodes in this periphery have few connections between them and are also only loosely coupled to the network's core. The core itself can again be subdivided into three major groupings, one comprising Germany, Austria and Greece, a second containing Spain, the UK, Ireland, Portugal and The Netherlands and a third group made up by Italy and France.

For the ITN the most significant split distinguishes mainly two large trading blocks, one centred around the US and the other one comprising most European countries. Further, a clique of former Soviet Republics and several small cliques, mostly composed of only two or three countries are found. Both, the US centred and the European cliques are not homogeneous and can be further subdivided. For instance, the US clique contains a South American subgroup (comprising Brazil, Argentina and Chile), and several Asian subgroups (for instance China-South Korea or Malaysia, Signgapore and Vietnam). The bulk of Europe is centred around the Germany,France,Italy and Switzerland group, but also The Netherlands and Belgium and Sweden and Norway or Spain and Portugal form distinct subgroups.

3   Summary and Conclusions

By interpreting a network in terms of a transport system and introducing the concept of maximum flows between nodes, we proposed a natural extension of 'global' network measures traditionally used to study and characterise binary networks. Some, like vertex strength and clustering coefficient, can be seen as straightforward generalisations from binary to weighted networks. Others, such as cluster size, path length, component sizes and betweenness centrality, provide information which would not be available, or would be scarcely significant, if applied to the binary discretisation of a weighted network.

The measures we introduced allow to study a system accounting not only for the presence of links between constituents, but also for the strength of their interaction. We demonstrated the approach by analysing two real world networks: the International Trade Network and the Network of Passenger Flows between EU member countries in 2004. The two example networks have different size, and are characterized by link weights which span over several orders of magnitude with very skewed link weight distributions. The measures we discussed allowed to discriminate between these networks and reference 'null' models and to automatically detect several known features of the real systems.

References

1. Albert, R., Barabási, A.-L.: Rev. Mod. Phys. 74, 47 (2002)
2. Newman, M.E.J.: SIAM Rev. 45(2), 167 (2003)
3. Bocaletti, S., Latora, V., Moreno, Y., Chavez, M., Hwang, D.-U.: Phys. Rep. 424, 175 (2006)
4. Barrat, A., Barthélemy, M., Pastor-Sarras, R., Vespignani, A.: Proc. Natl. Sci. U.S.A. 101, 3747 (2004)

5. Ahnert, S.E., Garlaschelli, D., Fink, T.M.A., Caldarelli, G.: Phys. Rev. E 76, 016101 (2007)

6. Onella, J.-P., Chakraborti, A., Kaski, K., Kertész, J., Kanto, A.: Phys. Rev. E 68, 056110 (2003)

7. Soromäki, K., Blech, M.L., Arnold, J., Glass, R.J., Beyeler, W.E.: Physica A 379, 317 (2007)

8. The data for the international trade network can be downloaded from, http://weber.ucsd.edu/~kgledits/exptradegdp.html We follow the procedure in [16] and reduce inconsistencies in the dataset by defining the link strength between countries as an average of reported imports/exports from the perspective of both countries

9. Newman, M.E.J.: Phys. Rev. E 64, 016131 (2001)

10. Almaas, A., et al.: Nature 427, 839x (2004)

11. Zhang, B., Horvath, S.: Stat. Appl. Gen. Mol. Bio. 4, Article 17 (2005)

12. Salwinski, L., et al.: Nucleic Acids Res. 32, D449 (2004)

13. Csermely, P.: Trends in Biochemical Sciences 29(7), 331 (2004)

14. Granovetter, M.: Am. J. Sociol. 73, 1360 (1973)

15. Onella, J.-P., Saramäki, J., Kertész, J., Kaski, K.: Phys. Rev. E 71, 065103R (2005)

16. Saramäki, J., Kivelä, M., Onella, J.-P., Kaski, K., Kertész, J.: e-print arXiv:cond-mat/0608670 (2006)

17. Newman, M.E.J.: Phys. Rev. E 70, 056131 (2004)

18. Dall' Asta, L., Barrat, A., Barthélmy, M., Vespignani, A.: J. Stat. Mech., P04006 (2006)

19. Holme, P., Park, S.M., Kim, B.J., Edling, C.R.: Physica A 373, 821 (2007)

20. Newman, M.E.J.: Phys. Rev. E 70, 056131 (2004)

21. Watts, D.J., Strogatz, S.H.: Nature 393, 440 (1998)

22. Girvan, M., Newman, M.E.J.: Proc. Natl. Sci. U.S.A. 99, 7821 (2002)

23. The data can be downloaded from, http://epp.eurostat.cec.eu.int

24. Newman, M.E.J.: Phys. Rev. Lett. 89, 208701 (2002)

25. Serrano, M.A., et al.: e-print ArXiv:cond-mat:0511035 (2005)

26. May, R.M., Levin, S.A., Sugihara, G.: Nature 451, 893 (2008)

An Emergence Principle for Complex Systems

Michel Cotsaftis

ECE, 37 Quai de Grenelle
mcot@ece.fr

"Beware, you who seek first and final principles, for you are trampling the
Garden of an angry God and He awaits you just beyond the last theorem"
Sister Miriam Godwinson

Abstract. From elementary system graph representation, systems are shown to
belong to only three states: simple, complicated, and complex. First two have
been studied over past centuries. Last one originates in existence of threshold
above which components interaction overtakes outside interaction, leading to
system self-organization which filters outer action, making it more robust with
emergence of new behaviour not predictable from components study. The
threshold value, expressed in terms of coupling system parameters, is verified to
recovers limits found in a broad range of domains in Physics and Mathematics,
giving explicit criterion for emergence in complex system. Application to man-
made systems concentrates on the balance between relative system isolation
when becoming complex and delegation of more "intelligence" in adequate
frame between new augmented system state and supervising operator. Entering
complexity state opens the possibility for the function to feedback onto the
structure, ie to mimic technically the early invention of Nature.

Keywords: Complex Systems, Emergence Criterion.

1 Introduction

Accumulation of recent observations on natural phenomena with high technical
development boosting the access to a wide range of new parameter indicates without
doubt the existence of phenomena not following main stream laws established from
patient analysis of natural phenomena over millennium long previous period. These
"classical" laws are concerning phenomena "reasonably" isolated over broad range in
size from galaxy to atoms when including quantum and relativistic improvements. In
the mean time, technology advance and observation accuracy drove the attention on
more complicated systems with always larger number of elements, for which previous
laws are not sufficient to represent correctly enough their behaviour. Such systems are
forming a new huge class in all scientific and technical human activities, and have
reached their own status by the corner of the millennium under the name of
"complex" systems. There is today a strong questioning about their origin and their
formation [1]. This has been addressed in a very pedestrian approach [2] based on

J. Zhou (Ed.): Complex 2009, Part I, LNICST 4, pp. 1105–1117, 2009.
© ICST Institute for Computer Sciences, Social Informatics and Telecommunications Engineering 2009

elementary source-sink model applied to the graph representing the aggregate of system components, showing that system structure falls into three different groups, simple, complicated and complex, with specific and explicit properties. The first two groups are usual ones approachable by the methods of scientific reductionism [3]. The third group, by its very global nature, is not reducible to the effect of its components [4], and requires some adjustment for being correctly handled, because now a key point is the way the system behaves under (or against) environment action. The system mainly self-organizes and develops a global reaction hiding details of specific component effect. A consequence is that the mechanistic notion of individual component "trajectory" pertaining to first two groups looses its meaning and should be replaced by more general "manifold" entity corresponding to accessible "invariants" under environment action. So there is emergence of new natural properties which will be discussed depending of complexity grade and which can be related to well known classes of observed phenomena. Within the pedestrian graph approach an explicit criterion for passing to complex state and to have emergence of new behaviour is given in terms of system coupling parameters and is recovering all previous expressions in Physics and in Mathematics. Advantages in application of complex structure to artificial man made systems are stressed.

2 System State Analysis

Let the graph with N nodes N_i representing a system with a finite number N of identified and separable components. There exists three types of vertices in between the N components i and outside sources e whenever an exchange exists between them.. System dynamics result from a combination of previous three different exchanges to which three characteristic fluxes can be associated for each system component the nature of which (power, information, chemical,..) unambiguously characterize system components status. First flux corresponds to "free" dynamics of i-th component p_{ii} along vertex V_{ii}, second one $p_{i,e}$ to transfer flux between outer source and system i-th component along vertex V_{ie}, and last one to inter components effects $p_{i,int} = \mathrm{Inf}_j\{p_{i,j}\}$, with $p_{i,j}$ the characteristic and oriented flux exchange between components i and j along vertex V_{ij}, see Fig. 1. For weak coupling $p_{ii} \gg p_{i,int}$ (case (**A**)), system dynamics are reducible to a set of almost independent one-component sub-systems, and system will be termed as a *simple* one. For strong outer coupling $p_{i,e}$ $\gg p_{i,int}, p_{ii}$ (case (**B**)), system dynamics can be decoupled (as other components action creates a weak coupling between them), at least locally, into a set of sub-systems controlled by as many exterior sources as there are components in the system because they can still be identified. The system can be termed as *complicated*. Finally for strong internal coupling $p_{i,int} \gg p_{ie}, p_{ii}$ (case(**C**)),system dynamics are now determined by components interaction satisfying the inequality, with fundamentally different outside action compared to case (**B**). Here internal interactions dominate and shield input tracking from outer source to component. So control action can only be a "global" one from other system components satisfying condition for case (**B**), so

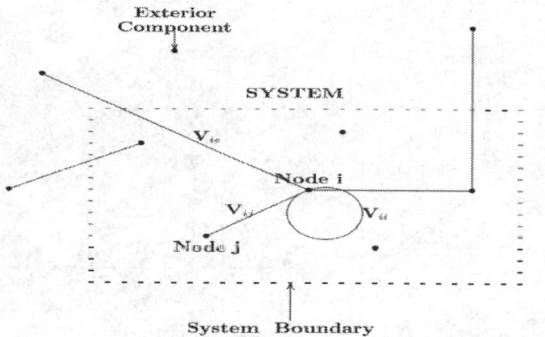

Fig. 1. Graph representation of system with its three exclusive types of vertices V_{ii}, V_{ie}, and V_{ij}

system dynamics are now also driven by internal action. For all system components passing in third state their control cannot as in case (**B**) be fixed only by outer source action because of stronger interaction effect dominating the dynamics of concerned components, and a *self-organization* takes place inside the system leading to an internal control replacing classical one from outside. So manipulating inputs with as many dof as in initial system is no longer possible because of the conflict with internal control. External system control dimension is thus *reduced*. The system will be termed as *complex* (from Latin *cum plexus* : tied up with). A very elementary test for determining if a system is passing to complex state is thus to verify that its control requires manipulation of less dof than in initial system with weaker internal interactions. So paradoxically most complex possible structure corresponds to totally autarchic system, which joins simple system definition by isolation. In fact this apparently contradictory statement is resulting from the very nature of internal interactions effect which reduces the number of invariants on which system trajectory takes place. Example is neutral gas particles for which their initial *6N* positions and velocities (the mechanical invariants of motion) are reduced to the only energy invariant (or temperature), justifying thermodynamic representation. Consequently, when increasing internal interactions, contrary to a complicated system which remains complicated from either side, a system is the less complicated seen from outside as it is more complex internally, on top of being less sensitive to outer action, a very useful property used very early by Nature and at the origin of Her evolution. In summary, exactly like there exists three states of matter (solid, liquid, gas), there are three states (simple, complicated, complex) for each system component. On a 3-d space, plotting the three values $\{p_{ii}, p_{i,e}, p_{i,int}\}$ for each system component gives a cluster of N points C_i the status of which is determined by their location with respect to boundaries of inequalities (**A**),(**B**),(**C**), see Fig. 2. Moreover, consequence of parameter variation can be analyzed, especially when crossing inequality (**C**), extremely important for control of man-made systems as this boundary is nowadays very often passed over with modern technology advance.

An immense literature exists about complexity, its definition and its properties covering an extremely wide range of domains from Philosophy to Technology [5], especially in recent years where its role has been "discovered" in many different

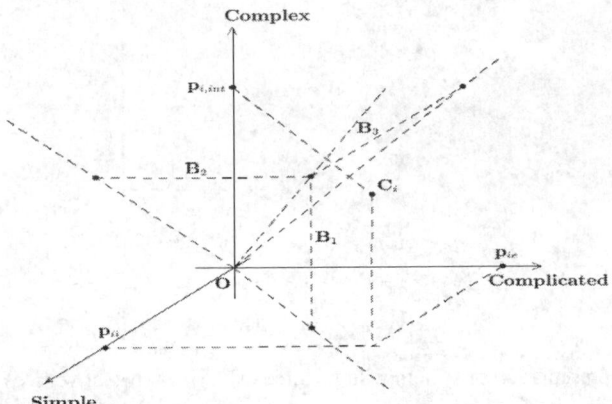

Fig. 2. System representation in [Simple-Complicated-Complex] state-space domain
- Inequality (**A**) is satisfied in tetrahedral domain (O,B_1,B_2)
- Inequality (**B**) is satisfied in tetrahedral domain (O,B_1,B_3)
- Inequality (**C**) is satisfied in tetrahedral domain (O,B_2,B_3)

fields such as networks now playing a crucial role with the ascent of Information Technologies. In the sequel emphasis is more modestly put only on more restricted complex state compared to complicated state as concerns action from outside environment (ie from control point of view) onto the system. It is intended to show from the very elementary picture above that most of important results in a broad range of domains from Mathematics to Physics can be easily recovered and that still unclear emergence phenomenon finds here a natural explanation.

3 Emergence

First the deep difference between the first two states and last complex one is in the possibility for the formers to split the system into as many independent one-component systems in a first approximation, which is impossible in the later where all interacting components have to be taken as a whole. In mathematical terms the consequence is that usual approximation methods developed for the first two states do not straightforwardly apply and have to be revised in order to handle the global aspect of the coordinated response of components in complex state, at the origin of important computer research on the problem. In specific situations, other elements also enter the description and it is interesting from description above to recover various situations observed and analytically studied for different parameters values. In general system state exhibits a mixed structure where some components are in one state and others in another one. Important examples are (weakly) inhomogeneous and continuous natural systems such as fluids with non zero gradients in a domain. Here, fluctuations are universally observed in a very large range of frequencies (roughly because the system has a very large number of components) the source of which is the free energy available in between this stationary equilibrium and complete (homogeneous) thermodynamic one, ie from the space gradients related to medium non homogeneity.

Splitting fluctuations into two groups depending on their wavelength compared to system characteristic gradient length, large wavelength ones excited in the medium are in complicated state and, because they are sensitive to boundary conditions, can be observed and possibly acted upon as such from outside, whereas fluctuations with small wavelength are in complex state and can only be considered globally. Under their strong interactions, and because they are much less sensitive to boundary conditions, they are loosing their phase and globally excite an outflux (usually called a *transport*) expressing non equilibrium system situation counteracting input flux responsible of medium non homogeneity. Clearly emergence of this transport is a direct consequence of components self interaction when crossing condition (**C**) which now takes place on a manifold in small wavelength component state space and constraints their motion to take place on it. So transports determination is an important element qualifying system behaviour and is an active research problem studied worldwide. This feature is observed for all natural (open) systems on the *dissipative* branch [6] exchanging (particles, chemical, energy) fluxes with outside environment. Evidently the channels by which internal energy sources are related to these fluxes are playing a privileged role because they regulate the influx ultimately available for the system and finally determine its *self-organized* state[7]. Dissipative systems only exist to the expense of these fluxes, and they evolve with parameter change – such as power input – along a set of neighbouring states determined by branching due to bifurcations where internal structure changes in compatibility with boundary conditions and by following the principle of largest stability. So the picture is a transport system governing flux exchanges guided by the bifurcation system which, as a pilot, fixes the structure along which these exchanges are taking place. Finding the branching pattern thus entirely defines possible system states and determines fluctuation spectrum. Branching is found as nontrivial solutions of variation equations deduced from general system dynamics equations. Despite clear identification of phenomena physical origin, their analysis is still in progress in many situations [8,9], and cleared up for fluctuations in deformable solid bodies[10]. Moreover, modification of system dynamics is the more important as non homogeneity gradient is steeper, with extreme case of living cell systems completely encapsulating within a filtering membrane (ie a steep gradient) a space domain where very specific "memory" DNA molecules are fixing the dynamics of inside system they control, with corresponding exchange across the membrane.

However, aside dissipative pattern followed by natural systems exhibiting components with relatively elementary features (charge, mass, geometrical structure, chemical activity, wavelength, frequency..), there exists cases where complex state occurs in systems with more sophisticated components, usually called "*agents*". Examples are herds of animals, insect colonies, living cell behaviour in organs and organisms, and population activity in an economy. In all cases, when observed from outside the systems are exhibiting relatively well defined behaviours but a very important element missing in previous analysis is the influence of the *goal* the systems are seemingly aiming at. Very often the components of these systems are searching through a collective action the satisfaction of properties they cannot reach alone, and to represent this situation the specific word "*emergence*" has also been coined [11]. The point is that it is now possible to return back to previous case and in a unified picture to envision the laws of Physics themselves as emerging phenomena.

For instance for an ensemble of neutral particles with hard ball interactions, and beyond the threshold of rarefied gas, (ie when the Knudsen number $K_n = \lambda/L$ is decreasing to 0 from the value 1, with λ the particle mean free path and L a characteristic length scale), the particle system is suddenly passing from a complicated to a complex state (due to the huge value of Avogadro number). Consecutive to overtaking of collective interaction by collisions (expressed by decrease of Knudsen number below the value *1* representing the limit of constraint (**C**) for neutral particle system), it could be said that there is emergence of a pressure and a temperature, which, from a point of view outside the gas, summarizes perfectly well the representative parameters (the invariants) describing it at this global level (the thermodynamic representation). Similarly at atomic level, after baryons are assembled from primitive quark particles below some threshold energy, protons and neutrons assemble in turn themselves below another lower energy threshold into ions with only mass and charge parameters, able to combine finally with electrons to create atoms. In all cases, it can be verified that emergence of new compound system does occur when boundary condition (**C**) is crossed when applied to each component of the system and at each interaction level. Interestingly, independent of the background system and of the vocabulary, it is easily verified that all systems exhibit first emergence of self organization out of which there appears a specific behaviour. In fact it is elementarily understandable from previous source-sink model that a key point is in the accuracy of modelling the components in complex state, as long as the resulting "invariants" which will grasp all system information for interaction with environment are directly depending on this modelling. This has been at the origin of a computer "blind" search where the agents are given properties and "emerging" behaviour is obtained in a bottom-up approach, sometimes in surprising compatibility with experimental observations [12], in parallel to theoretical analysis [13]. Finally it should be observed that the logical chain:

{stimuli/parameter change} → **{higher interactions between some system components}** →**{passage to complex state}** → **{system self organization}** → **{emergence of new behaviour}**

discussed here is nothing but the sequence leading to the final step of system evolution toward more independence, and which is the feedback of "*function*" onto "*structure*", a specific property of living organisms explaining their remarkable survival capability by structure modification.

4 Mathematical Representation of Complex Systems Emergence

The few basic previous examples from common sense observation illustrate the generality of the elements described above providing a unified base for complex system paradigm. From atomic nucleus to galaxy natural systems are seen to be constituted by aggregates of identifiable components (which, as already stressed, can be themselves, at each observation level, aggregates of smaller components) with well defined properties. These aggregates have been said to exhibit a complex behaviour when interaction between the components –or some of them– is overtaking their interactions with exterior environment. Similarly living beings are exhibiting the same behaviour, as observed with gregarious species, and in artificial man-made systems the same phenomenon is occurring when the overtaking conditions are satisfied. This is the case for high enough performance level systems because the components are then tightly packed, as for high torque compact electrical actuators. Despite an extremely large variety of possible situations there are few basic interactive processes leading to complex self-organization. First exist weak gradients natural systems discussed in previous paragraph and entering the more general class of *reducible systems* mainly because it can be shown that their complete dynamics including generated fluctuations can be reduced by projection onto {initial state plus large wavelength components} dynamics without small wavelength components dynamics (in complex state), now globally represented by transport coefficients modifying initial dynamics, see Fig. 3. Their mathematical analysis is still in progress in non Gaussian case, due to difficulty for specific small parameter ordering to analytically express transport coefficients despite their source is well identified [8,9,14].

More generally, a system may be in complete complex state, examples of which are atomic nucleus, herd of animals, and galaxies. Despite their very different space sizes, the systems exhibit always the same characteristic feature to finally depend on an extremely restricted number of parameters as compared to the aggregate of their initial components. Searching the way to extract directly the remaining ''control'' parameters of such systems from their dynamics is a fundamental issue which today motivates a huge research effort worldwide, especially in relation with information networking. Extensive analytical and numerical study has been developed for differential systems of generic form

$$\frac{dX}{dt} = A(t, X, v) + \lambda.F(t, X, u) + \mu.S(t, X) \tag{1}$$

where $X = col(X_1, X_2, .. X_n)$ is system state space, λ, μ are n-vector coupling parameters, and $A(.,.,.) : R_1^+ \times R_n \times R_p \to R_n$, $F(.,.,.) : R_1^+ \times R_n \times R_q \times U \to R_n$, $S(.,.) : R_1^+ \times R_n \to R_n$ are three specific $q_1 \times q_2$–matrix terms ($q_1, q_2 \le n$) corresponding to isolated free flight,

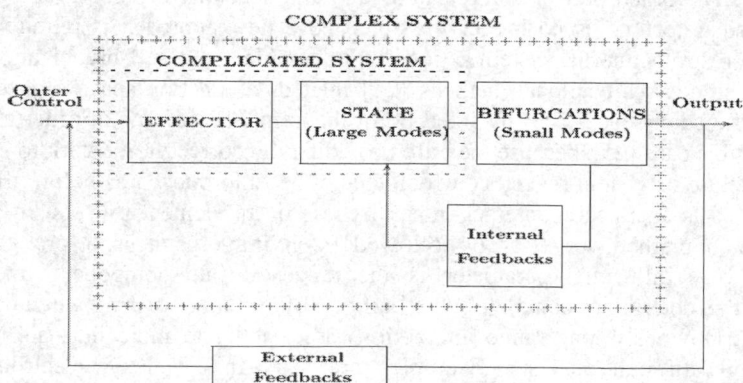

Fig. 3. Schematic block representation of (controlled) complex system

nonlinear internal interactions and source terms respectively (the linear and source terms in the right hand side of eq.(1) are here split apart to indicate their respective role). The other variables $v=v(X,t) \in \mathbf{R}_p$ and $u(X,t) \in \mathbf{U}$ in A and F function account more generally for possible feedback evolution of $X(t)$ onto their own dynamics as it often occurs in systems when splitting parameters into given and manipulated control ones. For fixed $u(.,.)$, $v(.,.)$, depending on the values of λ,μ components the system will be in simple, complicated or complex state described in §1. Increase in λ components moves the system into complex state, and eq.(1) transforms for very large λ into a singular system with small parameter in front of derivative amenable to asymptotic analysis [15]. For instance, 2–d Van der Pol system exhibits individual mode oscillations in state (**A**), driven oscillations in state (**B**). Under strong coupling between components when condition (**C**) is crossed relaxation oscillations are produced on a restricted (closed) 1–d curve representing the manifold on which complex state motion takes place. So even with two dof, complex state can occur, illustrating the fact that complexity is completely independent of complication with which it is very often confused.

More generally, it has been repeatedly observed, especially on systems close to Hamiltonian ones[16], that system representative point in n–dimensional state space follows a more and more chaotic trajectory when crossing bifurcation values and at the end fills up a complete domain[17]. Of course sensitivity is largest when the system exhibits resonances, ie is close to conservative, and adapted mathematical expansion methods have to be worked out[18]. In this case, it is easily verified that resonance overlapping condition[19] is nothing but application of condition (**C**). Because systems are basically non integrable[20], this is a direct evidence of increasing effect of internal interactions which reduce system dynamics to stay on attractor manifold of degree p < n, so that system dynamics are now layered on this manifold. This also expresses the fact that trajectories on the remaining n–p dimension space are becoming totally indistinguishable (from outside) when taken care of by internal interactions of n–p components going to complex state. So system trajectories reorganize here in equivalence classes which cannot be further split, a dual way to express the fact that there exists an invariant manifold on which system trajectories are lying. Continuing to control these components by regular previous control [21] worked out for complicated state and specially designed for tracking a prescribed trajectory, is no longer possible and a new approach is required which carefully respects internal system action due to complex state self-organization. More global methods of functional analysis [22] related to function space embedding in adapted function spaces [23] by fixed point theorem[24] are now in order as shown for reducible case [25], because they are providing the correct framework to grasp the new structure of system trajectory which cannot be fully tracked as before. Basically the method is again to counteract impreciseness in an element by robustness to its variation, a method very largely followed by living organisms. More generally, another very influential parameter is the range of inter-component interaction, because this determines completely the build up of system clustering when becoming complex. Obviously long range interactions are leading to more intricate response with more difficult analysis. Examples are stars in a galaxy, electromagnetic interactions between ions and electrons in a plasma, animals in a herd and social behaviour of human population in economic trading such as stock market with internet link. In all cases a new element is coming from the size of the neighbouring domain each system component is sensitive to, and implies a time extension to past neighbouring components trajectories. In this case, the resulting complex behaviour is more generally determined by interactive component effects over a past time interval and weighted according to their importance [2]. Finally one can summarize previous analysis by the universal.

Emergence Principle for Complex Systems: When interaction between some system components takes over by satisfying condition (**C**), they cluster into a subsystem the dynamics of which are only guided by the invariants of the generated manifold, and which are the only control parameters left by this (internal) reorganization.

This principle provides the way to express explicitly in mathematical terms the boundaries of domains in parameter space where complexity state is reached once a model of the system is given, and to fix its dynamical behaviour in new state by determination of its invariants.

5 Discussion and Conclusion

Systems exhibiting behaviours which do not fit with main stream scientific laws established from patient observations of Nature over past centuries have been repeatedly observed over last decades with the ascent of modern technology, where new natural and man made systems with very intricate structure implying a large number of heterogeneous components in strong interaction have been observed and developed. Application of usual laws is often unable to describe their dynamics, because they stay outside the domain of complicated multi component systems only covered by use of reductionism method. The main reason is in the overtaking of component interaction strength which dominates enough over other effects to force the system to close on itself and to manifest an internal self-organization responsible of its new behaviour. Differently said in elementary terms, the new paradigm is that "increasing interactions between components lead to their isolation" as easily verifiable. Such systems are termed as "complex", and their main feature is that components in complex state are internally ruled through this self-organization so that they are less sensitive to environment action. So natural complex systems are structurally more robust than complicated ones as evidenced by observation of living organisms, the most complex known ones. Analysis of complex systems dynamics shows the possibility of "emergence" of a new behaviour not included into the set of initial components behaviour as a direct consequence of self-organization opening the possibility for the "function" to feedback onto the "structure", as illustrated by development of living systems on Earth. The criterion for crossing "complexity" barrier has been explicitly expressed in terms of system coupling parameters by condition (**C**), which is verified to cover all known formulae in broad range of applications in Physics and Mathematics, thus providing a general emergence condition for a complex system.

Previous properties are of up-most importance when applied to man made industrial systems now appearing in open and global economy with an always increasing number of heterogeneous components to be operated all together for production of higher value objects. Previous centralized control structure cannot be maintained for keeping complete system mastery of larger number of elements going to complex state due to higher value of coupling parameters. In parallel, reduction of input control parameters by transforming the system into a (partially) complex one by clustering some components into bigger parts reduces system fragility. The industrial challenge civilisation is facing today justifies if any the needs to study and to create these complex systems [26]. On Fig. 2, this would mean to vary adequate parameters to move the representative points along complex axis in order to fix exactly new system status. In any case, internal non controlled dynamics are taken care of by system self organization resulting from passing to complex state, implying that precise trajectory control is now delegated to system. The challenging difficulty is that to comply with new structure, some "intelligence" has also to be delegated to the system, leading for the operator to a more supervisory position [27]. In present case, this is contribution to trajectory management by shifting usual (imposed) trajectory control to more elaborated task control [28], a way followed by all living creatures in their daily life to guarantee strong robustness while still keeping accuracy and preciseness. This illustrates the limited possibility of behaviours from laws of Physics because they are tightly linking *information flux* related to the described action to *power flux* implied in them. This opens on searching an adequate merging of information flux mastery from recent Information Technology development with power flux mastery resulting from classical long term mechanical development [29]. Though apparently loosing some hand on such systems, it has been surprisingly possible along this line to find explicit conditions in terms of system parameters expressing somewhat contradictory high preciseness (by asymptotic stability condition) and strong robustness (against unknown system and environment parts)[30]. In this way system dynamics are finally controlled and asymptotic stability can be demonstrated, but in general to the price of a not necessarily decreasing exponential asymptotic type.

References

1. Pattee, H.H.: Hierarchy Theory, the Challenge of Complex Systems. Braziller, New York (1973); Galbraith, J.: Designing Complex Organisations. Addison-Wesley, Reading (1973); Hayek, F.A.: The Theory of Complex Phenomena. In: Bunge, M. (ed.) The Critical Approach to Science and Philosophy, pp. 332–349. Collier McMillan, London (1964); Trappler (ed.): Power, Autonomy, Utopia: New Approaches towards Complex Systems. Plenum Press, New York (1986); Dyke, C.: The Evolutionary Dynamics of Complex Systems. OUP, New York (1988); Funtowicz, S., Ravetz, J.: Emergent Complex Systems. Futures 26, 568–582 (1994); Haken, H.: Information and Self–Organization: a Macroscopic Approach to Complex Systems. Springer, Heidelberg 1988)
2. Cotsaftis, M.: What Makes a System Complex: an Approach to Self-Organization and Emergence, Survey talk ECCS, Dresden, Germany, October 6-8, 2007 (2007), http://www.arxiv/nl/07060440
3. Laplace, P.S.: OEuvres Complètes, Part 6, Gauthiers–Villard, Paris (1878-1912); Lagrange J.L.: Mécanique Analytique, vol. I-II. Albert Blanchard, Paris (1788)

4. Aritotle: Physics, part I, chap. 9; Anderson P.W.: More is Different. Science 177, 393–396 (1972)
5. see reference [7] in ref [2] above
6. Prigogine, I., Nicolis, G.: Self–Organization in Non–Equilibrium Systems: From Dissipative Structures to Order through Fluctuations. J. Wiley and Sons, New–York (1977); Klimontovitch, Y.L.: Theory of Open Systems. Kluwer, Dordrecht (1995); Kleidon, A., Lorenz, R.D. (eds.): Non-equilibrium Thermodynamics and the Production of Entropy: Life, Earth, and Beyond. Springer, New–York (2005)
7. Kadomtsev, B.B.: Self Organization and Transport in Tokamak Plasma, Plasma Phys. and Nucl. Fus., 34(13), 1931–1938 (1992); Biebricher, C.K., Nicolis, G., Schuster, P.: Self–Organization in the Physico–Chemical and Life Sciences, vol.16546, EU Report (1995); Schweitzer, F.: Multi–Agent Approach to the Self–Organization of Networks. In: Reed–Tsochas, F., Johnson, N.F., Efstathiou, J. (eds.) Understanding and Managing Complex Agent–Based Dynamical Networks. World Scientific, Singapore (2007); Ebeling, W., Schweitzer, F.: Self-Organization, Active Brownian Dynamics, and Biological Applications. Nova Acta Leopoldina NF 88(332), 169–188 (2003)
8. Strand, P.I.: Predictive Simulations of Transport in Tokamaks, PhD Thesis, Dept. of Electromagnetics, Chalmers Univ. of Technology (1999)
9. Zakharov, V.E., L'vov, V.S., Falkovich, G.: Kolmogorov Spectra of Turbulence. Springer, New–York (1992); Klimontovich, Y.L.: Turbulent Motion and Structure of Chaos. Kluwer, Dordrecht (1991); Frisch, U.: Turbulence. Cambridge Univ. Press, Mass (1995)
10. Cotsaftis, M.: Lectures on Advanced Dynamics, Taiwan Univ., Taipeh, R.O.C (1993)
11. Fromm, J.: The Emergence of Complexity. Kassel Univ. Press (2004); Johnson, S.: Emergence. Penguin Books, New–York (2001)
12. Camazine, S., Deneubourg, J.L., Franks, N.R., Sneyd, J., Theraulaz, G., Bonabeau, E.: Self–Organization in Biological Systems. Princeton Univ. Press, Princeton (2002); Shen, W.-M., et al.: Hormone–Inspired Self–Organization and Distributed Control of Robotic Swarms. Autonomous Robots 17, 93–105 (2004); Bonabeau, E., Dorigo, M., Theraulaz, G.: Swarm Intelligence: from Natural to Artificial Systems. Oxford Univ. Press, New–York (1999)
13. Yates, F.E.(ed.) : Self–Organizing Systems: The Emergence of Order. Plenum Press (1987); Lewin, R.: Complexity–Life at the Edge of Chaos. Macmillan, Basingstoke (1993); Lerner, A.Y.(ed.) : Principles of Self–Organization. Mir, Moscow (1966); Nolfi, S., Floreano, D.: Evolutionary Robotics : the Biology, Intelligence and Technology of Self–Organizing Machines. The MIT Press, Cambridge (2000); Serra, R., Andretta, M., Compiani, M., Zanarini, G.: Introduction to the Physics of Complex Systems (the Mesoscopic Approach to Fluctuations, Nonlinearity and Self–Organization). Pergamon Press (1986)
14. Chandrasekhar, S.: Stochastic Problems in Physics and Astronomy. Rev. Mod. Phys. 15(1), 1–89 (1943); Risken, H.: The Fokker-Planck Equation, 2nd edn. Springer, Berlin (1996)
15. Grasman, J.: Asymptotic Methods for Relaxation Oscillations and Applications. Applied Math Sciences, vol. 63. Springer, New–York (1987)
16. Salamon, D.: The Kolmogorov–Arnold–Moser Theorem, ETH-Zürich (preprint) (1986); Arnold, V.I.: Mathematical Methods of Classical Mechanics, 2nd edn. Springer, Heidelberg (1989); Moser, J., Zehnder, E.: Lectures on Hamiltonian Systems. Lecture Notes, ITCP (1991)

17. Alliwood, K.T., Sauer, T.D., Yorke, J.A.: Chaos. An Introduction to Dynamical Systems. Springer, New–York (1997); Wiggins, S.: Chaotic Transport in Dynamical Systems. Springer, New–York (1991); Strogatz, S.H.: Nonlinear Dynamics and Chaos with Applications to Physics, Biology, Chemistry, and Engineering. Addison–Wesley, Reading (1994); Magnitskii, N.A., Vasilevich, S.V.: New Methods for Chaotic Dynamics. World Scientific Series on Nonlinear Science, Ser.A, Singapore (2006); Sagdeev, R.Z., Usikov, D.A., Zaslavsky, G.M.: Nonlinear Physics; from Pendulum to Turbulence and Chaos. Harwood Academic, New–York (1988)

18. Poincaré, H.: Les Méthodes Nouvelles de la Mécanique Céleste, vol. 3. Gauthier–Villars, Paris (1892–1899); Arnold, V.I., Koslov, V.V., Neishtadt, A.I.: Dynamic Systems III : Mathematical Aspects of Classical and Celestial Mechanics. Springer, Berlin (1993); Broer, H.W., Huitema, G.B., Sevryuk, M.B.: Quasi–Periodicity in Families of Dynamical Systems: Order amidst Chaos, LNM, vol. 1645. Springer, New–York (1996); Katok, A., Hasselblatt, B.: Introduction to the Modern Theory of Dynamical Systems, Cambridge Univ. Press. Mass. (1996); Bogoliubov, N.N., Mitropolskii, Yu.A., Samoilenko, A.M.: Methods of Accelerated Convergence in Nonlinear Mechanics. Springer, Berlin (1976)

19. Chirikov, B.V.: Statistical Properties of a Nonlinear String. Sov. Phys. Dokl. 1, 30 (1966); A Universal Instability of Many Dimensional Oscillator System. Phys. Rept. 52, 263–279 (1979); Lichtenberg, A.J., Lieberman, M.A.: Regular and Chaotic Dynamics, 2nd edn. Springer, New–York (1992)

20. Hirsch, M., Pugh, C., Shub, M.: Invariant Manifolds. Lecture Notes in Math. vol. 583. Springer, Berlin (1977); Tabor, M.: Chaos and Integrability in Nonlinear Dynamics, an Introduction. Wiley and Sons, New–York (1989); Goriely, A.: Integrability and non Integrability of Dynamical Systems. World Sci. Publ., Singapore (2001)

21. Lefschetz, S.: Stability of Nonlinear Control Systems. Academic Press, New–York (1965); Dullerud, G.E., Paganini, F.: A Course in Robust Control Theory : a Convex Approach. Springer, New–York (2000); Leonov, A.A., Ponomarenko, I.V., Smirnova, V.B.: Frequency Domain Methods for Nonlinear Analysis : Theory and Applications. World Scientific Publ., Singapore (1996); Astrom, K.J.: Control of Complex Systems. Springer, Berlin (2000); Grigorenko, I.: Optimal Control and Forecasting of Complex Dynamical Systems. World Scientific, Singapore (2006); Ng, G.W.: Application of Neural Networks to Adaptive Control of Nonlinear Systems. Research Studies Press, London (1997); Emelyanov S.V., Burovoi, A., Levada, F.Y.: Control of Indefinite Nonlinear Dynamic Systems: Induced Internal Feedback. Springer, London (1998); Fradkov, A.L., Miroshnik, I.V., Nikiforov, V.O.: Nonlinear and Adaptive Control of Complex Systems. Kluwer Acad. Publ., Dordrecht (1999); Krstic, M., Kanellakopoulos, I., Kokotovic, P.V.: Nonlinear and Adaptive Control Design. J. Wiley and Sons, New-York (1995); Francis, B.A., Tanenbaum, A.R. (eds.) : Feedback Control, Nonlinear Systems and Complexity. Springer, New-York (1995); Solodovnikov, V.V., Tumarkin, V.I.: Complexity Theory and the Design of Control Systems, Theory and Methods of Systems Analysis, vol. 28. Nauka, Moscow (1990); Wang, L.X.: Adaptive Fuzzy Systems and Control. Prentice Hall, Englewwod Cliffs (1994)

22. Deimling, K.: Nonlinear Functional Analysis. Springer, Berlin (1985); Krasnoselskii, M.A.: Asymptotics of Nonlinearities and Operator Equations. Birkhauser, Boston (1995); Vainberg, M.M.: Variational Methods for the Study of Nonlinear Operators. Holden–Day, San Francisco (1964); Zeidler, E.: Nonlinear Functional Analysis and its Applications, vol. I-III-IV. Springer, New–York (1985-1986-1988); Bendat, J.S.: Nonlinear Systems Techniques and Applications. J. Wiley and Sons, New–York (1998); Bobylev, N.A., Burman, Y.M., Korovin, S.K.: Nonlinear Analysis and Applications. Walter de Gruyter, Berlin (1994); Bogaesvski, V.N., Povzner, A.: Algebraic Methods in Nonlinear Perturbation Theory. Springer, New–York (1991); Hale, J.K., Verduyn Lunel, S.M.: Theory of Functional Differential Equations. Springer, New-York (1993)

23. Amerio, L., Prouse, G.: Almost–Periodic Functions and Functional Equations. Van Nostrand-Reinhold, New–York (1971); Mazja, V.G.: Sobolev Spaces. Springer, New–York (1985); Rao, M.M., Ren, Z.D.: Theory of Orlicz Spaces. Marcel Dekker, New–York (1991)

24. Jiang, T.H.: Fixed–Point Theory. Springer, New–York (1985); Zeidler, E.: Nonlinear Functional Analysis and its Applications, vol. I. Springer, New–York (1986); Burton T.A.: Stability by Fixed Point Theory for Functional Differential Equations. Dover Publ., New–York (2006)

25. Cotsaftis, M.: Recent Advances in Control of Complex Systems, Survey Lecture. In: Proc. ESDA 1996, Montpellier, France, ASME, vol. I, p. 1 (1996); Cotsaftis, M.: Comportement et Contrôle des Systèmes Complexes, Diderot, Paris (1997); Cotsaftis, M.: Popov Criterion Revisited for Other Nonlinear Systems. In: Proc. ISIC 2003 (International Symposium on Intelligent Control), October 5-8, Houston, Texas (2003)

26. Cotsaftis, M.: A Passage to Complex Systems. In: Proc. CoSSoM 2006, Toulouse, France, September 24-27 (2006); to be published in Springer Series "Understanding Complex Systems"

27. Cotsaftis, M.: Beyond Mechatronics, Toward Global Machine Intelligence. In: Proc. ICMT 2005, Kuala–Lumpur, December 6-8 (2005); Kosko, B.: Neural Networks and Fuzzy Systems: A Dynamical Systems Approach to Machine Intelligence. Prentice–Hall, Englewood Cliffs (1991)

28. Cotsaftis, M.: From Trajectory Control to Task Control – Emergence of Self Organization in Complex Systems. In: Aziz-Alaoui, M.A., Bertelle, C. (eds.) Emergent Properties in Natural and Artificial Dynamical Systems, pp. 3–22. Springer, Heidelberg (2006); also: On the Definition of Task Oriented Intelligent Control. In: Proc. ISIC 2002 Conf., Vancouver, October 27-30 (2002)

29. Cotsaftis, M.: Merging Information Technologies with Mechatronics – The Autonomous Intelligence Challenge. In: Proc. IEECON 2006, Paris, November 6-10 (2006)

30. Cotsaftis, M.: Robust Asymptotically Stable Control for Intelligent Unknown Mechatronic Systems. In: Proc. ICARCV 2006, Singapore, December 5-8 (2006)

An Effective Local Routing Strategy on the Communication Network

Yu-jian Li[1,2], Bing-hong Wang[1], Zheng-dong Xi[2], Chuan-yang Yin[3], Han-xin Yang[1], and Duo Sun[1]

[1] Department of Modern Physics, University of Science and Technology of China, Hefei 230026, P.R. China
jinzhili@mail.ustc.edu.cn, bhwang@ustc.edu.cn
[2] Department of Satellite Measurement and Control on Sea of China, Jiangyin 214400, P.R. China
[3] Nanjing University of Information Science and Technology, Nanjing 210000, P.R. China

Abstract. In this paper, we propose an effective routing strategy on the basis of the so-called nearest neighbor search strategy by introducing a preferential cut-off exponent K. We assume that the handling capacity of one vertex is proportional to its degree when the degree is smaller than K, and is a constant C_0 otherwise. It is found that by tuning the parameter α, the scale-free network capacity measured by the order parameter is considerably enhanced compared to the normal nearest-neighbor strategy. Traffic dynamics both near and far away from the critical generating rate R_c are discussed. Simulation results demonstrate that the optimal performance of the system corresponds to $\alpha = -0.5$. Due to the low cost of acquiring nearest-neighbor information and the strongly improved network capacity, our strategy may be useful and reasonable for the protocol designing of modern communication networks.

Keywords: complex networks, scale-free, local routing strategy.

1 Introduction

A variety of systems in nature can be described by complex networks and the most important statistical features of complex networks are the small-world effect and scale-free property [1,2,3]. It may serve as a very useful tool for understanding nature and our society. Since the discovery of some common interesting features of many real networks such as small-world phenomena by Strogatz and Watts [1] and Scale-free phenomena by Albert and Barabási, [2] processes of dynamics conducting on the network structure such as traffic congestion of information now have drew more and more attention from engineering and physical field [10,12,13,16], due to the importance of large communication networks such as WWW [4] and Internet [5] in modern society.

Many previous excellent works focus on the evolution of structure driven by the increment of traffic [6,7] and some explore how different topologies impact the traffic dynamics [10,11]. Some works [9] gave several models to mimic the traffic routing on complex networks by introducing randomly selected source as well as particles (packets) generating rate and destination of each particles [14,15,16]. Those models define

J. Zhou (Ed.): Complex 2009, Part I, LNICST 4, pp. 1118–1123, 2009.

the capacity of networks described by critical generating rate. At this critical rate, a continuous phase transition from free flow state to congested state occurs. In the free state, the numbers of created and delivered particles are balanced, leading to a steady state. While on the jammed state, the number of accumulated particles increases with time due to the limited delivering capacity or finite queue length of each vertex. We believe that the study on the network search is very important for traffic systems, for the existence of particles routing from origin to destination and communication cost is very meaningful.

A few previous studies [8,9] adopt the search strategies and the traffic processes on networks. In this paper, we present a traffic model in which particles are routed only based on local topological information with a single tunable parameter α. In order to maximize the nodes handling and delivering capacity of the networks which can be measured by an introduced order parameter η , the optimal α is found out. The dynamics right after the critical generating rate R_c exhibits some interesting properties independent of α, which indicates that although the system enters the jammed state, it possesses partial capacity for forwarding particles. Our model can be considered as a preferential walk among neighbor vertexes. We arrange the paper as follows: in the first section we describe the model in detail. then simulation results of traffic dynamics are provided in both the steady and congested states, A conclusion and discussion are given in the last section.

2   Traffic Model

Our traffic model is described as follows: at each time step, there are R particles generated in the system, with randomly chosen sources and destinations, and all vertexes can deliver at most C particles toward their destinations, which is one of the most interesting properties of the whole traffic network. The capacity of each vertex is set to be $C_i = k_i$ when the degree is smaller than a cut-off value K, and be C_0 otherwise. As a remark, there is difference between the capacity of network and vertexes. The capacity of the whole network is measured by the critical generating rate R_c at which a continuous phase transition will occur from free state to congestion. The free state refers to the balance between created particles and removed particles at the same time. When the system enters the jam state, it means particles continuously accumulate in the system and at last few particles can reach their destinations. In order to describe the critical point accurately, we use the order parameter [8,9]:

$$\eta(R) = \lim_{t \to \infty} \frac{C_i}{R} \frac{<\Delta N_p>}{\Delta t} \tag{1}$$

$\Delta N_p = N(t + \Delta t) - N(t)$ with $< \cdots >$ indicates average over time windows of width Δt and $Np(t)$ represents the number of data particles within the networks at time t, here. For $R < R_c$, $< \Delta N >= 0$ and $\eta = 0$, indicating that the system is in the free state with no traffic congestion. Otherwise for $R > R_c$, ηr, where r is a constant larger than zero, the system will collapse at last. To navigate particles, each vertex performs a local search among its neighbors. If a particle's destination is found in the searched area, it will be

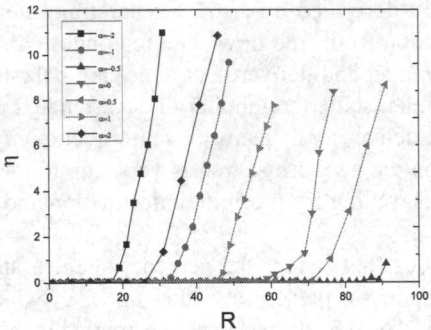

Fig. 1. The order parameter η versus R for BA network with different free parameter α. Other parameters are networks size $N = 3000$, $K = 200$, and $m = 3$.

delivered directly to its target, otherwise, it will be forwarded to a neighbor j of vertex i according to the probability:

$$\Pi_{i \to j} = \frac{k_j^{\alpha}}{\sum_l k_l^{\alpha}} \tag{2}$$

Here, the sum runs over the neighbors of vertex i on the searched area and α is an adjustable parameter studied by us in the next context. Once a particle reaches its destination, it will be canceled from the system.

As shown in Fig.1, the order parameter versus generating rate R by choosing different value of parameter α is displayed. It is easy to find that the capacity of the system is not alike for different α, thus, a natural question is addressed: what is the optimal value of α for making the network's capacity maximal in our model?

Many studies [1,2,3] indicate that many communication networks such as Internet are not homogeneous like random or regular networks. Barabási and Albert proposed a famous model (BA for short) called scale-free networks, [2] of which the degree distribution is in good accordance with modern communication networks, which has a power-law distribution $P(k) \propto k^{-r}$. Our study is based on the so-called BA network, we construct the network structure following the same method used in Ref. [2]: starting from m fully connected vertexes, a new vertex with m links is added to the existing graph at each time step according to the rule of preferential attachment i.e. the probability of being connected to an existing vertex is proportional to the degree of that vertex. Here, we choose $m = 5$ and network size $N = 1000$ fixed for simulations.

We should also note that the queue length of each vertex is assumed to be unlimited and the $FIFO$(First in First out) discipline is applied at each queue [8,9,15,16]. Another important rule called path iteration avoidance (PIA) is that a link between any pair of vertexes is not allowed to be visited more than twice by the same particle [8,9]. Without this rule, the capacity of the network is quite low due to many times' unnecessary visiting to the same links by the same particles, which does not exist in the real traffic systems. We note that this PIA routing algorithm does not damage the advantage of local routing strategy. If each particle records the links it has visited, the PIA can be easily performed. One can find that this rule does not need the global topological information.

Therefore, we think this rule is rational and can considerably improve the network capacity. With the development of science and technology, the handling capacity of the main central node can be set up large enough artificially, that is why we propose the cutoff value K. We hold that it is more practical and more reasonable.

3 Simulation Results

We have carried on the simulation under the definition of the model and the order parameter η versus generating rate R with choosing different value of parameter α is reported.

As shown in Fig. 1, one can see that, for all different α, η is approximately zero when R is small; it suddenly increases when R is larger than the critical point R_c. It is easy to find that the capacity of the system is not the same for different α. For the same η, when $\alpha = -0.5$, the R reaches its max. We can preliminarily determine that it is the best situation.

We also observed the handling capacity R_c for different α in the system, one can read from Fig.2 that the tolerance is the best when $\alpha=-0.5$. This is another strong evidence to show that $\alpha=-0.5$ is a perfect point.

As shown in Fig.3, when the K increases, the capacity of BA network measured by R_c considerably has the optimal performance at $K=200$. Although K is a variable parameter, the system always reaches its best case at $\alpha = -0.5$. We study the critical point R_c affected by the link density of BA network.

As shown in Fig.4, the increment of m considerably enhances the capacity of BA network measured by R_c due to the fact that with higher link density, particles can more easily find their target vertexes. From Fig.3 and Fig.4, we can know that the critical R_c reaches its max when $\alpha=-0.5$ at the same K or m.

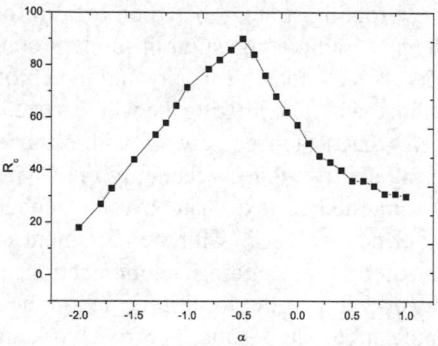

Fig. 2. The critical R_c versus α with network size $N = 3000$, $K = 200, m = 3$. The maximum of R_c corresponds to $\alpha = -0.5$ marked by a black solid line.

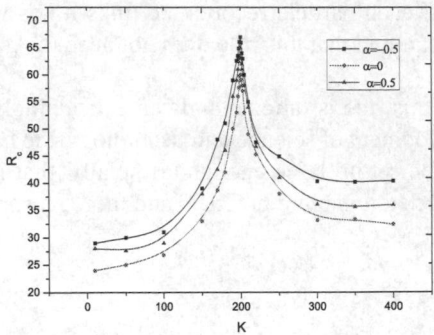

Fig. 3. (color online). The variance of R_c with the increasing of K with network size N = 3000, m = 3.

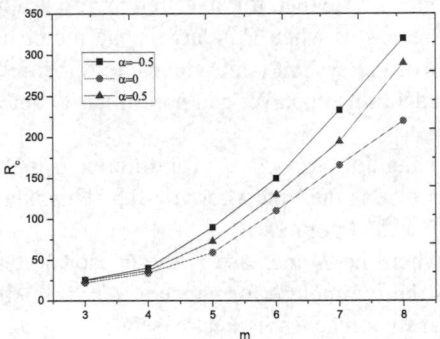

Fig. 4. (color online). The variance of R_c with the increasing of m with network size N = 3000, K = 200.

4   Conclusion

We have introduced a new routing strategy based on local information, trying to give a solution to the problem of traffic congestion in modern communication networks. Influenced by two factors of each node's capacity and the cutoff value K, the optimal parameter α=-0.5 is obtained with maximizing the whole system's capacity.

In addition, the property that scale-free network with occurrence of congestion still possesses partial delivering ability suggests that only improving processing ability of the minority of heavily congested vertexes can obviously enhance the capacity of the system. The variance of critical value R_c with the increment of m and K is also discussed. Our study may be useful for designing communication protocols for large scale-free communication networks due to the local information the strategy only based on and the simplicity for application. The results of current work may also shed some light on alleviating the congestion of modern technological networks.

Further work could be carried out, for the queue length of each vertex is infinite and the live time of a particle.

Acknowledgement

We thank Dr. WANG Wen-xu and LIU Jian-guo for useful discussions.

This work was Supported by the National Basic Research Programme of China under Grant No 2006CB705500, the National Natural Science Foundation of China under Grant Nos 60744003, 10635040, 10532060 and 10472116, the Special Research Funds for Theoretical Physics Frontier Problems under Grant Nos 10547004 and A0524701, the President Funding of Chinese Academy of Sciences, and the Specialized Research Fund for the Doctoral Programme of Higher Education of China, and Funded by The CAS Special Grant for Postgraduate Research, Innovation and Practice.

References

1. Watts, D.J., Strogatz, S.H.: Collective Dynamics of 'small-world' Networks. Nature 393, 409–410 (1998)
2. Barabási, A.L., Albert, R.: Emergence of Scaling in Random Networks. Science 286, 509–512 (1999)
3. Albert, R., Barabási, A.L.: Statistical Mechanics of Complex Networks. Rev. Mod. Phys. 74, 47–97 (2002)
4. Pastor-Satorras, R., Vázquez, A., Vespignani, A.: Dynamical and Correlation Properties of the Internet. Phys. Rev. Lett. 87, 258701 (2001)
5. Albert, R., Jeong, H., Barabási, A.L.: Diameter of the World Wide Web. Nature 401, 130–131 (1999)
6. Barrat, A., Barthélemy, M., Vespignani, A.: Weighted Evolving Networks: Coupling Topology and Weighted Dynamics. Phys. Rev. Lett. 92, 228701 (2004)
7. Wang, W.X., Wang, B.H., Hu, B., Yan, G., Ou, Q.: General Dynamics of Topology and Traffic on Weighted Technological Networks. Phys. Rev. Lett. 94, 188702 (2005)
8. Wang, W.X., Wang, B.H., Yin, C.Y., Xie, Y.B., Zhou, T.: Traffic Dynamics Based on Local Routing Protocol on Scale Free Network. Phys. Rev. E 74, 016101 (2006)
9. Yin, C.Y., Wang, B.H., Wang, W.X., Yan, G., Yang, H.J.: Traffic Dynamics Based on an Efficient Routing Strategy on Scale Free Networks. Eur. Phys. Jour. B 49, 205–211 (2006)
10. Singh, B.K., Gupte, N.: Congestion and De-congestion in a Communication Network. Phys. Rev. E 71, 055103(R) (2005)
11. Ashton, D.J., Jarrett, T.C., Johnson, N.F.: Effect of Congestion Costs on Shortest Paths through Complex Networks. Phys. Rev. Lett. 94, 058701 (2005)
12. Zhao, L., Park, K., Lai, Y.C.: Attack Vulnerability of Scale-free Networks Due to Cascading Breakdown. Phys. Rev. E 70, 035101(R) (2004)
13. Kim, B.J., Yoon, C.N., Han, S.K., Jeong, H.: Pathfinding Strategies in Scale-free Networks. Phys. Rev. E 65, 027103 (2002)
14. Zhao, L., Lai, Y.C., Park, K., Ye, N.: Onset of Traffic Congestion in Complex Networks. Phys. Rev. E 71, 026125 (2005)
15. Zhang, H., Liu, Z.H., Tang, M., Hui, P.M.: An Adaptive Routing Strategy for Packet Delivery in Complex Networks. Phys. Lett. A 364, 177–182 (2007)
16. Yang, H.X., Wang, W.X., Wu, Z.X, Wang, B.H.: Traffic dynamics in scale-free networks with limited packet-delivering capacity. Physica A 387, 6857–6862 (2008)

Average Consensus in Delayed Networks of Dynamic Agents with Impulsive Effects

Quanjun Wu[1], Lan Xiang[2], and Jin Zhou[1,*]

[1] Shanghai Institute of Applied Mathematics and Mechanics, Shanghai University,
Shanghai 200072, China
Jinzhousu@yahoo.com.cn
[2] Department of Physics, School of Science, Shanghai University,
Shanghai, 200444, China

Abstract. In this paper, the issues of average consensus in undirected delayed networks of dynamic agents with impulsive effects are investigated. The primary contribution of this paper is to propose the consensus schemes in undirected delayed networks of dynamic agents having impulsive effects as well as fixed, switching topology. Based on impulsive stability theory on delayed dynamical systems, we derive some simple sufficient conditions under which all the nodes in the network achieve average consensus globally exponentially. It is shown that average consensus in the networks is heavily dependent on impulsive effects of communication topology of the networks. Subsequently, two numerical examples illustrate and visualize the effectiveness and feasibility of our theoretical results.

Keywords: average consensus, undirected network, multi-agent systems, time-delays, impulsive effects.

1 Introduction

During the last few decades, distributed coordination in dynamic networks of multi-agents has attracted a great deal of attention in many fields such as biology, ecology, robotics, physics, etc., [1,2,3,4,5,6,7,8,9]. This is partly due to potential applications in many areas including cooperative control of unmanned air vehicles (UAVs), formation control, flocking, distributed sensor networks, attitude alignment of cluster of satellites, and congestion control in communication networks [8,10,11,12,13]. A critical problem in distributed coordinated control of multiple agents is to design appropriate protocols and algorithms such that all agents can reach an agreement regarding a certain quantity of interest that depends on the states of all agents. This problem is usually called the consensus problem.

Consensus problems have a long history in the field of computer science, and it has also been considered as the foundation of the distributed computing [14]. The study of consensus problems in groups of experts originated in management science and statistics in 1960s [15]. In the past decade, many researchers have investigated the consensus problems from various perspectives [6,8,10,16,17,18,19,20]: Vicsek et al. proposed a

---

* Corresponding author.

J. Zhou (Ed.): Complex 2009, Part I, LNICST 4, pp. 1124–1138, 2009.

simple model for phase transition of a group of self-driven particles and numerically demonstrated complex dynamics of the model [8]; Jadbabaie et al. attempted to provide a formal analysis of emergence of alignment in the simplified model of flocking proposed by Vicsek [16]; Saber et al. addressed the consensus problems under a variety of assumptions on the network topology (fixed or switching), presence or lack of communication delays, and directed or undirected network information flow [6]; Ren et al. extended the results of Jadbabaie et al. and gave some more relaxable conditions [19]; Moreau focused on the consensus problems under dynamically changing interaction topologies[18].

In many evolutionary systems there are two common phenomena: delay effects and impulsive effects [21,22,23,24]. Time delays often occurs in such systems as transportation and communication systems, chemical and metallurgical processes, environmental models and power networks. In many scenarios, networked systems can possess a dynamic topology that is time-varying due to node and link failures/creations, packet-loss [25], asynchronous consensus [26], state-dependence [27], formation reconfiguration, evolution, and flocking [7]. There has been increasing interest in the study of consensus problem in dynamic networks of multi-agents with time delays in the last several years. Among the existing research works, Earl and Strogatz proposed a stability criterion for a network of specific oscillators with time-delayed coupling, a necessary and sufficient consensus condition was established [28]. Olfati-Saber discussed average consensus problems in undirected networks with a common constant communication delay and fixed topology [6]. Moreau studied the case when the common constant delay affects only those variables that are actually being communicated between distinct agents in the network [18]. Sun discussed the average consensus problem in undirected networks of dynamic agents with fixed and switching topologies as well as multiple time-varying communication delays [35]. Lin studied the average-consensus problem in directed networks of agents with both switching topology and time-delays [30]. On the other hand, many evolutionary processes, particularly some biological systems such as biological neural networks and bursting rhythm models in pathology, as well as optimal control models in economics, frequency-modulated signal processing systems, and flying object motions, are characterized by abrupt changes of states at certain time instants. This is the familiar impulsive phenomena [31,32]. However, to our best knowledge, up to now just few works involved the consensus problems in delayed dynamic networks with impulsive effects. Therefore, as an interesting and challenging topic, this motivates the present investigation of consensus issue in dynamic networks of multi-agents associated with time delays and impulsive effects.

In this paper, we investigate average consensus problem in undirected delayed networks of dynamic agents with impulsive effects. The primary contribution of this paper is to propose the consensus schemes in undirected delayed networks of dynamic agents having impulsive effects as well as fixed, switching topology. Based on impulsive stability theory on delayed dynamical systems, we derive some simple sufficient conditions under which all the nodes in the network achieve average consensus globally exponentially. It is shown that average consensus in the networks is heavily dependent on impulsive effects of communication topology of the networks. Subsequently, two numerical examples illustrate and visualize the effectiveness and feasibility of our theoretical results.

An outline of this paper is as follows. In Section 2, some mathematical preliminaries are first prepared. Section 3 introduces the problem formulations with respect to average consensus problem in dynamic networks of multi-agents with time delays and impulsive effects. Section 4 deals with the average consensus problem in undirected delayed networks of dynamic agents having impulsive effects as well as fixed, switching topology. Some numerical simulations are presented in Section 5. Finally, some conclusions are drawn in Section 6.

2   Mathematical Preliminaries

Throughout this paper, the following notations and definitions will be used.

Let R denotes the set of real number, and $R^n = \underbrace{R \times R \times \cdots \times R}_{n}$, $R^{n \times n}$ is $n \times n$ the set of real matrices, $\text{diag}(\gamma_1, \cdots, \gamma_n) \in R^{n \times n}$ is the diagonal matrix with diagonal entries $\gamma_i \, (i = 1, \cdots, n)$. For $x = (x_1, \cdots, x_n)^T \in R^n$, $A = (a_{ij})_{n \times n} \in R^{n \times n}$, x^T denotes its transpose, we denote $|x| = (|x_1|, \cdots, |x_n|)^T$. The norm of the vector x is defined as $\|x\| = (x^T x)^{1/2}$. For $\psi : R \to R$, denote: $\psi(t^+) = \lim_{s \to 0^+} \psi(t + s)$, $\psi(t^-) = \lim_{s \to 0^-} \psi(t + s)$, $[\psi(t)]_\tau = \sup_{-\tau \le s \le 0} \{\psi(t + s)\}$, $[\psi(t)]_{\tau^-} = \sup_{-\tau \le s < 0} \{\psi(t + s)\}$. $PC([t_0 - \tau, t_0], R^n])$ denotes the set of all functions of the bounded variation and right-hand continuous on any compact subinterval of $[t_0 - \tau, t_0]$. Denote $\|\phi(t)\|_\tau = \sup_{-\tau \le s \le 0} \|\phi(t + s)\|$.

For the later use, the following inequality and the famous Halanay differential inequality on delayed impulsive dynamical systems are listed in the following:

Lemma 1. *(Park [36]). For any positive scalar ϵ and vectors x and y, the following inequality holds:*

$$x^T y + y^T x \le \epsilon x^T x + \epsilon^{-1} y^T y. \tag{1}$$

Lemma 2. *(Yang and Xu [34]). Suppose $p > q \ge 0$ and $u(t)$ satisfies scalar impulsive differential inequality:*

$$\begin{cases} D^+ u(t) \le -pu(t) + q[u(t)]_\tau, & t \ne t_k, \quad t \ge t_0 \\ u(t_k^+) \le b_k u(t_k^-) + d_k[u(t_k)]_{\tau^-}, & k \in N \\ u(t) = \phi(t), & t \in [t_0 - \tau, t_0] \end{cases} \tag{2}$$

where $u(t)$ is continuous at $t \ne t_k$, $t \ge t_0$, $u(t_k) = u(t_k^+)$, and $u(t_k^-)$ exists, $\phi \in PC$ with $n = 1$. Then

$$u(t) \le \left(\prod_{t_0 < t_k \le t} \theta_k \right) e^{-\lambda(t - t_0)} \|\phi(t_0)\|_\tau, \quad t \ge t_0 \tag{3}$$

where $\delta_k := \max\{1, |b_k| + |d_k| e^{\lambda \tau}\}$ and $\lambda > 0$ is a solution of the inequality $\lambda - p + q e^{\lambda \tau} \le 0$.

3 Problem Formulations

3.1 Graph Theory

Let $G = (\mathcal{V}, \mathcal{E}, \mathcal{A})$ be a weighted undirected graph of order n ($n \geq 2$) with the set of nodes $\mathcal{V} = \{v_1, \cdots, v_n\}$, set of edges $\mathcal{E} \subseteq \mathcal{V} \times \mathcal{V}$, and a symmetric weighted adjacency matrix $\mathcal{A} = [a_{ij}]$ with nonnegative adjacency elements a_{ij}. The node indexes belong to a finite index set $\mathcal{I} = \{1, 2, \cdots, n\}$. An edge of G is denoted by $e_{ij} = (v_i, v_j)$. The adjacency elements associated with the edges of the graph are positive, i.e., $e_{ij} \in \mathcal{E}$ if and only if $a_{ij} > 0$. Moreover, we assume $a_{ii} = 0$ for all $i \in \mathcal{I}$. The set of *neighbors* of the node v_i is denoted by $N_i = \{v_j \in \mathcal{V} : (v_i, v_j) \in \mathcal{E}\}$. An undirected graph is called *connected* if any two distinct nodes of the graph can be connected via a path that follows the edges of the graph.

3.2 Consensus Problems

Let $x_i \in R$ denote the value of the node v_i. We refer to $G_x = (G, x)$ with $x = (x_1, \cdots, x_n)^T$ as a *network* (or *algebraic graph*) with the value $x \in R^n$ and the *topology* (or *information flow*) G. The value of a node might represent physical quantities such as attitude, position, temperature, voltage, and so on. We say both the nodes v_i and v_j *agree* in a network if and only if $x_i = x_j$. We say the nodes of a network have reached a *consensus* if and only if $x_i = x_j$ for all $i, j \in \mathcal{I}, i \neq j$. Whenever the nodes of a network are all in agreement, the common value of all nodes is called the *group decision value*.

Suppose each node of a graph is a *dynamic agent* with dynamics

$$\dot{x}_i = f(x_i, u_i), \quad i \in \mathcal{I} \tag{4}$$

A *dynamic graph* (or *dynamic network*) is a dynamical system with a state (G, x) in which the value x evolves according to the *network dynamics* $\dot{x} = F(x, u)$. Here, $F(x, u)$ is the column-wise concatenation of the elements $f(x_i, u_i)$ for $1, \cdots, n$. In a dynamic network with switching topology, the information flow G is a discrete-state of the system that changes in time.

Let $\mathcal{X} : R^n \to R$ be a function of n variables $x_1, \cdots, x_n$ and $a = x(0)$ denote the initial state of the system. The $\mathcal{X}$-*consensus problem* in a dynamic graph is a distributed way to calculate $\mathcal{X}(a)$ by applying inputs u_i that only depend on the states of node v_i and its neighbors. We say a state feedback

$$u_i = k_i(x_{j_1}, \cdots, x_{j_{m_i}}) \tag{A}$$

is a *protocol* with topology G if the cluster $J_i = \{v_{j_1}, \cdots, v_{j_{m_i}}\}$ of nodes with indexes $j_1, \cdots, j_{m_i} \in \mathcal{I}$ satisfies the property $J_i \subseteq \{v_i\} \cup N_i$. In addition, if $|J_i| < n$ for all $i \in \mathcal{I}$, (A) is called a *distributed protocol*.

We say protocol (A) asymptotically solves the $\mathcal{X}$-consensus problem if and only if there exists an asymptotically stable equilibrium x^* of $\dot{x} = F(x, k(x))$ satisfying $x_i^* = \mathcal{X}(x(0))$ for all $i \in \mathcal{I}$. We are interested in distributed solutions of the $\mathcal{X}$-consensus problem in which no node is connected to all other nodes. The special case with $\mathcal{X}(x) = Ave(x) = \dfrac{1}{n}(\sum_{i=1}^{n} x_i)$, is called *average-consensus*.

Solving the average-consensus problem is an example of *distributed computation* of a linear function $\mathscr{X}(a) = Ave(a)$ using a network of dynamic systems (or integrators). This is a more challenging task than reaching a consensus with initial state a. Since an extra condition $x_i^* = \mathscr{X}(a)$, $\forall i \in \mathscr{I}$ has to be satisfied which relates the limiting state x^* of the system to the initial state a.

3.3   Model Formulations

In this paper, we are interested in discussing average consensus problem in undirected delayed networks of dynamic agents having impulsive effects as well as fixed, switching topology, where the information (from v_j to v_i) passes through edge (v_i, v_j) with the coupling time-delays $0 < \tau(t) \leq \tau$. To solve such a problem, we use the following consensus protocol:

$$u_i(t) = \sum_{v_j \in N_i} a_{ij}\big(x_j(t - \tau(t)) - x_i(t - \tau(t))\big)Dw_j(t), \tag{5}$$

where D denotes the distributional derivative, $w_i : J = [t_0, +\infty) \rightarrow R$ are functions of the bounded variations which are right-continuous on any compact subinterval of J. We remark that the model formulation given above implies that Dw_i represents the impulsive effect of switching topology in the dynamical network.

Under the consensus protocol (5), the system (4) can be described by the following measure differential equations:

$$Dx_i(t) = \sum_{v_j \in N_i} a_{ij}\big(x_j(t - \tau(t)) - x_i(t - \tau(t))\big)Dw_j(t). \tag{6}$$

Without loss of generality, we assume that

$$w_j(t) = t + \sum_{m=1}^{\infty} \mu_m H_m(t), \tag{7}$$

with discontinuity points

$$t_1 < t_2 < \cdots < t_m < \cdots, \qquad \lim_{m \to \infty} t_m = \infty,$$

where μ_m are constants, represents the strength of impulsive effects of connection between the jth node and the ith node at time t_m, and $H_m(t)$ are Heaviside functions defined by

$$H_m(t) = \begin{cases} 0, & t < t_m, \\ 1, & t \geq t_m. \end{cases} \tag{8}$$

It is easy to see that

$$Dw_j = 1 + \sum_{m=1}^{\infty} \mu_m \delta(t - t_m), \tag{9}$$

where $\delta(t)$ is the Dirac impulsive function. Clearly, if $\mu_k = 0$, then the model (6) becomes continuous consensus scheme with time delays [35].

$$\dot{x}_i(t) = \sum_{v_j \in N_i} a_{ij}\big(x_j(t - \tau(t)) - x_i(t - \tau(t))\big). \tag{10}$$

For the impulsive functional differential equation (6), its initial conditions are given by $x_i(t) = \phi_i(t) \in PC([t_0 - \tau, t_0], R^n)$. We always assume that Eq. (6) has a unique solution with respect to initial conditions [33,34].

Rewrite (6) in matrix form as

$$Dx(t) = -Lx(t - \tau(t))Dw(t), \tag{11}$$

where $L = [l_{ij}]$ is called *graph Laplacian* (Laplacian matrix or Laplacian) induced by the information flow G and is defined by

$$l_{ij} = \begin{cases} \sum_{k=1, k \neq i}^{n} a_{ik}, & j = i \\ -a_{ij}, & j \neq i \end{cases} \tag{12}$$

It is easy to see that the matrix L is given by

$$L = \mathcal{L}(G) = D - A \in R^{n \times n}, \tag{13}$$

where $A = [a_{ij}]_{n \times n}$, and $D = diag(d_1, \cdots, d_n) \in R^{n \times n}$, which is called as a degree matrix of the topology G, whose diagonal elements $d_i = \sum_{j \in N_i} a_{ij}$ for $i = 1, 2, \cdots, n$.

As indicated in [6], the topology G with the Laplacian matrix L is a connected undirected graph, then all eigenvalues but one simple eigenvalue at zero of L have positive real-parts. Furthermore, it always has a zero eigenvalue corresponding to a right eigenvector $\mathbf{1} = (1, , \cdots, 1)^T$. This means that $rank(L) \leq n - 1$.

Notice that $\mathbf{1}^T L = 0$. Thus, $\alpha = Ave(x)$ is an invariant quantity. The invariance of $Ave(x)$ allows decomposition of x according to the following equation:

$$x = \alpha \mathbf{1} + \delta, \tag{14}$$

where $\alpha = Ave(x)$ and $\delta = (\delta_1, \cdots, \delta_n)^T \in R^n$ satisfies $\mathbf{1}^T \delta = 0$. Here, we refer to δ as the (group) *disagreement vector*. The vector δ is orthogonal to $\mathbf{1}$ and belongs to an $(n - 1)$-dimensional subspace. Moreover, δ evolves according to the (group) *disagreement dynamics* given by

$$D\delta(t) = -L\delta(t - \tau(t))Dw(t). \tag{15}$$

In what follows, we will consider the average consensus problem of the in two cases: 1) networks with fixed topology; 2) networks with switching topology. We will prove that under appropriate conditions the system achieves average consensus globally exponentially.

4   Average Consensus in Delayed Networks with Impulsive Effects

4.1   Networks with Fixed Topology

Based on impulsive stability theory on delayed dynamical systems, the following sufficient condition for average consensus of the system (11) is established.

Theorem 1. *Let the eigenvalues of the Laplacian matrix L can be ordered as*

$$0 = \lambda_1(L) < \lambda_2(L) \leq \cdots \leq \lambda_n(L).$$

Assume that the topology of G with L is connected and the following conditions are satisfied for $m \in Z^+ = \{1, 2, \cdots, \infty\}$:

(A_1) *Denote $p = 2\lambda_2(L) - \lambda_n^2(L)\tau$, $q = \lambda_n^2(L)\tau$, such that $\tau < \dfrac{\lambda_2(L)}{\lambda_n^2(L)}$.*

(A_2) *Let $\lambda > 0$ and $\varepsilon > 0$ satisfy $\lambda - p + qe^{2\lambda\tau} \leq 0$, and*

$$\theta_m = 1 + \varepsilon + (1 + \varepsilon^{-1})\mu_m^2\lambda_n^2 e^{2\lambda\tau}, \quad \theta = \sup_{m \in z^+}\{\frac{\ln\theta_m}{t_m - t_{m-1}}\}.$$

such that $\theta < \lambda$.
Then the dynamical network (11) achieve average consensus globally exponentially. □

Proof. Now we rewrite Eq. (15) as

$$D\delta(t) = -L\delta(t - \tau(t))Dw(t) = \left[-L\delta(t) + L\int_{t-\tau(t)}^t \dot\delta(s)ds\right]Dw(t). \tag{16}$$

Let us construct the Lyapunov functional as the following:

$$V(t) = \frac{1}{2}\delta^T(t)\delta(t). \tag{17}$$

For $t \neq t_m$, we have

$$D^+V(t) = -\delta^T(t)L\delta(t) - \delta^T(t)L\int_{t-\tau(t)}^t L\delta(s - \tau(s))ds$$

$$\leq -\lambda_2(L)\delta^T(t)\delta(t) + \frac{1}{2}\lambda_n^2(L)\int_{t-\tau(t)}^t \left[\delta^T(t)\delta(t) + \delta^T(s - \tau(s))\delta(s - \tau(s))\right]ds$$

$$\leq -\lambda_2(L)\delta^T(t)\delta(t) + \lambda_n^2(L)\tau\left[V(t) + \sup_{t-2\tau\leq s\leq t}V(s)\right]$$

$$\leq -pV(t) + q\sup_{t-2\tau\leq s\leq t}V(s). \tag{18}$$

On the other hand, by using the properties of Dirac measure, we have

$$\delta(t_m) = \delta(t_m^-) - L\delta(t_m - \tau(t))\mu_m.$$

Then, by Lemma 1, we can get

$$V(t_m) = \frac{1}{2}|\delta(t_m)|^T|\delta(t_m)|$$

$$\leq \frac{1}{2}[|\delta(t_m^-)|^T|\delta(t_m^-)| + |\mu_m||\delta(t_m^-)|^T L|\delta(t_m - \tau(t))|$$

$$+ |\mu_m||\delta(t_m - \tau(t))|^T L|\delta(t_m^-)| + \mu_m^2|\delta(t_m - \tau(t))|^T L^2|\delta(t_m - \tau(t))|]$$

$$\leq \frac{1}{2}[(1 + \varepsilon)|\delta(t_m^-)|^T|\delta(t_m^-)| + (1 + \varepsilon^{-1})\mu_m^2\lambda_n^2|\delta(t_m - \tau(t))|^T|\delta(t_m - \tau(t))|]$$

$$\leq (1 + \varepsilon)V(t_m^-) + (1 + \varepsilon^{-1})\mu_m^2\lambda_n^2\sup_{-2\tau\leq s<0}V(t_m + s). \tag{19}$$

It follows from Lemma 2 that, for $t_{m-1} \leq t < t_m$, $m \in Z^+$, we have

$$V(t) \leq \theta_1 \cdots \theta_{m-1} e^{\lambda(t-t_0)} \sup_{-2\tau \leq s \leq 0} V(t_0 + s)$$

$$\leq e^{-(\lambda-\theta)(t-t_0)} \sup_{-2\tau \leq s \leq 0} V(t_0 + s). \tag{20}$$

Therefore, for all $t \geq t_0$,

$$V(t) \leq e^{-(\lambda-\theta)(t-t_0)} \sup_{-2\tau \leq s \leq 0} V(t_0 + s). \tag{21}$$

This completes the proof of Theorem 1.   □

4.2 Networks with Switching Topology

Since the nodes of the network are moving, it is not hard to imagine that some of the existing communication links can fail simply due to the existence of an obstacle between two agents. The opposite situation can arise where new links between nearby agents are created because the agents come to an effective range of detection with respect to each other. In terms of the network topology G, this means that certain number of edges are added or removed from the graph. Here, we are interested in investigating such a problem: for a *network with switching topology*, whether it is still possible to reach a consensus or not. In this case, the following hybrid system is considered:

$$Dx(t) = -L_k x(t - \tau(t)) Dw(t), \quad k = s(t) \in \mathscr{I}_0, \tag{22}$$

where $L_k = L(G_k)$ is the Laplacian of graph G_k, and $s(t) : [0, +\infty) \rightarrow \mathscr{I}_0 \subseteq \{1, \cdots, \frac{n(n-1)}{2}\}$ ($\frac{n(n-1)}{2}$ denotes the total number of all possible undirected graphs) is a switching signal that determines the communication topology G. If $s(t)$ is a constant function, then the corresponding topology is fixed.

Under arbitrary switching signal, $\alpha = Ave(x)$ is also an invariant quantity. This allows the decomposition of any solution $x(t)$ of the system (22) in the form Eq. (14). Therefore, the disagreement switching system induced by the system (22) takes the following form:

$$D\delta(t) = -L_k \delta(t - \tau(t)) Dw(t), \quad k = s(t) \in \mathscr{I}_0. \tag{23}$$

Theorem 2. *Let the eigenvalues of the Laplacian matricex L_k be ordered as*

$$0 = \lambda_1(L_k) < \lambda_2(L_k) \leq \cdots \leq \lambda_n(L_k).$$

and denote

$$\bar{\lambda}_2 = \min \lambda_2(L_k), \qquad \bar{\lambda}_n = \max \lambda_n(L_k).$$

Assume that the topology of G_k with L_k is connected, and the following conditions are satisfied for $m \in Z^+ = \{1, 2, \cdots, \infty\}$:

(A_1) *Denote* $p = 2\bar{\lambda}_2 - \bar{\lambda}_n^2 \tau$, $q = \bar{\lambda}_n^2 \tau$, *such that* $\tau < \dfrac{\bar{\lambda}_2}{\bar{\lambda}_n^2}$.

(A_2) *Let* $\lambda > 0$ *and* $\varepsilon > 0$ *satisfy* $\lambda - p + q e^{2\lambda\tau} \leq 0$, *and*

$$\theta_m = 1 + \varepsilon + (1 + \varepsilon^{-1})\mu_m^2 \bar{\lambda}_n^2 e^{2\lambda\tau}, \quad \theta = \sup_{m \in z^+}\{\frac{\ln \theta_m}{t_m - t_{m-1}}\}.$$

such that $\theta < \lambda$.

Then the dynamical network (22) achieve average consensus globally exponentially.

$\square$

Proof. The proof of Theorem 2 is similar to that of Theorem 1 by the same Lyapunov function, and hence it is omitted. The proof of Theorem 2 is completed. $\square$

5   Numerical Simulations

In this section, numerical simulations will be given to illustrate the theoretical results obtained in the previous section.

Example 1. Consider an undirected network with fixed topology G_a in Fig. 1. It is easy to see that G_a is a connected graph. For the case with fixed topology, we have

$$L_a = \begin{pmatrix} 2 & -1 & 0 & \cdots & -1 \\ -1 & 2 & -1 & \cdots & 0 \\ 0 & -1 & 2 & \cdots & 0 \\ \vdots & \vdots & \vdots & \ddots & \vdots \\ -1 & 0 & 0 & \cdots & 2 \end{pmatrix}$$

and

$$\lambda_2 = 0.3820, \qquad \lambda_{10} = 4, \qquad \frac{\lambda_2}{\lambda_{10}^2} = 0.0239.$$

By taking $\tau = 0.0200$, we obtain $p = 2\lambda_2 - \lambda_{10}^2 \tau = 0.4440$, $q = \lambda_{10}^2 \tau = 0.3200$.

For simplicity, we consider the equidistant impulsive interval $t_m - t_{m-1} = 0.5$, and $\mu_m = 0.0050$. By picking $\varepsilon = 0.01$, it is easy to find that $\theta_m = 1.0506$, and

$$\frac{\ln \theta_m}{t_m - t_{m-1}} = 0.0987 < \lambda = 0.1224,$$

where $\lambda = 0.1224$ is an unique solution of equation: $\lambda - p + q e^{2\lambda\tau} = 0$. It follows from Theorem 1 that the dynamical network (11) with the graph G_a achieve average

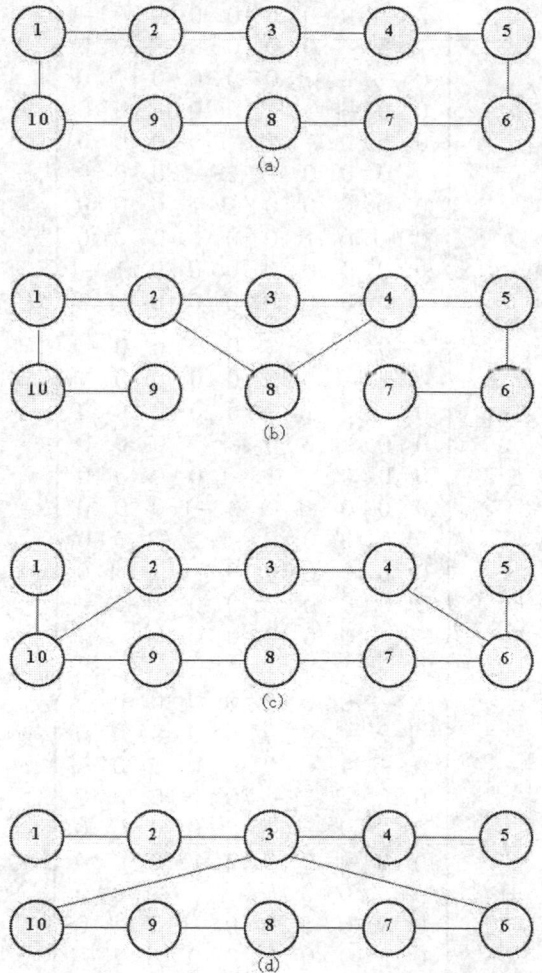

Fig. 1. Four examples of undirected graphs

consensus globally exponentially. Fig. 2 is the simulations results corresponding to this situations.

Example 2. Consider an undirected network with switching topology $\{G_a, G_b, G_c, G_d\}$ in Fig. 1. In Fig. 3, a finite state machine is shown with four states $\{G_a, G_b, G_c, G_d\}$, which represent the discrete-states of a network with switching topology and time-delays as a hybrid system. The hybrid system starts at the discrete state G_a, and switches every 0.2s to the next state according to the state machine as shown in Fig. 3. In this case, some of the existing communication links fail and some of them are created due to the moving of the agents. It is easy to see that the topologies G_b, G_c and G_d are all connected,

$$\mathbf{L_b} = \begin{pmatrix} 2 & -1 & 0 & 0 & 0 & 0 & 0 & 0 & 0 & -1 \\ -1 & 3 & -1 & 0 & 0 & 0 & 0 & -1 & 0 & 0 \\ 0 & -1 & 2 & -1 & 0 & 0 & 0 & 0 & 0 & 0 \\ 0 & 0 & -1 & 3 & -1 & 0 & 0 & -1 & 0 & 0 \\ 0 & 0 & 0 & -1 & 2 & -1 & 0 & 0 & 0 & 0 \\ 0 & 0 & 0 & 0 & -1 & 2 & -1 & 0 & 0 & 0 \\ 0 & 0 & 0 & 0 & 0 & -1 & 1 & 0 & 0 & 0 \\ 0 & -1 & 0 & -1 & 0 & 0 & 0 & 2 & 0 & 0 \\ 0 & 0 & 0 & 0 & 0 & 0 & 0 & 0 & 1 & -1 \\ -1 & 0 & 0 & 0 & 0 & 0 & 0 & 0 & -1 & 2 \end{pmatrix},$$

$$\mathbf{L_c} = \begin{pmatrix} 1 & 0 & 0 & 0 & 0 & 0 & 0 & 0 & 0 & -1 \\ 0 & 2 & -1 & 0 & 0 & 0 & 0 & 0 & 0 & -1 \\ 0 & -1 & 2 & -1 & 0 & 0 & 0 & 0 & 0 & 0 \\ 0 & 0 & -1 & 2 & 0 & -1 & 0 & 0 & 0 & 0 \\ 0 & 0 & 0 & 0 & 1 & -1 & 0 & 0 & 0 & 0 \\ 0 & 0 & 0 & -1 & -1 & 3 & -1 & 0 & 0 & 0 \\ 0 & 0 & 0 & 0 & 0 & -1 & 2 & -1 & 0 & 0 \\ 0 & 0 & 0 & 0 & 0 & 0 & -1 & 2 & -1 & 0 \\ 0 & 0 & 0 & 0 & 0 & 0 & 0 & -1 & 2 & -1 \\ -1 & -1 & 0 & 0 & 0 & 0 & 0 & 0 & -1 & 3 \end{pmatrix},$$

and

$$\mathbf{L_d} = \begin{pmatrix} 1 & -1 & 0 & 0 & 0 & 0 & 0 & 0 & 0 & 0 \\ -1 & 2 & -1 & 0 & 0 & 0 & 0 & 0 & 0 & 0 \\ 0 & -1 & 4 & -1 & 0 & -1 & 0 & 0 & 0 & -1 \\ 0 & 0 & -1 & 2 & -1 & 0 & 0 & 0 & 0 & 0 \\ 0 & 0 & 0 & -1 & 1 & 0 & 0 & 0 & 0 & 0 \\ 0 & 0 & -1 & 0 & 0 & 2 & -1 & 0 & 0 & 0 \\ 0 & 0 & 0 & 0 & 0 & -1 & 2 & -1 & 0 & 0 \\ 0 & 0 & 0 & 0 & 0 & 0 & -1 & 2 & -1 & 0 \\ 0 & 0 & 0 & 0 & 0 & 0 & 0 & -1 & 2 & -1 \\ 0 & 0 & -1 & 0 & 0 & 0 & 0 & 0 & -1 & 2 \end{pmatrix},$$

with

$$\lambda_2(L_b) = 0.1522, \quad \lambda_{10}(L_b) = 4.8260, \quad \lambda_2(L_c) = 0.3249,$$
$$\lambda_{10}(L_c) = 4.4812, \quad \lambda_2(L_d) = 0.3187, \quad \lambda_{10}(L_d) = 5.3234,$$

and so

$$\overline{\lambda}_2 = \min \lambda_2(L_k) = 0.1522, \quad \overline{\lambda}_{10} = \max \lambda_{10}(L_k) = 5.3234, \quad \frac{\overline{\lambda}_2}{\overline{\lambda}_{10}^2} = 0.0054.$$

By taking $\tau = 0.0050$, we obtain $p = 2\overline{\lambda}_2 - \overline{\lambda}_{10}^2\tau = 0.1627, q = \overline{\lambda}_{10}^2\tau = 0.1417$.

We also consider the equidistant impulsive interval $t_m - t_{m-1} = 0.8$, and $\mu_m = 0.0015$. By picking $\varepsilon = 0.01$, it is easy to find that $\theta_m = 1.0164$, and

$$\frac{\ln \theta_m}{t_m - t_{m-1}} = 0.0203 < \lambda = 0.0210,$$

where $\lambda = 0.0210$ is an unique solution of equation: $\lambda - p + q e^{2\lambda\tau} = 0$. From Theorem 2, we know that the dynamical network (22) with switching topology $\{G_a, G_b, G_c, G_d\}$ given in Fig. 3. achieve average consensus globally exponentially, whose the simulations is shown in Fig. 4.

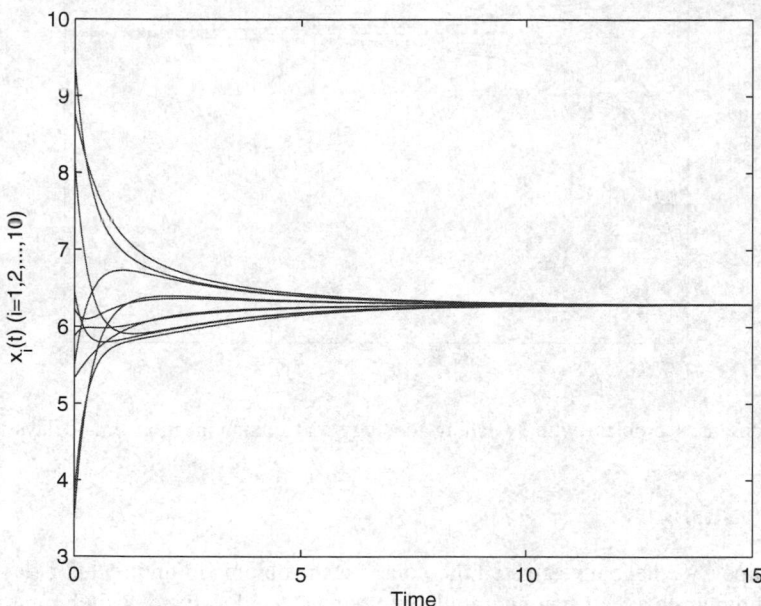

Fig. 2. Consensus problem with fixed topology and communication time-delays on graph G_a given in Fig. 1

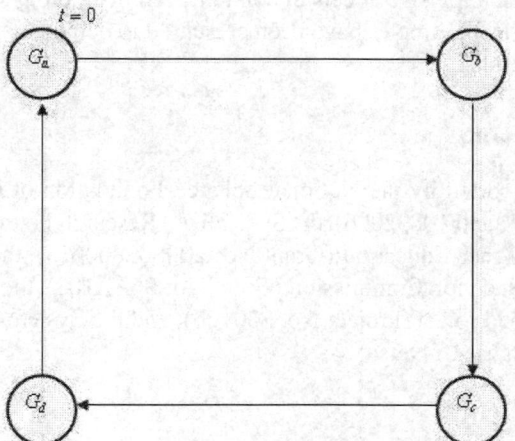

Fig. 3. Finite automation with four states representing the discrete-states of a network with switching topology and time-delays

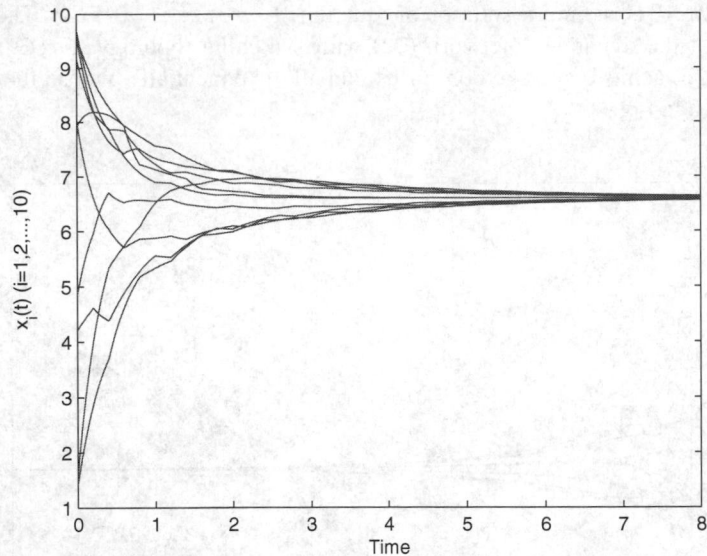

Fig. 4. Consensus problem with switching topology and communication time-delays given in Fig. 1

6   Conclusions

In this paper, we have investigated the consensus problems in undirected delayed networks of dynamic agents having impulsive effects as well as fixed, switching topology. Based on impulsive stability theory on delayed dynamical systems, we derive some simple sufficient conditions under which all the nodes in the network achieve average consensus globally exponentially. It is shown that average consensus in the networks is heavily dependent on impulsive effects of communication topology of the networks. To this end, two numerical examples have been presented to demonstrate the effectiveness of our theoretical results.

Acknowledgements

This work was supported by the National Science Foundation of China (Grant Nos. 10672094, 60474071 and 10832006), the Specialized Research Foundation for the Doctoral Program of Higher Education (Grant No. 200802800015), the Science Foundation of Shanghai Education Commission (Grant No. 06AZ101), the Shanghai Leading Academic Discipline Project (Project No. S30106), and the Systems Biology Research Foundation of Shanghai University.

References

1. Amritkar, R.E., Jalan, S.: Self-Organized and Driven Phase Synchronization Incoupled Map Networks. Physica A 321, 220–225 (2003)
2. Warburton, K., Lazarus, J.: Tendency-Distance Models of Social Cohesion in Animal Groups. J. Theor. Biol. 150(4), 473–488 (1991)

3. Breder, C.M.: Equations Descriptive of Fish Schools and Other Animal Aggregations. Ecology 35(3), 361–370 (1954)
4. Lin, Z., Francis, B., Maggiore, M.: Necessary and Sufficient Graphical Conditions for Formation Control of Unicycles. IEEE Trans. Autom. Control 50(1), 121–127 (2005)
5. Hong, Y., Hu, J., Gao, L.: Tracking Control for Multi-Agent Consensus with an Active Leader and Variable Topology. Automatica 42, 1177–1182 (2006)
6. Olfati-Saber, R., Murray, R.M.: Consensus Problems in Networks of Agents with Switching Topology and Time-Delays. IEEE Trans. Autom. Control 49(9), 1520–1533 (2004)
7. Olfati-Saber, R.: Flocking for Multi-Agent Dynamic Systems: Algorithms and Theory. IEEE Trans. Autom. Control 51(3), 401–420 (2006)
8. Vicsek, T., Czirok, A., Jacob, E.B., Cohen, I., Schochet, O.: Novel Type of Phase Transition in a System of Self-Driven Particles. Phys. Rev. Lett. 75(6), 1226–1229 (1995)
9. Czirok, A., Vicsek, T.: Collective Behavior of Interacting Self-Propelled Particles. Physica A 281, 17–29 (2000)
10. Fax, J.A., Murray, R.M.: Information Flow and Cooperative Control of Vehicle Formations. IEEE Trans. Automat. Control 49(9), 1465–1476 (2004)
11. Mu, S., Chu, T., Wang, L.: Coordinated Collective Motion in a Motile Particle Group with a Leader. Physica A 351, 211–226 (2005)
12. Toner, J., Tu, Y.: Flocks, Herds, and Schools: A Quantitative Theory of Flocking. Phys. Rev. E 58(4), 4828–4858 (1998)
13. Lawton, J.R., Beard, R.W.: Synchronized Multiple Spacecraft Rotations. Automatica 38, 1359–1364 (2002)
14. Lynch, N.A.: Distributed Algorithms. Morgan Kaufmann, San Mateo (1997)
15. DeGroot, M.H.: Reaching a Consensus. J. Am. Statist. Assoc. 69(345), 118–121 (1974)
16. Jadbabaie, A., Lin, J., Morse, A.S.: Coordination of Groups of Mobile Autonomous Agents Using Nearest Neighbor Rules. IEEE Trans. Autom. Control 48(6), 988–1001 (2003)
17. Lin, Z., Broucke, M., Francis, B.: Local Control Strategies for Groups of Mobile Autonomous Agents. IEEE Trans. Automat. Control 49(4), 622–629 (2004)
18. Moreau, L.: Stability of Multiagent Systems with Time-Dependent Communication Links. IEEE Trans. Automat. Control 50(2), 169–182 (2005)
19. Ren, W., Beard, R.W.: Consensus Seeking in Multiagent Systems under Dynamically Changing Interaction Topologies. IEEE Trans. Automat. Control 50(5), 655–661 (2005)
20. Xiao, L., Boyd, S.: Fast Linear Iterations for Distributed Averaging. Systems Control Lett. 53, 65–78 (2004)
21. Hale, J.K., Verduyn Lunel, S.M.: Introduction to Functional Differential Equations. Springer, New York (1993)
22. Zhou, J., Chen, T.P.: Synchronization in General Complex Delayed Dynamical Networks. IEEE Trans. Circ. Syst. I 53(3), 733–744 (2006)
23. Zhou, J., Xiang, L., Liu, Z.R.: Global Synchronization in General Complex Delayed Dynamical Networks and Its Applications. Physica A 382(2), 729–742 (2007)
24. Chen, G.R., Zhou, J., Liu, Z.R.: Global Synchronization of Coupled Delayed Neural Networks and Applications to Chaotic CNN Models. Int. J. Bifur & Chaos 14(7), 2229–2240 (2004)
25. Sinopoli, B., Schenato, L., Franceschetti, M., Poola, K., Jordan, M.I., Sastry, S.S.: Kalman Filtering with Intermittent Observations. IEEE Trans. Autom. Control 49(9), 1453–1464 (2004)
26. Hatano, Y., Mesbahi, M.: Agreement Over Random Networks. IEEE Trans. Autom. Control 50(11), 1867–1872 (2005)
27. Mesbahi, M.: On State-Dependent Dynamic Graphs and Their Controllability Properties. IEEE Trans. Autom. Control 50(3), 387–392 (2005)

28. Earl, M.G., Strogatz, S.H.: Synchronization in Oscillator Networks with Delayed Coupling: A Stability Criterion. Phys. Rev. E 67, 36204 (2003)
29. Sun, Y.G., Wang, L., Xie, G.: Average Consensus in Networks of Dynamic Agents with Switching Topologies and Multiple Time-Varying Delays. System Control Lett. 57, 175–183 (2008)
30. Lin, P., Jia, Y.: Average Consensus in Networks of Multi-Agents with both Switching Topology and Coupling Time-Delay. Physica A 387, 303–313 (2008)
31. Guan, Z.H., Liu, Y.Q., Wen, X.C.: Decentralized Stabilization of Singular and Time-Delay Large-Scale Control Systems with Impulsive Solutions. IEEE Trans. Autom. Control 40(8), 1437–1441 (1995)
32. Zhou, J., Xiang, L., Liu, Z.R.: Synchronization in Complex Delayed Dynamical Networks with Impulsive Effects. Physica A 384, 684–692 (2007)
33. Xu, J., Chung, K.W.: Effects of Time Delayed Position Feedback on a Van Der Pol-Duffing Oscillator. Physica D 180(1), 17–39 (2003)
34. Yang, Z.C., Xu, D.Y.: Stability Analysis of Delay Neural Networks with Impulsive Effects. IEEE Trans. Circuits Syst. II 52(8), 517–521 (2005)
35. Sun, Y.G., Wang, L., Xie, G.: Average Consensus in Directed Networks of Dynamic Agents with Time-Varying Communication Delays. In: 45th IEEE Conf. Decision and Control, pp. 3393–3398 (2006)
36. Park, J.H.: Synchronization of a Class of Chaotic Dynamic Systems with Controller Gain Variations. Chaos, Solitons and Fractals 27, 1279–1284 (2006)

Basic Notions and Models in Systems Science

Janos Korn

Visiting Academic, Middlesex University, UK
janos999@btinternet.com

Abstract. The development of the idea of seeing parts of the world as 'related objects' or the 'systemic view' and its relation to conventional science is briefly described. Concepts in the systemic view regarded as fundamental and their expression as linguistic and mathematical models which would turn this view into 'systems science', are introduced. Products are represented as sets and linguistic networks of ordered pairs. Semantic diagrams describe the dynamics of change. A case study to illustrate the basic notions and models is given.

Keywords: systemic view, linguistic modeling, sets of ordered pairs, sequences of predicate logic statements.

1 Introduction

People have been trying to create *symbolic structures* for expressing thoughts about aspects of the surrounding world of concrete objects including parts of their own bodies since ancient times. In particular, they have been using this kind of structures for the pursuance of knowledge to understand and possibly to change parts of the world.

Concepts formulated in the brain/mind assembly as a result of abstraction of observation of parts of the world are used for producing *descriptions, explanations and predictions* by means of 'things which stand for other things'. The history of human knowledge of the world may be seen as the history of development of *'things which stand for other things'*. Ancient and current methods of predictions like heated bones of selected animals, flight of birds, tarot cards, lines in the palm of hand or patterns of tea leaves at the bottom of a tea cup, are *things* which are claimed to stand for *other things* but there is *no* recognised *relation* between the two. There is no systematic correlation between the change of shape of a heated bone and the outcome of a forthcoming battle.

The idea of viewing parts of the world as *combinations of 'building blocks'* as models in terms of which parts of the world could be seen like Thales' water, Aristotle's suggestion of earth, water, air and fire, Democritos' 'atoms' or the modern 'binary digits', had been an advance. These notions were empirical concepts with relations to parts of the world for which they claimed to have stood as 'things which stand for other things'.

J. Zhou (Ed.): Complex 2009, Part I, LNICST 4, pp. 1139–1145, 2009.

Conventional science of physics, chemistry etc, the first organised and methodical body of 'things which stand for other things', views parts of the world in terms of groups of *quantitative properties*, the 'building blocks', usually classified into mechanical, electrical etc and arranged in mathematical relations of varying complexity. Its generalisations are based on regular recurrence of phenomena found mostly in the inanimate world. Pursuance of *truth* of views and the acceptance or rejection of a view through testing against experience, are marks of the scientific method. Irreversible thermodynamics, information theory, multidisciplinary, especially dynamic, purposive or control activities in technology and in animate, especially human scenarios together with design thinking are by and large *outside the scope* of conventional science. Conventional science is unhappy with dealing with more than one object as happens in network analysis, for example.

A part of the world called 'object' in static or dynamic states or events, can also be viewed as an ordered collection of 'related properties or related other objects' which is the *systemic view*, the view of complexity and hierarchy. The view is applicable to inanimate and animate, technical and human objects with *qualitative* as well as *quantitative properties* joined into *related* or *interacting wholes* such as relations like 'diameter -'is perpendicular to' - length', 'john – 'is father of' – mary ', 'john – 'tells' - tom (to cut the grass)' or 'engine – 'pulls' the long train (fast)'.

We may say that by and large:

Conventional science deals with 'Groups of quantifiable properties of objects organised into mathematical models to be exposed to *tests of experience*' in pursuance of reliable knowledge, and

Systems science deals with 'Groups of properties or objects organised into explicitly stated linguistic relations or interactions which can carry mathematical models, to be investigated for the occurrence of specific *outcomes*'. The aim of current work is to develop the details of this topic.

The relatively recent systemic view, perhaps began in earnest with the development of servomechanisms in the 2nd world war and has created immense interest. The result has been: A vast amount of verbal and written discussion on a variety of topics loosely if at all related to the systemic view or systems thinking presented at conferences and courses at university departments but not in schools (!) together with the appearance of a large number of books and papers. And the activity is going on unabated. All this is of great interest but the essentially *pervasive, empirical* and *indivisible* systemic view has remained by and large fragmented, speculative and a subject of discussion.

The 'founding fathers' of the 'systemic view' concentrated on the horizontal view or complexity and discarded conventional science *in its entirety* as inapplicable to describing this view. This, with hindsight, was a mistake since the approach or the methodology of conventional science appears to be not only applicable but needed if we want to turn the systemic view from a view into *systems science*. The search for general, fundamental notions behind the immense *diversity and variety* of the systems phenomenon has ceased after the initial enthusiasm in the 1950's to 1990's. Systems thinking has become fragmented into information systems, systems engineering, 'soft'/'hard' methods, systems dynamics, chaos theory and other interesting ideas [1], [2], [3], [4].

The objective of current work is to suggest what may be regarded as *fundamental notions* of the systemic view and the *symbolism* that is capable of expressing these notions in operational form so as to lead to models with *outcomes* which can be exposed to at least thought experiments. This comprehensive approach perhaps can turn the systemic view from being a view into systems science, can lead to a general design methodology including elicitation of requirements and to a more concrete view of organisations and management [5], [6], [7], [8], [9].

2 Basic Notions and Models

A part of the world in static state is viewed as *related* concrete properties or qualified objects with relations signaled by *relation indicators* of space (left), order (first), kinship (father), stative (to support) and passive voice of dynamic verbs (is kicked). A part of the world in dynamic state is viewed as *interacting* objects qualified by properties with interaction signaled by *dynamic verbs* like 'to kick'. Interaction in terms of physical power carries the appropriate *energy* and interaction in terms of influence carries relevant *information*. There is no 'change of state' without interaction. Only one property can be changed at a time by a *product* which is produced by a collection of interacting objects acting in accordance with an algorithm, called an *organisation*.

The starting point of modeling static and dynamic states is a story of a scenario in natural language, the primary model, which, due to linguistic complexities, is transcribed by linguistic analysis through meaning preserving transformations, into *homogeneous language* of one – and two – place, context dependent sentences. These sentences are regarded as *building blocks*. Static states are modeled by identifying these sentences with 'ordered pairs' leading to 'linguistic networks'. Dynamic states are modeled by turning similar sentences into 'semantic diagrams' from the topology of which sequences of 'predicate logic' statements qualified by mathematics if there is need, can be derived.

In general, linguistic modeling exhibits the *structure or topology of products* (through sets of ordered pairs and linguistic networks) and the *structure or topology of systems* (through semantic diagrams and sequences of logic statements). Elements of these structures can carry qualifiers with *uncertainties and mathematics*, aiding or hindering the appearance of *'emergent properties'* of products and *'final states as acquired properties'* of systems as outcomes.

3 A Case Study

The story of a scenario is as follows: 'A shopkeeper wants to increase his takings from customers who appear rather unhappy at present possibly because their level of satisfaction is low. This is the shopkeeper's impression and he is seeking ways and means to effect improvement which is in the interest of customers as well. His main line of merchandise is cheese, ham and tuna sandwiches which are delivered to the shop all mixed up. The shopkeeper thinks that easy access and selection of sandwiches may lead to improvement. A single assistant operating sequentially in purposive mode, is used to implement a scheme'.

The terms 'access' and 'selection' are abstract. They need to be expressed in concrete terms:

'Access' means that 'sandwiches' are *placed* on 'shelves' when the latter are *cleaned*.

'Selection' means that 'sandwiches' are to be *arranged* and *priced* according to their 'contents (cheese, ham or tuna)'.

'Priced' means that labels (cheese, £, ham, £, tuna, £) are *attached* to sandwiches.

We construct a set of ordered pairs or an array based on data in (1). There are 4 common nouns or objects: sw (sandwiches), sh (shelves), co (contents) and la (labels). We assign relation indicators as stative verbs and the passive voice of dynamic verbs to the objects i.e. introducing ordered pairs:

$$
\begin{aligned}
&i = 1 = \text{sandwiches, sw1 (are placed on, pl)}\\
&i = 2 = \text{shelves, sh (are clean, cl)}\\
&i = 3 = \text{sandwiches, sw2 (use to arrange, us)}\\
&i = 4 = \text{sandwiches, sw3 (carry, ca)}\\
&i = 5 = \text{contents, co (used to arrange, ar)}\\
&i = 6 = \text{labels, la (to mark prices, ma)}\,.
\end{aligned}
\tag{1}
$$

Here we have three 'sandwiches' as distinguished by the relations representing the roles which the same object can play in different scenarios.

(2) shows the set of ordered pairs which gives a large number of choices the number of which can be precisely calculated. A particular choice or subset is selected from considerations of *particular product* which defines the *conceptual boundary*

n_{11}	n_{12}	n_{13}	n_{14}	n_{15}	n_{16}
0	<u>sw1 pl sh</u>	sw1 pl sw2	sw1 pl sw3	sw1 pl co	sw1 pl la
n_{21}	n_{22}	n_{23}	n_{24}	n_{25}	n_{26}
0	<u>sh cl sh</u>	0	0	0	0
n_{31}	n_{32}	n_{33}	n_{34}	n_{35}	n_{36}
sw2 us sw1	sw2 us sh	0	sw2 us sw3	<u>sw2 us co</u>	sw2 us la
n_{41}	n_{42}	n_{43}	n_{44}	n_{45}	n_{46}
sw3 ca sw1	sw3 ca sh	sw3 ca sw2	0	sw3 ca co	<u>sw3 ca la</u>
n_{51}	n_{52}	n_{53}	n_{54}	n_{55}	n_{56}
<u>co ar sw1</u>	co ar sh	co ar sw2	co ar sw3	0	co ar la
n_{61}	n_{62}	n_{63}	n_{64}	n_{65}	n_{66}
la ma sw1	la ma sh	la ma sw2	la ma sw3	<u>la ma co</u>	0 .

$$\tag{2}$$

with an *emergent property* which, in this case, is labeled as, 'creator of easy access and selection of sandwiches'. (3) is a conceptual boundary and Fig. 1. shows the tree which corresponds to this subset.

$$\text{creator} \ldots = \prod_{i=1}^{i=6} (n_{12} \times n_{22} \times n_{35} \times n_{46} \times n_{51} \times n_{65}) \,. \tag{3}$$

where: $n_{12} = $ sw1 are placed on sh, $n_{22} = $ sh are clean, $n_{35} = $ sw2 use to arrange co, $n_{46} = $ sw3 carry la, $n_{51} = $ co used to arrange sw1, $n_{65} = $ la mark co.

We ignore those ordered pairs in (3) in which the first element is the second element in those ordered pairs in which the first element undergoes change of state, i.e. $n_{51} = $ co used to arrange sw1 and $n_{65} = $ la mark co. This step is *recognisable by computer software* and will affect the tree in Fig. 1. which is modified to that in Fig. 2.

The modified network in Fig. 2. will be isomorphic to the semantic diagram representing the dynamics of scenario which is shown in Fig. 3.

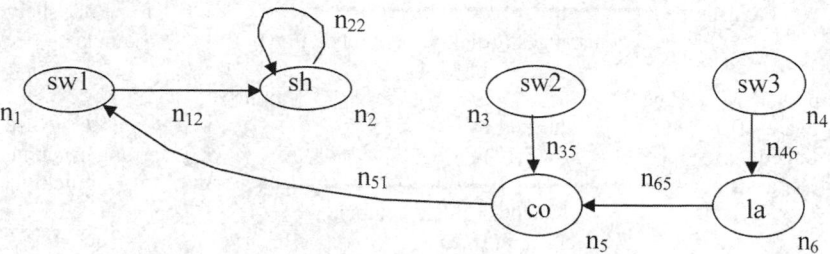

Fig. 1. Linguistic network of tree of subset

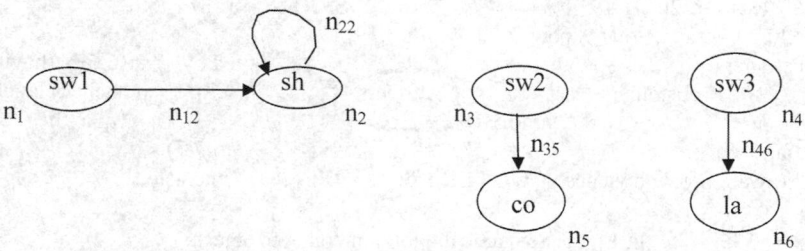

Fig. 2. Modified linguistic network from Fig. 1

A representative sample of predicate logic sequences in which the numerals designate objects in Fig. 3., is as follows:

For causal chain 1.
1/1 dp(15,15(I1),16(I2),...) ∧ ip(15,15) ∧ cp(19,15) → in(15,1)
1/2 in(15,1) ∧ ep(1,1) → ap(2,6)
and so on....

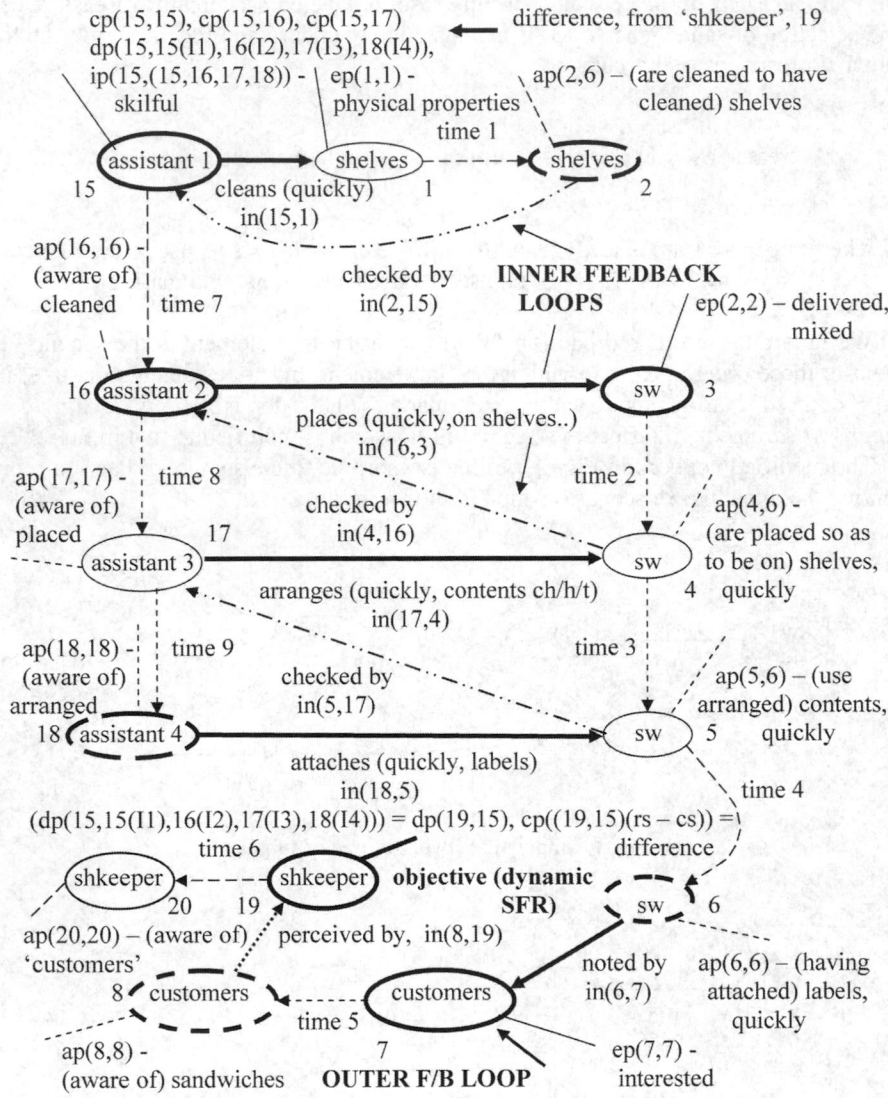

Fig. 3. Semantic diagram of concrete objects

4 Conclusions

We have introduced the basic notions and models which appear to be needed for an attempt at developing a comprehensive approach to viewing parts of the world as 'related objects' in more concrete terms. In particular, using these notions we aim at modeling human activity scenarios with their immense diversity, variety and uncertainties operating in purposive mode. We can perhaps say that purposive operations in animate activities are as common as gravity in nature [9].

The model of 'product' as in (2) offers a large number of choices one or more of which survive through selection by *evolution* or *design*. A product is constructed by an organisation and generates interaction to accomplish a change of state as shown in a semantic diagram such as Fig. 3. Dynamic linguistic modeling demonstrates how changes of state propagate in time towards *final state* of a changing object or outcome (customers). The large number of choices of product and elements of organisation is not shown. The example shown here is part of a *design* case study.

References

1. Checkland, P.: Systems Thinking Systems Practice. Wiley, Chichester (1982)
2. Jackson, M.C.: Systems Approaches to Management. Kluwer Academic, NY (2000)
3. Lane, D.C.: The Emergence and Use of Diagramming in Systems Dynamics. Systems Research and Behavioural Science 25(1), 3–23 (2008)
4. Barabasi, A.L.: Linked: The New Science of Networks. Perseus Publications, USA (2002)
5. Korn, J.: Reductionism in Systems Science. In: Proc. of UKSS Conf., Oxford, September 8-11 (2006a)
6. Korn, J.: Systemic View of Parts of the World. In: 50th Annual Conference, Sonoma State University, USA, July 9-14 (2006b)
7. Korn, J.: Linguistic Modeling of Scenarios. Scientific Inquiry 7(2), 193–210 (2006c)
8. Korn, J.: Systems View, Emergence and Complexity. Kybernetes 36(5/6), 776–790 (2007)
9. Korn, J.: Science and Design of Systems. Matador, Leicester (2009)

Bifurcation Phenomena of Opinion Dynamics in Complex Networks

Long Guo and Xu Cai

Complexity Science Center and Institute of Particle Physics,
Huazhong (Central China) Normal University, Wuhan, 430079, P.R. China
longkuo0314@gmail.com, xcai@mail.ccnu.edu.cn

Abstract. In this paper, we study the opinion dynamics of Improved Deffuant model (IDM), where the convergence parameter μ is a function of the opposite's degree K according to the celebrity effect, in small-world network (SWN) and scale-free network (SFN). Generically, the system undergoes a phase transition from the plurality state to the polarization state and to the consensus state as the confidence parameter ϵ increasing. Furthermore, the evolution of the steady opinion s_* as a function of ϵ, and the relation between the minority steady opinion s_*^{min} and the individual connectivity k also have been analyzed. Our present work shows the crucial role of the confidence parameter and the complex system topology in the opinion dynamics of IDM.

Keywords: opinion dynamics, complex networks topology, bifurcation phenomena.

1 Introduction

Our local society, which can be well modelled as complex networks, has its own structure depending on the geography, culture and history. Recently it has also been realized that many real social networks arising in society, such as networks of sexual relationships [1], collaborations between actors [2,3] and scientists [4,5], web-based social networks [6] , peer-to-peer social network [7], and the social network of a bulletin board system in a university [8] all share some universal characteristics such as the small-world effect, high clustering coefficient property and the power-law degree distribution. Those features affect the dynamics in society systems, especially the opinion dynamics in complex networks. Many natural and man-made networks have been successfully studied as a framework of several celebrated opinion models. Nevertheless, the understanding of the opinion dynamics in complex networks remains a challenge.

Social impact theory founded by Latané [9,10], was developed as a metatheo-retical framework for modelling situations where each individual is influenced by others around him/her to change his/her beliefs, attributes or behaviors. Based on social impact theory, there are two celebrated models about opinion dynamics that were proposed in recent years. One celebrated model is the binary opinion

J. Zhou (Ed.): Complex 2009, Part I, LNICST 4, pp. 1146–1153, 2009.

model that proposed by K. Sznajd-weron and J. Sznajd (S model) [11] to describe a simple mechanism of making up decisions in a closed community. In this model, the opinion of individual is a binary variable assuming the values +1 and -1 referring to two opposite opinions on a particular issue. The updating rules follow the principle of "united we stand, divided we fall". The other is the continuous model proposed by Deffuant et al (D model) [12]. In D model, the opinion s of individual can vary continuously between zero and one. Each agents selects randomly one of the other agents and checks first if an exchange of opinions makes sense. If the two opinions differ by less than ϵ ($0 < \epsilon < 1$), each opinion moves partly in the direction of the other, by amount $\mu \Delta s$, where Δs is the opinion difference and μ the convergence parameter ($0 < \mu \leq 0.5$); otherwise, the two refuse to discuss and no opinion is changed. The parameter ϵ is called confidence bound or confidence parameter. In our society, individual typically has a continuous opinion and always is influenced by his/her acquaintances or other external factors, such as advertisement, newspapers and broadcast, to change his/her opinion. Here, we pay most of our attention to the interaction between individuals and do not consider the influence of the external factors. On the other hand, each agent has his/her own influence of persuading others to agree with him/her, and also has his/her own ability to keep his/her opinion from changing. In our present work, we study the dynamics of continuous opinion of improved Deffuant model (IDM) with heterogeneous convergence parameters μ, which is a function of the opposite's degree k according to the celebrity effect, in complex networks.

The main goal of this paper is to study the opinion dynamics of IDM in complex networks with various confidence parameter ϵ. Generically, the system undergoes a phase transition from the plurality state to the polarization state to the consensus state as ϵ increasing, in the both small-world network (SWN) and scale-free network (SFN). Then, we focus on the evolution of the steady opinion s_* as the function of the confidence parameter ϵ in the both celebrated complex networks, and find that there exists a bifurcation diagram of the steady opinion s_* as ϵ increasing. Furthermore, we analyzed the relation between the minority steady opinion s_*^{min} and the individual connectivity k in the both SWN and SFN, and find that the process of opinion dynamics of IDM in SWN is complete different from that in SFN. Our present work reveals the dependence of the opinion dynamics of IDM on the confidence parameter ϵ between individuals and the complex system topology.

2   Improved Deffuant Model

Many real society systems can be well mapped to complex networks, which are sets of distinguishable nodes $i = 1, 2, \ldots, N$, connected by a fixed number of $l = 1, 2, \ldots, L$ indistinguishable edges. Those edges represent the different relationships among agents in society, such as friendship, collaboration, business, sexual and other interactions [13]. The network is represented by its adjacency matrix A, where $a_{ij} = 1$, if an edge connects nodes i and j and $a_{ij} = 0$, otherwise. There are no self-connections or multiple edges.

To realize our model simulation, we employ the celebrated small-world network (SWN) proposed by Watts and Strogatz [2] and scale-free network (SFN) proposed by Barabási and Albert [3] to study the opinion dynamics. The SWN is defined on a lattice consisting of N nodes arranged in a ring. Initially each node is connected to all of its neighbors up to some fixed range K to make the network have average coordination number $z = 2K$, and randomness is then introduced by rewiring edges between two randomly chosen nodes with rewiring probability ϕ. The random rewiring process introduces ϕNK long ranges which connect nodes that otherwise would be part of different neighborhoods. By varying ϕ one can closely monitor the transition between order ($\phi = 0$) and randomness ($\phi = 1$) [14]. And the scale-free network is built following the principle of growth and preferential attachment. The SFN of size N is generated starting from a randomly connected core of m_0 nodes and a set $U(0)$ of $(N - m_0)$ unconnected nodes. At each time step, a new node is chosen from $U(0)$ and linked to $m(m < m_0)$ other nodes with the probability of Π that a new node will be connected to node i depending on the degree k_i of node i, i.e, $\Pi(k_i) = k_i / \sum_j k_j$. Numerical simulations indicated that this network evolves into a scale-invariant state with the probability that a node has k edges following a power law with an exponent $\gamma = 3$ [14].

Many previous works about D model have considered the situation where the convergence parameters μ between pairwise agents are uniform [12,15,16]. However, in our society, we often change our opinion as the one of individual who is a famous expert about the particular issue according to the celebrity effect. In our present work, we assume that the larger the agent's connectivity is, the more famous expert the agent will be. Hence, the convergence parameters μ between pairwise agents are different, which is a function of the opposite's connectivity k.

We choose a pairwise agents i and j randomly at each time step. If the two opinions differ more than a fixed threshold parameter $\epsilon(0 < \epsilon < 1)$, called the confidence parameter, both opinions refuse to discuss and no opinion is changed. If, instead, $|s_i(t) - s_j(t)| < \epsilon$, then each opinion moves partly in the direction of the other as:

$$\begin{cases} s_i(t+1) = s_i(t) + \mu_j[s_j(t) - s_i(t)], \text{with prob.} p_i; \\ s_j(t+1) = s_j(t) + \mu_i[s_i(t) - s_j(t)], \text{with prob.} p_j. \end{cases} \quad (1)$$

where, $\mu_j = k_j/(2*k_{max})$ is the convergence parameter $(0 < \mu \le 1/2)$ that agent j interacts other agents and k_{max} is the largest connectivity degree in social complex system. The probability $p_j(= 1 - \mu_j)$ is the probability that agent j is persuaded to change his/her opinion, since each agent has the ability to keep his/her opinion from changing. The famous expert changes his/her opinion with smaller probability.

3   Results

We simulate the opinion dynamics of IDM, Eq. (1) in SWN and SFN of size $N = 1000$. The initial opinion of agent varies continuously from zero to one with a uniform distribution. All the results have been averaged over at least 100 realizations, with each running lasting for at least 2×10^4 updating steps. We choose about 400 different pairwise agents randomly at each updating step.

Generally, the system of opinion dynamics reaches a steady state from the plurality state to polarization state or to the consensus state as time elapses, which has also been found in many previous works about D model. Hence, the pictures of the evolution of opinions as a function of time steps t are not shown in our present work. Here, we focus on the dependence of the opinion dynamics of IDM on the confidence parameter ϵ and the complex system structure topology.

In Fig. 1 we represent the evolution of the steady opinion s_* as a function of confidence parameter ϵ both in SWN ($\times$) and SFN ($+$). We find that there exists bifurcation phenomena of the steady opinion s_* as the confidence parameter ϵ decreases from one to zero. Namely, the system undergoes a phase transition from the consensus state to the polarization state and then to the plurality state with ϵ decreasing. Here, the polarization state is defined as that the individuals can be divided into two or more camps according to their opinions. Each camp has its opinion that different from others obviously. The consensus state is defined as that all the individuals share the same opinion. On the other hand, we find that the steady opinion s_* of consensus state is around 0.5 in SWN and SFN. However, the fluctuation of steady opinion s_* in SFN is larger than that in SWN, probably caused by the topology structure of complex networks, which will be explained below. A detailed finite-size scaling analysis performed for both complex networks shows that the critical value of polarization and the critical value of consensus, (ϵ_p, ϵ_c), corresponds in SFN to $(0.21(4), 0.48(2))$, and in SWN to $(0.15(5), 0.40(3))$, accordingly, as shown in Fig. 1. From the bifurcation diagram of steady opinion s_* as the function of confidence parameter ϵ, the ability of polarization and consensus of SWN is much stronger than that of SFN. Namely, the ability of polarization and consensus depends on the heterogeneous property of complex networks, the more heterogeneous the complex network is, the weaker the ability of polarization and consensus of complex network will be.

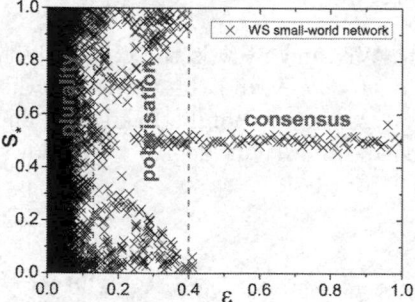 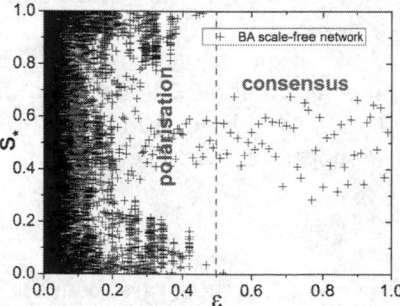

Fig. 1. (color online) Bifurcation diagram for the steady opinion s_* as a function of the confidence parameter ϵ in small-world network ($\times$ SWN) and in scale-free network ($+$ SFN) for one evolution case. The parameters of the two complex networks are: $N = 1000$, $\phi = 0.05$, $z = 18$, $m_0 = 10$, $m = 6$.

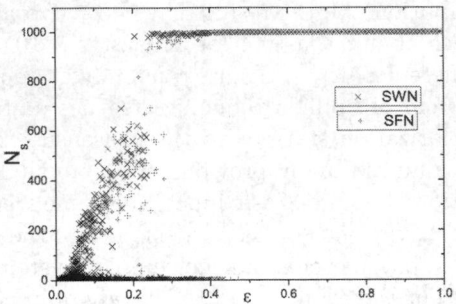

Fig. 2. (color online)Plots for the evolution of the N_{s_*} as a function of the confidence parameter ϵ in SFN (+) and in SWN (×). The network parameters are as in Fig. 1.

Further light can be shed on the dependence of the opinion dynamics on the complex network topology. In order to do this, we define the physical quantity N_{s_*} as the number of agents with the same steady opinion s_* and study the evolution of N_{s_*} as a function of ϵ in the both SWN and SFN.

In Fig. 2 we represent the evolution of the N_{s_*} as a function of the confidence parameter ϵ in SFN and SWN respectively. In the polarization region in the both complex networks, all individuals can be divided into more than one camps according to their opinions. Each camp has its opinion and size that is different from others. Here, we call the steady opinion s_* that a few individuals share, i.e., $N_{s_*} < 10$, the minority opinion s_*^{min}. Of course, there also exists the second-largest and the third-largest camps in the both complex networks, see the middle part of the picture in Fig. 2. The evolution of the largest and the second-largest camps as a function of confidence parameter ϵ of D model on adaptive networks has been analyzed by Balazs Kozma and Alain Barrat [17]. Surprisingly, we cannot find any signals to identify the different roles between the topology of SWN and that of SFN in opinion dynamics of IDM. To see this, we focus on the minority steady opinion s_*^{min} and study the relationship between the s_*^{min} and the connectivity degree k of the SWN and SFN, see the dots lie on the horizontal axis in Fig. 2.

As well known, the obvious difference of SWN and SFN is the connectivity degree distribution, the bell-form distribution to SWN and the power-law distribution to SFN accordingly. In order to analyze the relation between s_*^{min} and the connectivity degree k, we define the relative connectivity degree λ as follows,

$$\lambda = \frac{k}{k_{max}} \qquad (2)$$

where, k_{max} is the largest connectivity degree in complex network. The larger the relative connectivity degree λ of one agent is, the more famous the agent is, who plays the more important role in affecting others to change their opinions.

In Fig. 3 we represent the evolution of the relative connectivity degree λ as a function of the ϵ and the s_*^{min} in the polarization region in the both SWN and SFN. Along with the results in Fig. 1, we find that the steady opinion s_*

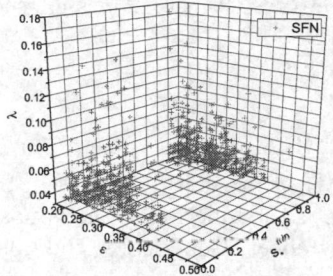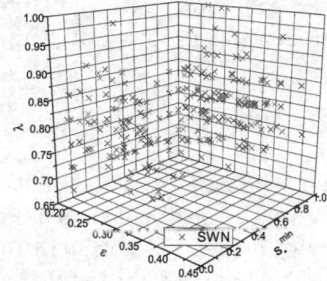

Fig. 3. (color online)Plots for the evolution of the relative degree λ as a function of confidence parameter ϵ and the minority steady opinion s_*^{min} in SFN (|) and SWN($\times$) when $N_{s_*} < 10$. The network parameters are as in Fig. 1.

is away from the middle opinion 0.5 in polarization region and is about 0.5 in the consensus state, which show the role of the compromise factor in our model algorithm. In the polarization region, there exists the steady opinions that smaller than 0.5 and more than 0.5 simultaneity in the system. The larger the confidence parameter ϵ, the farther the steady opinion will be away from 0.5.

Surprisingly, we also find that there exists a huge difference of the λ of agents with the minority steady opinion in the both celebrated complex networks. The relative connectivity degrees λ with the minority steady opinion s_*^{min} are smaller than 0.18 in SFN and larger than 0.65 in SWN respectively, which indicates that the process of the opinion formation in SFN is much different from that in SWN. As well known, the SFN following the algorithm of growth and preference attachment ia a disassortative complex network, i.e., the agents with higher connectivity, which plays the same role as hubs in the Internet, always are connected with those with smaller connectivity degree. Since the more difference between those agents' connectivity, agents with smaller connectivity are always persuaded to move their opinions enough in the direction of their nearest neighbor who has larger connectivity. Namely, the common people always change their opinions following the famous experts according to the celebrity effect. Note that the process of the polarization and consensus starts from the agents with highest connectivity and then to his nearest neighbors and last to all the agents in SFN, the agents with smaller connectivity away from all the agents who have higher connectivity will be separated as the minority with larger probability. On the other hand, the difference between agents' connectivity is much smaller due to the bell-form connectivity distribution in SWN. Hence, each pairwise agents with almost the same connectivity reaches the consensus opinion easily. Although agents change their opinions according to the celebrity effect in our present work, the agents who have higher connectivity degrees will be separated as the minority with larger probability. As time elapses and the interaction between individuals in SWN, the system will be divided into more than one camps and each camps share the same opinion firstly, then, larger camps can be made of those smaller camps. The difference process of opinion formation in the both

SFN and SWN, along with the larger fluctuation in SFN in Fig. 1, is caused by the heterogeneous property of complex network.

4   Conclusion

We propose an Improved Deffuant model (IDM) of opinion dynamics in SWN and in SFN, where the convergence parameter μ is the function of the opposite's connectivity degree k according to the celebrity effect. Generically, the opinion dynamics reaches the steady state (polarization state or consensus state, which is related to the confidence parameter ϵ) in the both celebrated complex networks. We find that the steady opinion s_* undergoes a bifurcation phenomenon as the confidence parameter ϵ increases, from the plurality state to the polarization state and to the consensus state. In order to show the effect of complex network topology, we pay most of our attention to the property of the agents with the same minority steady opinion s_*^{min} in the polarization region. We find that there also exists a bifurcation phenomena of the s_*^{min} as ϵ increases in SFN and in SWN, which is caused by the compromise factor in our model. Further, we find that a few agents who have smaller connectivity degree are persuaded easily as the minority with larger probability in SFN; otherwise, a few agents who have higher connectivity degree are persuaded easily as the minority with larger probability in SWN. All those results indicate that the process of the polarization and consensus of opinion dynamics in SFN is different from that in SWN, along with the fluctuation in SFN in Fig. 1, probably caused by the heterogeneous property of complex networks. Our present work opens new paths to understand the bifurcation phenomena of opinion formation in complex networks.

Acknowledgments

L.Guo thanks Prof. W. Li for helpful suggestions and comments. This work was supported by the National Natural Science Foundation of China under Grant No.s 70571027, 10635020 and by the Ministry of Education in China under Grant No. 306022.

References

1. Liljeros, F., Edling, C.R., Amaral, L.A.N., Stanley, H.E., Aberg, Y.: The web of human sexual contacts. Nature 411, 907 (2001)
2. Watts, D.J., Strogatz, S.H.: Collective dynamics of 'small-world' networks. Nature 393, 440 (1998)
3. Barabási, A.-L., Albert, R.: Emergence of Scaling in Random Networks. Science 286, 509 (1999)
4. Newman, M.E.J.: The structure of scientific collaboration networks. Proc. Natl. Acad. Sci. U.S.A. 98, 404 (2001)
5. Barabási, A.-L., Jeong, H., Néda, Z., Ravasz, E., Schubert, A., Vicsek, T.: Evolution of the social network of scientific collaborations. Physica A 311, 590 (2002)

6. Csányi, G., Szendrői, B.: Structure of a large social network. Phys. Rev. E 69, 036131 (2004)
7. Wang, F., Moreno, Y., Sun, Y.: Structure of peer-to-peer social networks. Phys. Rev. E 73, 036123 (2006)
8. Goh, K.-I., Eom, Y.-H., Jeong, H., Kahng, B., Kim, D.: Structure and evolution of online social relationships: Heterogeneity in unrestricted discussions. Phys. Rev. E 73, 066123 (2006)
9. Latané, B.: The psychology of social impact. American Psychologist 36, 343 (1981)
10. Latané, B., Wolf, S.: The social impact of majorities and minorities. Psychological Review 88, 438 (1981)
11. Sznajd-Weron, K., Sznajd, J.: Opinion evolution in closed community. Int. J. Mod. Phys. C 11, 1157 (2000)
12. Deffuant, G., Neau, D., Amblard, F., Weisbuch, G.: Mixing beliefs among inter acting agents. Adv. Complex Systems 3, 87 (2000)
13. Boccaletti, S., Latora, V., Moreno, Y., Chavez, M., Hwang, D.-U.: Complex networks: structure and dynamics. Phys. Rep. 424, 175 (2006)
14. Albert, R., Barabási, A.-L.: Statistical mechanics of complex networks. Rev. Mod. Phys. 74, 47 (2002)
15. Porfiri, M., Bollt, E.M., Stilwell, D.J.: Decline of minorities in stubborn societies. Eur. Phys. J. B 57, 481 (2007)
16. Weisbuch, G., Dedduant, G., Amblard, F.: Persuasion dynamics. Physica A 353, 555 (2005)
17. Kozma, B., Barrat, A.: Consensus formation on adaptive networks. Phys. Rev. E 77, 016102 (2008)

Community Detection of Time-Varying Mobile Social Networks

Shu-Yan Chan[1], Pan Hui[2], and Kuang Xu[1,3,*]

[1] Computer Laboratory, University of Cambridge,
15 JJ Thomson Avenue, CB3 0FD Cambridge, UK
[2] Deutsche Telekom Laboratories / TU Berlin,
Ernst-Reuter-Platz 7, 10587 Berlin, Germany
[3] Department of Electrical and Electronic Engineering,
The University of Hong Kong, Hong Kong
firstname.lastname@cl.cam.ac.uk,
Pan.Hui@telekom.de

Abstract. In this paper, we present our ongoing work on developing a framework for detecting time-varying communities on human mobile networks. We define the term *community* in environments where the mobility patterns and clustering behaviors of individuals vary in time. This work provides a method to describe, analyze, and compare the clustering behaviors of collections of mobile entities, and how they evolve over time.

1 Introduction

Categorizing mobile objects into communities has numerous potential applications. In Ecology, identifying the animals under observation with their social groups can enhance the study of hunting, mating and other behaviours in context. In online social network studies, grouping individuals into community can help to highlight interaction patterns and identify common attribute amongst individuals. Our target in this paper is human mobile networking.

Previous work has showed that identifying communities can help to improve message forwarding efficiency in Pocket Switched Network (PSN) [1], which is one kind of mobile ad hoc network with intermittent connectivity problem. The improvement is achieved by preferentially forwarding messages to devices that are in the same communities with the messages' destinations.

Our goal is to develop a community detection algorithm that requires little user interventions/adjustments once initialized, and can adapt to the changing and evolving networks. Our work consists of three phases:

1. We first lay a theoretical foundation and formulation for community detection on human mobility traces over a long period of time. This definition is also applicable to community detections in a distributed manner.
2. We evaluate a centralized community detection algorithm which matches our definition of community. (*work in progress*)

---

* This work begins when the authors were all in Cambridge.

J. Zhou (Ed.): Complex 2009, Part I, LNICST 4, pp. 1154–1159, 2009.

3. We develop and compare the performance of several distributed community detection algorithms. The results from the centralized algorithm serve as an upper bound on how close the communities detected by distributed algorithms can approach the definition.(*future work*)

2   Related Work

Evolution of communities has been well studied in the literature. For example, by utilizing temporal information explicitly, Berger-Wolf *et al.* [2] proposed a framework to identify communities and analyze their evolution in dynamic social networks. Tantipathananandh *et al.* in [3] further propsed a framework for finding communities in social networks that develop over time, and formulated it as a combinatorial optimization problem. They evaluated their algorithms by utilizing several synthetic and real-world datasets of social network. All the above work provides an insightful investigation on the time-varying characteristics in dynamic social networks, but our work differs from them since we focus on the clustering behavior of entities in mobile networks based on human-mobility patterns. By utilizing global time stamp, inter-contact time and contact-duration between mobile devices carried by human in their real lives, we aim at uncovering time-varying communities in dynamic mobile networks which will aid information dissemination [1] in human social life.

Community evolution has also been studied in on-line (Internet) social networks. In [4] [5], communities are identified within some time windows and then merged to reflect its evolution, and heuristics are proposed in those approaches to approximate the optimal solution. Backstrom *et al.* in [6] resorted to several large sources of on-line data which embeds explicit user-defined communities, finding that the inclination of an individual to join a community is affected crucially by the connecting structure of his friends. We do not focus on how the structural features influence the evolving communities in social networks, but study the time-varying communities in human-mobility networks by utilizing temporal information. The work similar to our study is in [7], which took into account of time variability of the information from mobile networks, as in [8], considering a community as a dense connected sub-graph over time, and a node as its member only when it attaches to this sub-graph in a series of time steps. Our study is more ambitious that by incorporating aging factor and history accumulator into the weighted temporal property, we pave the way to detect time-varying communities on mobile network in a distributed way and in real time.

3   Community Detection with Time-Varying Mobility Pattern

A satisfactory community detection algorithm on mobile network should require as few user settings as possible, e.g. a mobile device can be pre-programmed with the algorithm and then distributed to a user and the amount that the user need to fiddle with the algorithm/device's parameters should be minimal, even after the device has been running for a long time and the user may have changed his mobility pattern.

3.1   Definition of Community

Our intuition of a community is a clustering of entities that are "closely" linked to each other, either by direct linkage or by some "easily accessible" entities that can act as intermediates. An entity can belong to more than one community and communities can be hierarchical, such that within a closely connected community there can be sub-communities of which their members have even closer connections between each other.

In our work, we adapted the k-clique [9] community as the basis of our definition of community.

Within a given time period, the *Immediate Neighbours* of an entity n are the set of entities that have "heavy interactions" with n. Some of the Immediate Neighbours of n may be in the same community as n, but it is not necessarily for all of them to be in the same community, for n can be associated with multiple communities. *Immediate Community Neighbours* of an entity n in a community C is thus defined as the set of *Immediate Neighbours* of n that are also in community C.

The concept of "heavy interactions" is subjected to parametrization. In our work, we will explore the contact duration domain and leave other possibility for future works. Within the contact duration domain, the interaction between two entities can be classified as "heavy interaction" if their contact duration exceeds a certain threshold λ. It is the effect of varying the threshold that we wish to investigate.

We redefine k-Clique community in the following way, if entity n is in community C, then there exist at least $k-1$ other entities in it's immediate community neighbours set that all the k entities have "heavy interactions" (above λ) with each other (hence forming a $k-$clique in a graph). Two entities are in the same k-clique community, C, iff their two corresponding nodes are linked by at least one series of adjacent cliques (two k-cliques are adjacent if they share $k-1$ common nodes).

However, over a long period of time, community memberships may be broken or newly formed and communities may evolve. Therefore, if one wants to study the dynamic of communities, it is necessary to partition the trace into smaller time intervals, and applying the community detection algorithm to the data in each smaller time intervals. Ideally, the smaller time intervals should be chosen such that each only cover a period when there is little disturbance to the communities memberships.

However, there is no rigorous definition on what constitute a good partitioning of the time intervals, nor heuristic on how to find "good" partitioning. A "good" partitioning will reveal results that are either desirable (agree with expectation, perhaps inferred by prior knowledge or information) or insightful (reveals communities structures that are not previously known).

Varying λ, α(to be introduced later), and the time intervals might yield different set of communities on a given set of data, thus affecting the *usefulness* of the resulting communities, but there is no mathematical way to differentiate which set of results are better.

3.2 Algorithm

We seek to develop algorithms that can dynamically detect time-varying communities. Our criteria for the algorithm is that it can *adapt* to the change in interaction patterns between the entities and detect the change in communities structures.

Although the final goal in our work is to develop distributed algorithm that will run on pervasive computing devices, in this work-in-progress paper, we present a centralized version of the algorithm. The centralized version provides a upper bound on the communities detected by future distributed algorithms, base on the definition of community in this paper.

The (centralized) algorithm also runs in *real time*, that the communities it detects at any given time are computed solely base on the interaction histories of the entities and have no knowledge of their interactions in the future. This algorithm has its own application, in scenarios where global knowledge of all interactions between entities are accessible as they happen and community information are required on demand. Since we develop this algorithm to facilitate comparison of the distributed version and that it has its own potential applications, we put a bound on the resource it requires - instead of keeping the histories of all interactions between entities (which would have enabled the algorithm to search through all the histories when computing the present communities associations, at a significant increase of computation power and storage requirements), the algorithm keeps a summarized view of the histories (as the distributed version does).

For each pair of entities, there are four variables: a `history accumulator` which is initialized as 0, a current `tally`(of interaction) $tally_t$, an `aging factor` α and a `time interval setting`. The timeline in the environment clock is partitioned according to `time interval setting`. During each time interval, the interactions between the pair(e.g. the contact duration) are tallied. When the environment clock progress to the next time interval, the current tally is "aged" into the `history accumulator` according to the following formula, where the contact duration of the current time interval t is tallied as $tally_t$, w^{ij} represents the `history accumulator` between entities i and j(hereafter in this paper, we will use the graph theory terminologies *nodes* instead of entities, and the w^{ij} is also the *weight* between nodes i and j). w^{ij} takes into account of all the contact histories up to and including the previous time interval using the **Aging Formula**:

$$w_t^{ij} = \frac{w_{t-1}^{ij} * \alpha + tally_{t-1}}{\alpha + 1} \qquad (1)$$

and at the first time step, $w_0^{ij} = 0$. It will be shown in later section that the Aging Formula ages the weight in an exponential fashion.

The algorithm allows each pair of nodes to have their own `aging factor` α and `time interval setting`, and they can be updated in response to the change in interaction pattern. However, in this paper, we implement the centralized algorithm using a broadbrush approach that all nodes pairs use the same pre-set α value and use the same `time interval setting` and investigate the effect of changing the α value.

At any environment time, users can request the algorithm to compute for them the current communities structures. They need to specify the k-clique communities they want to detect by specifying the value of k and the contact threshold value, λ. The algorithm will then launch sub-algorithm to detect the k-clique communities. In this paper we implemented the standard CPM $k-clique$ detection algorithm [10].

3.3   Discussion and Preliminary Result

The algorithm needs to be able to *forget* old associations between nodes when they fade away, yet it also need to have some resilience to the temporary fluctuation. The contact history ages with an exponential decay: consider after time t_{end} that there is no further contact between node a and b, then it is straight forward to show that their `history accumulator`s age exponentially according to:

$$w_t^{ab} = w_{t_{end}}^{ab} * e^{-\gamma(t-t_{end})} \tag{2}$$

When the aging factor α is less than 1, the history ages rapidly (since for each new time step, the influence of the entire contact history before is less than that of the new contact statistic); however, if the aging factor is allowed to be set to be greater than 1, the resulting edge weights would be more resilience to temporary fluctuation as more emphasis is placed on the history than the current contact statistic, yet, over time, the influence of past contact counts would still fade away. This can therefore offset some of the impacts from setting a sub-optimal time step length while still allowing aging to take place.

Fig. 1 shows result of running our algorithm against an artificial mobility traces which is generated in such a way that there is a sudden change in mobility pattern between time units 4 and 5 on the x-axis. The y-axis denotes the classic Jaccard index [11] when the communities detected in one snapshot is compared to those detected in the previous snapshot. It is clearly shown that the algorithm managed to adapt to change in clustering behaviour occur at around time index 4 and 5 (the dip of lower similarity values around that time indicate a bigger change in detected communities) , with different settings of aging factor α. The figure also shows that the algorithm adapts slower with a smaller α value.

However, natural and social systems are inherently noisy, therefore, in future work we will experiment on relaxing the requirement that all pairs in a clique need to be linked. We would consider a group of nodes as $clique_{relax}$ if they are just missing a few links to form a proper clique (this is a more relax condition than those found in the study of clique community on weight network ,$CPMw$ [10]). In a clique, the distance between two nodes is 1 (direct connection), however, when one of the link is removed, the maximum distance between two nodes

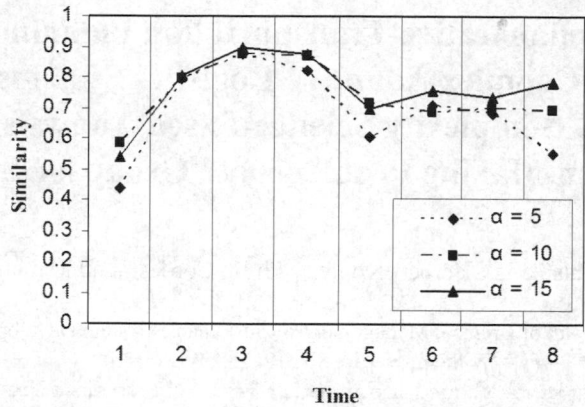

Fig. 1. The effect of community detection with different Aging factor (α)

in that clique becomes 2. If we restrict ourselves so that the maximum number of links that can be dropped to be less than $k - 1$, then we ensure that all nodes pairs are still at least indirectly connected and the maximum distance are still 2. We call such definition of k-clique as k-clique$_{relax}$. A community is equivalent to a maximal set of k-clique$_{relax}$, with edge weight higher than λ, that can be reached from each other via series of k-clique$_{relax}$ adjacency connections.

References

1. Hui, P., Crowcroft, J.: Bubble Rap: Forwarding in small world DTNs in ever decreasing circles. Technical Report UCAM-CL-TR-684, University of Cambridge, Computer Laboratory (2007)
2. Berger-Wolf, T.Y., Saia, J.: A framework for analysis of dynamic social networks In: KDD 2006 Philadelphia, USA (2006)
3. Tantipathananandh, C., et al.: A framework for community identification in dynamic social networks. In: KDD 2007, San Jose, California, USA (2007)
4. Chakrabarti, D., Kumar, R., Tomkins, A.: Evolutionary clustering. In: KDD 2006, Philadelphia, USA (2006)
5. Tantipathananandh, C., et al.: Structural and temporal analysis of the blogosphere through community factorization. In: KDD 2007, San Jose, California, USA (2007)
6. Backstrom, L., Huttenlocher, D., Kleinberg, J.: Group formation in large social networks: membership, growth, and evolution. In: KDD 2006, Philadelphia, USA (2006)
7. Scherrer, A., et al.: Description and simulation of dynamic mobility networks. Comput. Netw. 52(15) (2008)
8. Scherrer, A., et al.: Synchronization reveals topological scales in complex networks. Physical Review Letters 96 (2006)
9. Palla, G., et al.: Uncovering the overlapping community structure of complex networks in nature and society. Nature 435(7043), 814–818 (2005)
10. Farkas, I., Abel, D., Palla, G., Vicsek, T.: Weighted network modules. New Journal of Physics 9 180 (2007)
11. Jaccard, P.: Bulletin de la Societe Vaudoise des Sciences Naturelles 37, 547 (1901)

Collaborative Transportation Planning
in Complex Adaptive Logistics Systems:
A Complexity Science-Based Analysis
of Decision-Making Problems of "Groupage Systems"

Michael Hülsmann[1], Herbert Kopfer[2], Philip Cordes[1], and Melanie Bloos[2]

[1] University of Bremen, Management of Sustainable System Development,
Wilhelm-Herbst-Str. 12, 28359 Bremen
{michael.huelsmann,pcordes}@uni-bremen.de
http://www.wiwi.uni-bremen.de/mh/
[2] University of Bremen, Chair of Logistics,
Wilhelm-Herbst-Str. 5, 28359 Bremen
{kopfer,bloos}@uni-bremen.de
http://www.logistik.uni-bremen.de/

Abstract. This paper aims to analyze decision-making problems in Groupage Systems from a complexity-science perspective. Therefore, the idea of Complex Adaptive Logistics Systems (CALS) and its inherent organization principle of autonomous co-operation and control will be presented. Furthermore, Groupage systems as a way to implement collaborative transportation planning will be introduced and, in combination with the idea of CALS, resulting decision-making problems for so-called 'smart parts' in logistics systems will be deduced.

Keywords: Complex Adaptive Systems, Logistics, Collaboration, Groupage Systems, Decision-making Problems, Complexity Science, Autonomous Co-operation.

1 Introduction

Modern logistics has become more complex than ever before [e.g. 1,2,3]. Some reasons for this development can be observed on different basic levels of supply network systems. One reason is evident on the level of the system's elements: the management of logistics systems has to face an increasing number of agents which have to be controlled within such a system [2]. Another group of reasons can be found on the level of inter-relations: resulting from the rising number of agents more and more inter-relations between numerous and heterogeneous agents have been established [e.g. 4] – in the managerial dimension (e.g. recursive negotiations between opposing stakeholders) as well as in the informational and communicational dimension (e.g. integrated data exchange and warehousing) and in the dimension of material flow (e.g. atomization of goods and transportation means). Finally, some reasons may

J. Zhou (Ed.): Complex 2009, Part I, LNICST 4, pp. 1160–1166, 2009.

emerge on the level of characteristics, which may be represented by altering functionalities in logistics (e.g. Hülsmann & Grapp [5] describe the development from isolated basic corporate functions of transportation, storage, and handling up to a comprehensive concept of globally integrated, boundary spanning network processes of value creation). Therefore, one major question of logistics management today is, how to cope with the immanent and increasing complexity of supply systems [6].

An organizational principle that has recently been discussed as a capable approach to cope with complexity in logistics management is based on the idea of self-organization and is called autonomous co-operation and control [6]. It can be understood as *"(...) processes of decentralized decision-making in heterarchical structures. It presumes interacting elements in non-deterministic systems, which possess the capability and possibility to render decisions. The objective of Autonomous Control is the achievement of increased robustness and positive emergence of the total system due to distributed and flexible coping with dynamics and complexity."* [7]. With the implementation of the organizational principle in a large and diversified logistics structure like an international supply network (ISN) [e.g. 5] this structure intensifies its characteristics as so-called complex adaptive logistic systems (CALS) [2,3,4]. One example of such a system that already incorporates elements of autonomous co-operation and control is the concept of collaborative operational transportation planning in so-called "Groupage Systems" [8].

Hence, this paper intends to analyze Groupage Systems from a complexity-science perspective with regards to resulting decision-making problems. The results show that Groupage Systems are CALS and the limitations imposed on the system by its complexity are explained. Therefore, this paper comprises three major steps in its argumentation: First, it briefly outlines the idea of CALS, especially the concept of autonomy-driven co-operation and control; second, it describes the so-called "Groupage Systems" as a way to implement collaborative transportation planning; finally, it discusses under the postulates of complexity sciences-principles the decision-making problems of "Groupage Systems" in CALS.

2 Complex Adaptive Logistics Systems (CALS) – Autonomously Controlled International Supply Networks (ISN)

Recent research works on logistics systems observe a shift from linear supply chains to non-linear and complex networks [2,9]. One example for this are so-called international supply networks (ISN), which can be described as a consortium of companies involved in diverse organizational structures of different supply chains and competing with each other to a certain degree [e.g. 5,10]. In order to consider the developments of ISN to shift by trend towards complex networks, several authors took on the analogy of complex adaptive systems (CAS) as they have been described by e.g. Holland [11,12] or Kauffman [13]. Consequently, the term CALS was introduced to supply chain management [2,4,14].

CALS consist of a large number of elements as well as sub-systems. In order to sustain the logistics system's operational reliability and due to their interdependencies, it is necessary for them, to exchange resources (e.g. products, finances or information). In other words, the logistics system's entities are to a certain degree **interacting** with each other [2]. Their incentives to interact derive from their different endowment with these resources [4]. In consequence, the logistics system's operational reliability depends to a certain degree on its elements' **heterogeneity**. Furthermore, due to the interactions between the system's elements and the mentioned phenomenon of increasing interdependencies between them, they react and have therefore mutual influences on each other. In consequence, a logistics system **co-evolves** on the one hand with the evolution of its elements and on the other hand with its environment [3]. In addition to these different endowments with their resources, the elements have different goals as well as different rules, which, in turn, determine their behaviour, in order to reach these goals. Whereas human elements in logistics systems (e.g. management teams) are able to change these rules over time, and therefore **to learn**, this ability is at least arguable for non-human parts, like containers or single goods. However, recent developments in information and communication technologies (e.g. RFID or smart tags) make it possible, to extend this perspective [15] which leads to the development of the term 'smart parts' [4]. In consequence, this ability to learn enables the respective single entities to act **autonomously** and therefore to plan, decide, and act without any impact of an external entity [16]. Theoretically, this enables the logistics system to develop as well without any impact of an external control entity on a global basis. In other words, the system develops the ability of **self-organization** [2,3]. An essentiality of self-organizing systems is that they are located in so called **melting zones** [13], which means, that they do neither pass over the edge of chaos [17,18] nor the edge of order [19,20]. Within CALS there might be two ways of decision making to establish the ability to adapt the system's performance, strategies, organizational profiles, and resources to changing and diversifying environmental requirements. One is autonomous cooperation and control run by the agents of such a system; the other is centralized delegation, which creates a hierarchical structure among the agents. Because the latter has shown its limitations in coping with the complexity of large scale supply network structures, the organizational principle of self-organization shall here be applied to logistics via the concept of autonomous cooperation and control.

What are the supplementary major characteristics autonomous cooperation and control is described by? The idea of self-organization, *"(...) does not present an 'over aging paradigm', but there is a general overlapping of attributes such as autonomy, interaction and non-determinism (...)"* [21] as they can be found in the contributions of Von Foerster 1979 [22], Glansdorff and Prigogine 1971 [23], Haken 1973 [24], Maturana and Varela 1980 [25], Lorenz 1963 [26], Mandelbrot 1977 [27], and Bick 1973 [28]. In addition to autonomy and interaction as they have been already mentioned for CALS generally, further characteristics of autonomously controlled systems are decentralized decision-making, heterarchy and non-determinism. **Decentralized decision-making** describes the ability of a system's elements, to decide on their own about their next steps, without the need to consult a control entity, which is located on a hierarchically higher level. Consequently, an autonomous controlled system is characterized by **heterarchy**, which means, that all elements are equipped with the same (or respectively a similar level) of decision-making power. The combination of these characteristics leads in logical consequence to the impossibility of predicting future system states and therefore, to the system's **non-determinism** [7].

3 Groupage Systems

One approach to increase the degree of autonomous cooperation in CALS are so called Groupage Systems [8]. In Groupage Systems the options for transportation planning at each freight forwarder are extended by horizontal cooperation between several freight forwarders. Then, an additional mode order execution is the forwarding of some transportation orders to cooperating partners in order to achieve a better leveling of available capacity. This cooperation between several competing freight forwarders in Groupage Systems offers the freight forwarders more possibilities than the option of subcontracting only because the plans of the partners can be harmonized in order to improve capacity utilization. This exchange of orders is part of the collaboration and requires the incorporation of acquired partners' orders into the planning process. The Groupage System may include a mechanism for exchanging the transportation orders automatically and thus, for adjusting transport capacity across the partners [8].

Kopfer and Pankratz [8] discuss the modeling of Groupage Systems as Multi-Agent Systems. Multi-Agent Systems offer the possibility to model the decentralized planning situation, where each autonomous participant pursues individual rational objectives. Autonomous decision making in transport logistics can be modeled at various levels of detail ranging from freight forwarding agents as smallest autonomous units [29] via agents for each vehicle [30,31] to agents representing individual orders which strive for certain transports according to pre-specified rules [32]. However, the more detailed the degree of autonomy is, the more communication links between the individual agents are necessary to find a good solution.

4 Decision-Making Problems

Decision making problems occur whenever there is more than one possibility to achieve a certain goal, whereas decisions within a logistics system vary in their degree of complexity and predictability. Exemplary decision-making problems shall be outlined: In the initiating phase of transport collaboration [33], a decision has to be made on how many partners should participate, which will be guided by considerations on transaction costs and efficiency potential [8]. Following that, every involved decision making unit has to decide, whether to participate or not and to take into account, that participation includes the sharing of information with cooperation partners [34]. Furthermore, smart parts that are able to decide on their own about their next steps [4] render their decision based on local information [2], which can depart from information about the whole system's global optimum.

Additionally, in connection with the complex structures of ISN, problems like hyper-linking occur. Hyper-linking means that single agents are affected by the behavior of other agents – not only within a certain logistics system (e.g. warehouses and logistics service firms in a fruit supply chain), but also from different logistics systems (e.g. a certain fruit supply chain and meat supply chain use the same transportation means like container ships) as well as from non-logistics systems (e.g. financial or societal systems). Therefore, agents are highly interwoven with others [35] they might not even know. Each of these systems and their agents depend on the numerous, heterogeneous and dissimilar agent's activities and system performances (e.g. financial crisis). Therefore, the decision making of each single agent is influenced by the decision making of other agents. This might lead to manifold decision making dilemmas [1] and can lead to a complexity induced lock-in situation that results in suboptimal and dysfunctional decision making with a limited choice of possible decisions [36]. "Dysfunctional" reflects on the limited capability of a rational choice. The evident lack of information for the decision, as it is described by the problem of bounded rationality [37], refers to "suboptimal" [21].

5   Conclusions

Groupage Systems might be an appropriate and capable way to cope with a high degree of complexity in modern supply networks, which can be described as complex adaptive logistics systems (CALS). Groupage Systems are based on the concept of autonomous co-operation and control, which strives for the implementation of the idea of self-organization via the processes of decentralized decision-making in the heterarchical structures of logistics networks with numerous, heterogeneous, but equal agents. However, decision-making within Groupage Systems bears several problems, which limit the contributions of collaborative transportation planning to cope with complexity in large-scale supply systems such as hyper-linking. As an illustrative example the phenomenon of hyper-linking might lead to a higher risk of suboptimal and dysfunctional decision making, which can counterproductively affect the decision of partners to participate in the co-operation of Groupage Systems and to share their information with their partners in such a system. Therewith, future research requirements result on the one hand from the non-existence of a general optimum degree of collaboration regarding a minimized risk that emanates from decision making problems. This leads to the necessity, to analyze this question individual as the case arises. On the other hand the same problem is still unsolved for the optimal degree of its basic concept autonomous co-operation and its single degrees, like interaction or autonomy.

Acknowledgement. This research was supported by the German Research Foundation (DFG) as part of the Collaborative Research Centre 637 »Autonomous Cooperating Logistic Processes – A Paradigm Shift and its Limitations«.

References

1. Hülsmann, M., Berry, A.: Strategic management dilemmas: Its necessity in a world of diversity and change. In: Lundin, R., et al. (eds.) Proceedings of the SAM/IFSAM VIIth World Congress on Management in a World of Diversity and Change, 18 pages. Göteborg, Sweden (2004)
2. Surana, A., Kumara, S., Greaves, M., Raghavan, U.N.: Supply-chain networks: a complex adaptive systems perspective. International Journal of Production Research 43(20), 4235–4265 (2005)
3. Choi, T.Y., Dooley, K.J., Rungtusanatham, M.: Supply networks and complex adaptive systems: control versus emergence. Journal of Operations Management 19(3), 351–366 (2001)

4. Wycisk, C., McKelvey, B., Hülsmann, M.: 'Smart parts' logistics systems as complex adaptive systems. International Journal of Physical Distribution and Logistics Management 38(2), 108–125 (2008)
5. Hülsmann, M., Grapp, J.: Autonomous Cooperation in International-Supply-Networks – The Need for a Shift from Centralized Planning to Decentralized Decision Making in Logistic Processes. In: Pawar, K.S. et al. (eds.) Proceedings of the 10th International Symposium on Logistics (10th ISL). Loughborough, United Kingdom, pp. 243–249 (2005)
6. Hülsmann, M., Scholz-Reiter, B., Freitag, M., Wycisk, C., De Beer, C.: Autonomous Cooperation as a Method to cope with Complexity and Dynamics? – A Simulation based Analyses and Measurement Concept Approach. In: Bar-Yam, Y. (ed.) Proceedings of the International Conference on Complex Systems (ICCS 2006), Boston, MA, USA, web-publication, 8 pages (2006)
7. Windt, K., Hülsmann, M.: Changing Paradigms in Logistics. In: Hülsmann, M., Katja, W. (eds.) Understanding Autonomous Cooperation & Control: The Impact of Autonomy on Management, Information, Communication, and Material Flow, pp. 1–12. Springer, Berlin (2007)
8. Kopfer, H., Pankratz, G.: Das Groupage-Problem kooperierender Verkehrsträger. In: Kall, P., Lüthi, H.-J. (eds.) Proceedings of OR 1998. Springer, Heidelberg (1999)
9. Mason, R.B.: The external environment's effect on management and strategy - a complexity theory approach. Management Decision 45(1), 10–28 (2007)
10. Lambert, D.M., Cooper, M.C., Pagh, J.D.: Supply Chain Management: Implementation Issues and Research Opportunities. The International Journal of Logistics Management 9(2), 1–19 (1998)
11. Holland, J.H.: The global economy as an adaptive system. In: Anderson, P.W., Arrow, K.J., Pines, D. (eds.) The Economy as an Evolving Complex System, vol. V, pp. 117–124. Addison-Wesley, Reading (1988)
12. Holland, J.H.: Complex Adaptive Systems and Spontaneous Emergence. In: Quadrio Curzio, A., Fortis, M. (eds.) Complexity and Industrial Clusters, pp. 25–34. Physica-Verl., Heidelberg (2002)
13. Kauffman, S.A.: The Origins of Order: Self-Organization and Selection in Evolution. Oxford Univ. Press, New York (1993)
14. Choi, T.Y., Dooley, K.J., Rungtusanatham, M.: Supply networks and complex adaptive systems: control versus emergence. Journal of Operations Management 19(3), 351–366 (2001)
15. Spekman, R.E., Sweeney, P.J.: RFID: from concept to implementation. International Journal of Physical Distribution & Logistics Management 36(10), 736–754 (2006)
16. Kappler, E.: Autonomie. In: Frese, E. (ed.) Handwörterbuch der Organisation, 3rd edn., pp. 272–280. Poeschel, Stuttgart (1992)
17. Langton, C.G.: Artificial Life. Addison-Wesley, Reading (1989)
18. Lewin, R.: Complexity. University of Chicago Press, Chicago (1992)
19. Bérnard, H.: Les Tourbillons Cellulaires dans une Nappe Liquide Transportant de la Chaleur par Convection en Régime Permanent. Annales de Chimie et de Physique 23, 62–144 (1901)
20. Prigogine, I.: An Introduction to Thermodynamics of Irreversible Processes. Thomas, Springfield (1955)
21. Hülsmann, M., Wycisk, C.: Unlocking Organizations through Autonomous Cooperation - Applied and Evaluated Principles of Self-Organization in Business Structures. In: Proceedings of the 21st EGOS Colloquium, Berlin, webpublication, 25 pages (2005)

22. von Foerster, H.: Cybernetics of Cybernetics. In: Krippendorff, K. (ed.) Communication and Control in Society, pp. 5–8. Gordon and Breach Science Publishers, New York (1979)
23. Glansdorff, P., Prigogine, I.: Thermodynamic theory of structure, stability and fluctuations. Wiley, New York (1971)
24. Haken, H.: Synergetics: cooperative phenomena in multi-component systems. In: Symposium on Synergetics, 30 April - 6 May 1972. Schloß Elmau, Stuttgart (1973)
25. Maturana, H.R., Varela, F.: Autopoiesis and cognition: the realization of living. Reidel, Dordrecht (1980)
26. Lorenz, E.N.: Deterministic Nonperiodic Flow. Journal of the Atmospheric Sciences 20, 130–141 (1963)
27. Mandelbrot, B.: The Fractal Geometry of Nature. Freeman, New York (1977)
28. Bick, H.: Population dynamics of Protozoa associated with the decay of organic materials in fresh water. American Zoologist 13(1), 149–160 (1973)
29. Gomber, P., Schmidt, C., Weinhardt, C.: Elektronische Märkte für die dezentrale Transportplanung. Wirtschaftsinformatik 39, 137–145 (1997)
30. Sandholm, T.: An Implementation of the Contract Net Protocol Based on Marginal Cost Calculations. In: Proceedings of the 11th National Conference on Artificial Intelligence (AAAI-93), Washington DC, pp. 256–263 (1993)
31. Fischer, K., Müller, J.P., Pischel, M., Schier, D.: A Model for Cooperative Transportation Scheduling. In: Lesser, V. (ed.) Proceedings of the First International Conference on Multiagent Systems (ICMAS 1995), pp. 109–116. MIT Press, Cambridge (1995)
32. Utecht, T.: Kooperatives Problemlösen in Workstationclustern. Verlag für Wissenschaft und Forschung, Berlin (1997)
33. Minner, S.: Modellgestützte Ermittlung und Verteilung von Kooperationsvorteilen in der Logstik. In: Spengler, T., Voss, S., Kopfer, H. (eds.) Logistik Management. Prozesse, Systeme, Ausbildung, pp. 111–132. Physika-Verlag, Heidelberg (2004)
34. Krajewska, M.A., Kopfer, H.: Collaborating freight forwarding enterprises. Request allocation and profit sharing. OR Spectrum 28, 301–317 (2006)
35. Tapscott, D.: Creating value in the network economy. Harvard Business School Publication, Boston (1999)
36. Schreyögg, G., Sydow, J., Koch, J.: Organisatorische Pfade – Von der Pfadabhängigkeit zur Pfadkreation? In: Schreyögg, G., Sydow, J. (eds.) Strategische Prozesse und Pfade, Managementforschung 13, Gabler, Wiesbaden (2003)
37. Simon, H.A.: Theories of Bounded Rationality. In: McGuire, C.B., Radner, R. (eds.) Decision and Organization, pp. 161–172. North-Holland Publ., Amsterdam (1972)

Classification Based on the Optimal
K-Associated Network

Alneu A. Lopes, João R. Bertini Jr., Robson Motta, and Liang Zhao

University of São Paulo, Institute of Mathematics and Computer Science,
Av. Trabalhador São Carlense 400, São Carlos, Brazil
{alneu,bertini,rmotta,zhao}@icmc.usp.br

Abstract. In this paper, we propose a new graph-based classifier which uses a special network, referred to as optimal K-*associated network*, for modeling data. The K-associated network is capable of representing (dis)similarity relationships among data samples and data classes. Here, we describe the main properties of the K-associated network as well as the classification algorithm based on it. Experimental evaluation indicates that the model based on an optimal K-associated network captures topological structure of the training data leading to good results on the classification task particularly for noisy data.

Keywords: Complex Network, Data Mining, Data Classification, Network formation.

1 Introduction

Complex networks are today a unifying topic in complex systems. Such theory not only had provided a deeper understanding of complex systems, such as biological and social networks, but also gave rise a new approach for modeling such structures. Recently, triggered by some discoveries [1], [2], the theory of complex networks had spread to many branch of sciences. Results obtained so far ranged from microbiology to economy or epidemic analysis to traffic planning (see [3],[4],[5],[6] for instance). Although complex networks theory have been used in a large number of areas there are still plenty of tasks that could be potentially helped by such theory, such as Data Mining tasks. In this scenario complex networks can improve the representation of data from the traditional attribute-value paradigm to a richer relational representation.

Data mining tasks, in general, can be divided in two categories, predictive and descriptive mining [7] and two respective major branches are data classification and clustering. In data classification, each data instance has an associated label that characterizes it in a group named class. The objective is to predict the classification of a new pattern based on a dataset with patterns already classified, used as training set. Some examples of classification problem are medical diagnosis, credit assignment, market prediction [8]. In clustering tasks the data does not have an associated label and the objective in this case is to describe the patterns

J. Zhou (Ed.): Complex 2009, Part I, LNICST 4, pp. 1167–1177, 2009.

formed by some groups of the dataset. Some common application is clustering are gene categorization, Web documents classification, market research [9].

Throughout recent years, graph-based clustering algorithms have received great attention and have been widely studied (see [10],[11],[12]). This interest is mostly justified due to some advantages the network approach provides. Among the most important ones, graph representation (i) can capture topological underline structure of similarity relation among data leading to interesting approach for dealing with clustering problem or community detection; (ii) promotes hierarchical representation of cluster and; (iii) enables detection of clusters with arbitrary shapes. In graph-based algorithms, each node of the graph represents a data point and the edges the similarity between them.

There exists various ways of connecting the nodes in a network (see [10] for instance), but usually the basic idea is that the probability of connection between two vertices are proportional to the similarity between them. Stated in this way, it is clear that close groups of instances tend to be heavier linked together than the rest of the data. Hence, it is straightforward the usage of community detection algorithms [13] to reveal similarity in data which in fact has been widely used clustering problems [14],[15].

This wide usage of complex network in clustering problems does not reflect its use in classification tasks. Motivated by the richness of graph representation as well as the new methods provided by complex networks theory, this paper presents a classification method based on complex networks. Taking into consideration that graph-based classification has not been much explored in the literature, this work is an effort toward this direction. In what follows will be developed a network-based approach for dealing with classification tasks. The obtained results indicate a potentially new approach for dealing with classification problems especially in the presence of noise.

The remainder of the paper is organized as follows: Section 2 presents the three parts for generating a network that will be used by the classifier: 1) the K-associated network - how the network is built from data; 2) how to extract the purity measure for a given component and 3) how to obtain the network that maximizes the component's purity, called optimal network. In Section 3 the classifier that uses the network described in Section 2 is derived. Section 4 presents some results of using our classifier in ten knowledge domain, as well as a comparison with two other well know algorithms. Finally, Section 5 concludes the paper.

2   The Graph-Based Model

In this section we present a new network based on similarity relations among data and on its classes, referred to as K-associated Network. The main properties of such network are presented, as well as the algorithm to build this network from training data and the usage of this model in the classification task. First of all given a vector-based dataset it is necessary a method for converting it to a network. Once the network is obtained, the purity measure for each component of the network is computed. This measure is detailed next.

Note that the K-associated network depends on the parameter K, and as will become clear ahead, each value of K will generate a network with different number of components with different purity. In a classification context however, it is desirable components with minimum of noise, i.e. with high purity. The idea is to overcome the parameter restriction for creating a single network by creating various networks with different K, keeping, along this process the components with highest purity, obtaining an optimal network.

2.1 The K-Associated Network

A network built from data that supposes to represent the training set must inherit the main characteristic of the data. In this network, each pattern in the training set is mapped to a node in the network and the linkages must be done in order to preserve some desirable similarity relations.

In a K-associated network, among the K nearest neighbors of a given vertex, the connections will be established between the given vertex and those within the same class. As this process is done for all vertices in the network, in many cases a vertex v_i is already connected to vertex v_j (supposing v_i and v_j belongs to the same class) when vertex v_j selects vertex v_i to connect, it is possible that v_i also selects v_j, resulting in two undirected connections between vertices v_i and v_j. This is justified by the fact that if v_j is in the K-neighborhood of v_i it would establish a connection but this connection does not mean that vi belongs to the K-neighborhood of v_j since this not necessarily happens. So when v_j is found in the K-neighborhood of v_i a connection is established no matter they were already connected. This particular way of wiring the network is fundamental for the properties defined ahead. At the end of this process, for any K, there will be at least as many components as the number of classes. Notice also that the number of components decreases monotonically to the number of classes as K increases.

In a formal way, the resulting K-associated network $A = (V, E)$ consists of a set of labeled vertices V and a set of edges E between them, where an edge e_{ij} connects vertex v_i with vertex v_j if $class(v_i) = class(v_j)$ and v_j belongs to the K nearest neighbors of v_i. Where $class(v_i)$ stands for the class associated with vertex v_i.

Let $knn(v_i)$ be the set of the K nearest neighbors of vertex v_i by a given similarity measure. The neighborhood N_{K_i} for a vertex v_i is defined as its immediately connected K-neighbors as follows: $N_{K_i} = \{e_{ij} \in E | v_j \in class(v_i)$ and $v_j \in knn(v_i)\}$. The degree g_i of a vertex v_i is defined as the number of vertices, $|N_{K_i}|$, in its neighborhood N_{K_i}.

Following the above definitions, we can notice:

1. If $class(v_i) = class(v_j)$ and $v_j \in knn(v_i)$ e $v_i \in knn(v_j)$ both vertices (e_{ij} and e_{ji}) are considered in the network, and in the calculation of g_i.
2. One can easily check that (see Fig. 2) the maximum degree of v_i is $2K$, that occurs when there are only vertices with the same label in the K-neighbourhood of v_i in the component. This maximum degree can be achieved only when the component of v_i has more than K vertices.

3. In a component, only vertices which belong to a same label can be linked. Thus each component is associated to only one label.

Figure 1 illustrates a bi-dimensional representation of a toy dataset with 10 examples from black label and 5 from white label, and its correspondent 1,3, and 5-associated networks.

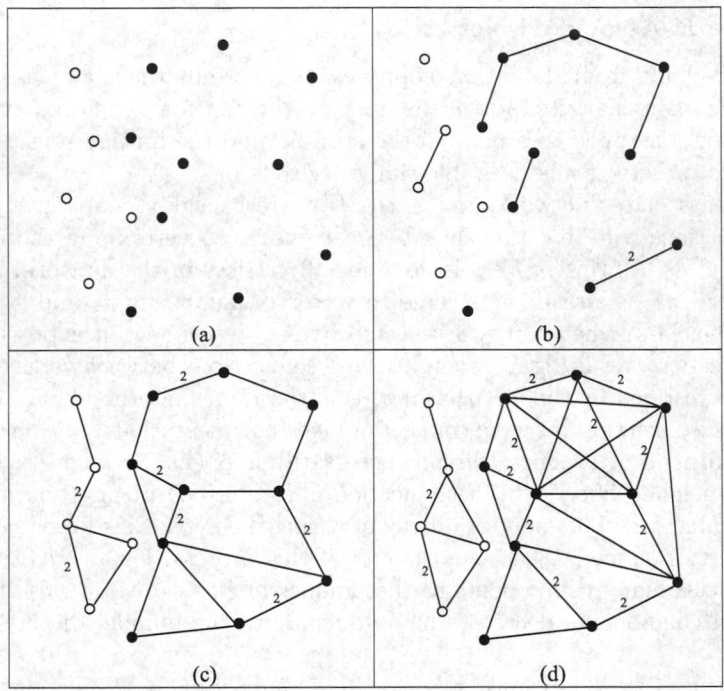

Fig. 1. (a) 2D representation of the dataset. (b) (c) and (d) are 1, 3 and 5-associated correspondent networks, respectively. Notice that edges between vertices can represent more than one connection.

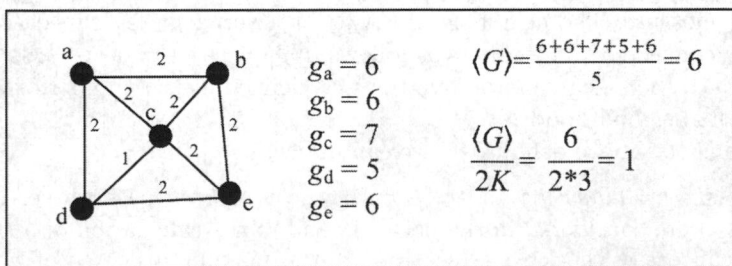

$g_a = 6$
$g_b = 6$
$g_c = 7$
$g_d = 5$
$g_e = 6$

$\langle G \rangle = \dfrac{6+6+7+5+6}{5} = 6$

$\dfrac{\langle G \rangle}{2K} = \dfrac{6}{2*3} = 1$

Fig. 2. An example of a "pure" component with 5 vertices and $K = 3$

2.2   The Purity Measure

The method of generation of the K-associated network improves the representation of the training set and enables some interesting calculation such as the measure of purity. Such measure uses network topologies to quantify how intertwined are the nodes of different class in the network.

Let g_i be the degree of vertex v_i, N the number of patterns in the training set, K the number of neighbors used in the construction of the network. Consider the ratio $y_i/2K$, this relation corresponds to the fraction of links between the vertex v_i and vertices in its own component. This ratio varies between 0 and 1, inclusively. Hence, the total of links between N_c vertices in a component C is given by eq. (1).

$$|E_c| = \frac{1}{2} \sum_{i=1}^{N_c} g_i = \frac{N_c}{2} \sum_{i=1}^{N_c} \frac{g_i}{N_c} = \frac{N_c}{2} \langle G_c \rangle \tag{1}$$

Now, if we consider that for each vertex v_i there are $(2K - g_i)$ edges "rejected" (they would link v_i to vertices of another component), the total number of such rejected edges is given by,

$$|-E_c| = -\frac{1}{2} \sum_{i=1}^{N_c} (2K - g_i) = K \left(\sum_{i=1}^{N_c} 1 - \sum_{i=1}^{N_c} \frac{g_i}{2K} \right) \tag{2}$$

$$= K \left(N_c - \frac{N_c}{2K} \langle G_c \rangle \right) = N_c \left(K - \frac{1}{2} \langle G_c \rangle \right).$$

Such total can be seen as the number of vertices that would link vertices with different components (classes), in a network in which any vertex was linked to its K neighbors.

Thus, the probability of edges between vertices in the same component C (intra-component links) is given by,

$$P_i = \frac{\frac{N_c \langle G_c \rangle}{2}}{\frac{N_c \langle G_c \rangle}{2} + \frac{N_c (2K - \langle G_c \rangle)}{2}} = \frac{\langle G_c \rangle}{2K}. \tag{3}$$

In the equation (3) $P_i = 1$ when there are only vertices with the same label in the K-neighborhood of every vi in the component, see Fig. 2. Thus, $\langle G_c \rangle / 2K$ ratio can be seen as a measure of "purity" of the region of the component C.

We can also compute the probability of links between vertices from distinct components (which do not exist in a K-associated network, but such probability can be understood as a measure of "impurity" in the region of the component).

$$P_i = \frac{N_c (2K - \langle G_c \rangle)}{N_c \langle G_c \rangle + N_c (2K - \langle G_c \rangle)} = \frac{2K - \langle G_c \rangle}{2K} \tag{4}$$

We have demonstrated that the ratio $\langle G_c \rangle / 2K$ expresses the probability of intra-component connection. Thus it expresses the purity in the region of a component. This assumption can be also empirically evaluated. In Fig. 3 is represented the ratio (averaged in 10 runs) for five artificial datasets with 250 vertices.

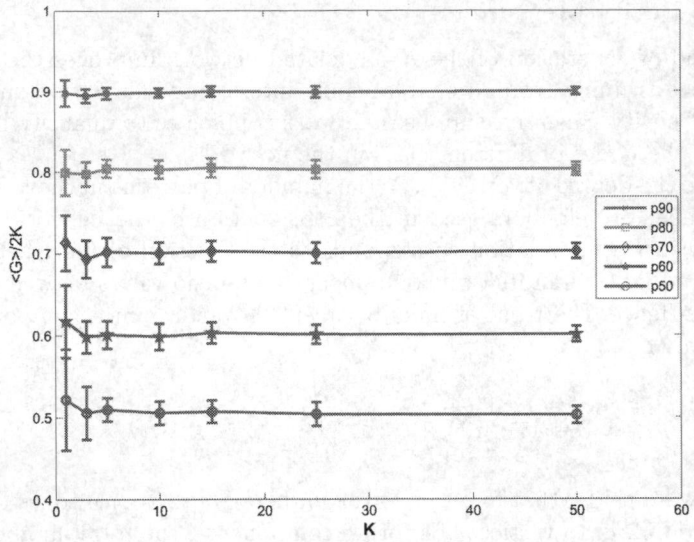

Fig. 3. The average $\langle G \rangle / 2K$ for 5 K-associated networks from datasets with 90, 80, 70, 60, and 50% of purity in the region of the component

The datasets, referred to as p90, p80, p70, p60, p50, respectively, were built using a normal distribution with 90.0, 80.0, 70.0, 60.0, and 50.0% of "purity". This experiment shows that $\langle G \rangle / 2K$ is a good approximation of the purity of the component.

2.3 Optimal Network

In the process described so far, each K yields a network; clearly some networks will have better components than others, according to the notion of purity. In general, rarely a network obtained with a unique value of K will have all the best components among all others components in all K possible networks. Bearing this in mind, a suggestive idea is to obtain a network with the best organization of data into components independently of a unique K. For doing this, the idea is to vary K keeping the best components found. This process will result in a network called optimum network, with components formed by distinct values of K.

The idea of the optimum network based only in the purity of components has some drawbacks. As purity does not consider the length of the component it tends to favor the small ones, hence it is not a good gauge for constructing a network that must optimally represent the data. The quantity to be maximized though is not purity itself but a slightly variant that considers the size of the components. An intuitive way for overcoming this problem is multiply purity by the number of vertices for a given component, as stated in eq. (5). However the measure W incurs another problem. Now, very large components with low purity may have advantage over smaller ones with high purity. To solve this problem a new constraint that relates size and purity in an indirect way is added to the equation.

$$W_j = \frac{\sum_{i=1}^{N_j} g_i}{2K_j} \qquad and \qquad \langle G_c \rangle > K \qquad (5)$$

Where W_j is the new measure for component C_j, g_i is the degree of vertex v_i, N_j is the number of patterns in component C_j, K_j is the number of neighbors used in the construction of component j.

Notice that the purity measure will still be used, after the optimum network has been obtained, in the classification process. The optimum network is the final structure obtained through this process. This network can be viewed as the result of a supervised learning process and will be used in the classification process as exposed in the next section. The algorithms 1 and 2 detail the optimum K-associated model generation from data.

Algorithm 1. Optimum K-associated Model

Input:

$V = \{v_i, ..., v_n\}$	set of vertices - examples
D	matrix of distance between vertices
$L = \{label(v_i), ..., label(v_n)\}$	set of labels
K_{max}	number of iterations

Output

$LC_{best} = \{(C_i(V_i; E_i); P_i)\};$ best components' network

Algorithm

$K = 1;$
$LC_{best} = Kac(V, D, L, K);$
$For(K = 2; K \le K_{max}; K = K + 1)$
    $LC_K = Kac(V, D, L, K);$
        For each component C_{K_i} in LC_K
            Determines correspondent components $j\{C_{(K-1)j}\}$ in LC_{K-1}
            If $(W(C_{K_i}) \ge W(C_{(K-1)j})$ for any j and $\langle G_i \rangle > K$
               $LC_{best} = LC_{best} - \{C_{(K-1)j}\} \cup \{C_{K_i}\};$

Return (LC_{best})

In this algorithm, the optimum network is firstly set as the 1-associated network ($K = 1$). As K increases, different components from $(K-1)$-associated network can be merged into just one component C in the current K-associated network. If this new component has a better measure W_c, and $\langle G_c \rangle > K$ it replaces the corresponding components of the previous $(K-1)$-associated networks (LC_{best}). The experiments have shown that in few iterations (at most 5) the number of components converges to the number of classes. The following algorithm details the construction of a K-associated network from data.

Algorithm 2. K-associated Network from data (Kac)

Input:

$V = \{v_i, ..., v_n\}$	set of vertices - examples
D	matrix of distance between vertices
$L = \{label(v_i), ..., label(v_n)\}$	set of labels
$K = \{1, ..., n\}$	number of nearest neighbors to be used

Output

$$LC = \{(C_i(V_i; E_i); P_i)\} \qquad \text{set of components and purity}$$

Algorithm

$C = \emptyset$;
For each vertex v_i in V
$\qquad N_{k_i} = \{v_j | v_j = knn(i) \text{ and } label(v_j) == label(v_i)\}$;
$\qquad E = E \cup \{e_{ij} | v_i \in N_{K_i} \text{ and } v_j \in N_{K_i}\}$;
$\qquad$ For each connected sub-graph C_i from $A(V; E)$
$\qquad\qquad$ Compute the purity of the component (P_{C_i})

Return $(LC = \{(C_i(V_i; E_i); P_i)\})$

This algorithm builds the K-associated network for desired K and return a list of its components with respective purity.

3 Non-parametric K-Associated Classifier

The objective is to derive a non-parametric classifier that uses the optimal K-associated network as model from training data to accurately classify new patterns. As stated before this structure stores the best components of data found through a large range of K. The component purity can be seen as a priori of the data represented in the component. Since each component contains vertices (instances) from only one class, we can compute the probability of a new instance to belong to given class by computing the probability of this instance to belong to the components of the same class. Before presenting details on this classifier some notation must be introduced.

Typically a training pattern x_i is represented by $x_i = (x_{i1}, x_{i2}, ..., x_{ip}, \omega_i)$, which x_i represents the i-th training pattern with ω_i its associated class, in a M-class problem $\Omega = \{\omega_1, \omega_2, ..., \omega_M\}$. In the same way, a new pattern is defined as $y_j = (y_{j1}, y_{j2}, ..., y_{jp})$, excepted that now the class ω_j associated with the new pattern y_j must be estimated. Consider also the set of components of the optimum network $C = \{C_1, ..., C_R\}$, where R is the number of components and $R \geq M$.

According to Bayes theory [16] the posteriori probability of a new instance y_i to belong to the component C_j given the neighbors N_{K_i} of y_i that belongs to the component C_j is,

$$P(y \in C_j | N_{K_i}) = \frac{P(N_{K_i}|C_j) P(C_j)}{P(N_{K_i})}. \tag{6}$$

It is important to bear in mind that each component C_i came from a particular K-associated network. Hence, the neighborhood N_{K_i} must considers this particular K.

As purity scores individually how pure is each component, the normalized purity acts as priori probability,

$$P(C_j) = \frac{g_j}{\sum_{i=1}^{M} g_i}. \tag{7}$$

Probability of having N_{K_i} connections, among the K_j possible, to component C_j, is

$$P(N_{K_i}|C_j) = \frac{\#\{e_{N_k} \in C_j\}}{K_j}. \tag{8}$$

Probability of N_{K_i} connections is given by eq. (9).

$$P(N_{K_i}) = \sum_{i=1}^{M} P(N_{K_i}|C_i) P(C_i) \tag{9}$$

As in many cases there are more components than classes, according to Bayes optimal classifier, it is necessary to sum the posteriori probability that correspond to a common class. So the posteriori probability of the new instance to belong to a given class is given by eq. (10).

$$P(y|\omega_i) = \sum_{C_j = \omega_j} P(y \in C_j | N_j) \tag{10}$$

Finally the greatest values between the found *posteriori* probabilities reflect the most probable class to assign for the new instance, according to eq. (11).

$$\varphi(y) = arg\ max\{P(y|\omega_1), ..., P(y|\omega_M)\} \tag{11}$$

where $\varphi(y)$ stands for the class attributed for instance y.

4   Experiments and Results

This section presents and discusses the results of using the proposed algorithm and the two well known multiclass classification algorithms, the K-nearest neighbor, for tree frequently used values of K (1, 3, and 5) and the decision tree C4.5 algorithm. The tests were carried out learning from nine multiclass knowledge domain data taken from the UCI-Repository [17]. Each of the algorithms was implemented in Java and the results were obtained through 10-fold stratified cross-validation process. Table 1 presents the test averaged under 10 runs followed by its standard deviation. From these ten datasets, we produce new datasets with noise changing the classes in 5 and 10% of the training data. The results carried out on these noise data are also showed in the Table 1.

Table 1. Comparison results through nine knowledge domains

Domain	Proposed Algorithm	C4.5	K-NN K=1	K-NN K=3	K-NN K=5
Yeast	98.2±0.79	**98.9±7.8**	98.8±0.8	98.5±0.9	98.3±1.0
Yeast (5%)	85.2±2.7	81.3±3.2	80.5±1.7	86.9±2.6	**88.6±3.1**
Yeast (10%)	**78.5±2.1**	66.5±1.5	68.6±4.4	74.3±3.6	76.9±2.9
Tae	**63.6±15.1**	61.6±11.9	63.3±11.7	41.9±11.7	43.6±11.5
Tae (5%)	50.3±14.5	51.7±18.5	**55.9±12.8**	34.6±13.1	39.3±13.5
Tae (10%)	47.1±14.2	**48.9±14.4**	47.0±11.8	36.1±12.1	31.2±10.5
Zoo	**97.1±4.7**	93.1±9.5	96.2±7.2	93.4±9.6	88.0±11.9
Zoo (5%)	**86.1±8.5**	82.2±13.2	78.6±12.6	82.7±12.4	79.2±10.4
Zoo (10%)	**72.2±16.8**	63.2±15.9	63.0±15.1	71.3±12.9	71.5±14.5
Image	74.3±8.1	**79.5±7.5**	74.9±8.9	74.4±8.7	70.4±8.1
Image (5%)	60.9±12.0	**66.2±9.4**	61.3±12.1	64.1±11.4	64.6±9.3
Image (10%)	**56.3±11.3**	54.7±10.8	54.2±10.9	54.9±10.3	54.2±10.8
Wine	88.8±6.9	**90.9±6.8**	83.9±7.4	80.3± 8.5	83.1±8.6
Wine (5%)	**76.5±8.9**	74.3±10.7	72.6±10.1	69.9±10.2	75.1±10.1
Wine (10%)	64.6±10.7	64.1±10.9	59.3±10.3	61.7±12.3	**65.1±9.7**
Iris	**98.0±3.2**	94.6±6.1	97.8±3.6	98.1±3.7	97.8±3.5
Iris (5%)	**88.6±8.9**	82.0±6.3	83.9±9.1	85.7±7.6	87.6±8.7
Iris (10%)	**81.3±8.2**	70.6±8.4	71.9±10.5	78.3±10.2	80.8± 11.0
Glass	66.8±9.3	64.9±5.7	**73.3±8.3**	70.1±13.9	68.7±9.1
Glass (5%)	**64.0±8.6**	58.4±9.6	57.6±9.8	63.2±11.9	59.1±10.2
Glass (10%)	**57.7±9.2**	49.0±11.4	56.4±9.9	54.2±10.2	54.8±9.4
E.coli	**97.6±2.6**	96.5±4.3	97.6±2.8	97.0±3.2	96.8±2.9
E.coli (5%)	**85.4±6.6**	80.4±6.4	79.8±6.3	83.6±5.7	84.2±6.7
E.coli (10%)	**74.4±6.1**	71.2±5.8	62.1±7.6	72.2±7.5	74.1±7.7
Balance	94.2±3.4	89.9 4.2	**96.9±1.9**	94.7±2.9	95.6±2.4
Balance (5%)	80.3±3.9	78.8±3.9	82.3±4.1	82.8±4.2	**84.8±4.1**
Balance (10%)	72.9±5.6	61.9±4.3	66.3±5.4	73.6±5.5	**75.9±5.0**

The proposed classifier had better performance on 15 of 27 datasets; the C4.5 had better performance on 4 of 27; the KNN for (K=1, 3, and 5) had better performance on 4 of 27, 0 of 27, and 4 of 27, respectively. As one could expect, the KNN with K=5 has better performance than KNN with K=1 for noise data, but poorer results in the other cases.

5   Conclusions

This paper presented an effort considering complex networks for dealing with classification problems. Previously unseen in complex network literature, this paper proved that it is possible to use complex networks not only to clustering problems but also to classification tasks. The results, although considering little datasets and few comparison, indicates a favorable scenario for our approach; particulary for noisy data. Future work expects some more comparisons to better validate our method.

References

1. Watts, D.J., Strogatz, S.H.: Collective Dynamics of 'Small-World' Networks. Nature 393, 440–442 (1998)
2. Albert, R., Jeong, H., Barabási, A.-L.: Diameter of the World Wide Web. Nature 401, 130–131 (1999)
3. Newman, M.E.J.: The Structure and Function of Complex Networks. SIAM Review 45(2), 167–256 (2003)
4. Albert, R., Barabási, A.-L.: Statistical Mechanics of Complex Networks. Review of Modern Physics 74, 47–97 (2002)
5. Bornholdt, S., Schuster, H.G.: Handbook of Graphs and Networks: From the Genome to the Internet. Wiley-vch, Weinheim (2003)
6. Dorogovtsev, S.N., Mendes, J.F.F.: Evolution of Networks: From Biological Nets to the Internet and WWW. Oxford University Press, Oxford (2003)
7. Han, J., Kamber, M.: Data Mining: Concepts and Techniques. Morgan Kaufmann, San Francisco (2006)
8. Duda, R.O., Hart, P.E., Stork, D.G.: Pattern Classification. John Wiley & Sons, Inc., Chichester (2001)
9. Berkhin, P.: Survey of Clustering Data Mining Techniques. Technical report, Accrue Software (2002)
10. Schaeffer, S.E.: Graph Clustering. Computer Science Review 1, 27–34 (2007)
11. Karypis, G., Han, E.-H., Kumar, V.: Chameleon: Hierarchical Clustering using Dynamic Modeling. IEEE Computer 32(8), 68–75 (1999)
12. Guha, S., Rastogi, R., Shim, K.: CURE: An Efficient Clustering Algorithm for Large Databases. In: Proc. of 1998 ACM-SIGMOD Int. Conf. on Management of Data, pp. 73–84 (1998)
13. Newman, M.E.J., Girvan, M.: Finding and Evaluating Community Structure in Networks. Physical Review E 69, 026113(1-15) (2004)
14. Danon, L., Duch, J., Arenas, A., Dáz-Guilera, A.: Comparing Community Structure Identification. Journal of Statistical Mechanics: Theory and Experiment, P09008(1-10) (2005)
15. Hopcroft, J., Khan, O., Kulis, B., Selman, B.: Tracking Evolving Communities in Large Networks. Publications of the National Academy of Sciences USA 101(1), 5249–5253 (2004)
16. Hastie, T., Tibshirani, R., Friedman, J.: The Elements of Statistical Learning: Data Mining, Inference and Prediction. Springer, Heidelberg (2001)
17. Asuncion, A., Newman, D.J.: UCI Machine Learning Repository. University of California, School of Information and Computer Science, Irvine, CA (2007), http://www.ics.uci.edu/~mlearn/MLRepository.html

Characterizing the Structural Complexity of Real-World Complex Networks

Jun Wang and Gregory Provan

Department of Computer Science, University College Cork, Ireland
{jw8,g.provan}@cs.ucc.ie

Abstract. Although recent research has shown that the complexity of a network depends on its structural organization, which is linked to the functional constraints the network must satisfy, there is still no systematic study on how to distinguish topological structure and measure the corresponding structural complexity of complex networks. In this paper, we propose the first consistent framework for distinguishing and measuring the structural complexity of real-world complex networks. In terms of the smallest d of the dK model with high-order constraints necessary for fitting real networks, we can classify real-world networks into different structural complexity levels. We demonstrate the approach by measuring and classifying a variety of real-world networks, including biological and technological networks, small-world and non-small-world networks, and spatial and non-spatial networks.

Keywords: Complex Networks, Structural Complexity, Random Graph Generators.

1 Introduction

Heavy-tailed or scaling degree distributions, found in many real-world networks (including a variety of social networks, biological systems and technological systems) [1], have been posited as a "universal class" of such complex systems. However, recent research has challenged the arguments that such distributions are special and signify a common architecture, independent of the system's functional properties or domain role [2,3]. Although people realized that the complexity of a network depends on its structural organization, which is linked to the functional constraints the network has to satisfy [4], there is still no systematic study on how to distinguish topological structure and measure the corresponding structural complexity of various real-world networks. The structural complexity measures of complex networks have been discussed before [5,6]. However, there is little consistency among the proposed measures, and most analyses are based on very small graphs with only a few nodes [5,6]. More importantly, prior work [5,6] only showed that real-world networks are "complex" in the sense that different topological features deviate from classic ER random graphs or simple structures

J. Zhou (Ed.): Complex 2009, Part I, LNICST 4, pp. 1178–1189, 2009.

like regular lattices [1]. In contrast, we try to finely distinguish structural complexity among real-world networks.

One approach to characterize the structure of real-world networks is to compare them to "appropriate" null models. Appropriate null models include random network ensembles with some of the statistical features being present in the real-world network under investigation. The classic ER ensemble is the simplest example of the so-called "maximally random" graphs [1], and the only constraint is the average degree of the real network. The deviation of data collected on real-world networks from the predictions of the ER model triggered interest in more advanced random network models [1], because it implied that those graphs were not created just by joining vertices at random, but required the existence of additional constraints. The classic random graph model can be naturally extended to define network ensembles that have other high-order topological characteristics in common with a real network [7,4].

Recently, a dK-random graph model was proposed to specify all degree correlations within d-sized subgraphs of a given network [7]. The $1K$-distribution defines a family of $1K$-graphs which reproduce the original graph's node degree distribution, and is equal to generating the widely-used generalized random graph (GRG) model. $2K$-graphs reproduce the joint degree distribution, the $2K$-distribution, of the original graph. $3K$-graphs consider interconnectivity among triples of nodes, and so forth. Generally, the set of $(d + 1)$K-graphs is a subset of dK-graphs. In other words, larger values of d capture increasingly complex properties of the original graph and further constrain the number of possible graphs, so any specified topology metric we can define on a real network will eventually be captured by dK-graphs with a sufficiently large d. However, the computational complexity of generating dK-graphs increases exponentially in d. One main concern with dK-graphs is how fast the dK model converges toward the real network. So for creating realistic but "random" ensembles, it is important to find the smallest d which can match the real network with sufficient fidelity in terms of the specified topology metrics of specific applications. Since the smallest d in the dK-graphs determines the number of constraints, as well as the computational complexity necessary for fitting the real network, we use the smallest value of d as an indicator of the level of structural complexity of the real network.

We also need a set of graph metrics to evaluate the fidelity of generated random graphs, and a wide range of topological metrics have been proposed recently [7]. Not all topology metrics are mutually independent: some either fully define others, or significantly narrow down the spectrum of their possible values [7]. Therefore, identifying the underlying principles of such definitive metrics reduces the number of topology characteristics that models must reproduce. The dK-distributions themselves present one possible approach to constructing a family of such simple metrics which define all others. Recent research showed that the $2K$-distribution, the joint degree distribution, appears to play a central role in determining a wide range of other existing topological properties [7]. The s-Metric[1] is a scalar summary statistic of the joint degree distribution, and potentially unifies many aspects of complex networks, because it is closely related to betweenness, and linearly related to graph assortativity [3]. Obviously, the s-Metric is a succinct but rich topology metric. In addition, shortest paths play an important role in transport and communication within a network. A measure of the typical separation in the network is given by the characteristic path length, defined as the mean of shortest lengths over all pairs of nodes [1]. The above two metrics help to characterize the topological structures of various real-world complex networks, and are effectively applied to experimental analyses in this paper.

This paper makes two main contributions. First, based on the dK model analysis, we propose the first consistent framework for distinguishing and measuring structural complexity of real-world complex networks. The approach can be applied to complex networks with different topologies in any application domain.

Second, we demonstrate our analyses on a variety of real-world complex networks, and classify them into different levels of structural complexity. This provides the first clear classification of real-world networks in terms of their structural complexity. In our analyses, we surprisingly found that a wide range of complex networks, including electronic circuits, transportation systems, brain and neuronal systems, and protein interaction networks (PINs), have the same level of structural complexity and can be matched well by the simple $1K$ (GRG) model. We argue that these networks have a common set of explicit or implicit geometric constraints. The router-level Internet and transcription regulatory networks (TRNs) show higher structural complexity, and at least the $2K$ model is necessary for fitting them. Recent research showed that highly complicated technological and economic constraints have big impacts on shaping the topological structure of the Internet. It will be very interesting to study why the topological structure of the TRN is much more complex than that of the PIN. We also surprisingly found that a pulp mill system has a very high level of structural complexity, which cannot even be captured by the $3K$ model. We think that different types of devices and complicated interfaces between them lead to the high level of structural complexity of the system.

We organize the remainder of the document as follows. Section 2 introduces some related work. Section 3 applies the dK model analysis to a variety of real-world complex networks and classifies them into different levels of structural complexity. Finally, section 4 summarizes our work.

2   Related Work

This level of structural complexity based on the dK model is consistent with the concept of *entropy* in statistical mechanics. In Bianconi's definition [4], the entropy of a network ensemble under specific constraints is proportional to the logarithm of the number of networks belonging to the ensemble. The complexity

---

[1] The s-Metric of a graph G is defined as $s(G) = \sum_{edge(i,j)} d_i d_j$, where (i,j) ranges over the edges in the graph, and d_i and d_j are the degrees of the node i and j respectively.

of a given ensemble of networks increases as the number of networks in the ensemble decreases. As we add further constraints that a desired ensemble is to have in common with a given real network, we effectively consider ensembles with decreasing cardinality. Consequently a network ensemble of high complexity corresponds to a small variability of the networks in the ensemble. We expect that a very complex network belongs to an ensemble of functionally equivalent networks of small entropy. A larger value of d reflects higher complexity and smaller entropy.

Real-world networks generally are classified based on specific topological properties, such as different (power-law or exponential) degree distributions [8], or power-law exponents of the betweenness centrality distributions [9]. In contrast, our approach classifies real-world networks from a new dimension using the corresponding random graph models necessary for fitting original networks.

3   Analyzing Structural Complexity of Real-World Complex Networks

In this section we analyze a variety of real-world complex networks and classify them into the corresponding levels of structural complexity. Figure 1 shows a general view of our results, and the details are discussed in the following sections.

3.1   Networks with $1K$ Complexity

We surprisingly found that the simple $1K$ (GRG) model, which is independent of any domain-specific growth process and only reproduces the degree distribution [1], can closely match the topologies of a variety of technological and biological networks, as listed in Table 1. In this paper, the Markov-chain Monte Carlo (MCMC) switching algorithm [10,11,7] has been used to implement the $1K$, $2K$ and $3K$ models, and to generate experimental data in order to reduce statistical variance [2].

All networks listed in Table 1 have highly heterogeneous heavy-tailed degree distribution [17,18,13,14,15,16]. Except the core S. Cerevisiae protein interaction network [16], all networks are spatial networks which occupy some physical space, such that their nodes occupy a position in two- or three-dimensional Euclidean space, and their edges are real physical connections [1]. It is not surprising that the topology of spatial networks is strongly constrained by their geographical embedding. All these man-made and biological spatial networks share a common planning principle: wire cost optimization over the entire network [19,13,20].

Wire cost optimization is obvious and natural in transportation planning. In circuit design, wire length has been treated as the prime parameter for performance evaluation since it has a direct impact on several important design

---

[2] The MCMC switching algorithm generates uniform sample of graphs having the dK-distribution, while remaining unbiased (random) with respect to all other properties. However, this results in non-uniform sampling of graphs with different values of properties that are not fully defined by the dK-distribution. In this sense, the graphs generated by the dK model are the *maximally random* graphs [11,7].

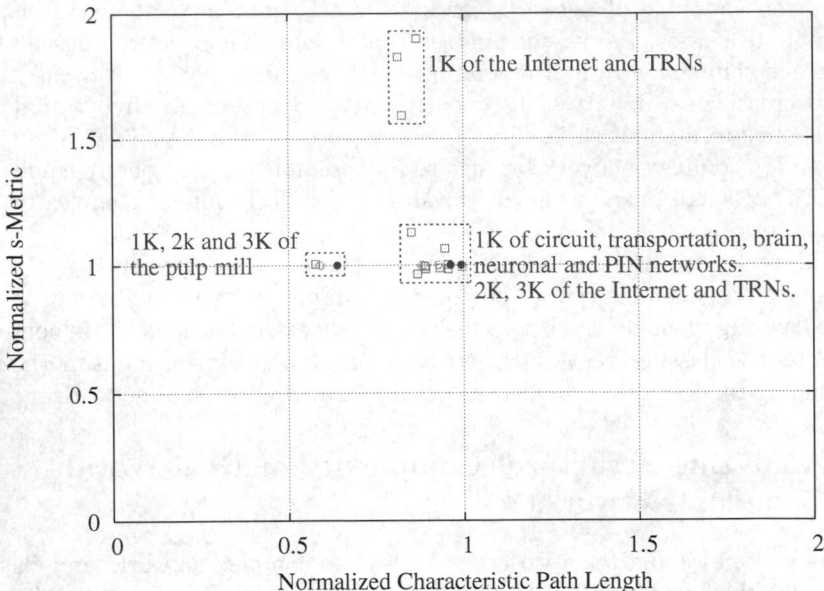

Fig. 1. The plot of the normalized values of characteristic path length and s-Metric of $1K$ (squares), $2K$ (open circles) and $3K$ (solid circles) models of a variety of real-world networks discussed in this paper. The values are normalized by those of real networks, so the coordinate $(1, 1)$ means perfect fitting of the real networks.

Table 1. Comparing topology metrics for the real-world networks (the digital circuits in ISCAS-85 benchmark suite [12], the German highway network (Autobahn)[13], the Chinese airport network [14], the giant component of the anatomical network of the human cerebral cortex using cortical thickness measurements from magnetic resonance images [15], the Macaque cortical connectivity network within one hemisphere [13], the C. elegans neuronal networks [13], the giant component of the core S. Cerevisiae protein interaction network [16]) and the corresponding $1K$ model. All values of random graphs are averaged over 100 graphs respectively.

Network	Characteristic path length		s-Metric	
	real	$1K$	real	$1K$
Circuit C432	4.53	4.33 ± 0.05	6986	6875.99 ± 143.46
Circuit C499	4.65	4.4 ± 0.06	9848	10491.57 ± 306.78
German highway	19.42	17.33 ± 0.63	8025	7904.1 ± 61.45
Chinese airport	2.07	2.06 ± 0.01	1728592	1716900.08 ± 3647.17
Human Brain	3.05	2.65 ± 0.05	3957	3819.35 ± 65.61
Macaque Brain	1.78	1.70 ± 0.08	2368861	2375055.27 ± 4136.29
C. elegans(local)	2.52	2.35 ± 0.08	127622	126103.72 ± 591.41
C. elegans(global)	2.64	2.35 ± 0.06	916807	911946.68 ± 9739.35
S. Cerevisiae PIN	5.26	4.48 ± 0.01	749149	846220.16 ± 11800.32

parameters [20]. In the brain and neuronal systems, energy is consumed in establishing fibre tracts between neurons, and in propagating action potentials over these fibres. Thus, the total cost of all wires should be kept as low as possible [21]. Although the exact origin of the wiring cost is not completely known, the farther apart two neurons are, the more costly is the connection between them [21].

In addition, the graph distance (the number of hops between nodes) also has an important impact on functions of spatial networks. For instance, graph distance can capture another driving force underlying circuit design, timing, where it is important to reduce the delay of signal transmission among components. Similarly, minimizing the average number of processing steps–that is, reducing the number of intermediate transmission steps in neural integration pathways– has several functional advantages [21]. Too many transfer flights are bothering in an air journey, but most road travelers look for routes that are short in terms of miles, and the number of legs is often considered less important. The graph distance can be characterized by the characteristic path length.

Recent research showed that an optimization model (OPT) trading-off the wire cost and the graph distance can capture the topologies of specific spatial networks, like the electronic circuits [18], and under appropriate parameters, a preferential attachment model with spatial constraints (SPA) can generate small-world network structure close to that of networks generated by the OPT model [18,22]. The SPA model and its extension can closely match the topologies of electronic circuits, the brain networks and airport networks [18]. Our experiments also showed that the OPT model with setting of strong preference for reducing wire cost can also capture the topology of the highway network. Naturally, the highway network is not a small-world network, as the characteristic path length is twice as large as for comparable ER models [13]. All other spatial networks listed in Table 1 are small-world graphs [17,18,13,14,15,16]. So the $1K$ model can match both small-world networks and non-small-world networks well.

The $1K$ model itself is independent of any system growth process, but the degree distributions of the above spatial networks are shaped and constrained under domain-specific spatial constraints. Actually, the parameters corresponding to spatial constraints in the SPA and OPT model can be tuned to generate diverse degree distributions. The degree distributions of the above networks implicitly reflect some spatial constraints with various strength shaping the network structures. Maybe that's the reason why the $1K$ model can closely match these spatial networks.

In addition to our results in Table 1, there is also a lot of other solid evidence that the $1K$ model closely captures the topologies of the PINs. Przulj et al. proposed a "Stickiness Index" model for the a series of PINs and showed that it outperforms other models in terms of a range of topology metrics including relative subgraph frequency [23]. Actually, the "Stickiness Index" model is a stochastic implementation of the $1K$ model proposed by Chung et al., in which the connection between nodes i and j is chosen independently with probability p_{ij}, with p_{ij} proportional to the product of the degree of i and j [24].

This approach is convenient for theoretical analysis, since rigorous proofs for a random graph with exact degree sequences is rather complicated and usually requires additional "smoothing" conditions because of the dependency among the edges [24]. This stochastic implementation and the MCMC switching implementation are "basically asymptotically equivalent, subject to bounding error estimates" [3]. Ivanic et al. recently also analyzed a series of PINs and found the so-called "degree-weighted behavior" that the probability of an interaction between two proteins is generally proportional to the numerical product of their individual interacting partners, or degrees [25]. The "degree-weighted behavior" is consistent with the definition of the stochastic $1K$ model. They found that the degree-weighted behavior is manifested throughout the PINs studied, except for the high-degree, or hub, interaction areas. Their finding is also consistent with our results of the s-Metric in Table 1, in which the s-Metric of the corresponding $1K$ model is about 12% higher than that of the S. Cerevisiae PIN. But the probabilities of interaction between the hubs are still high, and these hubs are separated by very few links, so the discrepancy of s-Metric data of the S. Cerevisiae PIN in Table 1 is only about 12%. Ivanic et al. further proposed a degree-conserving degree-weighted (DCDW) model [25], which actually is a matching implementation of the $1K$ model and has only very small deviations from the MCMC switching implementation [11], and showed that this model can closely capture the PINs in terms of a series of topological properties. Friedel et al. showed that PINs are in general most similar to uncorrelated networks, which are implemented by the MCMC switching $1K$ model, with regard to degree correlations and all other network properties considered [26].

The PINs are different from the above spatial networks because they are not explicitly embedded in any observable physical space. However, Przulj et al. showed that a random geometric model can accurately capture the PIN structures in terms of relative subgraph frequency [27]. Higham et al. [28] pushed the research further by exploiting the fact that the geometric property can be tested for directly. They applied a algorithm, which has been verified in the sense that it successfully rediscovers the geometric structure in artificially constructed geometric networks, to a series of publicly available PINs of various organisms, and indicated that geometric effects are present. Testing on a high-confidence yeast data set produced a very strong indication of geometric structure. Overall, the results add support to the hypothesis that PINs have a geometric structure. Serrano et al. discussed the hidden variables formalism, taking as hidden variables nodes's coordinates in a metric space [29]. Each two nodes are located at a certain hidden metric distance, and connected with a probability, which relates the network topology to the underlying metric space. This probability depends on the metric distance [29]. It seems that hidden metric spaces do exist for the PINs, and implicit geometric constraints play an important role in shaping the observed PIN topologies.

According to the above analyses, we think that the explicit or implicit geometric constraints are probable underlying driving forces shared by all the above networks with $1K$ complexity. For these networks, reproducing only the $1K$ constraints can also closely fit the $2K$ statistics represented by the s-Metrics.

3.2   Networks with $2K$ Complexity

In this section, we show additional complex networks which also have power law degree distributions but need higher-order statistics to capture their structures. As shown in Table 2, the $1K$ model is not sufficient to match the topologies of the router-level Internet [3,7] and the transcriptional regulatory networks of E. Coli [30].

Table 2. Comparing topology metrics for the real-world networks (the router-level topology of the Internet of a single ISP (HOT) [3,7], the giant components of two E. coli transcriptional regulatory networks, collected by Shen-Orr. et al. and Ma et al. [30], respectively, and the pulp mill [31]), and the corresponding $1K$, $2K$, $3K$ models. All values of random graphs are averaged over 100 graphs respectively.

Network	Characteristic path length				s-Metric			
	real	$1K$	$2K$	$3K$	real	$1K$	$2K$	$3K$
Internet	6.81	5.91 ±0.17	6.33 ±0.13	6.55 ±0.13	28442	54023.71 ±4437.59	28442	28442
TRN(Shen-Orr)	4.83	3.99 ±0.06	4.28 ±0.06	4.65 ±0.05	26621	42402.61 ±1782.63	26621	26621
TRN(Ma)	3.99	3.25 ±0.02	3.51 ±0.01	3.96 ±0.01	1301244	2375893.92 ±43876.86	1301244	1301244
Pulp Mill	11.62	6.71 ±0.13	6.87 ±0.14	7.43 ±0.16	3629	3647.44 ±54.88	3629	3629

The router-level Internet shown in Table 2 has an s-Metric value much lower than that of the corresponding $1K$ model, so the organizing principles of the Internet are completely different from the networks with $1K$ complexity listed in Table 1. The s-Metric is linearly related to the network assortivity coefficient, and a relatively lower s-Metric value means a relatively disassortive connectivity pattern in which high-degree nodes are less likely to be connected with each other [3]. The router-level Internet is also a spatial network, but it is subject to more complicated technological and economic constraints [3], and has much more complex topology structure. In general, a router can have a few high bandwidth connections or many low bandwidth connections, because limits in technology fundamentally preclude the possibility of high-degree, high-bandwidth routers [3]. The high-end backbone routers in the network core have only a few high-speed and long-haul connections, and edge routers (in the "last mile") are typically slower overall, but have many low-speed connections. So, for the router-level Internet, high-degree nodes can exist, but are found only within local networks at the far periphery of the network, and would not appear anywhere close to the backbone [3]. This pattern can result in high performance (traffic flow) and robustness to failures [3]. In contrast, in the networks shown in Table 1, the high-degree nodes are more likely to connect to each other and appear in the cores of the networks, because these networks have high s-Metric values close to those of the corresponding $1K$ models [3].

As we mentioned before, the joint degree distribution can determine a wide range of other important topological properties, except clustering [3,7]. The $2K$ model reproduces the joint degree distribution, so the s-Metric, which is scalar summary statistics of the joint degree distribution, can be completely matched when $d \geq 2$. As shown in Table 2, by increasing d the dK-random graphs constantly converge toward the real networks. Actually, in additional to the topology metrics listed in Table 2, clustering coefficients and 3-node motifs [1], which depend on interconnectivity among tripes of nodes, can be completely matched when $d = 3$, and 4-nodes motifs can be completely matched when $d = 4$. But model selection has to make trade-offs between fidelity and complexity according to domain requirements. The implementations of $1K$ and $2K$ models are relatively simple, but when $d \geq 3$ the implementations become much more complex due to the increasing number of non-isomorphic simple connected graph of size d [7]. As shown in Table 2 and Figure 1, from $1K$ to $2K$ the fidelity improves dramatically, but there is only relatively mild improvement from $2K$ to $3K$. Mahadevan et al. also found that the $d = 2$ case is sufficient for most practical purposes for the Internet topology [7].

Furthermore, we analyzed two widely-used TRNs listed in Table 2. The TRNs are directed networks where a transcription factor positively or negatively regulates the RNA transcription of the controlled protein. In this paper, we mainly focus on general organizational principles of networks, so we ignore the direction of links in the TRNs and treat them as undirected graphs. But all methods in our analyses can be easily applied to directed graphs as well. The structures of the TRNs have patterns similar to the Internet: links between high-degree nodes are systematically suppressed, whereas those between high-degree nodes and low-degree nodes are favored, so as shown in Table 2, they naturally have much lower s-Metric values than those of the $1K$ model. Maslov et al. also quantified correlations between connectivities of interacting nodes in the TRN of the yeast S. cerevisiae and compared them to the $1K$ model, and their empirical results showed the disassortive pattern similar to that in the TRNs of E. Coli we analyzed [10]. It is feasible that molecular networks in a living cell have organized themselves in an interaction pattern that is both robust and specific. Topologically, the specificity of different functional modules can be enhanced by limiting interactions between hubs and suppressing the average connectivity of their neighbors. This effect decreases the likelihood of cross talk between different functional modules of the cell, and increases the overall robustness of a network by localizing effects of deleterious perturbations [10]. Similarly, the $2K$ model captures the structures of the TRNs listed in Table 2 and Figure 1 much better than does the $1K$ model, and setting $d = 3$ only improves fitting mildly. Some researchers conjectured that it appears likely that the $3K$ model will be sufficient for self-organized small-world graphs in general [7].

Maslov et al. also claimed that they found a similar disassortative pattern in the yeast PIN [10], but as we discussed before, many recent studies showed no such disassortative correlation between node degrees in yeast for high-confidence interaction sets, and the opposite results discovered by Maslov et al. may be explained by a bias in the yeast–two hybrid system which might artificially increase negative degree correlations [26]. Molecular networks guide the biochemistry of a living cell on multiple levels: its metabolic and signaling pathways are shaped by the network of interacting proteins, whose production, in turn, is controlled by the genetic regulatory network, so it will be very interesting to study why these two tightly-related molecular networks have completely different topological structures.

3.3 Networks with Higher Complexity

We studied a real pulp mill benchmark model developed by Castro and Doyle [31], which consists of modular representations of unit operations in a complete pulp mill. The benchmark can be used for studying several process-system tasks, including modeling, control, estimation and fault diagnosis [31]. In the pulp mill, the major units of operation are: a digester, pulp washers, oxygen tower, storage vessels, bleaching towers, evaporators, recovery boiler, smelt dissolving tank, clarifiers, slaker, causticizers and lime kiln [31]. There are also many valves, which are used to connect components in and between various key units.

The system structure has big impacts on a series of test and control tasks in engineering systems. For example, the complexity of specific diagnosis algorithms only depends on the system topology [32,17]. We analyzed physical connections between the fundamental components, and studied the corresponding topology of the whole pulp mill system. The degree distribution of the pulp mill follows a power law as well. However, as shown in Table 2 and Figure 1, the pulp mill is a non-small-world network, and even the $3K$ model highly deviates from the pulp mill and the corresponding characteristic path length can only reach about 64% of the real network.

Although the router-level Internet and electronic circuits we analyzed are also highly-engineered complex systems under specific design principles, their elementary components and connection interfaces are relatively homogeneous. In digital circuits the components are only different types of basic logic gates, and in the router-level Internet the components are only routers with various speeds. Different types of logic gates in circuits and routers in the Internet can be easily connected with each other, respectively. But in the pulp mill, the components are diverse heterogeneous devices, and only specific types of devices, which are functionally related and have compatible interfaces, can be connected with each other. For an complex engineering system like the pulp mill, it is largely impossible to fit any nontrivial network structure while ignoring domain-specific constraints [3], and the random graph generators seems not suitable for capturing topologies of this kind of complex systems since the cost of the dK-distribution representation and resulting computational complexity will be too high for practical applications when $d > 3$.

4 Summary

This article describes a consistent framework for distinguishing and measuring structural complexity of real-world complex networks. As shown in Figure 1, the experiments show that our approach can clearly distinguish the underlying structure of various real-world complex networks, and convincingly classify these networks from a new dimension.

We can apply the approach to measure structural complexity of more real-world networks, and the measured results can provide useful guidance on synthetic benchmark model generation for various simulation tasks. As shown in this paper, for complex systems with relatively low structural complexity, we can generate realistic (high-fidelity) but "random" benchmark models [17] with computationally efficient and simple random graph generators. In contrast, random graph generators cannot feasibly synthesize complex systems classified into high-complexity levels, due to the corresponding high computational cost and small ensemble-size generated.

References

1. Boccaletti, S., Latora, V., Moreno, Y., Chavez, M., Hwang, D.U.: Complex networks: Structure and dynamics. Physics Reports 424(4-5), 175–308 (2006)
2. Keller, E.F.: Revisiting "scale-free" networks. Bioessays 27(10), 1060–1068 (2005)
3. Li, L., Doyle, J.C., Willinger, W.: Towards a theory of scale-free graphs: Definition, properties, and implications. Internet Mathematics 2(4), 431–523 (2006)
4. Bianconi, G.: The entropy of randomized network ensembles. Europhysics Letters 81(2), 28005 (2008)
5. Claussen, J.C.: Offdiagonal complexity: A computationally quick complexity measure for graphs and networks. Physica A: Statistical Mechanics and its Applications 375(1), 365–373 (2007)
6. Kim, J., Wilhelm, T.: What is a complex graph? Physica A: Statistical Mechanics and its Applications 387(11), 2637–2652 (2008)
7. Mahadevan, P., Krioukov, D.V., Fall, K.R., Vahdat, A.: Systematic topology analysis and generation using degree correlations. In: SIGCOMM, pp. 135–146 (2006)
8. Amaral, L.A., Scala, A., Barthelemy, M., Stanley, H.E.: Classes of small-world networks. Proc. Natl. Acad. Sci. USA 97(21), 11149–11152 (2000)
9. Goh, K.I., Oh, E., Jeong, H., Kahng, B., Kim, D.: Classification of scale-free networks. Proc. Natl. Acad. Sci. USA 99(20), 12583–12588 (2002)
10. Maslov, S., Sneppen, K.: Specificity and stability in topology of protein networks. Science 296, 910–913 (2002)
11. Milo, R., Kashtan, N., Itzkovitz, S., Newman, M.E.J., Alon, U.: On the uniform generation of random graphs with prescribed degree sequences (2004)
12. Hansen, M.C., Yalcin, H., Hayes, J.P.: Unveiling the iscas-85 benchmarks: A case study in reverse engineering. IEEE Des. Test 16(3), 72–80 (1999)
13. Kaiser, M., Hilgetag, C.C.: Spatial growth of real-world networks. Phys. Rev. E 69(3), 036103 (2004)
14. Li, W., Cai, X.: Statistical analysis of airport network of china. Phys. Rev. E 69(4), 046106 (2004)

15. He, Y., Chen, Z.J.J., Evans, A.C.C.: Small-world anatomical networks in the human brain revealed by cortical thickness from MRI. Cereb Cortex (2007)
16. Hormozdiari, F., Berenbrink, P., Przulj, N., Sahinalp, S.C.C.: Not all scale-free networks are born equal: The role of the seed graph in ppi network evolution. PLoS Comput. Biol. 3(7) (2007)
17. Wang, J., Provan, G.M.: Generating application-specific benchmark models for complex systems. In: AAAI, pp. 566–571 (2008)
18. Wang, J., Provan, G.M.: Topological analysis of specific spatial complex networks. Advances in Complex Systems (in press)
19. Costa, L., Kaiser, M., Hilgetag, C.: Predicting the connectivity of primate cortical networks from topological and spatial node properties. BMC Systems Biology 1, 16 (2007)
20. Dambre, J.: Prediction of interconnect properties for digital circuit design and technology exploration. Ph.D. dissertation: Ghent University, Faculty of Engineering (2003)
21. Kaiser, M.: Brain architecture: a design for natural computation. Philosophical Transactions of the Royal Society A 365, 3033–3045 (2007)
22. Barthlemy, M.: Crossover from scale-free to spatial networks. Europhysics Letters 63, 915–921 (2003)
23. Przulj, N., Higham, D.J.: Modelling protein-protein interaction networks via a stickiness index. J. R. Soc. Interface 3(10), 711–716 (2006)
24. Chung, F., Lu, L.: The average distances in a random graph with given expected degrees. Internet Math. 1, 91–113 (2003)
25. Ivanic, J., Wallqvist, A., Reifman, J.: Probing the extent of randomness in protein interaction networks. PLoS Comput. Biol. 4(7), e1000114+ (2008)
26. Friedel, C.C., Zimmer, R.: Influence of degree correlations on network structure and stability in protein-protein interaction networks. BMC Bioinformatics 8, 297+ (2007)
27. Przulj, N., Corneil, D.G., Jurisica, I.: Modeling interactome: scale-free or geometric? Bioinformatics 20(18), 3508–3515 (2004)
28. Higham, D.J.J., Rasajski, M., Przulj, N.: Fitting a geometric graph to a protein-protein interaction network. Bioinformatics (2008)
29. Serrano, A.M., Krioukov, D., Boguna, M.: Self-similarity of complex networks and hidden metric spaces. Physical Review Letters 100, 078701 (2008)
30. Ma, H., Kumar, B., Ditges, U., Gunzer, F., Buer, J., Zeng, A.P.: An extended transcriptional regulatory network of escherichia coli and analysis of its hierarchical structure and network motifs. Nucleic acids research 32, 6643 (2004)
31. Castro, J.J., Doyle III, F.J.: A pulp mill benchmark problem for control: Problem description. J. Proc. Cont. 14, 17–29 (2004)
32. Provan, G.M., Wang, J.: Automated benchmark model generators for model-based diagnostic inference. In: IJCAI, pp. 513–518 (2007)